STOCKMAN'S HANDBOOK DIGEST

(Animal Agriculture Series)

First Edition

INTERSTATE PUBLISHERS, INC.
Danville, Illinois

Printed in the United States of America.

Editions:
First 1992

Order from
Interstate Publishers, Inc.
510 North Vermilion Street, P.O. Box 50
Danville, IL 61834-0050
Phone: (800) 843-4774
FAX: (217) 446-9706

Library of Congress Catalog Card No. 90-85249

ISBN 0-8134-2896-3

STOCKMAN'S HANDBOOK DIGEST

by

M. E. Ensminger, B.S., M.A., Ph.D.

Dr. M. E. Ensminger is President of Agriservices Foundation, a nonprofit foundation serving world agriculture. Also, he is Adjunct Professor, California State University-Fresno; Adjunct Professor, The University of Arizona-Tucson; Distinguished Professor, University of Wisconsin-River Falls; Collaborator, U.S. Department of Agriculture; and Honorary Professor, Huazhong Agriculture College-Wuhan, People's Republic of China. Dr. Ensminger (1) grew up on a Missouri farm; (2) completed B.S. and M.S degrees at the University of Missouri, and the Ph.D. at the University of Minnesota; (3) served on the staffs of the University of Massachusetts, the University of Minnesota, and Washington State University; and (4) served as Consultant, General Electric Company, Nucleonics Department (Atomic Energy Commission).

Dr. Ensminger is the author of 21 widely used books that are translated into several languages and used throughout the world.

Among Dr. Ensminger's honors and awards are: Distinguished Teacher Award, American Society of Animal Science; Washington State University named and dedicated the *Ensminger Beef Cattle Research Center*, in recognition of his contributions to the University; and an oil portrait of him was placed in the 300-year-old gallery of the famed Saddle and Sirloin Club, which is recognized as the highest honor that can be bestowed on anyone in the livestock industry.

Other books by M. E. Ensminger
available from Interstate Publishers, Inc.

Animal Science
Animal Science Digest
Beef Cattle Science
Dairy Cattle Science
Horses and Horsemanship
Poultry Science
Sheep and Goat Science
The Stockman's Handbook
Swine Science

To
the memory of
my beloved brother,
Dr. Doug Ensminger,
world famed rural sociologist
who dedicated his life to his roots—
rural people/rural poor

PREFACE TO *STOCKMAN'S HANDBOOK DIGEST*

Stockman's Handbook Digest puts it all together! It's several books rolled into one.

There are 21 sections in *Stockman's Handbook Digest*. It opens with an important and a unique 22-page section on careers. Then, in order, a section is devoted to each of the following: animal behavior and environment, business aspects, breeding, feeding, pasture and range forages, hay and crop residues, silage/haylage/high-moisture grain, management, buildings and equipment, animal health/disease prevention/parasite control, selecting and judging livestock, fitting and showing livestock, marketing livestock and milk, meat and milk, wool and mohair, law on the livestock farm, breeds and breed registry associations, where to go for help, feed composition tables, and weights and measures.

Stockman's Handbook Digest correlates and applies the art and science of livestock production, marketing, and processing of all species of four-footed animals. It's the answer book for teachers, students, farmers/ranchers, counselors, and people of all walks of life engaged in the field of agriculture, directly or indirectly.

Today's teachers, students, farmers/ranchers, and counselors, are sophisticated—they want to know "how" and "why." To them, knowledge is not solely for the privileged few, but for all who seek it. This work fills this need.

Stockman's Handbook Digest is 21st century new. It covers everything from genes to genetic engineering; from pastures to patented animals; from parasites to pollution. Also, better to portray modern technology, illustrations have been used throughout the book, including a beautiful 8-page colored picture section.

Stockman's Handbook Digest is concise, but complete. But, I am willing to let its readers and posterity evaluate the book.

As an author, I have always used a team approach; I rely on many people for help. Special appreciation is expressed to the following persons who contributed richly to *Stockman's Handbook Digest*. Audrey H. Ensminger, who shepherded it from my Missouri hieroglyphics to camera ready and who designed the cover; Dr. Lawrence A. Duewer, Agricultural Economist, USDA, ERS, who provided many of the statistical facts and figures; Professor Richard F. Johnson, Department of Animal Sciences and Industry, California Polytechnic State University, San Luis Obispo, CA, who prepared most of the line drawings; Joan Wright, who typed the manuscript and correlated much of the processing; Margo Williams, who did the art and pasteup work; Randall and Susan Rapp, who set the type; and Jean Nelson and Deanna Ross, who did the proofreading. Additionally, a host of individuals, associations, and companies provided pictures, served as reviewers or made other notable contributions, which are gratefully acknowledged at appropriate places throughout the book.

M. E. Ensminger

Clovis, California
1992

Coming down the lane! Bitterroot Stock Farm. (Courtesy, Ernst Peterson, Hamilton, MT)

Contents

Animal behavior. (Courtesy, American Saddlebred Horse Assn., Lexington, KY)

Sustainable agriculture—terracing. (Courtesy, Allis-Chalmers, Milwaukee, WI)

Livestock Farming/Ranching In The 21st Century

Below: Business aspects. (Courtesy, Babcock Swine, Inc., Plainview, MN)

Below: Two sets of identical twins developed from split embryos. (Courtesy, Colorado State University, Fort Collins, CO)

Right: Pasture/range. (American Hereford Assn., Kansas City, MO)

Below: Cattle feedlot. (Courtesy, Union Pacific R.R.)

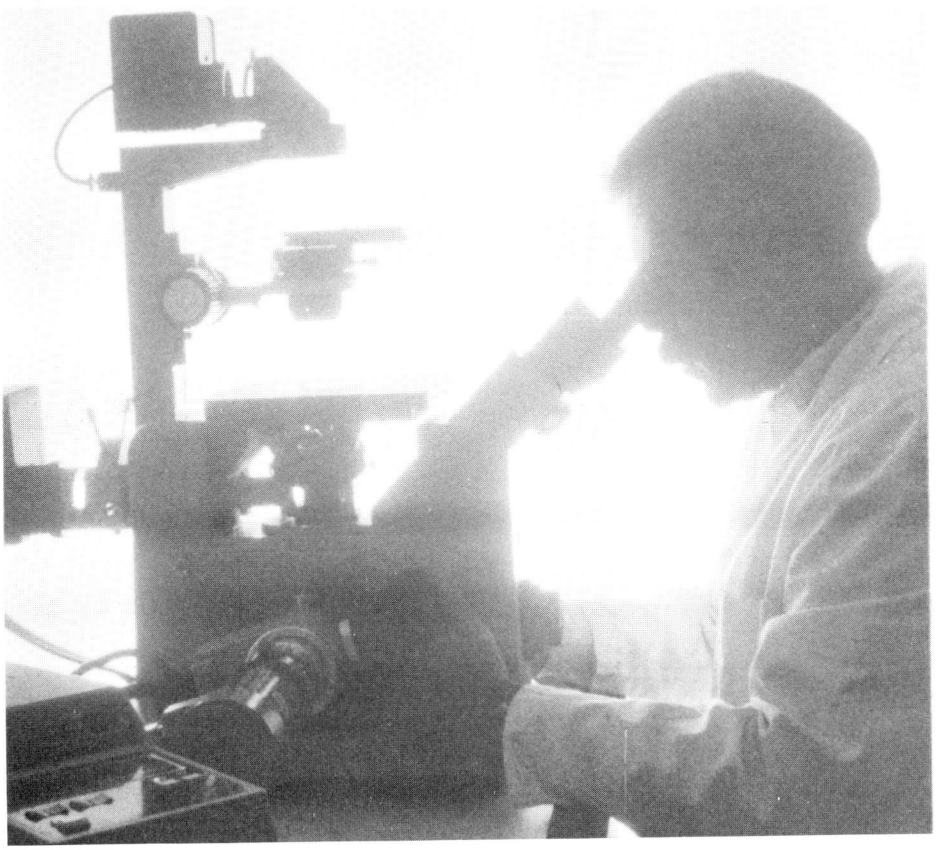

Looking into your future! (Courtesy, University of Idaho, Moscow)

CAREERS

Contents *Page*

Fig. C–1. Campus scene. (Courtesy, University of Idaho, Moscow)

Ask a dozen people how they chose their current occupations and you will get a dozen different answers. Some followed in their parents' footsteps. Others drifted there. Still others were pushed. A few discovered occupations of interest to them. Some had to take what was available. Everyone ended up somewhere!

But, do you have any control over where you end up? Can you really choose your own road, follow it, and end up where it leads? Well, maybe not entirely, because many unexpected things do happen. Still, most people have considerable control over their future. If they really want something badly enough, train for it, and work for it, the chances are that they will succeed in getting to where they want to go. Such a choice and pursuit is known as a career.

Career is defined as a course of continued progress in the work life of a person. Webster adds: *It is a profession for which one undergoes special training and which is undertaken as a permanent calling.*

WHY GO TO SCHOOL?

Practical young people graduate from high school and college for four primary reasons:

1. Because it takes trained people to operate high-tech; which is already here in the form of such things as computer-

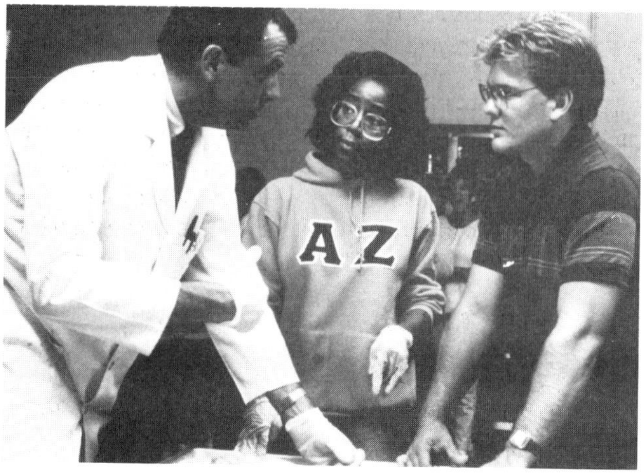

Fig. C–2. Professor training students to operate high-tech. (Courtesy, North Carolina State University-Raleigh)

controlled growth environments, robots that harvest crops, gene transfers that make plants and animals resistant to disease, and laser-guided machinery.

2. Because in many cases a diploma is requisite to getting a position. There are fewer unskilled agricultural jobs; more positions for managers, technicians, and scientists.

3. Because it increases earning power. As a result, they can enjoy a higher standard of living and more of the finer things of life. Indeed, a college education is a big key to financial success (see Fig. C–3). But remember that a job is more than a paycheck! It reflects who you are, your success in life, and your position in the community. It also affects your well-being. Remember, too, that when work is play and play is work, the job changes from an 8-hour per day *task* to an *exciting career.*

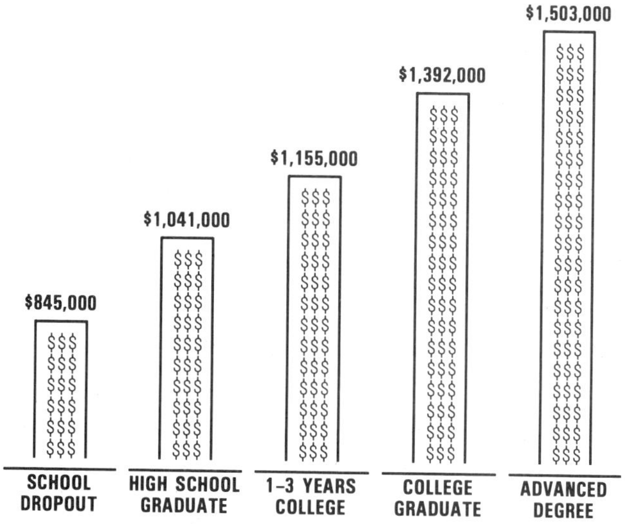

Fig. C–3. In a lifetime, and on the average, this shows what you may expect to earn. Prepared by the author from the most current data available: *Lifetime Earning Estimates for Men and Women in the United States*, U.S. Bureau of the Census (*Current Population Reports*, Series P–60, Number 139, page 3). The data are for 18-year-old, year-round, full-time, male workers, for 1978–80 and are expressed in 1981 dollars. In future years, lifetime earnings will be affected by (1) supply of, and demand for, the educated labor force, and (2) inflation/value of the dollar. In the decades to come, it is expected that lifetime earnings of each group will be higher, and that the gap between the least and most educated will grow wider.

4. Because most people like to do some good in the world. Certainly everyone wants to make more money; and there's no doubt that an education is most helpful from this standpoint. But an education also enhances those things of the spirit and intellect that do not wear dollar marks—the pure enjoyment from living and the enlarged contribution to society. Generally speaking, those who command respect as leaders, and whose lives are fullest and most productive, are college graduates.

WHY AGRICULTURE?

Agriculture is the nation's largest industry. The following facts and figures point up the enormity and importance of American agriculture:

1. Total farm assets exceed $800 billion, which is equal to about 40% of the total capital assets of all manufacturing corporations in the United States.
2. More than 21 million people are employed in some phase of agriculture; they are either producing, processing, or marketing farm commodities, or they are supplying goods used on the farm. About one job out of every five in the U.S. is farm/food related.
3. A hundred years ago, half of the nation's population was required on farms to feed themselves and the other half. In 1820, each farm worker supplied farm products for 4 people (self and 3 others). By 1940, one farm worker produced enough for 11 people (self and 10 others). Today one farm worker supplies enough agricultural products for 95 people (self and 94 others), more than 8 times the 1940 productivity—thereby freeing 94 people to produce such luxuries as autos, refrigerators, TV sets, and a host of other goods and services for modern American living.
4. In recent years, the production from about one-fourth of U.S. cropland has moved into export markets. So, farm exports make a most important contribution to the U.S. balance of trade.
5. Farmers are big buyers of goods produced by other industries. Annually, they normally spend over $6 billion for tractors, trucks, cars, and machinery. Each year, they spend another $9.5 billion to fuel, lubricate, and maintain this fleet. Additionally, farmers make annual expenditures of about $19 billion for feed and seed and nearly $6 billion for fertilizers and lime.

ABOUT THE WORLD OF WORK

The average person will work more than 2,000 hours each year throughout a lifetime. Moreover, most people will work from the time they are age 20 until they are 60 to 65 years old. This is a very long time. So, it is very important that young people familiarize themselves with the world of work before entering into it.

WHAT IS WORK?

Someone once said that the world is full of two kinds of people: "Those willing to work; and those willing to let them work." This makes work sound like a dirty word. It is not. Everything that the human race has achieved has been accomplished through work. Civilization depends upon it. Survival requires it. But work should not be drudgery!

Isn't it odd that so few people seem to *play* for a living? Most folks work for a living, but play for fun. When high school students join the marching band, we think of them as playing. If some of them form a band and play for dances and get paid for it, the *playing* becomes work. Even though they are working, they may enjoy the work, which is as it should be.

With the proper choice of a career, along with the proper attitude, *work is fun, and fun is work.*

When choosing a career, the following two steps are suggested:

1. Self-analysis, or self-appraisal.
2. Evaluating occupations.

SELF-ANALYSIS, OR SELF-APPRAISAL

Self-analysis, or self-appraisal, is an inventory of your personal qualities for the world of work.

Right off, you should recognize that you are a unique and distinct individual. There is only one person in the world like you. They threw the mold away after making you. You differ from your classmates in ability, talent, interest, attitude, and ambition.

You possess individual strengths and weaknesses. You do some things well and easily *(ability)*; you do not do other things well—and they are a chore *(limitations)*. At the moment, you are what you are. But, fortunately, interests can change and skills can be improved. So, study yourself; know your strengths and weaknesses. Choose and build your career around your strengths, interests, and abilities. Strive to overcome your weaknesses, and plan a career in which your weaknesses will be minimized. For success, your occupation should be satisfying; the job should fit you as an individual and make the maximum use of your talents.

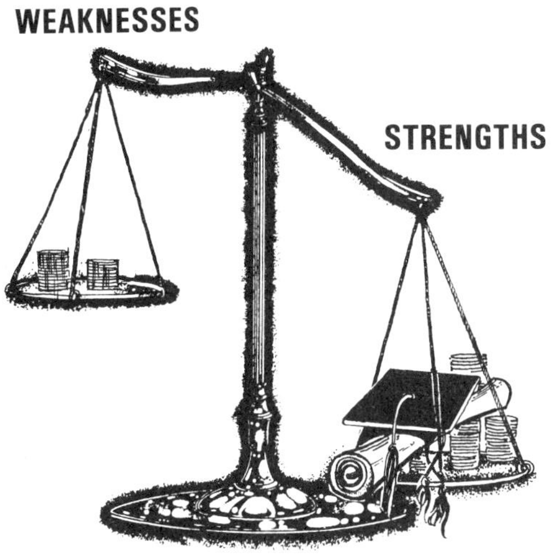

Fig. C–4. Weigh your strengths and weaknesses.

Analyze yourself realistically as to what you really are. You know yourself better than anyone else. Interested persons—parents, relatives, friends, teachers, and guidance counselors—can give valuable assistance and guidance, but they cannot decide for you. *You must make your own decisions.* Thus, a self-analysis or self-appraisal is very important when planning a career.

The following 10 personal qualities should be considered in your self-analysis:

1. Ability.
2. Talents.
3. Interest.
4. Physical makeup.
5. Attitude and values.
6. Self-concept.
7. Training and experience.
8. Educational aspirations.
9. How others see you.
10. Willingness to change.

Appraising yourself for a career is complex, but it can be done—and it is very necessary. The task can be simplified by a systematic approach.

Ask yourself the following straightforward questions, and give your honest answers:

1. What are my abilities?
2. What special talents do I possess?
3. What are my special interests?
4. What is my physical makeup, and what are my handicaps?
5. What are my attitudes and values?
6. How do I see myself; what is my self-concept?
7. What is my training and previous experience?
8. What are my educational plans for the future?
9. How do others see me; what kind of personality do I have?
10. Am I willing to change?

If you make a good occupational choice, it won't be work. Rather, you will be one of those fortunate persons who find deep satisfaction in earning your daily bread.

Remember, too, that there is a right person for every job, and that every job is as honorable as you make it. So, you should seek the right occupation for you—a job that you are good at, and that you will enjoy doing all your life.

For success now and in the decades to come, employees must have the most skills ever. Young people need to be well prepared for the *Biotechnological Age* which we are now entering.

EVALUATING OCCUPATIONS

Up front, and for a successful career, you should evaluate the job on the following basis:

Not what the job can do for me,
but what I can do for the job.

Note well: Such things as starting salary, raises, hours, and fringe benefis should never be lead questions.

If you contribute richly to the job, it will contribute richly to you. Also, remember that the greatest fringe benefit of all is the kind of people with whom you will be associated, and the opportunity that the job provides for your professional growth and development.

With the right occupational choice, and with superior job performance on your part, you may expect the following benefits:

1. **Livelihood.** People work to provide themselves and their families with the essentials of life—food, clothing, and shelter. If you do a good job, most "bosses" will happily accord merited salary increases; so, it is best that you not ask for a raise. *Your job value should determine your wage.*

2. **Human relationship.** The opportunity to be associated, and work, with top people cannot be evaluated in dollars—it is priceless.

3. **Personal development.** Your job should provide an opportunity to learn and grow intellectually and professionally. It should provide an opportunity to develop new skills and learn new things. Your work should give you an opportunity to grow and reach your highest potential.

4. **Service.** Your job should provide an opportunity for service; an opportunity for you to serve humankind and do some good in the world.

5. **Security.** Your job will be as good as you make it. It follows, that you can make your own security.

6. **Success.** Employers are eager to employ persons who like a challenge, who will work hard, and who will succeed.

CAREERS IN THE WORLD OF AGRICULTURE

Career opportunities abound in agriculture; and there will always be an agriculture. The U.S. Department of Agriculture reports that more than 48,000 jobs are being created annually for college graduates with expertise in agriculture and related industries. Current projections indicate that, to the year 2000 and beyond, the annual demand for university graduates with training in agriculture will exceed the suppy by 12 to 15%. The livestock industry and related fields are major components of the U.S. economy; hence, career opportunities in the world of agriculture are excellent for both men and women trained in animal science.

Rather than attempt the impossible task of listing all the different kinds of jobs available in agriculture, the author has categorized them to correspond to the sections in this book.

Stockman's Handbook Digest covers the whole gamut of agriculture; from genes to genetic engineering, from pastures to patented animals, from parasites to pollution. It also portrays the world of work of the subjects covered, section by section. Logically, this section (chapter) on *Careers* follows the same order as the subjects listed in the contents of the book; so, it is not necessary to repeat the narrative in the careers section. Instead, the careers presentation which follows is a pictorial portrayal; and the reader is referred to the corresponding narraitve section for details. For example, the first careers presentation portrays "Careers in Animal Behavior and Environment"; then, Section 1, pp. 1 to 26, tells about animal behavior and environment in narrative form. The second careers presentation portrays "Careers in Business Aspects"; and Section 2, pp. 27 to 66, tells about business aspects in narrative form. So, it goes from section to section, from beginning to end of the book.

CAREERS IN ANIMAL BEHAVIOR AND ENVIRONMENT

(Also see Section 1)

Fig. C-5. Horse behavior. (Courtesy, USDA)

Fig. C-6. Swine behavior. (Courtesy, College of Agriculture, Kansas State University, Manhattan)

Fig. C-7. People need pets and pets need people. (Courtesy, University of New Hampshire, Penacook)

Fig. C-8. Environment/weather stress. (Courtesy, USDA)

Fig. C-9. Sustainable agriculture—contour strip cropping. (Courtesy, USDA, Soil Conservation Service)

Fig. C-10. Manure, the polluting barnyard centerpiece of olden times. Note chickens on the manure pile. (Courtesy, USDA)

CAREERS IN BUSINESS ASPECTS

(Also see Section 2)

Fig. C–11. Student using computer technology at Michigan State University to determine the value of feeds. (Courtesy, Michigan State University, East Lansing)

Fig. C–12. Students in animal science computer lab at Kansas State University. (Courtesy, College of Agriculture, Kansas State University, Manhattan)

Fig. C–13. Trained personnel is needed to operate a totally computerized feed facility like this unit. (Courtesy, Farr Feeders, Inc., Greeley, CO)

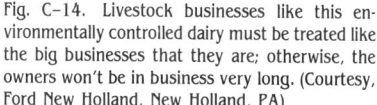

Fig. C–14. Livestock businesses like this environmentally controlled dairy must be treated like the big businesses that they are; otherwise, the owners won't be in business very long. (Courtesy, Ford New Holland, New Holland, PA)

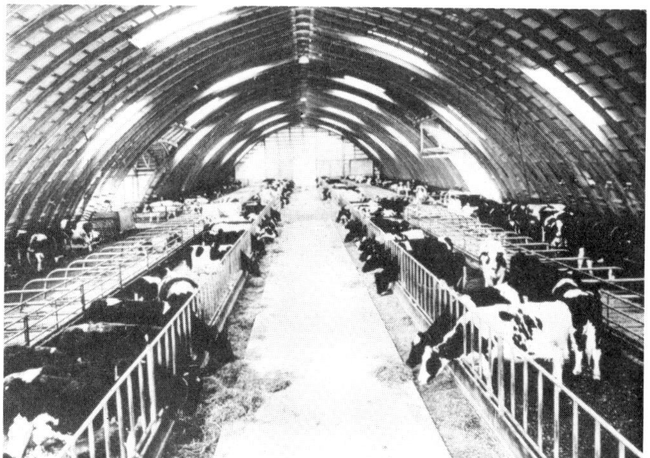

CAREERS IN ANIMAL BREEDING

(Also see Section 3)

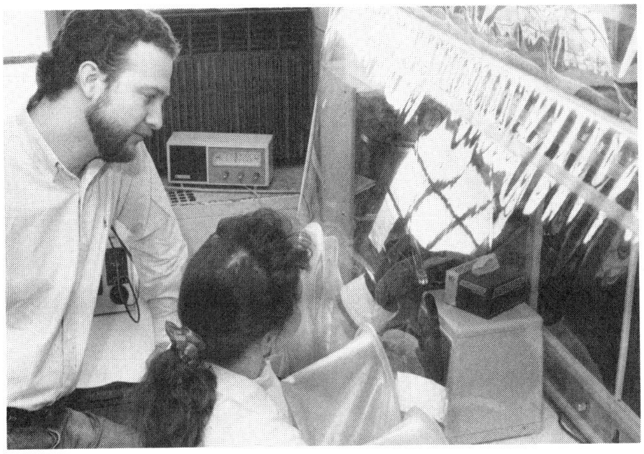

Fig. C–15. Dr. Bryan White is shown overseeing a student who is performing a gene transfer in an anaerobic chamber. (Courtesy, University of Illinois, Urbana)

Fig. C–16. Students pregnancy testing a ewe by ultrasonic scanning. (Courtesy, North Carolina State University, Raleigh)

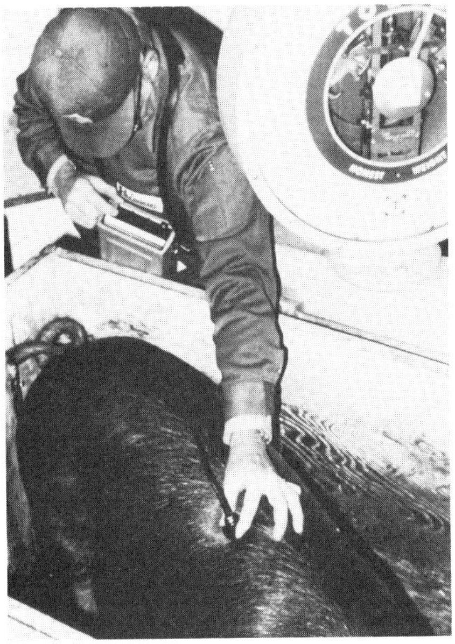

Fig. C–17. Ultrasonic measurement of backfat on a hog. (Courtesy, Farmland Industries, Inc., Kansas City, MO)

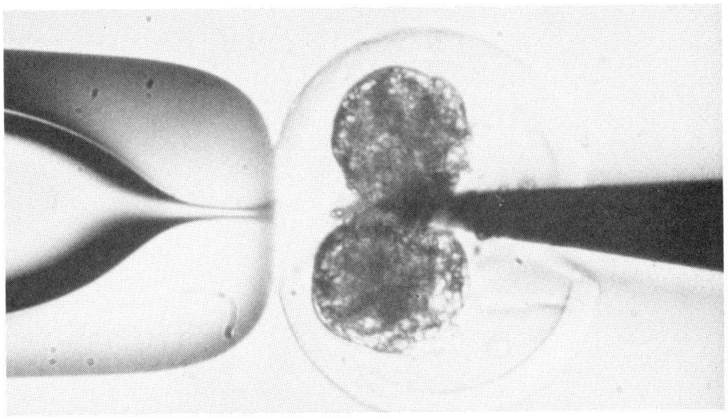

Fig. C–18. Microsurgical bisection of a 7-day bovine embryo to produce identical twins. (Courtesy, Dr. George Seidel, Colorado State University, Fort Collins)

Fig. C–19. Cloned Brangus bulls, genetically identical, produced by the same embryo. (Courtesy, Granada Land & Cattle Co., Wheelock, TX)

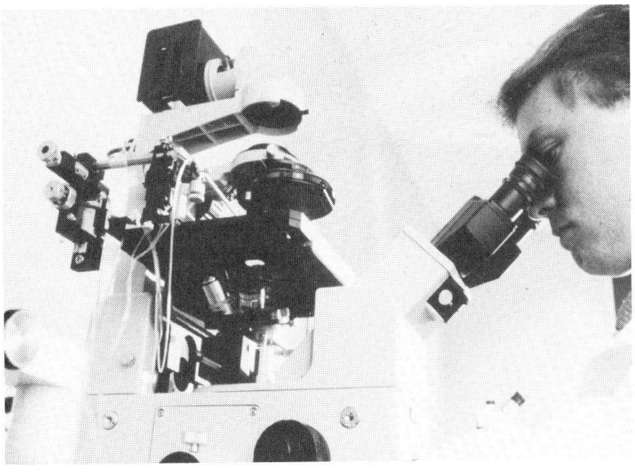

Fig. C–20. Granada is involved in nuclear transfer, recombinant animal hormones, embryo sexing, and genetic engineering. (Courtesy, Granada Land & Cattle Co., Wheelock, TX)

CAREERS IN FEEDS AND FEEDING

(Also see Section 4)

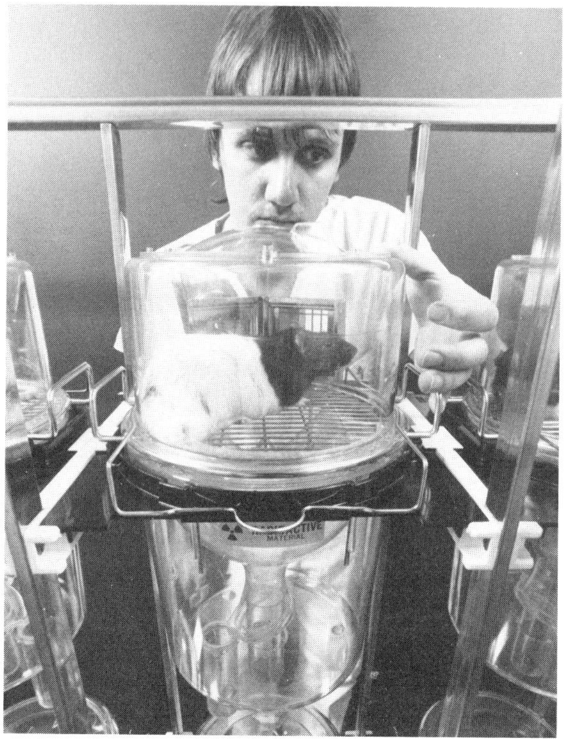

Fig. C-21. A USDA researcher is shown examining a rat in a metabolic chamber; tests conducted in search of copper deficiencies, suspected to be a leading cause of heart disease. (Courtesy, USDA)

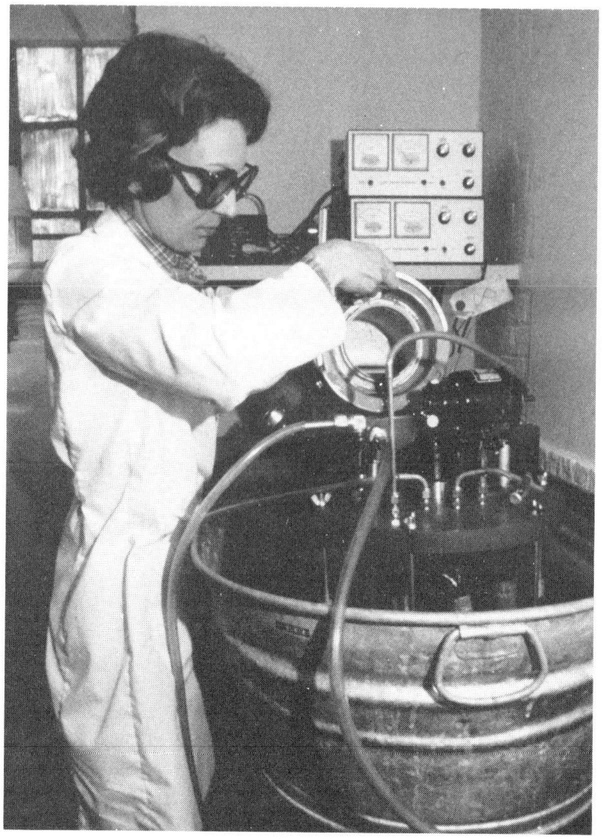

Fig. C-22. Artificial rumen used to study basic rumen fermentation. (Courtesy, University of Georgia, Athens)

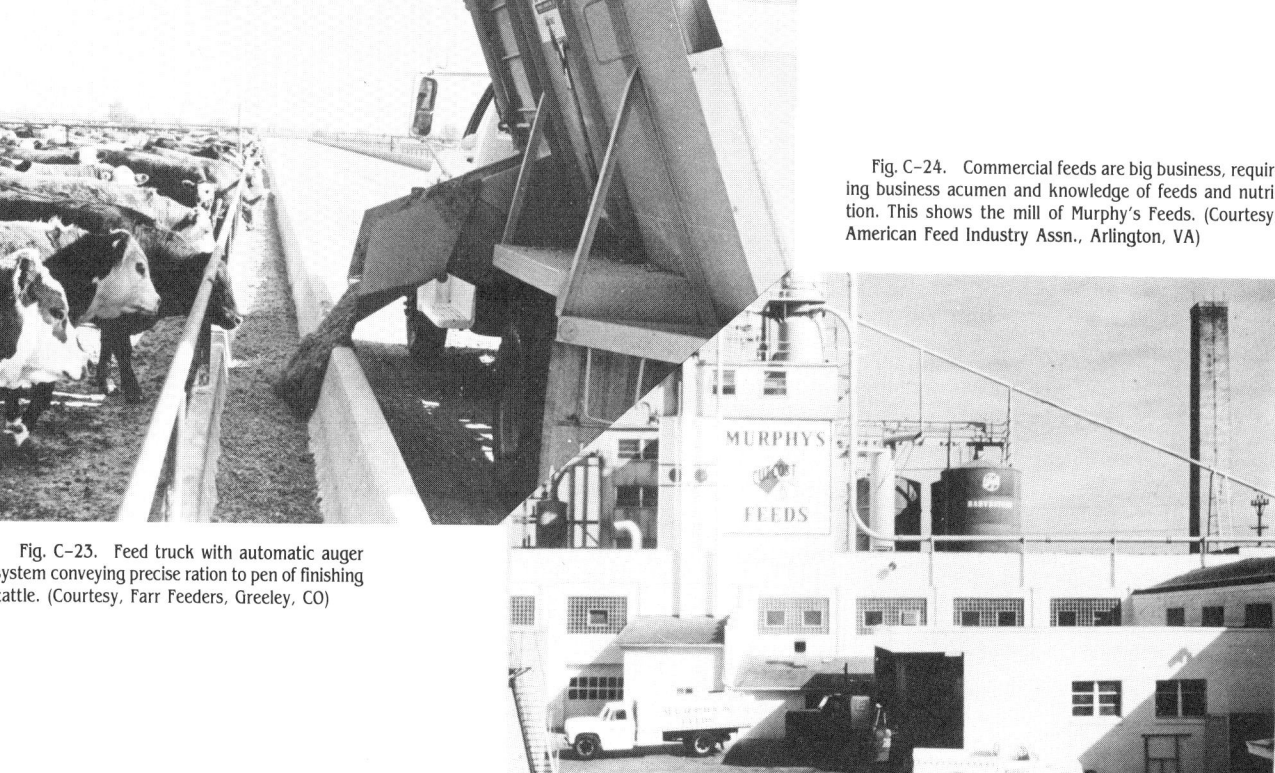

Fig. C-23. Feed truck with automatic auger system conveying precise ration to pen of finishing cattle. (Courtesy, Farr Feeders, Greeley, CO)

Fig. C-24. Commercial feeds are big business, requiring business acumen and knowledge of feeds and nutrition. This shows the mill of Murphy's Feeds. (Courtesy, American Feed Industry Assn., Arlington, VA)

CAREERS IN PASTURES AND RANGE FORAGES
(Also see Section 5)

Fig. C–25. Two ranchers checking seedfill. Seed harvest is a potential bonus crop on some rangelands. (Courtesy, H. Dietz, USDA-SCS, Fort Worth, TX)

Fig. C–26. Rotating cattle from one dryland pasture to another, on Lelos Hubbard Ranch, Caribou, SD. (Courtesy, USDA-SCS)

Fig. C–27. Holstein cows on pasture. (Courtesy, National Milk Producers Federation, Arlington, VA)

Fig. C–28. On the range! Sheep are unsurpassed in converting what would otherwise be wasteland into meat and fiber. (Courtesy, USDA)

Fig. C–29. Recreation on the range. Trail rides are popular and a source of income for ranchers.

Fig. C–30. Elk on the range, in winter snow.

CAREERS IN MAKING HAY AND UTILIZING CROP RESIDUES

(Also see Section 6)

Fig. C-31. Del Thomas, Germain's field representative (left), and Richard Rasmussen evaluate leafiness and fine stems of alfalfa hay, which produced 10 tons per acre during the season. (Courtesy, Germain Seed Co.)

Fig. C-32. Making round bales of hay. (Courtesy, Ford New Holland, New Holland, PA)

Fig. C-33. Come snow, sleet, or rain, cattle must eat. This may require feeding hay on the snow, as shown here. (Courtesy, USDA)

Fig. C-34. Cornstalk residue being harvested for feed. (Courtesy, Fox Equipment Co.)

CAREERS IN SILAGE/HAYLAGE/HIGH-MOISTURE GRAIN

(Also see Section 7)

Fig. C-35. Upright silos on a dairy farm. (Courtesy, Harvestore Systems, A. O. Smith Harvestore Products, DeKalb, IL)

Fig. C-36. Field chopping haylage into a covered forage wagon. (Courtesy, Ford New Holland, New Holland, PA)

Fig. C-37. Dairy cows feeding on grass silage. (Courtesy, The American Jersey Cattle Club, Reynoldsburg, OH)

Fig. C-38. High-moisture corn being removed (for feeding) from a 60,000 ton concrete-lined trench storage pit. (Courtesy, Farr Feeders, Greeley, CO)

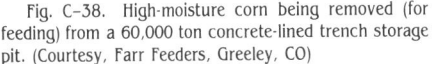

CAREERS IN ANIMAL MANAGEMENT

(Also see Section 8)

Fig. C–39. Professor and student discussing management of the lactating herd. (Courtesy, Pennsylvania State University, University Park)

Fig. C–40. Students discussing the ear notching system used on this pig. (Courtesy, North Carolina State University, Raleigh)

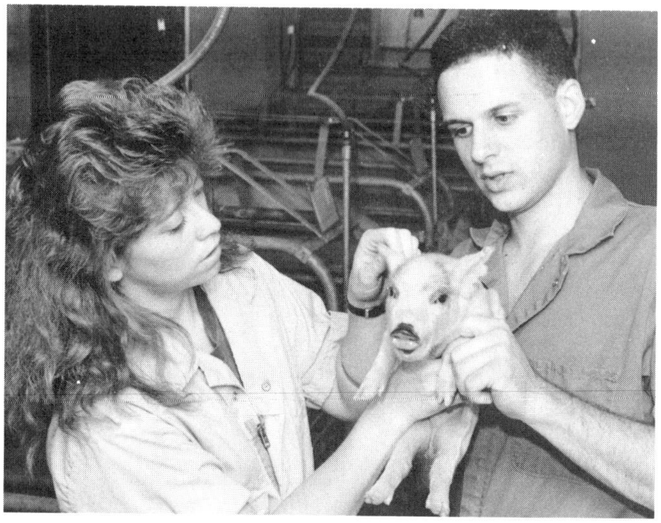

Fig. C–41. The cattle roundup; branding, dehorning, and castrating. (Courtesy, Iowa Beef Processors, Dakota City, NE)

Fig. C–42. Manure management—fertilizing the fields. (Courtesy, Gehl Company, West Bend, IN)

CAREERS IN BUILDINGS AND EQUIPMENT

(Also see Section 9)

Fig. C-43. Students receiving instruction in designing farm machinery. Such instruction is very important in today's highly mechanized agriculture. (Courtesy, Purdue University, West Lafayette, IN)

Fig. C-44. Three John Deere machines harvesting wheat in eastern Washington. (Courtesy, H. Fisher, Oakesdale, WA)

Fig. C-45. An attractive farm headquarters. (Courtesy, Photomaps, Belleville, KS)

Fig. C-46. Hayloader tossing a bale into the wagon behind. (Courtesy, Ford New Holland, New Holland, PA)

Fig. C-47. Environmentally controlled dairy barn, with free stall arrangement. (Courtesy, Babson Bros., Oak Brook, IL)

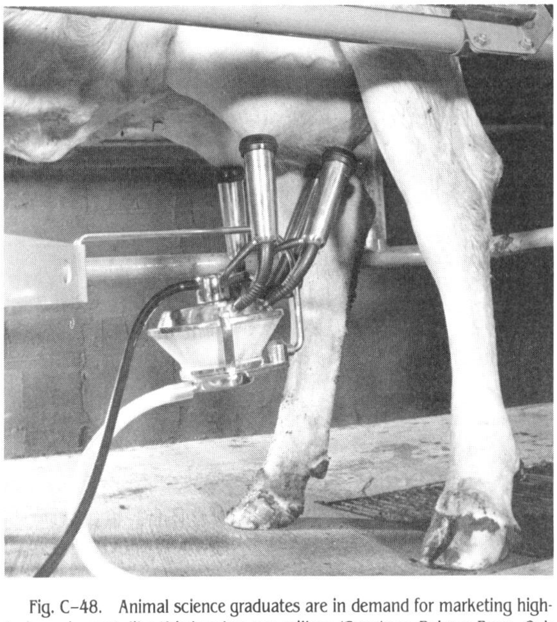

Fig. C-48. Animal science graduates are in demand for marketing high-tech equipment, like this breaker cup milker. (Courtesy, Babson Bros., Oak Brook, IL)

CAREERS IN ANIMAL HEALTH/DISEASE PREVENTION/PARASITE CONTROL

(Also see Section 10)

Fig. C–49. Students at Cal Poly learning by doing. (Courtesy, California Polytechnic State University, San Luis Obispo, CA)

Fig. C–50. Students at North Carolina State University treating an animal. (Courtesy, North Carolina State University, Raleigh)

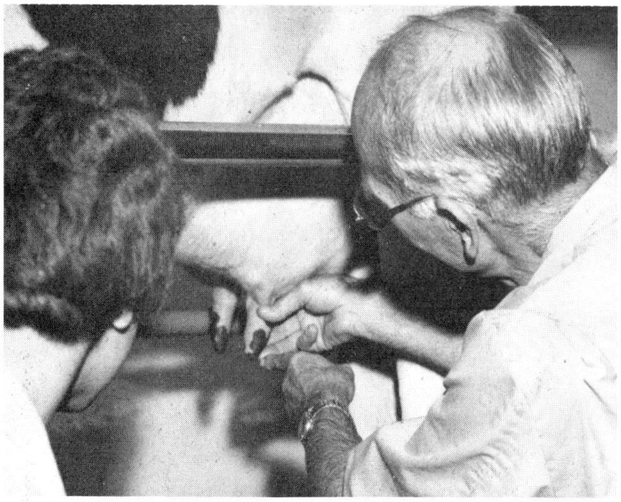

Fig. C–51. Dr. John K. Winkler, DVM, School of Veterinary Medicine, Auburn University, Auburn, AL, shown discussing an udder problem with students. (Courtesy, Dr. Winkler)

Fig. C–52. "Rosette" formation of dairy cattle, a bovine behavioral trait exemplified when seeking protection from certain insects. (Courtesy, Livestock Insect Laboratory, Beltsville, MD)

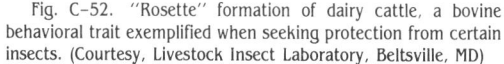

CAREERS IN SELECTING, FITTING, SHOWING, AND JUDGING LIVESTOCK

(Also see Sections 11 and 12)

Fig. C-53. Students in the beginning animal science class at the University of Wyoming must fit and show an animal. These students are fulfilling this requirement by fitting and showing beef cattle. (Courtesy, University of Wyoming, Laramie)

Fig. C-54. FFA students engaged in a swine judging contest sponsored by the University of Missouri-Columbia. (Courtesy, University of Missouri-Columbia)

Fig. C-55. Brahman heifers being judged at the San Antonio Livestock Exposition. (Courtesy, San Antonio Livestock Exposition, San Antonio, TX)

Fig. C-56. Junior Jersey show in the All-American show at the North American International Livestock Exposition, Louisville, KY. (Courtesy, The American Jersey Cattle Club, Reynoldsburg, OH)

Fig. C-57. Beginning animal science students at the University of Wyoming must fit and show an animal. These students are fulfilling this requirement by fitting and showing sheep. (Courtesy, University of Wyoming, Laramie)

Fig. C-58. Draft horse class at the National Western Stock Show & Rodeo. (Courtesy, National Western Stock Show & Rodeo, Denver, CO)

CAREERS IN MARKETING LIVESTOCK AND MILK

(Also see Section 13)

Fig. C-59. Successful marketers know their products. This shows Dr. F. McKeith, University of Illinois, telling about meat cuts. (Courtesy, National Live Stock and Meat Board, Chicago, IL)

Fig. C-60. Trucking livestock to market. (Courtesy, Howard Barnes Livestock Trucking, Fayetteville, AR)

Fig. C-61. Hectic futures trading action at the Chicago Mercantile Exchange. (Courtesy, Chicago Mercantile Exchange, Chicago, IL)

Fig. C-62. Professor explaining lactation to students. (Courtesy, North Carolina State University, Raleigh)

Fig. C-63. Bulk milk truck transporting milk from dairy farm to processing plant. (Courtesy, USDA)

Fig. C-64. Milk in a supermarket. (Photo by A. H. Ensminger)

CAREERS IN MEAT AND MILK

(Also see Section 14)

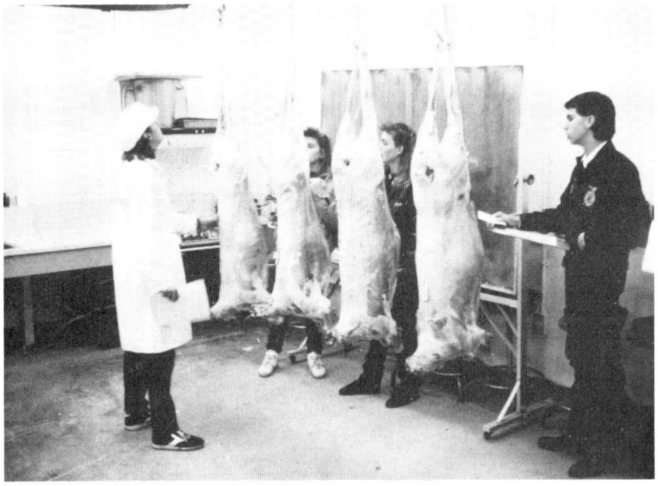

Fig. C-65. FFA students placing a class of lamb carcasses at the annual Agricultural Sciences Field Day contest, sponsored by the University of California-Davis. (Courtesy, University of California-Davis)

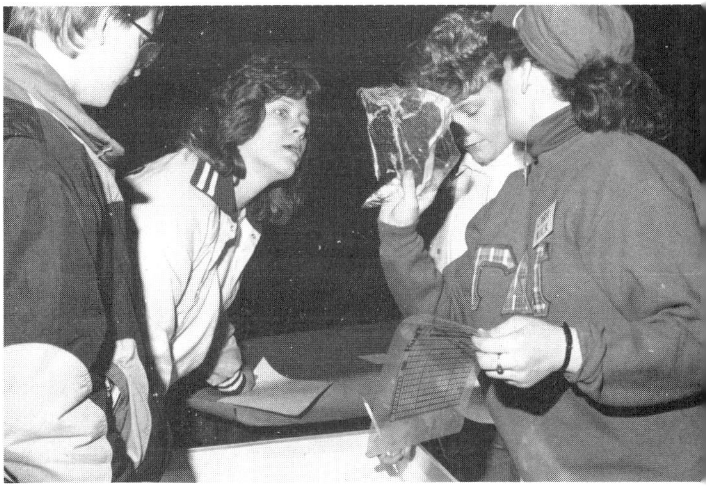

Fig. C-66. College meats judging team at work at Kansas State University. (Courtesy, Kansas State University, Manhattan)

Fig. C-67. Meat on the table—a pork loin; the end result of production, marketing, slaughtering, processing, and distribution. (Courtesy, National Live Stock and Meat Board, Chicago, IL)

Fig. C-68. *Right:* A trio of cheeses—Parmesan, Provolone, and Cheddar. (Courtesy, United Dairy Industry Assn., Rosemont, IL)

Fig. C-69. Ice cream—a favorite forever. (Courtesy, American Dairy Assn., Rosemont, IL)

Fig. C-70. Milk on the table; the end result of production, marketing, processing, and distribution. (Courtesy, USDA)

CAREERS IN WOOL AND MOHAIR

(Also see Section 15)

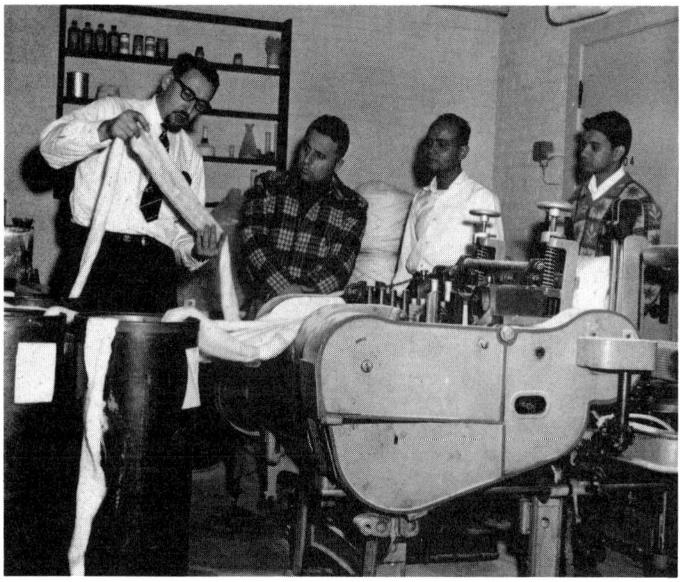

Fig. C–71. Wool scoring laboratory at the University of Wyoming. (Courtesy, University of Wyoming, Laramie)

Fig. C–72. Made of wool. Lady's double-knit pant suit, with sweater; man's sweater. (Courtesy, The Wool Bureau, New York, NY)

Fig. C–73. Chair upholstered with mohair fabric. (Courtesy, Mohair Council of America, San Angelo, TX)

Fig. C–74. Men's jacket made of mohair. (Courtesy, Mohair Council of America, San Angelo, TX)

CAREERS IN WHERE TO GO FOR HELP (INFORMATION)

(Also see Section 18)

Fig. C-75. Two 4-H Club members learn about using a microscope from Dr. M. Smith, University of Missouri. (Courtesy, Dr. M. Smith, University of Missouri, Columbia)

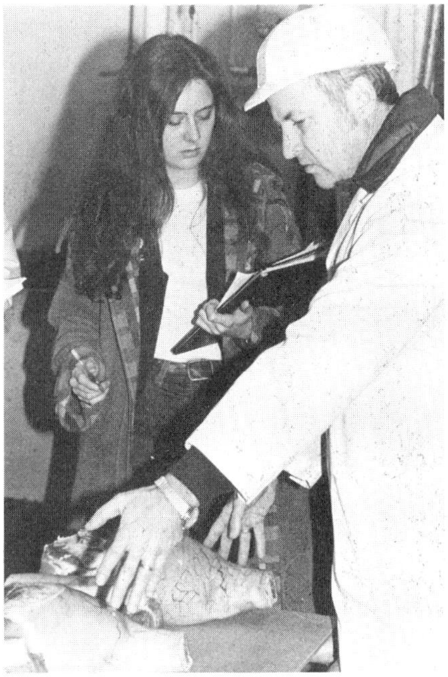

Fig. C-76. *Right:* Dr. W. Moody, University of Kentucky, giving a demonstration of different types of fresh hams. (Courtesy, National Live Stock and Meat Board, Chicago, IL)

Fig. C-77. An agricultural extension service worker giving instruction on various beef cookery methods. (Courtesy, National Live Stock and Meat Board, Chicago, IL)

Fig. C-78. A college-sponsored adult education meeting for farmers. (Courtesy, American Society of Animal Science, Champaign-Urbana, IL)

Fig. C-79. Mary Adolf, National Live Stock and Meat Board, giving a cooking demonstration before a food service audience. (Courtesy, National Live Stock and Meat Board, Chicago, IL)

Fig. C-80. Dr. W. T. (Dub) Berry, Agricultural Consultant, The Helming Group, Overland Park, KS.

CAREERS FOR THE 21st CENTURY—AND BEYOND

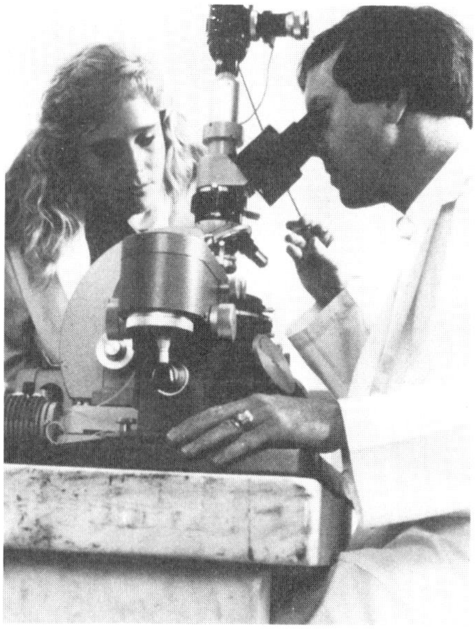

Fig. C–81. *Left:* Professor and student searching the future—your future. (Courtesy, North Carolina State University, Raleigh)

Fig. C–82. *Right:* Preparing for the future—your future. (Courtesy, Pennsylvania State University, University Park, PA)

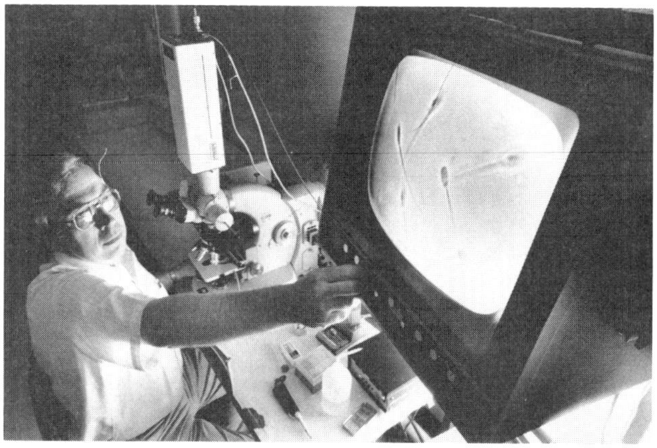

Fig. C–83. Sex control will come! This shows Dr. L. Johnson, USDA Animal Pathologist, doing sex control research. When sex control is perfected, farmers will be able to select the sex of their animals. For example, a dairy producer will be able to produce all heifer calves for replacements and eliminate unwanted bull calves. (Courtesy, USDA)

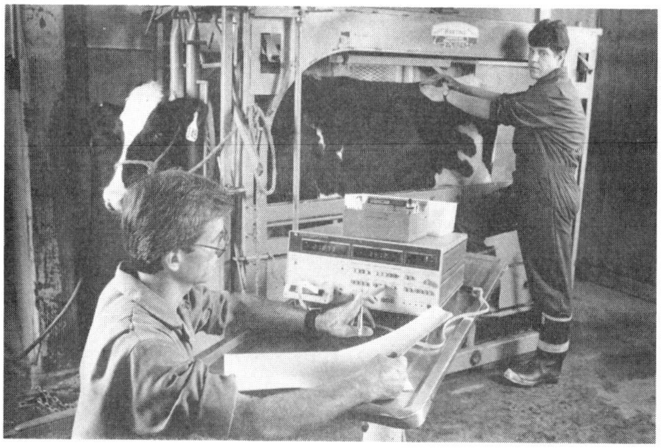

Fig. C–84. This shows two USDA biologists using an electronic sensor to test when this cow will give birth. Electronic sensing technology is now being perfected to measure animal weight, carcass composition, body temperature, fertility, animal identification, and detection and diagnosis of disease. (Courtesy, USDA)

Fig. C–85. Careers will be global! This shows Dr. D. W. Andrews, California Polytechnic State University, with a group of students from Africa. (Courtesy, California Polytechnic State University, San Luis Obispo)

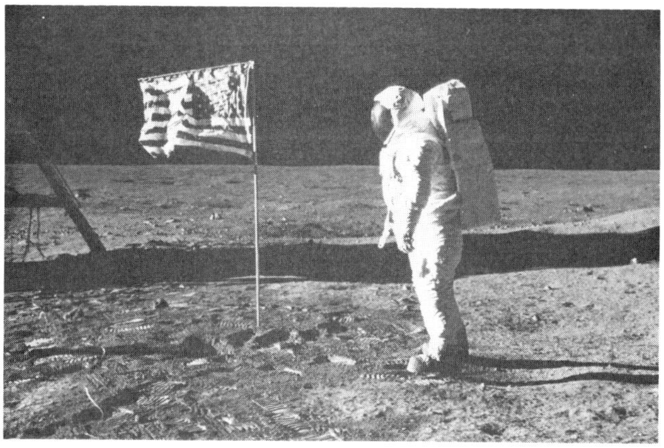

Fig. C–86. The sky is without limit for careers in animal science in the 21st century—and beyond, for those who are there first with the most of the best. This shows Astronaut E. E. Aldrin, Jr. on the moon on July 20, 1969. (Courtesy, Manned Spacecraft Center, Houston, TX)

SUCCESS STORIES ABOUND; AMONG THEM, THE FOLLOWING:

Fig. C–87. W. H. "Bill" Stuart, Jr., a graduate of Washington State University, Pullman, with a B.S. degree in Animal Science followed by graduating with an MBA degree from the Wharton School of Finance and Commerce, University of Pennsylvania, now a successful owner of several animal related businesses headquartered at Bartow, FL.

Fig. C–88. Bill Bennett, Jr., who studied Animal Science at Washington State University, Pullman, presently a leading Hereford breeder of America.

Fig. C–89. *Left:* D. E. McGlothlin, Manager, Horse Division, a graduate of Colorado State University, Fort Collins, with B.S. and M.S. degrees. *Right:* J. C. Harris, owner, Harris Farms, a graduate of the University of California-Davis. *Center:* The Thoroughbred mare, *Mistress Gay*. Harris Farms, Coalinga, CA, is one of the most diverse, unique, and successful farms in America. It consists of 20,000 acres of crops, 400 Thoroughbred breeding and racing horses, a 100,000 cattle feedlot, a modern meat packing plant, a large restaurant,a country store, and a hotel.

Fig. C–90. *Below:* Dr. N. Brown received his B.S. degree in agricultural education, his master's degree in educational administration, and his doctorate in vocational education. Dr. Brown is president and programming officer of the W. K. Kellogg Foundation, Battle Creek, MI. (Courtesy, Michigan State University, East Lansing)

NEW FRONTIERS IN THE 21st CENTURY

The past is prologue! The future is now! Genetic wizardry by gene splicing is giving rise to a major scientific revolution called biotechnology and spawning many new developments exceeding our fondest dreams. Biotechnology will involve every facet of animal production from breeding and feeding to the finished product, including the genetic makeup of animals and the feeds they eat; the digestion, physiology, stress tolerance, disease resistance, and efficiency of production of animals; the composition, quality, and quantity of products produced; along with the production of large quantities of drugs and chemicals. While some aspects of biotechnology are decades away from commercial production, others are near, and still others are here now.

But advanced technology calls for advanced animal adaptation, welfare, and environmental control. We need to breed and select animals adapted to an artificially-made environment—animals that not only survive, but thrive, under the conditions in which they are kept. We need to heed the warnings of endangered animals, endangered people, and an endangered planet—presaged by increased pollution, the greenhouse effect, acid rain, depletion of the ozone layer, and destruction of rain forests.

Biotechnology will be the key to unlocking vast improvement per animal, per feed unit, and per dollar investment. The speed with which innovations are developed will be limited only by people.

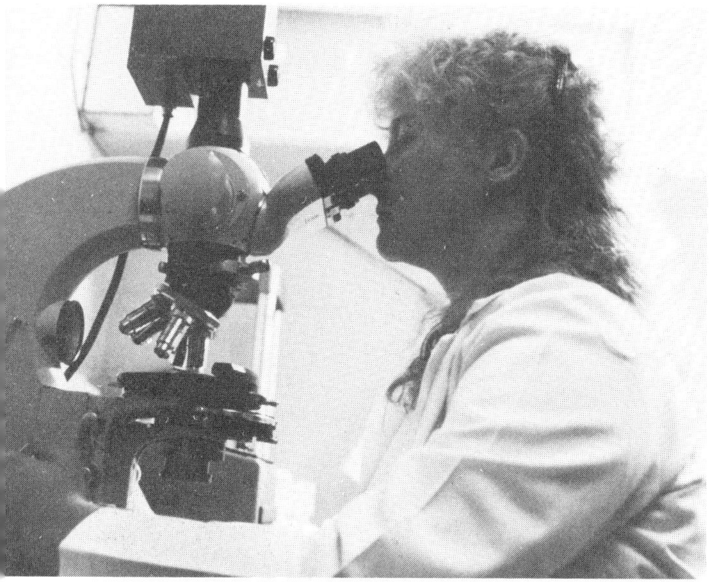

Fig. C-91. This shows a student at Pennsylvania State University searching for new frontiers. (Courtesy, The Pennsylvania State University, University Park)

WORLD WITHOUT END—WITH ANIMALS

Global food demand is rising significantly. World population will grow from over 5 billion now to more than 6 billion soon after the year 2000. Thus, early in the 21st century, there will be another billion mouths to feed.

Fig. C-92. FFA students placing a class of Suffolk ewes at the annual Agricultural Sciences Field Day contest, sponsored by the University of California-Davis. (Courtesy, University of California-Davis)

World food consumption is determined by population and the amount eaten per person. It is expected to double over the next three decades, led by greater per capita consumption linked to rising incomes, changing tastes, and improved food supplies in the developing countries.

Practicality dictates that in the years ahead a hungry world will meet its increased food needs through having plants and animals play complementary roles—with animal products complementing the deficiencies of plant products. Grazing land is highly efficient in the capture of solar energy—requiring little energy for a high return. However, grass does not store the energy in a form available to humans. It follows that ruminants which can utilize grass not suitable for human consumption and convert it into meat and milk are essential.

For a world without end, the developing countries need a massive infusion of research, technology, and education, with emphasis on plant-animal relationships. Other approaches serve only to prolong and aggravate the current disparities.

FOR MORE INFORMATION, CONTACT:

1. The Animal Science Department at the Land Grant or State University.
2. The American Society of Animal Science, 309 West Clark St., Champaign, IL 61820.
3. The local County Extension Agent or FFA Instructor.
4. The career offices in local colleges.
5. The state employment service.
6. The public libraries. Among other references, see (a) *Dictionary of Occupational Titles*, U.S. Government Printing Office, 1977, reprinted 1988; (b) *Employment Opportunities for College Graduates in the Food and Agricultural Sciences*, 1986 by K. Jane Coulter, Marge Stanton, and Allan D. Goecker; and (c) *Technical Addendum to Employment Opportunities for College Graduates in the Food and Agricultural Sciences*, by Kyle Jane Coulter and Marge Stanton, July, 1986. The latter two publications were prepared by, and are distributed by, Higher Education Programs, U.S. Department of Agriculture, Washington, DC.

Cattle exhibiting shelter-seeking behavior—in a ravine, and facing away from the storm. (Courtesy, American Hereford Assn., Kansas City, MO)

SECTION 1

ANIMAL BEHAVIOR AND ENVIRONMENT

Contents — Page

Fig. 1–1. Gregarious (flocking) behavior exemplified by primitive Soay sheep, an unimproved race of sheep on the uninhabited island of Soay, northwest of Scotland. (Courtesy, National Wool Growers' Assn., Salt Lake City, UT)

Today, there is great interest in animal behavior and environment. Those who grew up around animals and dealt with them in practical ways already have accumulated substantial workaday knowledge about their reaction to certain stimuli or their environment. But those who are less familiar with them may need to acquaint themselves with animal behavior in an unnatural environment, better to produce and care for them, and in order to recognize the signs when all is not well in the barnyard. Whether we come from farms or are city-bred, the principles and application of animal behavior and environment depend on understanding.

This section is presented for the purpose of bridging the gap between something old and something new in animal behavior and environment.

ANIMAL BEHAVIOR

Animal behavior is the reaction of animals to certain stimuli, or the manner in which they react to their environment. Through the years, behavior has received less attention than the quantity and quality of the meat, milk, eggs, fiber, and power produced by animals. But modern breeding, feeding, and management have brought renewed interest in behavior, especially as a factor in obtaining maximum production and efficiency. With the restriction, or confinement, of herds and flocks, many abnormal behaviors evolved to plague those who raise them, including cannibalism, loss of appetite, stereotyped movements, poor parental care, over-aggressiveness, dullness, degenerate sexual behavior, tail biting, cribbing, and a host of other behavior disorders. Confinement has not only limited space, but it has interfered with the habitat and social organization to which the species was adapted and best suited. This has been due to a genetic time lag; livestock producers have altered the environment faster than the genetic makeup of animals.

HOW ANIMALS BEHAVE—BEHAVIORAL SYSTEMS

People have always had to know something about the behavior of the animals around them. It required knowledge of basic behavior patterns to capture, confine, and herd animals, as did breeding, feeding, watering, and sheltering them. Without this understanding, domestication would have failed and animals would not have survived.

Animals behave differently, according to species. Also, some behavioral systems or patterns are better developed in certain species than in others. Ingestive and sexual behavior systems have been most extensively studied because of their importance commercially. Nevertheless, most animals exhibit the following nine general functions or behavioral systems, each of which is summarized in Table 1–1.

1. Agonistic behavior (combat).
2. Allelomimetic behavior.
3. Gregarious behavior.
4. Care-giving and care-seeking (mother-young) behavior.
5. Eliminative behavior.
6. Ingestive behavior (eating and drinking).
7. Investigative behavior.
8. Sexual behavior.
9. Shelter-seeking behavior.

TABLE 1-1
HOW ANIMALS BEHAVE

AGONISTIC BEHAVIOR (COMBAT)

This type of behavior includes fighting, flight, and other related reactions associated with conflict. Among all species of farm mammals, males are more likely to fight than females. Nevertheless, females may exhibit fighting behavior under certain conditions. Castrated males are usually quite passive, which indicates that hormones (especially testosterone) are involved in this type of behavior. Thus, farmers have for centuries used castration as a means of producing docile males, particularly cattle, swine, and horses.

Bulls, rams, boars, and stallions that are run together from a very young age seldom fight. Perhaps they have already settled their social rank. On the other hand, bringing together sexually mature strange males of these species almost always results in a fight. The intensity of fighting depends upon the tenacity of the two combatants. Although fighting rarely results in death, it usually continues until one gives up.

Cattle (Beef and Dairy)

In combat, bulls, paw the ground and bellow, followed by putting their heads together and butting.

Although young bulls raised together will seldom fight, a group of bulls may single out one individual and ride him to death, unless he is removed from the group.

Bringing together sexually mature strange bulls almost always results in a fight. Also, it is noteworthy that breeds of cattle differ in their agonistic behavior.

There is the hazard that bulls will be stifled as a result of fighting; hence, conditions that result in combat should be minimized.

Under range conditions, it is common for large numbers of bulls to be run together with a herd of cows. Even though many different bulls of different ages are included in the herd, fighting among them seldom occurs. Outside of the breeding season, as in the fall of the year, it is not uncommon to see bulls congregated together on the range, away from the cow herd.

Sheep

Rams fight by backing off and charging at each other headlong. The fight generally continues until one ram gives up, usually after both combatants have bloody noses.

Swine

When strange boars are brought together, some fighting ensues. Sows and barrows will also fight, but they do not exhibit the jaw-clicking and saliva-producing (champing) characteristics of fighting boars. A sow will try to bite, whereas a boar will slash his opponent with his tusks.

When strange boars are first penned together, they smell one another and begin to circle as they "size up" each other. They frequently strut shoulder to shoulder with the hair on their crest bristled, ears cocked, and head raised in an alert, threatening position. In a serious encounter, the combatants utter deep-throated barking grunts and champ throughout the fight. As the fighting becomes intense, each boar repeatedly thrusts his head and neck sideways and upward, with his jaws open and his teeth bared. If the boars have tusks, slashes are usually inflicted on the shoulders of each other.

Fighting boars await the opportunity to discontinue shoulder contact and to nip at the ears or the neck and front legs. Sometimes, they even charge the side of the opponent with their mouth wide open.

Fighting may continue for as long as an hour, or it may end very quickly. In any event, it will continue until the dominant boar is satisfied and the loser retreats, with the winner biting and slashing him as he scampers away. A fight on a summer day may end in the death of one or both combatants due to heat exhaustion.

Fig. 1-2 Agonistic (combat) behavior exhibited by two pigs (Courtesy, J. V. Craig, Department of Animal Sciences and Industry. Kansas State University, Manhattan)

Horses

Bringing together sexually mature, strange stallions for the first time almost always results in a vicious fight. Stallions fight by biting, kicking, and striking. Generally they fight head to head and most of the biting is done on the neck, shoulders and front legs. Although fighting rarely results in death, it usually continues until one gives up—battle scarred by teeth and hoof marks.

Fighting among mares is less vicious than between stallions. Body biting and kicking are used as a means of establishing social order. Geldings may fight much like mares.

Jacks are unusually vicious fighters. They rely on their teeth, rather than kicking. Sometimes wild jacks killed a rival by cutting his windpipe or jugular vein. Also, it is reported that the dominant jack occasionally castrated the weaker jacks with his teeth.

Agonistic behavior is of practical importance when strange horses are first put together. One way or another,a social order must be established. Hence, there is always the potential of injury until rank is settled. Also, agonistic behavior may create a potentially dangerous situation to both horses and riders in group riding. To reduce the hazard of such accidents, all horses should be spaced well apart when standing or moving.

Wild bands of horses and bands of domestic horses on the range range behave very much alike. The stallions have keen sight, hearing, and smell; and each stallion leader is very good at protecting his harem, which usually includes 10 to 20 mares. When frightened or facing danger, the stallion warns his band with snorting and restless movements and takes his place ready for battle if necessary.

ALLELOMIMETIC BEHAVIOR

Allelomimetic behavior is mutual mimicking behavior. Thus, when one member of a group does something, another tends to do the same thing; and because others are doing it, the original individual continues.

In the wild state, this trait was advantageous in detecting the enemy, and in providing protection therefrom. In wolves and coyotes, this behavior is important in attacks on prey, since a pack working together is much more likely to be successful than an individual working alone.

Under domestication, animals are usually protected from predators. Nevertheless, the allelomimetic behavior still has important consequences.

Cattle (Beef and Dairy)

Cows moving across a pasture toward a milking barn often display allelomimetic behavior. One cow starts toward the barn, and the others follow. Because the rest of the herd is following, the first cow proceeds on.

Fig. 1-3. Allelomimetic behavior exhibited by three steers. Because of stimulating and competing with one another, there is usually higher per steer feed consumption among a group of steers than by one steer alone. (Courtesy, American Feed Industry Assn., Arlington, VA)

(Continued)

TABLE 1-1 *(Continued)*

ALLELOMIMETIC BEHAVIOR *(Continued)*

Because of stimulating and competing with each other, there is usually higher per steer feed consumption among a group of steers than by one steer alone. Thus one steer penned alone may eat "X" pounds of feed per day. However, when he is placed with other steers, his intake may be "X + Y" pounds. But, of course, the feed consumption advantage can be nullified when the animals are placed together too closely, with the result that the agonistic behavior comes into play.

Sheep

Sheep walk, run, graze, and bed down together.

Sheep graze when they observe others in adjacent paddocks doing so.

Swine

Swine exhibit allelomimetic behavior in their eating habits. Thus, when one pig eats, there is a tendency for the rest to join it. As a result, pigs in a group usually average higher feed consumption than one pig alone.

Horses

A timid horse will follow behind a pack, in order not to be left behind.

When kept alone, high-strung racehorses may become nervous and fail to eat properly. To alleviate this situation a companion—known as a mascot, is often provided. All sorts of mascots are used—a goat, a sheep, a chicken, a duck, or a pony.

CARE-GIVING AND CARE-SEEKING (MOTHER-YOUNG) BEHAVIOR

The care-giving behavior is largely confined to females among domestic animals, where it is usually described as "maternal"; the care-seeking behavior is normal for young animals. This behavior begins shortly after birth and extends until the young are weaned. Care-giving and care-seeking vary widely among different species of farm animals.

Cattle (Beef and Dairy)

Nature ordained that cows seek isolation at calving time. So, where possible, they'll hide out.

Following birth, the care-giving behavior of the new mother becomes evident almost immediately. She gets up and begins to dry her newborn calf by licking it. Simultaneously, some cows "talk" to their newborn. They may become quite concerned and nervous as their "baby" first attempts to stand, takes a few footsteps—and falters. Aided by its mother's licking, and encouraged by her "talking" eventually the calf makes it to its unsteady feet and commences to search for a teat.

A newborn calf cannot see too well, but it can smell, touch, and taste. It associates everything that is good and that cares for it with its mother. This is the beginning of herd instinct.

If on pasture, the new mother usually hides her calf. During the first day or two, the calf sleeps a great deal, while the mother grazes nearby. But a mother takes great pains not to disclose the hiding place of her calf. At intervals, she returns to feed it. If it is necessary for her to leave her calf in order to get water or supplemental feed, she does not tarry much along the way. Frequently, where there are a number of newborn calves, the cows "baby-sit" for each other. Part of the cows will leave for feed or water, but one or two will remain behind and guard all the calves. Then, when the first cows to leave have returned, the "baby-sitters" will take their turn and depart. In this manner, there are older cows with the calves at all times.

When a calf in hiding is approached by a human, it will usually lie as close to the ground as possible, without any movement except for its eyes. If picked up, and if scared, it may bawl (cry) for its mother. If the mother hears the call, she will come running—often ready to fight. Frequently, other cows in the vicinity, especially if they have calves of their own, may join in the response. If the disturbed calf runs away, it will return to the area after the danger has passed.

By the time the calf is two days old, the mother wanders more extensively, with the calf at her side. Soon, they rejoin the herd.

Recognition between mother and calf is by smell (olfactory), sight (visual), and sound (auditory). Cows usually sniff their calves after being away for a time; and the calf recognizes its mother's call. The attachment of the mother to her calf is very strong. However, the calf accepts separation with less stress.

If a calf is stillborn, or dies soon after birth, some cows will leave the place where the fetus lies, never to return. Others may return to their dead calves at frequent intervals over a period of several days, smelling it and mooing gently.

Beef calves are normally weaned at about 7 months of age. The bond between cows and calves is very considerable, with the result that the separation is a traumatic experience. Thus, both mothers and calves bawl, often in unison, for 2 to 3 days. In all cases, however, the weaning separation should be complete and final, preferably with no opportunity for the calf to see or hear its dam again. In no case should the cows and calves be turned together once the separation has been made, for it will only prolong the weaning process, and it may cause digestive disorders in the calf.

Dairy calves are normally removed from their mothers when they are from 1 to 4 days of age, with the result that the tie between the mothers and offspring is soon severed.

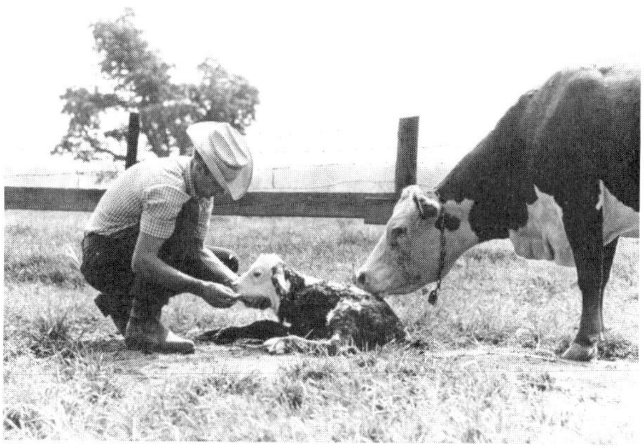

Fig. 1-4. Care-giving behavior evidenced by a Hereford cow as she licks her newborn calf. (Courtesy, USDA)

Sheep and Goats

After parturition, th ewe licks the newborn lamb, removing the moisture and placental membranes. The lamb soon staggers to its feet and makes awkward efforts to find a teat to nurse. Quite often a very weak lamb will have to be held to the teat. Normally, lambs suckle in a standing or kneeling position. While suckling, they wiggle their tails from side to side. The mother-young bond in sheep is very strong; the ewe becomes attached to her offspring, and the lamb develops an attachment to its mother. Although ewes are normally timid and easily frightened, they will defend their young even if the attacker is formidable. It is noteworthy, too, that sheep will accept and suckle orphan goats (kids), and vice versa (interspecies rearing).

Swine

The sow is very protective of her pigs, especially if they squeal. She goes toward the intruder with mouth open and emits a series of sharp, barking grunts in rapid succession. She continues to mother her pigs until they are weaned, but after 2 to 3 days' separation she loses interest in them. If pigs are left with the sow for 3 or 4 months she will usually wean them herself. Sows will readily accept pigs from another litter, provided the transfer is made the first day or two following farrowing. Exchanging of pigs among sows, in order to even out the size litters, is a common practice in herds where many sows are farrowing about the same time.

Some nervous sows eat their pigs during or immediately after farrowing. If this trait is observed, all pigs, both live and dead, along with the placental membranes should be removed as soon as possible, before the sow has an opportunity to eat them. Once the sow has acquired a taste for flesh, she may develop a permanent pig-eating habit. Usually, such nervous sows calm down following farrowing, after which their pigs may be returned to them and they will express normal protective behavior.

(Continued)

TABLE 1-1 *(Continued)*

CARE-GIVING AND CARE-SEEKING BEHAVIOR *(Continued)*

Horses

Mares show much the same maternal behavior toward their young as is exhibited by females of other species of farm animals. Thus, a mare calls her foal with a neigh or whinny and exhibits nervousness and distress when her young is disturbed. When mares are separated from their foals, such as sometimes happens when they are worked or taken away for rebreeding, there is usually a noisy exchange of whinnying between mother and foal when they are put back together again and the foal is allowed to nurse.

It is noteworthy that a mare will devote as much attention and affection to a mule colt—a hybrid (ass × horse)—as she will to a horse foal.

ELIMINATIVE BEHAVIOR

In recent years, elimination has become a most important phenomenon, and pollution has become a dirty word. Nevertheless, nature ordained that if animals eat, they must eliminate.

A full understanding of eliminative behavior will make for improved animal building design and give a big assist in handling manure. Right off, it should be recognized that the eliminative behavior in farm animals tends to follow the general pattern of their wild ancestors; but it can be influenced by the method of management.

Cattle (Beef and Dairy)

Cattle deposit their feces in a random fashion. Although cows can defecate while walking, with the result that their feces are scattered, generally they manage to deposit their "chips" in neat piles. Most cows hump up to urinate, whereas bulls are inclined to stand squarely on all "fours."

Sheep

The eliminative behavior in sheep is very similar to that of cattle. However, ewes usually assume a squat position when they urinate, and their feces are relatively dry and pelleted.

Swine

If given an opportunity, pigs are of very clean habits. They like to keep their bedding area clean and dry. Hence, they usually deposit their feces in a corner of the pen, away from the sleeping quarters. Modern methods of raising pigs in restricted quarters, which are often overcrowded, has disturbed their natural eliminative patterns.

Horses

Horses tend to deposit their feces in certain locations, such as along well-traveled paths, like those leading to waterholes. Hence, if given the opportunity, they often return to these locations for defecation.

Fig. 1-5. Eliminative behavior was ordained by nature, but it can be modified. This shows ewes and lambs on expanded metal floor with slots through which the feces and urine pass to a pit below. (Courtesy, University of Illinois, Urbana)

Stallions are much more prone to deposit their droppings on the same old mound than mares or geldings. Mares and geldings are inclined to use the border of their defecating area, with the result that they enlarge it each time.

There is disagreement whether a feral or range stallion marks the outside boundary of his home range with his feces, thereby staking it out for himself and his harem of mares, with most horse owners believing and most ethologists disbelieving.

GREGARIOUS BEHAVIOR

Gregarious behavior refers to the flocking or herding instinct of certain species. It is closely related to allelomimetic behavior. If animals imitate each other, they must stay together. If they stay together as a mobile group, they must use allelomimetic behavior to do so. All such behavior arises out of the process of social attachment.

Gregarious behavior differs among species.

Cattle (Beef and Dairy)

Cattle tend to roam in groups of various sizes when a large herd is placed on a pasture or range. However, there is usually considerable space between the members of the herd. Moreover, on close observation it is evident that there are several small groups within a herd, each ranging from 3 to 5 head.

Sheep and Goats

The gregarious, or flocking, instinct is particularly strong in sheep. Moreover, it is more evident in some breeds than in others. The Merino, and animals carrying Merino breeding, are noted for their flocking instinct. This makes it possible to herd them on the range.

It is noteworthy that the gregarious instinct of sheep diminishes to some extent when they are placed within fenced holdings, instead of herded. As a result, those who handle western range bands do not try to switch back and forth from fenced range to herding, for the reason that the band becomes unmanageable from the standpoint of herding once they have been in a fenced holding for an extended period of time.

Fig. 1-6. Gregarious (flocking) behavior exemplified in a range band of ewes. (Courtesy, Utah State University, Logan)

Packers use the gregarious instinct of sheep by having an old goat, appropriately called a "Judas," lead sheep to slaughter. A well-trained Judas will lead group after group of sheep to slaughter all day long.

(Continued)

TABLE 1–1 *(Continued)*

GREGARIOUS BEHAVIOR *(Continued)*

Swine

In the wild state, swine roved through the forest in herds. Usually these wild groups consisted of 1 to 4 females, along with their young of the year and their yearlings. Adult males join these groups during mating season, but range separately the rest of the year.

Under domestication, swine retain their gregarious nature. However, caretakers have altered it a great deal. Today, hogs are usually confined to a very limited area. Also, under domestication, they have lost most of their ferocity and are usually gentle and easily handled.

Horses

In the wild state, horses ran in bands; thus, they were gregarious by nature. These bands seldom consisted of more than 40 animals; and always there was a stallion in each group.

Under domestication, horses show definite preferences for their herdmates; they will even avoid certain horses in the herd. In the draft horse era, animals that were worked together usually stayed together when they were turned to pasture.

INGESTIVE BEHAVIOR (EATING AND DRINKING)

This type of behavior includes eating and drinking; hence, it is characteristic of animals of all species and all ages. It is very important because animals cannot live without feed and water.

Rumination is the act of chewing the cud, characteristic of herbivorous animals with split hoofs—cattle, sheep, and goats. It involves regurgitation of ingesta from the reticulo-rumen, swallowing of regurgitated liquids, remastication of the solids accompanied by reinsalivation, and reswallowing of the bolus.

The first ingestive behavior trait, common to all young mammals, is suckling.

Each species has its own particular method of ingesting feed.

Cattle (Beef and Dairy)

The natural feeding (grazing) position of cattle is heads down. In this position, they produce more saliva; and saliva aids digestion.

Cattle wrap their tongues around grass, then jerk their heads forward so that the vegetation is cut by the lower teeth. (There are no upper incisor teeth; only the thick hard dental pad.)

Rumination occupies about 8 hours of the cow's time each day. (In addition, the harvesting or grazing time may take another 8 hours. This means that a cow may work a 16-hour day.)

The Iowa Station reported that steers in lots on self-feeders spent 12 hours per day lying down, and this time was unaffected by shelter or season.

Rumination has an important effect on the amount of feed the animal can utilize. Feed particle size must be reduced to allow passage of the material from the rumen. It follows that high-quality forages require much less rechewing and pass out of the rumen at a faster rate; hence, they allow a cow to eat more.

Sheep and Goats

Sheep graze very much like cattle, but their cleft upper lip allows them to graze vegetation closer to the ground. As in cattle, the incisors are in the lower jaw only.

Goats can graze like cattle and sheep, but they are very fond of browse—the young shoots of shrubs and trees.

Swine

Swine possess teeth in the upper and lower jaws; hence, they bite off grass or take a mouthful of grain, then chew and swallow it.

Fig. 1–7. Ingestive (eating) behavior begins with nursing, soon after birth. (Courtesy, Maple Leaf Mills Ltd., Toronto, Canada)

Pigs have a single stomach, whereas ruminants have a four-compartment stomach.

By nature, pigs love to root. If given the opportunity, they will stick their noses into the ground and lift forward and upward, moving earth out of the way and exposing earthworms, grubs, and roots.

Horses

The mobile upper lip of the horse is used in gathering grass and other feed, in the same way that a cow uses her tongue or an elephant uses its trunk.

The long, strong, and roughened teeth of the horse are well suited to grinding common feeds.

The horse is somewhat between ruminants and monogastrics, primarily due to the large "blind gut," which is the seat of considerable bacterial action.

INVESTIGATIVE BEHAVIOR

All animals are curious and have a tendency to explore their environment. Investigation takes place through seeing, hearing, smelling, tasting, and touching. Whenever an animal is introduced into a new area, its first reaction is to explore it. Experienced caretakers recognize that it is important to allow animals time for investigation before attempting to work them, either when they are placed in new quarters or when new animals are introduced into the herd.

Cattle (Beef and Dairy)

If they are not afraid, cattle investigate a strange object at close range. They proceed toward it with their ears pointed forward and their eyes focused directly upon it. As they approach the object, they sniff and their nostrils quiver. When they reach the object, sniffing is replaced by licking; and if the object is small and pliable, they may chew it or even swallow it.

Cattle exhibit investigative behavior when placed in a new pasture or in a new barn. As a result, if there is an open gate in a pasture or a hole in the fence, they usually find it, then proceed to explore the new area.

Calves are generally more curious than older cattle. Perhaps this is due to the fact that older animals have seen more objects, with the result that fewer things are new or strange to them.

Sheep and Goats

Sheep and goats investigate strange objects and quarters much like cattle. They also approach objects in the same heads-up, ears forward, and eyes-fixed manner. However, sheep are much more timid than goats or cattle, with the result that they usually turn and run if the object moves or if something frightens them.

(Continued)

TABLE 1-1 *(Continued)*

INVESTIGATIVE BEHAVIOR *(Continued)*

Swine

Pigs are curious. When a strange person approaches a herd of hogs, an alarm, or "woof," is sounded and the animals scatter—scampering as fast as they can for a short distance. In the meantime, if the intruder remains stationary, either standing or sitting, the pigs invariably return to investigate by smelling, rooting, and nibbling. Of course, when pigs are placed in confinement, they have little area to investigate.

Horses

Foals are more curious than older horses. Young equines spend much of their time looking at and sniffing objects in their pastures or stalls. As the foal grows older, it may exhibit fear of certain objects. At this stage, it may even move away from its caretaker. When this happens, the caretaker should never run after the foal. Rather, stand still; very soon, the foal's curiosity will get the best of it, and it will return. A mare frequently becomes nervous as she watches her offspring investigate, fearful that it may get hurt in the process.

Fig. 1-8. *Left:* Investigative behavior displayed by young kids (goats). (Courtesy, *Dairy Goat Guide*, Waterloo, WI)

SEXUAL BEHAVIOR

Reproduction is the first and most important requisite of livestock breeding. Without young being born and born alive, the other economic traits are of academic interest only. Thus, it is important that all those who breed animals should have a working knowledge of sexual behavior.

Sexual behavior involves courtship and mating. It is largely controlled by hormones, although males that are castrated after reaching sexual maturity (which, among farm animals, are known as stags) usually retain considerable sex drive and exhibit sexual behavior. This suggests that psychological, or learned, as well as hormonal factors may be involved in sexual behavior.

Each animal species has a special pattern of sexual behavior. As a result, interspecies matings do not often occur. There are two notable exceptions, however; The best-known cross between animal species is the mule, a hybrid, which is a cross between the horse and the ass. Also, when sheep and goats are confined, they readily mate with each other, although such matings are never fertile.

Males in most species of farm animals detect females in heat by sight or smell. Also, it is noteworthy that courtship is more intense on pasture or range than under confinement, and that captivity has the effect of producing many distortions of sexual behavior compared to wild animals. Perhaps this explains the high percentage foal crop of wild bands of mares where conception and foaling rates of 90% or better were commonplace, in comparison with the average 50% foaling rate under domestication.

Today, livestock producers are attempting to control the sex life of animals, by bringing about ovulation at the time of choice of the owner, rather than of the female.

Fig. 1-9. Sexual behavior displayed by a Texas Longhorn bull and cow. (From an original painting by artist Tom Phillips, 3333 17th St., San Francisco, CA)

Cattle (Beef and Dairy)

Experienced cattle producers can usually detect in-heat cows through one or more of the following characteristic symptoms: (1) restlessness; (2) mounting other cows, and standing to be mounted by another cow (standing heat appears to be the best single indicator of the proper time to breed); (3) a noticeable swelling of the labia of the vulva; (4) an inflamed appearance about the lips of the vulva; (5) frequent urination; (6) switching and raising the tail; and (7) a mucous discharge. A day or two following estrus, a bloody discharge is sometimes seen. Dry cows and heifers usually show a noticeable swelling or enlargement of the udder during estrus, whereas in lactating cows a rather sharp decrease in milk production is often encountered. When kept alone, some cows become restless, walk the fence, and bawl when they are in heat. Some may even jump the fence, or go through it, as they attempt to find a bull.

A bull can often detect a cow that's coming in heat 24 to 48 hours before she will mate, at which time he will remain in her company. Courtship of the bull consists of following the in-heat cow; licking and smelling the external genitalia, with the head extended horizontally and the lip upcurled; with the chin and throat resting on the cow's rump.

Sheep

Unlike other farm animals, the ewe shows few visible external indications of heat. The acceptance of the ram (or of a teaser with an apron) is the best method of detection. Ovulation seems to occur late in the heat period usually from about 24 to 30 hours after the onset of estrus.

In sheep, the display of sexual behavior of the male is more elaborate than that of the female. Typically, the ram responds to the urination of a ewe in estrus by sniffing the urine, then extending his head with lips upcurled. He sticks his tongue in and out

(Continued)

TABLE 1-1 *(Continued)*

SEXUAL BEHAVIOR *(Continued)*

of his mouth as he follows the ewe, noses her external genitalia, and rubs along her side biting her wool. A characteristic part of the sexual display, or teasing, by the ram is the raising and lowering of one front leg in a stiff-legged striking motion.

Swine

The external signs of heat in the sow are restless activity; swelling or enlargement of, and discharge from, the vulva (although these signs are not always present); mounting of other sows; frequent urination; and occasional loud grunting.

The boar often nudges the sow or gilt around the head or in the flanks with his head and nose and emits a courting song. He will then attempt to mount her.

Horses

The signs of estrus in the mare are (1) the relaxation of the external genitalia; (2) frequent urination in small quantities; (3) the teasing of other mares; (4) the apparent desire for company; (5) a slight mucous discharge from the vulva; (6) allowing the stallion to smell and bite her; (7) spreading the hind legs; and (8) lifting the tail sideways. But many mares are shy breeders. Thus, when there is any question about a mare's being in season, she should be tried with the stallion. When possible, it is usually good business to regularly present mares to the teaser every day or every other day as the breeding season approaches. A systematic plan of this sort will save much time and trouble.

The courtship of the stallion is characterized by neighing and smelling the external genitalia of the mare, followed by extended head and upcurled lip; and pinching the mare in the croup area with his teeth.

SHELTER-SEEKING BEHAVIOR

All species of animals seek shelter—protection from the sun, wind, rain and snow, insects, and predators.

Cattle (Beef and Dairy)

Cattle are not as sensitive to extremes in temperature—heat and cold—as are swine. Nevertheless, they do seek shelter under natural conditions—this may consist of hills, valleys, timber, and other natural windbreaks; or they may even group closely together.

Cattle seem to be able to sense the coming of a storm, at which time they may race about and "act up." During a severe rain or snow storm, they turn their rear ends to the storm and tend to drift away from the direction of the wind. By contrast, bison (buffalo) face a storm head on.

During the hot summer months, cattle seek either shade or a waterhole during the heat of the day. Then, they graze in the cool of the evening or early morning. There are well known breed differences in tolerance to heat. Brahman cattle can withstand more heat than the European breeds, whereas the heat tolerance of the Santa Gertrudis is intermediate.

Sheep

Sheep seek shelter by moving into barns or under trees, by huddling together to keep off flies, by crowding together in extremely cold weather, and by pawing the ground and lying down. Like cattle, during a severe storm they turn their rear ends towards the wind.

When there is no shelter, there is danger of sheep massing together and smothering during a very severe storm.

Swine

Hogs are very sensitive to extremes of heat and cold; hence, shelter seeking is a very important trait with them. It is particularly important that swine be provided with shade during hot weather so that they may avoid the direct rays of the sun, because they do not possess an adequate cooling mechanism. In hot weather, hogs will wallow in water if given the opportunity.

Fig. 1-10. Shelter-seeking behavior exhibited by a band of sheep, making use of a tree shelter during a blizzard on the range. (Courtesy, Charles Belden, Pitchfork, WY)

When they are hot, hogs pant rapidly and sleep stretched out full length—so as to expose the maximum body surface to the air. During cold weather, swine sleep curled up and huddled together, thereby exposing minimal body surface to the air.

Horses

Horses are not very sensitive to either heat or cold. When not confined in cold areas, they develop a shaggy coat of hair in the wintertime and seek shelter from storms under trees and in the valleys. They paw to get their feed supply when the ground is covered with snow. Like cattle, horses face away from the direction of a severe storm.

SOCIAL RELATIONSHIPS

Social behavior may be defined as any behavior caused by or affecting another animal, usually of the same species, but also in some cases, of another species.

Social organization may be defined as an aggregation of individuals into a fairly well integrated and self-consistent group in which the unity is based upon the interdependence of the separate organisms and upon their responses to one another.

The social structure and infrastructure in herds and flocks are of great practical importance. Livestock producers should be knowledgeable relative to the social relationships of each species with which they work. Then, if this social relationship is disturbed and/or modified under intensive, confined conditions, they will be better able to feed, care, and manage the animals with maximum consideration accorded to both economy of production and animal welfare.

SOCIAL ORDER (DOMINANCE)

Within most groups of farm animals of the same species, there is a well-organized social rank. When we restrict or confine them and force them into spaces that bring them within the natural, individual distance that has been established (the distance between each other when moving as a herd or flock), we immediately create stress throughout the herd or flock. Thereupon, the dominants have to pay more attention to maintaining their dominance. They have to be more aggressive in their reactions. The subservients become far more nervous, and their nervousness spreads throughout the herd.

Fig. 1-11. Dominance. This shows a dominant cow attacking the neck of a subordinate. The latter submits and avoids a fight.

Once the social rank order is established, it results in a peaceful coexistence of the herd or flock. Thereafter, when the dominant one merely threatens, the subordinate animal submits and avoids conflict. Of course, there are some pairs that fight every time they chance to meet. Also, if strange animals are introduced into such a group, social disorganization results in the outbreak of new fighting, as a new social rank order is established.

Social rank among farm animals exists, but does not affect production adversely, so long as they are on pasture or range, and if there is plenty of feed and water. But it becomes of very great importance when animals are placed in confinement. When cows are moved into winter quarters, social dominance decrees that replacement heifers be sorted out and fed separately, that young bulls be cared for in separate quarters, and that old cows with poor teeth be fed separately; otherwise, these animals will not get enough feed.

Social rank becomes of importance when a group of animals is fed in confinement; and it becomes doubly important if limited feeding is practiced. Under such circumstances, the dominant individuals crowd the subordinate ones away from the feed bunk, with the result that the subordinates may go hungry. This happens both in feedlot cattle and in breeding cattle being wintered.

Several factors influence social rank; among them, (1) age—both young animals and those that are senile rank toward the bottom; (2) early experience—once a subordinate in a particular relationship, usually always a subordinate; (3) weight and size; and (4) aggressiveness or timidity.

In feedlots and other confinement operations, social facilitation is of great practical importance. Since dominance tends to conflict or interfere with social facilitation, dominant animals should be sorted out and, if possible, grouped together. Of course, they will fight it out until a new social order is established. In the meantime, both feed efficiency and gains will suffer. But, as a result of removing the dominants, the feed intake of the rest of the animals will be improved, followed by greater feed conversion efficiency and profit. Among the more settled animals, social facilitation will become more evident. After the dominants have been removed, the rest of the animals will settle down into a new hierarchy, but within the limits of their dominance. Their interaction or social facilitation will be far more likely to have a calming effect on this group, to both the economic and practical advantages of the operator.

Dominance and subordination are not inherited as such; they are developed by experience. Rather, the capacity to fight (agonistic behavior) is inherited, and, in turn, this determines dominance and subordination. Hence, when combat has been bred into a herd, such herds never have the same settled appearance and docility that is desired of high-producing animals under intensive management.

LEADER-FOLLOWER

The leader is the animal that is frequently at the head of a moving column and often initiates a new activity; the other animals in the group are followers. "Followership" appears to be a fairly strong phenomenon in many species, as most animals resist being left behind.

If the lead animal can be controlled, generally the remainder of the group can be moved easily.

Leader-follower relationships are particularly strong in sheep, where lambs follow their mothers from birth. In a naturally formed flock of sheep, the oldest ewes lead, followed immediately by their young lambs. Each is followed less closely by her descendants, with the females followed by their own lambs. Thus, the leader in the flock is usually the oldest ewe with the largest number of descendants. This type of leadership is broken up in flocks where unrelated animals are brought together.

Domesticated groups of horses also exhibit the leader-follower relationship. But the leadership may be shared between the stallion and the dominant mare, although some activities may be initiated by other animals in the herd. When the latter happens, usually the activator soon pauses and the leader proceeds to lead.

INTERSPECIES RELATIONSHIPS

Social relationships are normally formed between members of the same species. However, they can be developed between two different species. In domestication this tendency is important (1) because it permits several species to be kept together in the same pasture or corral, and (2) because of the close relationship between caretakers and animals. Such interspecies relationships can be produced artificially, generally by taking advantage of the maternal instinct of females and using them as foster mothers.

All sorts of bizarre interspecies relationships have been arranged—including cows raising pigs, bitches (dogs) raising pigs, rabbits, and cats; and cats raising mice.

Interspecies relationship is being used to protect sheep from predators. By raising puppies, young llamas, or young donkeys with sheep, at maturity they become their protectors (guards).

PEOPLE-ANIMAL RELATIONSHIPS

Fig. 1–12. *Left:* People-animal relationship displayed by a miniature donkey and its young admirer. (Courtesy, the Langfelds of Danby Farm, Omaha, NE)

Social relationships can also be transferred to human beings. Thus, animal caretakers usually form care-dependency relationships with the animals under their care. This is particularly true with pets—horses, dogs, and cats—and with single housed animals such as calves.

People need pets and pets need people! Both groups desire to love and be loved. This relationship is especially valuable for children, shut-in, handicapped, and elderly people. The Delta Society, a movement ably spearheaded by Leo K. Bustad, DVM, Ph.D., a former student of the author of whom he is very proud, is contributing richly to the happiness and well-being of people through furthering the human-animal bond and animal-facilitated therapy.

NORMAL ANIMAL BEHAVIOR

The producer needs to be familiar with the behavior norms of animals in order to detect and treat abnormal situations—especially illness. Many sicknesses are first suspected because of some change in behavior—loss of appetite (anorexia); listlessness; labored breathing; posture; reluctance or unusual movement; persistent rubbing or licking; and altered social behavior, such as one animal leaving the herd or flock and going off by itself—these are among the useful diagnostic tools.

Normal behavior in sleep should be recognized, especially since it differs widely among species.

Table 1–2 is a summary of normal animal behavior.

TABLE 1–2
NORMAL ANIMAL BEHAVIOR

HEALTH

Some of the signs of good health are:

1. Contentment.
2. Alertness.
3. Eating with relish and cudding by ruminants.
4. Sleek coat and pliable and elastic skin.
5. Bright eyes and pink membranes.
6. Normal feces and urine.
7. Normal temperature, pulse rate, and breathing rate.

Fig. 1–13. *Health*, enhanced by this dairy calf being raised in a clean, well-bedded, individual pen. (Courtesy, Maddox Dairy, Riverdale, CA)

Cattle (Beef and Dairy)

Normal rectal temperature:
Average, 101.5°F
Range, 100.4–102.8°F

Normal pulse rate:
60–70/min.

Normal breathing rate:
10–30/min.

Swine

Normal rectal temperature:
Average, 102.6°F
Range, 102–103.6°F

Normal pulse rate:
60–80/min.

Normal breathing rate:
8–13/min.

Sheep

Normal rectal temperature:
Average, 102.3°F
Range, 100.9–103.8°F

Normal pulse rate:
70–80/min.

Normal breathing rate:
12–20/min.

Horses

Normal rectal temperature:
Average, 100.5°F
Range, 99–100.8°F

Normal pulse rate:
32–44/min.

Normal breathing rate:
8–16/min.

(Continued)

TABLE 1-2 *(Continued)*

SIGHT

The eyes of most animals are on the side of the head (the cat is an exception). This gives them an orbital, or panoramic, view—to the front, to the side, and to the back—virtually at the same time. Also, this is a rounded, or globular, type of vision. This leads to a different interpretation than that of the binocular type of vision of people.

Cattle (Beef and Dairy)

The wide-set eyes of cattle enable them to have a large panoramic field of vision, even to the extent of seeing everything around them, with slight head movements. Only what is immediately behind their hindquarters is outside their field of view.

Sheep

Members of the flock maintain contact with each other largely through vision. As a flock grazes, each individual throws up its head at intervals, presumably to respond to the position of other members.

Swine

Swine, with their large snouts, have more efficient scent direction than sight.

Horses

In its natural habitat, the adult horse keeps a sharp lookout for its enemies, even while grazing. It is rare to see all members of a herd lying down together, one horse is almost always on the lookout.

Horses have monocular vision; that is, each eye is independent of the other and can see different pictures. This gives them a panoramic view—to the sides, the front, and the back—virtually at the same time. When a horse wants to see an object very clearly, it will face the object and use both eyes in a binocular manner. By contrast, humans have binocular vision and see the same picture with both eyes.

The lens of the horse's eyes is nonelastic; but the retina is arranged on a slope, the bottom part being nearer the lens than the top part. Thus, in order to focus on objects at different distances, the horse has to raise or lower its head so that the image is brought on to that part of the retina at the correct distance to achieve a sharp image.

Because of its monocular vision, it is difficult for a horse to judge distance accurately. In its evolution, it was more important that the horse see a wide area around it as it watched for predators than to judge distance. With domesticated horses, however, being able to judge distances is very important in certain types of performance. Thus, a rope horse must accurately judge the distance between itself and the animal it is following; a barrel racing horse must accurately judge the distance to the barrel as it prepares for the turn; and a jumping horse must accurately determine distance to the jump in order to select the take-off point, and it must determine the height and spread of the jump. Of course, top performing horses used for these purposes possess the ability to learn to judge distances; and they receive expert training.

Also, it is noteworthy that the horse has good vision in darkness. It is not as good as a cat's night vision, but it is considerably better than that of people. Thus, a horse may be ridden at night with reasonable safety, particularly if it is familiar with the area.

Fig. 1-14. *Sight*; a barrel racing horse must accurately judge the distance to the barrel as it turns. (Courtesy, McLaughlin Photography, Morrison, CO)

SLEEP

Normal behavior in sleep should be recognized, especially since it differs widely among species.

Cattle (Beef and Dairy)

Cattle typically lie on their stomachs or tilt to one side, with the fore limbs folded under the body; one hind limb extends forward, while the other protrudes toward the outside. Although cattle rest in this manner, they do not sleep in the sense that the term connotes. While lying down, they do shut their eyes for short periods of time. Beef bulls sometimes assume a sitting position.

Fig. 1-15. *Sleep*, a placid pasture scene. Some dairy cows are sleeping, others are chewing their cuds, and others are standing. (Courtesy, D. R. Rush, Berkeley, CA)

Calves commonly spend up to one-half hour at a time with their heads turned back in the flank position.

Sheep

The normal sleeping posture of sheep is on the stomach, but tilted to the side with one front leg folded under the body and the other extended forward. Usually the head is turned to one side and the eyes are closed. Although sheep are usually inactive about half of the day, as with cattle there is considerable debate as to whether they actually sleep. Certainly, sheep do not enter the state of deep sleep that exists in horses, dogs, and cats.

Swine

The resting position of swine varies according to temperature—in the summer, they sleep stretched full length; in cold weather, they sleep curled up. In any event, pigs sleep soundly—they even snore.

Horses

The horse rests and sleeps standing up. This is made possible by a system of ligaments, which do not get tired like muscles, and which take the weight off muscles during rest. Sometimes horses lie down in the sun, apparently to expose the body to warmth.

In contrast to cattle and sheep which sleep very little, the horse may sleep soundly for as much as 7 hours out of each 24 hours, mostly during the warmest part of the day. But not all of the 7 hours of sleep are taken at one time; rather, it is short and irregular, depending on the degree of hunger and the climatic conditions.

ABNORMAL ANIMAL BEHAVIOR

Abnormal behaviors of domestic animals are not fully understood. As with human behavior disorders, more research work is needed. However, we have learned from studies of captured wild animals that when the amount and quality, including variability, of the surroundings of an animal are reduced, there is increased probability that abnormal behaviors will develop. Also, it is recognized that confinement of animals makes for lack of space; this often leads to unfavorable changes in habitat and social interactions for which the species have become adapted and best suited over thousands of years of evolution. Among the abnormal behaviors that frequently develop with domestic animals in confinement are those summarized in Table 1-3.

TABLE 1-3
ABNORMAL ANIMAL BEHAVIOR

Abnormal behavior in animals develops where there is a combination of confinement, excess stimulation and forced production with a lack of opportunity to adapt to the situation. For example, homosexual behavior is common among all species where adult mammals of one sex are confined together.

Cattle (Beef and Dairy)

There are inherited differences in the temperaments of cattle. Nevertheless, constant stress can change the temperament of an animal, just as it can in people. Thus, when a bull is kept for hand mating in a corral by which the cow herd passes each day, cows in heat, or coming in heat, stimulate his sexual behavior. Since he cannot respond naturally through coitus, he becomes a mean bull. Thus, the "mean bull" complex is an example of abnormal behavior in cattle.

Milk cows may kick because they are in pain or frightened, or because they have been mistreated.

Cattle may develop pica (consumption of dirt, hair, bones, and/or feces) due to boredom, nutritional deficiencies, or physiological stress.

Sheep

As a consultant for the Atomic Energy Commission, the author had an opportunity to observe sheep that were kept in confinement, generation after generation, for 20 years. These animals developed a "wool-eating" habit. They didn't inflict special harm, for they only took small nibbles of wool from each other. Since these sheep were getting the most complete diet that science knew how to formulate, the only conclusion was that the wool-eating habit came about as a result of the unnatural confinement.

Swine

Fig. 1-16. Tail biting, abnormal behavior in swine. (Photo by J. C. Allen and Sons, West Lafayette, IN)

Tail biting accompanies close confinement. It results when pigs are prevented from rooting, nibbling, and chewing—it follows when the pig's normal behavior pattern is disturbed.

Docking is the best way in which to stop pigs from tail biting. Swine producers have tried all sorts of things to prevent tail biting. Some have substituted other materials for the pigs to bite, such as rubber tires or chains hung near the pigpen. Others have tried spraying the tails with distasteful chemicals. But none of these methods work very well.

Newly weaned pigs may nuzzle their penmates abdomens, mimicking early stages of nursing. Persistent nuzzling may cause ulcers and destruction of tissue.

Horses

Fig. 1-17. A cribber (wind-sucker, or stump-sucker) in action. This is the vice of biting or setting the teeth against some object, such as a post or manger, while sucking in air. (Courtesy, Dr. George H. Waring, Department of Zoology, Southern Illinois University, Carbondale)

Few animals have undergone such drastic change through evolution as equines. Little *Eohippus* (the dawn horse of 58 million years ago) was a denizen of the swamp. Later, through evolution, the horse became a creature of the prairie. Even though their natural habitat shifted during this long predomestication period, until people confined them they gleaned the feeds provided by nature. Inevitably, this occupied their time and provided exercise. But domestication and confinement to stalls wrought many changes—changes which spawned abnormal behaviors, including balking, bolting feed, cribbing, halter pulling, kicking, tail rubbing, weaving, wood chewing (pica), backing, rearing, shying, striking with the front feet, a tendency to run away, and objection to harnessing, saddling, and grooming, Many of these vices originate with incompetent handling; nevertheless, they may be difficult to cope with or to correct. This is especially true in older animals.

BREEDING FOR ADAPTATION

The wide variety of livestock in different parts of the world reflects a continuous process of natural and artificial selection which has resulted in the survival of animals well adapted to climate and other environmental factors. Changes in the physical structure of the species is dependent upon (1) the ability of animals to mutate and/or respond to selection pressure (natural or artificial), and (2) the effect of environmental pressure on the animal which results in a survival of the fittest. Among the examples of adaptation to environment are haired sheep (devoid of wool) in desert areas, fat-tailed sheep in arid zones. *Bos indicus* (Zebu) types of cattle in tropical areas, and *Bos taurus* cattle in temperate zones. Such adaptations relate to survival of the animals, but they do not necessarily entail maximum productivity of food. European cattle usually have much higher yields of milk and propensities for rapid growth than the breeds native to Africa or India. It is understandable, therefore, why there have been many attempts to introduce improved European livestock into countries in which the productivity of native stock is low. But there are many problems in breed replacement, with the result that a large number of experimental introductions of new breeds have not been successful. Tropical Africa provides an example. Because of disease problems, poor resistance to high temperatures, and limited feed supplies, many of the attempts made by former colonial powers to improve the output of native stock by replacing them with the European breeds failed. Breed replacement or a crossbreeding system might seem to be a simple panacea for low productivity. However, unless associated with special provisions for subsequent importation of breeding stock and simultaneous improvement of the nutritional, parasitological, disease, and husbandry environment of the crossbreds, it is not likely to succeed.

Fig. 1-18. Katahdin hair sheep bred for adaptation to a hot climate. There are more than 100 million hair sheep in the world, 90 million of which are in Africa. (Courtesy, Laura Callan, Perryville, AR)

Selection should be from among animals kept in an environment similar to that in which it is expected that their offspring shall perform—this requisite applies to animals brought in from another herd, either foundation or replacement animals. For example, animals that are going into a range herd should be selected from animals handled under range conditions, rather than from among stall-fed animals.

Animals can be changed through heredity and selection. For example, in Israel, which has one of the highest average milk yields per cow of any nation in the world, the individual distance between cows approaches contiguousness in some herds; this is due to Israel's having selected intensively for docility for 25 years. In other words, the animals are literally touching each other, with no agonistic or dominance-type response. This allows them to concentrate their animals even more than they had previously, thereby giving them a higher productivity per unit area. The only problem reported by Israel is that estrus, or heat, in animals in close proximity is difficult to detect.

Another example of breeding and selecting for a behavioral characteristic is milk production. When milkers first modified the milking technique and went from hand milking to machine milking, producers assumed the necessity of hand stripping. In recent years, many producers have given up hand stripping and selected cows capable of producing large quantities of milk by being milked by machine, without either stimulation before milking or stripping afterwards. This has been a selection for a behavioral characteristic—a low threshold to the milking stimulus. As a result, today a very large number of cows, especially within the Holstein-Friesian breed, will let down their milk effectively with no other stimulation than having the milking machine applied.

CONTROLLING ANIMAL BEHAVIOR

Many abnormal behaviors can be controlled rather easily, even though the cause is not removed; among them, those which follow.

- **Tail biting**—Tail biting accompanies close confinement. It results when pigs are prevented from rooting, nibbling, and chewing—from disturbing the pig's normal behavior pattern.

Docking is the best way to stop pigs from tail biting. Swine producers have tried all sorts of things to keep pigs from tail biting. Some have substituted other materials for the pigs to bite, such as rubber tires or chains hung near the pigpens. Others have tried spraying the tails with distasteful chemicals. But none of these methods works very well.

It is recommended that tail docking be a part of the regular management program, with the tails docked at the same time that the needle teeth are cut, when the pigs are about 3 days old. The side-cutting type pliers will work for both jobs, but tails will bleed less when they're cut with a dull blade. Emasculators and poultry debeakers also work well. To dock the tail, clean it first, then cut it ½ to ¾ in. from its base, lifting it gently so as not to stretch the skin. The skin won't heal over the end bone as rapidly if you pull the tail away from the body. Don't cut the tail shorter than ½ in. because it will make for excess bleeding and slow healing.

- **Bolting feed**—Horses that eat too rapidly are said to be bolting their feed. It can be lessened by spreading the concentrate thinly over the bottom of a large grain box, so that the horse cannot get a large mouthful; or by placing in the grain box a few smooth stones about the size of baseballs, so that the horse has to work to get feed.

• **Eating bedding**—Sometimes gluttonous animals eat their bedding. This is undesirable because (1) most bedding materials are low in nutritional value, and (2) feces-soiled bedding adds to the parasite problem. The problem can be alleviated by muzzling the horse.

• **Wood chewing**—Wood chewing is a common abnormal behavior in horses. In the final analysis, there is only one foolproof way to prevent wood chewing; to have no wood on which they can chew—to use metal, plastic, or other similar materials, for fences and barns. Of course, this isn't always practical. But wood chewing can be lessened, although it cannot be entirely prevented, by one or more of the following practices:

1. Stepping up the exercise.
2. Feeding three times a day, rather than twice a day, even though the total daily feed allowance remains the same.
3. Spreading out the feed in a larger feed container, and/or placing a few large stones about the size of a baseball in the feed container.
4. Providing 2 to 4 lb of straw or coarse grass hay per animal per day, thereby giving the horse something to nibble on during its spare time.

ANIMAL ENVIRONMENT

Fig. 1-19. Environmentally controlled dairy barn, well insulated and ventilated (note fan). (Courtesy, Babson Bros. Co., Oak Brook, IL)

Environment may be defined as all the conditions, circumstances, and influences surrounding and affecting the growth, development, and production of animals. The most important influences in the environment are the feed and quarters (space and shelter).

The branch of science concerned with the relation of living things to their environment and to one another is known as ecology.

Through the years, the domesticated animals best suited to a particular environment survived, and those that were poorly adapted either moved to a more favorable environment or perished. During the past two centuries, livestock producers have made great strides in the selection and propagation of animals suited to a particular environment, and during the past 50 years they have made progress in modifying the environment for the benefit of their animals and themselves.

It is becoming increasingly difficult to define environment, because scientists continue to discover important new environmental factors. Primitive people recognized that the sun and fire provided both heat and light, that body heat could be conserved by draping the body with animal skins, and that trees and caves provided protection from the weather. Today, it is recognized that these, along with a host of other environmental factors, affect animals and people.

The keepers of herds and flocks were little concerned with the effect of environment on animals so long as they grazed on pastures or ranges. But rising feed, land, and labor costs, along with the concentration of animals into smaller spaces, changed all this. Today, most layers and broilers are on litter floors. Turkeys are shifting rapidly from range to confinement. Water is important for ducks, but even with ducks the trend is toward higher population densities and more confinement. Many swine are raised partially or totally in confinement; and confinement production is increasing with beef cattle, dairy cattle, and sheep.

Among animals, environmental control involves space requirements, light, air temperature, relative humidity, air velocity, wet bedding, ammonia buildup, dust, odors, and manure disposal, along with proper feed and water. Control or modification of these factors offers possibilities for improving animal performance. Although there is still much to be learned about environmental control, the gap between awareness and application is becoming smaller. Research on animal environment has lagged, primarily because it requires a melding of several disciplines—nutrition, physiology, genetics, engineering, and climatology. Those engaged in such studies are known as ecologists.

In the present era, pollution control is the first and most important requisite in locating a new livestock establishment, or in continuing an old one. The location should be such as to avoid (1) the neighbors complaining about odors, insects, and dust; and (2) pollution of surface and underground water. Without knowledge of animal behavior, or without pollution control, no amount of capital, native intelligence, and sweat will make for a successful livestock enterprise.

HOW ENVIRONMENT AFFECTS ANIMALS

Heredity has already made its contribution at the time of fertilization, but environment works ceaselessly away until death. Among the environmental factors affecting animals are the following:

1. Feed.
2. Water.
3. Weather.
4. Facilities.
5. Health.
6. Stress.

FEED/ENVIRONMENTAL INTERACTIONS

Animals may be affected by either (1) too little or too much feed, (2) rations that are too low in one or more nutrients, (3) an imbalance between certain nutrients, or (4) objection to the physical form of the ration—for example, it may be ground too finely.

Forced production (such as growth, milk products, and racing 2-year-old horses) and the feeding of forages and grains which are often produced on leached and depleted soils have created many problems in nutrition. These conditions have been further aggravated through the increased confinement of animals, many animals being confined to stalls or lots all or a large part of the year. Under these unnatural conditions, nutritional diseases and ailments have become increasingly common.

Also, nutritional reproductive failures plague livestock operations. Generally speaking, energy supply tends to be more limiting than protein in reproduction. The level and kind of feed before and after parturition will determine how many females will show heat and conceive. After giving birth, feed requirements increase tremendously because of milk production; hence, a female suckling young needs approximately 50% greater feed allowance than during the pregnancy period. Otherwise, she will suffer a serious loss in weight, and she may fail to come in heat and conceive. This basic fact, along with other pertinent findings, was confirmed by researchers at the Montana Agricultural Experiment Station. Based on 12 years research at the Havre and Miles City Stations, they concluded that beef cattle size and milk production should be tailored to fit the environment. Big size and more milk are not better unless the range forage supply is better. The best size cow is one that fits the range conditions. Small cows do best on poor range because they can usually get 100% of their daily feed requirement for maintenance and milk production, whereas big cows on a poor range are borderline hungry all the time. Also, cows that give a lot of milk must have a good range, otherwise, they are stressed by lack of feed; and their fertility rate and calf crops drop. So, cow size and milk production should match their environment.

WATER/ENVIRONMENTAL INTERACTIONS

Animals can survive for a longer period without feed than without water. Water is one of the largest constituents in the animal body, ranging from 40% in very fat, mature animals to 80% in newborn animals. Deficits or excesses of more than a few percent of the total body water are incompatible with health, and large deficits of about 20% of the body weight lead to death.

The total water requirement of animals varies primarily with the weather (temperature and humidity); feed (kind and amount); the species, age, and weight of animal; and the physiological state. The need for water increases with increased intakes of protein and salt, and with increased milk production of lactating animals. Water quality is also important, especially with respect to the content of salts and toxic compounds.

WEATHER/ENVIRONMENTAL INTERACTIONS

Webster defines weather as a state of the atmosphere with respect to heat or cold, wetness or dryness, calm or storm, clearness and cloudiness.

Extreme weather can cause wide fluctuations in animal performance. The difference in weather impact from one year to the next, and between areas of the country, causes difficulty in making a realistic analysis of buildings and management techniques used to reduce weather stress.

During hot weather, feedlot cattle "peak" their eating during early morning and again during the evening hours—when it is cool. In cool weather, they eat more during midday than when it is hot. The feeder should sense these changes in cattle eating habits and program their feeding accordingly. Also, cattle eat more following a bad storm or a hot spell. At such times, the bunk may be "slick" for 2 to 3 hours and the cattle may line up waiting to be fed. When this happens, the ration should be increased. Also, by going to a higher roughage ration at these times, acidosis and laminitis can be minimized. The ability to recognize the "sign language of animals" and to change the feeding program accordingly is responsible for the oft-quoted statement that "the eye of the master fattens the cattle."

The maintenance requirements of animals increase as temperature, humidity, and air movements depart from the comfort zone. Likewise, the heat loss from animals is affected by these three items.

The research data clearly show that winter shelters and summer shades improve production and feed efficiency. The issue is clouded only because the additional costs incurred by shelters have frequently exceeded the benefits gained by the improved performance, particularly in those areas with less severe weather and climate.

The maintenance requirements of animals increases as temperature, humidity, and air movement depart from the comfort zone. Likewise, the heat loss from animals is affected by these three factors. Animals adapt to weather as follows:

- **In cold weather**, the heating mechanisms are employed, including (1) increased insulation from growth of hair and more subcutaneous fat; (2) increase in thyroid activity; (3) seeking protective shelter and warming solar radiations (the animals sun themselves); (4) huddling together; (5) consumption of more feed, which increases the heat increment and warms animals; and (6) increasing activity. The most important animal body heating mechanisms are amount of feed consumed and body activity, which are also evidenced in people. For example, after skiing in bitter cold weather, a skier feels comfortable after eating a beefsteak; and during a marathon race, a runner may feel quite warm when the temperature is near freezing (30°F).

- **In hot weather**, the cooling mechanisms are employed, including (1) moisture vaporization (from the skin and lungs), (2) avoidance of the heating solar radiation (the animals seek shade), (3) depression of thyroid activity, and (4) loafing (including lessening the production of meat, milk, and eggs, since they increase heat production).

THERMONEUTRAL ZONE (COMFORT ZONE)

Fig. 1–20 and the definitions that follow are pertinent to an understanding of thermal zones.

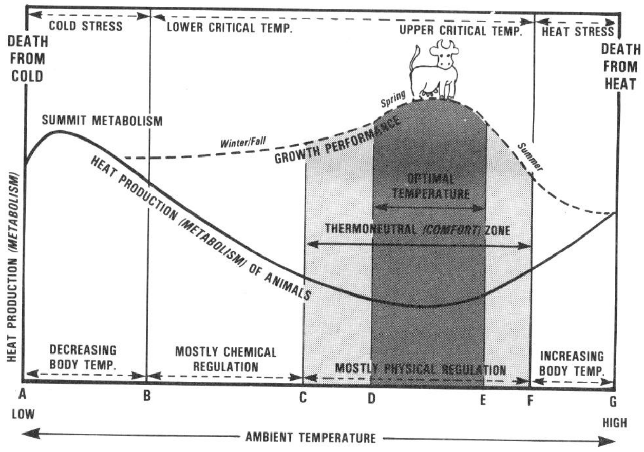

Fig. 1–20. Diagram showing (1) the influence of thermal zones and temperature on homeotherms (warm-blooded animals), and (2) the peak of milk yields in the spring, followed by the summer slump due to high (hot) summer temperature and lignification of forage.

In Fig., 1–20 *heat production (metabolism)* is plotted against *ambient temperature* to depict the relationship between chemical and physical heat regulation. Note, too, the broad range of accommodation to low (cool) temperatures in contrast to the restricted range of accommodation to high (warm) temperatures. Definitions of terms pertaining to Fig. 1–20 follow.

• ***Thermoneutral (comfort) zone*** *(C to F) is the range in temperature within which the animal may perform with little discomfort, and in which physical temperature regulation is employed.*

• ***Optimum temperature*** *(D to E) is the temperature at which the animal responds most favorably, as determined by maximum production (gains, milk, wool, work, eggs) and feed efficiency.*

• ***Lower critical temperature*** *(B) is the low point of the cold temperature beyond which the animal cannot maintain normal body temperature.* The chemical temperature regulation is employed in the zone below C. When the environmental temperature reaches below point B, the chemical-regulation mechanism is no longer able to cope with cold, and the body temperature drops, followed by death. The French physiologist, Giaja, used the term *summit metabolism* (maximum sustained heat production) to indicate the point beyond which a decrease in ambient temperature causes the homeothermic mechanisms to break down, resulting in a decline in both heat production and body temperature and eventually death of the animal.

• ***Upper critical temperature*** *(F) is the high point on the range of the comfort zone, beyond which animals are heat stressed and physical regulation comes into play to cool them.*

The cow produces the maximum yield of milk during the spring when the temperature is optimum (D to E), and the minimum yield in the summer when it is hot (F to G).

The comfort zone, optimum temperature, and both upper and lower critical temperatures vary with different species, breeds, ages, body sizes, physiological and production status, acclimatizations, feed consumed (kind and amount), the activity of the animal, and the opportunity for evaporative cooling.

The temperature varies according to age, too. For example, the comfort zone of newborn lambs is 75 to 80°F, whereas the comfort zone of mature sheep is 45 to 75°F.

Animals that consume large quantities of roughage or high-protein feeds produce more heat during digestion; hence, they have a different critical temperature than the same animals fed a high-concentrate, moderate-protein ration. Because of this, experienced cattle feeders decrease the roughage and increase the concentrate of finishing cattle during the hot summer months.

Stresses of both high and low temperatures are increased with high humidity. The cooling effect of evaporating sweat is minimized and the respired air has less of a cooling effect. As humidity of the air increases, discomfort at any temperature, and nutrient utilization, decrease proportionately.

Air movement (wind) results in body heat being removed at a more rapid rate than when there is no wind. In warm weather, air movement may make the animal more comfortable, but in cold weather it adds to the stress temperature. At low temperatures, the nutrients required to maintain the body temperature are increased as the wind velocity increases. In addition to the wind, a drafty condition where the wind passes through small openings directly onto some portion or all of the animal body will usually be more detrimental to comfort and nutrient utilization than the wind itself.

ADAPTATION, ACCLIMATION, ACCLIMATIZATION, AND HABITUATION OF SPECIES/BREEDS TO THE ENVIRONMENT

Every discipline has developed its own vocabulary. The study of adaptation/environment is no exception. So, the following definitions are pertinent to a discussion of this subject:

Adaptation refers to the adjustment of animals to changes in their environment.

Acclimation refers to the short(term (over days or weeks) response of animals to their immediate environment.

Acclimatization refers to evolutionary changes of a species to a changed environment which may be passed on to succeeding generations.

Habituation is the act or process of making animals familiar with, or accustomed to, a new environment through use or experience.

Species differences in response to environmental factors result primarily from the kind of thermoregulatory mechanism provided by nature, such as type of coat (hair, wool, feathers), and sweat glands. Thus, hogs, which have a light coat of hair, are very sensitive to extremes of heat and cold. On the other hand, nature gave cattle an assist through growing more hair for winter and shedding hair for summer, with the result that they can withstand higher and lower temperatures than hogs. The long-haired, shaggy yak of Tibet and the wooly Scotch Highland cattle of Scotland are as cold tolerant as the arctic-dwelling caribou and reindeer.

From time to time, American buffalo *(bison bison)* and domestic beef cattle *(Bos taurus)* have been crossed to obtain a more hardy beast than cattle. The most publicized early work of this type was the development of the Cattalo (bison × domestic cattle), the initial cross for which was made at the Dominion Experiment Station, at Scott, Saskatchewan, in Canada, in 1915.

Fig. 1–21. Cattalo (¼ buffalo, ¾ domestic cattle) cow. The initial Cattalo breeding experiment was started by the Dominion Experimental Station, Scott, Saskatchewan, Canada, in 1915. The foundation herd consisted of 16 female and 4 male hybrids (Courtesy, Research Station, Canada Department of Agriculture, Lethbridge, Alberta, Canada)

Male fertility and female reproductive rate have remained a problem in Cattalo. Although unquestionably hardy, bison × cattle crosses can be outperformed in nearly all environments by the currently available cattle breeds or crosses; and management procedures.

Also, there are breed differences, which make it possible to select animals well adapted to specific environments. Thus, the breeds of cattle that originated in the British Isles and Northern Europe are cold tolerant, whereas the Indian-evolved Zebu, or Brahman, cattle are heat tolerant. The long-fibered Black-faced Highland sheep are cold tolerant; the haired sheep are suited to hot, desert areas; and the fat-tailed sheep are adapted to arid conditions. The Shetland Pony, native to the Shetland Isles, no more than 400 miles from the Arctic Circle, evolved in the rigors of the northland climate and on sparse vegetation, which imparted that hardiness for which the stocky breed is famed. The long-legged donkey is adapted to hot, desert areas.

In recent years, attempts have been made to combine the heat tolerance characteristics of tropical breeds with the high productive capacity of European stock. The best known of these planned beef breeds is Santa Gertrudis, developed on the famed King Ranch of Texas, in the early 1900s, which carry approximately ⅝ Shorthorn and ⅜ Brahman breeding.

ENVIRONMENTALLY CONTROLLED BUILDINGS

With the shift to confinement structures and high-density production operations, building design became more critical, with consideration given to air temperature, relative humidity, air velocity, wet bedding, dust, light, ammonia buildup, odors, and space requirements.

Environmentally controlled buildings are costly to construct, but they make for the ultimate in animal comfort, health, and efficiency of feed utilization. Also, they lend themselves to automation, which results in a saving in labor; and, because of minimizing space requirements, they effect a saving in land cost. Today, environmental control is rather common in poultry and swine housing, and it is on the increase with other classes of livestock—especially dairy cattle.

In hot climates, increased use is being made of shades for the purpose of enhancing animal comfort and minimizing the maintenance requirements. Also, studies with lactating dairy cows reveal that putting only the head in an air-conditioned chamber, with the rest of the body left exposed to the heat, will increase production and feed efficiency.

Before an environmental system can be designed for animals, it is important to know their (1) heat production, (2) vapor production, and (3) space requirements. This information is presented in Section 9, Buildings and Equipment, of this book; hence, the reader is referred thereto.

ARTIFICIAL LIGHTING

The number of hours of light in the day affects the initiation of the normal breeding season of ewes and mares, both of which are seasonal breeders. It is noteworthy, too, that the reproductive function in poultry and migratory fowl is regulated by the length of daylight.

The ratio of hours of daylight to darkness throughout the year acts on nerves in the region of the pituitary gland, and stimulates or inhibits the release of the follicle-stimulating hormone (FSH). Lengthening the daylight hours activates the pituitary, and causes it to release increasing amounts of the FSH which stimulates ovarian function. Thus, sometime after the daylight period increases, the estrous cycle begins in ewes and mares.

Artificial lighting will accomplish the same thing as daylight; hence, it may be used to alter the estrous cycle in both ewes and mares.

- **Sheep**—Normally, ewes come in heat during the late summer or early fall, though there is both an area and a breed difference. The breeding season is usually restricted to about four months.

Ewes generally begin cycling when the number of daylight hours drops below 14. This is the reason that most breeds of sheep come into heat during the fall months. To initiate estrus, however, it appears that the shorter days must be preceded by longer days.

• **Horses**—Normally, the natural breeding season of mares begins in March and extends to late July or August.

Artificial lighting of broodmares enables breeders to bring mares in season about 6 weeks earlier than normal. By the use of the artificial light technique, a mare that would normally conceive on March 15 may get in foal sometime in January. By avoiding the necessity of skipping a year due to late breeding, this technique may actually result in obtaining two additional foals during the lifetime of a mare.

The procedure consists in using a 200-watt light bulb in a box stall so as to extend the hours of light to 16 hours daily. By beginning the light treatment of mares about December 1, they may be bred the latter part of January.

Slight adjustments in the schedule will need to be made in different locations, depending upon the sunrise and sunset times of the particular area.

FACILITIES/ENVIRONMENTAL INTERACTIONS

Optimum facility environments can only provide the means for animals to express their full genetic potential or production, but they do not compensate for poor management, health problems, or improper rations.

Research has shown that animals are more productive and feed-efficient when raised in an ideal environment. The primary reason for having facilities, therefore, is to modify the environment. Proper barns and other shelters, shades, sprinklers, insulation, ventilation, heating, air conditioning, and lighting can be used to approach the desired environment. Also, increasing attention needs to be given to other stress sources such as space requirements, and the grouping of animals as affected by class, age, size, and sex.

The principal scientific and practical criteria for decision making relative to the facilities for animals in modern, intensive operations is the productivity and cost of production of animals , which can be achieved only by healthy animals under minimal stress. So, the investment in environmental control facilities is usually balanced against the expected increased returns.

Temperature, humidity, and ventilation recommendations for different classes of livestock are given in Section 9 of this book, Table 9–8, "Recommended Environmental Conditions for Animals." This table will be helpful in obtaining a satisfactory environment in confinement livestock buildings, which require careful planning and design.

In recent years, there has been a trend to modify the environmental control facilities as much as possible; among such modifications designed for maximum animal comfort and efficiency of production are fans, floors, lights, shades, sprinklers/sprayers/foggers, ventilation, wallows, and windbreaks.

HEALTH ENVIRONMENTAL INTERACTIONS

Health is the state of complete well-being, and not merely the absence of disease.

Environment embraces the forces and conditions, both physical and biological, that (1) surround animals, and (2) interact with heredity to determine behavior, growth, and development.

Disease is defined as any departure from the state of health.

Parasites are organisms living in, on, or at the expense of another living organism.

Feed, air quality, lighting, noise, other animals, and weather are among the many factors that constitute an animal's environment. Extremes or alterations in the environment may subject an animal to stress; and stress may affect health and lead to more diseases and parasites.

The importance of good animal health is underscored by the following statistics: It is estimated that the animal diseases and parasites in the United States (1) decrease animal productivity by 15 to 20%; and (2) make for annual losses equivalent to 15% of the annual cash farm income from marketing livestock and products, on which basis the estimated livestock losses for 1987 totalled $11.4 billion. Further, there is evidence that nutrition has some involvement in 85% of the veterinary cases. In the developing countries, diseases and parasites take an even greater toll—they decrease animal productivity by 30 to 40%.

Some important health/environmental interactions not covered elsewhere in this book are discussed in the sections that follow.

STRESS/ENVIRONMENTAL INTERACTIONS

Stress is the nonspecific response of the body to any demand.

As used herein, stress indicates an environmental condition that is adverse to an animal's well-being, either external (nutritional, weather, social) or internal (disease, parasites).

Stresses of many kinds affect animals; among them, cold stress, heat stress, drafts, poor ventilation, excitement, presence of strangers, fatigue, mixing animals, number of animals together, space, changing corral and corral mates, weaning, previous nutrition, hunger, thirst, poor sanitation, disease, parasites, surgical operations, injury, and management.

Race and show horses are always under stress; and the greater the speed and the more tired they become, the greater the stress. Also, the greater the stress, the more exacting the nutritive requirements. Thus, the ration of race and show horses should be scientifically formulated.

Animals can be prepared, or adapted, to the environment, in such a manner as to reduce stress. For example, if calves are properly *preconditioned* (started on feed, vaccinated, treated for parasites, etc.) prior to weaning, the stress of subsequent weaning and movement to a feedlot will be minimized.

In the life of an animal, some stresses are normal, and they may even be beneficial—they can stimulate favorable action on the part of an individual. Thus, we need to differentiate between stress and distress. Distress—not being able to adapt—is responsible for harmful effects. The trick is to manage stress so that it doesn't become distress and cause damage and to recognize the warning signals of distress. For example, Texas Agricultural Experiment Station workers recently reported that added vitamin C, in either the feed or water, may reduce many of the health hazards associated with various kinds of stress to chickens such as hot weather, interaction with other birds in crowded conditions, and exposure to diseases.

The principal criteria used to evaluate, or measure, the well-being or stress of people are: increased blood pressure, increased muscle tension, body temperature, rapid heart rate, rapid breathing, and altered endocrine gland function. In the whole scheme, the nervous system and endocrine system are intimately involved in the response to stress and the effects of stress.

The principal criteria used to evaluate, or measure, the well-being or stress of animals are: growth rate or production, efficiency of feed use, efficiency of reproduction, body temperature, pulse rate, breathing rate, mortality, and morbidity. Other signs of animal well-being, any departure from which constitutes a warning signal, are: contentment, alertness, eating with relish (and cudding by ruminants), sleek coat and pliable and elastic skin, bright eyes and pink eye membranes, and normal feces and urine.

Stress is unavoidable. Wild animals were often subjected to great stress; there were no caretakers to modify their weather, often their range was overgrazed, and sometimes malnutrition, predators, diseases, and parasites took a tremendous toll.

Domestic animals are subjected to different stresses than their wild ancestors, especially to more restricted areas and greater animal density. However, in order to be profitable, their stresses must be minimal.

POLLUTION OF THE ENVIRONMENT

We must ever be mindful that life, beauty, wealth, and progress depend upon how wisely man uses nature's gifts—the soil, the water, the air, the minerals, and the plant and animal life.

Pollution is the issue of the decade. Anything that defiles, desecrates, or makes impure or unclean the surroundings pollutes the environment and can have a detrimental effect on animal health and performance. Thus, gases, odorous vapors, and dust particles from animal wastes (feces and urine) in buildings directly affect the quality of the environment. Muddy lots and stray electrical voltage may also pollute the environment. For healthy and productive animals, each of these pollutants must be maintained at an acceptable level. Among the most troublesome animals pollutants are: dust, manure, muddy lots, pests and pesticides, and stray voltage.

Fig. 1-22. Cattle feeding has gone modern. *Upper:* Pollution from a muddy cattle feedlot. *Lower:* Pollution control, with the runoff from the cattle feedlot flowing into a basin. (Lower picture, courtesy, USDA, Soil Conservation Service.)

DUST

Dust may be defined as a mixture of small particles of different sizes of dry matter.

Dust is a contributing factor to both animal and human health, especially with respect to respiratory diseases. Thus, it should be considered a significant contaminant that adversely affects environmental quality of animal houses and feedlots.

Dust may be present in significant amounts both inside and outside buildings.

Cattle and sheep feedlot dust is both organic, from excreta, and inorganic, from soil. Sprinkling and increased animal density (resulting in more moisture from urine) are the most effective methods of preventing feedlot dust in dry climates or during dry seasons. In swine houses and horse barns, most of the dust comes from the feed. In poultry houses, the dust contains a considerable amount of feather and skin debris, along with particles from the feed and litter. Dust in animal buildings may also carry microbes, gases, and vapors.

MANURE

In animal agriculture, we need to give particular attention to the pollution caused by manure. One cow produces as much waste as 16 humans. Hence, with 20,000 steers in a feedlot, the disposal problem is equal to a city of 320,000 people. In addition to being used as fertilizer, manure is now being recycled as livestock feed and serving as a source of energy (methane gas).

Of course, there is no one best manure management system for all situations. But, one way or another, science and technology must evolve with ways of disposing of 1.5 billion tons of manure annually; and this must be accomplished without polluting streams or the atmosphere or being offensive to neighbors.

If not managed properly, animals may produce the following pollutants in troublesome quantities: manure, gases/odors, dust, and flies/other insects. Also, they may pollute water supplies.

• **Precautions when using manure as a fertilizer on the land**—The following precautions should be observed when using manure as a fertilizer:

1. Avoid applying closer than 100 ft to waterways, streams, lakes, wells, springs, or ponds.
2. Do not apply where downward movement of water is not good, or where irrigation water is very salty or inadequate to move salts down.
3. Incorporate (preferably by plowing or discing) manure into the soil as quickly as possible after application. This will maximize nutrient conservation, reduce odors, and minimize runoff pollution.
4. Distribute the waste as uniformly as possible in the area to be covered.
5. Irrigate thoroughly to leach excess salts below the root zone.
6. Allow about a month after irrigation before planting, to enable soil microorganisms to begin decomposition of manure.
7. Minimize odor problems by—
    a. Spreading raw manure frequently, especially during the summer.
    b. Spreading early in the day as the air is warming up, rather than late in the day when the air is cooling.
    c. Spreading only on days when the wind is not blowing toward populated areas.

Also, see Section 8, under heading of "Manure."

MUDDY LOTS

Muddy lots often plague livestock producers, especially during the winter months. Mud increases scours and other diseases in newborn animals and reduces production and feed efficiency in older animals.

California Agricultural Experiment Station studies show that mud can reduce finishing cattle gains by as much as 10 to 35%, and increase the feed required per pound of gain by a like amount. Thus, it is important that the problem be minimized, especially in high rainfall areas. Good drainage is the first essential. This should be assured at the time the lot is located and constructed.

Mounds 6 to 12 ft high, preferably perpendicular to the feed bunk, will provide finishing cattle a dry place on which to lie. Concrete aprons 10 × 12 ft wide and sloping 1 in. per foot along the bunk will provide them with solid footing on which to stand and feed. Also, lessening of cattle density during the winter months—fewer animals per lot—is an effective method of controlling the mud problem. Thus, many feedlots plan to feed fewer cattle during the muddy season.

To cope with the mud and alleviate calf scours, cow-calf operators should move the cows to a clean pasture during the calving season. If no pasture is available, dirt mounds and a course straw bedding are recommended.

PESTS AND PESTICIDES

Although science and technology have been the great multipliers in increasing our food supply, potential food supplies are still destroyed by the ravages of pests. For example, in the high-rainfall belt of tropical Africa, the dreaded tsetse fly has kept large areas out of agricultural production. It is estimated that if *trypanosomiasis*, a disease borne by the tsetse fly, were brought under control the Savannah pastures of the tsetse fly infested area would carry a cattle population of 140 million head, which is more than the cattle population of the United States.

Fig. 1-23. Face-fly avoidance behavior of dairy cows, showing rosette-like formation with heads directed medially. (Courtesy, E. Schmidtmann, Livestock Insects Laboratory, Agricultural Environment Quality Institute, Beltsville, MD)

• **Pesticides**—*A pesticide is a substance that is used to control pests.* Pesticides are an integral part of modern agricultural production and contribute greatly to the quality of food, clothing, and forest products we enjoy. Also, they protect our health from disease and vermin. Pesticides have been condemned, however, for polluting the environment, and in some cases for posing human health hazards. Unfortunately, opinions relative to pesticides tend to be polarized. A report by the National Research Council summarized the situation as follows:

> Users of pesticides fear that they will be regulated to the point where pests cannot be effectively controlled, with the concomitant losses of food while opponents of the use of pesticides fear that people are being poisoned and that irreversible damage is being done to the environment.[1]

No pest control system is perfect; and new pests keep evolving. So, research and development on a wide variety of fronts should be continued. We need to develop safer and more effective pesticides, both chemical and nonchemical. In the meantime, there is need for prudence and patience.

STRAY VOLTAGE

Stray electrical voltage has caused serious problems on many dairy farms—affecting animal behavior and lowering milk production, although it may affect other animal species also. Contrary to popular belief, stray voltage is not new; it is as old as electricity itself. However, it has become a problem on many farms recently for two reasons: (1) There is more electrical load on today's farms; and (2) in the last 20 years we have used more equipment grounding for safety purposes.

Stray voltage is excessive voltage between two animal contact points. The conditions that cause stray voltage are, electrically, quite simple: If sufficient voltage is present, it may force a current through any available conductor, including a cow's body. Cows are good conductors because of their body design (the length from mouth to front and rear legs); cows bridge the gaps between electrically grounded objects and "true earth." The cow doesn't feel the voltage as such; she feels the tingling current running through her body.

People seldom feel the current for several reasons. Usually, caretakers wear rubber-soled shoes when in the barn, whereas the bare-footed cow stands on concrete that is often wet. Also, humans have only two legs instead of four like the cow, and human's legs touch the floor near the same vicinity.

- **Sources of stray voltage**—Any electrical condition which creates large enough voltage between two animal contact points may create a stray voltage problem. The source of stray voltage may be either "on-farm" or "off-farm."

On-farm voltage problems stem from defective equipment, faulty wiring, bad connections, or having several 120-volt motors on the same line. On-farm stray voltage can be minimized by maintaining good electrical wiring systems that meet the requirements of the National Electric Code. Also, properly balanced 120-volt circuits and conversion of large 120-volt meters to 240 volts will reduce the effect of secondary neutral voltage drops at the farm service entrance.

Off-farm voltage comes onto the farm through the electrical supplier's lines. Voltage will vary with the load and the natural grounding ability of the area. As usage increases, so may stray voltage. Heavier loads are seen at milking time and in the fall when grain dryers may be running on many farms.

- **Signs of stray voltage**—One or more of the following signs may indicate that stray voltage exists in a dairy:

1. Cows reluctant to enter the parlor.
2. Cows nervous in parlor.
3. Uneven milk let-down and milk-out.
4. Increased mastitis.
5. Reduced feed intake in the parlor.
6. Reluctance to drink water.
7. Lowered milk production.

But detection of stray voltage is not easy! Other factors such as mistreatment of animals, milking machine problems, disease, sanitation, and nutritional disorders can create problems which manifest themselves in the seven symptoms mentioned above.

- **Use voltmeter to monitor voltage**—The only sure method to determine if significant stray voltage is present is to have a qualified person perform a stray voltage survey, using approved equipment and monitoring the voltage through one, and preferably two, milkings. Point to point measurements between cow and contact points will determine if the voltage is actually getting to the cow. Generally, stray voltage is not constant throughout the day; so, readings should be taken over a long period.

POLLUTION POTENTIAL OF GRAZING LANDS

Little pollution potential exists from pasture systems with low animal densities or numbers, or where pastures are rotated. So, except for high-density pasture systems involving a number of animals, pollution is no problem. Nevertheless, some environmentalists have centered their attack on the grazing of public lands.

Grazing influences the environment on federal lands. Under poor range management, the environment is affected adversely; under good range management, such as exists on most ranges today, grazing actually improves the environment.

Eating of plant materials by animals is a natural process in earthly and aquatic systems. Thus, the coming of the white colonists to what is now the United States, along with the introduction of domestic animals, did not constitute an entirely

[1] *Pesticide Decision Making*, Vol. VII of the Analytical Studies for the U.S. EPA, NRC, National Academy Press, 1978, pp. 14–15.

new component in the environment. Rather, domestic animals replaced, or added to, the wild animals that were already there.

The environmental effects of grazing depend upon the kind of range, the intensity of grazing, and the kind of management employed to control livestock on the range. it is generally recognized that unregulated heavy grazing results in loss of desirable forage plants, increased runoff and erosion, and other indications of range deterioration. On the other hand, planned seasonal grazing and controlled animal distribution foster rapid vegetational growth. Most grazing experiments have shown that ranges may be improved more rapidly under proper grazing management than with no grazing at all.

There is no evidence that well-managed grazing of domestic livestock is incompatible with a high-quality environment. But there is ample evidence that managed grazing by livestock enhances certain uses and that poor management detracts from them. Properly managed grazing is a reasonable and beneficial use of the range.

Ecologists tell us that good range management will support more wildlife than the wilderness. This explains why big game numbers on federal lands have increased during recent years, and why wildlife production is an increasingly important use of rangelands.

Indeed, ranges actually improve while being properly utilized by domestic livestock. The benefits which accrue to the range include increased vegetation cover, improved plant species composition, improved soil fertility and soil structure, and greater yield of high-quality water. When sheep and cattle go, rank underbrush takes over, and fire becomes a real hazard.

Both upland birds and big game animals are benefited by grazing that promotes good cover for mating sites and enhanced food supply and other habitat requirements.

On ranges with mixed types of vegetation, herbaceous species increase and browse species decline when grazed only by game. The converse is true when cattle graze the land. The combined grazing by two groups of animals maintains a better balance of browse species—preferred by game animals, and of herbaceous species, preferred by cattle and sheep.

Heavy livestock grazing is beneficial to irrigated pastures used by geese and other migratory waterfowl. Unless the vegetation is closely cropped, these areas are unattractive to the birds.

Thus, livestock grazing of the public lands is contributing to improved wildlife habitat conditions and increased numbers of game animals. Range development programs, particularly livestock water developments, have made more public land usable by game animals and is partly responsible for the vast increase in game numbers over the years.

On many grass-shrub ranges, livestock grazing reduces the danger of fire by preventing buildup of dry grass, which is highly flammable.

Grazing systems and manipulation of vegetation can create contrast in vegetation color and pattern, thereby improving the aesthetic value of the landscape. Also, the livestock industry is traditional to the West; hence, a well-managed range with its cattle herd and roundup, or with its sheep camp, has recreational values. Indeed, cattle and sheep on the landscape are pleasing to tourists who come to view the "old West."

Ranges properly grazed by hoofed animals produce safe water. Counts of fecal coliform organisms, as indicators of water pollution by warm-blooded animals, relate more closely to the quantity of the fecal material than to the kind of animal. Investigations have shown that the count of harmful bacteria in streams is no greater in areas grazed by livestock than in areas grazed by wild animals alone, and that modern livestock grazing has little effect upon the chemical and physical quality of the water.

It is noteworthy, too, that few western ranges are ever in a stable, natural condition, whether or not they are grazed by domestic animals. Rather, most of them are in a stage of vegetational development following disturbances by such phenomena as drought, flood, avalanche, frost, or fire. Also, cyclical phenomena, such as large numbers of deer, rodent epidemics, or insect plagues, temporarily change the natural ecosystems. Thus, an absolutely stable rangeland is seldom attained or maintained.

Significantly, the greatest diversity of animal and plant species and the highest rates of reproduction occur when the landscape supports many stages of ecosystem development. Fire, grazing, and drought stimulate plants and animals to new growth. Each stage of vegetational development is more productive of certain animal species than of others.

Finally, in an era of food shortages, the contribution of properly managed federal lands in terms of food and fiber production need to be recognized. More and more grains will be used for direct human consumption. As a result, there will be an increased reliance on ranges for meat and wool production. It just makes sense to preserve all the natural food and fiber that we can. Remember that petroleum is not needed to make wool. Remember, too, that cattle and sheep are completely recyclable. It takes thousand of years to create coal, oil, and natural gas; and when they're gone, they're gone forever. But animals produce a new crop each year and perpetuate themselves through their offspring.

Approximately 261 million acres of federal land are administered for livestock grazing. In 1980, lands in the 11 western states administered by the Bureau of Land Management and the U.S. Forest Service provided grazing all or part of the year for 5,981,980 head of all classes of livestock—cattle, horses, sheep, and goats.

Both livestock producers and environmentalists need to recognize (1) that forage is a renewable natural resource, which regrows each year and is wasted unless it is utilized annually; (2) that grazing on federal rangelands helps to keep the natural environmental systems active and productive; (3) that we cannot allow overgrazing by domestic livestock, bison, deer, or wild horses; and (4) that grazing must be scientifically controlled and responsive to the needs of all users.

Indeed, it may be said that the influence of people on, and the use of, the environment will determine how well we live—and how long we live.

POLLUTION LAWS AND REGULATIONS

Invoking an old law (the Refuse Act of 1899, which gave the Corps of Engineers control over runoff or seepage into any stream which flows into navigable waters), the U.S. Environmental Protection Agency (EPA) launched a program to control water pollution by requiring that all cattle feedlots which had 1,000 head or more the previous year must apply for a permit by July 1, 1971. The states followed suit; although differing their regulations, all of them increased legal pressures for clean water and air. Then followed the Federal Water Pollution Control Act Amendments, enacted by Congress in 1971, charging the EPA with developing a broad national program to eliminate water pollution.

Owners/operators of animal feeding facilities with more than 1,000 animal units must apply. Animal units are computed as follows: multiply number of slaughter and feeder cattle by 1.0; multiply number of mature dairy cattle by 1.4; multiply swine weighing over 55 lb by 0.4; multiply number of sheep by 0.1; and multiply number of horses by 2.0. (See Table 1–4, footnote 1, for what constitutes 1,000 animal units.)

TABLE 1–4
SUMMARY OF REGULATIONS

Feedlots with 1,000 or More Animal Units[1]	Feedlots with Less than 1,000 but with 300 or More Animal Units[2]	Feedlots with Less than 300 Animal Units
Permit required for all feedlots with discharges[3] of pollutants.	Permit required if feedlot— 1. Discharges[3] pollutants through an unnatural conveyance, or 2. Discharges[3] pollutants into waters passing through or coming into direct contact with animals in the confined area. Feedlots subject to case-by-case designation requiring an individual permit only after on-site inspection and notice to the owner or operator.	No permit required unless— 1. Feedlot discharges pollutants through an unnatural conveyance, or 2. Feedlot discharges pollutants into waters passing through or coming into direct contact with the animals in the confined area, and 3. After on-site inspection, written notice is transmitted to the owner or operator.

[1]More than 1,000 feeder or slaughter cattle, 700 mature dairy cows (milked or dry), 2,500 swine weighing over 55 lb *(24.9 kg)*, 500 horses, 10,000 sheep or lambs, 55,000 turkeys, 100,000 laying hens or broilers with continuous overflow watering, 30,000 laying hens or broilers with liquid manure handling, 5,000 ducks; or any combination of these animals adding up to 1,000 animal units.

[2]More than 300 slaughter or feeder cattle, 200 mature dairy cows (milked or dry), 750 swine weighing over 55 lb *(24.9 kg)*, 150 horses, 3,000 sheep, 16,500 turkeys, 30,000 laying hens or broilers with continuous overflow watering, 9,000 laying hens or broilers with liquid manure handling, 1,500 ducks; or any combination of these animals adding up to 300 animal units.

[3]Feedlot not subject to requirement to obtain permit if discharge occurs only in the event of a 25-year, 24-hour storm event.

SUSTAINABLE AGRICULTURE

Fig. 1–24. Contour farming—a field of soybeans. (Courtesy, USDA, Soil Conservation Service)

Endangered species—and more! Today, it is endangered planet, endangered people and animals, and endangered agriculture. Among the deluge of warnings of environmental catastrophes are:

- Pollution-caused warming of the atmosphere, known as the *greenhouse effect,* threatening weather changes that could render large areas of the planet unproductive and uninhabitable.
- Toxic and radioactive wastes and dumped garbage that could poison drinking water and despoil the land.
- Chemical pollution that is depleting the atmosphere's protective ozone layer.
- Slashing and burning of tropical rain forests, driving thousands of species to extinction, increasing the amount of carbon dioxide in the atmosphere, and contributing to the greenhouse effect that warms the earth.

Is ¼ lb of hamburger worth ½ ton of Brazil's rain forest? Is 67 sq ft of rain forest (an area about the size of one small kitchen) too much to pay for 1 hamburger? Should we form cattle pastures to produce hamburgers in the Amazon, or should we retain the rain forest and the natural environment? These and other similar questions are being asked too little and too late to preserve much of the great tropical rain forest of the Amazon and its environment. It took nature thousands of years to form the rain forest, but it took a mere 25 years for people to destroy much of it. And when a rain forest is gone, it is gone forever![2]

[2]Uhl, C. and G. Parker, "Is a One-Quarter Pound Hamburger Worth a Half-Ton of Rain Forest?," *Interciencir,* 1986. Sept.-Oct., Vol. II, No. 5, p. 213.

Although less dramatic, the Amazon rain forest story has been, or is being, repeated all over the world in the form of the greenhouse effect, toxicities, polluted streams, and/or other harbingers of threats to our environment. Too long we have managed our nonrenewable resources like there is no tomorrow! Now, the situation is being righted. World-wide, environmental quality and economic efficiency are in vogue. In the United States, this movement is called *Sustainable Agriculture.*

Sustainable agriculture is often described as farming that is ecologically sound and economically viable. It may be high or low input, large scale or small scale, a single crop or diversified farm, and use either organic or conventional inputs and practices. Obviously, the actual practices will differ from farm to farm. A definition follows.

Sustainable agriculture is farming with reduced off-farm purchased inputs of pesticides, herbicides and fertilizers, along with reduced negative impact on natural resources and improved environmental quality and economic efficiency, while producing and distributing abundant, nutritious, affordable, high-quality foods and fibers for America and world markets.

The development of improved crops, cropping systems, irrigation, farm management, and marketing will be needed to make farms more profitable and sustainable. Typically, such farms will rely more on biological resources and management than on nonrenewable inputs of energy and chemicals. The foundation of a sustainable farm system is a comprehensive understanding of the land, the farm resources and operations, and potential short- and long-term markets.

Many of the practices advocated under sustainable agriculture are not new; they involve such timeless agricultural practices as soil erosion control, the protection of groundwater, the use of legumes as a source of nitrogen, biological insect and weed control, and the use of pastures as a primary feed source.

ANIMAL WELFARE/ANIMAL RIGHTS

In recent years, the behavior and environment of animals in confinement have come under increased scrutiny of animal welfare/animal rights groups all over the world. For example, in 1987 Sweden passed legislation designed (1) to phase out layer cages as soon as a viable alternative can be found; (2) to discontinue the use of sow stalls and farrowing crates; (3) to provide more space and straw bedding for slaughter hogs; and (4) to forbid the use of genetic engineering, growth hormones, and other drugs on farm animals except for veterinary therapy. Also, the law provides for fining and imprisoning violators.

Animals welfarists see many modern practices as unnatural, and not conducive to the welfare of animals. In general they construe animal welfare as the well-being, health, and happiness of animals; and they believe that certain intensive production systems are cruel and should be outlawed. The animal rightists go further; they maintain that humans are animals, too, and that all animals should be accorded the same moral protection. They contend that animals have essential physical and behavioral requirements, which, if denied, lead to privation, stress, and suffering; and they conclude that all animals have the right to live.

Livestock producers know that the abuse of animals in intensive/confinement systems leads to lowered production and income—a case in which decency and profits are on the same side of the ledger. They recognize that husbandry that reduces labor and housing costs often results in physical and social conditions that increase animal problems. Nevertheless, means of reducing behavioral and environmental stress are needed so that decreased labor and housing costs are not offset by losses in productivity. The welfarists/rightists counter with the claim that the evaluation of animal welfare must be based on more than productivity; they believe that there should be behavioral, physiological, and environmental evidence of well being, too. And so the arguments go!

But wild animals are often more severely stressed than domesticated animals. They didn't have caretakers to store feed for winter or to irrigate during droughts; to provide protection against storms, extreme temperatures, and predators; and to control diseases and parasites. Often survival was grim business. In America, the entire horse population died out during the Pleistocene Epoch. Fossil remains prove that members of the horse family roamed the plains of America (especially the area that is now known as the Great Plains of the United States) during most of tertiary period, beginning about 58 million years ago. Yet no horses were present on this continent when Columbus discovered America in 1492. Why they perished, only a few thousand years before, is still one of the unexplained mysteries. As the disappearance was so complete and so sudden, many scientists believe that it must have been caused by some contagious disease or some fatal parasite. Others feel that perhaps it was due to multiple causes, including (1) climatic changes, (2) competition, and/or failure to adapt. Regardless of why horses disappeared, it is known that conditions were favorable to them at the time of their reestablishment by the Spanish conquistadores about 500 years ago.

To all animal caretakers, the principles and application of animal behavior and environment depend on understanding; and on recognizing that they should provide as comfortable an environment as feasible for their animals, for both humanitarian and economic reasons. This requires that attention be paid to environmental factors that influence the behavioral welfare of their animals as well as their physical comfort, with emphasis on the two most important influences of all in animal behavior and environment—feed and confinement.

Animal welfare issues tend to increase with urbanization. Moreover, fewer and fewer urbanites have farm backgrounds. As a result, the animal welfare gap between town and country widens. Also, both the news media and the legislators are increasingly from urban centers. It follows that the urban views that are propounded will have greater and greater impact in the years ahead.

FOOD SAFETY AND DIET/HEALTH CONCERNS

Many food safety and diet/health concerns are unwarranted. American consumers are prone to over-react to rumors relative to their food. They care little about what they put on their backs, but they are greatly concerned about what goes into their stomachs.

America's food supply is the safest in the world! Nevertheless, there is need for constant vigilance and improvement, especially in animal products which are subject to all the hazards of other foods (spoilage, pesticides, toxicities), plus being capable of transmitting, or serving as passive carriers, of certain diseases to humans.

In colonial times, the livestock producer slaughtered animals and processed meats, milked the cows, and gathered the eggs; then, delivered the products door-to-door to urban customers. If the products were not acceptable (spoiled meat, sour milk, cracked eggs), the matter was resolved quickly and on the spot, or the producer lost a customer. Today, the public expects the livestock team—farmers, processors, and retailers—to provide wholesome and safe products free from disease agents, toxic substances, and pesticide and drug residues.

Uptake of pesticides by animals, leading to residues in animal products, can result either from direct application of pesticides to animals or from animals ingesting feeds carrying pesticide residues. Drug residues are caused by (1) producers failing to withdraw drugs from livestock far enough in advance of marketing products; (2) contaminated feed storage, mixing, and handling equipment; and/or (3) the wastes (feces and urine) of treated animals coming in contact with untreated animals. *Reading and following the directions on the label is the key to safe pesticide and drug use.*

Dr. C. Everett Koop, former U.S. Surgeon General, is the authority for the following statement:

> People who are worried about pesticides fail to recognize that cancer rates in the United States have dropped remarkably over the past 40 years. During this period of time, stomach cancer has dropped more than 75%, and rectal cancer has dropped more than 65%. The only cancer going up today is cigarette-induced lung cancer.

Dr. Koop continued:

> The same hysteria that sometimes accompanies food safety issues also has affected how Americans look at diet and health. However, one major issue—cholesterol—is waning, as consumers realize how much it has been oversold. The cholesterol bubble has been pricked and is slowly deflating. While cholesterol is a risk factor in coronary heart disease, scientists consider other risk factors, such as smoking, hypertension, and heredity, to be much more significant than cholesterol. Because cholesterol is manufactured in the body naturally, diet does not have the direct relationship to blood levels of cholesterol that many mislead laymen assume.[3]

Because the welfare of the nation is dependent upon the health of its people, animal (and other) products are carefully monitored by various government agencies to assure consumers that they are wholesome and safe; and because of recognizing the importance of consumers in the safety of their food, the private sector may do additional testing. The agencies most responsible for this important work are:

1. The U.S. Department of Health and Human Services, including the following agencies: The Center for Disease Control, the Food and Drug Administration (FDA), and the National Institute of Health.
2. The U.S. Department of Agriculture, including the following agencies: the Agricultural Research Service, the Animal and Plant Health Inspection Service, the Cooperative State Research Service, the Federal Extension Service, the Labeling and Registration Section, and the Veterinary Service Division.
3. State and local government agencies.
4. International organizations engaged in health and/or nutrition activities, including the World Health Organization (WHO), and Food and Agriculture Organization (FAO).
5. Private industry groups such as the National Livestock and Meat Board and the National Dairy Council.
6. Professional organizations, including dentists, dietitians, doctors, health educators, nurses, and public health workers.
7. Food processors and retailers. For example, in 1988 Lucky Stores, Inc., one of the largest food handlers in California, initiated a testing program in cooperation with the California Department of Food and Agriculture (CDFA) to have the CDFA check their produce for pesticide residues on a regular basis, as a way of assuring their customers that the produce that they buy in Lucky Stores is completely safe. Some food handlers are using private laboratories to conduct similar tests.

CONSERVE ENERGY

Fossil fuels—the stored photosynthesis of previous millennia—are like a bank account. There is nothing wrong with drawing upon either of them, but neither is inexhaustible. It is highly imprudent not to be aware of big withdrawals and not to cover them. Within a short span of a few years, the world made the transition from a positive energy balance based upon the capture of the energy of the sun via green plants, crops, and forests to an imbalance, or even a negative balance, by resorting primarily to the bank of trapped sun energy of fossil fuels that had accumulated over millions of years. Currently, the global use of energy resources is increasing at an alarming rate.

Fig. 1–25. An Oriental rice peasant, using animal power (water buffalo), expends only 1 calorie of energy to produce 50 calories of food. By comparison, the average U.S. farmer, using mechanical power (tractors), expends 2.5 calories of fuel energy to produce 1 calorie of food.

[3] Koop, Dr. C. Everett, *Gulf Coast Cattleman*, Vol. 55, No. 9, Dec. 1989, p. 9.

Everyone knows that cars and airplanes are powered by fuel. But how many people realize that modern, mechanized food production in the developed countries requires an extra input of fuel, which is mostly of fossil origin? How many people know that the developed nations have used fossil fuels to supplement the natural energy that comes directly from the sun on a day-to-day basis, and that they have grown dependent upon them?

Of course, the direct input of fuel into food production is of rather recent origin. It all began in a very small way about 1840, when fuel-powered ships transported fertilizer (guano, and later bone meal) from South America to Europe. Then, after 1910, transportation vehicles relied almost exclusively on fossil fuels. But the direct use of fossil fuels in agriculture started with the manufacturing of chemical fertilizer beginning about 1922—little more than 60 years ago. Following closely in period of time, the tractor was substituted for horses, mules, and oxen—eventually almost completely replacing them.

In addition to food production on the farm, there are two other important steps in the food line as it moves from the producer to the consumer; namely, processing and marketing, both of which require bigger energy inputs than to produce the food on the farm (see Table 1-5).

Table 1-5 points up the increasing drain that modern food production is putting on the energy supply. In 1990, U.S. farms put in 2.8 calories of fuel per calorie of food grown, 3.1 times more than the on-farm energy input in 1940.

TABLE 1-5
MODERN FOOD PRODUCTION IS INEFFICIENT IN ENERGY UTILIZATION—THE STORY FROM PRODUCER TO CONSUMER[1]

Year	On the Farm	Food Processing	Marketing and Home Cooking	Total/ Person/ Year
1940[2]				
Million kcal . .	0.9	2.2	2.1	5.2
Percent	18.0	42.0	40.0	100.0
1990[3]				
Million kcal . .	2.8	5.7	4.6	13.1[4]
Percent	21.4	43.5	35.1	100.0
Increase, times 1940-1990	3.1	2.6	2.2	2.5

[1]Energy in million kcal used per capita to produce 1 million kcal of food in the U.S.

[2]Values from Borgstrom, G., "The Price of a Tractor," *Ceres*, FAO of the U.N., Rome, Italy, Nov.-Dec. 1974, p. 18, Table 3.

[3]Author's estimate based on several reports detailing trends in energy usage.

[4]This means that in 1990 it required 13.1 million kcal to produce 1 million kcal of food for each person, a daily consumption of 2,740 kcal (1,000,000 ÷ 365 = 2,740)

Table 1-5 also shows that, in the United States in 1990, a total of 13.1 calories were used in the production, food processing, and marketing-cooking for every calorie of food consumed, with a percentage distribution of the total cost of energy at each step from producer to consumer as follows: on the farm 21.4%; food processing, 43.5%; and marketing and home cooking, 35.1%. In 1940, it took only 5.2 calories—somewhat less than half the 1990 figure—to get 1 calorie of food on the table. It's noteworthy, too, that more energy is required for food processing and marketing-home cooking than for growing the product; and that, from 1940 to 1990, the on-the-farm energy requirement increased by 3.1 times, in comparison with an increase of 2.6 and 2.2 times for each of the other steps—processing and marketing/home cooking.

Prior to the advent of machines and fuel in crop production, 1 calorie of energy input on the farm produced about 16 calories of food energy. Today, on the average, U.S. farms put in about 2.8 calories of fuel per calorie of food grown; hence, to produce a daily intake of 3,000 calories of edible food from cultivated crops may require 8,400 calories of energy from fossil fuels—an exhaustible source. It's more surprising yet—and thought-provoking—to know that, even today in the poorer or developing countries, it takes only 1 calorie to produce each 10 calories of food consumed. The Oriental wet rice peasant uses only 1 unit of energy to produce 50 units of food energy. This gives the Orientals a favorable position among the major powers as the energy crisis worsens.

The following additional points are pertinent to any energy conservation program:

1. **Photosynthesis fixes energy**—Photosynthesis is by far the most important energy-producing process. But currently only about 1% of the solar energy falling on an area is fixed by photosynthesis; and only 5% of this captured energy is fixed in a form suitable as food for humans. Thus, (a) the manipulation of plants for increased efficiency of solar energy conversion, and (b) the conversion of a greater percentage of total energy fixed as chemical energy in plants (the other 95%) into a form available to humans would appear to hold great promise in solving the future food problems of the world.

2. **Animals step up energy**—The increase in the energy level through animal products—through animals consuming the photosynthetic energy in crops—almost equals the energy subsidies at each of the two steps after the product leaves the farm (in food processing and marketing-home cooking—see Table 1-5).

3. **Crop residues contain energy**—Crop residues left in the field, above or below the soil surface, may well constitute four to five times more energy than is harvested. Increasingly, this potential source of added feed, organic fertilizer, and energy will be utilized in the future.

4. **Increased yields; the law of diminishing returns in energy**—Modern intensive farming has markedly increased crop yields per acre and per man-hour—by as much as 50- to 100-fold. But this has been done at the cost of large inputs of fuel (including electricity).

For a surprising number of modern cropping systems, a 10- to 50-fold increase in the energy output merely doubles or triples the food energy. Substantial expenditures fail to produce corresponding increases in yields. Thus, the law of diminishing return prevails.

High petroleum costs have spurred a search for other energy sources and for means of conserving energy. Higher productivity of the agriculture of tomorrow must be achieved through ingenious approaches in order to reverse the present lopsided energy balance. In obtaining increased food yields, we must consider the use of energy to produce energy. We must remember that photosynthesis is by far the most important energy-producing process; indeed, that it is the only basic food manufacturing process in the world. We must remember, too, that grazing animals do not require fuel outside of their own body use to harvest the energy and other nutrients of grass (solar energy converted into chemical energy by grass), a renewable resource.

The headquarters is the center of all farm/ranch business. This shows the attractive ranch headquarters of Martin Jorgensen, noted cattleman, near Ideal, South Dakota.

SECTION

2

BUSINESS ASPECTS

Contents	Page

Fig. 2-1. The attractive Maddox Dairy headquarters, Riverdale, California. The herd consists of more than 3,600 lactating cows with a rolling herd average of 20,850 lb milk and 3.72% fat on 3× daily milking. (Photo by A. H. Ensminger)

Agriculture, with assets totaling $813.1 billion in 1987, (1) ranks as the nation's biggest single industry, and (2) has a value equivalent to 40% of all manufacturing corporations in the United States. Moreover, it is destined to get bigger and more complicated.

From 1935 to 1988, within a span of 53 years, the number of farms decreased from 6.8 million to 2.2 million and the size of farms increased from 154.8 acres to 463 acres.[1] Thus, within 53 years, nearly 68% of the farms disappeared from American agriculture and the average size of farms tripled. With this transition, herds, flocks, and feedlots became bigger.

In 1952, each farm worker supplied enough food and fiber for 17 persons, including self; by 1988, each farm worker supplied enough food and fiber for 93 persons, including self.

The above trends to bigness will continue.

The business and management aspects of animal production will be increasingly important in the future. Changes in the type of business organization and in financial management will come. More capital will be required, more money will be borrowed, competent managers will be in demand, better and more complete records will be necessary, futures trading will increase, and livestock operators will become more knowledgeable relative to tax management, estate planning, and liability.

[1]The Census definition of a farm is as follows: "A place that sells $1,000 a year in ag products."

The net result will be that those engaged in the business of agriculture must treat it as the big business that it is and become more sophisticated and efficient; otherwise, they won't be in business very long.

The successful stock raisers of the future will possess business acumen as well as superior livestock know-how. Also, they will need operations that are large enough to provide their families an adequate standard of living and generate enough capital to keep expanding. Since profit margins will likely decline still further, there will be greater stress on business and financial management skills.

Generating both equity and debt capital, or risk and borrowed capital, will be one of the main concerns of the future livestock producer. The large investment, plus the need to keep competitive by utilizing new and usually expensive technological advances, will cause capital to be very important.

To obtain capital, several things will be necessary. The producer-business person will have to prepare (1) profit and loss statements to show that the operation is profitable, (2) financial statements to show that progress is being made, and (3) cash-flow projections to show loan repayability. Then, and then only, will the borrower be ready to go looking for money.

Skill in capital budgeting and analyzing alternative investment opportunities will be needed to see that the limited capital is invested where payoff will be the greatest. Producers will also have to exercise budget and cost controls of their businesses. Skill in building sound credit will be needed.

The greatest payoffs in the future are likely to accrue to those producers who improve their skills in business and financial management.

BUSINESS ORGANIZATION

Big land-livestock operations will get bigger, demanding more and more capital and top management. In the years ahead, many investors will become part owners in land and livestock, much as they now do through corporate stocks and bonds; and they will leave the management of the holdings to the professionals. Such an arrangement will also make it possible to (1) diversify in countries and types of investments—in different areas of the United States, and in Australia, Canada, South America, and other areas where there are vast acreages of rangeland with great potential for improvement that can be secured at reasonable prices; (2) minimize risks of loss from droughts and local depressions; (3) obtain for investors, big and little, the benefits that accrue to bigness, such as lower investment per animal unit, and lower feed and labor costs; (4) furnish recreational and vacation areas on farms and ranches of which they are part owner; and (5) provide know-how, continuity, and able management.

Part ownership in land and livestock investment companies affords a modern way in which to spread investments and minimize risks—as is done with stocks and bonds, grain and livestock futures, and syndicated sires.

TYPES OF BUSINESS ORGANIZATION

The success of today's farming is very dependent on the type of business organization. No one type of organization is superior under all circumstances; rather, each situation must be considered individually. The size of the operation, the family situation, the enterprises, the objectives—all these, and more, are important in determining the best way in which to organize the farming business.

Three major types of business organizations are commonly found among farming enterprises: (1) the sole proprietorship, (2) the partnership, and (3) the corporation. Among the factors which should be considered when deciding which business form best fits a given set of circumstances are the following:

1. Which type of organization is most likely to be looked upon favorably from the standpoint of more credit and capital?
2. How much capital will be required of each individual involved?
3. Are there tax advantages to be gained from the business organization?
4. Is expansion of the business feasible and facilitated?
5. Which type of organization reduces risks and liability most?
6. Which type of organization can be terminated most easily and readily?
7. Which type of ownership provides for the most continuity and ease of transfer?
8. What costs for legal and accounting fees are involved, in setting up the organization and in the preparation of the annual reports required by law?
9. Who will manage the business?

Most agricultural enterprises are operated as sole proprietorships, not necessarily because this is the best type of organization, but with no effort to form some other type of organization it naturally results. Both the partnership and the corporation, which require special planning and effort to bring about, are well suited to the operation of large livestock establishments.

ACQUIRING A FARM OR RANCH

Acquiring a suitable farm is a big problem, particularly for beginners. They must compete for available farms with established farmers as well as with other beginners. Many established farmers need more land to enlarge their operations. Others move during the year, getting a better or more suitable farm. Some simply move to a new locality for personal reasons.

Whatever the motive for acquiring a farm or ranch, some buyers will be happy with their purchases. Others will rue the day they made the decision. How well the farm or ranch was bought, and how it was bought, may make the difference. Since most farmers buy only one farm and operate it for a lifetime, it follows that the vast majority of them lack expertise relative to the procedures to follow when acquiring a farm or ranch. The guidelines that follow have been prepared to fill this need.

Fig. 2–2. Acquiring a farm/ranch ranks second in importance only to acquiring a spouse! Some buyers will be happy with their purchase, whereas others will rue the day that they made the decision. This shows an attractive farm headquarters. (Courtesy, *Livestock Breeder Journal*, Macon GA)

WAYS TO ACQUIRE A FARM OR RANCH

Farms or ranches may be acquired through gift or inheritance, by marriage, by renting or leasing, or by purchasing.

1. **Gift or inheritance.** Many farms and ranches are inherited and stay in the family for several generations. Certainly, inheriting a farm or ranch is a real advantage to anyone desiring to stay in the business and having the know-how and ability to operate the enterprise. Farm and ranch land and livestock (as well as corporate stock, U.S. savings bonds, mutual funds, money, whole life and annuity life insurance, and real estate) can be transferred to relatives and friends, with certain tax savings, provided certain well-established rules are observed.

The basic rule in giving land or animals to another member of the family is that the donor (person giving the property) must give up control.

Income producing property can be given to relatives or friends, and the income will be taxed to the person receiving the gift (donee). But the property must actually be transferred. Any strings attached to the ownership whereby the donor can have control over the property or get it back at some future time will not meet the requirements of the law. The basic elements are:

a. There must be intention to make a gift.

b. Transfer of legal title and control.

c. The donee accepts the gift.

d. No consideration (money or property) to be exchanged for the property.

Persons expecting to inherit a farm or ranch, should become fully cognizant of the inheritance tax laws. Also, it is most important that they have clear title to the land. If the title of ownership of a farm or ranch is left unsettled through two or three generations, the value of the land may almost be expended in settling the estate.

Where gift or inheritance money or property are involved, always seek the advice of a tax accountant and/or tax attorney.

2. **Marry it.** Although young people wanting to enter farming or ranching are frequently admonished, facetiously, that they should "marry for love, but love a person with plenty of money," there is more than a little bit of truth in the advice.

In some countries, the social orders call for the parents to arrange the marriages of sons and daughters, primarily to keep them within the same class strata, thereby not dividing property with those who "have not." However, this method is fast giving way to the new social order, which, like that in the United States, results in marriages between individuals of vastly different amounts of wealth.

3. **Rent or lease.** The main advantage in renting over buying is that less capital is required and less financial risk is involved. The main disadvantages are insecurity of tenure and that the farming enterprise may be limited in size or kind because the landowner is reluctant to make needed additional investments in buildings and facilities. These disadvantages can be minimized, and sometimes eliminated, by a suitable lease—an agreement between landlord and tenant under which the farm or ranch is rented and operated. Such a lease should always be in writing.

The most common types of farm and ranch leases are:

a. **Cash leases.** This is a good type of lease for (1) the small farm or where the owner lives at a distance, and (2) a tenant who has adequate livestock, equipment, and working capital. It encourages livestock farming because all of the crop can easily be fed on the farm. Also, it is simple, with little chance for controversy.

There are two types of cash leases: (1) that type in which a fixed rent per acre is agreed upon when the lease is drawn, and (2) that type in which the rent is adjusted to prices of farm products which prevail during the lease year. Under the second plan, the land owner bears part of the risk of price changes; however, it is difficult to keep cash rent in line with farm product prices. If product prices are used as a basis for rent changes, the products, markets, and dates should be specified.

The owner may prefer a cash lease because (1) the amount paid is definite, and (2) it requires less supervision. On the other hand, it may not always be desirable from the standpoint of the owner because (1) it generally makes for lower income, (2) it gives less control of the farm, and (3) it is difficult to collect rent if crops fail.

The tenant may prefer a cash lease because (1) it will make for more profit if the tenant is a successful manager, (2) it makes for more independence in the operation, and (3) it makes for more profit in the good years.

b. **Livestock share lease.** Livestock share leases vary considerably. But most of them provide for 50–50 ownership of the herd or flock; 50–50 sharing of the costs of production—especially feed and veterinary expenses; and 50–50 division of the income from the sale of the animals. Buildings are generally a cost borne by the landowner. Labor is the responsibility of the operator.

A livestock share lease fits the tenant who wants to raise livestock, but cannot finance a program. It is especially suited where tenant and owner get along well and where the owner can make a good contribution in management.

In order for this type of lease to work best, the owner should live close to the farm, and either give it personal attention or arrange for adequate management help such as can be provided through a professional farm management service.

The owner may prefer a livestock share lease because (1) it encourages more livestock and more manure, (2) low-quality crops can be utilized more easily, (3) active interest in management is retained, and (4) it generally makes for more profits.

The tenant may prefer a livestock share lease because (1) the risk is less since rent is based on net income on the farm, (2) it requires less tenant capital, (3) the owner may be more willing to make improvements, and (4) experience can be gained from the guidance of a successful owner.

A careful determination of lease provisions and putting them in writing will result in a lease that is more equitable to both tenant and landowner, and will avert later misunderstandings and friction between the two parties. Standard lease forms are available, so the detailed provisions of a lease need not be spelled out in this book.

Renting or leasing might be a desirable way to start in the livestock business even if funds are available for purchase. This is particularly true where there is an option-to-buy clause. This gives the renter an opportunity to study the ranch more carefully and gain additional management experience without committing his/her entire assets.

4. **Purchase.** Purchase of a livestock farm or ranch has the advantages of security of tenure and freedom to make management decisions. Earnings from the operator's equity capital may be added to labor and management earnings for living expenses, reinvestment in the business, or other uses. Also, the value of the land may rise over a period of time. On the other hand, ownership may involve substantial indebtedness. Also, risks of financial loss are greater than in renting.

Of course, ownership brings with it financial responsibility that is both greater and longer lasting than the financial responsibility which renting entails. Few persons buy more than one farm in a lifetime; moves are time-consuming and expensive.

Some part-time farmers, working at nonfarm jobs, use their off-farm income to move gradually into full-time farming. They use their initial savings to make a down payment on a small farm and to buy enough livestock and equipment to permit limited farming or ranching operations for the first few years. The off-farm income makes them better credit risks for lenders than if they were wholly dependent on farm earnings. They can continue to borrow to build up their farm business to a point where it will support their family and pay off previous loans. Such a gradual shift into full-time farming can usually be made with less sacrifice in family living standards and better chances of eventual success than an abrupt change to full-time farming.

Generally speaking, purchase of land is either by (a) land purchase contract, or (b) mortgage contract.

Purchase of land by use of land purchase contract has become more important in recent years. These contracts allow the use of lower buyer down payments (usually from nothing to 29%) with the balance paid over a long period of years in annual payments. For the buyer, this offers a way in which to acquire land without having to make a big down payment. For the seller, it has certain capital gain tax advantages and it usually attracts more prospective buyers and makes for a higher sale price. In order to qualify for special treatment on capital gains for federal income tax purposes, the seller must not receive more than 30% of the purchase price in the year of sale.

Mortgage contracts differ from land purchase contracts in the following ways: (1) They are of longer duration—usually 20 to 30 years, or up to 40 years, whereas land purchase contracts are commonly for 10 years or less; (2) the law provides for specified grace periods after default in payments before the seller can foreclose the mortgage; and (3) larger down payments are normally required—frequently 40 to 50% of the purchase price.

FACTORS TO CONSIDER WHEN BUYING A FARM OR RANCH

Many factors in addition to worth should be considered in the selection of a farm or ranch; among them, the following, each of which is discussed in a section which follows: (1) deeded vs government leased land; (2) water and water rights; (3) mineral rights; (4) oil and gas leases; (5) timber; (6) easements and property lines; and (7) risks. Although space limitations will not permit a discussion of them, the following factors should also be considered: (1) area and climate; (2) soil and topography; (3) improvements; (4) wind direction, windbreaks, and natural shelters; (5) service facilities, communities and markets; (6) expansion possibilities; and (7) proximity to factories or cities, with the possible advantages of (a) off-farm employment, and (b) big-city attractions; and the likely disadvantages of (c) air and noise pollution, and (d) limited expansion possibilities.

CHOOSING THE CLASS OF ANIMALS

Under some conditions a hog farmer may have one neighbor who is a cattle feeder, a second who operates a dairy, a third who keeps a sizable farm flock of sheep, a fourth who produces light horses for recreation and sport, and a fifth whose chief source of income is from poultry. All may be successful and satisfied with their respective livestock enterprises. This indicates, therefore, that several types of livestock farming may be nearly equally well adapted to an area or region. This means that the selection of the dominant type of livestock enterprise should be analyzed from the standpoint of the individual farm or ranch.

Usually a combination of several factors suggests the livestock enterprise or enterprises best adapted to a particular farm or ranch. One of these factors is the labor requirement. Available feeds and market outlets for animals and animal products are also very important. Table 2-1 may be of assistance in arriving at a decision as to the kind or kinds of livestock best suited to the individual farm or ranch.

Fig. 2-3. Choosing the class of animals is important. Sheep are adapted to rough terrain grazing. But experienced sheep herders (shepherds) are difficult to come by, and predators often inflict heavy losses on sheep.

TABLE 2-1
ANIMAL LABOR REQUIREMENTS

Animal or Product	Work Hours/Hundredweight Production[1] 1935–39	Work Hours/Hundredweight Production[1] 1982–86	No. Head Cared for in 1990 by One Worker in the Most Efficient Operations[2]	Characteristics, Requisites, and/or Conditions Under Which They Fit
	(hours)	(hours)		
Beef cattle	4.2	0.9	Cow-calf (brood cows) . . . . 300–500 Feedlot cattle 2,500	Plenty of roughages (pasture and hay); moderate capital available on intermediate and long-term basis. Plenty of grain or cheap by-product feeds; strong finances; adequate know-how in buying, feeding, and marketing cattle.
Milk cows	3.4	0.2	Milk cows 150	Plenty of labor; intensive farming; suitable outlet for milk.
Sheep	6.1	1.0[2]	Farm flock ewes . . . . 1,000 Range ewes . . . . 1,000–2,000 Feedlot lambs 7,500–10,000	Plenty of roughages (pasture and hay); moderate capital. Plenty of grain or cheap by-product feeds; strong finances; adequate know-how of buying, feeding, and marketing.
Hogs	3.2	0.3	Sows 200	Feed grains abundant; good sanitation; where capital is limited.

[1]Source: Data on "work hours/hundredweight production," except sheep, from *Agricultural Statistics 1988*, USDA, p. 395, Table 569.
[2]Estimates by M. E. Ensminger.

CAPITAL REQUIREMENTS

Those thinking of becoming livestock operators inevitably ask: "How much money will it take, and what can I make?"

It takes a lot of capital to own and/or operate a modern livestock establishment. In 1987, U.S. farm assets—investments in land, improvements, machinery, equipment, animals, feed, and supplies—totalled $813.1 billion, while farm debt totaled $153.3 billion. Thus, in the aggregate, farmers had 81.2% equity in their businesses and 18.8% borrowed money (debts). The balance sheet of U.S. farming from 1970 through 1987 is shown in Fig. 2-4.

Perhaps agriculturists have been too conservative, for it is estimated that one-fourth to one-third of American farmers could profit from the use of more credit in their operations.

In 1987, the average U.S. farmer or rancher had more than $370,000 invested in land, machinery, livestock, working capital, and farm buildings. Some were much larger; invest-

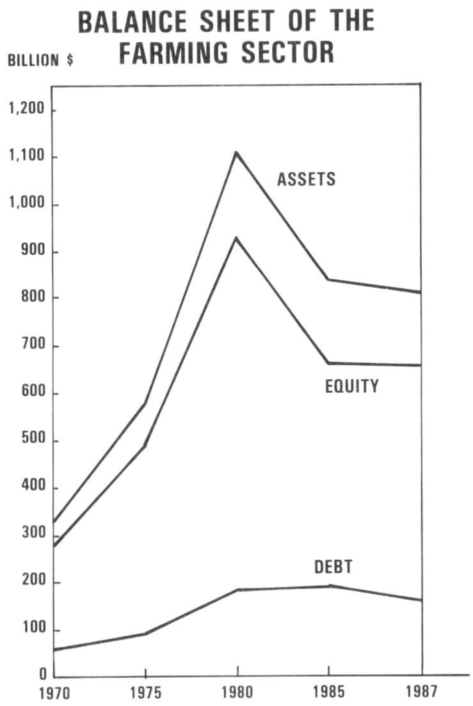

Fig. 2-4. Balance sheet of U.S. farming, showing (1) assets, (2) debts, and (3) equities. (Source: *Statistical Abstracts of the United States 1989*, U.S. Department of Commerce, p. 632, No. 1089)

ments of more than $1,000,000 per farm or ranch were not uncommon.

Another statistic which points up the enormity of capital needs is that it takes about $17.58 in farm assets to produce $1 of net farm income.[2]

Capital needs are determined primarily by (1) size of operation, (2) kind of animal enterprise, (3) ownership vs contract operation, and (4) location.

• **Size**—A commercial dairy with 5,000 lactating cows requires far more capital than a family owned and operated dairy consisting of 50 lactating cows, although the per cow capital needs are generally less with increased size.

• **Kind of animal enterprise**—The kind of animal enterprise is a factor. Thus, far more capital is needed to own and operate a milk producing dairy than to raise dairy heifer replacements. Also, it requires far more capital for a light horse establishment than for a ewe-lamb operation.

• **Ownership vs contract operation**—Ownership necessitates that the operator have, or borrow, all of the capital, whereas contract producers generally operate on the contractor's capital. Thus, contract producers of feeder pigs may own the buildings and the land that they use, but the contractor may furnish the breeding animals and the feed.

• **Location**—Some locations require more investment per animal unit than others. Table 2-2 shows the capital requirements in two different areas of Nebraska: (1) the Sandhills ranching area—a reputation cattle area; and (2) the general farming area of eastern Nebraska. Thus, based on these figures, the capital requirement for a 500-cow outfit in the Nebraska Sandhills would be $2,000,000 (500 × $4,000); while a 500-cow crop and livestock farm in eastern Nebraska would require a capital investment of $1,527,500 (500 × $3,055).

TABLE 2-2
ESTIMATED CAPITAL REQUIREMENT ON A PER COW BASIS IN TWO AREAS OF NEBRASKA[1]

	Sandhills Ranches in Nebraska		Crop and Livestock Farms in Eastern Nebraska	
Cow		$ 515		$ 515
Heifer replacement	(1/8)	85		85
Bull (share)	(1 to 25)	50		50
Pastureland	(15-16 acres)	2,800	(3-4 acres)	2,000
Hayland	(1¼-1¾ acres)	500	(½ acre)	325
Buildings and equipment		50		80
Total		$4,000		$3,055

[1]Estimates submitted by NE Beef and Agricultural Economics Specialists: Guyer and Jose. Data for 1981-82.

CATTLE FEEDLOT CAPITAL REQUIREMENTS

Cattle feeders need to know the size investment that they need, and can justify, in cattle feeding facilities. A nomograph may be used for this purpose. It can give a quick preliminary idea of cost, gross profit, returns, and investment relationships. But a nomograph should not replace more detailed figuring which should precede all major investment decisions. Also, one should realize that a nomograph will give erroneous and misleading information unless based upon accurate and realistic cost and return data from the problem at hand.

• **Cost of feedlot**—Before constructing a feedlot, cost must be considered for two reasons: (1) capital must be secured; and (2) cost must be amortized. The usual basis of computing cost is on a "per animal unit capacity." This will run about the same whether calves or yearlings are involved, because per unit capacity must consider carrying the animals to market time.

The area affects cost from the standpoint of shelter requirements and land values. Thus, because of the necessity for winter protection and shelters, feedlot costs are higher in the northern tier of states than in the South. Land values are higher in California than most areas of the United States, with the result that land costs become a factor.

Size of feedlot affects per animal cost. Most studies reveal that investment savings do accrue to the larger feedlots. Thus, the cost per animal usually decreases up to about 10,000-head

[2]Based on 1987 figures, when farm assets were $813.1 billion and net farm income was $46.264 billion ($813.1 ÷ $46.26 = $17.58). Source, *Statistical Abstracts of the United States 1989*, Department of Commerce, p. 636, No. 1099.

capacity, then it increases slightly with larger lots. The slightly higher cost per head capacity of the larger lots appears to be due to duplication in equipment and the tendency to become more highly mechanized and elaborate.

An open lot without shelter is the cheapest type of feedlot construction. In the Southern Plains area, where the weather is mild and shelters are unnecessary, investment costs range from $100 to $125 per head of capacity.[3]

Housing increases costs, and the more elaborate the housing the greater the cost. The author's consensus showed the costs given in Table 2-3 relative to three types of cattle feedlot facilities.[4]

TABLE 2-3
COSTS OF THREE TYPES OF CATTLE FEEDLOT FACILITIES

Type of Facility	Sq Ft per Animal	Cost per Animal Capacity
		($)
Open shed	20	140
Cold confinement	17	185
Warm confinement (heated)	17	300

GUIDELINES RELATIVE TO FACILITY AND EQUIPMENT COSTS

Overinvestment in facilities is a mistake. Some livestock producers are prone to invest more in facilities and equipment than reasonably can be expected to make a satisfactory return; others invest too much in feed mills. Sometimes small cattle feeders fail to recognize that it may cost half as much to mechanize to feed 500 head as it costs to mechanize to feed 2,000 head.

Fig. 2-5. Will this hay windrowing equipment pay for itself? Two guidelines relative to facility and equipment costs are presented in this section. (Courtesy, Sperry New Holland, New Holland, PA)

[3]Estimates made by the author

[4]*Ibid.*

In order to lessen overinvestment by the uninformed, guidelines are useful. Here are some:

1. **Guideline No. 1**—*The break-even point on how much you can afford to invest in equipment to replace hired labor can be arrived at by the following formula:*

$$\frac{\textbf{Annual saving in hired labor from new equipment}}{\textbf{(divide by) .15}} = \textbf{amount you can afford to invest.}$$

Example:
If hired labor costs $20,000 per year, this becomes—

$$\frac{\$20{,}000}{.15} = \textbf{\$133,333, the break-even point on new equipment}$$

Since labor costs are going up faster than machinery and equipment costs, it may be good business to exceed this limitation under some circumstances. Nevertheless, the break-even point, $133,333 in this case, is probably the maximum expenditure that can be economically justified at the time.

2. **Guideline No. 2**—*The break-even point on new facility-equipment costs is five times the annual salary of each person replaced.*

Assuming an annual cost plus operation of power machinery and equipment equal to 20% of new cost, the break-even point to justify replacement of one hired laborer is as follows:

Example:

If annual cost of one hired laborer is:[5]	The break-even point on new investment
$10,000 (20%) × 5	$ 50,000
15,000 (20%) × 5	75,000
20,000 (20%) × 5	100,000

Assume that the new cost of added equipment comes to $20,000, that the annual cost is 20% of this amount, and that the new equipment would save one hour of labor per day for 6 months of the year. Here's how to figure the value of labor to justify an expenditure of $20,000 for this item:

$20,000 (new cost) × 20% = $4,000
4,000 (annual ownership use cost) ÷ 80 hours (labor saved) = $22.22 per hour.

So, if labor costs less than $22.22 per hour, you probably shouldn't buy the new item.

FINANCING THE LIVESTOCK BUSINESS

A big livestock operation necessitates both big money and knowledge of financing.

• **Credit**—*Credit may be defined as belief in the truth of a statement, or in the sincerity of a person.* In farming and ranching, or in any other business transaction, credit means

[5]This is assuming that the productivity of workers at different salaries is the same, which may or may not be the case.

confidence that people will take care of their future obligations. Credit is the lifeblood of the livestock business. Without it, few large operations would be possible; for not many people are able to provide all the capital that they need.

Most commercial lenders have guides and standards that set upper limits on the amount they will lend. Usually, to get credit on a mortgage for buying a farm, the borrower is expected to make a down payment of 40 to 50% of the purchase price. Lenders usually will make loans on livestock and on new machinery for up to 80% of the purchase price.

Total farm investment in land, buildings, livestock, and equipment has increased 4.8 times in 31 years, rising from $170 billion in 1956 to $813 billion in 1987. Farm debts have increased even more—they are 8.2 times larger than 31 years ago. The amount of debt owed by farmers and ranchers has risen from $18.8 billion in 1956 to $153.3 billion in 1987, and the trend shows no signs of letting up.

• **Sources of loans**—Hand in hand with getting the right kind of loan, it is important that the best available source of the loan be secured. Table 2–4 shows the primary sources of the three main kinds of loans. It is noteworthy that banks, merchants, and individuals provide 80% of the farm credit.

TABLE 2–4
PRINCIPAL SOURCES OF THREE MAIN KINDS OF FARM LOANS

	Kind of Loan		
Credit Source	**Long Term**	**Intermediate Term**	**Short Term**
Commercial banks	X	X	X
Dealers and merchants		X	X
Farm mortgage companies	X		
Farmers Home Administration	X	X	X
Federal Land Bank Associations	X	X	
Individual lenders	X	X	X
Insurance companies	X		
Production Credit Associations		X	X

• **Types of credit or loans**—Getting the needed credit through the right kind of loan is an important part of sound financial farm management. The following three general types of agricultural credit are available, based on length of life and type of collateral needed:

1. **Short-term loans.** This type of loan is made for operating expenses and is usually for one year or less. It is used for the purchase of feeders or birds, feed, seed, fertilizer, gasoline, and family living expenses. Security, such as a chattel mortgage on the feeders, birds, or crop, may be required by the lender; and the loan is repaid when the animals or crop are sold.

2. **Intermediate-term loans.** These loans are used to buy equipment and breeding stock, for making land improvements, and for remodeling existing buildings. They are paid back in 1 to 7 years. Generally, they are secured by a chattel mortgage on livestock and machinery.

3. **Long-term loans.** These loans are secured by mortgage on real estate and are used to buy land or make major improvements to farmland and buildings or to finance construction of new buildings. They may be for as long as 40 years. Usually they are paid off in regular annual or semiannual payments. The best sources for long-term loans are: an insurance company, the Farm Credit Bank, the Farmers Home Administration, or an individual.

CREDIT SOURCES

Table 2–5 and Fig. 2–6 show where farmers borrow, the amount of loans from each source, and the percent of the total held by each type of lender.

TABLE 2–5
WHERE FARMERS BORROW (1989)[1]

Type and Source of Loan	Amount of Loan	Percent of Total
	(million $)	(%)
Real estate mortgage loans:		
Farm Credit System	26,059	35.0
Individuals and others	15,751	21.2
Commercial banks	15,263	20.5
Insurance companies	8,852	11.9
Farmers Home Administration	8,421	11.3
Total	74,346	100.0
Nonreal estate loans:		
Commercial banks	28,595	41.9
Farmers Home Administration	11,792	17.3
Individuals and others	11,760	17.2
Farm Credit System	9,120	13.4
Commodity Credit Corp.	7,000	10.2
Total	68,267	100.0
Total loans	142,613	
Percent real estate	52.1	
Percent nonreal estate	47.9	

[1]Data provided in a personal communication to the author from George D. Irwin, Deputy Director, Office of Financial Analysis, Farm Credit Administration, McLean, VA.

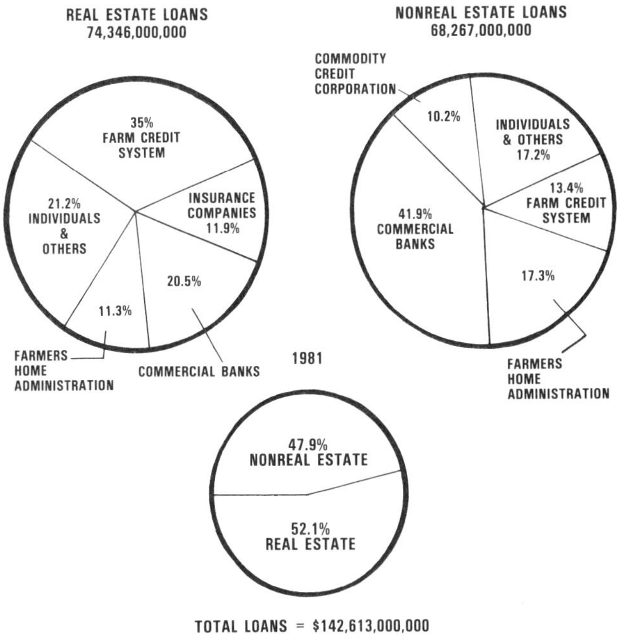

Fig. 2–6. Where farmers borrow (1989). (Source: Farm Credit Administration, Washington, DC)

But, agricultural financing is changing, and it will continue to change even more in the years ahead. Today, farmers are tapping the vast supply of farm equity or risk capital that is constantly seeking investment opportunities—nonfarm equity capital is being used in agriculture.

Some time or other most farmers and ranchers find it necessary to borrow money to buy land; to construct buildings and other improvements; to purchase equipment, seed, and livestock; and/or to pay for seasonal labor. They should know something, therefore, about the lending organizations available to them in order that they may determine which one will best serve their needs. The leading sources of farm credit are summarized in Table 2-6.

TABLE 2-6
MAJOR SOURCES OF CREDIT, AND THE CHARACTERISTICS OF EACH[1]

Lenders	Sources of Funds	Limitations on Agricultural Lending	Loans Offered to Farmers	Comments
Commercial banks	Demand and time deposits, bank stock, retained earnings. Funds from correspondent banks, the Federal Reserve, and participation with PCAs or FmHA.	Commercial banks may prefer more profitable alternatives, such as installment loans. Also, legal reserve laws limit the volume of deposits available for loans.	(1) Short-term operating loans repayable within 1 yr. (2) Intermediate-term loans, repayable in 1 to 7 yr. (3) Some banks make long-term or real estate loans (7 to 25 yr).	A financial statement is required by bank examiners. Some commercial banks have special agricultural representatives who are qualified to assist the borrower in many ways.
Farm Credit System, which embraces the following 3 federal lending units, all under the supervision of the Farm Credit Administration, an independent federal agency:	Sale of its bonds and discount notes in the private money market.			The Federal Land Banks and PCAs supply nearly ⅓ of the credit used by farmers, and the banks for cooperatives provide nearly ⅔ of the borrowed capital used by farmer cooperatives.
1. Farm Credit Banks (FCB) Loans to farmers are made or serviced by one of four types of local associations: (a) Federal Land Bank Assn. (b) Federal Land Credit Assn. (c) Production Credit Assn. (d) Agricultural Credit Assn. The type of credit varies by area and by length of loan.	Sell bonds publicly on the national money markets. FCBs also draw on some money available from capital stock and retained earnings. The associations borrow from, or act as agents for, the FCBs.	To borrow, you must buy stock in the local association equal to the lesser of $1,000 or 2% of the loan amount. Congress currently limits a real estate loan to 85% or less of the appraised property value. Farmers can also get loans for farm related services, such as rural home construction.	Real estate loans, amortized over periods ranging up to 40 years. FCBs secure short-term and intermediate-term loans for improvements by mortgaging the improved real estate. Only short-term and intermediate-term loans.	Farm Credit Banks and Associations are actually farmers cooperatives. Mortgage loans are on first mortgage only. When making such loans, consideration is given to market value of the real estate, plus the income and the management ability of the borrower. The short- and intermediate-term loan limit varies with individual cases, but it can be up to 100% of cost. However, 70–30 is most common, with the borrower providing 30% margin.
2. Banks for Cooperatives				Banks for Cooperatives provide the majority of financing for the nation's farm supply, marketing, and business service cooperatives.
Farmers Home Administration (FmHA)	Congressional appropriations and from emergency and revolving funds. FmHA also guarantees loans from other lenders.	Eligible only to farmers who can't obtain credit elsewhere. Legal maximums: $200,000—direct operating, $400,000—guaranteed operating loans, $200,000 direct real estate, $300,000 guaranteed real estate loans.	Short-, intermediate-, and long-term, FmHA provides supervision for its loans. The guaranteed loan program is receiving priority.	Applicants must meet loan eligibility requirements. Where a natural disaster has occurred, under the Emergency Loan Program, a farmer may borrow up to 80% of the loss, but not to exceed $500,000.
Federal Agricultural Mortgage Corporation (Farmer Mac)	It provides a way for lenders to pool groups of qualifying first mortgage loans to sell to investors.	Source of funds to all types of lenders who follow Farmer Mac standards.		Farmer Mac was created by Congress in 1987; and it became operational in 1990.
Individuals	Personal loans, made by one individual to another.	Compared to the other sources, individuals frequently offer lower interest rates and down payments. However, the repayment period may be shorter.	There's an infinite variety of conditions and interest rates. Range all the way from real estate contracts or mortgages to personal unsecured loans.	One disadvantage of a loan from an individual is that the arrangement may be complicated by the lender's death unless adequate provision has been made for this eventuality.
Insurance Companies	Premiums received on insurance policies. They also draw funds from reserves held to pay insurance claims, and from capital and retained earnings.	Prefer long-term loans. Higher yielding, more secure investment opportunities lure money away from agricultural lending. May limit to selected geographic areas.	Typically, insurance companies limit themselves to real estate loans up to 30 to 40 years, and improvement loans which they secure through real estate mortgages.	Generally insurance companies will make loans up to 60% of the appraised value of the farm or up to 50% of the sale value.
Merchants and Dealers	Borrow from lending institutions, capital and retained earnings. Their supplier, distributor, or manufacturer may extend similar credit to them.	Limited credit, because they must keep a certain level of liquidity. Rates are higher than other lenders. Also limited by their supplier and other sources of capital.	Open account or sales contract. Short-term loans on equipment and machinery, intermediate-term loans on equipment. May charge add-on interest. Some offer discounts if you pay cash.	It must be realized that merchants and dealers extend credit to farmers and ranchers primarily for the purpose of promoting the sale of products and services, and that their profits come from both sales and interest.

[1]This table was authoritatively reviewed by the following: LaVerne Ausman, Administrator, Farmers Home Administration, Washington, DC, and George D. Irwin, Deputy Director, Office of Financial Analysis, Farm Credit Administration, McLean, VA.

HELPFUL HINTS FOR BUILDING AND MAINTAINING A GOOD CREDIT RATING

Livestock operators who wish to build up and maintain good credit are admonished to do the following:

1. **Keep credit in one place, or in a few places.** Generally, lenders frown upon *split financing.* Shop around for a creditor who (a) is able, willing and interested in extending the kind and amount of credit needed, and (b) will lend at a reasonable rate of interest, then stay with you.
2. **Get the right kind of credit.** Don't use short-term credit to finance long-term improvements or other capital investments. Also, use the credit for the purpose intended.
3. **Be frank with the lender.** Be completely open and aboveboard. Mutual confidence and esteem should prevail between borrower and lender.
4. **Keep complete and accurate records.** Complete and accurate records should be kept by enterprises. By knowing the cost of doing business, decision making can be on a sound basis.
5. **Keep annual inventory.** Take an annual inventory for the purpose of showing progress made during the year.
6. **Repay loans when due.** Borrowers should work out a repayment schedule on each loan, then meet payments when due. Sale proceeds should be promptly applied on loans.
7. **Plan ahead.** Analyze the next year's operation and project ahead.

BORROW MONEY TO MAKE MONEY

Livestock producers should never borrow money unless they are reasonably certain that it will make or save money. With this in mind, borrowers should ask, "How much should I borrow?" rather than, "How much will you lend me?"

Fig. 2-7. Borrow money to make money! Would this free-stall dairy barn, equipped with automatic aisle scraper, make money on a new dairy farm? (Courtesy, Ohio Agricultural Research and Development Center, Wooster, OH)

CALCULATING INTEREST

The charge for the use of money is called interest. The basic charge is strongly influenced by the following:

1. The *basic cost* of money in the money market.
2. The *servicing costs* of making, handling, collecting, and keeping necessary records on loans.
3. The *risk* of loss.

Interest rates vary among lenders and can be quoted and applied in several different ways. The quoted rate is not always the basis for proper comparison and analysis of credit costs. Even though several lenders may quote the same interest rate, the effective or simple annual rate of interest may vary widely. The more common procedures for determining the actual annual interest rate, or the equivalent of simple interest on the unpaid balance, follow.

1. **Simple or true annual interest on the unpaid balance.** A $1,200 note payable at maturity (12 months) with 12% interest:

Interest paid .... $.12 \times \$1{,}200 = \$144$

Average use of the money .... $1,200 for the entire year

Actual rate of interest .... $\dfrac{\$144 \text{ (interest)}}{\$1{,}200 \text{ (used for 1 year)}} = 12\%$

2. **Installment loan (with interest on unpaid balance).**[6] A $1,200 note payable in 12 monthly installments with 12% interest on the unpaid balance:

Interest paid ranges from:

First month .... $\dfrac{.12 \times \$1{,}200}{12} = \$12$

to

Twelfth month .... $\dfrac{.12 \times 100}{12} = \$1$

Total for 12 months is $78

Average use of the money ranges from $1,200 for the first month down to $100 for the twelfth month, an average of $650 for 12 months.

Effective rate of interest .... $\dfrac{\$\ 78}{\$650} = 12\%$

3. **Add-on installment loan (with interest on face amount).** A $1,200 note payable in 12 monthly installments with 12% interest on face amount of loan:

Interest paid .... $.12 \times \$1{,}200 = \$144$

[6]This method is used for amortized loans.

Average use of the money ranges from $1,200 for the first month down to $100 for the twelfth month, an average of $650 for 12 months.

$$\text{Effective rate of interest} \ldots\ldots\ldots\ldots \frac{\$144}{\$650} = 22.15\%$$

4. **Points and interest.** Some lenders now charge *points*. A point is 1% of the face value of the loan. Thus, if 4 points are being charged on a $1,200 loan, $48 dollars will be deducted and the borrower will receive only $1,152. But the borrower will have to repay the full $1,200. Obviously, this means that the actual interest rate will be more than the stated rate. But how much more?

Assume that a $1,200 loan is for 1 year and the annual rate of interest is 12%. Then the payment by the borrower of 4 points would make the actual interest rate as follows:

Interest	$.12 \times \$1,200 = \$144$
Average use of money	$1,152 for one year
Effective rate of interest	$\frac{\$144 \text{ (interest)}}{\$1,152 \text{ (used for 1 year)}} = 12.5\%$

5. **If interest is not stated, use this formula to determine the effective annual interest rate:**

Effective rate of interest =

$$\frac{\text{Number of payment periods} \times 2 \text{ in 1 year}^{7} \times \text{Finance charges}^{8}}{\text{Balance owed}^{9} \times \text{Number of payments in contract plus 1}}$$

For example, a store advertises a refrigerator for $500. It can be purchased on the installment plan for $80 down and monthly payments of $35 for 12 months. What is the actual rate of interest if you buy on the time payment plan?

Effective rate of interest =

$$\frac{2 \times 12 \times \$35}{\$420 \times (12 + 1)} = \frac{\$\ 840}{\$5,460} = 15.4\%$$

[7]Regardless of the total number of payments to be made, use 12 if the payments are monthly, use 6 if payments are every other month, or use 2 if payments are semiannual.

[8]Use either the time payment price less the cash price, or the amount you pay the lender less the amount you received if negotiating for a loan.

[9]Use cash price less down payment or, if negotiating for a loan, the amount you receive.

MANAGER

Fig. 2-8. Cattle feedlot managers must wear many hats—business acumen, labor relations, feeds and nutrition, additives, animal health, pollution control, marketing, and consumer demands. (Courtesy, Colorado State University, Ft. Collins)

According to Webster, *a manager is one who conducts business affairs with economy; and management is the act, or art, of managing, handling, controlling or directing.*

Three major ingredients are essential to success in the livestock business: (1) good livestock, (2) good feeding, and (3) good management. A manager can make or break any livestock enterprise. Unfortunately, this fact was long overlooked, primarily because the accent was on scientific findings, automation, and new products.

Management gives point and purpose to everything else. The skill of the manager materially affects how well animals are bought and sold, the quality of the animals, the health of the animals, the results of the ration, the stress of the stock, efficiency of production, the performance of labor, the public relations of the establishment, and even the expression of the genetic potential. Indeed, a livestock manager must wear many hats—and all of them must be worn well.

The bigger and the more complicated the operation, the more competent the management required. This point merits emphasis because, currently, (1) bigness is a sign of the times, and (2) the most common method of attempting to bail out of an unprofitable business venture is to increase its size. Although it's easier to achieve efficiency of equipment, labor, purchases, and marketing in big operations, bigness alone will not make for greater efficiency, as some owners have discovered to their sorrow, and others will experience. Management is still the key to success. When in financial trouble, owners should have no illusions on this point.

In manufacturing and commerce, the importance and scarcity of top managers are generally recognized and reflected in the salaries paid to persons in such positions. Unfortunately, agriculture as a whole has lagged; and altogether too many owners still subscribe to the philosophy that the way to make money out of the livestock business is to hire a manager cheaply, with the result that they usually get what they pay for—a "cheap" manager.

TRAITS OF A GOOD LIVESTOCK MANAGER

There are established bases for evaluating many articles of trade including animals, hay, and grain. They are graded according to well-defined standards. Additionally, feeds are chemically analyzed and feeding trials conducted with them. But no such standard or system of evaluation has evolved for livestock managers, despite their acknowledged importance.

The author has prepared the Livestock Manager Checklist, given in Table 2–7, which (1) employers may find useful when selecting or evaluating a manager, (2) managers may apply to themselves for self-improvement purposes, and (3) students may use for guidance as they prepare themselves for management positions. No attempt has been made to assign a percentage score to each trait, because this will vary among livestock establishments. Rather, it is hoped that this checklist will serve as a useful guide (1) to the traits of a good manager, and (2) to what the boss wants.

TABLE 2–7
LIVESTOCK MANAGER CHECKLIST

- ☐ **Character**
  Has absolute sincerity, honesty, integrity, and loyalty; is ethical.
- ☐ **Industry**
  Has enthusiasm, initiative, and aggressiveness; is willing to work, work, work.
- ☐ **Ability**
  Has livestock know-how and experience, business acumen—including ability systematically to arrive at the financial aspects and convert this information into sound and timely management decisions—knowledge of how to automate and cut costs, common sense, organization, imagination, and growth potential.
- ☐ **Plans**
  Sets goals; prepares organization chart and job description; plans work and works plans.
- ☐ **Analyzes**
  Identifies the problem, determines the pros and cons, then comes to a decision.
- ☐ **Courage**
  Has the courage to accept responsibility, to innovate, and to keep on keeping on.
- ☐ **Promptness and Dependability**
  Is a self-starter; has "T.N.T." which means doing it "today, not tomorrow."
- ☐ **Leadership**
  Stimulates subordinates and delegates responsibility.
- ☐ **Personality**
  Is cheerful; not a complainer.

ORGANIZATION CHART AND JOB DESCRIPTION

It is important that workers know to whom they are responsible and for what they are responsible; and the bigger and more complex the operation, the more important this becomes. This should be written down in an organization chart and a job description. A sample Organization Chart is given in Fig. 2–9, and a sample Job Description is given in Table 2–8.

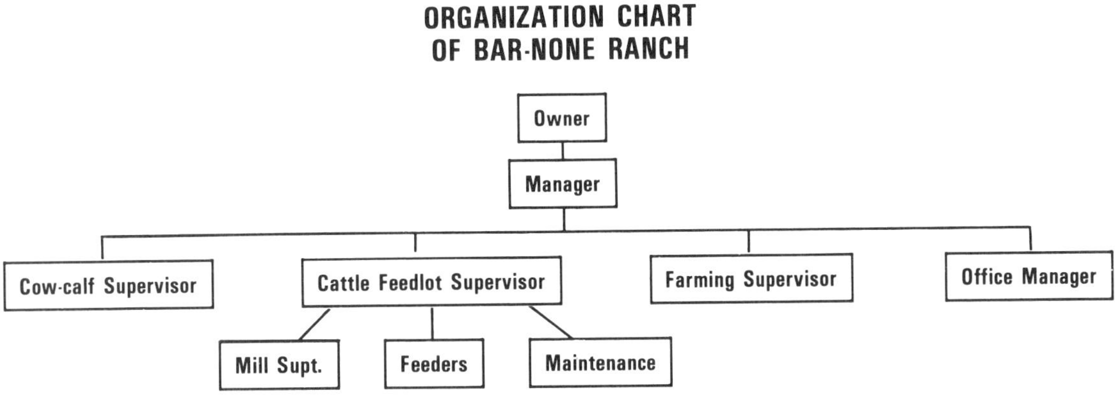

Fig. 2–9. A suggested farm-ranch Organization Chart.

TABLE 2–8
SUGGESTED JOB DESCRIPTIONS OF BAR-NONE RANCH

Owner	Manager	Cattle Feedlot Supervisor	Cattle Feeder No. 1
Responsible for: 1. Selecting management. 2. Making policy decisions. 3. Borrowing capital. 4. (List others)	Responsible for: 1. Supervising all staff. 2. Preparing proposed long-time plan. 3. Budgets. 4. (List others)	Responsible for: 1. Directing feedlot staff. 2. Buying and selling cattle. 3. Processing incoming cattle. 4. Animal health. 5. Feedlot rations. 6. (List others)	Responsible for: 1. Morning and evening feedings. 2. Cleaning water troughs. 3. (List others)

AN INCENTIVE BASIS FOR THE HELP

Big farms and ranches must rely on hired labor, all or in part. Good help—the kind that everyone wants—is hard to come by; it's scarce, in strong demand, and difficult to keep. Moreover, the farm labor situation is going to become more difficult in the years ahead. There is need, therefore, for some system that will (1) give a big assist in getting and holding top-flight help, and (2) cut costs and boost profits. An incentive basis that makes hired help partners-in-profit is the answer.

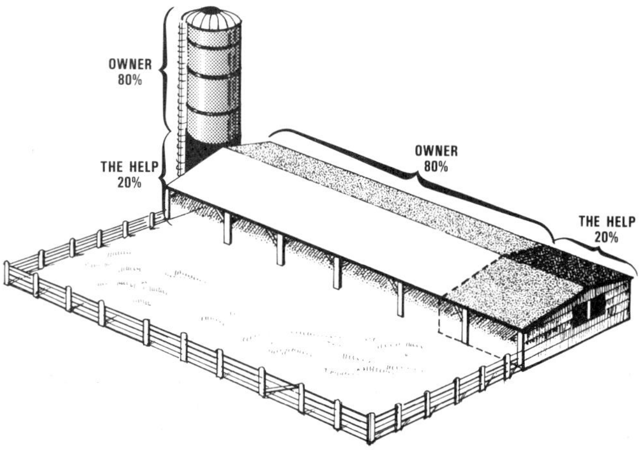

Fig. 2–10. A good incentive basis makes hired help partners-in-profit.

Many manufacturers have long had an incentive basis. Executives are frequently accorded stock option privileges, through which they prosper as the business prospers. Common laborers may receive bonuses based on piecework or quotas (number of units or pounds produced). Also, most factory workers get overtime pay and have group insurance and a retirement plan. A few industries have a true profit-sharing arrangement based on net profit as such, a specified percentage of which is divided among employees. No two systems are alike. Yet, each is designed to pay more for labor, provided labor improves production and efficiency. In this way, both owners and laborers benefit from better performance.

Family owned and family operated farms have a built-in incentive basis; there is pride of ownership, and all members of the family are fully cognizant that they prosper as the business prospers.

Many different incentive plans can be, and are, used. There is no best one for all operations.

Note well: The various plans given in Table 2–9 and in the narrative that follows are for the purpose of showing how different incentives work; and *not* as indicators of amounts. The type and amount of incentive chosen, along with the provisions, should be tailored to fit the specific organization, with consideration given to the kind and size of the operation, the extent of the owner's supervision, the present and projected productivity levels, mechanization, and other factors.

For most livestock operations, the author favors a "production sharing and prevailing price" type of incentive (see Table 2–9).

HOW MUCH INCENTIVE PAY?

After (1) reaching a decision to go on an incentive basis, and (2) deciding on the kind of incentive, it is necessary to arrive at how much to pay. Here are some guidelines that may be helpful in determining this:

1. Pay the going base, or guaranteed, salary; then add the incentive pay above this.

2. Determine the total stipend (the base salary plus incentive) to which you are willing to go.

3. Before making any offers, always check the plan on paper to see (a) how it would have worked out in past years based on your records, and (b) how it will work out as you achieve the future projected production.

Let's take the following example:

> A supervisor of a 500-cow herd is now producing an average of 400 pounds of calf weaned per cow bred. The supervisor is receiving a base salary of $1,000 per month, plus house, garden, and 600 lb of dressed beef per year. The owner prefers a "production sharing and prevailing price" type of incentive.

Step by step, here is the procedure for arriving at an incentive arrangement based on increased production:

1. By checking with local sources, it is determined that the present salary of $1,000 per month plus extras is the going wage; and,of course, the supervisor receives this regardless of what the year's calf production or price turns out to be—it's guaranteed.

2. A study of the cow-calf records reveals that with a little extra care on the part of the supervisor—particularly in pregnancy testing, at calving time, and in rotating pastures—the average weaning weight of calf per cow bred can be boosted enough to permit paying the supervisor $1,300 per month, or $300 per month more than the current wage. That's $3,600 per year. This can be fitted into the incentive plan.

3. An average increase of 50 lb of calf weaned per cow bred at 60¢ per pound would mean $30 per cow, or $15,000, on a 500-cow herd. With an 80:20 split between owner and manager, the supervisor would get $3,000, or $250 per month.

TABLE 2-9
INCENTIVE PLANS FOR LIVESTOCK ESTABLISHMENTS

Types of Incentives	Pertinent Provisions of Some Known Incentive Systems in Use	Advantages	Disadvantages	Comments
Bonuses	A flat, arbitrary bonus; at Chirstmastime, year-end, quarterly, or other intervals. A tenure bonus such as (1) 5 to 10% of the base wage or 2 to 4 weeks' additional salary paid at Christmastime or year-end, (2) 2 to 4 weeks vacation with pay, depending on length and quality of service or (3) $10 to $20/week set aside and to be paid if employee stays on the job a specified time.	It's simple and direct.	Not very effective in increasing production and profits.	
Equity-building plan	Employee is allowed to own a certain number of animals. In breeding operations, these are usually fed without charge.	It imparts pride of ownership to the employee.	The hazard that the owner may feel that employees accord their animals preferential treatment; suspected if not proved.	
Production sharing	$2 to $6/calf weaned, $1/cwt on gain of feeder cattle; 50¢ to $2/head on fed cattle marketed; $1 to $2/pig marketed above 7 pigs/litter; 50¢ to $1/lamb weaned; $1/cwt of gain on lambs fed; 20¢ to $1/head on fed lambs marketed; so much per day for meeting certain levels of milk production/cow (for example, a 20¢ bonus/cow/month for 45.0 to 51.9 lb milk/cow/day; and 70¢ for 52.0 to 58.9 lb), with the bonus graduated upward with higher production in order to reflect the difficulty of increasing milk production in a herd where milk yield is already high.	It's an effective way to achieve higher production.	Net returns may suffer. For example, a higher rate of gain than is economical may be achieved by feeding stockers more concentrated and expensive feeds than are practical. This can be alleviated by (1) specifying the ration and (2) setting an upper limit on the gains to which the incentive will apply. If a high performance level already exists, further gains or improvements may be hard to come by.	Incentive payments for production above certain levels—for example, above 450 lb calf weaned/cow bred—are more effective than paying for all units produced.
Profit sharing 1. Percent of gross income 2. Percent of net income	 1% or 2% of the gross. 10% up to 20% of the net after deducting all costs.	Net income sharing works better for managers, supervisors, and other administrators than for common laborers because fewer hazards are involved to opening the books to them. It's an effective way to get hired help to cut costs. It's a good plan for a hustler.	Percent of gross does not impart cost of production consciousness. Both (1) percent of gross income and (2) percent of net income expose the books to workers, who may not understand accounting principles. This can lead to suspicion and distrust. Controversy may arise (1) over accounting procedures—for example, from the standpoint of the owner a fast tax write-off may be desirable on new equipment, but this reduces the net shared with the worker, and (2) because some owners are prone to overbuild and over-equip, thereby decreasing net.	There must be prior agreement on what constitutes gross or net receipts, as the case may be, and how it is figured.
Production sharing and prevailing price	*Cow-calf, ewe-lamb, or sow operation:* Basis (1) percent offspring weaned, and (2) weaning weight (which means pounds offspring weaned/female bred). *Finishing cattle, or finishing hogs:* Basis (1) pounds feed/lb gain, and (2) daily rate of gain. *Horse breeding establishment:* Basis (1) percent foal crop weaned, and (2) price of yearlings. In each of the above, establish break-even point(s), then split profit(s) beyond this point(s) basis (1) 80% (owner) and 20% (help), or (2) use escalator arrangement, giving help greater percentage as profits rise. With a dairy, the following production-sharing basis will work: *Dairy:* Basis (1) udder health (for example, 20¢/cow/month for a somatic cell count of 500,000 to 600,000, graduated upward to 50¢/cow/month for a somatic cell count of 400,000 or less); or, (2) calving interval (for example, $20/month for a herd calving interval of 13.0 to 13.4 months, graduated upward to $50 bonus/month for a calving interval of 12.5 months or under).	It embraces the best features of both production sharing and profit sharing, without the major disadvantages of each. It (1) encourages high productivity and likely profits, (2) is tied in with prevailing prices, (3) does not necessitate opening the books, and (4) is flexible—it can be split between owner and employee on any basis desired, and the production part can be adapted to a sliding scale or escalator arrangement—for example, the incentive basis can be higher for the ¼ lb feedlot gain made in excess of 2¾ lb than for a ¼ lb gain in excess of 2¼ lb	It is a bit more complicated than some other plans, and it requires more complete records.	When properly done, and all factors considered, this is the most satisfactory incentive basis for livestock enterprises.

REQUISITES OF AN INCENTIVE BASIS

Owners who have not previously had experience with an incentive basis are admonished not to start with any plan until they are sure of both their plan and their help. Also, it is well to start with a simple plan; then change can be made to a more inclusive and sophisticated plan after experience is acquired.

Regardless of the incentive plan adopted for a specific operation, it should encompass the following essential features:

1. A good owner (or manager) and good workers. No incentive basis can overcome a poor manager. A good manager must be a good supervisor and fair to the help. Also, on big establishments, the manager must prepare a written organiza-

tion chart and job description so the employees know (a) to whom they are responsible, and (b) for what they are responsible. Likewise, no incentive basis can spur employees who are not able, interested, and/or willing. This necessitates that they be selected with special care where they will be on an incentive basis. Hence, the three—good owner (manager), good employees, and good incentive—go hand in hand.

2. It must be fair to both employer and employees.

3. It must be based on and make for mutual trust and esteem.

4. It must compensate for extra performance, rather than substitute for a reasonable base salary and other considerations (house, utilities, and certain provisions).

5. It must be simple, direct, and as easily understood as possible.

6. It should compensate all members of the team—from cowhands to manager on a cow-calf outfit, from feeders and feed processors to manager in a cattle feedlot, and from milkers to supervisors in a dairy.

7. It must be put in writing, so that there will be no misunderstanding. For example, if some production-sharing plan is used in a cattle feedlot, it should stipulate the ration (or who is responsible for ration formulation), the maximum gain of stocker cattle, and the grade to which finishing cattle are to be carried. On a cow-calf outfit, it should stipulate the ration, the culling of cows, and other pertinent factors.

8. It is preferable, although not essential, that workers receive incentive payments (a) at rather frequent intervals, rather than annually, and (b) immediately after accomplishing the extra performance.

9. It should give the hired help a certain amount of responsibility, from the wise exercise of which they will benefit through the incentive arrangement.

10. It must be backed up by good records; otherwise, there is nothing on which to base incentive payments.

11. It should be a two-way street. If employees are compensated for superior performance, they should be penalized (or, under most circumstances, fired) for poor performance. It serves no useful purpose to reward the unwilling, the incompetent, and the stupid. For example, no overtime pay should be given to an employee who must work longer because of slowness or in order to correct mistakes of his/her own making. Likewise, if the reasonable break-even point on a cow-calf operation is an average of a 400-lb calf weaned per cow bred, and this production level is not reached because of obvious neglect (for example not being on the job at calving time), the employee(s) should be penalized (or fired).

INDIRECT INCENTIVES

Normally, we think of incentives as monetary in nature—as direct payments or bonuses for extra production or efficiency. However, there are other ways of encouraging employees to do a better job. The latter are known as indirect incentives. Among them are (1) good wages; (2) good labor relations; (3) adequate house plus such privileges as the use of the farm truck or car, payment of electric bill, use of a swimming pool, hunting and fishing, use of a horse, and furnishing meat, milk, and eggs; (4) good buildings and equipment; (5) vacation time with pay, time off, sick leave; (6) group health; (7) security; (8) the opportunity for self-improvement that can accrue from working for a top person; (9) the right to invest in the business; (10) an all-expense-paid trip to a short course, show, or convention; and (11) year-end bonus for staying all year. These indirect incentives will be accorded to the help of more and more establishments, especially the big ones.

FARM RECORDS AND ACCOUNTS

Modern farming is more than a job; it is a business. Therefore, it should be conducted in a businesslike manner. This means that there should be adequate records and accounts.

Fig. 2–11. Modern farming/ranching is big business, which calls for good records and accounts. (Courtesy, *Holstein World*, Sandy Creek, NY)

WHY KEEP RECORDS?

There are many reasons for keeping good records, the most important of which follow:

1. **To provide profit and progress indicators.** Production records on livestock and crops are profit indicators, and a way to measure progress.

2. **To guide changes in enterprises.** Farm and ranch records should provide information from which the farm business may be analyzed, with its strong and its weak points ascertained. From the facts thus determined, the manager may adjust current operations and develop a more profitable organization. The enterprise should be above average before the owner borrows to expand it. Is the ranch too small? Is it more profitable to sell weaners or yearlings? Should the farm or ranch produce or buy hay?

3. **To provide a net worth statement.** Farm and ranch records should provide a net worth statement, showing financial progress during the year.

4. **To serve as a guide for current income and expenses.** Records of cost and returns of previous years are very valuable as guides, and as a means of spotting trouble. Items which deviate substantially from the historical record should be studied with care.

5. **To obtain needed credit.** Lenders need certain basic information in order to evaluate the soundness of a loan request. The financial record will show the borrower's current financial position, potential ahead, and liability to others.

Also, livestock producers must realize that they are competing with many other users of credit, including retail merchants, manufacturers, home buyers, and professional people. Many of the borrowers can and do provide the lender with profit and loss statements and net worth statements, prepared by a CPA. Also, they usually submit (a) annual budgets , and (b) cash flow projections, showing when money is needed, and showing loan repayability. As livestock producers increase the amount of borrowed capital in their operations, they, too, will be required to furnish adequate records and budgets if they are to compete successfully for the available capital.

6. **To save on income taxes.** Keeping good and adequate records usually makes for a savings in income taxes.

7. **To provide for continuity of the business.** "Barn door" and "memory" records are insufficient. They mitigate against continuity of the business. The sudden passing of a manager places a severe stress on a business even under the most favorable circumstances. However, a good set of records gives a big assist to those who must take over during such times.

8. **To keep an historical record.** Good and complete historical records are needed for future reference purposes.

TYPES OF RECORDS

In general, the functions enumerated under the earlier section headed "Why Keep Records?" can be met by the following types of records:

1. **Annual inventory.** The annual inventory is the most valuable record that a farmer can keep. It should include a list and value of real estate, livestock, equipment, feed, supplies, and all other property, including cash on hand, notes, bills receivable, and growing crops. Also, it should include a list of mortgages, notes, and bills payable. It shows farmers what they own and what they owe; whether they are getting ahead or going behind. The following pointers may be helpful relative to the annual inventory.

a. **Time to take inventory.** The inventory should be taken at the beginning of the account year; usually this means December 31 or January 1.

b. **Proper and complete listing.** It is important that each item be properly and separately listed.

c. **Method of arriving at inventory values.** It is difficult to set up any hard and fast rule to follow in estimating values when taking inventories. Perhaps the following guides are as good as any:

(1) **Real estate.** Estimating the value of farm real estate is, without doubt, the most difficult of all. It is suggested that the farmer use either (a) the cost of the farm, (b) the present sale value of the farm, or (c) the capitalized rent value according to its productive ability with an average operator.

(2) **Buildings.** Buildings are generally inventoried on the basis of cost less observed depreciation and obsolescence. Once the original value of a building is arrived at, it is usually best to take depreciation on a straight line basis by dividing the original value by the estimated life in terms of years. Usually 4% or more depreciation is charged off each year for income tax purposes.

(3) **Livestock.** Animals are usually not too difficult to inventory because there are generally sufficient current sales to serve as a reliable estimate of value.

(4) **Machinery.** The inventory value of machinery is usually arrived at by one of two methods: (a) the original cost less a reasonable allowance for depreciation each year, or (b) the probable price that it would bring at a well-attended auction.

Under conditions of ordinary wear and reasonable care, it can be assumed that the general run of farm machinery (except trucks and autos) will last about 10 years. Thus, with new machinery, the annual depreciation will be the original cost divided by 10.

(5) **Feed and supplies.** The value of feed and supplies can be based on market price.

Two further points are important. Whatever method is used in arriving at inventory value (a) should be followed at both the beginning and the end of the year, and (b) should reflect the operator's opinion of the value of the property involved.

2. **Daily and weekly report.** A farm or ranch manager should keep a daily record, like the daily-weekly report shown in Fig. 2–12 (next page). It takes little time to keep it. Nevertheless, a certain time should be set aside daily for this purpose, thereby assuring that it will be kept—and that it will be kept accurately. For the manager, such a report provides an invaluable record of the day-by-day operations. For the owner, it is a quick and easy way to keep informed. This record should be filed where it can be referred to as needed.

"BAR-NONE RANCH," FOR WEEK BEGINNING ____________ (month, day, year), Prepared by ____________

	Monday	Tuesday	Wednesday	Thursday	Friday	Saturday	Sunday	Comments for Day or Week
WEATHER Rain or snow, inch Temp.—high low								
LABOR Accident or sick Who, what, how Changes								
EQUIPMENT BREAKDOWN Kind and make Cause Cost to repair								
ANIMAL HEALTH PROGRAM (indicate cow-calf, swine, etc.) Vaccination Treatment								
ANIMAL LOSSES No. Kind Cause								
ANIMALS RECEIVED No. Kind From Price or custom feed								
ANIMALS SOLD No. Kind To Price								
CROP HARVESTED Kind Field No. Acreage Yield/acre								
VISITORS Name and address								

Fig. 2-12. A good type of daily and weekly report.

3. **Record of livestock and crop production.** A record of the production and sale of animals and their products, and of the yield of crops, is most important, for the success of the farm depends upon production. Such records are important profit indicators and help in analyzing the farm business. They may be few or many, depending upon the wishes of the operator and the class of livestock. The production records of a cow-calf operator, for example, will usually include percent calf crop dropped, percent calf crop weaned, weaning weight of calves at seven months, the pounds of beef per cow or per acre, the pounds of feed required to produce a pound of beef, and death losses.

4. **Record of receipts and expenses.** Such a record is essential to any type of well-managed business. To be most useful, these entries should not only record the amount of the transaction, but should give the source of the income or the purpose of the expense, as the case may be. In other words, they should show the farmer from what sources the income is derived and for what it is spent.

The following kinds and arrangements of farm record books are commonly used for recording receipts and expenditures:

a. Those that devote a separate page to each enterprise; that is, a separate page is used for the beef cattle finishing enterprise, another for swine, still another for sheep, and so on.

b. Those that provide for a record of receipts and expenses on the same page, using one column for receipts and another for expenses. This type is easy to keep, but very difficult to analyze from the standpoint of any particular enterprise.

c. Those that combine the features of both 1 and 2 above. The latter are more difficult to keep than the others, and may be confusing to the person keeping the record.

Household and personal accounts should be kept, but should be handled entirely separate from farm accounts because they are not farming expenses as such.

5. **Financial records.** The name of the game is *profit.* Thus, it is necessary that production be translated into dollars and weighed against costs involved in achieving that production.

Many different kinds of financial records are used by livestock producers, among the most common ones are the following: annual inventory, budgets, income, expenses, cash flow, depreciation schedule, annual net income, enterprise accounts, profit and loss (P&L) statement, and net worth statement.

KIND OF RECORD AND ACCOUNT BOOK

Farmers can make their own record book by simply ruling off the pages of a bound notebook to fit their specific needs, but the saving is negligible. Instead, it is recommended that they obtain a copy of a farm record book prepared for and adapted to their area. Such a book may usually be obtained at a nominal cost from the agricultural economics department of each state college of agriculture. Also, certain commercial companies distribute very acceptable farm record and account books at no cost.

At the outset, it should be recognized that a farm record should be easy to keep and should give the information desired to make a valuable analysis of the business.

Most farm record and account books contain simple and specific instructions relative to their use.

WHO SHALL KEEP THE RECORDS?

The records may be kept either by someone in the farm or ranch business, or by someone hired to perform this service—a professional.

• **Farm or ranch help**—Very frequently, farm or ranch records are kept by a member of the family. With adequate time and a bit of training, this should result in good records.

Most farmers are not good record keepers, primarily because the operation of an ordinary farm requires large amounts of physical labor. As a result, most farmers and ranchers are physically exhausted at the end of the day's work and have neither the time nor the ambition to record in at least four different places each transaction which occurred during the day, as would be necessary of the usual double-entry bookkeeping system.

• **A professional**—There are, of course, individuals and firms who make a business of keeping farm or ranch records. Usually, they are accountants or farm management specialists.

HOW SHALL RECORDS BE KEPT

Records may be kept either by hand or by computer. Accurate and up-to-the-minute records and controls have taken on increasing importance in all agriculture, including the livestock business, as the investment required to engage in farming and ranching has risen and profit margins have narrowed. Today's successful farmers and ranchers must have, and use, as complete records as any other business. Also, records must be kept current; it no longer suffices merely to know the bank balance at the end of the year.

• **By hand**—The hand system can be used, but is slow and tedious. This service is usually performed by a member of the family or by an accountant living in the community. Nevertheless, after learning what is wanted, a first-rate accounting system can be kept by hand.

• **By computer**—Big and complex agricultural enterprises have outgrown hand record keeping. It's too time-consuming, with the result that it doesn't allow management enough time for planning and decision making. Additionally, it does not permit an all-at-once consideration of the complex interrelationships which affect the economic success of the business. This has prompted a new computer technique known as linear programming.

Whether an owner-operator should keep the records or hire a professional, and whether records should be kept by hand or by computer, each individual must decide. For the most part, the decision should be based on weighing the usefulness of the information each provides against the cost of obtaining it.

SUMMARIZING AND ANALYZING THE RECORDS

At the end of the year, the second or closing inventory should be taken, using the same method as was followed in taking the initial inventory. The final summary should then be made, following which the records should be analyzed. In the latter connection, farmers should remember that the purpose of the analysis is not to prove that they have or have not been prosperous. They probably know the answer to this question already. Rather, the analysis should show actual conditions on the farm and point out ways in which these conditions may be improved.

Although farmers and ranchers can summarize and analyze their own records, there are many advantages in having the services of a specialist for this purpose. such a specialist is in a better position to make a "cold" appraisal without prejudice, and to compare enterprises with those of other similar operators. Thus, the specialist may discover that, in comparison with other operators, the hogs on a given farm are requiring too much feed to make a hundred pounds of gain, or that the steer feeding enterprise is much less profitable than others have experienced. The local county agent can either render or recommend such specialized assistance. In some areas, it may consist in joining a cooperative farm record group or engaging the services of a consultant; in some states, such service is provided by the state agricultural college.

ANALYZING A LIVESTOCK BUSINESS—IS IT PROFITABLE?

Most people are in business to make money—and livestock producers are people. In some areas, particularly near cities and where the population is dense, land values may appreciate so as to be a very considerable profit factor. Also, a tax angle may be important. But neither of these should be counted upon. The livestock operation should make a reasonable return on the investment; otherwise, the owner should not be in the business.

For land and livestock to be profitable, they should yield a return sufficient to the owner to (1) meet the interest payment on the investment, (2) retire a reasonable portion of the loan, and (3) provide satisfactory management return. But there is no more reason large land and livestock holdings should be debt-free than there is for General Motors, or any other big corporation, to be debt-free.

Livestock owners or managers need to analyze their business—to determine how well they are doing. With big operations, it is no longer possible to base such an analysis on the bank balance statement at the end of the year. In the first place, once per year is not frequent enough, for it is possible to go broke, without really knowing it, in that period of time. Secondly, a balance statement gives no basis for analyzing an operation—for ferreting out its strengths and weaknesses. In large cattle, lamb, and hog feeding operations, it is strongly recommended that progress be charted by means of monthly or quarterly closings of financial records.

Also, producers must not only compete with other producers down the road, but they must compete with themselves—with their respective records last year and the year before. They must work ceaselessly at making progress, of improving the end product, and lowering costs of production.

To analyze a livestock business, two things are essential: (1) good records; and (2) yardsticks, or profit indicators, with which to measure an operation.

Profit indicators are gauges for measuring the primary factors contributing to profit. In order for producers to determine how well they're doing, they must be able to compare their own operations with something else; for example, (1) their own historical five-year average, (2) the average for the U.S. or for their particular area, or (3) the top 5%. The author favors the latter, for high goals have a tendency to spur superior achievement.

Space limitations will not permit presentation in this book of all conceivable profit indicators for all classes of livestock. Hence, only two—cow-calf profit indicators and cattle feedlot profit indicators—will follow. However, these can readily be adapted to dairy, sheep, swine, and horse operations.

Like most profit indicators, the ones given in Table 2-10 and Table 2-11 are not perfect. But they will serve as useful guides. Also, on some establishments, there may be reason for adding or deleting some of the indicators; and this can be done. The important thing is that each operation has adequate profit indicators, and that these be applied as frequently as possible; in a cattle feedlot, for example, this may be done monthly with some indicators.

COW-CALF PROFIT INDICATORS

Fig. 2-13. Is the cow-calf operation profitable? Three profit indicators will provide the answer: (1) capital investment in land and improvements per cow, (2) percent calf crop weaned, and (3) weaning weight. (Courtesy, USDA)

Many factors determine the profitableness of a cow-calf enterprise. Certainly, a favorable per animal unit capital investment in land and improvements is a first requisite. Additionally, percent calf crop weaned and weaning weight are exceedingly important, as shown in Table 2-10.

TABLE 2-10
SIZE AND WEIGHT OF CALF CROP WEANED ARE IMPORTANT

	Yearly Operating Cost—$300 per Cow-Calf Weights				
Calf Crop	**450 Lb**	**425 Lb**	**400 Lb**	**375 Lb**	**350 Lb**
(%)					
90	405 74.1¢	382 78.5¢	360 83.3¢	337 89.0¢	315 95.2¢
85	382 78.5¢	361 83.1¢	340 88.2¢	319 94.0¢	298 $1.01
80	360 83.3¢	340 88.2¢	320 93.7¢	300 $1.00	280 $1.07
75	337 89.0¢	319 94.0¢	300 $1.00	281 $1.07	263 $1.14
	Break-even point @ 88-89 cents/lb.				

In Table 2–10, it is assumed that the yearly operating cost is $300 per cow. Then, the effect of size and weight of calf crop is computed. As shown, with a 90% calf crop and an average weaning weight per cow bred of 405 lb (450 × 90% = 405), a selling price of 74.1¢/lb will meet the break-even cost of $300. With a 75% calf crop, because of fewer pounds per cow bred (450 × 75% = 337 lb), the calves would have to bring 89.0¢/lb in order to break even.

Table 2–11 will serve as a yardstick for determining (1) how you stack up with the nation's (a) average, and (b) top 5% of cow-calf operators; and (2) where you are falling down in your cow-calf enterprise.

It is noteworthy that Table 2–11 reveals that the top 5% operators have a higher investment/animal unit in land and improvements than their average counterparts. Obviously, better operators have better land and improvements; their savings are made in the handling of the herd. It's not unlike selecting a ration—where it's net returns, rather than cost per ton, that counts.

No claim is made that Table 2–11 is perfect. Admittedly, there are wide area differences, and no two farms or ranches are alike. Also, there are seasonal differences; for example, a drought will materially affect weaning weight of calves. Yet, Table 2–11 will serve as a useful guide.

TABLE 2–11
COW-CALF PROFIT INDICATORS[1]

	Average for U.S. Cow-Calf Operations			
	Commercial		Purebred	
	Average	Top 5%	Average	Top 5%
Investment unit (one mature cow) in land and improvements (real estate) . . . . . . ($)	2,000	2,250	2,500	3,000
Percent calf crop dropped (based on no. cows bred) . . . . (%)	88	93	89	94
Percent of calf crop weaned (based on no. cows bred) . . . . (%)	81	89	82	91
Weaning weight of calf at 7 mo. . . . . (lb)	425	525	450	550
Age and longevity of cows:				
Age when removed from herd . . . . (yr)	9	10	9.5	10.5
No. calves produced in lifetime of cow	6.0	8.0	7.0	8.5
Labor/cow/year . . . . (hr)	10.0	9.5	16.0	13.5
Net return per cow to management[2] . . . . ($)	27.25	41.50	38.00	95.00

[1]This is a consensus (or judgment) arrived at from the following knowledgeable sources: National Cattlemen's Association *(Cattle-Fax)*, heads of university animal science departments, and agricultural consultants. A consensus was resorted to for the reasons that (1) no extensive, nationwide, scientific sampling of cattle producers on these matters has ever been made; (2) there was considerable variation in the figures obtained from the various sources; and (3) this informatioin is much needed by producers and those who counsel with them. No claim is made relative to the scientific accuracy of the data; rather, it is presented (1) because it is the best information of its kind presently available on a nationwide basis, and (2) with the hope that it will stimulate needed research along these lines.

[2]Net return to management after deducting from gross receipts all costs, including depreciation on machinery, buildings and cattle, and interest on investment.

BUDGETS IN THE LIVESTOCK BUSINESS

A budget is a projection of records and accounts and a plan for organizing and operating ahead for a specified period of time. A short-time budget is usually for one year, whereas a long-time budget is for a period of years. The principal value of a farm budget is that it provides a working plan through which the operation can be coordinated. Changes in prices, droughts, and other factors make adjustments necessary. But these adjustments are more simply and wisely made if there is a written budget to use as a reference.

HOW TO SET UP A BUDGET

It's unimportant whether a printed form (of which there are many good ones) is used or one made up on an ordinary ruled 8½ in. × 11 in. sheet placed sidewise. The important things are: (1) that a budget is kept, (2) that it be on a monthly basis, and (3) that the operator be "comfortable" with whatever form or system is used.

No budget is perfect. But it should be as good an estimate as can be made—despite the fact that it will be affected by such things as droughts, diseases, markets, and many other unpredictables.

A simple, easily kept, and adequate budget can be prepared by using the following three types of forms:

1. Annual cash expense budget (see Table 2–12).
2. Annual cash income budget (see Table 2–13).
3. Annual cash expense and income budget—cash flow (see Table 2–14).

(See pages 48 and 49 for Tables 2–12, 2–13, and 2–14.)

The annual cash expense budget (Table 2–12) should show the monthly breakdown of various recurring items—everything except the initial loan and capital improvements. It includes labor, feed, supplies, fertilizer, taxes, interest, utilities, etc.

The annual cash income budget (Table 2–13) is just what the name implies—an estimated cash income by months.

The annual cash expense and income budget (Table 2–14) is a cash flow budget, obtained from the first two forms. It's a money "flow" summary by months. From this, it can be ascertained when, and how much, money will need to be borrowed, and the length of the loan along with a repayment schedule. It makes it possible to avoid tying up capital unnecessarily, and to avoid unnecessary interest.

TABLE 2–12
ANNUAL CASH EXPENSE BUDGET[1]

______________________________ for 19 ________
(name of farm or ranch)

Item	Total	Jan.	Feb.	Mar.	Apr.	May	June	July	Aug.	Sept.	Oct.	Nov.	Dec.
Labor hired													
Feed purchased													
Gas, fuel, grease													
Taxes													
Insurance													
Interest													
Utilities													
etc.													
etc.													
etc.													
Total													

[1]The Annual Cash Expense Budget should show the monthly breakdown of various recurring items—everything except the initial loan and capital improvements. It includes labor, feed, supplies, fertilizer, taxes, interest, utilities, etc.

TABLE 2–13
ANNUAL CASH INCOME BUDGET[1]

______________________________ for 19 ________
(name of farm or ranch)

Item	Total	Jan.	Feb.	Mar.	Apr.	May	June	July	Aug.	Sept.	Oct.	Nov.	Dec.
500 steers													
430 bu oats													
etc.													
etc.													
etc.													
Total													

[1]The Annual Cash Income Budget is just what the name implies—an estimated cash income by months.

TABLE 2-14
ANNUAL CASH EXPENSE AND INCOME BUDGET (Cash Flow)[1]

__________________________________ **for 19** ____________

(name of farm or ranch)

Item	Total	Jan.	Feb.	Mar.	Apr.	May	June	July	Aug.	Sept.	Oct.	Nov.	Dec.
Gross income	25,670					1,000	1,000						
Gross expense	13,910					575	2,405						
Difference	11,760					425	1,405						
Surplus (+) or Deficit (−)	+					+	−						

[1]The Annual Cash Expense and Income Budget is a cash flow budget, obtained from the first two forms. It is a money "flow" summary by months. From this can be ascertained when, and how much, money will need to be borrowed, the length of the loan and a suitable repayment schedule. It makes it possible to avoid tying up capital unnecessarily, and to avoid unnecessary interest.

HOW TO FIGURE NET INCOME

Table 2–14 shows a gross income statement. There are other expenses that must be taken care of before net profit is determined; namely:

1. **Depreciation on buildings and equipment.** It is suggested that the "useful life" of buildings and equipment be as follows, with depreciation accordingly: buildings, 15 years; and machinery and equipment, 5 years. These depreciation figures are in compliance with the Economic Recovery Tax Act of 1981, Accelerated Cost Recovery System (ACRS).

2. **Interest on owner's money invested in farm and equipment.** This should be computed at the going rate in the area, say 12%.

Here's an example of how the above works: Let's assume that on a given farm there was a gross income of $200,000 and a gross expense of $125,000, or a surplus of $75,000. Let's further assume that there are $60,000 worth of machinery, $60,000 worth of buildings, and $200,000 of the owner's money invested in farm and equipment. Let's further assume that buildings are being depreciated in 15 years and machinery in 5 years. Here is the result:

Gross profit				$75,000
Depreciation—				
Machinery:	$ 60,000 @	20% =	$12,000	
Buildings:	$ 60,000 @	6.67% =	4,002	
			$16,002	
Interest:	$200,000 @	12% =	24,000	
				40,002
Return to labor and management				$34,998

Some people prefer to measure management in terms of return on invested capital, and not wages. This approach may be accomplished by paying management wages first, then figuring return on investment.

ENTERPRISE ACCOUNTS

Where a cattle enterprise is diversified (for example, a farm or ranch having a cow-calf operation, a feedlot, and crops), enterprise accounts should be kept—in this case three different accounts for three different enterprises. The reason for keeping enterprise accounts are:

1. It makes it possible to determine which enterprises have been most profitable, and which least profitable.
2. It makes it possible to compare a given enterprise with competing enterprises of like kind, from the standpoint of ascertaining comparative performance.
3. It makes it possible to determine the profitableness of an enterprise at the margin (the last unit of production). This will give an indication as to whether to increase the size of a certain enterprise at the expense of an alternative existing enterprise when both enterprises are profitable in total.

COMPUTERS IN THE LIVESTOCK BUSINESS

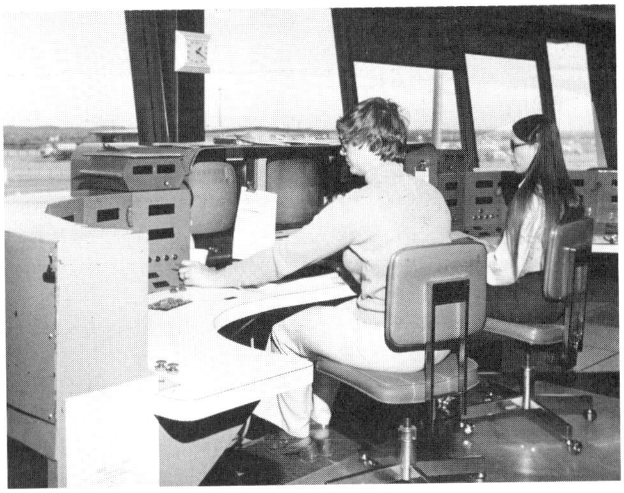

Fig. 2–14. Most modern farms, ranches, and agribusinesses are computerized. This shows trained personnel operating a totally computerized feed facility. (Courtesy, Union Pacific Railroad, Omaha, NE)

The computer technique known as linear programming is similar to budgeting, in that it compares several plans simultaneously and chooses from among them the one likely to yield the highest returns. It is a way in which to analyze a great mass of data and consider many alternatives. It is not a managerial genie, nor will it replace decision-making managers. However, it is a modern and effective tool in the present age, when just a few dollars per head or per acre can spell the difference between profit and loss.

There is hardly any limit to what computers can do if fed the proper information. Among the difficult questions that they can answer for a specific farm or ranch are:

1. **How is the entire operation doing so far?** It is possible to obtain quarterly or monthly progress reports, often making it possible to spot trouble before it's too late.

2. **What farm enterprises are making money; which ones are freeloading or losing?** By keeping records by enterprises—cow-calf, cattle feedlot, hogs, wheat, corn, etc.—it is possible to determine strengths and weaknesses, then either to rectify the situation or to shift labor and capital to a more profitable operation. Through *enterprise analysis*, some operators have discovered that one part of the farm business may earn $10 or more per hour for labor and management, whereas another may earn only $2 per hour, and still another may lose money.

3. **Is each enterprise yielding maximum returns?** By having profit, or performance, indicators in each enterprise (see Table 2–11), it is possible to compare these (a) with the historical average of the same farm or ranch, or (b) with the same indicators of other farms or ranches.

4. **How does this ranch stack up with its competition?** Without revealing names, the computing center (local, state, area, or national) can determine how a given ranch compares with others—either the average, or the top (say 5%).

5. **How to plan ahead?** By using projected prices and costs, computers can show what moves to make for the future; they can be a powerful planning tool. They can be used in determining when to plant, when to schedule farm machine use, etc.

6. **How can income taxes be cut to the legal minimum?** By keeping an accurate record of expenses and figuring depreciation accurately, computers make for a saving in income taxes on most farms and ranches.

7. **What are the *least cost* and *highest net returns* rations?** Instruction on how to balance a ration by computer is given in this book in Section 4, Balanced Ration, How to Balance Rations, Computer Method. Hence, the reader is referred thereto.

For providing answers to the above questions and many more, computer accounting costs an average of about 1% of the gross farm income. By comparison, it is noteworthy that city businesses pay double this amount.

There are three requisites for linear programming a farm or ranch; namely:

1. Access to a computer.
2. Computer know-how, so as to set the program up properly and be able to analyze and interpret the results.
3. Good records.

The pioneering computer services available to farmers and ranchers were operated by universities, trade associations, and the government, with most of them being on an experimental basis. Subsequently, others have entered the field, including commercial data processing firms, banks, machinery companies, feed and fertilizer companies, and farm suppliers. They are using it as a "service sell," as a replacement for the days of the "hard sell."

Programmed farming is here to stay, and it will increase. Space limitations will not permit elucidation in this book of the role of computers in each kind of livestock operation. Only one kind—cow-calf—will follow. However, the same principles can be adapted to any other livestock operation.

BUSINESS ASPECTS OF CATTLE FEEDING[10]

Cattle feeders conduct a big business; so, as is true in other businesses, they need to make a reasonable profit for the use of their capital, labor, and management. To this end, their business aspects must be sophisticated and efficient; they must—

1. Develop an economic plan for feeding cattle for a particular market.
2. Develop the most practical and economical size unit for the area.
3. Know the market specifications for weight and grade.
4. Make continuous break-even price projections in order to guide the purchase of feeder cattle and feed.
5. Buy feeder cattle of the right size and quality, and at the right price.
6. Sell the cattle to the best advantage.
7. Integrate when possible and practical.
8. Feed cattle to weight and grade.
9. Evaluate performance.
10. Maintain a feedlot and cattle financing program that provides adequate equity for a sustained operation.
11. Know how and when to use cattle and feed futures trading and options. (See Section on "Livestock Futures Trading and Options.")

Of course, the above 11 points represent a great oversimplification of a complex business, but they do clearly set forth the main requisites for profitable cattle feeding. Anyone who wishes to make money feeding cattle must have expertise in these 11 areas.

Business aspects must balance with technical factors—feed, additives, crossbreds, pollution control, etc. It is important, therefore, that feeders and those who counsel with them be thoroughly grounded in each of these areas.

---

[10]The entire section on "Business Aspects of Cattle Feeding" was authoritatively reviewed by the following: Ron Baker, President, C & B Livestock, Inc., Hermiston, OR; and Lee H. Stampe, General Manager, American Cattle Feeders, Clovis, NM.

Other factors than the cost of feed, amount of gain, and price of slaughter cattle affect the price that a feeder can afford to pay for feeder cattle; among them, are the following:

1. Condition of the cattle. Thin cattle, if in good health, will make faster gains than fleshy cattle.

2. Growthy cattle—cattle that are big framed and on the rangy order—make better gains than the little, compact kind; and they may be carried to heavier weights. Feeder cattle backed by production records enhance visual appraisal.

3. Younger, lighter weight cattle tend to make more efficient gains.

4. Cattle of known, superior ancestry with gaining ability are worth more.

5. Higher grade cattle are worth more. This is so because better grades generally bring a higher selling price, and, therefore, a higher price is obtained on their gains made in the feedlot.

6. The higher the cost of feed, the greater the necessary margin between the cost of feeder cattle and the selling price of finished animals. This is so because of the high cost of gains as compared to their selling price.

7. The longer the feeding period and the greater the gains necessary to get the cattle in a finished condition, the greater the necessary margin. This is due to the fact that gains made in the feedlot generally rise in cost with longer feeding periods.

8. Feeder steers are generally worth more than heifers. This is because they generally gain faster, require less feed, and bring from $0.50 to $1.50 per cwt more than heifers when finished. Additionally, there is no pregnancy problem. However, the use of more exotic crossbreeding and MGA has narrowed this spread.

9. Good crossbreds will make 2 to 4% more rapid and efficient gains than the average of the parent breeds.

MARGIN

The cost of feed and the cost of cattle are the two major capital expenses in any cattle feeding venture. Likewise, these same two factors are the most important ones in determining profits and losses.

Generally speaking, about 80% of the cost of finishing cattle (exclusive of the initial purchase price for animals) is for feed. Another 6% is usually absorbed by interest on the purchase price of the cattle. Then labor costs, taxes, purchasing and marketing charges, shrinkage losses, and death losses (about 1 to 2%) make up most of the remaining expenses.

A positive margin exists when feeder cattle cost less than finished cattle. A negative margin exists when feeder cattle cost more than finished cattle. Cattle feeders will pay more for feeder cattle than they expect to receive for them as finished animals when there appears to be a favorable margin on the gain in weight. That is, when they can sell the gain in weight for considerably more than it cost to produce it.

Profits in cattle feeding come from two different kinds of margins—price margin and feeding margin.

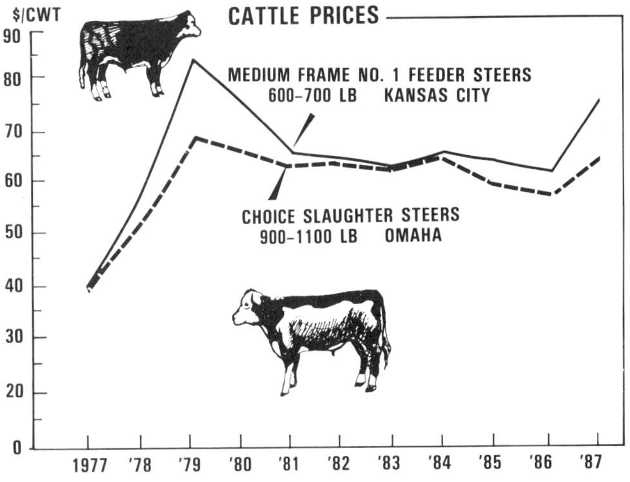

Fig. 2-15. A negative cattle feeder's margin existed until 1983-84; that is, feeder steers cost more than the price of slaughter steers. (Source: *Livestock and Meat Statistics*, 1984-88, USDA, Stat. Bul. No. 784, pp. 189-190)

Price margin is the difference between the cost per cwt of the feeder animals and the selling price per cwt of the same animal when finished.

For example, if a feeder pays $66 per cwt for a 600-lb steer and sells him for $63 per cwt, the price margin is a negative $3. This means that the cattle feeder would take an $18 loss on the original 600 lb bought.

Feeding margin is the difference between the cost of putting on 100 lb of gain and the selling price per cwt of the same animal when finished. Thus, if it costs $50 per cwt to put gain on yearling steers, and if a cattle feeder could sell cattle for $63 per cwt, the feeding margin would be $13 per cwt. Assuming a market weight of 1,000 lb, or of 400 lb gain, the producer could expect to make about $52 on feeding margin.

The amount a cattle feeder makes as a result of a good feeding margin can more than offset the losses accruing from a negative price margin, but it doesn't always work that way. It depends on many different things—the selling price, the cost of gain, the price paid for feeder animals, and other factors.

In the example just cited, the feeding margin amounted to $52 per animal. This is not to suggest, however, that cattle feeders should always put more gain on yearling steers. How much gain a cattle feeder should put on depends upon the kind of feeding program being followed, the kind and condition of the feeder cattle when they go into the lot, rate of gain, and several other factors. Research and experiences have clearly demonstrated that costs of gain go up pretty fast if cattle are fed much beyond Choice slaughter grade.

The principles of profits from price margin and feeding margin apply to feeder calves, also. But the relative importance of price margin vs feeding margin is not quite the same for calves as for yearlings. Let us analyze the situation further: The feeder who is feeding yearlings buys 600 lb of the 1,000 lb finally sold. Getting cattle bought right is pretty important. If the feeder pays $1 per cwt too much for feeders, that takes $6 off the potential profit from price margin and from total profits. On the other hand, the feeder who buys calves is more interested in costs of gain and feeding margin than in price margin because about 60% of the weight sold is from the gain put on in the feedlot. Thus, if the feeder pays $1 per cwt too much for a 400-lb feeder calf, it hurts, but not quite so much—$4 compared to $6 per cwt per head.

If a farmer-feeder just manages to balance gains from feeding margin with losses from price margin, this does not necessarily mean that cattle should not be fed. Actually, the feeder isn't in too bad shape. The feeder is getting paid market price for feed, a going wage for labor, around 8% on capital invested, and enough to cover all fixed costs like depreciation, taxes, etc., on the lot and equipment. With the commercial feeders, it's another story. They usually buy most of their feed, and they operate on borrowed capital. Thus, to stay in business, they must turn in a profit over and above these costs.

CUSTOM (CONTRACT) FEEDING[11]

Custom cattle feeding is the feeding of cattle for a fee, usually without taking ownership of the animals.

Contract feeding is not new. It made rapid development after 1929, and there was much of it during the severe drought of 1934. From this time to World War II, contract feeding decreased in importance—a decline attributed to improved feed conditions on the western range, higher prices for feeder animals, and the availability of more credit through federal and private loan agencies.

Custom cattle feeding as we know it today paralleled the development of commercial feedlots. California pioneered in it; thence it spread to Arizona, the Northwest, and other areas of the West, Nebraska, Texas, the Oklahoma Panhandle, and western Kansas. Even today, these are the principal custom feeding areas. Custom feeding provided a means of financing the rapid growth of cattle numbers needed to utilize the highly mechanized, large volume feeding operations. Individuals who did not have sufficient capital to build and operate a feedlot large enough to perform economically could acquire the necessary capital and volume by custom feeding cattle for others. In this way, part of the burden of providing capital for efficient operation of a feedlot was shifted to outside interests.

Capital requirements, periods of severe economic conditions (like scarce money and high interest), times of depressed feeder cattle prices, and adverse pasture conditions caused custom feeding to grow following World War II. These same forces, along with the need for high occupancy (full feedlots) and increased integration, have resulted in further expansion of custom feeding.

Most custom feeders have developed large, highly mechanized, and very efficient plants. Usually, they have on their staffs, or retain as consultants, highly trained nutritionists who are charged with the responsibility of formulating rations and of obtaining maximum gains and feed efficiency at the lowest possible cost. Through custom feeding, they sell the use of their facilities, services, and know-how to cattle owners, usually with profit to each party.

The Packers and Stockyards Administration, of the U.S. Department of Agriculture, ruled that, effective July 1, 1974, (1) packers could not own, operate, or control *custom* feedlots; and (2) *custom* feedlot owners could not own, operate, or control packing plants. This action was taken in order to avoid monopolistic conditions. This ruling does not prohibit packers from feeding their own cattle for their own slaughter needs.

The proportion of custom fed cattle to cattle owned by the feedlot varies (1) in period of time—it increases in times of financial stress (when cattle feeding is not profitable, money is scarce, and interest is high); (2) according to area—for example, there is more custom feeding in California than Colorado; (3) according to size of feedlot—generally speaking, the larger the feedlot, the greater the percentage of custom feeding. Some feedlots do not do any custom feeding whatsoever; others are almost wholly on a custom basis; but most lots have part of each. Feedlots that do both—those in the dual role of custom feeding and owning cattle—vary in the proportion of cattle in each category, but most of them seem to prefer about ⅔ custom fed cattle and ⅓ ownership. It's a good bread-and-butter division; in times when fed cattle lose money, such a feedlot has sufficient assured income to pay its bills.

The ownership of custom fed cattle is diverse. It includes (1) cow-calf operators (farmers and ranchers) who wish to retain ownership of the cattle that they produce through the feedlot phase; (2) packers; and (3) investors, including limited partnerships, corporations, cattle buyers, cattle dealers, and others.

STOCKER AND GROWER CONTRACTS[12]

Hand in hand with the development of big feedlots and year-round feeding came the need for an assured supply of feeder cattle of the desired kind on a continuous basis. To meet this need, more and more feedlots have turned to contractual arrangements with stocker growers, with numerous kinds of contracts. Usually, the cattle are owned by the feedlot, most of which are large and in a stronger financial position than the majority of stocker growers. The two most common kinds of contracts are based on either (1) a fixed cost for the gain, or (2) an agreed feed cost plus an extra charge for labor and lot rental. Usually, there is provision for adjusting for death loss. Such contracts should always be in writing, with all provisions, including weighing conditions, spelled out.

Although the use of stocker and grower contracts has increased in recent years, the concept is not new. Many of the owners of the lush bluestem pastures of the flint hills of Kansas have long grown out stockers on a contract basis. Today, many of the cattle grazed in this area are owned by ranchers;

[11]This section was authoritatively reviewed by, and helpful suggestions were received from the following: Dr. W. F. Williams, Department of Agricultural Economics, Texas Tech University, Lubbock, and Mr. R. R. Baker, President, C & B Livestock, Inc., Hermiston, OR)

[12]This section was authoritatively reviewed by the following: Dr. R. M. Koch, U.S. Meat Animal Research Center, Clay Center, NE; and Mr. A. Solomon, cattleman, Humboldt, KS.

some are owned by feedlots; and still others are owned by investors from all walks of life. The common practice is to contract the cattle to a cattle feedlot from 1 to 3 months ahead of delivery, with the buyer making a down payment of approximately 10% of the projected delivery price, at the time of signing the contract.

Fig. 2-16. Stocker cattle on bluestem pasture in the famed flint hills of Kansas. (Courtesy, A. Solomon, cattleman, Humboldt, KS)

Today, many corn farmers in the fertile irrigated area in the vicinity of Greeley, Colorado, make corn silage and feed cattle on a contract basis to stockers owned by one of several large feedlots in the vicinity. Stocker cattle are also being grown under contract on the wheat pastures of Kansas, Oklahoma, and Texas; on hay and other roughages in the irrigated valleys of the West; and on sorghum silage and stalk fields throughout the Southwest.

HOG PRODUCTION CONTRACTS

A hog contract is an agreement between two or more persons to do or refrain from doing certain things. In recent years, swine producers have shown increasing interest in contract hog production due to (1) the high cost of capital, (2) the difficulty of many producers in obtaining adequate financing, and (3) the desire to forego the possibility of large profits for the assurance of more reliable returns. In the late 1980s, an estimated 8 to 10% of U.S. hogs were under some kind of production contract, with a much smaller number under a marketing contract.

An overview of the most common types of contract used in the swine industry follows.

TYPES OF HOG PRODUCTION CONTRACTS

Investors, feed dealers, farmers and others are often interested in producing hogs, but are unwilling or unable to provide the necessary labor, facilities, and equipment. Some producers have also found contract production to be an effective method of expanding their operations. So, these entrepreneurs find producers who are willing to furnish the labor and equipment in exchange for a fixed wage or share of the profits. The resulting contracts vary considerably in form and responsibility of each party involved.

The more popular contracts provide for fixed payment, direct feeding, or profit sharing. These contracts are most commonly used for feeder pig production and hog finishing.

- **Fixed (guaranteed) payments**—These contracts guarantee the producer a fixed payment per head, usually along with provision for bonuses and penalties.

Under the fixed (guaranteed) payment, the contractor generally supplies the pigs, feed, veterinary services and medication, transportation, management assistance, and marketing. The producer normally provides the buildings and equipment, labor, utilities, and the necessary insurance. Most fixed payment contracts also provide for bonuses for keeping death losses low and feed efficiency high, and for penalties for high death losses and unmarketable animals.

- **Direct feeding**—In this type of contract, a feed dealer or cooperative contracts with producers to finish hogs. The contractor's objective is to increase feed sales and secure a reliable feed outlet.

Typically, the contractor provides the feed, directs the feeding program, and provides some management assistance. The producer agrees to purchase all feed and related services from the contractor and is responsible for all production costs. Upon sale of the hogs, the producer receives all proceeds less any outstanding balance owed to the contractor.

- **Profit sharing**—In a profit-sharing contract, the contractor and producer divide the profit on an agreed basis, such as 50–50, 60–40, depending upon who provides the majority of inputs and their value.

Normally, the contractor purchases the pigs and is responsible for all feed, veterinary expenses, transportation, and marketing costs. Over the duration of the contract, the contractor's costs are charged to the account. Then, this account balance is subtracted from the sale proceeds to determine the profit. The contractor's return depends upon the profit made on the sale of hogs and the gain received from the markup on feed, pigs, and supplies. Typically, the producer provides the facilities, labor, utilities, and insurance for a portion of the profit. The producer is usually guaranteed a minimum amount per head as long as death losses are below a set percentage.

FEEDER PIG PRODUCTION CONTRACTS

The following feeder pig production options are available:

1. **Option 1.** The contractor provides the breeding stock and contracts with the feeder pig producer to provide feeder pigs based on production criteria, such as pigs weaned per litter, with bonuses and docks based on target level. Most of the risk is retained by the producer.
2. **Option 2.** The contractor with a finishing operation provides the breeding stock, feed, and management assistance, and pays the feeder pig producer a flat fee for each pig. The fee will vary according to pig weight and current production costs. Most of the risk is assumed by the contractor.

3. **Option 3.** The contractor provides breeding stock, feed, facilities, and veterinary costs. The producer provides labor, utilities, maintenance, and manure handling. The producer is paid a fee for each pig produced and a monthly fee for each sow and boar maintained. This option is suited for owners who no longer want to be actively involved, but who have a good manager with limited cash willing to take over.

4. **Option 4.** This is a shared revenue program, with the percentage of gross sales based on inputs and services provided, and risks borne by each participant. For example, the feed dealer would receive a percentage of gross sales based on feed dealer inputs; the breeding stock supplier would receive a percentage; the management firm that supplied consultants and computerized records would receive a percentage; and the producer who supplied facilities, veterinary care, utilities, labor, and insurance would receive a percentage.

FARROW-TO-FINISH CONTRACTS

Most farrow-to-finish programs are set up on a percentage basis to reflect the relative amount of input supplied by each person or firm, with the following options available.

1. **Option 1.** Based on input costs, the following participants may share in a percentage of gross sales: The breeding stock supplier; the feed and medications supplier; the management consultant and computerized record services; and the producer-supplier of facilities, labor, veterinary care, utilities, and insurance.

2. **Option 2.** The current hog inventory is purchased by a limited partnership, which supplies sow replacements. Each of the following contract participants receives a percentage of the proceeds when hogs are marketed: The feed and medications supplier; the management agency-supplier of production and marketing guidance; and the producer-supplier of facilities, labor, utilities, veterinary costs, repairs, and manure disposal. The remaining percentage is split between the limited partnership and the general partner who manages the partnership.

BREEDING STOCK LEASING

Under a breeding stock lease, the contractor furnishes the producer with breeding age gilts and/or boars. The rent paid by the producer for the breeding stock may be either a specified number of pigs or an equivalent amount of money at designated times. The popularity and use of breeding stock leases has declined in recent years.

MARKETING CONTRACTS

Marketing contracts are of two kinds: (1) market hog contracts, and (2) feeder pig contracts.

- **Market hog contract**—A market hog contract is a forward sale between a buyer (normally a meat packer or a marketing agent) and a seller (normally a producer), in which the producer agrees to sell, at a specified date, a specified number of hogs to a buyer at a certain price. Normally, the following terms are detailed in a forward marketing contract: (1) the quantity, with the minimum ranging from 5,000 lb to 30,000 lb; (2) the date and location of delivery; (3) acceptable weights and grades, including premiums and discounts; (4) a description of the pricing mechanism (some contracts now price hogs on a grade and yield basis); (5) provisions for non-deliverable hogs and unacceptable carcasses; (6) provisions outlining the credit requirements of the seller and inspection of the hogs by the buyer; and (7) provisions for breach of contract.

  Under a forward sale contract, the producer retains all risks of production, other than selling price. The producer uses the forward sale contract to reduce the risk of price fluctuations and to lock in an acceptable profit. But a forward sale contract may also cause the producer to miss out on greater profits if prices rise. Sometimes, a minimum price (a floor price) is used for hogs, in which the buyer guarantees the seller a minimum price (a floor price), with the seller receiving whichever price is higher at market time—the floor price, or the market price.

- **Feeder pig contracts**—Typically, feeder pig marketing contracts are between a marketing agency, often a cooperative, and a pig producer, in which the marketing agency agrees to market the pigs of a producer for a fee. A feeder pig marketing contract may contain the following provisions: (1) the producer agrees to market exclusively through the agency; and (2) the marketing agency specifies management practices and weight of feeder pigs at marketing. Essentially, producers are hiring market expertise through feeder pig contracts.

CHARACTERISTICS OF A GOOD HOG CONTRACT

A contract is no better than the parties back of it. Moreover, it should be fair to both parties. Additionally, a good contract meets the following requisites:

1. It is in writing, and it is clear and concise.
2. It clearly defines the rights and responsibilities of both parties.
3. It contains the following: the names of both parties, the number of pigs involved, the duration of the contract, the time and method of payment, and who should supply certain inputs.

HORSE CONTRACTS[13]

Today, most horse enterprises are owned and operated as businesses, rather than as hobbies. In the conduct of horse businesses, the following types of contracts may be involved: (1) syndicated horses, (2) stallion breeding contracts, and (3) boarding agreements.

---

[13]In the preparation of this entire section, the author benefited from the expertise of the following noted horse specialists: D. E. McGlothlin, Manager, Horse Division, Harris Farms, Coalinga, CA; and R. Vacca, General Manager, Washington Thoroughbred Breeders Assn., Seattle, WA.

SYNDICATED HORSES

Reduced to simple terms, a syndicated horse is one that is owned by several people. Most commonly, it's a stallion, although an expensive yearling or broodmare is sometimes syndicated. Also, any number of people can form a syndicate. However, there is a tendency to use the term *partnership* where two to four owners are involved, and to confine the word *syndicate* to a larger group of owners. Recently, racing syndicates have become fashionable at virtually every level; partnerships and syndicates are being formed to purchase modestly priced horses. For example, five persons may form a syndicate with each contributing $2,500 to buy a $10,000 yearling or racehorse.

Fig. 2-17. The syndicated Arabian stallion **Aladdin*, syndicated for $6.3 million. (Courtesy, International Arabian Horse Assn., Denver, CO)

Each member of the syndicate owns a certain number of shares, depending on how much he/she purchased or contributed. It's much like a stock market investor, who may own one or several shares in General Electric, IBM, or some other company. Sometimes one person may own as much as half interest in a horse. Occasionally, half shares are sold.

Generally speaking, the number of shares in a stallion is limited to the number of mares that may reasonably be bred to him in one season—usually 30 to 35, with Thoroughbred stallions.

WHY AND HOW HORSE OWNERS SYNDICATE

The owner of a stallion that has raced successfully usually has the opportunity to choose between (1) continuing as sole owner of the horse, and standing him for service privately or publicly, or (2) syndicating him. In recent years, more and more owners of top stallions have elected to syndicate. The most common reasons for so doing are:

1. The stallion owner does not have a breeding farm or an extensive band of broodmares.

2. The owner believes that the stallion under consideration may not nick well with many of his/her mares; or perhaps the stallion is closely related to the mares.

3. The owner has need for immediate income.

4. Syndicating spreads the risk, should the stallion get injured or die, or prove unsuccessful as a sire.

The owner may arrange the syndication, usually with competent legal advice; or, if preferred, the syndication can be turned over to a professional manager, who will generally take a free share as an organization fee.

The following pointers are pertinent to successful syndication of stallions:

1. **Check fertility.** Before syndicating, it is a good idea to check the fertility by test-mating to a coldblood (draft) mare. Of course, if the stallion is still racing, and has not been retired to stud, this is impossible.

When a test-mating is made on any stallion that is to be syndicated, it is essential that each potential purchaser of a share receive documented results of any test.

2. **Establish a stud fee.** A common rule of thumb is that each syndicate share is worth 4 times the stud fee. Hence, if it is decided that the stallion under consideration will command a $10,000-stud fee, each share would be worth $40,000. If 30 shares are involved, the horse would have a value of $1,200,000 for syndication purposes.

3. **Determine time of payment.** In most cases, payment is due upon the signing of the syndicate contract, although some contracts (a) allow 30, 60, or 90 days; or (b) provide that the price of a share may be paid on the installment plan over a 2- or 3-year period.

4. **Insurance.** Generally, the syndicate does not carry mortality insurance. So, persons purchasing shares are responsible for insuring their shares.

5. **Put it in writing.** Syndication agreements should be clear, detailed, and in writing. In addition to identifying the horse, the agreement should state (a) the shareholders' proportionate interest (say $\frac{1}{32}$); (b) the breeding rights of a shareholder (for example, the right to breed one mare per season to the horse, so long as he is in good health and able to breed); (c) the method of distributing services by lot, should it be necessary to limit the number of mares bred during any given

season; (d) the method of disposing of, and the price to charge for, any extra services (over and above one per share, for example) during a given season; (e) the place where the horse shall stand, or how such determination will be made (usually by majority vote of the shareholders); and (f) how other policy matters not covered in the agreement will be determined (usually by majority vote).

Generally, such routine matters as the feed, care, and health of the stallion, and the scheduling of mares are left to the discretion of the syndicate manager, at a stated fee per month, with each shareholder billed proportionate to his/her number of shares. The manager also handles the promotion and advertising, insurance, and unusual veterinary expenses, as stipulated by the syndicate, with the costs prorated among its members.

Normally, shareholders can barter their breeding service to another stallion. However, they cannot sell their shares without prior approval of the manager and giving the other shareholders the right to buy it at the price offered; and, normally, this same stipulation applies to the sale of a service during any season.

Also, provision is usually made for sale of the horse should the majority of the shareholders so desire, with them also determining, at the time of sale, the price and whether sale shall be at private treaty or auction. Further, the contract usually provides for "pensioning," or otherwise disposing of, a sire should he become sterile or be overtaken by old age before dying.

In short, a syndicate agreement, like any good legal contract, attempts to spell out every foreseeable contingency that may arise during the stud's career, and to arrange for majority vote of the shareholders to settle any unforeseen contingencies.

The Nashua Syndicate Agreement, used by the late Mr. Leslie B. Combs II, Spendthrift Farm, Lexington, Kentucky, who probably contributed more than any other person to stallion syndication, follows. It is also noteworthy that, when Mr. Combs syndicated *Nashua*, he sold all but one share of a total of 32, each at $39,100 (for a total of $1,251,200), over the telephone in one afternoon; and the only reason that the one share was not sold until the next morning was that he couldn't reach one of his regular clients on the telephone that afternoon.

STALLION BREEDING CONTRACT

Stallion breeding contracts should always be in writing; and the higher the stud fee, the more important it is that good business methods prevail. Neither verbal agreements nor barn door records will suffice.

From a legal standpoint, a stallion breeding contract is binding to the parties whose signatures are affixed thereto. Thus, it is important that the contract be carefully read and fully understood before signing.

In addition to the provisions made in the example Stallion Service and Boarding Agreement presented here, and most other similar contracts, the author suggests that the following matters be covered in the stallion breeding contract:

1. **Facts about the mare.** There should be record of the mare's temperament; thereby lessening danger to her, to the stallion, and to the personnel. Also, historical information should be included about the mare's breeding record and peculiarities, and her health—preferably with the health record provided by the veterinarian who has looked after her.

2. **Some management understanding.** The parties to the contract should reach an understanding relative to the mare's veterinary care, parasite control, seasonal injections, foot trimming, etc., and then put it in writing.

3. **An incentive basis.** Generally, stallion owners guarantee a live foal, which means that the foal must stand and nurse; otherwise, the stud fee is either refunded or not collected, according to the stipulations. Of course, it is in the best interests of both parties that a strong, healthy foal be born. One well-known Quarter Horse establishment reports that their records reveal that of all mares settled during a particular 3-year period, 19% of them subsequently either resorbed or aborted feti, or the foal or mare died. Further, their investigation of these situations showed that the vast majority of these losses could have been averted by better care and management. They found many things wrong—ranging from racing mares in foal to turning them to pastures where there was insufficient feed. To alleviate many, if not most, of these losses—losses that accrue after the mare has been examined and pronounced safe in foal, then taken away from the stallion owner's premises—the author suggests that an incentive basis be incorporated in the stallion breeding contract. For example, the stallion owner might agree to reduce the stud fee (1) by 10, 15, or 20% (state which), provided a live foal is born; or (2) by 25 to 33⅓% provided the mare owner's veterinarian certifies that the mare is safe in foal 30 days after being removed from the place where bred, with payment made at that time and based on conception rather than on birth of a live foal.

BOARDING AGREEMENT

Today's tough zoning laws and antipollution campaigns are making it increasingly difficult to keep horses in towns and suburban areas. As a result, more and more horses are being stabled and cared for in boarding establishments out in the country, to which owners commute. This prompts the need for an agreement.

Boarding agreements should always be in writing, rather than verbal handshake agreements. From a legal standpoint, a boarding agreement is binding to the parties whose signatures are affixed thereto. Thus, it is important that the agreement be carefully filled out, read, and fully understood before signing. A sample boarding agreement is given in Fig. 2-18.

BOARDING AGREEMENT

(To be executed in duplicate; one copy to be retained by each party.)

This agreement made and entered into by and between ______________ (owner of horse) ______________ (address)

hereinafter designated "Horse Owner," and ______________ (owner of stable) ______________ (address) hereinafter designated, "Stable Owner." This agreement covers the horse described as follows:

______________ (name) (sex) ______ (age) ______ (color) ______________

I. *Stable owner agrees to—*

1. Keep the horse in a stall and/or paddock described as follows:

2. Feed, water, and care for the horse in a good and husbandlike manner; feeding horse as follows:

Kind of Feed	Amount of Feed		
	Morning	Noon	Night
	(lbs)	*(lbs)*	*(lbs)*

3. Perform the following additional services:

   a. *Grooming (specify):* ______________

   b. *Exercising (specify):* ______________

   c. *Parasite treatments (specify):* ______________

   d. *Others (list):* ______________

____

____

II. *Horse Owner agrees to—*

1. Make all arrangements for the periodic shoeing of the horse, and assume the cost thereof. Any exception to this shoeing arrangement shall be in the space that follows:

2. Pay Stable Owner (a) for the foregoing facilities, feed, and services the sum of $______________ per month, payable on the ______________ day of each month in advance, and (b) for drugs and medications, cost, the first of each month following invoicing.

3. Entitlement of Stable Owner to a lien against the boarded horse for the value of services rendered, and to enforce said lien according to the appropriate laws of the state, *provided* (a) Stable Owner performs the services herein specified, and (b) Horse Owner fails to make a scheduled payment.

III. *Horse Owner and Stable Owner mutually agree that—*

1. In the event the horse shall require the services of a Veterinarian, Stable Owner will immediately contact the Horse Owner. In the event Horse Owner cannot be reached, Stable Owner is hereby authorized, as agent for Horse Owner, (a) to call Dr. ______________ DVM; and should Dr. ______________ be unavailable, (b) to call any other licensed veterinarian of choice. All fees charged by said veterinarian shall be the sole and exclusive responsibility of the Horse Owner, with no liability whatsoever on the part of Stable Owner for such fees.

2. This document constitutes the entire agreement between the parties and there are no other agreements between them except as noted below.

______________ (Signature of Horse Owner) ______________ (date)

______________ (Signature of Stable Owner) ______________ (date)

Fig. 2-18. Boarding Agreement.

FUTURES TRADING[14]

Fig. 2–19. Futures trading on the Chicago Mercantile Exchange (CME) floor. (Courtesy, Chicago Mercantile Exchange, Chicago, IL)

Futures trading is not new. It is a well-accepted, century-old procedure used in many commodities, for managing risk, protecting profits, stabilizing prices, and smoothing out the flow of merchandise. For example, it has long been an integral part of the grain industry; grain elevators, flour millers, feed manufacturers, and others have used it to protect themselves against losses due to price fluctuations. Also, a number of livestock products—hides, tallow, frozen pork bellies, and hams—were traded on the futures market before the advent of beef futures. Many of these operators prefer to forego the possibility of making a high speculative profit in favor of earning a normal margin or service charge through efficient operation of their business. They look to futures markets to provide (1) an insurance medium in the marketing field, and (2) the facilities and machinery for underwriting price risks.

A commodity exchange is a place where buyers and sellers meet on an organized market and transact business on paper, without the physical presence of the commodity. The exchange neither buys nor sells; rather, it provides the facilities, establishes rules, serves as a clearinghouse, holds the margin money deposited by both buyers and sellers, and guarantees delivery on all contracts. Buyers and sellers are represented by brokerage firms.

The unique characteristics of futures markets is that trading is in terms of contracts to deliver or to take delivery, rather than on the immediate transfer of the physical commodity. In practice, however, very few contracts are held until the delivery date. The vast majority of them are cancelled by offsetting transactions made before the delivery date.

Many cow-calf raisers have long forward contracted their calves for future delivery without the medium of an exchange. They contract to sell and deliver to a buyer a certain number and kind of calves at an agreed upon price and place. Hence, the risk of loss from a decrease in price after the contract is shifted to the buyer; and, by the same token, the seller foregoes

[14]This entire section of beef futures was authoritatively reviewed by and helpful suggestions were received from J. Graham, Director Commodity Marketing and Education, Chicago Mercantile Exchange, Chicago, IL; and R. E. Sheldon, Manager-Agriculture Group, Economic Analysis and Planning Dept., Chicago Board of Trade, Chicago, IL.

the possibility of a price rise. In reality, such contracting is a form of futures trading. Unlike futures trading on an exchange, however, actual delivery of the cattle is a must. Also, such privately arranged contracts are not always available, the terms may not be acceptable, and the only recourse to default on the contract is a lawsuit. By contrast, futures contracts are readily available and easily offset.

LIVESTOCK FUTURES TRADING

Livestock futures trading is relatively new. It was not until November 1964 that trading in live cattle futures opened on the Chicago Mercantile Exchange. Trading in live hogs began 15 months later. Since then, livestock futures trading has grown enormously. Between 1965 and 1989, the number of cattle futures contracts increased from 59,219 to 4,265,710. Between 1966 and 1989, the number of hog futures contracts increased from 8,063 to 2,008,750.

FUTURES TRADING IN FINISHED CATTLE, FEEDER CATTLE, AND FEED

The three big uncertainties in the cattle feeding business, any one of which can cause a cattle feeder to suffer heavy losses, are prices of (1) feeder cattle, (2) feed, and (3) finished cattle. Through futures contracts, cattle feeders can now hedge all three. In advance of feeding, they can lock in the price of feeder cattle, feed, and finished cattle.

This discussion is devoted primarily to live (slaughter) beef cattle futures as they apply to cattle feedlot operators, because it is the highest risk phase of the cattle business, as well as the least flexible. Unless feeders contract ahead, they have no assurance of what their finished cattle will bring when they are ready to go. Moreover, there is little flexibility in market time, for the reason that excess finish is costly and unwanted by the consumer. As a result of this uncertainty of market price, and in realization of the high risks involved, sleepless nights are rather commonplace among cattle feeders; they find it difficult to concentrate on the business at hand—the efficient feeding and management of cattle. Live (slaughter) beef cattle futures provide a means through which a cattle feeder can fix the selling price before the cattle are ready to be marketed.

The second major item of the triumvirate making for uncertainties in cattle feeding is the price of feeder cattle. Only by contracting ahead can cattle feeders be sure of the price that they will have to pay when ready to lay in feeder cattle. For many years, a fairly effective, albeit unorganized, cash contracting system has been operating relative to feeder cattle. Feeder cattle futures now offer, on an organized basis, a method for cattle feeders to lock in the price of feeder cattle well ahead of taking delivery, thereby alleviating possible heavy losses due to sharp price rises of feeder cattle. Without feeder cattle, a feedlot is not in business. Yet, much of the overhead cost for facilities and staff continues. Hence, a full feedlot is important. Cow-calf raisers—the producers of feeder cattle—have more flexibility, and are less dependent on contracting ahead, than cattle feeders. If the feeder cattle market isn't good, they can hold their calf crop for a time; they may even carry them over for another year—to the yearling stage. Also, rather than accept what they consider to be unfavorable prices for their stockers, they can have them custom fed, or they can feed them out themselves. By retaining ownership for a longer period of time, they increase the probability of being able to price their cattle at a profit. Certainly, there are risks in the cow-calf business, but, in comparison with cattle feeding, there is more flexibility, and the timing is not so exacting.

Since feed represents such a large proportion of the cost of feeding cattle (amounting to approximately 80% of the costs exclusive of the purchase price of the feeder cattle), it is wise to set the price months in advance when possible. Usually, feed can be bought most advantageously at harvest time. thus, cattle feedlot owners who have adequate storage and finances generally buy their main feed ingredients at that time. By so doing, they can project with reasonable accuracy what it will cost to feed cattle. Corn and soybean meal futures permit the cattle feeder to accomplish the same thing without actually taking delivery on the feed and incurring storage costs and risks of physical deterioration. The cattle feeder can use such futures to protect against increases in feed prices.

WHAT CONSTITUTES A FUTURES CATTLE CONTRACT?

A futures contract is a standardized, legally binding transaction in which the seller promises to make delivery of a specified quantity and type of a commodity at a specified location(s) during a specified future month. The buying and selling are done through a third party (the exchange clearing member) so that the buyer and seller remain anonymous; the validity of the contract is guaranteed by a reputable and well-financed exchange clearing member; and either buyer or seller can readily liquidate their positions by simply offsetting sale or purchase.

The Chicago Mercantile Exchange specifications of finished cattle and feeder cattle contracts follow:

- **Specifications for a live (slaughter) cattle contract are**—Delivery and acceptance of 40,000 lb of USDA yield grade 1, 2, 3, or 4 Choice grade steers (approximately 37 head), within the weight range of 1,050 to 1,125.5 lb, and yielding 62%, or within the weight range of 1,125.6 to 1,200 lb, and yielding 63%; stated discounts and tolerances including substitutions in estimated grade, weight, yield, fat thickness, and other details; and delivery to: Greeley, Colorado; Sioux City, Iowa; Dodge City, Kansas; Omaha, Nebraska; and Amarillo, Texas.

- **Specifications for a feeder cattle contract are**—44,000 lb of feeder steers averaging 600–800 lb (approximately 65 head) consisting of steers that will grade 60–80% Choice when fed to slaughter weight. The contract, on the last trading day, is cash settled to the United States Feeder Steer Price (USFSP) as determined by Cattle-Fax.

COMMISSION FEES AND MARGIN REQUIREMENTS ON CATTLE CONTRACTS

The commission fee on all futures contracts covering both purchase and sale (called a round turn) is negotiable between the brokerage firm and customer. In 1990, the minimum hedge margin on live cattle was $500, and the speculative margin was $700. On feeders, the speculative margin was $700 and the hedge margin was $500. The margin deposit may be increased by the broker if the value of the contract should change unfavorably.

EXAMPLES OF FUTURES HEDGING STRATEGIES

Examples of futures hedging by (1) a cattle feeder, (2) a cow-calf producer, (3) a packer, and (4) a cattle feeder using a long hedge to protect the price of feeder cattle replacements at the time of forward contracting, finished cattle follow. These illustrate hedging procedures, although it must be borne in mind that in actual application the hedges may not work out as perfectly as these.

• **Example 1: A cattle feeder hedging to lock in price (see Table 2–15)**—It is now November, and the cattle feeder has just purchased feeder cattle to place in the feedlot. Based on past experience the feeder is quite confident that these cattle should be ready for market the following April. Through good record keeping, the feeder is also quite confident that production (including labor) and marketing costs should be about $67.50/cwt.

TABLE 2–15
EXAMPLE OF A CATTLE FEEDER USING A SHORT HEDGE TO LOCK IN A PRICE

Cash Market		Futures Market		Basis
	per cwt		per cwt	per cwt
Nov. 15:				
Expects to receive in April	$71.45	Sells April futures at	$72.45	–$1.00 Expected
April 10:				
Sells cattle on cash market at	$67.35	Buys April futures at	$68.85	–$1.50 Actual
Futures gain	$ 3.60			
Realized price	$70.95	Gain	$ 3.60	Loss $0.50

The cattle owner-feeder decides to hedge the cattle with the April futures contract which at the time is selling for $72.45/cwt. The feeder has also estimated the basis will be about $1.00/cwt in April. So, this figure is subtracted from the April futures price resulting in a localized futures price of $71.45/cwt or an estimated $3.95/cwt profit; hence, the feeder sells April futures.

The cattle feeder sold finished cattle on the cash market for $67.35/cwt which, after subtracting production costs of $67.50/cwt, gives a loss of $0.15/cwt. However, the feeder realized a profit of $3.60/cwt on the futures transactions, so that the total profit was $3.45/cwt.

This example illustrates what could happen on a declining market. The feeder still showed a profit, even though the cattle had to be sold in the cash market for a price lower than production costs because this loss was offset by a larger profit in the futures market. This is true because, as the cash price declined, the futures prices also declined.

If, however, the cash and futures prices had risen, the cattle feeder still could have made a profit, this time in the cash market. But because of a loss in the futures market, the total profit would have been less than had the feeder not hedged. Nevertheless, the feeder still received the price protection desired, which was the main purpose in hedging.

DELIVERY AGAINST THE CONTRACT

Although very few contracts, usually fewer than 3%, are consummated by actual delivery of the commodity, a hedger should consider delivery as one alternative, particularly when the cash and futures prices are out of line with each other. However, due consideration must be given to the costs of delivering or receiving delivery, since such costs may be of such magnitude as to offset the differences between the cash and futures prices.

It is not the function of the futures market to provide an alternative source of supply nor an alternate means of disposal of surplus commodities. The purpose of delivery is merely to serve a safeguard to be used when all else fails.

FACTS ABOUT FUTURES CONTRACTS

A cardinal feature of any workable futures contract—whether it be steers, grain, or any other commodity—is that there shall be maintained a solid connection with the commodity; that is, cash and futures must be tied together.

Any contract held until maturity must be delivered or settled to an index of cash prices. This keeps the futures price in line with the cash price at the livestock market.

During the delivery month the cash and futures markets tend to come together at the point of delivery. If this were not so, traders would quickly take advantage of the situation. For example, if prior to the termination of trading on August live cattle futures, the price of U.S. slaughter steers on the terminal market was $5/cwt below August futures, traders could buy cattle and sell futures, then deliver on the contract for a profit of $5/cwt (less marketing and brokerage fees).

LIVESTOCK OPTIONS[15]

Livestock options are much newer than futures and put an interesting new twist on the concept of hedging. Live cattle options began trading on October 31, 1984, and in 1989 there were 4,265,710 *put* and *call* options traded. Feeder cattle options started on January 9, 1987, and in 1989 there was a total of 702,438 trades.

Options are much more like insurance than futures. In exchange for a relatively small premium payment, the hedge buyer

[15]This section on *Livestock Options* was prepared especially for this book by J. Graham, Director, Commodity Marketing and Education, Chicago Mercantile Exchange, Chicago, IL; and adapted by the author.

on an option receives protection against adverse price moves, but is able to take advantage of favorable price moves. Unlike futures, where a producer is locked in at a price and subject to margin calls, options have no margins and margin calls.

The specifications for options are the same as for futures. In fact these are options on the underlying futures. The owner of a put or call option has the *right* to take a position at any time in the futures contract. The only difference is that live cattle options expire (as of February 1991) on the first Friday of the delivery month. The feeder cattle options, since the contract is cash settled, expire on the same day as the futures. There are two types of options; *puts* and *calls*. Puts give the buyer the right to sell futures at a fixed price and increase in value if prices fall. Calls give the buyer the right to buy futures at a fixed price and increase in value if prices rise.

This fixed price is also known as the *strike price* or exercise price. The Chicago Mercantile Exchange sets the strike prices at $2, even-numbered intervals, e.g., $70, $72, $74, etc. The buyer of an option gets to choose the strike price. The cost or *premium* of the option at each strike price will be different. Premium costs normally vary throughout the trading day.

A cattle feeder would most likely be a buyer of a live cattle put option, against cattle that will be sold in the cash market at a later date. A put hedge will give the cattle feeder a minimum (or floor) selling price, if price levels fall. Since the risk is limited to the premium paid, the cattle feeder would have no maximum price (less the premium lost), if price levels go higher.

A cattle feeder may also be a buyer of a *Feeder Cattle Call Option* to protect the cost of feeders that it is planned to buy. Buying a call could set a price ceiling, if prices rise, but no price floor, if prices fall. Once again, the rise is limited to the premium paid.

A rancher, or stocker operator, most likely would find *Feeder Cattle Put Options* useful in setting floor prices. There are many other option strategies, besides just buying puts and buying calls, which are beyond the scope of this section.

EXAMPLES OF OPTION PRICING

• **Example 1: Buying a Live Cattle Put Option**—A producer purchases a February 74 put option at $2/cwt to price a group of cattle. At the time, February live cattle futures are at 74.75/cwt. Estimated basis for the end of January is $1/cwt. The producer's estimated minimum selling price would be the 74 strike, minus the premium of $2, and the estimated basis of $1, which would equal $71/cwt. Let's take a look at what happens in late January if the market goes up, stays about the same, or goes down.

At the end of January, the cattle are ready for market:

If Feb. Futures Are ($)	Value of Put ($)	A Put Net Gain or Loss ($)	B Local Cash Sale ($)	C Net Realized Price ($)
		← (A + B = C) →		
84	0	-2.00	83	81
74	0	-2.00	73	71
64	10	8.00	63	71

As can be seen, when the futures price drops below the put strike price, the minimum selling price or insurance kicks in and protects the floor that was established when the 74 live cattle put was purchased. Should the market go higher, the increase less the cost of the premium will be realized while enjoying protection from a price drop.

A long hedger is one (such as a feedlot operator, a backgrounder, or a stocker operator) who needs a commodity at some point in the future and seeks to forward price the anticipated purchase. Again, choosing a particular hedging strategy depends upon the level of protection desired.

• **Example 2: Buying a Feeder Cattle Call Option**—A January 84 call option is purchased at $2.55/cwt to protect the purchase price of feeder cattle that will be needed in January. At the same time, January feeder cattle futures are at $85.50/cwt. Estimated basis for the end of January is +$3. The estimated maximum purchase price would be the 84 strike price, plus the premium of $2.55, plus the estimated basis of +$3 or a total of $89.55/cwt. Let's take a look at what happens in late January if the market goes up, stays the same, or goes down.

At the end of January when feeder cattle are purchased for feeding:

If Jan. Futures Are ($)	A Local Cash Purchase ($)	Value of Call ($)	B Call Gain or Loss ($)	C Net Realized Price ($)
	←	(A – B = C)	→	
94	97	10	7.45	89.55
84	87	0	-2.55	89.55
74	77	0	-2.55	79.55

Should the market rise between the time the 84 call was purchased and the time the animals were actually purchased, the feeder cattle cost would be limited to the ceiling price created. In this example an in-the-money call was purchased, thus benefiting from any price increase. Should the market fall after the 84 call was purchased, there is still, benefit from a lower feeder cattle purchase price (although the purchaser would be out the premium paid for the call).

TAX MANAGEMENT AND REPORTING[16]

Good tax management and reporting consists in complying with the law, but in paying no more tax than is required. It is the duty of revenue agents to see that taxpayers pay the correct amount, and it is the business of taxpayers to make sure that they do not pay more than is required. From both standpoints, it is important that farmers and ranchers should familiarize themselves with as many of the tax laws and regulations as possible.

[16]This section and the Estate Planning were prepared by the author's son, John J. Ensminger, LL.M., an attorney specializing in taxes and estate planning.

Fig. 2-20. The late John Wayne, co-owner of 26-Bar Ranch, is shown here with a group of his Hereford females on summer range near Springerville, Arizona. John Wayne operated his ranch as a serious business, from which he derived much pleasure. (Courtesy, *American Hereford Journal*, Kansas City, MO)

The cardinal principles of good tax management are: (1) maintenance of adequate records, and (2) conduct of business affairs to the end that the tax required is no greater than necessary. Good tax management and good farm management do not necessarily go hand in hand, and may sometimes be in conflict. When the latter condition prevails, the advantages of one must be balanced against the disadvantages of the other to the end that there shall be the greatest net return.

It is recognized that tax matters constitute a highly specialized and complex field, and each farm or ranch will need separate considerations in appropriate planning. The recent rounds of federal tax legislation have made significant changes in the procedures livestock producers must use in accounting, as well as in their approaches to financial and estate planning. More than ever, it is important that they consult competent professionals before embarking upon any business operation involving livestock. It is noteworthy that, if a livestock producer's return is to be audited, under the recently enacted Taxpayer Bill of Rights, the taxpayer is entitled to be represented at the audit by a representative. Though the IRS can require the taxpayer's attendance with a special summons, this is not likely to be used at the initial meeting.

Increasingly, as local governments must make up for decreased federal support, local tax matters become more important in planning; this also makes consultation with a specialist knowledgeable in state and local tax law, crucial for effective management.

SELECT THE BEST METHOD OF ACCOUNTING

Farmers and ranchers may report on either the *cash receipts and disbursements (cash basis)* or the *accrual basis*. Also, they can use a combination of the two bases, a ***hybrid*** method of reporting. Thus, the accrual basis can be used for livestock, and the cash basis otherwise. While most taxpayers who produce, buy or sell merchandise must use inventories, and thus must use the accrual method, farmers must only do so to the extent that they are subject to inventory accounting requirements. If a return has never been filed, the choice is made on the first return.

If the accrual method is used, annual inventories must be kept, with taxes determined from increases in inventory and deductions given for decreases. Having elected a method, the farmer or rancher will not generally be allowed to change methods without the approval of the IRS. This approval will not be granted unless the IRS and the taxpayer agree to the terms and conditions under which the change will be effected. With the general trend towards requiring accrual accounting, there is likely to be more resistance towards shifting from accrual to cash accounting than the reverse.

A description of each system follows.

CASH BASIS

Under this system, farm income includes all cash or value of merchandise or other property received during the tax year. It includes all receipts from the sale of items produced on the farm and profits from the sales of items that have been sold. It does not include proceeds from sales if the proceeds were not actually available during the tax year.

Allowable deductions include those business expenses incurred that were actually paid during the year, and depreciation on depreciable items.

1. **Drought relief.** After 1987, a farmer using the cash method who is forced to sell livestock as a result of drought conditions can defer income on the excess sales to the following year. The former limitations by which this treatment was restricted to certain animals were lifted by the Technical and Miscellaneous Revenue Act of 1988. The deferral of income is available only if the livestock would not have been sold but for the drought and the drought conditions resulted in the area being eligible for federal assistance.

2. **Feed purchases.** Purchasing feed for use in subsequent years by cash basis producers is an effective means of reducing current income, but the IRS will contest the deduction if (a) the payment is, in fact, a deposit, (b) there was no business purpose for it, or (c) it distorts income. Acceptable business purposes include guaranteeing prices and making certain of supply.

3. **Prepayments.** In 1986, Congress added a provision concerning *farming syndicates* which generally limited deductions

for feed, seed, fertilizer or similar supplies to the taxable year in which the supplies are actually used or consumed. A farming syndicate is a partnership or other enterprise (but not a C corporation) engaged in the business of farming, including the feeding, training and management of animals, if at any time any interest requiring federal or state registration has been offered for sale in the enterprise. It also includes an entity where more than 35% of the losses during any period are allocable to limited partners or limited entrepreneurs.

The IRS has indicated that a farming syndicate can include a general or limited partnership, a sole proprietorship involving an agency relationship created by a management contract, a trust, a common trust fund and an S corporation. In 1988, Congress determined that the effect of the provision requiring farming syndicates to expense feed in the year consumed was unnecessary as such entities must use the accrual method of accounting in any case.

Of more general application is a provision stating that, to the extent that prepaid farming supplies (feed, seed, fertilizer or similar supplies) exceed 50% of other deductible farming expenses, the excess amounts can only be deducted as consumed. This includes interest and depreciation. A farmer whose principal residence is on the farm, whose principal occupation is farming, or who is a member of such a farmer's family, is excepted from this provision if the excess supplies are attributable to extraordinary circumstances. Such an individual is also excepted if he/she does not have excess prepaid farm supplies based on the prior three years' operations.

4. **Deferred payment contracts.** It is sometimes desirable to delay recognition of income from a sale of animals until a subsequent year. Though the IRS is prone to finding that income which has been earned has, in fact, been constructively received by a taxpayer, and thus includable in his/her income, there are certain procedures which can be used to delay recognition. The general procedure for such a deferral involves the receipt of the funds by a middle party, but not an agent. The livestock producer should not receive cash equivalents, such as negotiable notes or securities or letters of credit. If the farmer/rancher is to receive funds directly from the buyer, the contract should require a deferral of income. The contract should specify the terms under which the producer will receive payment. In the typical escrow agreement, there must be conditions, enforceable by both buyer and seller, which preclude the producer from receiving payment until a subsequent date. The producer cannot receive any present beneficial interest from the receipt of funds by the escrowee.

ACCRUAL BASIS

This system requires the keeping of complete annual inventories. Tax is paid on all income earned during the taxable year, regardless of whether payment was actually received, and on increases of inventory values of livestock, crops, feed, produce, etc., at the end of the year as compared with the beginning of the year. All expenses incurred during the year's business are deducted from gross income regardless of whether payment is actually made, and deductions are made for any decrease in inventory values of livestock, etc., during the year.

Four methods of inventorying are available to the accrual basis farmer or rancher.

1. **Cost.** Inventory items are valued at the actual cost of producing or purchasing them.
2. **The lower of cost or market value.** The comparison is made separately for each item in the inventory, not for the entire inventory. The entire stock should *not* be valued at cost and then at market, with the lower selected.
3. **Farm price.** Each item, raised or purchased, is valued at its market price less estimated direct cost of disposition. This method must be used for the entire inventory, except that livestock may be inventoried by the next method.
4. **Unit livestock price.** Animals are classified according to kind and age, and a standard unit price is used for each animal within a class. All raised livestock must be included in inventory under this method. Unit prices must reflect any costs required to be capitalized under the uniform capitalization rules. This method is usually chosen by many large operations. Producers using the unit-livestock method are permitted to elect a simplified production method for determining costs required to be capitalized.

The third and fourth methods are unique to farmers and ranchers.

- **Corporations and partnerships**—The Tax Reform Act of 1986 required most C corporations, partnerships in which one of the partners is a C corporation and tax shelters to use the accrual method of accounting. An exception was made for farming businesses (unless they are tax shelters). Another exception applies to taxable entities with average annual gross receipts of less than $5 million. S corporations (in which there is generally no corporate-level tax) and partnerships can continue to use the cash method.

For purposes of the farming exception to the accrual-method requirement, a farming business is defined as cultivation of land or the raising or harvesting of any agricultural or horticultural commodity. This includes the raising, shearing, feeding, caring for, training or management of animals. Also included is the raising or harvesting of timber and ornamental trees. The exception does not include processing of commodities or products beyond normal agricultural production. Thus, a meat processing plant is not within the farming exception and must use the accrual method.

- **Capitalization of inventory costs**—The Revenue Act of 1987 reduced the ability of many businesses to deduct expenses currently by requiring that a number of expenses associated with the production of inventory be capitalized. This applies to the direct costs of producing the inventory, as well as to certain indirect costs which are determined to be allocable to the inventory. An exception allowing current deductions for expenses was provided for taxpayers with gross receipts of $10 million or less. Also, an exception was provided for any plant or animal which is produced in a farming business and which has a preproductive period of two years or less. In 1988, Congress extended the exception so that only farms under the accrual method must use uniform capitalization. Because of the flux in this area, livestock producers who decided on an accounting method prior to 1988 should have their tax advisors reconsider these choices.

DISTINGUISH CAPITAL GAINS FROM ORDINARY INCOME

As this book goes to press, there is no longer (and not yet) a difference in the tax rates applied to ordinary income and capital gains. However, political considerations are such that a differential appears probable in the near future. If this should be the case, income reported as capital gains will be taxed at a lower rate. Thus, livestock held for sale in inventory, and livestock held for breeding purposes, may produce different tax effects when sold, even if the sale prices are the same. Nevertheless, developments in this area will have to come about before planning can be discussed.

SET UP DEPRECIATION SCHEDULES PROPERLY

Depreciation is estimated operating expense covering wear, tear, exhaustion, and obsolescence of property used in a farm business.

Depreciation may be taken on all farm buildings (except the livestock producer's personal residence), and on everything from grain elevators to horse clippers, including tile drains, water systems, fences, machinery and equipment.

Those who file returns on a cash basis may also take depreciation on dairy cattle, breeding and work stock which were purchased, but they cannot take depreciation on livestock they raised because all costs of raising are deducted as operating expenses. On the accrual basis, depreciation may be taken on purchased animals that are not included in inventory.

Taxpayers should list each building, and each piece of machinery on which depreciation is to be computed on the depreciation schedule. Such items as cows and small implements may be grouped together, but such groupings should be derived from totaling of a detailed individual list kept current in a permanent farm record book.

Depreciation is not available for inventory, which would include livestock held for sale to customers. After 1986, depreciable property is placed in specific classes. Because the period over which property is amortized affects the overall tax revenues. Congress has tended to lengthen recovery periods as a means of increasing tax collections without *raising* taxes.

1. **Three-year property.** This includes horses that are more than 12 years old when placed in service (racehorses more than 2 years old when placed in service).
2. **Five-year property.** This includes automobiles and light-general purpose trucks, certain technological equipment and research and experimentation property.
3. **Seven-year property.** This includes breeding and work horses, 12 years or younger, and any horse not in any other category.
4. **Ten-year property.** Horticultural or single-purpose agricultural structures were originally recovered over 7 years, but after 1988 have a 10-year recovery period. A companion requirement limits recovery on such items to the 150% declining balance method (discussed below). Orchards, groves and vineyards placed in service after 1988 are depreciated on a straight-line basis over 10 years.
5. **Fifteen-year property.** This includes equipment used for two-way exchange of voice and data communications.
6. **27.5-year property.** This covers residential rental property.
7. **31.5-year property.** This covers nonresidential real property. This will include most farm buildings.

For property in the 3-, 5-, 7-, and 10-year classes, depreciation was, prior to 1989, calculated on the double declining balance method, switching to the straight-line method at the time where depreciation is maximized. For property in the 15- and 20-year classes, the 150% declining balance method is used. For the 27.5- and 31.5-year classes, the straight-line method is used. However, for personal property (i.e., nonreal property) placed in service in a farming business after 1988, the 150% declining balance must be used regardless of the recovery period.

A horse owner can elect to depreciate a 2-year-old race horse under the straight-line method provided the election is made for all property in the same class. Once the election is made it is irrevocable. Special provisions apply to property which is not placed in service at the beginning of the year. If property depreciated under certain methods is sold, the gain will be characterized as ordinary income, a factor which may become relevant if differential capital gains rates are reintroduced.

For purchased animals, the price paid will generally determine the amount which can be depreciated. Inherited or gift animals can be depreciated. However, their value may have to be established by a qualified appraiser, if the IRS contests the taxpayer's valuation.

AVOID OPERATING THE BUSINESS AS A HOBBY

If an activity is not engaged in for profit, deductions are generally not available for the conduct of the activity except to the extent of income from it. (Actually, because the expenses of a hobby are limited to the extent of 2% of adjusted gross income, such expenses will not be fully deductible.) This requirement has often been applied when the IRS determines that a livestock operation is actually a hobby. Though the problem will generally not apply to full-time livestock producers, others who devote a smaller amount of their time to an operation may find their activity is classified by the IRS as a hobby.

The general presumption for activities is that if an activity is profitable for 3 of the 5 consecutive years before the year being audited, it will be presumed to be engaged in for profit. Recognizing that horse operations often depend on the success of a rare horse, in such an operation Congress has allowed the activity to be presumed to be engaged in for profit if only 2 of 7 years are profitable. The IRS has indicated that an activity cannot be considered as engaged in for profit until there is a profit year.

Horse owners can delay the determination of whether a horse operation is engaged in for profit until the seventh taxable year of the activity. This election also keeps open the statute of limitations for those years.

In determining whether a livestock operation is a business or a hobby, the IRS will examine the following factors:

1. The manner in which the operator carries on the activity. The more businesslike the conduct of the activity, the more likely it is to be recognized as a business.
2. The expertise of the operator and the employees. A study of the industry and of other successful operations indicates a profit-making approach.
3. Time and effort spent in carrying out the operation. The more time the owner devotes to the activity as a business and not as a recreational pursuit, the more likely the Service will find that the operation is a business.
4. The expectation that assets used in the activity will appreciate in value.
5. Prior successes of the livestock producer.
6. The operation's history of income and losses.
7. Occasional profits.
8. Financial status of the livestock producer.
9. Elements of recreation or pleasure.

ESTATE PLANNING

Human nature being what it is, most livestock producers shy away from suggestions that someone help plan the disposition of their property and other assets after they are gone. Also, they have a long-standing distrust of lawyers, legal terms, and trusts; and to them the subject of taxes on death seldom makes for pleasant conversation.

If a farmer has prepared a valid will, or placed the property in joint tenancy, the estate will be distributed as intended. If not, it goes to the heirs, according to the laws governing intestate (without a will) succession. The heirs are those persons whom the law appoints to succeed to the property in the event of intestacy, and are not necessarily the persons to whom the farmer would want to leave the property. These laws vary somewhat from state to state.

If no plans are made, estate taxes and settlement costs often run considerably higher than if proper estate planning is done. Today, livestock business is big business; many have well over $1 million invested in land, animals and equipment. Thus, it is not a satisfying thought to one who has worked hard to build and maintain a good livestock establishment during their lifetime to feel that the heirs will have to sell the facilities and animals to raise enough cash to pay estate and inheritance taxes. Therefore, livestock producers should go to an estate planning specialist—a lawyer or company specializing in this work, or the trust department of a commercial bank.

• **Consult a professional.** The preparation of wills, trusts, redemption agreements (if the farm is incorporated), partnership agreements, etc., requires consideration of the effects of federal and state tax law, as well as state law governing the various potential arrangements. Consequently, it is strongly advised that competent professionals be consulted in order to achieve an effective and cost-saving estate plan.

WILLS

A will is a set of instructions drawn up by or for an individual which details how he/she wishes the estate to be handled after death.

Despite the importance of a will in distributing property in keeping with the individual's wishes, about 50% of farmers and ranchers pass away without having written a will. This means that state law determines property distribution in such cases.

Every farmer/rancher should have a will. By so doing, (1) the property will be distributed in keeping with their wishes, (2) they can name the executor of the estate, and (3) sizable tax savings can be made by the way in which the property is distributed. Because technical and legal rules govern the preparation, validity, and execution of a will, it should be drawn up by an attorney. Wills can and should be changed and updated from time to time. This can be done either by (1) a properly drawn-up codicil (formal amendment to a will), or (2) a completely new will which revokes the old one.

The same attorney should prepare both the husband's and wife's wills so that a common disaster clause can be incorporated and the estate planning of each can be coordinated.

TRUSTS

A trust is a written agreement by which an owner of property (the trustor) transfers title to a trustee for the benefit of persons called beneficiaries. Both real and personal property may be placed in trust.

The trustee may be an individual(s), bank, or corporation, or a combination of two or three of these. Management skill should be considered carefully in choosing a trustee.

A trust can continue for any period of time set by the owner—for a lifetime, until the youngest child reaches age 21, etc. If the trust extends beyond a lifetime, there are limitations which should be explained by an attorney.

KINDS OF TRUSTS

Basically, there are two kinds of trusts, the *living* and the *testamentary*. The living or *inter vivos* trust is in essence an agreement between the trustor and the trustee and may be revocable or irrevocable.

The *revocable trust* can be terminated or altered; under it the trustor is concerned about the here and now, rather than only the hereafter. The trustor continues to make decisions, and can call off the whole arrangement (it's revocable) if it doesn't work out as expected. The revocable trust offers no special estate tax advantage; the assets of a revocable trust are included in the estate of the deceased creating the trust. However, it can be written in such a manner as to reduce substantially the estate taxes of the beneficiaries. Also, the revocable trust will eliminate the cost of probate—costs which may include executor's fees, attorney's fees, court costs, and appraisal fees.

The *irrevocable trust* cannot be amended, altered, revoked, or terminated. Under an irrevocable trust, the trustor must be willing to part with the trust property forever (irrevocably) and have nothing further to do with it and its administration. However, the irrevocable trust has many favorable aspects in estate planning; it will reduce estate taxes in both the estate of the trustor and the estate(s) of the life beneficiaries, and it avoids probate.

The *testamentary trust* is so-called because it is established under the provisions of the trustor's last will and testament. The testamentary trust does not become effective until after death of the trustor, followed by probate. there is no tax saving in the trustor's estate. However, the trust may be drafted to save estate taxes in the estates of the beneficiaries. A testamentary trust is useful when the heirs are minors or inexperienced in money matters.

PARTNERSHIP CONTRACT

Another logical step in the transfer of property is a partnership contract between the parents and their heir(s) recorded in accordance with law. Appropriate counsel should be consulted in the preparation of such an agreement. Because of recently added "estate freeze" provisions, this approach to control of a farm does not provide the estate tax savings it could previously.

LIVESTOCK INSURANCE[17]

Fig. 2-21. The ownership of a fine animal, like this Duroc boar for which the purchaser paid $6,200, constitutes a risk. Unless the owner is in a strong financial position, insurance should be carried. (Courtesy, United Duroc Swine Registry, Peoria, IL)

[17]This section was accorded the authoritative review of Mr. Frank Harding, and Mr. Duncan Alexander, American Live Stock Insurance Company, 200 South Fourth Street, Geneva, IL.

The ownership of a fine animal constitutes a risk; which means that there is a chance of financial loss. Unless the owner is in sufficiently strong financial position to assume this risk, the animal should be insured.

When insuring animals, the 10 pertinent points that follow should be considered, according to the American Live Stock Insurance Company:

1. Livestock mortality insurance is written for the purpose of protecting the actual investment of the livestock owner, not potential gain or profit.

2. A mortality policy cannot be construed in any way as a maintenance coverage; it does not include veterinarian or similar expenses.

3. Indemnity is payable only as a result of death loss.

4. Mortality coverage does not indemnify an insured against loss of an animal's ability to perform the functions for which it is kept.

5. Death from natural or accidental causes is included but mandatory slaughter by governmental authority or decree, or for expediency is not included.

6. The basis for valuing an animal should be actual sales price or fair and conservative appraisal by competent judges when no actual sales transaction has taken place. These values shall be subjected to acceptance by the Company.

7. Mortality insurance is renewable only on evidence or reinsurability, both as to physical condition and market value.

8. Cancellation may only be effected by the insured, or by the company on notice given in conformation with whatever existing laws govern for the address of the insured as shown on policy. Short rate basis if ordered by insured and *pro rata* basis if by company.

9. Policies may not be transferred from one insured to another unless agreed to through endorsement by company, nor may cover be switched from one animal to another unless agreed to by company.

10. Application subject to acceptance by company.

Special stipulations and rates apply to (a) fire, lightning, windstorm and transportation losses only, (b) castration of colts and setting tails, (c) air and ocean transportation, and (d) group insurance. For information relative to these, the owner should see a livestock insurance agent.

In order to obtain insurance, the following information is generally required: Name, registry number, ear tag or tattoo number (markings on horse), breed, sex, date of birth, amount to be insured for and period of insurance required, and a statement of health examination (made not more than five days prior to insuring) by an approved federal or state veterinarian to the effect, "that the animal(s) (referring to it by name) is at the time of applying for insurance in a state of good physical health and condition."

Livestock Farming/Ranching In The 21st Century

Animal Behavior

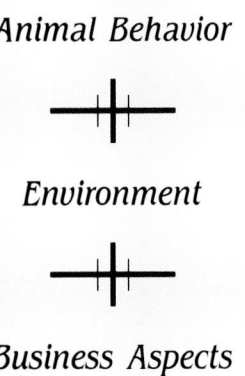

Environment

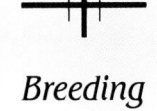

Business Aspects

Breeding

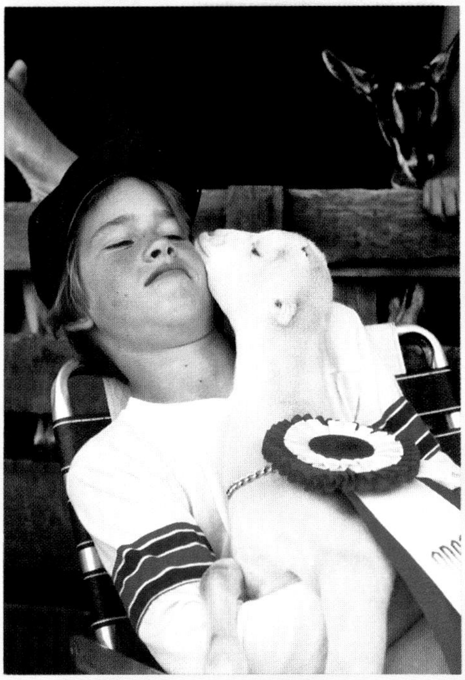

Plate 1. People need pets, and pets need people. (Courtesy, Cindy L. Schneider, Sacred Heart, MN)

Plate 2. Sheep displaying the gregarious, or flocking, instinct. (Courtesy, Texas A&M Extension Service, San Angelo)

Plate 3. A beautiful environment, enhanced by windbreaks, on a ranch in North Dakota. (Courtesy, USDA Soil Conservation Service)

Plate 4. Farming is big business, and it will get bigger, demanding more capital and top management. (Photo by J. C. Allen & Son, West Lafayette, IN)

Plate 5. *Right:* Records that are up-to-the-minute and accurate are of increasing importance in the livestock business. (Courtesy, Babcock Genetics, Rochester, MN)

Plate 6. *Below:* Futuristic animal breeding! This shows seven genetically identical Brangus calves cloned from a single embryo. (Courtesy, Granada Genetics, Inc., Wheelock, TX)

Livestock Farming/Ranching In The 21st Century

Feeding

Pasture & Range Forages

Hay

Silage

Plate 7. *Left:* Finishing cattle on feed. (Courtesy, American Angus Assn., St. Joseph, MO)

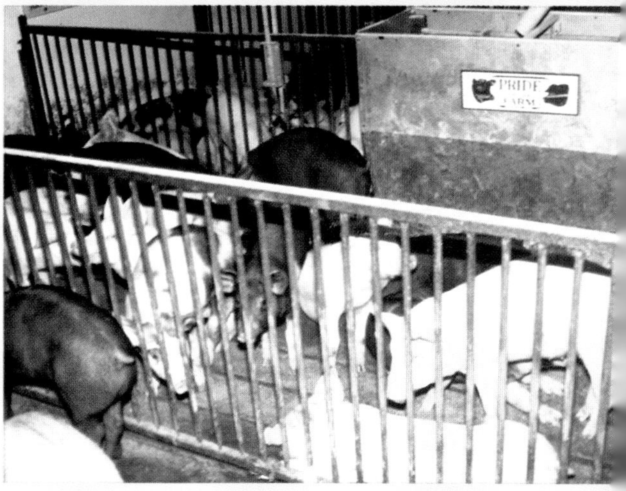

Plate 8. Growing-finishing swine self-fed on slotted floor. (Courtesy, Iowa State University, Ames)

Plate 9. Guernsey cows on pasture. (Courtesy, American Guernsey Assn., Reynoldsburg, OH)

Plate 10. Round hay baler in operation. (Courtesy, Case IH, Racine, WI)

Plate 11. *Below:* Harvesting corn for silage with a self-propelled forage harvester. (Courtesy, Deere & Company, Moline, IL)

Plate 12. *Below:* Upright (tower) silos. (Courtesy, Milk Marketing, Inc., Strongsville, OH)

Livestock Farming/Ranching In The 21st Century

Management

Buildings & Equipment

Animal Health

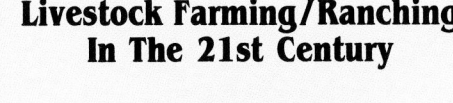

Plate 13. Animals provide manure (fertilizer) for the fields. This shows a manure spreader in operation. (Courtesy, Ford New Holland, New Holland, PA)

Plate 14. *Right:* Burning is the oldest brush control method. It is effective, but sometimes hazardous. Also, air pollution must be considered. (Courtesy, K. C. McDaniel, New Mexico State University, Las Cruces)

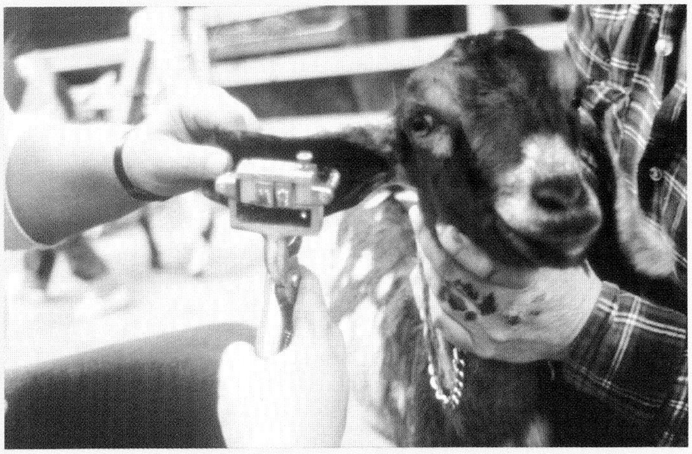

Plate 15. Each animal should be individually identified. This shows a Nubian dairy goat being tattooed. (Courtesy, University of Delaware, Newark)

Plate 16. Properly designed, constructed, and arranged buildings and equipment increase animal production and efficiency. (Courtesy, Milk Marketing, Inc., Strongsville, OH)

Plate 17. *Below:* Say oink! Veterinarian shown treating a pig. (Courtesy, USDA)

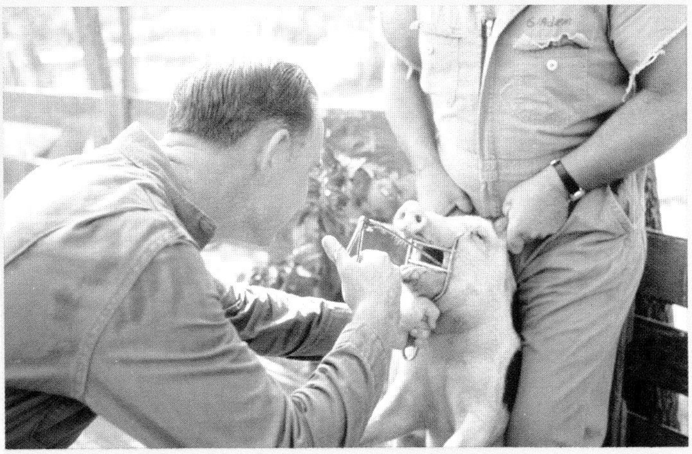

Plate 18. *Below:* A swimming pool just for horses. Some trainers favor swimming horses for conditioning, especially where there are minor leg weaknesses. (Courtesy, El Rancho, Murrieta, CA)

Livestock Farming/Ranching In The 21st Century

Selecting & Judging — *Fitting & Showing* — *Marketing*

Plate 19. Three-year-old Guernsey cows being judged at the North American International Livestock Exposition, Louisville, KY. (Courtesy, American Guernsey Assn., Reynoldsburg, OH)

Plate 20. Hampshire boars being judged at the Winter Type Conference. (Courtesy, Hampshire Swine Registry, Peoria, IL)

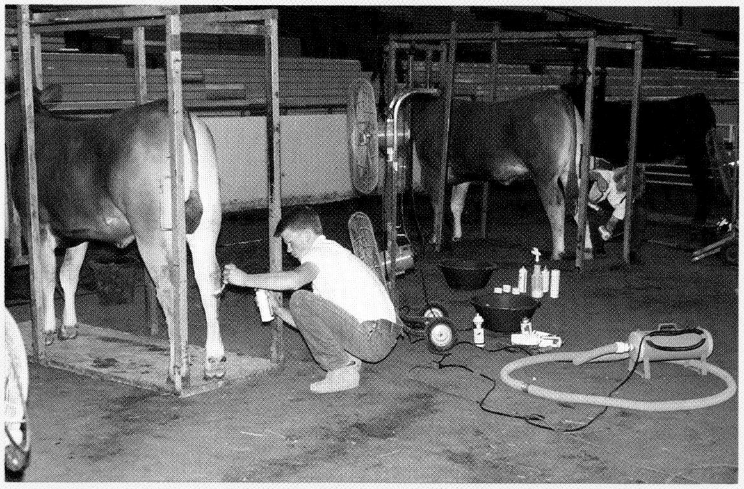

Plate 21. Boning the legs (1) to impart an illusion of bigger bone, and (2) to make them appear more correctly set. (Courtesy, *Limousin World*, Yukon, OK)

Plate 22. *Right:* Tracy Molsby showing a half-Arabian in a Western Pleasure Class. (Courtesy, International Arabian Horse Assn., Denver, CO)

Plate 23. *Below:* Auction selling. This shows a Rambouillet ram being sold at auction in the National Ram Sale in Salt Lake City, UT. (Courtesy, American Sheep Industry Assn., Englewood, CO)

Plate 24. *Below:* A view of the trading floor on the Chicago Mercantile Exchange. (Courtesy, Chicago Mercantile Exchange, Chicago, IL)

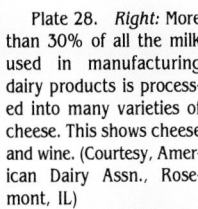

Livestock Farming/Ranching In The 21st Century

Meat & Milk — *Wool & Mohair*

Plate 25. Beef burgers. (Courtesy, National Live Stock and Meat Board, Chicago, IL)

Plate 26. T-bone steaks. (Courtesy, National Live Stock and Meat Board, Chicago, IL)

Plate 27. *Left:* Milk is good—and good for you. (Courtesy, University of Maryland, College Park)

Plate 28. *Right:* More than 30% of all the milk used in manufacturing dairy products is processed into many varieties of cheese. This shows cheese and wine. (Courtesy, American Dairy Assn., Rosemont, IL)

Plate 29. *Below:* An attractive and fashionable red wool dress accented with gold buttons. (Courtesy, The Wool Bureau, New York, NY)

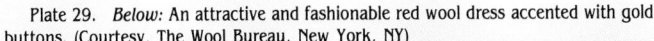

Plate 30. *Below:* Billies in bluebonnets. This shows Angora bucks (producers of mohair) in Texas bluebonnets. (Courtesy, Mohair Council of America, San Angelo, TX)

Livestock Farming/Ranching In The 21st Century

Plate 31. Livestock farming and ranching is big business, and it will get bigger. Hence, it needs the protection of laws and the counsel of attorneys. (Courtesy, Harvestore Systems, DeKalb, IL)

Plate 32. *Above:* Mammoth (or American Standard) Jack. (Courtesy, Standard Jack and Jennet Registry of America, Elk Horn, KY)

Law

Breeds

Feed Compositions

Plate 33. *Below:* Miniature Horse pulling a cart. The Miniature Horse cannot exceed 34 in. at the withers. (Photo by A. H. Ensminger)

Plate 34. *Right:* Duroc pigs. (Courtesy, *National Hog Farmer*, Minneapolis, MN)

Plate 35. *Below:* Feed composition includes the determination of energy value in the laboratory. (Courtesy, Georgia Experiment Station, Experiment, GA)

Plate 36. *Below:* Measuring thickness of soybean flakes. (Courtesy, American Soybean Assn., St. Louis, MO)

Livestock Farming/Ranching In The 21st Century

Animals Contribute—

Food

Clothing

Recreation

Plate 37. *Above:* Meat on the table—a tenderloin steak. (Courtesy, National Live Stock and Meat Board, Chicago, IL)

Plate 38. *Right:* A fancy dessert with whipped cream. (Courtesy, USDA)

Plate 39. *Left:* Made of wool. This shows a ladies' wool plaid jacket and wool pants, and a man's dark red, sleeveless sweater. (Courtesy, The Wool Bureau, New York, NY)

Plate 40. *Below:* A polo match, which is played by four mounted players on each team. (Courtesy, Kentucky Horse Park, Lexington, KY)

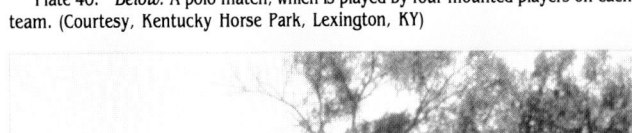

Plate 41. *Below:* Harness race. (Courtesy, United States Trotting Assn., Columbus, OH)

Plate 42. *Below:* Multiple use of the western range includes provision for wild game such as these buffalo (bison). (Courtesy, H. Dietz, USDA, Soil Conservation Service, Fort Worth, TX)

Livestock Farming/Ranching In The 21st Century

Plate 43. The signs of a sustainable, productive, profitable agriculture. (Courtesy, The American Morgan Horse Assn., Shelburne, VT)

The Goal In The Decades To Come:

A world of sustainable animal agriculture, with animal behavior, pollution control, protection of the environment, food safety, and technology fully integrated and in harmony.

Plate 44. A mountain meadow, with nature and animals in harmony. (Photo by J. C. Allen & Son, West Lafayette, IN)

Plate 45. *Left:* At its best, an animal agriculture enhances the cycle of nature by manure being applied to the land and decomposed by microbes. (Courtesy, Gehl Company, West Bend, IN)

Plate 46. *Right:* Many of the practices advocated in a sustainable agriculture are not new; among them, the use of pasture as a primary feed source. (Courtesy, American Shorthorn Assn., Omaha, NE)

Plate 47. *Below:* Pollution control calls for a world that is mindful that life, beauty, wealth, and progress depend upon how wisely people use nature's gifts—the soil, the water, the air, the minerals, and the plant and animal life.

Plate 48. *Below:* The primary challenges to the year 2000 and beyond will be (1) feeding the hungry, and (2) protecting the environment. (Courtesy, *Sheep Breeder*, Columbia, MO)

Progeny from one superior donor female produced through embryo transfer. (Courtesy, Granada Land & Cattle Co., Wheelock, TX)

SECTION

3

BREEDING

Contents *Page*

Contents *Page*

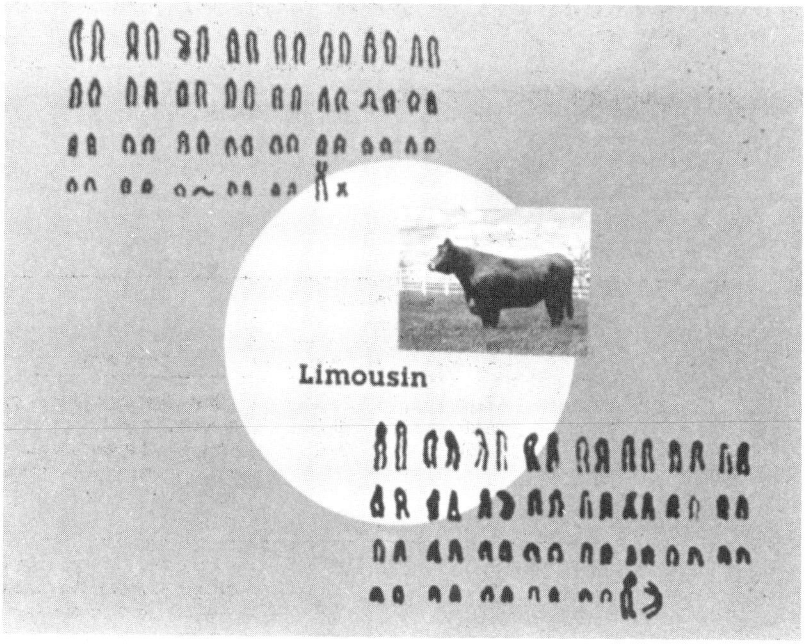

Fig. 3-1. Scientists can now read and manipulate the genetic code—and tailor-make animals. (Courtesy, Department of Animal Science, University of Missouri, Columbia)

Something new has been added to breeding! It is called biotechnology. Scientists can now read and manipulate the genetic code, directly and precisely. Snipping and splicing genes (genetic engineering) is spawning new animals exceeding our fondest dreams. Biotechnology is reshaping every facet of animal production from breeding to finished product, including genetic makeup, physiology, stress tolerance, disease resistance, feed efficiency, and the quality and quantity of products produced. But advanced technology calls for advanced animal adaptation, behavior, environmental control, and care, and for keepers of the herds and flocks with the tomorrow mind instead of the yesterday mind. This chapter is designed to fill this need.

PUBERTY, HEAT, AND GESTATION PERIODS OF ANIMALS

The normal age of puberty, and the normal heat and gestation periods of animals are given in Table 3-1.

TABLE 3-1
PUBERTY, HEAT, AND GESTATION PERIODS OF ANIMALS

Class of Animal	Age of Puberty	Duration of Heat		Interval of Heat		Gestation Period	
		Range	Average	Range	Average	Range	Average
	(mos.)			(days)	(days)	(days)	(days)
Cattle	8–12	6–30 hr	16–20 hr	19–23	21	278–288	283
Sheep	5–7	20–42 hr	30 hr	14–20	16–17	144–152	148
Goats	4–8	20–80 hr	36–48 hr	12–25	20	140–160	151
Swine	4–7	1–5 days	2–3 days	18–24	21	98–124	114
Horses	12–15	1–37 days	4–6 days	10–37	21	310–370	336

FLUSHING AND CONDITIONING

Flushing is that practice of feeding cows, ewes, and sows more generously 2 to 3 weeks before breeding. This may be accomplished by grain feeding, or cows and ewes may be turned on more lush pasture or range. Under most circumstances, the following amounts per head daily of suitable concentrate are added to the ration that the females were receiving prior to flushing: cows, 2 to 5 lb; ewes, 1 to 2 lb; and sows, about 2 lb. Immediately after breeding, females should be returned to normal rations.

Although it is not likely that all the benefits ascribed to flushing will be fully realized under all conditions, the general feeling persists that the practice will result in (1) more eggs being shed, (2) the females coming in heat more promptly, (3) more certain and prompt conception—with the young arriving more nearly at the same time, and (4) a 15 to 30% increase in lamb and pig crops.

Fat cows, ewes, and sows can best be conditioned for breeding by increasing the exercise.

In mares, preparation for breeding is known as *conditioning.* Usually it involves exercising plus emulating spring conditions through blanketing and exercising.

MATING TABLE

Table 3–2, Mating Table, gives pertinent information relative to the use of sires of different classes of livestock, including considerations that should be given to age and method of mating.

TABLE 3–2
MATING TABLE

Class of Animal	Age	No. of Females Bred/Year: Hand-mating	No. of Females Bred/Year: Pasture-mating	Comments
Bull	Yearling 2-year-old 3-year-old or over	10–12 25–30 40–50	8–10 20–25 25–40	Most western ranchers use 1 bull to about 25 cows. A bull should remain a vigorous and reliable breeder up to 10 years or older; up to 6 to 7 years under range conditions.
Ram	Lamb	20–25		Lambs should be used in hand-mating only. Unless well grown, and under close supervision of an experienced sheep producer, they should not be used at all.
	Yearling or older	50–75	35–60	Most range operators use 1 ram to 25 to 35 ewes. A ram should remain a vigorous and reliable breeder up to 6 to 8 years of age.
Boar	8 to 12 months of age Yearling or older	24 50	12 35–40	Boar pigs should be limited to 1 service/day; older boars to 2 services/day. A boar should remain a vigorous and reliable breeder up to 6 to 8 years of age. Under hand-mating, 2 services are recommended; the first mating on gilts should be on the first day of estrus and the first mating on sows on the second day of estrus, with a second mating following the first by 24 hours in each case.
Stallion[1]	2-year-old 3-year-old 4-year-old Mature horse Over 18 years old	10–15 20–40 40–60 50–70 20–40	Preferably no pasture mating unless the stallion is prepared for same and certain precautions are taken.	Limit the 2-year-old to 2 to 3 services/week, the 3-year-old to 1 service/day, and the 4-year-old or over to 2 services/day. A stallion should remain a vigorous and reliable breeder up to 20 to 25 years of age.

[1]There are breed differences. Thus, when first entering stud duty, the average 3-year-old Thoroughbred should be limited to 15 to 20 mares per season, whereas a Standardbred of the same age may breed 20 to 30 mares; and the 4- or 5-year-old Thoroughbred should be limited to 25 to 30 mares, whereas a Standardbred of the same age may breed 30 to 40 mares. Mature stallions of the draft breeds may and do breed up to 100 mares in a season.

GESTATION TABLE

The producer who has information relative to breeding dates can easily estimate parturition dates from Table 3–3, Gestation Table.

TABLE 3–3
GESTATION TABLE

Date Bred	Cow 283 Days (date due)	Ewe 148 Days (date due)	Sow 114 Days (date due)	Mare 336 Days (date due)
Jan. 1	Oct. 11	May 29	April 25	Dec. 3
Jan. 6	Oct. 16	June 3	April 30	Dec. 8
Jan. 11	Oct. 21	June 8	May 5	Dec. 13
Jan. 16	Oct. 26	June 13	May 10	Dec. 18
Jan. 21	Oct. 31	June 18	May 15	Dec. 23
Jan. 26	Nov. 5	June 23	May 20	Dec. 28
Jan. 31	Nov. 10	June 28	May 25	Jan. 2
Feb. 5	Nov. 15	July 3	May 30	Jan. 7
Feb. 10	Nov. 20	July 8	June 4	Jan. 12
Feb. 15	Nov. 25	July 13	June 9	Jan. 17
Feb. 20	Nov. 30	July 18	June 14	Jan. 22
Feb. 25	Dec. 5	July 23	June 19	Jan. 27
Mar. 2	Dec. 10	July 28	June 24	Feb. 1
Mar. 7	Dec. 15	Aug. 2	June 29	Feb. 6
Mar. 12	Dec. 20	Aug. 7	July 4	Feb. 11
Mar. 17	Dec. 25	Aug. 12	July 9	Feb. 16
Mar. 22	Dec. 30	Aug. 17	July 14	Feb. 21
Mar. 27	Jan. 4	Aug. 22	July 19	Feb. 26
April 1	Jan. 9	Aug. 27	July 24	Mar. 3
April 6	Jan. 14	Sept. 1	July 29	Mar. 8
April 11	Jan. 19	Sept. 6	Aug. 3	Mar. 13
April 16	Jan. 24	Sept. 11	Aug. 8	Mar. 18
April 21	Jan. 29	Sept. 16	Aug. 13	Mar. 23
April 26	Feb. 3	Sept. 21	Aug. 18	Mar. 28
May 1	Feb. 8	Sept. 26	Aug. 23	April 2
May 6	Feb. 13	Oct. 1	Aug. 28	April 7
May 11	Feb. 18	Oct. 6	Sept. 2	April 12
May 16	Feb. 23	Oct. 11	Sept. 7	April 17
May 21	Feb. 28	Oct. 16	Sept. 12	April 22
May 26	Mar. 5	Oct. 21	Sept. 17	April 27
May 31	Mar. 10	Oct. 26	Sept. 22	May 2
June 5	Mar. 15	Oct. 31	Sept. 27	May 7
June 10	Mar. 20	Nov. 5	Oct. 2	May 12
June 15	Mar. 25	Nov. 10	Oct. 7	May 17
June 20	Mar. 30	Nov. 15	Oct. 12	May 22
June 25	April 4	Nov. 20	Oct. 17	May 27
June 30	April 9	Nov. 25	Oct. 22	June 1

Date Bred	Cow 283 Days (date due)	Ewe 148 Days (date due)	Sow 114 Days (date due)	Mare 336 Days (date due)
July 5	April 14	Nov. 30	Oct. 27	June 6
July 10	April 19	Dec. 5	Nov. 1	June 11
July 15	April 24	Dec. 10	Nov. 6	June 16
July 20	April 29	Dec. 15	Nov. 11	June 21
July 25	May 4	Dec. 20	Nov. 16	June 26
July 30	May 9	Dec. 25	Nov. 21	July 1
Aug. 4	May 14	Dec. 30	Nov. 26	July 6
Aug. 9	May 19	Jan. 4	Nov. 31	July 11
Aug. 14	May 24	Jan. 9	Dec. 6	July 16
Aug. 19	May 29	Jan. 14	Dec. 11	July 21
Aug. 24	June 3	Jan. 19	Dec. 16	July 26
Aug. 29	June 8	Jan. 24	Dec. 21	July 31
Sept. 3	June 13	Jan. 29	Dec. 26	Aug. 5
Sept. 8	June 18	Feb. 3	Dec. 31	Aug. 10
Sept. 13	June 23	Feb. 8	Jan. 5	Aug. 15
Sept. 18	June 28	Feb. 13	Jan. 10	Aug. 20
Sept. 23	July 3	Feb. 18	Jan. 15	Aug. 25
Sept. 28	July 8	Feb. 23	Jan. 20	Aug. 30
Oct. 3	July 13	Feb. 28	Jan. 25	Sept. 4
Oct. 8	July 18	Mar. 5	Jan. 30	Sept. 9
Oct. 13	July 23	Mar. 10	Feb. 4	Sept. 14
Oct. 18	July 28	Mar. 15	Feb. 9	Sept. 19
Oct. 23	Aug. 2	Mar. 20	Feb. 14	Sept. 24
Oct. 28	Aug. 7	Mar. 25	Feb. 19	Sept. 29
Nov. 2	Aug. 12	Mar. 30	Feb. 24	Oct. 4
Nov. 7	Aug. 17	April 4	Mar. 1	Oct. 9
Nov. 12	Aug. 22	April 9	Mar. 6	Oct. 14
Nov. 17	Aug. 27	April 14	Mar. 11	Oct. 19
Nov. 22	Sept. 1	April 19	Mar. 16	Oct. 24
Nov. 27	Sept. 6	April 24	Mar. 21	Oct. 29
Dec. 2	Sept. 11	April 29	Mar. 26	Nov. 3
Dec. 7	Sept. 16	May 4	Mar. 31	Nov. 8
Dec. 12	Sept. 21	May 9	April 5	Nov. 13
Dec. 17	Sept. 26	May 14	April 10	Nov. 18
Dec. 22	Oct. 1	May 19	April 15	Nov. 23
Dec. 27	Oct. 6	May 24	April 20	Nov. 28

COLOR INHERITANCE IN SHORTHORNS

Fig. 3–2 shows how color in Shorthorn cattle is inherited.

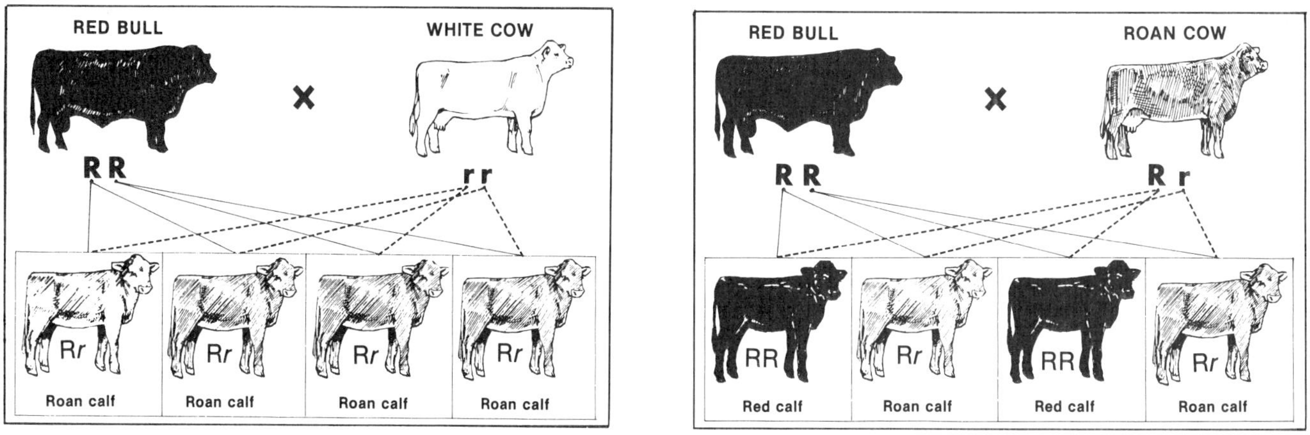

Fig. 3–2. Diagrammatic illustration of the inheritance of color in Shorthorn cattle. Red × white matings (left) in Shorthorn cattle usually produce roan offspring; whereas red × roan matings (right) produce ½ red offspring and ½ roan offspring.

DEHORNING WITH POLLED BULLS

Fig. 3-3 shows how horned cattle may be dehorned through the use of polled bulls.

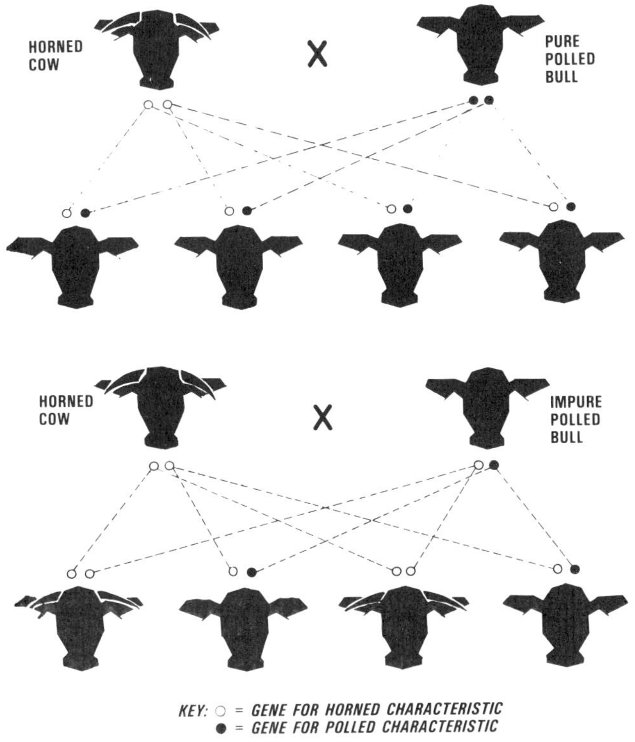

Fig. 3-3. Diagrammatic illustration of the inheritance of horns in cattle. If a bull that is pure or homozygous for the polled character is mated with a number of horned females, all of the calves will be polled; whereas if a bull that is impure or heterozygous for the polled character is mated with a number of horned females, only half of the calves will, on the average, be polled. (Drawing by R. F. Johnson)

DWARFISM IN CATTLE

Beginning about 1940, a disturbing condition known as dwarfism appeared in increasing frequency among beef cattle, probably in all breeds. There are several different types of dwarfism, of which the short-headed, short-legged, pot-bellied dwarf—commonly referred to as the snorter dwarf—is the most frequent. The discussion which follows applies specifically to snorter dwarfism.

There is complete agreement among scientists (1) that the dwarf condition is of genetic origin, and (2) that it is inherited as a simple autosomal recessive (the word *autosomal* merely means that it is not carried on the sex chromosomes), and conditioned by at least two pairs of modifying genes. Thus, the birth of a dwarf calf identifies both the sire and the dam as carriers of the dwarf gene.

- **Conditions prevailing in dwarf-afflicted herds**—One or the other of the conditions (or perhaps both conditions) shown in Figs. 3-4 and 3-5 prevail in any herd of cattle in which dwarf-carrying animals are being used. Thus, 100 offspring from matings of carrier bulls × noncarrier cows will, on the average, possess the following genetic picture from the standpoint of dwarfism:[1]

50—carriers, although not dwarfs
50—noncarriers and nondwarfs
100—total

Likewise, 100 offspring from matings of carrier bulls × carrier cows will, on the average, possess the following genetic picture from the standpoint of dwarfism:[2]

25—dwarfs
50—carriers, although not dwarfs
25—noncarriers and nondwarfs
100—total

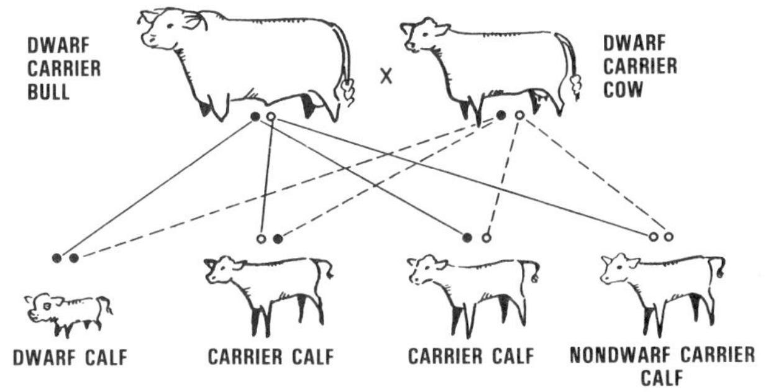

Fig. 3-4. Diagrammatic illustration of the inheritance of the most common kind (short-headed or snorter type) of dwarfism, showing what to expect when a carrier (heterozygous) bull(s) is mated to a noncarrier (homozygous normal) cow(s); or the sexes may be reversed. As shown, carrier × noncarrier matings will, *on the average*, produce calves of which (1) 50% are carriers, although not dwarfs, and (2) 50% are noncarriers, and nondwarfs. Unfortunately, the two groups look alike and cannot be detected by sight. (Drawing by R. F. Johnson)

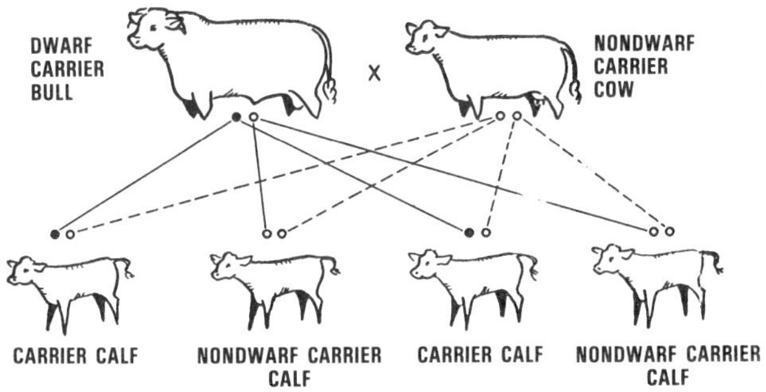

Fig. 3-5. Diagrammatic illustration of the inheritance of the most common kind (short-headed or snorter type) of dwarfism, showing what to expect when a carrier (heterozygous) bull(s) is mated to a carrier (heterozygous) cow(s). As shown, carrier × carrier matings, will, *on the average*, produce calves of which (1) 25% are dwarfs, (2) 50% are carriers, although not dwarfs, and (3) 25% are noncarriers, and nondwarfs. Unfortunately, only the dwarfs can be detected by sight; the 2 nondwarf groups look alike and cannot be distinguished by sight. (Drawing by R. F. Johnson)

---

[1]All ratios are averages based on large numbers; thus, they may not apply to any given herd.

[2]Ibid.

DOUBLE MUSCLING (MUSCULAR HYPERTROPHY)

Double muscling refers to cattle characterized by bulging muscles of the shoulder and thigh, a very rounded rear end (as viewed from the side), a wide but shallow body throughout, appearance of intermuscular grooves, and fine bones.

Double muscling is really a misnomer. Likewise, the scientific name, muscular hypertrophy, is incorrect because it implies increased size of fibers in each muscle, which is not the case. Rather, it has been shown that double muscled cattle have more fibers, not larger fibers.

Double muscling is a genetically controlled character. It appears to be caused by a single recessive gene, which tends to be masked by the dominant gene in the heterozygous carriers. Other examples of a character controlled by one pair of genes are: polledness and hornedness, and dwarfism. The genetics, therefore, are relatively simple. Since each animal has two genes for such characters, all cattle can be classified as follows:

DM DM—Homozygous normal; two dominant normal genes—a normal animal.

DM dm—Heterozygous; one dominant normal gene (DM) which tends to cover up the one recessive gene (dm)—these are called carriers. This cover-up is not complete; hence, there is a tendency toward double muscling.

dm dm—Homozygous recessive; two recessive double muscling genes—a double muscled animal.

• **Appearance of homozygous double muscled (dm dm) cattle**—Obviously, the problem is to determine if any animal is DM dm (a carrier), rather then DM DM (a normal animal). There are two ways to do this: (1) appearance, and (2) breeding tests. Detection by appearance is not 100% sure, but the experienced observer doesn't make many mistakes.

• **Appearance of carrier (DM dm) cattle**—Generally speaking, homozygous, double muscled animals can be identified. But it isn't easy to pick out the heterozygotes—the carriers of the double muscle gene, due to the wide variation in expression. Some of the carriers look quite normal, others look like homozygous double muscled animals, and still others are intermediate between these two extremes. Also, identity is further complicated because few, if any, double muscled animals show all of the characteristics. Nevertheless, carriers are characterized by general overall trim appearance, thicker quarter with bulging, thicker round, and a higher tailhead setting than normal animals.

• **Double muscled cattle—good or bad?**—Double muscled Piedmontese (Piedmont) cattle mean to the Italian cattle industry what broad breasted turkeys and Cornish cross broilers mean to the U.S. poultry industry. All are meat producers *par excellence.*

Piedmontese cattle are the most popular breed in Italy. The fact that 80% of the bulls are double muscled to some degree indicates that the character responds to selection. Knowledgeable Piedmontese cattle breeders in Italy told the author that they expect the following results in their Piedmontese breeding programs (see Fig. 3–6):

1. Phenotypically normal heifers mated to double muscled bulls will produce 80% double muscled bull calves.
2. By culling out the first calf heifers whose bull calves from the above mating were not double muscled, then mating only proved heterozygotes (or carriers) to double muscled bulls, 95% double muscled bull progeny will be produced.

Fig. 3–6 A

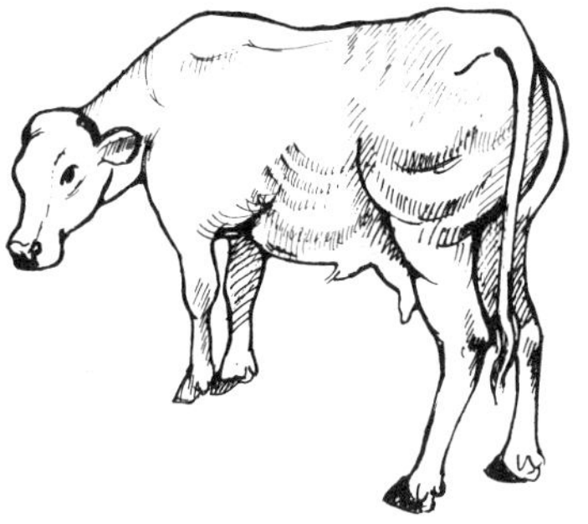

Fig. 3–6 B

Fig. 3–6 C

Alternate Breeding Programs:

No. 1: Piedmontese bull—double muscled; homozygous	× (bred to)	Piedmontese cow—phenotypically normal	= (will produce)	80% double muscled bull calves
No. 2: Piedmontese bull—double muscled; homozygous	× (bred to)	Piedmontese cow—a proven heterozygote (or carrier)	= (will produce)	95% double muscled bull calves

Fig. 3–6 A-B-C. Double muscling predictability in Piedmontese (Piedmont) cattle, in Italy.

SEX DETERMINATION

On the average, and when considering a large population, approximately equal numbers of males and females are born in all common species of animals. To be sure, many notable exceptions can be found in individual herds. The history of the Washington State University Angus herd, for example, reads like a storybook. The entire herd was built up from one foundation female purchased in 1910. She produced 7 daughters, in turn, her first daughter produced 6 females. Most remarkable yet and extremely fortunate from the standpoint of building up the cow herd, it was 4 years before any bull calves were dropped in the herd.

Such unusual examples often erroneously lead livestock producers to think that something peculiar to the treatment or management of a particular herd resulted in a preponderance of males or females, as the case may be. In brief, through such examples, the breeder may get the impression that variation in the sex ratio is not random but that it is under the control of some unknown and mysterious influence. Under such conditions, it can be readily understood why the field of sex control is a fertile one for fraudulent operators. Certainly, any foolproof method of controlling sex would have tremendous commercial possibilities. For example, a cattle producer wishing to build up a herd could then secure a high percentage of heifer calves. On the other hand, the commercial producer would then elect to produce only enough heifers for replacement purposes. From an economic standpoint, the commercial producer would want a preponderance of bull calves for the reason that commercial steers sell for a higher price than do commercial heifers.

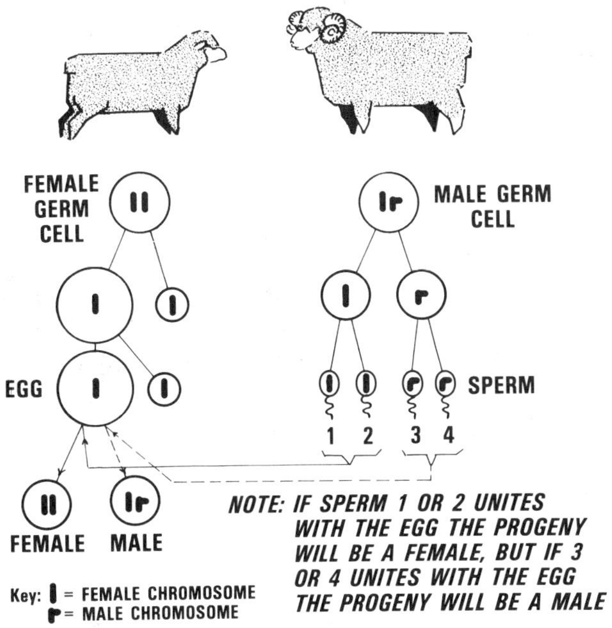
FEMALE GERM CELL
MALE GERM CELL
EGG
SPERM
1 2 3 4
FEMALE
MALE
NOTE: IF SPERM 1 OR 2 UNITES WITH THE EGG THE PROGENY WILL BE A FEMALE, BUT IF 3 OR 4 UNITES WITH THE EGG THE PROGENY WILL BE A MALE
Key: = FEMALE CHROMOSOME
= MALE CHROMOSOME

Fig. 3-7. Diagrammatic illustration of the mechanism of sex determination in farm animals, showing how sex is determined by the chromosomal makeup of the individual. The female has a pair of like sex chromosomes, whereas the male has a pair of unlike sex chromosomes. Thus, if an egg and a sperm of like sex chromosomal makeup unite, the offspring will be a female; whereas if an egg and sperm of unlike sex chromosomal makeup unite, the offspring will be a male. (Drawing by R. F. Johnson)

The most widely accepted theory of sex determination at the present time is that sex is determined by the chromosomal makeup of the individual. One particular pair of chromosomes is called the sex chromosomes. In farm animals, the female has a pair of similar chromosomes (usually called X chromosomes), whereas the male has a pair of unlike sex chromosomes (usually called X and Y chromosomes). In the bird, this condition is reversed, the females having the unlike pair and the male having the like pair.

The pairs of sex chromosomes separate out when the germ cells are formed. Thus, each of the ova, or eggs, produced by the cow contains the X chromosomes; whereas the sperm of the bull are of two types, one half containing the X chromosome and the other half the Y chromosome. Since on the average the eggs and sperm unite at random, it can be understood that half of the progeny will contain the chromosomal makeup XX (females) and the other half, XY (males).[3]

Many unsuccessful attempts have been made to control sex, including the electrophoretic, mechanical, and chemical methods of separation of the two types of sperm cells. Obviously, some method of controlling sex of offspring would have tremendous significance in the livestock field. However, to date, no practical solution has been found. Until there is adequate experimental evidence, any method or theory that purports to control sex should be regarded with skepticism. Of course, research is being continued because the stakes are high if any workable method can be found.

PREGNANCY TESTING

The absence of heat is not always a sign of pregnancy, but a positive diagnosis can be made. Among the advantages of early pregnancy detection in all species are the following:

1. It gives early warning of breeding troubles, such as infertile males and cystic ovaries of females, and makes it possible to accord special care or treatment.

2. It makes it possible to detect shy breeders and females that show signs of heat even when well advanced in gestation.

3. It makes it possible to rebreed nonpregnant, feed-wasting females.

4. It allows for the separation and grouping of females—as pregnant, and nonpregnant, which is requisite to proper nutrition and husbandry.

5. It makes for more effective use of facilities, including providing adequate facilities for parturition.

6. It makes it possible to guarantee pregnancy of females that are for sale.

---

[3]The scientists' symbols for the male and female, respectively, are: ♂ (the sacred shield and spear of Mars, the Roman god of war) and ♀ (the looking glass of Venus, the Roman goddess of love and beauty).

Several methods for pregnancy determination have been developed and used experimentally or semipractically. The most common tests for each species follows.

• **Pregnancy testing cows**—Cows are commonly pregnancy tested by rectal palpation. By about the second month in heifers and the third month in cows, the uterus becomes enlarged, especially in the pregnant horn, and drops into the abdominal cavity. An experienced technician can ascertain this sign of pregnancy by *feeling with the gloved hand through the rectum wall.* Application of this method depends upon the recognition of changes in tone, size, and location of the uterine horns and changes in the uterine arteries.

This cow pregnancy test is popular because it affords early diagnosis, and there is little hazard when performed by experienced operators. It is recommended that cows be pregnancy tested by this method about 2 months after the bulls have been removed. With convenient facilities—corrals and squeeze—an experienced operator can pregnancy test 800, or more, cows per day.

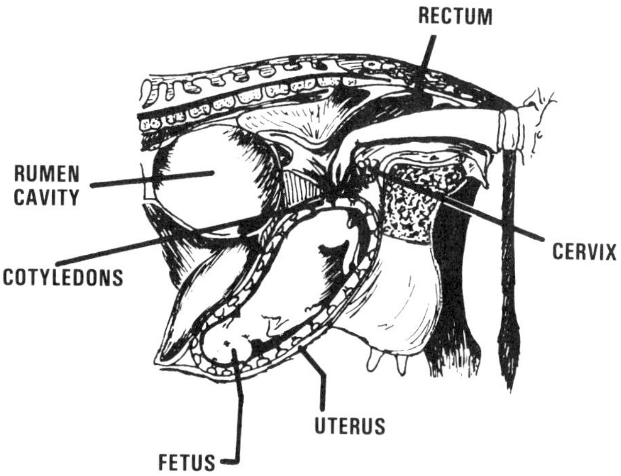

Fig. 3-8. Rectal method for determining pregnancy in the cow.

• **Pregnancy testing ewes**—Barren ewes can no longer return feed costs for their wool production. Moreover, returns from lambs are of increasing importance to the sheep industry; today, about ¾ of the income is from the lamb, and only ¼ from wool. These factors make pregnancy diagnosis of great economic importance. Detection of multiple fetuses would also be of significance, because feed costs could then be lowered on ewes carrying only one lamb.

Several methods for pregnancy testing have been developed and used experimentally or practically in recent years; among them, the following: (1) rectal-abdominal palpation, (2) ultrasonic scanning, and (3) intrarectal Doppler technique.

Fig. 3-9. The intrarectal Doppler technique for pregnancy determination in ewes. It detects the fetal heartbeat, which is 130 to 160 beats per minute compared to the ewe's 90 to 110 beats per minute. (Courtesy, Sheepman Supply Co., Barboursville, VA)

• **Pregnancy testing sows**—With ultrasonic detectors, pregnancy diagnosis of sows is a reality. Producers can determine with a 90 to 95% accuracy the number of females that have settled. These detectors are most accurate and yield the best return per dollar invested when they are used between 30 and 45 days after mating.

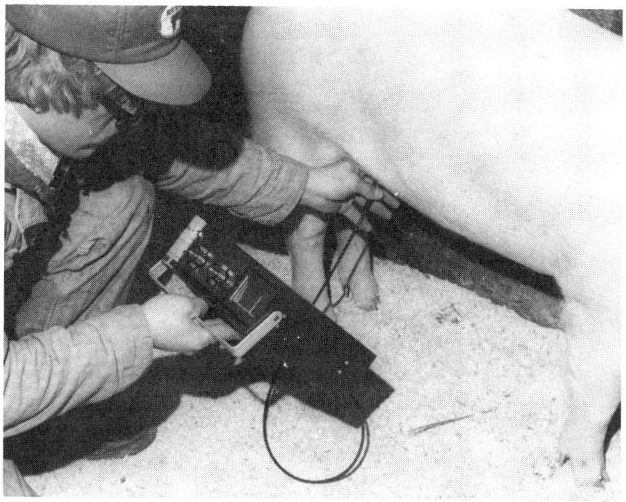

Fig. 3-10. Ultrasonic testing for pregnancy. (Courtesy, International Livestock Improvement Services Corp., Ames, IA)

The principle of ultrasonic pregnancy detectors is an ultrasonic echo from fluid in the uterus. Uterine fluid increases rapidly following conception and reaches detectable levels 25 to 30 days after breeding. It remains detectable for 80 to 90 days after breeding, following which the mass of pigs in the uterus exceeds the fluid content.

Among the several advantages of early pregnancy detection in sows and gilts are (1) it makes it possible to cull or rebreed nonpregnant, feed-wasting females; (2) it allows closer grouping of a number of sows for a farrowing period; (3) it gives early warning of breeding troubles, such as infertile boars and cystic ovaries of sows; (4) it enables producers to make more effective use of their breeding facilities and to plan more adequately for farrowing, nursing, and finishing; and (5) it makes it possible to guarantee pregnancy in females that are for sale.

• **Pregnancy testing mares**—In order to produce as high a percentage of foals as possible and to have them arrive at the time desired, the horse breeder should be familiar with the signs of and tests for pregnancy. This is doubly important when it is recognized that a great many mares may either be shy breeders or show signs of heat even when well advanced in gestation. The signs and tests of pregnancy follow:

1. **The cessation of the heat period.** One of the simplest determinations of pregnancy is the cessation of the heat period—the mare does not exhibit any signs of heat 18 to 20 days after her last ovulation. This may be difficult to determine as well as misleading. Some mares will continue to exhibit the characteristic heat symptoms when in foal; sometimes, they show such pronounced signs of heat that they are given the service of a stallion, which may result in abortion. Other mares may not cycle due to follicular or corpora luteal abnormalities or have silent heat periods in which external signs of estrus are not evident. Because of these situations, other methods of pregnancy determination are commonly used.

2. **Rectal palpation.** The most widely used method of pregnancy determination is by rectal palpation. An experienced technician can determine pregnancy (or barrenness) of mares at 98 to 100% accuracy by feeling with the hand through the rectal wall. Normally, the test is made 43 to 45 days after breeding, but pregnancy in maiden mares often can be detected within 35 to 40 days following conception. When performed by an experienced person, the manual test is quite reliable.

3. **Blood tests.** Blood tests, involving blood samples taken 20 to 120 days following breeding, give an indication of pregnancy.

4. **Ultrasonography.** In recent years, ultrasonography has increased in use in pregnancy determination. This technique can be used to obtain a visual image of the mare's reproductive tract, and thus detection of pregnancy, before palpation is normally performed. Ultrasonography can detect pregnancies as early as 10 days postovulation.

The ultrasound can also be used to detect follicular development, to detect estrus, and to detect reproductive abnormalities.

FREEMARTIN HEIFERS

Sterile heifers that are born twin with a bull are known as freemartins. This condition prevails in about 85% twin births when a calf of each sex is involved. The fetal circulations fuse, and the male hormone gets into the circulation of the unborn female where they interfere with the normal development of sex and modify the female embryo in the direction of the male. In approximately 15% of twin births of unlike sexes, fusion of the circulation does not occur, and the animal is normal and fertile.

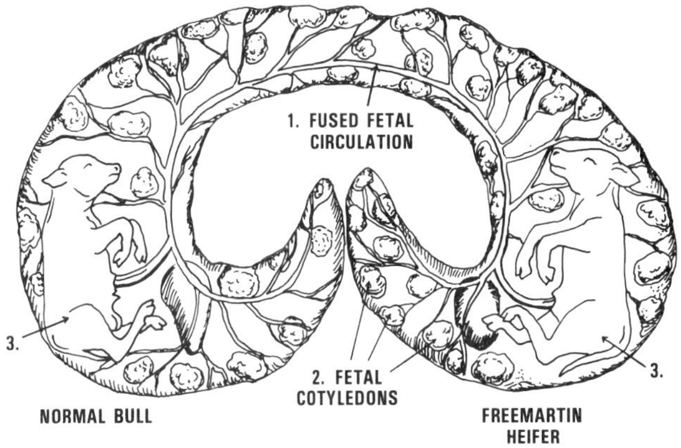

Fig. 3-11. Diagram showing fused fetal circulation of twin calves of opposite sex. Note (1) the fetal circulation of the male fused with that of the female, (2) fetal cotyledon free yolk sac, and (3) normal bull on the left and freemartin heifer on the right. (Source, *Physiology of Reproduction*, by Marshall; courtesy, the publisher, Longmans, Green and Co., Ltd., London, England)

Since only about 15% of such heifers are fertile, it is usually best to assume that they are sterile and market them, unless (1) an experienced person determined at the time of birth that their circulatory systems were not fused, (2) an examination of the vagina reveals that the animal is normal (in freemartin heifers, the vagina is usually about one-third normal length), or (3) skin-grafting[4] or blood-typing[5] techniques show that they are not freemartins and that they may, therefore, be regarded as reproductively normal.

PRODUCTION TESTING

Four bases of selection are available to the livestock breeder; namely, (1) selection based on type of individual, (2) selection based on pedigree, (3) selection based on show-ring winnings, and (4) selection based on production testing. Breeders have always followed one or all of the first three bases, and, therefore, need no introduction to them; now they are placing increasing emphasis on production testing.

---

[4]Billingham, R. E., and G. H. Lampkin, J., *Embryol. Exp. Morph.*, Vol. 5, Part 4, Dec. 1957, pp. 351–367.

[5]Stormont, C., *Journal of Animal Science*, Vol. 13, No. 1, Feb. 1954, pp. 94–98.

Production testing embraces both (1) performance testing (sometimes called individual merit testing), and (2) progeny testing. The distinction between and the relationship of these terms are set forth in the following definitions:

1. **Performance testing.** The practice of evaluating and selecting animals on the basis of their performance or individual merit.

2. **Progeny testing.** The practice of selecting animals on the basis of the merit of their progeny.

3. **Production testing.** A more inclusive term, including performance testing and/or progeny testing.

Production testing involves the taking of accurate records rather than casual observations. Also, in order to be most effective, the accompanying selection must be based on characteristics of economic importance and high heritability, and an objective measure or "yardstick" (such as pounds, inches, etc.) should be placed upon each of the traits to be measured. Finally, those breeding animals that fail to meet the high standards set forth must be removed from the herd promptly and unflinchingly.

PRODUCTION TESTING BEEF CATTLE[6]

Production testing is now an accepted beef cattle improvement tool, which may be used by purebred and commercial breeders alike.

The traits of economic value include those which contribute to either productive efficiency or desirability of product, the major ones of which are: fertility (reproductive efficiency), mothering and nursing ability, rate of gain, efficiency of feed utilization, longevity, and carcass merit (meatiness and quality).

The performance of cattle on the farm or ranch or in the feedlot, and their value on the rail (in the cooler), varies widely. For example, in the same feedlot and on the same ration, some cattle gain only 1 lb daily while others exceed 3 lb. Although this difference is disturbing, the fact that such variations do exist, and the further fact that such traits are, to a considerable extent, inherited differences, makes it possible to select and improve. Research has shown that when cattle are kept under nearly uniform conditions and their performance records are adjusted for known environmental differences, genetically superior animals can be identified.

Thus, production testing is the systematic measurement of differences in traits, the recording of the measurements, and the use of records in selection. It is a selection tool that will increase the rate of genetic improvement in individual herds, and eventually in the breed and in the total cattle population.

The rate of improvement in a herd, breed, and population is dependent on (1) the percentage of observed differences between animals that is due to heredity (heritability), (2) the difference between selected individuals and the average of the herd or group from which they come (selection differential), (3) the genetic association among traits upon which selection is based (genetic correlations), and (4) the average age of parents when the offspring are born (generation interval).

It should be noted that production testing will not increase the production of the animal that is tested. That is, keeping records does nothing to change the genetics of the test animal.

Production records are primarily useful for comparing cattle that are handled alike in a herd, and are not reliable for estimating differences among herds or among groups treated differently within a herd. This is so because large environmental differences due to location, management, and nutrition are likely to exist between herds or between management groups within a herd. It is not possible to adjust accurately for these differences.

ECONOMICALLY IMPORTANT TRAITS AND THEIR HERITABILITY

Research has revealed a high degree of heritability for several economically important traits in beef cattle. By heritability is meant the amount of variation in a trait, such as weaning weight, which is inherited or bred-in. This is usually expressed in a percentage; for example, birth weight has a heritability of 40%. The balance, or 60% (100 − 40 = 60) is due to environment—such things as feed, climate, and disease. The important thing to remember is that the heritable portion is passed on by parents to their offspring—the environmental portion is not.

This means that if the heritability of a trait is high, marked progress can be made through selection. On the other hand, if most of the improvement in an economically important trait is due to environment, the heritability of that character is low and little progress toward improvement can be made through selection. Nevertheless, the economic value of some traits may be great enough for them to receive emphasis in a breeding program even though their heritabilities may be low; fertility is an example of such a trait. Even though the heritability of many characters is disappointingly small, it is gratifying to know that progress from selection is cumulative and permanent.

Table 3–4 lists the economically important traits in beef cattle and gives their estimated heritability.

[6]This section was authoritatively reviewed by the following person who has great expertise in production testing beef cattle: M. Jorgensen, Jorgensen Ranches, Ideal, SD.

TABLE 3-4
ECONOMICALLY IMPORTANT TRAITS IN BEEF CATTLE, AND THEIR HERITABILITY

Economically Important Characters	Approximate Heritability of Characters[1]	Comments
	(%)	
Calving interval (fertility)	10	Fertility is economically the most important trait in beef cattle. Without a calf being born, and born alive, cattle are self-eliminating.
Birth weight	40	Birth weight is associated with calving survival. Also, it has a positive correlation of .39 with growth rate. Selecting for increased birth weight is generally avoided because of likely increased calving difficulty.
Weaning weight	30	Heavy weaning weight is important because: 1. It is indicative of the milking ability of the cow. 2. Gains made before weaning are cheaper than those made after weaning. 3. Those who sell calves at weaning usually make more profit due to the heavier weight available to sell.
Cow maternal ability	40	Mothering ability is important in beef cows, because it contributes to calf survival and weaning weight.
Feedlot gain	45	Daily rate of gain is important because: 1. It is highly correlated with efficiency of gain. 2. It makes for a shorter time in reaching market weight and condition, thereby effecting a saving in labor and making for a more rapid turnover in capital.
Pasture gain	30	Most beef animals spend a good part of their lives on grass; hence, pasture gain is important.
Efficiency of gain	40	Efficiency of feed conversion is expressed as pounds of feed intake per 100 lb of gain. It is seldom measured in performance and progeny tests because of the difficulty in securing feed intake records on individual animals. However, with the development of various types of electronic devices to facilitate such measurements, a number of central testing stations are now securing data on feed efficiency. A positive relationship exists between rate and efficiency of gain, so selection for rapid rate of gain does give some automatic selection for efficiency of gains. However, more rapid genetic progress can be made in improving efficiency of gain by selecting directly for efficiency of feed utilization, because some genes related to efficiency of gain are not related to rapid gains and would be "missed" in selecting only for correlated trait of rapid gain. With beef production becoming so highly competitive, cost factors have become so important to profit and success that an increasing number of livestock producers will be measuring and selecting directly for feed efficiency in the years ahead.
Final feedlot weight	60	Final feedlot weight is usually referred to as *weight per day of age.* It is generally computed at one year of age or at the end of the performance test. It is probably the most important measurement of the estimated value of a beef bull. It is composed of birth weight, weaning weight, and postweaning gain.
Conformation score:		This score should be based on skeletal soundness and indications of carcass desirability. Structural soundness, especially of the feet and legs, is most important in breeding animals.
1. Weaning	25	
2. Slaughter	40	
Carcass traits		Quality of product and quantity of edible portion are the basic factors of carcass merit. Where breeding animals are involved, and are not to be slaughtered, carcass quality may be evaluated by either (1) ultrasonic measurements, or (2) the K^{40} counter. Ultrasonic can be used to measure rib eye area and outside fat cover. The K^{40} counter evaluates the entire animal; it provides an effective method of measuring the total content of the live animal.
1. Carcass grade	40	High carcass grade is important because it helps determine selling price and is related to the juiciness and palatability of the meat.
2. Rib eye area	70	The rib eye (the large muscle which lies in the angle of the rib and vertebra) is indicative of the bred-in muscling of the entire carcass. Thus, a large area of rib eye is much sought.
3. Tenderness	60	Warner-Bratzier shear test and taste panel test are recommended as methods of measuring tenderness.
4. Fat thickness	45	Fat thickness is taken at the twelfth rib.
Cancer eye susceptibility	30	There is indication that susceptibility to cancer eye is hereditary.

[1]Gregory, K. E., *Beef Cattle Breeding*, Ag. Info. Bull. No. 286, Agricultural Research Service, USDA, 1969.

RECORD FORMS

A prerequisite for any production data is that each animal be positively identified—by means of ear notches, ear tags, or tattoos. For purebred breeders, who must use a system of animal identification anyway, this does not constitute an additional detail. But the taking of weights and grades does require additional time and labor—an expenditure which is highly worthwhile, however.

In order not to be burdensome, the record forms should be relatively simple. Also, they should be in a form that will permit easy summarization—for example, the record of one cow should be on one sheet if possible. Suggested record forms are shown in Figs. 3-12, 3-13a, and 3-13b (see pp. 78-79).

Information on the productivity of *close relatives* (the sire and the dam and the brothers and sisters) can supplement that on the animal itself and thus be a distinct aid in selection. The production records of more distant relatives are of little significance, because, individually, due to the sampling nature of inheritance, they contribute only a few genes to an animal many generations removed.

GET OF SIRE RECORD

Calf Crop for Year of ______________________

Sex of Get[1] ______________________

Owner and Address ______________________

Sire's Name ______________ Reg. No. __________

Date of Birth ______________________

Herd No. of Calf	Date of Birth	Calf Data							Yearling Data					Dam Data				Remarks
		Weaning Date	Weaning Age in Days	Weight in Lb	Daily Gain from Birth Weight,Lb	Adj. 205-Day Weaning Weight,Lb	Weanling Weight Ratio[2]	Conformation Score	Date Weighed	Weight, Lb	Wt. Adj. to days	Yr. Wt. Ratio[2]	Conformation Score	Herd No.	Age This Year	Mature Weight, Lb	Conformation Score	
Totals																		
Averages																		

[1]One sheet should be used to record all the bull calves and another sheet to record all the heifer calves by the same sire.

[2]Ratio calculated as follows:

$$\frac{\text{Individual record}}{\text{Av. of all calves on same farm and same season}} \times 100$$

Fig. 3-12. Get of Sire Record Form.

INDIVIDUAL COW RECORD

Tattoo, Horn Brand, and/or Neck Chain No. ______________

Name ______________ Reg. No. ______________

Sire ______________

Dam ______________

Bred by ______________ Birth Date ______________

Purchased from ______________ Birth Wt., Lb ______________

Address ______________ Weaning Wt., Lb ______ Age ______ Conf. Score ______

Purchase Date ______ Price, $ ______ Yearling Wt., Lb ______ Age ______ Conf. Score ______

Disposition ______ Price, $ ______ Two Year Wt., Lb ______ Age ______ Conf. Score ______

Av. Daily Gain Weaning to 1 yr., Lb ______________

Reason for Disposal ______________ Feed Efficiency ______ lb feed/100 lb gain

Date ______________ Temperament ______________

Faults & Abnormalities ______________

PRODUCE OF DAM RECORD

Calf Data												Yearling Data					Production Testing				
Birth Date	Sex	Tattoo	Sire	Birth Wt., Lb	Vigor at Birth[1]	Weaning Age Days	Weaning Wt., Lb: Act.	Weaning Wt., Lb: 205 day Adj.	Weaning Wt. Ratio[2]	Weaning Cond.	Conf. Score	Date	Yr.,Wt., Lb.: Adj.	Yr.,Wt., Lb.: —Days	Yearling Wt. Ratio[2]	Conf. Score[1]	Days on Feed	Av. Daily Gain, Lb	Gain Ratio[2]	Lb Feed /100 Lb Gain	Disposition, Price Remarks

[1]0=dead at birth; 1=definitely undersized at birth; 2=unthrifty, definite indications of disorders; 3=moderately thrifty, slight indications of disorders; 4=thrifty, no signs of disorders, dry hair coat; 5=thrifty, no signs of disorders, sleek hair coat; 6=very large, healthy, and vigorous

[2]Ratio calculated as follows:

$$\frac{\text{Individual record}}{\text{Av. of all calves on same farm and same season}} \times 100$$

Fig. 3-13a. Individual Cow Record (see Fig 3-13b for reverse side of record form).

IMMUNIZATION AND TEST RECORD

Immunizations					Health Tests							Remarks
Date[1]	Blklg.	M. Edema	Bangs	Misc.	TB-Bangs	Johnes	Lepto.	Anaplas.	Vib.	Trich.	Misc.	

[1] Indicate vaccinations by check in appropriate column opposite date given; indicate test results by P (positive), N (negative), or S (suspect) opposite date of test.

GENERAL INFORMATION

Record all facts pertinent to the history of this cow, viz.: veterinary treatment (except immunizations), udder condition, mothering instinct, calving peculiarities, etc.

Date	Remarks

Fig. 3–13b. Individual Cow Record. This is the reverse side of the record form shown in Fig. 3–13a.

MILESTONES IN PRODUCTION TESTING BEEF CATTLE

Fig. 3–14. The goal of production testing beef cattle is the genetic improvement of individual animals, herds, breeds, and the total beef cattle population. As a yearling, this purebred Simbrah bull had a ribeye of 17.8 sq in., measured by ultrasound. (Courtesy, Granada, Wheelock, TX)

The primary goal of production testing beef cattle, which has remained unchanged through the years, is—

> To increase the rate of genetic improvement of individual animals and herds, followed by improvement of breeds and the total cattle population.

In pursuing this goal, the measurement techniques, amount of data, method of evaluation, and terminology have greatly changed along the way. Two recent milestones in the production testing of beef cattle follow.

EXPECTED PROGENY DIFFERENCE (EPD)/ NATIONAL SIRE EVALUATIONS (NSE)

In most cases, the primary desire in genetic selection is an accurate estimate of comparative progeny performance. The most recent concept for predicting progeny performance is *Expected Progeny Difference*, or *EPD*. The first use of EPD came through *National Sire Evaluations (NSE)* conducted by some breed associations. These programs involved a *sire summary* comparing sires across an entire breed. Comparisons were made possible only if sires had progeny in more than one contemporary group where other sires were represented. Widespread use of AI, especially in new breeds available in the United States beginning in the late 1960s, made these comparisons possible. The first National Sire Summary, published by one of these registry associations in 1972, compared 13 sires.

However, only sires with adequate progeny in more than one comparison group were included in these programs. Also, some incorrect assumptions reduced their accuracy. One of these assumptions was that the sires being evaluated were not genetically related, which was not always true. Another assumption was that the breed trait-average did not change from one year to the next, which is false. Still another assumption in NSE analyses was that sires are mated randomly; that is, to females of equal genetic merit, which may or may not be true, depending upon the intent of the particular breeder. Finally, it was assumed that progeny were not selected or culled before all records were collected. Here again, this is a false assumption; in most operations some animals for which weaning information is available are not maintained for further collection of performance data. This results in a changed contemporary group average which is the basis for all comparisons.

NATIONAL CATTLE EVALUATION (NCE)

Finally, refined mathematical techniques and additional data resulted in the *National Cattle Evaluation (NCE)* where all animals, or even planned matings, in a breed can be more accurately compared than with progeny NSE. All major breed associations now have such programs. The limitations and incorrect assumptions of NSE have been addressed by NCE as follows:

a. The new values combine records of relatives, the individual's performance, and/or progeny. This results in improved accuracy.

b. An adjustment is made for differences in genetic merit of females to which a bull is mated.

c. A separate analysis is performed to distinguish direct and maternal genetics.

d. Trait correlation, or the known relationship between various performance traits, is used to improve estimates. This is possible by utilizing breed analyses which have established numerical relationships between various performance characteristics.

e. Animals with no records of any kind for some characteristics can be estimated from trait correlation.

f. The breed genetic change in performance over time is considered.

g. There are no more *Reference Sires* as such. Rather, any bull with progeny in more than one contemporary group is, in effect, a reference.

h. Accuracy is now expressed by all breeds from 0 to 1, with the larger number meaning higher accuracy.

Until now, information in a performance pedigree has been of primary use for within-herd comparisons. But new techniques and new analyses have been implemented so that values in a performance pedigree are now comparable to those in the National Sire Summary; that is, these values are comparable across the entire breed. In fact, estimates can be made of Expected Progeny Difference of individuals that would result from a possible mating, i.e., an EPD can be estimated for an individual before a mating is even made.

BEEF IMPROVEMENT FEDERATION (BIF) GUIDELINES[7]

Fig. 3-15. A linebred Angus bull from a performance program in the 36th year. (Courtesy, M. Jorgensen, Jorgensen Ranches, Ideal, SD)

Performance testing offers those engaged in beef production a way of measuring heritable differences among animals so producers can select those individuals which are expected to transmit superior performance to their offspring. But, for maximum value, there is need for uniformity in measuring, recording, and evaluating beef cattle performance data. The latter role has been filled by the Beef Improvement Federation, and is detailed in *BIF Guidelines for Uniform Beef Improvement Programs,* sixth edition, 1990. Some pertinent points from this valuable, but voluminous, report are briefed in the sections that follow.

It is noteworthy that, among its several guidelines, the BIF recommends the pelvic and live animal measurements which follow.

• **Pelvic measurements**—Many producers today are interested in using pelvic measurements as a management tool to assist in reducing the incidence and severity of calving difficulty. Many factors are associated with calving difficulty, in-

[7]The Beef Improvement Federation headquarters at Oklahoma State University, Stillwater.

cluding small first calf heifers, large calves, male calves, small pelvic size of heifer, long gestation, condition score of cow and abnormal presentation. Research indicates that a disproportion between the calf size (birth weight) and female birth canal (pelvic area, see Fig. 3–16) can be a big contributor to calving difficulty.

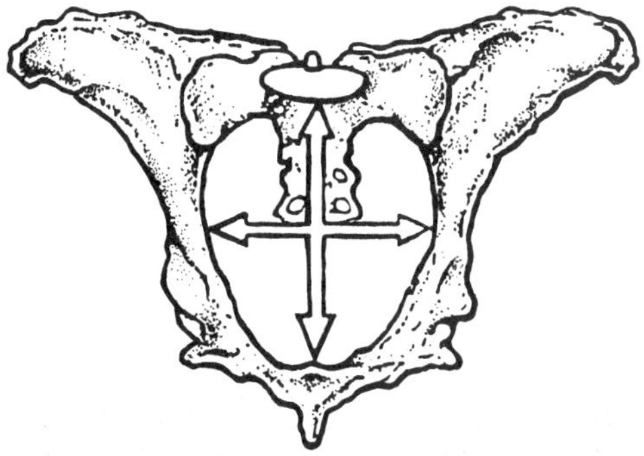

Fig. 3–16. Vertical and horizontal measurements are obtained to determine pelvic area.

a. **Measurement of heifers.** Since a large majority of the calving difficulty occurs in first calf heifers (calving at 22 to 28 months of age), many producers take pelvic measurements in their heifers at 12 months of age.

b. **Measurement of bulls.** To increase the pelvic area of their heifers, many producers are selecting bulls with a larger pelvic area. Pelvic size has been shown to be 60% heritable, indicating that selection for large pelvic size in bulls should result in increasing pelvic size in females.

• **Live animal**—Evaluation of live animals takes into consideration any measurements or subjective evaluations that help describe an animal. For example, evaluation involves physical examination of bulls to include penis, rectal examination, and scrotum (including scrotal circumference).

Some other common measurements of cattle include: backfat, pelvic size, height at the shoulder, height at the hip, and length of body.

In recent years, measurements for height have become a descriptive supplement to many herd testing programs. Adjusted weights and weight ratios accompanied by linear measurements for height have added another dimension to evaluating the fat-lean ratio of an individual animal in a performance program.

The recommended point for linear measurement for height is to a point directly over the hooks (Fig 3–17). This measurement is adjusted to relatively logical production end points at 205 days and 365 days (with the BIF ranges currently used for adjusted weights).

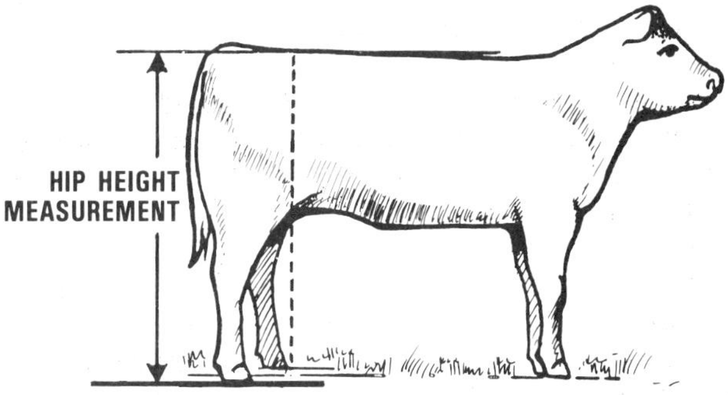

Fig. 3–17. Height measurement.

PRODUCTION TESTING DAIRY CATTLE[8]

Production records are the most important management tool in dairying. Almost every decision regarding the dairy herd is based on them—how much to feed, when to turn cows dry, which cows to cull, the level of herd health, and which bull to use. Records necessitate that each cow be individually identified, and that milk and butterfat production records be kept. (In the future, protein production records will likely be necessary.)

ECONOMICALLY IMPORTANT TRAITS AND THEIR HERITABILITY

The economically important characters in dairy cattle are those which contribute to the production of milk, more abundantly and efficiently. Variation and heritability provide the potential for improvement; and the various testing programs provide means of measuring traits.

Heritability is a measure of additive genetic variation. It is a measure of accuracy when choosing parents. To the producer, it determines how much of that for which selection is made is achieved. When heritability is low, there will be many errors when choosing parents and progress will be slow. When heritability is high, progress will be rapid. Heritability of milk production is $\pm 25\%$ which is moderate. But progress will be much faster if producers performance test all cows (DHI) and progeny test all bulls (DHI).

[8]This entire section was authoritatively reviewed by the following: L. Curkendall, General Manager, California Dairy Herd Improvement Assn., Fresno, CA; Dr. W. D. Gilson, Extension Dairy Specialist, College of Agriculture, The University of Georgia, Athens; T. Sawyer, Diamond S Ranch, Registered Holstein-Friesian Cattle, Waterford, CA; and Dr. D. J. Schingoethe, Dairy Science Department, College of Agriculture and Biological Sciences, South Dakota State University, Brookings.

The following example will serve to illustrate the importance and application of heritability estimates: If heifers from a herd of cows averaging 15,000 lb of milk were selected from 18,000-lb dams, and by sires equal to these dams, it would be important to know how much of this 3000-lb superiority in the parents would show up in the offspring. The heritability of milk production is 25%, which indicates that only 25% of the superiority of the selected parents is genetic and will appear in the daughters; 25% of 3,000 is 750 lb of milk. Adding this 750 to 15,000, the estimate of 15,750-lb daughters results from mating 18,000-lb parents that originated in a 15,000-lb herd. Heritability estimates for several traits in dairy cattle are given in Table 3–5.

TABLE 3–5
HERITABILITY ESTIMATES

Trait	Heritability (%)
Production Traits	
Milk Production	25
Percent Fat	50
Percent Protein	50
Percent SNF	50
Feedlot Gain	45
Physical Traits	
Stature	40
Udder Support	20
Legs and Feet	15
Management Traits	
Milking Speed	25
Birth Weight	40
Temperament	40
Fertility	05

ALTERNATE DHI TESTING PLANS

The National Cooperative Dairy Herd Improvement (DHI) Program is a voluntary cooperative effort to improve the level and efficiency of milk production and increase dairy profits. It involves milk producers, local and state DHI organizations, extension services of land grant colleges and universities, and the U.S. Department of Agriculture.

The U.S. Department of Agriculture (Animal Science Research Division) aids in conducting and distributing results of the sire evaluation phase of the DHI Program. It also coordinates, furnishes materials, provides statistical information, analyzes data, and researches various aspects of the program.

State and local Dairy Herd Improvement Associations (DHIA) conduct the program among producers, working through the Cooperative Extension Service in cooperation with the Federal Extension Service and Animal Science Research Service of the USDA.

Dairy cattle record-keeping plans may be either *official* or *unofficial*, with alternate choices under each grouping.

OFFICIAL DHI TESTING

This includes the Standard Dairy Herd Improvement Association (DHIA) and the Dairy Herd Improvement Registry (DHIR). Since both programs are official, a supervisor tests herds 4 to 12 times annually. Records from both programs are used in proving dairy sires.

Dairy Herd Improvement Association (DHIA)

This program, first adopted in 1926, is the most complete of all dairy production and record plans. More than half of the cows in the United States on production test are on this program. Both registered and grade cows can be enrolled.

In this program, a supervisor or tester employed by the local or state testing association visits the herd 1 day each month. The tester identifies all cows in the herd; weighs and takes representative samples of the milk from all animals in the herd for 2 consecutive milkings (3 milkings on herds 3-times-daily-milking); then combines the milk samples and sends them to a central testing laboratory for analyses of components such as butterfat, protein, and somatic cell count (SCC). Records are obtained on an individual cow basis on monthly and accumulative records for milk, fat (the latter in pounds and percentage), and protein; amount and cost of feed, and income over feed cost; breeding dates, calving dates, dry dates, and other factors affecting productivity; and in some testing associations, somatic cells or the California Mastitis Test (CMT) is made as an aid in monitoring udder health.

The above information is fed into a computer, programmed to provide monthly summaries of (1) individual cows, and (2) the herd; and this information is sent to each producer.

Alternate AM-PM Test

Dairy producers have additional options which can reduce the cost and still maintain official records. The supervisor can weigh two milkings and sample only one of the milkings, or only one milking may be weighed and sampled. A milking time record is required if only one milking is weighed and sampled. The milking which is weighed and sampled is alternated from month to month.

The primary advantages of these programs is the lower cost and the reduced supervisor time. They also reduce the amount of disruption in the parlor by reducing the sampling time. These programs are especially attractive to larger dairy herds. Supervisors may also find these programs beneficial since they allow additional herds to be tested.

Dairy Herd Improvement Registry (DHIR)

This is the Standard DHIA record *plus* added requirements to satisfy the needs of breed associations. Among the latter, are *surprise tests*, made when the milk production of certain cows exceeds the breed average or another specified amount.

Only registered dairy cows are eligible for DHIR records. The production records of herds enrolled in DHIR are transferred to the respective breed registries for official recording.

UNOFFICIAL DHI TESTING

These unofficial record plans are designed to aid within herd management at minimal cost. Among such plans are: Owner Sampler (OS), Weigh-A-Day-A-Month (WADAM), Milk Only Record (MOR), and Alternate AM-PM.

Owner-Sampler Records (OS)

Under the owner-sampler plan, the owner, rather than a supervisor, weighs and samples the milk. The samples are then tested at a central laboratory.

Weigh-A-Day-A-Month (WADAM)

In this program, the owner weighs the milk from each cow one day each month, and enters the weight and feeding information on the form provided. The information and forms are mailed to the supervisor, or a central office, where calculations are completed, following which summaries are returned to the owner.

Milk Only Records (MOR)

This testing plan, which originated in North Carolina, is receiving considerable attention, especially in the southeastern states. These records involve milk weights recorded by the DHIA supervisor, without fat determination.

ADJUSTMENT FACTORS

It is frequently desirable to compare the performance of individuals or groups of animals. To do so, it is necessary to adjust all records to a comparable basis. For this purpose, adjustment factors have been developed for each breed for (1) length of lactation, (2) the number of milkings per day, (3) age and month of calving, and (4) fat content of milk. These four adjustments are important for comparing milk and fat of cows in different environmental conditions. Each of these factors will be discussed. At the outset, however, the following point is pertinent: Although adjustment factors are usually necessary in order to get two or more records to a common basis, it is recognized that records that are comparable without factors are more reliable.

GENETIC EVALUATIONS

Genetically, the sire and the dam contribute equally to their offspring. However, the breeding value of a dairy sire can be more accurately determined than that of a cow because he will have many more offspring. Thus, it is through the selection of sires that the major portion of progress in genetic improvement is made.

Generally speaking, the dairy producer has three sources from which to choose a herd sire: (1) artificial insemination (AI), (2) purchase from another breeder, or (3) raise from within the herd.

Sires should be selected with a primary emphasis on production. Components have varying degrees of importance depending upon the marketplace. Sire selection should be adjusted with consideration given to these variations. Sires may also be selected based on pedigree, type (conformation), and family or bloodlines. Selection for calving ease may be important when breeding heifers.

USDA-DHIA SIRE SUMMARIES

Sire evaluations are published twice yearly in January and July. The U.S. Department of Agriculture calculates the production summaries and the respective breed organizations calculate the type summaries. The National Association of Animal Breeders (NAAB) also calculates a calving ease summary for the Holstein breed. Selected portions of the summary are published in a variety of magazines. The complete production summary may also be obtained from the Animal Improvement Programs Laboratory, U.S. Department of Agriculture, Agricultural Research Service, Beltsville, MD 20705. The complete summary (production and type) for each breed can be obtained from the respective breed organizations.

A sire summary is compiled when 10 or more daughters of a bull have lactation data and herdmate records reported. (Some sire summaries report only sires having 20 or more production tested daughters.) The production records used in USDA-DHIA sire summaries consist of lactation records of 305-days or less, standardized to twice daily (2×) milking, mature equivalent (ME) basis.

The dairy registry associations also publish sire summaries. Some of these are useful for type (conformation) information, which is not included in USDA-DHIA summaries.

Genetic evaluations are also conducted on cows twice each year using the same procedures as the sire summaries. The evaluations provide a means of objectively evaluating the genetic merit of females. This increases the amount of genetic progress which can be achieved by allowing accurate selection of both parents.

PREDICTED TRANSMITTING ABILITY

Predicted transmitting ability (PTA) is an estimate of the amount of superiority (improvement) or inferiority an animal will transmit to its offspring. It is the most accurate measure available of an animal's genetic ability. For the purpose of illustration, the following Predicted Transmitting Ability for a bull owned by Tri-State Breeders Cooperative is taken from the January, 1990 USDA Sire Summary:

	Predicted Transmitting Ability							
		Fat				Protein		
Name of Bull	Milk	%	lb	$$	%	lb	$$	CY$$
Zeilland Fast Future	+2,767	-.09	+79	+170	-.06	+73	+297	+282

Daughters from *Zeilland Fast Future* would be expected to produce 2,767 lb more milk, 79 lb more fat, and 73 lb more protein than daughters from breed average bulls. They would also be expected to produce milk which tests .09% and .06% lower for fat and protein respectively.

The USDA-DHIA sire summary also includes three economic indexes. They are PTA$$, PTA Protein $$, and PTA CY$$. The indexes are based on the respective PTA values, the average U.S. milk price for 3.5% milk and the average fat and protein differentials. The indexes provide a means of selecting for components as well as production simultaneously with little additional effort. Sires are ranked on the basis of their PTA Protein $$. Sires in the 99th percentile are in the top 1% while bulls in the 50th percentile are in the top half of the active AI sires.

TABLE 3-6
LEVEL OF PRODUCTION AND PREDICTED DIFFERENCE FOR SEVERAL BULLS

Number of Daughters	Avg. Milk Production of Daughters (M.E.-305-2X)	Predicted Transmitting Ability
	(lb)	(lb of protein)
43	20,929	+1916
73	21,074	+1746
42	23,213	+1571
304	20,406	+1215
53	20,637	+ 834
43	21,398	+ 740
947	19,399	+ 492
20	24,073	+ 483
648	19,739	+ 127

If the bull, whose daughters produced the most milk was selected (24,073 lb of milk), the expected production of his future daughters would only be 483 lb greater than cows sired by breed average bulls. However, if the bull with the second highest daughter average was selected (23,213 lb), his future daughters would be expected to produce 1,571 lb more milk than cows sired by breed average bulls. This is a difference of over 1,000 lb more milk per lactation for the sire whose daughters averaged over 850 lb less milk per lactation. Future daughters of the top bull for PTA would be expected to produce over 1,400 lb more milk even though his daughters have averaged producing over 3,100 lb less (see Table 3–6).

RELIABILITY

Reliability (REL) provides an indication of the confidence that can be placed in the genetic evaluation. It is a measure of the amount of information available.

Reliability is unrelated to the transmitting ability of the individual. It simply provides a measure of how sure we are that the PTA reflects the individual's true transmitting ability. The higher the reliability, the greater the confidence we have in the evaluation. Conversely, the lower the reliability; the less confidence we should have in the evaluation. Reliability increases as the number of offspring increases, the number of herds represented increases, and the amount of information available on each offspring increases.

Table 3–7 illustrates the value of reliability when evaluating bulls with essentially the same predicted transmitting abilities. The true transmitting ability of the first sire listed has a 98% probability of being between +1,517 and +1,675 lb while the last sire has a 52% probability of being between +1,199 and +1,975 lb.

TABLE 3-7
VARIATION IN RELIABILITY AND THE CONFIDENCE INTERVAL WITH NEARLY IDENTICAL PREDICTED TRANSMITTING ABILITIES FOR MILK

No. of Daughters	No. of Herds	Predicted Transmitting Ability	Reliability	Confidence Interval
		(lb)	(%)	
947	608	1,596	98	79
170	118	1,580	90	177
79	63	1,592	85	217
68	49	1,571	80	250
62	47	1,574	77	268
47	17	1,582	73	291
32	21	1,583	70	306
17	14	1,599	61	349
12	10	1,587	52	388

The higher the reliability, the more confidence one can place in genetic evaluation. Reliability increases with the number of progeny, the number of lactations per daughter, and the number of herds in which daughters are located. We can see from the example that the PTA of the first bull is quite accurate while the PTA of the second bull is subject to much wider variation. The latter sire however has the potential to be superior to the first sire by 300 lb or more.

SELECTING THE DAIRY BULL

Predicted transmitting ability in dollars is an excellent tool to use in ranking bulls. It provides an indication of the income which can be expected from the daughters of each bull. The value is based on the PTA for milk, the PTA for fat, and the national average milk price and differential. Thus, this ranking includes both milk and fat production as well as the economics.

Another method of ranking bulls is by the PTA$$ Protein. This value includes the PTA for protein and the average national protein differential in the formula. The USDA uses this value to rank bulls. Herds which are paid on the basis of milk components (fat and protein) will find this to be the best system for ranking bulls. It results in the selection of bulls which more nearly reflects the market than a ranking based on components alone.

Reliability indicates the confidence which may be placed in the PTA. Bulls with high reliability can be used with the assurance that their proof will not vary greatly. Those bulls with reliabilities below 75% should be used with more caution. The selection of several bulls with low reliabilities, guards against the overuse of a single bull whose proof drops significantly. It should be realized that all of the daughters of a bull will not produce up to expectations; however, some will exceed expectations. The average of the group will be very near the anticipated levels.

Conformation or type traits can also be included in the selection program. The rationale behind the selection for type is that cows with better type are more desirable animals for milk production. It is also believed that they stay in the herd longer. The only type traits which have consistently been shown to have relationship to longevity are the udder traits. Feet and leg traits may be related to longevity in some situations.

For herd improvement, a dairy producer should choose—

1. Bulls with the highest PTA value.

2. Bulls with high PTA values that also have high reliability values (narrow confidence interval).

3. Several bulls with high PTA values when the reliability value is below 75%.

4. Bulls with a low percent difficult births for use in breeding heifers.

PRODUCTION TESTING SHEEP[9]

In sheep, as with other classes of livestock, production testing (performance and progeny testing) is used for the following purposes:

1. To aid in the selection of both male and female flock replacements.

2. To provide an accurate basis for culling flocks.

3. To aid in the promotion and sale of breeding stock.

The amount of improvement in sheep that can be achieved through production testing depends on the following:

1. The accuracy in measuring a trait, and the use made of the record obtained.

2. The selection pressure applied. This is limited by (a) reproductive rate, (b) the number or percentage of animals that need to be saved for flock replacements, and (c) the number of traits being selected for simultaneously. Most of the selection pressure and improvement will result from the selection of rams.

3. The variability in the trait or traits being selected. The greater the variability, the more rapid the improvement.

4. The heritability of the trait for which selection is being made.

5. The emphasis on weight or number of lambs weaned per year by the dam.

ON-FARM VS CENTRAL TEST STATION

From a practical standpoint, some traits can be evaluated only on the farm; among them, fertility, prolificacy, longevity, and all traits of lambs measured before weaning.

Postweaning traits may be measured either on the farm or at a central test station. The **advantages** of central testing include:

1. It facilitates the accurate measurement of some traits that are difficult or impossible to measure on the farm, such as feed efficiency and ultrasonic estimate of fat thickness.

2. Comparisons of rams between flocks are possible, *provided* pretest management differences are minimal so that such comparisons are valid.

3. Greater standardization of the test environment may be possible than on the farm.

4. There is greater confidence in the results of a test conducted by an independent agency.

The **disadvantages** of central testing include:

1. The impossibility of completely alleviating pretest differences in flock management, rations, disease and parasite control, etc.

2. The relatively high cost.

ECONOMICALLY IMPORTANT TRAITS AND THEIR HERITABILITY

Two basic principles of animal breeding are:

1. The more traits included as criteria for selection the slower the progress for any one of them.

2. A trait will not respond to selection unless variation for the particular trait exists in the flock, and unless the trait is heritable.

---

[9]This entire section was authoritatively reviewed by Dr. C. E. Terrill, USDA retired, Beltsville, MD.

To the above may be added the observation that more measurements mean more time and cost. Thus, only those traits should be considered which contribute to net income, which display variation in the flock, and which are heritable.

Table 3–8 lists the economically important traits in sheep and gives their heritability.

TABLE 3–8
ECONOMICALLY IMPORTANT TRAITS IN SHEEP AND THEIR HERITABILITY

Economically Important Characters	Approximate Heritability of Characters	Comments
	(%)	
Multiple births	15	Where adequate feeds are available, twin lambs are desirable because (1) they greatly increase the weight of lambs sold per ewe, and (2) the annual maintenance requirement of ewes is not far different, whether they are producing twins or singles. Australian workers have increased twinning rate in Merino sheep by 2.3% per year by selection. In New Zealand, the Romney has responded to selection for twinning by increasing 1.1% per year.
Birth weight of lambs	30	The larger lambs at birth are generally more vigorous and make faster gains.
Weaning weight:		Heavy weaning weights are especially important in those areas where cost of production is largely on a per head rather than on a per pound basis, such as the western range.
1. 60 days of age	10	
2. 100 days of age	30	
Rate of gain	30	Preweaning rate of gain, or growth rate, is largely a reflection of the milk production of the ewe. It is affected by twinning, sex of lamb, and age of ewe. Postweaning rate of gain is a reflection of inherent growth potential of the individual. It is also positively correlated with mature size. Growth rate is economically important for three reasons: (1) It is highly associated with feed efficiency—rapid growth is efficient growth; (2) rapid growth allows for the sale of a larger amount of product; and (3) it makes for a shorter time in reaching market weight and condition, thus effecting a saving in labor, making for less exposure to risk and disease, and allowing for more rapid turnover in capital. Postweaning growth rate can be measured effectively by average daily gain, either on the farm or in the central station.
Type score:		Type can include any or all of the following: (1) characteristics that influence an animal's ability to live and perform in its environment—such as feet and legs, teeth, and udder; (2) traits that indicate meatiness; and/or (3) breed type. Type is a factor in determining today's market values. Yet, type within itself—unsupported by performance records for other traits—will not likely be sufficient to ensure high selling prices in the future. The determination of optimum type, the evaluation of it, and the use made of the information, should remain the responsibility of the individual breeder.
1. Weanling	10	
2. Yearling	40	
Finish or condition at weaning	17	Finish at weaning is largely determined by available feed and is not highly heritable. Yet it is most important because milk-fat lambs suitable for slaughter at weaning time almost always bring more per pound than thinner lambs that are sold as feeders. For the range area as a whole, about 25% of the lambs lack sufficient finish for slaughter at weaning time.
Wrinkles or skin folds:		Sheep with smooth bodies are preferred. Wrinkled sheep are difficult to shear, and lack fiber uniformity.
1. Neck folds (weanling)	39	
2. Body folds (yearling)	40	
Face covering	56	Wool-blind ewes do not graze well, require more labor if they are clipped around the eyes, and wean fewer pounds of lamb. At the Western Sheep Breeding Laboratory, ewes with open faces produced 11 lb more lamb per ewe bred than those with covered faces.
Fleece weight:		Clean fleece weight is most important, for the fiber is far more important than the materials scoured from grease wool. However, scouring a whole fleece, or even a sample, requires much time and equipment; hence, it likely can only be justified in the selection of stud rams in purebred flocks of fine- and medium-wool sheep. Since there is a close correlation between clean fleece weight and grease fleece weight, grease fleece weight will suffice under most circumstances.
1. Grease weight	38	
2. Clean weight	40	
Staple length:		Fiber length is important because it is a major factor in determining fleece weight and grade.
1. Weanling	39	
2. Yearling	47	
Fleece grade	35	The grade of fleece—which is based primarily on fiber diameter, but with consideration given to length, also—is important because it determines the use and price of wool.
Fat thickness over loin eye	23	Fat thickness is a measure of meatiness; excess fat results in an increase in fat trim and a decrease in percent lean cuts.
Loin eye area	53	Loin eye area is a good indicator of muscling.
Carcass weight/day of age	22	This trait is moderately heritable.
Carcass grade	12	High carcass grade is important because it determines eating and selling qualities.
Carcass length	31	Long carcasses are usually meaty carcasses.

RECORD FORMS

A carefully planned and executed on-the-farm recording scheme for lamb production is the first step in a sheep production testing program. Records should be in understandable form; and, most important, proper use of the records should be made in selection and culling.

Fig 3–18 shows an individual ewe or ram record form which will meet the needs of most herds and flocks.

SHEEP

INDIVIDUAL EWE OR RAM RECORD FLOCK NO. ______

Breed ______ Reg. No. ______ Ear Nick ______ Tattoo ______ Birth Date ______

Type of birth (Single, Twin) ______ Date ______

Bred by ______ Temperament ______ (Gentle, nervous)

Sire ______

Bought from ______

Address ______ Face Covering[6] ______ (As a lamb)

Date Purchased ______

Type, Weaned[1] ______ Date ______ Face Covering[6] ______ (As yearling)

Type, Yearling[1] ______ Date ______

Dam ______

Back[2] ______ Rump[3] ______ Leg[4] ______ Disposed to ______ Date ______

Defects & Abnormalities[5] ______ Why Disposed[11] ______

LAMBS (Use one line for each lamb for ewe's offspring; use one line for the average of a ram's progeny for each year).

Date of birth	Ear nick and No.	Vigor at birth	Type of birth[7]	Type of rearing[8]	Sex	Birth Wt.	Defects and abnormalities[5]	Sire	Milking ability - ewe[9]	Weaning age, days	120 Day Weight	Weaning condition[10]	Weaning type[1]	Disposition[11] or remarks

[1] Trueness to breed appearance and desired mutton conformation:
"1" Excellent; "2" Good; "3" Medium; "4" Fair; "5" Poor.

[2] Straightness, strength, and spring of rib; width.
"1" Excellent; "2" Good; "3" Medium; "4" Fair; "5" Poor.

[3] Width and levelness: 1-2-3-4-5 as above.

[4] Plumpness of thigh: 1-2-3-4-5 as above.

[5] Including overshot or undershot jaw, scurs, black fiber, etc.

[6] "1" Not covered beyond poll; "2" Covered to eyes; "3" Covered slightly below eyes, but open faced; "4" Covered partially below eyes, but not subject to wool blindness; "5" Face covered and subject to wool blindness.

[7] S—Single; T—Twin; Tr—Triplet.

[8] S—Single; T—Twin; Tr—Triplet; Gr—Grafted on foster mother and give her number.

[9] Good, medium, poor.

[10] Condition or degree of fatness:
"1" Excellent; "2" Good; "3" Medium; "4" Fair; "5" Poor.

[11] Cause of death, reason for disposal, kept for breeding purposes, whom sold to.

Fig. 3–18a. Individual Ewe or Ram Record Form. (See Fig. 3–18b, p. 88, for reverse side of record form.)

PREWEANING LAMB PERFORMANCE TEST

The following preweaning lamb performance test program is recommended for use in all purebred or commercial farm flocks:

Step 1: Record data—*Minimum data* should include (a) identification—lamb ear tag number; (b) sire number; (c) dam number; (d) age of dam, in years, at lambing time; (e) birth date of lamb; (f) sex of lamb; (g) type of birth—single, twin, triplet, and (h) how reared—single, twin, triplet.

Optional data may include (a) whether or not lamb was creep fed; (b) slaughter grade; (c) 200-day yearling body weight; and (d) grease fleece weight and staple length to nearest $\frac{1}{10}$ in.

WEIGHT RECORD OF EWE OR RAM

Date	Age	Weight	Condition[1]	Remarks[2]

REMARKS

(For example: bad udder, poor mother, aborted, veterinary treatment and nature of ailment.)

Date: Remarks:

FLEECE
(Use one line for each year)

Length Side[3]	Fineness[4]			Date of Shearing	Days Growth	Grease Weight	Per cent of Yield	Clean Wt.	Color of Skin	Purity[5]	Remarks About Fleece
	Shoulders	Side	Thigh								

1. Condition or degree of fatness: "1" Excellent; "2" Good; "3" Medium; "4" Fair; "5" Poor.
2. Factors affecting weight: e.g. just shorn, soon lamb, etc.
3. Length of staple, middle at side, to nearest 0.2 cm., just before shearing.
4. Numerical grade as determined by U.S.D.A. samples, just before shearing.
5. Kemp, black fibers, etc.

Fig. 3-18b. Individual Ewe or Ram Record Form (reverse side of Fig. 3-18a, p. 86).

Step 2: Wean and weigh—In advance, decide on weaning age—usually 90, 120, or 140 days. Wean and weigh as near to the intended age as possible.

Step 3: Type score—Even though the heritability of type, or conformation, is low at weaning (10%), it is a factor in determining today's market values; hence, all performance records should be augmented by type scores at weaning time. Also, and even more important because of the higher heritability (40%), all animals retained for breeding purposes should be type scored as yearlings.

Type can include any or all of the following: (1) characteristics that influence an animal's ability to live and perform in its environment—such as feet and legs, teeth, udder, and lethals and sublethals; (2) traits that indicate meatiness; and/or (3) breed type.

Also type score may include face cover score, wrinkle score, and record of the presence or absence of scurs or horns.

The determination of optimum type, the evaluation of it, and use made of it, should remain the responsibility of the individual breeder.

Step 4: Adjust for certain environmental factors—Adjust records for certain environmental factors such as age, sex, type of birth and rearing, and age of dam.

Weaning weights of lambs within a flock may be adjusted to 90, 120, or 140 days of age by finding the weight per day of age, then multiplying by the standardized age desired. Thus, the following formula may be used to provide the estimated 120-day weight of lambs:

$$\textbf{Adjusted 120-day weight (lb)} = \frac{\textbf{Actual weaning weight}}{\textbf{Actual days of age}} \times 120$$

Additional adjustment factors are given in Table 3–9.

TABLE 3–9
ADJUSTMENT FACTORS[1]

	Age of Dam		
	1 Year Old	2 Years Old or Over 6 Years Old	3 to 6 Years Old
Ewe Lamb			
Single	1.22	1.09	1.00
Twin—raised as twin	1.33	1.20	1.11
Twin—raised as single	1.28	1.14	1.05
Triplet—raised as triplet	1.46	1.33	1.22
Triplet—raised as twin	1.42	1.28	1.17
Triplet—raised as single	1.36	1.21	1.11
Wether			
Single	1.19	1.06	0.97
Twin—raised as twin	1.30	1.17	1.08
Twin—raised as single	1.25	1.11	1.02
Triplet—raised as triplet	1.43	1.30	1.19
Triplet—raised as twin	1.39	1.25	1.14
Triplet—raised as single	1.33	1.18	1.08
Ram Lamb			
Single	1.11	0.98	0.89
Twin—raised as twin	1.22	1.09	1.00
Twin—raised as single	1.17	1.03	0.94
Triplet—raised as triplet	1.35	1.22	1.11
Triplet—raised as twin	1.31	1.17	1.06
Triplet—raised as single	1.25	1.10	1.00

Multiply 90-, 120-, or 140-day weight by the appropriate factor.

Example: To find the adjusted 120-day weight of a twin-born and reared ram lamb from a 2-year-old ewe that weighed 90 lb at 110 days of age, make the following calculations:

$$\frac{90 \text{ lb}}{110 \text{ days (age)}} = .82 \text{ lb} \times 120 = 98 \text{ lb} \times 1.09 \text{ (adjustment factor)} = 107 \text{ lb}$$

The adjusted 120-day weight of the lamb would be 107 lb.

[1]Scott, G., *The Sheepman's Production Handbook*, 2nd ed., Sheep Industry Development Program, May 1975, p. 27.

Step 5: Cull—Cull the lambs that fail to measure up in the preweaning performance test.

POSTWEANING LAMB PERFORMANCE TEST

The postweaning lamb performance test may be conducted either on the farm or at a central test station, as follows:

Step 1: Postweaning feed test—Determine the growth rate by placing weaned lambs on a uniform feeding test of 42 to 90 days. Make selection on the basis of growth rate during the feeding test.

Step 2: Cull—Cull the lambs that were poor gainers in the postweaning lamb performance test.

EWE PRODUCTION TEST

Reproductive traits in ewes should be measured as part of the on-the-farm performance record scheme. For this purpose, the record form shown in Figs. 3–18a and b will suffice for most herds and flocks.

The following ewe performance test program is recommended for use in all purebred flocks:

Step 1: Record data—*Minimum data* should include (a) ewe number; (b) sire number; (c) dam number; (d) age of dam in years; (e) birth date of ewe; (f) type of birth of ewe—single, twin, triplet; (g) how reared—single, twin, triplet, artificially; (h) number of lambs born; (i) number of lambs weaned; and (j) adjusted weight of lambs weaned.

Step 2: Type score yearlings—At the yearling stage, type score and cull rigidly (type score as a yearling is 40% heritable).

Step 3: Evaluate fleece—Record the shearing date, grease fleece weight, staple length to nearest $\frac{1}{10}$ in., and fleece grade.

Step 4: Compute productivity for each ewe in the flock as—

a. *Lamb productivity per ewe,* in total adjusted weight of lambs produced. It is intended that this should put strong emphasis on twinning.

b. *Ewe combined productivity score,* by combining both lamb and wool as follows:

Ewe combined productivity score = Total pounds of lamb produced (using adjusted weights) + 3 times the 12-month wool total.

c. *Ewe productivity weight ratio or index,* to show the performance of an individual in relation to the average of all animals of the same group. Thus, the average productivity (total lamb and/or wool production) of the flock would be considered 100%. Individual ewes can then be rated based on their production above or below the flock average. Thus, a ewe producing 20% more lamb and/or wool than the average would have a productivity index of 120.

PROGENY TESTING RAMS

Progeny testing of rams can be very reliable if carefully planned and executed. For a valid progeny test, ewes must be assigned at random to the rams being tested, and sufficient ewes must be allotted to each ram to allow an accurate test. Progeny testing for rate of gain and carcass merit, for example, requires a minimum of 10 ewes per ram.

Progeny tests can be slow and expensive. Thus, if rams are tested as yearlings, they will be two-and-a-half years old when the test is complete. Of course, testing ram lambs would speed up the process. Fortunately, research shows that selecting sires on the basis of their own growth rate (selecting them on the basis of their own performance test) will result in more than half as much economic gain as selecting them on the basis of progeny test. Selecting on the mother's performance (number or weight of lambs weaned per year) may be even better because the generation length for rams may be reduced from 3 to 1. It would appear, therefore, that the added gain from progeny testing, in comparison with performance testing, is not sufficient to justify progeny testing of most rams that are to be used in natural service. If artificial insemination with frozen semen becomes practicable in sheep, progeny testing will become more important.

The following procedure is recommended for progeny testing rams:

Step 1: Record individual data—Record the following for each ram being progeny tested: (a) ram number and breed; (b) sire number; (c) dam number; (d) age of dam in years; (e) birth date of ram; (f) type of birth—single, twin, triplet; (g) how reared—single, twin, triplet; (h) lambs born from dam per year; and (i) adjusted 90-, 120-, or 140-day weight of lambs from dame per ewe year (weight per day of age to 200 days of age, or to 12 to 16 months of age, is desirable; but, of course, this won't be available where ram lambs are being tested).

Step 2: Postweaning feed test—Wean prospective stud rams at 60 to 90 days of age and place them on uniform feed test for 42 to 90 days. The heritability of growth rate is increased as maternal influence is decreased; hence, selection for growth should be based on the growth rate during the postweaning feeding test.

Step 3: Select top ram lambs—Select ram lambs that have excelled in both preweaning and postweaning performance.

If further testing—progeny testing—is not to be made, select future sires with high postweaning growth rate, that are well muscled and have a minimum of fat, and that are out of ewes producing multiple births with maximum growth rate.

Where elite stud rams are desired, they should be progeny tested—proceed to steps 4, 5, and 6.

Step 4: Mate to randomly chosen ewes—Mate each ram lamb to a minimum group of 10 randomly chosen ewes.

Step 5: Test lambs—Wean the lambs early (60 to 90 days), feed on uniform test for approximately 90 days, and slaughter.

Step 6: Use best rams as yearlings—The progeny test for gain and carcass merit can be computed in time to select and use the top progeny tested rams in their yearling breeding season.

Step 7: Evaluate fleece—Record the shearing date, grease fleece weight, staple length to nearest 1/10 in., and fleece grade.

Step 8: Type score—At the yearling stage, type score and cull rigidly (for type score as a yearling is 40% heritable).

Step 9: Cull—Cull those rams that fail to measure up in the progeny test.

PRODUCTION TESTING SWINE[10]

Fig. 3-19. The world's first swine production testing station, established in 1907 at Elsesminde, Denmark. (Courtesy, Danish Embassy, Washington, DC)

[10]This entire section was authoritatively reviewed by the following: D. D. Anderson, American Yorkshire Club, West Lafayette, IN; and Dr. C. J. Christians, Secretary Treasurer, National Swine Improvement Federation, University of Minnesota, St. Paul.

The effectiveness of swine selection can be increased, provided it is based upon carefully taken records rather than upon casual observation. Naturally, it would be illogical to expect upstanding, narrow-bodied, shallow sows and boars to beget meaty barrows that would be market toppers. Breeding animals of acceptable meat type can only transmit these qualities unfailingly to all their offspring when they themselves have been rendered relatively homozygous or pure for the necessary genes—a process that can be gradually accomplished through judgment by the eye method, but which can be made more rapid and certain through securing and intelligently using production records.

ECONOMICALLY IMPORTANT TRAITS AND THEIR HERITABILITY

That swine show variations in economically important traits is generally recognized. The problem is to measure these differences from the standpoint of discovering the most desirable genes and then increasing their concentration and, at the same time, to purge the herd of the less desirable characters.

Production testing begins with the birth of the litter. Females should be selected and culled continually during the growing period. Structurally sound, fast-growing gilts with reasonable fat cover should be chosen as potential breeding animals. At about 180 to 200 lb, these gilts should be weighed and probed for backfat thickness, with both values adjusted to a 230-lb basis. If desired, backfat thickness may be determined by use of the backfat probe or a scan device. Details relative to the 3 common methods of making backfat thickness determinations are given in a later section headed, "Meatiness; Measuring Backfat."

Except for backfat thickness, carcass traits can be measured and appraised only after slaughter. So, after selecting the replacements gilts, a representative group of the remaining animals not used for breeding purposes should be slaughtered, with carcass measurements made. The latter data should be considered when making final decisions in selecting herd replacements.

Table 3–10 lists the economically important traits in swine and gives their estimated heritability.

TABLE 3–10
ECONOMICALLY IMPORTANT TRAITS IN SWINE AND THEIR HERITABILITY[1]

Economically Important Characters	Approximate Heritability of Characters[2]	Comments
	(%)	
Litter size at birth	15	On the average, a sow will have consumed a total of ¾ to 1 ton of feed during the period between breeding and the date her litter is weaned. Thus, if this quantity of feed must be charged against a litter of 4 or 5 pigs, the chance of eventual profit is small.
Litter size at weaning	12	Although greatly influenced by herdsmanship, litter survival to weaning is a measure of the mothering ability of the sow.
Birth weight of pigs	5	Very light pigs usually lack vigor.
Litter weight at weaning	15	Weaning weight is important, for it has been shown that the pigs that are heaviest at weaning time reach market weight more quickly. The low heritability of this factor indicates that it is largely a function of the nursing ability of the sow rather than genetic.
Daily rate of gain from weaning to marketing	40	Daily rate of gain from weaning to marketing is important because (1) it is highly correlated with efficiency of gain, and (2) it makes for a shorter time in reaching market weight and condition, thus effecting a saving in labor, making for less exposure to risk and disease, and allowing for a more rapid turnover in capital.
Days to 230 lb	35	Rate of gain and fat deposition may be correlated to some degree. Thus, one should not let this be the only factor upon which selection is based.
Efficiency of feed utilization	30	Where convenient, accurate litter feed records should be kept, for the most profitable animals generally require less feed to make 100 lb of gain.
Conformation score	29	This heritability figure is likely to be considerably higher in a herd of low quality.
Carcass characteristics:		
1. Length	60	Carcass length is perhaps the most highly heritable trait in hogs. This accounts for the rapid shifts that frequently have been observed; for example, in changing from chuffy to rangy hogs.
2. Backfat thickness	40	The probe, lean meter, or ultrasonic equipment can be used to measure backfat thickness on prospective breeding animals.
3. Loin muscle area	50	Loin area is an indication of muscling or red meat.
4. Predicted pounds lean (predicted % lean)	58	The aim is to maximize the percentage lean in the carcass.
5. Percent lean cuts, based on carcass weight	50	A high yield of lean cuts means trimmable fat and more edible meat.

[1]These heritability estimates apply to within herd and within breed variations. Variations between breeds are much higher in heritability than the variations within.

[2]The rest is due to environment. The heritability figures given herein are averages based on large numbers; thus, some variation from these may be expected in individual herds.

RECORD FORMS

A prerequisite for any swine production test data is that each animal be positively identified—by means of ear notches. For purebred breeders, who must use a system of animal identification anyway, this does not constitute an additional detail. But the taking of weights and scoring does require additional time and labor—an expenditure which is highly worthwhile, however.

In order not to be burdensome, the record form should be relatively simple. Figs. 3-20 and 3-21 (next page) are recommended forms.

SWINE

Individual Sow Record

Breed ______ Name and registration no. ______

Date farrowed ______ Identification ______ (ear notch, tattoo)

Bred by ______ (Name and address)

Sow's pedigree: ______ (Sire) { ______ / ______ }

______ (Dam) { ______ / ______ }

Record of litter of which the sow was a member:

No. alive in litter ______ No. of pigs weaned ______

Weaning wt. at ______ (fill in) days of age:

Her own wt. ______ Avg. 21 day wt. of litter ______

Litter mate carcass record, if any:

No. carcasses ______; avg. back fat ______ (in.); loin eye ______ (sq in.); length ______ (in.)

Number of teats ______

Production Record of Sow

Litter no.	1	2	3	4	5	6	7	8
Sire								
No. services								
Farrowing data:								
Date								
Temperament of sow: (Gentle; nervous; cross)								
No. pigs born: Alive								
Dead								
Mummies								
Total								
Avg. birth weight								
No. functioning teats								
Weaning data: Age								
No. weaned								
Avg. 21 day weaning wt.								
Offspring saved for breeding: No. gilts								
No. boars								

Disposal of Sow

Date ______ Reasons ______

Sold to ______ (Name and address)

Price $ ______

Fig. 3-20. Individual Sow Record Form.

SWINE

Litter Record

Breed ______________________ Litter No. ______________________ (notch, tattoo)

Data on Dam:

Pedigree: ______________________ (name, reg. no., and ear notch) { ______________ (Sire) ______________ (Dam)

Birth date ______________________ (date and year)

Litter mate carcass data, if any:

No. carcasses __________; Avg. back fat __________ (in.); loin eye __________ (sq in.); length __________ (in.)

Sow's __________ (1st, 2nd, etc.) Litter.

Data on Sire:

Pedigree: ______________________ (name, reg. no., and ear notch) { ______________ (Sire) ______________ (Dam)

Birth date ______________________ (date and year)

Litter mate carcass data, if any:

No. carcasses __________; Avg. back fat __________ (in.); loin eye __________ (sq in.); length __________ (in.)

Date of birth ______________________

No. pigs born:

Alive ______________________

Dead ______________________

Mummies ______________________

Total ______________________

No. pigs weaned ______________

Health Services:

Date vaccinated ______________________

Date erysipelas vaccinated ______________________

Date wormed ______________________

Other, including iron pills or shots (list) ______________________

____

____

Individual Pig Record

Pig's No.	Sex	No. Teats	Birth Wt.	Off Color Markings	Defects & Abnormalities	Weaning Wt. _____ days (fill in)	Date Castrated	Date & Cause of Death	Days to 230 lb	Backfat Thickness	Disposal Date & To Whom	Remarks

Fig. 3-21. Litter Record Form.

MEATINESS; MEASURING BACKFAT

The importance of backfat as a measure of meatiness becomes apparent when it is realized that each additional 0.1 in. of fat in a 160-lb carcass results in a 1.5% (2 lb) increase in fat trim and a decrease of about 5% (7 lb) in percent lean cuts. This is reflected in the marketplace. Also, it requires less feed to produce a pound of lean gain than a pound of fat gain.

Thickness of backfat, which has a heritability of 50%, has long been recognized as an important measure of meatiness in hogs. For many years, visual appraisal was the only method of estimating backfat thickness on live animals. However, even the most skilled are oftentimes wrong in their visual measurements.

Today, mechanical methods are available and may be used by producers in determining backfat on live hogs; namely, the probe or scan devices. Each of these methods requires hog restraint. The probe may be used with the hog restrained by a snare.

The probe and scanning devices were developed for the purpose of obtaining objective measures of backfat.

• **Probing**—Fig. 3–22 shows the probing sites on the live hog. These three locations correspond to the three locations where backfat determinations are made on the carcass.

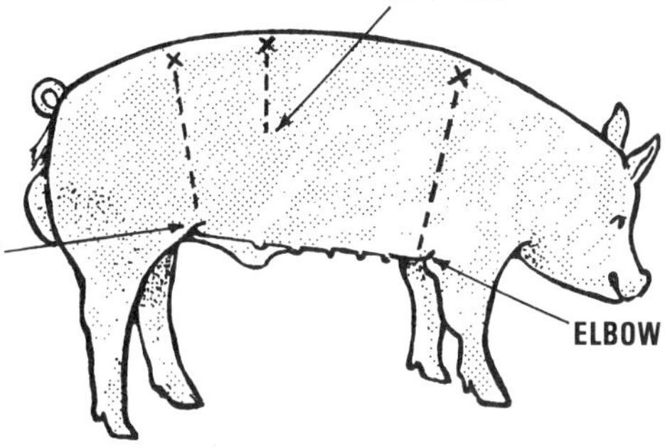

Fig. 3–22. Probe each hog at three locations: (1) midpoint of shoulder above elbow; (2) middle of back where last rib joins the vertebrae; and (3) rump, straight above the stifle joint.

The only equipment needed for probing is a snare to restrain the hog, a sharp knife or scalpel blade, and a narrow 6-in. metal ruler with $\frac{1}{10}$-in. gradations. The steps and techniques in probing are:

1. Wrap the knife or scalpel with tape about $\frac{3}{8}$ in. from the tip (in order to keep the blade from going too deep).
2. Restrain the hog with a nose snare.
3. Jab the knife through the skin at a right angle to the hog's body.
4. Insert the probe in the cut and slant it so that it points toward the center of the hog's body.
5. Force the probe through the fat down to the loin muscle. When the probe reaches the loin muscle, a firm resistance will be noted.
6. Push the clip on the probe down to the skin line. Remove the ruler and read the measurement.

The backfat thickness should be adjusted to a 230-lb basis.

Where pinpoint accuracy is not considered essential, measuring backfat at a single probe site is suggested. The recommended site where one probe only is taken is the last rib, located at a distance approximately four fingers wide behind the shoulder-probing location and 1 in. of the midline of the back.

• **Sonoray (ultrasonic)**—The sonoray machine employs the *pulse echo* technique. Basically, this is the generation of very short bursts of high-frequency (nondestructive, inaudible) sound into the animal, detecting the reflection of the pulses, and measuring the elapsed time between introduction of the sound pulse into the animal and the return of the reflected pulse. When the machine is properly calibrated, the fat depth can be read directly from the scale.

The sonoray backfat reading is made at the last rib, usually at the midline of the back. However, the reading may also be made 2 in. from the midline.

The sonoray may also be used to estimate loin eye area at the tenth rib.

Any one of these three mechanical methods for determining backfat thickness of the live animal is a valuable adjunct to visual appraisal and scales in the selection of meaty-type breeding animals and the production of high quality pork carcasses. Hence, backfat measurement should be used in the selection program of both purebreds and commercial producers.

GUIDELINES FOR UNIFORM SWINE IMPROVEMENT PROGRAMS

In swine, as in other animal species, there is need for uniform procedures for objectively measuring and recording swine performance data.

This need is being met by the National Swine Improvement Federation (NSIF), which, in cooperation with the Science and Education Administration (SEA-Extension and SEA-Agricultural Research), U.S. Department of Agriculture, and National Pork Producers Council, sponsored the development of the *Guidelines for Uniform Swine Improvement Programs,* 1988.

The *Guidelines* outline procedures for measuring and recording swine performance. They also strive to achieve greater uniformity of terminology and methods of measuring performance traits. This is important in accomplishing rapid and accurate communication to foster cooperation among all segments of the swine industry in compiling and using performance records.

Economic traits of swine include those that contribute to both productive efficiency and desirability of product. Growth rate, feed efficiency, reproductive efficiency, and carcass merit are the economic traits of greatest importance. Performance testing offers those engaged in swine production a way of measuring heritable differences among their animals in order to select parents that are expected to transmit their superior performance to their offspring.

PRODUCTION TESTING HORSES

The breeders of racehorses have always followed a program for mating animals of proved performance on the track. For example, it is noteworthy that the first breed register which appeared in 1791—known as "An Introduction to the General Stud Book"—recorded the pedigrees of all the Thoroughbred horses winning important races. In a similar way, the Standardbred horse—which is an American creation—takes its name from the fact that, in its early history, animals were required to trot a mile in 2 minutes and 30 seconds, or to pace a mile in 2 minutes and 25 seconds, before they could be considered as eligible for registry. The chief aim, therefore, of the early-day breeders of racehorses was to record the pedigree of outstanding performers rather than all members of the breed.

The simplest type of progeny testing in horses consists of the average record of merit of an individual stallion's or mare's offspring. Thus, the offspring of Thoroughbred or Standardbred animals bred for racing may be tested by timing on the track. Less satisfactory tests for saddle horses and harness horses have been devised. However, it is conceivable that actual exhibiting on the tanbark in the great horse shows of the country may be an acceptable criterion for saddle- and harness-bred animals. Also, the dynamometer might conceivably be used for testing animals of draft horse breeding, although it has not been so used in the past.

HERITABILITY OF PERFORMANCE

Relatively little scientific work has been done on the heritability of performance of horses—on the genetics of working ability, racing ability, cutting ability, jumping ability, etc. As a result, the horse is the last of the farm animals to which the science of genetics has been added to the art of breeding. Nevertheless, horse breeders have selected for performance. For example, the Thoroughbred horse has been selected and bred for speed and stamina for 300 years. Because more often than not the "best" horse wins, the breeding of the best to the best has resulted in improvement in the track performance of the Thoroughbred horse over the centuries.

The underlying genetic principle which determines the success of mating the best to the best is based upon the assumption that phenotypes of the best for a given trait, such as speed, are due to simple *additive*-type genes without regard to family relationships. On the other hand, when a breeder plans matings on the basis of a nick, family or pedigree relationships receive careful consideration.

The underlying genetic principle in making an outcross is that the members of the unrelated strains of families will bring together genes which will act in a complementary fashion to produce hybrid vigor in the offspring for the traits desired.

Differences in the performance ability (working, racing, cutting, jumping) are due to two major forces—heredity and environment. Success in selecting superior breeding animals for each of these traits depends entirely upon how accurately we are able to partition the differences in performance capacity of horses into causes due to the environment and causes due to heredity.

The important environmental factors in determining the overall performance of horses are nutrition (both prenatal and postnatal), health care, quality of training, ability of the rider or driver, and injuries.

An important genetic principle is that traits as such are not inherited. Rather, what is inherited is the ability to respond to a given set of environmental conditions in order to produce a trait with a measurable effect.

The key to continued genetic improvement in the performance of a horse, such as the racing capacity of the Thoroughbred, rests essentially on two factors: (1) the magnitude of the heritable component (additive genes) of performance (racing) capacity, and (2) the accuracy with which the breeder can identify those individuals which are truly genetically superior to their contemporaries.

Essentially, the breeding value of a horse is the fraction of the differences that will be transmitted to the progeny. The most straightforward measure of this is the heritability of the trait. It follows that an estimate of heritability of a trait is one of the most important considerations in formulating an effective program of improvement through breeding. Reliable estimates on the heritability of performance traits in horses are limited in comparison with other species. Nevertheless, further and important knowledge has been accumulated in recent years. Some heritability estimates follow:

- **Working ability**—In most countries, work horses, as distinct from light horses (sporting breeds), still make up the bulk of the population. In France, for example, only 15% of the horse population consists of the sporting breeds.

The main measure of the working ability in a horse is pulling power. This performance trait has been estimated to have a heritability of 26%.[11]

- **Racing ability**—Racing performance can be measured in different ways: by purses earned, time per unit distance, handicap weight, or *Timeform* ratings or other year-end handicaps. In a 1971 study, the Texas Agricultural Experiment Station[12] determined the racing ability of individual horses through a computer comparison of the number of lengths (one length = 8 ft) the horse would win or lose to other horses in a typical race. For this unique study, each horse was given a rating called the *Performance Rate*. Theoretically, it was assumed that the average horse would have a Performance Rate of zero. Then, in an average race, a horse with a Performance Rate of +12 would, theoretically, finish 12 lengths in front of the average horse. Likewise, a horse whose Performance Rate was −12 would, theoretically, finish 12 lengths behind the average horse and 24 lengths behind one with a +12 Performance Rate.

The Texas Station study included all 3-year-olds which raced on North American tracks in 1971. It involved 6,458 fillies and 7,113 colts and geldings, which were sired by 3,228 different stallions. Statistical analysis of the data showed that racing ability is about 40% heritable. This means that, on the average, about 40% of the difference in racing superiority of one horse

[11]Cunningham, Professor E. P., Head of Animal Breeding Genetics, Dublin University, Ireland. "Equine Genetics," *The Blood-Horse*, Oct. 6, 1975, p. 4210.

[12]Kieffer, Dr. N. M., Geneticist, Texas A & M University, College Station, "Heritability of Racing Ability," *The Blood-Horse*, Oct. 13, 1975, p. 4292.

over another is due to differences in heredity. The remaining 60% is due to difference in environment—nutrition, state of health, and abilities of trainers and jockeys.

So, after nearly three centuries of selection for speed and stamina, it should still be possible to improve the racing performance of Thoroughbred horses through the selection of superior stock for future parents.

• **Cutting ability**—Based on a study made by the Texas Station, the cutting ability of horses is less than 10% heritable. Obviously, training is most important in determining cutting ability.

• **Jumping ability**—Based on a study of steeplechase results in France, involving 3,500 progeny of 326 stallions, the heritability of jumping was estimated to be 15%.[13]

Although the heritability estimates of performance traits in horses reported above are disturbingly low, genes are a permanent, transmissable investment, whereas environmental factors are not. When buying horses, therefore, it is important to know whether you are buying desirable genes or superior environment.

RECORD FORMS

Figs. 3-23 and 3-24 show record forms that will be useful on most breeding establishments.

[13]Cunningham, Professor E. P., Head of Animal Breeding and Genetics, Dublin University, Ireland, "Equine Genetics," *The Blood-Horse,* Oct. 6, 1975, p. 4210.

HORSE
INDIVIDUAL LIFETIME BROODMARE RECORD

Name of mare ______________________

Number or other identity ______________________

Birth date ______________________

Show or performance record ______________________

____

____

Temperament ______________________
(gentle, nervous, cross)

PHOTO

Bred by ______________________
(name and address)

Purchased: from ______________________
(name and address)

Date ____________ Price ____________

Disposal: Sold to ______________________
(name and address)

Date ____________ Price ____________

Remarks ______________________

Production Record of Mares

Year	Sire of foal	Birth date of foal	Temperament of mare at foaling (gentle, nervous, cross)	Foaling (normal, requiring assistance, retained placenta)	Vigor of foal at birth (deformities)	Sex of foal	Identity of foal	Date foal was weaned	Score of foal: Suckling or Weanling	Score of foal: Yearling	Score of foal: 2-year-old	Score of foal: 3-year-old	Disposal of foal: Sold to: (name and address)	Date	Price	Reasons	Remarks

Fig. 3-23a. Individual Lifetime Broodmare Record. (See Fig. 3-23b for reverse side of record form.)

HEALTH RECORD

Date	Immunization: Encephalomyelitis	Immunization: Tetanus	Immunization: Abortion	Immunization:	Immunization:	Type of parasite treatment	Other veterinary treatment	Remarks

Fig. 3-23b. Individual Lifetime Broodmare Record (reverse side of Fig. 3-23a).

HORSE
INDIVIDUAL YEARLY STALLION BREEDING RECORD

Name of stallion ______________________

Number or other identity ______________________

Birth date ______________________

Show or performance record ______________________

____

____

PHOTO

For breeding year of ______________________

For foaling year of ______________________

Total number of services ______________________

No. services/conception ______________________

Mares in Foal to Stallion

Name of Mare	Dates mare was bred				Date foaled	Vigor of foal at birth	Sex of foal	Disposal of foal				Remarks
								Sold to: (name and address)	Date	Price	Reasons	

Fig. 3-24a. Individual Yearly Stallion Breeding Record. (See Fig. 3-24b for reverse side of record form.)

HEALTH RECORD

Date	Immunization							Type of parasite treatment	Semen test	Veterinary treatment	Remarks
	Encephalomyelitis	Tetanus	Other								

Fig. 3-24b. Individual Yearly Stallion Breeding Record (reverse side of Fig. 3-24a).

CROSSBREEDING

Crossbreeding is the mating of animals of different breeds. In a broad sense, crossbreeding also includes the mating of purebred sires of one breed with high-grade females of another breed.

Today, there is great interest in crossbreeding, and increased research is under way on the subject. Crossbreeding is being used by livestock producers to (1) increase productivity over straightbreds because of the resulting hybrid vigor or heterosis, just as is being done by commercial corn and poultry producers; (2) produce commercial animals with a desired combination of traits not available in any one breed; and (3) produce foundation stock for developing new breeds.

The motivating forces back of increased crossbreeding in farm animals are (1) more artificial insemination, thereby simplifying the rotation of sires of different breeds; and (2) the necessity for producers to become more efficient in order to meet their competition, both from within their respective industries and from without.

Crossbreeding will play an increasing role in the production of market animals in the future, because it offers the several advantages discussed in the sections which follow.

HYBRID VIGOR OR HETEROSIS

Heterosis, or hybrid vigor, is the name given to the biological phenomenon which causes crossbreds to outproduce the average of their parents. For numerous traits, the performance of the cross is superior to the average of the parental breeds. This phenomenon has been well known for years and has been used in many breeding programs. The production of hybrid seed corn by developing inbred lines and then crossing them is probably the most important attempt by scientists to take advantage of hybrid vigor. Also, heterosis is being used extensively in commercial swine, sheep, layer and broiler production today; an estimated 80% of market hogs, market lambs, and layers are crossbreds, and 95% of broilers are crosses.

The genetic explanation for the hybrid's extra vigor is basically the same, whether it be cattle, hogs, sheep, layers, broilers, hybrid corn, hybrid sorghum, or whatnot. Heterosis is produced by the fact that the dominant gene of a parent is usually more favorable than its recessive partner. When the genetic groups differ in the frequency of genes they have and dominance exists, then heterosis will be produced.

Heterosis is measured by the amount the crossbred offspring exceed the average of the two parent breeds or inbred lines for a particular trait, using the following formula for any one trait:

$$\frac{\text{Crossbred average} \ \text{(minus)} \ \text{Purebred average}}{\text{Purebred average}} \times 100 = \text{Percent hybrid vigor}$$

Thus, if the average of the 2 parent populations for weaning weight of calves at 205 days of age is 400 lb and the average of their crossbred offspring is 420 lb, application of the above formula shows that the amount of heterosis is 20 lb, or 5%.

Traits high in heritability—like tenderness of rib eye in cattle—respond consistently to selection, but show little response in hybrid vigor. Traits low in heritability—like mothering ability, calving interval, and conception rate—usually show good response in hybrid vigor.

The level of hybrid vigor for all traits depends on the breeds crossed. The greater the genetic difference between two breeds, the greater the hybrid vigor expected. The genetic difference between a British breed (*Bos taurus*) and a breed of Indian origin (*Bos indicus*) is greater than the difference between one British breed and another British breed.

COMPLEMENTARY

Complementary refers to the advantage of a cross over another cross or over a purebred, resulting from the manner in which two or more characters combine or complement each other. It is a matching of breeds so that they complement each other, the objective being to get the desirable traits of each. Thus, in a crossbreeding program, breeds that complement each other should be selected, thereby maximizing the desirable traits and minimizing the undesirable traits. Since breeds which are selected because they tend to express a maximum of some trait (e.g., high daily gain) will have some undesirable traits (e.g., large mature cow size and high maintenance cost, different breeds must be selected for different purposes. A well-known example of breed complementation for improving overall carcass desirability in the market animal is the Angus × Charolais cross, combining the higher carcass grade of the Angus with the higher cutability of the Charolais.

INTRODUCE NEW GENES QUICKLY

Crossbreeding provides a way in which to introduce new and desired genes quickly—at a faster rate than can be achieved by selection within a breed. A practical example of this sort is the introduction of new genes for milk production in a beef herd by crossing a dairy breed with a beef breed, then selecting females from within the crossbred foundation for the future cow herd.

GET HYBRID VIGOR EXPRESSED IN THE FEMALE

Except for a two-breed cross, crossbreeding offers an opportunity to have hybrid vigor expressed in breeding females. This is most important in the cow herd where it results in increased fertility, survivability of the calves, milk production, growth rate of calves, and longevity of the cows—all factors that mean more profit to the cattle producer.

CROSSBREEDING BEEF CATTLE

Fig. 3-25. Hereford cows with black baldie calves—Hereford × Angus crossbreds. (Courtesy, Oklahoma State University, Stillwater)

Without a planned breeding program, crossbreeding will almost inevitably end up with (1) a motley collection of females and progeny varying in type and color, and (2) minimum benefits from hybrid vigor or heterosis.

"Where do I go from here?" This is the question that many cattle breeders frequently ask, almost frantically, after having heifers of breeding age sired by exotic bulls. Others get worried when they notice that calves out of their crossbred cows aren't doing so well as the first-cross calves. Of course, what these breeders really want to know is how they can maintain satisfactory hybrid vigor (heterosis) in animals when a herd is on a continuous crossbreeding program. They want to know how they can maintain 15 to 25% greater total efficiency in the crossbreds than the average of their parents; in production rate, calf livability, growth rate, and feed conversion.

Several different systems of crossbreeding may be used. Among them are the following:

1. **Two-breed cross.** This consists of mating purebred bulls to purebred or high-grade cows of another breed. An example would be using Angus bulls on Hereford cows, to give crossbred Angus × Hereford offspring—black baldies. This system of crossing has been used with success by cattle breeders for many years.

In the 2-breed cross, only the calves are crossbred—the breeding of the sires and dams remains the same. Hence, the 2-breed cross imparts hybrid vigor only in the calf. On the average, it gives about an 8 to 10% increase in pounds of calf weaned per cow bred, plus another 2 to 3% advantage in rate of gain in the feedlot. In order to follow the 2-breed cross indefinitely, the purebred females must be replaced with other purebreds sooner or later. They may either be purchased from other breeders, or breeders may want to produce their own purebred heifers within their own herds.

The 2-breed cross is relatively simple. However, it has one major deficiency; it does not make use of the crossbred cow.

2. **Two-breed backcross or crisscross.** This system involves the use of bulls of breed A on cows of breed B, then backcrossing the progeny of bulls to either breed A or B. The rotation is accomplished by using bulls of the breed least related to the particular set of cows. For example, if Charolais bulls are mated to Hereford cows, the crossbred Charolais × Hereford heifers could be retained and bred to either a Charolais or a Hereford bull. If Hereford bulls were used, the calves produced would be ¼ Charolais and ¾ Hereford. Later, if the heifers of this breeding are saved, they should be bred to a Charolais bull. The 2-breed backcross results in about 67% of the maximum heterosis being attained in the crossbred calves. But since crossbred cows are used, overall performance should be a little better in pounds of calf weaned per cow bred than in the 2-breed cross.

TWO-BREED BACKCROSS OR CRISSCROSS

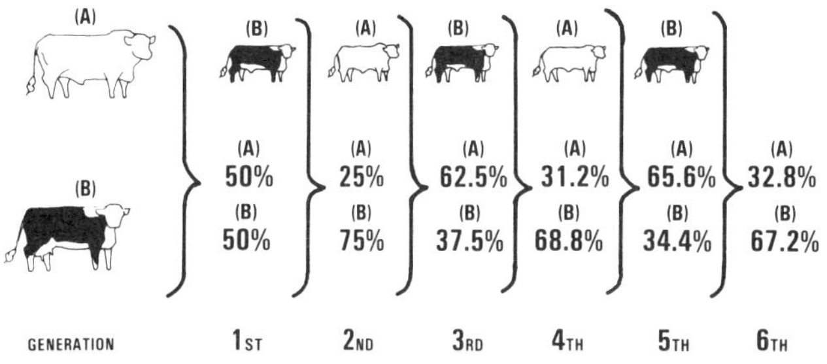

Fig. 3-26. Two-breed backcross or crisscross: Charolais bull × Hereford cow, thence female offspring bred to Hereford bull, thence female offspring bred to Charolais bull.

3. **Three-breed rotation cross.** This system calls for the selection of 3 breeds (e.g., breeds A, B, and C, which might represent Herefords, Brahmans, and Charolais), possessing the combination of maternal, growth, and carcass traits desired in the crossbred cows and the slaughter cattle produced. Crossbred females, selected for growth rate, are retained for breeding and bred to a purebred bull of one of the 3 breeds. Each new generation of crossbred females is retained for breeding and mated to a purebred bull until bulls of all 3 breeds have been used in rotation. Thus, such a system would operate as follows:

Mate the existing B cow herd continuously to bulls of breed A; select crossbred heifers for growth rate and mate them continuously to bulls of breed C; mate the selected C (AB) females to bulls of breed B. After the rotation of bulls from the 3 breeds is completed, the rotation of purebred sires begins all over again. Thus, mate the selected B × (ABC) females to bulls of breed A.

Continue the same system indefinitely, always selecting the best performing crossbred females to be mated to the breed of sire in the program to which they are least related.

In addition to the genetic advantages of this system, commercial producers select their own replacements; hence, the only outside cattle purchases are production tested bulls. The major disadvantage is that after the first four years it is necessary to maintain bulls of all three breeds simultaneously (unless AI is used).

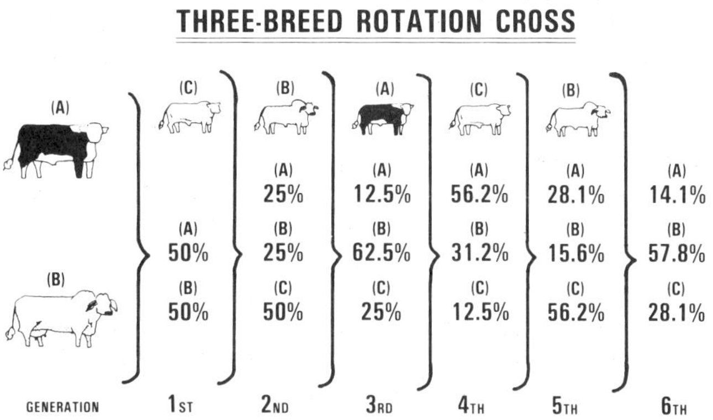

Fig. 3-27 Three-breed rotation cross; Hereford bull × Brahman female, thence female offspring bred to Charolais bull, thence female offspring bred to Brahman bull.

A three-way rotation system results in about 87% of the maximum heterosis being attained.

4. **Three-breed fixed or static cross (terminal cross).** In this system, crossbred cows from a 2-breed cross (F_1s) are used as females and are mated to a bull of a third breed. All offspring from this cross are sold. When replacement females are needed, crossbred cows are purchased. In this way, crossbred calves with a fixed percentage of inheritance from three breeds are always produced.

In addition to realizing 100% of the maximum heterosis of each calf crop, this system allows the selection of maternal breeds to go into the production of the crossbred female and the selection of growthy breeds having desirable carcasses for the terminal cross sire breed. It allows the breeds to be used for their strong points without regard to some of their weaker points. A breeder can tailor-make the crossbred market animal, putting together in one animal desirable traits of several breeds. Such specification is not possible in the rotational system because all breeds contribute to maternal performance and calf performance.

The mechanics of this system consist in selecting three breeds for crossbreeding—two breeds (A and B) that will produce crossbred cows with outstanding maternal characteristics for fertility, milking ability, mothering ability, and adaptation. Select a third breed (breed C) with rapid, efficient, postweaning muscle growth rate. Breed C would be considered a terminal sire breed. All crossbred progeny of bull C are marketed for slaughter.

The problem with this system is the acquisition of production tested, crossbred (F_1) heifers for replacements in such a program, since all the three-way crosses are marketed. The system is perpetuated by having specialized multipliers produce crossbred (F_1) replacement females. Small operators (those with under 100 cows) might well use such a system where heifers are purchased along with the bulls. Large operators might well produce their own F_1 heifers in a specialized portion of their herd. Purebred breeders (seed stock breeders) would supply production selected terminal sires which the commercial producer would purchase for such a program.

Four or more breeds may be used in a rotation crossbreeding system if the commercial producer so desires. However, the maximum hybrid vigor is usually realized with the 3-breed cross.

Also, it is noteworthy that all of these crossbreeding systems rely upon the use of purebred bulls. Additionally, the 2-breed cross relies on the use of purebred females and the 3-breed fixed or static cross relies on purebred females to produce F_1 heifers necessary for the program.

Before going into a long-range crossbreeding program, the owner should know what is involved and what to expect. Plans should be developed before committing all available cattle and resources to a crossbreeding program. Consideration should be given to size of herd, markets, number of pastures, natural vs AI breeding, availability of breeding stock, etc.

Crossbreeding is no magic or cure-all, but it will give a powerful assist to the pocketbook if properly used. Also—and this point bears emphasis—sound management and sound selection of breeding stock based on performance, potential carcass characteristics, and overall productivity are just as important in crossbreeding as in any other breeding program.

All crossbreeding programs involve some animal identification system so that (1) growth rate of heifers may be determined, with selection of replacements made on this basis; and (2) where more than one breed is involved, the cow herd can be sorted for assignment to specific sire breeds for mating. Unless AI is used, it is necessary to maintain bulls of whatever breeds are involved, along with separate breeding pastures for each sire breed. Also, it should be recognized that where bulls of two or three different breeds are used, the crossbred slaughter progeny will vary considerably in performance and carcass traits, because they will be sired by bulls of different breeds and be produced by cows of divergent breed backgrounds.

CROSSBREEDING DAIRY CATTLE

The crossing of dairy breeds produces hybrid vigor or heterosis, just as it does in meat animals. It results in more live calves, improved livability, and increased early growth. Yet, crossbreeding for heterosis has been little used in dairy cattle breeding programs in the United States. This prompts the question: Why has crossbreeding been more widely used in beef cattle, sheep, and swine than in dairy cattle? The answer: There is one great difference between a dairy cow and females of meat species. A dairy cow is both a breeding animal and a producing animal—she must produce both calves and milk. In meat animals, however, it is practical to keep females primarily for breeding purposes—to produce feeder calves, feeder lambs, or feeder pigs. Thus, it is expected that most U.S. dairy producers will continue to rely upon selection based on genetic variation to make genetic improvements in their herds.

CROSSBREEDING SHEEP

Fig. 3-28. Rambouillet × Lincoln crossbred yearling rams.

Although the common systems used in breeding sheep are not unlike those applied to other classes of farm animals, there appears to be more crossbreeding among sheep because of (1) the fact that sheep are called upon to produce two products, lambs and wool; (2) the many diverse conditions under which they are produced; and (3) the conviction on the part of many sheep producers that the hybrid vigor of crossbreeding accounts for increased vigor and livability in the lamb crop. Crossbreeding, therefore, is extensively followed in commercial sheep production on the western range. The ewe bands are predominantly of Rambouillet extraction; whereas, for market lamb production, Suffolk or Hampshire rams are generally used. The Rambouillet ewe bands are desired because of their (1) gregarious or flocking instinct, (2) great hardiness, and (3) superior shearing qualities. On the other hand, lambs of this breeding are not so desirable for market lambs. Thus, mutton-type rams are used in order to get large, fast growing lambs that will attain a good market finish on milk and range vegetation or that can be readily sold to go into feedlots for further finishing. As black-faced crossbred lambs of this type are not suitable as flock replacements, both ewe and wether lambs are marketed. Replacement females are obtained by (1) outright purchase from a breeder who has used white-faced rams (Rambouillets, Columbias, Targhees, or Panamas) for purposes of raising animals for sale as replacements; (2) using white-faced rams on the band every third year and retaining the ewe lambs (some breeders with several bands simply use certain bands for producing lambs for replacement purposes); or (3) using both white-faced and black-faced rams simultaneously on the same ewe band. In the last type of program, the better white-faced ewe lambs—which are easily recognized as the offspring of the white-faced rams—are selected out for breeding purposes.

As can be readily surmised, crossbreeding in sheep does make for a considerable problem from the standpoint of producing or purchasing replacement animals. Also, it often makes the ram problem a difficult one. This practice, however, was born of necessity, there being few or no existing breeds or types possessing all the desirable features needed. In recent years, considerable effort has been made toward developing breeds of sheep better adapted to the needs, with the hope of alleviating the necessity of crossbreeding. The Columbia, Targhee, and Panama breeds evolved out of this need.

In addition to the crossbreeding common to the western range, most hothouse lambs are produced through using this system of breeding. Usually grade Merino or Dorset ewes are topped with a Southdown ram. Ewes of this extraction will breed out of season, and they are excellent milkers; and Southdown rams impart to their progeny the ultimate in early maturity and mutton type. Crossbreeding has also gained in popularity in Kentucky where crossbred Hampshire-Rambouillet ewes are frequently bred to Southdown rams for the production of grass-fat lambs.

Studies by the U.S. Department of Agriculture show that purebred ewes crossed with purebred rams of another breed raised 2 more lambs per 100 ewes than purebred ewes bred to rams of the same breed. Also, the lambs averaged 6 lb heavier at weaning.

From the above cross, the first cross ewe lambs bred to purebred rams of a third breed raised 14 more lambs per 100 ewes that were 10 lb heavier at weaning than those of the purebred breeds in the cross.

The ewe lambs from the above (containing blood of 3 breeds) crossed with a purebred ram of a fourth breed raised 27 more lambs per 100 ewes that were 7 lb heavier at weaning than those of the purebred breeds in the cross.

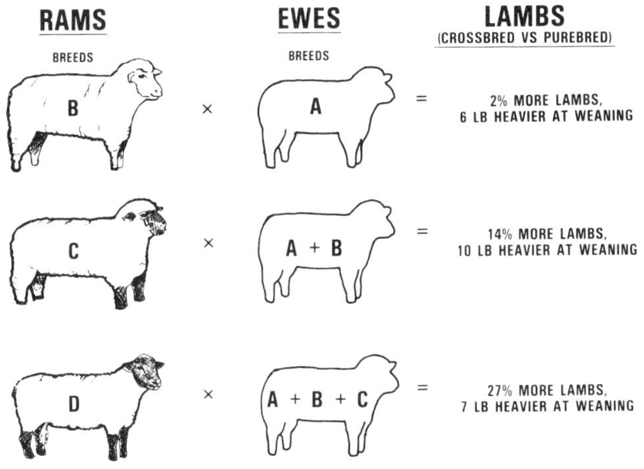

Fig. 3-29. What the USDA sheep crossbreeding studies showed.

In addition to more and heavier lambs per ewe, crossbreeding enables the commercial sheep raiser to benefit from the desirable characteristics of one breed—such as long life, flocking instinct, and wool production—and excellent body conformation and rapid growth of another breed.

Thus, systematic crossbreeding as a mating system for sheep production will continue to provide heterosis, or hybrid vigor, for some very important traits with which sheep producers are concerned. For example, it appears that an adapted crossbred ewe is more fertile than a straightbred ewe. But if such crossbred ewes are used and mated to rams of an unrelated breed for maximum commercial production, a minimum of three breeds must be involved. The breeds that are crossed to produce the crossbred ewe (the F_1) that will be retained as the producer in the flock should possess those characteristics which contribute to making a highly productive female—such as reproductive efficiency, milk production, maternal instincts, and wool quantity and quality. In turn, these crossbred ewes should be bred to one of the ram breeds. The ram breeds would be mated to the crossbred ewes as a terminal cross, to produce market lambs. Such ram breeds should possess growthiness, carcass quality, sexual aggressiveness, and fertility. An example of a crossbreeding program would be:

1. **Foundation ewe breeds.** Rambouillet, Merino, Columbia, Corriedale, Targhee. Two of these breeds to be selected and crossed to produce F_1 females.

2. **F_1 females.** The F_1 females resulting from the above cross to be bred to Suffolk or Hampshire rams as a terminal cross. All lambs to be marketed.

CROSSBREEDING SWINE

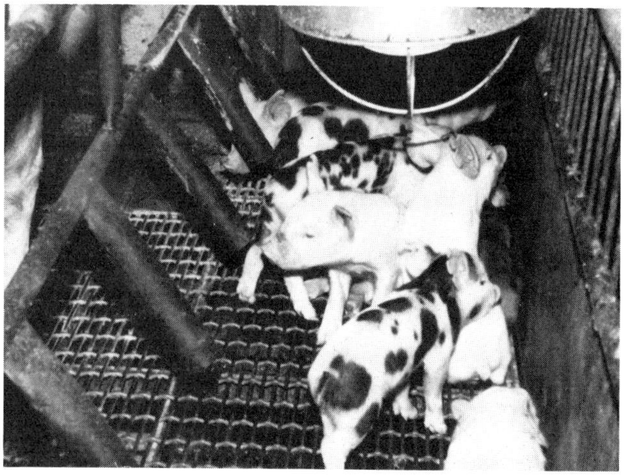

Fig. 3-30. Crossbred pigs at 1 week of age. (Courtesy, Dr. E. J. Stevermer, Iowa State University, Ames)

Crossbreeding has been widely used in swine because it makes for increased production and profit. At this time, about 80 to 85% of all commercial hogs are crossbreds.

The mating of individuals from different breeds produces heterosis or hybrid vigor, a condition in which the offspring are superior in certain traits to the average of their parents. In a swine crossbreeding program, heterosis results in larger litters farrowed and weaned, stronger pigs at birth, and faster growth. However, heterosis gives little or no improvement in feed efficiency and carcass merit. The latter traits must be obtained through the selection of parents.

In order to obtain the full benefits of crossbreeding, the producer should follow a systematic program in selecting crossbred replacement gilts and purebred boars. Superior crossbred gilts should be mated to rapid-growing, muscular boars from families with proven performance and carcass quality.

The three common crossbreeding systems followed in swine are:

1. **Two-breed cross.** This consists in mating a purebred boar to purebred or high-grade sows of another breed; for example, a Yorkshire boar and Hampshire sows.

Where this system is held to first crosses only, the breeder is faced with the problems of sooner or later breeding the females back to a purebred boar of the same breed in order to secure replacement females. Under these conditions, the breeder is prone to make little or no selection and to keep all of the females for replacement purposes. In such a program, it is usually found that the producer does well to maintain the quality of the female herd.

2. **Two-breed backcross or crisscross.**—This system, which uses 2 breeds alternately, is recommended where good individuals of only 2 breeds are available. Boars of 2 different breeds (breeds that complement each other) are used in alternate generations. Selected crossbred gilts, produced by mating sows of breed A to a boar of breed B, are bred back to a boar belonging to one of the parent breeds (boar of breed A, for example). Then, selected offspring from this mating are next bred back to a boar of breed B. Crossbred gilts and sows are always mated to the boar of the breed farthest away in their pedigree.

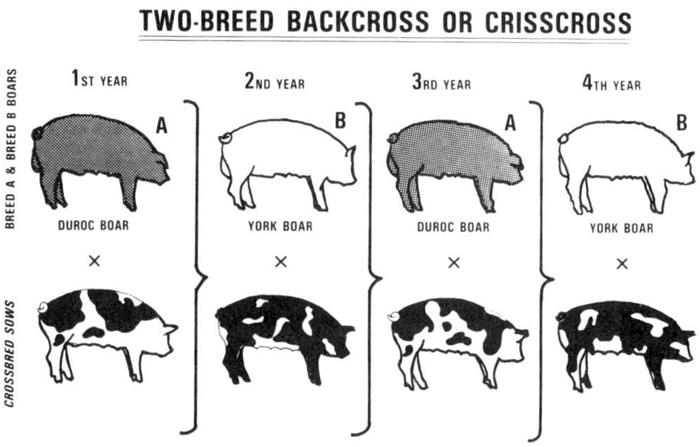

Fig. 3-31. A two-breed backcross or crisscross using only two breeds of boars in alternate generations. Crossbred gilts are retained and mated to boars of the same breed as the grandsire of the dam's side.

3. **Three-breed rotation cross.** The 3-breed rotation cross is perhaps the most widely used system of crossbreeding in swine. In this system, first cross gilts are mated to a boar of a third breed. The program is then continued through rotating the sires among the three breeds.

Some swine producers follow a 4-breed, or even a 5-breed system of crossbreeding. It is noteworthy, however, that the optimum amount of hybrid vigor is attained with the 3-breed cross, and that a 4-breed or 5-breed cross merely retains that level in later generations. Contrary to the belief of some swine producers, using a 3-breed or 4-breed rotation as described above does not cause a decline in the level of heterosis after several generations *provided* purebred boars are always used.

Before starting a crossbreeding program, the swine producer should know what is involved and what to expect. Also, the producer should realize that sound management and the selection of superior breeding stock on performance, potential carcass characteristics, and overall productivity are requisites for success. Moreover, the breeds used should complement each other.

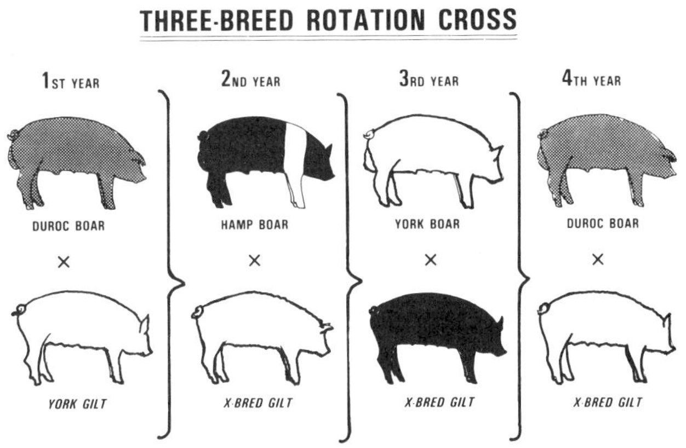

Fig. 3-32. A three-breed rotation cross, using Duroc, Hampshire, and Yorkshire breeds.

CROSSBREEDING HORSES

The greatest use of equine crossbreeding in this country was in the production of mules, prior to the mechanization of American agriculture. The mule, representing a cross between a jack (male of the ass family) and the mare (female of the horse family), is the best-known hybrid in the United States. The resulting offspring of the reciprocal cross of the stallion mated to a jennet is known as a hinny.

Horse breeders have crossed certain breeds to obtain desired colors or color patterns. Arabians are sometimes crossed on other breeds to produce horses with great endurance and stamina. The infusion of some coldblood (draft breeding) is sometimes relied upon to secure hunters of greater size—of the weight-carrying variety.

According to the rules of the American Horse Shows Association, harness show ponies may be of any breed or combination of breeds; the only requisite is that they must be under 12–2 hands in height. Three breeds produce animals that qualify under this category—Shetland, Welsh, and Hackney. Heavy harness ponies are, as the name indicates, miniature heavy harness horses—they're under 14–2 hands. Generally, they are either purebred Hackneys, or predominantly of Hackney breeding.

Although crossbreeding of horses has a place for the purposes indicated above, purebreeding will continue to control the destiny of further improvement in horses and furnish the desired homozygosity and uniformity which many horse breeders insist is a part of the art of producing better horses.

SYSTEMS OF SELECTION

Hand in hand with the breeding system—with production testing, or crossbreeding—the breeder needs to follow a system of selection which will result in maximum total progress over a period of several years or animal generations. The three common systems are:

1. **Tandem selection.**—This refers to that system in which there is selection for only one trait at a time until the desired improvement in that particular trait is reached, following which selection is made for another trait, etc. This system makes it possible to achieve rapid improvement in the trait for which selection is being practiced, but it has two major disadvantages: (a) usually it is not possible to select for one trait only, and (b) generally income is dependent on several traits.

Tandem selection is recommended only in those rare herds and flocks where one character only is primarily in need of improvement; for example, where a certain flock of fine-wool sheep needs improving primarily in staple length.

2. **Establishing minimum standards for each character, and selecting simultaneously but independently for each character.** This system, in which several of the most important characters are selected for simultaneously, is without doubt the most common system of selection. It involves establishing minimum standards for each character and culling animals which fall below these standards. For example, it might be decided to cull all beef calves having a gain ratio under 110. Of course, the minimum standards may have to vary from year to year if environmental factors change markedly (for example, if calves average light at weaning time due to a severe drought and poor pasture).

The chief weakness of this system is that an individual may be culled because of being faulty in one character only, even though it is well nigh ideal otherwise.

3. **Selection index.** Selection indexes combine all important traits into one overall value or index. Theoretically, a selection index provides a more desirable way in which to select for several traits than either (a) the tandem method, or (b) the method of establishing minimum standards for each character and selecting simultaneously but independently for each character.

Selection indexes are designed to accomplish the following:

a. To give emphasis to the different traits in keeping with their relative importance; for example, to give emphasis to the mutton and wool qualities of sheep in keeping with the economic importance of each.

b. To balance the strong points against the weak points of each animal.

c. To obtain an overall total score for each animal, following which all animals can be ranked from best to poorest.

d. To assure a constant and objective degree of emphasis on each trait being considered, without any shifting of ideals from year to year.

e. To provide a convenient way in which to correct for environmental effects, such as type of birth (single or twin), age of dam, etc.

Despite their acknowledged virtues, selection indexes are not perfect. Among their weaknesses are the following:

a. Their use may result in covering up or masking certain bad faults or defects, when they are overbalanced by strong points.

b. They may not adequately allow for year to year differences as genetic changes may be confounded or confused with environmental changes.

c. Their accuracy is dependent upon (1) the correct evaluation of the net worth of the economic traits considered, (2) the correctness of the estimate of heritability of the traits, and (3) the genetic correlation between the traits; and these estimates are often difficult to make.

In practice, a well designed selection index, including items measuring both performance and quality or market type, will serve as a good guide in making selections. It should, however, be supplemented with careful observation on individual animals in order to eliminate those which have severe defects not adequately covered in the index. Caution should be used, however, in departing from the index unless there is overwhelming reason for doing so.

HEAT DETECTION METHODS AND DEVICES

The problem of heat detection becomes more important as herds get larger, good hired help is more difficult to come by, cows produce more milk, and animal value increases.

Under ordinary farm conditions, caretakers miss an estimated 25 to 50% of the heat periods. On the average, a missed heat period prolongs the calving interval by 30 to 40 days and means a loss of more than $40 in a dairy herd and $20 in a beef herd. Some owners pay their employees a bonus for catching a cow in heat. For these reasons, cattle breeders are interested in heat detection methods. Among them are the following:

1. **Chin-Ball Marker.** This device was developed in New Zealand. It is similar to a ball-point pen attached to a halter under the chin of a surgically modified teaser bull, often called a *Gomer.* (One of the first ranches in North America to use the Chin-Ball Marker gave this name to the bull on which it was used.) During preservice sex play, it is usual for a bull to place his head over the shoulders, back, and rump of the cow. This causes a smearing of the colored ink from the ball-point onto the cow.

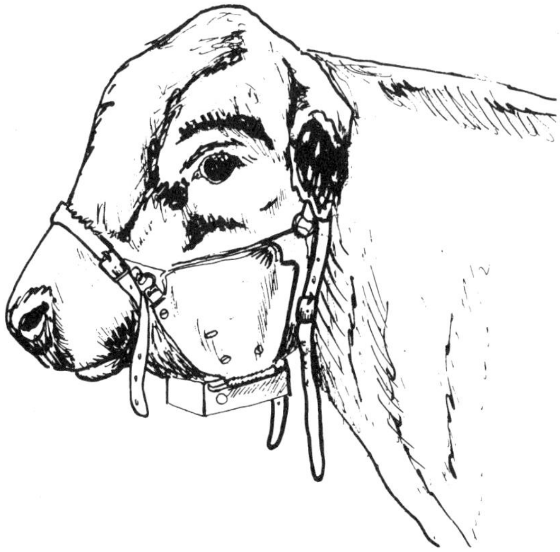

Fig. 3-33. Chin-Ball marking device. (Courtesy, American Breeders Service, De Forest, WI)

One filling of the stainless steel container is sufficient to mark 15 to 20 cows. Experience indicates that one Gomer bull can work approximately 80 cows. In large pastures and in larger sized herds, it is best to have two bulls.

This method of heat detection is a most dependable management tool.

2. **Heat-Mount Detector.** The heat-mount detector is a 2 × 4½ in. fabric base to which is attached a white plastic capsule. Inside the capsule is a small plastic tube containing red dye. The tube is constructed so the dye is released slowly by moderate pressure. When enough dye is released from the tube (after about 4 to 5 seconds of pressure), it spreads over the inner lining of the capsule, causing it to turn red.

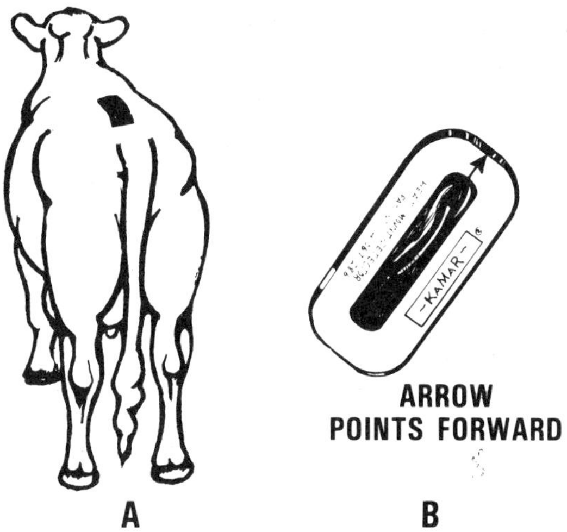

Fig. 3-34. Heat-Mount Detector. "A" shows the device in place on a cow's back. "B" is an enlarged top view of the device before activation.

The detector relies on the natural bovine instinct of "bulling" or mounting during estrus. The pressure from the brisket of a mounting animal causes the dye to be released and the detector to turn red. If the cow does not stand for the mounting animal, there will not be enough pressure to release the dye and turn the detector red. This device has resulted in catching 95% of the heat periods.

3. **Pen-O-Block.** The Pen-O-Block is a plastic tube placed within the bull's sheath and held in place with a stainless steel pin. The bull can detect cows in heat and mount them in a normal way, but the device mechanically prevents him from making contact with the cow.

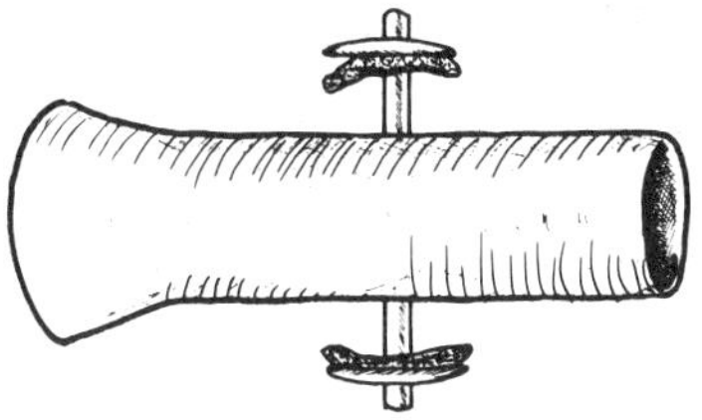

Fig. 3-35. Pen-O-Block mating device. (Courtesy, American Breeders, De Forest, WI)

The Pen-O-Block consists of a white plastic tube, the pin or cannula, two washers, and a cotter pin. The device is inserted within the bull's sheath and held in place by the cannula. The procedure is best carried out by a veterinarian, as it requires skill.

The **advantages** of the Pen-O-Block are (a) that venereal diseases cannot be transmitted; and (b) that the Pen-O-Block can be removed anytime, following which the bull can be used in natural service. The **disadvantage** is that the bull may lose his sex drive.

4. **Vasectomized bull; deviated penis and sheath.** The vasectomized bull has been used as a heat detector for many years. The vasectomy operation, which should be performed by a veterinarian, involves removing part of the vas deferens from the spermatic cord. The **advantages** of the vasectomized bull are: the blood and nerve supply of the spermatic cord are left intact; the testicles and penis function normally, but the transport of spermatozoa to the urethra is blocked; and the sexual activity of the bull remains unaltered. The **disadvantage** is that the bull continues to copulate; hence, he can transmit venereal diseases from one cow to another.

The deviated penis and sheath results from a surgical technique that transplants the penis and sheath from their normal position to the folds of the flank. The **advantage** to this method is that it permits normal erection, but does not allow copulation. The **disadvantage** is that the bull may become frustrated and lose his sex drive.

5. **Other heat detection methods and devices.** Other methods that are used to detect cows in heat, with varying degrees of success are: (a) twice daily turnout of cows that are continually confined to stalls, (b) hormone-treated steers or females, (c) a colored paste or chalk on the tail head or rump of open cows, and (d) electronic monitoring.

Properly used, these aids will improve heat detection. They are by no means replacements for visual heat detection; nor will they solve all the problems in breeding a beef or dairy herd artificially.

- **Heat detection of sheep**—The acceptance of the ram, or of a teaser with an apron, is the best method of detection. Also, breeders commonly keep a breeding record of ewes in heat (a) by using a marking harness (breeding harness) on the ram, for which different colored crayons are available; or (2) by smearing the breast of the ram and the area between his forelegs with a colored paste.

- **Heat detection of swine**—Sows that are in heat can be detected by their mounting of other sows, frequent urination, and occasional loud grunting.

- **Heat detection of horses**—The most reliable detection of mares in heat consists of trying them with a stallion or teaser.

ESTROUS CYCLE MANIPULATION

Fig. 3-36. When the estrous cycle is manipulated properly, females produce a more uniform group of offspring. (Courtesy, The Upjohn Co., Kalamazoo, MI)

Planned parenthood is not new. It has long been practiced among females of all species, women included. Livestock producers have long tampered with the breeding and parturition season that was common in the wild state. Prior to domestication, animals brought forth their young in the fields and glens, inhibited only by age and feed, and influenced somewhat by seasons. But breeders changed all this—even without the use of hormones. Sly controls have been exercised over breeding for a very long time. For example, farm flock owners controlled reproduction in chickens by the simple act of putting eggs under an old setting hen—unless she hid out. Today, modern poultry producers regulate chicken hatchings by controlling when, and how many, eggs go into the incubator. It is more difficult to accomplish the same thing in four-footed animals.

The motivating reasons back of estrous cycle manipulation vary somewhat by species. Horse breeders, especially those who race or show, want their mares to foal as soon after January 1 as possible, because a horse's age is computed on a January 1 basis, regardless of how late in the year it may have been born. Sheep and swine breeders strive for two crops of offspring per year, and for multiple births. Purebred cattle breeders who show, plan their breeding programs to take maximum advantage of show classifications; commercial cattle producers are concerned with weather and feed supply; and dairy producers want the largest flow of milk at a time when the product is likely to bring the highest price. Also, cattle producers recognize that controlled estrus would greatly facilitate both artificial insemination and ova transplantation.

Livestock breeders have altered nature's way in farm animals (a) by confining the male at certain times, or hand mating; (2) by emulating spring conditions—through providing better feed, shelter, and/or blankets when breeding at other times of the year; (3) by flushing—through feeding females more liberally 2 to 3 weeks ahead of the breeding season; and (4) by artificially controlling the hours of light per day.

ARTIFICIAL LIGHTING

The number of hours of light in the day affects the initiation of the normal breeding season of ewes and mares, both of which are seasonal breeders. It is noteworthy, too, that the reproductive function in poultry and migratory fowl is regulated by the length of daylight.

The ratio of hours of daylight to darkness through the year acts on the nerves in the region of the pituitary gland, and stimulates or inhibits the release of the follicle-stimulating hormone (FSH). Lengthening the daylight hours activates the pituitary, and causes it to release increasing amounts of FSH which stimulates ovarian function. Thus, some time after the daylight period increases, the estrous cycle begins in both ewes and mares.

Artificial lighting will accomplish the same thing as daylight; hence, it may be used to alter the estrous cycle in both ewes and mares.

- **Sheep**—Normally, ewes come in heat during the late summer or early fall, though there is both an area and a breed difference. The breeding season is usually restricted to about four months.

Ewes generally begin cycling when the number of daylight hours drops below 14. This is the reason that most breeds of sheep come into heat during the fall months. To initiate estrus, however, it appears that the shorter days must be preceded by longer days.

- **Horses**—Normally, the natural breeding season of mares begins in March and extends to late July or August.

Artificial lighting of broodmares enables breeders to bring mares in season about 6 weeks earlier than normal. By the use of the artificial light technique, a mare that would normally conceive on March 15 may get in foal some time in January. By avoiding the necessity of skipping a year due to late breeding, this technique may actually result in obtaining two additional foals during the lifetime of the mare.

The procedure consists of using a 200-watt light bulb in a box stall so as to extend the hours of light to 16 hours daily. Where a large band of mares is involved, corral lighting may be used. A general rule is that if a newspaper can be easily read in any corner of the stall or corral, the lighting is sufficient. By beginning the light treatment about December 1, they may be bred the latter part of January. Slight adjustments in the schedule will need to be made in different locations, depending upon the sunrise and sunset times of the particular area.

HORMONAL CONTROL OF HEAT

Many different drugs have been administered (either orally, by injection, or by implantation) in an attempt to control the estrous cycle of females, with progestagens, prostaglandins, and human chorionic gonadotropin heading the list. These products are available under different trade names, such as Syncro-Mate B (SMB).

Researchers in both colleges and industries are in general agreement that hormone controlled estrus synchronization will work, and that it offers promise of good returns when properly used in a well-managed herd. Scientists also realize that we do not know all the answers—that further research work is necessary. Nevertheless, it appears that planned parenthood is here to stay—that its wide use only awaits getting the technique perfected and lowering costs, both of which will come. In the meantime, livestock breeders are admonished to keep abreast of developments and to rely on well-informed advisers.

INDUCED CALVING (SHORTENED GESTATION)

Instead of "letting nature take its course," scientists are now artificially shortening gestation. The objectives of induced early calving are: (1) lowering birth weight of calves, thereby lessening parturition difficulty; (2) predicting calving dates in order to pool labor and concentrate watching; and (3) gaining a longer period from calving until rebreeding.

ARTIFICIAL INSEMINATION

Artificial insemination is, by definition, *the deposition of spermatozoa in the female genitalia by artificial rather than by natural means.* Legend has it that this method had its origin in 1322, at which time an Arab chieftain used artificial methods to impregnate a prized mare with semen stealthily collected by night from the sheath of a beautiful stallion belonging to an enemy tribe. However, the first scientific research relative to the artificial insemination of domestic animals was conducted with dogs by the Italian physiologist, Lazarro Spallanzani, in 1780.

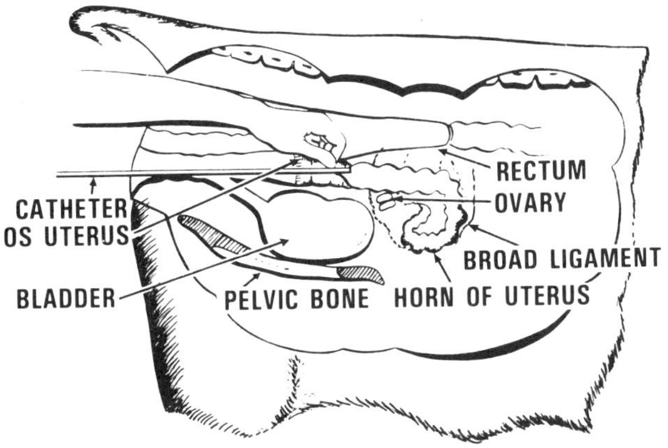

Fig. 3-37. Insemination of the cow. The cervix is grasped *per rectum* and the inseminating tube is carefully worked into and through the cervical canal. (Courtesy, Dr. H. A. Herman, Secretary/Treasurer, Association of Missouri Dairy Organizations, Colombia, MO)

Artificial insemination has been practiced successfully in the United States for more than 40 years. In 1988, more than 65% of the nation's dairy cattle were artificially inseminated. Also, as beef cattle breeders achieve increased genetic progress through artificial insemination, AI of beef cattle will increase.

ADVANTAGES OF ARTIFICIAL INSEMINATION

Some of the **advantages** of artificial insemination are:

1. It increases the use of outstanding sires.
2. It alleviates the danger and bother of keeping a sire.
3. It makes it possible to overcome some physical handicaps of mating.
4. It makes it possible to use a sire that is not alive at the time.
5. It lessens sire costs.
6. It reduces the likelihood of costly delays through using sterile sires.
7. It helps to control disease.
8. It makes it feasible to prove more sires.
9. It creates large families of animals.
10. It increases pride of ownership.
11. It increases profits.

LIMITATIONS OF ARTIFICIAL INSEMINATION

Like many other wonderful techniques, artificial insemination is not without its limitations. A full understanding of such limitations, however, will merely accentuate and extend its usefulness. Some of the **limitations** of artificial insemination are:

1. It must conform to physiological principals.
2. It requires skilled technicians.
3. It necessitates considerable capital to initiate and operate an artificial insemination organization.
4. It may accentuate the damage of a poor sire.
5. It may restrict the sire market.
6. It may increase the spread of disease.
7. It may be subject to certain abuses.

Of course, with skilled workers performing the techniques required in artificial insemination, there usually is more check on the operations and perhaps less likelihood of dishonesty than when only the owner is involved, such as is usually the situation with natural service.

COLLECTION AND HANDLING OF SEMEN

The method of collecting semen should be reasonably adapted to the males of the species, should be easy for the operator to use, and should permit the collection of a sample of normally ejaculated semen free from contamination with dirt, bacteria, or excess secretions.

There are four recognized methods of collecting semen: (1) using an artificial vagina; (2) using an electro-ejaculator or electrical stimulation (a necessity for crippled bulls that are unable to mount); (3) massaging the accessory genital organs (seminal vesicles and ampulla) per rectum; and (4) collecting the semen from the vagina that has just been bred by natural service.

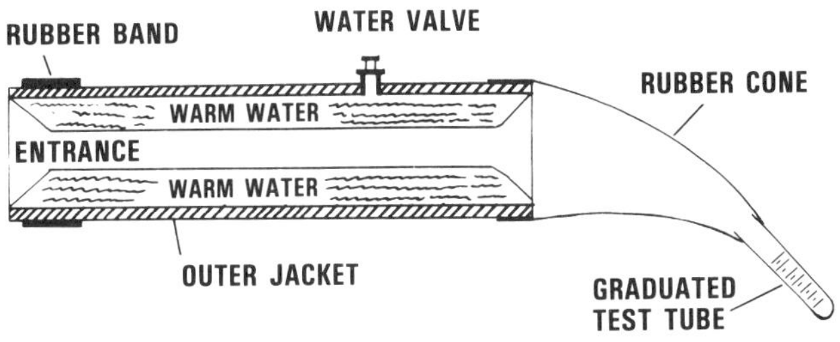

Fig. 3-38. Diagrammatic artificial vagina for the collection of bull semen. (Drawing by R. F. Johnson)

VOLUME OF SEMEN/CONCENTRATION OF SPERM

The volume of semen ejaculated at one service and the concentration of sperm vary according to species and individuals. Table 3-11 gives average figures by classes of farm animals.

TABLE 3-11
SEMEN VOLUME AND SPERM CONCENTRATION OF FARM ANIMALS

Class of Animal	Avg. Volume of Semen per Ejaculate	Avg. Concentration of Sperm	No. Females That Can Be Bred per Ejaculate
	(ml)	(millions/ml)	
Bull	5-6	800-1,200	300-500
Ram	1	800-4,000	40-100
Boar	200-300	25-1,000	15-25
Stallion	50-150	30-800	8-12

As a rule, the smaller ejaculates of high sperm concentration, such as found in bull and ram semen, are the most suitable for artificial insemination; for they can be diluted into a large volume with special dilutes that retain the lift of the sperm to a higher degree than the natural seminal fluid. It is possible, however, to fractionate boar ejaculate during collection so as to obtain the sperm-rich portion apart from the large volume of aspermic seminal fluids.

In cattle, sperm numbers are of more importance than semen volume. In swine, however, any volume below 50 ml seems to decrease conception rate.

SEMEN EXTENDERS

Addition of extenders to freshly collected semen is almost imperative because (1) they provide needed volume; and (2) they exert a beneficial effect on the sperm.

A recent survey showed that about half of the bull studs in America use the yolk-citrate extender and most of the remainder use boiled milk. Other extenders are (1) glucose plus tartrate, sulfate, or phosphate salts (developed by Milovanov, a Russian); (2) modified Ringer solution (developed by Bonnier and Trulsson, which is considered the best of the extenders for rooster semen); (3) skim milk; (4) dry skim milk powder; (5) homogenized whole milk; (6) synthetic pabulum (developed by Phillips and Spitzer); (7) Krebs solution (developed by Lardy and Phillips, for boar semen); (8) IVT (developed by Illinois); (9) Tris buffered extenders; (10) saline solution extenders (for short-time durations), (11) Caprogen; and (12) certain commercially prepared extenders.

STORAGE AND HANDLING OF SEMEN

Fig. 3–39. Transferring semen from shipping tank to regular storage tank. (Photo by T. O. and J. Jaggers, Leitchfield, KY)

Semen may be used as (1) fresh liquid semen, or (2) frozen semen.

• **Fresh liquid semen**—If semen is to be used within 1 or 2 hours following collection, it may be kept at room temperature. For longer storage (1 to 4 days) as liquid semen, and delayed used, it should be properly diluted and gradually cooled (avoiding temperature shock) and stored at a temperature of 35 to 40°F.

Fresh semen is extensively used in the artificial insemination of sheep and swine. Since fresh stallion semen does not store well beyond 24 hours and boar semen not beyond 48 hours, shipment of these is impractical at the present time. It is noteworthy that the greatest use of liquid semen stored at ambient temperature is made in New Zealand. Liquid semen is used to inseminate most of the cattle on AI because the main breeding season is short. Of the approximately 1 million head of cows inseminated annually, 90% are bred from mid-September to December, during the lush grazing season of their spring months. New Zealand workers have developed an ambient temperature storage extender by gassing the citrate-buffered egg yolk extender with nitrogen and adding capronic acid. The extender is called *caprogen.*

• **Frozen semen**—In 1949, British scientists reported that the addition of glycerine to semen diluters permitted them to freeze certain semen at temperatures much below zero (they used dry ice to freeze at a temperature of –79°C or –110°F), and still retain a high degree of fertility following thawing.

Today, semen is being frozen for storage in liquid nitrogen at –196°C, or –320°F. Liquid nitrogen, which is the fourth coldest known substance, is the universally used refrigerant in the United States because uniform temperatures may be maintained for long periods of time and the method is more convenient for shipping and storing frozen semen. Dry ice at –101°F isn't cold enough, and mechanical refrigerators aren't reliable.

Frozen semen is currently being used successfully for cattle, goats, and horses; but, to date, it has not been perfected for sheep and swine.

Frozen bull semen has been stored for more than 30 years and conception obtained. There is a small reduction in fertility of semen stored long periods, but it is usual to find semen stored one year or longer in routine use.

Frozen semen, refrigerated with liquid nitrogen, can be shipped to all parts of the world. The thawing of the semen is accomplished by placing the vial in a container of water containing thawing ice (34–38°F) immediately prior to insemination.

Frozen semen is potentially the most valuable breeding technique yet known. Through it, the following may be achieved:

1. The usefulness of outstanding sires can be extended far beyond their lifetime; also, it ensures the proven sire should he die.

2. Outstanding sires can be used, nationwide and worldwide.

3. A multiherd progeny test can be completed at a much earlier age.

4. A stock of semen can be built up while waiting for progeny record assessment.

5. Long-term storage of semen lessens semen wastage, and facilitates long-distance transport.

6. Semen from valuable sires may be fully utilized.

7. A herd owner can usually obtain the sire of choice at any time.

• **Custom freezing**—On a fee basis, many established bull studs collect, freeze, process, store, and dispense semen from bulls owned by private breeders.

Such semen is generally used to service cattle belonging to the owner of the sire. More than likely, this practice will increase in the future.

PACKAGING SEMEN

The world trend is toward packaging and freezing semen in concentrated form. Among the methods are ampules, pellets, straws, shell-freezing, and lyophilization of semen.

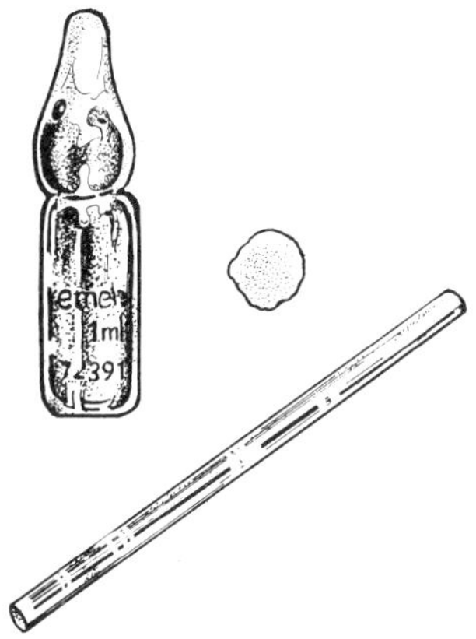

Fig. 3-40. The three most common methods of packaging and storing frozen semen (left to right): ampule, pellet, and straw.

• **Ampule**—This is a glass container with enough semen for only one service. Usually, each ampule contains 0.5 to 1 ml of diluted semen. In comparison with straws, ampules take much more time to thaw. Also, in going (1) from ampule, (2) to breeding tube, (3) to cow, there is loss of some sperm in semen which clings to the ampule and breeding tube.

• **Pellet**—The pelleting of semen was developed by the Japanese, in 1962. Pellets are thawed in physiological saline (0.9% NaCl). Pellets are a concentrated method of storing semen, but they present problems in identification, automation, and sanitation.

• **Straws**—Plastic straws range from about 2½ to 5 in. in length and commonly hold about 0.25 to 0.5 ml of semen. In comparison with glass ampules, plastic straws require less storage space, yield a higher recovery rate of motile sperm when thawed, and result in fewer sperm being lost in the insemination process (e.g., fewer sperm adhere to the container). Today, 95% of all units of frozen semen marketed in the United States are in straws.

• **Shell-freezing**—This method was developed by Graham of Minnesota in 1966. It results in a higher percentage of live sperm than other methods of packaging; the dead and abnormal sperm are removed by passing the semen through a glass-fiber filter.

• **Lyophilization of semen**—The goal in the use of this technique is the preservation of semen in the dry state at low temperature. Progress along these lines is encouraging, but the process is still in the experimental stage.

WHEN TO BREED

A female is fertile only when an egg is present which can be fertilized. Moreover, an egg can live for only a short time after being shed from an ovary unless it is fertilized. The optimal time for insemination is in advance of the time of ovulation, which varies according to species.

A cow doesn't shed her egg from the ovary until about 10 hours after the close of standing heat, and the egg lives only 6 to 10 hours. Thus, for optimal results from insemination, cows should be bred in the latter two-thirds of heat or within a few hours after having gone out of heat; roughly, this means that a cow should be bred within a 24-hour period after she is first noticed in standing heat (see Fig. 3-41). To accomplish this, it is recommended that cows first observed in standing heat in the morning be bred during the afternoon or evening of the same day, and that those observed in heat in the evening be bred the next morning.

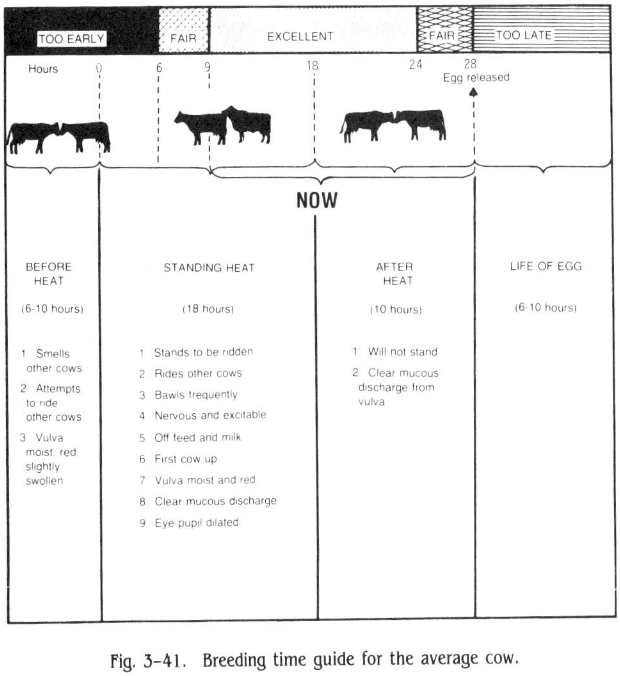

Fig. 3-41. Breeding time guide for the average cow.

Mares normally ovulate 24 to 48 hours before the termination of the heat period, ewes near the end of heat, and sows about 30 to 40 hours after the beginning of the heat period. Thus, a mare with a 5-day heat period (4 to 6 days is considered normal for a mare) should be inseminated on the third day and at least every other day thereafter for the duration of the heat. Ewes should be inseminated during the last day of heat, and sows 12 to 24 hours after the beginning of heat. In sows, a second service 12 to 24 hours later increases conception and litter size.

INSEMINATION OF THE FEMALE

Cleanliness during all insemination manipulations is essential and is a crucial point for success or failure. This applies to the instruments used, to the hands of the operator, and to the animals.

The inseminating tube should be passed just through the cervix, stopping at the location where the cervix ends and the uterus begins. Precisely at this point the semen should be deposited—slowly (see Fig. 3-42). The most desirable region of deposition has not yet been fully established for all farm animals. In sows, the uterus would be the place of deposition, as it is in natural service.

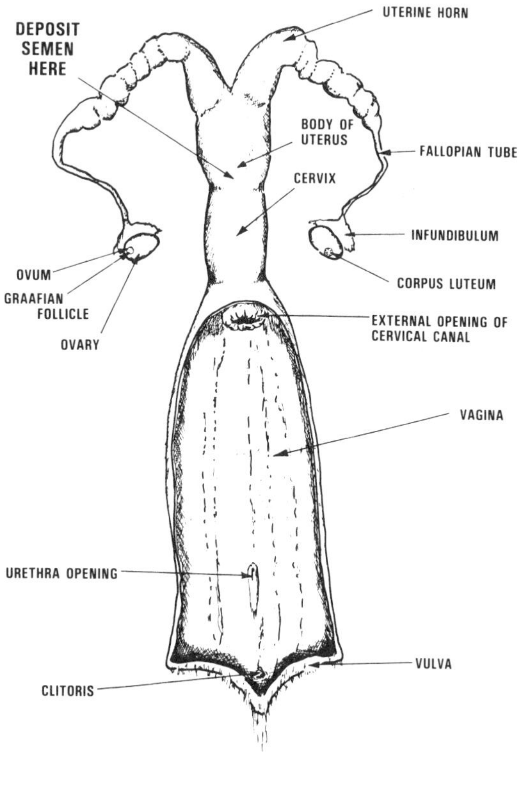

Fig. 3-42. Where to deposit semen.

The amount of extended semen required varies with the species and the individual concerned. Sows are usually inseminated with 50 to 100 ml of extended sperm, mares 20 to 30, cows 0.5 to 1.5, and ewes 0.1 to 0.2.

PRACTICAL APPLICATION

Today, artificial insemination is more extensively practiced with dairy cattle and beef cattle than with any other class of farm animals. In 1938, when the program first began in America, only 7, 359 cows and 646 herds were bred by this means in organized groups in the United States. During 1988, a total of 17,965 dairy units of semen, and a total of 1,025 beef units of beef semen were marketed in the United States. *Note well*: Currently, most reporting is in units of semen, not in number of cows bred. (It takes 1½ to 2 units per cow.)

In 1988, artificial insemination programs involved 65% of milking cattle and more than 90% of the dairy herds in the United States.

In the future, there will be increases in the use of AI in beef cattle, and to a lesser extent in sheep, swine, horses, and milk goats (a special section devoted to each of the first four classes of farm animals follows).

Based on present knowledge, gained through research and practical observation, it may be concluded that livestock producers can make artificial insemination more successful through the following:

1. Give the female a reasonable rest following parturition and before rebreeding; in cows this should be 50 to 60 days, and in sows 35 to 49 days.
2. Keep records of heat periods and note irregularities.
3. Watch carefully for heat signs, especially at the approximate time.
4. Where an association is involved, notify the insemination technician promptly when an animal comes in heat.
5. Avoid breeding diseased females or females showing cloudy mucus. The latter condition indicates an infection somewhere in the reproductive tract.
6. Have the veterinarian examine females that have been bred three times without conception or that show other reproductive abnormalities.
7. Follow a proper nutrition program at all times.

BULL COSTS VS AI COSTS

With the increase in artificial insemination in recent years, a frequently asked question is: What's the cost of AI vs natural service?

Many cost figures seen in various publications show bull costs by natural service ranging from \$6 to \$12 per cow. The Nebraska Station reported that the cost of keeping a bull is much higher than this, even if the bull is depreciated over a 4-year period (see Table 3-12).

TABLE 3-12
ANNUAL COSTS OF OWNING AND MAINTAINING A BULL[1]

	1981-82 Prices			
	If Depreciated Over 2 Years		If Depreciated Over 4 Years	
	Total Costs	Direct Costs	Total Costs	Direct Costs
Hay, 1.5 tons	$ 75.00	$ 75.00	$ 75.00	$ 75.00
Winter pasture	25.00	—	25.00	—
Summer pasture	108.00	—	108.00	—
Salt and mineral	4.00	4.00	4.00	4.00
Veterinary and medicine	14.00	14.00	14.00	14.00
Death loss	6.00	6.00	6.00	6.00
Depreciation ($1,500 purchase price, $840 selling price)	330.00	330.00	165.00	165.00
Interest on bull (12%)	140.00	140.00	140.00	140.00
Interest on feed and operating expense (12%)	6.00	6.00	6.00	6.00
Labor (10 hr)	55.00	—	55.00	—
Use of buildings and equipment	9.00	3.00	9.00	3.00
Miscellaneous expense	6.00	6.00	6.00	6.00
Total	$778.20	$584.00	$613.00	$419.00
30 cows	26.00	19.50	20.50	14.00

[1]Estimates submitted by NE Beef and Ag Economics Specialists: Guyer & Jose.

As shown in Table 3-12, one of the largest bull cost items is depreciation. These figures are based on a purchase price of $1,500 and a selling price of $840. On this basis, bull depreciation alone would amount to more than $10 per cow. Direct costs, excluding pasture, amount to $13.97 per cow (depending on how long the bull is kept), and $14.87 to $21.27 per cow if the charge for summer pasture is included.

Artificial insemination eliminates the expense and problems associated with maintaining bulls. But of course, it involves some different expenses, primarily for the cost of semen and the added labor.

Based on a study of 37 commercial ranches in Wyoming, the Wyoming Experiment Station reported $1.87 greater cost per calf from AI than from natural service.[14] However, the AI-sired calves had a $7.05 per head greater value than calves sired by natural service, leaving $5.18 net per calf in favor of AI. This comparison included all charges related to breeding the cow herds.

SUPEROVULATION

Superovulation consists of injecting the female with drugs which cause the larger follicles (each of which contains one oocyte, or egg) to mature, rupture, and release the egg.

Since the commercial employment of superovulation is limited primarily to cattle at the present time, the discussion that follows will pertain to this class of animals. However, with proper adaptation, the same principles and techniques will work with females of all species.

[14]Stevens, D. M. and T. Mohr, *Artificial Insemination of Range Cattle in Wyoming: An Economic Analysis,* Wyo. Bull. 496, 1969.

The bull is capable of producing from several thousand to millions of sperm daily whereas the cow normally produces one ovum (occasionally two ova) every 17 to 21 days. Now it is possible, through the administration of hormones, to obtain several ova (5 to 50) from a cow at one estrous cycle. It is also feasible to obtain a larger number of eggs from very young heifer calves, by injection of hormones.

Eggs which are shed from the ovaries are stored in large follicles. The basic principle of superovulation is to stimulate extensive follicular development through the use of a hormone preparation, given intramuscularly or subcutaneously, with follicle-stimulating hormone (FSH) activity. The most common sources of such a hormone are pregnant mares' serum (PMSG) and FSH extracts from pituitaries of slaughtered animals. Many animals so treated will come into estrus about five days after initiation of treatment and ovulate, through release of their own luteinizing hormone (LH). However, to help assure that multiple ovulations occur, the ovulating LH from pituitaries or from chronic gonadotropin (HCG) is injected. The multiple ovulations occur at about the same time the cow would have normally ovulated one egg (21 days after the previous ovulation).

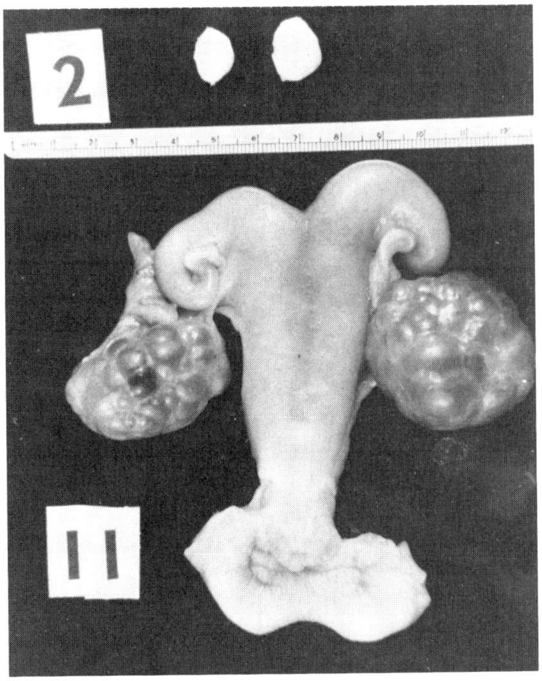

Fig. 3-43. Superovulation of the calf, **Calf No. 2**—Ovaries of a control 4-month-old calf. The ovaries contain thousands of dormant Graafian follicles. **Calf No. 11**—Ovaries and genital organs of a superovulated 4-month-old calf. The calf was not showing any signs of heat, since it had not reached sexual maturity. However, it was injected with 10 units of the gonadotropic hormone Vetrophin on each of 3 successive days and slaughtered 5 days after the last injection. The 2 ovaries contained 97 ripe follicles. Note the size of the superovulated ovaries in relation to the immature uterus (Courtesy, Washington State University, Pullman)

Studies have shown that FSH should be administered twice daily over a period of 4 to 5 days. PMSG has a longer biological life than FSH and a single subcutaneous injection is normally used. Then, 5 or 6 days after the original FSH or PMSG shot, LH or HCG is given intravenously.

The heifers should ovulate by the seventh day after starting hormone treatment.

Since ovulation occurs over a period of time, not all the eggs are fertilized unless the donor is inseminated repeatedly. A yield of four or five good fertilized eggs per donor is about average.

Of course, the real economic value of superovulation lies in the successful transfer of excess eggs from more valuable donor cows to less valuable recipient cows.

EMBRYO TRANSFER

Embryo transfer is the placing of an embryo into the lumen of the oviduct or uterus.

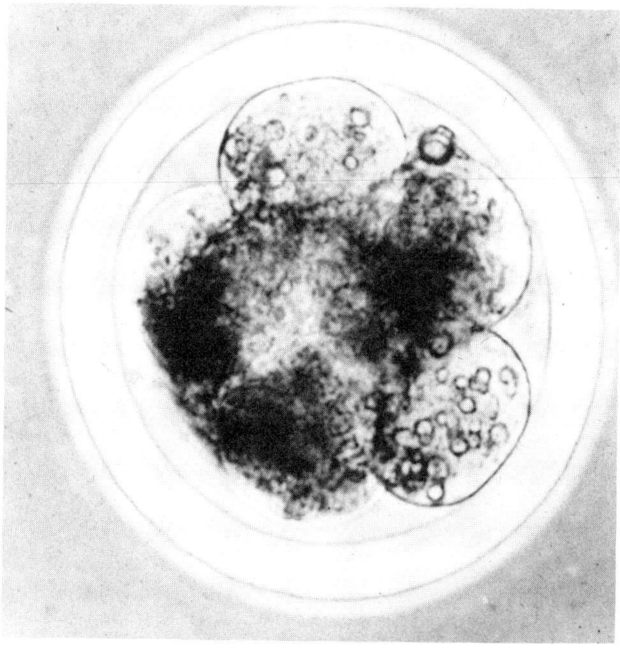

Fig. 3-44. Fertilized ovum in dividing stage, ready to be taken from donor cow's oviduct and transferred to recipient cow. (Courtesy, A. H. J. Rajamannan, International Cryo-Biological Services, Inc., St. Paul, MN)

Artificial insemination has given a means for the widespread distribution of desirable genes via the sperm. Similar genetic selection through high-quality females has, however, been limited since, normally, one cow will produce one calf per year and the average number of offspring per female will seldom exceed five in a lifetime. Out of the latter arose the idea that a marked increase in the production of offspring from desirable cows might be effected by superovulation, followed by transfer of the fertilized ova to less desirable cows, with the latter serving as host-mothers or foster-mothers to the developing embryo.

Embryo transfer in cattle was developed as a result of research done by Jim Rowson at Cambridge, England, in the early 1950s. The earliest work was done with sheep, then cattle and pigs.

The first commercial embryo transfers in the United States were done in the early 1970s. Initially embryos were recovered from valuable donors and transferred into recipient animals, using surgical procedures. During the mid-1970s, nonsurgical methods for the recovery of embryos were developed. In the late 1970s, nonsurgical transfers grew in popularity.

A major factor responsible for the growth of embryo transfer is the success that has been achieved in producing offspring from valuable donor cows. Many donors have produced more than 20 calves, and a few more than 50 calves, within a year. On the average, 8 good embryos are recovered from normal donors. However, about ⅓ of all recoveries result in fewer than 3 embryos, and about ⅓ of them produce 8 or more embryos. It is noteworthy, too, that commercial embryo transfer services report up to 65% embryo transfer conception rates.

Embryo transfer is a 7-step process as follows: (1) synchronize heat cycles of donor and recipient cows; (2) obtain a large number of ova from the donor cow by giving her a drug so that she superovulates; (3) breed donor cow (AI or natural); (4) collect ova from donor cow, 5 days after breeding (see Fig. 3-45); (5) examine eggs, make sure that they are normal and fertilized; (6) prepare foster mothers, by synchronizing (usually by hormone control) their ovulation with the donor; and (7) transfer eggs to recipient (see Fig. 3-46 for all seven steps).

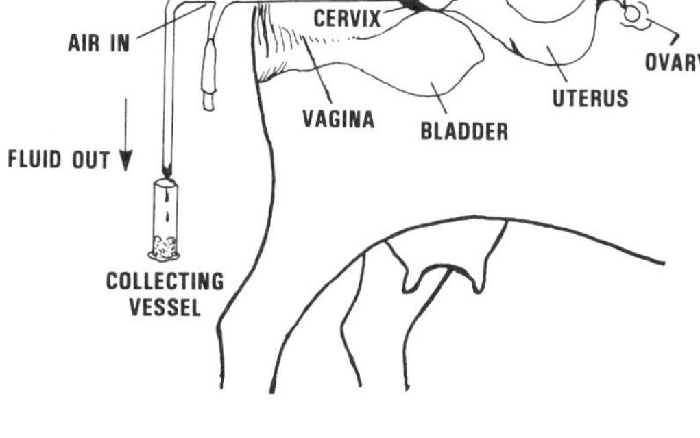

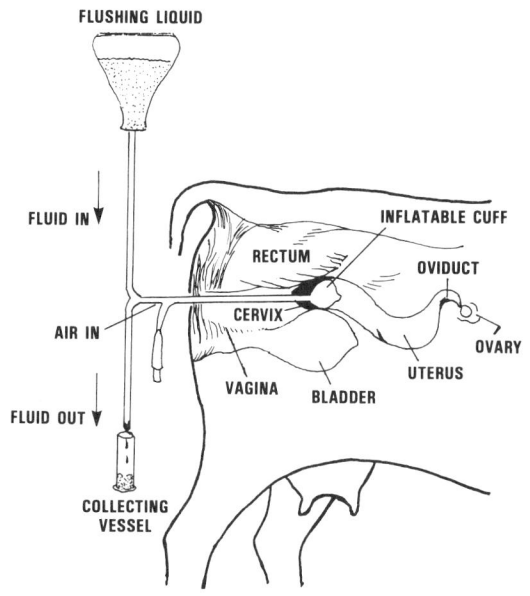

Fig. 3-45. Diagram showing the technique of nonsurgical embryo recovery. The collection procedure involves filling the uterus with fluid through a series of flexible tubes inserted in the uterus through the cervix. (Source: Wright, R. W., "The State of the Art of Cattle Embryo Transfer," *Beef Cattle Science Handbook*, Vol. 18, edited by M. E. Ensminger and published by Agriservices Foundation, 1981, p. 478, Fig. 1)

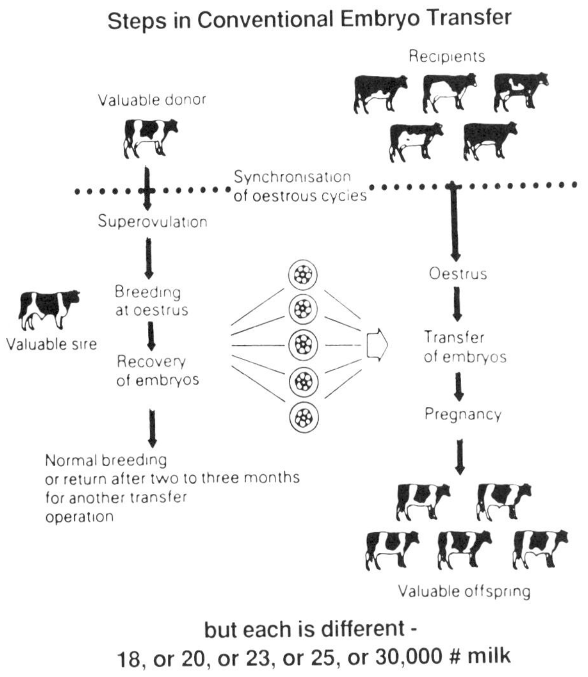

Fig. 3-46. Steps in conventional embryo transfer. (Prepared by Dr. R. Walton, American Breeders Service, a division of W. R. Grace & Co., De Forest, WI)

Pregnancy in the recipients can be diagnosed in about 30 days. Full-term pregnancies result in full sibs (brothers and sisters) with the genetic traits of the donor cow and the bull to which she was bred. Recipients have no genetic influence on the calves they carry—they merely serve as incubators.

The following **advantages** accrue from embryo transfer:

1. Up to 50 calves may be obtained from a valuable cow during a year's time.

2. The rate of progress in genetic improvement is speeded up, because of the increased number of progeny from valuable cows.

3. Valuable cows that produce normal ova but fail to conceive due to some hormonal or anatomic defects need not be culled because of sterility; such animals may be used as donors for supplying ova for transplantation.

4. Embryo transfer offers a practical method through which to import and export cattle genetics, with three primary advantages: (a) reduced transportation costs, (b) reduced disease hazard, and (c) easy adaptation to the climate and environment in which the animal is born.

5. Heifers may be effectively progeny tested at an early age. If large numbers of fertilized eggs could be procured from calves and transplanted to sexually mature recipients, the generation time of cattle could be reduced by one year or more.

6. It is possible to produce calves of the beef breed of preference from dairy cows.

The following three advancements in bovine embryo transfer technology were reported in the early 1980s:

- **The one-step freezing process**—This is a new method of freezing and thawing embryos in a single straw container. This process enables embryos to be implanted in a recipient cow in much the same manner as breeding a cow by artificial insemination.

- **The embryo splitting or cloning procedures**—If cloning is defined as asexual reproduction accomplished by replicating genetic material, then embryo splitting is a form of cloning. Embryo splitting is what the designation implies. After recovery from donor cows at approximately 7 days following insemination, the embryos are actually cut in half and each half is implanted into a recipient cow to produce twins from one recipient. Splitting embryos doubles the number of potential offspring from embryos recovered from donor females. Also, ½ of a split embryo can be held in reserve by freezing, then brought out later if progeny from the first half-embryo possess superior characteristics. In 1984, the Louisiana Agricultural Experiment Station reported on the first ever calves produced from an embryo split four ways.

- **Embryo sexing**—Experienced technicians report 90% accuracy in embryo sexing, thereby permitting embryos of known sex to be transferred.

An estimated 40,000 embryo transfer calves were born in the United States in 1983. Transfers cost up to $2,000 when the commercial transfer operator provided recipient cows; and as low as $400 to $500 per pregnancy when the breeder supplied the recipient.

In summary, it may be said that embryo transfer can be done very successfully by skilled teams. But the high cost of present techniques limits the application to the most elite stock. With more research, techniques will become more efficient, simple, and economical, and embryo transfer will be more widely used.

GENETIC ENGINEERING

Building upon the present knowledge and understanding of genes, and the nucleic acids DNA and RNA, scientists are now making such ideas as genetic engineering and cloning into realities.

The recent development of gene-splicing (also known as recombinant DNA) ushered in a new era of genetic engineering—with all its promise and possible peril.

On May 23, 1977, scientists at the University of California, San Francisco reported a major breakthrough as a result of altering genes—turning ordinary bacteria into factories capable of producing insulin, a valuable hormone previously extracted at slaughter from pigs, sheep, and cattle, so essential to the survival of 1.8 million diabetics. The feat opened the door to further genetic engineering or splicing—the transferring of a gene from one individual to another. Already, this genetic wizardry has been used in transplanting into bacteria (and recently into yeast cells) genes responsible for many critical biochemicals in addition to insulin; among them, endorphin, somatotrophin, interferon, and vaccines.

Genetic manipulations to create new forms of life make biologists custodians of great power. Despite different schools of thought, scare headlines, and political hearings, molecular biologists will continue recombinant DNA studies, with reasonable restraints, and work ceaselessly away at making the world a better place in which to live (see Fig. 3-47).

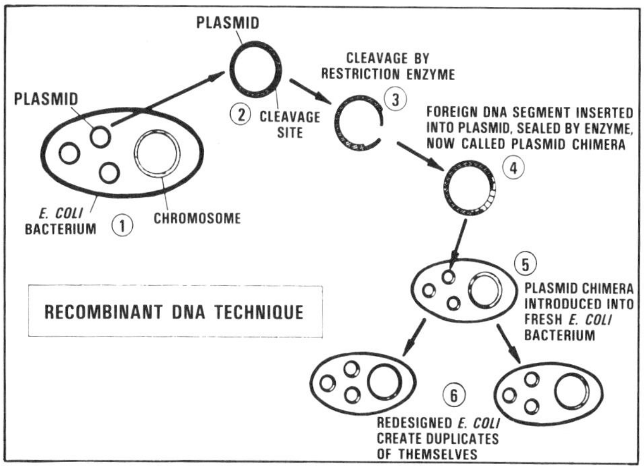

Fig. 3-47. Redesigning *E. coli*, common bacteria of animal and human intestines. The steps:

1. The scientist places the bacterium in a test tube with a detergent. This dissolves the microbe's outer membrane, causing its DNA strands to spill out.
2. The plasmids (the closed loops), which are genetic particles found in bacteria, are separated from the chromosomal DNA in a centrifuge.
3. The plasmids are placed in a solution with a chemical catalyst called a restriction enzyme, which cuts through the plasmids' DNA strips at specific points.
4. The opened plasmid loops are then mixed in a solution with genes—also removed by the use of restriction enzymes—from the DNA of a plant, animal, bacterium, or virus. In the solution is another enzyme called a DNA ligase, which cements the foreign gene into place in the opening of the plasmids. These new loops of DNA are called plasmid chimeras because, like the chimera—the mythical lion-goat-serpent after which they are named—they contain the components of more than one organism.
5. The chimeras are placed in a cold solution of calcium chloride containing normal *E. coli* bacteria. Then the solution is suddenly heated, at which time the membranes of the *E. coli* become permeable, allowing the plasmid chimeras to pass through and become a part of the microbe's new genetic structure.
6. When the redesigned *E. coli* reproduce, they create duplicates of themselves, new plasmids—and DNA sequences—and all.

CLONING

Cloning of an animal is the production of an exact genetic copy. In a technical sense, identical twins are clones; they are derived from a single cell, as a result of the embryo splitting early in development to yield what is essentially two carbon copies.

The exciting and much sought technological breakthrough in the cloning of mammals involves the manipulation of embryos.

The dream of cloning is based on the following two pieces of scientific evidence:

1. With few exceptions, all cells in the body of an animal appear to contain the same genetic information. This information is contained in the DNA, a molecule that is located in a sac inside cells called the nucleus. Thus, within an animal, the DNA sequence in the nucleus of a liver cell is identical to that in a skin cell. These cells differ in appearance and function because they make use of different parts of the genetic information, not because the total amount of information differs. Further, all of these cells have the genetic information that was present in the one-cell embryo that developed in the animal. Therefore, if the nucleus of any of these cells were used to replace the genetic information in any one-cell embryo, an exact genetic copy of the animal whose cells donated the nucleus would develop. With such an approach, thousands of cloned copies could be made.

2. The second piece of scientific evidence is that nuclear transplantation experiments have been done successfully with several species of animals, especially frogs and fish, which have the big advantage of their eggs being thousands of times larger than mammalian eggs.

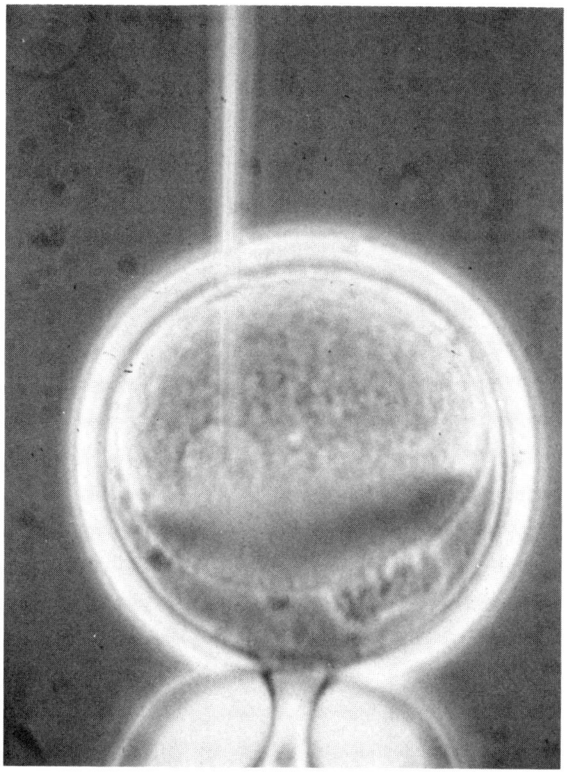

Fig. 3-48. Microinjection of DNA into cow egg to produce transgenic cattle. (Courtesy, Dr. R. Walton, American Breeders Service, Inc., De Forest, WI)

Historically, research and development in cloning has passed through the following stages, in order and period of time:

1. Identical twin calves were produced by microsurgically splitting embryos, then transferring half embryos to recipients.

2. A bull calf was born as a result of using the laboratory culturing technique (*in vitro*) of maturing an egg, fertilizing the egg *in vitro*, then transferring the fertilized egg to a surrogate mother.

Nuclei have been take from 16-cell bovine embryos and placed in one-cell bovine eggs whose nuclei had already been removed. The new one-cell embryos were matured and transferred into recipient cows, which subsequently gave birth to two cloned female calves.

During the late 1980s, the most advanced cloning procedure consisted of flushing the embryo out of the donor cow at day five of its development (at the 32-cell stage); followed by putting the embryo under a microscope and manually removing one of the cells, then freezing the remaining 31-cell embryo (much like semen is frozen) and putting it away until an order is received. Next, the technician takes an unfertilized egg that has been flushed from a low-grade donor cow, removes the nucleus from it, inserts the borrowed cell into the egg, patches up the hole with a short zap of electricity, and the cloned embryo divided day after day and develops into a genetically identical duplicate of the heifer that results from the 31-cell embryo.

The goal ahead is (1) to let embryos grow to the 32-cell stage, then split them all into 32 more embryos each, resulting in 1,024 (32 × 32) genetically identical copies of the same pedigree—*the exact same*; (2) to let 31 of the embryos grow up, split apart the 32nd, and keep carrying out the chain as long as desired; and the first animal born would be identical to the last one—*identical*. The goal ahead is illustrated in Fig. 3-49.

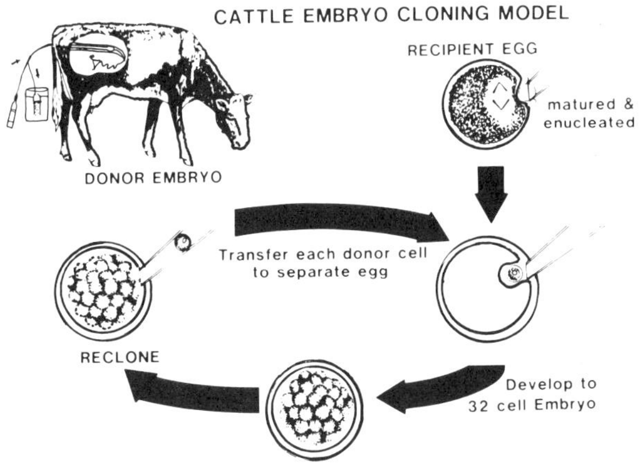

Fig. 3-49. Cattle embryo cloning model. (Prepared by Dr. R. Walton, American Breeders Service, Inc., De Forest, WI)

BLOOD TYPING

Blood typing was developed at the University of Wisconsin during the decade 1940–50. It involves a study of the components of the blood, which are inherited according to strict genetic rules that have been established in the research laboratory. By determining the genetic *markers* in each sample and then applying the rules of inheritance, parentage can be determined. To qualify as the offspring of a given male and female, an animal must not possess any genetic markers not present in its alleged parents. If it does, it constitutes grounds for illegitimacy.

Blood typing is used for the following purposes:

- **To verify parentage**—The test is used in instances where the offspring may bear some unusual color or markings or carry some undesirable recessive characteristic. It may also be used to verify a registration certificate. Through blood typing parentage can be verified with 90% accuracy.[15] Although this means that 10% of the cases cannot be settled, it is not possible to do any better in human blood typing.

- **To determine which of two sires**—When a female has been served by two or more males during one breeding season, blood typing can identify the sire.

- **To provide a permanent blood type record for identification purposes**—Two samples of blood are required for each animal to be studied; and the samples must be taken in tubes and in keeping with detailed instructions provided by the laboratory. In parentage cases, this calls for blood samples from the offspring and both parents; in paternity cases, samples must be taken from the offspring, the dam, and all the sires.

- **To detect fertile heifers born co-twin with bulls**—About 15% of all heifers born twin with a bull are potentially fertile; the other 85% are sterile, or freemartins. A cattle producer need not wait until such heifers reach breeding age in order to ascertain their potentialities. Instead, blood samples can be submitted from each of the twins (the bull and the heifer) to a service-typing laboratory, along with a request for a diagnosis. If the bull and heifer have *like* blood types (except possible differences in the J system), the heifer is diagnosed as a freemartin and nonbreeder. If the bull and the heifer have unlike blood types (except possible differences in the J system alone), the heifer is diagnosed as potentially fertile.

The basis for this remarkable method of diagnosing the breeding potentialities of a heifer born twin with a bull goes back to the early events in the embryology of cattle twins. In about 85% of the cattle embryos, some of the chorionic blood vessels become anastomosed or joined together. This results in a communal blood vascular system. Hence, the twins come to share each other's blood-forming tissues. As a result, they have like blood types.

---

[15]In a personal communication to the author, Dr. C. Stormont, Professor Emeritus, University of California at Davis and owner of Stormont Laboratories, Inc., 1237 Beamer St., Suite D, Woodland, CA 95695, reported that they have been able to solve approximately 91% of all the parentage cases.

- **To substitute for fingerprinting**—Much attention is now being given to the idea of utilizing blood typing as a positive means of identification of stolen animals, through proving their parentage.

A list of the blood typing laboratories in the United States and Canada follows.

In the United States

Cattle Blood Typing Laboratory, Ohio State University, Columbus, OH 43210 (cattle blood typing only)

Department of Veterinary Sciences, University of Kentucky, 102 Animal Pathology Building, Lexington, KY 40546 (horse blood typing only)

ImmGen, Inc., P.O. Box 10135, Rt. 4, Box 539C, College Station, TX 77843 (cattle blood typing only)

Serology Laboratory, School of Veterinary Medicine, University of California, Davis, CA 95616 (both cattle and horse blood typing)

Stormont Laboratories, Inc., 1237 E. Beamer St., Suite D, Woodland, CA 95695 (provides a variety of blood typing services involving several species: bison, cat, cattle, dog, horse)

In Canada

Bovine Blood Testing Laboratory, Saskatchewan Research Council, 15 Innovation Blvd., Saskatoon, Saskatchewan, S7N 2X8 (cattle blood typing only)

Mann Equitest, Inc., 550 McAdam Road, Mississauga, Ontario L4Z 1P1 (horse blood typing only)

PATENTED ANIMALS

In April 1987, the U.S. Patent and Trademark Office (PTO) ruled that patents could be issued on genetically engineered animals. Subsequently, the PTO (1) put the new animals on a par with mechanical inventions by decreeing that livestock producers must pay fees to those who patent genetically altered animals, and (2) ruled that livestock producers must pay royalties to the patent holder on each generation of patented animals for the life of the patent—which may be as long as 17 years.

Stud fees and one-time payments for animals have always been a part of livestock farming, but farmers have the rights to, and complete control over, subsequent generations without any additional payments.

In 1988, the U.S. Patent Office granted a patent to Harvard University for a genetically engineered mouse, involving developing a way to add cancer genes to the embryo of mice, thus making them more likely to get cancer thereby facilitating cancer experiments.

Farmers and ranchers face a dilemma: Generally speaking, they want to reap the benefits of genetically improved livestock, but they do not want to pay royalties on each succeeding generation.

Standardbred stallion in action. The name *Standardbred* is derived from the fact that, beginning in 1879, eligibility for registration was based on a performance test—animals had to trot the mile in 2:30 or pace it in 2:25. (Courtesy, United States Trotting Assn., Columbus, OH)

Feed is the most important factor in the environment of animals. This shows hay being fed to a commercial cow herd on a winter range. Today, most commercial feeding operations are mechanized—with tractors and trucks. But horses and sleds, like this one, are occasionally used in deep snow and inaccessible areas. (Courtesy, Union Pacific Railroad)

SECTION

4

FEEDING

Contents *Page*

Contents *Page*

Contents *Page*

PART I
GENERAL LIVESTOCK FEEDING

Fig. 4-1. This central computerized system in a feed mill controls all feed ingredient distribution and mixing. (Courtesy, National Broiler Council, Washington, DC)

Animals inherit certain genetic possibilities, but how well these potentialities develop depends upon the environment to which they are subjected; and the most important influence in the environment is the feed. In turn, all feeds come directly or indirectly from plants which have their tops in the sun and their roots in the soil. Thus, we have the cycle as a whole—from the soil, through the plant, thence to the animal and back to the soil again.

Pastures and dry forages produced on fertile soils, together with a multitide of grains and by-product feeds constitute the basis of successful livestock production. Fig. 4-2 shows the relative importance of the principal U.S. livestock feeds.

The primary purpose of the 763 million (1988–89 figures) U.S. animals, including poultry, is to convert feeds into food, clothing, power, and recreation. About two-thirds of the feed consumed by U.S. livestock is not suited for human consumption. In this category are hay, pasture, coarse forages (such as straws, fodders, etc.); certain grains; such by-products as are obtained from mills, packinghouses, and food-processing plants; and damaged grains and foods; and garbage. These are converted into meat, milk, eggs, and fiber.

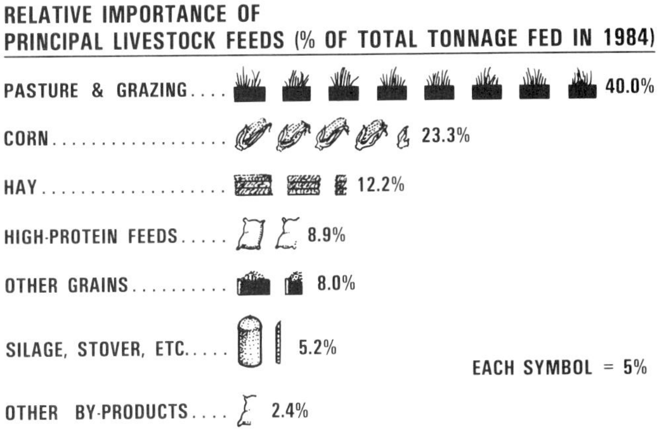

Fig. 4-2. These principal livestock feeds are converted into meat, eggs, milk, and wool. (Source of data: USDA, Economic Research Service)

Also, feeding is important from an an economic standpoint; it is the major item of expense in producing livestock. For example, feed accounts for approximately the following proportions of the cost of livestock production: finishing cattle, 70%; milk production, 55%; feedlot lambs, 50%; and pork, 65 to 75%.

This section is a capsule presentation of livestock feeding. The first part of the section is devoted to general feeding information and recommendations pertinent to all classes of livestock. This is followed by a section devoted to the feeding of each class of livestock—beef cattle (including finishing cattle), dairy cattle, sheep/goats, swine, and horses.

TYPES OF DIGESTIVE SYSTEMS AND KINDS OF FEEDS EATEN

There are major differences in the anatomy and physiology of the organs of the digestive tract of different animal species. These differences are of great nutritional significance, as they affect both the nature of the digestion processes and the kind of feed that can be utilized. Based on type of digestive system, animals may be grouped as (1) monogastric (simple-stomached), (2) polygastric (ruminants), or (3) pseudo-ruminants (those with functional ceca).

Based on the kind of feed eaten, animals are classed as follows:

1. **Herbivores.** The vegetarians; they depend entirely upon plants for their feed supply.

2. **Carnivores.** The flesh eaters; they feed almost entirely upon the flesh of other animals.

3. **Omnivores.** The consumers of both plants and flesh.

An understanding of specific differences in types of digestive systems and kind of feed eaten is essential to intelligent feeding. This information is presented in Fig. 4–3.

Monogastric animals, like hogs, must eat a large percentage of grains and other concentrates and depend almost entirely on digestive enzymes to break down these compounds. Ruminants, such as cattle and sheep, with their four stomach compartments and the help of microorganisms, can subsist largely, or entirely, on bulky, high-fiber forages which, because of their low energy per unit weight of dry matter, must be consumed in large quantities to supply their nutrient needs. The horse, because of its greatly enlarged cecum and large intestine, can utilize quantities of roughage intermediate between simple-stomached and ruminant animals.

Group 1—Polygastric (ruminants)

Animal	*Class of Food*
Cow	Herbivore
Sheep	Herbivore
Goat	Herbivore

Group 2—Monogastric (simple stomach)

Animal	*Class of Food*
Pig	Omnivore
Dog	Carnivore
Monkey	Omnivore
Human	Omnivore

Group 3—Pseudo-ruminants (functional cecum)

Animal	*Class of Food*
Horse	Herbivore
Rabbit	Herbivore
Guinea pig	Herbivore
Hamster	Omnivore

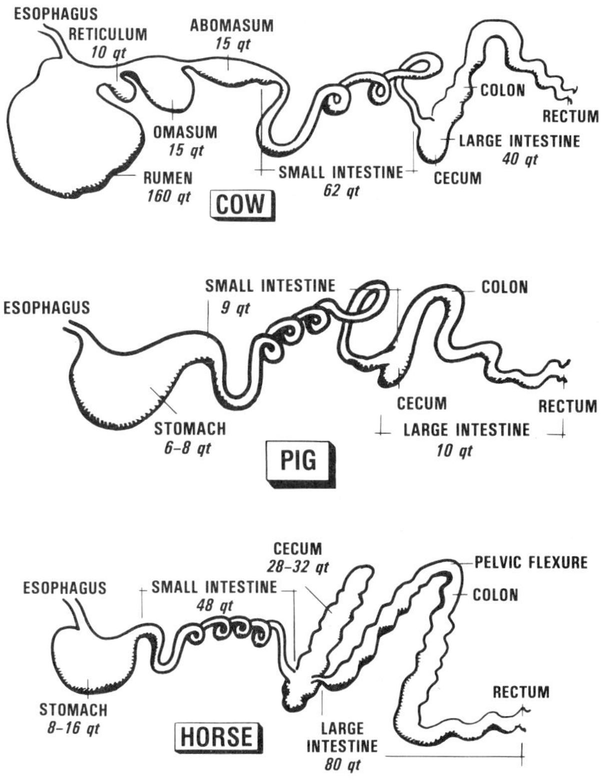

Fig. 4–3. Classification of animals according to type of digestive system and class of feed, with schematic diagram of digestive tracts of cow, pig, and horse.

NUTRITIVE NEEDS

Nutrients are the chemical substances found in feed materials that can be used, and are necessary for the maintenance, production, and health of animals. The chief classes of nutrient substances are carbohydrates, fats, proteins, minerals, vitamins, and water. Nutrients are needed by the animal in definite amounts, with the quantities varying according to the class and age of animal, and the purpose for which it is being fed. A deficiency in a nutrient can be, and often is, a limiting factor in animal production.

The economical production of animal products is dependent upon (1) meeting the total nutritional requirements of the animal, and (2) knowing both the nutritional requirements of the animal and the nutritional value of the feeds.

ENERGY NEEDS

Energy is required for practically all life processes—for the action of the heart, maintenance of blood pressure and muscle tone, transmission of nerve impulses, ion transport across membranes, reabsorption in the kidneys, protein and fat synthesis, the secretion of milk, and the production of eggs and wool.

It is common knowledge that a ration must contain carbohydrates, fats, and proteins. Although each of these has specific functions in maintaining a normal body, they can all be used to provide energy for maintenance, for work, or for finishing. From the standpoint of supplying the normal energy needs of animals, however, the carbohydrates are by far the most important, more of them being consumed than any other compound, whereas the fats are next in importance for energy purposes.

A deficiency of energy is manifested by slow or stunted growth, body tissue losses, and/or lowered production of meat, milk, eggs, or fiber, rather than by specific signs such as those which characterize many mineral and vitamin deficiencies. For this reason, energy deficiencies often go undetected and unrectified for extended periods of time.

PROTEIN NEEDS

Proteins are complex organic compounds made up chiefly of amino acids, which are present in characteristic proportions for each specific protein. This nutrient always contains carbon, hydrogen, oxygen, and nitrogen, and in addition it usually contains sulfur and frequently phosphorus. Proteins are essentially in all plant and animal life as components of the active protoplasm of each living cell.

Crude protein refers to all the nitrogenous compounds in a feed. It is determined by finding the nitrogen content and multiplying the result by 6.25. The nitrogen content of protein averages about 16% ($100 \div .16 = 6.25$).

Animals of all ages and kinds require adequate amounts of protein of suitable quality; for maintenance, growth, finishing, reproduction, work, and wool production. Of course, the protein requirements for growth, reproduction, and lactation are the greatest and most critical.

In addition to supplying an adequate quantity of proteins, it is essential that the character of proteins be thoroughly understood. Proteins are very complex compounds with each molecule made up of hundreds of thousands of amino acids combined with each other. The amino acids, of which some 23 are known, are sometimes referred to as the building stones of proteins. Certain of these amino acids can be made by the animal's body to satisfy its needs. Others cannot be formed fast enough to supply the body's needs and, therefore, are known

Fig. 4-4. Lysine made the difference! Big pig (B) received high-lysine corn. Little pig (C), a littermate, was fed a lysine-deficient ration. (Courtesy, Cornell University, Ithaca, NY)

as essential (or indispensable) amino acids. These must be supplied in the feed. Thus, rations that furnish an insufficient amount of any of the essential amino acids are said to have proteins of poor quality, whereas those which provide the proportions of the various necessary amino acids are said to supply proteins of good quality.

In general, animal proteins are superior to plant proteins for monogastric animals (including man) because they are better balanced in the essential amino acids. For example, zein (a corn protein) is an incomplete plant protein. It is deficient in the essential amino acids lysine and tryptophan. On the other hand, animal proteins are excellent sources of lysine, and many of them (especially milk and eggs) are abundant in tryptophan.

The necessity of each amino acid in the ration of the experimental rat has been thoroughly tested, but less is known about the requirements of large animals or even the human. According to our present knowledge, based largely on work with the rat, the following division of amino acids as essential and nonessential seems proper:

Essential (indispensable)	Nonessential (dispensable)
Arginine	Alanine
Histidine	Asparagine
Isoleucine	Aspartic acid
Leucine	Cysteine
Lysine	Cystine
Methionine (may be replaced in part by cystine)	Glutamic acid
	Glutamine
Phenylalanine	Glycine
Threonine	Hydroxyproline
Tryptophan	Proline
Valine	Serine
	Tyrosine

Arginine is regarded as essential for animals, whereas it is not for humans; most young mammals cannot synthesize it in sufficient amounts to meet their needs for growth.

In practical animal nutrition, the amino acids most likely to be deficient are lysine, methionine, and tryptophan. This stems from the fact that cereal grains, which are primary energy feeds, are quite low in these amino acids. So, it follows that rations based on a high percentage of these grains usually require supplementation with proteins which contain higher levels of these amino acids.

Fortunately, the amino-acid content of proteins from various sources varies. Thus, the deficiencies of one protein may be improved by combining it with another, and the mixture of the two proteins often will have a higher feeding value than either one alone. It is for this reason that a considerable variety of feeds in the ration is usually recommended.

The feed proteins are broken down into amino acids by digestion. They are absorbed and distributed by the bloodstream to the body cells, which rebuild these amino acids into body protein.

MINERAL NEEDS

Minerals are the inorganic elements of animals and plants, determined by burning off the organic matter and weighing the residue, which is called ash.

Animal bodies contain small amounts—only 2 to 5%—of inorganic elements, called minerals. But these constituents play an important role in animal nutrition. In addition to furnishing structural material for bones and teeth, as constituents of the soft tissues, the blood, the fluids of the body, and certain of the secretions, they regulate many of the vital processes.

Eighteen mineral elements are known to be required by at least some animal species. They can be divided into two groups based on the quantity required in the ration.

- **Major or macrominerals**—These elements are required in amounts ranging from a few tenths of a gram to one or more grams per day.

- **Trace or microminerals**—These elements are required in minute quantities, ranging from a millionth of a gram (microgram) to a thousandth of a gram (milligram) per day.

The terms—major/macromineral and trace/micromineral—do not imply any lesser role for the latter group; rather, they represent quantity designation based on the amounts needed by animals.

The two groups follow:

Major/Macrominerals	Trace/Microminerals	
Salt (sodium & chlorine, NaCl)	Chromium (Cr)	Molybdenum (Mo)
Calcium (Ca)	Cobalt (Co)	Selenium (Se)
Phosphorus (P)	Copper (Cu)	Silicon (Si)
Magnesium (Mg)	Fluorine (F)	Zinc (Zn)
Potassium (K)	Iodine (I)	
Sulfur (S)	Iron (Fe)	
	Manganese (Mn)	

The general functions of minerals are as follows:

1. Give rigidity and strength to the skeletal structure.

2. Serve as constituents of the organic compounds, such as protein and lipid, which make up the muscles, organs, blood cells, and other soft tissues of the body.

3. Activate enzyme systems.

4. Control fluid balance—osmotic pressure and excretion.

5. Regulate acid-base balance.

6. Exert characteristic effects on the irritability of muscles and nerves.

7. Engage in mineral-vitamin relationships.

A summary of each mineral is presented in Table 4-1, Animal Mineral Chart, with the minerals listed alphabetically within each of two classifications: (1) Major/Macrominerals, or (2) Trace/Microminerals.

TABLE 4–1
ANIMAL MINERAL CHART

Mineral	Major Functions	Some Deficiency Symptoms	Major Interrelationships; Toxicities	Good Sources for Animals	Comments
Major or Macrominerals					
Salt (NaCl)	Salt serves as both a condiment and a nutrient. Sodium and chlorine help maintain osmotic pressure in body cells, upon which depends the transfer of nutrients to the cells and the removal of waste materials. Sodium is associated with muscle contraction and is important in making bile, which aids in the digestion of fats and carbohydrates. Chlorine is required for the formation of hydrochloric acid in the gastric juice so vital to protein digestion.	Reduced growth and efficiency of feed utilization in growing animals; reduced milk production and weight loss in adults. Lowered reproduction (infertility in males, and delayed sexual maturity in females). Craving for sodium, evidenced by such things as drinking urine. In laying hens, a deficiency of sodium results in lowered production, loss of weight, and cannibalism. Chicks on chlorine-deficient diet exhibit nervous symptoms induced by sudden noise.	Salt toxicity, which is accentuated with restriction of water intake, readily occurs in nonruminants. It is characterized by a staggering gait, blindness,and other nervous disorders. Excess Na results in hypertension. Excess Cl is not likely.	Salt: free choice, or added to the ration at a level of 0.25 to 0.50%.	In practice, Na and Cl are supplied together as common salt. The body's requirement for Cl is approximately half that of Na. The body contains approximately 0.2% sodium.
Calcium (Ca)	Bone and teeth formation and maintenance; nerve function; muscle contraction; blood coagulation; cell permeability. Essential for milk production and for formation of eggshell in poultry.	Rickets in young. Osteomalacia in adults. Tetany (hypocalcemia). Milk fever in dairy cows is the classical example of Ca tetany. Hens: Thin-shelled eggs, drop in egg production, and lowered hatchability.	Calcium-phosphorus ratio is important. For non-ruminants, it should be 1 to 2 parts Ca to 1 part P. For ruminants, it may be anywhere from 1:1 to 7:1. Vitamin D is involved. If adequate vitamin D is present, the ratio of calcium to phosphorus is less important. Excess Ca reduces the absorption and utilization of Zn. In swine, this causes parakeratosis. Excess Mg decreases Ca absorption, replaces Ca in the bone and increases Ca excretion.	Oystershells. Limestone. Dicalcium phosphate. Defluorinated phosphate. Protein supplements of animal origin, legume forages, and rape. Milk. Bone meal.	Only 20 to 30% of the calcium in the average ration is absorbed from the intestinal track and taken into the bloodstream. Over 70% of the ash of the body consists of Ca and P Approximately 99% of the Ca in the body is present in the bones and teeth.
Phosphorus (P)	Bone and teeth formation and maintenance; a component of phospholipids which are important in lipid transport and metabolism and cell-membrane structure. Milk secretion. In energy metabolism. A component of RNA and DNA, the vital cellular constituents required for protein synthesis. A constituent of several enzyme systems.	Rickets in young. Osteomalacia in adults. Depraved appetite (pica), but this is not specific for phosphorus deficiency. Breeding problems. Hens: Reduced egg production in poultry.	Ratio of Ca-P is important; somewhere between 1 to 2 parts of Ca to 1 part of P. Sufficient vitamin D is necessary for P assimilation and utilization. Excess Ca and Mg cause decrease in P absorption. In ruminants, excess P in relation to Ca is likely to cause calculi.	Monosodium phosphate. Diammonium phosphate. Dicalcium phosphate. Defluorinated phosphate. Bone meal. Most cereal grains and their by-products (notably wheat bran) are high in P.	Phosphorus is more efficiently absorbed than calcium; about 70% of the ingested phosphorus is absorbed. Approximately 80% of the P of the body is present in the bones and teeth. Excess P may result in lameness and spontaneous fracture of long bones. High P has a laxative effect.

(Continued)

Fig. 4–5. Soft bones because of a calcium/phosphorus deficiency. (Courtesy, Dean R. S. Sugg, School of Veterinary Medicine, Auburn University, Auburn, AL)

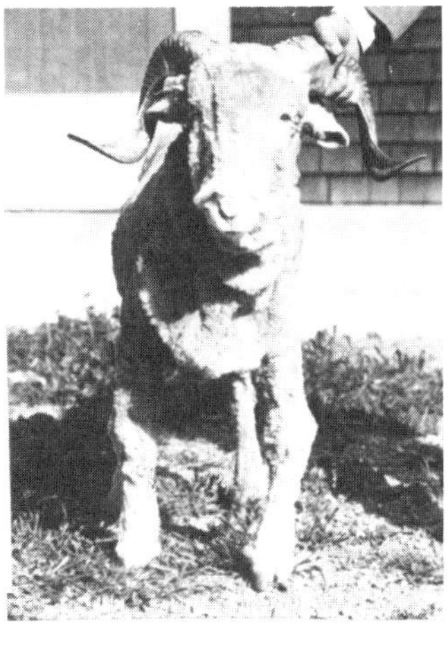

Fig. 4–6. Rickets in sheep produced from insufficient calcium or phosphorus or an imbalance of the two. (Courtesy, California Ag., Exp. Sta.)

TABLE 4–1 *(Continued)*

Mineral	Major Functions	Some Deficiency Symptoms	Major Interrelationships; Toxicities	Good Sources for Animals	Comments
Magnesium (Mg)	Essential for normal skeletal development; as a constituent of bones and teeth; enzyme activator, primarily in glycolytic system; involved in activating certain peptidases in protein digestion; relaxes nerve impulses; serves as a ruminant alkalizer and buffer.	Vasodilation, with resulting reduction in blood presure (manifested outwardly by a flushing of the skin). Hyperirritability. Tetany (grass tetany, or grass staggers) characterized by loss of appetite (anorexia), hyperemia, convulsions, and death.	Excess of Mg upsets Ca and P metabolism. Mg toxicity has not been demonstrated.	Magnesium sulfate or oxide, mixed with salt or small amount of feed	Deficiencies of Mg may be encountered with suckling calves and pigs.
Potassium (K)	Major cation intracellular fluid where it is involved in osmotic pressure and acid-base balance. Relaxes the heart muscle. Involved in secretion of insulin, in enzyme reactions of the phosphorylation of creatine, in carbohydrate metabolism, and in protein synthesis.	Growth retardation, unsteady gait, general muscle weakness, pica, diarrhea, distended abdomen, emaciation, enlargement of the heart and kidneys, followed by death.	Magnesium deficiency results in failure to retain potassium; hence, it may lead to K deficiency. Excessive levels of potassium interfere with magnesium absorption. Excessive use of salt depletes the body's potassium.	Potassium chloride, kelp. Molasses (beet and cane), beet tops. Roughages usually contain ample potassium.	Potassium deficiency may occur in drylot finishing cattle or sheep on a high-concentration ration.
Sulfur (S)	Required as a component of sulfur-containing amino acids, cystine and methionine. As a component of biotin, sulfur is important in lipid metabolism. As a component of thiamin and insulin, it is important in carbohydrate metabolism. As a component of coenzyme A, it is important in energy metabolism. As a component of hair, wool, and feathers.	Retarded growth, primarily due to not meeting the sulfur amino acid requirement for protein synthesis. Sheep fed nonprotein N to replace protein without S supplementation show reduced wool growth (wool contains approximately 4% sulfur).	Sulfur is related to the amino acids, cystine and methionine, and to biotin, thiamin, and coenzyme A (see column to left, "Major Functions"). Sulfur toxicity is not a practical problem.	Nonruminants should be provided sulfur-containing proteins. Ruminants and horses may be provided sulfur in protein, as elemental sulfur or as sulfate sulfur.	The body contains approximately 0.15% sulfur. Sulfur requirements are primarily those involving amino acid nutrition. Ruminants fed urea as a source of protein nitrogen may benefit from supplemental sulfur.
Trace or microminerals					
Chromium (Cr)	In glucose metabolism. Activator of certain enzymes. Stabilizer of nucleic acids. Stimulation of the synthesis of fatty acids and cholesterol in the liver.	Impaired glucose tolerance. Disturbance of lipid and protein metabolism.	Diets high in carbohydrates may cause the supply of GTF-chromium to be depleted.	There is no evidence that practical animal rations need to be supplemented with Cr.	The importance of Cr in glucose metabolism of other animals (other than the rat) and humans has not been established to date.
Cobalt (Co)	As a component of vitamin B–12. Rumen microorganisms use Co for the synthesis of vitamin B–12 and the growth of rumen bacteria.	Deficiency of Co in cattle and sheep produces symptoms similar to a deficiency of vitamin B–12. Ruminants grazing in Co-deficient areas show loss of appetite, reduced growth, and loss in body weight, followed by emaciation, anemia, and eventually death. Frequently a depraved appetite is noted. The disease called *salt sick* in Florida is due to Co deficiency associated with Cu deficiency. In different parts of the world, Co deficiency is known as Denmark disease, coast disease, enzootic marasmus, bush sickness, wasting disease, Nakuritis, and pining disease.	Related to vitamin B–12. Cobalt toxicity is not likely.	Cobaltized mineral mixture made by adding Co at rate of 0.2 oz/100 lb of salt as cobalt chloride, cobalt sulfate, cobalt oxide, or cobalt carbonate. Also, several good Co-containing commerical minerals are on the market. Grazing animals may be given pellets composed of cobalt oxide and iron administered orally with a balling gun. The pellets lodge in the rumen and are gradually dissolved over a period of months. Poultry by-product meal, soybean meal, meat meal, rice bran, and blackstrap molasses.	The Co content of the leaves of the catalpa tree is regarded as a good indicator of the adequacy of cobalt in an area. Co-deficient areas have been reported in Australia, western Canada, and in the U.S. in the states of Florida, Michigan, Wisconsin, Massachusetts, New Hampshire, Pennsylvania, and New York.

(Continued)

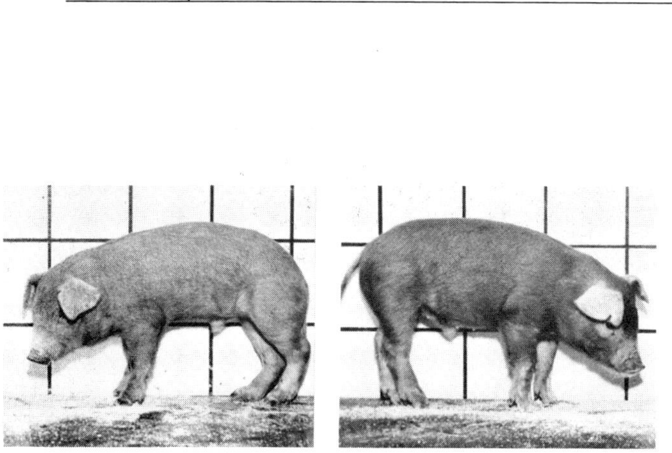

Fig. 4–7. Magnesium made the difference! (Courtesy, Purdue University, Lafayette, IN)

Fig. 4–8. Cobalt made the difference! (Courtesy, Michigan State University, East Lansing)

TABLE 4-1 *(Continued)*

Mineral	Major Functions	Some Deficiency Symptoms	Major Interrelationships; Toxicities	Good Sources for Animals	Comments
Copper (Cu)	Along with iron and vitamin B-12, copper is necessary for hemoglobin formation, although it forms no part of the hemoglobin molecule (or red blood cells). Essential in enzyme systems, hair development and pigmentation, bone development, reproduction, and lactation.	Fading hair coat; light wool growth and straight, hairlike fibers, known as steely wool. A condition called *swayback* (enzootic ataxia) in newborn lambs. Lameness, swelling of joints, and fragility of bones. Nutritional anemia, commonly called *salt sick*.	Copper is involved in iron metabolism. An excess of molybdenum in the presence of sulfate causes a condition which can be cured by administering copper. Excess copper is toxic; it accumulates in the liver, and death may result. In high molybdenum areas, the Cu level for horses and cattle should be about 5 times higher than normal.	Trace mineralized salt containing copper sulfate or copper carbonate. Cane or blackstrap molasses, liver meal, brewers' grains, and gluten feed and meal.	A variable store of copper is located in the liver and spleen. Milk is low in Cu; hence, young animals raised almost exclusively on milk may develop anemia. Copper deficiencies are common in Australia and New Zealand,, and in southern U.S.
Fluorine (F)	Necessary for sound bones and teeth.	Excesses of fluorine are of more concern than deficiencies in livestock production.	Large amounts of calcium, aluminum, or fat will lower the absorption rate of fluorine. High dietary Ca depresses F uptake of bone. F is a cumulative poison; hence, the toxic effects may not be noticed for some time. High levels result in enlarged bones; softening, mottling, and irregular wear of the teeth; roughened hair coat; delayed maturity; and less efficient utilization of feed.	No need to supplement livestock with fluorine has been demonstrated. Should such supplementation be necessary, 1 ppm in the drinking water should suffice.	Fluorine in excess of 20 to 40 ppm of the dry matter of the diet (depending on the species of animal, age, and rate of production) may show a progressive severe toxicity.
Iodine (I)	Needed by the thyroid gland for making thyroxin, an iodine-containing hormone which controls the rate of body metabolism or heat production.	Goiter (big-neck) in humans, calves, lambs, and kids; stillbirths and weak young; hairless pigs; woolless lambs at birth. There is no satisfactory treatment for animals that have developed pronounced I-deficiency symptoms. Iodine deficiency in young animals is called cretinism. In adults it is known as myxedema.	Feeds of the cabbage family contain goitrogens, which interefere with the use of thyroxin and may produce goiter. Long-term chronic intake of large amounts of I reduces thyroid uptake of I. Marked species differences exist in tolerance to high intakes of I.	Stabilized iodized salt containing 0.01% potassium iodide (0.0076%I). Calcium iodate. Ethylenediamine dihydriodide (EDDI). Whey, marine by-products, poultry by-products, blackstrap molasses, and meat meal.	Enlargement of the thyroid gland (goiter) is nature's way of trying to make enough thyroxin (an I-containing hormone) when there is insufficient I in the feed. Mature animal body contains less than 0.00004% I. I deficiencies are worldwide. In the U.S., the Northwest, the Pacific Coast, and the Great Lakes regions are goiter areas.
Iron (Fe)	Iron is a constituent of hemoglobin, the iron-containing compound that transports oxygen. Also, iron plays a role in cellular oxidations, being a component of certain enzymes concerned with oxygen transfer.	Fe-deficiency anemia, characterized by smaller than normal number of red cells and less than normal amount of hemoglobin.	Iron is related to hemoglobin. Cu is required for proper Fe metabolism. Pyridoxine deficiency decreases the absorption of Fe. Too much iron may be deleterious—interfering with phosphorus absorption by forming an insoluble phosphate.	Ferrous sulfate administered orally, or iron dextran infection. Leafy portions of plants, meat by-products, legume seeds, cereal grains, and cane molasses. Trace mineralized salt.	The body contains only about 0.004% iron. Thus, a mature human contains only about 1/10 ounce of this mineral. Iron is stored in the liver, spleen, and kidneys. Young animals are born with a store of iron. But milk is low in iron. So, when young animals are continued on milk for a long time, particularly under confined conditions and with little or no supplemental feed, nutritional anemia will likely develop.

(Continued)

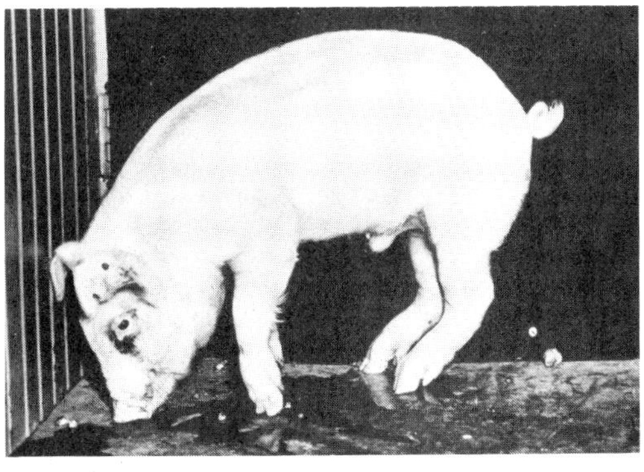

Fig. 4-9. Copper deficiency. Note the drawing-under of the rear legs and crookedness of the forelegs. (Courtesy, Hormel Institute, Austin, MN)

Fig. 4-10. Suckling pig with nutritional anemia, caused by a lack of iron, characterized by swollen condition about the head and paleness of the mucous membranes. (Courtesy, College of Veterinary Medicine, University of Illinois, Urbana)

TABLE 4-1 *(Continued)*

Mineral	Major Functions	Some Deficiency Symptoms	Major Interrelationships; Toxicities	Good Sources for Animals	Comments
Manganese (Mn)	Essential for normal bone formation (as a component of the organic matrix), and growth of other connective tissues. Blood clotting. Insulin action. Activator of enzyme systems in the metabolism of carbohydrates, fats, proteins, and nucleic acids.	Poor growth. Lameness, shortening and bowing of the legs, and enlarged joints. *Knuckling over* in calves. In pigs, crooked legs and enlarged hocks. Impaired reproduction (testicular degeneration of males; defective ovulation of females). Slipped tendons (perosis) in poultry.	Excess Ca and P decreases absorption. Mn is not toxic in moderate excesses.	Trace mineralized salt containing 0.25% manganese (or more). Rice, wheat, hays, blackstrap molasses, cottonseed hulls.	The manganese content of plants is dependent on soil content. Plants grown on alkali soils may be abnormally low in manganese.
Molybdenum (Mo)	As a component of three different enzyme systems involved in the metabolism of carbohydrates, fats, proteins, sulfur-containing amino acids, nucleic acids, and iron. As a component of the enzyme xanthine oxidase especially important in poultry for uric acid formation. Stimulates action of rumen organisms.	Toxic levels of Mo are of greater practical concern than deficiencies.	Molybdenum utilization is reduced by excess copper, sulfate, and tungsten. Mo is related to uric acid formation in poultry and microbial action in ruminants. Mo as a toxic mineral affects cattle and sheep grazing pastures grown on soils high in Mo content. Toxic levels of Mo interfere with copper metabolism; hence, increase copper requirements.	No Mo supplementation of normal rations is necessary.	Mo toxicity results in severe scours and loss of condition.
Selenium (Se)	Not completely known. But involved in vitamin E absorption and/or retention. Also, a required nutrient in its own right. Se prevents degeneration and fibrosis of the pancreas in chicks. Component of the enzyme glutathione peroxidase, which protects against oxidation of polyunsaturated fatty acids. Protects tissue against certain poisonous substances, such as arsenic, cadmium, and mercury. Interrelation with vitamin E.	Nutritional muscular dystrophy, called *white muscle disease* in calves and *stiff lamb disease* in lambs. Exudative diathesis in poultry. Liver necrosis in pigs.	Selenium is closely related to vitamin E and the sulfur-containing amino acids. Animals consuming forage or grain produced on seleniferous soils develop blind staggers or alkali disease, characterized by emaciation, loss of hair, soreness and sloughing of hooves, lameness, anemia, excess salivation, grinding of the teeth, blindness, paralysis, and death. In poultry, egg production and hatchability are reduced and deformities are common, including lack of eyes and deformed wings and feet.	Sodium selenate, sodium selenite. Marine by-products, cereal grains, wheat by-products, and plants grown on selenium-rich soils.	In 1987, FDA approved a maximum of 0.3 ppm selenium in complete feed for all classes of animals.
Silicon (Si)	Necessary for normal growth and skeletal development of the chick and rat.	Deficiency in chicks and rats results in retarded growth and skeleton deformities, especially in the skull.	From a practical standpoint, adverse effects of high Si intake, rather than Si deficiency, appear to be of concern.	One of most abundant elements on earth. Present in large amounts in soils and plants.	On purified diets, the addition of Si has increased the growth rate of chicks and rats.
Zinc (Zn)	Zinc is needed in normal skin, bones, hair, and feathers. Zinc is a component of several enzyme systems, including peptidases and carbonic anhydrase. Also, Zn is required for normal protein synthesis and metabolism and is a component of insulin. Zinc imparts bloom to the hair coat.	Loss of appetite and stunted growth. Poor hair or feather development; slipping of wool. Rough and thickened skin in swine, known as parakeratosis.	Excess Ca reduces the absorption and utilization of Zn, precipitating parakeratosis in swine. Excess Zn interferes with Cu metabolism and may cause anemia.	Zinc carbonate. Zinc sulfate. Fish meal. Corn gluten feed and meal. Poultry by-products. Distillers' solubles.	Zinc availability is affected adversely by phytates.

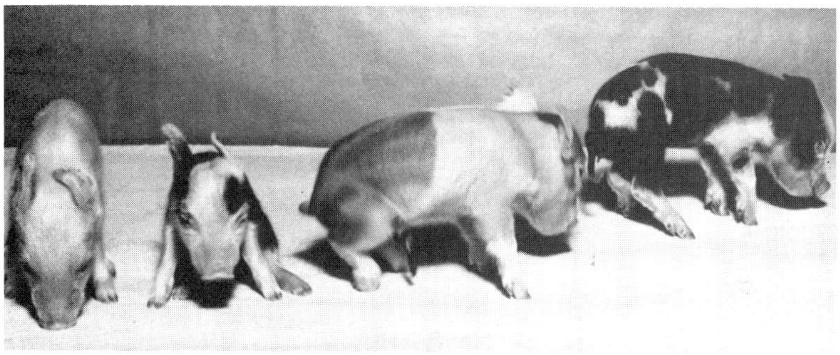

Fig. 4-11. Manganese deficiency. Litter from a sow that was fed 0.5 ppm of manganese. The pigs showed weakness and poor sense of balance at birth. (Courtesy, Purdue University, Lafayette, IN)

Fig. 4-12. Zinc deficiency. Calf showing loss of hair on legs and severe scaliness, cracking, and thickening of the skin. (Courtesy, University of Georgia, Athens)

VITAMIN NEEDS

Vitamins are substances that are required in minute amounts by one or more animal species for normal growth, production, reproduction, and/or health.

The omission of a single vitamin from the diet of a species that requires it will produce specific deficiency symptoms. Many of the vitamins function as coenzymes (metabolic catalysts); others have no such role, but perform certain essential functions.

Many phenomena of vitamin nutrition are related to solubility—vitamins are soluble in either fat or water. Consequently, it is important that both nutritionists and producers be well informed about solubility differences in vitamins and make use of such differences in programs and practices. Based on solubility, vitamins may be grouped as follows:

The Fat-Soluble Vitamins	**The Water-Soluble Vitamins**
Vitamin A	Biotin
Vitamin D	Choline
Vitamin E	Folacin (folic acid)
Vitamin K	Inositol
	Niacin (nicotinic acid, nicotinamide)
	Pantothenic acid (vitamin B–3)
	Para-aminobenzoic acid (PABA)
	Riboflavin (vitamin B–2)
	Thiamin (vitamin B–1)
	Vitamin B–6 (pyridoxine, pyridoxal, pyridoxamine)
	Vitamin B–12 (cobalamins)
	Vitamin C (ascorbic acid, dehydroascorbic acid)

It is noteworthy that vitamin C is the only member of the water-soluble group that is not a member of the B family.

The two groups of vitamins exhibit the following differences that distinguish them both chemically and biologically.

The fat-soluble vitamins contain only carbon, hydrogen, and oxygen, whereas the water-soluble B vitamins contain these three elements plus nitrogen and occasionally sulfur.

Vitamins originate primarily in plant tissues; with the exceptions of vitamins C and D, they are present in the animal tissues only if an animal consumes feed containing them or harbors microorganisms that synthesize them. Fat-soluble vitamins can occur in plant tissue in the form of a provitamin (or precursor of a vitamin), which can be converted into a vitamin in the animal body. Also, the B vitamins are universally distributed in all living tissues, whereas the fat-soluble vitamins are completely absent from some.

The fat-soluble vitamins are stored in appreciable quantities in the body, whereas the water-soluble vitamins are not. Any of the fat-soluble vitamins can be stored wherever fat is deposited; and the greater the intake, the greater the storage. By contrast, the water-soluble B vitamins are not stored in any appreciable amount. Moreover, the large amounts of water which pass through the body daily tend to carry out the water-soluble vitamins, thereby depleting the supply. Hence, they should be supplied in the diet on a daily basis. However, because all living cells contain all the B vitamins, and because the body conserves nutrients that are in short supply by using them only in vital reactions, deficiency symptoms do not appear immediately following their removal from the diet.

Each of the vitamins is as much a distinct chemical compound as is cane sugar, for example. Even when added to the diet in very small amounts, vitamins are extraordinarily potent.

Table 4–2 (next page) contains a list of the 16 vitamins, the existence of which is undisputed with the vitamins listed alphabetically within each of two classifications: (1) fat-soluble vitamins, and (2) water-soluble vitamins.

Fig. 4–13. Effect of vitamin A deficiency on reproduction. *Upper:* This heifer received a vitamin A deficient, but otherwise complete, ration. She became night blind, aborted, and retained the placenta. *Lower:* This heifer received the same ration plus vitamin A. She produced a normal, vigorous calf. (Courtesy, California Ag. Exp. Sta., Davis, CA)

TABLE 4-2
ANIMAL VITAMIN CHART

Name of Vitamin	Animals Most Affected	Functions	Some Deficiency Symptoms	Good Sources for Animals	Comments
Fat-soluble Vitamins					
Vitamin A	Affects all farm animals, including poultry.	Bone growth. Night vision (formation of visual purple in the eye). Prevents xerophthalmia. Essential for body growth, bone growth, and normal tooth development. Epithelial tissue maintenance—respiratory, urogenital and digestive tracts, and the skin.	Stunted growth or loss in weight and loss of appetite, xerophthalmia (an eye disease), night blindness, nervous incoordination as shown by a staggering gait, unsound teeth, rough, dry skin, and sterility in males and females or young which are born weak or dead. Chicks: Wobbly gait. Hens: Reduced egg production and hatchability.	Vitamin A can be provided as the synthetic vitamin or as its precursor, carotene. Rich sources of carotene follow: Green, leafy hays, not over 1 year old. Grass silages. Lush, green pastures. Yellow corn. Green and yellow peas. Fish oils. Carrots. Whole milk. Dehydrated alfalfa meal.	Vitamin A is found only in animals; plants contain the precursor, carotene. Animals are able to store considerable vitamin A, but because of their greater requirements and less storage, young animals suffer from a deficiency much sooner than those that are mature. Both carotene and vitamin A are readily destroyed by oxidation, thus resulting in considerable losses in processing and storing (as in making or storing of hay).
Vitamin D	Affects all farm animals, including poultry.	Aids in the assimilation and utilization of calcium and phosphorus and necessary in normal bone development of animals, including the bone of the fetus. Promotes sound teeth.	Rickets in young. Osteomalacia in adults. Tetany, characterized by muscle twitching, convulsions, and low serum calcium. Chicks: Reduced growth, soft bones (rickets), leg deformities. Hens: Poor eggshells and lowered hatchability.	Vitamin D_2 (irradiated ergosterol), the plant form. Vitamin D_3, the animal form. Sunlight. Sun-cured hays. Cod and certain other fish-liver oils. Irradiated yeast.	Most mammals can use either D_2 or D_3, but birds require vitamin D_3. When animals are exposed sufficiently to direct sunlight, the ultraviolet light in the sunlight penetrates the skin and produces vitaim D from traces of certain cholesterols in the tissues. Tissue storage is very limited. The vitamin D requirement is less when a proper balance of calcium and phosphorus exists.
Vitamin E	Calves, sheep, horses, poultry, rats, and perhaps certain other animals.	Antitoxidant. A sparer of selenium. As an essential factor for the integrity of red blood cells. Essential in cellular respiration, primarily in heart and skeletal muscle tissue. Regulator in the synthesis of DNA, vitamin C, and coenzyme Q.	Muscular dystrophy (stiff-lamb disease in lambs and white muscle disease in calves). Reproductive failure. Steatitis. Chicks: Encephalomalacia (crazy chick disease). Hens: Poor hatchability.	Alpha-tocopherol. Rice polishings. Wheat germ meal. Alfalfa leaf meal. Green grass. Early cut hay.	Vitamin E is widely distributed in all natural feeds. Utilization of vitamin E is dependent on adequate selenium.
Vitamin K	All species, but ruminants have the advantage of microbial synthesis.	Essential for blood clotting.	Prolonged blood clotting time, generalized hemorrhages, and death in severe cases.	Menadione (vitamin K_3). Green pastures. Well-cured hays. Fish meal. In general, this factor is widely distributed in normal farm rations. Also, all classes of farm animals synthesize it.	Vitamin K has definite value in human therapy where clotting of the blood is impaired due to a deficiency of the vitamin. Menadione is widely used commercially as a source of vitamin K. Well-known antagonists of vitamin K are dicoumarol and warfarin.

(Continued)

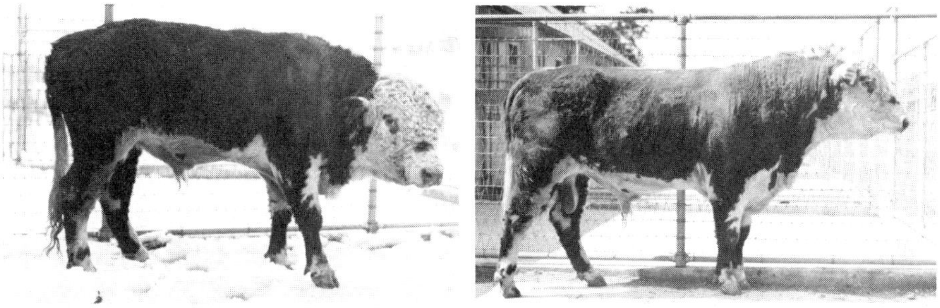

Fig. 4-14. Same bull before (left) and after (right) vitamin A (carotene) feeding. At the left, the bull shows advanced stages of Vitamin A deficiency—note the dejected appearance and rough hair coat. At the right, the bull shows a general improvement in appearance and male characteristics following vitamin A feeding. (Courtesy, USDA)

Fig. 4-15. Advanced case of rickets caused by a deficiency of vitamin D. The pig was fed indoors, without exposure to sunlight. Because of leg abnormalities, it was unable to walk. Later, the pig responded to vitamin D. (Courtesy, University of Saskatchewan, Saskatoon, Canada)

TABLE 4-2 *Continued*

Name of Vitamin	Animals Most Affected	Functions	Some Deficiency Symptoms	Good Sources for Animals	Comments
Water-soluble Vitamins					
Biotin	Required by all species.	Biotin is required in many reactions in the metabolism of carbohydrates, fats, and proteins. Biotin serves as a coenzyme for transferring CO_2 from one compound to another. Biotin also serves as a coenzyme for deamination (removal of NH_2) of amino acids for the production of energy.	Pigs exhibit spasticity of the hind legs, cracks in the feet, and a dermatitis. There is also lowered efficiency of feed utilization. Chicks and turkey poults show dermatitis and perosis. In hens, hatchability is severely reduced. In mink, biotin deficiency makes for abnormal fur.	Synthetic biotin. Alfalfa leaf meal (dehy.) Rice polishings. Yeast. Distillers' solubles. Safflower meal. Cottonseed meal. Blackstrap molasses.	Ordinary farm rations probably contain ample biotin, or farm animals synthesize all they need. Biotin is rendered unavailable by raw egg white.
Choline	Swine, rats, and poultry.	Choline is involved in the prevention of fatty livers, in transmitting nerve impulses, and in the metabolism of fat.	Poor growth and fatty livers in most species. In chickens and turkeys, slipped tendon (perosis). In swine, abnormal gait in growing pigs and reproductive failure in adult females.	Choline chloride or choline dihydrogen. Rice polishings. Soybean lecithin. Wheat germ. Yeast. Rapeseed (canola) meal. Poultry by-products. Fish meal.	With a high-protein diet, enough choline is synthesized from certain precursors and amino acids. Deficiency symptoms are more readily obtained as the protein content is lowered.
Folacin (folic acid)	All animals and birds may be affected.	Folacin enzymes are responsible for the following important functions: (1) the formation of purines and pyrimidines; (2) the formation of heme; (3) the interconversion of the amino acid serene to the amino acid glycine; (4) the formation of the amino acids tyrosine from phenylalanine and glutamic acid from histidine; (5) the formation of the amino acid methionine from hemocystene; (6) the synthesis of choline from ethanolamine; and (7) the conversion of nicotinamide to N-methylnicotinamide.	Macrocytic anemia (of young) and macrocytic anemia (of pregnancy). In chicks, retarded growth and depigmentation of colored feathers. In humans and dogs, a sore, red, smooth tongue, disturbance of the digestive tract, and poor growth.	Synthetic folacin. Wheat germ. Yeast. Soybean meal. Alfalfa hay. Cottonseed meal. Linseed meal.	Folic acid is widely distributed in both plants and animals. It was given this name because of the abundance of the factor in plant leaves.
Inositol	Chicks, fish, swine, guinea pigs, hamsters, rats, and mice.	Not known. But it appears to aid in the metabolism of fats and helps reduce blood cholesterol. In combination with choline, it prevents hardening of arteries and protects the heart. As a precursor of phosphoinosities, which is found in various tissues.	Not demonstrated in animals.	Synthetic inositol. Wheat germ. Yeast. Liver meal. Citrus meal. Blackstrap molasses.	Widely distributed in animal feeds. Synthesized in intestines.
Niacin (Nicotinic Acid, Nicotinamide)	It is a dietary essential of pigs, chickens, monkeys, and humans. Apparently synthesized in the digestive tract of ruminants (sheep and cattle) and the horse.	Constituent of two coenzymes, which are necessary in cell respiration; in the release of energy from carbohydrates, fats, and protein; and in biological oxidation-reduction systems.	Reduced growth and appetite. Swine exhibit diarrhea, vomiting, dermatitis, unthriftiness, and ulcerated intestine. Chicks show poor feathering, scaly dermatitis, and sometimes, a "spectacled eye." Dogs show a darkening of the tongue (black tongue) and mouth lesions. Humans develop pellagra characterized by a bright red tongue, mouth lesions, anorexia, and nausea.	Synthetic niacin. Rice polishings. Yeast. Rice bran. Marine by-products. Liver meal. Animal by-products. Green alfalfa is a fair source.	Niacin is the most stable of the B-complex vitamins. Niacin present in most cereal grains is not available to the pig and other simple-stomached animals. Niacin can be synthesized in the body from surplus tryptophan. Mature ruminants do not need dietary niacin under most conditions because of synthesis of rumen microflora.

(Continued)

Fig. 4-16. Biotin made the difference! (Courtesy, Washington State University, Pullman)

Fig. 4-17. Choline made the difference! (Courtesy, University of Illinois, Champagne-Urbana)

TABLE 4–2 *Continued*

Name of Vitamin	Animals Most Affected	Functions	Some Deficiency Symptoms	Good Sources for Animals	Comments
Pantothenic Acid (vitamin B–3)	Rats, dogs, pigs, chickens, and turkeys. Synthesized in rumen of cow and sheep; perhaps the horse also synthesizes it.	Component (1) of coenzyme A, required for energy metabolism; and (2) of acyl carrier protein (ACP). ACP, along with CoA, is required by the cells in the biosynthesis of fatty acids.	All species exhibit reduced growth, loss of hair, and enteritis. Pigs develop "goose-stepping" gait. Chicks show dermatitis and embryonic death. Dogs vomit and show fatty infiltration of liver. Mature ruminants synthesize pantothenic acid in rumen. Signs of deficiency in calves are rough coat, dermatitis, anorexia, and loss of hair around eyes.	Calcium pantothenate. Rice polishings. Yeast. Safflower meal. Whey. Fish solubles. Blackstrap molasses. Alfalfa meal.	Grain is very deficient in pantothenic acid. Of all B vitamins, it is most likely to be deficient under drylot conditions. Pantothenic acid is commonly added to commercial swine and poultry rations.
Para-aminobenzoic Acid (PABA)	Essential growth factor for certain microorganisms.	For higher animals, PABA functions as an essential part of the folacin molecule.	Not demonstrated in animals.	Synthetic para-aminobenzoic acid. Lecithin. Wheat germ. Yeast. Fish meal. Soybean meal. Peanut meal. Blackstrap molasses.	Abundantly synthesized in intestines.
Riboflavin (vitamin B–2)	Thought to be required by all animals, but deficiency symptoms not observed in ruminants, perhaps due to rumen synthesis. Deficiency symptoms noted in poultry, swine, and horses.	Promotes growth and functions in the body as a constituent of several enzyme systems and as such is important in carbohydrate, fatty acid, and amino acid metabolism.	Retarded growth in most species, with a wide variety of other symptoms somewhat variable with the species. Periodic ophthalmia (moon blindness) in horses; reproductive failure in the sow, and slow growth, anemia, diarrhea, unthrifty appearance, eye opacities, and an abnormal gait in the young pig; and curled toe paralysis in birds.	Synthetic riboflavin. Yeast. Skim milk. Whey. Liver meal. Alfalfa hay. Grass (immature/green). Poultry by-product meal.	Grains are poor source of riboflavin. Many common rations are borderline or deficient in riboflavin, especially swine and poultry rations. Riboflavin is destroyed by light or heat.
Thiamin (vitamin B–1)	All animals must have a dietary source, unless there is rumen synthesis, as in cattle and sheep.	As a coenzyme in energy metabolism. In the functioning of the peripheral nerves. Maintains (1) normal appetite, (2) tone of the muscles, and (3) healthy mental attitude.	Reduction in appetite (anorexia) and loss in weight. Cardiovascular disturbances. Beriberi (in humans). Lowered body temperature. Chicks: Polyneuritis (retraction of the head). Hens: Lowered egg production.	Thiamin hydrochloride. Thiamin mononitrate. Rice polishings. Wheat germ meal. Yeast. Rice bran. Wheat and wheat by-products. Cottonseed meal.	Fats exhibit a thiamin-sparing effect.
Vitamin B–6 (pyridoxine, pyridoxal, pyridoxamine)	B–6 is a dietary essential for the rat, pig, chick, and dog. It is synthesized in the rumen of cattle and sheep and perhaps in the cecum of the horse; thus, no deficiency symptoms in these species have been reported.	As coenzyme in protein and nitrogen metabolism. Involved in red blood cell formation and in absorption of amino acids. Involved in carbohydrate and fat metabolism. Involved in clinical problems, including (1) anemia that is iron-resistant, (2) kidney stones, and (3) physiological demands of pregnancy.	All species exhibit convulsions. Pigs show anorexia, poor growth, and convulsions. Chicks show retarded growth and abnormal feathering. Hens show lowered egg laying and hatchability. Rats develop a specific dermatitis.	Safflower meal. Fish solubles. Pasture (green). Meat meal and tankage. Wheat and wheat by-products. Alfalfa hay.	Normally, animal rations are not lacking in vitamin B–6.

(Continued)

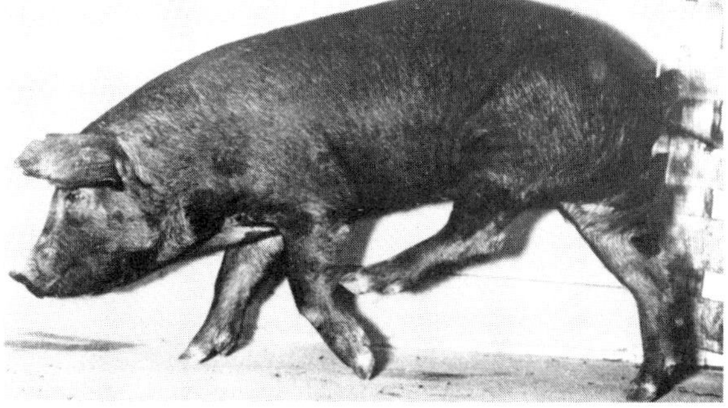

Fig. 4–18. Pantothenic acid deficiency. Note locomotor incoordination (goose stepping), which was produced by feeding a corn-soybean meal ration low in pantothenic acid. (Courtesy, Michigan State University, East Lansing)

Fig. 4–19. Thiamin deficiency in littermate pigs. *Right:* Pig received no thiamin. *Left:* Pig received the equivalent of 2 mg thiamin/100 lb liveweight. Otherwise, their diets were the same. (Courtesy, USDA)

TABLE 4-2 *Continued*

Name of Vitamin	Animals Most Affected	Functions	Some Deficiency Symptoms	Good Sources for Animals	Comments
Vitamin B-12 (cobalamins)	Swine, rats, poultry, and humans. Ruminants synthesize B-12 unless cobalt is deficient.	Vitamin B-12 functions in two enzyme forms: coenzyme B-12 and Methyl B-12. Also, the role of B-12 is closely related to other vitamins.	All animals show retarded growth. Pigs show uncoordinated hind leg movements; and there is reproductive failure in sows. Eggs from B-12-deficient hens fail to hatch.	Synthetic B-12. Protein supplements of animal origin. Fermentation products.	B-12 is apt to be lacking in swine and breeder poultry rations.
Vitamin C (ascorbic acid, dehydroascorbic acid)	Dietary need is limited to humans, guinea pigs, monkeys, fur-eating bats, and bulbul birds. Probably required by other species but synthesized in the body.	Collagen formation. Metabolism of the amino acids, tyrosine and tryptophan. Absorption and movement of iron. Metabolism of fats and lipids, and cholesterol control. Sound teeth and bones. Strong capillary walls and healthy blood vessels. Metabolism of folic acid. As a general antioxidant. Requirement increases in periods of stress.	Scurvy; swollen, bleeding and ulcerated gums, loosening of teeth, and weak bones.	Vitamin C (ascorbic acid) Acerola cherry. Rose hips. Citrus pulp. Well-cured hay. Green pasture.	Ordinary farm rations and body synthesis provide adequate vitamin C.

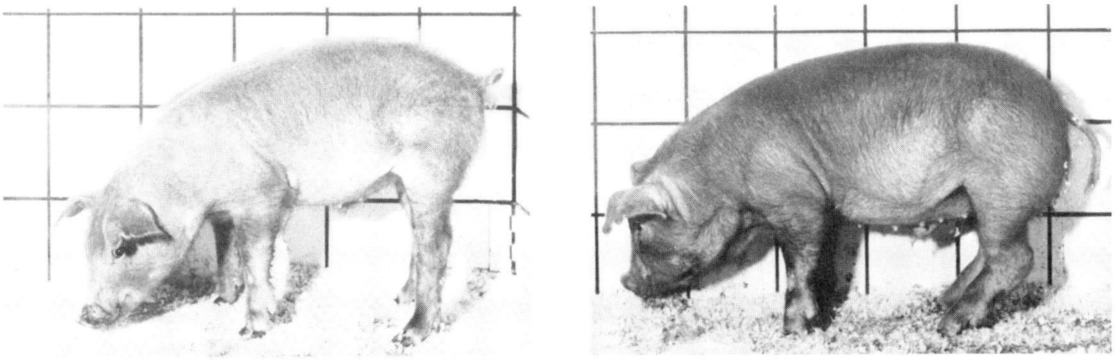

Fig. 4-20. Vitamin B-12 deficiency. *Left:* Pig deficient in vitamin B-12. Note rough hair coat and dermatitis. *Right:* Control pig. (Courtesy, Iowa State University, Ames)

UNIDENTIFIED FACTORS

In addition to the vitamins listed in Table 4-2, certain unidentified or unknown factors are important in animal nutrition. They are referred to as *unidentified* or *unknown* because they have not yet been isolated or synthesized in the laboratory. Nevertheless, rich sources of these factors and their effects have been well established. There is evidence that the unknown factors exist in dried whey, marine and packinghouse by-products, distillers' solubles, antibiotic fermentation residues, alfalfa meal, and certain green forages. There is also evidence that at least one unknown hatchability factor is in fish solubles and green forage. Most of the unidentified factor sources are added to the diet at a level of 1 to 3%.

WATER NEEDS

Water is one of the most vital of all nutrients. In fact, animals can survive for a longer period without feed than they can without water. Fortunately, under most conditions, it can be readily provided in abundance and at little cost. In addition to what animals drink, water is found in all feeds, ranging from about 10% in air-dry feeds to over 80% in fresh green forage.

Surplus water is excreted from the body, principally in the urine, and to a slight extent in the perspiration, feces, and water vapor from the lungs.

The specific water requirements of each class of animals will receive further consideration in the sections devoted to the respective species. In general, however, under practical conditions, the needs for water can best be taken care of by allowing the animals free access to plenty of clean, fresh water at all times.

FEEDS

Feed (or feedstuff) is any ingredient, or material, fed to animals for purposes of sustaining them. Most feedstuffs provide one or more nutrients, but nonnutritive products may be fed for such purposes as providing flavor, color, or other factors related to palatability, adding bulk, or preserving feeds.

A wide variety of feedstuffs can be, and are, used for animal feeding throughout the world. More than 2,000 different products have been classified as animal feeds, not counting varietal, grade, and stage of maturity differences. However, as shown in Table 4–3, relatively few of these products make up the great bulk of the U.S. feed supply.

TABLE 4–3
ANIMAL FEEDS CONSUMED IN THE UNITED STATES[1]

	Acreage Harvested	Used for Feed	Yield Per Acre
	(1,000 acres)	(1,000 tons)	(tons)
Hay (all):	60,748	149,302	2.46
Alfalfa	25,535	84,794	3.32
All other hay	35,215	64,508	1.83
Silage:			
Corn	5,829	84,468	14.5
Sorghum	424	5,157	12.2
Grains:		(1,000 bu)	(bu)
Corn	59,208	7,072,073	119.4
Sorghum	10,604	739,249	69.7
Barley	10,057	529,530	52.7
Oats	6,925	374,000	54.0
High-protein:		(1,000 metric tons)	
Oilseed meals:			
Soybean		19,410	
Cottonseed		1,429	
Sunflower		390	
Canola		205	
Linseed		120	
Peanut		100	
Animal proteins:			
Tankage and meat meal		2,471	
Fishmeal and solubles		481	
Milk products		364	
Grain protein feeds:			
Gluten feed and meal		1,321	
Distillers' dried grains		1,046	
Brewers' dried grains		120	
Miscellaneous feeds:			
Wheat millfeeds		5,659	
Molasses		1,605	
Fats and oils		888	
Dried and molasses beet pulp		662	
Alfalfa meal		554	
Rice millfeeds		548	
Urea[2]		265	
Miscellaneous by-product feeds		1,267	

[1]USDA sources. 1987 data.
[2]Estimate for the feed year 1986–87 from *Feed Management*, June, 1987, p. 18.

CLASSES OF FEEDS

The number of feedstuffs is so great that it is impossible to cover each of them in this book. Rather, we shall classify them, then discuss their nutritional properties by groups. For convenience, the commonly used feeds are herein classified as (1) roughages, (2) concentrates, (3) by-product feeds, (4) protein supplements, (5) minerals, (6) vitamins, (7) special feeds, and (8) additives, implants, and injections.

Table 4–4 shows that roughages account for 61.7% of all U.S. livestock feeds, and that concentrates account for 38.3%. Of course, the proportion of roughages to concentrate consumption varies widely according to relative price and the class of animal. As shown in Table 4–4, sheep and goats head the list of roughage consumers, with 93.8% of their total feed coming therefrom, including pasture. Beef cattle obtain 84.5% of their feed from roughages. Swine consume only a negligible amount of roughage.

TABLE 4–4
PERCENTAGE OF FEED FOR DIFFERENT CLASSES OF U.S. LIVESTOCK DERIVED FROM (1) ROUGHAGES, INCLUDING PASTURE; AND (2) CONCENTRATES[1]

Class of Animal	Roughages	Concentrates
	(%)	(%)
Sheep and goats	93.8	6.2
Beef cattle	84.5	15.5
Horses and mules	73.0	27.0
Dairy cattle	58.7	41.3
Swine	4.3	95.7
All livestock	61.7	38.3

[1]USDA, Economic Research Service. Data for the feed year 1983–1984.

ROUGHAGES

Roughages are bulky feeds that are low in weight per unit of volume, contain more than 18% crude fiber, and are low in energy. They are the natural feeds of all herbivorous animals, including ruminants and horses. Although swine can survive solely on roughages, productivity is too low to be economical.

In the sections that follow, the following groups of roughages will be discussed: pasture, hay, crop residue, silage, haylage (low-moisture silage), green chop, and other roughages.

- **Pasture**—*A pasture is an area of land on which there is a growth of forage that animals may graze.* Pasture and rangeland account for 29.2% of the total land area of the United States, including Alaska and Hawaii.

Broadly speaking, all U.S. pastures may be classified as either (1) seeded pastures, or (2) native pastures. Although no sharp line of demarcation exists between the two groups, seeded pastures include those which either receive more than approximately 20 in. of rainfall annually or are irrigated. They are the seeded (cultivated) pastures of the Corn Belt, the South, the East, and the irrigated areas, and smaller and scattered moderate-to-high rainfall areas throughout the West. The native pastures include those range pastures which receive less than 20 in. of rainfall annually.

Pasture may be further classified as—

1. **Permanent pastures.** Those which, with proper care, last for many years. They are most commonly found on land that cannot be used profitably for cultivated crops, mainly because of topography, moisture, or fertility. The vast majority of the farms of the United States have one or more permanent pastures, and most range areas come under this classification.

2. **Semipermanent or rotation pastures.** Those that are used as a part of the established crop rotation. These are seeded pastures that are generally used for 2 to 7 years before plowing.

3. **Temporary and supplemental pastures.** Those that are used for a short period, usually annuals, such as Sudan grass, sorghum, millet, rye, barley, wheat, oats, rape, or soybeans. They are generally seeded for the purpose of providing supplemental grazing during the season when the regular permanent or rotation pastures are relatively unproductive.

(Also see Section 5, Pasture and Range Forages.)

• **Hay**—*Hay is forage harvested during the growing period and preserved for drying and subsequent use.* It is the most important harvested roughage for U.S. livestock, and ranks third among all livestock feeds, being exceeded only by pasture and corn (see Fig. 4–2). The importance of the nation's crop is further attested to by the fact that more than 60 million acres of hay, producing more than 140 million tons, worth more than $9 billion, are harvested annually.

Hay varies more in nutritive value than any other feed, primarily because of (1) differences in the crop from which it is made, (2) stage of cutting, (3) handling, and (4) possible weather damage during curing. Average quality hay will run 25 to 35% crude fiber and 45 to 55% TDN.

Hays are made from legumes, grasses, or cereal crops. In terms of total tonnage produced annually, alfalfa accounts for approximately 40% of the nation's hay production. Many different kinds of hay make up the other 60% of the country's hay supply; among them, cereal hays made from oats, barley, wheat, and rye, and grass hays made from Bermuda grass, prairie grass, redtop, Johnson grass, orchard grass, and timothy.

Hay is primarily a cattle, sheep, and horse feed, although dehydrated alfalfa may be included in swine and poultry rations.

(Also see Section 6, Hay and Crop Residues.)

• **Crop Residues**—*Crop residues are the portions of crops that are normally left in the field following harvest.* Among such crop residues are: corn stalks and husklage, sorghum stalks, soybean refuse, small grain straws and chaff, and legume and grass seed straws. Crop residues must be fed to the right class of animals, and they must be properly supplemented.

Of all the crop residues, the residue of corn is produced in greatest abundance in the United States and offers the greatest potential for expansion in animal numbers. Corn usually produces an amount of residue equal to the quantity of the grain produced. So, the more than 200 million tons of corn grain produced each year results in more than 200 million tons of corn residue, an amount approximately equal to all other kinds of crop residue combined. Each year, that's more than 200 million tons of potential cow feed, enough to winter more than 150 million dry pregnant cows.

Mature cows are physiologically well adapted to utilize crop residues. Such feeds will meet the daily energy (TDN) needs of dry pregnant cows, but they are slightly deficient in protein, and low in phosphorus and carotene.

(Also see Section 6, Hay and Crop Residues)

• **Silage**—*Silage is fermented forage plants.*

Silage making is one of the 3 common methods of utilizing forage crops, the other 2 methods being pasturing and haymaking. Pasturing is the least expensive of the 3 methods, but it is seasonal in nature. In the spring and early summer, forage plants generally grow faster than they can be utilized by normal grazing, and become dormant in cold weather.

The importance of silage in this country is evidenced by the fact that about 120 million tons of it are made annually.

Most silage in the United States is made from either corn or sorghum, with corn silage far in the lead—over 16 times as much corn as sorghum silage is made. At the present time, it is estimated that 65% of the nation's silage is made from corn and sorghum and 35% from grasses, legumes, and other feeds. Among the other feeds made into silage are small grains, waste from food processing (sweet corn, green beans, and green peas), root crops, and various vegetable residues.

Silage is primarily a beef and dairy feed, where it is used as part or the only roughage in the ration. It is also a good sheep feed. Sometimes it is fed to brood sows. Very little silage is fed to horses.

From 2½ to 3 lb of silage are required to replace 1 lb of hay, due to the lower dry matter content (usually only 25 to 35%) of the silage.

(Also see Section 7, Silage/Haylage/High-Moisture Grain.)

• **Haylage (low-moisture silage)**—*Haylage is made from grass and/or legume that is wilted to 40 to 60% moisture content before ensiling.* Properly made haylage has a pleasant aroma and is a palatable, high-quality feed. Animals usually receive more dry matter and net feed value in haylage than in silage made from the same cut.

Haylage is easy to prepare and preserve in a gas-type silo where air is excluded. But it can be made in a conventional silo provided certain precautions designed to keep out the air are taken.

Haylage is growing in popularity, especially as a dairy feed. Its nutritive value depends on the stage of the growth of the crop when cut and the percentage of dry matter in it.

(Also see Section 7, Silage/Haylage/High-Moisture Grain.)

• **Green chop (soilage)**—*Green chop, or soilage is fresh herbage that is cut and chopped in the field, then transported and fed to animals in confinement.* Legumes, Sudan grass, and corn are sometimes used in this manner. With tall growing crops, 50% more feed value may be realized from a given area than can be obtained by any other method of harvesting. However, green chop requires special equipment and harvesting every day. Also, there are harvesting problems in wet weather, and there is an inevitable change in feed quality as the season progresses.

Most green chop is fed to lactating dairy cows, usually in combination with hay or silage because the total intake tends to be greater.

(Also see Section 5, Pasture and Range Forages)

• **Other roughages**—Among the other roughages (other than pasture, hays, crop residue, silage, haylage, and green chop) used for livestock are cottonseed hulls, corncobs, sawdust and other wood products, oat hulls, beet tops, root crops, peanut hay, newspapers, and a host of others. When properly (1) used for the right species and class of animal, (2) combined with a high-quality legume roughage, and/or (3) supplemented with the necessary protein, minerals, and vitamins, all of them are excellent feeds. Availability, costs, and results should be the determining factors in their use, just as the economics of the situation should determine the use of any other feed ingredient.

CONCENTRATES

Concentrates are feeds that are high in energy and low in fiber (under 18%).

Many different kinds of concentrate feeds can be, and are, used as animal feeds. Availability and price are the two most important factors determining the choice of concentrates. Consideration of the latter factor—price—necessitates that feeders be keen students of values. They must change the formulations of their rations in keeping with comparative feed prices. Fig. 4–21 shows the tonnage of feed concentrates fed to U.S. livestock and poultry from 1977 to 1987.

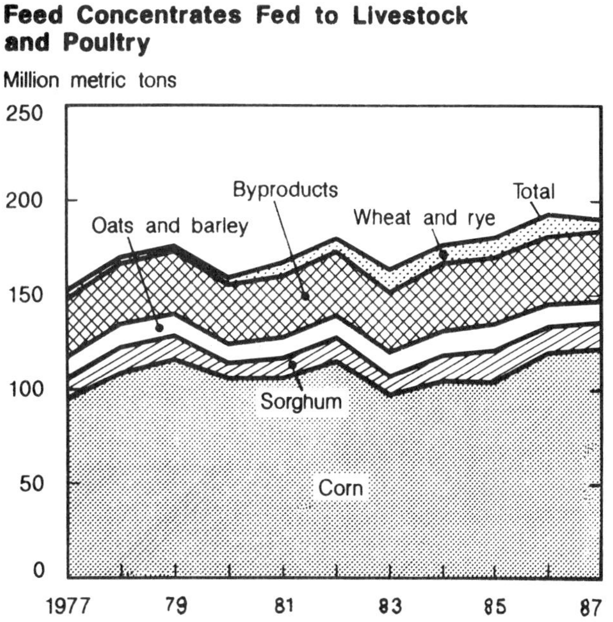

Fig. 4–21. This shows the concentrates, including by-product feeds (oilseed meals, animal protein feeds, and milk by-products only), fed to U.S. livestock and poultry, 1977–87. This figure shows (1) that corn is by far the most important livestock feed, (2) that by-products rank second, and (3) that wheat is of minor importance as a livestock feed. (From: 1988 *Handbook of Agricultural Charts*, USDA, p. 97)

Corn is the most common grain fed to livestock; about 9 times as much corn as sorghum is fed in the United States. It is palatable and rich in the energy-producing carbohydrates and fats (80% TDN), and low in fiber. Also, corn is easily stored, with only moisture and carotene being lost over a period of time. Corn-and-cob meal (consisting of 20 to 25% cob) is excellent for finishing cattle. However, corn has certain very definite limitations—it is low in protein (and in the essential amino acids lysine, methionine, and tryptophan) and calcium.

The grain sorghums are assuming an increasingly important role in livestock feeding, particularly in the fringe areas of the Corn Belt, and in the South and Southwest where moisture conditions are less favorable. New and high-yielding varieties have been developed and have become popular. As a result, more and more grain sorghums are being fed to livestock. The chemical composition of sorghum (milo) is similar to corn except that the protein content is generally higher and more variable. Its feeding value is greatly enhanced by proper processing.

Although corn and sorghum are by far the most common feed grains, such grains as barley, rye, oats, wheat, and triticale, are used in many sections of the United States and Canada. The small grains are excellent when properly prepared and used.

BY-PRODUCT FEEDS

By-product feeds are concentrates and roughages other than the primary products from animal and plant processing and from industrial manufacturing.

Innumerable by-products—both roughages and concentrates—from plant and animal processing, and from manufacturing—are standard and valuable livestock feeds; among them, the following: milling by-products from the cereal grains and oilseeds, root crops (cull potatoes and by-products of potato processing, turnips, mangels, swedes, fodder beets, carrots, and parsnips), dried beet pulp, and beet tops (from sugar beet processing), distillery and brewing by-products, unused bakery products, and by-products from numerous fruits and nuts. (Also, see section on "Protein Supplements," which follows.)

As is true of any ration ingredient, the requisites to effective and profitable use of each by-product feed are (1) that it be bought at a favorable price, nutritive composition considered; (2) that its proximate composition be known, and that it be incorporated in a balanced ration; (3) that it be palatable and consumed in adequate quantity; and (4) that it not adversely affect carcass quality, particularly from the standpoint of harmful chemical residues from pesticides applied to crops. Generally speaking, the use of by-product feeds calls for ingenuity and experience in handling them, special knowledge relative to their nutritive qualities and use in balanced rations, and relatively high labor costs. As a result, many feeders are not interested in using them, whereas others find it a lucrative business.

PROTEIN SUPPLEMENTS

Protein supplements are feedstuffs that contain more than 20% protein or protein equivalent. At least 23 amino acids have been identified and may occur in combinations to form an almost limitless number of proteins.

High-protein feeds are usually named and classified according to their origin and method of processing. On the basis of origin, they are usually grouped into two general categories as follows:

1. **Animal proteins.** Animal protein supplements are derived from inedible tissues from meat-packing or rendering plants, from surplus milk or milk products, and from marine sources. They include proteins from meat, fish, poultry, eggs, milk, and their products. With hogs and chickens, one of these protein sources was formerly a must. With the discovery and general availability of vitamin B-12, high-protein feeds of animal origin became less essential for swine and poultry.

2. **Plant proteins.** This group includes the common oilseed by-products—soybean meal, cottonseed meal, linseed meal, peanut meal, safflower meal, sunflower meal, rapeseed meal (canola meal), and coconut (or copra) meal. They vary in protein content and feeding value, depending on the seed from which they are produced, the amount of hull and/or seed coat included, and the method of oil extraction used.

In addition to the oilseed meals, numerous good commercially manufactured protein supplements are available. Usually, they are prepared for a particular class of livestock. They are generally blends of animal and vegetable protein ingredients, with urea added for ruminants. They may also include minerals, vitamins, and/or antibiotics.

Fig. 4–22 shows the kinds and quantities of high-protein feeds fed to U.S. livestock.

High-Protein Feed Use

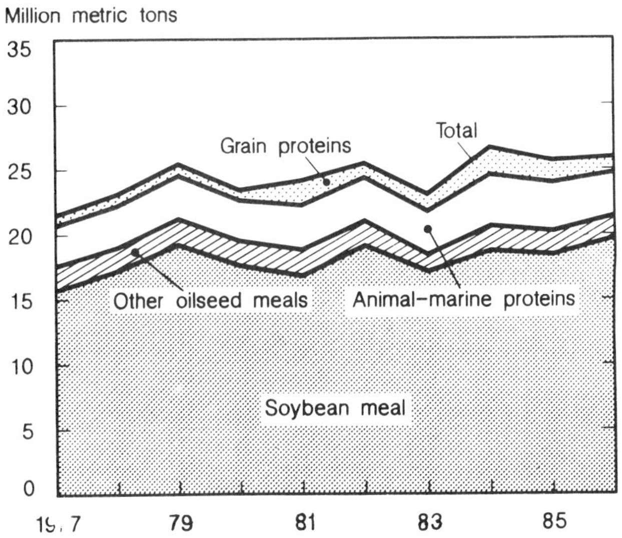

Fig. 4–22. This shows the high-protein feeds fed to U.S. livestock, 1977–86, in 44% protein soybean meal equivalent. The following explanation is pertinent: Grain proteins include gluten feed and meal, and brewers' and distillers' dried grains; animal-marine proteins include tankage, meat meal, marine by-products, and milk products; other oilseed meals include cottonseed, linseed, peanut, sunflower, and copra. This figure shows that soybean meal is by far the most important high-protein feed. (From 1988 *Handbook of Agricultural Charts*, USDA, p. 97)

NONPROTEIN NITROGEN (NPN) SOURCES

Certain nonprotein nitrogen sources may be substituted for all or much of the supplemental protein required in most ruminant rations, provided such rations are adequate in minerals and readily available carbohydrates. Among such products are urea, ammoniated molasses, ammoniated beet pulp, ammoniated cottonseed meal, ammoniated citrus pulp, and ammoniated rice hulls.

The rumen microorganisms—which are a low form of plant life and are able to use inorganic compounds much like plants utilize chemical fertilizers—build proteins of high quality in their cells from sources of inorganic nitrogen that nonruminants cannot use. Since the life-span of these microorganisms is short, further on in the digestive tract the ruminant digests the bacteria and obtains good protein therefrom. In ruminant nutrition, therefore, even such nonprotein sources of nitrogen as urea and ammonia have a protein replacement value. An exception is the very young ruminant in which the rumen and its ability to synthesize are not yet well developed.

• **Urea**—Urea is a white, crystalline, odorless, nonprotein nitrogen compound of the formula N_2H_4CO. It is manufactured in chemical plants that produce anhydrous ammonia by fixing some of the nitrogen of the air, some of the ammonia gas is combined with gaseous carbon dioxide to produce the white crystalline solid urea which is quite stable. In addition to feeds, urea is used as a fertilizer, either dry or in solutions, and in making plastics. Also, it is noteworthy that urea occurs as the principal end product of nitrogen metabolism in nearly all mammals; it is found in the urine of all farm animals and humans.

Approximately 265,000 metric tons of urea are fed annually in the United States, as a source of protein for cattle, sheep, and goats.

Urea may constitute up to one-third of the total protein of the ration of ruminants, provided additional energy is added in the form of molasses or grain to compensate for the lack of energy in the urea, in order to feed properly the rumen bacteria. By total protein is meant the protein intake of the entire ration—including forage, grain, and protein supplements.

• **Slow-release urea products**—Several products in which urea is bound in a slow-release complex have been developed in recent years; among them, urea combined with starch from grain, and urea combined with the sugars in molasses through heat and chemical treatment. These products are designed to decrease the solubility of urea in the rumen and thereby slow the release of ammonia. Slow ammonia release, or a more uniform ammonia level in the rumen throughout the day, is desirable, especially for urea used in low-energy rations. Additionally, there should be less danger of urea toxicity from overconsumption with slow-release products.

Common guidelines relative to the use of urea for cattle are given in Table 4–5.

TABLE 4–5
COMMON GUIDELINES TO THE USE OF UREA FOR CATTLE

	For Finishing Cattle	For Grower (stocker) Cattle	For Wintering Pregnant and Lactating Cows
Percent of total protein in ration from urea .... (%)	33⅓	25.0	25.0
Maximum urea/animal/day .... (lb)	0.22 *(100 g)*	0.15 *(68 g)*	–
Percent of urea, by weight, of total air-dry feed consumed .... (%)	1.0	1.0	1.0
Percent of urea, by weight, of concentrate mix (grain plus protein supplement)[1] .... (%)	2.0–3.0	3.0	3.0
Percent of urea, by weight, of the protein supplement .... (%)	20–30[2]	10.0[3]	10.0
Percent of supplemental nitrogen in high-protein supplement from urea[4] .... (%)	60–90[5]	30.0	30.0
Pounds of urea added/ton of corn silage at ensiling time[6] .... (lb)	10.0 *(4.5 kg)*	10.0 *(4.5 kg)*	10.0 *(4.5 kg)*

[1]Feed intake may be depressed if over 1% is used. Yet, many beef producers are successfully using 2%.

[2]High-urea supplements are best fed in complete mixed rations, which are *thoroughly* mixed. *Supplements containing 20–30% urea require extreme caution when being hand-fed.*

[3]A protein supplement containing 10% urea provides 28.1% of the protein equivalent (281% × .10) from nonprotein nitrogen.

[4]This means that as much as 60–90% of the protein value of the supplement may come from nonprotein sources. However, because such a supplement will constitute only 2–5% of the total ration fed, the first rule of thumb given in Table 4–8 still applies; namely, only ¼–⅓ of the total protein in the ration will be supplied from a nonprotein source.

[5]In a feedlot ration, this may be equivalent to 25–40% of the total nitrogen from all sources.

[6]On a dry matter basis, corn silage ensiled at the well-dented stage contains about 8% protein. The addition of 10 lb of urea per ton (or *5 kg/1,000 kg*) of silage increases the protein content from 8–13%. However, there is loss of flexibility in feeding such a ration, and the rate of gain will be less than can be secured from higher energy, more dense rations. Also, it is extremely important that the urea be well mixed in the silage; otherwise, there is hazard of toxicity.

• **Single-celled protein (SCP)**—*Single-cell protein (SCP) is protein obtained from single-cell organisms, such as yeast, bacteria, fungi, and algae, that have been grown on specially prepared growth media.*

Some single-celled protein types can be useful sources of protein and vitamins for animal feeding. The safety of these feeds depends on the organisms selected, the quality of substrate used, and the conditions of growth. Of course, yeast and bacteria have been used for centuries in the baking, brewing, and distilling industries, in making cheeses and other fermented foods, and in storage and preservation of foods.

Dried brewers' yeast, a residue from the brewing industry, and Torula yeast, resulting from the fermentation of wood residues and other cellulose sources, have long been used as animal feeds.

Bacteria grow faster than yeasts under favorable conditions, doubling their mass in a matter of minutes, rather than hours. Dried bacterial cells contain at least 55% protein.

Algae, single-celled plants which contain about 50% protein on a dry basis, offer an attractive possibility as a protein source. They synthesize proteins by the use of solar energy. Preliminary results with cultivated freshwater algae indicate that they will produce about 10 times as much protein per unit of land area as soybeans.

Considerable research is in progress to convert manure from poultry and other animals through bacterial fermentation into animal protein feed. This recycling process could produce much protein and help solve a pollution problem.

MINERAL SUPPLEMENTS

Mineral supplements are rich sources of one or more of the inorganic elements needed to perform certain essential body functions.

When livestock are fed a mixed feed, totally or in part, the needed minerals are generally incorporated into the ration in keeping with known requirements. This is usually accomplished by adding 0.25 to 0.50% trace mineralized salt to the total ration, plus calcium and phosphorus (and any other minerals that are in short supply) as needed to balance the ration. In addition, where the lower level of salt (0.25%) is incorporated in the ration, trace mineralized salt is usually provided free choice.

Where animals are fed an unmixed ration or are on pasture, minerals may be provided as follows:

1. **Where animals are on liberal grain feeding.** Provide free access to a 2-compartment mineral box, with (a) trace mineralized salt in one side, and (b) in the other side, a mixture of ⅓ trace mineralized salt (salt included for purposes of palatability), ⅓ defluorinated phosphate or steamed bone meal, and ⅓ ground limestone or oystershell flour.

2. **Where animals are primarily on roughage (pasture, hay, and/or silage).** Provide free access to a 2-compartment mineral box, with (a) trace mineralized salt in one side, and (b) in the other side, a mixture of ⅓ trace mineralized salt (salt included for purposes of palatability), and ⅔ defluorinated phosphate or steamed bone meal.

As noted, no limestone or oystershell flour (source of calcium only) is needed in the latter mix, because forages are generally more deficient in phosphorus than in calcium.

VITAMIN SUPPLEMENTS

Vitamin supplements are rich synthetic or natural feed sources of one or more of the complex organic compounds, called vitamins, that are required in minute amounts by animals for normal growth, production, reproduction, and/or health.

Formerly, a wide variety of feed ingredients were added to livestock rations for their vitamin content. But it was found that the vitamin concentration of feedstuffs varied tremendously, being affected by plant species and part (leaf, stalk, or seed), harvesting, storing, and processing. Generally speaking, vitamins are easily destroyed by heat, sunlight, oxidation, and mold growth. So, today, nutritionists rely on vitamin supplements, which in many cases are chemically pure sources that need to be used only in very minute amounts. In modern feed formulation, premixes often represent the common sense approach to providing vitamins.

For adult ruminants, vitamins A, D, and E are of concern, with A being the one most likely to be deficient. Under ordinary circumstances, ruminants synthesize adequate B vitamins, and vitamins C and K. Unless they are kept indoors, they usually receive sufficient exposure from direct sunlight to meet their needs for vitamin D.

Because of the greater prevalence of confinement feeding, along with limited gastrointestinal synthesis, swine are more apt to suffer from vitamin deficiencies than ruminants. Under practical conditions, special consideration should be given to the need for supplementing swine rations with the following vitamins: A, D, E, riboflavin, niacin, pantothenic acid, B-12, and choline.

SPECIAL FEEDS

Among the special feeds used by livestock producers are colostrum, milk replacers, fats and oils, and molasses.

• **Colostrum**—*Colostrum is the first milk secreted by mammalian females following parturition.*

Newborn mammals are unable to produce antibodies within their own bodies for some time after birth; they acquire these antibodies from their mothers either while in the uterus before birth or through colostrum after birth. Newborn calves, lambs, kids, pigs, and foals do not acquire passive immunity while *in utero*; so, the transfer of immunoglobulins via colostrum is of special importance to them. However, they should receive colostrum within the first 15 minutes to 4 hours after birth if they are to acquire passive immunity. When newborn animals fail to nurse naturally, colostrum may be given from a rubber nippled bottle. Some successful dairy producers use a specially designed tube to force colostrum into the abomasum of newborn calves.

After about 24 hours following birth, gut closure occurs, following which the newborn animal digests these proteins, which then lose their immunization properties. Apparently, this results from the newborn not being able to absorb the large protein molecule.

Surplus colostrum from all species can be frozen and stored for a period of 1 year or longer without losing its antibody value. Then, it may be thawed, warmed to about 100°F and fed as needed.

Orphan calves may receive colostrum from any fresh cow or from frozen colostrum that has been stored. Also, calves can absorb antibodies in ewe or mare colostrum, but certain diseases are species specific; hence, colostrum from another species may not afford the desired protection.

• **Milk replacers**—*Milk replacers are formulated feeds designed to replace the mother's milk of young mammals during the critical, early suckling or milk-feeding stage of life.* Milk replacers are available for calves, lambs, pigs, foals, and other animals.

Although scientists have not yet learned how to formulate a synthetic product that will alleviate the necessity of colostrum, in certain other respects they have been able to improve upon nature's product, milk. For example, it has long been known that milk is deficient in iron and copper, thus resulting in anemia in suckling young if proper precautions are not taken. In addition to correcting these deficiencies, milk replacers are fortified with minerals, vitamins, and antibiotics.

A good commercial milk replacer should contain (1) 22–25% protein, preferably derived from milk products (whey, casein, or nonfat dry milk, although about 25% of the milk protein can be replaced by a modified soybean protein); (2) 10 to 20% fat (preferably derived from animal fats, although homogenized soy lecithin is a very acceptable fat source), the higher fat level tends to reduce the severity of diarrhea and provide additional energy for growth; (3) carbohydrates from lactose (milk sugar) and dextrose, and *not* from starch and sucrose (table sugar); and (4) the essential minerals and vitamins. Acidified milk replacers for free-choice calf feeding systems, which were developed and tested in Europe, are gaining increased acceptance in large U.S. commercial operations.

From the standpoint of livestock producers, synthetic milk is of interest in raising orphaned or early-weaned animals of each class of livestock. For the dairy producers, it is generally more profitable to sell the whole milk and purchase a high-quality milk replacer for the young calves. Also, it is a valuable adjunct in certain disease control programs, especially those diseases that may be transmitted from dam to offspring; and, in some cases, it makes it practical to retain in production those valuable females which, due to injury or disease to the udder, cannot suckle their young.

• **Fats and oils**—Feeding of fats was promoted in an effort to find a profitable outlet for surplus packinghouse and rendering plant fats. For the most part fats were formerly used for soapmaking, but they are not used extensively in detergents. Thus, with the rise in the use of detergents in recent years, they became a "drug" on the market.

In 1987, a total of 888,000 metric tons of fats and oils were used in animal feeds in the United States.

Animal and vegetable fats seem to be equally effective additions to rations; thus, selection should be determined solely by comparative price. Ordinarily, animal fats are much cheaper than such vegetable oils as soybean oil or cottonseed oil. Vegetable oils are generally priced out of the animal feed market, for use in margarine, paint, and other industrial uses.

Several different fat products are used as animal feed; among them, acidulated soap stock (foots), tallows, greases (white and yellow), blended feeding fat, house grease, brown grease, sewer grease, and modified yellow grease. Each of them should be bought by specifications and guarantees.

Fat serves the following three practical functions when added to livestock rations:

1. It increases the caloric density of the ration.
2. It controls dust.
3. It lessens the wear and tear on feed-mixing equipment.

If the price is favorable, fat may be added to rations at the following levels: for swine and poultry, 5 to 10%; and for cattle, 2 to 6%. Higher levels of fat usually result in drastically lowered feed consumption. When fed at the levels recommended above, the energy value of fat is approximately 2¼ times that of the grains. When corn is the major source of grain, fat additions can be expected to be less useful than with the small grains. This is understandable when it is realized that corn contains approximately 4% fat as compared to 1 to 1½ for the other feed grains.

Higher levels of fat than indicated above may be used for young ruminants in milk replacers; depending on the purpose, replacers may contain 15 to 30% added fat.

• **Molasses**—Molasses (including cane or blackstrap, beet, citrus, and wood molasses) is extensively used as a livestock feed. 1,605,000 metric tons are used for animal feeds in the United States, annually (see Table 4-3).

Cane and beet molasses are by-products of the manufacture of sugar from sugarcane and sugar beets, respectively. Citrus molasses is produced from the juice of citrus wastes. Wood molasses is a by-product of the manufacture of paper, fiberboard, and pure cellulose from wood; it's an extract from the more soluble carbohydrates and minerals of the wood material. Cane or blackstrap molasses is by far the most extensively used type.

When used at levels of 10 to 15% of the ration, molasses has about three-fourths the energy value of corn. However, molasses has added value as an appetizer, to reduce dustiness of a ration, as a binder for pelleting, to stimulate rumen microbial activity, and as a source of unidentifiable factors. Also, cane molasses is a good source of certain minerals.

Brix is a term used to express molasses quality, as reflected by the relative level of sugar present. It is arrived at by first determining specific gravity. Then by use of conversion tables, the degrees Brix, or level of sucrose present is obtained.

The different types of molasses may also be available in dehydrated form.

ADDITIVES, IMPLANTS, AND INJECTIONS

More than 1,000 drug products are approved by the Food and Drug Administration (FDA) for use by livestock and poultry producers. This includes additives, implants, and injectables, along with other drugs that are used to fight disease and protect animals from infections. Two other statistics which point up the important role of drugs in animal production are: (1) 8 out of every 10 animals raised for food in the United States receive some drugs during their lifetime; and (2) chemicals that regulate growth, modify the rumen's activity, and/or improve feed efficiency increase U.S. meat, milk, and egg production approximately 15% each year. Used properly, these drugs enable livestock producers to provide safe and wholesome meat, eggs, and milk to consumers at lower costs than would otherwise be possible. Used improperly, however, these drugs can be hazardous to consumers.

Consumers are little bothered about what goes on their backs, but they are much concerned about what goes in their stomachs. While they enjoy the price and supply benefits of modern food production technology, they want to be assured of the safety of the food they eat.

Thus, livestock producers have the task of choosing the right drug(s) to maximize rate and efficiency of production, while, at the same time, observing FDA regulations and protecting the consumer. Under such circumstances, they should carefully analyze all the information presented by each company in support of its product. Also, they should study the results of unbiased experimental work, as reported in both scientific literature and popular articles; and they should sound out reliable users of the product. Finally, in the United States, they must comply with FDA regulations.

FEED SUBSTITUTIONS

Successful livestock producers are keen students of values. They recognize that feeds of similar nutritive properties can and should be interchanged in the ration as price relationships warrant, thus making it possible at all times to obtain a balanced ration at the lowest cost.

In arriving at feed substitutions, two primary factors besides cost, chemical composition, and feeding value should be considered—namely, palatability and product quality.

Feed substitutions should be based on the class and age of animal and quality of feed, together with experience and experiments.

FEED PROCESSING

Feed is the major cost in animal production. Hence,it is economically important that it be processed in such a manner as to make for maximum efficiency (1) in handling, from a mechanical standpoint; and (2) in feed efficiency, from the animal standpoint.

Feed preparation can influence the nutritive value of a feed. For example, fine grinding and pelleting of forages tend to increase rate of passage through the gut, which lowers fiber digestibility. However, overall animal response to pelleted forages is usually increased over the same forage fed in long or chopped form, because the slightly lower digestibility is more than offset by increased feed consumption.

Generally speaking, the higher the level of feeding and the greater the production desired, the more important proper feed preparation becomes. This is so because (1) the higher the level of feeding, the more selective animals become in their eating habits; and (2) in ruminants, digestibility decreases as level of feeding increases, primarily because the feed does not remain in the digestive tract long enough for maximum effect of the various digestive processes.

Most of the recent technology in feed preparation has been with feedlot cattle. It came in with the development of large commercial feedlots. But much of it is applicable to all ruminants. Feed preparation for swine and poultry has remained relatively simple as compared with the variety of methods available and in use for ruminant feeds. The major change in horse feed preparation has been the increased use of all-pelleted rations (hay and grain combined).

CONCENTRATE PROCESSING METHODS

Several concentrate processing methods have evolved. Some are physical, others are chemical; some are dry processing, others are wet processing.

It is recognized that any grouping of processing methods cannot be precise, for two or more processing treatments may be involved in a feed; for example, in making pellets, grinding is followed by adding heat and moisture, then pressure. Despite some overlapping, the author evolved with the following classification of grain processing methods:

- **Mechanical alterations**
  - Dehulling
  - Extruding (gelatinization)
  - Grinding
  - Rolling
  - Dry rolling (cracking, crushing)
  - Steam rolling (crimping, steam crimping)
- **Heat treatments**
  - Dry heat processing
    - Micronizing
    - Popping—Jet-sploding
    - Roasting
  - Moist heat processing
    - Cooking
    - Exploding
  - Flaking
    - Steam Flaking
    - Pressure Flaking
  - Pelleting
    - Crumbling
- **Moisture alterations**
  - Bran mash
  - Drying (dehydration)
  - High-moisture grain (early harvested)
  - Reconstituted grain
  - Watered feeds
- **Blocks**
- **Liquid Supplements**
- **Fermenting**
- **Hydroponics (sprouted grain)**
- **Unprocessed (whole) corn**

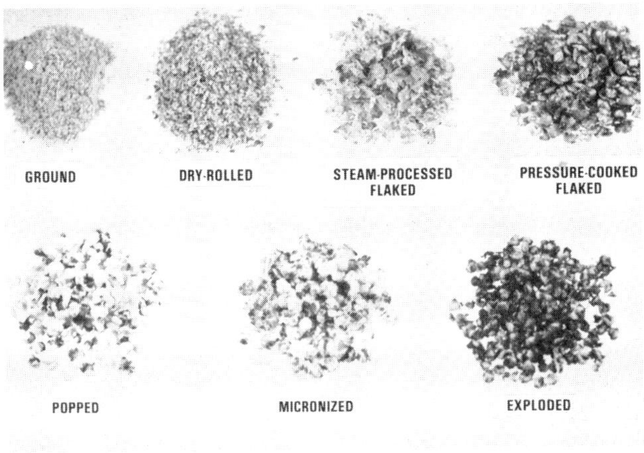

Fig. 4-23. Grain sorghum processed by several different methods. (Courtesy, Dept. of Animal Science, University of Arizona, Tucson)

FORAGE PROCESSING METHODS

The common methods of forage preparation are chopping, grinding, shredding, cubing (wafering), drying, ensiling, and pelleting.

MISCELLANEOUS PROCESSING METHODS

There is hardly any limit to the number of processing methods—some old, others new. Some preserve quality, others increase consumption and lessen labor, and still others change the chemical composition and feeding value. In addition to the processing methods already covered, several miscellaneous, but important, methods are discussed in the sections that follow.

- **Ammoniation**—Ammonium salts and anhydrous ammonia (gas or liquid) have been used for ammoniating feeds that contain high levels of carbohydrates and low levels of nitrogen. Among such ammoniated feeds are: citrus pulp, beet pulp, molasses, sugarcane bagasse, and rice hulls. Also, low quality roughages may be ammoniated.

- **Animal waste (manure) processing**—Animal waste (manure) has nutritive value for ruminants because these animals are capable of utilizing nonprotein nitrogen and fiber. So, proper processing is important.

Broiler and layer litter have been successfully used as an ingredient of cattle feed for many years. However, wastes from all species may be, and are, used. Among the methods employed to process animal wastes prior to feeding are: deep-stacking, ensiling (fermentation), dehydration, and pelleting. The two most common and practical methods of processing are:

1. **Deep-stacking.** In this method,the litter is deep-stacked for several weeks, during which it generates temperatures of 160°F or higher, which render it free of any potentially pathogenic microorganisms that might be present (Pathogenic bacteria do not grow at temperatures over 80°F, and they are

killed at 145°F in a matter of minutes.) It follows that there have been no documented animal health problems associated with feeding broiler or layer litter processed in this manner.

2. **Ensiling (fermentation).** Ensiling is a controlled fermentation process during which carbohydrates in the mixture are converted to lactic, acetic, and other acids. Once sufficient acids are produced, bacterial action ceases and the ensilage is stable. During the fermentation, heat is generated, thereby diminishing the hazard from certain pathogenic organisms that might be present.

Dehydration and pelleting of animal wastes are excellent processing methods as such. However, current energy costs make them uneconomical.

In December 1980, the U.S. Food and Drug Administration published a document leaving regulation of feeding animal waste to the individual states.

• **Irradiation**—For many years, it was known that both ultraviolet light (from the sun) and cod-liver oil had identical effects in healing of rickets. In 1924, Steenbock of Wisconsin and Hess of Columbia University, independently announced that certain food materials could be made antirachitic by exposing them to ultraviolet light.

Upon irradiation, ergosterol, a plant sterol, yields ergocalciferol, commonly known as vitamin D_2.

The ultraviolet radiation in sunlight serves as a source of radiant energy necessary to convert 7-dehydrocholesterol (an animal sterol stored beneath the skin surface) into biologically active vitamin D_3.

Vitamin D_2, the plant form of the vitamin, and vitamin D_3, the animal form, have the same antirachitic value for the rat, dog, pig, ruminant, and human, but vitamin D_3 is more active for poultry.

Sun-cured hay is a good natural source of vitamin D for four-footed animals, but not for poultry. Fortified fish oils and irradiated sterols are good sources of vitamin D for both four-footed animals and poultry.

• **Molasses added**—Molasses (including cane or blackstrap, beet, citrus, wood and starch molasses) is extensively used as a livestock feed. When used at levels of 5 to 15% of the ration, it has about ¾ the energy value of corn. However, molasses has added value as an appetizer, to reduce dustiness of a ration, as a binder for pelleting, to stimulate rumen microbial activity, and as a source of unidentified factors. Also, cane molasses is a good source of certain trace minerals.

In hot, humid areas, molasses should be limited to 5% of the ration; otherwise, mold may develop. Where mustiness is a problem, it may be controlled by adding calcium propionate to the feed according to the manufacturer's directions.

• **Organic Acids**—The proper use of organic acids provides another way in which to preserve high-moisture grains. The organic acid treatment involves the application of 1 to 1½% acid (i.e. propionic, acetic, formic, ammonium isobutyric, etc.) at time of harvest, followed by storage in a pile.

Acid treatment inhibits the growth of molds and bacteria. Research has shown that propionic acid alone or a mixture of 75% acetic and 25% propionic acid are quite effective. Acetic acid should not be used alone. Limited research indicates that sodium propionate, formalin, ammonium isobutyrate and citric acids have been successful, as well as combinations of propionic acid and formic acid or formalin.

Experimental studies indicate that acid-treated grain has approximately the same feeding value as high-moisture grain stored in an oxygen-limiting silo. Also, it alleviates the cost of drying. Thus, organic acid treatment of grain may be a practical way in which to preserve high-moisture grains.

• **Preservatives**—*A preservative is a material added at the time of mixing or storing to enhance the keeping qualities of a feed. A brief description of hay and silage preservatives follows:*

1. **Hay preservatives.** Preservatives are available commercially which can be applied to hay. Usually the directions (1) recommend the addition of 1 to 3 lb of these products for each ton of damp hay, and (2) claim that there will be no heating or molding.

More experimental work is needed relative to chemical hay preservatives. But, available data indicate that propionic acid is the hay preservative of choice. Missouri workers report that most hay (28% moisture at baling) treated with an organic acid was equivalent in digestion to dry hay, whereas hay baled moist and not treated had significantly lower digestibility than treated or control hays.

Anhydrous ammonia is one of the most recent materials being studied as a hay preservative. In Indiana trials, applying this material at the rate of 1.0% to hay baled at 30% moisture successfully prevented molding, heating, and quality deterioration.

2. **Silage preservatives**—Two types of additives have generally been used in silage making: (1) feed additives, and (2) chemical additives.

Feed additives supply a readily available source of carbohydrates for bacterial fermentation of the silage. Some feed additives, such as corn-and-cob meal, when mixed with high-moisture forages, also absorb water and help to reduce runoff. When used as preservatives, approximately 75 to 85% of the feed nutrients added may be recovered as feed.

A large number of chemical additives have been used in silage making, with variable results.

• **Self-feeding governors**—The commonly used self-feeding governors are (1) bulky fibrous feeds; (2) salt-feed or fat-feed mixtures; (3) fat content of block; and (4) liquid supplements.

1. **Bulky, fibrous feeds.** Bulk can be used as a self-feeding governor. This consists in adding to the bulkiness of the ration, such as can be achieved by increasing the amount of chopped hay and lessening the concentrate. Actually, this is a way in which to lower the energy content of the ration. Since an

animal can hold only so much, it is an effective control of feed intake.

2. **Salt-feed mixtures.** The practice of using salt as a governor to limit feed consumption of pasture or range has been used for a very long time. It was ushered in as a labor-saving device for cattle and sheep in inaccessible and rough areas. Today, salt-feed mixtures are used in either meal or block form.

3. **Fat content of block.** Since animals tend to eat until a certain caloric intake is reached, they consume less total weight when fed high-fat rations. Thus, pounds of feed consumed can be governed by the amount of fat in a block. It is noteworthy, too, that fat serves as a needed feed nutrient, whereas consuming more salt than required (as happens when a salt-feed mixture is used) makes for a waste of salt.

4. **Liquid supplements.** When self-fed, the consumption of liquid supplements is generally controlled by (1) the use of a lick tank, and/or (2) incorporating in the formulation phosphoric acid, beet solubles, and/or citrus peel liquor.

• **Slow-release and rumen bypass treatments**—Two feed processing techniques—slow-release nonprotein nitrogen, and rumen bypass protein—are designed to delay digestion.

1. **Slow-release nonprotein nitrogen.** Among the slow-release nonprotein nitrogen products that liberate nitrogen slowly are a combination of urea and gelatinized starch, and urea combined with gelatinized corn.

2. **Rumen bypass.** This refers to bypass protein (also known as protected or escaped protein) in feed that escapes digestion in the rumen and passes into the lower digestive tract where it is digested and absorbed. Feed processors have developed treatments through which the bypass proteins in certain feeds can be increased; among them, heat and pressure treatment, treatment with tannins, treatment with formaldehyde or other aldehydes, lipid (fat) treatment, complexing with bentonite clay, use of amino acid analogs, increasing microbial metabolism in the rumen, and adding ionophores.

• **Treatment of high-cellulose feeds**—High feed prices and more stringent burning regulations have spurred research to find a practical method of improving the feeding value of several high-cellulose products, such as rice, wheat, barley and oat straws; bagasse; tree bark; corncobs; gin trash; newspaper; and seed hulls.

In their natural state, these products make poor feedstuffs because lignin or silica, or a combination of the two, (1) encrust the energy-rich carbohydrates, cellulose, and hemicellulose; and (2) keep the microbes in the ruminant's stomach from breaking them down to release energy.

The answer to this problem lies in some treatment that opens up the fibers enough to permit increased digestion in the rumen. Several methods of chemical and/or physical treatment are being investigated; among them, alkali treatment, ammoniation, hydrogen peroxide treatment, and high pressure steam.

• **Complete (all-in-one) rations**—Most experiments and experiences have not shown any difference between mixed rations and the feeding of roughage and concentrates separately insofar as efficiency and production are concerned. However, a mixed ration has the following advantages:

1. It makes for greater efficiency in feeding and lessens the sorting at the feed bunk.

2. When the roughage is relatively unpalatable, a mixed ration forces consumption.

3. When it is desired to limit concentrate consumption, mixing with the roughage is desirable.

4. A mixed ration makes it easier to get animals on full feed.

Thus, each feeder must decide on the matter of mixed feed vs feeding roughage and concentrate separately, with relative costs and other factors considered. Most large cattle and sheep feedlots use completely mixed rations. Also, the trend is toward complete feeds for both dairy cows and swine, primarily because such complete feeds (1) lend themselves better to automation, and (2) provide better control of nutrient intake.

All-pelleted rations (grain and forage combined). Increasingly, complete pelleted rations are being used for horses, swine, and fish. Among the virtues ascribed to all-pelleted rations are (1) they prevent selective eating—if properly formulated, each mouthful is a balanced diet; (2) they alleviate waste; (3) they eliminate dust (thereby lessening heaves in horses); (4) they lessen labor and equipment; (5) they lessen storage; and (6) they facilitate automation.

CHOICE OF PROCESSING METHOD

The choice of a processing method is highly dependent on the feedstuff to be fed. It is clear that a given processing technique may be very desirable for one grain, but quite detrimental to another. Corn may be fed without any processing, but not milo. Pressure treating appears to be desirable for milk, but harmful to wheat.

Comparison of grain processing techniques are difficult because there are a number of interactions between processing technique and roughage level or type of ration fed. For example, data from Ohio State University have shown that whole shelled corn was superior to crimped corn in very low-roughage rations, whereas crimped corn was clearly superior in high-roughage rations.

FEED PROCESSING TABLE

Table 4-6 is a summary of pertinent information relative to the preparation of feeds for each class of livestock.

TABLE 4-6
PREPARATION OF FEEDS

Class of Animal	Concentrates	Forages	Comments
Beef cattle	Extruding, flaking, micronizing, popping, roasting, or high-moisture grain—with choice determined by cost—are preferable, especially for full-fed animals on a high-grain ration. But such equipment is costly to purchase and operate; hence, a large-volume operation is required to cover fixed costs. Dry or steam roll or grind coarsely for most beef cattle, especially those not full fed high-grain rations and those in smaller operations. On high-concentrate rations (those with 80% or more concentrate), whole corn need not be processed. Grain (except for very hard seeds) need not be processed for calves under 6 months of age, for young calves masticate feed thoroughly. Cubes (large pellets) preferred for feeding on pasture or range. Professional caretakers often cook feed (especially barley) for show cattle to increase palatability.	Long hay is satisfactory for most cattle other than commercial feedlot operations. Chopped (2" length), cubed (wafered), or pelleted forage should be used (1) in commercial feedlots or when quality of hay is poor; and (2) in all cattle operations from standpoints of ease of handling and lessening wastage. Shredding fodders and stovers (corn or milo) makes them easier to handle and lessens waste.	Fine grinding grain increases incidence of hyperkeratosis (ruminal parakeratosis) in feedlot cattle. Dry or steam rolling or coarse grinding of grains are of about equal value for most beef cattle. Either method is just as satisfactory as more expensive methods (like flaking) when grain intake is relatively low. Chopping or pelleting low-quality hay is more advantageous than chopping high-quality hay. Finely ground hay not recommended, as it decreases digestibility.
Dairy cattle	Grinding is the simplest and the most widely used grain processing method for dairy cattle. Cracking, steam rolling, and pelleting are the other popular procedures. Exploding, extruding, flaking, micronizing, popping, roasting, or high-moisture grain are preferable for high-producing lactating cows, but are not widely used. Dry or steam roll, or grind, for all but high-producing, lactating cows. Feed grain whole to calves under 6 mo. of age.	Long hay or cubes. Cubes lend themselves to automation; and lower milk fat percentage only slightly, if at all.	Butterfat is depressed unless the ration contains some threshold level of coarse material. Finely ground or pelleted roughage will result in reduced rumen acetate production and lower milk fat percentage.
Sheep and goats	Processing grains not necessary unless seeds are hard (like sorghum or millet) or the teeth are poor. Hard seeds (like sorghum) may be prepared by exploding, extruding, flaking, micronizing, popping, roasting, or high-moisture grain, with cost determining the choice. Pellets are increasingly being used by lamb feeders. Cubes or pellets preferred for feeding on pasture or range. Professional shepherds prefer flaked grain for show sheep, as the ration is lighter and there are fewer digestive disturbances.	Chop (2" in length), pellet, or cube. Many lamb feeders are using all-pelleted rations (hay and grain combined).	Sheep and goats masticate grain more thoroughly than cattle, with the result that feed preparation for them is of less value than for cattle. A high incidence of parakeratosis—a degeneration of the rumen papilla—appears to result from feeding pellets, especially when low forage-high concentrate pellets are used. Hence, breeding sheep should not be fed pellets for extended periods without any long or chopped forage.
Swine	Corn, barley, grain sorghum, and oats should be finely ground for swine. Medium to coarse grinding is best for wheat, because fine grinding makes it pasty and less palatable. Pelleting corn-soybean rations generally improves feed utilization and increases rate of gain by at least 4–5%. Cook Irish potatoes, beans, soybeans, and garbage. Cooking (except for the feeds listed above), soaking, or fermenting are not of value when swine are on full feed. Liquid and paste feeding give inconsistent results in feed consumption and rate of gain; hence, they should be evaluated on the basis of a mechanical means of dispensing feed. However, slop (slurry or gruel) is desirable for early weaned pigs, and perhaps for pigs being fitted for show or sale. High-moisture corn does not result in any improvement of efficiency for swine; hence, the value of high-moisture corn as compared to regular corn should be computed on a dry matter basis.	Alfalfa (or other legume) that is to be incorporated in mixed feeds should be ground. Rations containing considerable amounts of fiber are improved by pelleting because of increased consumption, improved carbohydrate digestibility, and reduced sorting and wastage compared to meal rations.	Fine grinding will cause some bridging in self-feeders. Also, finely ground feed is associated with increased incidence of stomach ulcers in swine.
Horses	Flaking is the preferred method of grain preparation for horses; it makes for a light ration and few digestive disturbances. For horses with good teeth, the value of oats is increased only 5% by processing.	Either feed long hay or an all-pelleted ration (grain and hay combined).	Cubes (wafers) sometimes cause horses to choke. Horses are very sensitive to moldy feed. Horses should not be fed dusty feed, because of the hazard of heaves.

EVALUATING FEEDSTUFFS

Some feeds are more valuable than others; hence, measures of their relative usefulness are important. Among such methods of evaluating feeds are the following:

1. Physical evaluation.
2. Cost per unit of nutrients.
3. Proximate analysis.
4. Digestion (or metabolism) trials.
5. Feeding trials.
6. Other feed evaluations.

PHYSICAL EVALUATION

In order to produce or buy superior feeds, producers need to know what constitutes feed quality, and how to recognize it. They need to be familiar with those recognizable characteristics of feeds which indicate high palatability and nutrient content. If in doubt, the animals will tell them, for they like and thrive on high-quality feed.

The physical evaluation of feedstuffs, especially forages, is based largely on eye and smell appeal. Does it look good and smell good? The easily recognizable characteristics of hay of high feeding value are:

1. It is made from plants cut at an early stage of maturity, thus assuring the maximum content of protein, minerals, and vitamins, and the highest digestibility.
2. It is leafy, thus giving assurance of high protein content.
3. It is bright green in color, thus indicating proper curing, a high carotene or provitamin A content, and palatability.
4. It is free from foreign material, such as weeds and stubble.
5. It is free from mold and dust.
6. It is fine stemmed and pliable—not coarse, stiff, and woody.
7. It has a pleasing fragrant aroma; it "smells good enough to eat."

COST PER UNIT OF NUTRIENTS

One method of arriving at the best buy in feeds is to compute and compare the cost per unit of nutrients, based on feed composition. Where chemical analysis of a specific feed is not available, feed composition tables may be used. Thus, feed composition tables may serve as a basis of feed purchasing and merchandising, as well as for ration formulation.

The use of the cost per unit of nutrients method can best be illustrated by the examples that follow:

If 44% protein (crude) soybean meal is selling at $9.88 per 100 lb whereas 35% protein (crude) linseed meal sells for $6.25 per 100 lb, which is the better buy? Divide $9.88 by 44 to get 22.5¢ per pound of crude protein for the soybean meal. Then divide $6.25 by 35 and get 17.9¢ per pound of crude protein for the linseed meal. Thus, at these prices linseed meal is the better buy—by 4.6¢ (22.5 − 17.9 = 4.6) per pound of crude protein.

When buying energy feed, one can compare the cost per pound of total digestible nutrients (TDN). For example, if corn is priced at $3.63 per 100 lb and has a TDN of 91% divide $3.63 by 91 and the result is 3.99¢ per pound of TDN. If milo with 86% TDN sells for $3.25 per 100 lb, divide $3.25 by 86, and the price is 3.78¢ per pound of TDN. Thus, milo would be the better buy by 0.21¢ (3.99 − 3.78 = 0.21) per pound of TDN.

Of course, it is recognized that many other factors affect the actual feeding value of each feed, such as (1) palatability, (2) grade of feed, (3) preparation of feed, (4) ingredients with which each feed is combined, and (5) quantities of each feed fed. It follows that, from the standpoint of the producer, the most important measurement of a feed's usefulness is in terms of *net returns* rather than cost per bag or cost per ton.

PROXIMATE ANALYSIS

For more than 100 years, feeds have been analyzed by a method developed by two scientists, Henneberg and Stohmann, at the Weende Experiment Station in Germany. This method is called the proximate analysis, or the Weende System of feed analysis. Feeds are evaluated in terms of 6 components: (1) moisture, (2) ash, (3) crude protein, (4) ether extract, (5) crude fiber, and (6) nitrogen-free extract (see Table 4–7).

TABLE 4–7
THE FRACTIONS OF PROXIMATE ANALYSIS

Fraction	Procedure[1]	Major Components
1. Moisture (dry matter).	Heat sample to constant weight at temperature just above boiling point of water. Loss in weight equals water.	Water and any volatile compounds (100% − H_2O = DM%).
2. Ash (mineral matter).	Burn at 500° to 600°C for 2 hours.	Mineral elements.
3. Crude protein (protein averages 16% N; hence, N × 6.25 = crude protein).	Determine nitrogen by Kjeldahl process.	Proteins, amino acids, nonprotein nitrogen.
4. Ether extract (fat).	Extraction with diethyl ether.	Fats, oils, waxes, resins, pigments.
5. Crude fiber (CF).[2]	Residue after boiling in weak acid and weak alkali.	Cellulose, hemicellulose, lignin.
6. Nitrogen-free extract (NFE).[2]	Remainder; *i.e.*, 100 minus sum of the other fractions.	Starch, sugars, some cellulose, hemicellulose, and lignin.

[1]Each procedure can be applied to a separate sample, of standard weight, of the feedstuff to be analyzed; or a single sample can be used to determine dry matter, crude fat, and crude fiber. In the latter case, separate samples would be run for ash and crude protein.

[2]Carbohydrates (CHO = CF + NFE).

Today, feeds are being analyzed routinely through highly sophisticated chemical procedures. Many agricultural experiment stations, as well as most large feed companies, have facilities to analyze feeds for both the prevention and diagnosis of nutritional problems.

A chemical analysis gives a solid foundation on which to start in the evaluation of feeds. Thus, feed composition tables serve as a basis for ration formulation and for feed purchasing and merchandising. Commercially prepared feeds are required by state law to be labeled with a list of ingredients and a guaranteed analysis. Although state laws vary slightly, most of them require that the feed label (tag) show in percent the minimum crude protein and fat; and maximum crude fiber and ash. Some feed labels also include maximum salt, minimum TDN, and/or minimum calcium and phosphorus. These figures are the buyer's assurance that the feed contains the minimal amounts of the higher cost items—protein and fat; and not more than the stipulated amounts of the lower cost, and less valuable, items—the crude fiber and ash.

In addition to proximate analysis, methods are available for assaying individual vitamins—biological assays for some, chemical determinations for others.

DIGESTION (OR METABOLISM) TRIAL

Animals are not able to extract all the nutrients present in feeds. The actual value of ingested nutrients is dependent upon the use which the body is able to make of them. The first consideration here is digestibility, since undigested nutrients do not get into the body proper.

A digestion trial is made by determining the percentage of each nutrient in the feed through chemical analysis; giving the feed to the test animal for a preliminary period (usually 7 to 10 days for ruminants), so that all residues of former feeds will pass out of the digestive tract; giving weighed amounts of the feed during the test period (7 to 10 days for ruminants); collecting, weighing, and analyzing the feces; determining the difference between the amount of the nutrient fed and the amount found in the feces; and computing the percentage of each nutrient digested. The latter figure is known as the *digestion coefficient* for that nutrient in the feed.

Various techniques and equipment may be used to make the fecal collections; among them; a specially designed digestion stall (see Fig. 4–24); collection harness and bag; markers (such as carmine, ferric oxide, chromic oxide, or soot), fed with the ration at the beginning and the end of the collection period; and indicators of inert reference subject.

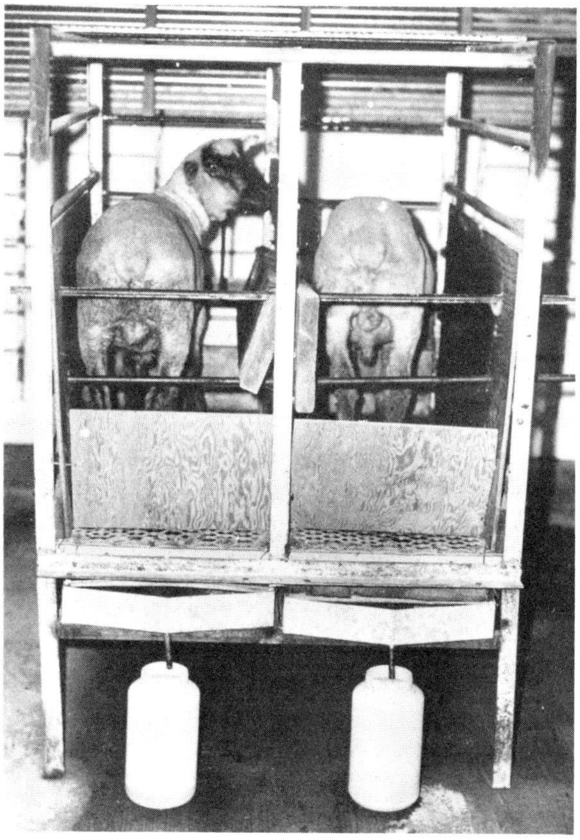

Fig. 4–24. Metabolism stalls for digestibility studies with sheep. Note the pans at the bottom for collecting feces and the containers on the floor for collecting urine. (Courtesy, Irrigated Agriculture Research and Extension Center, Washington State University, Pullman)

FEEDING TRIALS

Each method of evaluating feedstuffs, discussed earlier in this chapter, has a place and is valuable. But none of them takes into consideration all the factors which determine the true value of any feed for a particular class of livestock. The ''court of last judgment'' for determining the true value of a feedstuff is the animal. How well do the animals eat the ''stuff''? How does it affect their health and well-being? How are they producing? Answers to these questions call for feeding the ingredient or ration under controlled conditions to the particular class of livestock.

OTHER FEED EVALUATIONS

In addition to being nutritionally complete, the following feed requirements are important:

1. **Palatability.** If they don't eat it, they won't produce; and if they don't eat enough, feed efficiency will be poor. The relationship of feed consumption to feed efficiency becomes clear when it is realized that the maintenance requirement of an animal producing at a low rate represents a much greater percent of the total feed required than for an animal producing at a more rapid rate.

2. **Variety.** Some variety in the ration is desirable, particularly from the standpoints of assuring (a) increased palatability, and (b) balance of nutrients—for example, all the essential amino acids.

3. **Digestive disturbances.** Bloat, colic, scours, and constipation are the bane of all feeders. The choice of feeds can give a big assist in minimizing such disturbances. For example, bloat in cattle and colic in horses can be lessened by avoiding lush or frosted pastures; scours can be lessened by proper feeding; and constipation can be corrected by feeding alfalfa, wheat bran, linseed meal, and molasses.

4. **Bulk.** The amount of bulk in the ration will vary. Ruminants can consume bulkier feeds than monogastric animals; the younger the animal, the less bulk; and the higher the production desired, the less bulky the ration. Also, the relative cost of feeds—concentrates vs roughages—will influence the relative amount of bulk in the ration.

5. **Cost.** Cost is important. But even more important is net returns; hence, it may well be said that it is net returns rather than cost per ton, or per bag, that counts.

6. **Poisonous plants and feeds.** Poisonous plants and feeds should be avoided. The livestock producer should know the poisonous plants common to the area, and avoid them. Also, the following poisons should be avoided: prussic acid, hydrocyanic acid, ergot, scabbed grain, smut on grain, spoiled or moldy feed, botulism, aflatoxins, selenium poisoning, nitrate poisoning, lead poisoning, and mercury poisoning.

MEASURING AND EXPRESSING ENERGY VALUE OF FEEDSTUFFS

One nutrient cannot be considered as more important than another, because all nutrients must be present in adequate amounts if efficient production is to be maintained. Yet, historically, feedstuffs have been compared or evaluated primarily on their ability to supply energy to animals. This is understandable because (1) energy is required in larger amounts than any other nutrient, (2) energy is most often the limiting factor in livestock production, and (3) energy is the major cost associated with feeding animals.

Our understanding of energy metabolism has increased through the years. With this added knowledge, changes have come in both the methods and terms used to express the energy value of feeds.

The methods of measuring the energy value of feedstuffs currently employed in the United States are:

1. Total digestible nutrient (TDN)
2. Calorie System, including—
   a. Gross energy (GE)
   b. Digestible energy (DE)
   c. Metabolizable energy (ME)
   d. Net energy (NE)

Each system has its advantages and advocates. Also, both the difficulty in determining energy values of feeds according to the different systems of measurement and the accuracy of the results increase in the order that they are listed above. Nevertheless, more and more feedstuffs are being evaluated in calories, with net energy being the method of choice.

Although net energy is the most precise measure known of the real value of a feed, it is very difficult to determine. It requires either (1) the measurement of all forms of energy loss, or (2) the actual amount of energy retained by the animal or produced as a useful product. The first determination may be accomplished by balance methods for determining energy, and the second by the comparative slaughter technique.

TOTAL DIGESTIBLE NUTRIENT (TDN) SYSTEM

Total Digestible Nutrients (TDN) is the sum of the digestible protein, fiber, nitrogen-free extract, and fat × 2.25. It has been the most extensively used measure for energy in the United States.

Back of TDN values are the following steps:

1. **Digestibility**—The digestibility of a particular feed for a specific species is determined by a digestion trial.

2. **Computation of digestible nutrients**—Digestible nutrients are computed by multiplying the percentage of each nutrient in the feed (protein, fiber, nitrogen-free extract (NFE), and fat) by its digestion coefficient. The result is expressed as digestible protein, digestible fiber, digestible NFE, and digestible fat. For example if dent corn contains 8.9% protein of which 77% is digestible, the percent of digestible protein is 6.9.

3. **Computation of total digestible nutrients (TDN)**—The TDN is computed by using the following formula:

$$\% \text{ TDN} = \% \text{ DCP} + \% \text{ DCF} + \% \text{ DNFE} + (\% \text{ DEE} \times 2.25)$$

where DCP = digestible crude protein; DCF = digestible crude fiber; DNFE = digestible nitrogen-free extract; and DEE = digestible ether extract.

TDN is ordinarily expressed as a percent of the ration or in units of weight (lb or kg), not as a caloric figure.

The main **advantage** of the TDN system is that it has been used for a very long time and many people are acquainted with it.

The main **disadvantages** of the TDN system are:

1. It is really a misnomer, because TDN is not an actual total of the digestible nutrients in a feed. It does not include the digestible mineral matter (such as salt, limestone, and defluorinated phosphate—all of which are digestible); and the digestible fat is multiplied by the factor 2.25 before being included in the TDN figure, because its energy value is higher than carbohydrates and protein. As a result of multiplying fat by the factor 2.25, feeds high in fat will sometimes exceed 100 in percentage TDN (a pure fat with a coefficient of digestibility of 100% would have a theoretical TDN value of 225%—100% × 2.25).

2. It is an empirical formula based upon chemical determinations that are not related to actual metabolism of the animal.

3. It is expressed as a percent or in weight (lb or kg), whereas energy is expressed in calories.

4. It takes into consideration only digestive losses; it does not take into account other important losses, such as losses in the urine, gases, and increased heat production (heat increment).

5. It overevaluates roughages in relation to concentrates when fed for high rates of production, due to the higher heat loss per pound of TDN in high-fiber feeds.

Because of these several limitations, in the United States the TDN system is gradually being replaced by other energy evaluation systems, particularly net energy. However, due to the voluminous TDN data on many feeds and long-standing tradition, it will continue to be used by many people for a long time to come.

CALORIE SYSTEM

Calories are used to express the energy value of feedstuffs. *One calorie* (always written with a small "c") *is the amount of heat required to raise the temperature of 1 g of water 1°C* (precisely from 14.5°C to 15.5°C).

To measure this heat energy, an instrument known as the bomb calorimeter is used, in which the feed (or other substance) to be tested is placed and burned in the presence of oxygen.

Through various digestive and metabolic processes, much of the energy in feed is dissipated as it passes through the animal's digestive system. About 60% of the total combustible energy in grain and about 80% of the total combustible energy in roughage is lost as feces, urine, gases, and heat. These losses are illustrated in Fig. 4–25.

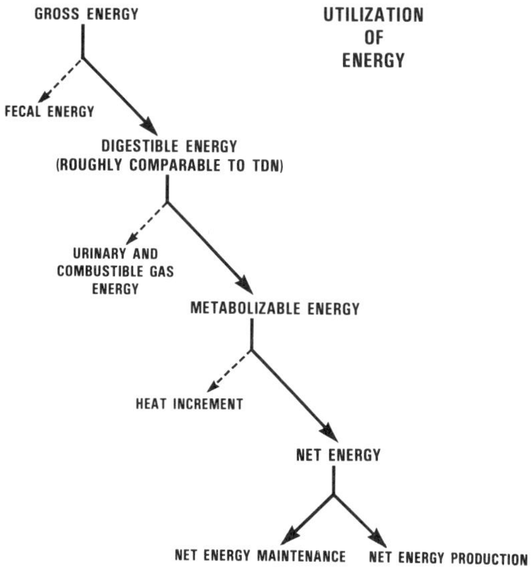

Fig. 4–25. Utilization of energy.

As shown in Fig. 4–25, energy losses occur in the digestion and metabolism of feed. Measures that are used to express animal requirements and the energy content of feeds differ primarily in the digestive and metabolic losses that are included in their determination. Thus, the following terms are used to express the energy value of feeds:

- **Gross energy (GE)**—*Gross energy represents the total combustible energy in a feedstuff.* It does not differ greatly among feeds, except for those high in fat. For example, 1 lb of corn-cobs contains about the same amount of GE as 1 lb of shelled corn. Therefore, GE does little to describe the useful energy in feeds for finishing animals.

Fig. 4–25a. Bomb calorimeter for the determination of gross energy (caloric content). (Courtesy, Parr Instrument Co., Moline, IL)

- **Digestible energy (DE)**—*Digestible energy is that portion of the GE in a feed that is not excreted in the feces.*

- **Metabolizable energy (ME)**—*Metabolizable energy represents that portion of the GE that is not lost in the feces, urine, and gas.* Although ME more accurately describes the useful energy in the feed than does GE or DE, it does not take into account the energy lost as heat.

- **Net energy (NE)**—*Net energy represents the energy fraction in a feed that is left after the fecal, urinary, gas, and heat losses are deducted from the GE.* The net energy, because of its greater accuracy, is being used increasingly in ration formulations, especially in computerized formulations for large operations.

Although net energy is a more precise measure of the real value of the feed than other energy values, it is much more difficult to determine.

Two systems of net energy evaluation are presently being used. Lofgreen and Garrett[1] developed a system whereby the net energy requirements are listed as dictated by physiological functions—for example, net energy for maintenance (NE_m) and net energy for gain (NE_g). Also, Moe and Flatt[2] developed a net energy system that compares the physiological function to that of lactation through the use of regression analysis. This value, $NE_{lactation}$, is applicable for all physiological functions.

FEEDING STANDARDS

Feeding standards are tables showing the amounts of one or more nutrients needed by different species of animals for different purposes, such as growth, finishing, and lactation. They serve as guides in balancing rations and feeding practices. Most feeding standards are expressed in (1) quantities of nutrients required per day, and/or (2) percent of the ration; the first type being used for animals given exact quantities of a ration, and the second type used when animals are fed free choice (*ad libitum*).

Today, the most widely used feeding standards in the United States are those published for each animal species by the National Research Council (NRC) of the National Academy of Sciences. These requirements were adapted for, and presented in, *Feeds & Nutrition* and *Feeds & Nutrition Digest*, books by M. E. Ensminger, *et al.*, published by Ensminger Publishing Company; hence, the reader is referred thereto. In England, similar standards are issued by the Agricultural Research Council (ARC). Other countries have similar bodies which make recommendations on the nutritive requirements of animals.

---

[1]Lofgreen, G. P., and W. N. Garrett, "A System for Expressing Net Energy Requirements and Feed Values for Growing and Finishing Beef Cattle," *Journal of Animal Science*, Vol. 27, 1968, p. 793.

[2]Moe, P. W., and W. P. Flatt, "Net Energy of Feedstuffs for Lactation," *Journal of Dairy Science*, Vol. 52, 1969, p. 928.

BALANCED RATIONS

To supply all the needs—for maintenance, growth, finishing, reproduction, lactation, work (or running), and/or wool—the different classes of animals must receive sufficient feed to furnish the necessary quantity of energy (carbohydrates and fats), proteins, minerals, vitamins, and water. Perhaps under certain conditions feed additives may be desirable, although it is not likely that they are essential. A ration that meets all these needs is said to be balanced. More specifically, by definition, *a balanced ration is one which provides an animal the proper proportions and amounts of all the required nutrients for a period of 24 hours.*[3]

When in confinement, animals have access only to the feeds provided by the caretaker. This points up the importance of balanced rations.

Several suggested rations for different classes of livestock are given in this section, in each of the parts devoted to a species. Generally these rations will suffice, but it is recognized that rations should vary with conditions, and that many times they should be formulated to meet the conditions of a specific farm or ranch, or to meet the practices common to an area.

Also, good livestock producers should know how to balance rations. Then, if the occasion demands, they can do it. Perhaps of even greater importance, they will then be able more intelligently to select and buy rations with informed appraisal; to check on how well their manufacturer, dealer, or consultant is meeting their needs; and to evaluate the results.

HOW TO BALANCE A RATION

Ration formulation consists of combining feeds to make a ration that will be eaten in the amount needed to supply the daily nutrient requirements of the animal. This may be accomplished by the methods presented later in this chapter, but first the following pointers are necessary:

1. In computing rations, more than simple arithmetic should be considered, for no set figures can substitute for experience. Compounding rations is both an art and a science—the art comes from animal know-how and experience, and keen observation; the science is largely founded on chemistry, physiology, and bacteriology. Both are essential for success.

2. Before attempting to balance a ration, the following major points should be considered:

   a. **Availability and cost of the different feed ingredients.** Preferably, cost of ingredients should be based on delivery after processing—because delivery and processing costs are quite variable.

   b. **Moisture content.** When considering costs and balancing rations, feeds should be placed on a comparable moisture basis; usually, an air-dry basis, or 10% moisture content, is used. This is especially important in the case of high-moisture grain or silage.

   c. **Composition of the feeds under consideration.** Feed composition tables (*book values*), or average analysis, should be considered as good guides, but not precise, because of wide variations in the composition of feeds. For example, the protein and moisture contents of sorghum, hay, and silages are quite variable. Whenever possible, especially with large operations, it is best to take a representative sample of each major feed ingredient and have a chemical analysis made of it for the more common constituents—protein, fat, fiber, nitrogen-free extract, and moisture; and often calcium, phosphorus, and carotene. Such ingredients as oil meals and prepared supplements, which must meet specific standards, need not be analyzed so often, except as quality-control measures.

   Feed compositions of the most commonly used feeds are presented in Section 19, Feed Composition Tables, of this book.

   d. **The nutrient allowances.** This should be known for the particular class of animals for which a ration is to be formulated. Also, it must be recognized that nutrient requirements and allowances must be changed from time to time, as a result of new experimental findings.

3. In addition to providing a proper quantity of feed and to meeting the protein and energy requirements, a well-balanced and satisfactory ration should be:

   a. Palatable and digestible.

   b. Economical. Generally speaking, this calls for the maximum use of feeds available in the area, especially forages.

   c. Adequate in protein content, but not higher than is actually needed. Generally speaking, medium and high-protein feeds are in scarcer supply and higher in price than high-energy feeds.

   d. Well fortified with the needed minerals, or free access to suitable minerals should be provided; but mineral imbalances should be avoided.

   e. Well fortified with the needed vitamins.

   f. So formulated, where ruminants are involved, as to nourish the billions of bacteria in the paunch in order that there will be satisfactory (1) digestion of roughages, (2) utilization of lower quality and cheaper proteins and other nitrogenous products (thus, it is possible to use urea to constitute up to one-third of the total protein of the ration of ruminants, provided care is taken to supply enough carbohydrates and other nutrients to assure adequate nutrition for rumen bacteria), and (3) synthesis of B vitamins.

   This means that rumen microorganisms must be supplied adequate (1) energy, including small amounts of readily available energy such as sugars or starches; (2) ammonia-

[3]Although Webster defines the noun *ration* as "the amount of food supplied to an animal for a definite period, usually for a day," to most livestock producers the word implies the feeds fed to an animal or animals, without limitation to the time in which they are consumed. The author accedes to the common usage of the word, rather than to dictionary correctness.

bearing ingredients such as proteins, urea, and ammonium salts; (3) major minerals, especially sodium, potassium, and phosphorus; (4) cobalt and possibly other trace minerals; and (5) unidentified factors found in certain natural feeds rich in protein or nonprotein nitrogenous constituents.

g. One that will enhance, rather than impair, the quality of the product (meat, milk, eggs, or wool/mohair) produced.

4. In addition to considering changes in availability of feeds and feed prices, ration formulation should be altered at stages to correspond to changes in weight and productivity of animals.

The above points are pertinent to the balancing of rations, regardless of the mechanics of computation used.

METHODS OF FORMULATING RATIONS

In the sections that follow, four different methods of ration formulation are presented: (1) the square method, (2) the trial-and-error method, (3) the net energy method, and (4) the computer method. Despite the sometimes confusing mechanics of each system, if done properly, the end result of all four methods is the same—a ration that provides the desired allowance of nutrients in correct proportions economically (or at least cost), but more important, so as to achieve the greatest net returns—for it is net profit, rather than cost, that counts. Since feed usually represents the greatest cost item in livestock production, the importance of balanced rations is evident.

An exercise in ration formulation follows for purposes of illustrating the application of each of these four methods:

1. **Square method,** applied to a swine ration.
2. **Trial-and-error method,** applied to a lactating cow ration.
3. **Net energy method,** applied to a cattle finishing ration.
4. **Computer method,** applied to a layer ration.

SQUARE (OR PEARSON SQUARE) METHOD

The square method is a simple, direct, and easy way in which to figure proportions between two ingredients. It permits quick substitution of feed ingredients in keeping with market fluctuations, without disturbing the protein content.

In balancing rations by the square method, it is recognized that one specific nutrient alone receives major consideration. Correctly speaking, therefore, it is a method of balancing one nutrient requirement, with no consideration given to the other nutritive requirements.

To compute rations by the square method, or by any other method, it is first necessary to have available both feeding standards and feed composition tables. These may be obtained from *Feeds & Nutrition*, by Ensminger, *et al.*, or from the nutrient requirement publications of the National Academy of Sciences, of which there are separate publications for each species of animal.

The following example shows how to use the square method in formulating a swine ration:

Example. *A swine producer has 40-lb pigs to which it is desired to feed a 16% protein ration until they reach 120-lb weight. Corn containing 8.9% protein is on hand. A 36% protein supplement, which is reinforced with minerals and vitamins, can be bought. What percent of the ration should consist of corn and of the 36% protein supplement?*

Step by step, the procedure in balancing this ration is as follows:

1. Draw a square, and place the number 16 (desired protein level) in the center.

2. At the upper left-hand corner of the square, write *protein supplement* and its protein content (36); at the lower left-hand corner, write *corn* and its protein content (8.9).

3. Subtract diagonally across the square (the smaller number from the larger number), and record the difference at the corners on the right-hand side (36 − 16 = 20; 16 − 8.9 = 7.1). The number at the upper right-hand corner gives the parts of protein supplement by weight, and the number at the lower right-hand corner gives the parts of corn by weight to make a ration with 16% protein.

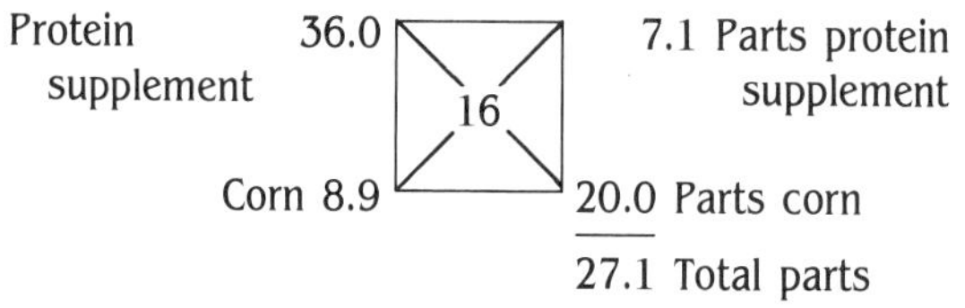

4. To determine what percent of the ration would be corn, divide the parts of corn by the total parts and multiply by 100: 20.0 ÷ 27.1 × 100 = 73.8% corn. The remainder, 26.2%, would be supplement.

TRIAL-AND-ERROR METHOD

In the example that follows, the trial-and-error method is used, with consideration given to energy and protein. Also, crude protein rather than digestible protein is used because (1) this is what feed manufacturers want to know as they plan feed formulas, and (2) this is what livestock producers see on the tag when they purchase feed. In most mixed feeds, approximately 80% of the total protein is digestible.

Example. *Let's assume that a dairy producer has a 1,433-lb cow producing 65 lb of milk testing 4.0% fat. The producer is feeding 14 lb of alfalfa hay and 40 lb of corn silage per day. Corn, oats, and soybean meal are available. What concentrate mix should the producer use to meet the needs of this lactating cow, from the standpoint of energy and protein?*

The available feeds have approximately the following composition (as-fed basis):

	TDN	Crude Protein
	(%)	(%)
Alfalfa hay, all analyses	51.0	16.0
Corn silage, all analyses	18.0	2.2
Corn, all analyses	80.0	9.9
Oats, all analyses	69.0	11.9
Soybean meal, solv extd, 44%	76.0	44.4

Here are the steps in balancing this ration:

Step 1. The daily TDN and crude protein requirements of this cow (1,433 lb body weight, 65 lb of milk testing 4% fat) are:

Requirements of cow for—

	TDN	Crude Protein	
	(lb)	(lb)	*(g)*
Maintenance	9.94	0.94	*428*
Milk production	20.9	5.87	*2,665*
Total	30.84	6.81	*3,093*

Step 2. The forage (14 lb alfalfa hay, 40 lb corn silage) is supplying:

	TDN	Crude Protein
	(lb)	(lb)
Alfalfa hay, 14 lb	7.14	2.24
Corn silage, 40 lb	7.20	0.88
Total from forage	14.34	3.12

Step 3. Remainder, to be supplied by concentrate:

TDN	Crude Protein
(lb)	(lb)
16.5	3.69

Step 4. Let's try out (that's why it is called the *trial-and-error method*) a grain mix of 700 lb corn, 280 lb oats, 10 lb monosodium phosphate, and 10 lb salt, and determine the amounts of TDN and crude protein in 1,000 lb of the grain mix:

	TDN	Crude Protein
	(lb)	(lb)
Corn, 700 lb	560.0	69.3
Oats, 280 lb	193.2	33.3
Monosodium phosphate, 10 lb	—	—
Salt, 10 lb	—	—
Total	753.2	102.6
or in percent	75.3%	10.3%

Step 5. Divide the TDN needed from concentrate (16.5 lb) by the percent TDN in the mixture (75.3%). Thus, feeding 21.9 lb of the concentrate will meet the energy needs.

Step 6. Will this level of grain mix (21.9 lb) also meet the crude protein needs? By multiplying the pounds of concentrate mixture by the percent crude protein (21.9 × 10.3%), we find that the proposed concentrate would supply 2.26 lb of crude protein, whereas 3.69 lb are needed. Therefore, a high-protein supplement must be substituted for some of the homegrown grain.

Step 7. Let's substitute 175 lb of soybean meal for 175 lb of corn. Hence, the concentrate mix as now proposed will consist of:

	TDN	Crude Protein
	(lb)	(lb)
Corn, 525 lb	420.0	52.0
Oats, 280 lb	193.2	33.3
Soybean meal, 175 lb	133.0	77.7
Monosodium phosphate, 10 lb	—	—
Salt, 10 lb	—	—
Total	746.2	163.0
or in percent	74.6%	16.3%

Step 8. By referring back to Step 3, we can divide the pounds of TDN and crude protein needed from the concentrate, by the percentage of TDN and crude protein found in the grain mix in Step 7. We find that 16.5 ÷ .746 = 22.1 lb needed to supply 16.5 lb TDN; and 3.69 ÷ .163 = 22.63 lb needed to supply 3.69 lb crude protein. Thus, we find that the following ration will supply the needed TDN (with a slight overage) and crude protein for a 1,433-lb lactating cow producing 65 lb of milk testing 4% fat:

	TDN	Crude Protein
	(lb)	(lb)
Alfalfa hay, 14 lb	7.1	2.2
Corn silage, 40 lb	7.2	0.9
Concentrate (Steps 7 & 8), 22.63 lb	16.9	3.7
Total	31.2	6.8

In many sections of the country, especially in grain-deficient areas and on highly specialized dairies where little or no grain is grown, the dairy producer may find it most economical to purchase a commercial dairy feed to augment the roughage that is being fed.

NET ENERGY METHOD

In order to apply the net energy method to the feeding of livestock, the following net energy values must be available:

1. A table showing the net energy requirements of the particular class of animal.

2. A table showing the nutrient composition of feeds, with the net energy of each feed partitioned into energy used for body maintenance and for gain; thus, the net energy values in megacalories (Mcal) per unit (lb or kg) are needed for each feed for maintenance (NE_m) and for gain (NE_g).

These net energy values may be obtained from *Feeds & Nutrition*, by Ensminger, *et al.*, and from the nutrient requirement publications of the National Academy of Sciences, of which there are separate publications for each species of animals.

The two examples that follow will show how to apply the net energy method. In the first example, net energy values of feeds are used to calculate the number of pounds of a given ration that a steer would need to consume to make a specified daily gain. In the second example, net energy is used to predict average daily gain based on consuming a certain number of pounds of a specified ration. Bear in mind that the ration in both cases (in these examples, the ration in Table 4–8) must be balanced for protein, minerals, and vitamins, in order for these net energy values to have validity for calculating daily consumption and predicting average daily gain.

TABLE 4–8
RATION FOR FINISHING CATTLE

Ration Ingredient			Composition of Ingredients (as-fed basis) NE_m[1]		Ration Supplies NE_m[1]	Composition of Ingredients (as-fed basis) NE_g[2]		Ration Supplies NE_g[2]
	(lb)	*(kg)*	(Mcal/lb)[3]	*(Mcal/kg)*[3]	(Mcal)[3]	(Mcal/lb)[3]	*(Mcal/kg)*[3]	(Mcal)[3]
Shelled corn, all analyses	68.60	*31.14*	0.86	*1.90*	59.00[4]	0.59	*1.89*	40.47[5]
Soybean meal (solvent), 44%	4.00	*1.82*	0.79	*1.74*	3.16	0.53	*1.17*	2.12
Alfalfa hay (mid-bloom)	27.00	*12.26*	0.52	*1.15*	14.04	0.28	*0.62*	7.56
Salt	0.40	*0.18*	—	—	—	—	—	—
Total	100.00	*45.4*	—	—	76.20	—	—	50.15

[1]NE_m = net energy for maintenance. [2]NE_g = net energy for gain. [3]Mcal stands for megacalorie. [4]68.60 lb × 0.86 = 59.00 [5]68.60 lb × 0.59 Mcal = 40.47

Example 1. *Using net energy values to calculate the number of pounds of the ration that must be consumed to produce a specific gain*—How many calories would a 770–lb medium-frame steer calf need to consume to gain 2.6 lb daily?

Step 1. Calculate the net energy for maintenance (NE_m) and gain (NE_g) values for a pound of the ration shown in Table 4–8.

By referring to a table of Composition of Feeds, it is determined that 1 lb of the Table 4–8 ration supplies 0.7620 megacalories of net energy for maintenance (Mcal NE_m) and 0.5015 megacalories of net energy for gain (Mcal NE_g).

Step 2. By referring to a table of Net Energy Requirements, we find that the requirement for a 770–lb medium-frame steer calf to gain 2.6 lb daily is as follows:

	Mcal / day
NE_m	6.24
NE_g	5.50

Step 3. Pounds of feed to meet the daily maintenance requirement:

6.24 Mcal ÷ .7620 Mcal = 8.19 lb

Step 4. Pounds of feed to meet the requirement for 2.6 lb daily gain:

5.50 Mcal ÷ .5015 Mcal = 10.97 lb

Step 5. Total pounds of feed that the steer calf must eat daily to gain 2.6 lb:

8.19 lb + 10.97 lb = 19.16 lb

Example 2. *Using net energy to predict the average daily gain of a 770–lb medium-frame steer calf that is consuming a certain number of pounds of a specified ration*—Let's assume that we have a 770–lb steer that is consuming 18 lb of the ration shown in Table 4–8. What daily gain should be expected?

Step 1. Pounds of feed to meet the daily maintenance requirement = 8.19 lb (see prior example).

Step 2. Pounds of feed left for gain:

18 lb – 8.19 lb = 9.81 lb

Step 3. Mcal of NE_g supplied by remaining feed:

9.81 lb × .5015 Mcal = 4.92 Mcal

Step 4. Daily gain expected from 4.92 Mcal of NE_g. (Value obtained from a table of Net Energy Requirements. In this case, *Feeds & Nutrition*, p. 698, Table 19–3.)

4.51 Mcal produces 2.2 lb gain

Therefore, 4.92 Mcal will produce 2.40 lb daily gain

$$\left(\frac{4.92\ (2.2)}{4.51}\right)$$

COMPUTER METHODS[4]

Most large livestock establishments and feed companies now use computers in ration formulation. Also, many of the state universities, through their Federal-State Extension Services, are offering ration balancing computer services to farmers within their respective states on a charge basis. Consulting nutritionists are available throughout the United States and provide computer services, as well as other services. With the recent advent of the low cost personal computer, this powerful technology is available to almost everyone.

Despite their sophistication, there is nothing magical or mysterious about the use of computers in ration balancing. Their primary advantages are accuracy and speed of computation. In addition, computer programs (software) used in ration balancing provide a means of organizing needed information in a logical and systematic manner. The computer should be viewed as an extension of the knowledge and skills of the formulator.

At this time, there is no "push-button" system of feed formulation available. The degree of success realized is very dependent on the management of data put into the computer, and on the evaluation of the resulting formulations that the computer generates. In the hands of experienced users, the computer enables the producer and nutritionist to be more precise in carrying out ration formulation.

Two basic approaches to ration formulation are practiced with computers:

1. Trial-and-error formulation.
2. Linear programming (LP).

• **Trial-and-error formulation with the computer**—For a discussion of the trial-and-error method of ration balancing, see the earlier discussion headed "Trial-and-Error Method." Many ration balancing software programs written for the computer allow for trial-and-error ration balancing. Feed mill nutritionists frequently use this technique to enter into the computer rations that are given to them by other nutritionists or by a producer. The objective in this case is to confirm the nutrient values for the ration based on the specific ingredients used by the feed manufacturer. In many cases, these rations are not to be altered without permission. In other cases, the number of ingredients for a specific ration may be limited so that the trial-and-error technique is just as fast as using linear programming to arrive at the desired nutrient levels in the ration.

Note: It does not take specialized computer software to use the trial-and-error method. Spreadsheet (or Financial Spreadsheet) programs, for instance, organize data into rows and columns. Information, such as nutrient values for a feedstuff, may be entered into data cells (see Fig. 4–26). Simple and complex arithmetic operations can be controlled by the user to the extent that rather large trial-and-error method rations can be programmed and run.

Spreadsheets have been developed with specific microcomputers in mind; and there are a great number of them on the market.

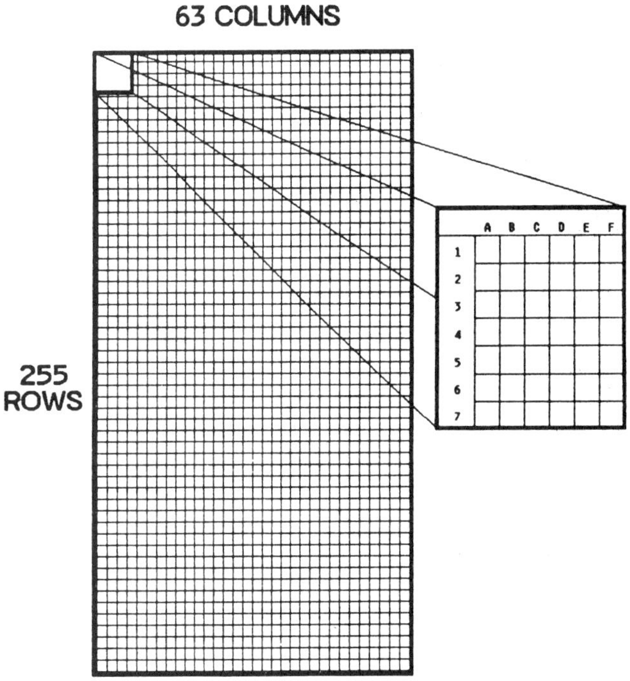

Fig. 4–26. Graphic representation of a spreadsheet (From Lane, R. J. and T. L. Cross, *Spreadsheet Applications for Animal Nutrition and Feeding*, Reston Publishing Co., Inc., Reston, VA, 1985).

• **Linear programming (LP)**—The most common technique for computer formulation of rations is the linear programming (LP) technique. At times, this is referred to as *least cost* ration formulation. This designation results from the fact that most LP techniques for ration formulation have as their objective *minimization of cost*. A few LP programs are in use that solve for *maximization of income over feed costs*. Regardless, the livestock producer and nutritionist should always keep in mind that maximizing net profit is the only true objective of most ration formulations. A skilled user of the LP system will control ration quality by writing specifications that lead to rations that will maximize profit.

Briefly described, the LP program is a mathematical technique in which a large number of simultaneous equations are solved in such a way as to meet the minimum and maximum levels of nutrients and levels of feedstuffs specified by the user at the lowest possible cost. It is not necessary to understand the inner workings of the computer program to use LP, though it does take experience to use it to good advantage and to avoid certain pitfalls. The most common pitfalls are incorrectly entered

[4]The section on "Computer Methods" was prepared by L. M. Larsen, Ph.D., Consultant, Nutri-Systems, 426 E. Shields, Fresno, CA 93704.

or missing data and the specification of minimums and maximums that cannot be met with the feedstuffs available. The latter is called an infeasible solution. When an infeasible solution is encountered, the user must determine (1) if this is due to incorrect or missing data, or (2) if the specifications must be relaxed.

Note: Important additional items to consider when creating ration specifications are upper limits on the use of nonprotein nitrogen (or urea) and limits on the usage of feed additives, such as drugs, feed flavors, and the like.

Fig. 4–27 illustrates, by means of a worksheet, a logical method of organizing the restrictions for a ration.

- **Procedure for use of linear programming (LP)**—Before using the LP approach to ration formulation, the user should become familiar with the specific software package to be used. It is also desirable to study the LP technique as applied to feed formulation. After users are familiar with LP and their computer software, they are ready to begin using the computer for ration formulation by LP. It must first be understood that all data entered into the computer is directed to files. In most cases, these files are located on disks, or perhaps on tapes. Currently, most computers use keyboards and CRT (cathode ray tube) displays for entry of data. The necessary data files are generally created in steps as follows:

1. **Enter names of available feed ingredients, and the cost of each.** It is necessary that all of the available feeds be listed along with the unit cost. It matters little if the formulator uses cost per ton, cost per hundred weight (cwt), or cost per pound, but the same method of cost input must be used for all feeds. The computer software may call for a specific form for entering costs.

2. **Enter nutrient values for feeds.** Tables of feed composition using average or typical values may be used, but, because of the wide variation in the composition of feeds, a chemical analysis of a representative sample of each lot of feed is more precise and should be used if available. This is especially true of forages, in which composition may be affected in a major way by cultural conditions and stage of maturity.

3. **Enter ration specifications.** Ration specifications are generally broken into two parts: (1) Nutrient limits, and (2) Ingredient limits. In each case, the formulator specifies either a lower limit and/or an upper limit for each item. If no specification for the particular item is desired, it may be specified as zero (0) or left blank, depending on the circumstances. It is also appropriate to list feedstuffs available, but not currently on hand (with an upper limit of zero). Most LP solutions will then tell the user the highest cost at which such feeds would enter the solution if allowed to do so. Ratios between nutrients (such as a calcium/phosphorus ratio) or feedstuffs (corn/barley ratio) may also be specified in most LP software packages. The experienced formulator usually deals with palatability or feedstuff quality considerations by setting an upper limit on the amounts of problem feeds or a lower limit on feeds that contribute a positive quality to the ration. Nonnutritive attributes, such as bulk density, may also be programmed into the LP system. The LP technique is a very flexible and powerful ration balancing tool.

LEAST-COST FORMULATION WORKSHEET

Specifications	Ingredient A	Ingredient B	Ingredient C	Restrictions
Cost				Minimize
Total weight				1,000 lb
Crude protein				133 lb
Digestible protein				100 lb
Ether extract				25 to 80 lb
Net energy lactation				900 Mcal
Calcium				5 to 10 lb
Phosphorus				7 lb
Vitamin A equivalent				35,000
Vitamin D				60,000
Limits on ingredients				
Minimum				
Maximum				

Fig. 4–27. Sample worksheet for a least-cost formulation. The first column lists the specifications. The various feedstuffs to be considered are listed in the succeeding columns with their respective costs and nutritive values. The last column lists the restrictions desired on the final formulation.

4. **Submit all of the above information to the matrix building and solving portion of the LP software package.** Matrix building and solving are generally accomplished automatically by the computer software once the specifications have been entered into the computer. Mathematically, the procedure involves the solution of a complex algebraic problem, with an answer being derived in seconds or minutes. Using the LP program, the computer produces a mix that will meet the desired specifications at the lowest possible cost.

5. **Examine the solution provided by the computer software.** The end result should be feasible, both from a mathematical standpoint and from a nutritional standpoint. The feedstuff mixture should be acceptable to the animals for which it is intended. In most cases the first solution provided to the user is not acceptable. Repeat runs may be necessary to obtain the best solution.

Fig. 4–28 is a computer printout of an LP solution. It is a computer printout of a ration for lactating cows formulated with the use of the National Research Council (NRC) prediction equations.

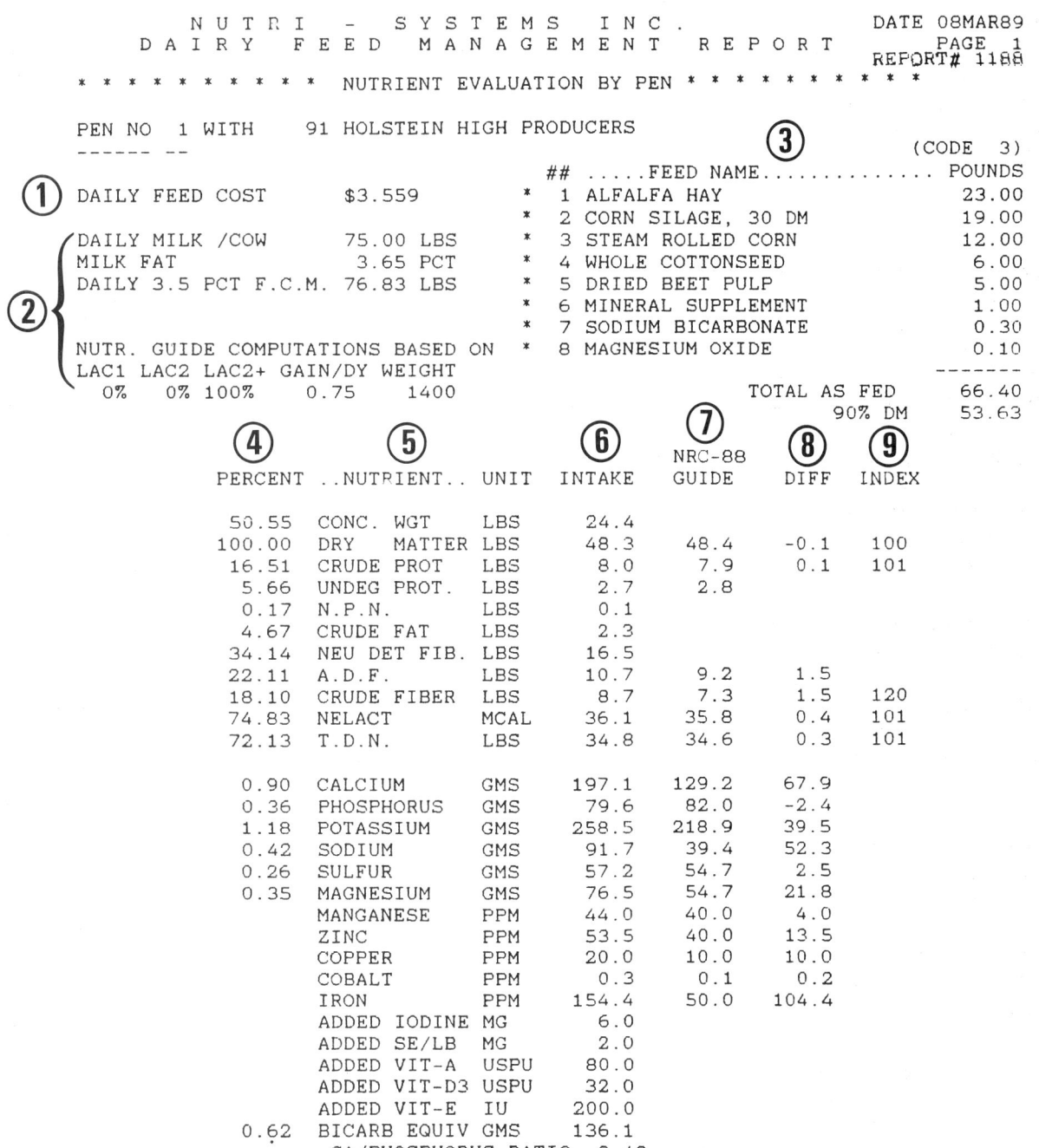

NUTRI - SYSTEMS INC. DATE 08MAR89
DAIRY FEED MANAGEMENT REPORT PAGE 1
REPORT# 1188

* * * * * * * * * * NUTRIENT EVALUATION BY PEN * * * * * * * * * *

PEN NO 1 WITH 91 HOLSTEIN HIGH PRODUCERS

(1) DAILY FEED COST $3.559

(2)
DAILY MILK /COW 75.00 LBS
MILK FAT 3.65 PCT
DAILY 3.5 PCT F.C.M. 76.83 LBS

NUTR. GUIDE COMPUTATIONS BASED ON
LAC1 LAC2 LAC2+ GAIN/DY WEIGHT
0% 0% 100% 0.75 1400

(3) (CODE 3)

##	.....FEED NAME..............	POUNDS
* 1	ALFALFA HAY	23.00
* 2	CORN SILAGE, 30 DM	19.00
* 3	STEAM ROLLED CORN	12.00
* 4	WHOLE COTTONSEED	6.00
* 5	DRIED BEET PULP	5.00
* 6	MINERAL SUPPLEMENT	1.00
* 7	SODIUM BICARBONATE	0.30
* 8	MAGNESIUM OXIDE	0.10
	TOTAL AS FED	66.40
	90% DM	53.63

(4) PERCENT	(5) ..NUTRIENT..	UNIT	(6) INTAKE	(7) NRC-88 GUIDE	(8) DIFF	(9) INDEX
50.55	CONC. WGT	LBS	24.4			
100.00	DRY MATTER	LBS	48.3	48.4	-0.1	100
16.51	CRUDE PROT	LBS	8.0	7.9	0.1	101
5.66	UNDEG PROT.	LBS	2.7	2.8		
0.17	N.P.N.	LBS	0.1			
4.67	CRUDE FAT	LBS	2.3			
34.14	NEU DET FIB.	LBS	16.5			
22.11	A.D.F.	LBS	10.7	9.2	1.5	
18.10	CRUDE FIBER	LBS	8.7	7.3	1.5	120
74.83	NELACT	MCAL	36.1	35.8	0.4	101
72.13	T.D.N.	LBS	34.8	34.6	0.3	101
0.90	CALCIUM	GMS	197.1	129.2	67.9	
0.36	PHOSPHORUS	GMS	79.6	82.0	-2.4	
1.18	POTASSIUM	GMS	258.5	218.9	39.5	
0.42	SODIUM	GMS	91.7	39.4	52.3	
0.26	SULFUR	GMS	57.2	54.7	2.5	
0.35	MAGNESIUM	GMS	76.5	54.7	21.8	
	MANGANESE	PPM	44.0	40.0	4.0	
	ZINC	PPM	53.5	40.0	13.5	
	COPPER	PPM	20.0	10.0	10.0	
	COBALT	PPM	0.3	0.1	0.2	
	IRON	PPM	154.4	50.0	104.4	
	ADDED IODINE	MG	6.0			
	ADDED SE/LB	MG	2.0			
	ADDED VIT-A	USPU	80.0			
	ADDED VIT-D3	USPU	32.0			
	ADDED VIT-E	IU	200.0			
0.62	BICARB EQUIV	GMS	136.1			

CA/PHOSPHORUS RATIO 2.48

Fig. 4-28. Lactating dairy ration comparing nutrient intake (column 6) with the NRC requirement (column 7).

An explanation of the numbered sections (columns) of Fig. 4-28 follows:

1. Feed cost per cow per day.
2. Input data used for computing the requirements.
3. Feedstuff amounts given in pounds per cow per day.
4. Nutrient composition of the ration on a 100% dry matter basis.
5. Nutrient names and units.
6. Daily intake of nutrients per cow per day.
7. The NRC requirements computed from the prediction equations.
8. Deficiencies (negative values) or excesses (positive values). These values are computed by subtracting the requirements from the daily intake figures.
9. A computed index for dry matter, crude protein, NE_{lc}, and TDN. This value is computed as the percentage that nutrient intake is of the nutrient requirement.

Changes in ingredient costs, in ingredient availability, and in the needs of animals dictate the need for reprocessing the ration. The good formulator monitors all these items on a regular basis. It is also critical to evaluate the feeding results to confirm that production goals and cost objectives are being met with the ration. Computers don't feed animals—people do!

COMMERCIAL FEEDS

Commercial feeds are just what the term implies—instead of being farm mixed, these feeds are mixed by commercial feed manufacturers who specialize in the business. In 1988, a total of 103.1 million tons of primary feeds (complete feeds) were manufactured in the United States, and an additional 10 million tons of secondary feeds (supplements) were produced; making for a total of 113.1 million tons of commercial feeds. Primary feed is that which is mixed from individual ingredients, sometimes with the addition of a premix at a rate of less than 100 lb per ton of finished feed. Secondary feed is that which is mixed with one or more ingredients and a formula feed supplement (which is a primary feed); normally, the supplement is used at a rate of 300 lb or more per ton of finished feed, depending upon the protein content of the supplement and the percentage of protein content desired in the finished feed. The breakdown, percentagewise, by classes of livestock for which primary (complete) commercial feeds were used in 1988 follows: poultry, 44.5%; beef and sheep, 17.6%; dairy, 16.9%; hogs, 14.2%, and all other, 6.8%.[5]

Fig. 4-29. A new and modern commercial feed mill of slip-form concrete construction. (Courtesy, Pennfield Corporation, Lancaster, PA)

The commercial feed manufacturer has the distinct advantages of (1) purchasing feed in quantity lots, making possible price advantages; (2) economical and controlled mixing; (3) the hiring of scientifically trained personnel for use in determining the rations; and (4) quality control. Most livestock producers have neither the know-how nor the quantity of business to provide these services on their own. Because of these several advantages, commercial feeds are finding a place of increasing importance in livestock feeding. Also, it is to the everlasting credit of reputable feed dealers that they have been good teachers, often getting livestock producers started in feeding balanced rations.

Numerous types of commercial feeds, ranging from additives to complete rations, are on the market, with most of them designed for the specific species, age, or need. Among them, are complete rations (including hay for ruminants and pseudo-ruminants), concentrates, pelleted or cubed forages, protein supplements (with or without reinforcements of minerals and/or vitamins), mineral and/or vitamin supplements, additives (antibiotics, hormones, etc.), milk replacers, starters, young stock rations, fitting rations, rations for different levels of production—for the idle (like dry cow rations) to the forced producers, and medicated feeds.

• **How to select a commercial feed**—There is a difference in commercial feeds! That is, there is a difference from the standpoint of what producers can purchase with their feed dollars. Enlightened operators will know how to determine what constitutes the best in commercial feeds for their specific needs. They will not rely solely on how the feed looks and smells or on the feed salesperson. The most important factors to consider or look for in buying a commercial feed are:

1. Buy from a reputable manufacturer.
2. Select feeds for specific needs.
3. Study the feed tag.
4. Use flexible formulas because they are usually the best buy.

• **State commercial feed laws**—Nearly all states have laws regulating the sale of commercial feeds. These benefit both producers and reputable feed manufacturers. In most states the laws require that every brand of commercial feed sold in the state be licensed, and that the chemical composition be guaranteed.

Samples of each commercial feed are taken each year, and analyzed chemically in the state's laboratory to determine if manufacturers lived up to their guarantees. Additionally, skilled microscopists examine the sample to ascertain that the ingredients present are the same as those guaranteed. Flagrant violations on the latter point may be prosecuted.

[5]Source: American Feed Manufacturers Assn.

Results of these examinations are generally published, annually, by the state department in charge of such regulatory work. Usually, the publication of the guarantee alongside any "short-changing" is sufficient to cause the manufacturer promptly to rectify the situation, for such public information soon becomes known to both users and competitors.

HOME-MIXED VS COMMERCIAL FEEDS

The value of farm-grown grains—plus the cost of ingredients which need to be purchased in order to balance the ration, and the cost of grinding and mixing—as compared to the cost of commercial ready-mixed feeds laid down on the farm, should determine whether it is best to mix feeds at home or depend on ready-mixed feeds.

Feeders have the following options for purchasing and preparing feeds:

1. Purchase of a commercially prepared, complete feed.
2. Purchase of a commercially prepared protein supplement (likely reinforced with minerals and vitamins), which may be blended with local or homegrown grain.
3. Purchase of a commercially prepared mineral-vitamin premix which may be mixed with an oil meal, and then blended with local or homegrown grain.
4. Purchase of individual ingredients (including minerals and vitamins) and mixing the feed from the ground up.

In summary, it may be said that there now exist two good alternative sources of most feeds and rations—home mixed or commercial—and the able manager will choose wisely between them.

PART II
FEEDING BEEF CATTLE

Fig. 4-30. Pastures and other roughages are the very foundation of successful beef production. (Courtesy, American Angus Assn., St. Joseph, MO)

The feeding of beef cattle constitutes the greatest single cost item of their production. It is important, therefore, that the feeding practices be as satisfactory as possible.

Pastures and other roughages, preferably with a maximum of the former, are the very foundation of successful beef cattle production. In fact, it may be said that the principal function of beef cattle is to harvest vast acreages of forages, and, with or without supplementation, to convert these feeds into more nutritious and palatable products for human consumption. It is estimated (1) that 84.5% of the total feed of beef cattle is derived from roughages (see Table 4–4), and (2) that 31% of the land area of the continental United States is grazed (pastured) all or part of the year, with much of this area utilized by beef cattle. If produced on well-fertilized soils, green grass and well-cured, green, leafy hay can supply all of the nutrient requirements of beef cattle except the need for common salt and whatever energy-rich feeds may be necessary for additional conditioning or drylot finishing.

NUTRITIVE NEEDS OF BEEF CATTLE

In recent years, the introduction of crossbreeding and the exotic breeds has produced faster-gaining calves, later-maturing cattle, and heavier-milking cows. Hand in hand with this development, scarce and high-priced grains, compared to roughages, have caused feeder cattle to be carried to heavier weights on grass and other roughages before going into feedlots, then grain fed for a shorter period than formerly. Also, more and more heifers are being bred to calve as 2-year-olds. In this revision, provisions have been made for the nutritive needs created by these changes.

ENERGY

Carbohydrates, which constitute about 75% of all the dry matter of plants, are the chief source of energy in cattle feeds. Next to carbohydrates, fats are important as energy sources.

A relatively large portion of the feeds consumed by beef cattle is used in meeting the energy needs, regardless of whether the animals are merely being maintained (as in wintering) or fed for growth, finishing, or reproduction.

After the energy needs for body maintenance have been met, any surplus energy may be used for growth, finishing, reproduction, or lactation. When cattle are finished at early ages, growth and finishing are in most instances simultaneous, and, therefore, not easily separated.

Through bacterial action in the rumen, cattle are able to utilize a considerable portion of roughages as sources of energy. Yet it must be realized that with extremely bulky rations, the animal cannot consume sufficient quantities to produce the maximum amount of fat. For this reason, finishing rations contain a considerable proportion of concentrated feeds, mostly cereal grains. On the other hand, when the energy requirements are primarily for maintenance, roughages are usually the most economical sources of energy for beef cattle.

- **Symptoms of energy deficiency (underfeeding)**—Lack of energy is accompanied by an inevitable loss in body weight and condition; and, varying with the degree of underfeeding, there may be a slowing or cessation of growth (including the skeletal growth), failure to conceive, and increased mortality. Low feed intake also commonly results in increased deaths from toxic plants and from lowered resistance to parasites and diseases.

PROTEIN

The protein allowance for beef cattle, regardless of age or system of production, should be ample to replace the daily breakdown of the tissues of the body including the growth of hair, horns, and hoofs. In general, the protein needs are greatest for the growth of the young calf and for the gestating-lactating cow.

- **Symptoms of protein deficiency**—Depressed appetite is the primary symptom of protein deficiency in beef cattle rations. Depressed appetite may, in turn, lead to an inadequate intake of energy; hence, protein deficiency and energy deficiency often occur together.

Other symptoms of protein deficiency are loss of weight, poor growth, irregular or delayed estrus, and reduced milk production.

MINERALS

The functions, deficiency symptoms, interrelationships/toxicities, and sources of minerals for beef cattle, which are similar to those of other animal species, are presented in Table 4–1, Animal Mineral Chart.

Table 4–9 presents the NRC mineral requirements and maximum tolerable levels for beef cattle.

Needed minerals may be incorporated in beef cattle rations or in the water. In addition, it is recommended that all classes and ages of cattle be allowed free access to a two-compartment mineral box, with (1) salt (iodized salt in iodine-deficient areas) in one side, and (2) a suitable mineral mixture in the other side. Free-choice feeding is in the nature of cheap insurance with the animals consuming the minerals if they are needed.

TABLE 4-9
MINERAL REQUIREMENTS AND MAXIMUM TOLERABLE LEVELS FOR BEEF CATTLE[1]

Mineral	Requirements: Suggested Value	Requirements: Range[2]	Maximum Tolerable Level[3]	Mineral	Requirements: Suggested Value	Requirements: Range[2]	Maximum Tolerable Level[3]
Calcium .... (%)	—	Breeding 0.15–0.43 Growing/finishing 0.17–1.38	2	Phosphorus .... (%)	—	Breeding 0.15–0.28 Growing/finishing 0.15–0.53	1
Cobalt .... (ppm)	0.10	0.07–0.11	5	Potassium .... (%)	0.65	0.5–0.7	3
Copper .... (ppm)	8	4–10	115	Selenium .... (ppm)	0.20	0.05–0.30	2
Iodine .... (ppm)	0.5	0.20–2.0	50	Sodium .... (%)	0.08	0.06–0.10	10[4]
Iron .... (ppm)	50	50–100	1,000	Chlorine .... (%)	—	—	—
Magnesium .... (%)	0.10	0.05–0.25	0.40	Sulfur .... (%)	0.10	0.08–0.15	0.40
Manganese .... (ppm)	40	20–50	1,000	Zinc .... (ppm)	30	20–40	500
Molybdenum .... (ppm)	—	6					

[1]Adapted from *Nutrient Requirements of Beef Cattle,* 6th revised edition, National Research Council—National Academy of Sciences, 1984, p. 43.

[2]The listing of a range in which requirements are likely to be met recognizes that requirements for most minerals are affected by a variety of dietary and animal factors (body weight, sex, rate of gain). Thus, it may be better to evaluate rations based on a range of mineral requirements and for content of interfering substances than to meet a specific dietary value.

[3]From National Research Council (1980). Maximum tolerable levels are given on the basis of the ration dry matter.

[4]10% sodium chloride.

When buying and home mixing minerals, or when buying commercial mineral mixes, the cattle producers should first determine their needs, based on (1) available feeds, (2) area (for example, the Northern Great Plains and the Southwest are phosphorus-deficient areas), and (3) the age and reproduction status (pregnancy and lactation make a difference) of the animals for which the mineral mix is intended.

Cattle pastured on native grass should be offered a free-choice mineral mix consisting of 40% dicalcium phosphate or bone meal and 60% trace mineralized salt. Minerals that are self-fed on pastures or in corrals should be in boxes protected from the weather.

Salt should always be available on a free-choice basis in addition to whatever mineral mix is provided.

VITAMINS

The functions, deficiency symptoms, and good sources of vitamins for beef cattle, which are similar to those of other animal species, are presented in Table 4-2, Animal Vitamin Chart.

Table 4-10 presents the National Research Council vitamin requirements of beef cattle.

Vitamin deficiencies in cattle may occur as a result of lack of availability of vitamins or because of the presence of antimetabolites. Both are important concepts. For example, analyses show corn to be adequate in niacin. Yet, due either to an antimetabolite or unavailability, there may be niacin deficiencies when corn is fed—deficiencies which can be remedied by niacin supplementation.

TABLE 4-10
VITAMIN REQUIREMENTS OF BEEF CATTLE
(IN AMOUNT PER KILOGRAM OF DRY RATION)[1]

Nutrient	Growing and Finishing Steers and Heifers	Dry Pregnant Cows	Breeding Bulls and Lactating Cows
Vitamin A activity ....IU[2]	2,200	2,800	3,900
Vitamin D ....IU	275	275	275
Vitamin E ....IU	15–60	—	15–60

[1]From: *Nutrient Requirements of Beef Cattle,* Sixth Revised Edition, National Research Council, National Academy of Sciences, Washington, DC, 1984. Requirements given in IU/kg may be converted to IU/lb by dividing by 2.2.

[2]May be vitamin A or provitamin A equivalent.

WATER

Beef cattle should have an abundant supply of water before them at all times. Mature cattle will consume an average of about 11 gal of water per head daily, with younger animals requiring proportionately less. The water requirement is influenced by several factors, including rate and composition of gain, pregnancy, lactation, activity, type of ration, feed intake, and environmental temperature.

GROWTH STIMULANTS AND IMPLANTS FOR BEEF CATTLE

Table 4-11 (next page) summarizes the growth stimulants and implants that are presently available and can be used for beef cattle. All of these products have been shown to improve gain and feed efficiency significantly.

TABLE 4-11
GROWTH STIMULANTS AND IMPLANTS FOR CATTLE[1]

Additive	Method of Adminis-tering	Dosage	Increase in Daily Rate of Gain	Increase in Feed Effi-ciency	Effect on Carcass Quality	Other Comments	Withdrawal Period Prior to Slaughter
			Finishing Steers				
Antibiotic	Oral.	10 mg/100 lb body wt. daily; or 70 to 75 mg/head daily.	6%	4%	Improves carcass quality slightly; more fat deposition and marbling. Decreases liver and rumen condemnations.	Antibiotics will also reduce the disease level. More effective with high-roughage rations than with high-concentrate rations.	No withdrawal required.
Bovatec (lasalocid)	Oral.	150 to 360 mg/day. 250 to 360 mg/day.	 8%	8% 5%	No effect.	Alters rumen fermentation similar to monensin.	No withdrawal required.
Rumensin (monensin)	Oral.	50–360 mg/head/day, drylot. 50–200 mg/head/day, pasture.		10%	No effect.	Not a hormone. It results in more propionic acid and less butyric and acetic acids; hence, more energy.	No withdrawal required.
Compudose	Implant.	24 mg estradiol.	10–15%	5–10%	No effect.	Only one implant. Effective for 200 days.	No withdrawal required.
Finaplex	Implant.	140 mg.	7%	7%		The active ingredient is trenbolene acetate (TBA). It is a synthetic analog of testosterone. It gives an additional response when used in combination with an estrogen.	
Ralgro (Zeranol)	Implant.	36 mg resorcyclic acid lactone.	10%	5–10%	No effect.	Nonestrogenic.	65 days.
Steer-oid	Implant.	200 mg progesterone, 20 mg estradiol.	10–15%	5–10%	No effect.	Effective period of 140 days.	No withdrawal.
Synovex-S (for steers)	Implant.	200 mg progesterone, 20 mg estradiol benzoate.	10–15%	5–10%	No effect.	Effective period of 90 to 120 days.	No withdrawal required.
			Finishing Heifers				
Antibiotic	Oral.	10 mg/100 lb body wt. daily; or 70 to 75 mg/head daily.	6%	4%	Improves carcass quality slightly; more fat deposition and marbling.	Antibiotics will also reduce the disease level. More effective with high-roughage than with high-concentrate rations.	No withdrawal required.
Bovatec (lasalocid)	Oral.	150 to 360 mg/day.	5%	8%	No effect.	Alters rumen fermentation similar to monensin.	No withdrawal.
MGA	Oral.	0.25 to 0.50 mg daily melengestrol acetate.	10%	6%	MGA will lower the incidence of estrus in heifers and increase rate and efficiency of gain. It is not effective with pregnant heifers.	MGA is effective for heifers, but not for steers.	48 hours.
Rumensin	Oral.	50–360 mg/head/day, drylot. 50–200 mg/head/day, pasture.		15%	No effect.	Not a hormone. It results in more propionic acid and less butyric and acetic acids; hence, more energy.	No withdrawal required.
Heifer-oid	Implant.	Follow label directions.	10%	5–10%	No effect.	Effective period of 140 days. For heifers over 400 lb.	No withdrawal.
Ralgro (zeranol)	Implant.	36 mg resorcyclic acid lactone.	10%	5–10%	No effect.	Nonestrogenic.	65 days.
Synovex-H (for heifers)	Implant.	200 mg testosterone propionate. 20 mg estradiol benzoate.	10%	5–10%	No effect.	Recommended for use in heifers during last 60 to 150 days of the finishing period.	No withdrawal required.
			Suckling Calves				
Antibiotic	Oral (in creep feed).	15 to 20 mg/100 lb body wt. daily.	6%	4%		Antibiotics will also reduce the disease level. Administer in creep feed.	No withdrawal required. **Note well**: If fed at level of 350 mg or over/day, 48-hour withdrawal required.
Ralgro (zeranol)	Implant.	36 mg resorcyclic acid lactone.	10%	5–10%	No effect.	Nonestrogenic.	65 days.
Synovex-C	Implant.	110 mg.	5–10%		No effect.	For calves of either sex.	No withdrawal required.

[1] *CAUTIONS:* FDA regulations are subject to change. Always follow the manufacturer's directions on the use of these products.

In considering the stimulants and implants listed in Table 4–11, it should be noted that there is no evidence to indicate that the use of these products can or will alleviate the need for vigilant sanitation, improved nutrition, and superior management. Also, the benefits of each one must be weighed against its cost.

RATIONS FOR BEEF CATTLE[6]

Some general rules of feeding may be given, but it must be remembered that "the eye of the expert feeder fattens the cattle."

Table 4–12 presents rations for different classes and ages of beef cattle. Variations can and should be made in the rations used. The feeder should give consideration to (1) the supply of homegrown feeds, (2) the availability and price of purchased feeds, (3) the class and age of cattle, (4) the health and condition of the animals, and (5) the length of the grazing season.

In using Table 4–12, it is to be recognized that feeds of similar nutritive properties can and should be interchanged as price relationships warrant. Thus, (1) the cereal grains may consist of corn, barley, wheat, oats, and/or sorghum; (2) the protein supplement may consist of soybean, cottonseed, peanut, sunflower, and/or linseed meal; (3) the roughage may include many varieties of hays and silages, and (4) a vast array of by-product feeds may be utilized.

FEEDING BROOD COWS

Feed affects total profit and cow productivity. It accounts for 65 to 70% of the total cost of keeping cows, and it exerts a powerful influence on cow fertility and calf weaning weight—the two biggest success factors in the cattle business.

• **Nutritional needs**—Cows should be fed according to their needs. It is impossible to feed the herd properly where calving occurs the year around, or when dry pregnant cows, replacement heifers, and cows nursing calves are run together. This point becomes apparent from the following nutritional differences of (1) cows nursing calves, and milking well; (2) yearling heifers, last 3-4 months of pregnancy; and (3) dry mature cows, last 2-3 months of pregnancy. Additional pertinent information about nutritional needs follows:

1. The nutritional needs of cows nursing calves are higher and more critical than those (a) of yearling heifers the last 3-4 months of pregnancy, or (b) of dry mature cows the last 2-3 months of pregnancy in total feed consumed, in energy and protein of the ration, and in calcium and phosphorus. After a cow calves, her energy needs jump about 50%, her protein needs double, and her calcium and phosphorus needs triple.

2. The nutritional needs of yearling heifers the last 3-4 months of pregnancy are higher than those of dry mature cows the last 2-3 months of pregnancy in energy and protein, and in calcium and phosphorus.

3. The nutritional needs are affected by weight; they increase with weight.

Heavy grain feeding is uneconomical and unnecessary for the beef breeding herd. The nutrient allowances should be adequate merely to provide for maintenance, growth (if the animals are immature), and reproduction. Fortunately, these needs can largely be met through the feeding of roughages.

• **Nutritional reproductive failure**—Since producers largely determine their own destiny when it comes to feeding, it is important that they know the causes of nutritional reproductive failure and how to rectify them.

A review of the literature clearly points to 3 important reproductive difficulties: (1) the small number of cows in heat and bred the first 21 days of the breeding season, (2) the low conception rate at first service, and (3) the excessive calf losses at birth or within the first 2 weeks of age. Also, it is noteworthy that each of the causes is more marked in young cows (first-calf heifers) than in mature cows.

Fig. 4-31. Being born and born alive is the most important requisite in the cow-calf business. (Courtesy, American Polled Hereford Assn., Kansas City, MO)

Research throughout the country gives ample evidence that the real cause of most beef cow reproductive failures is a deficiency of one or more nutrients just before and immediately following calving—nutritive deficiencies during that critical 100-day period when life begins—a deficiency of energy, protein, minerals, and/or vitamins.

---

[6]Insofar as possible, these rations were computed from the requirements as reported by the National Research Council and applied by the author.

TABLE
DAILY RATIONS
(As-Fed

Suggested Rations With all rations and for all classes and ages of cattle, provide free access in separate containers to (1) salt (iodized salt in iodine-deficient areas), and (2) a suitable mineral mixture.	Wintering Mature Pregnant Beef Breeding Cows (avg. wt. 1,100 lb or ***499 kg***)		Wintering Mature Lactating Beef Breeding Cows (avg. wt. 1,100 lb or ***499 kg***)		Wintering Replacement Heifers (weighing 400–500 lb or ***181–227 kg*** start of wintering)	
	Per Day		Per Day		Per Day	
	(lb)	*(kg)*	(lb)	*(kg)*	(lb)	*(kg)*
1. Legume hay or grass-legume mixed hay, good quality	18–20	*8.2–9.1*	30	*13.6*	13–15[3]	*5.9–6.8*[3]
Grain	—	—	—	—	2–3	*0.91–1.36*
Protein supplement	—	—	—	—	—	—
2. Grass hay or other nonlegume dry roughage	18–20	*8.2–9.1*	24–26	*10.9–11.8*	12–18[3]	*5.4–8.2*[3]
Grain	—	—	2	*0.91*	2½–4½	*1.13–2.04*
Protein Supplement	½–1	*0.23–0.45*	3	*1.36*	1¼–1½	*0.57–0.68*
3. Legume hay or grass-legume mixed hay, good quality	7–11	*3.2–5.0*	26–28	*11.8–12.7*	8–12[3]	*3.6–5.4*[3]
Grass hay or other nonlegume dry roughage	9–11	*4.1–5.0*	—	—	4-6	*1.8–2.7*
Grain	—	—	1	*0.45*	2½–4	*1.13–1.81*
Protein supplement	—	—	1	*0.45*	½–1	*0.23–0.54*
4. Corn or sorghum silage	50–55	*22.7–25*	55	*25*	25–40	*11.3–18.2*
Grain	—	—	2	*0.91*	—	—
Protein supplement	0–½	*0–0.23*	3	*1.36*	1½–1¾	*0.68–0.79*
5. Grass silage, half or more legume	50	*22.7*	50	*22.7*	25–40	*11.3–18.2*
Grain	—	—	4	*1.81*	3–4	*1.36–1.81*
Protein supplement	—	—	—	—	½	*0.23*
6. Silage (corn or sorghum silage fed with legume hay or legume silage fed with grass hay)	35	*15.9*	40	*18.1*	15–30	*6.8–13.6*
Hay	5–6	*2.3–2.7*	10	*4.5*	3–4	*1.4–1.8*
Grain	—	—	—	—	1–2	*0.45–0.91*
Protein supplement	0–½	*0–0.23*	—	—	½–1	*0.22–0.45*

[1]If stocker calves are late or the roughage is fair to poor quality, it may be desirable to add 2–4 lb (*0.91–1.81 kg*) of grain per head daily. If farm scales are available, monthly weights may be used as the criterion for grain feeding. Keep in mind that calves should gain ¾–1 lb (*0.34–0.45 kg*) daily.

[2]In general, the experienced feeder plans that cattle on full feed shall consume (1) feeds in amounts (daily: air-dry basis) equal to about 2.5–3.0% of their liveweight, (2) 70–90% concentrates, and (3) a minimum of 2–4 lb (*0.9–1.8 kg*) roughage for each 100 lb (*45 kg*) liveweight. In areas where roughage is more abundant and comparatively cheaper than grain, the proportions of roughage to grain should be somewhat higher than indicated. In computing roughage consumption, 3 lb (*1.36 kg*) of silage are considered equivalent to 1 lb (*0.45 kg*) of hay.

Based on a review of literature, the author concluded as follows:

1. Energy is more important than protein in reproduction.
2. Beef cows receiving inadequate energy reproduce at a low level.
3. Phosphorus supplementation of cows on range areas deficient in phosphorus increases the calf crop.
4. Administering additional vitamin A to heifers grazing dry forage increases the calf crop.
5. The level and kind of feed before and after calving will determine how many cows will show heat—and conceive. After calving, feed requirements increase tremendously because of milk production; hence, when a cow is suckling a calf, she needs approximately 50% greater feed allowance than during the pregnancy period. Otherwise, she will suffer a serious loss in weight and fail to come in heat and conceive.
6. Cows in average condition should gain a minimum of 100 lb during the pregnancy period, followed by a gain of ½ to ¾ lb daily after calving and extending through the breeding season. If they are on the thin side at calving time, they should gain 1½ to 2 lb daily after they drop calves. This calls for 7 to 12 lb of TDN daily before calving (which can be provided by feeding 14 to 22 lb of average quality hay), and 10 to 17 lb of TDN after calving (which can be provided by feeding 14 to 28 lb of hay plus 4 lb grain), with the lactating requirement dependent on both cow weight and milking ability. Additionally, there must be adequate protein, minerals, and vitamins.

• **Crop residues and winter pastures**—Two requisites are important in wintering the cow herd; (1) bringing them through the winter in proper condition for calving, and (2) keeping feed costs to the minimum consistent with nutritional demands. Meeting these requirements has prompted increased use of crop residues and winter pastures for brood cows. As the ever-increasing human population of the world consumes a higher proportion of grains and seeds, and their by-products, directly, cattle will utilize increasing amounts of crop residues and pastures and a minimum of products suitable for human consumption. Thus, more and more farmers with crops will include

4-12
FOR BEEF CATTLE
Basis)

Wintering Stocker Calves Roughed Through Winter and Grazed the Following Summer. Fed for winter gain of ¾-1 lb (*0.34-0.45 kg*) per head daily (weighing 400-500 lb or *181-227 kg* start of wintering)[1]		Finishing Calves in Drylot, Generally in Winter (weighing 400-500 lb or *181-227 kg* start of feeding and 750-850 lb or *340-386 kg* at marketing)[2]		Wintering Yearlings; Roughed Through the Winter, and Generally Pasture Finished the Following Summer. Fed for winter gains of 1-1¼ lb or *0.45-0.57 kg* per head daily (weighing about 600 lb or *227 kg* start of wintering)		Finishing Yearlings in Drylot, Generally in Winter (weighing about 600 lb or *272 kg* start of feeding, and 900-1,050 lb or *409-477 kg* at marketing)[2]		Finishing Long-yearling Steers in Drylot Generally in Winter (weighing about 850 lb or *386 kg* start of feeding and 1,000-1,100 lb or *454-499 kg* at marketing)[2]	
Per Day		Per Day		Per Day		Per Day		Per Day	
(lb)	(*kg*)	(lb)	(*kg*)	(lb)	(*kg*)	(lb)	(*kg*)	(lb)	(*kg*)
12-18[3]	*5.4-8.2*	4-6	*1.8-2.7*	16-24	*7.2-10.9*	4-8	*1.8-3.6*	6-12	*2.7-5.4*
—	—	12-15	*5.4-6.8*	—	—	15-19½	*6.80-8.8*	16-22	*7.2-10.0*
—	—	1-1½	*0.45-0.68*	—	—	1-1½	*0.45-0.68*	—	—
12-18[3]	*5.4-8.2*	4-5	*1.8-2.3*	16-24	*7.2-10.9*	4-8	*1.8-3.6*	6-12	*2.7-5.4*
—	—	12-15	*5.4-6.8*	—	—	15-20	*6.8-9.1*	16½-22¾	*7.5-10.3*
¼-1½	*0.57-0.68*	1¾-2	*0.79-0.91*	1½-1¾	*0.68-0.79*	1½-2½	*0.68-1.1*	1½-1¾	*0.68-0.79*
8-12[3]	*5.4-8.2*	2-3	*0.91-1.36*	6-8	*2.7-3.6*	2-4	*0.91-1.81*	3-6	*1.4-2.7*
4-6	*1.8-2.7*	2-3	*0.91-1.36*	10-16	*4.5-7.2*	2-4	*0.91-1.81*	3-6	*1.4-2.7*
—	—	12-15	*5.4-6.8*	—	—	15-19¾	*6.8-9.0*	16-22	*7.2-10.0*
¼-1	*0.11-0.45*	1½-1¾	*0.68-0.79*	1-1½	*0.45-0.68*	1¼-1¾	*0.57-0.79*	½-¾	*0.23-0.34*
25-40	*11.3-18.1*	6-16	*2.7-7.3*	40-55	*18.2-24.9*	6-25	*2.7-11.3*	6-35	*2.7-5.9*
—	—	8-12	*3.6-5.4*	—	—	11-16	*5.0-7.3*	15-21	*6.8-9.5*
1-1¼	*0.45-0.57*	2	*0.91*	1¼-1½	*0.57-0.68*	2	*0.91*	1¼-1½	*0.57-0.68*
25-40	*11.3-18.1*	6-16	*2.7-7.3*	40-55	*18.1-24.9*	6-25	*2.7-11.3*	6-35	*2.7-15.9*
2-3	*0.91-1.36*	8-12	*3.6-5.4*	4-5	*1.8-2.3*	11-16	*5.0-7.3*	15-21	*6.8-9.5*
½	*0.23*	1-2	*0.45-0.91*	½	*0.23*	1-1½	*0.45-0.68*	1	*0.45*
15-30	*6.8-13.6*	3-8	*1.4-3.6*	20-35	*9.1-15.9*	3-15	*1.4-6.8*	3-15	*1.4-6.8*
3-4	*1.4-1.8*	1-3	*0.45-1.4*	7	*3.2*	1-4	*0.45-1.8*	1-7	*0.45-3.2*
1-2	*0.45-0.91*	8-12	*2.6-5.4*	—	—	11-16	*5.0-7.2*	15-21	*6.8-9.5*
½	*0.23*	1-2	*0.45-0.91*	½-¾	*0.23-0.34*	1-1¾	*0.45-0.79*	1-1¼	*0.45-0.57*

[3]With calves (both replacement heifers and stockers) an extra 2 lb (*0.91 kg*) of hay daily, over and above requirements, are herewith indicated to allow for wastage. Practical operators generally feed stemmy or other hay left over by calves to the cow herd.

Fig. 4-32. Cows grazing cornstalks. (Courtesy, Iowa State University, Ames)

a beef herd in their operations and realize a fair return from feeds which would otherwise be wasted.

a. **Crop residues.** *Crop residues are the parts of forages that remain after harvesting a grain or seed crop.* Among such crop residues are cornstalks and husklage, sorghum stalks, soybean refuse, small grain straws and chaff, and legume and grass seed straws. Crop residues left in the field, above or below the soil surface, may well constitute 4 to 5 times more energy than is harvested. This potential source of added feed, organic fertilizer, and energy will be increasingly utilized in the future.

b. **Winter pastures.** Where feasible, winter pasture offers cattle producers a means of reducing costs. By accumulating the feed in the field, rather than harvesting, storing, and handling the forage, the cost and labor of winter feeding can be substantially reduced. Also, cost of bedding and manure hauling can be eliminated.

Tall fescue is used as a winter pasture in the area to which it is adapted—Missouri, Illinois, Indiana, and Ohio. Usually, the new regrowth is baled in late June into round bales and left in the field. The round bales shed rain and snow and, together with the regrowth, make excellent late fall and winter grazing. Experience shows that field-stored forage has adequate quality to maintain beef cows in good condition.

• **Pasture and range cattle supplementation**—Improved range should be the first goal of cattle producers, without using supplemental feeding as a substitute for good grass or as a crutch for poor range. Instead, the two—good range and proper supplemental feeding—go hand in hand.

Where dried grass cured on the stalk is grazed, or where insufficient pasture is available—perhaps due to drought or overstocking—supplemental feeding is necessary. Also, supplemental feeding is a way in which to extend the grazing season, both early and late.

There is no one best and most practical pasture or range supplement for any and all conditions. Many different feeds may be, and are, used; among them (1) ranch or locally produced hay, (2) alfalfa pellets or cubes, with or without fortification, and (3) supplements of various kinds.

Also, farmers and ranchers can lessen the labor attendant to the daily feeding of a pasture or range supplement by (1) hand feeding cubes at intervals, rather than daily, (2) use of protein blocks, (3) use of liquid protein supplements, or (4) self-feeding salt-feed mixtures. Where these feeding systems do not result in the neglect of the herd, there is no effect upon the health and weight of the cows, percent calf crop, or weaning weight of calves.

1. **Range cubes or pellets.** Traditionally, cattle have been supplemented either once or twice daily on pasture or range, with the cubes scattered on the ground.

Fig. 4–33. Range cubes fed on pasture or range. Many cattle producers prefer this method of supplementation, primarily for reasons of convenience and reducing losses from wind blowing. (Courtesy, Ralston Purina Company, St. Louis, MO)

Urea-containing supplements, particularly those containing high levels of urea, should not be fed at intervals on the range because (1) range forages are relatively low in energy, and (2) urea is extremely soluble and its nitrogen becomes available very quickly in the rumen. Where nonprotein nitrogen is used in a range cube or pellet, a slow-release product is safest.

A suggested formulation without urea for pasture/range cubes or pellets is given in Table 4–13.

TABLE 4–13
RANGE CUBE OR PELLET, WITHOUT UREA (AS-FED BASIS)

Ingredient	Percent	Per Ton
	(%)	(lb)
Soybean meal, 44% (or cottonseed meal)	72.7	1,454
Alfalfa meal, 15%	15.0	300
Molasses (sugarcane)	8.5	170
Dicalcium phosphate, or equivalent	2.0	40
Salt	1.0	20
Trace minerals[1]	.5	10
Vitamin A[2] (30,000 IU/g potency)	.3	6
Total	100.0	2,000
Proximate analysis:	(%)	
Crude protein	35.9	
Fat	1.2	
Fiber	8.3	
Calcium	1.01	
Phosphorus	.9	
TDN	68.7	

[1]Trace minerals should be in keeping with Table 4–9. Generally, trace minerals can best be provided by a mineral or feed company, rather than home mixed.

[2]In low-sunshine areas, also add 6 million IU of vitamin D/ton of finished feed.

2. **Hand feeding at intervals, rather than daily.** Based on a four-year study done at the Texas Station, plus observations and experiences, the author recommends feeding nonurea range supplements twice weekly, allocating in each of the two feedings one-half as much supplement as would have been fed in a week on a daily feeding basis.

Protein cubes may be scattered on the ground—two or three times a week. This offers a method of checking the animals because they are attracted by the sight or sound of the vehicle when they know that there is something to eat.

Twice-weekly feeding has two distinct advantages over the use of salt-feed mixes: (1) It alleviates the cost of using excess salt, which has no nutritive value when so used; and (2) it forces inspection of the herd two times per week, which is as infrequent as is desirable.

3. **Protein blocks.** Protein blocks are just what the designation implies. They are compressed protein blocks, generally weighing from 50 to 500 lb each.

Blocks may be placed in grazing areas where cattle have frequent access to them, with one block provided to 15 cows. Intake will vary with the feed supply and the type of block. Generally, it is planned to limit feed consumption to about 2 lb per head per day by hardness of the block and salt and/or fat content.

4. **Liquid protein supplements.** Liquid supplement in a *lick tank* can be offered free choice. This is a convenient and satisfactory way in which to supply protein, energy, and other nutrients, so long as the cattle do not consume more than they need.

5. **Self-feeding salt-feed mixture.** The practice of using salt as a governor to limit feed consumption on pasture or range has been around a very long time. It was ushered in as a labor-saving device for cattle and sheep in inaccessible and rough areas.

The proportion of salt to feed may vary anywhere from 5 to 40% (with 30 to 33⅓% salt content being most common), with the actual intake of feed supplements limited to 1 to 2½ lb daily. By varying the proportion of salt in the mixture, it is possible to hold the consumption of feed supplement to any level desired. In some range areas, a reduction of the salt level from 33⅓ to 24% will increase consumption by about 50%. When a liberal feeding of grain on pasture is desired, 5% salt may be sufficient.

Two suggested salt-meal supplements (salt-cottonseed or soybean meal, 41%; *do not pellet*) follow:

Ingredient	Salt-Meal Mix No. 1	Salt-Meal Mix No. 2
	(lb)	(lb)
Salt	665	499
Meal (either 41% cottonseed or soybean meal)	1,331	1,497
Vitamin A (30,000 IU/g)	4	4
	2,000	2,000
Consumption level:	approx. 1½ lb daily	approx. 2 lb daily
Guarantee:		
Crude protein	min. 27%	min. 30%
Salt	max. 35%	max. 27%
Vitamin A	24,000 IU/lb	24,000 IU/lb

• **Rations for brood cows**—Suggested wintering rations for (1) mature pregnant beef cows, and (2) mature lactating beef cows are given in Table 4–12, pp. 160 and 161.

FEEDING BULLS

Frequently, little thought is given to the feeding and management of bulls except during the breeding season. Instead, the feeding program for herd bulls should be such as to keep them in a thrifty, vigorous condition at all times. They should neither be overfitted nor in thin, run-down condition. Also, exercise is necessary for the normal well-being of the bull.

The feeding and management of bulls differ according to age and condition. For this reason, sale bulls, young bulls, and mature bulls should be fed separately and differently.

Fig. 4–34. Hereford bull and cows on good pasture.

• **Feeding sale bulls**—Bull sales are generally held in late winter and early spring, at which time mostly yearling and 2-year-old bulls are sold. In order to attract buyers, they have usually been grain fed since calfhood. Most bull buyers—especially commercial producers in rougher range areas—would rather have their new bulls in less than fitted sale condition. They find that such bulls are more fertile and more apt to range with the cows when turned to pasture during the breeding season.

• **Feeding young bulls**—To achieve proper development, young bulls should gain at least 2½ lb daily from weaning to 12 to 15 months of age. This will necessitate a daily feed allowance equal to about 2½% of their body weight, with a ration comprised of 50% or more concentrate. From 15 months to 3 years old, they should make a daily gain of 2 to 2¼ lb and receive a feed allowance equal to 2 to 2¼% of their body weight, with the proportion of roughage increased after the first year.

During the breeding season, young bulls should be fed a grain ration consistent with pasture quality and number of cows to be bred in order to promote proper growth and development. Drought, overpasturing, and poor quality pastures are situations in which grain supplementation is particularly needed. Heavy service and poor pasture with no supplemental feeding may shorten the breeding career of a young bull.

After the breeding season, yearling bulls generally need 5 to 6 lb of grain along with good roughage.

• **Feeding mature bulls**—Winter is the proper time to condition bulls for the next breeding season. Bulls that have been running on pasture with the cows are likely to be thin; thus, they require sufficient concentrate to put them in proper flesh. Mature bulls will consume a daily amount of feeds equal to 1½ to 3% of their liveweight, depending upon condition and individuality.

Mature bulls should be fed all the legume hay they will eat plus 3 to 5 lb of ground or rolled grain and 1 lb of a 32% protein supplement (or equivalent) per head per day. Also, free access to a suitable mineral mixture should be provided. About

60 days before the bulls are turned out with the cows, the concentrate allowance should be increased by 25 to 50% with the amount of the increase determined by the condition of the bulls.

Mature herd bulls need no additional feed when running with the cow herd on good summer pasture.

FEEDING CALVES

Cattle producers as a whole, have lagged in applying much of what we know about feeding and managing calves. They're inclined to let mother cows and mother nature fend for the calves. Indeed, more good proven practices, based on both successful operations and research, need to be put to use in feeding and handling calves.

• **Feeding orphan and multiple birth calves**—Occasionally a cow dies during or immediately after parturition, leaving an orphan calf to be raised. Also, there are times when cows fail to give a sufficient quantity of milk for the newborn calf. Sometimes, there are multiple births.

If there are only a few orphans, usually they can be grafted onto other cows (or adopted)—either cows that have lost their calves or that give sufficient milk to raise two calves. When such calves cannot be grafted, they must be raised by artificial methods—without a cow.

A suggested feeding program for calves is given in this section, Part III—Feeding Dairy Cattle; and six starter rations are presented in the same section, in Tables 4–20 and 4–21 (see p. 179).

• **Feeding early weaned calves**—*Early weaning refers to the practice of weaning calves earlier than the usual weaning age of about 7 months, usually within the range of 35 days to 5 months of age.* Although it is not common practice among U.S. beef producers, dairy producers have been weaning 3-day-old calves for years. Also, early weaning has long been an integral part of many of the beef programs of Europe.

Currently, there is much interest in early weaning because (1) it fits into a drylot cow-calf management system, and (2) it can give a big assist in getting females, especially 2-year-old heifers, to rebreed in a short period of time.

From 35 days of age on, early-weaned calves can be fed any good starter ration. Six such rations are given in this section, in Part III—Feeding Dairy Cattle, in Tables 4–20 and 4–21. Of course, the starter ration should be made available to the calves well ahead of weaning in order that they will be accustomed to it, thereby avoiding any setback.

• **Creep feeding calves**—*Creep feeding is the supplementation of calves while they are nursing their dams.* It increases weaning weight. The basis for this response is related to the lactation curve of beef cows, the increasing nutrient requirements of the calf during the nursing period, and the decline in feed quality and quantity typical of most pastures or ranges which support the cows and calves during lactation. Studies reveal that milk production of dairy cows increases up to the fourth to sixth month following freshening, then declines gradually. By contrast, maximum milk production of beef cows occurs during the first two months after calving, then declines.

To fill the *hungry-calf gap*—the nutrient requirements over and above that provided by 13 lb of milk—would require the consumption of 50 lb of green grass daily. Of course, that's a physical impossibility, because a 500-lb calf simply cannot hold that much bulk. So the best way to fill the *hungry-calf gap* is to creep feed with concentrate mixes.

1. **The creep.** *A creep is an enclosure or feeder for feeding purposes which is accessible to the calves but through which the cows cannot pass.* It allows for the feeding of the calves but not their dams.

Fig. 4–35. Movable calf creep (note runners), with openings that will permit the calves to enter and keep the cows out. (Courtesy, *Livestock Breeder Journal*, Macon, GA)

2. **Creep rations; feeding directions.** Creep-fed calves need special rations. They are bovine babies; and they are both in forced production and finishing. The calf's body is expected simultaneously to lay on fat and grow in protein tissues and skeleton. Consequently, their ration requirements are for feed high in protein; rich in readily available energy; fortified with vitamins, minerals, and unidentified factors; and with all the nutrients in proper balance. Also, the ration must be very palatable. This calls for an exacting ration. To meet these needs, more and more producers are finding it practical to buy a commercially prepared complete creep feed, or a well-fortified and highly concentrated supplement to add to locally available feeds, rather than purchase individual ingredients and mix from the ground up.

Table 4–14 shows a creep ration, formulated by the author, that has been widely and successfully used. A simple, yet very satisfactory, creep ration may be made by grinding and pelleting 75% alfalfa and 25% cereal grain.

TABLE 4–14
CALF CREEP RATION #1[1] (AS-FED BASIS)

Ingredient	Precent	Per Ton	
	(%)	(lb)	*(kg)*
Oats	39.60	800.0	363.2
Corn #2	14.80	300.0	136.2
Barley	8.90	177.5	80.7
Wheat bran	9.90	200.0	90.8
Dried molasses beet pulp	9.90	200.0	90.8
Soybean meal, 44%	9.90	200.0	90.8
Molasses	4.90	100.0	45.4
Salt	.50	10.0	4.5
Dicalcium phosphate	.50	10.0	4.5
Trace minerals[2]	.04	1.0	0.45
Vitamin A (30,000 IU/g)	.06	1.5	0.68
Total	100.00	2,000.0	907.2
Proximate analysis:	(%)		
Crude protein	14.30		
Fat	3.20		
Fiber	8.30		
Calcium	.32		
Phosphorus	.50		
TDN	69.60		

[1]*Feed preparation:* Preferably ⅛– or 3/16–in. pellets. Otherwise, steam roll and flake grains, or grind grains coarsely.

[2]See Table 4–9 for recommended trace mineral levels. Follow manufacturer's directions.

Calves will consume approximately 500 lb of creep feed per head from one month of age to weaning. In years of lush pasture, it will be less; in dry years, more.

FEEDING REPLACEMENT HEIFERS

The feed and management program of replacement heifers will have a lifelong effect on their productivity. It will determine how young they may be bred, whether they calve early or late, whether they are good milkers or poor milkers, the weaning weight of their calves, and how long they remain in the herd. Also, feed accounts for 40 to 70% of the cost of raising replacement heifers; hence, it is important to know whether it is possible to effect savings on feed and during the growing period without affecting reproduction adversely. It is even more important to know whether their performance as adult animals can be enhanced by proper nutrition and management.

The pregnancy requirements of replacement heifers are really not too great. The body of an 80-lb newborn calf contains only about 12 lb of protein, 3.0 lb of fat, and 3.6 lb of mineral matter. But the lactation requirements are much more rigorous. If a 2-year-old heifer gives her calf an average of 1¾ gal of milk per day over a seven-month suckling period, she will produce in that milk a total of 93 lb of protein, 107 lb of fat, 133 lb of sugar, and 20 lb of minerals.

Fig. 4–36. Replacement heifers on pasture. (Courtesy, E. Peterson, Hamilton, MT)

Hence, the comparison: 12 lb of protein in the fetus vs 93 lb in the milk during the suckling period. This means that nearly 8 times more protein is required for 7 months of lactation than for 9 months of pregnancy.

Also, when breeding yearlings to calve as 2-year-olds, producers should be aware that nature has ordained that the growth of the fetus, and the lactation which follows, shall take priority over the maternal requirements. Hence, when there is a nutritive deficiency, the young mother's body will be deprived, or even stunted, before the developing fetus or milk production will be materially affected.

- **Rations for replacement heifers**—In season, good pasture plus mineral supplements fed free choice will meet the nutrient requirements for proper growth and development of heifers.

On the winter range, when dry forage is of low quality, and sometimes not too abundant, 1 to 2 lb of a protein supplement should be provided in the form of cubes, blocks, meal-salt, or liquid. When consumed at the intended level, the supplement should contain sufficient vitamin A to meet the requirements. Mineral supplements should also be provided, preferably free-choice.

Where winter grazing is not available, heifers must be drylotted and fed a complete ration. Sufficient nutrients should be provided to meet the requirements and to keep heifers in a thrifty condition, neither too fat nor too thin.

Suggested rations for wintering replacement heifers are given in Table 4–12, pp. 160 and 161.

Replacement heifers should be fed rather liberally—more so than stocker cattle which are being grown for the feedlot, to the end that they will acquire most of their growth and development before calving. With limited feeding, they will not have enough weight for age to breed when they are 15 months old; and it is best not to have them calve until they are 30 months of age.

• **Separate heifers by age**—The nutritive requirements for heifers differ according to body weight and expected daily gain. Consequently, the recommended ration for a 500-lb heifer calf differs from that of an 800- to 900-lb bred heifer. It is important, therefore, that replacement heifers be separated by ages for wintering, with coming yearlings in one group and coming twos in another.

• **Feeding and managing 2-year-old heifers**—From the above, it may be concluded that more producers can, and should, breed yearling heifers to calve as 2-year-olds.

Of course, the below-average breeder—the person who has lightweight, poorly developed heifers, and who wouldn't think of staying up nights and having cold, numb fingers while being attendant to a heifer and a newborn calf—should take another year and stick to calving out 3-year-olds. But progressive, commercial cattle producers should calve out more 2-year-olds from the standpoint of cutting production costs and increasing profits.

FEEDING STOCKERS

For a stocker operation to be profitable, the grower must be ever aware of the following reasons back of it, then feed stockers accordingly: (1) to provide a supply of the kind of cattle desired by finishing lots at the time needed; (2) to utilize the roughages and other low-cost feeds; and (3) to "cheapen down" the cattle.

Because of the very nature of the operation, the successful feeding of the stockers requires the maximum of economy consistent with normal growth and development. This necessitates cheap feed—either pasture or range grazing or such cheap harvested roughage as hay, straw, fodder, and silage. In general, the winter feeds for stockers consist of the less desirable and less marketable roughages. It is important, therefore, that the high-roughage rations of young stockers be properly supplemented from the standpoints of proteins, minerals, and vitamins.

Of course, too small gains may be unprofitable to the grower. Besides, young animals can be stunted. To make maximum growth without fattening—just to maintain condition—calves of the British breeds and crossbreds should gain 1.25 lb daily, and yearlings should gain 0.9 lb daily.

Table 4–12, pp. 160 and 161, contains recommended daily rations for stocker calves and stocker yearlings. Variations can and should be made in the rations used. The grower should give consideration to (1) the supply of homegrown feeds, (2) the availability and price of purchased feeds, (3) the class and age of cattle, (4) the health and condition of animals, and (5) the kind of feeder cattle in demand by feedlots.

The following points are pertinent to feeding stocker cattle and should be kept in mind:

1. **Recommended nutrient allowances.** Where grower rations are formulated on the basis of percentage of nutrients in the ration, the following allowances are recommended:

Protein

For up to 1.5 lb daily gain 10.0%
For 1.5 lb daily gain or more 10.5%

Calcium and Phosphorus

For up to 500 lb liveweight 0.30%
For over 500 lb liveweight 0.25%

Vitamin A

Air-dry feed (10% moisture) . . 800 to 1,000 IU per lb
. . . . 10,000 IU daily per head

2. **Level of wintering.** The level of wintering stockers affects the gains in the next stage. Thus, calves gaining the most during the winter make the least gains on pasture the following summer.

Calves wintered to gain 1.0 lb daily make satisfactory summer pasture gains. This level is recommended for calves to be grazed season-long the following summer, provided the same ownership is retained all the way through. A daily gain of 1 to 2 lb during the winter is usually desirable if calves (1) are to be sold in the spring, (2) will be on full feed 2 to 3 months after going to grass, (3) will be receiving a limited feed of grain when on grass, or (4) are replacement heifers that are to be bred at 13 to 15 months of age.

Since yearlings are not growing as rapidly as calves, they may be fed for smaller gains than calves, and yet show comparable condition. Thus, for maximum growth without fattening (for just holding their condition) calves should gain approximately 1.25 lb daily, whereas yearlings need to gain only 0.9 lb daily.

• **Preconditioning**—*Preconditioning is a way of preparing the calf to withstand the stress and rigors of leaving its mother, learning to eat new kinds of feed, and shipping from the farm or ranch to the feedlot.* To the cow-calf producer, it is a program of management, nutrition, and immunization. It, along with improved breeding based on production testing, is the trademark of the producer of feeder calves. To the feedlot operator, preconditioning is a way in which to prepare calves to fit into the program and to minimize costly and unnecessary procedures.

Changed environment; excitement of sorting, loading, and shipping; long periods without feed and/or water; movement through one or more assembly points; change of feed; and exposure to disease—all add up to *fatigue, stress, shrink,* and *lowered disease resistance.*

The term *preconditioning* generally consists of the following practices being conducted on the farm or ranch of origin and certified to by a licensed veterinarian: weaned; bunk broke and water tank or fountain trained; castrated and dehorned; vaccinated for IBR, PI–3, BVD, and *Haemophilus somnus;* and, depending on local conditions, additional vaccinations may be required for blackleg and malignant edema, and for brucellosis of heifers.

It is important that the program be written down, adhered to rigidly, then certified to by both the owner and the veterinarian. The producer should take the lead in developing such a program, but the counsel of the veterinarian and potential buyers should be sought.

Preconditioning is often confused with handling newly arrived feedlot cattle, and backgrounding. This is understandable because all three of them are important phases between weaning and finishing. Yet, each of them is distinct and separate.

• **Compensatory growth**—*Compensatory growth is increased growth rate in one time period as a result of growth restriction imposed during an earlier time period.*

It is common practice for stocker cattle to be *roughed through* the winter as cheaply as possible, with limited daily gains. Then, in the spring, the animals are turned to lush pasture or put in a feedlot and fed a high-energy ration. Animals so managed exhibit the phenomenon of *compensatory growth;* that is, on the high-energy ration they gain faster and more efficiently than similar cattle which were fed more liberally during the wintering period. Feedlot operators were quick to sense this situation, and to take advantage of it. This is the chief reason for the popularity of Okie-type cattle. Usually, they are animals whose growth has been held back to less than their genetic potential. When fed more liberally, they exhibit a surge in growth rate and feed efficiency. Large compensatory growth usually indicates that someone (the stocker operator) has lost money while someone else (the feeder) has made money. It is noteworthy that Holsteins and the larger exotics should never be handled so as to exhibit compensatory gains. If they're held back in the winter, they're too heavy when they finish.

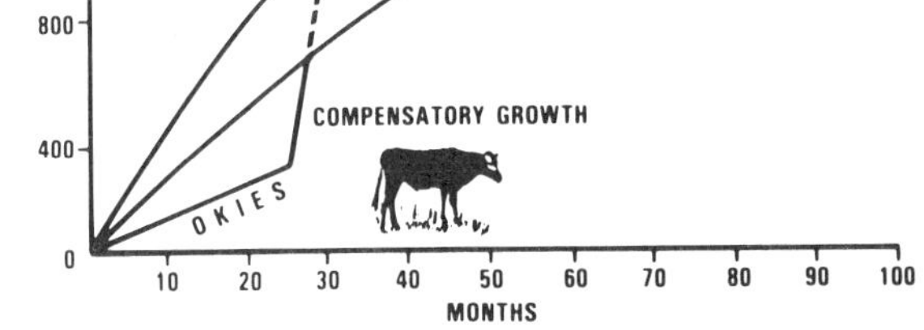

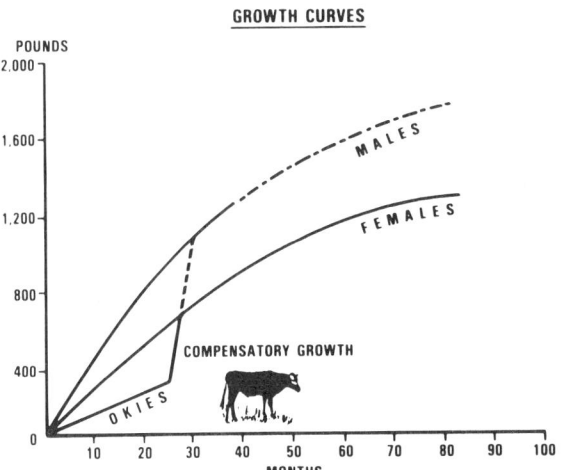

Fig. 4–37. The two curved lines show the normal growth of beef cattle under proper environment. Note that females grow most rapidly from birth to 10 months and males from birth to 15 months. Okie-type cattle grow more slowly during the first part of their lives when their environment is poor and they are under stress. When put on feedlot rations, their growth curve is steep. This rapid, economical gain is called *compensatory growth.*

FEEDLOT FINISHING (FATTENING) OF CATTLE

Feedlot finishing refers to feeding cattle in a restricted area, with the feed conveyed to the animals; and it may involve either (1) an open pen or feedlot, or (2) confinement (sheltered) feeding.

The major nutritional requirements of finishing cattle are: energy, protein, minerals, vitamins, and water.

About 75% of the cost of finishing cattle, exclusive of the purchase price of the feeders, is feedstuffs—grain, hay, silage, and miscellaneous wastes and by-products. The greatest need is for energy. Of course, net profit depends on how much of that energy can be converted to pounds of gain—and how efficiently.

Fig. 4–38. Open pen feedlot finishing of cattle. (Courtesy, Union Pacific Railroad Co., Omaha, NE)

• **Ration formulation**—Feedlot cattle have access only to the rations provided by the caretaker. It is important, therefore, that cattle feedlot rations be balanced, and that they make for maximum net returns.

In addition to considering changes in availability of feeds and feed prices, ration formulation should be altered at stages to correspond to weight increases in the cattle.

Some suggested rations that may serve as useful guides are given in Table 4–12 (pp, 160 and 161).

• **Feeds**—The growth of the cattle-feeding industry of America has gone hand in hand with the production and feeding of more grains and by-product feeds. Such feeds are high in energy and low in fiber. Hence, their availability and price influence the extent, location, and type of feeding program. Roughages, which require relatively more energy to digest and metabolize than grains, are used at low levels in most finishing rations. However, they are important in grower programs or in warm-up rations and they are even more important for maintenance of the breeding herd.

• **Pasture finishing cattle**—When grains are scarce and high in price, more cattle are grass finished. But, because young cattle grow and do not reach market finish under usual pasture conditions, it is impossible to finish them at early ages and light weights without either supplemental feeding on pasture and/or lot finishing at the end of the grazing season.

Generally speaking, no cheaper method of harvesting forage crops has been devised than is afforded by harvesting directly by grazing animals. Moreover, even most seeded pastures last several years; thus, seeding cost may be distributed over the entire period. Naturally, the cash income to be derived from pastures will vary from year to year and from place to place depending upon such factors as market price levels, class of animals, soil, season, and the use of adapted varieties.

When cattle are finished on pasture, any one of the following systems may be employed:

1. Finishing on pastures alone—no concentrates being fed.
2. Limited grain allowance during the entire pasture period.
3. Full feeding during the entire pasture period.
4. Full or limited grain feeding on pasture following the period of peak pasture growth.
5. Short feeding (60 to 120 days) in the feedlot at the end of the pasture period.

The system of pasture finishing that will be decided upon will depend upon the age of the cattle, the quality of the pasture, the price of concentrates, the rapidity of gains desired, and the market conditions.

FEEDING SHOW AND SALE CATTLE

Fig. 4-39. Brett Nelson, Valley City, North Dakota, showing a heifer in the National Junior Polled Hereford show. (Courtesy, American Polled Hereford Assn., Kansas City, MO)

All animals intended for show or sale purposes, including both breeding animals and steers, must be placed in proper condition. To accomplish this, a suitable ration must be selected and the animal or animals must be fed with care over a sufficiently long period. The rations listed in Table 4-15 have been used by successful fitters. They are higher in protein content than rations normally used in commercial-finishing operations, but most experienced fitters feel that by such means they get more bloom. In general, when show animals are being force-fed on any one of these concentrate mixtures, experienced caretakers prefer to feed a grass hay or a grass-legume mixed hay to a straight legume, because of the laxative effect and possible bloat hazard of the latter.

TABLE 4-15
FITTING RATIONS FOR SHOW AND SALE CATTLE
(As-Fed Basis)

Rations 1 to 5 are bulky. They are recommended for use (1) by the inexperienced feeder, and (2) in starting prospective show animals on feed.

Rations 6 to 11 are less bulky and higher in energy. They are recommended for use (1) by the experienced feeder, and (2) during the latter part of the fitting period.

Ration	(lb)	*(kg)*
Ration No. 1		
Rolled barley	50	*22.7*
Crushed oats	20	*9.1*
Wheat bran	20	*9.1*
Protein supplement[1]	10	*4.5*
Ration No. 2		
Rolled barley	30	*13.6*
Flaked corn	20	*9.1*
Crushed oats	20	*9.1*
Wheat bran	20	*9.1*
Protein supplement[1]	10	*4.5*
Ration No. 3		
Flaked corn	40	*18.1*
Crushed oats	30	*13.6*
Wheat bran	20	*9.1*
Protein supplement[1]	10	*4.5*
Ration No. 4		
Crushed oats	30	*13.6*
Rolled barley	30	*13.6*
Wheat bran	20	*9.1*
Flaked corn	10	*4.5*
Protein supplement[1]	10	*4.5*
Ration No. 5		
Flaked corn	55	*25.0*
Crushed oats	30	*13.6*
Protein supplement[1]	15	*6.8*
Ration No. 6		
Flaked corn or sorghum	50	22.7
Rolled barley	40	*18.1*
Protein supplement[1]	10	*4.5*
Ration No. 7		
Flaked corn	55	*25.0*
Crushed oats	20	*9.1*
Dried beet pulp	10	*4.5*
Protein supplement[1]	15	*6.8*
Ration No. 8		
Flaked corn	40	*18.1*
Rolled barley	20	*9.1*
Crushed oats	10	*4.5*
Dried beet pulp	10	*4.5*
Wheat bran	10	*4.5*
Protein supplement[1]	10	*4.5*
Ration No. 9		
Crushed oats	25	*11.3*
Rolled barley	20	*9.1*
Rolled wheat	20	*9.1*
Flaked corn	20	*9.1*
Wheat bran	10	*4.5*
Protein supplement[1]	5	*2.3*
Ration No. 10		
Rolled barley	35	*15.9*
Crushed oats	20	*9.1*
Rolled wheat	20	*9.1*
Dry beet pulp	15	*6.8*
Protein supplement[1]	10	*4.5*
Ration No. 11		
Rolled barley	20	*9.1*
Flaked corn	20	*9.1*
Crushed oats	20	*9.1*
Whole barley (dry wt. basis but cooked before feeding)	13	*5.9*
Commercial supplement	8	*3.6*
Linseed meal	8	*3.6*
Wheat bran	6	*2.7*
Beet pulp, dried molasses	4	*1.8*
Salt	1	*.5*

[1]The protein supplement may consist of linseed, soybean, cottonseed, or peanut meal. With most caretakers, linseed meal is the preferred protein supplement. It gives the animal a sleek hair coat and a pliable hide. Because it is a laxative feed, however, caution should be used in feeding it. Although it is true that an animal getting good clover or alfalfa hay needs less protein supplement than does one eating nonleguminous roughage, it is not possible to supply all the needed protein with hay and still get enough grain into young animals to finish them quickly.

Ration 11, which the author has used extensively in fitting show steers, is prepared as follows: The whole barley is processed by (1) adding water in the proportion of 2 to 2½ gal to each gallon of dry barley, and (2) cooking until the kernels are thoroughly swelled and can be easily squashed between the thumb and forefinger. Then it is mixed with the balance of the ration in about the proportions (on a dry basis) indicated. Each steer also receives 4 lb daily of a supplement high in milk by-products. As the animal approaches show finish, the ration is changed by decreasing the rolled barley to 7 lb and increasing the rolled oats by 5 lb and the wheat bran by 2 lb.

• **Rules of feeding show and sale cattle**—The selection of the fitting ration should be made largely on the basis of availability and price of feeds and the results obtained. Other important points in compounding the show ration and feeding the animal are:

1. Use care in getting the animal on feed.
2. When on full feed, the average animal will eat from 1½ to 2½ lb of grain for each 100 lb of liveweight. Feed only as much grain as the animal will clean up in ½ to 1 hour's time.
3. Most fitters prefer flaked grains in fitting rations. But they may be coarsely ground or crushed.
4. Provide needed minerals.
5. If the droppings are too thin or there is scouring, (a) cut down on the grain allowance, and (b) clean up the quarters.
6. The palatability of the ration may be enhanced by adding blackstrap molasses. Make it by diluting one-half to one pint of molasses with an equal volume of water and mixing it with each grain ration just before feeding. Although blackstrap molasses is preferable, beet molasses is satisfactory.
7. Satisfactory milk replacers, which most caretakers like to include in the fitting ration, are now on the market. These products should be used in keeping with manufacturer's directions.

PART III
FEEDING DAIRY CATTLE

Fig. 4-40. Open-air, free-stall housing of dairy cows, with milking parlor and maternity barn in the background. (Courtesy, Babson Bros., Oak Brook, IL)

Feed, more than any other one factor, determines the productivity and profitability of dairy cows. Within a herd, approximately 25% of the difference in milk production between cows is due to heredity; the remaining 75% is determined by environmental factors, with feed making up the largest portion. Feed accounts for about 55% (with a range from 45 to 65%) of the cost of milk production. Therefore, a good feeding program is necessary for profitable milk production.

It costs more to feed high producers than low producers. But high producers generally return more net income over feed cost than low producers. Fig. 4–41 shows how income over feed cost improves as production per cow increases.

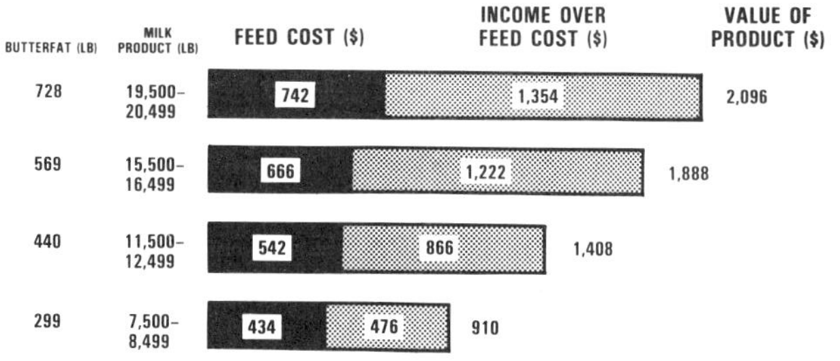

Fig. 4–41. It costs more to feed high-producing cows—but it pays. The reason: feed and overhead costs for maintenance are practically the same, regardless of level of production.

NUTRITIVE NEEDS OF DAIRY CATTLE

A nutrient is any substance that aids in the support of life. The first consideration in any dairy feeding program, therefore, is to determine the nutritive need. Dairy cattle require nutrients for growth, body maintenance, pregnancy or reproduction, and milk production

The nutritive needs for growth, body maintenance, and pregnancy generally are provided for before milk production can take place in quantity. For this reason, it does not pay to underfeed. A cow produces more economically when worked near full capacity.

• **National Research Council nutrition requirements of dairy cattle**—The nutritive requirements for dairy cattle have been established by the National Academy of Sciences—NRC (*Nutrient Requirements of Dairy Cattle,* sixth revised edition, update 1989).

These requirements were adapted for, and are presented in, *Feeds & Nutrition* and *Feeds & Nutrition Digest,* books by M. E. Ensminger, *et al.,* published by the Ensminger Publishing Company; hence, the reader is referred thereto. In using these tables, cognizance should be taken of the fact that the nutritive requirements given in them do not allow for any margin of safety; that is, they do not provide for animal differences, feed differences, and losses of certain nutrients in storage. Accordingly, in the formulation of rations, certain margins of safety should be provided.

Fig. 4–42. Dairy cows consuming a complete ration. (Courtesy, J. Tappan, Higley, AZ)

In their feed compositions, the NRC committee assumed an average increase of 4% per unit of dry matter intake above maintenance in calculating NE_{lc} values for feed ingredients, or an average discount of 8% based on their assumption that lactating cows are fed at 3X maintenance. These values may be obtained from *Nutrient Requirements of Dairy Cattle,* Sixth Revised Edition, 1988, or from *Feeds & Nutrition* or *Feeds & Nutrition Digest* of which Dr. Ensminger is the senior author.

ENERGY

Lack of energy is the most common deficiency of dairy rations. Cows cannot produce milk at peak levels if their rations are too low in energy.

Most of the energy required is supplied by carbohydrates and fats in forage and grain. All cows, except low-producing ones—those producing less than 15 to 20 lb of milk per day, need some grain if they are to yield at top levels.

The NRC energy requirements for dairy cattle are expressed as digestible energy (DE), metabolizable energy (ME), net energy for maintenance (NE_m), net energy for body gain (NE_g), net energy for lactation (NE_{lc}), and total digestible nutrients (TDN). Separate net energy values for each maintenance (NE_m) and gain (NE_g) are given because animals use energy for maintenance more efficiently than for growth. However, the efficiency of energy use by lactating cows for maintenance, pregnancy, and milk production is similar; so, only one energy value, net energy for lactation (NE_{lc}), is used for these functions.

The energy value of a feed may be separated into: (1) the losses that occur in digestion and metabolism, and (2) the net energy (NE) that is available to the animal for maintenance and production. The total energy in feed, which is determined by complete oxidation (burning) of the feedstuff and measurement of the heat produced, is known as *gross energy* and is expressed as calories. Common feedstuffs are similar in gross energy content, but differ in feeding value because of variations of digestibility. About 60% of the total energy in grain and 80% of the total energy in roughage is lost in feces, urine, gases, and heat.

The energy requirement for maintaining a lactating cow is affected by a number of factors, especially the following: (1) *body size*—the larger the animal, the higher the maintenance energy requirement; (2) *activity*—to support grazing activity, the maintenance allowance may be increased by 10% on good pasture and up to 20% on poor pasture; and (3) *cold temperature*—under severe winter conditions without access to dry shelter, the maintenance feed allowance may be increased up to 8%. Also, during the first lactation, when a heifer is still growing, her energy needs are about 20% greater than a mature cow; and during the second lactation, her energy needs are 10% greater than a mature cow. The energy requirement for gestation is about 30% of that required for maintenance alone, with most of the increase during the last 8 weeks of pregnancy.

Fig. 4-43. It takes 3 steers 18 months' time (for each steer) to store as much mineral in their bodies as 1 cow produces in milk in 1 year. And the cow remains alive to do it all over again!

PROTEIN

Protein is essential for dairy cattle maintenance, growth, milk production, and the development of the fetus. Also, it is required for the formulation of enzymes and certain hormones that control or regulate chemical reactions in the body. The protein requirement is really a requirement for amino acids.

The protein composition of feeds, and the protein requirements of dairy cattle, may be expressed as crude protein, digestible protein, degraded intake protein, undegraded intake protein, and/or nonprotein nitrogen (NPN).

The amount of protein needed in the total ration of lactating cows is determined primarily by the amount of milk produced. Milk is a rich source of high-quality protein; so, as milk production increases, a substantial amount of dietary protein is necessary. Thus, a high-producing 1,320-lb cow yielding 88 lb of 3.5% protein milk daily secretes 3.08 lb of milk protein. A deficiency of protein results in lowered milk production and may depress the protein content of milk. Excess protein usually results in high cost rations.

The amount of protein needed in the concentrate mix depends on the kind and quality of forage fed. As the amount of legume increases, the percentage of protein in the concentrate can be lowered. For most lactating cows, the total ration (forage plus grains and protein and energy supplements) should have 19% crude protein during the first ⅓ of lactation, lowered to 14% in midlactation and 12% during the dry period.

MINERALS

The functions, deficiency symptoms, interrelationships/toxicities, and sources of minerals for dairy cattle, which are similar to those of other animal species, are presented in Table 4–1, Animal Mineral Chart.

Milk contains about 0.7% minerals. Thus, one cow producing 15,000 lb of milk gives 105 lb of mineral per year. Additionally, a milk cow needs minerals for body maintenance, for development of the unborn calf, and for growth if she is a young cow.

Dairy cattle of all ages and stages of production are more apt to suffer from a lack of phosphorus in their feed than from a deficiency of any other mineral element. Generally speaking, the calcium-phosphorus ratio of the total ration should not be wider than 2:1.

It is also good business to guard against any trace mineral deficiencies by providing cobalt, copper, iodine, iron, manganese, molybdenum, selenium, and zinc. These trace minerals may be provided in the mineral mix, in trace-mineralized salt, or in the ration itself.

Salt and other minerals may be added to the concentrate mix, usually at the rate of about 1% salt and 1% other minerals. Even so, they should always be available free choice.

VITAMINS

The functions, deficiency symptoms, and good sources of vitamins for dairy cattle, which are similar to those of other animal species, are presented in Table 4–2, Animal Vitamin Chart.

Dairy cattle, like other animals, require vitamins. Of the known vitamins, only A and D, and perhaps E under certain conditions, are likely to be lacking in the average dairy ration. Vitamin K and the B-complex vitamins are synthesized by the body tissues.

WATER

Large amounts of water are essential if a cow is to produce to her maximum capacity. Cows drink an average of 100 to 200 lb of water per day, with heavy producers drinking up to 300 lb per day. The amount of water a cow will drink depends on her size and milk yield, the temperature and relative humidity of the air, the temperature of the water, and the amount of moisture in her feed.

RATIONS

Dairy producers must put together the available feeds so as to achieve the most profitable production. At its best, developing a dairy ration involves combining the art and the science of feeding. For small herds, individual animal response may be satisfactory. With large commercial herds, formulating rations must be more precise, because small costs per cow become large costs when multiplied by many cows. Yet, the most sophisticated computer must be augmented by the good judgment of the manager if the rations are to be successful in meeting the nutrient needs of individual cows and of the herd as a whole. Producers must always keep in mind that the best formula on paper is not always the best feed. A feed is of no value if it is not actually consumed.

Also, there should be a specific ration for every need—for lactating cows, dry cows, calves, replacement heifers, dairy beef, and show or sale animals.

Fig. 4–44. A complete (all-in-one) ration being fed to lactating cows at Arizona Dairy Co., Higley, Arizona. (Courtesy, J. Tappan, Arizona Dairy Co., Higley, AZ)

FEEDING LACTATING COWS

Few animal stresses are as great as those involved in the production of a large volume of milk. For each gallon of milk produced, 400 to 500 gal of blood must pass through the udder. Thus, if a cow is producing 10 gal (86 lb) of milk daily, 15 to 20 tons of blood course through the udder each 24 hours. This 10 gal of milk contains more than 3 lb of fat, more than 3 lb of protein, more than 4 lb of lactose (milk sugar), and more than ½ lb of minerals. All these must be supplied in the ration over and above the nutrients needed for the body processes, wastes, and energy to sustain the whole operation.

Also, producers realize greatest profits from feeding when cows convert the maximum proportion of their feed into milk. The nutrient requirements for production depend primarily on the amount and composition of the milk.

Additional considerations in feeding lactating cows include palatability of the ration; physical form, protein and mineral content of concentrates; proportion of concentrate to roughage; relative prices of ingredients; voluntary feed intake; and frequency and regularity of feeding. Thus, the proper feeding of lactating cows necessitates that producers have sufficient knowledge relative to basic nutrient requirements and principles to plan an efficient feeding program, and the experience and management ability to apply it.

Dry matter consumption is very important in feeding dairy cows. The best ration formulation on paper will not make for profitable production if the cows either fail to eat it or are given insufficient amounts of it. Also, high-producing cows must consume very large amounts of a balanced ration if they are to produce to their maximum.

- **Fiber**—Fiber is important in dairy rations. Excessive fiber levels limit intake and energy concentrations, while a shortage of fiber reduces rumen digestibility and milk fat test.

The amount of fiber to include in the ration of dairy cattle is influenced by the body condition and level of production of the animal, the type of fiber, the particle size, the amount of total DM consumed and its bulk density, the buffering capacity of the forage, the frequency of feeding, and the economics. Lactating cows that are fed to produce large amounts of milk, or young animals that are fed to achieve rapid growth, should receive more energy and less fiber than lower producing animals. Forages that are finely ground (processed to small particle size) are rapidly consumed and fermented in the rumen, which reduces (a) the animal's chewing time, (b) ruminal fluid, (c) the acetate-to-propionate ratio in ruminal fluid. The result of these effects is a depression in milk fat percentage. Chopped alfalfa should average about ¼ in. in length to maintain a normal milk fat percentage. Feed factors, such as small particle size of forages, that reduce the pH of ruminal fluid, decrease the number and activity of fiber-degrading bacteria and cause a depression in fiber degradation. Feeding an insufficient amount of fiber or feeding forages that have a poor buffering capacity in the rumen may have undesirable effects on rumen fermentation, fiber degradation, and milk fat percentage that are similar to those caused by reducing the particle size of the forage.

So, the general recommendation is that lactating dairy cows should receive at least one-third of the total ration dry matter as long hay or as its DM equivalent in medium-to-coarse chopped silage or other forage. A minimum of 5 lb of forage dry matter measuring 1 to 2 in. in length will meet the fiber need of most lactating cows.

The values for NDF and ADF are more accurate measures of the fiber component of feeds than are values for crude fiber. Yet, because both chemical and physical properties of feeds are involved in determining fiber quality and the energy value of feeds, there is currently no one fiber analysis that can accurately predict fiber quality and energy values for all feeds. NDF content is negatively correlated with dry matter intake and apparent digestibility of forages, but it is positively correlated with chewing time. ADF is more negatively correlated with apparent digestibility than is NDF. NDF and bulk density are positively related, which may explain the negative relationship between dry matter intake and the NDF content of the ration. According to University of Wisconsin researchers, NDF is a better predictor than ADF of dairy cow feed intake and milk production.

The optimum amount of NDF and ADF to include in the ration varies with the level of milk production and the type of forage that is fed to dairy cattle. A minimum of 21% of ADF and 28% of NDF is recommended for cows during the first 3 weeks of lactation. During times of high milk production, however, ADF and NDF contents of the ration are usually reduced to 19 and 25%, respectively, so that adequate dietary energy can be included to meet the cow's requirement. The ADF and NDF contents of the ration should be increased in later lactation to help prevent milk fat depression and because less energy is required for milk production. Seventy-five percent of the NDF in the ration should be supplied as forage.

THUMB RULES FOR FEEDING LACTATING COWS

The feed requirements of lactating cows are significantly influenced by the volume and composition of the milk that they produce. Although knowledge of the nutrient requirements of the animals and of the composition of feeds is essential in order to feed properly, the ability of the cows to consume sufficient volume of the feed complicates adequate feeding. The two sections that follow give some thumb rules relative to the amount and kind of forage and the amount and kind of concentrate to feed.

- **Amount and kind of forage to feed**—The common thumb rules for forage feeding of lactating cows follow.

1. **Forage dry matter and intake.** The forage should constitute a minimum of 40% of the total dry matter of the ration and account for an intake of approximately 1.5% of the body weight daily.
2. **Acid detergent fiber (ADF).** The ADF should constitute 19% of the ration dry matter, increased to 21% during the first 3 weeks of lactation.
3. **Neutral detergent fiber (NDF).** The NDF should constitute 25% of the ration dry matter, increased to 28% during the first 3 weeks of lactation.
4. **Hay consumption.** If good quality hay only is fed, a cow will eat about 3 lb per 100 lb of body weight.
5. **Silage.** Depending on the moisture content, 2.5 to 4.5 lb of silage are equal to (and may replace) 1 lb of hay; the lower feeding value of silage is due to its high moisture content—hay runs 10 to 15% moisture, whereas silage runs 65 to 75% moisture.
6. **Hay/grain equivalent.** It takes about 3 lb of good hay to supply the same amount of usable energy as 2 lb of grain.
7. **Pasture (grass) consumption.** Cows will consume 100 to 200 lb of pasture per day; since pasture normally contains 70 to 85% moisture, that's 15 to 60 lb of dry matter per day.
8. **Yearly hay consumption.** Except for cows fed high grain rations, it takes 5 to 6 tons of hay (or an equivalent amount in dry matter from pasture or silage) to feed 1 cow for 1 year.
9. **Forage:concentrate ratio.** If forage is very high in quality, cows will eat more of it, with the result that the grain requirement will be lessened. However, over and above meeting the minimum requirement, the proportion of forage to concentrate should be determined primarily by the economics of the situation—that is, it should be decided on the basis of the relative price of available forage and concentrate, the milk production, and the net returns.

- **Amount and kind of concentrate (grain) to feed**—The common thumb rules for concentrate feeding of dairy cows follow.

1. **Amount of concentrate (grain).** The concentrate should constitute a maximum of 60% of the total dry matter of the ration and account for an intake of not to exceed 2.3% of the body weight daily. Table 4–16 can be used as a guide for feeding concentrate according to milk production.

TABLE 4–16
AMOUNT OF CONCENTRATE (GRAIN)
TO FEED BY PERIODS (1,400 LB *(636 KG)* COW, 4% MILK)[1]

	Milk Production Ability of the Cow[2]			
Average Daily 1st Period .. (lb)	**50**	**60**	**80**	**90–100**
Average Daily 1st Period . (kg)	***23***	***27***	***36***	***41–45***
Lactation Total .... (lb)	**10,000**	**12,000**	**15,000**	**18,000**
Lactation Total .... (kg)	***4,540***	***5,448***	***6,810***	***8,172***
Phase of Lactation	**Grain to Milk Ratio**			
1 (1st 10 weeks)	1:4	1:3	1:3	1:2.5
2 (2nd 10 weeks)	1:4	1:3	1:3	1:3
3 (last 24 weeks)	1:4	1:4	1:2.5	1:2.5
	Daily	**Daily**	**Daily**	**Daily**
4 (dry, 6–8 weeks) .... (lb)	0–4	0–4	0–4	0–6
(dry, 6–8 weeks) .... (kg)	*0–1.8*	*0–1.8*	*0–1.8*	*0–2.7*
Total grain (approximate) .... (lb)	3,000	4,000	5,000	6,000
***Total grain** (approximate) .... (kg)*	*1,362*	*1,816*	*2,270*	*2,724*

[1]Adapted by the author from: Linn, J. G., M. F. Hutjens, W. T. Howard, L. H. Kilmer, and D. E. Otterby, *Feeding the Dairy Herd,* Cooperative Extension Services, Universities of Illinois, Iowa State, Minnesota, and Wisconsin, 1988, p.26, Table 18.

[2]Ratios based on 100% dry matter basis, grain containing 80 Mcal, and forage 60 Mcal of NE_{lc} per 100 lb *(45 kg)*.

2. **Amount and kind of protein.** Feed protein according to requirements (19% in early lactation, decreased thereafter according to milk production). A low rumen degradable protein source is recommended for high-producing cows in early lactation. Limit urea to 0.4 lb per day, and preferably to 0.2 lb per day, in phases 1 and 2.
3. **Added fat.** In addition to the fat present in natural feedstuffs, lactating cows may be fed 1 to 1½ lb of *added fat* per day; which translates into about 6% added fat to the concentrate ration, or 3% added fat to the total mixed ration (grain and forage combined). Fats in oilseeds (soybeans or whole cottonseed) should be considered as added fat. When feeding added fat, increase the calcium to 0.9 to 1.0%, the magnesium to 0.3%, and the acid detergent fiber to 20%.

4. **Salt.** Include 1% salt in the concentrate mix, or 0.5% salt in the total ration (concentrate and forage combined); which will provide for a salt intake of 2 to 3 oz per cow per day.

5. **Calcium/phosphorus and trace minerals.** A calcium/phosphorus mineral source should constitute 1 to 2% of the grain mix, or be fed at a rate of 1 oz per 10 lb of milk. Trace minerals should be incorporated in the ration or self-fed in trace mineralized salt to meet the requirements.

6. **Vitamins.** Vitamins A, D, and E should be added to the ration to meet the requirements.

RATIONS FOR LACTATING COWS

Not all dairy producers, in the United States or in other countries, who home-mix feeds balance rations (1) on the basis of chemical analyses of their feed ingredients, (2) with the use of computers, or (3) by combining the concentrates and forage into a complete ration. For these producers, Table 4–17, Feed Mixing Guide for Lactating Cows, may serve as a useful guide. It shows how ingredients partitioned into four approximate

TABLE 4–17
FEED MIXING GUIDE FOR LACTATING COWS (AS-FED BASIS)[1]

Note: *This shows how ingredients of 4 protein levels may be combined to make different concentrate mixes of approximate protein content to match 3 different qualities of roughages.*

(1) Suggested Grain Mix, Based on Kind of Roughage Available	(2) Low Protein (under 12%) Ingredients		(3) Low-Medium Protein (12–18%) Ingredients		(4) Medium-High Protein (18–28%) Ingredients		(5) High Protein (over 32%) Ingredients	
Feeds	(% protein) Barley, all analyses ...... 11.7 Beet pulp w/molasses, dried . 9.3 Corn-and-cob meal .... 7.8 Corn #2 .... 8.9 Dairy feed, 12% .... 12.0 Hominy feed .... 10.3 Molasses, cane* .... 4.3 Oats, all analyses .... 11.9 Rye, all analyses* .... 12.0 Sorghum (milo) .... 10.1 Wheat, all analyses 13.1		(% protein) Dairy feed, 16% .... 16.0 Wheat bran .... 15.5 Wheat middlings 16.4		(% protein) Brewers' dried grains* .... 27.3 Copra (coconut) meal ..... 21.3 Corn gluten feed .... 23.0 Dairy feed, 18–24% .... 18–24 Distillers' dried grains ..... 27.3 Malt sprouts .... 22.9 Peas, field* 23.2		(% protein) Dairy feed, 32–34% .. 32–24 Corn gluten meal ..... 60.8 Cottonseed meal* ..... 41.2 Linseed meal .... 35.7 Peanut meal .... 49.0 Soybean meal 44.4	
	(lb)	*(kg)*	(lb)	*(kg)*	(lb)	*(kg)*	(lb)	*(kg)*
Excellent roughage—High protein forage, 18%: (1) legume, or (2) legume and nonlegume mixed forages of *high quality;* consisting of dry forages and/or silage.								
Mix No. 1	1,000	*454*						
Mix No. 2	900	*409*					100	*45*
Mix No. 3	800	*363*			200	*91*		
Mix No. 4	850	*386*	100	*45*			50	*23*
Medium roughage—Medium protein forage, 15–17%: (1) legume, or (2) legume and nonlegume mixed forages of *medium quality;* consisting of dry forages and/or silage.								
Mix No. 5	800	363					200	*91*
Mix No. 6	650	*295*			350	*159*		
Mix No. 7	700	*318*	100	*45*	100	*45*	100	*45*
Mix No. 8	Straight 16% dairy feed, or ½ Mix No. 9 and ½ 16% dairy feed							
Poor roughage—Low protein forage, under 14%: nonlegume forage; consisting of dry forages and/or silage.								
Mix No. 9	700	*318*	300	*136*				
Mix No. 10	600	*272*			200	*91*	200	*91*
Mix No. 11	600	*272*	100	*45*	100	*45*	200	*91*
Mix No. 12	500	*227*					500	*227*

[1]The protein compositions in columns 2 to 5 were obtained from Section 19, Composition of Feeds.

Comments:

Add—To all rations (1) 1% iodized or trace-mineralized salt; (2) 1% steamed bone meal, dicalcium phosphate, or the equivalent (use monosodium phosphate or a high-phosphorus commercial mineral where alfalfa is fed liberally); (3) 1,000 IU of vitamin A/lb *(2,205 IU of vitamin A/kg)* of concentrate and, unless cows are in sunlight, add 150 IU of vitamin D/lb *(331 IU of vitamin D/kg)* of concentrate.

***Limitations**—Wheat, not more than 50% of the ration; dried molasses beet pulp, 20%; molasses, 15%; peas and brewers' dried grains, 30%; rye, 10%; and cottonseed meal, 20% of the mix for calves, but as needed for mature cows.

protein levels (columns 2, 3, 4, and 5) may be combined to make concentrates suitable for feeding with three different qualities of roughages—excellent, medium, and poor.

Variations can and should be made in the rations listed in Table 4–17. Producers should give consideration to the supply of homegrown feeds, and to the availability and price of ingredients. Feeds of similar nutritive properties can and should be interchanged as price relationships warrant. Thus, the cereal grains may consist of corn, barley, wheat, oats, and/or sorghum; the protein supplements may consist of soybean, cottonseed, peanut and/or linseed meal; and a vast array of by-product feeds may be utilized.

Here is how to use Table 4–17: Let's assume that a producer has (1) medium quality forage, and (2) both low- and medium-high protein (columns 2 and 4) ingredients from which to choose. How many pounds each of the low- and medium-high protein ingredients will be required in a 1,000-lb concentrate mix? Step by step, here is the answer:

1. Look under "Medium roughage—medium protein forage" 15–17% (column to the left).

2. Mix No. 6, containing 650 lb of low protein ingredients (under column 2: under 12% ingredients) and 350 lb of medium-high protein ingredients (column 4: 18 to 28% protein), will meet the needs. The concentrates may be chosen from among those listed at the top of the respective columns of Table 4–17—the low protein concentrates from column 2 (under 12%) and the medium-high protein concentrates from column 4 (18 to 28%).

• **How to balance a dairy ration**—Complete instructions on how to balance a ration (including an example of balancing a dairy ration) are given in Part I of this section, under the heading "How to Balance a Ration"; hence, the reader is referred thereto.

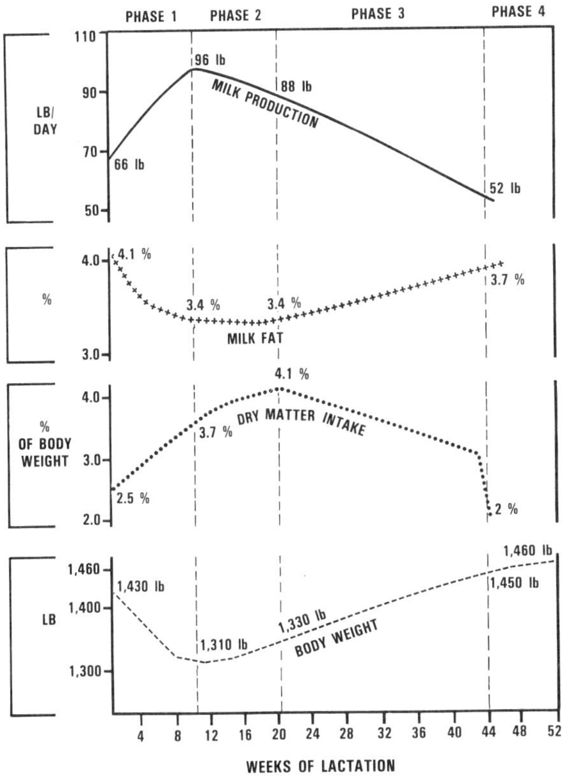

Fig. 4–45. Lactation cycle phases with corresponding changes in milk production, milk fat percentage, dry matter intake, and body weight. (Source: Linn, J. G., M. F. Hutjens, W. T. Howard, L. H. Kilmer, and D. E. Otterby, *Feeding the Dairy Herd*, Cooperative Extension Services, Universities of Illinois, Iowa State, Minnesota, and Wisconsin, 1988, p. 15, Fig. 6)

FEEDING SYSTEMS

Traditional individual feeding of lactating cows in stanchioned barns or milking parlors is giving way to new feeding systems. Although the newer methods are not as effective as feeding cows individually, they are much more economical than feeding all cows in the herd the same amount of grain, regardless of production. Additionally, they make for considerable saving in labor and facilities.

• **Phase feeding**—*Phase feeding is a feeding program that is divided into periods based on milk production, milk fat percentage, feed intake, and body weight.* Fig. 4–45 illustrates the shape and relationship of curves for milk production, fat percentage, dry matter intake, and body weight. Based on these curves, four distinct feeding phases of lactating cows can be identified.

Producers should formulate rations to match each of these phases in order to optimize milk yield, minimize metabolic disorders, increase longevity, and increase profits. The four phases are:

1. **Phase 1, early lactation, 0 to 70 days postpartum.** During this period, milk production increases rapidly, peaking at 4 to 6 weeks after calving. But feed intake does not keep pace with nutrient needs (especially energy needs) for milk production, so body tissues are mobilized to meet these needs. During this phase, adjusting the cow to the lactation ration is an important management practice. After calving, the grain should be increased by 1 to 1.5 lb daily to meet increased nutrient demands and minimize off-feed problems and acidosis.

2. **Phase 2, peak dry matter intake, second 10 weeks postpartum.** During this phase, cows should be fed to maintain peak milk production as long as possible. Feed intake is near maximum and can supply nutrient needs. Cows should be maintaining weight or making slight gains.

3. **Phase 3, mid- to late-lactation, 140 to 305 days postpartum.** This is the easiest phase to manage. During this period, milk production is declining, the cow is pregnant, and nutrient intake will easily meet or exceed requirements. The level of grain feeding should be adequate to meet production requirements, and to begin to replace body weight lost during early lactation. Lactating cows require less feed to replace a pound of body tissue than dry cows; hence, it is more efficient

to have cows gain body weight near the end of lactation than during the dry period. Young cows should receive additional nutrients for growth; 2-year-old heifers should receive 20% more than for maintenance, and 3-year-olds should receive 10% more.

4. **Phase 4, dry period, 45 to 60 days before parturition.**

The example rations presented in Table 4–18 are suitable for the three lactating phases.

TABLE 4–18
EXAMPLE RATIONS
FOR VARIOUS MILK PRODUCTION PHASES,
1,350 LB *(613 KG)* COW, 3.8% FAT TEST[1]

Item	Phase 1		Phase 2		Phase 3	
Milk (lb/day)	90		80		50	
Milk *(kg/day)*	*40.9*		*36.3*		*22.7*	
DM intake[2] (lb/day)	49		51		38	
DM intake[2] *(kg/day)*	*22.2*		*23.2*		*17.3*	
	As-Fed					
Ration 1	(lb/day)	*(kg/day)*	(lb/day)	*(kg/day)*	(lb/day)	*(kg/day)*
Alfalfa hay (88% DM), 140 RFV, 20% crude protein	28	*12.71*	34	*15.44*	27	*12.26*
Corn-oats[3]	21	*9.53*	24	*10.90*	16	*7.26*
Soybean meal, 44%	5.0	*2.27*				
Dical, 18% phosphorus	0.5	*0.23*	0.45	*0.20*	0.30	*0.14*
Salt, vitamins, trace mineralized	0.30	*0.14*	0.25	*0.11*	0.25	*0.11*
Weight change	−1.5	*−0.68*	—	—	+.5	*+.23*
Ration 2 (corn silage limit fed)						
Alfalfa hay, 140 RFV, 20% CP	19	*8.63*	34	*15.44*	23	*10.44*
Corn silage (35% DM)	25	*11.35*	25	*11.35*	25	*11.35*
Corn-oats	18	*8.17*	12	*5.45*	10	*4.54*
Soybean meal, 44%	7.5	*3.41*	0.3	*0.14*	—	—
Dical, 18% phosphrous	0.45	*0.20*	0.50	*0.23*	0.3	*0.14*
Salt, vitamins, trace mineralized	0.30	*0.14*	0.25	*0.11*	0.25	*0.11*
Weight change	−1.2	*−0.54*	—	—	+.5	*+.23*
Ration 3 (hay limit fed)[4]						
Alf-grass hay, 113 RFV, 16% CP	10	*4.54*	10	*4.54*	10	*4.54*
Corn silage	41	*18.61*	70	*31.78*	57	*25.88*
Corn-oats	16	7.26	11	*4.99*	6	*2.72*
Soybean meal, 44%	11.5	*5.22*	8.2	*3.72*	4.5	*2.04*
Dical, 18% phosphorus	0.40	*0.18*	0.30	*0.14*	0.25	*0.11*
Limestone	0.40	*0.18*	0.30	*0.14*	0.15	*0.07*
Salt, vitamins, trace mineralized	0.30	*0.14*	0.25	*0.11*	0.25	*0.11*
Weight change	−1.4	*−0.64*	+.7	*+.32*	+.5	*+.23*
Ration 4						
Alf-grass hay, 113 RFV, 16% CP	23	*10.44*	32	*14.53*	24	*10.90*
Corn-oats	22	*9.99*	22	*9.99*	19	*8.63*
Soybean meal, 44%	8.5	*3.86*	3.5	*1.59*	1.1	*0.50*
Dical, 18% phosphorus	0.45	*0.20*	0.40	*0.18*	0.25	*0.11*
Limestone	0.20	*0.09*				
Salt, vitamins, trace mineralized	0.30	*0.14*	0.25	*0.11*	0.25	*0.11*
Weight change	−1.9	*−0.86*	—	—	+.5	*+.23*

[1]*Source:* Linn, J. G., M. F. Hutjens, W. T. Howard, L. H. Kilmer, and D. E. Otterby, *Feeding the Dairy Herd,* Cooperative Extension Services, Universities of Illinois, Iowa State, Minnesota, and Wisconsin, 1988, p. 16, Table 6.

[2]Estimated average intake during the phase.

[3]85% corn–15% oats mix.

[4]Feed amounts may have to be limited during phase 2 and 3 to avoid over-conditioning.

The example ration presented in Table 4–19 is suitable for dry cows.

TABLE 4–19
EXAMPLE DRY COW RATIONS, 1,400 LB *(636 KG)* DRY COW[1]

Forage	As-Fed	
	(lb/day)	*(kg/day)*
Grass forage		
Orchard grass hay, 12% crude protein	25.0	*11.35*
Corn	3.0	*1.36*
Soybean meal	0.5	*0.23*
Limestone	0.15	*0.07*
Trace mineralized salt and vitamins	0.1	*0.05*
Limited legume forage[2]		
Alfalfa hay, RFV 140, 20% crude protein	12.0	*5.45*
Corn silage	43.0	*19.52*
Monosodium phosphate	0.1	*0.05*
Trace mineralized salt and vitamins	0.1	*0.05*
Limited corn silage		
Alfalfa-grass hay, RFV 113, 16% crude protein	21	*9.53*
Corn silage	20	*9.08*
Dicalcium phosphate	0.1	*0.05*
Trace mineralized salt and vitamins	0.1	*0.05*

[1]*Source:* Linn, J. G., M. F. Hutjens, W. T. Howard, L. H. Kilmer, and D. E. Otterby, *Feeding the Dairy Herd,* Cooperative Extension Services, Universities of Illinois, Iowa State, Minnesota, and Wisconsin, 1988, p. 17, Table 7.

[2]Ration contains excess energy as formulated and may over-condition cows in some situations.

• **Challenge feeding (lead feeding)**—*Challenge feeding, or lead feeding, refers to feeding the lactating cow so that she is challenged to reach her peak (summit) production level early in lactation.*

Because of the strong relationship between peak (summit) milk yield and the total milk production for the entire lactation period, emphasis should be placed on attaining maximum yield between weeks 3 and 8.

For every 5 lb increase in the summit peak, it is estimated that rolling herd average will increase 1,000 lb.

• **Corral (group) feeding**—Individual feeding of lactating cows has largely given way to mechanized group feeding. The latter was developed for convenience and saving of labor, rather than for improved animal well-being or feed efficiency. Today, lactating herds with several hundred cows are common; and some herds number several thousand. In order to design a nutritional program for such large numbers that can be adapted to the specific needs of the cows, they are separated into groups according to production (and, therefore, nutritional needs).

When producers decide to go to group feeding, they must decide on the number of groups into which to divide the herd. To answer this question, consideration should be given to the following: (1) herd size; (2) types and costs of available feeds; (3) current type of housing, feeding, and milking system; and (4) overall economic integration of the operation—for example, labor, machinery, etc.

In large herds (more than 250 milking cows), a commonly used system is one in which a minimum of 5 groups are established: (1) high-production cows (about 90 lb of milk/head/day), (2) medium-production cows (about 65 lb of

Fig. 4-46. Brown Swiss cows in a corral system of feeding. (Courtesy, The Brown Swiss Cattle Breeders Assn., Beloit, WI)

milk/head/day), (3) low-production cows (about 45 lb of milk/head/day), (4) dry cows, and (5) first calf heifers. More groups are desirable in very large herds if corrals and facilities are available. Because of feeding and social considerations, a maximum of 100 cows per group is advisable. With this program, there can be a maximum of two moves during the lactation cycle. In many cases, only one move is necessary; and in a few cases, no moves are required. This system allows each group to be fed according to need. The high-producing groups should be fed the highest quality ingredients at maximum levels. The middle-producing cows should be fed in such a way as to reduce feed costs, increase butterfat test, improve rumen function, and promote lactation persistency. The same holds true for the low-producing cows as for the medium producers except that considerable care must be exercised to avoid excessive fattening.

One of the problems inherent in group feeding concerns the behavioral adaptation of a newly introduced cow to a group. Group acceptance—*pecking order*—can pose occasional problems with a new cow, but the magnitude of the problem is usually not very great. One means of reducing this is to move several cows into a new group at the same time and just before feeding, rather than individually.

When group feeding programs are followed, grain is seldom fed in the milking parlor. This is commonly referred to as corral or bunk feeding since feeding generally takes place in bunks along the fenceline of the corrals or pens. Studies have demonstrated that cows fed their grain as a group in a common manger do as well as those fed individually in the milking parlor, but some cows may not always come into the parlor as easily when there is no grain to attract them. Some producers offer a minimum amount of feed in the parlor and the remaining amount in the corral with good success. The high producers seem to be more aggressive than the low producers; hence, they usually eat more when group fed.

Group feeding can be easily adapted to the use of complete feeds when the concentrates, roughages, and supplements are mixed into one feed rather than being fed separately. Some producers who use complete feeds prefer to feed dried roughages—especially long-stemmed hay—separately in order to enhance stimulation of the rumen and to facilitate mixing, because long hay does not lend itself to mixing in a mixer.

- **Example rations for group feeding**—Example rations for group feeding mature lactating cows at each of 3 milk production phases are given in Table 4-18; and example rations suitable for feeding dry cows are presented in Table 4-19.

When group feeding, first-calf heifers should be handled in a separate group and fed for both milk production and growth. Their nutrient requirements for milk production are similar to the requirements of their older counterparts producing milk at the same level, but, because of their growth, they should receive about 20% more nutrients than are required for maintenance.

Although variations can and should be made in Tables 4-18 and 4-19 rations, they are excellent guides. When milk yields and/or body weights differ from those used in these tables, a suitable computer program or hand calculations should be employed to obtain amounts to be fed, and to give consideration to cost of alternative feeds.

FEEDING ON PASTURE

Problems of milk production are at a minimum during the early pasture season, when plant growth is lush. However, when the weather gets hot, it is a different story; there is the period known as the *summer slump*. High temperatures actually affect pasture growth more than the well-being of the cows. Many dairy producers have discontinued pasture grazing for two reasons: (1) It is difficult to keep milk production uniform when cows are on pasture, because of changing temperatures and pasture growth; and (2) with larger herds, it is not possible to have sufficient pastures in close proximity to headquarters.

Fig. 4-47. Guernsey cows on pasture. (Courtesy, American Guernsey Cattle Club)

FEEDING DRY COWS

Dry cows have three important jobs: (1) recovering from a heavy milk producing period and resting the mammary glands, (2) developing the unborn calf (more than half the fetal growth occurs during the last two months of lactation), and (3) storing up body reserves for the next milking period. This necessitates that they be properly fed.

Toward the end of the dry period, many successful producers follow a program of challenge or lead feeding. They reach a feeding level of 1 to 1½ lb of grain to each 100 lb of liveweight about one week before freshening, and continuing at this rate right up to freshening. This precalf feeding gets the rumen, and the cow's appetite and eating habits, adjusted to liberal feeding before freshening. Also, a cow freshening in good condition starts off better and maintains a higher level of production; her milk is usually higher in total solids; and the incidence of milk fever and ketosis is usually reduced.

FEEDING AT CALVING TIME

Until recent years, cows were not fed much grain prior to freshening; and, immediately after calving. They were given all the roughage they wanted but the grain allowance was sharply reduced. Today, dairy producers feed appreciably more grain to cows prior to calving—from 12 to 20 lb per day before freshening; and feed is not withheld at calving. Instead, on the first day after calving, cows are fed the same amount of grain that they were used to before calving, followed by an increase of 2 to 3 lb per day according to the cow's appetite. The experienced caretaker is in the best position to determine how much, and what, to feed each individual cow at calving time.

FEEDING SHOW AND SALE ANIMALS

Dairy animals intended for show or sale should be fed so as to achieve a certain amount of finish or bloom, but they should not be too fat. Linseed meal, beet pulp, oats, barley, and wheat bran are popular feeds in a fitting and showing ration. Likewise, good roughages are always very important.

FEEDING DAIRY CALVES

One of the most important phases of dairy production is that of feeding and managing the dairy calves raised for replacement purposes. Statistics reveal that more than 20% of the dairy calves die of sickness or disease before reaching maturity. With good management, many of these losses may be reduced to 3 to 5%. Many of these deaths are caused by faulty nutrition and/or poor housing and management.

A carefully planned and executed feeding program is necessary to produce growthy, vigorous, and healthy calves. The following feeding program is recommended:

Day 1	Dam's colostrum
Day 2	Dam's colostrum
Day 3	Dam's colostrum
Day 4	Liquid feed of choice, introduce starter and water
Day 5 to weaning . . .	Continue feeding program
Weaning to 12 weeks	Starter (up to 5 lb daily), introduce forage

Since milk is the primary product of dairy production, it is necessary to switch the young calf to cheaper feeds as expeditiously as possible. At the same time, it is important that the diet promote good health, growth, and development. Four feeds that are routinely fed to calves are (1) colostrum, (2) milk, (3) milk replacers, and (4) calf starters.

• **Colostrum**—*Colostrum is the milk which is high in antibodies, and which is secreted by cows, and other mammalian females, for the first few days following parturition.*

Colostrum (either dam's colostrum or mixed colostrum from first milking of older cows) should be fed to calves as soon after birth as possible (ideally within 15 minutes and certainly within 4 hours) to protect against disease. Some successful producers remove calves from their mothers *wet* and without nursing, so as to minimize infection from nursing. Then, as soon as possible, they are offered colostrum from a nippled bottle. If they fail to nurse naturally, the colostrum is hand-fed, using a specially designed tube to force it into the abomasum.

Surplus colostrum can be frozen and stored for a period of 1 year or longer without losing its antibody value. It may be thawed, warmed to about 100°F, and fed as needed.

• **Milk replacers**—Milk replacers vary in quality, so the buyer/user should study the feed tag. The best milk replacers contain 22% protein, all derived from milk products—skim milk powder, buttermilk powder, dried whole whey, de-lactosed whey, casein, and/or milk albumen. Chemically modified soy protein, soy isolates, and soy concentrates are good, but as plant proteins they are less digestible than milk protein. Meat solubles, fish protein concentrate, distillers' dried solubles, brewers' dried yeast, oat flour, and wheat flour are inferior as protein sources in milk replacers.

A good milk replacer powder should contain a minimum of 15% fat, and it may contain more than 20%. The higher fat level tends to reduce the severity of diarrhea and produce additional energy for growth. Good quality animal fats are preferable to most vegetable fats. However, soy lecithin, especially when homogenized, is an acceptable fat source and improves mixing qualities of the replacer.

• **Calf starters**—A high quality palatable calf starter should be offered when the calf is 4 days old, and not later than 10 to 12 days of age. The best starters are high in energy, contain 16 to 18% protein (20% if calves are weaned before 4 weeks of age), and are free of excessive fines. To encourage consumption, starters should consist of whole, coarsely ground, cracked, or rolled grains. Up to 5% molasses improves palatability and minimizes fines and dust. Whole grains, especially oats, can be fed with starter rations to calves up to 3 months of age. Calf starters should be fed until calves are about 12 weeks of age, with intake limited to 5 to 7 lb per calf daily.

Fig. 4–48. Group-fed dairy calves. (Courtesy, Holstein-Friesian Assn. of America, Brattleboro, VT)

Many good commercial starters are on the market. Also, calf starters may be home-mixed. Table 4–20 presents examples of some good grain calf starters.

TABLE 4–20
GRAIN STARTER RATIONS FOR CALVES[1]

	Grain Starters[2]		
	1	2	3
Ingredients (air dry basis)			
Corn (cracked or coarse ground) ...... (%)	50	30	
Ear corn (coarse ground) .... (%)			50
Oats (rolled or crushed) .... (%)	22	18	
Barley (rolled or coarse ground) ...... (%)		20	21
Wheat bran .... (%)		8	
Soybean meal .... (%)	20	16	21
Molasses .... (%)	5	5	5
Dicalcium phosphate .... (%)	0.5	0.5	0.5
Limestone .... (%)	1.5	1.5	1.5
Trace mineralized salt and vitamins ... (%)	1	1	1
Composition (dry matter basis)			
ADF .... (%)	7.0	6.9	9.1
Crude protein .... (%)	18.1	18.0	18.4
TDN .... (%)	80.0	78.8	78.0
Calcium (Ca) .... (%)	0.80	0.80	0.82
Phosphorus (P) .... (%)	0.48	0.56	0.47
Vitamin A .... (IU/lb)	1,000	1,000	1,000
Vitamin A .... *(IU/kg)*	*2,205*	*2,205*	*2,205*
Vitamin D .... (IU/lb)	150	150	150
Vitamin D .... *(IU/kg)*	*331*	*331*	*331*
Vitamin E .... (IU/lb)	11	11	11
Vitamin E .... *(IU/kg)*	*24*	*24*	*24*

[1]*Source:* Linn, J. G., M. F. Hutjens, W. T. Howard, L. H. Kilmer, and D. E. Otterby, *Feeding the Dairy Herd,* Cooperative Extension Services, Universities of Illinois, Iowa State, Minnesota, and Wisconsin, 1988, p. 20, Table 11.

[2]Hay may be offered free choice with grain starters.

While calves may begin nibbling on good quality hay as early as 5 to 10 days of age, it is not necessary to feed forage before 8 to 10 weeks of age. If the housing and management system makes it inconvenient to provide forage, it may be desirable to incorporate a forage factor (more fiber) in the starter ration. Table 4–21 presents examples of suitable rations for calves not receiving hay or silage. Corn silage or pasture should not be fed before 3 months of age because of their high moisture content which can limit intake and growth. Low moisture haylage is acceptable if it is kept fresh.

TABLE 4–21
COMPLETE STARTER RATIONS FOR CALVES[1]

	Complete Starters		
	1	2	3
Ingredients (air dry basis)			
Corn (cracked or coarse ground) ...... (%)	40	25	30
Oats (rolled or crushed) .... (%)	14.5	8	18
Beet pulp .... (%)		25	25
Alfalfa hay (ground) .... (%)		10	
Corn cobs (ground) .... (%)	15		
Soybean meal .... (%)	23	18	20
Molasses .... (%)	5	5	5
Dried whey .... (%)		7	
Dicalcium phosphate .... (%)	0.5	0.5	0.5
Limestone .... (%)	1	0.5	0.5
Trace mineralized salt and vitamins ... (%)	1	1	1
Composition (dry matter basis)			
ADF .... (%)	13.3	15.8	14.2
Crude protein .... (%)	18.3	18.0	18.2
TDN .... (%)	75.5	78.0	79.4
Calcium (Ca) .... (%)	0.63	0.72	0.58
Phosphorus (P) .... (%)	0.45	0.44	0.43
Vitamin A .... (IU/lb)	1,000	1,000	1,000
Vitamin A .... *(IU/kg)*	*2,205*	*2,205*	*2,205*
Vitamin D .... (IU/lb)	150	150	150
Vitamin D .... *(IU/kg)*	*331*	*331*	*331*
Vitamin E .... (IU/lb)	11	11	11
Vitamin E .... *(IU/kg)*	*24*	*24*	*24*

[1]*Source:* Linn, J. G., M. F. Hutjens, W. T. Howard, L. H. Kilmer, and D. E. Otterby, *Feeding the Dairy Herd,* Cooperative Extension Services, Universities of Illinois, Iowa State, Minnesota, and Wisconsin, 1988, p. 20, Table 11.

• **Amount and method of feeding, frequency of feeding, and age of weaning**—Calves may be separated from their dams at birth, or within 12 to 24 hours after birth. In any case, they should receive their dam's colostrum for the first 3 days of life, following which they may be shifted to a liquid feed of the feeder's choice.

In order to obtain proper growth, calves must be provided adequate dry matter. For an 80- to 100-lb calf, this calls for 1 lb of dry matter (solids) daily from milk, surplus colostrum, or milk replacer, from birth to weaning at 4 weeks.

Milk or milk replacer may be fed by open pail, by nipple feeding from a pail or bottle, or by automated feeding equipment. Each method of feeding is satisfactory, provided it is accompanied by cleanliness and sanitation.

Most calf raisers feed twice daily. Most producers wean calves between 4 and 8 weeks of age.

FEEDING DAIRY HEIFER REPLACEMENTS

Fig. 4–49. Replacement heifers—bunk-fed in confinement. (Courtesy, Holstein-Friesian Assn. of America, Brattleboro, VT)

Between weaning and calving (12 weeks to 2-year-olds), the nutrition of heifer replacements is often neglected. At its best, the feeding and management program during this period involves 3 distinct phases: (1) weaning (about 12 weeks of age) to 1 year; (2) 1 year to 2 months before calving at 2 years; and (3) 2 months before calving to calving.

A suitable ration for each of these three stages is given in Table 4–22.

TABLE 4–22
RATIONS FOR LARGE BREED DAIRY HEIFERS OF DIFFERENT WEIGHTS[1]

Weight and Age (lb)	(kg)	Rate of Gain (lb/day)	(kg/day)	Ration (As-fed)	(lb)	(kg)	Percent (%)
400	*182*	1.7	*0.8*	Alfalfa-grass hay, 113 RFV,			
Weaning (about 12 weeks) to 1 year of age.				16% CP	7.0	*3.2*	
				Grain mix, 14.5% CP	4.5	*2.0*	
				coarse ground corn			83.1
				soybean meal			15.5
				trace mineral salt			0.5
				dicalcium phosphate			0.8
				vitamin premix			0.1
700	*318*	1.7	*0.8*	Alfalfa-grass hay, 113 RFV,			
1 year to 2 months before calving at 2 years of age.				16% CP	12.5	*5.7*	
				Grain mix	6.0	*2.7*	
				ground ear corn			95.9
				soybean meal			2.5
				trace mineral salt			0.6
				dicalcium phosphate			0.9
				vitamin premix			0.1
1,000	*454*	1.7	*0.8*	Orchardgrass hay	10.0	*4.5*	
2 months before calving at 2 years of age.				Corn silage	38.0	*17.3*	
				Supplement	2.0	*0.9*	
				soybean meal			89.0
				trace mineral salt			3.0
				limestone			7.5
				vitamin premix			0.5

[1]*Source:* Linn, J. G., M. F. Hutjens, W. T. Howard, L. H. Kilmer, and D. E. Otterby, *Feeding the Dairy Herd,* Cooperative Extension Services, Universities of Illinois, Iowa State, Minnesota, and Wisconsin, 1988, pp. 20–21, Table 12.

Table 4–23 shows desirable weights for first breeding at 15 months of age along with weights for other age categories.

TABLE 4–23
NORMAL HEART GIRTH MEASUREMENT AND WEIGHT OF CALVES AND HEIFERS DURING THE GROWING PERIOD[1]

Age in Months	Holstein (in.)	Holstein (lb)	Ayrshire (in.)	Ayrshire (lb)	Guernsey (in.)	Guernsey (lb)	Jersey (in.)	Jersey (lb)
Birth	31	96	29½	72	29	66	24½	56
1	33½	118	32	98	31½	90	29½	72
2	37	161	35½	132	34½	122	32½	102
4	43½	272	42¾	236	41¼	217	38¼	181
6	50	396	48¼	340	47	304	44½	277
12	62½	714	59	583	58¼	549	56½	520
15	65¼	805	63	703	61¾	640	59	585
18	68½	912	66	781	65	727	61½	660
21	71½	1,025	68½	885	67½	816	64	740

[1]Body weight for Holsteins and Jerseys from *USDA Tech. Bull. 1098 and 1099.* Heart girth measurements for these weights taken from *Res. Bull. 194* (1960). Nebraska Ag. Exp. Sta. Weights and heart girth measurements for Ayrshires and Guernseys calculated from data furnished by Professor H. P. Davis, University of Nebraska, Lincoln.

FEEDING DAIRY BULLS

Bull calves raised for breeding purposes should be fed and handled much the same as heifers. But, since they grow slightly faster than heifers, they should receive somewhat more feed than heifers of the same age.

Older bulls should be kept in thrifty, vigorous condition, but they should not be permitted to become too fat. Mature bulls can be fed the same grain ration as the lactating cows. Depending on the quality of the roughage, usually about ½ lb of grain per 100 lb of body weight will suffice for the mature bull. Also, individual differences must be considered, for some bulls are easier keepers than others.

PART IV
FEEDING SHEEP AND GOATS

Fig. 4–50. In the northern (snowbelt) area of the U.S., sheep are fed ½ to 1 lb grain daily, in addition to the roughage allowance, during the winter months. (Photo by J. C. Allen and Son, West Lafayette, IN)

Sheep consume a higher proportion of forages than any other class of livestock, it being estimated that 94% of the total feed supply of the U.S. sheep production is derived from roughages. They are naturally adapted to grazing on pastures and ranges which supply a variety of forage plants, and they thrive best on forage that is short and fine rather than high and coarse. Although sheep will eat considerable quantities of weeds and brush, they prefer choice grasses and legumes.

Except at lambing season, sheep seldom receive much grain. In the northern latitudes, farm-flock ewes are frequently given from ½ to 1 lb daily of a grain ration in addition to the roughage allowance from about 6 weeks before lambing to the time that they are turned to spring pasture. Higher levels of grain are fed during the suckling period than during gestation. Many of the farm flocks of the South and range bands of the Southwest, however, are kept in good thrifty condition, and the lambs are raised to the marketing stages, without the feeding of any grain. In still other areas, the ewes are fed only during periods of deep snows or extended droughts. The range bands in the colder regions of the West are normally fed alfalfa hay and grain during the period of about 3 to 4 weeks that they are confined to the lambing camp.

In general, for practical reasons, the ration of ewes should consist of as nearly year-round pastures as possible, with well-cured hay and other forages available the balance of the year, plus a limited grain allowance under certain conditions. Good quality sun-cured hay and lush pastures will not only provide most of the necessary proteins, but they are excellent sources of most of the minerals and vitamins, also.

NUTRITIVE NEEDS OF SHEEP AND GOATS

As with other classes of livestock, the nutritive needs of sheep and goats have been established by the National Academy of Sciences, and may be classified as (1) energy, (2) protein, (3) minerals, (4) vitamins, and (5) water.

The nutritive requirements are the values considered necessary for maintenance, optimum production, and prevention of any signs of nutritional deficiency.

ENERGY

Fig. 4-51. The energy needs of sheep are largely met through the consumption of hay and pasture. (Courtesy, Ralston Purina Company, St. Louis, MO)

Inadequate amounts of feed may result from overgrazing, droughts, snow covering the feed, or from a low dry matter content of lush, washy feeds. Also, poorly digested low-quality forage leads to reduced feed intake.

The energy needs of sheep are largely met through the consumption and digestion of roughages—pasture, hay, and silage. Grains, such as corn, barley, milo, wheat, and oats, are used to raise the energy level of the ration during periods when supplementation is necessary. In general, sheep subsist on an even higher proportion of roughages to concentrates than do beef cattle, and this applies to finishing lambs. The bacterial action in the paunch of the sheep efficiently converts roughages into suitable sources of energy.

It is generally recognized that the energy requirements of sheep are affected by size, age, pregnancy, lactation, growth, and protein content of the ration. It is also affected by environment, shearing, and sex.

- **Symptoms of energy deficiency**—An energy deficiency is characterized by slowing and cessation of growth, loss of weight, reduced fertility or reproductive failure, lowered milk production and shortened lactation period, reduced quantity and quality of wool (including breaks in the fiber), lowered resistance to infection with internal parasites, and increased mortality.

PROTEIN

Sheep need protein, as do other classes of animals, for maintenance, growth, reproduction, and finishing. Additionally, sheep need protein for the production of wool—a protein product.

Green pastures and legume hays (alfalfa, clover, soybeans, lespedeza, etc.) are excellent and practical sources of proteins for sheep in most areas. Where the ranges are bleached and dry for an extended period, or legume hays cannot be produced for winter feeding, however, it may be desirable to provide sheep with such protein-rich supplements as soybean meal, cottonseed meal, linseed meal, canola meal, peanut meal, sunflower meal, or a commercial protein supplement, at the rate of about ¼ to ⅓ lb per ewe per day.

The protein requirements of sheep are affected by growth, pregnancy, lactation, mature size, weight for age, body condition, rate of gain, and protein-energy ratio. Though correspondingly less because of their smaller body size and lower milk production, the protein requirements of ewes nursing lambs are much like those of lactating cows.

- **Symptoms of protein deficiency**—A protein deficiency is characterized by reduced appetite, lowered feed intake, and poor feed efficiency. In turn, this makes for poor growth, poor muscular development, loss of weight, reduced reproductive efficiency, and reduced wool production. Under extreme conditions, there are severe digestive disturbances, nutritional anemia, and edema.

MINERALS

The functions, deficiency symptoms, interrelationships/toxicities, and sources of minerals for sheep, which are similar to other animal species, are presented in Table 4-1 Animal Mineral Chart.

Fig. 4-52. Lamb fed a ration deficient in phosphorus. Note the knock-kneed conformation. (Courtesy, University of Idaho)

Tables 4-24 and 4-25 present the macromineral and micromineral requirements, respectively, of the NRC-National Academy of Sciences. Where known, the toxic levels of the microminerals are given in Table 4-25, also.

TABLE 4-24
MACROMINERAL REQUIREMENTS OF SHEEP (PERCENTAGE OF RATION)[1]

Nutrient	Requirement	
	As-fed[2]	Moisture-free
	(%)	(%)
Sodium	**0.08-0.16**	0.09-0.18
Chlorine	—	—
Calcium	**0.18-0.74**	0.20-0.82
Phosphorus	**0.14-0.34**	0.16-0.38
Magnesium	**0.11-0.16**	0.12-0.18
Potassium	**0.45-0.72**	0.50-0.80
Sulfur	**0.13-0.23**	0.14-0.26

[1]Adapted by the author from *Nutrient Requirements of Sheep,* Sixth Revised Edition, NRC-National Academy of Sciences, 1985, p. 48.

[2]As-fed was calculated using 90% dry matter (moisture-free).

TABLE 4-25
MICROMINERAL REQUIREMENTS OF SHEEP AND MAXIMUM TOLERABLE LEVELS (PPM OR MG/KG OF RATION)[1]

Nutrient	Requirement		Maximum Tolerable Level	
	As-fed[2]	Moisture-free	As-fed	Moisture-free
	(ppm or mg/kg)	(ppm or mg/kg)	(ppm or mg/kg)	(ppm or mg/kg)
Cobalt	**0.09-0.18**	0.1-0.2	**9**	10
Copper	**6-10**	7-11[3]	**23**	25[4]
Fluorine	—	—	**54-135**	60-150
Iodine	**0.09-0.72**	0.10-0.80[5]	**45**	50
Iron	**27-45**	30-50	**450**	500
Manganese	**18-36**	20-40	**900**	1,000
Molybdenum	**0.45**	0.5	**9**	10[4]
Selenium	**0.09-0.18**	0.1-0.2	**1.8**	2
Zinc	**18-30**	20-33	**675**	750

[1]Adapted by the author from *Nutrient Requirements of Sheep*, Sixth Revised Edition, NRC-National Academy of Sciences, 1985, p. 50.

[2]As-fed was calculated using 90% dry matter (moisture-free).

[3]Requirement when dietary Mo concentrations are <1 mg/kg DM.

[4]Lower levels may be toxic under some circumstances.

[5]High level for pregnancy and lactation in rations not containing goitrogens; should be increased if rations contain goitrogens.

VITAMINS

Many phenomena of vitamin nutrition are related to solubility—vitamins are soluble in either fat or water. Consequently, it is important that both nutritionists and sheep producers be well informed about solubility differences in vitamins and make use of such differences in programs and practices. Thus, in the discussion that follows, vitamins are grouped as either (1) fat-soluble vitamins, or (2) water-soluble vitamins.

The fat-soluble vitamins are vitamin A (carotene), vitamin D, vitamin E, and vitamin K.

All of the water-soluble vitamins except vitamin C are known as B vitamins. These are *not* stored.

Fig. 4-53. Lamb with stiff-lamb disease, caused by a deficiency of vitamin E. (Courtesy, Cornell University, Ithaca, NY)

The functions, deficiency symptoms, and good sources of vitamins for sheep, which are similar to those of other animal species, are presented in Table 4–2, Animal Vitamin Chart.

Table 4–26 presents the NRC-National Academy of Sciences vitamin E requirements of growing-finishing lambs and the suggested amounts of alpha-tocopherol acetate to add to the rations to provide 100% of the requirements.

TABLE 4–26
VITAMIN E REQUIREMENTS OF GROWING-FINISHING LAMBS AND SUGGESTED LEVELS OF FEED FORTIFICATION TO PROVIDE 100% OF REQUIREMENTS[1]

Body Weight		Alpha-Tocopherol Acetate			Feed Intake Per Lamb		Amount of Vitamin E Added to Concentrate			Amount of Vitamin E Added to Protein Supplement[2]		
(lb)	(*kg*)	(mg/lamb/day)[3]	(mg/lb ration)	(*mg/kg ration*)	(lb)	(*kg*)	(mg/lb)	(*mg/kg*)	(mg/ton)	(mg/lb)	(*mg/kg*)	(mg/ton)
22	*10*	5.0	44	*20*	0.50	*0.23*	9.1	*20*	18,200	133	*60*	120,000
44	*20*	10.0	44	*20*	1.00	*0.45*	9.1	*20*	18,200	60	*133*	120,000
66	*30*	15.0	33	*15*	2.10	*0.96*	6.8	*15*	13,600	45	*100*	90,000
88	*40*	20.0	33	*15*	2.86	*1.30*	6.8	*15*	13,600	45	*100*	90,000
110	*50*	25.0	33	*15*	3.50	*1.60*	6.8	*15*	13,600	45	*100*	90,000

[1]Adapted by the author from *Nutrient Requirements of Sheep,* Sixth Revised Edition, NRC-National Academy of Sciences, 1985, p. 51.

[2]Assumes the concentrate diet contains 15% protein supplement.

[3]Rounded values based on approximate diet intake containing recommended vitamin E levels.

WATER

Sheep get water by drinking, and from snow, dew, and feed. The amount of water that sheep voluntarily consume is affected by temperature, rainfall, snow and dew covering, age, breed, stage of production, number of lambs carried, wool covering, respiratory rate, frequency of watering, kind and amount of feed, and exercise. On the average, mature animals consume approximately a gallon of water per day, whereas feeder lambs require about half this amount. However, sheep may go for weeks without drinking water when foraging on grasses and other feeds of high moisture content. This condition often prevails on desert ranges in the early spring and on many of the mountain ranges during the summer months.

Fig. 4–54. Sheep watering at the snow melter. (Courtesy, USDA, Soil Conservation Service)

FEED ADDITIVES

Table 4–27 summarizes the growth stimulants that are presently available and can be used. All of these products have been shown to improve gain and feed efficiency of sheep. The information presented in Table 4–27 is the most recent available. But feed additives and implants do change from time to time; new products are developed, and sometimes old products are banned by the Food and Drug Administration. So, those using additives should always confer with local authorities and read and follow manufacturer's label directions for more complete details on the use of a specific drug or combination of drugs.

Antibiotics may improve performance when added to creep and lamb-finishing rations. Chlortetracycline and oxytetracycline are especially effective. The response to antibiotics seems to be affected by differences in management and the amount of stress to which the lambs are subjected. There is some evidence that antibiotics reduce the incidence of enterotoxemia.

TABLE 4-27
SHEEP FEED ADDITIVES AND IMPLANTS

Type of Additive	Method of Administering	Dosage	Effect On: Daily Rate of Gain	Feed Efficiency	Carcass Quality	Comments
			(% increase)	(% increase)		
Antibiotics (chlortetracycline and oxytetracycline)	Feeding (oral)	Aureomycin (chlortetracycline) 10 to 25 mg/lb of feed. Terramycin (oxytetracycline) 5 to 10 mg/lb of feed.	Range: 0–31 Average: 11	Range: 4–27 Average: 10	No effect to slight improvement.	Antibiotics (especially chlortetracycline and oxytetracycline) may improve performance when added to creep and lamb finishing rations. Response to antibiotics varies markedly according to differences in management and degree of stress to which lambs are subjected. There is some evidence that antibiotics reduce the incidence of enterotoxemia.
Bovatec (lasalocid)	Feeding (oral)	10–15 mg/lb complete feed, fed at rate of 15–70 mg lasalocid /day.	Range: 0–20 Average: 6–8	Range: 5–15 Average: 8–10	No effect.	Bovatec is an ionophore. In addition to increasing rate of gain and feed efficiency, Bovatec reduces rumen protein degradation and increases the amount of bypass protein. Greatest response is obtained where coccidiosis is a problem, for which purpose Bovatec was initially approved by FDA.
Ralgro (zeranol)	Implant	12 mg/head	Range: 0–25 Average: 10	Average: 6		Do not implant animals within 40 days of slaughter. Do not implant breeding animals.

FEEDING BREEDING EWES

Success in the sheep business is largely measured by the percentage lamb crop raised and the pounds of lamb marketed per ewe. The most important factor affecting these criteria is the feed of the ewe. Also, the yearly feed of the ewe represents about 50% of all production costs. For purposes of convenience, the feeding of ewes will be discussed under the following headings: (1) drylot (confinement) feeding, (2) flushing ewes, (3) feeding pregnant ewes, (4) feeding at lambing time, (5) feeding lactating ewes, and (6) feeding ewes in accelerated lambing.

Fig. 4-55. Polled Dorset ewes with a Polled Dorset ram, owned by Riverwood Farms, Powell, Ohio. (Courtesy, *Sheep Breeder and Sheepman*, Columbia, MO)

• **Drylot (confinement) feeding**—The vast majority of the nation's sheep utilize pasture in season. However, some ewe-lamb producers are drylotting all or part of the year. So, now there are two alternatives, and the producer may choose between the two.

The **advantages** of drylot (confinement) production are:

1. The virtual elimination of losses from predators.
2. Freedom from the most harmful internal parasites.
3. Lowering of the energy requirement due to limited activity.
4. The opportunity to feed ewes according to their productivity and nutrient requirements rather than their appetites.
5. It results in more rapid gains, in lambs reaching market weight at an earlier age, and in improved carcass grade.

The **disadvantages** are:

1. A higher initial capital investment, especially in buildings and equipment.
2. It requires superior management.
3. All nutritive requirements must be met.
4. External parasites and contagious diseases may be increased.
5. Animal manure disposal and bedding costs will be greater.

The Table 4-31 rations are satisfactory for ewes in confinement, and the Table 4-29 and 4-30 (p. 189) rations are excellent for creep feeding lambs raised in confinement.

• **Flushing ewes**—*Flushing is the practice of conditioning or having thin ewes gain in weight just prior to breeding.* Its purpose is to increase the ovulation rate and, consequently, the lambing rate.

• **Feeding pregnant ewes**—Ewes should gain in weight during the entire period of pregnancy, making a total gain of 20 to 30 lb for the period. They should enter the nursing period with some reserve flesh, because the lactation requirements are much more rigorous than those of the gestation period.

Fig. 4–56. Pregnant ewes in excellent condition. (Courtesy, American Hampshire Sheep Assn., Ashlan, MO)

After the ewes are bred, they should have access to pastures as long as they are available and open. When the ground is firm, winter pasture or range, stalk, or stubble fields may be pastured to advantage. Green rye or wheat pastures furnish a very succulent feed and valuable exercise for the flock. Where winter pastures are either unavailable or inadequate, supplemental feeds must be provided. The most satisfactory forage is a good-quality legume hay—alfalfa, clover, lespedeza, or soybeans. The sheep producer, however, often seems to find it difficult to grow such roughage at satisfactory prices. Where grass hay, such as native hays or timothy, is used, every effort should be made to cut it at an early stage of maturity and to have it properly cured. Even then, a protein supplement should be provided, together with suitable minerals. Because of the known value of legumes from the standpoint of quality of proteins, minerals, and vitamins and the fact that grass hays are not recognized as too desirable for sheep, every effort should be made to supply at least a third of a good-quality legume roughage to pregnant ewes. A 150-lb ewe will eat about 4 lb of hay daily. In order to prevent waste and protect the wool from chaff and hay seeds, suitable racks should be provided.

• **Feeding at lambing time**—As lambing time approaches, or immediately after lambing, each ewe should be placed in an individual holding or lambing pen. At this time, the grain allowance should be materially reduced, but dry roughage may be fed free choice, when it is certain that it is of good quality and palatable. Usually, some five to seven days should elapse before ewes are placed on full feed following parturition. In general, feeds of a bulky and laxative nature should be provided during the first few days. A mixture of equal parts of oats and wheat bran is excellent. Soon after lambing, the ewe should be given water with the chill removed but should not be allowed to gorge.

Fig. 4–57. Polypay ewe and her newborn triplet lambs in a lambing pen. (Courtesy, American Polypay Sheep Assn., Sidney, MT)

• **Feeding lactating ewes**—Following lambing, the feed allowance of the ewe should be increased according to her capacity and needs.

In general, it is considered good practice to feed lactating ewes rather liberally, for lambs make the most economical gains when suckling. It is a good plan to separate the ewes with twins from those with singles, giving the former more liberal rations or the benefit of the better pastures or ranges. In fact, some large sheep operators find this practice so advisable that they regularly separate out the twin bands.

Though varying somewhat with the size and condition of the ewe and whether there are twins or a single, an adequate ration for a lactating ewe may consist of approximately 4 lb of high-quality alfalfa hay plus 1 to 2 lb of grain daily. If neither a legume hay nor legume silage is available, a protein supplement should be included in the grain ration.

As soon as the spring pasture season has arrived, the use of harvested feeds should be discontinued, being both uneconomical and unnecessary.

• **Feeding ewes in accelerated lambing**—*Accelerated lambing involves ewe lambs dropping their first lambs at 1 year of age, and lambing at intervals of 6 to 8 months thereafter.*

Ewe lambs that are to be bred so that they lamb at 12 months of age should be liberally fed (1) from birth, using one of the creep rations given in Tables 4–29 or 4–30 (p. 189); and (2) during pregnancy and lactation, using one of the suggested rations in Table 4–31.

FEEDING RAMS

Fig. 4–58. Polled Dorset ram in strong breeding condition, owned by Riverwood Farms, Powell, OK.

The rams should be fed so as to remain in vigorous, active breeding condition. In general, rams should be fed the same kind of feed as ewes but in slightly larger quantities. They need a generous allowance of relatively high-quality feed just before and during the breeding season, when pasture is not available. During the balance of the year, pasture is usually adequate when available; otherwise, the ration may be comparable to that of ewes.

FEEDING RANGE SHEEP

Range sheep are normally maintained on winter grazing areas, with or without supplemental feeds, as long as possible. Usually these ranges are located at the lower altitudes and the vegetation consists of rather mature and bleached grasses or brush and browse. When the vegetation is sparse or covered by deep snow, supplemental feeds of hays, preferably alfalfa, some other legume, or concentrates are provided. Often protein supplements in the form of pellets or cubes are used, for these may be scattered about the feeding grounds, neither being blown away nor difficult for the sheep to find. Usually such expensive protein supplements are fed only when native grass hays are being utilized, high-quality alfalfa not requiring a protein supplement.

- **Nutrient deficiencies of range forage**—Mature, weathered native range grass is almost always deficient in protein—being

Fig. 4-59. Hunger, due to lack of feed, is the most common deficiency on the western ranges of the U.S. (Courtesy, Amos Dee Jones Ranches of Roswell and Tatum, NM)

as low as 3% or less. Protein-leaching losses due to fall and winter rains may range from 37 to 73%.

Phosphorus deficiencies are rather common among range sheep, but calcium deficiency is seldom encountered.

Of the vitamins, vitamin A is most likely to be deficient in range forage, because dry, bleached range grass is very low in carotene (the precursor of vitamin A).

- **Range supplements**—Four suggested range supplements, ranging from high to low protein, are given in Table 4–28 (see next page).

Also, producers can lessen the labor attendant to the daily feeding of a pasture or range supplement by (1) using protein blocks, or (2) self-feeding salt-feed mixtures.

Where salt is used for the purpose of governing consumption, the proportion of salt to feed may vary anywhere from 5 to 40% (with 30 to 33⅓% salt content being most common).

- **Rate of supplemental feeding**—The normal range of supplementation for sheep is ¼ to ½ lb per head per day. Rates above ½ lb approach a level that will result in reduced intake of range forage. Where range vegetation is so short as to require supplementation in excess of ½ lb per head per day, consideration should be given either to moving the sheep into drylot or to moving them to a better grazing area.

Some managers divide their sheep according to age, condition, and twins vs single lambs. Of course, this is facilitated where there are several bands. By so doing, it is possible (1) to give the animals that require the highest level of nutrition the best pasture or range, and/or (2) to supplement according to need.

TABLE 4-28
FORMULAS FOR RANGE SHEEP SUPPLEMENTS[1]

Feed[2]	Recommended Level of Protein: High	Medium-High	Medium-Low	Low
	(%)	(%)	(%)	(%)
Barley, grain or corn, dent yellow, grain, grade 2 US, minimum 54 lb (*24.5 kg*)/bu	5	40	75	65
Beet, sugar, molasses, or sugar cane molasses, 48% invert sugar, minimum 79.5° Brix	5	5	5	5
Cottonseed with some hulls, solvent extracted, ground, minimum 41% protein, maximum 14% fiber, minimum 0.5% fat (cottonseed meal)	66	36	–	16
Soybean, seeds, solvent extracted, ground, maximum 7% fiber, 44% protein (soybean meal)	10	10	10	10
Urea, technical, 282% protein equivalent	–	–	5	–
Alfalfa, aerial parts, dehydrated, ground, minimum 17% protein or alfalfa, hay, sun-cured, early bloom	10	5	–	–
Vitamin A (IU/lb)	–	1,818	3,636	3,636
Vitamin A (IU/kg)	–	*4,000*	*8,000*	*8,000*
Calcium phosphate, monobasic, commercial	1	1	2	1
Sodium phosphate, monobasic, technical	2	2	2	2
Salt or trace mineralized salt	1	1	1	1
Total	100	100	100	100

Composition[3]	High: As-Fed[4]	High: *M-F*	Medium-High: As-Fed[4]	Medium-High: *M-F*	Medium-Low: As-Fed[4]	Medium-Low: *M-F*	Low: As-Fed[4]	Low: *M-F*
Digestible energy (Mcal/lb)	**1.4**	1.5	**1.4**	1.5	**1.4**	1.5	**1.3**	1.4
Digestible energy (Mcal/kg)	***3.0***	*3.3*	***3.0***	*3.3*	***3.0***	*3.3*	***2.8***	*3.1*
Protein (N × 6.25) (%)	**30.4**	33.8	**21.9**	24.3	**23.6**	26.2	**15.9**	17.7
Phosphorus (%)	**1.8**	2.0	**1.4**	1.5	**0.8**	0.9	**1.1**	1.2
Carotene (mg/lb)	**9.0**	10.0	**4.1**	4.5	–	–	–	–
Carotene (mg/kg)	***19.8***	*22.0*	***9.0***	*10.0*	–	–	–	–
Vitamin A (IU/lb)	–	–	**1,636.0**	1,818.0	**3,273.0**	3,636.0	**3,273.0**	3,636.0
Vitamin A (IU/kg)	–	–	***3,600.0***	*4,000.0*	***7,200.0***	*8,000.0*	***7,200.0***	*8,000.0*
Rate of feeding (lb/day)	**0.20–0.40**	0.22–0.44	**0.20–0.40**	0.22–0.44	**0.20–0.40**[5]	0.22–0.44[5]	**0.20–0.40**[5]	0.22–0.44[5]
Rate of feeding (kg/day)	***0.09–0.18***	*0.1–0.2*	***0.09–0.18***	*0.1–0.2*	***0.09–0.18***[5]	*0.1–0.2*[5]	***0.09–0.18***[5]	*0.1–0.2*[5]

[1]Adapted by the author from *Nutrient Requirements of Sheep*, Sixth Revised Edition, NRC-National Academy of Sciences, 1985, p. 52, Table 11.

[2]Feeds mixed and fed in meal or pellet form.

[3]Molasses and alfalfa hay, sun-cured, early bloom not included.

[4]Estimated 90% dry matter.

[5]In emergency situations, up to 1.1 lb (*0.5 kg*) may be fed.

FEEDING GROWING-FINISHING LAMBS

Fig. 4-60. Growing-finishing lambs at market weight. (Courtesy, *The Dorset Journal*.

The growing-finishing stage of lambs refers to that period extending from birth to weaning at 4 to 6 months of age. At no other period in the life of the sheep is the promotion of growth and prevention of disease so important.

Where succulent pastures are available, most practical sheep producers, including producers with both farm flocks and range bands, consider that a combination of such green forage plus the ewe's milk is ample. In fact, lambs are unique among farm animals, inasmuch as they may be marketed at top prices off grass. Although young cattle may be sold off grass without having any other feed, they will usually fail to get sufficiently fat to bring top prices.

- **Early weaning**—*Early weaning refers to the practice of weaning lambs earlier than usual—to weaning at 6 to 8 weeks of age or earlier.* There is much interest in early weaning because of lambing out of season, multiple births, and more than one lamb crop per year.

An early-weaned lamb ration should meet the following specifications: contain a minimum of 16% crude protein; be fortified with supplemental iron if the lambs are raised on slotted floors; and have a calcium:phosphorus ratio of at least 1:1 (2:1 if urinary calculi has been experienced).

Milk replacers containing approximately 30% fat and 24% protein have been used successfully in feeding lambs receiving colostrum and weaned at one day of age. Replacers with reduced lactose content (from 42 to 27% on a dry matter basis) give improved performance. The milk is fed (1) cold at 36 to 40°F, rather than warm, to reduce overeating and bacterial contamination, and (2) free choice. From the beginning, lambs are offered a very palatable solid feed in addition to the milk. The milk replacer is discontinued when the lambs are eating sufficient quantities of the dry feed, usually at 21 to 35 days of age.

TABLE 4-30
SOME SIMPLE CREEP RATIONS (AS-FED BASIS)[1]

	Unpelleted		Pelleted	
	First 2 Months	2 Months to Market	First 2 Months	2 Months to Market
	(%)	(%)	(%)	(%)
Ground corn	49	89	64	59
Crushed oats	30	—	—	—
Soybean meal	20	10	20	10
Limestone	1.0	1.0	1.0	1.0
Trace mineral salt	.5	.5	.5	.5
Alfalfa	—	—	10	25
Molasses	—	—	5	5

[1]The addition of 0.25 to 0.50% ammonium chloride will minimize urinary calculi.

Feeding Directions:

Same as presented with Table 4-29; so, see the latter.

• **Creep feeding**—*The practice of supplemental feeding of nursing lambs in a separate enclosure away from their dams is known as creep feeding.* Lambs will usually consume some creep feed at 10 to 14 days of age.

Creep rations can either be hand-fed or self-fed. Many sheep producers hand-feed until the lambs begin to eat regularly, then self-feed from this point on.

Suggested creep rations are given in Tables 4-29 and 4-30.

TABLE 4-29
SOME EXCELLENT CREEP RATIONS (AS-FED BASIS)[1]

	Unpelleted		Pelleted	
	First 2 Months	2 Months to Market	First 2 Months	2 Months to Market
	(%)	(%)	(%)	(%)
Ground corn	80	60	40	50
Ground oats		20	15	—
Soybean meal	20	10	20	10
Alfalfa hay	—	—	10	35
Bran		10	10	10
Molasses	—	—	5	5
Trace mineral salt	.5	.5	.5	.5
Limestone	1.0	1.0	1.0	1.0
Antibiotic (mg/lb)[2]	50	20	50	15
Vitamin A (IU/lb)	1,000	1,000	1,000	1,000
Vitamin D (IU/lb)	200	200	200	200
Vitamin E (mg/lb)	20	20	20	20

[1]The addition of 0.25 to 0.50% ammonium chloride will minimize urinary calculi.

[2]Chlortetracycline (Aureomycin) or oxytetracycline (Terramycin).

Feeding Directions:

1. Lambs should be started on creep feed about 10 days after birth. Although they will not consume significant amounts of feed until 3-4 weeks of age, the small amounts consumed at earlier ages are critical for establishing both rumen function and the habit of eating.

2. Feed high quality legume hay in a separate rack. Feed hay and creep ration twice daily to keep them fresh.

3. The amount of creep feed consumed by lambs 2 to 6 weeks of age is affected by the palatability of the ration (ration composition and ration form) and the location and environment of the creep area. A well-bedded, well-lighted area located close to where the ewes congregate is preferred.

• **Feeding orphan ("bummer") lambs**—Sheep producers estimate that about 10% of their lamb crop dies from starvation during the first week after birth. Some starvation results from newborn lambs sucking the scrotum and/or navel of another lamb. But most starved lambs are orphans (bummers) resulting from (1) the mother dying at lambing, (2) rejection by the mother, (3) the mother not being able to suckle the lamb because of mastitis or some similar problem, or (4) multiple births beyond the ewe's nursing capacity. Whatever the cause, the most satisfactory arrangement for the orphan is to provide a foster mother—to transfer (graft) the lamb to another ewe. The alternate to a foster mother arrangement is artificial rearing.

Fig. 4-61. Orphan lambs being self-fed milk replacer from a multiple-nipple container. (Courtesy, Ralston Purina Co., St. Louis, MO)

Observance of the following principles and practices will increase the chances of raising orphan lambs artificially:

1. Give colostrum.

2. Inject orphans with (1) vitamins A, D, and E, (2) iron-dextran, and (3) selenium in selenium-deficient areas.

3. Use milk replacer.

4. Provide a good starter feed and water from day one. A good starter ration follows:

Lamb Starter Rations:

Ingredients	%
Soybean meal (49% CP)	40.0
Ground corn	27.0
Alfalfa meal	15.0
Dextrose (corn sugar)	10.0
Fat (*e.g.* vegetable oil)	5.0
Limestone	2.0
Trace mineral salt	0.7
Vitamin premix	0.3
Total	100.0

Once the lambs have fully adjusted to the starter ration, they can be slowly switched onto the regular creep or grower ration. (See Tables 4-29 and 4-30, previous page.)

5. Maximize sanitation, observation, and TLC.

Fig. 4-62. Lambs finishing on winter wheat in western Kansas. (Courtesy, Rufus F. Cox, Kansas State University)

In Kansas, Oklahoma, Nebraska, and Texas, thousands of lambs are finished primarily by fall pasturing of the wheat fields. In the Pacific Northwest, a limited number of lambs are finished by gleaning pea stubble.

• **Drylot feeding**—Drylot feeding is, as the name indicates, feeding under restricted conditions. This may be either (1) shelter or barn feeding, or (2) open-yard feeding.

FEEDING FINISHING (FATTENING) LAMBS[7]

The primary objective of the sheep producer is that of producing milk-fat lambs suitable for slaughter at weaning time. Only when pasture is inadequate are lambs sold via the feeder route. Almost all feeder lambs come from the range area. Some range areas produce only a small percentage of lambs which are classed as feeders, whereas in other areas almost all the lambs must be sold as feeders because the vegetation is not sufficient to promote rapid growth and finishing. It is estimated that, for the range area as a whole, an average of at least 50% of all lambs produced in one year receive additional feed after they are removed from the range prior to slaughter.

• **Field finishing**—This method of finishing lambs is somewhat comparable to the pasture finishing of cattle, except that a greater variety of feeds is used by lamb feeders.

Throughout the Corn Belt, feeder lambs are usually used as scavengers during the early part of the field feeding process. Frequently, the lambs are pastured in the stubble fields or on the meadow until all these feeds are consumed, after which they are turned into corn fields.

Fig. 4-63. Drylot feeding of lambs.

FEED ALLOWANCE AND SOME SUGGESTED RATIONS FOR SHEEP

Sheep rations vary with the section of the country, depending chiefly on available local feeds. Fortunately, many feeds of similar nutritive properties can be interchanged in the ration as price relationships warrant. This makes it possible at all times to obtain a balanced ration at the lowest cost.

[7]Many helpful suggestions for this section were received from the following authority: Dr. Clair Acord, Ph.D., Utah Wool Growers Assn., Salt Lake City, UT.

Except at lambing time or when emergencies occur as a result of drought or inclement weather, western bands receive little supplemental feed. Even with farm flocks, a minimum of grain is fed to breeding animals. Grain feeding usually is limited to the latter part of gestation and to the lactation period prior to turning to pasture.

Tables 4-31 and 4-32 (see next page for table 4-32) contain some rations that have been used by successful sheep operators in various sections of the country.

TABLE 4-31
DAILY RATIONS FOR BREEDING EWES AT VARIOUS STAGES OF PRODUCTION

Ration No.	Moisture Basis[1] A-F (as-fed) M-F (moisture-free)	Hay[2]		Corn Silage		Haylage		Corn Straw		Stover (Stalks)		Grain[3]		Protein Supplement[4]	
		(lb)	*(kg)*	(lb)	*(kg)*	(lb)	*(kg)*	(lb)	*(kg)*	(lb)	*(kg)*	(lb)	*(kg)*	(lb)	*(kg)*
Maintenance															
1	**A-F**	**3.0**	***1.4***												
	M-F	3.3	1.5												
2	**A-F**			**6.0**	***2.7***									**0.20**	***0.09***
	M-F			6.7	*3.0*									0.22	*0.10*
3	**A-F**					**6.0**	***2.7***								
	M-F					6.7	*3.0*								
4	**A-F**							**3.0**	***1.4***					**0.40**	***0.18***
	M-F							3.3	*1.5*					0.44	*0.20*
Gestation, early (first 15 weeks)															
1	**A-F**	**3.5**	***1.6***												
	M-F	3.9	*1.8*												
2	**A-F**	**2.0**	***0.9***									**1.0**	***0.45***		
	M-F	2.2	*1.1*									1.1	0.49		
3	**A-F**	**1.8**	***0.8***									**0.6**	***0.27***	**0.20**	***0.09***
	M-F	2.0	*0.9*									0.7	*0.31*	0.22	*0.10*
4	**A-F**			**8.0**	***3.6***									**0.20**	***0.09***
	M-F			8.9	*4.0*									0.22	*0.10*
5	**A-F**					**7.0**	***3.2***					**0.20**	***0.09***		
	M-F					7.8	*3.5*					0.22	*0.10*		
6	**A-F**	**2.0**	***0.9***							**2.0**	***0.9***	**0.5**	***0.23***		
	M-F	2.2	*1.0*							2.2	*1.0*	0.6	*0.27*		
7	**A-F**	**1.0**	***0.45***							**2.0**	***0.9***	**0.5**	***0.23***	**0.30**	***0.14***
	M-F	1.1	*0.49*							2.2	*1.0*	0.6	*0.27*	0.33	*0.15*
Gestation, late (last 4 weeks): Add 0.5–1.0 lb (*0.23–0.45 kg*) grain per ewe daily to any of the above rations.															
Lactation															
1	**A-F**	**4.0**	***1.8***									**2.0–3.0**	***0.9–1.4***	–	–
	M-F	4.4	*2.0*									2.2–3.3	*1.1–1.5*	–	–
2	**A-F**			**10.0**	***4.5***							**1.5**	***0.7***	**0.25**	***0.11***
	M-F			11.1	*5.0*							1.7	*0.8*	0.28	*0.13*
3	**A-F**	**1.0**	***0.45***	**8.0**	***3.6***							**1.5**	***0.7***	**0.20**	***0.09***
	M-F	1.1	*0.49*	8.9	*4.0*							1.7	*0.8*	0.22	*0.10*
4	**A-F**					**8.9**	***3.6***					**2.0–3.0**	***0.9–1.4***		
	M-F					8.9	*4.0*					2.2–3.3	*1.1–1.4*		

[1]As-fed was calculated using an average figure of 90% dry matter. When using silages, roots, and other wet feeds, these feeds should be converted to a moisture-free basis and the ration calculated using the moisture-free data.

[2]Alfalfa hay, midbloom, preferred.

[3]Grain may consist of corn, barley, wheat, oats, and/or grain sorghum.

[4]Protein supplement may consist of soybean, cottonseed, linseed, sunflower, safflower, or rapeseed meal.

Feeding Directions:

1. These rations are formulated to meet the requirements of a 154 lb (*70 kg*) ewe in average condition, and are designed for hand-feeding. The daily feed allowance can be increased or decreased, depending on the actual size of the ewe and the body condition.

2. Some of these rations are deficient in calcium and/or phosphorus; therefore, a supplement containing 50% trace mineral salt (for sheep) and 50% dicalcium phosphate should be fed free choice. The consumption of 0.05 lb (*0.02 kg*) per sheep per day of this mixture will provide the amounts of calcium and phosphorus needed for maintenance and the first 15 weeks of gestation; and 0.10 lb (*0.05 kg*)/day will provide the needed Ca and P for late gestation and lactation. Vitamins A and E should be added to the salt-mineral mix when sheep are fed the wheat straw and corn stover rations.

3. Ewes should gain 15 to 25 lb (*6.8 to 11.4 kg*) during gestation. During early gestation (first 15 weeks), they should gain 0.05 lb (*0.02 kg*)/day. During late gestation (last 4 weeks), they should gain 0.5 lb (*0.23 kg*)/day. If, during the last 4 weeks of pregnancy, ewes are fed 0.5 to 1.0 lb (*0.23 to 0.45 kg*) grain per head daily and gain 8 to 15 lb (*3.6 to 6.8 kg*), ketosis (lambing paralysis) can be prevented almost entirely.

4. During maintenance and early gestation, each ewe should have 14 in. (*36 cm*) of bunk feed space. In late gestation and during lactation, bunk space should be increased to 15 to 18 in. (*38 to 46 cm*).

TABLE 4–32
GROWING–FINISHING RATIONS FOR LAMBS[1]

Ingredient	Moisture Basis[2] A-F (as-fed) M-F (moisture-free)	Rations Using Corn/Alfalfa Hay/Soybean Meal					Rations Using Milo/Cottonseed Hulls/Cottonseed Meal				
		1	2	3	4	5	1	2	3	4	5
		(%)	(%)	(%)	(%)	(%)	(%)	(%)	(%)	(%)	(%)
Corn grain (dent yellow)		31.0	41.5	51.7	63.0	73.3	—	—	—	—	—
Sorghum grain (milo)		—	—	—	—	—	19.5	32.7	46.2	60.7	73.7
Alfalfa hay (mature)		55.0	45.0	35.0	25.0	15.0	15.0	15.0	15.0	15.0	15.0
Cottonseed hulls		—	—	—	—	—	40.0	30.0	20.0	10.0	—
Soybean meal (solvent 44% CP)		7.0	6.5	6.0	5.5	5.0	—	—	—	—	—
Cottonseed meal (solvent 41% CP)		—	—	—	—	—	17.5	14.0	10.5	7.0	4.0
Molasses (cane)		6.0	6.0	6.0	5.0	5.0	6.0	6.0	6.0	5.0	5.0
Calcium carbonate		—	—	.3	.5	.7	1.0	1.3	1.3	1.3	1.3
Trace mineral salt (sheep)		.5	.5	.5	.5	.5	.5	.5	.5	.5	.5
Ammonium chloride		.5	.5	.5	.5	.5	.5	.5	.5	.5	.5
Nutritional content											
Dry matter (%)	**A-F**	**87.5**	**87.5**	**87.5**	**87.6**	**87.6**	**89.0**	**88.9**	**88.7**	**88.7**	**88.5**
	M-F	100	100	100	100	100	100	100	100	100	100
TDN (%)	**A-F**	**75.3**	**80.1**	**84.6**	**89.2**	**93.1**	**67.9**	**72.3**	**77.2**	**82.0**	**86.8**
	M-F	65.9	70.1	74.0	78.1	82.0	60.4	64.3	68.5	72.7	76.8
Net energy for maintenance (Mcal/lb)	**A-F**	**.80**	**.86**	**.91**	**.98**	**1.04**	**.71**	**.76**	**.82**	**.88**	**.94**
	M-F	.70	.75	.80	.86	.91	.63	.68	.73	.78	.83
Net energy for maintenance (Mcal/kg)	***A-F***	***.36***	***.39***	***.41***	***.44***	***.47***	***.32***	***.35***	***.37***	***.40***	***.43***
	M-F	*.32*	*.34*	*.36*	*.39*	*.41*	*.29*	*.32*	*.33*	[illegible]	*.38*
Net energy for gain (Mcal/lb)	**A-F**	**.40**	**.47**	**.53**	**.59**	**.66**	**.33**	**.38**	**.45**	**.52**	**.59**
	M-F	.35	.41	.46	.52	.58	.29	.34	.40	.46	.52
Net energy for gain (Mcal/kg)	***A-F***	***.18***	***.21***	***.24***	***.27***	***.30***	***.15***	***.17***	***.20***	***.24***	***.27***
	M-F	*.16*	*.19*	*.21*	*.24*	*.26*	*.13*	*.15*	*.18*	*.21*	*.24*
Crude protein (%)	**A-F**	**17.1**	**16.6**	**16.0**	**15.4**	**14.8**	**17.0**	**16.3**	**15.8**	**15.2**	**14.8**
	M-F	15.0	14.5	14.0	13.5	13.0	15.1	14.5	14.0	13.5	13.1
Protein bypass (%)	**A-F**	**42.3**	**44.1**	**45.8**	**47.9**	**49.7**	**44.6**	**47.6**	**51.0**	**54.7**	**57.9**
	M-F	37.0	38.6	40.1	42.0	43.5	39.7	42.3	45.2	48.5	51.2
Calcium (%)	**A-F**	**.86**	**.71**	**.70**	**.67**	**.66**	**.85**	**.96**	**.94**	**.91**	**.89**
	M-F	.75	.62	.61	.59	.58	.76	.85	.83	.81	.79
Phosphorus (%)	**A-F**	**.29**	**.30**	**.31**	**.31**	**.32**	**.37**	**.37**	**.36**	**.36**	**.36**
	M-F	.25	.26	.27	.27	.28	.33	.33	.32	.32	.32

[1]Adapted by the author from *The Sheepman's Production Handbook,* published by the Sheep Industry Development Program, Inc., Denver, CO, 1986, p. N-44, Table 13.

[2]As-fed was calculated using an average figure of 90% dry matter. When using silages, roots, and other wet feeds, these feeds should be converted to a moisture-free basis and the ration calculated using the moisture-free data.

Feeding Directions:

1. These rations can be fed once daily in troughs or bunks if there is capacity for a day's feed. They can also be self-fed if the feeders are designed to handle such feed without bridging.
2. Offering lambs a good quality hay for 1–3 days along with rations 1 or 2 (provided free choice) can be used to start lambs on feed.
3. About 3 in. (*7.6 cm*) of self-feeder or trough space must be provided per lamb for self-feeding and about 12 in. (*30.5 cm*) if hand-fed.
4. Gradually adapt the lambs to the higher energy rations by allowing 4–7 days on a ration before switching to the ration with the next higher energy level.
5. Complete mixing to prepare a uniform ration is important.
6. Lambs must not be allowed to be without feed even for a short period of time.
7. The mineral and vitamin mixture given in Table 4–33 may replace the trace mineral salt in all Table 4–32 rations.

The rations in Table 4–32 are nutritionally adequate and balanced with respect to Ca:P and N:S ratios. The mineral and vitamin mixture given in Table 4–33 may replace the trace mineral salt in all the Table 4–32 rations.

The rations in Table 4–32 are nutritionally adequate and balanced with respect to Ca:P and N:S ratios. The mineral and vitamin mixture given in Table 4–33 may replace the trace mineral salt in all the Table 4–32 rations.

TABLE 4–33
MINERAL AND VITAMIN MIXTURE FOR LAMB RATIONS[1]

Ingredient	Lb/Ton	*Kg/Ton*	Contribution to Complete Ration[2]
Salt, plain fine mixing	1,729.613	*785.244*	.43% salt
Sulfur, elemental[3]	200.00	*90.80*	.05% S
Cobalt carbonate ($CaCO_3$)	0.087	*0.039*	.1 ppm Co
Ethylenediamine dihydro-iodide (EDDI)	0.100	*0.045*	.2 ppm I
Manganese oxide (MnO)	10.300	*4.676*	20 ppm Mn
Zinc oxide (ZnO)	10.300	*4.676*	20 ppm Zn
Vitamin A[4]	17.6	*8.0*	600 IU/lb
Vitamin E[5]	32.0	*14.5*	10 IU/lb

[1]Adapted by the author from *The Sheepman's Production Handbook*, published by the Sheep Industry Development Program, Inc., Denver, CO, 1986, p. N-45, Table 14.

[2]Contribution to the complete ration when 10 lb (*4.5 kg*) of the mineral and vitamin mixture is added to 1 ton of complete lamb ration.

[3]In complete rations containing ammonium sulfate instead of ammonium chloride for prevention of urinary calculi, sulfur should not be added to the mineral and vitamin premix (.5% NH_4SO_4 contributes .12% S to the ration).

[4]Contains 13,607,700 IU of vitamin A per pound.

[5]Contains 125,000 IU of vitamin E per pound.

FITTING RATIONS FOR SHEEP

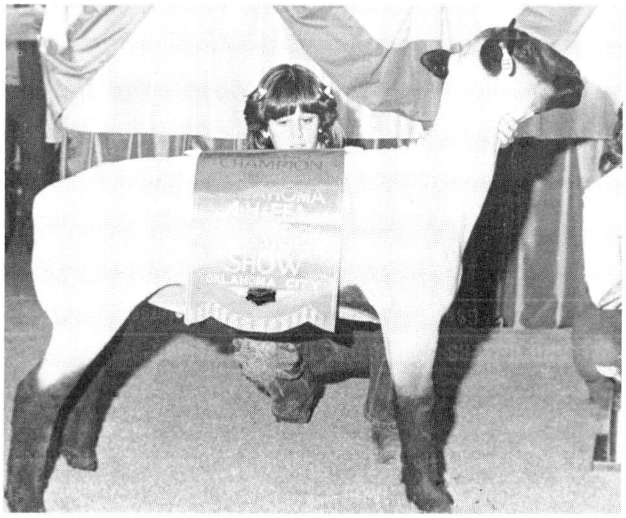

Fig. 4–64. A beautifully fitted lamb, Champion in the Omaha, Nebraska 4-H and FFA show. (Courtesy, American Hampshire Sheep Assn., Ashland, MO)

In addition to being reasonably economical (mostly home-grown) and well-balanced, the ration for show and sale sheep must be palatable. Many feed combinations meet these specifications. The ration selected is usually determined by (1) the availability and price of feed in the area, and (2) the preference and judgment of the feeder.

Some suggested grain fitting rations are given in Tables 4–34 and 4–35. To each of these grain rations, good quality roughage should be added.

TABLE 4–34
FITTING CONCENTRATE MIX FOR SHOW LAMBS

Ingredient	%
Cracked corn	50
Whole or rolled oats	35
Soybean meal	10
Molasses	4
Mineral (limestone/sheep salt-mineral mix)	1
Total	100

TABLE 4–35
RATIONS FOR FITTING YEARLING AND MATURE SHEEP

Ingredient	Ration Number							
	1		2		3		4	
	(lb)	*(kg)*	(lb)	*(kg)*	(lb)	*(kg)*	(lb)	*(kg)*
Barley, rolled	—	—	40	*18.2*	—	—	10	*4.5*
Corn, cracked	—	—	—	—	40	*18.2*	—	—
Oats, rolled	50	*22.7*	40	*18.2*	40	*18.2*	60	*27.2*
Peas (split)	40	*18.2*	—	—	—	—	10	*4.5*
Protein supplement[1]	—	—	10	*4.5*	10	*4.5*	10	*4.5*
Wheat bran	10	*4.5*	10	*4.5*	10	*4.5*	10	*4.5*
Total	100	*45.4*	100	*45.4*	100	*45.4*	100	*45.4*

[1]Cottonseed, linseed, rapeseed (canola), soybean, and/or sunflower meal.

Lambs (either creep fed or weaned lambs that are being fitted for show) will eat about 2½ lb of Table 4–34 fitting concentrate, per head daily when on full feed.

Yearlings will eat about 3 lb per head daily of one of the rations shown in Table 4–35 when on full feed, whereas mature sheep will eat about 3½ lb of one of these rations.

Exhibitors prefer to feed steam rolled oats and barley and nutted (pea-sized) old process, linseed meal.[8] Corn is usually cracked or coarsely ground, and peas are split or cracked. When pastures are not available, alfalfa is the most popular hay. But any good legume is quite satisfactory.

In fitting animals for show or sale, most successful shepherds feed a limited quantity of sliced carrots, cabbage, mangels (stock beets), rutabagas (swedes), or turnips. These succulent feeds are highly relished by sheep and appear to help their digestion and general thrift.

The following points also are pertinent in feeding sheep:

1. All classes and ages of sheep should be allowed free access to a double compartment mineral box, with loose salt in one compartment and a mixture of ⅓ salt and ⅔ steamed bone meal, or other suitable mineral, in the other.
2. Unless grains are unusually hard, they need not be ground for sheep. The animals prefer to do their own grinding, and the feeds are no more effectively utilized when ground.

[8]Among experienced shepherds, old-process linseed meal is especially popular for fine-wool sheep because of its conditioning effect on the fleece.

FEEDING GOATS

Fig. 4-65. Angora goats on pasture. (Courtesy, Mohair Council of America, San Angelo, TX)

Worldwide, there are over 450 million goats; and there are over 2 million goats in the United States. These goats provide three products: milk, meat, and mohair. As with other livestock, the feed costs comprise a large share of the costs of production. Moreover, with goats the feed needs differ with regard to the final product. The end product also dictates management practices which affect the nutritional needs. For example, Angora (mohair) and Spanish or meat-type goats are raised primarily on rangelands; whereas, dairy goats in high production require well-balanced rations high in protein and energy, similar to dairy cows. Nevertheless, some principles and practices are applicable to all goats.

FEEDING ANGORA GOATS AND SPANISH (MEAT) GOATS

The vast majority of the goats in the United States belong to the mohair-bearing Angora breed. Although there are more than 1.2 million head of these strange-looking, heavy-coated creatures in this country, few people outside the Angora goat districts know what they look like. Texas alone accounts for over 90% of these goats in the United States, due largely to the large areas of rugged grazing land, which is well adapted to utilization by this species.

The Spanish goat, the primary meat type goat, is of uncertain origin. In all likelihood, many Spanish goats contain infusions of one or more dairy goat breeds. The meat from the kids of this type of gat has tradtionally been called *cabrito*—Spanish for little goat. Much like its cousin, the Angora, the Spanish goat is raised primarily in the range areas of the United States. Because both are raised under similar conditions, their feeding is discussed together.

Available browse and forages will satisfy many of the nutritive needs of goats that are raised on ranges, but, for maximum performance, it is advisable to provide supplemental feed when range conditions become adverse. Twenty percent protein range cubes are a popular supplement. Also, shelled corn can be fed on the ground to goats, with very little waste. Usually, about 0.25 to 1.0 lb of supplement per head per day is adequate in the winter or during dry periods when green feed is scarce. Some examples of concentrate supplements for range goats are found in Table 4-36.

TABLE 4-36
CONCENTRATE SUPPLEMENTS FOR RANGE GOATS[1]

Ingredients	Supplement A 20% Protein	Supplement B 30% Protein	Supplement C 40% Protein
	← (%) →		
Corn or soghum	82	58	25
Cottonseed meal or soybean meal	14	37	70
Urea	2	3	3
Dicalcium phosphate	2	2	2
Vitamin A supplement	—[2]	—[2]	—[2]
Proximate analysis:			
TDN (%)	75	72	70
DE (Mcal/lb)	1.50	1.45	1.40
Crude protein (%)	20	30	40
Digestible protein (%)	16	24	32
Phosphorus (%)	0.55	0.65	0.77

[1]Adapted by the author from *Nutrient Requirements of the Angora Goat* by J. Z. Huston, M. Shelton, and W. C. Ellis, Tex. Ag. Exp. Sta. Bull. B-1105, March, 1971.

[2]Sufficient supplement to provide 2,500 IU of vitamin A per lb of feed (5 million IU/ton of total concentrate supplement mix).

• **Feeding does**—On good range in the spring, mature dry does will consume enough feed to satisfy all their nutrient demands except salt and phosphorus. During lactation, they may need 0.5 to 0.75 lb of a supplement of type A as given in Table 4-36. (The total ration, supplement plus range grass, will usually run about 11 to 12% protein.)

During the summer and early fall, the quality of range feed is reduced and a higher protein supplement, such as B (Table 4-36), should be provided at the rate of 1 lb for each 10 to 20 does. Immature does (yearling and those not fully developed) should be provided 1 lb of supplement for each 5 does. If the range is particularly poor, double the amount of supplement.

In late fall and winter, ranges tend to be at their lowest nutritive value. Poor ranges require supplemental feeding—Supplement C, Table 4-36—at levels of 1 lb for each 3 to 5 mature does and 1 lb for each 1 to 3 yearling or underdeveloped does. These supplements should take care of the needs of late pregnancy and of lactation. For ranges with new growth, supplements should follow the recommendations given above for good spring range.

Does with more than two kids should be given 25 to 50% more of the supplement during lactation than is recommended for does with singles or twins. Does which have a history of giving birth to more than one kid should be fed at a high rate of supplementation at least 3 weeks before expected parturition (kidding).

Additional supplementation of 0.25 to 0.33 lb grain or range cubes should be fed to does 1 to 2 weeks prior to turning the bucks in for the breeding season. This added feed (called flushing) improves conception rate by having does in a positive nutrient balance during breeding. When flushing is to be used, feed should be increased gradually. Likewise, at the end of the breeding season, the feed allowance should be reduced gradually, so as to avoid upsetting the animals' appetites and digestive tracts.

• **Feeding kids (nursing and weaned)**—As long as kids are receiving adequate amounts of milk from their mothers, they do very well on good range. Additional supplementation, however, makes for more rapid growth and better prepares them for breeding or market. One pound of supplement for each 2½ to 3 kids should be provided. Older and larger kids may have their supplement reduced to 1 lb daily for each 5 kids if the range conditions are good. When range is poor, the grain and supplement should be increased to provide 1.00 to 1.33 lb of grain daily per kid. In addition, kids should have access to good-quality hay.

FEEDING DAIRY GOATS

A good dairy doe will average 5 lb of milk, or more, per day over a lactation period of 10 months, while superior animals will average 10 lb or more. The highest official milk production on record in the United States was made by a Toggenburg doe that produced 5,750 lb of milk in 305 days. The butterfat record is held by an Alpine doe that produced 249 lb in 305 days.

Milk production in dairy goats parallels that in dairy cattle, only on smaller scale. Thus, the general nutrient requirements for dairy cattle apply to dairy goats.

Some other suggested rations for lactating dairy goats are found in Table 4–37.

TABLE 4–37
SUGGESTED RATIONS FOR LACTATING DAIRY GOATS[1]

	Ration[2]							
	1		2		3		4	
Ingredient	Alfalfa[3]		Alfalfa[3]		Legume and Grass[3]		Legume and Grass[3]	
	(lb)	*(kg)*	(lb)	*(kg)*	(lb)	*(kg)*		
Hay, free-choice								
Concentrate								
Cornmeal	800	*362.9*	—	—	350	*158.8*	—	—
Ground oats	400	*181.4*	400	*181.4*	200	*90.7*	250	*113.4*
Wheat bran	500	*226.8*	500	*226.8*	700	*317.5*	600	*272.2*
Linseed meal	200	*90.7*	200	*90.7*	—	—	—	—
Cottonseed meal	100	*45.4*	—	—	—	—	150	*68.0*
Gluten feed	—	—	200	*90.7*	750	*340.2*	—	—
Brewers' grains	—	—	—	—	—	—	600	*272.2*
Hominy	—	—	700	*317.5*	—	—	—	—
Corn-and-cob meal	—	—	—	—	—	—	400	*181.4*
Total	2,000	*907.2*	2,000	*907.2*	2,000	*907.2*	2,000	*907.2*

[1]Adapted by the author from *Dairy Goat Management*, Rutgers University Ext. Bull., 334.

[2]Provide minerals free-choice. A mixture of 50% trace mineralized salt and 50% dicalcium phosphate is a suitable mineral.

[3]The type of hay fed free choice.

• **Feeding the dairy doe**—Table 4–38 provides some suggested guidelines for feeding concentrates (grains and grain mixes) to lactating does, depending upon their daily milk production and the fat content of their milk.

TABLE 4–38
SUGGESTED AMOUNTS OF CONCENTRATE TO FEED LACTATING DOES[1]

Daily Milk Production[2]		Percentage of Fat in Milk											
		3.0		3.5		4.0		4.5		5.0		5.5	
(lb)	*(kg)*	(lb)	*(kg)*	(lb)	*(kg)*	(lb)	*(kg)*	(lb)	*(kg)*	(lb)	*(kg)*	(lb)	*(kg)*
1	*0.45*	0.5	*0.23*	1.0	*0.45*	1.0	*0.45*	1.0	*0.45*	1.0	*0.45*	1.0	*0.45*
2	*0.91*	1.0	*0.45*	1.5	*0.68*	1.5	*0.68*	1.5	*0.68*	1.5	*0.68*	1.5	*0.68*
3	*1.36*	1.5	*0.68*	1.5	*0.68*	2.0	*0.91*	2.0	*0.91*	2.0	*0.91*	2.0	*0.91*
4	*1.81*	2.0	*0.91*	2.0	*0.91*	2.0	*0.91*	2.5	*1.13*	2.5	*1.13*	2.5	*1.13*
5	*2.27*	2.5	*1.13*	2.5	*1.13*	2.5	*1.13*	3.0	*1.36*	3.0	*1.36*	3.0	*1.36*
6	*2.72*	2.5	*1.13*	3.0	*1.36*	3.0	*1.36*	3.5	*1.59*	3.5	*1.59*	3.5	*1.59*
7	*3.18*	3.0	*1.36*	3.5	*1.59*	3.5	*1.59*	4.0	*1.81*	4.0	*1.81*	4.0	*1.81*
8	*3.63*	3.5	*1.59*	4.0	*1.81*	4.0	*1.81*	4.5	*2.04*	4.5	*2.04*	5.0	*2.27*
9	*4.08*	4.0	*1.81*	4.0	*1.81*	4.5	*2.04*	5.0	*2.27*	5.0	*2.27*	5.5	*2.49*
10	*4.54*	4.0	*1.81*	4.5	*2.04*	5.0	*2.27*	5.0	*2.27*	5.5	*2.49*	6.0	*2.72*
11	*4.99*	4.5	*2.04*	5.0	*2.27*	5.5	*2.49*	5.5	*2.49*	6.0	*2.72*	6.5	*2.95*
12[3]	*5.44*	5.0	*2.27*	5.5	*2.49*	6.0	*2.72*	6.0	*2.72*	6.5	*2.95*	7.0	*3.18*

[1]Suggested with average-quality hay fed free-choice. Less grain can be used when high-quality forage is fed, and more hay may be needed with low-quality forage.

[2]One pt = 1.075 lb; 1 qt = 2.15 lb; 1 gal = 8.60 lb.

[3]Does producing 12 lb or more of milk may be fed to appetite.

PART V
FEEDING SWINE

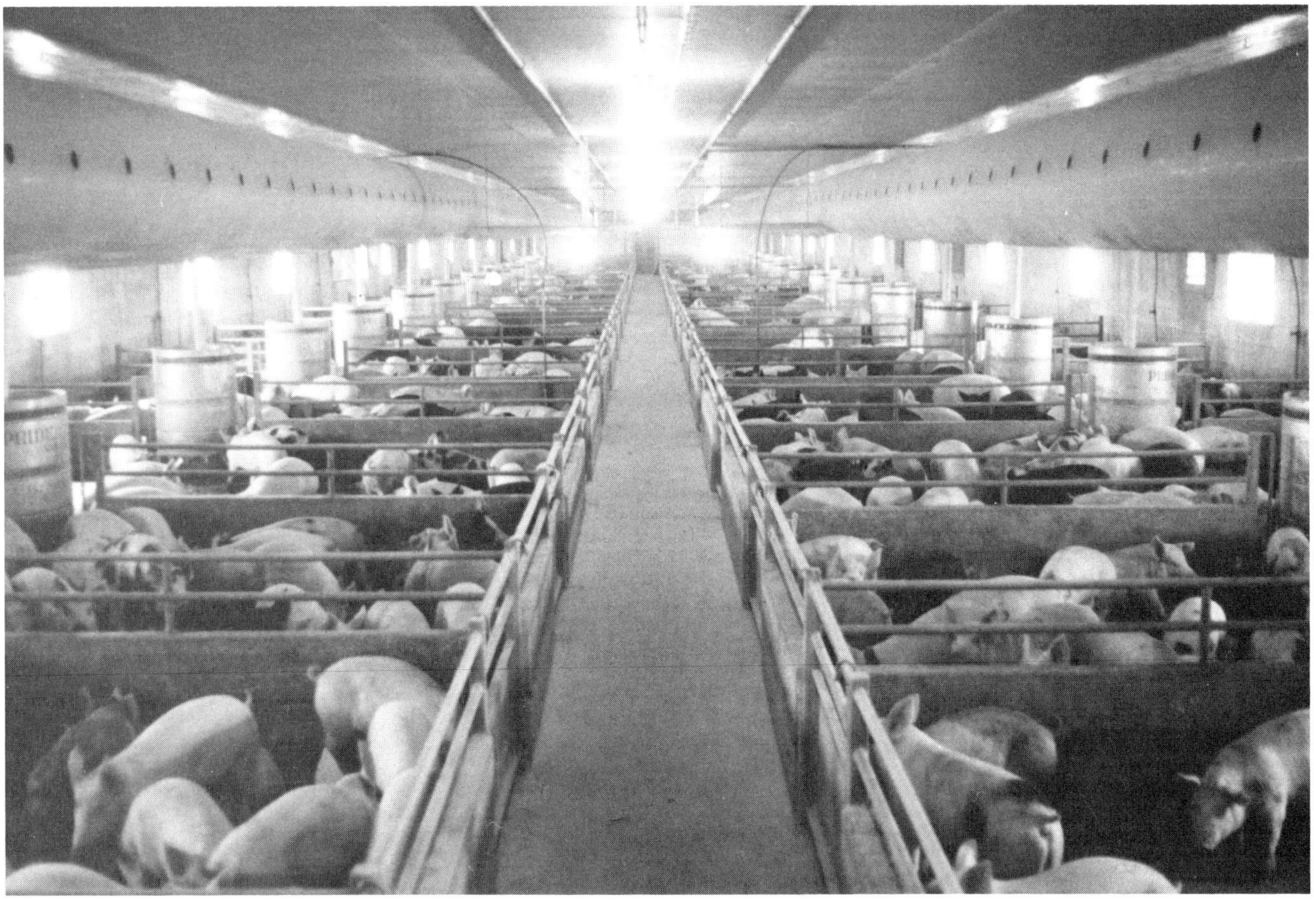

Fig. 4–66. Environmentally controlled hog finishing building. This unit holds 1,200 hogs and has a partially slotted floor and scraping gutter manure-handling system. (Courtesy, *National Hog Farmer*, St. Paul, MN)

By 1990, more than 60% of the market hogs in the United States were raised from farrow to finish in some type of confinement system—ranging all the way from simple shelters to environmentally controlled pig palaces. As a result of this confinement, domestic swine have less choice in their selection of feed than any other class of four-footed animals. For the most part they are able to consume only what the caretaker provides. This consists largely of concentrated feeds with only a small proportion of roughage. These conditions are made more critical because hogs grow much faster in proportion to their body weight than the larger farm animals, and they produce young at an earlier age. Thus, a knowledge of the nutritional needs of swine is especially important.

Extensive surveys indicate that about 25 to 30% of all pigs farrowed fail to live to weaning age. Although these heavy losses are due to many and variable factors, certainly nutritional deficiencies play a major role.

Knowledge of feeding swine is also important from an economic standpoint, because feed accounts for approximately 65 to 75% of the total cost of producing pork.

NUTRITIVE NEEDS OF SWINE

Swine differ in the kinds and amounts of nutrients needed. The need is influenced by age, function, disease level, nutrient interaction, environment, etc. It has been established that the pig has a requirement for over 40 individual nutrients. Fortunately, not all of them are of practical concern. It should be emphasized that the pig has a requirement for nutrients and not for particular ingredients.

A swine producer has a wide variety of ingredients from which to choose in formulating a ration. Each ingredient may contain several nutrients in varying amounts. Because ingredients vary in price, and in amount and quality of nutrients contained, judgment must be exercised in the choice made.

The nutrient present in the largest amount determines how an ingredient is classified. Thus, soybean meal is classified as a protein supplement because it is high in protein content. There are five main categories of nutrients: (1) energy (carbohydrates and fats), (2) protein (amino acids), (3) minerals, (4) vitamins, and (5) water. Their functions may be described as energy producing, structural, or regulatory.

No one nutrient is more important than another—and all are essential. Each nutrient has one or more particular and specific functions to perform in the body. If the nutrient is not supplied by the ration in proper amounts, the functions (growth, reproduction, lactation, etc.) will be impaired. Since the modern swine producer is interested in maximum performance, it follows that the input of nutrients must be ample to bring this about.

• **Recommended allowances**—The recommended allowances of the various important nutrients that should be included in ration formulation for optimum performance are given in Table 4–39.

Sometimes nutritional requirements, or standards, like those in Table 4–39, impart the erroneous impression that such figures are absolute, final and unchangeable. Nothing could be further from the truth. Rather, Table 4–39 figures are guides, prepared by swine specialists of Iowa State University, the Land Grant University in the leading swine producing state in the nation.

In using Table 4–39, the following pertinent points should be recognized:

1. Feedstuffs produced in various parts of the country vary in nutritive value.

2. The environment in which pigs are produced can modify the requirements.

3. Animals bred for high performance have nutritional needs that are quite different from average performers.

TABLE 4–39
RECOMMENDED NUTRIENT ALLOWANCES FOR SWINE[1,2,3]

	Sows, Gilts and Boars				Young Pigs		Grower-Finisher Pigs	
	Pregestation, Breeding, and Gestation			Lactation	Prestarter, Nursing to 12 lb	Starter, Creep to 40 lb	Grower, Replacements 40–120 lb	Finisher 120–240 lb
	3 lb/day	4 lb/day	5 lb/day					
Protein, amino acids:								
Protein[4] (%)	13.00	12.00	11.00	13.00	20–24	18–20	15–17	13–15
Lysine[4] (%)	0.60	0.45	0.35	0.60	1.40	1.15	0.75–0.85	0.60–0.70
Threonine (%)	0.40	0.30	0.24	0.45	0.80	0.70	0.50	0.40
Tryptophan (%)	0.13	0.10	0.08	0.12	0.20	0.18	0.13	0.11
Major or Macrominerals:								
Salt (NaCl) (%)	0.50	0.40	0.30	0.50	0.25	0.25	0.25	0.25
Calcium (Ca) (%)	1.00	0.75	0.60	0.75	0.90	0.80	0.60	0.55
Phosphorus (P) (%)	0.75	0.60	0.50	0.60	0.70	0.65	0.50	0.45
Trace or Microminerals, added:[5]								
Copper (Cu) (ppm)	7	5	4	5	8	8	4	2
Iodine (I) (ppm)	0.20	0.14	0.11	0.14	0.14	0.14	0.07	0.04
Iron (Fe) (ppm)	100	80	60	80	100	100	50	25
Manganese (Mn) (ppm)	13	10	8	10	4	4	2	1
Selenium (Se) (ppm)	0.20	0.15	0.12	0.15	0.30	0.30	0.15	0.08
Zinc (Zn) (ppm)	65	50	40	50	100	100	50	25
Vitamins, added:[5]								
Vitamin A (IU/lb)	2,500	2,000	1,500	2,000	2,000	2,000	1,000	500
Vitamin D (IU/lb)	250	200	150	200	200	200	100	50
Vitamin E (IU/lb)	13	10	8	10	10	10	5	2.5
Biotin[6] (mg/lb)	0.13	0.10	0.08	0.10	0	0	0	0
Niacin (Nicotinic Acid, Nicotinamide) (mg/lb)	7	5	4	5	10	10	5	2.5
Pantothenic Acid (Vitamin B-3) (mg/lb)	8	6	5	6	8	8	4	2
Riboflavin (Vitamin B-2) (mg/lb)	3	2	1.5	2	2	2	1.5	0.8
Vitamin B-12 (Cobalamins) (mcg/lb)	10	7.5	6	7.5	10	10	5	2.5
Feed additives[7] (g/ton)	0–300	0–300	0–300	100–300	100–300	100–300	0–100	0–50

[1]Adapted by the author from *Life Cycle Swine Nutrition,* Iowa State University, Ames, PM–489, June, 1988.

[2]The nutrient allowances are suggested for maximum performance, not as minimum requirements. They are based on research work with natural feedstuffs and have been found to give satisfactory results. Trace mineral and vitamin levels listed should be added to the ration in addition to those occurring in natural feedstuffs.

[3]To convert lb to kg, divide by 2.2. To convert IU/lb to IU/kg, multiply by 2.2. To convert mg/lb to mg/kg, multiply by 2.2. To convert mcg/lb to mcg/kg, multiply by 2.2. To convert g/ton (short) to g/ton (metric), divide by 0.907.

[4]Sow protein recommendations are based on corn-soybean meal rations. Other feedstuffs may require more protein to meet the amino acid requirement. Protein and lysine ranges for growing and finishing hogs allow for least cost formulation per unit of gain.

[5]Trace mineral and vitamin recommendations for finishing pigs are 50% of grower values. To convert trace or microminerals from ppm to mg/lb, divide by 2.2.

[6]Biotin additions are not needed in corn-soybean meal based rations.

[7]The feed additives may be antibiotics, arsenicals or other chemotherapeutics or combinations. Levels and combinations used and stage of production for which they are used must comply with Food and Drug Administration regulations. High levels for sows may be beneficial just before and after breeding and at farrowing. They are not recommended during the entire gestation-lactation period unless specific diseases are present. The feed additive and the level used during growing and finishing phases should be primarily for growth promotion and improvement of feed efficiency.

ENERGY

Energy is the body's fuel supply. Every movement and activity of the pig's life involves the expenditure of fuel—energy for breathing, heart action, digestion, muscular movement, as well as heat to keep the body warm. If more energy is consumed than necessary to carry on vital functions, the excess is stored as body fat. In fact, this is what is done in finishing hogs. More energy is eaten than is needed for growth and body maintenance, with the result that the animal lays down fatty tissue with the excess.

The main nutrients supplying energy are carbohydrates. Thus, the kind of carbohydrate a feed contains determines its value as a source of energy for the pig. Cereal grains are widely used in swine feeding because of their very high NFE (60–70%) and low crude fiber content.

Another group of energy nutrients is the fats and oils. Fat, which is abundant in such common hog feeds as peanuts and soybeans, is a very concentrated source of fuel. It supplies approximately 2.25 times as much metabolizable energy as an equal weight of carbohydrates. Therefore, a feed high in fat, or a ration containing added fat, is much higher in energy value than a feed or ration low in fat. It is emphasized, however, that liberal qualities of either soybeans or peanuts will produce soft pork.

Although roughages are a good source of energy for ruminants, because of their bulky nature and the restricted size of the digestive tract of hogs in comparison with ruminants, only limited quantities of them are contained in normal swine rations. Roughages (pastures and ground legumes and hay) are added to swine rations because of their protein, minerals, and vitamins, rather than for energy purposes.

• **Symptoms of energy deficiency**—These are slow or interrupted growth, lowered reproduction, and offspring dead or weak at birth.

PROTEIN

While carbohydrates and fats are the principal sources of energy, proteins supply the building materials from which body tissue and many body regulators, such as enzymes and hormones, are made. Each protein is made up of several nitrogen compounds called amino acids.

The pig has a specific requirement for each of the essential amino acids. (See Part I of this section under the heading "Protein Needs" for a list of amino acids.) Since they are needed for the formation of every new cell, the need is most critical when growth is rapid. This makes ration formulation for the young pig very important, because the protein provided at this time must supply the amino acids for muscle growth (lean meat), internal organs, blood, bone, and all other parts associated with growth and development.

Quality of protein is a term used to describe the amino acid balance of proteins. A protein is said to be of good quality when it contains all the essential amino acids in proper proportions and amounts, and to be of poor quality when it is deficient in either content or balance of essential amino acids. From this it is evident that the usefulness of a protein source depends upon its amino acid composition, because the real need of the pig is for amino acids and not for protein as such.

Although it is common practice to refer to *percent protein* in a ration, this term has little significance in swine nutrition unless there is information about the amino acids present. For swine, quality is just as important as quantity. It is possible for pigs to perform better on a 12% protein ration, well-balanced for amino acids, than on a 16% protein ration having a poor amino acid balance.

• **Symptoms of protein (amino acid) deficiency**—These are reduced feed intake, stunted growth, poor hair and skin condition, and lowered reproduction.

MINERALS

Of all common farm animals, the pig is most likely to suffer from mineral deficiencies.

The functions, deficiency symptoms, interrelationships/toxicities, and sources of minerals for swine, which are similar to those of other animal species, are presented in Table 4–1, Animal Mineral Chart.

Table 4–39 presents recommended mineral allowances for swine.

Salt (sodium chloride), calcium, phosphorus, magnesium, potassium, and sulfur are the supplemental minerals needed in largest quantities by the pig. These are called *major* or *macrominerals.* Other minerals are required in small amounts and are known as *trace* or microminerals. The latter include cobalt, copper, iodine, iron, manganese, selenium, and zinc. Although minerals constitute a small percentage of the swine ration, their importance to the health and well-being of the pig cannot be minimized.

• **Method of feeding the mineral supplement**—Generally, minerals are incorporated in the ration. However, they may be provided by giving free access to a suitable mineral mix.

VITAMINS

Because of the greater prevalence of confinement feeding, swine are more likely to suffer from vitamin deficiencies than any other class of four-footed animals.

Vitamins are classified into two groups—fat-soluble and water-soluble. The body can store reserves of the fat-soluble vitamins for a considerable period of time. But stores of the water-soluble vitamins are depleted quite rapidly.

The *fat-soluble vitamins* of practical importance for swine are vitamins A, D, and E. Vitamin K may be of concern under some circumstances.

The *water-soluble vitamins* most likely to be deficient in swine rations are: biotin, niacin, pantothenic acid, riboflavin, and vitamin B-12.

The functions, deficiency symptoms, and good sources of vitamins for swine, which are similar to those of other animal species, are presented in Table 4–2, Animal Vitamin Chart.

Table 4–39 presents the recommended vitamin allowances for swine.

• **Unidentified factors**—Some unidentified factor or factors may, under certain circumstances, be involved in securing optimum results during the critical periods (early growth and gestation-lactation). Sources of the unknown factor or factors are: distillers' dried solubles, fish solubles, dried whey, grass juice concentrate, soil, alfalfa meal, brewers' dried yeast, liver, and pasture.

WATER

Water is so common that it is seldom thought of as a nutrient. However, it is the largest single part of nearly all living things. The body of a baby pig is about three-fourths water.

In general, swine will consume ¼ to ⅓ gal. of water for every pound of dry feed. The higher the temperature, the greater the water consumption. It is preferable that swine have access to automatic waterers, with cool, clean water available at all times. Otherwise, they should be hand watered at least twice daily. During winter, the drinking water should not be permitted to fall below 40°F.

FEEDS FOR SWINE

Throughout the world, swine are raised on a great variety of feeds, including numerous by-products. Except when on pasture or when ground dry forages are incorporated in the ration, they eat relatively little roughage; only 4.3% of the total feed consumed by swine in the United States is derived from roughages (see Table 4–4).

Although corn is the chief concentrate fed to swine, the agriculture of the 50 states is very diverse, and the diet of the pig is readily adapted to the feeds produced locally. A similar adaptation in feeding practices is found in other countries. Thus, in most sections of the world, swine are fed predominantly on homegrown feeds. Ireland depends largely upon potatoes and dairy by-products; the swine industry of Denmark has been built up to augment the dairy industry, with milk and whey supplementing homegrown and imported cereals (mostly barley); in East and West Germany, the pig is fed on such crops as potatoes, sugar beets, and green forage; and in China, the pig is primarily a scavenger, competing very little for grains suitable for human consumption.

ENERGY FEEDS

Carbohydrates and fats may be classed together as energy feeds for swine. Three essential fatty acids—linoleic, linolenic, and arachidonic acid—are required by swine, but cereal grains contain adequate quantities of fat to meet the fatty acid requirements. The ingredients commonly used as sources of energy are corn, barley, sorghum, wheat, and fats and oils. Full replacement of corn with barley, oats, sorghum, or wheat will decrease dressing percentage by approximately 1% and decrease backfat by about 0.1 in.

In this country, corn and swine production have always gone hand in hand. Normally, about one-fourth of the U.S. corn crop is fed to hogs. Because of the dominant position of corn as a swine feed, it is herein singled out for further elucidation.

In spite of its virtues, corn alone will not keep pigs alive. It contains 7 to 9% protein, but the protein is deficient in some of the essential amino acids required by the weaning pig, especially lysine and tryptophan (see Fig. 4–67). It is also so deficient in calcium and other minerals, and so inadequate in vitamin content, that pigs will die if they are limited to a ration containing only corn. So, corn must be supplemented with a protein that makes up its amino acid deficiencies. Equally important are the needed minerals and vitamins. When properly supplemented, corn is an excellent energy feed for all classes of swine.

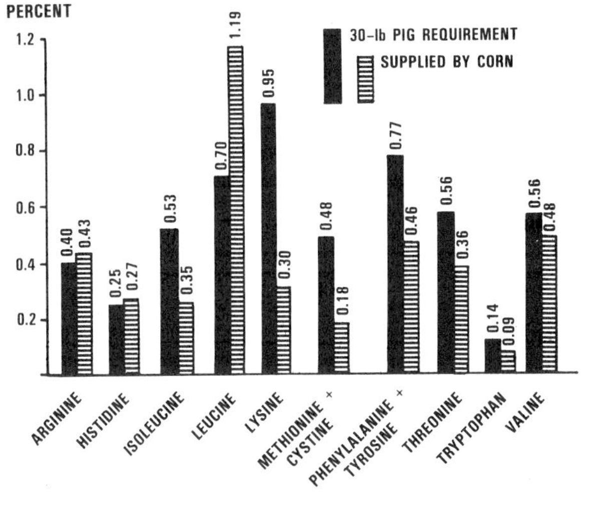

Fig. 4–67. Hogs cannot live by corn alone! This shows the amino acid content of corn (in %) in comparison with the ration requirements (in %) of a 30-lb pig. Note that corn is especially deficient in lysine and tryptophan. (Sources of data: Amino acid ration requirements (in % as-fed) of a 30-lb pig from *Nutrient Requirements of Swine*, 9th rev. ed., National Research Council, National Academy Press, 1988, p. 50, Table 5–1.)

PROTEIN FEEDS

Protein is made up of nitrogenous compounds called amino acids. Protein feeds vary in the kind and amount of amino acids they contain. During the digestion process, the protein in feed is broken down into the various amino acids and the pig recombines them into the kind of protein needed for muscle development, repair of worn-out tissue, etc. Thus, the real need of the pig is for amino acids, not protein as such.

The pig can synthesize some of the amino acids, with the result that they are not required in the diet. However, 10 of the amino acids are termed *essential,* because the body cannot manufacture them in sufficient quantity to permit maximum growth and performance. It is important that ingredients rich in the essential amino acids be used in formulating the ration.

Soybean meal is by far the leading high-protein supplement of the United States. About 75% total high-protein animal feed (including both oilseed meals and animal proteins) consists of soybean oil.

PASTURES

Fig. 4–68. Pigs on pasture. Although more than 60% of the market hogs in the U.S. are raised in confinement, the use of pasture is still a viable alternative. (Courtesy, National Spotted Swine Record, Inc., Bainbridge, IN)

Pasture was formerly thought to be an absolute essential for a successful swine operation. However, in recent years producing hogs in confinement has become a reality because of vastly improved rations, along with greater disease and parasite control. But it is still possible to utilize large amounts of forage effectively for the breeding herd.

Over much of the country, alfalfa, the clovers (principally ladino, red, alsike, and sweet clover), rape, and oat mixtures make excellent swine pastures. Other plants which find use as hog pastures in certain areas and under certain conditions include bluegrass, bromegrass, orchardgrass, lespedeza, carpetgrass, rye, wheat, soybeans, cowpeas, field beans, sorghum, and Sudangrass.

ADDITIVES

Certain additives have become standard ingredients of swine rations, especially for pigs from birth to market weight. They are not nutrients as such; hence, they should not be considered as dietary essentials. Although many different additives are used in swine rations, antibiotics and sulfas are most common.

• **Antibiotics**—Antibiotics are widely used as feed additives to stimulate growth, improve feed efficiency, secure uniformity of performance, and control infections. The response secured from their use depends on (1) the age of the pig, (2) the sanitary conditions, (3) the level fed, (4) the health and environment of the animal, (5) the type of ration, and (6) the season of the year. When fed to young, unthrifty pigs, antibiotics have increased growth rate by over 200%. For growing-finishing hogs under good sanitary conditions, antibiotics generally result in about 10% faster gains on 5% less feed. Pigs up to 100 lb weight give the greatest response to antibiotic feeding. Experimental results have been inconsistent relative to the value of antibiotics in brood sow rations, but it appears that breeding herds with a high disease level may show a favorable response.

In addition to antibiotics, certain other antimicrobial compounds can be used as feed additives, either (1) at low levels for promotion of growth and improvement in feed efficiency, or (2) for treatment and prevention of disease. Among these are nitrofurans, sulfonamides, copper, and arsenicals. Such compounds, alone or in combination with other antimicrobial compounds, should be used only at approved levels and for the specific purpose for which they are authorized.

CAUTION: Toxic residues in meat, milk, and/or eggs may result from the improper use of certain antimicrobial compounds, pesticides, tranquilizers, and other chemicals. So, follow the directions of the manufacturer.

• **Sulfonamides (sulfas)**—*Sulfonamides are organic compounds with bacterial and growth promotant properties similar to those of the antibiotics.*

Of the more than 5,000 sulfa compounds that have been synthesized, only two—sulfamethazine and sulfathiazole—are approved for use in swine feeds. Two additional sulfa drugs, sulfamerazine and sulfapyridine, are approved, along with sulfamethazine and sulfathiazole, for use as water medicants for swine.

When used properly and withdrawn at the specified time before marketing, sulfa drugs have been shown to be safe.

• **Porcine somatotropin (PST) for swine**—Porcine somatotropin (PST) is the scientific name for the growth hormone in swine. It is normally produced by the anterior pituitary at the base of the brain. It stimulates protein synthesis and growth in most tissues of the body, and it causes breakdown of fat deposits in adipose tissue.

Experiments have consistently shown that PST can make the following great leaps forward in the swine industry: 15–20%

higher average daily gain, 20–30% improvement in feed conversion efficiency, 10–15% improvement in muscle mass, and 25–30% less backfat.

Before PST can be used, it must have FDA approval. Then its acceptance and use will be determined by a practical method of administering the product, and producer and consumer acceptance.

CREEP FEEDING PIGS

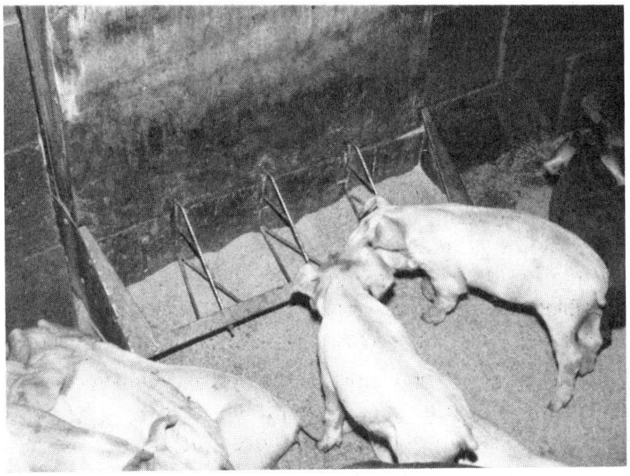

Fig. 4-69. Fig. 4-96. Creed fed pigs, feeding from a small self-feeder. (Courtesy, Watt Publishing Co., Mount Morris, IL)

Baby pigs should have access to a creep feed beginning at 7 to 10 days of age. Commercial milk replacers (prestarters) and starters are readily available, or farm-mixed creep rations can be used (see Table 4–45 (p. 206) for suggested prestarter rations, and see Table 4–46 (p. 206) for suggested starter rations). A prestarter ration is normally fed to pigs weaned before 3 weeks of age and/or until they weigh approximately 12 lb, after which they can be switched to a starter ration. Also, a prestarter ration may be used for orphan pigs, when extreme disease outbreaks occur, or when the sow fails to produce sufficient milk. They should be fed a starter ration until they weigh about 40 lb. Pigs should receive a total of 3 to 5 lb of milk replacer (prestarter), after which they should be switched to starter until they weigh about 40 lb.

WEANING PIGS

The optimum age to wean pigs varies considerably, depending on nutritional programs, facilities, environment, health, and management. The average age at weaning for pigs in the United States is about 4 weeks, with a range of 2 to 7 weeks. The earlier pigs are weaned, the better the required feeding and management practices.

For best results, the guidelines given in Table 4–40 should be observed when planning for early weaning (at 3 to 4 weeks of age).

TABLE 4–40
GUIDELINES TO SUCCESSFUL EARLY WEANING[1]

Guideline	Age in Weeks				
	1	2	3	4	5
Minimum pig weight (lb)	5	9	12	15	21
Nursery temperature at pig level (°F)	85	85	83	81	79
Minimum floor space per pig (sq ft)[2]	3	3	3	3	3
Maximum number of pigs per linear foot of feeding space	5	5	5	5	5
Maximum number of pigs per nipple waterer[3]	8	8	8	8	8
Maximum number of pigs per group	10	10	10	15	25

[1]See Section 20, "Weights and Measures" for conversion of U.S. customary to metric.

[2]The figures given herein are for solid floors. On slotted floors, this may be lowered 2 sq ft per pig from 1 to 5 weeks of age.

[3]Where bowls are used instead of nipples, there should be one bowl for each 12 pigs.

FEEDING GROWING-FINISHING PIGS

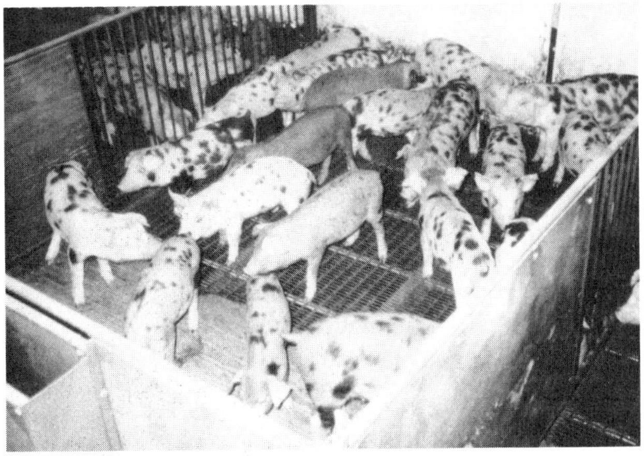

Fig. 4–70. Growing-finishing pigs self-fed on a partially slotted floor. (Courtesy, McLean County Hog Service, Inc., LeRoy, IL)

In the practical swine enterprise, growing-finishing generally refers to that period from weaning to market weight of about 240 lb. Because hogs are finished at an early age, the process really consists of both growing and finishing. In a general way, there are two methods of finishing hogs for market: (1) full feeding all the time until the animals attain market weight, and (2) limited feeding early in the period, with full feeding the last 60 to 70 days of the period before marketing.

When on full feed, finishing pigs will consume 5 to 6 lb of feed daily per 100 lb liveweight up to 120 lb in weight. From 120 lb to a finished weight of 240 lb, pigs on full feed consume about 4 lb of feed daily for each 100 lb of liveweight. With good feed and management, about 700 lb of feed are required to produce 200 lb of gain during the growing-finishing period from 40 lb to 240 lb weight. This means that 3.5 lb of feed are required to produce 1 lb of gain during the growing-finishing stage, from 40 lb weight to 240 lb weight (700 ÷ 200 = 3.5 lb).

Suggested growing-finishing rations are given in Table 4–47 (p. 207). Note that provision is made for meeting the nutritional needs at two different stages of growth—40 to 120 lb, and 121 to 240 lb—with three different rations suggested for each stage.

FEEDING PROSPECTIVE BREEDING GILTS

Prospective breeding gilts should be kept from getting too fat. Meat-type animals can usually be left on a high-energy ration until they reach 175 to 200 lb without becoming too fat. It is neither necessary nor desirable that females intended for breeding purposes carry the same degree of finish as market animals. After selecting replacement gilts, they should be fed as follows:

1. Give about 5 lb per day through their second heat period.

2. Flush—full feed—after the second heat period until breeding on the third heat period.

3. After breeding, limit the feed intake to 3 to 5 lb per day. Overfeeding during gestation can cause embryonic death and thus decrease litter size.

FEEDING BOARS

The feed requirements of the herd boar are abut the same as those of a female of equal weight. He should always be kept in thrifty, vigorous condition and virile. In no case should boars be overfat, nor should they be in a thin run-down condition. Normally, the following feed allowances will suffice: for boars weighing 120 to 150 lb, 6 to 9 lb of feed daily; for mature boars, 5 to 7 lb of feed daily. A more liberal ration must be provided in the wintertime and when the sire is in heavy service. The feed allowance should be varied with the age, development, temperament, breeding demands, and roughage consumed.

Boars and pregestating/gestating sows may be fed the same rations. Rations, formulated for feeding 4 lb per head daily are presented in Table 4-43, p. 205.

FEEDING BROOD SOWS

The nutrition of brood sows is critical, for it may materially affect conception, reproduction, and lactation. Proper feeding of sows should begin with replacement gilts and continue through each stage of the breeding cycle—finishing, gestation, farrowing, and lactation.

• **Flushing sows**—*The practice of conditioning or having the sows gain in weight just prior to breeding is known as flushing.* The purpose of flushing is to increase the number of ova shed during estrus. About 10 to 14 days prior to expected breeding, the female should be fed a ration that will make for gains of 1 to 1¼ lb per day. Generally 6 to 8 lb per day of a high-energy, 14 to 16% protein feed that is well balanced in minerals and vitamins, is adequate. Immediately after breeding, the females should be put back on limited feeding. Continuation of high level of feeding after breeding will result in a higher embryo mortality.

• **Gestation period**—The nutrients fed the pregnant gilt or sow must first take care of the usual maintenance needs. If the gilt is not fully mature, nutrients are required for both maternal growth and growth of the fetus. Quality and quantity of proteins, minerals, and vitamins become particularly important in the ration of young pregnant gilts, for their requirements are much greater and more exacting than those of the mature sow.

It is important that the condition of dry sows should be regulated so that they are neither too fat nor too thin at farrowing time. Overly fat sows may have difficulty in farrowing and give birth to weak or dead pigs. Sows that are too thin at farrowing tend to become suckled down during lactation. Thus, one way or another, limited feeding is a must for gestating gilts and sows. This may be accomplished by any one of the following feeding systems (these feeding systems are further detailed later in this chapter in the section headed "Feeding Systems").

1. By adding sufficient bulk.

2. By interval feeding.

3. By group hand feeding.

4. By individual feeding.

Suggested gestation rations, formulated for feeding 4 lb per head daily are given in Table 4-43, p. 205.

• **Farrowing time**—It is considered good practice to feed lightly and with bulky laxative feeds from 4 to 5 days before and after farrowing. Wheat bran or oats may constitute half of the limited ration, and a small amount of linseed meal may be added.

The sow may be watered at frequent intervals before or after farrowing, but in no event should she be allowed to gorge. It is also a good plan to take the chill off the water in the wintertime.

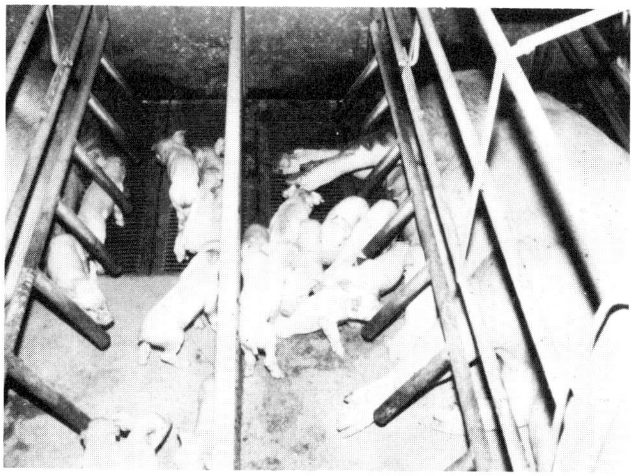

Fig. 4-71. Life for pigs generally begins with the protection of the farrowing crate. (Courtesy, Iowa State University, Ames)

• **Lactation period**—The nutritive requirements of a lactating sow are more rigorous than those during gestation. They are very similar to those of a milk cow, except they are more exacting relative to the quality of proteins and the B vitamins because of the absence of rumen synthesis in the pig. A good lactating sow will produce an average of 10 to 15 lb of milk daily during the suckling period. A sow's milk is also richer than cow's milk in all nutrients, especially in fat. Thus, sows suckling litters need a liberal allowance of concentrates rich in protein, calcium, phosphorus, and vitamins.

Suggested lactation rations are given in Table 4–44 (p. 205).

FEEDING ORPHAN PIGS

There is no replacement for the sow's colostrum. If the newborn pig does not receive colostrum, it has a lesser chance for survival. An orphan pig can obtain colostrum by being placed with another sow (a foster sow) that has just farrowed. If no such sow is available, the orphan can be fed a commercial milk replacer, several good ones of which are on the market. A homemade milk replacer can be prepared by mixing the following ingredients:

1 quart milk
1 pint half-and-half
1 raw egg
Oral, water-soluble antibiotic

Portions of this mixture should be warmed to 98 to 100°F and fed about every 3 hours, with each pig receiving about ¼ cup per feeding. The use of a shallow pan for feeding is recommended. Immersing the baby pig's nose in the milk a few times will result in its drinking readily. It is extremely important that all feeding utensils be kept clean and sanitary; otherwise, scouring will occur.

Orphan pigs can be fed a dry 22 to 23% crude protein prestarter from 5 to 7 days of age until about 2 to 3 weeks of age (see Table 4–45, p. 206). At this time, they can be switched to a 20 to 21% crude protein pig starter (see Table 4–46, p. 206).

CORN-HOG RATIO

The corn-hog ratio refers to the number of bushels of corn required to be equivalent in value to 100 lb of live hogs at local markets, based on average prices received by farmers for corn and hogs. During the 14-year period 1973 through 1987, the corn-hog ratio averaged 19.3. This means that the price relationship was such that 19.3 bushels of corn equalled in value 100 lb of hogs.

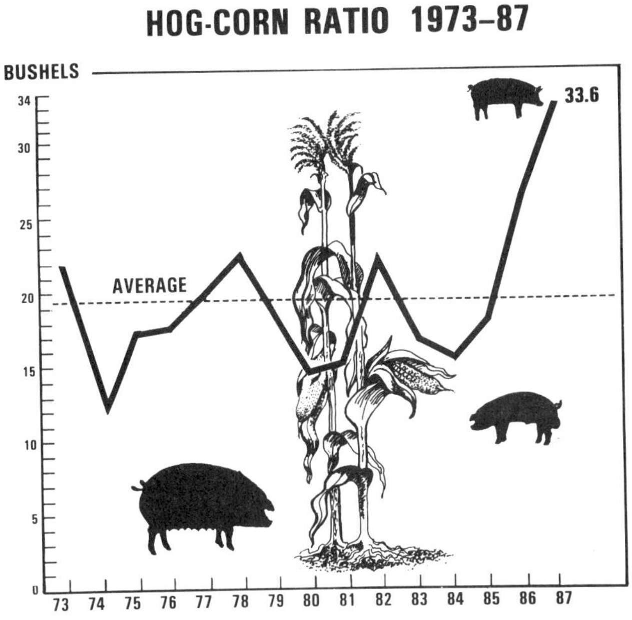

Fig. 4–72. The corn-hog ratio, 1973 to 1987. As shown, it averaged 19.3. (Based on data from *Agricultural Statistics, 1988*, USDA, p. 279, Table 414.

A high corn-hog ratio—one above 19.3 in recent years—indicates cheap corn and high-priced hogs and likely profit to the producer—conditions that stimulate more breeding and more feeding to heavier weights. On the other hand, a low ratio, one which is below 19.3, means high-priced corn and low-priced hogs—conditions that result in less breeding and feeding of swine.

It is noteworthy that the corn-hog ratio averaged 12.6 for the period 1930 to 1981, whereas it averaged 19.3 for the period 1973 to 1987.

FEED ALLOWANCES AND SOME SUGGESTED RATIONS FOR SWINE

In most instances, the nutrient allowances should be higher than the minimum requirements established by the national Research Council. This is desirable to obtain maximum performance and reduce the risk of nutrient deficiencies that might occur because of the differences in ingredient quality, environment, health, genetics, and performance of individual animals.

Following the selection of feed ingredients, they may be fed to swine in either combined (complete) form or cafeteria style, and they may be purchased commercially or home mixed.

• **Recommended nutrient allowances of the swine specialists at Iowa State University**—In this chapter, the author presents the nutrient allowances, along with premixes and rations, of the swine specialists of Iowa State University, the Land Grant University in the state of Iowa, the leading swine state in the nation.

The recommended allowances and rations formulated by the swine specialists at Iowa State University consist of two premixes (Tables 4–41 and 4–42) and five complete ration tables (Tables 4–43, –44, –45, –46, and –47. This makes for a popular and convenient method of building swine rations by blending the proper premix with the main ingredients of the ration.

Normally, premixes should be purchased from a commercial company, because they have much better quality control and mixing facilities to handle the small quantities of minerals and/or vitamins required.

TABLE 4–41
COMPOSITION AND ANALYSIS OF TRACE MINERAL PREMIX[1]

Element	Source[2]	Amount		Percent in Premix	Parts Per Million When Added to a Complete Ration At The Following Pounds Per Ton:			
					2	3	4	5
		(lb)	*(kg)*	(%)	(ppm)	(ppm)	(ppm)	(ppm)
Copper (Cu)	Copper sulfate	1.500	*0.681*	0.38	4	6	8	10
Iodine (I)	Potassium iodide[3]	0.010	*0.005*	0.008	0.08	0.11	0.15	0.19
Iron (Fe)	Ferrous sulfate	25.000	*11.350*	5.03	50	75	101	126
Manganese (Mn)	Manganese sulfate	2.500	*1.135*	0.57	6	9	11	14
Selenium (Se)	Sodium selenite[3]	0.025	*0.011*	0.011	0.11	0.17	0.23	0.29
Zinc (Zn)	Zinc sulfate	25.000	*11.350*	5.68	57	85	114	142
	Carrier	45.965	*20.868*					
	Total	100.000	*45.400*					

[1]Adapted by the author from *Life Cycle Swine Nutrition,* Iowa State University, Ames, PM-489, June, 1988.

[2]Other sources of trace minerals may be substituted. Iodine may be omitted if iodized salt is used.

[3]Iodine and selenium probably will be added in a separate premix form.

TABLE 4–42
COMPOSITION OF VITAMIN PREMIX[1 2]

Vitamins	Amount	Units Per Pound of Complete Ration When Added At The Following Pounds Per Ton:				
		Unit	3	5	8	10
Essential[3]						
Vitamin A (million IU)	5.0	(IU)	750.00	1,250.00	2,000.00	2,500.00
Vitamin D (million IU)	0.6	(IU)	90.00	150.00	240.00	300.00
Vitamin E (thousand IU)	26.0	(IU)	3.90	6.50	10.40	13.00
Niacin (nicotinic acid, nicotinamide) (g)	25.0	(mg)	3.75	6.25	10.00	12.50
d-Pantothenic acid (vitamin B-3) (g)	20.0	(mg)	3.00	5.00	8.00	10.00
Riboflavin (vitamin B-2) (g)	6.0	(mg)	0.90	1.50	2.40	3.00
Vitamin B-12 (cobalamins) (mg)	25.0	(mcg)	3.75	6.25	10.00	12.50
Optional[4]						
Biotin (g)	0.3	(mg)	0.05	0.08	0.12	0.15
Menadione (source of vitamin K) (g)	4.0	(mg)	0.60	1.00	1.60	2.00
Carrier	?					
Total[5] (lb)	10.0					

[1]Adapted by the author from *Life Cycle Swine Nutrition,* Iowa State University, Ames, PM-489, June, 1988.

[2]A feed additive may be included in the vitamin premix.

[3]Most natural feedstuffs contain very little vitamin D or B-12. The amount of provitamin A (beta-carotene) in feedstuffs will depend on processing and storage, while niacin in most grains is relatively unavailable for swine. Riboflavin and pantothenic acid in natural feedstuffs can meet part of the requirement.

[4]Supplemental biotin is not necessary with corn-soybean meal based rations. It should be included in sow rations based on other grains. The vitamin K requirement is normally met by the level present in natural feedstuffs and by intestinal synthesis. A hemorrhagic or bleeding syndrome has been diagnosed which is probably due to a vitamin K antimetabolite. The antimetabolite is thought to be produced by mold occurring in one or more of the ration ingredients. When this has occurred, adding menadione has been helpful in preventing or overcoming the problem.

Suggested rations are shown for each of the following classes of swine:

Table 4–43, Pregestation, Breeding, and Gestation Rations—for boars, sows, or gilts fed 4 lb per day.

Table 4–44, Lactation Rations.

Table 4–45, Prestarter Rations—for baby pigs before 3 weeks of age.

Table 4–46, Starter Rations.

Table 4–47, Grower-Finisher Rations—for 40 to 240 lb.

As previously indicated, the rations shown in Tables 4–43 to 4–47 call for the use of the mineral and vitamin premixes shown in Tables 4–41 to 4–42.

TABLE 4–43
PREGESTATION, BREEDING, AND GESTATION RATIONS (FOR BOARS, SOWS, OR GILTS FED 4 LB *(1.8 KG)* PER DAY)[1 2]

Ingredient	Ration #							
	1		2		3		4	
	(lb)	*(kg)*	(lb)	*(kg)*	(lb)	*(kg)*	(lb)	*(kg)*
Corn, yellow (8.4% protein)[3]	1,769	*803*	1,692	*768*	1,767	*802*	1,692	*768*
Soybean meal, solvent extracted (44.0% protein)[4]	160	*72*	140	*64*	90	*41*	68	*31*
Alfalfa meal, dehydrated (17.0% protein)			100	*45*			100	*45*
Meat and bone meal (50.0% protein)					100	*45*	100	*45*
Dicalcium phosphate	36	*16*	36	*16*	16	*7*	16	*7*
Limestone	15	*7*	12	*5*	7	*3*	4	*2*
Iodized salt	8	*4*	8	*4*	8	*4*	8	*4*
Trace mineral premix (Table 4–41)	4	*2*	4	*2*	4	*2*	4	*2*
Vitamin premix (Table 4–42)	8	*4*	8	*4*	8	*4*	8	*4*
Feed additives[5]								
Total	2,000	*908*	2,000	*908*	2,000	*908*	2,000	*908*
Calculated analyses:								
Metabolizable energy (kcal/lb)	1,493		1,457		1,492		1,456	
Metabolizable energy *(kcal/kg)*	*3,285*		*3,205*		*3,282*		*3,203*	
Protein (%)	10.95		11.04		11.90		11.95	
Lysine (%)	0.45		0.45		0.46		0.46	
Threonine (%)	0.41		0.41		0.43		0.43	
Tryptophan (%)	0.11		0.12		0.10		0.11	
Calcium (Ca) (%)	0.75		0.76		0.75		0.76	
Phosphorus (P) (%)	0.60		0.60		0.60		0.60	

[1]Adapted by the author from *Life Cycle Swine Nutrition,* Iowa State University, Ames, PM–489, June, 1988.

[2]These rations can be used for gilts on pasture during gestation since they require 3 to 4 lb *(1.4 to 1.8 kg)* of feed daily. These rations can also be used for interval-fed sows or gilts if the average daily intake is approximately 4 lb *(1.8 kg)*.

[3]Ground oats can replace corn up to 20% of the total ration. Ground milo, wheat, or barley can replace the corn.

[4]To replace 44% soybean meal with 47% soybean meal or whole soybeans, use the following ratios:

Each 100 lb *(45 kg)* SBM (44%) = 93 lb *(42 kg)* SBM (47%) + 7 lb *(3 kg)* corn.

Each 100 lb *(45 kg)* SBM (44%) + 35 lb *(16 kg)* corn = 135 lb *(61 kg)* whole soybeans.

[5]Feed additives are not generally recommended during gestation or for gilts during the developer period after selection unless specific disease problems exist. High levels (100–300 g/ton) may be beneficial when fed 2 weeks before and after breeding and 2 weeks before farrowing.

TABLE 4–44
LACTATION RATIONS[1 2]

Ingredient	Ration #							
	1		2		3		4	
	(lb)	*(kg)*	(lb)	*(kg)*	(lb)	*(kg)*	(lb)	*(kg)*
Corn, yellow (8.4% protein)	1,658	*752*	1,545	*701*	1,580	*717*	1,469	*667*
Soybean meal, solvent extracted (44.0% protein)[3]	270	*122*	283	*128*	177	*80*	260	*118*
Fat or oil source			100	*45*				
Meat and bone meal (50.0% protein)					100	*45*		
Oats (11.5% protein)							200	*90*
Beet pulp (8.0% protein)					100	*45*		
Dicalcium phosphate	35	*16*	35	*16*	15	*7*	33	*15*
Limestone	15	*7*	15	*7*	6	*3*	16	*7*
Iodized salt	10	*5*	10	*5*	10	*5*	10	*5*
Trace mineral premix (Table 4–41)	4	*2*	4	*2*	4	*2*	4	*2*
Vitamin premix (Table 4–42)	8	*4*	8	*4*	8	*4*	8	*4*
Feed additives[4]								
Total	2,000	*908*	2,000	*908*	2,000	*908*	2,000	*908*
Calculated analyses:								
Metabolizable energy (kcal/lb)	1,487		1,409		1,468		1,457	
Metabolizable energy *(kcal/kg)*	*3,271*		*3,100*		*3,230*		*3,205*	
Protein (%)	12.90		12.72		13.43		13.04	
Lysine (%)	0.60		0.60		0.60		0.60	
Threonine (%)	0.49		0.49		0.48		0.49	
Tryptophan (%)	0.14		0.14		0.13		0.15	
Calcium (Ca) (%)	0.75		0.76		0.76		0.76	
Phosphorus (P) (%)	0.61		0.60		0.60		0.60	

[1]Adapted by the author from *Life Cycle Swine Nutrition,* Iowa State University, Ames, PM–489, June, 1988.

[2]These rations may be limit-fed from a few days before farrowing and full-fed during lactation.

[3]To replace 44% soybean meal with 47% soybean meal or whole soybeans, use the following ratios:

Each 100 lb *(45 kg)* SBM (44%) = 93 lb *(42 kg)* SBM (47%) + 7 lb *(3 kg)* corn.

Each 100 lb *(45 kg)* SBM (44%) + 35 lb *(16 kg)* corn = 135 lb *(61 kg)* whole soybeans.

[4]High levels of feed additives (100–300 g/ton) may be beneficial when fed 2 weeks before and after breeding and 2 weeks before farrowing.

TABLE 4-45
PRESTARTER RATIONS (FOR BABY PIGS BEFORE 3 WEEKS OF AGE)[1 2]

Ingredient	Ration # 1 (lb)	1 *(kg)*	2 (lb)	2 *(kg)*	3 (lb)	3 *(kg)*	4 (lb)	4 *(kg)*	5 (lb)	5 *(kg)*
Corn, yellow (8.4% protein)	917	*416*	878	*399*	866	*393*	621	*282*	774	*351*
Soybean meal, solvent extracted (44.0% protein)					635	*288*				
Soybean meal, solvent extracted, dehulled (47.0% protein)	680	*309*	580	*263*			580	*263*	725	*329*
Oat groats (dehulled oats) (16.0% protein)							200	*91*		
Skim milk, dried (33.0% protein)	100	*45*	100	*45*	200	*91*	200	*91*		
Whey, dried (12.0% protein)	200	*91*	300	*136*	200	*91*	200	*91*	400	*182*
Fish meal, menhaden (62.0% protein)			50	*23*						
Sugar							100	*45*		
Fat or oil source	40	*18*	40	*18*	40	*18*	40	*18*	40	*18*
Dicalcium phosphate	26	*12*	19	*9*	25	*12*	25	*12*	25	*12*
Limestone	19	*9*	15	*7*	16	*7*	16	*7*	18	*8*
Iodized salt	5	*2*	5	*2*	5	*2*	5	*2*	5	*2*
Trace mineral premix (Table 4-41)	5	*2*	5	*2*	5	*2*	5	*2*	5	*2*
Vitamin premix (Table 4-42)	8	*4*	8	*4*	8	*4*	8	*4*	8	*4*
Feed additives[3]	← 100 to 300 g per ton →									
Total	2,000	*908*	2,000	*908*	2,000	*908*	2,000	*908*	2,000	*908*
Calculated analyses:										
Metabolizable energy (kcal/lb)	1,458		1,459		1,441		1,387		1,441	
Metabolizable energy *(kcal/kg)*	*3,208*		*3,210*		*3,170*		*3,051*		*3,170*	
Protein (%)	22.68		22.32		22.11		22.34		22.64	
Lysine (%)	1.40		1.40		1.40		1.40		1.40	
Threonine (%)	0.94		0.95		0.94		0.93		0.96	
Tryptophan (%)	0.29		0.28		0.29		0.29		0.30	
Calcium (Ca) (%)	0.91		0.91		0.91		0.90		0.91	
Phosphorus (P) (%)	0.70		0.70		0.70		0.71		0.70	

[1]Adapted by the author from *Life Cycle Swine Nutrition,* Iowa State University, Ames, PM-489, June, 1988.

[2]The prestarter ration is normally fed in only limited amounts. It should be fed to pigs weaned before 3 weeks of age until they reach approximately 12 lb *(5.4 kg)*. They can then be switched to a starter ration. These are good rations for orphan pigs, when extreme disease outbreaks (TGE) occur, or when the sow fails to produce sufficient milk.

[3]The feed additive may be part of the vitamin premix, or if a separate premix, it should replace an equal amount of corn.

TABLE 4-46
PIG STARTER RATIONS[1 2]

Ingredient	Ration # 1 (lb)	1 *(kg)*	2 (lb)	2 *(kg)*	3 (lb)	3 *(kg)*	4 (lb)	4 *(kg)*	5 (lb)	5 *(kg)*
Corn, yellow (8.4% protein)	1,072	*487*	939	*426*	1,032	*468*	1,273	*578*	865	*393*
Soybean meal, solvent extracted (44.0% protein)					610	*277*	689	*313*		
Soybean meal, solvent extracted, dehulled (47.0% protein)	570	*259*	580	*263*						
Soybeans, full-fat, cooked (37.0% protein)									800	*363*
Oat groats, dehulled (16.0% protein)			200	*91*						
Whey, dried (12.0% protein)	300	*136*	200	*91*	300	*136*			300	*136*
Fat or oil source			20	*9*						
Dicalcium phosphate	24	*11*	25	*12*	25	*12*	3	*1*	2	*1*
Limestone	16	*7*	18	*8*	15	*7*	17	*8*	15	*7*
Iodized salt	5	*2*	5	*2*	5	*2*	5	*2*	5	*2*
Trace mineral premix (Table 4-41)	5	*2*	5	*2*	5	*2*	5	*2*	5	*2*
Vitamin premix (Table 4-42)	8	*4*	8	*4*	8	*4*	8	*4*	8	*4*
Feed additives[3]	← 100 to 300 g per ton →									
Total	2,000	*908*	2,000	*908*	2,000	*908*	2,000	*908*	2,000	*908*
Calculated analyses:										
Metabolizable energy (kcal/lb)	1,483		1,472		1,460		1,473		1,525	
Metabolizable energy *(kcal/kg)*	*3,263*		*3,238*		*3,212*		*3,241*		*3,355*	
Protein (%)	19.70		20.37		19.55		20.39		20.13	
Lysine (%)	1.15		1.16		1.15		1.15		1.20	
Threonine (%)	0.82		0.82		0.82		0.81		0.85	
Tryptophan (%)	0.25		0.25		0.26		0.26		0.28	
Calcium (Ca) (%)	0.80		0.81		0.81		0.80		0.84	
Phosphorus (P) (%)	0.65		0.65		0.65		0.65		0.68	

[1]Adapted by the author from *Life Cycle Swine Nutrition,* Iowa State University, Ames, PM-489, June, 1988.

[2]The pig starter ration can be used as a creep ration before weaning and fed after weaning until the pigs reach approximately 40 lb *(18 kg)*. They can then be switched to a grower-finisher ration.

[3]The feed additive may be part of the vitamin premix, or if a separate premix, it should replace an equal amount of corn.

TABLE 4-47

GROWER-FINISHER RATIONS (FOR PIGS FROM 40 TO 240 LB *(18 TO 109 KG)*)[1,2,3,4]

Ingredient	For Pigs 40–120 Lb *(18–54 Kg)*						For Pigs 121–240 Lb *(55–109 Kg)*[5]					
	1		2		3		4		5		6	
	(lb)	(lb)	(lb)	(lb)	(lb)	(lb)	(lb)	(lb)	(lb)	(lb)	(lb)	(lb)
Corn, yellow (8.4% protein)[6,7]	1,571	1,497	1,422	1,343	1,561	1,492	1,692	1,613	1,575	1,481	1,694	1,620
Soybean meal, solvent extracted (44.0% protein)[8]	380	455			310	380	265	345			190	264
Soybeans, full-fat, cooked (37.0% protein)[9]			530	610					380	475		
Meat and bone meal (50.0% protein)					110	110					100	100
Dicalcium phosphate	21	19	19	18			18	17	20	20		
Limestone	15	16	16	16	6	5	15	15	15	14	6	6
Iodized salt	5	5	5	5	5	5	5	5	5	5	5	5
Trace mineral premix (Table 4-41)	3	3	3	3	3	3	2	2	2	2	2	2
Vitamin premix (Table 4-42)	5	5	5	5	5	5	3	3	3	3	3	3
Feed additives[10]	◄—— 0 to 100 g per ton ——►						◄—— 0 to 50 g per ton ——►					
Total	2,000	2,000	2,000	2,000	2,000	2,000	2,000	2,000	2,000	2,000	2,000	2,000
Calculated analyses:												
Metabolizable energy (kcal/lb)	1,500	1,497	1,543	1,547	1,498	1,495	1,510	1,507	1,538	1,543	1,508	1,505
Metabolizable energy *(kcal/kg)*	*3,300*	*3,293*	*3,395*	*3,403*	*3,296*	*3,289*	*3,322*	*3,315*	*3,384*	*3,395*	*3,318*	*3,311*
Protein (%)	14.96	16.30	15.78	16.93	16.13	17.38	12.94	14.36	13.65	15.01	13.79	15.11
Lysine (%)	0.75	0.85	0.81	0.90	0.77	0.86	0.60	0.70	0.65	0.76	0.60	0.70
Threonine (%)	0.58	0.63	0.61	0.66	0.60	0.65	0.49	0.55	0.52	0.58	0.51	0.56
Tryptophan (%)	0.17	0.20	0.20	0.21	0.17	0.19	0.14	0.17	0.16	0.18	0.13	0.15
Calcium (Ca) (%)	0.60	0.60	0.61	0.61	0.61	0.60	0.55	0.55	0.59	0.58	0.55	0.56
Phosphorus (P) (%)	0.50	0.50	0.51	0.51	0.51	0.53	0.46	0.46	0.49	0.51	0.47	0.49

[1]Adapted by the author from *Life Cycle Swine Nutrition,* Iowa State University, Ames, PM-489, June, 1988.

[2]Feed the ration with the higher level of soybean meal (lower level of corn) to lighter pigs in each group and decrease the soybean meal (increase the corn) until you reach the lower level as pig weights increase. If preferred, one level of protein can be fed from 40 to 240 lb *(18 to 109 kg)* with similar results as with varying the levels. To accomplish this, use the lower protein formulations from rations 1, 2, or 3 (for example, in ration No. 1 use 1571 lb *(713 kg)* of corn and 380 lb *(173 kg)* of soybean meal).

[3]If barrows and gilts are separated, use the higher range for soybean meal for the gilts and the lower range for the barrows.

[4]To convert lb to kg, divide by 2.2. To convert g/ton (short) to g/ton (metric), divide by 0.907.

[5]For potential replacement gilts, the level of dicalcium phosphate should be increased by 10 lb *(4.5 kg)* per ton. This will provide a minimum dietary level of 0.67% calcium and 0.55% phosphorus.

[6]Ground milo, wheat, or barley can replace the ground corn. Ground oats can replace corn up to 20% of the total ration.

[7]If the ration is to be pelleted, 25 to 50 lb *(11 to 23 kg)* of molasses or binder can replace an equal amount of corn.

[8]To replace 44% soybean meal with 47% soybean meal or synthetic lysine, use the following ratios:
Each 100 lb *(45 kg)* SBM (44%) = 93 lb *(42 kg)* SBM (47%) + 7 lb *(3 kg)* corn.
Each 100 lb *(45 kg)* SBM (44%) = 3 lb *(1.4 kg)* 98% lysine hydrochloride + 1 lb *(0.45 kg)* dicalcium phosphate + 96 lb *(43.6 kg)* corn.

[9]The fat content of whole soybeans increases the energy content of the ration. For maximum utilization of the ration, the protein content has been increased to maintain a similar energy-to-protein ratio.

[10]The feed additive may be part of the vitamin premix, or if it is a separate premix, it should replace an equal amount of corn.

Fig. 4-73. Market hogs self-fed on a concrete floor. (Courtesy, Gehl Company, West Bend, WI)

• **Complete commercial protein supplements**—Also, commercial supplements (usually a combined protein-mineral-vitamin-additive supplement) may be bought and mixed with the locally available grain. Table 4–48 shows the proportion of grain containing 8.4% protein to mix with supplements ranging from 30 to 45% protein to obtain finished rations ranging from 10 to 18% protein content.

Where commercial supplements are bought to use with farm-grown grains, they may be utilized in the following ways:

1. Mixed with ground, farm-grown grain in the approximate amounts shown in Table 4–48 to make a complete ration (see the Table 4–48 footnote example for mixing directions).
2. Self-fed in separate feeders, with the ground or whole grain also being self-fed in separate self-feeders.
3. Hand-fed; with the supplement and the grain each being hand-fed in the proportions recommended in Table 4–48.

TABLE 4–48
GRAIN AND SUPPLEMENT COMBINATIONS (POUNDS)
NEEDED TO FORMULATE RATIONS OF DIFFERENT PROTEIN LEVELS (GRAIN CONTAINING 8.4% PROTEIN)[1,2,3]

Protein in Supplement		Percent Protein in Total Ration								
		10	11	12	13	14	15	16	17	18
(%)		(lb)	(lb)	(lb)	(lb)	(lb)	(lb)	(lb)	(lb)	(lb)
30	Grain	1,852	1,759	1,667	1,574	1,481	1,389	1,296	1,204	1,111
	Supplement	148	241	333	426	519	611	704	796	889
31	Grain	1,858	1,770	1,681	1,593	1,504	1,416	1,327	1,239	1,150
	Supplement	142	230	319	407	496	584	673	761	850
32	Grain	1,864	1,780	1,695	1,610	1,525	1,441	1,356	1,271	1,186
	Supplement	136	220	305	390	475	559	644	729	814
33	Grain	1,870	1,789	1,707	1,626	1,545	1,463	1,382	1,301	1,220
	Supplement	130	211	293	374	455	537	618	699	780
34	Grain	1,875	1,797	1,719	1,641	1,563	1,484	1,406	1,328	1,250
	Supplement	125	203	281	359	438	516	594	672	750
35	Grain	1,880	1,805	1,729	1,654	1,579	1,504	1,429	1,353	1,278
	Supplement	120	195	271	346	421	496	571	647	722
36	Grain	1,884	1,812	1,739	1,667	1,594	1,522	1,449	1,377	1,304
	Supplement	116	188	261	333	406	478	551	623	696
37	Grain	1,888	1,818	1,748	1,678	1,608	1,538	1,469	1,399	1,329
	Supplement	112	182	252	322	392	462	531	601	671
38	Grain	1,892	1,824	1,757	1,689	1,622	1,554	1,486	1,419	1,351
	Supplement	108	176	243	311	378	446	514	581	649
39	Grain	1,895	1,830	1,765	1,699	1,634	1,569	1,503	1,438	1,373
	Supplement	105	170	235	301	366	431	497	562	627
40	Grain	1,899	1,835	1,772	1,709	1,646	1,582	1,519	1,456	1,392
	Supplement	101	165	228	291	354	418	481	544	608
41	Grain	1,902	1,840	1,779	1,718	1,656	1,595	1,534	1,472	1,411
	Supplement	98	160	221	282	344	405	466	528	589
42	Grain	1,905	1,845	1,786	1,726	1,667	1,607	1,548	1,488	1,429
	Supplement	95	155	214	274	333	393	452	512	571
43	Grain	1,908	1,850	1,792	1,734	1,676	1,618	1,561	1,503	1,445
	Supplement	92	150	208	266	324	382	439	497	555
44	Grain	1,910	1,854	1,798	1,742	1,685	1,629	1,573	1,517	1,461
	Supplement	90	146	202	258	315	371	427	483	539
45	Grain	1,913	1,858	1,803	1,749	1,694	1,639	1,585	1,530	1,475
	Supplement	87	142	197	251	306	361	415	470	525

[1]Adapted by the author from *Life Cycle Swine Nutrition*, Iowa State University, Ames, PM-489, June, 1988. To convert lb to kg, divide by 2.2.

[2]The grain common to the area may be substituted in Table 4-48, with the 8.4% protein content changed in keeping with the protein content of the grain used, and the proportions of grain and supplement adjusted to obtain the desired percent protein in the total ration.

[3]**Example**: In order to obtain a total ration with 15% protein, each 2,000 lb *(908 kg)* of feed should contain 1,389 lb *(631 kg)* of the 8.4% protein grain and 611 lb *(277 kg)* of the 30% supplement.

• **Pointers in formulating rations and feeding swine**—In formulating rations and in feeding swine, the following points are noteworthy:

1. Feeds of similar nutritive properties can be interchanged in the ration as price relationships warrant.
2. If wheat, barley, oats, or grain sorghum is used instead of corn as the grain in a ration, the protein supplement may be slightly reduced.
3. When proteins of animal origin predominate, adequate mineral protection can be obtained by allowing hogs free access to a 2-compartment box or self-feeder with (a) salt (trace mineralized) in one side, and (b) a mixture of ⅓ salt (salt added for purposes of palatability) and ⅔ monosodium phosphate or other phosphorus supplement, in the other side. When supplements of plant origin constitute most of the source of proteins, add a third compartment to the mineral box and place

in it a mixture of ⅓ salt (trace mineralized) and ⅔ ground limestone or oystershell flour.

4. When hogs are not exposed to sunlight or when dehydrated alfalfa meal is fed, vitamin D should be added in keeping with the recommended allowances (see Table 4–39).

5. Where the ration consists chiefly of white corn, barley, wheat, oats, rye, kafir, or by-products of these grains, there may be a deficiency of vitamin A (see Table 4–39 for recommended allowances).

6. Except for gestating sows and boars of breeding age, hogs are generally self-fed. All of the ingredients may be mixed together and placed in the same self-feeder or the grain may be placed in one self-feeder (or compartment) and the protein supplements (including any ground alfalfa) in another. If the (a) cereal grains and (b) protein supplements (including ground alfalfa) are hand-fed, the grain and supplement should be fed separately, in the proportions indicated in the suggested rations.

7. An exception should be made to the cafeteria-style feeding when the grain ration consists of barley, oats, rye, or kafir. These feeds are higher in protein content than corn, and for this reason are generally fed as a mixed ration. Otherwise, the pigs will often eat more protein supplement than is necessary to balance the ration. Likewise, when corn is fed as the grain, sometimes such protein supplements as (a) roasted soybeans, (b) soybean meal, and (c) peanut meal are too palatable to be fed separately from the corn, especially if the corn is not of good quality.

8. Full-fed finishing hogs will consume 4 to 5 lb of feed daily per 100 lb liveweight until they weigh 100 lb. They will eat 3 to 4 lb daily per 100 lb weight from this stage until marketing.

• **Fitting rations for show and sale swine**—Any of the rations listed in Tables 4–43 or 4–47 for the respective classes and ages of swine are suitable for use in fitting show animals of similar classification. Because of the high cost of labor, the recent trend has been toward self-feeding both young breeding animals and market barrows and gilts that are being fitted for show. Many of them are left on self-feeders right up to show time, others are hand-fed only during the last month or two of the fitting period. However, most experienced exhibitors feel that they can get superior bloom and condition by either (1) hand-feeding, or (2) using a combination of hand-feeding and self-feeding (hand-feeding twice daily and allowing free access to a self-feeder). When hand-feeding, they also prefer mixing the ration with skim milk, buttermilk, or condensed buttermilk, and feeding the entire ration in the form of a slop.

Adding milk to a ration that is already properly balanced makes for a higher protein content than necessary. On the other hand, most experienced caretakers prefer using rations of higher protein content for fitting purposes. They feel they get more bloom that way. In general, however, when skim milk or buttermilk is used in slop feeding, the protein feeds of the ration may be reduced by one-half without harm to the animal.

In fitting show barrows, it may be necessary to decrease or discontinue slop feeding 2 to 4 weeks before the show to avoid paunchiness and lowering of the dressing percentage.

When oatmeal (oat groats, rolled hulled oats) is not too high priced, many successful hog caretakers replace up to 50% of the grain (corn, wheat, barley, oats, and/or sorghum) in the ration with oatmeal. They do this especially when fitting hogs—both breeding animals and barrows—in the younger age groups. Oatmeal is highly palatable, lighter, and less fattening than corn.

Suitable minerals and vitamins should always be provided.

FEEDING SYSTEMS

The choice of the feeding system(s) and the choice of the ration(s) must go hand in hand. For example, if the grain and the protein supplements are to be self-fed in separate feeders or compartments, it is important that they be of equal palatability; otherwise, pigs will consume too much of one and too little of the other. A summary of each of the common feeding systems follows.

• **Complete self-fed rations**—The trend is toward the use of complete, self-fed rations for baby pigs and growing-finishing hogs, because, in comparison with free-choice feeding, they (1) lend themselves better to automation, (2) provide better control of nutrient intake, and (3) result in faster gains.

Complete rations may be formulated either by "building from the ground up" (by adding each ingredient, one by one), or by mixing a complete supplement, a base mix, or a premix with ground grain.

• **Floor or drop feeding**—Floor or drop feeding is particularly suited to the controlled feeding of growing-finishing swine or the breeding herd. Feeding in the sleeping area encourages cleanliness, since pigs are less inclined to defecate where they eat. Feed wastage is reduced to a minimum when the animals do not have more feed available than they will consume at one eating. Even though automated, restricted feeding requires close attention, because the daily feed intake of pigs is affected by the weather.

• **Free-choice**—Grain and protein supplements may be fed separately and free-choice. Generally, pigs fed free-choice rations in separate feeders or compartments will not make as uniform or as fast gains as pigs fed a complete mixed ration. The free-choice system requires more supervision, as the palatability of the grain or the protein supplement may vary and the pigs will then overeat or undereat the supplement or the grain. There is very little, if any, difference in economy of gain between feeding a free-choice or a complete ground mixed ration. Free-choice feeding may be the best feeding system for the small producer who does not have mixing equipment.

• **Liquid feeding**—Liquid feeding usually involves mixing predetermined amounts of feed and water prior to, or at the time of, feeding. When properly used, this method can practically eliminate feed dust in the feeding area and minimize wastage. Ratios of feed and water can be varied to produce a free-flowing liquid or a thick paste. In some cases, feed is automatically dropped into the water in the feed trough. Research has shown no difference in the rate of gain of pigs full-fed on liquid or dry feeds. Neither does liquid feeding have any effect on dressing percentage, carcass measurements, or carcass quality. However, pigs full-fed liquid rations generally require more feed per pound of gain than pigs full-fed dry rations.

• **Limit feeding**—With gestating sows, limit feeding to 4 to 6 lb per head daily is a must in order to keep them from getting too fat. Overly fat sows have difficulty in farrowing and give birth to weak or dead pigs. With growing-finishing pigs, it is a way in which to increase slightly the proportion of lean to fat in the carcass. A discussion of limit feeding of (1) gilts and sows, and (2) growing-finishing pigs follows.

1. **Gilts and sows.** Replacement gilts should be started on a limited feeding program at 180 to 200 lb; and all gestating sows and gilts should be limit fed. Limit feeding may be accomplished by any one of the following methods:

a. **By feeding bulky, fibrous feeds**, such as silage, haylage, or alfalfa, with such feed constituting at least one-third of the ration. Actually, this is a way in which to lower the energy content of the ration. Although bulky feeds will hold the weight down, they usually do not lower feed cost.

b. **By interval feeding**, in which gilts or sows are turned to self-feeders for 2 to 8 hours every second or third day. Under this system, gilts will usually eat around 12 lb of feed at a time (or an average of 4 lb per day) and older sows will consume around 15 lb (or an average of 5 lb per day). The amount of feed consumed in interval feeding may be controlled either (1) by varying the interval, from every other day to twice a week, (2) by varying the length of time that the gilts and sows are left on the self-feeders (from 2 to 8 hours), or (3) by hand-feeding.

c. **By group hand-feeding** a limited ration to several sows. This is apt to result in the "bossy" sows getting too much and the timid sows getting too little. This problem can be partially alleviated by feeding over a large area.

d. **By individual feeding** in either individual stalls or in tie stalls, tethered by a neck collar or belt.

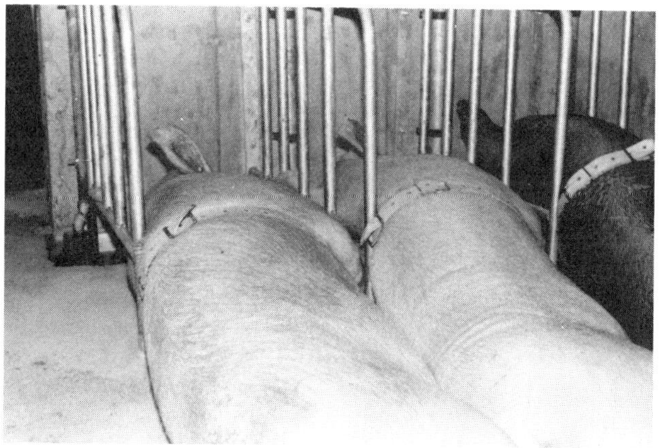

Fig. 4–74. Individually fed sows, comfortably tethered. (Courtesy, Lone Willow Genetics, Roanoke, IL)

2. **Growing-finishing pigs.** Sometimes growing-finishing pigs are limit-fed in order to produce leaner carcasses. Usually, it is started when pigs weigh around 100 lb and feed is limited to about 85 to 95% of what pigs of comparable age consume when self-fed. Limit feeding of market hogs results in slower gains, increased labor, and more mechanization. Thus, unless sufficient premium is paid for the modestly leaner carcasses, it cannot be justified.

• **Pelleted complete rations**—The use of pelleted complete rations for growing-finishing hogs will increase the average daily gain by 2 to 5% and improve the feed efficiency by approximately 5 to 10%. Thus, when a complete ration is purchased, buying a pelleted feed may be more economical than buying a meal. But the advantage of pelleting will usually not be sufficient to offset the cost of hauling grain to the mill and having a pelleted ration made. Also, pellet machines are costly; hence, the purchase of such equipment cannot be justified with the volume of hogs handled by most swine producers.

FEED REQUIRED TO PRODUCE A POUND OF MARKET HOG

Nationally, it has been estimated that it requires 4.0 lb of feed to produce 1 lb of on-foot hog (live) from birth to market weight, exclusive of the feed required by sows and boars to produce pigs. This is high. But remember that 25 to 30% of all pigs farrowed die before weaning. Remember too, that many swine producers are inefficient.

Table 4–49 shows realistic goals for well-managed swine operations.

TABLE 4–49
ESTIMATED FEED REQUIRED TO PRODUCE 240-LB MARKET PIG[1]

Stage of Production	Feed Required per 240-Lb Market Pig
	(lb)
Sow gestation ration (includes pregestation and breeding)	110
Boar ration	8
Lactation ration	45
Starter ration (creep to 40 lb)	54
Grower-finisher ration (40 to 240 lb)	700
Total, lb	917

Per 100 lb of pork produced $\frac{917}{240} \times 100 = 382$ **lb**

[1]See Section 20, "Weights and Measures" for conversion of U.S. customary to metric.

The values given in Table 4–49 are estimates based on realistic standards for apportioning the quantities of sow and boar feed to each pig and the feed conversion normally attained during the starter and grower-finisher periods. Although these data do not provide for pig deaths after weaning, normal milling losses, and feed wastage, it is assumed that these losses are not considered in the pounds of hog produced. Data obtained in commercial herds where accurate records have been kept indicate that 382 lb of feed per 200 lb live hog produced is a realistic goal for a practical swine operation. However, to achieve this level of efficiency, a sound feeding and management program must be followed, including limit feeding of pregnant sows, high conception rates, large litters weaned, early weaning and rebreeding, low death losses, minimal disease problems, balanced rations, and minimal feed wastage.

PART VI
FEEDING HORSES

Fig. 4–75. Getting acquainted! Feed is the most important influence in the environment of the horse. (Photo by E. Peterson, Hamilton, MT)

Feed is the most important influence in the environment of the horse. Unless horses are fed properly, their maximum potential in reproduction, growth, body form, speed, endurance, style, and attractiveness cannot be achieved.

The following conditions make it imperative that the nutrition of horses be the best that science and technology can devise:

1. **Confinement.** Many horses are kept in stables or corrals most of the time.

2. **Fitting yearlings.** When forcing young equines, it is important to their development and soundness that the ration be nutritionally balanced.

3. **Racing two-year-olds.** In the United States, we race more two-year-olds than any other nation in the world; our richest races are for them. If the nutrient content of the ration is not adequate, there is bound to be more breakdown on the track than with older horses—this is costly.

4. **Stress.** Stress is affected by excitement, temperament, fatigue, number of horses together, previous nutrition, breed, age, and management. Race and show horses are always under stress; and the more tired they are and the greater the speed, the greater the stress. Thus, the ration for race and show horses should be scientifically formulated, rather than based on fads, foibles, and trade secrets. The greater the stress the more exacting the nutritive requirements.

5. **Horses are unique.** They differ from other farm animals from the standpoint of use and should not be fed the same feeds. They have greater value; are kept for recreation, sport, and work; are fed for a longer life of usefulness; have a smaller digestive tract; should not carry surplus weight; and are fed for nerve, mettle, animation, and character of muscle.

Also, feed constitutes the greatest single item in the horse business.

NUTRITIVE NEEDS OF HORSES

The proper nutrition of horses is a major factor in determining their efficiency and years of service.

The various nutritive needs of the horse are: (1) energy (carbohydrates and fats), (2) protein, (3) minerals, (4) vitamins, and (5) water.

• **Recommended nutrient allowances**—Presently available information indicates that the recommended nutrient allowances given in Table 4–50 will meet the minimum requirements for horses and provide reasonable margins of safety.

TABLE 4–50
RECOMMENDED NUTRIENT ALLOWANCES FOR HORSES (TOTAL RATION/AS-FED BASIS)[1] (See footnotes at end of table.)

	Mature Horses (Consuming 25 lb feed/horse/day. Idle horses require less feed and/or consume more roughage than heavily worked horses or lactating mares.)					Young Horses, Based on Mature Weight 1,000 lb				
	Idle Horses/ Light Work/ Moderate Work (1,000 lb Wt.)	Heavy Training/ Heavy Work (1,000 lb Wt.)	Stallions in Breeding Season (1,000 lb Wt.)	Mares, Last 90 Days Gestation (1,000 lb Wt.)	Mares, Peak of Lactation (1,000 lb Wt.)	Creep Feed (250 lb Body Wt/11 lb Feed Daily)	Weanlings (450 lb Body Wt/12 lb Feed Daily)	Yearlings (650 lb Body Wt/13 lb Feed Daily)	2-Yr-Olds & 3-Yr-Olds (800 lb Body Wt/14 lb Feed Daily)	2-Yr-Olds in Light Training (800 lb Body Wt/15 lb Feed Daily)
Digestible Energy:										
TDN[2] .... (%)	55	62.50	75	62.50	75	75	75	70	60	65
Mcal per .... (lb)	0.8	1.2	1.0	0.90	1.10	1.25	1.25	1.15	1.00	1.10
Mcal per .... *(kg)*[3]	*1.80*	*2.55*	*2.15*	*2.0*	*2.35*	*2.60*	*2.60*	*2.50*	*2.20*	*2.40*
Crude Protein .... (%)	9.0	11.0	14.0	13.0	14.0	18.0	16.0	14.0	13.0	13.0
Lysine .... (%)	0.25	0.36	0.30	0.32	0.41	0.54	0.55	0.48	0.38	0.41
Major or Macrominerals:										
Salt .... (%)	0.75	0.75	0.75	0.75	0.75	0.75	0.75	0.75	0.75	0.75
Calcium .... (%)	0.21	0.31	0.26	0.29	0.47	0.62	0.55	0.40	0.28	0.31
Phosphorus .... (%)	0.15	0.23	0.19	0.30	0.30	0.34	0.30	0.22	0.15	0.17
Magnesium .... (%)	0.08	0.12	0.10	0.10	0.09	0.07	0.07	0.07	0.08	0.09
Potassium .... (%)	0.27	0.39	0.33	0.33	0.38	0.27	0.27	0.27	0.27	0.29
Sulfur .... (%)	0.15	0.15	0.15	0.15	0.15	0.15	0.15	0.15	0.15	0.15
Trace or Microminerals:										
Cobalt .... (ppm)[4]	0.11	0.11	0.11	0.11	0.11	0.11	0.11	0.11	0.11	0.11
Copper .... (ppm)	25	25	25	25	30	40	40	30	25	25
Iodine .... (ppm)	0.11	0.11	0.11	0.11	0.11	0.11	0.11	0.11	0.11	0.11
Iron .... (ppm)	40	60	90	90	90	90	80	60	60	60
Manganese .... (ppm)	46	46	46	46	46	46	46	46	46	46
Selenium .... (ppm)	0.11	0.11	0.11	0.11	0.11	0.11	0.11	0.11	0.11	0.11
Zinc .... (ppm)	80	90	90	100	100	100	100	100	90	90
	(/lb)	**(/lb)**	**(/lb)**	**(/lb)**	**(/lb)**	**(/lb)**	**(/lb)**	**(/lb)**	**(/lb)**	**(/lb)**
Fat-soluble Vitamins in Feed:										
Vitamin A .... (IU)	1,045	1,045	1,045	1,569	1,569	1,045	1,045	1,045	1,045	1,045
Vitamin D .... (IU)	156	156	156	314	314	419	419	419	419	419
Vitamin E .... (IU)	26	41	41	41	41	41	41	41	41	41
Vitamin K .... (mg)	0.32	0.32	0.32	0.32	0.32	0.30	0.30	0.30	0.30	0.30
Water-soluble Vitamins in Feed:										
Biotin .... (mg)	0.1	0.1	0.1	0.1	0.1	0.1	0.1	0.1	0.1	0.1
Choline .... (mg)	20	30	30	30	30	62.5	62.5	62.5	62.5	62.5
Folacin .... (mg)	0.8	1.2	1.2	1.2	1.2	3.0	3.0	3.0	3.0	3.0
Niacin .... (mg)	10	20.8	10	10	10	10	10	10	10	10
Pantothenic acid .... (mg)	10	20.8	10	10	10	10	10	10	10	10
Riboflavin .... (mg)	1.6	1.6	1.6	1.6	1.6	1.6	1.6	1.6	1.6	1.6
Thiamin (B-1) .... (mg)	1.57	2.61	1.57	1.57	1.57	1.57	1.57	1.57	1.57	1.57
Vitamin B-6 .... (mg)	1.0	1.0	1.0	1.0	1.0	0.5	0.5	0.5	0.5	0.5
Vitamin B-12 .... (mg)	0.005	0.006	0.006	0.006	0.006	0.007	0.007	0.007	0.007	0.007
Vitamin C (Ascorbic acid) .... (mg)	2.4	4.0	4.0	4.0	4.0	3.75	3.75	3.75	3.75	3.75

[1]Where hay is fed separately, double the amounts shown in this table should be added to the concentrate.

[2]1 lb TDN = 2 Mcal or 2,000 Kcal.

[3]1 kg = 2.2 lb or 1,000 g.

[4]1 ppm (parts per million) = 1 mg/kg.

ENERGY

A lack of energy intake may cause slow and stunted growth in foals and loss of weight, poor condition, and excessive fatigue in mature horses. Excess energy may result in obese horses, which are more susceptible to stress and founder and have lowered reproductive efficiency and decreased longevity.

Generally, increased energy for horses is met by increasing the grain and decreasing the roughage.

The fiber of growing pasture grass, fresh or dried, is more digestible than the fiber of most hay. Likewise, the fiber of early cut hay is more digestible than that of hay cut in the late bloom or seed stages. The difference is due to both chemical and physical structure, especially the presence of certain encrusting substances (notably lignin) which are deposited in the cell wall with age.

Fig. 4-76. When racing, horses may require up to 100 times more energy than at rest. (Courtesy, Turf Paradise, Phoenix, AZ)

Young equines and working (or running) horses must have rations in which a large part of the carbohydrate content of the ration is low in fiber, and in the form of nitrogen-free extract.

PROTEIN

Horses of all ages and kinds require adequate amounts of protein of suitable quality for maintenance, growth, finishing, reproduction, and work. Of course, the protein requirements for growth and reproduction are the greatest and most critical.

The extent to which the horse's ration is supplemented with protein depends on the age of the horse and on the quality of the forage fed. Growing or lactating animals require somewhat more protein than horses that are idle, gestating, or working. Also, grass hays are generally low in quality and quantity of proteins and require more supplementation than legumes.

A deficiency of proteins in the horse may result in the following deficiency symptoms; depressed appetite, poor growth, loss of weight, reduced milk production, irregular estrus, lowered foal crops, loss of condition, and lack of stamina.

MINERALS

Eighteen mineral elements are known to be required by at least some animal species. They can be divided into the following two groups based on the relative amounts needed in the ration:

Major or Macrominerals	Trace or Microminerals
Calcium (Ca)	Iodine (I)
Phosphorus (P)	Manganese (Mn)
Sodium (Na)	Iron (Fe)
Chlorine (Cl)	Zinc (Zn)
Potassium (K)	Copper (Cu)
Magnesium (Mg)	Molybdenum (Mo)
Sulfur (S)	Fluorine (F)
	Chromium (Cr)
	Selenium (Se)
	Silicon (Si)
	Cobalt (Co)

Approximately 70% of the mineral content of the horse's body consists of calcium and phosphorus. About 99% of the calcium and over 80% of the phosphorus are found in the bones and teeth.

The typical horse ration of grass hay and farm grains is usually deficient in calcium, but adequate in phosphorus. Also, salt is almost always deficient; and many horse rations do not contain sufficient iodine and certain other trace elements. Thus, horses usually need special mineral supplements. But they should not be fed either more or less minerals than needed. Also, it is recognized that mineral allowances given with the ration or in a mineral mix should vary according to the mineral content of the soil on which feeds are grown.

The proper development of the bone is particularly important in the horse, as evidenced by the stress and strain on the skeletal structure of the racehorse, especially when racing the 2-year-old. Since the greatest development of the skeleton takes place in the young, growing animal, it is evident that adequate minerals must be provided at an early age if the bone is to remain sound.

At this time, the cause of the increase in the incidence of bone diseases among horses is not entirely clear. However, it appears that the major factors are: (1) rapid growth and excess weight, (2) injury to the epiphysis, (3) nutritional imbalances, (4) genetic predisposition, (5) limited forced exercise, (6) exercise on hard ground, and (7) faulty conformation.

The functions, deficiency symptoms, interrelationships/toxicities, and sources of minerals for horses, which are similar to those of other animal species, are presented in Table 4-1, Animal Mineral Chart.

Table 4-50 presents recommended mineral allowances for horses.

VITAMINS[9]

Certain vitamins are necessary for the growth, development, health, and reproduction of horses. Deficiencies of vitamins A and D are sometimes encountered. Also, indications are that

[9]In this section, when reference is made to a National Academy of Sciences recommendation, this implies the following source: *Nutrient Requirements of Horses*, No. 6, 5th rev. ed., National Academy of Sciences, 1989.

vitamin E and some of the B vitamins are required by horses. Further, it is recognized that single, uncomplicated vitamin deficiencies are the exception rather than the rule.

High-quality, leafy, green forages plus plenty of sunshine generally give horses most of the vitamins they need. Horses get carotene (which they can convert to vitamin A) and riboflavin from green pasture and green hay not over a year old, and they get vitamin D from sunlight and sun-cured hay. If plenty of green forage and sunlight are not available, the caretaker should get the advice of a nutritionist or veterinarian on the use of vitamin additives to the feed.

The functions, deficiency symptoms, and good sources of vitamins for horses, which are similar to those of other animal species, are presented in Table 4–2, Animal Vitamin Chart.

Table 4–50 lists the vitamins most commonly involved in horse nutrition, and gives the recommended allowances of each of them. Although there is no evidence of deficiencies of certain vitamins, it is possible that more of them may be destroyed or used by horses during stress or strain than can be obtained through normal feeds or synthesized by the intestinal microflora of the horse; hence, adding them to the ration may assure maximum performance.

The fat-soluble vitamins, which are stored in the body in appreciable quantities, are vitamin A (carotene), vitamin D, vitamin E, and vitamin K.

The large amounts of water which pass through the horse's body daily tend to carry out the water-soluble vitamins, thereby depleting the supply. Thus, they must be supplied in the horse's ration on a day-to-day basis. All of the water-soluble vitamins except C are known as B vitamins.

Vitamins of the B-complex, particularly biotin, choline, folacin (folic acid), niacin, pantothenic acid, riboflavin, thiamin (B-1), vitamin B-6 (pyridoxine) and vitamin B-12 may be essential, especially for (1) young horses before the synthesis of the B-complex vitamins by the microflora begins, and (2) horses that are under stress, as in racing and showing.

• **Vitamin imbalances**—Experiments have shown that the amounts needed of certain vitamins may be affected by the supply of another vitamin or of some other nutritive essential. Also, it is known that excess fortification of the horse's ration with certain vitamins may prove more detrimental than helpful. Thus, caretakers should avoid harmful imbalances; they should provide vitamins on the basis of recommended allowances. Also, when fortifying with vitamins, consideration should be given to the vitamins provided by the ingredients of the normal ration, for it is the total composition of the feed that counts.

• **Unidentified factors**—Since the U.S. foal crop is only around 50%, and since horses under stress (racing, showing, etc.) frequently become temperamental in their eating habits, it is obvious that there is room for improvement in the ration somewhere along the line. Perhaps unidentified factors are involved.

Unidentified factors include those vitamins which the chemist has not yet isolated and identified. For this reason, they are sometimes referred to as the vitamins of the future. There is mounting evidence of the importance of unidentified factors for animals, including humans. Among other things, they lower the incidence of ulcers in humans and swine. For horses, they appear to increase growth and improve feed efficiency and breeding performance when added to rations thought to be complete with regard to known nutrients. The anatomical and physiological mechanism of the digestive system of the horse, plus the stresses and strains to which modern horses are subjected, would indicate the wisdom of adding unidentified factor sources to the ration of the horse. Unidentified factors appear to be of special importance during breeding, gestation, lactation, and growth.

Three highly regarded unidentified factor sources are dried whey product, corn fermentation solubles, and dehydrated alfalfa meal.

WATER

Water is one of the most vital of all nutrients. In fact, horses can survive for a longer period without feed than they can without water. The loss of 10% body water will result in disorders; the loss of 20% body water will cause death. But, fortunately, under ordinary conditions water can be readily provided and at little cost.

Water is one of the largest single constituents of the animal body, varying in amount with condition and age. The younger the animal, the more water it contains. Also, the fatter the animal, the lower the water content. Thus, as an animal matures, it requires proportionately less water on a weight basis, because it consumes less feed per unit of weight and the water content of the body is being replaced by fat.

Surplus water is excreted from the body, principally in the urine, and to a slight extent in the perspiration, feces, and water vapor from the lungs.

The average horse will drink 10 to 12 gal. of water daily, the amount varying according to the weather, amount of work done, (sweating), rations fed, and size of horse.

Free access to water is desirable. When this is not possible, horses should be watered at approximately the same times daily. Opinions vary among caretakers as to the proper times and method of watering horses. All agree, however, that regularity and frequency are desirable. Most caretakers agree that water may be given before, during, and after feeding.

Frequent, small waterings between feedings are desirable during warm weather or when the animal is being put to hard use. Do not allow a horse to drink heavily when it is hot, because it may founder; and do not allow a horse to drink heavily just before being put to work.

Automatic waterers are the modern way to provide clean, fresh water at all times—as nature intended. Also, frequent but small waterings prevent gorging. All waterers should have drains for easy cleaning, and should be heated to 40 to 45°F during the winter months in cold regions. Waterers should be available in both stalls and corrals.

FEEDS FOR HORSES

More than one kind of hay makes for appetite appeal. In season, any good pasture can replace part or all of the hay unless work or training conditions make substitution impractical.

Good quality oats and timothy hay always have been considered standard feeds for light horses. However, feeds of similar nutritive properties can be interchanged in the ration as price relationships warrant; among them, the grains—oats, corn, barley, wheat, and sorghum; the protein supplements—linseed meal, soybean meal, and cottonseed meal; and hays of many varieties. Feed substitution makes it possible to obtain a balanced ration at lowest cost.

During the winter months, it is well to add a few sliced carrots to the ration, an occasional bran mash, or a small amount of linseed meal. Also a bran mash or linseed meal may be used to regulate the bowels.

The proportion of concentrates must be increased and the roughages decreased as energy needs rise with the greater amount, severity, or speed of work. A horse that works at a trot needs considerably more feed than one that works at a walk. For this reason, riding horses in medium to light use require somewhat less grain and more hay in proportion to body weight than light horses that are racing. Also, from an aesthetic standpoint, large, paunchy stomachs are objectionable on horses that are used for recreation and sport.

In addition to making for a nutritionally complete ration, the following factors should be considered when choosing horse feeds: cost, palatability, preparation, variety, bulk, and laxativeness.

SPECIAL FEEDS AND ADDITIVES

Special horse feeds may be needed from time to time for promoting growth of young stock, preventing disease,or imparting bloom and attractiveness.

- **Antibiotics**—Certain antibiotics, at stipulated levels, are approved by the FDA for growth promotion and for the improvement of feed efficiency of young equines up to one year of age. Unless there is a disease level, however, there is no evidence to warrant the continuous feeding of antibiotics to mature horses. Such practice may even be harmful, Hence, where antibiotics are needed for therapeutic purposes, it is best to seek the advice of a veterinarian.

It appears that antibiotics may be especially helpful for young foals which suffer setbacks from infections, digestive disturbances, inclement weather, and other stress factors. Also, horses may benefit from antibiotics (1) when being transported from one location to another—for example, when being moved to a new show or track; (2) when there is a low disease level in the herd; or (3) when mares are foaling.

When added to feed, the level of antibiotics should be in keeping with the directions of the manufacturer and with the Food and Drug Administration regulations.

- **Bloom-imparting feeds**—Bloom or gloss is important in horses. But sometimes they lack this desired quality—their hair is dull and dry. Feeding a well-balanced ration will usually rectify this situation. Also feeding the following products will make for an attractive, shiny coat:

1. **Corn oil or safflower oil.** Feed at the rate of 2 oz (2 Tbsp) per horse twice a day.
2. **Whole flaxseed soaked.** Put a handful of whole flaxseed in a teacup, cover it with water, let it stand overnight, then pour it over the morning feed. Repeat twice each week.

Unless the horse is afflicted with lice, mange, or some other ailment, either of the above treatments will impart bloom or gloss to the coat.

- **Lysine**—Cornell University reported that the addition of lysine to the diet of growing horses increased weight gains, feed consumption, and feed efficiency.

In recognition that lysine is the first limiting amino acid of horses and is thus an indicator of the quality of protein which horses require, the recommended lysine allowance for horses is given in Table 4–50.

- **Milk by-products**—The superior nutritive values of milk by-products are due to their high-quality proteins, vitamins, a good mineral balance, and the beneficial effect of the milk sugar, lactose. In addition, these products are palatable and highly digestible. They are an ideal feed for young equines and for balancing out the deficiencies of the cereal grains. Most foal rations contain one or more milk by-products, primarily dried skim milk, with some dried whey and dried buttermilk included at times. The chief limitation to their wider use is price.

- **Milk replacer**—As indicated by the name, a milk replacer is a replacement for milk. Such replacers generally contain the following composition: animal or vegetable fat, 17–20%; crude soybean lecithin, 1–2%; skimmed milk solids, 78–82% (10–15% dried whey powder can be included in place of an equivalent amount of skimmed milk solids); plus fortification with minerals and vitamins.

Foals suckling their dams generally develop satisfactorily up to weaning time. But the most critical period in the entire life of a horse is that space from weaning time (about six months of age) until one year of age. This is especially so in the case of the young horse being fitted for shows or sales, where condition is so important. Thus, where valuable weanlings or yearlings are to shown or sold, the use of a milk replacer may be practical.

RATIONS

Table 4–51 (see next page) contains some suggested rations for different classes of horses. This is merely intended as a general guide. The feeder should give consideration to (1) the quality, availability, and cost of feeds; (2) the character and severity of the work; and (3) the age and individuality of the animal.

TABLE 4–51
LIGHT HORSE FEEDING GUIDE[1]

Age, Sex, and Use	Daily Allowance	Kind of Hay	Suggested Grain Rations: Rations No. 1 (lb)	Rations No. 2 (lb)	Rations No. 3 (lb)
Stallions in breeding season (weighing 900 to 1,400 lb)	¾ to 1½ lb grain per 100 lb body weight, together with a quantity of hay within same range.	Grass-legume mixed; or ⅓ to ½ legume hay, with remainder grass hay.	Oats ... 55 Wheat ... 20 Wheat bran ... 20 Linseed meal ... 5	Corn ... 35 Oats ... 35 Wheat ... 15 Wheat bran ... 15	Oats ... 100
Pregant mares (weighing 900 to 1,400 lb)	¾ to 1½ lb grain per 100 lb body weight, together with a quantity of hay within the same range.	Grass-legume mixed; or ⅓ to ½ legume hay, with remainder grass hay (straight grass hay may be used first half of pregnancy).	Oats ... 80 Wheat bran ... 20	Barley ... 45 Oats ... 45 Wheat bran ... 10	Oats ... 95 Linseed meal ... 5
Foals before weaning (weighing 100 to 350 lb with projected mature weights of 900 to 1,400 lb)	½ to ¾ grain per 100 lb body weight, together with a quantity of hay within same range.	Legume hay.	Oats ... 50 Wheat bran ... 40 Linseed meal ... 10 Rations balanced on basis of following assumption: Mares of mature weights of 600, 800, 1,000, and 1,200 lb may produce 36, 42, 44 and 49 lb of milk daily.	Oats ... 30 Barley ... 30 Wheat bran ... 30 Linseed meal ... 10	Oats ... 80 Wheat bran ... 20
Weanlings (weighing 350 to 450 lb)	**1** to 1½ lb grain and 1½ to 2 lb hay per 100 lb body weight.	Grass-legume mixed; or ½ legume hay, with remainder grass hay.	Oats ... 30 Barley ... 30 Wheat bran ... 30 Linseed meal ... 10	Oats ... 70 Wheat bran ... 15 Linseed meal ... 15	Oats ... 80 Linseed meal ... 20
Yearlings, second summer (weighing 450 to 700 lb)	Good, luxuriant pastures. (If in training for other reasons without access to pastures, the ration should be intermediate between the adjacent upper and lower groups.)				
Yearlings, or rising 2-year-olds, second winter (weighing 700 to 1,000 lb)	½ to 1 lb grain and 1 to 1½ lb hay per 100 lb body weight.	Grass-legume mixed; or ⅓ to ½ legume hay, with remainder grass hay.	Oats ... 80 Wheat bran ... 20	Barley ... 35 Oats ... 35 Bran ... 15 Linseed meal ... 15	Oats ... 100
Light horses at work; riding, driving, and racing (weighing 900 to 1,400 lb)	Hard use—1¼ to 1⅓ lb grain and 1 to 1¼ lb hay per 100 lb body weight. Medium use—¾ to 1 lb grain and 1 to 1¼ lb hay per 100 lb body weight. Light use—⅖ to ½ lb grain and 1¼ to 1½ lb hay per 100 lb body weight.	Grass hay.	Oats ... 100	Oats ... 70 Corn ... 30	Oats ... 70 Barley ... 30
Mature idle horses; stallions, mares, and geldings (weighing 900 to 1,400 lb)	**1½** to 1¾ lb hay per 100 lb body weight.	Pasture in season; or grass-legume mixed hay.	(With grass hay, and ¾ lb of	a high-protein supplement daily.)	

[1]With all rations and for all classes and ages of horses, provide free access to a mineral box as follows: (1) *Where the pasture or hay is primarily grass*, use a mixture containing 2 parts of calcium to 1 part of phosphorus; and (2) *where the pasture or hay is primarily a legume*, use a mixture containing 1 part of calcium to 1 part of phosphorus. To each of these mixes, add ⅓ salt (trace mineralized) to improve acceptability. If preferred, a good commercial mineral may be used. Self-feed salt separately.

AMOUNT TO FEED

When given all the feed that they will consume, mature horses will generally eat an amount equivalent to about 2.5% of their body weight. Growing foals and lactating mares eat more heartily—they will consume up to 3% of their body weight.

Because the horse has a rather limited digestive capacity, the amount of concentrates must be increased and the roughage decreased when the energy needs rise with greater amount, severity, or speed of work. The following are general guides for the daily ration of horses under usual conditions.

1. **For horses at light work** (1 to 3 hours per day of riding or driving) allow ⅔ to ½ lb of grain and 1¼ to 1½ lb of hay per day per 100 lb liveweight.
2. **For horses at medium work** (3 to 5 hours per day of riding or driving), allow ¾ to 1 lb of grain and 1 to 1¼ of hay per 100 lb of liveweight.
3. **For horses at hard work** (5 to 8 hours per day of riding or driving), allow about 1¼ to 1⅓ lb of grain and 1 to 1¼ lb of hay per 100 lb of liveweight.

GENERAL FEEDING RULES

In addition to the guides already mentioned, observance of the following general rules will help avoid some of the common difficulties:

1. Know the approximate weight and age of each animal.

2. Feed by weight of feed, not by volume (volume as determined by a coffee can or marked bucket). Horses do not require a certain volume of feed; rather, they require a certain weight of nutrients based on their body weights.

3. Avoid sudden changes in the ration.

4. Never feed moldy, musty, dusty, or frozen feed.

5. Feed regularly. Horses anticipate their feed.

6. Look for problems at feeding time; don't just dump the feed and run. Look for injuries and abnormalities.

7. Check the feces. Any change in quantity, odor, color, or composition may presage trouble.

8. Inspect the feedbox frequently to see if the horse goes off feed. Feed refusal means (1) the horse was over-fed, (2) something is wrong with the feed, or (3) the horse is sick.

9. Keep the feed and water containers clean. Scrub them periodically to insure proper sanitation.

10. Do not overfeed. Some horses suffer from obesity, while others suffer from deficiency. Fat horses not receiving adequate exercise are predisposed to colic and founder. An old Arab proverb cautions: "The greatest enemies of horses are fat and rest."

11. Force aggressive eaters to slow down. Some horses may bolt their feed when fed in deep narrow feed boxes. Their eating may be slowed by scattering the feed in a larger box, or by placing large round stones, bricks, or salt blocks in the feed container.

12. Accord timid eaters solitude to eat. Feed them where it is quiet and they will not be disturbed.

13. Do not feed from the hand; this can lead to *nibbling.*

14. Exercise stalled horses daily. It improves their appetite, digestion, and overall well being. This may be accomplished by riding, longeing, walking, ponying, swimming, or treadmilling.

15. Avoid excessive exercise (to the point of fatigue and stress), rough treatment, noise, and excitement.

16. Do not feed concentrates 1 hour before or within 1 hour after hard work.

17. Feed horses as individuals; consider their likes and temperaments. Learn the peculiarities and desires of each animal because each one is different.

18. Gradually decrease the condition of horses that have been fitted for show or sale. Many caretakers accomplish this difficult task, and yet retain strong vigorous animals, by cutting down gradually on the feed and increasing the exercise.

19. Prevent wood chewing. This habit usually results from boredom, lack of exercise, lack of adequate roughage, or lack of phosphorus; so, alleviate the causes.

20. Make certain that the horse's teeth are sound.

21. Know the signs of a well-fed, healthy horse, any departure from which constitutes a warning signal.

FEEDING PLEASURE HORSES

It is difficult to feed pleasure horses properly because their exercise is often irregular. Sometimes they are used moderately; at other times they are idle; at still other times they are worked hard over the weekend or on a trail ride.

Fig. 4-77. Pleasure horses—on a trail ride. (Courtesy, *The Western Horseman*, Colorado Springs, CO)

Most horses used for pleasure are worked lightly, perhaps 1 to 3 hours of riding per day. Others are worked medium hard, as when ridden 3 to 5 hours per day. Still others are worked very hard, as when raced or when ridden 5 to 8 hours per day. The recommended daily feed allowance per 100 lb body weight of pleasure horses in light, medium, and hard use follows:

Lb Daily/100 Lb Weight of Horse	Light Use	Medium Use	Hard Use
Hay	1¼–1½	1–1¼	1–1¼
Grain	⅖–½	¾–1	1¼–1⅓

As shown above, the roughage content of the ration decreases and the concentrate content increases as the amount of work increases. This is because the digestibility and the efficiency of conversion are greater for high-energy concentrates than for roughages.

FEEDING HORSES IN TRAINING

Horses in heavy training for specific purposes—such as training for racing, cutting, roping, jumping, or hunting—have a higher nutritional requirement than most pleasure horses. And the younger the animal in training, the higher the level of nutrition needed in order to develop and maintain sound legs and build a strong frame and body. Therefore, the level of work, the temperament of the individual, and the age of the horse determine the nutritional needs. For this reason, horses in training should be fed as individuals.

Horses in training will eat about 1½ lb of grain and 1 lb of hay per 100-lb liveweight.

FEEDING RACEHORSES

Racehorses are equine athletes whose nutritive requirements are the most exacting, but the most poorly met, of all animals.

Fig. 4–78. A harness horse—racing. (Courtesy, U.S. Trotting Assn., Columbus, OH)

High strung and highly stressed racehorses need special rations just as human athletes do—and for the same reasons; and, the younger the age, the more acute the need. This calls for rations rich in protein, rich in readily available energy, fortified with vitamins, minerals, and unidentified factors—and with all nutrients in proper balance.

A racehorse is asked to develop a large amount of horsepower in a period of 1 to 3 minutes. The oxidations that occur in a racehorse's body are at a higher pitch than in an idle horse, and therefore, more vitamins are required.

Also, racehorses are the *prima donnas* of the equine world; most of them are temperamental, and no two of them can be fed alike. They vary in rapidity of eating, in the quantity of feed that they will consume, in the proportion of concentrate to roughage that they will take, and in the response to different caretakers. Thus, for best results, they must be fed as individuals.

During the racing season, the hay of a racehorse should be limited to 7 or 8 lb, whereas the concentrate allowance may range up to 16 lb. Heavy roughage eaters may have to be muzzled, to keep them from eating their bedding. A bran mash is commonly fed once a week.

FEEDING BROODMARES

The following pointers are pertinent to feeding a broodmare properly:

1. Condition the mare for breeding by providing adequate and proper feed and the right amount of exercise prior to the breeding season.

2. See that adequate proteins, minerals, and vitamins are available during the last third of pregnancy when the fetus grows most rapidly.

3. Feed and water with care immediately before and after foaling.

4. Provide adequate nutrition during lactation, because the requirements during this period are more rigorous than the requirements during pregnancy.

5. Make sure that young growing mares receive adequate nutrients; otherwise, the fetus will not develop properly or the dam will not produce milk except at the expense of her body tissue.

FEEDING STALLIONS

Fig. 4–79. *GAIPARADA+++*. Triple National Champion Arabian stallion, owned by Gainey Fountainhead Arabians, Santa Ynez, CA. (Courtesy, Gainey Arabians, Santa Ynez, CA)

The ration exerts a powerful effect on sperm production and semen quality. Successful breeders adhere to the following stallion feeding rules:

1. Feed a balanced ration, giving particular attention to proteins, minerals, and vitamins.

2. Regulate the feed allowance because the stallion can become infertile if he gets too fat. Also, increase the exercise when the stallion is not a sure breeder.

3. Provide pasture in season as a source of both nutrients and exercise.

FEEDING FOALS

Fig. 4-80. Creep feeder for weanlings. (Courtesy, University of Kentucky, Lexington)

The following pointers are pertinent to proper feeding of foals:

1. Start on feed early, which means at 10 days to 3 weeks of age. Rolled oats and wheat bran, to which a little brown sugar has been added, is especially palatable as a starting ration. Crushed or ground oats, cracked or ground corn, wheat bran, and a little linseed meal may be provided later with good result; or, a good commercial ration may be fed if desired and available.

2. Use a scientifically formulated ration even though it seems expensive, for usually it will represent a wise investment. Supplemental feeding also affords a convenient way in which to improve upon milk, by reinforcing it with certain minerals (for example, milk is low in iron and copper), vitamins, and additives.

Provide a good hay, preferably a legume, or pasture, in addition to its grain ration.

3. Allow about ½ lb of grain daily per 100 lb of body weight at 4 to 5 weeks of age. This ration should be increased by weaning time to about ¾ lb or more per 100 lb of body weight. The exact amount of the ration will vary with the individual, the type of feed, and the development desired.

4. Obtain growth with durability and soundness, which calls for expert care and particular emphasis on the kind of ration, feed allowance, and exercise.

5. Simplify weaning and setback by feeding foals so that they rely less upon their mothers.

FEEDING WEANLINGS

No great setback or disturbances will be encountered at weaning time provided that the foals have developed a certain independence from proper grain feedings during the suckling period. Generally, weanlings should receive 1 to 1½ lb of grain and 1½ to 2 lb of hay daily per each 100 lb liveweight. The amount of feed will vary somewhat with the individuality of the animal, the quality of roughage, available pastures, the price of feeds, and whether the weanling is being developed for show, race, or sale. Naturally, animals being developed for early use or sale should be fed more liberally, although it is equally important to retain clean, sound joints, legs, and feet—a condition which cannot be obtained so easily in heavily fitted animals.

Because of the rapid development of bone and muscle in weanlings, it is important that, in addition to ample quantity of feed, the ration also provides quality of proteins, and adequate minerals and vitamins.

FEEDING YEARLINGS

If young animals have been fed and cared for so that they are well grown and thrifty as yearlings, usually little difficulty will be experienced at any later date.

When on pasture, yearlings that are being grown for show or sale should receive grain in addition to grass. They should be confined to their stalls in the daytime during the hot days and turned out at night (because of not being exposed to sunshine, adequate vitamin D must be provided). This point needs to be emphasized when forced development is desired; for, good as pastures may be, they are roughages rather than concentrates.

The winter feeding program for the rising 2-year-old should be such as to produce plenty of bone and muscle rather than fat. From ½ to 1 lb of grain and 1 to 1½ lb of hay should be fed for each 100 lb of liveweight. The quantity will vary with the quality of the roughage, the individuality of the animal, and the use for which the animal is produced. In producing for sale, more liberal feeding may be economical. Access to salt and to a mineral mixture should be incorporated in the ration. An abundance of fresh, pure water must be available.

FEEDING TWO- AND THREE-YEAR-OLDS

Except for the fact that the 2- and 3-year-olds will be larger, and, therefore, will require more feed, a description of their proper care and management would be merely a repetition of the principles that have already been discussed for the yearling.

FITTING FOR SHOW OR SALE

Each year, many horses are fitted for shows or sales. In both cases, a fattening process is involved, but exercise is doubly essential.

For horses that are being fitted for shows, the conditioning process is also a matter of hardening, and the horses are used daily in harness or under saddle. Regardless of whether a sale or a show is the major objective, fleshing should be obtained without sacrificing action or soundness or without causing filling of the legs and hocks.

In fattening horses, the animals should be brought to a full feed rather gradually, until the ration reaches a maximum of about 2 lb of grain daily for each 100 lb of liveweight. When on full feed, horses make surprising gains. Daily weight gains of 4 to 5 lb are not uncommon. Such animals soon become fat, sleek, and attractive. This is probably the basis for the statement that "fat will cover up a multitude of sins in a horse."

Although exercise is desirable from the standpoint of keeping the animals sound, it is estimated that such activity decreases the daily rate of gains by as much as 20%. Because of the greater cost of gains and the expense involved in bringing about forced exercise, most feeders of sale horses limit the exercise to that obtained naturally from running in a paddock.

In comparison with finishing cattle or sheep, there is more risk in fattening horses. Heavily fed horses kept in idleness are likely to become blemished and injured through playfulness, and there are more sicknesses among liberally fed horses than in other classes of stock handled in a similar manner.

In fitting show horses, the finish must remain firm and hard, the action superb, and the soundness unquestioned. Thus, they must be carefully fed, groomed, and exercised to bring them to proper bloom.

Persons fitting and selling yearlings or younger animals may feed a palatable milk replacer or commercial feed to advantage.

PART VII
NUTRITIONAL DISEASES AND AILMENTS

More animals (and people) throughout the world suffer from hunger—from just plain lack of sufficient feed—than from the lack of a specific nutrient (or nutrients); therefore, it is recognized that nutritional deficiencies may be brought about either by (1) too little feed, or (2) rations that are too low in one or more nutrients.

Also, forced production (such as very high milk yields and finishing animals at early ages) and the feeding of forages and grains which are often produced on leached and depleted soils have created many problems in nutrition. This condition has been further aggravated through the increased confinement of stock, many animals being confined to lots or buildings all or a large part of the year. Under these unnatural conditions, nutritional diseases and ailments have become increasingly common.

Although the cause, prevention, and treatment of most of these nutritional diseases and ailments are known, they continue to reduce profits in the livestock industry simply because the available knowledge is not put into practice. Moreover, those widespread nutritional deficiencies which are not of sufficient proportions to produce clear-cut deficiency symptoms cause even greater economic losses because they go unnoticed and unrectified. Table 4–52 contains a summary of the important nutritional diseases and ailments affecting animals.

TABLE 4–52
NUTRITIONAL DISEASES AND AILMENTS

ACETONEMIA (See KETOSIS.)

ACIDOSIS (LACTIC ACIDOSIS)

Species Affected: Cattle, especially feedlot cattle; and sheep, especially feedlot lambs.

Cause: Acidosis is caused by an increase in lactic acid-producing bacteria and the rapid production of lactic acid. It commonly occurs when there is a sudden shift from a high-roughage to a high-concentrate ration. However, cattle maintained on high-energy rations may be in a marginal state of acidosis due to the formation of lactic acid by the rumen flora. Thus, ingredient changes, poor mixing of grain in the ration, or faulty feeding can produce acute acidosis.

Symptoms and Signs (or age group most affected): Marginal acidosis is characterized by poor performance and inconsistent feed ingestion. If ingredient changes or erratic feeding persist, acute acidosis may result, creating laminitis—and eventually "ski shoe" cattle (founder). In severe cases, the rumen becomes immobilized, followed by increased pulse and respiration rate, variable rectal temperature, sunken eyes, loss of dermal elasticity (dehydration), staggering, coma, and death.

Distribution and Losses Caused By: Acidosis occurs wherever beef or dairy cattle and lambs are fed, especially when consuming high-concentrate rations.

The annual loss from acidosis has been estimated at about 1% of the production.

Treatment: Different treatments have been used with varying degrees of success; among them: (1) removing rumen contents and replacement by contents of an animal on a normal ration; (2) feeding a high level of an antibiotic to suppress lactic acid-producing bacteria; (3) drenching (or intravenous injection) with a solution of sodium bicarbonate to restore the acid-base balance; (4) administering intramuscularly antihistamines and cortical steroids daily for each of several days to help prevent intoxication and laminitis; and/or (5) backing the cattle down on both amount and kind of feed (lessening the total amount of the ration, and returning to a higher forage mix).

Control: Acidosis is best controled by (1) avoiding accidental access of cattle to large amounts of concentrates, (2) changing gradually and stepwise from a low to a high proportion of concentrate in the ration, and (3) adding buffer salts, such as sodium bicarbonate, to the ration.

(Continued)

TABLE 4–52 *(Continued)*

ACIDOSIS *(Continued)*

Prevention: Prevention consists of starting animals on a high-roughage ration and gradually reducing the roughage and increasing the grain; avoiding erratic feeding; and avoiding abrupt ration changes.

Remarks: A feedlot history of deliberate or accidental starting of animals on high-energy feeds, or of sudden ration changes, helps establish the correct diagnosis.

A rapid field test can be used to diagnose and differentiate between rumen acidosis and urea poisoning. Samples can be collected by stomach tube or postmortem collection. In general, a rumen content pH of 5.0 or less is indicative of rumen acidosis; a rumen content pH greater than 7.5 is indicative of urea or NPN toxicosis.

ANEMIA, NUTRITIONAL

Species Affected: All warm-blooded animals, including humans.

Cause: Commonly an iron deficiency, but it may be caused by a deficiency of copper, cobalt, and/or certain vitamins.

The baby pig is born with a total of about 40 mg of iron in the body. With an iron requirement of about 7 mg daily, it is apparent that without supplemental iron, body stores will not last very long.

Sow's milk is a good source of all nutrients the baby pig is known to require with the exception of iron.

Symptoms and Signs (or age group most affected): Loss of appetite, progressive emaciation, and death.

Most prevalent in suckling young.

Pigs show listlessness, rough hair coat, wrinkled skins, drooping ears and tails, pale membranes around the mouth and eyes, labored breathing, and a swollen condition about the head and shoulders.

Distribution and Losses Caused By: Worldwide.

Losses consist of slow and inefficient gains, and deaths.

Treatment: Provide sources of the nutrient or nutrients, the deficiency of which is known to cause the condition. If iron deficiency is indicated, iron may be given by injection, in organic combination (iron dextran).

Control: When nutritional anemia is encountered, it can usually be brought under control by supplying dietary sources of the deficient nutrient(s).

Fig. 4–81. Anemia, caused by an iron deficiency, characterized by listlessness, rough hair coat, and wrinkled skin. (Courtesy, University of Florida, Gainesville)

Prevention: Supply dietary sources of iron, copper, cobalt, and certain vitamins (especially folacin, riboflavin, and vitamin B–6).

Keep confinement of suckling animals to a minimum and provide dry feeds at an early age.

Anemia in pigs can be prevented by providing supplemental iron in one of the following forms:

1. Inject intramuscularly 100 to 200 mg of iron from iron dextran into baby pigs at 2 to 3 days of age. If pigs remain in confinement and do not have access to creep feed at an early age, a second injection at 2 to 3 weeks of age is desirable. Injection is the method of choice, for it assures that every pig receives its requirement.
2. Orally administer iron dextran in a liquid or a solid preparation. To ensure daily intake by all pigs, it is important to have a preparation that is palatable and readily consumed. Also, placement of the oral preparation at the right location in the creep area is most important.
3. Give the pigs iron tablets or paste at 2 to 3 days of age. Repeat the treatment every 7 to 10 days until the pigs are eating the creep ration adequately. If pills are given, it is important to see that the pigs swallow them and not spit them out.
4. Place clean soil in the farrowing pen daily. Soil should not be contaminated with parasite eggs and other disease organisms. Iron sulfate can be sprinkled over the soil.
5. Swab sow's udder daily with a solution of 1 lb ferrous sulfate dissolved in 1 gal of warm water.
6. Provide pigs with access to a creep feed by the time they are 10 days old.

Remarks: Anemia is a condition in which the blood is either deficient in quality or quantity. (A deficient quality refers to a deficiency in hemoglobin and/or red cells.)

Levels of iron in dry feed are generally believed to be ample, since feeds contain 40 to 400 mg/lb.

AZOTURIA (HEMOGLOBINURIA, MONDAY MORNING DISEASE, BLACKWATER)

Species Affected: Horses.

Cause: Sudden exercise, following a day or two of rest during which time the horse has been on full feed, resulting in partial spasm or "tie-up." Azoturia is caused by an abnormal amount of glycogen being stored in the muscle. As the glycogen breaks down, lactic acid is formed, which builds up in the muscle causing severe muscle destruction and the release of myoglobin which manifests itself as partial spasm or "tie-up" and wine-colored urine.

Symptoms and Signs (or age group most affected): Symptoms usually develop 15 to 60 minutes after the beginning of exercise. Azoturia is characterized by profuse sweating, elevated temperature and pulse, wine-colored urine (caused by the release of myoglobin—the red pigment in muscle tissue), tight (cramping) and sore loin hindquarter muscles—they're "tied up" due to semi-paralysis, stiff gait, reluctance to move due to pain, and knuckling over of the hind pasterns. Finally, the animal may assume a sitting position and, eventually, fall prostrate on its side. The breath and urine may have a peculiar odor.

Distribution and Losses Caused By: Worldwide, but the disease is seldom seen in horses at pasture and rarely in horses at constant work.

(Continued)

TABLE 4–52 *(Continued)*

AZOTURIA *(Continued)*

Treatment: Absolute rest and quiet. While awaiting the veterinarian, apply heated cloths or blankets, or hot-water bottles to the swollen and hardened muscles, *but don't try to move the horse—don't take the horse back to the barn.* Keep it on its feet if possible, even if you have to use a sling.

The veterinarian should determine treatment. In mild cases, treatment may consist of the use of a tranquilizer or a sedative. In severe cases, the veterinarian may use (1) muscle relaxers or (2) sodium bicarbonate in solution to readjust the acid balance in the muscles.

Control: When trouble is encountered, decrease the concentrate ration and increase the exercise on idle days.

Prevention: Restrict the grain ration, increase good quality roughage, and provide daily exercise when the animal is idle. Give a wet bran mash the evening before an idle day or turn the idle horses to pasture.

Some believe that a diuretic (a drug which will increase the flow of urine) will prevent the tie-up syndrome. This is a common treatment of racehorses.

Others feel that increased B vitamins will prevent the lactic acid buildup.

Remarks: The chances of recovery are good for horses that remain standing, are not forced to move after the signs are noticed, and whose pulse returns to normal within 24 hours.

Azoturia and colic have some similar symptoms; hence, there is danger of misdiagnosis and the wrong treatment. Walking, a standard part of colic treatment, is the worst thing to do when a horse has azoturia.

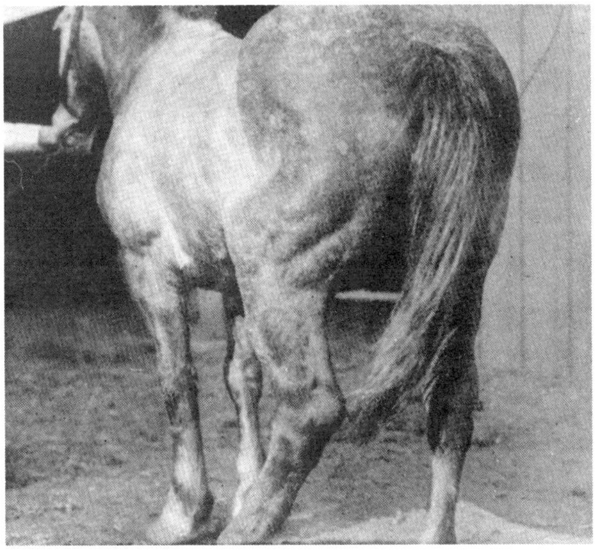

Fig. 4–82. Horse with Azoturia, evidencing sore loin and hindquarters—it is "tied up." (Courtesy, Pitman-Moore, Indianapolis, IN)

BLOAT—FEEDLOT

Species Affected: All ruminants.

Cause: Bloat is an excessive accumulation of gas in the rumen and reticulum of ruminants. High-concentrate rations, especially when finely ground, increase numbers of slime-producing bacteria in rumen. Slime traps fermentation gas and produces bloat. Both frothy and free gas bloat occur in feedlot bloat.

Genetic tendency or physiological abnormality.

Symptoms and Signs (or age group most affected): Symptoms same as pasture bloat (see "Bloat—Pasture" which follows).

Occurs when cattle or sheep have been fed high-concentrate, low-roughage rations for approximately 60 days or longer.

Distribution and Losses Caused By: A survey of Kansas feedlots showed the following losses from bloat: 0.1% died of bloat; 0.2% bloated severely; and 0.6% bloated mildly to moderately, with animal performance affected adversely.

Treatment: Reduce intraruminal pressure as quickly as possible. This may be done by means of a large stomach tube, although this method is usually disappointing in foamy bloat.

Administer a defoaming agent immediately, such as (1) 1 pint of corn oil, peanut oil, or soybean oil; or (2) poloxalene administered according to the manufacturer's directions.

As a last resort, a trocar and cannula can be inserted on the left side of the animal at the center of a triangle formed by the backbone, the hipbone, and the last rib. The trocar is removed, but the cannula should stay in place until all gases have dissipated. If a trocar and cannula are not available, a knife may be used in emergencies.

Control: If feasible, increase proportion of nonlegume roughage in ration. However, good-quality legume hay may increase incidence of feedlot bloat. In this case, poloxalene or oxytetracycline are effective preventives when used according to manufacturers' directions.

Prevention: (1) Use poloxalene (Bloat Guard) or oxytetracycline (Terramycin and Neo-Terramycin) according to manufacturers' directions; and (2) proper management.

Remarks: Feedlot bloat may occur during any month of the year; however, it is more common during hot, humid weather.

Two products are cleared by FDA for bloat control; namely, poloxalene (trade name, Bloat Guard) and oxytetracycline (trade names, Terramycin and Neo-Terramycin).

BLOAT—PASTURE (LEGUME BLOAT)

Species Affected: All ruminants.

Cause: Bloat is caused by the inability of the animal to get rid of ruminal gas. Lack of scabrous (rough) material in the rumen to stimulate eructation (belching), along with the formation of heavy foam bubbles, seems to be the main cause of pasture bloat. Pasture bloat is most common on immature, rapidly growing legumes and on wheat pasture. Pasture bloat is a frothy bloat caused by interaction of several factors—plant, animal, and microbial. Soluble plant proteins and the presence of saponins play a prominent role in permitting stable foam formation.

Heavy applications of urea fertilizer on pastures may also induce bloat.

Animals that will bloat on any feed are known as chronic bloaters. These animals, in which there may be a genetic tendency, are unable to eructate (belch) fermentation gases because of some physiological abnormality.

(Continued)

TABLE 4–52 *(Continued)*

BLOAT-PASTURE *(Continued)*

Symptoms and Signs (or age group most affected): First observed as a distention of the paunch on the left side in front of the hipbone. This is followed by distention of the right side, protrusion of the anus, respiratory distress, cyanosis (bluish coloration) of the tongue, struggling, and death if not treated. The entire period of time from when a ruminant enters a pasture until death occurs can be as short as a half hour.

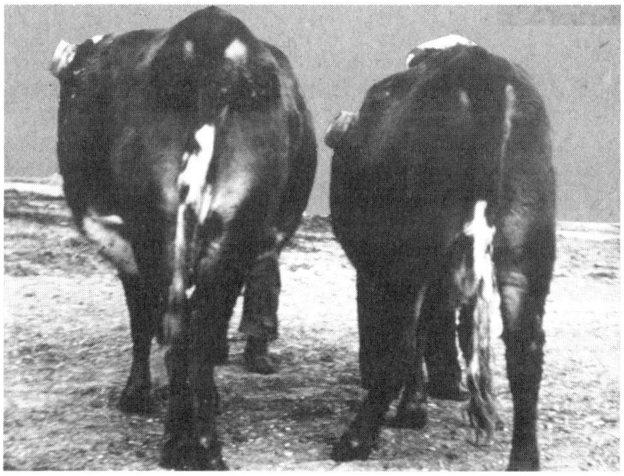

Fig. 4–83. Identical twins. *Left:* Bloated animal showing distention of the paunch on the left side in front of the hipbone. *Right:* Twin mate showing no bloating. (Courtesy, Kansas State University, Manhattan)

Distribution and Losses Caused By: Widespread, although some areas appear to have more bloat than others.

It often results in death.

Bloat causes annual losses in beef and dairy cattle of more than $100 million from reduced weight gain and lower milk production.

Treatment: Time permitting, severe cases of bloat should be treated by a veterinarian. The use of a stomach tube, carefully inserted, is usually very helpful in eliminating gases. Puncturing of the paunch with a trocar and cannula should be a last resort. A knife may be used in emergencies.

Mild cases may be home treated by (1) keeping the animal on its feet and moving; and (2) drenching cattle either with (a) 1 pint of corn oil, or soybean oil; or (b) 1–2 oz poloxalene.

Control: When there is a high incidence of bloat, it may be desirable to change the feed.

Where legume bloat is encountered, use poloxalene (Bloat Guard), oxytetracycline (antibiotic), or polyoxyethylene (23) lauryl ether (Laureth–23/Enproal Bloat Blox), according to the respective manufacturers' directions.

Prevention: The incidence is lessened by (1) avoiding straight legume pastures and immature legumes, (2) feeding a coarse grass hay prior to turning onto lush pasture, (3) feeding dry forage along with pasture, (4) avoiding a rapid fill from an empty start, (5) keeping animals continuously on pasture after they are once turned out, (6) keeping salt and water conveniently accessible at all times, (7) avoiding frosted pastures, or (8) using poloxalene (Bloat Guard), oxytetracycline (Terramycin or Neo-Terramycin), or Laureth–23 (Enproal Bloat Blox) according to manufacturers' directions, including placing blocks containing these antifoaming agents in various parts of the pasture.

Remarks: Legume or cereal pastures, or alfalfa hay, appear to be associated with a higher incidence of bloat than any other feeds.

Legume pastures are particularly hazardous when immature, when moist, after a light rain or dew.

COBALT DEFICIENCY (See SALT SICK.)

COLIC

Species Affected: Horses.

Cause: Internal parasites are the number one cause of colic; additional causes are improper feeding, working, or watering. There are more than 70 different things that can cause colic.

Symptoms and Signs (or age group most affected): Severe pain, usually in the abdomen; and depending on the type of colic, other symptoms are: the horse looking at his belly, distended abdomen, increased intestinal rumbling, violent rolling and pawing, profuse sweating, constipation, and refusal of feed and water.

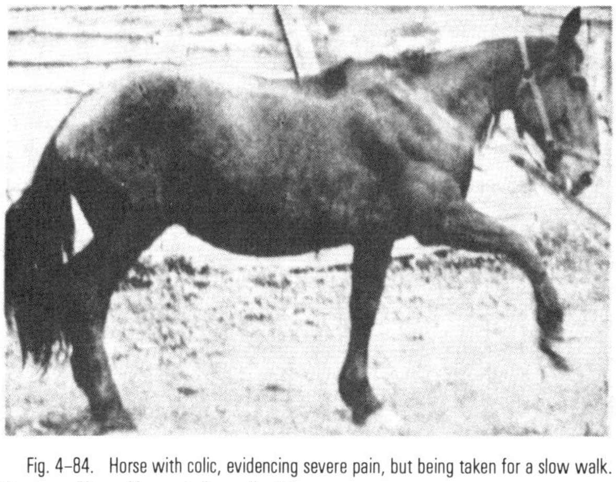

Fig. 4–84. Horse with colic, evidencing severe pain, but being taken for a slow walk. (Courtesy, Pitman-Moore, Indianapolis, IN)

Distribution and Losses Caused By: Worldwide.

Colic is the most common ailment among horses and is the leading cause of death. Livestock insurance companies report about ⅓ of all deaths of insured horses can be attributed to colic.

Treatment: Call a veterinarian. To avoid danger of inflicting self-injury, (1) place the animal in a large, well-bedded stable, or (2) take the animal for a slow walk. Do not give the horse any type of drug, unless so advised by the veterinarian when telephoned. Painkillers may cover up symptoms which are vital for the veterinarian in making an accurate diagnosis.

Depending on the diagnosis, the veterinarian may use one or more of the following: sedatives; laxatives, such as mineral oil; drugs; or surgery. The surgeon may avoid recurrence of twists and displacements of the horse's colon by attaching it to other organs or the abdominal wall, thereby deliberately creating adhesions which prevent further twisting.

In the late 1980s, Colorado State University scientists developed a "scorecard" method of evaluating colic (1) to help differentiate between a colicking horse that needs surgery and one that should be treated medically, and (2) to predict how likely a horse in each category is to survive.

Control: Follow a good management program, including parasite control.

Feed, work, and water horses properly.

Prevention: Proper feeding (including adequate roughage), working, watering, and parasite control.

Remarks: The word colic is not specific. There are many syndromes that can result in colic, and not all of them are gastrointestinal. For example, blood worms can cause colic due to damaging the walls of blood vessels, and mares with uterine torsions exhibit colic pain as do horses with urinary stones in the bladder. So, the first thing is to determine what is causing the animal to exhibit colic symptoms.

(Continued)

TABLE 4–52 *(Continued)*

ENTEROTOXEMIA (OVEREATING DISEASE, PULPY-KIDNEY DISEASE)

Species Affected: Sheep; less frequently goats, and rarely cattle.

Cause: *Clostridium perfringens* type D. However, predisposing factors are essential; the most common of these are overconsumption of high-energy feeds, an abundant milk supply, and lush pastures. Under such conditions, the *Clostridium perfringens* bacteria grow rapidly and produce a powerful toxin.

Symptoms and Signs (or age group most affected): Sudden death; frequently a lamb is found dead in the field or feedlot without having shown any previous signs of illness. Quite often it is the biggest lamb with the biggest appetite. The disease develops rapidly; and the animal becomes weaker and weaker and shows nervous disturbances such as circling, butting, or throwing the head from side to side or backwards. Finally, the animal collapses and may go into convulsions before dying. Enterotoxemia can be confirmed by laboratory tests if necropsy is performed shortly after death.

Distribution and Losses Caused By: Worldwide.

The death rate is a minimum of 1%, with an average of 3 to 4% in unvaccinated feedlot lambs. In explosive outbreaks, losses range from 10 to 40%.

Treatment: None.

Control: The method of control depends on the age of the lambs, the frequency with which the disease occurs, and the method of husbandry.

When an outbreak in feeder lambs occurs, for several days (1) increase the amount of roughage in the ration, and (2) add 200 g of chlortetracycline per ton of feed.

When an outbreak in nursing lambs occurs, injection of all susceptible lambs with enterotoxemia antiserum will provide protection for about 14 to 21 days, at which time the lambs can be vaccinated.

Prevention: Along with proper feeding, vaccinate. Ewes should be vaccinated with type C and D toxoid. Lambs should be vaccinated with type D only.

Nursing lambs can be protected by vaccinating the ewes for enterotoxemia. For best results, previously unvaccinated ewes should be vaccinated with type C and D toxoid twice before lambing. The 2 doses should be spaced at least 1 month apart, with the second dose given 2 to 4 weeks before start of lambing. Ewes which have been vaccinated previously need only one booster shot before lambing. Vaccinating the ewes prior to lambing ensures that the lambs will receive colostral protection for 2 to 3 weeks, following which the lambs should be vaccinated at 4 to 6 weeks of age.

Feedlot lambs can be protected by giving them one dose of type D toxoid soon after their arrival in the feedlot. It takes about 10 days after vaccination for immunity to develop. Sometimes revaccination with the toxoid or bacterin (a booster shot) is required 2 to 4 weeks following the first vaccination.

Remarks: The disease is caused by bacteria, but it is triggered by high-energy rations or excellent pastures.

FESCUE FOOT (FESCUE TOXICOSIS)

Species Affected: Cattle, sheep, and horses.

Cause: A fungus (endophyte), *Acremonium coenophialum,* which lives in the leaves, stems, and seeds of tall fescue, without adversely affecting the fescue plant.

Symptoms and Signs (or age group most affected): The symptoms vary. Some animals show no apparent lameness, whereas others show varying degrees of sloughing (necrosis) on the ends of their tails. Mild fescue toxicosis is characterized by poor conception rates, low pasture gains, and depressed milk production.

The most common symptoms in horses, in the order of their occurrence, are a decrease or absence of milk production, prolonged gestation, abortion, and thickened placenta.

Fescue toxicity is more common in animals suffering from malnutrition and/or parasitism.

Distribution and Losses Caused By: Fescue foot has occurred in the U.S., Australia, New Zealand, and Italy. In the U.S., it has been reported in California, Colorado, Florida, Kentucky, Missouri, and Tennessee.

Tall fescue is currently grown on about 35 million acres in the U.S.

The Mississippi Station reports that studies extending over several years show that each 10% fungus infection in a fescue pasture will lower the daily gains of cattle by about 10%.

University of Kentucky researchers report that dairy cows grazing 70% infected fescue produced an average of 11.2 lb less milk per day than cows grazing noninfected fescue.

Treatment: There is no effective medication. Cattle usually recover if removed from fescue pasture or fescue hay.

Control: Until, and unless, scientists find a way to remove the toxic factor(s) from fescue, the best control consists of good management, proper nutrition of animals, early detection of symptoms, and/or destroying toxic pastures and reseeding with endophyte-free seed.

Prevention: The seeding of fungus-free fescue seed is the best way to prevent fescue foot. Also, interseeding alfalfa, clovers, or other grasses into fescue stands dilutes the amount of fescue eaten and helps to reduce the toxic effects.

Also, tall fescue selections low in, or free of, this endophytic fungus are evolving.

In areas where fescue toxicity is a problem, gestating mares should be removed from fescue pasture the last 2 or 3 months of pregnancy.

Remarks: Most cases of fescue toxicity occur among cattle that graze pure stands of fescue during late fall and winter; and most toxic stands of fescue pasture are several years old.

FOUNDER (LAMINITIS)

Species Affected: Horses, cattle, sheep, goats.

Cause: A variety of causes have been recognized, including (1) overeating and too rapid increase in the ration (grain founder), (2) digestive disturbances (enterotoxemia), (3) retained afterbirth (foal founder), (4) lush pastures (grass founder), and (5) concussion (road founder).

Symptoms and Signs (or age group most affected): Extreme pain, fever (103°–106°F), and reluctance to move—the animal appears to be "walking on eggs." If neglected, it causes an acute or chronic degeneration of the joining of the sensitive and insensitive laminae of the foot; and, if the degeneration is severe, the coffin bone may rotate and come through the bottom of the foot.

Distribution and Losses Caused By: Worldwide.

Actual death losses from founder are not very frequent, but animal usefulness may be severely affected.

Treatment: There is no widely accepted, standard method of treating founder. If known, the condition(s) that caused the problem should be alleviated. Treatment of acute horse founder usually involves one or more of the following procedures and medications:

1. *Mineral oil.* To aid passage of the excessive feed consumed and prevent further absorption of lactic acid and endotoxin into the bloodstream. One gallon should be given via stomach tube. **Note well**: A purgative should not be employed in cases involving pneumonia or parturient laminitis of mares.
2. *Analgesics (pain-killers).* To obtain pain relief.
3. *Injectable antihistamines.* To provide anti-inflammatory effects.
4. *Antibiotics.* To combat the subsequent formation of endotoxin which destroys tissue.
5. *Sodium bicarbonate (baking soda).* To neutralize the acidic toxicity of these inflammatory products.
6. *Temporarily deadening the nerve supply to the feet.* To alleviate pain and assist in restoring blood supply by allowing the horse to be walked.

(Continued)

TABLE 4–52 *(Continued)*

FOUNDER *(Continued)*

7. *Water soaks.* To stimulate and massage the blood supply to the feet in an effort to open up previously constricted blood vessels, but there is disagreement as to whether the water should be warm or cold.

8. *Wraps applied to the affected feet.* To cushion the painful sole soreness which is evident in the toe region due to coffin bone rotation following breakdown of the laminae.

Other treatments for founder that are sometimes used, with varying degrees of reported success, are cortisones; and methionine, a sulfur containing amino acid.

Note well: Due to the complexity of laminitis and its far-reaching effect on many of the horse's internal systems, the treatments described above may not be effective in controlling founder. If the disease is not diagnosed and treated early enough, the coffin bone rotation and damage to the associated hoof wall structure can become irreversible.

Treatment of chronic founder consists of attempting to restore the normal alignment of the rotated coffin bone by lowering the heels, removing excess toe, and protecting the dropped sole. This may be accomplished by a competent farrier through proper trimming and perhaps by using leather pads or a steel-plate shoe. Also, soft acrylic plastics are sometimes applied to the sole to replace the injured hoof area and help realign the rotated coffin bone. Because of the tendency to refounder, weight control is extremely important and overfeeding should be avoided.

Control: Alleviate the causes.

Prevention: Avoid (1) overeating, (2) overdrinking (especially when hot), and/or (3) inflammation of the uterus following parturition.

Veterinary attention should be given if mares retain the after-birth longer than 12 hours.

Careful management practices related to grain feeding will prevent many cases of founder in cattle, sheep, and goats.

Remarks: Unless foundered animals are quite valuable, it is usually desirable to dispose of them following a case of severe founder.

Swine do not founder because they can unload their stomachs by vomiting.

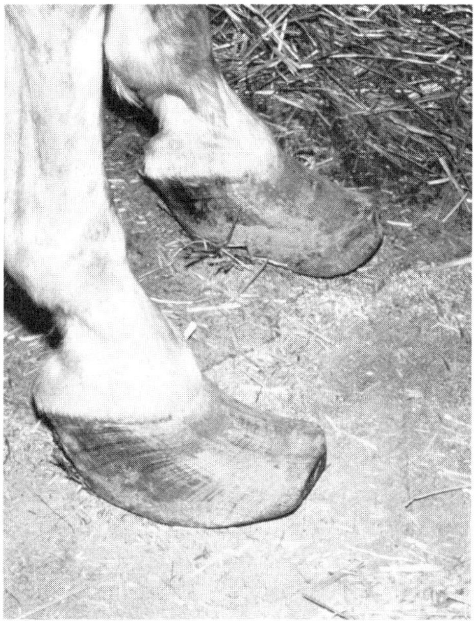

Fig. 4–85. "Snowshoe feet" of a horse due to chronic founder. (Courtesy, Colorado State University, Ft. Collins)

GOITER (See IODINE DEFICIENCY.)

GRASS TETANY (HYPOMAGNESEMIC TETANY, GRASS STAGGERS, WINTER TETANY. In Europe, it's called FOG FEVER.)

Species Affected: Cattle (beef and dairy; in the U.S., tetany is more common in beef herds than in dairy herds). Sometimes sheep and goats, in which the disease occurs under essentially the same conditions and has the same clinical signs as in cattle.

Cause: Grass tetany is a nutritional disease caused by an inadequate level of magnesium (Mg) in the blood. It most commonly occurs among lactating animals grazing rapidly growing, lush spring pastures containing less than 0.2% magnesium and more than 3% potassium and 4% nitrogen (25% protein). Forage that is high in potassium and nitrogen should have a magnesium content of at least 0.25%. Such low magnesium pastures are most commonly encountered during the first two weeks of the pasture season, although somewhat later in the season outbreaks have been reported during rainy and foggy weather. Sometimes tetany is a problem when cattle are allowed to overgraze a field, then moved abruptly to a field of new lush growth. Small grain pastures (wheat/rye/oats/barley) are especially troublesome. Also, the disease may occur when animals are fed poor quality hay, straw, or corn stover—feeds that are low in magnesium. It is not common on legume pasture or in animals wintered on legume hay. (Legumes may contain twice the magnesium concentration of grasses grown on the same soil.)

Several factors adversely influence magnesium metabolism in cattle and may "trigger" grass tetany; among them, drastic fluctuations in spring temperatures, prolonged cloudy weather, organic acid content of plants, hormonal status of the animal, level of higher fatty acids in plants, energy intake of the animal, and additional stress—such as a dog chasing animals, parasites, or a cold rain.

Grass tetany is most likely to occur on pasture plants grown on soils that are low in available magnesium and high in available potassium. If calcium is low as well as magnesium, the hazard of tetany is even greater. Many state soil-testing laboratories provide information on the danger of tetany on pastures, and can recommend corrective fertilization or dolomitic liming (which contains magnesium). Also, the historical record of grass tetany in an area or on a specific pasture is important.

Symptoms and Signs (or age group most affected): The initial signs of magnesium deficiency include nervousness, attentive ears, and decreased milk yield; signs which, to the experienced and observing caretaker, indicate the need for immediate preventive measures—before the animals become sick tomorrow. In more severe cases, affected animals may avoid the rest of the herd, walk with a stiff gait, lose their appetite, and urinate frequently. They are nervous, have staring eyes, and keep their head and ears in an erect position. Also, they stagger; have a twitching skin, especially on the face, ears, and flanks; and lie down and get up frequently. Animals may be irritable and behave aggressively; they may even charge or fight persons in the immediate area. After a time (as long as 2 to 3 days), extreme excitement and violent convulsions may develop. Animals lie flat on their sides, the fore legs pedal periodically, saliva flows freely, breathing is labored, and the heart pounds rapidly. If treatment is not given at this stage, animals usually die during or after a convulsion. The various symptoms of animals suffering from grass tetany indicate that the nervous system controlling both voluntary and involuntary muscles is affected. Quite often, clinical signs are not observed, and the only evidence is a dead animal.

Older cows are more susceptible to grass tetany than those with their first or second calves, because of lowered magnesium stores and decreased absorption efficiency. Also, the disease is most likely to strike beef cows during early lactation, especially those with high levels of milk production. Dry cows and bulls are seldom affected.

Normal plasma magnesium levels range from 1.8 to 2.0 mg/100 ml; values below 1.0 to 1.2 mg/100 ml are indicative of magnesium deficiency. However, not all cattle with low plasma magnesium develop tetany. Also, plasma magnesium levels in affected animals may return to almost normal during the convulsive stage. So, diagnosis cannot be based on blood tests alone. Since the kidneys apparently start conserving magnesium when the serum level reaches about 1.8 mg/100 ml, one of the better diagnostic aids to indicate grass tetany is low urinary magnesium.

(Continued)

TABLE 4–52 *(Continued)*

GRASS TETANY *(Continued)*

Distribution and Losses Caused By: Grass tetany is a worldwide problem, with occurrence sporadic and unpredictable for any given area. It is generally considered to be the leading cause of cattle deaths in the U.S., killing an estimated 1 to 3% of the cattle in temperate regions.

Treatment: Treatment of tetany cases can be successful if given early and without excessive handling of the affected animals. Chance of recovery is slight if treatment is delayed 8–12 hours; so, call the veterinarian immediately.

Under range conditions, 200 cubic centimeters (cc) of a sterile, saturated solution of magnesium sulfate (Epsom Salts) injected under the animal's skin (inject only 50 cc at any one place on the animal) places a high level of magnesium in the blood in 15 minutes.

Some veterinarians use intravenous injections of chloral hydrate or magnesium sulfate to calm excited animals, then follow with a calcium-magnesium gluconate solution. If the animal again goes into convulsions, a second dose of calcium-magnesium gluconate solution may be required. Intravenous injections should be administered slowly (allow about 15 minutes for a 500 cc bottle) by a trained person because there is a danger of heart failure if they are given too rapidly.

An enema of 60 g (*2 oz*) of magnesium chloride ($MgCl_2 \cdot 6H_2O$) in 10 oz of water is helpful. The enema may be given with an esophageal or oral calf feeder with the probe inserted 10 in. into the anus. Magnesium is absorbed through the walls of the large intestine and the lower bowel.

Oral administration of magnesium to sick animals, in place of intravenous injections or enemas, has not been effective because too much time is required for the magnesium to reach that part of the GI tract where it can be absorbed.

Herd treatment of the animals that are not down may involve adding magnesium sulfate (Epsom Salts) or magnesium acetate or chloride to the drinking water. Some diarrhea may occur, but this is no reason for concern. To be effective, the treated tanks should be the only source of drinking water. **Note well**: Production will be lowered by this treatment due to lowered consumption of water.

Follow-up treatment may involve removing all animals from the tetany-producing pasture and feeding alfalfa hay (plus concentrates if necessary). Additionally, each animal should consume 30 g of magnesium daily for 1 to 2 weeks, preferably through a highly palatable supplement; force-feeding should be resorted to if necessary.

Cattle that get tetany are likely to get it again later in the season or in later years; they are usually the high producers.

Note well: "Downer cows" should be turned daily—and more frequently if possible. (Also see Downer Cow Syndrome.)

- **Toxicity**—Magnesium toxicity does not occur in cattle fed normal rations. Supplemental levels of magnesium of 170 to 350 g have resulted in deleterious effects. Maximum tolerable levels have been established as 0.4% of the ration by the National Research Council. Feeding toxic levels has resulted in anorexia (loss of appetite), reduced performance, and occasional diarrhea. Also, cattle experiencing toxicity may exhibit lack of reflexes and respiration depression.

Control: Commonly used feedstuffs vary widely in magnesium concentration and availability. The magnesium content of most cereal grains runs between 0.12 and 0.18%. Protein supplements of animal origin are low in magnesium, while those of plant origin usually contain 0.3 to 0.6%. Fat is not as beneficial as carbohydrate as a source of energy under tetany conditions, as fat tends to tie up calcium and magnesium in the digestive tract, rendering them less available to the animal. The magnesium content of forages varies greatly; normally, the legumes contain more than the grasses. Magnesium availability increases with plant maturity. Magnesium fertilization usually increases plant magnesium content. The inclusion of energy or protein supplements in high-magnesium mineral supplements will help to overcome palatability problems.

Prevention: Prevention of grass tetany is always preferred to treatment. Prevention consists of providing magnesium daily throughout the high-risk period, because very little of it is stored in the body. **Note well**: Crash feeding programs begun after tetany appears in a herd are usually not adequate to stop the disease. A magnesium supplement should be started 30 days before grass tetany is usually observed in the area in order to get the animals accustomed to it. Since magnesium oxide or sulfate are not very palatable, cattle may not consume sufficient of them.

Meeting the magnesium requirements of beef cows calls for providing 10 g of magnesium daily for the dry cows, and 20 to 25 g daily for cows suckling calves. For dairy cows, 30 g of magnesium per day is recommended. For calves, 4 to 8 g per day is needed, depending on their ages.

Lactating ewes and does, just after parturition, which is the most tetany-susceptible period, should receive about 3 g of magnesium per day.

High levels of aluminum, potassium, phosphorus, or calcium decrease the efficiency of magnesium absorption and/or utilization; so, in areas where the levels of these elements are high, the magnesium allowance should be increased to overcome their antagonistic effect.

Normally, animals on pasture during the summer and early fall months receive an adequate supply of magnesium from the grasses on which they feed. However, during the late fall, winter, and spring months, many pastures are magnesium deficient. To prevent grass tetany during these months, cattle, sheep, and goats on pasture should receive a magnesium-rich feed in addition to pasture and/or have ready access to a magnesium mineral supplement.

One of the following high-magnesium feeds is commonly used:

1. Alfalfa, or other legume hay; 20 lb of average alfalfa hay will provide 30 g of magnesium. Additionally, the legume hay provides increased energy.
2. For self-feeding (with adequate available water and forage containing 10% protein), 65% ground grain (corn, barley, or grain sorghum), 20% magnesium oxide, and 15% iodized salt. Since early green grass is high in protein, a supplement containing cereal grain is preferred to a supplement containing an oilseed protein. When consumed at the rate of ½ lb/head/day, such a mix will provide 27 g of magnesium daily.
3. For self-feeding (with adequate available water and low-protein forage), 65% cottonseed meal or soybean meal and 35% magnesium oxide. When consumed at the rate of ⅓ lb/head/day, such a mix will provide 31 g of magnesium daily.
4. For hand-feeding of supplements, use (2) or (3) above, but omit the salt in (2). With the salt omitted, each ½ lb of mix No. 2 will provide 32 g of magnesium daily.
5. A liquid molasses supplement fortified with 4% magnesium sulfate[1] (80 lb of magnesium sulfate per ton of liquid molasses). When consumed at the rate of 2 lb/head/day, this will provide 7.2 g of magnesium.

Several good sources of inorganic magnesium may be used to supplement cattle; among them, those listed in the table that follows:

Magnesium Content of Various Magnesium Salts

Name	% Mg	Pounds of Mineral Required to Supply the Same Amount of Mg as 30 lb of MgO
Magnesium oxide (Magnesia) (available in light and heavy grades; heavy is more stable and easier to mix)	60.32	30.0
Magnesium hydroxide	41.69	43.4
Magnesium carbonate (Magnesite)	28.8	62.8
Magnesium carbonate hydroxide	27.0	67.0
Magnesium sulfate (Epsom Salts)	20.2	89.6
Potassium magnesium sulfate (Langbelinite)	11.6	156.0
Magnesium acetate (Cromosan)	11.34	159.6

[1]Although magnesium sulfate contains only 20% Mg, it is commonly used as a magnesium additive to a molasses supplement because of its solubility. Magnesium oxide will not remain dispersed in liquid feed without the aid of suspending agents.

(Continued)

TABLE 4–52 *(Continued)*

GRASS TETANY *(Continued)*

Magnesium from dolomitic limestone is less readily available to cattle, sheep, and goats than some other salts. Any of the following mineral supplements are satisfactory:

1. A mineral mix made by mixing $\frac{1}{3}$ each magnesium oxide, iodized salt, and either soybean oil meal or cottonseed meal. This mix should be made available as the only mineral. Each $\frac{1}{3}$ lb of this mix will provide 30 g of magnesium.

2. A mineral mix made by mixing 30% magnesium oxide, 30% iodized salt, 30% bone meal, and 10% dried molasses. Each $\frac{1}{3}$ lb of this mix will provide 27 g of magnesium.

3. A mineral mix made by mixing $\frac{2}{3}$ ($66\frac{2}{3}\%$) magnesium oxide and $\frac{1}{3}$ ($33\frac{1}{3}\%$) salt as the only source of salt is effective. Each $\frac{1}{6}$ lb of this mix will provide 30 g of magnesium.

4. A commercial high-magnesium supplement, in blocks or mineral-salt mixtures which usually contain molasses, grain, and/or some other material to make them more palatable to animals. Generally, these are formulated to be fed at the rate of $\frac{1}{2}$ to $1\frac{1}{2}$ lb/head/day, and to provide 10 to 15 g of magnesium daily.

Note well: Blocks or mineral mixes usually give the best results if no additional salt is provided. Since the desire for salt varies among animals and with seasons, a high-salt mixture may not provide the required level of magnesium consumption. It should be emphasized that cattle must consume adequate magnesium on a regular basis. When using a supplement, one should pay particular attention to (1) the percentage of magnesium it contains and (2) the daily intake of the supplement. From these two factors the daily intake of magnesium can be determined and compared to the animal's requirement. The intake should be checked frequently as magnesium salts are generally unpalatable. The intake may be increased by adding grain or cottonseed meal/soybean meal.

On farms and ranches with a history of grass tetany, free-choice feeding of a mineral mix to insure a daily intake of 25 g of magnesium, with about half the daily intake coming from the natural magnesium in the pasture or other feed and half from the magnesium-containing mineral supplement, will usually provide protection from the development of grass tetany.

In high-risk situations—such as cows near calving grazing lush spring grass, highly fertilized with nitrogen or potassium, or both—the total magnesium requirement should be provided in the supplement. The reasons: (1) cows near calving are approaching lactation, when the magnesium demands are the highest; (2) nitrogen and potassium are antagonistic to magnesium; and (3) magnesium availability in early spring grass is low—besides, cattle cannot consume enough such grass to meet their energy requirements. A free-choice or hand-fed grain or protein-mineral supplement, providing 35 g of magnesium daily, will usually prevent the occurrence of grass tetany in such high-risk situations. If the magnesium supplement is self-fed, it should be located for easy access to the cattle, especially near water and shade where cattle tend to congregate and loiter. If the magnesium supplement is hand-fed, it is important that all animals have access to feeder space.

Where grass tetany is particularly troublesome, as an additional preventive measure, consideration should be given to (1) applying magnesium fertilizer and dolomitic limestone, or (2) dusting the pastures with magnesium oxide (MgO). Also, generally speaking, such pastures may be grazed without hazard by steers. But, before fertilizing or dusting, or shifting from a cow-calf program to a steer program, the counsel and advice of the local Farm Advisor/County Extension Agent should be sought.

Remarks: Affected animals may be aggressive on getting up. So, watch out!

IODINE DEFICIENCY (GOITER, BIG NECK)

Species Affected: All farm animals and humans.

Cause: A failure of the body to obtain sufficient iodine from which the thyroid gland can form thyroxin (an iodine-containing compound).

Symptoms and Signs (or age group most affected): Goiter (big neck, which is a swelling under the chin) is the most characteristic symptom of iodine deficiency in calves, lambs, kids, and humans. Also, there may be reproductive failure and weak offspring that fail to survive. Pigs may be born hairless and show edema of the shoulders and neck. Foals may be born weak.

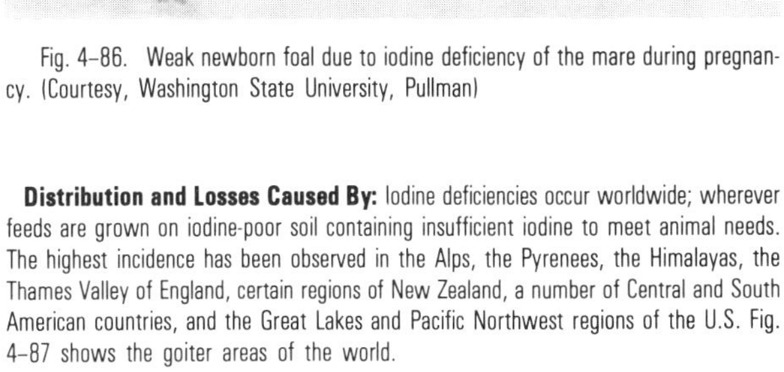

Fig. 4–86. Weak newborn foal due to iodine deficiency of the mare during pregnancy. (Courtesy, Washington State University, Pullman)

Distribution and Losses Caused By: Iodine deficiencies occur worldwide; wherever feeds are grown on iodine-poor soil containing insufficient iodine to meet animal needs. The highest incidence has been observed in the Alps, the Pyrenees, the Himalayas, the Thames Valley of England, certain regions of New Zealand, a number of Central and South American countries, and the Great Lakes and Pacific Northwest regions of the U.S. Fig. 4–87 shows the goiter areas of the world.

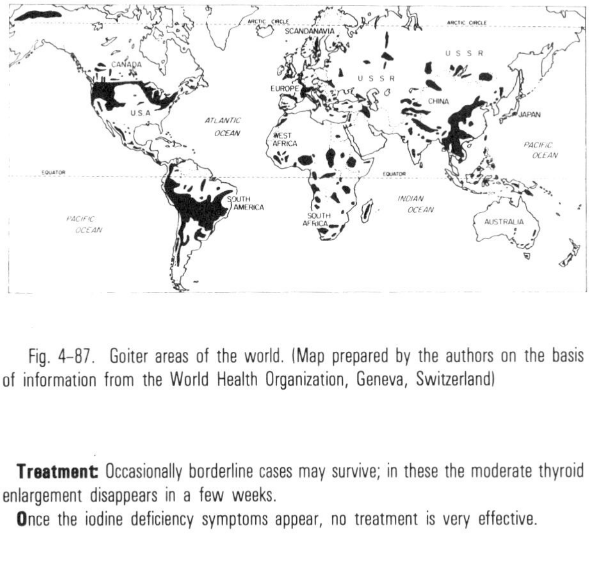

Fig. 4–87. Goiter areas of the world. (Map prepared by the authors on the basis of information from the World Health Organization, Geneva, Switzerland)

Treatment: Occasionally borderline cases may survive; in these the moderate thyroid enlargement disappears in a few weeks.

Once the iodine deficiency symptoms appear, no treatment is very effective.

Control: At the first signs of iodine deficiency, stabilized iodized salt should be fed to all farm animals.

Prevention: In iodine-deficient areas, feed stabilized iodized salt containing 0.01% potassium iodide to all farm animals throughout the year.

Remarks: The enlarged thyroid gland (goiter) is nature's way of attempting to make sufficient thyroid hormone, thyroxin, under conditions when an iodine deficiency exists.

Mares fed excess iodine (48 mg or more) during late gestation will produce foals with hyperplastic goiter. Some mares will also develop goiter.

(Continued)

TABLE 4–52 *(Continued)*

KETOSIS (ALSO KNOWN AS ACETONEMIA IN CATTLE AND PREGNANCY DISEASE IN SHEEP)

Species Affected: Cattle, sheep, goats.

Cause: A metabolic disorder of nutritional origin, characterized by hypoglycemia (low blood sugar). If the increased nutrient requirements are not met by more feed during the high-demand periods (in cows, 1–6 weeks after calving; in ewes, 2 weeks before lambing), the animal must draw on body fat reserves. If this is done too rapidly, and without adequate carbohydrates in the ration, ketosis follows.

Symptoms and Signs (or age group most affected): In cows, ketosis or acetonemia is usually observed 2 to 6 weeks after calving. Affected animals show loss of appetite and condition, a marked decline in milk production, and the production of a peculiar sweetish chloroform-like odor of acetone that may be present in the milk and urine and pervade the barn. A positive diagnosis can be made by testing the milk or urine for the presence of ketones.

In ewes and goats, ketosis or pregnancy disease generally strikes during the last 2 weeks of pregnancy. Usually, affected ewes are carrying twins or triplets. Symptoms include going off feed suddenly, grinding of teeth, dullness, weakness, frequent urination, trembling when exercised, and blindness—with the final stage being complete collapse, followed by death in 90% of the cases. In dairy goats, lactation ketosis, which is similar to the ketosis of dairy cows, may be observed in high milk producers following kidding.

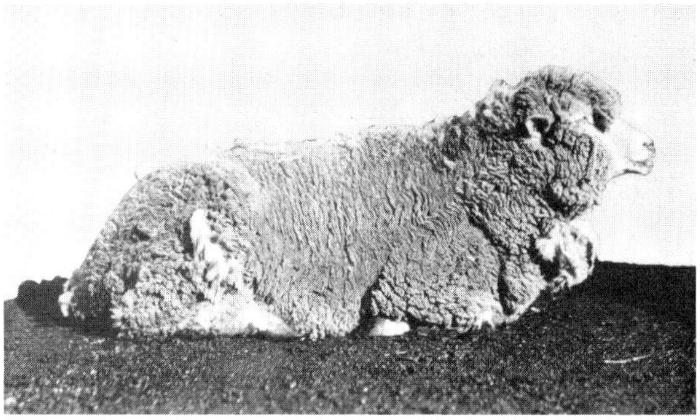

Fig. 4–88. Ewe with ketosis, or pregnancy disease. (Courtesy, College of Veterinary Medicine, University of Illinois, Champaign-Urbana)

Distribution and Losses Caused By: Worldwide.

Ketosis or acetonemia affects cattle throughout the U.S.

Ketosis or pregnancy disease in sheep affects farm flocks more than range bands, the losses in the former sometimes being as high as 25%.

Treatment: *Cattle:* ½–1 lb of either propylene glycol or sodium propionate daily, with the dose divided into 2 treatments for 5–10 days. Put treatment in grain if cow is eating; otherwise, give as drench.

Intravenous injection of glucose solution and glucocorticoids (to increase blood sugar levels temporarily) as well as the oral administration of propylene glycol. Numerous other treatments are sometimes used.

Sheep and goats before parturition: 3 to 4 oz of propylene glycol, given orally 3 times daily. Cesarean section early in the course of the disease usually leads to recovery and, if near term, the offspring may be saved.

Dairy goats after kidding: 6 to 8 oz of propylene glycol, given orally twice daily. Severe cases may be aided by intravenous injections of 50% dextrose solution. Cortocosteroid injection may be used in conjunction with either propylene glycol or dextrose solution in does that have kidded.

Control: *Cows:* Maintain relatively high-energy intake before calving; increase energy intake substantially after calving.

Ewes: Avoid obesity in early pregnancy. Feed grains rather liberally the last 6 weeks of pregnancy.

Prevention: *Cows:* The incidence of ketosis can be lessened by (1) avoiding excessively fat cows at calving; (2) increasing the level of concentrates gradually after calving; (3) feeding good-quality hay in preference to high-silage rations after calving, and avoiding abrupt changes in roughage; (4) feeding adequate proteins, minerals, and vitamins; and (5) providing comfort, exercise, and ventilation. In problem herds, feeding ¼ lb daily of propylene glycol or sodium propionate may be helpful.

Sheep and goats: Feed more hay and ½ to 1 lb of grain beginning a month before parturition. Good management is important, too, including exercise, freedom from parasites, and avoiding stress.

Remarks: The clinical findings are similar in the case of affected cattle and sheep, but it usually strikes ewes just before lambing, whereas cows are usually affected within the first 2–6 weeks after calving.

MILK FEVER (PARTURIENT PARESIS, HYPOCALCEMIA)

Species Affected: Cattle, sheep, goats.

Milk fever, a metabolic disease, is similar in cows, ewes, and does. So, the discussion that follows pertaining to cows also applies to sheep and goats, with treatment and prevention adjusted for size of animal.

Cause: Low blood calcium concentration. The name *milk fever* is a misnomer, because the animal does not have a fever.

Initiation of lactation places a severe strain on the calcium balance of the cow due to the amount of calcium secreted in the milk. All cows are slightly hypocalcemic at the time of calving, but some become so hypocalcemic that clinical signs of milk fever develop.

Symptoms and Signs (or age group most affected): Commonly occurs within 3 days after calving and in high-producing cows. Rarely occurs at first calving. First symptoms are loss of appetite, constipation, and general depression. This is followed by nervousness and finally collapse and complete loss of consciousness. The head is usually turned back toward the flank.

The incidence of the disease increases with the age of the cow and is highest in Guernsey and Jersey breeds.

Distribution and Losses Caused By: A common widespread disease of dairy cows. It is estimated that more than 8% of all dairy cows are stricken by milk fever; occasionally, with up to 80% affected in a single herd.

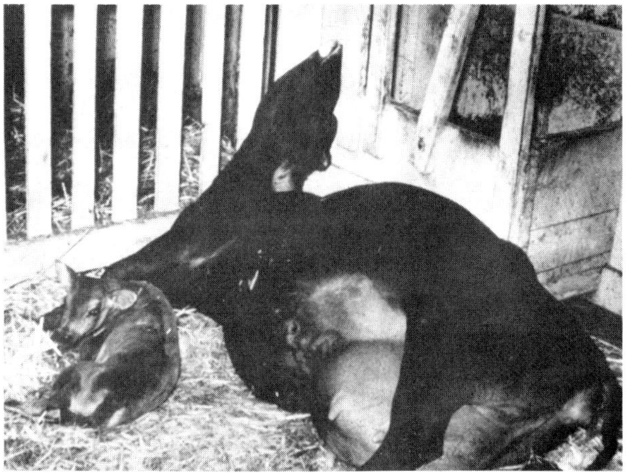

Fig. 4–89. Milk fever in Jersey cow, showing characteristic position of head—turned back over shoulder. (Courtesy, Washington State University, Pullman)

(Continued)

TABLE 4–52 *(Continued)*

MILK FEVER *(Continued)*

Losses are not great, although untreated animals are likely to die.

Causes estimated average annual losses in dairy cattle (including milk) of $100 million.

Treatment: Milk fever should be regarded as an emergency and the affected animal treated as soon as possible.

The standard treatment is intravenous infusion of calcium borogluconate as soon as the first signs appear. *CAUTION:* Overdoses of calcium salts can result in acute heart damage; so, dose level should be carefully calculated, and intravenous administration should be performed slowly.

Cows that are already down will usually stand up within 1 to 2 hours following treatment.

Control: (See Prevention.)

Prevention: Each of the following measures will lessen the incidence of milk fever:

1. *Low calcium during the dry period.* Feeding low calcium (less than 100 g/day)—high phosphorus (more than 40 g/day) rations during the dry period is important. High calcium levels in dry cow rations aggravate the problem. So, feeding a low calcium ration (less than 0.1 lb/day) before calving has shown promise of preventing milk fever.

2. *Calcium shock treatment.* Feed low calcium, high phosphorus rations containing only 15 to 20 g of calcium per day for 2 weeks prior to the expected calving date, rather than a restricted, but somewhat higher, calcium intake during the entire dry period. This creates a mild calcium deficiency which stimulates production of the biologically active form of vitamin D in the animal's body. In turn, this form of vitamin D stimulates the bone and gut to supply more calcium and phosphorus. As a result, when the greater demand for calcium and phosphorus occurs at calving, the bone and gut are already activated and are able to meet the increased demands for calcium. Thus, milk fever is avoided.

3. *Calcium-phosphorus ratio and amounts.* Balancing cow rations to contain 0.5% calcium and 0.25% phosphorus on a dry matter basis will limit the incidence of milk fever.

4. *High vitamin D.* Feeding massive doses of 20 million I.U. of vitamin D/cow/day starting about 5 days before calving and continuing through the first day postpartum, with a maximum dosage period of 7 days, has been effective in controlling milk fever. However, difficulty in predicting calving dates accurately has reduced the effectivemess of this treatment under practical conditions.

5. *Avoid excessive fatness.* Excessive fatness, or any other conditions that reduce feed intake at calving, tends to cause more milk fever.

Remarks: The name *milk fever* is a misnomer, because the disease is not accompanied by fever, the temperature really being below normal.

PREGNANCY DISEASE IN SHEEP (See Ketosis.)

RICKETS

Species Affected: All young farm animals and young humans.

Cause: Lack of either calcium, phosphorus, or vitamin D; or an incorrect ratio of the 2 minerals.

In housed animals, vitamin D deficiency is not uncommon; grazing animals are more likely to be phosphorus-deficient.

Symptoms and Signs (or age group most affected): Enlargement of the knee and hock joints, and the animal may exhibit great pain when moving. Irregular bulges (beaded ribs) at juncture of ribs with breastbone, and bowed legs.

Rickets is a disease of young animals—calves, foals, pigs, lambs, kids, pups, and chicks.

Poultry: Bones of growing birds become soft and rubbery.

Distribution and Losses Caused By: Worldwide.

It is seldom fatal, but it can be severely debilitating and economically disastrous.

Treatment: If the disease has not advanced too far, treatment may be successful by supplying adequate amounts of vitamin D, calcium, and phosphorus, and/or adjusting the ratio of calcium to phosphorus.

Control: Control of rickets is usually achieved by providing a balanced ration, with special consideration given to calcium, phosphorus, and vitamin D.

Prevention: Provide (1) sufficient calcium, phosphorus, and vitamin D, and (2) a correct ratio of the 2 minerals. Vitamin D_3, rather than D_2, is required by the chicken.

Fig. 4–90. Rickets (advanced case) caused by a deficiency of vitamin D. The pig was fed indoors, without exposure to sunlight. Because of leg abnormalities, it was unable to walk. (Courtesy, University of Saskatchewan, Saskatoon, Saskatchewan, Canada)

Remarks: Rickets is characterized by a failure of growing bone to ossify, or harden, properly.

Hens fed rations deficient in vitamin D lay eggs with progressively thinner shells until production ceases.

(Continued)

TABLE 4–52 *(Continued)*

SALT SICK (COBALT DEFICIENCY)

Species Affected: Cattle, sheep, goats.

Cause: Cobalt deficiency. In Florida, cobalt deficiency is associated with copper deficiency.

Symptoms and Signs (or age group most affected): Loss of appetite, emaciation, depraved appetite, scaliness of skin, rough hair coat, listlessness, and lack of thrift.

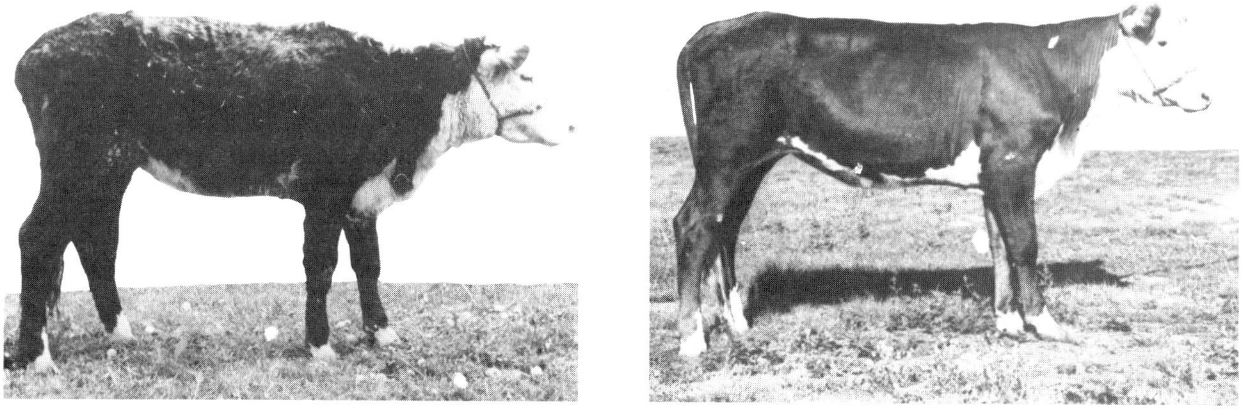

Fig. 4–91. Cobalt deficiency. *Left:* A heifer suffering from cobalt deficiency. Anemia, loss of appetite, and roughness of hair coat characterize the malady. *Right:* Illustrates the remarkable recovery in the same animal brought about by the administration of cobalt. (Courtesy, Michigan State University, East Lansing)

Distribution and Losses Caused By: Cobalt deficiency is widespread. In different parts of the world, it is known as *Denmark disease, coast disease, enzootic marasmus, bush sickness, wasting disease, Nakuritis,* and *pining disease.* Fig. 4–92 shows cobalt-deficient areas of the U.S.

Treatment: Provide 0.2–0.5 oz cobalt salt/100 lb of salt—or feed a suitable trace mineral supplement. Injection of cobalt salts is not satisfactory, since ruminal action is needed to form vitamin B–12, the active form of cobalt.

Control: Provide adequate cobalt in the ration—about 0.1 ppm. Deficiency symptoms appear when the level drops to the range of 0.04 to 0.07 ppm or lower.

Prevention: Mix 0.2–0.5 oz of cobalt chloride, cobalt sulfate, or cobalt carbonate/100 lb of either (1) salt, or (2) whatever mineral mix is being used.

Remarks: Cobalt is needed especially for rumen microbial synthesis of vitamin B–12. Nonruminants must be fed preformed vitamin B–12.

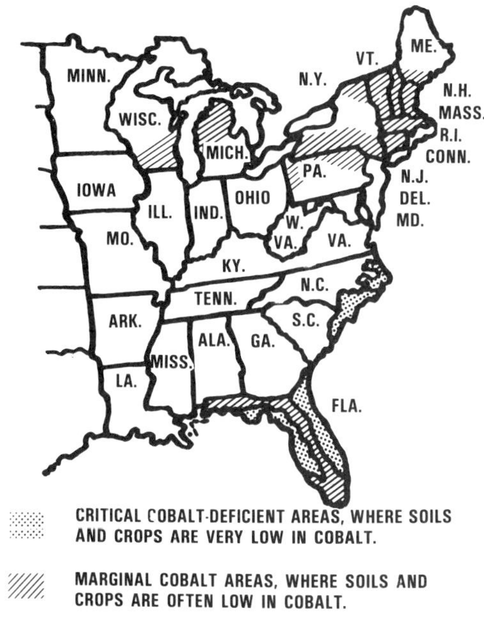

Fig. 4–92. Cobalt-deficient areas in eastern U.S., resulting from its deficiency in the soil and thus in the herbage produced thereon.

STIFF-LAMB DISEASE (WHITE MUSCLE DISEASE) (See White Muscle Disease.)

URINARY CALCULI (GRAVEL, STONES, WATER BELLY, UROLITHIASIS)

Species Affected: Cattle, sheep, goats, horses, mink, and humans.

Cause: The precipitation of various salts, usually inorganic, in the urine, frequently associated with rations high in cereal grains or grazing on the silica-rich soils of the northwest plains of Canada and the U.S. However, not all causative factors are known.

Experiments and experiences have shown a higher incidence of urinary calculi when there is (1) a high potassium intake, (2) a high phosphorus-low calcium ratio (from the standpoint of preventing urinary calculi, the Ca:P ratio should be about 2:1), (3) a high-silica content in the ration, or a high proportion of high-silica grains and forages, such as native grasses, wheat straw, sugar beet leaves or pulp, sorghums, and cottonseed meal. High dosages of diethylstilbestrol or a deficiency of vitamin A may be contributing factors.

Symptoms and Signs (or age group most affected): Frequent attempts to urinate, dribbling or stoppage of the urine, pain and renal colic.

Usually only males affected; females are able to pass the concretions.

Bladder may rupture, with death following. Otherwise, uremic poisoning may set in.

Urinary calculi is one of the most important diseases in feedlot cattle and sheep, particularly in steers and wethers on full feed.

Distribution and Losses Caused By: Worldwide. The economic loss may be considerable, since calculi formation frequently comes near the end of the feeding period.

Affected animals seldom recover completely.

(Continued)

TABLE 4–52 *(Continued)*

URINARY CALCULI (GRAVEL, STONES, WATER BELLY, UROLITHIASIS)

Treatment: (1) Add ammonium chloride at the rate of 1 oz (lambs) or 1¼–1½ oz (cattle) per head daily—or 50–60% more ammonium sulfate; (2) increase the phosphorus content of the ration (pasture) so that it equals the calcium content (by adding monosodium phosphate); (3) increase salt content of ration to 3 to 4% so as to increase water consumption (too much salt may lower feed intake); (4) incorporate 20% alfalfa in the ration; (5) administer muscle relaxants to help the passage of calculi from the bladder; or (6) surgically remove the calculi.

In cattle, surgical removal of the calculi is the most effective treatment, with the stone(s) removed at the point of blockage. In steers, the urethra may be bisected and brought to the outside of the body to bypass the constricted portion of the tract. After a short time to eliminate any tissue residue of urine, such animals are marketable.

In sheep, amputation of the urethral process is simple and allows the immediate passage of urine.

In horses, bladder calculi must be removed surgically.

Careful observation of susceptible animals by an experienced person several times daily will allow early detection and more successful treatment.

Control: If severe outbreaks of urinary calculi occur in finishing steers or lambs, it is usually well to dispose of them if they are carrying acceptable finish.

Increase water consumption by including 3 to 4% sodium chloride (salt) in the ration.

The addition of a broad-spectrum antibiotic to the ration has been useful in controlling urinary calculi in some cases.

Prevention: The basis of prevention is identification of the chemical composition of the calculi so that appropriate steps can be taken to reduce the concentration of the particular chemical in the urine by increasing urine volume, eliminating infection, changing urine pH, or altering the metabolism with drugs.

Good feed and management appear to lessen the incidence.

Delayed castration (castration of bull calves at 4–5 mo. of age) and high-salt rations for feedlot cattle (1–3% salt in the grain ration, using the upper limits in the winter months) in order to induce more water consumption are effective preventive measures.

Avoid (1) a high potassium or phosphorus intake, (2) an incorrect Ca:P ratio, or (3) an excessive amount of beet pulp or grain sorghum in the ration.

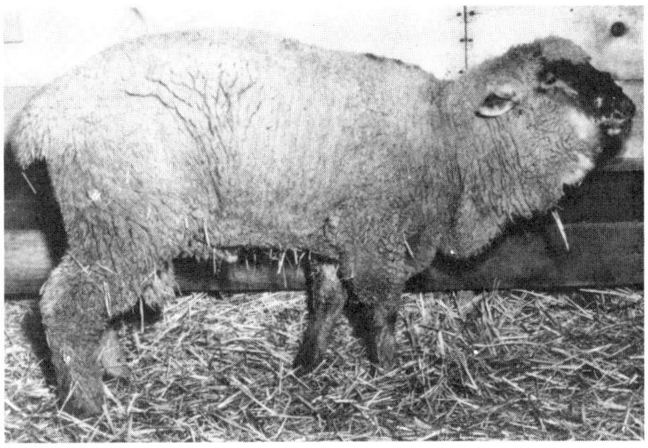

Fig. 4-93. Lamb suffering from urinary calculi. (Courtesy, Washington State University, Pullman)

Remarks: Calculi are stonelike concretions in the urinary tract which almost always originate in the kidneys. These stones block the passage of urine, resulting in the condition commonly referred to as *water belly.*

The mineral deposits may be of variable sizes, shapes, and composition. In cattle, the phosphatic type predominates under feedlot conditions and the silicate type occurs most frequently in range cattle.

According to researchers at Canada's Lethbridge, Alberta, research station, the incidence of calculi in calves grazing native pasture is 10 times higher than in calves grazing Russian wild ryegrass.

WHITE MUSCLE DISEASE (MUSCULAR DYSTROPHY; in sheep, STIFF-LAMB DISEASE)

Species Affected: Calves, lambs, and foals. In lambs, it is commonly referred to as stiff-lamb disease.

Cause: Selenium deficiency, due to the continuous consumption of a ration containing less than 0.02 ppm selenium.

Symptoms and Signs (or age group most affected): In calves, white muscle disease is characterized by lameness or inability to stand, and heart failure. It most commonly affects calves 2 to 4 months of age.

In lambs, the symptoms and signs are: A stiff, stilted way of moving, chiefly in the hind legs, although the front legs and shoulders may be involved. The back is usually humped or "roached." Lambs that live are usually stunted. Young, rapidly growing lambs are especially susceptible.

It seems that more calves than lambs or foals develop heart damage, which may be fatal, especially if subjected to unusual exercise. Affected calves and lambs show similar pathological lesions—whitish areas or streaks in the heart and other muscles.

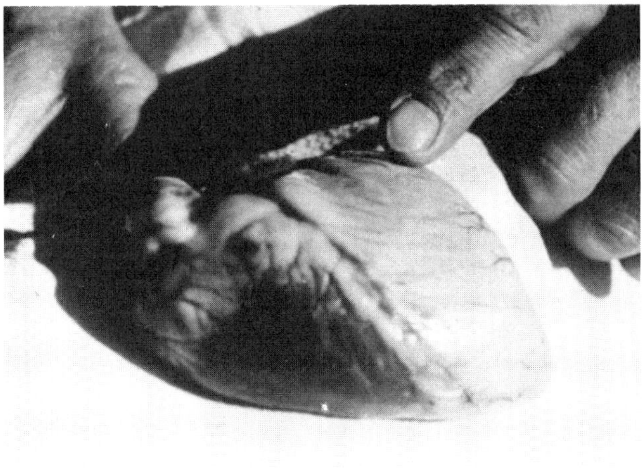

Fig. 4-94. White muscle disease in a calf. *Left:* Shows the generalized weakness of muscles, lameness, and difficulty in locomotion of an afflicted calf. Calf is about 3 months old. *Right:* Shows abnormal white areas in the heart muscles of a 6-week-old calf afflicted with white muscle disease. (Courtesy, Oregon State University, Corvallis)

(Continued)

TABLE 4–52 *(Continued)*

WHITE MUSCLE DISEASE *(Continued)*

Distribution and Losses Caused By: Geographically, white muscle disease has been reported in Australia, Canada, Finland, Italy, Japan, New Zealand, Norway, Scotland, South Africa, Sweden, U.S., U.S.S.R., and Yugoslavia.

In the U.S., the disease is widely distributed, but the severity is greatest on the two coasts.

Economic losses result from the deaths of severely affected calves and lambs, the unthriftiness of survivors, and the cost of preventive programs.

Death losses range up to 50%, with an average of 15%; and the mortality of untreated animals may reach 80%.

Treatment: Affected animals should receive early treatment—the intramuscular injection of sodium selenite/vitamin E in aqueous solution at the rate of 0.25 mg Se per pound of body weight. This may be repeated in 2 weeks, but should not exceed 4 doses. **Note well**: Federal law restricts injectable Se to the order of a licensed veterinarian. Do not use within 30 days of slaughter.

Control: Control consists in meeting the selenium requirements by (1) supplementing the ration of cows or ewes during the last ⅓ of pregnancy and the first part of lactation with selenium in the form of sodium selenite at the rate of 0.3 ppm dry matter; or (2) injecting intramuscularly each cow or ewe 1 month before parturition with approved levels of selenium/vitamin E preparation.

Prevention: Add selenium to the ration. In 1987, FDA approved the addition of 0.3 ppm selenium to the complete feed of cattle, sheep, swine, and poultry.

Also, mineral mixes containing selenium are available for free-choice feeding in areas of known selenium deficiency.

Remarks: White muscle disease is most common (1) in rapidly growing calves and lambs, and (2) in calves and lambs on lush pastures on selenium-deficient soils.

Because of muscle failure, severely affected young animals die from starvation or heart failure.

White muscle disease cannot be produced in calves on vitamin E-free rations unless the rations are high in unsaturated fats.

Complete, mixed ration—with the proportion of hay and concentrate controlled—being conveyed by a self-unloading truck into a fenceline bunk. (Courtesy, Butler Manufacturing Company, Garden City, KS)

"All flesh is grass." (Isaiah 40:6)

SECTION

5

PASTURE AND RANGE FORAGES[1]

By M. E. Ensminger
with Harland Dietz, USDA, Soil Conservation Service, Fort Worth, Texas

Contents *Page*

---

[1]Grateful appreciation is expressed to the following authorities who reviewed, and made helpful suggestions relative to the entire Section 5: C. W. Carter, Range Conservationist, USDA, Soil Conservation Service, Fort Worth TX; H. DeGarmo, National Range Conservationist, USDA, Soil Conservation Service, Washington, DC; and N. Hobson, Director, USDA, Soil Conservation Service, Portland, OR.

Grass is the largest and most remunerative crop in the United States, and the cornerstone of successful livestock production. Also, and most important, no method of harvesting has been devised which is as cheap as that which can be accomplished by grazing animals.

As the ever increasing human population of the world consumes a higher proportion of grains and seeds directly, there will be increased reliance on grass for meat, milk, and wool production. In this connection, it is noteworthy that petroleum is not needed to make wool, and that animals do not require fossil fuels to graze the land and recover the energy that is manufactured and stored in the grass. Noteworthy, too, is the fact that the grass-animal system is completely recyclable; it is driven by solar energy and a new crop is produced each year—thus the system is perpetual. But it takes thousands of years to create coal, oil, and natural gas; and when these energy sources are gone, they are gone forever.

It is also noteworthy that soil conservation—erosion control—is accomplished more effectively by pasture crops than any other means. The American Forage and Grassland Council is authority for the statement that, "A good stand of grasses and legumes is over 300 times as effective in saving soil, and six times as effective in reducing water runoff, as a clean tilled crop on the same kind of land."

Grassland agriculture, better than any other type of agriculture, will continue in the face of economic and social changes to conserve the land and ensure a food supply of the desired quantity, variety, and quality. At its best, it calls for an interdisciplinary approach—for knowledge and application of soil, plant, and animal sciences. This joint focus characterizes all the great livestock and forage producing areas of the world.

From the standpoint of organization, this chapter is presented in three parts: Part I—Pasture; Part II—The Western Range; and Part III—Multiple Use/Conservation of Land.

PART I—PASTURE

Fig. 5-1. Pasture and range, the cornerstone of successful livestock production, worldwide.

The term *pasture* is of Latin origin, from the word *pastus*, meaning *an area of land on which there is a growth of forage that animals may graze.* Additional definitions of pasture-related terms follow.

Pasture or tame pasture are terms that have evolved to mean grazinglands that are planted primarily to introduced species and receive periodic culture treatments such as tillage, fertilization, mowing, weed control, and irrigation.

Rangeland, which is of American origin, *refers to lands on which the native (range) vegetation is predominantly grasses, grasslike plants, forbs, or shrubs suitable for grazing or browsing.*

Grazingland includes both pasture and rangeland; it denotes all land on which animals are grazed, with the exception of annually seeded row crops. It includes some forest understories.

Forage, refers to leaves, stems, and other portions of plants in a fresh, dried, or ensiled state, which is fed to livestock through grazing or as hay or silage.

Grazing refers to the process by which animals harvest their own feed (grass, legume, browse, and/or forbs) from its growing environment.

Table 5-1 shows the types of grazingland, ownership, and total land area in the U.S. (excluding Alaska). Note (1) that 45.5% of the total land area is used for grazing, (2) that rangeland and pastureland account for 38.9%, (3) that 65.6% of the rangeland is non federally owned (mostly private) and 34.4% federally owned, and (4) that essentially all pastureland is privately owned.

TABLE 5-1
TYPES OF GRAZINGLAND, OWNERSHIP, AND TOTAL LAND AREA IN THE U.S. (EXCLUDING ALASKA)[1]

Type of Land/ Ownership	Total Area
	(1,000,000 acres)
Total land area	1,940
Rangeland:	
Federal	214
Nonfederal[2]	408
Pastureland:	
Nonfederal	132
Forestland grazed:	
Federal	62
Nonfederal	66
Total grazinglands	882

[1]Source: USDA-NRI.

[2]Includes state-owned and Indian lands.

Pastures vary greatly in both quality and production of forage, depending on type (plant species), soil, growing conditions, and stage of maturity. Mature grasses, especially those that are leached and bleached, are low in palatability, digestibility, protein, and carotene, and in some of the minerals. Grasses are usually adequate in calcium, magnesium, and potassium, but they are apt to be borderline or deficient in phosphorus, and they may be low in some of the trace minerals.

Also, pastures vary in quality and quantity as a result of grazing management. Too close grazing during the growing season will lower yields and stand persistence. All pasture plants, whether grass, legume, native, or introduced, manufacture food in their green leaves through photosynthesis. Removing too many leaves, either through overgrazing or frequent mowing, will eventually reduce the vigor and stamina of plants and lower both forage production and livestock performance. Progressive operators tailor their programs and consider the basic requirements of both livestock and forage.

For many farmers and ranchers, pasture management is both the biggest problem and the greatest opportunity for improvement in operational efficiency. Many have spent years, even lifetimes, increasing annual production and reproduction efficiency 20 to 30% through genetic improvement and selection. At the same time, these farmers and ranchers have increased forage production 30 to 100% by applying available technology in pasture and range management.

As opportunities in forage production are better known, pasture management should no longer be taken for granted. Most grazing land can be improved by management with use of such practices as scientifically controlled grazing, brush and weed control, seeding new and better varieties of grasses and legumes, and by fertilizing and the supplemental feeding of animals.

Again and again, scientists and practical farmers and ranchers have demonstrated that the following goals in pasture production are well within the realm of possibility:

- To produce higher yields of palatable and nutritious forage.
- To extend the grazing season from as early in the spring to as late in the fall as possible.
- To provide a fairly uniform supply of feed throughout the entire season.

At the outset, it should be recognized that no one plant embodies all the desirable characteristics necessary to meet the above goals. None of them will grow year-round, or during extremely cold or dry weather. Each of them has a period of peak growth, part of which must be conserved for periods of little growth. Consequently, the progressive producer will find it desirable (1) to grow more than one species; (2) to plan pastures for each season of the year; and (3) to have a reserve, or a supplemental, feeding plan for drought or other emergencies. In general, a combination of permanent, rotation, and temporary pastures—accompanied by scientific management—will best achieve these ends.

CLASSES OF PASTURE

Broadly speaking, all pastures in the United States may be classified as either (1) seeded pastures (pastureland) or (2) native pastures (rangeland) (see Fig. 5-2). Although no sharp line of demarcation exists between the two groups in area of occurrence, seeded pastures are generally established in areas where moisture conditions are favorable. This is usually where rainfall exceeds 20 in. annually or where irrigated. They are the seeded (cultivated) pastures of the Corn Belt, the South, the East, and the irrigated areas, and smaller and scattered moderate to high rainfall areas throughout the West.

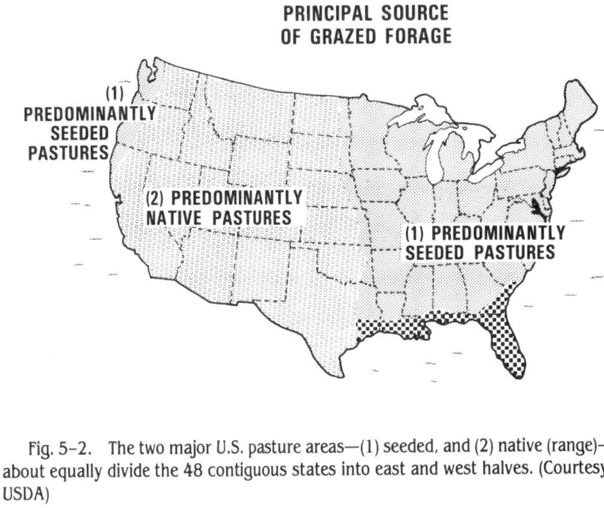

Fig. 5-2. The two major U.S. pasture areas—(1) seeded, and (2) native (range)—about equally divide the 48 contiguous states into east and west halves. (Courtesy, USDA)

The native (range) pastures include those natural grasslands found primarily west of the Mississippi River, although a small portion is found along the Gulf Coast states from Louisiana to Florida. Their vegetative cover consists of adapted plants developed by natural selection that have existed in the area for many years. Rangeland pastures and management are discussed in more detail later in this section under Part II—The Western Range.

Seeded pastures may be further classified as follows:

1. **Permanent pastures.** Those seeded to permanent grasses which, with proper care, last for many years. The vast majority of the farms of the United States have one or more permanent pastures.
2. **Semipermanent or rotation pastures.** Those that are used as part of the established crop rotation. These are seeded pastures that are generally used for 2 to 7 years before plowing.
3. **Temporary and supplemental pastures.** Those that are used for a short period; and they are usually annuals, such as Sudangrass, sorghum, millet, rye, barley, wheat, oats, ryegrass, arrowleaf clover, crimson clover, or rape. They are generally seeded for the purpose of providing supplemental grazing during the season when the permanent or rotation pastures are relatively unproductive.

Pasture plants are classed as (1) grasses, (2) legumes, (3) browse, and (4) forbs. The definition of each of these terms follows:

1. **Grass.** *Botanically, any plant of the family Gramineae.* In grassland agriculture, grass refers to the forage species of *Gramineae* when either grown alone or with a legume.
2. **Legume.** *Plants, such as alfalfa and the clovers, that generally have the ability to obtain nitrogen through bacteria that live in their roots are known as legumes.* The nitrogen fixation aspect of legumes will be of increasing interest as energy sources become more scarce and costly.
3. **Browse.** *The edible parts of woody vegetation, such as leaves, stems, and twigs from bushes, shrubs, or trees.*
4. **Forbs.** *Nongrasslike broadleaf herbs.* They are generally called weeds by farmers and ranchers but actually some of the most palatable plants to livestock are forbs.

The foregoing groups, especially the grasses, are further divided into either cool-season or warm-season, depending on period of major growth:

1. **Cool-season.** *Plants that grow rapidly during spring, often using most of the available soil moisture and nitrogen.* They are nearly dormant during summer, but growth begins early in the fall as cool temperatures return.
2. **Warm-weather.** *Plants that begin growing later and make most rapid growth during the late spring and summer months.*

Each kind of plant has its advantages in a forage program in extending the grazing season and meeting nutritional needs of livestock.

PASTURES FOR CATTLE, SHEEP, SWINE, AND HORSES

The economic importance of pastures for beef cattle, dairy cattle, sheep, and horses has been demonstrated in many experiments and on thousands of livestock farms and ranches.

Brood cows normally obtain practically all their feed from pasture. In the southern half of the United States, year-round grazing generally can be achieved. Even on northern Cornbelt farms cows graze pastures at least 200 days of the year. While producing nearly two-thirds of the total feed supply for the whole year, the cost of pasture is only about one-third of the feed bill; hence, pasture is a good buy.

As larger numbers of dairy cattle are concentrated on smaller acreages, and as milk production per cow increases, dairy producers depend less on pasture and more on other feeds.

Sheep are efficient grazers. They are able to utilize the various grasses, legumes, weeds, herbs, and browse that grow on millions of acres of cultivated and uncultivated lands in this and other countries.

Pastures for swine have decreased with confinement production. By 1990, 60% of the market hogs in the United States were raised from farrow to finish in some type of confinement, and only 4.3% of the feed for swine was derived from forage. Yet, it is possible to utilize large amounts of pasture effectively for the swine breeding herd.

Although suburban horse owners are finding it increasingly difficult to provide pasture, horses compete with cattle and sheep for grazing areas.[2]

Further facts pertinent to pasture for beef cattle, dairy cattle, sheep, swine, and horses follow:

1. **Beef cattle.** It is estimated that 84.5% of the total feed supply of all U.S. beef cattle is derived from forage (see Table 4-4, p. 132); in season, this means pasture. Some fertile pastures alone will produce 200 to 400 lb of beef per acre annually (in weight of calves weaned in added weight of older cattle).

Generally speaking, cattle can be produced more cheaply on pasture than in the drylot because (a) less labor is required for the animals to do their own harvesting; (b) grass is the cheapest of all roughages; (c) less expensive protein supplement is required; (d) the animals scatter their own droppings, which aids in maintaining fertility and alleviates hauling it; and (e) fewer buildings and less equipment are necessary.

---

[2]The pasture recommendations made in this section for cattle and sheep are equally applicable to horses, except for Sudangrass and sorghum-Sudangrass hybrids, which may cause cystitis—a fatal inflammation of the bladder of horses. Hence, Sudangrass should not be grazed by horses.

2. **Dairy cattle.** It is estimated that 58.7% of the nation's milk is produced from forages (see Table 4–4, p. 132). Good pasture alone will provide cows with sufficient nutrients for body maintenance and for the production of 20 lb to more than 40 lb of milk daily, depending on the quality of forage.

As larger numbers of lactating cows are concentrated on smaller acreages and as milk production per cow increases, dairy producers depend less on pasture and more on other feeds. A major reason for this is the inability of high-producing cows to consume enough feed to supply their energy requirements when pasture is their main feed source. The physical form and volume of pasture fill the rumen to capacity before nutrient needs of high producers are fulfilled. However, pastures continue to be practical in smaller herds and for heifers and dry cows in both large and small herds.

The general trend among dairy producers, however, is a gradual reduction in use of pasture and increased dependence on stored feeds and green chop as herd size and milk production per cow increase.

3. **Sheep.** It is estimated that 93.8% of the total feed supply of all U.S. sheep is derived from forage (see Table 4–4, p. 132); for the most part, this is from pasture. No other class of farm animals is so well adapted to the utilization of maximum quantities of pasture as sheep. They are unique in that the vast majority of the young are marketed as milk-fed animals directly off grass.

4. **Swine.** Prior to 1950, pastures were considered essential for successful swine production. Subsequently, the importance of pastures for swine declined with increased knowledge of nutrition and escalated land and labor costs. These forces resulted in more and more confinement rearing, accompanied by increased labor efficiency and less use of pastures.

Today, only 4.3% of U.S. swine feed is derived from forage, including pasture (see Table 4–4, p. 132). Yet, hogs, especially gestating sows, will often yield greater return from an acre of good pasture than any other class of farm animal.

Good pasture can reduce the concentrate requirements for the breeding herd by 75%, for sows and pigs by 20%, and for growing-finishing pigs by 15%.

The most common pasture forages used in U.S. swine production are alfalfa and ladino pastures. Additional pastures include other clovers, lespedeza, birdsfoot trefoil, rape, winter rye, and certain grasses.

5. **Horses.** When idle, horses do well on pasture and other forages as the only feed. Even when working, they can use some pasture, with the amount depending on the degree and severity of the work. most pleasure horses are kept in pasture paddocks when not in use, and they receive supplemental grain and hay when at work.

Do not use Sudangrass for horses.

ADAPTED AND/OR COMMON GRASSES AND LEGUMES OF THE U.S.A.

Establishment of specific grasses or grass-legume mixtures will vary from area to area, according to differences in soil, temperature, and rainfall. Fig. 5–3 shows the 10 generally recognized U.S. pasture areas; and Chart 5–1 (see next page) shows the better adapted and/or most common grasses and legumes for consideration when planning a pasture program in each of these areas.

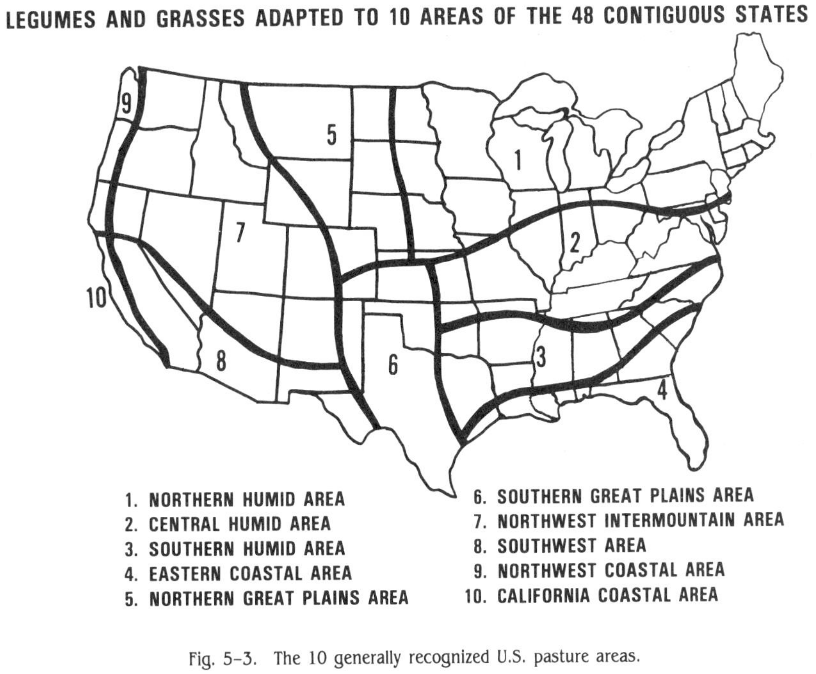

Fig. 5–3. The 10 generally recognized U.S. pasture areas.

CHART 5-1
ADAPTED GRASSES AND LEGUMES (INCLUDING BROWSE AND FORBS) FOR CATTLE, SHEEP, AND HORSE PASTURES, BY 10 GEOGRAPHICAL AREAS OF THE UNITED STATES (SEE FIG. 5-11 FOR GEOGRAPHICAL AREAS)[1]

	Areas of the United States									
	1	2	3	4	5	6	7	8	9	10
Grasses, shrubs, forbs:										
Bahiagrass			X	X						
Bermudagrass		X	X	X		X		X		X
Bluegrass, big							X		X	
Bluegrass, Kentucky	X	X	X		X		X		X	
Bluestem, big	X	X	X	X	X	X				
Bluestem, Caucasian		X	X			X				
Bluestem, little	X	X	X	X	X	X				
Bluestem, sand	X	X			X	X				
Bristlegrass, plains						X		X		
Bromegrass, meadow					X		X	X	X	
Bromegrass, smooth	X	X			X		X	X	X	X
Buckwheat (wild)								X		
Buffalograss					X	X				
Buffelgrass						X				
Canarygrass, reed	X	X					X		X	
Cottontop, Arizona								X		
Curly mesquite						X		X		
Dallisgrass			X	X						
Digitgrass, pangola			X	X						
Dropseed, sand						X		X		
Fescue, tall	X	X	X				X	X	X	X
Foxtail, creeping							X		X	
Galleta						X		X		
Gamagrass, eastern	X	X	X	X	X	X				
Grama, black								X		
Grama, blue					X	X	X	X		
Grama, sideoats	X	X	X	X	X	X	X	X		
Hardinggrass						X			X	X
Indiangrass	X	X			X	X				
Indianwheat								X		
Johnsongrass			X	X		X				
Kleingrass						X				
Koleagrass, Perla									X	X
Limpograss				X						
Lovegrass, Lehmann							X	X		
Lovegrass, sand	X	X			X	X				
Lovegrass, weeping			X			X				
Maidencane				X						
Millet	X	X	X	X		X				
Muhly, spike							X	X		
Needle-and-thread	X				X					
Needlegrass, green	X				X					
Oatgrass, tall									X	
Oats	X	X	X	X	X	X		X	X	X
Orchardgrass	X	X	X	X	X		X	X	X	X
Paragrass				X						
Pearlmillet		X	X	X		X				
Redtop	X						X		X	
Rescuegrass			X	X					X	
Rhodesgrass			X	X						
Ricegrass, Indian							X	X		
Rye	X	X	X	X	X	X		X	X	X
Ryegrass, annual		X	X	X		X			X	
Ryegrass, perennial	X	X	X						X	
Sacaton, alkali						X	X	X		
Sage, pitchers	X	X	X		X	X				
Saltbrush, fourwing					X	X	X	X		
Sorghum-Sudan hybrids	X	X	X	X	X	X	X			
Stargrass			X							
Sudangrass	X	X	X	X	X	X	X	X	X	X
Sunflower, Maximilian	X	X	X		X	X				
Switchgrass	X	X	X		X	X				
Three-awn				X	X	X	X	X		
Timothy	X	X					X		X	
Tobosa grass						X				
Wheat	X	X	X	X	X	X	X	X	X	X
Wheatgrass, bluebunch					X		X		X	
Wheatgrass, crested					X		X			
Wheatgrass, intermediate	X				X		X		X	X
Wheatgrass, pubescent					X	X	X			X
Wheatgrass, tall	X				X	X	X	X	X	
Wheatgrass, western	X	X			X	X	X		X	
Wild-rye, basin					X		X			
Wild-rye, Canada	X	X	X		X	X	X			
Wild-rye, Russian					X		X			
Winterfat (white sage)								X		
Wintergrass, Texas						X				
Legumes:										
Alfalfa (lucerne)	X	X	X	X	X	X	X	X	X	X
Alyceclover			X	X						
Black medic (yellow trefoil)			X			X		X		
Bur-clover		X						X		X
Clover, alsike	X	X			X		X	X	X	
Clover, arrowleaf		X	X							
Clover, crimson		X	X							
Clover, Hubam (white sweet clover)	X	X						X	X	
Clover, Kura	X	X			X		X			X
Clover, Ladino	X	X	X	X			X	X	X	X
Clover, prairie						X		X		
Clover, red	X	X	X	X			X	X	X	
Clover, strawberry					X		X	X		X
Clover, subterranean			X	X				X	X	X
Clover, white	X	X	X	X			X	X	X	X
Cowpeas			X	X						
Crown vetch	X	X								
Flat pea			X	X			X			
Hairy indigo				X						
Lespedeza (annual)		X	X	X						
Lespedeza (perennial, sericea)		X	X	X						
Milk vetch, cicer	X				X		X			
Peas, field									X	
Pea shrub								X		
Prairie clover, purple	X	X	X		X	X	X			
Ratany								X		
Soybeans	X	X	X	X			X			
Sweet clover, white	X	X						X	X	
Sweet clover, yellow	X	X			X	X	X	X	X	
Trefoil, birdsfoot	X	X	X				X		X	X
Velvet bean			X	X						
Vetch		X	X	X	X	X		X	X	

In using Fig. 5–3 and Chart 5–1, bear in mind that many species of forages have wide geographic adaptation, but varieties often have rather specific adaptation. Thus, alfalfa, for example, is represented by many varieties which give this species adaptability to nearly all states. Variety then, within species, makes many forages adapted to widely varying climate and geographic areas. Also, some species have very narrow adaptation ranges, but may be important within that limited area. Some species or varieties have adaptations to specific soil conditions, such as salinity, heavy clay, or high water tables.

For more specific and individual farm recommendations, farmers and ranchers are urged to seek the advice of local authorities, including the county agricultural agent and district conservationist, or to write to their state agricultural college.

ESTABLISHMENT AND MANAGEMENT OF SUBHUMID, HUMID, AND IRRIGATED PASTURES

This section, and the subsections under it, has reference to the seeded or cultivated pastures that occur in areas which either receive above approximately 20 in. of rainfall annually or are irrigated. This includes pastures of the Corn Belt, the South, the East, and the irrigated valleys and smaller, scattered moderate-to-high-rainfall areas throughout the West.

Fig. 5–4. Contented lactating cows on pasture. (Courtesy, Holstein-Friesian Assn. of America, Brattleboro, VT)

ESTABLISHING A NEW PASTURE

The following practices are usually adhered to when planning forage programs and in successfully establishing a new pasture in the subhumid, humid, and irrigated areas:

1. **The species and cultivars are selected.** The selection of species and cultivars to seed should receive first priority. One species or a specific combination best fits the needs of a certain climate, soil, and intended use. A single species may be adequate for one pasture while several species may be best for another. Chart 5–1 gives the general recommendations of competent specialists residing in different areas of the United States. For more specific recommendations for a particular farm or ranch, the livestock producer should consult such local authorities as the county extension agent, district conservationist, or successful neighbors. Often new varieties have been released that may be proving superior to standard varieties in production, palatability, digestibility, or disease resistance.

2. **The soil is tested and limed/fertilized.** Following testing of the soil, it should be limed and/or fertilized according to the needs. In humid regions where periodic application of lime is needed, it is best to work lime into the soil considerably in advance of seeding. Phosphorus levels are especially critical when establishing a new seeding. When the soil is deficient in nitrogen, the application of a small amount of nitrogen at seeding time may be helpful.

3. **High-quality seed is purchased.** The seed should be of good quality, of high germination and purity as indicated on the tag, and free of noxious weeds. Also, proof of origin is of prime importance when an imported variety is obtained. Certified seed carries more assurance of being high quality than noncertified seed, and gives proof of its origin much as a registration certificate does for a purebred animal.

Pure Live Seed (PLS) has become the standard in seeding recommendations. It is also advisable to purchase seed on a PLS basis. PLS can be determined by % PLS = % germination × % purity × 100. These items should be found on all seed tags.

4. **Scarified legume seed is used.** In the purchase of certain legume seed, it is important that it be scarified, which breaks the seed coat and allows faster moisture penetration—thus assuring quicker and more uniform germination and a better stand the first year.

5. **Legume seed is inoculated.** Since most legumes can use nitrogen from the air provided they are inoculated with the proper bacteria, it is important that legume seed be inoculated. This is traditionally done at planting time by coating the seed with a water-based slurry consisting of a commercial preparation of bacteria. The expiration date is given on the container. It is important that the seed not be inoculated more than a few hours before seeding because these nitrogen-fixing bacteria are easily killed by drying, heat, sunlight, chemical seed treatment, or by direct contact with acid fertilizers.

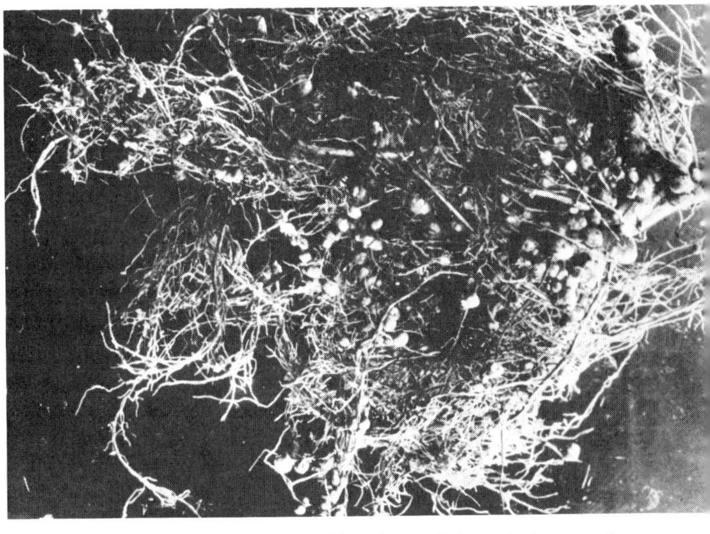

Fig. 5–5. Nodules (small bumps) containing nitrogen-fixing bacteria on a well-inoculated soybean root. (Courtesy, USDA)

On fields or pastures where the legume species has been grown regularly for a number of years, inoculation of seed is not necessary because the soil contains a sufficiently large population of nodule bacteria to fulfill crop requirements.

6. **A good seedbed is prepared.** A good seedbed is free from weeds, fine-textured, firm, and moist.

Weeds are usually destroyed by growing row crops or a small grain the year preceding seeding to pasture and by cultivating frequently following the harvesting of this crop.

There are many different ways in which to prepare a good seedbed. Perhaps as good a method as any consists in (a) plowing as far in advance of seeding as possible, (b) discing, (c) harrowing one or more times to level the field and smooth the surface, and (d) cultipacking or rolling. A properly prepared seedbed should be so firm that one barely leaves a footprint when walking across it; the firmer the better from the standpoint of moisture conservation of small seeds.

7. **The seeding operation is timed and carried out properly.** The seeding time will vary, being determined primarily by the area and by the species or mixture used. Cool-season species should be seeded early enough, in the spring or fall, for seedling development so they can withstand the stress of hot summers and/or avoid winter injury. Conversely, warm-season species may rot before germination if planted before critical soil temperatures are reached in the spring.

The actual seeding operation may be (a) by broadcast, with a whirlwind seeder or by hand; (b) by cultipacker, consisting of two corrugated rollers with seed-metering boxes; or (c) by drilling with one of several types of conventional seeders over a band of fertilizer $\frac{1}{5}$ to $\frac{3}{5}$ in. deep. Drilling is the preferred method, for it ensures more uniform placement of seeds in both depth and amount of seed per acre and results in a more uniform stand. When broadcast methods of seeding are used, rates should be increased 50 to 100% above the amount recommended for drilling.

Seeding rates are dependent on many factors. Normally, only about a third of sown seed produces seedlings, and only half of the seedlings survive the first year. So, high seeding rates appear to be justified. Moreover, high seeding rates reduce weed invasion and increase yields the first year.

Since most grass and legume seeds are very small, they should not be covered deeply. A good rule is that they should not be covered more than 4 to 5 times the diameter of the seed; usually this means not more than $\frac{1}{6}$ to $\frac{1}{2}$ in. deep.

8. **A companion (nurse crop) or a preemergence herbicide may be used.** The value of planting a "companion" or nurse crop—usually consisting of annuals or short-lived perennials—with new seed crops varies with the intended crop, soils, and season. The advantages are: (a) it furnishes a crop of value while the new seeding is being established, (b) it lessens erosion, and (c) it reduces the weed population. The disadvantages are: (a) it may retard the growth of the seedlings for whose protection it is grown and delay establishment, and (b) it may rob the new seedlings of so much moisture and light that it kills them during dry spells unless the companion crop is harvested early as pasture, hay, or silage.

Some perennial grass species are extremely fragile when in the early seedling stage. Establishment is often enhanced by protection and shade afforded either by a nurse crop or the noncompetitive stubble remaining from a previous crop.

Preemergence herbicides may replace the traditional companion crop, especially where a single species is being established. Usually these result in higher forage yields in the seeding year than can be obtained when a companion crop is used.

RENOVATING OLDER PASTURES

Renovation is the rejuvenation of existing pastures. Pastures are renovated in order to regain production potential by (a) reestablishing the desired mix of species; (2) eliminating persistent, undesirable weeds and/or brush; and (3) breaking up soil compaction to improve soil aeration, water infiltration, and root development.

Renovation is accomplished either by partial or complete destruction of the existing sod. The method will vary from one area to another depending on (1) stand and species, (2) soils and topography, and (3) degree or stage of degeneration. Pastures may be renovated by the following methods:

1. **Fertilizing and overseeding.** Often on pastures where a fair, but unproductive, stand of desired plants exist, renovation may simply consist of specially prescribed fertilizer applications, and liming if needed. Occasionally, this may be accompanied with weed and brush control treatments. This condition represents large areas in the south humid and subhumid regions and the occurrence is strangely influenced by the ratio of fertilizer costs and prices received for livestock.

Fig. 5-6. Power-drive seeder for fast, economical pasture and range renovation. (Courtesy, Deere & Co., Moline, IL)

2. **Complete seedbed preparation with reseeding.** Complete seedbed preparation followed by reseeding has been widely used in accomplishing the objective of renovation. This is especially effective where perennial, weedy grasses have invaded the sod and are otherwise difficult to control. This method facilitates the working of lime and fertilizers into the surface

soil and is generally superior to other methods in establishment of legumes such as alfalfa. Occasionally, the use of an intervening crop for one or two years may be desirable in reducing populations of undesirable plants; and a cash crop may help defray renovation costs.

3. **Sod-seeding or overseeding (no-till seeding).** In humid regions, reestablishment of desirable plants, especially legumes, into existing pasture sod is often preferred to complete seedbed preparation. The basic requirements for successful sod-seeding are: the partial (strip or rows) or complete destruction of existing plants by use of herbicides and close grazing, thereby reducing competition; liming when necessary; fertilizing; proper seed placement; and good pasture management following seeding.

FACTORS AFFECTING VALUE OF PASTURE

Many factors affect the value of pasture, including (1) soil and fertilizer, (2) plant species, (3) stage of maturity, (4) rate of growth and season of year, and (5) grazing.

• **Soil and fertilizer**—Soil and fertilizer affect the growth and composition of pasture crops. Many experiments have been conducted to determine the effect of soil fertility and fertilizer application on pasture. A brief summary of the benefits that generally accrue from pasture fertilization follows:

1. **Increased yields.** the chief benefit to accrue from applying fertilizer to pasture is an increase in yield.

2. **Increased proportion of legumes.** In grass-legume pastures, proper fertilizing can influence the proportion of legumes; in turn, this increases the protein, calcium, phosphorus, and vitamin content of the mixture. Generally legume-grass pastures with about 50% legumes do not require N fertilization. However, on such pastures it is important to maintain adequate levels of lime, P, and K.

3. **Extended grazing season.** Properly fertilized pasture plants grow over a longer period of time than those on infertile soils; they begin growth earlier in the spring, and they continue growth later in the fall.

4. **Increase protein and palatability.** The protein content of young, immature nonlegume pasture is increased appreciably by nitrogenous fertilization, unless of course, there is already plenty of nitrogen in the soil. This increase may be sufficient to add materially to the palatability and the feeding value of grass pasture.

5. **Increased calcium and phosphorus.** Calcium-deficient soils affect pastures in two ways: (a) the percentage of calcium in nonlegume crops is considerably reduced, and (b) the legume crop, if present, will not thrive.

On phosphorus-deficient soils, the phosphorus content of grasses may drop so low that phosphorus deficiency in animals is produced, unless a phosphorus supplement is provided. According to research workers at the Texas Station, pasture forage having less than 0.15% phosphorus on a dry matter basis will not supply enough of the element to meet safely the requirements of beef cattle grazed thereon without supplemental feed. The phosphorus content of legumes is less affected by a deficiency of soil phosphorus than that of nonlegumes. However, most legumes do not thrive or yield well on phosphorus-deficient soils.

• **Plant species**—Plant species affect the feeding value of pasture. Generally speaking, legumes contain a higher percentage of protein and calcium than nonlegumes. Also, there are marked differences between different kinds of pasture plants as growth advances. For example, bromegrass retains its palatability and nutritive value over a longer period than most grasses. By contrast, reed canarygrass is readily eaten when young, but becomes woody, high in alkaloids, and unpalatable with maturity. Most pasture legumes retain their palatability and nutritive value as they mature better than most grasses. An exception to the latter rule is *lespedeza sericea,* which becomes bitter and distasteful with maturity due to the accumulation of tannin in the plants. However, plant breeders have developed *sericea* that is low in tannin, thereby overcoming this problem to some degree.

• **Stage of maturity**—There are great differences in nutritive value between young, immature pasture and the same plants when they are mature or even at the usual hay stage. An example of these wide differences is shown in Fig. 5–7.

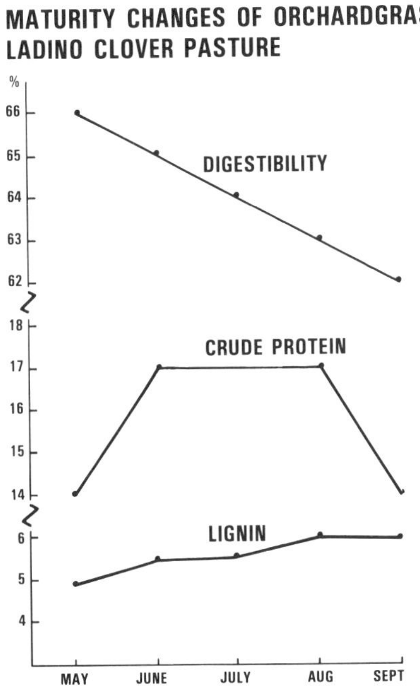

Fig. 5–7. Changes in digestibility, crude protein, and lignin content of orchardgrass (*Dactylis glomerata*)/ladino clover (*Trifolium repens*) pasture in Tennessee, during a 5-year period, from May to September. **Note well:** Digestibility decreased with maturity; it decreased steadily from 66% in May to 62% in September. Protein decreased with maturity; it dropped from 17% in June to 14% in September. Lignin increased with maturity; it increased from 4.8% in May to 5.8% in September. (Source of data: Lane, C. D., K. M. Barth, and J. B. McLaren, *Journal of Animal Science,* Vol. 34, No. 2, 1972, p. 351)

Stage of maturity affects pasture composition as follows:

1. **Protein decreases with maturity.** Young plants are much richer in protein, on a dry basis, than the same plants at maturity. Young actively growing grasses may run 16 to 20% protein content, or better; and young legumes, such as alfalfa and clover, are even higher—20 to 25% protein or more. The protein content of grasses decreases greatly as they head out. Very mature dormant grasses may contain as little as 2% protein.

The protein content of legumes decreases with maturity, but to a lesser extent than in grasses.

The above underscores the need for producers to take advantage of the high-protein content of young pastures when formulating rations so as to avoid overfeeding costly protein supplements.

2. **Fiber and lignin increase with maturity.** The percentage of fiber and lignin in plants influences both palatability and digestibility. Although there are inherent differences among plant species as well as among plants of the same species, young plants and regrowth are always lower in fiber and lignin than mature plants. As a result, they are more tender and digestible. On a dry matter basis, new growth in a grass legume pasture will range up to 68% total digestible nutrients in comparison with 51% at the normal hay stage. In humid areas, the nutrient levels of some grasses is leached rapidly following maturity and the digestibility and nutrient value is comparable to poor grade straw.

To a considerable degree, the decrease in digestibility as plants mature is due to an increase in lignification, which lowers digestibility, even by ruminants.

3. **Calcium and phosphorus decrease with maturity.**The calcium content of plants decreases with maturity. However, the percentage change in calcium is much less than occurs in phosphorus.

Young grasses or legumes grown on phosphorus-rich soils usually contain 0.25%, or more, phosphorus. Although phosphorus in pasture decreases with maturity, there is generally plenty left for livestock, unless (a) it is produced on phosphorus-deficient soils, or (b) it is left to cure on the stalk, which makes for weathering and bleaching. Early spring grass (April) may contain five times more phosphorus than leached and weathered grass in the winter (March).

4. **Vitamin value decreases with maturity.** Actively growing green parts of plants are high in carotene. Likewise, such plants are usually rich in most of the B complex vitamins, in vitamin E, in ascorbic acid, and in certain unidentified factors. But the content of vitamins, especially of carotene, decreases as plants mature. When pasture plants are left to cure on the stalk, and to leach and weather, carotene disappears rapidly.

• **Rate of growth and season of year**—Rapidly growing grass is usually rich in protein and in other nutrients. It is important, therefore, that pasture plants be properly fertilized and, to the extent possible, managed so that they keep growing, and that they be prevented from heading out. This can be accomplished by mowing, but it is more practical when done with livestock through intensive grazing programs.

Grass is usually higher in protein and other nutrients early in the spring than later in the season. If plant growth is sharply checked in the summer—due to drought, hot weather, and/or lack of available plant food—the protein content and the digestibility will be lower than that of grass at the same stage of maturity earlier in the season.

If pasture resumes growth after the fall rains come, it may be nearly as high in protein and other nutrients as spring growth.

It is also noteworthy that all plants are most digestible and most preferred during periods of rapid growth. Recognizing this, producers are establishing warm-season grass pastures to obtain higher livestock productivity during the normal midsummer growth slump in areas where cool-season grasses predominate.

• **Grazing**—When pastures are grazed closely throughout the season, the total yield of dry matter is usually 30 to 50% less than when they are allowed to grow to the normal hay stage. This is due to the smaller leaf surface and lowered photosynthesis. This explains why rotational, strip, and green chop grazing yield more than close continuous grazing.

The total effect of frequent grazing will depend on the kind of plants. The yield of tall-growing plants—such as timothy, orchardgrass, alfalfa, and the erect clovers—can be reduced much more than that of low-growing spreading plants, such as bluegrass, Bermudagrass, and white clover.

In contrast to the lowering of the yield of dry matter, frequent grazing usually results in greater total production of protein for the season than when the crop is cut for hay. Also, because immature plants are lower in fiber and more digestible than mature plants, the yield of total digestible nutrients is not reduced as much by frequent grazing as the dry matter yield—dry matter production is lowered by 30 to 50%, whereas digestibility is lowered only by 25 to 40%.

Also, it is noteworthy that when allowed to graze selectively in an extensive grazing program animals pick and choose the leaves and finer parts of stems, which are more tender and more nutritious, and reject the courser, stemmy parts. Thus, the portion consumed under such circumstances will always differ appreciably from the chemical composition of the entire plant.

PASTURE MANAGEMENT

Many good pastures have been carefully established only to be lost in succeeding years through poor management. Efficient and profitable pasture management in the subhumid, humid, and irrigated areas involves the following practices:

1. **Controlled grazing.** Nothing contributes more to good pasture management than controlled (proper) grazing. At its best, it embraces the following:

a. **Protecting first-year seedings.** First-year seedings should be grazed lightly or not at all in order that they may get a good start. Where practical, instead of grazing, it is preferable to mow a new seeding about 3 in. above the ground and to utilize it as hay or silage, provided there is sufficient growth to so justify.

b. **Shifting the location of salt and water.** Where portable salt containers are used, more uniform grazing may be obtained simply by the practice of shifting the location of the salt to less grazed areas of the pasture. Where possible and practical, water locations should be well distributed.

c. **Fencing into smaller units.** Development of more and smaller pastures and employing rotation grazing programs permits greater control of when and where livestock graze. Many more intensive systems are designed with stationary salt and water locations, while the area being grazed may change every 3 to 4 days, even daily in some cases. Various forms of electric fencing are being used in a number of different pasture arrangements, all designed to facilitate grazing greater numbers of animals on an area for shorter periods of time (short duration grazing).

d. **Deferred spring grazing.** When possible allow 6 to 8 in. of growth before turning out to pasture in the spring, thus giving grass a needed start. This is not always practicable on year-round pasture programs. However, when two or more pastures are available, early spring grazing can be rotated to the benefit of plants in all pastures.

e. **Avoiding close late-fall grazing.** Pastures that are closely grazed late in the fall start growing late in the spring. Plants should be allowed to replenish root reserves prior to going dormant. With most pastures, 3 to 5 in. of growth should be left for winter cover. An exception to this close grazing rule should be made where winter annual clovers are to be seeded, or are expected to volunteer, especially on Bermudagrass or bahiagrass pastures. Under such circumstances, close grazing or mowing of the Bermudagrass or bahiagrass is recommended. Likewise, in the Pacific Northwest, perennial ryegrass must be grazed hard in the summer in order for subterranean clover seeds to germinate and become established in the autumn.

f. **Avoiding overgrazing.** Never graze more closely than 2 to 3 in. during the pasture season. Continued close grazing reduces the yield, weakens the plants, encourages weeds to invade, and increases runoff and soil erosion. The use of temporary and supplemental pastures, such as Sudan, may ''spell off'' regular pastures through seasons of drought and other pasture shortages and thus alleviate overgrazing.

g. **Avoid undergrazing.** Undergrazing seeded pastures should be avoided, because (1) rank growth is less palatable and of lower nutritive value; (2) tall-growing grasses may drive out desirable low-growing plants such as white clover due to shading; and (3) weeds, brush, and coarse grasses are more apt to gain a foothold when the pasture is grazed insufficiently. It is a good rule, therefore, to graze the pasture fairly close at least once a year.

2. **Clipping pastures and controlling weeds.** Pastures should be clipped as necessary to control competing weeds (and brush) and to get rid of uneaten clumps and other unpalatable coarse growth left after incomplete grazing. Good grazing management will reduce the amount of clipping needed. Pastures that are grazed continuously may be clipped at or just preceding the usual haymaking time; rotated pastures may be clipped at the close of the grazing period. Weeds and brush may also be controlled by chemicals and burning.

It is noteworthy that mowing can be expensive ($3 to $8 per acre) and should be applied only when results are clearly beneficial. Clipping solely for cosmetic reasons should be critically evaluated.

3. **Topdressing.** Like animals, for best results, grasses and legumes must be fed properly throughout a lifetime. It is not sufficient that they be fertilized (and limed if necessary) only at or prior to seeding time. In addition, in most areas it is desirable and profitable to topdress pastures with fertilizer annually, and, at less frequent intervals, with reinforced manure and lime. Such treatments should be based on soil tests, and are usually applied in the spring or fall.

4. **Scattering droppings.** The dropping should be scattered at the end of each grazing season in order to prevent animals from leaving ungrazed clumps and to distribute the droppings over a larger area. This can best be done by the use of a brush harrow or a chain harrow.

When animals are concentrated under intensive systems such as strip grazing, there is a tendency for droppings to be broken up by hoof action and the need for harrowing is reduced.

5. **Grazing by more than one kind of animal.** Grazing by two or more species of animals makes for more uniform pasture utilization and fewer weeds and parasites, provided the area is not overstocked. Different kinds of livestock have different habits of grazing; they show preference for different plants and they graze to different heights. For example, sheep consume shorter and finer forages and more forbs than cattle.

6. **Irrigating where practical and feasible.** Where irrigation is practical and feasible, it alleviates the necessity of depending on natural precipitation.

EXTENDING THE GRAZING SEASON

In practically all United States pasture areas, the grazing season can be extended by grazing earlier in the spring and later in the fall/winter, thereby lessening the amount of stored feed and supplemental protein needed for winter.

Fig. 5-8 illustrates in graphic form the growth period of each of the common pasture plants of area 2 (see Fig. 5-3 for areas), the northern part of the humid South. As shown, by selecting the proper combination of crops, forage budgets concerning each month of the year can be designed. A similar chart for each area of the country can be developed.

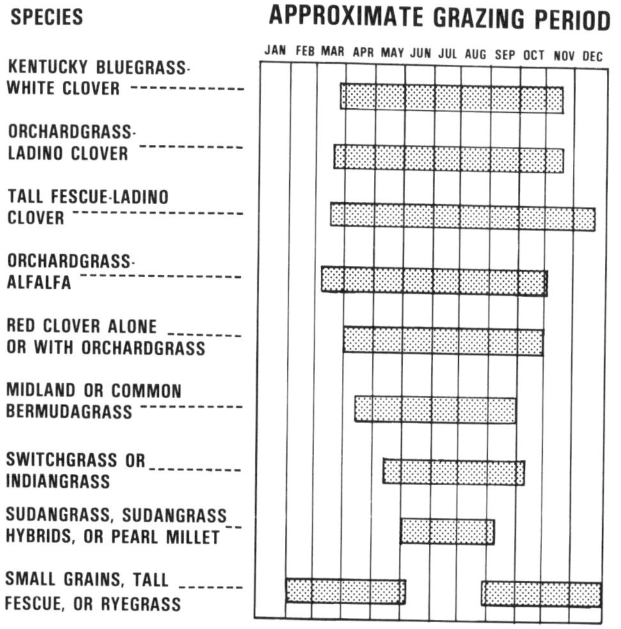

Fig. 5-8. Approximate grazing period of common pasture crops in area 2 (see Fig. 5-11 for areas), the northern part of the humid South. As shown, by selecting the proper crops a near year-round grazing can be achieved.

In addition to lengthening the grazing season through the selection of species, it may be extended as follows:

1. **By obtaining earlier spring pastures.** This can be accomplished by avoiding grazing late in the fall and by applying nitrogen fertilizer in the fall or early spring. Nitrogen fertilizers will often stimulate the growth of grass so that it will be ready for grazing 10 days to 2 weeks earlier than unfertilized pastures.

2. **By saving fall growth for winter grazing.** For example, tall fescue that has been cut for winter feed can be used to stockpile regrowth in August through October for grazing from November into the winter. Also, often the regrowth after hay harvest may be utilized.

3. **By using crop residues.** Following harvest, cornstalks will provide 2 months of winter grazing for dry cows and ewes. In addition to extending the grazing season, this utilizes feed which would otherwise be wasted or unmarketable.

SEEDED PASTURE GRAZING SYSTEMS

Several systems of grazing management have been successfully applied to pastures. Generally speaking, the more intensive the system of management on such pasture, the higher the yield of forage and of livestock products.

It is noteworthy that pasture grazing systems have been changed/adapted by both researchers and farmers; and with such changes/adaptations, they have been given different names. Nevertheless, under whatever name, the basic type of rotation grazing, intensive grazing, creep grazing, strip grazing, and green chop are covered in the sections that follow.

(Also, see the section on Range Grazing Systems in this chapter. The principles involved in grazing seeded pastures and western ranges are similar. However, the application differs because western range pastures are generally much larger and have lower rainfall.)

CONTINUOUS GRAZING

The name identifies the practice. *Continuous grazing is the uninterrupted grazing of a specific pasture by livestock throughout the year or grazing season.* It can be successful provided moderate stocking is practiced, with some adjustment in animal numbers to reduce the severity of under- or overgrazing.

The **advantages** of continuous grazing as compared to rotational grazing are (1) lower costs for fencing and watering facilities, (2) fewer management decisions when animals are not moved from pasture to pasture, and (3) often slightly better individual animal performance when younger animals, such as stocker steers or heifers, are grazed.

The **limitations** of continuous grazing are (1) animal numbers are seldom flexible; (2) pastures must be stocked lighter than desired when forage growth is maximal to avoid overgrazing during periods of minimal forage growth; (3) animals selectively graze some species in preference to others and return to graze the regrowth of the same plant, thus selectively reducing plant vigor; and (4) livestock often show preference to grazing certain portions of pastures, resulting in uneven fertilization.

ROTATION GRAZING

Rotation grazing is that system in which two or more pastures are grazed and rested in a planned sequence. In this system, pastures are divided into two or more pastures, with the objective to develop a grazing program where major forage species are harvested and then provided a period of rest enabling the plants to regrow and remain thrifty and vigorous.

Rotation grazing involves the concept of time as a management variable for either the grazing period or the regrowth interval of each pasture. Duration of grazing and rest generally are governed by herbage growth rate, which depends primarily on the time of year, moisture, fertility, and species. The number of animals grazing each system may be fixed or variable.

Experimental studies comparing continuous and rotation grazing have given inconsistent results. In general, however, rotation grazing has proven beneficial to cow-calf operations and continuous grazing is favored for stocker programs.

The **advantages** of a rotation grazing system are:

1. It permits the farmer to match grazing more adequately to the growth habit of the forage species, condition of the pasture, and animal needs than does continuous grazing.
2. It improves stand persistence and production. Plants are given recovery periods during the growing season for more or less unhampered development of tillers and leaves. This is essential to replenish root reserves. This system of grazing enables the tall-growing legumes and grasses to survive.
3. It increases carrying capacity. Greater amounts of feed nutrients can be removed in the form of herbage with reduced losses due to trampling, fouling, and herbage death and decay.
4. It encourages equalization of grazing. It helps prevent overgrazing and undergrazing, and results in maintaining a better balance of the legumes and grasses. Also, both the palatable and the inferior species are grazed more nearly the same.
5. It often provides more nutritious herbage since the herbage is at the most ideal pasture stage. It will be high in protein and low in crude fiber.
6. It helps prevent the grasses from heading out. This is done by concentrating grazing animals or by mowing when animals are shifted to new pastures. This allows new growth to come back uniformly and keeps it more palatable.
7. It helps control livestock parasites, especially intestinal worms. Life cycles of worms can be broken by proper planning of grazing and rest periods.
8. It makes it more convenient to harvest surplus forages as hay or silage.

The **limitations** of rotation grazing are:

1. It requires a higher input of capital and management than continuous grazing.
2. There is a continous day-to-day decline in the quality of the available forage, especially on the more intensive systems. At first turn-on, animals have access to leafy, high-quality forage, but the quality of the forage gets poorer and poorer during the grazing period.

INTENSIVE GRAZING

Several ingenious intensive grazing systems have evolved. All of them are designed to provide and harvest the maximum of high quality forage; to utilize the highest quality pastures for the highest producing animals; and to increase profits.

These systems were first described and used in Europe by Voisin and have since been installed to some extent throughout the United States and much of the world. Many designs are utilized for intensive short duration grazing systems. Basically, they fall into either a rectangular design or a circular or wagon wheel design.

- **Conventional (rectangular) systems**—The rectangular system generally is a series of small pastures, usually of equal size or production, that are fenced in a grid arrangement. Water and salt may be located in each of the pastures or a single source may be used with cattle gaining access by a lane or alley.

- **Cell or wagon wheel system**—This system was introduced and popularized in the United States by Alan Savory. The pasture fences radiate out from a central hub, giving it the appearance of a wheel, thus the name. Water, salt and mineral, and working pens are usually located at the hub. As livestock return to the hub daily, they can easily be moved from one pasture to the next.

- **First and second grazers**—This short duration grazing system involves two herds: first grazers and second grazers. It calls for using the best quality pastures for dairy cows, cows suckling calves, ewes with twin lambs, does with kids, or mares with foals. Here is how it may work with diary cows: High producing lactating cows, which have a high energy requirement, may be first grazers; they are allowed to graze the higher-quality (leafy) portion of pasture No. 1, following which they are moved to pasture No. 2—a fresh pasture. Dry cows, which have a low energy requirement, may be second grazers; they are turned into pasture No. 1 immediately following the removal of the high producers. This progression is continued through all pastures, then the cycle is repeated. Also, it may involve three or more groups of animals. The groups may consist of any animal species or class. For example, the first grazers may consist of beef cows and suckling calves, and the second grazers of replacement heifers; or the first grazers may be ewes with twin lambs, and the second grazers gestating ewes.

The chief **advantage** of the system of first and second grazers is the enhanced productivity of the first grazers. The main **limitations** are the necessity of maintaining (1) two groups of animals of different productivity levels, and (2) balanced stocking rates and pasture sizes.

- **Intensive early season stocking**—This grazing system calls for heavy stocking (perhaps twice the average summer carrying capacity of the pasture) in the spring and early summer, when the pasture is of highest quality and the most productive. Here is how it works with stocker cattle: Double the normal stocking rate in the spring and early summer. Then, around July 1, either sell half or more of the herd or place them in a feedlot.

The **advantages** of this system are (1) more pounds of beef per acre, (2) lower interest charges, because of owning the cattle for a shorter period of time, and (3) higher net returns. The main **limitation** is the lack of flexibility relative to removal of half the herd; they must either be sold or moved into the feedlot as scheduled—in mid-summer.

CREEP GRAZING

Creep grazing is a system of grazing nursing young on a high-quality pasture(s) (a grass-legume mixture, all legumes, or high-quality annuals) separated from their dams. This system may be used for cows and calves, ewes and lambs, goats and kids, and mares and foals. Creep grazing may be accomplished as follows:

1. Allowing young to forward graze ahead of their dams, then following with their dams later. This is similar in principle to "first and second grazers," with the young having access to the choicest most succulent pasture(s) without competition from their dams. It may be accomplished by confining the dams and young in one field for a period, but allowing the young to enter the next choice-quality pasture(s) through a creep opening large enough for the young but small enough to keep the dams out. The dams are kept on each pasture in the rotation until forage is utilized to the desired level, then moved to the new pasture. This same pattern is contained through all pastures in rotation.

2. Keeping the dams and young on a base pasture, and providing an additional creep pasture for the young. The dams and young are kept on a base pasture (or pastures if rotational). In addition the young are given access to a high quality creep pasture(s) through a creep opening.

Limited studies in creep grazing calves indicate that as much as one-half pound extra weight gain per day may be obtained by creep grazing.

STRIP GRAZING

In this system, animals are allowed access to a strip which may be large enough for several days of grazing or small enough for ½ to 1 day of grazing. Heavy stocking rates of upwards to 50 animal units per acre are used by fencing each strip with movable electric fences both in front and behind the grazing livestock. The **advantages** claimed for this method are:

1. Increased utilization of herbage, with wastage reduced to 10 to 20%.

2. Increased meat and milk yields per acre up to 25%.

3. Improved stability of meat and milk yield because the nutritive value of the pasturage consumed is quite constant.

4. Improved utilization of the available forage. Less herbage is soiled by dung, urine, and treading. Under strip grazing, animals are quieter and settle down quickly for steady grazing rather than roaming about and tramping forage.

5. Increased animal units maintained on a given area, although individual animal productivity may not be increased.

GREEN CHOP

Fig. 5-9. Harvesting alfalfa as green chop. (Courtesy, Sperry New Holland, New Holland, PA)

Green chop is fresh herbage that is cut and chopped in the field, then transported and fed to animals in confinement.

Green chop, which is also called soilage or zero grazing, consists of growing a succession of forage crops, harvesting them with mechanized equipment, and hauling the green feed to the animals rather than allowing the animals to harvest their own forage. Historically, cutting green forage and hauling it to animals developed where land and forage were scarce and labor was plentiful. Under this system of feeding, each animal unit has to be supplied with upwards of 150 lb of green forage daily, depending upon its succulence.

Green chop minimizes the loss of moisture, color, nutrients, and wastage. Alfalfa, ladino clover, orchardgrass, bromegrass, grass-legume mixtures, Sudangrass, corn, sorghum, soybeans, and cereal grains are sometimes used in this manner. With tall-growing crops, more feed value may be realized from a given area than can be obtained by conventional pasturing. However, green chop requires special equipment and harvesting every day. Also, there are harvesting problems in wet weather.

Most green chop is fed to lactating dairy cows, usually in combination with hay or silage because the total intake tends to be greater. Green chop has increased with herd size, with more intensive forms of dairying, with drylotting of cows, and with high grain prices. Also, the use of green chop has been facilitated by the greater mechanization present on larger and more modern dairy farms.

IRRIGATED PASTURES

Well managed irrigated pastures can enhance the flexibility and add to the stability of livestock forage programs. Throughout the western United States, irrigated pastures have been used successfully to improve carrying capacities, reduce feed costs, lengthen the grazing season, improve gains and milk production, and improve breeding efficiency.

Irrigated pastures provide forage of high quality at a relatively low cost, often on land unsuitable for other crops. Both perennial and annual irrigated pastures are important feed crops.

Successful pasture irrigation involves special decision making relative to (1) irrigation—the method, frequency, and amount of irrigating, and the removal of excess water; and (2) the kind and amount of fertilizer.

Fig. 5-10. Cows and calves on irrigated pasture. Note the sprinkler irrigation equipment in the distance. (Courtesy, American Hereford Assn., Kansas City, MO)

- **Method of water application**—Two basic methods of irrigating pastures are practiced: (1) flood, and (2) sprinkler. The choice of the method for any given pasture should be determined by soil type, topography, water supply, and funds available for irrigation development.

The efficiency of flood irrigation can be improved with the use of borders. The border-flood method is adapted where a large head of water is available and the land is level or requires only minor movement. When properly used, there is no runoff and efficiency of water utilization is high.

Sprinkler irrigation may be preferable to the flood method (1) on land that is not level enough for surface irrigation or where the cost of leveling would be prohibitive; (2) on soils of variable texture where the amount and frequency of application can be adjusted to the water-holding capacity of the soil; or (3) where water cost is high or the supply of water is limited. Sprinklers are on the increase throughout the United States, with laborsaving, center-pivot and wheel move systems making hand-move systems obsolete.

- **Frequency and amount of irrigation**—It is recommended that in many parts of the United States, especially the west, irrigation water must be applied when it is available, not necessarily when it is desired, and often it may not be available for part of the season. Such restriction may severely limit pasture production. However, when possible and practical, water should be applied: (1) at a rate and frequency to maintain good soil moisture throughout the root profile; (2) immediately following grazing where grazing is rotated; (3) according to the consumptive use rate of the major species; and (4) relative to the content of soluble salt, as soils high in salt may be flushed while irrigation water high in salts should be used sparingly.

Many pasture plants, especially the clovers, are shallow-rooted and require more frequent and lighter irrigations than deep-rooted plants. The Washington Station reports that the highest yields per acre from an orchardgrass-ladino clover pasture can be obtained with a summer irrigation frequency of 7 to 11 days, rather than at less frequent intervals, and that more frequent irrigations also give the highest proportion of clover. In important irrigated areas, county extension agents and district conservationists are usually knowledgeable relative to the proper time to irrigate specific crops in the particular area; hence, they should be consulted when developing a schedule.

- **Excess water**—Excess water in irrigated pastures is caused by either (1) overirrigation and the inability of the excess water to drain from the soil, or (2) subsurface drainage from adjacent and higher land. Allowing excess water to stand on pasture can drown desirable plants, with the resulting area growing up in weeds. Also, standing water is a breeding ground for insects. Surface drains are necessary to remove excess irrigation. Drainage is particularly important wherever there is danger of salt accumulation. Deep drainage ditches spaced at proper intervals help to remove these excess salts and control the level of the water table. In some areas, drainage ditches must be augmented by tile drainage in order to keep the salt content below levels that are harmful to plants.

- **Plant species**—Selection of species for establishment of irrigated pastures should be dependent on (a) adaptation to the general area, (2) water availability, (3) soils, (4) salinity problems, and (5) forage needs.

- **Fertilization**—Irrigated pastures require high soil fertility to be productive. The kind and amount of fertilizer should be determined by the level of the productivity desired, and the role of the legumes in the mixture. Production levels of irrigated pastures are increased more by N fertilizer than by other fertilizer elements, with responses also obtained from P and K where soils are deficient in these elements. Nitrogen stimulates grass growth, whereas P increases the legume component. Nitrogen can be supplied by either fertilizer or inoculated legumes. When legumes are a major component of pasture, economic returns from applied N, measured in increased animal production, may not be obtained.

- **Grazing**—Although continuous grazing has been used effectively in some locations, the potential benefits from irrigated pastures in the West are of such magnitude that some form or rotation grazing should be employed.

PART II—THE WESTERN RANGE[3]

Fig. 5-11. The western range. (Courtesy, USDA, Soil Conservation Service)

Rangelands are native pastures. As a resource, they occupy the largest expanse of land area in the United States utilized for livestock production. As defined by the Society for Range Management, *rangelands are lands on which the potential (climax) vegetation is made up of grasses, forbs, and shrubs suitable for grazing.* This includes natural grasslands, shrublands, savannahs, many wetlands and marshes, and some desert and tundra areas.

Rangelands have multiple uses; thus, they are important to many people other than livestock producers. In addition to producing livestock and livestock products, these lands (1) provide habitat for many species of animals and birds, (2) serve as extensive watersheds and are important for water yield, (3) provide the setting for many forms of recreation—including hunting, fishing, dude ranching, hiking, and horseback riding, and (4) produce wood products and minerals in some locations.

Because of the extent of rangelands and the magnitude of the range livestock industry, and the fact that it is a highly specialized type of operation, considerable discussion will be devoted to the range area and its care and management.

RANGE AREA

There are approximately 622 million acres of rangeland in the United States, excluding Alaska (Table 5-1). About 214 million acres, or 34% is federal land. The bulk of rangelands occurs in the intermountain portions of the 11 western states. Nearly 408 million acres of rangeland, or 66%, is nonfederal lands, mostly privately owned.

Various geographical divisions are assumed in referring to the range area. Sometimes reference is made to the *17 range states,* embracing a land area of approximately 1.16 billion acres. At other times this larger division is broken down chiefly on the basis of topography into (1) the *11 western states* (Arizona, California, Colorado, Idaho, Montana, Nevada, New Mexico, Oregon, Utah, Washington, and Wyoming); and (2) the *Great Plains States* (the 6 states of Kansas, Nebraska, North Dakota, Oklahoma, South Dakota, and Texas). Specific references to the Great Plains Area includes, in addition to the 6 Great Plains States, that portion of Colorado, Montana, New Mexico, and Wyoming lying east of the Rocky Mountains.

---

[4]Grateful appreciation is expressed to the following authorities who reviewed, and made helpful suggestions relative to Part II—The Western Range: M. Pefoll, Director, Bureau of Land Management, U.S. Department of the Interior, Washington, DC; and R. M. Williamson, Director, Range Management, Forest Service, USDA, Washington, DC.

Since the major portion of rangeland is located in the West, it is often referred to as the *Western Range.* Actually, rangelands occur in 25 of the 50 states. All states west of the Mississippi River, with the exception of Iowa, contain rangelands. They also occur in the east, extending through the Gulf Coast states from Louisiana to Florida. These eastern rangelands, although minor in extent, have high production potentials and are important to local economies.

Moreover, variation in climate, soils, and topography in the range country is accompanied by a great diversity in the kind and production of vegetation, and in the use made of it.

Most rangeland areas are suitable for year-round livestock operations, although supplements in forage or nutrients may be required at different seasons. In parts of the intermountain area, rangelands are grazed at different times of the year and herds and flocks migrate with the season, moving to higher elevations in summer and returning to lower elevations in winter.

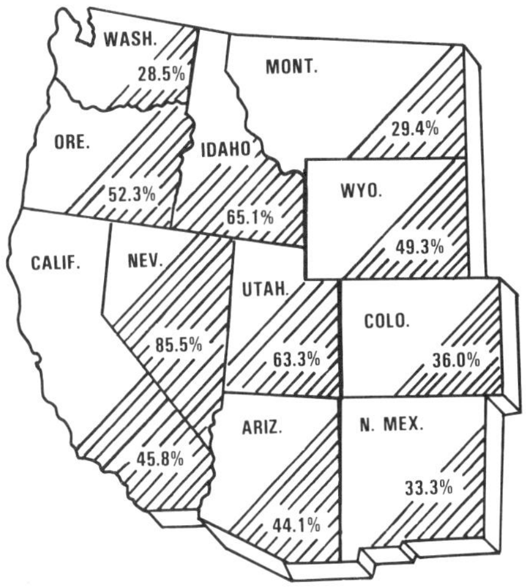

Fig. 5-12. A map showing the 11 western states and the proportion of land in each of these states that is owned by the U.S. Government. (Source: *Public Land Statistics,* 1984, Vol. 169, August 1985, p. 10)

Although some operations rely solely on range for forage, in most locations rangelands are used in combination with cultivated pastures and cropland forages in varying proportions.

There are striking differences in the nature and use of rangelands when compared to seeded pastures and in the principles and techniques required for management. From the standpoint of vegetation and utilization by livestock, ranges differ significantly from cultivated pastures as follows:

1. **They are less productive.** Generally their production capacity is lower. This is as one would expect for they are largely made up of land not suitable for cultivation; they include soils that are shallow or rocky, have alkali or salt accumulations, and/or occur on steep topography. Also, plant growth on rangeland frequently is limited by low and undependable rainfall (even drought).

However, it should be noted that some range areas are highly productive with yields often comparable or even exceeding that of cultivated pastures. These are largely restricted to the coastal marshes, the tall grass prairies of the eastern Great Plains, and the scattered areas with natural sub-irrigation and wetland meadows.

2. **They may deteriorate without showing detectable signs.** Tame pastures, as a rule require yearly, even seasonal, attention if the desired sod and production are to be maintained. By contrast, rangelands vary little from year to year, and the changes are subtle. The mute signs of a deteriorating range often go unnoticed by the untrained.

Range vegetation consists of a mixture of plants and each species varies in palatability, nutritive value and production ability. Grazing animals select the most palatable plants first. Thus, unless management is practiced, these plants will be overgrazed, weakened and replaced by lower value plants. Poor management, ever so slight, continued over decades, will result in good forage plants being almost completely replaced by plants of lower value.

3. **They are more difficult to restore when depleted.** Once a range becomes depleted it is a slow and often expensive process to rebuild it. Especially in more arid zones, plowing and drilling are impracticable and costs for brush control are frequently prohibitive. Thus, very often the only feasible way of restoring a range is to stock it conservatively and carefully manage the location and intensity of grazing.

4. **They must be managed differently.** Management on rangelands is conceptually the management of natural ecosystems, and, to the extent possible, is done through natural means. Native vegetation is encouraged to perpetuate, largely through livestock control and correct grazing management. This is done without the use of cultivation, commercial fertilizers, soil amendments, or periodic renovation, as commonly prescribed on seeded (tame) pasturelands.

RANGE ECOSYSTEMS

Because progress in range improvement is slow and gradual, management must become a "way of life" for the western rancher. Effective management is best accomplished when the principles of range ecology are well understood and incorporated into a planned program of grazing management.

CLIMAX PLANT COMMUNITIES

Through centuries of natural selection, native plant communities developed in conformity with the environment and, therefore, the plants are superbly adapted to the climate and soils on which they grow. The highest order of plants (most productive) that grow on an area is referred to as climax or the potential plant community. When well managed, this mixture of plants, although fluctuating some with weather cycles, can be perpetuated indefinitely. This is exemplified in the Flint

Hills region of Kansas and the Sand Hills of Nebraska. It is noteworthy that since climax plant communities developed with grazing they are more productive when properly grazed than when not grazed at all.

RANGE DETERIORATION

The stability of a climax plant community is altered when improperly managed. When livestock selectively overgraze, preferred plants become weakened, lose their ability to compete, and gradually decrease in abundance. They are replaced by the spread of less palatable and more shallow rooted plants. Degradation toward poorer and less productive plants can continue until (1) the original plants are completely replaced by communities of low value grasses, weeds, and woody shrubs; or (2) the range is left partially denuded and subject to erosion.

PLANT SUCCESSION

When causes of deterioration are detected in time and corrected, the cycle can be reversed. Plants that declined in amount because of overgrazing will, in time, increase with proper management. This sequence is referred to as *plant succession.*

To describe the amount of change that has occurred in the vegetation, four classes of range condition are recognized—excellent, good, fair, and poor. Excellent, the highest ranking, is representation of the potential community, production is near maximum, desired plants abound, and conditions are optimum for insoak of rainwater. Range classified as poor condition has less than one-fourth of the desired plants, forage production is only a fraction of the potential, soil may be exposed, and weeds and brush usually have invaded. Good and fair classifications are intermediate.

RANGE MANAGEMENT CONSIDERATIONS

Good range management may be achieved if an inventory or analysis is made of the forage resources, followed by a sound plan of management based upon the analysis. Consideration should be given to such factors as proper stocking rates, safe degree of plant use, season of use, kind of livestock, conditions and trend of forage, system of grazing, improvements needed, etc.

STOCKING RATE

The key to planning for successful long-term operations on rangeland lies in making (1) a reliable determination of the condition and trend of the range vegetation, (2) a realistic estimate of the production and grazing capacity for each pasture, and (3) a stocking rate with the flexibility to adjust with variations in weather cycles and to accomplish objectives in range improvement.

Fig. 5-13. Overgrazed vs properly grazed range, separated only by the fenceline. Note that the properly grazed range shows a vigorous stand of bunchgrass, whereas the bunchgrass has been practically destroyed on the other range. (Courtesy, USDA, Soil Conservation Service)

Of course the actual carrying capacity for any range unit may vary widely from year to year and within a given season, depending largely on rainfall patterns. For this reason, stocking should be adjusted as nearly as possible to the forage produced each year, or set at a constant rate conservative enough to assure safe management of the key forage plants. Greatest flexibility is possible on operations grazing stocker steers or heifers as all or part can be readily marketed or placed in feed lots in order to relieve grazing pressures.

Recognition must also be given to the fact that animals do not graze uniformly over a range unit, that certain areas are more attractive to them. Consequently, some areas are repeatedly grazed while others may go practically unused. Cattle prefer to graze the flatter portions, and tend to congregate along creek bottoms, on ridge tops, and around water and shade. Sheep generally prefer hillsides. If herded, sheep can be moved rather uniformly over a range. If not herded they tend to congregate in the portion of the pasture in the direction of the prevailing wind and spot graze much like cattle.

Grazing capacity determinations are relatively complex and require careful study over a period of several years. They are arrived at most simply and accurately by observing changes in plant cover. If it is observed that the best plants are being overgrazed, then numbers of animals or season of use should be reduced; conversely, if excessive forage remains at the end of the grazing season, numbers can be slowly increased until a balance is struck. In arriving at grazing capacity, it is generally wise to seek assistance from qualified range technicians.

PROPER USE

To measure effectiveness of stocking rates on range vegetation, the degree of utilization of desired plant species must be determined. The following rule of thumb applied to key species can be used effectively: "Use half and save half and the half you save will grow bigger and better." The rule refers to half of the plant by weight or volume and not height. For example: bluebunch wheatgrass, a grass common in the northwest, will have 50% of its volume remaining when grazed to a stubble height of 4 in. to 5 in.

The 50% rule assumes that sufficient leaf surface is present to provide for plant growth and reproduction. The scientific basis for this is one of the great miracles of nature.

SEASON OF USE

A common goal of management for both cattle and sheep is that there shall be as nearly year-round grazing as possible and that both the animals and range shall thrive. In some areas, especially in the southwestern Great Plains areas, these conditions are met without necessitating migration of animals. The winter climate is mild, and the native forages cure well on the stalk, thus providing nutritious dry feed at times when green vegetation is not available. Generally speaking, however, most of the cattle and sheep from such areas are marketed via the feeder route rather than as grass-fat slaughter animals.

In mountainous areas, the most desirable management, both from the standpoint of the animals and the vegetation, consists of the proper seasonal use of the range. There is wide variety in the customs and requirements for seasonal use because of the spread in climate, topography, and vegetative types included. As a result, rangelands are usually grazed at different times of the year. Herds and flocks move to higher elevations in summer and return to lower ranges and/or pastureland for winter.

Because a range band of sheep can be moved and herded on unenclosed areas with greater ease than a herd of cattle and because investigations in range livestock management have been conducted more extensively with sheep, greater seasonal use of ranges is made with sheep.

Despite the value of yearlong grazing, it is recognized that the prevalence of severe winters in some parts of the West preclude winter grazing except to a limited degree, and stock must be fed during at least a part of the winter season. Where these conditions prevail, cattle and sheep are usually wintered in the irrigated valleys, close to the feed supply, especially a supply of alfalfa or meadow hay.

Some pertinent points in determining the proper season to use the range follow:

1. **Elevation.** Generally speaking, vegetative development is delayed 10 to 15 days by each 1,000-ft increase in elevation. Also, severe storms occur later in the spring and earlier in the fall at higher altitudes than at lower ones.

2. **Availability of water.** Although great accomplishments have been made in pond, spring, and well development to provide water for livestock use, there are still areas in the United States where water supplies are deficient. Grazing in these areas is restricted to seasonal runoff and is, therefore, intermittent at best.

3. **Poisonous plants grow early.** Most poisonous plants are very early growers and cause their greatest damage when animals are turned out too early. Larkspur, which affects cattle, and death camas, which affects sheep and cattle, are two examples. Poisoning losses from these two plants are usually negligible if stock are detained until the best forage plants have made suitable growth.

4. **Winter range should be saved.** If stock are allowed to remain on winter ranges too long after spring growth begins, the next winter's feed will be reduced, because the forage produced on these ranges grows mainly during spring and early summer.

KIND OF LIVESTOCK

Sheep and cattle share in the utilization of the western range. In fact, some ranges are simultaneously grazed by these two kinds of animals. This dual system of grazing is practical and beneficial provided the grazing capacity for each is properly adjusted so that the major forage plants are properly used, and that, at intervals, a careful determination is made of condition and trend of soil and forage. Some ranges, especially in Texas, are grazed by three classes of animals—cattle, sheep, and goats—with the goats controlling the brush without adversely affecting environment.

When sheep and/or goats are added to a cattle range (or *vice versa*), the increased numbers should not result in a total which exceeds the previous animal units of a single species by more than 10%; otherwise, overgrazing will likely result.

Fig. 5-14. Cattle on the range. (Courtesy, Bureau of Land Management, U.S. Department of the Interior)

Actually, economic factors—often unrelated to range characteristics—probably have the greatest influence on the selection and popularity of kinds of livestock. The kind which the operator feels will return the greatest net profit is selected, and the choice changes with changing times. Nevertheless, range characteristics may be so specific as to favor one kind of

Fig. 5-15. Sheep on the range. (Courtesy, Ralston Purina Co., St. Louis, MO)

livestock to the point that other kinds would be produced under handicap. Among such range characteristics which should be considered in choosing the kind of livestock are:

1. **Poisonous plants.** The presence of certain poisonous plant species may limit the use of the range to one kind or another of livestock. Thus, larkspur is a serious menace to cattle, but normally sheep are not affected by it. On the other hand, generally cattle may safely graze lupine-infested ranges, which are sometimes extremely dangerous to sheep. Many other examples of selective poisoning could be cited.

2. **Topography.** Cattle prefer level to gently rolling topography, whereas sheep and goats are better adapted than cattle to steep, rocky, or bushy ranges. The latter seem to have a natural instinct for climbing, and, through the efforts of the herder, they can be encouraged to graze the more difficult terrain. In addition, because of greater ease in herding, and moving about on unfenced public domain, sheep are trailed about more than cattle, thus more effectively utilizing seasonal ranges.

3. **Water.** Sheep and goats are much better adapted than cattle to more poorly watered ranges, because they can go for longer periods without water. Also, sheep utilize snow as a sole source of water more satisfactorily than cattle; therefore, sheep use range dependent on snow for water more efficiently than cattle.

4. **Vegetative cover.** In general, sheep do not utilize tall-growing grasses as effectively as cattle. Sheep and goats are weed eaters and browsers; and goats probably do better than sheep on a straight diet of browse. Horses are more selective than any other kind of livestock; they prefer grasses, although they will eat small amounts of other kinds of forage. Hogs do best on acorns, pods of certain leguminous shrubs, roots, and other concentrated feeds found on the range only during limited seasons and in certain areas, principally in the Southeast and Southwest.

5. **Predators, insects, and diseases.** Coyotes are serious predators of sheep, but bother cattle very little, comparatively speaking. Thus, heavy concentrations of coyotes, or other sheep-killing predators, may make sheep raising unprofitable, but present less serious problems to cattle production. The presence of certain insects and diseases may also become factors in the selection of the best suited kind of livestock.

6. **Big game population.** Deer compete more directly with sheep, and elk with cattle.

Theoretically, the most efficient use of most ranges can be made by two or more kinds of livestock grazing at the same time; by *common use* or *dual use.* The most popular combination is that of cattle and sheep. Destructive grazing often results therefrom, however, because common use requires much more critical grazing management than grazing by one kind of livestock only. This is so primarily because the most popular parts of the range—waterholes, creek bottom meadows, ridge tops—are the most preferred by all kinds of animals; and, in addition, many of the most valuable forage plants are preferred by both cattle and sheep. As a result, only very careful management can prevent the destruction of these most valuable range areas. In brief, a range unit cannot support its full quota of cattle in addition to its total capacity for sheep. Rather, a studied adjustment should be made to fit the particular unit, based on topography, water distribution, class of forage, and other considerations.

RANGE GRAZING SYSTEMS

Ranges may be grazed continuously throughout the entire grazing season without rest, or the area may be subdivided and the pasture grazed rotationally with alternating periods of grazing and rest.

Theoretically, based on how plants grow and how animals graze, rotation grazing systems should improve the vigor and productivity of desirable plants, prevent the invasion of undesirable plants, and increase the carrying capacity and rate of gain of animals. In practice, however, all the benefits ascribed to rotation grazing may not accrue; perhaps due to human failure in management, or because there is not as much difference between the results of continuous and rotation grazing as had been thought. However, rotational systems usually make for increased plant vigor and carrying capacity, whereas individual animal gains has usually been in favor of continuous grazing.

Range grazing systems vary somewhat from unit to unit, depending on kind and class of livestock, kind and type of forage, mixture of range sites, time and amount of rainfall, pasture and corral layout, available water supply, the condition of the range, the long-time goals for improvement, the prevailing economics, and the time necessary and available to supervise and conduct the operation.

Some of the basic range grazing systems, of which there are many variations and adaptations from ranch to ranch, follow:

Continous Grazing
Rotation Grazing
- Deferred rotation grazing systems
  1. Two pastures—one herd system
  2. Three pastures—one herd system
  3. Four pastures—three herds
- Short duration grazing systems
  1. Conventional (rectangular) grazing system
  2. Savory (or cell) grazing system

Intensive Early-Season Stocking

CONTINUOUS GRAZING

Continuous grazing is the simplest and most common grazing system of the western ranges; and varying the number of animals allowed to graze pasture is the most commonly used means for grazing management. In comparison with rotation systems, continuous grazing requires less fence, water, development, less labor in moving animals and fixing fences, and less knowledge of livestock and range management. Also, continuous grazing may be more suitable and practical when used in conjunction with a seasonal range, than a complicated rotation system.

The major **disadvantages** of continuous grazing are: poor flexibility; lower stocking rate; less animal gains per acre; poorer livestock distribution on the range caused by animals concentrating around water, bedding grounds, and feed grounds, and overgrazing such areas; and less opportunity to use such improvement practices as burning, brush control, and livestock management.

ROTATION GRAZING

Rotation grazing is a system in which pastures are grazed and rested in a planned sequence. It gives the more desirable plants a chance to regrow, compete, and multiply, thus gradually increasing the number and production of high quality plants.

The objective of any rotation grazing system are to favor the growth and survival of desired plants; to obtain greater use of the less palatable plants; and to improve range conditions. The improved range increases livestock production, improves the habitat of wildlife, reduces erosion, and conserves water.

The two main types of rotation grazing systems are deferred rotation grazing and short duration grazing.

DEFERRED ROTATION GRAZING SYSTEMS

In deferred rotation grazing systems, the range is usually divided into two to four units. There are different ways in which to apply deferred rotation grazing, with the following three basic systems, or some variations therefrom, most common:

1. **Two pastures—one herd system (switch-back system).** With this system one herd of livestock is rotated between two pastures. Each pasture is grazed or rested at a different time during the 2-year period required to complete the grazing cycle.

2. **Three pasture—one herd system.** This system is similar to the two pastures-one herd system, except the herd is moved through three pastures instead of two. In any one year, one pasture may be used during the growing season; the second pasture may be used at a later stage of vegetative maturity, such as seed ripe; and the third pasture may be rested and not grazed by livestock. The length of each grazing period may be as short as 30 days or as long as 90 days. This sequence is rotated among years. By treating a unit in this manner, the entire area receives the equivalent of a year-long rest.

3. **Four pastures—three herd system (the Merrill system).** In Texas, and in much of the Southwest, ranges may be grazed throughout the year. Under these circumstances, a different type of rotation grazing system should be considered than where grazing is not year-round.

Where 4 pastures are available, or can be arranged, a 3-herds system is popular, with each pasture grazed 12 months and rested 4 months. This system is summarized in chart form (see Table 5–2).

Here is how the system outlined in Table 5–2 works:

1. Divide the grazing area into 4 units.
2. Rest pasture 1 for 4 months. Place all livestock on the remaining 3 pastures (2, 3, and 4), with little or no attempt to force animals to graze undesirable plants. Plants are not set back while being grazed, with the result that they make maximum improvement when they are rested.
3. At the end of 4 months, move livestock from pasture 2 (which has been grazed 4 months) to pasture 1 (which has been rested 4 months). For the next 4 months, rest pasture 2 and graze pastures 1, 3 and 4)
4. The next 4 months, graze pastures 2, 4, and 1, and rest pasture 3.
5. The next 4 months, graze pasture 3, 1, and 2, and rest pasture 4.

Note that the rest period of each pasture will begin 4 months later each succeeding year until in 4 years all pastures will have been rested 1 complete year.

TABLE 5–2
FOUR PASTURES–THREE HERDS SYSTEM

Year and Season				Pastures			
				1	2	3	4
1990:							
Mar.	Apr.	May	June	Rest	Graze	Graze	Graze
July	Aug.	Sept.	Oct.	Graze	Rest	Graze	Graze
1991:							
Nov.	Dec.	Jan.	Feb.	Graze	Graze	Rest	Graze
Mar.	Apr.	May	June	Graze	Graze	Graze	Rest
July	Aug.	Sept.	Oct.	Rest	Graze	Graze	Graze
1992:							
Nov.	Dec.	Jan.	Feb.	Graze	Rest	Graze	Graze
Mar.	Apr.	May	June	Graze	Graze	Rest	Graze
July	Aug.	Sept.	Oct.	Graze	Graze	Graze	Rest
1993:							
Nov.	Dec.	Jan.	Feb.	Rest	Graze	Graze	Graze
Mar.	Apr.	May	June	Graze	Rest	Graze	Graze
July	Aug.	Sept.	Oct.	Graze	Graze	Rest	Graze
1994:							
Nov.	Dec.	Jan.	Feb.	Graze	Graze	Graze	Rest

The Texas station reports that, in comparison with conventional yearlong grazing on the same area, the "4 pasture/3 herds system" results in greater livestock gains and 25% increase in carrying capacity.

SHORT DURATION GRAZING SYSTEMS

Short duration grazing, as the name implies, employs frequent movement of animals, with the speed of the rotation adjusted according to the growth rate of the plants. During the peak of the growing season, animals are moved at shorter intervals, with longer intervals during the remainder of the year when plant growth slows. In practice, a pasture may be grazed 3 days and rested 20 days in one part of the year and grazed 20 days and rested 60 days in another season. The short duration technique uses rest periods within the growing season in order to restore plant vigor. This system will usually give more rapid range improvement than deferred rotation grazing.

1. **Conventional (rectangular) grazing system.** This system involves the use of conventional, rectangular pastures, along with a 16 to 18 ft alley for cattle to get water and minerals and move from one pasture to another. The principle and practices (grazing time, resting time, central watering, and easy movement of cattle between pastures) for both the rectangular and cell systems were first described by Andre Voison of France in the early 1950s. The only difference is the layout and design; the cell system uses the wagon wheel design, whereas the conventional (rectangular) system uses the design identified by the name—rectangular pastures.

2. **Savory (cell) grazing system.** This system is named after Allan Savory, who originated and popularized it in the United States. Ideally, grazing is from 1 to 5 days, and resting is from 30 to 60 days. It usually involves 12 or more pastures, and generally, although not always, the pastures are arranged as a grazing cell, with pastures formed by fence lines fanning out from the hublike spokes on a wagon wheel, and with the water, minerals, and handling facilities at the hub. When the animals come to the center, or hub, for water and minerals, they can be moved, between pastures by opening and closing gates. Producers using the Savory system generally use electric fences in order to reduce costs. The Savory system is designed to lessen movement stress. However, it can add to nutritional stress if (1) stock are held too long in pastures, or (2) if livestock numbers and forage production are out of balance.

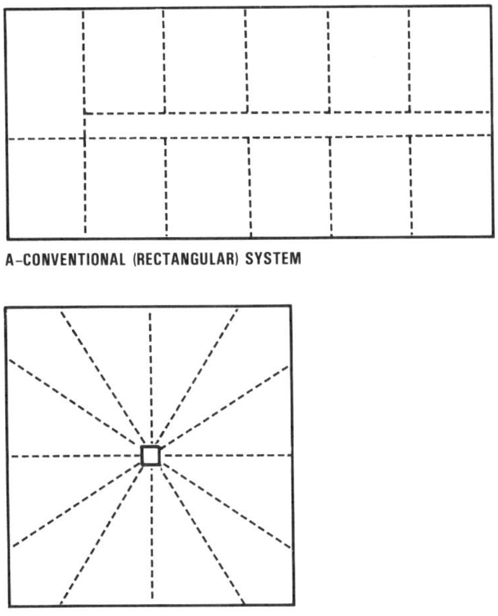

Fig. 5-16. Two basic layouts (designs) for short duration grazing: A, the conventional (rectangular) system; and B, the Savory (cell) system.

INTENSIVE EARLY SEASON STOCKING

This grazing system was developed in the Flint Hills of Kansas where large numbers of stocker steers are custom grazed annually. In this area, nearly three-fourths of the total annual production of the bluestem grasses is produced during the 3-month period from mid-April to mid-July. Traditionally, livestock gains are highest during this same period and then taper off the remainder of the season.

This system was designed to take advantage of the high rates in daily gains by doubling the normal stocking for three months. The pastures are emptied in mid-July by selling the cattle or moving to feedlots. Pasture grasses then have the remainder of the summer to regrow and regain vigor for the next year's cycle.

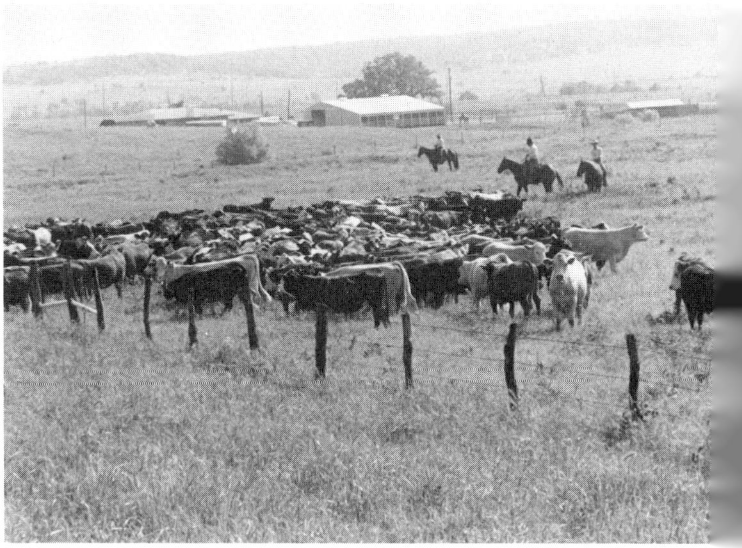

Fig. 5-17. Stocker cattle on bluestem pasture in the famed Flint Hills of Kansas, the area in which intensive early season stocking was developed. (Courtesy, A. Solomon, Humboldt, KS)

In this tall grass area greater steer gains are also obtained when pastures are burned. This has averaged nearly 30 lb per animal seasonally over unburned pastures. Prescribed burning, as a rule, is incorporated into this system.

The **advantages** of this system are (1) more pounds of beef per acre, (2) lower interest charges, because of owning the cattle for a shorter period of time, and (3) higher net returns. The main **limitation** is the lack of flexibility relative to removal of the herd; they must either be sold or moved into the feedlot as scheduled—in midsummer—or overgrazing and damage to the grasses will result.

RANGE IMPROVEMENT METHODS

The warning signals of a range that is on the downgrade and that is in need of improvement are:

1. Desirable plants are being overgrazed. Leaf removal is excessive.
2. Weakened vitality of the important plants as shown by reduced height and yield in period favorable for growth.
3. Desirable forage plants "going out" and being replaced by undesirable ones; the number of young, inferior plant species increasing.
4. Thinning of perennial grass cover, with the grass tufts breaking down and dying; and an increase in annual plants and perennial weeds. The greater the severity of overgrazing, the more rapidly this process takes place.
5. Excessive trampling damage.
6. Increased soil erosion, by wind and/or water.

When the warning signs of range deterioration are heeded and corrective action taken in time, rangelands can usually be restored to their original plant diversity and production levels.

Programs for improvement involve comprehensive inventorying of and planning to determine a sound course of action to overcome the problems and reverse the trend. Depending on the extent and severity of the problems, a number of practices or methods can be employed. Bear in mind that seldom is there a quick, easy fix. Generally, changes in range vegetation are gradual and subtle. Range plants spread vegetatively by rhizomes, tillers, and stolons, and by seed. Usually, initial stages in improvement will be obtained through vegetative means.

CONSERVATIVE STOCKING

Basically, improvement is accomplished by controlled stocking and manipulating livestock grazing to assure that the desired plant species are being properly used—that is, not more than 50% of the leaves are being utilized. Usually, improvement can be accelerated by employing one or more of the following practices: (1) rotation-deferred grazing; (2) rest-rotation grazing; (3) a lighter continuous grazing load; or (4) a shorter season of use, such as may be accomplished by using supplemental pastures.

DISTRIBUTION OF ANIMALS ON THE RANGE

Next to the proper rate of stocking and proper seasonal use, distribution of the animals on the range is the most important feature in range management. Proper distribution of animals is reflected in more even utilization of the forage. This assignment is more difficult with cattle than with sheep, especially on rough mountainous land. Cattle have a strong tendency to utilize the flatter areas and to congregate around watering places. Also, sheep are usually herded.

The ultimate in distributing livestock is accomplished by herding. When allowed to fend for themselves, livestock are selective and may repeatedly graze favored areas. Even when stocking rates are seemingly adequate for a given pasture, it is not uncommon for large areas to be overgrazed and other areas only lightly grazed. The result is lower production and plant composition changes on the one hand and wasted forage on the other.

Better distribution of livestock on the range may be accomplished through one or more of the following methods:

1. **Developing water.** Proper location of water sources—ponds, springs, wells, windmills, pipelines, tanks, or troughs—will help correct distribution, especially where pastures are large. Under ideal conditions the travel distances in rough country should not exceed ½ mile for cattle or 1 to 2 miles for sheep and goats; and in level country, 1 to 1½ miles for cattle and 2 to 3 miles for sheep and goats.

2. **Locating salt and minerals.** Seldom do livestock lick salt and then proceed to water for a drink. As a rule salt will be taken independently and, therefore, it's placement can be used to influence livestock grazing patterns. Whether locations are of a permanent nature or moved repeatedly, placement should be strategically located at a distance from water sources to encourage maximum coverage by livestock.

3. **Fencing smaller pastures.** Most distribution problems are primarily associated with terrain and size of pastures. Fencing into smaller pastures and, when possible, aligning new fences to separate areas with extreme differences in topography or in production, benefits distribution. In mountainous areas open-ended drift fences are used to alter the normal movement of livestock and to guide them into remote or undergrazed areas.

Innovations in electric fencing have alleviated high costs and stimulated many operators to crossfence into a number of pasture arrangements to facilitate various systems of grazing.

4. **Rotation grazing.** Most systems of rotation grazing result in better livestock distribution and more equitable use of plant species than obtained with continuous grazing. When livestock are concentrated into large herds and pastures are grazed for shorter time periods, livestock tend to distribute more and are less inclined to establish routine grazing patterns. The greater the grazing intensity over the shortest period of time, the more effective distribution becomes on a pasture.

5. **Building trails or walkways.** Construction of trails or roads into rough mountainous terrain may enhance livestock movements into those areas. At times even construction of dams across deep, narrow ravines will serve as a bridge allowing livestock access to areas previously not available.

Cattle grazing in coastal marshes tend to congregate near the natural ridges on firm soils. Construction of low, narrow earthen walkways provide additional "high ground" and encourages livestock to extend further into the marsh areas previously ungrazed.

6. **Controlling brush.** As a rule cattle prefer grazing prairie areas to areas with dense stands of woody shrubs and trees. Even narrow bands of brush may serve as a barrier and influence grazing patterns. Brush control either totally or in strips can improve grazing in these problem situations.

7. **Burning pastures.** Although effects last only 1 to 2 years, prescribed burning is beneficial in removing accumulations of weathered grass from previous years. This reduces species selection, erases the effects of old grazing patterns, and attracts livestock into areas previously avoided.

8. **Controlling pests.** Livestock pests such as flies, grubs, and ticks cause animals to congregate and seek protection. Timely pest control measures can aid livestock distribution to some degree.

RANGE RESEEDING

Fig. 5-18. Range seeding of previously cleared chaparral area for improved pasture and wildlife habitat. (Photo by G. F. Murphy. Courtesy, R. M. Timm, Superintendent and Extension Wildlife Specialist, University of California-Davis, Hopland Field Station, Hopland, CA)

Range seeding has been used to reestablish native plants on rangelands that have deteriorated through mismanagement, and on marginal or abandoned croplands being converted to rangelands.

When improper management and overstocking have seriously reduced the quality of the forage and the grazing capacity of the range, some method of reseeding may be the only logical alternative.

Where considerable of the better forage plants remain, natural regeneration is preferred. The latter is accomplished by managing the grazing season so as to favor the propagation of the remaining desirable native species and by controlling low-value brush and other competing vegetation. Often the recovery process can be speeded up by controlling most undesirable vegetation through the use of herbicides. This process has been especially successful on rangeland dominated by sagebrush in the West and by mesquite in the Southwest, and where a residual stand of native grasses is present as an understory. There are also other weed types which respond well to chemical treatment.

When most of the desirable plants have been destroyed and the soils are suitable, artificial reseeding is advocated—even though it is expensive and subject to failure.

It is recognized (1) that only a relatively small proportion of the more arid western range can be seeded because of the soil and moisture limitations, and (2) that seeding is not a satisfactory substitute for good management practices needed to prevent further destruction of forage plants.

In the Great Plains area, millions of acres have been seeded to rangeland mixtures during the past 30 years. This has been largely through federally supported programs, including the Soil Bank Program, Conservation Reserve Programs, and the Great Plains Program. Experience gained through these efforts has greatly improved rangeland revegetation technologies.

For successful seeding, the following rules are usually adhered to:

1. Select a site where success can reasonably be achieved. Rainfall, soil, topography, and other site factors must be suitable for seeding. Otherwise the practice is foredoomed.

2. Prepare a firm, weed-free seedbed. Light soils receiving limited rainfall are notoriously difficult to firm into a satisfactory seedbed. Also, seedlings of perennial grasses commonly used in range reseeding are weak competitors with weeds. In many Great Plains areas seeding into a noncompetitive crop stubble such as provided by grain sorghum or forage sorghum is needed to insure success. This standing residue is beneficial in that it (1) conserves moisture, (2) provides shade and protection to the new seedlings, and (3) reduces weed competition.

3. Select a species or mixture of species best adapted to the climate and other features of the area being planted and fitted to the seasonal forage needs of the particular ranch.

4. Use Pure Live Seed (PLS). The use of PLS has become standard in buying and selling seeds of native plants. All seeding recommendations are made on this basis. It removes guesswork when determining the quality and, therefore, the value of a seed lot; and it eliminates paying for pieces of stems, leaves, and other impurities. PLS is determined by the formula: % PLS = % germination × % pure seed divided by 100.

5. Plant just prior to the season of most dependable moisture and growing conditions.

6. Cover the seed, but do not get it too deep. The best depth for most range grasses is ¼ to ¾ in.; depth regulators on drills are advisable.

7. Protect the seeding from grazing use until the plants have become well established. This may require from 8 months to 3 years—depending on species planted, weather conditions, etc.

8. Fence the seeding or otherwise provide a means of managing it. Without adequate control, animals congregate on the new seedings and destroy them through overuse. Even when the stand is well established livestock may prefer to graze seeded areas.

When available, certified seed of known varieties or cultivars should be planted. There are many variances in adaptation within a native species. For example: Switchgrass grows naturally in nearly all of the eastern two-thirds of the United States. However, seed collected in Texas would not be well adapted to conditions in Nebraska or Iowa. As a result, nearly 15 different cultivars of switchgrass are available, each with a specific geographical area of adaptation.

Occasionally, when moisture and temperature are optimum, native ranges will produce a crop of seed that can be harvested by combine. Such harvest is beneficial as a (1) seed supply for planting other needed areas on a given ranch, or (2) cash crop when sold to area seed dealers. As a rule of thumb, seed from native stands should not be planted beyond a radius of 150 miles from the harvest site.

BRUSH AND WEED CONTROL

Brush and weeds compete with both native and introduced forage species for water, nutrients, and light. Brush is a primary problem on large portions of U.S. rangelands. An estimated 320 million acres of grazing lands are predominantly brush, with mesquite, juniper, sagebrush, and oak the major problems.

In many areas, brush has invaded grasslands and thickened to the point that only scant forage is produced. Efforts to reclaim these lands, initially, depends on some form of control. Numerous methods have been used, each with varying success depending on (1) the growth habits of the problem species, (2) weather conditions during and following treatment, (3) soils, especially depth and texture, (4) topography, and (5) grazing management following treatment. Often the most effective program is one that integrates a combination of control methods over a period of years.

Costs of the various treatment methods should be carefully evaluated with anticipated returns. Most treatment measures are expensive and to become profitable, control effectiveness must extend over a period of several years.

Major brush problems can generally be controlled by one, or a combination, of the following methods.

1. **Chemical methods.** This involves applying select herbicides to the soil or the plant by ground equipment or by aircraft. Use of herbicides is desirable in that the soil surface is left undisturbed and chances for erosion are minimized. Success depends on applying the right herbicide at the right rate and at the right time. The county extension agent or other specialist should be consulted for the latest recommendations, and the directions on the label of the product being used should be followed.
2. **Mechanical methods.** Most forms of mechanical control involve the use of heavy equipment and, therefore, are usually expensive. However, when brush stands are dense and other control methods prove ineffective, mechanical treatment can be justified. Mowing, axing, grubbing, roller chopping, root plowing, chaining, and bulldozing have each proven quite effective, depending on the species of brush and the terrain.
3. **Prescribed burning.** Fire is the oldest method of controlling unwanted woody plants, and it is still used. Prescribed burning refers to burning that is done at the end of the dormant season or at the time the desirable grasses are beginning spring growth; this suppresses certain undesirable plants. In addition to doing prescribed burning at the proper time, the following factors should be considered: fuel for the fire, favorable weather conditions, safety, and air pollution. Prescribed burning can be dangerous and safety precautions cannot be overemphasized.
4. **Proper grazing management.** Proper grazing management avoids undergrazing and assures healthy, vigorous forage plants. This results in a healthy ecosystem, thereby preventing many brush problems.

Weeds—the unwanted plants of pastures and ranges—are undesirable because (1) they compete with higher quality plants which might otherwise grow; (2) they are often less palatable and less well consumed than desirable species, with the result they are undergrazed; (3) they may impart undesirable flavors and quality to the food product produced by the grazing livestock; (4) they often contain toxic substances which may impair the usefulness of, or even kill, the animals which consume them, and which may cause abortion or birth deformities in calves, lambs, or foals; (5) they may inflict mechanical injury; (6) they may shade the desirable plants and interfere with their establishment and maintenance in the stand; and (7) they usually give an unsightly appearance to the pasture.

Weeds may be controlled in the following ways, singly or in combination: (1) biologically, through the use of insects, diseases, and other organisms, including grazing animals, as control agents; (2) roguing, or individual weeding, if the plants are scattered and not found in large numbers; (3) mowing the plants at a time when they are most susceptible; and (4) chemically, by treating either the plants or the soil to prevent seed germination or to kill the plant itself. Each of these methods has certain advantages over the others. The first step in any weed control program consists of identifying the nature of the weed problem. Next, using this information, along with a specialist's advice, one can determine how weeds can be controlled. Many counties have a weed control program, with provisions to help by spraying roadways, fencelines, and even concentrated spots in pastures if the problem has the potential to spread to adjoining pastures.

- **Biological control of brush and weeds**—Interest in the biological control of brush and weeds is increasing. The classical example of the biological control of an undesirable species was the practical elimination of prickly pear cactus (*Opuntia* spp) on 30,000,000 acres of grazing land in Queensland, Australia, by using a moth introduced from Argentina in 1925. In the United States, insects have been used to control St. Johnswort, lantana, puncturevine, and knapweed. The use of sheep and goats is also a form of biological control. Sheep will eat leafy spurge, tansy ragwort, tall larkspur, and spotted knapweed. Goats are effective for keeping some species of chaparral and other brushes under control.

To be effective, brush control must be followed by proper grazing management. During the first growing season following treatment, grazing should be deferred/limited in order to allow desirable grasses to become established. Reseeding of these areas may be necessary where a natural seed source of the desirable forage plant is not available.

(Also see Section 8, under the heading of "Weed and Brush Control.")

Fig. 5–19. Brush control! This shows an Angora goat browsing high up on a live oak tree. (Courtesy, Texas A&M University Agricultural Research Station, Sonora, TX)

RANGE NUTRIENT DEFICIENCIES

Nutrient deficiencies are rather common on the range. Many soils are deficient in certain nutrients, which affect the plants, and, in turn, the animals feeding on them. During droughts, and early or late in the season, forage may be in short supply, limiting energy and other nutrients. Later in the season, they become leached or bleached—they increase in fiber and decrease in protein, phosphorus, and carotene.

It should be noted that animals on rangelands are privileged to a wide range of selectivity in picking and choosing different plants and different plant parts. Research has shown that grazing animals consistently harvest a diet of forage much higher in chemical composition than that of an entire plant or group of plants.

• **Energy deficiencies**—Hunger, due to lack of feed, is the most common deficiency of range livestock. This will, of course, depend on the type and level of management. The most important requirement is sufficient feed for body maintenance. Over and above this, surplus energy is used for growth or fattening.

With bulky, low-quality roughages—such as many range grasses cured on the stalk, animals cannot consume sufficient quantities to meet energy needs. The younger the animal, the more acute the problem. Under these circumstances, the low-energy intake is met by breaking down of body tissue. This results in loss of weight and condition and lack of growth. Therefore, most range livestock are supplemented to maintain weight or minimize weight loss until green forage is again available.

In breeding animals, low-energy intake affects reproduction. Cows take longer to come in heat and require more services per conception, thus reducing the calf crop. Also, calves born from energy-deficient cows are lightweight at birth. Supplemental feeding is the practical way to eliminate energy deficiencies.

• **Protein deficiencies**—The protein intake of beef cattle must be adequate to develop muscle (meat) and replace worn body tissues. The protein need is most critical in young calves and in gestating-lactating cows.

Mature, weathered native range plants are almost always deficient in protein—sometimes containing as little as 3% (see Fig. 5–20). A deficiency of protein results in depressed appetite, poor growth, loss of weight, reduced milk production, irregular heat periods, and lowered calf crops.

Because protein supplements ordinarily cost more per ton than grains, the temptation is to feed too little of them. When grazing mature, weathered forage, cows should be fed a protein supplement.

• **Mineral deficiencies**—Phosphorus deficiencies are rather common on the range (see Fig. 5–21). A severe phosphorus deficiency results in depraved appetite, emaciation, retarded growth and development, failure to breed regularly, lowered calf crop, lowered milk production, and high death losses. On rangelands these deficiencies are corrected by supplementing livestock rather than attempting to amend the soil.

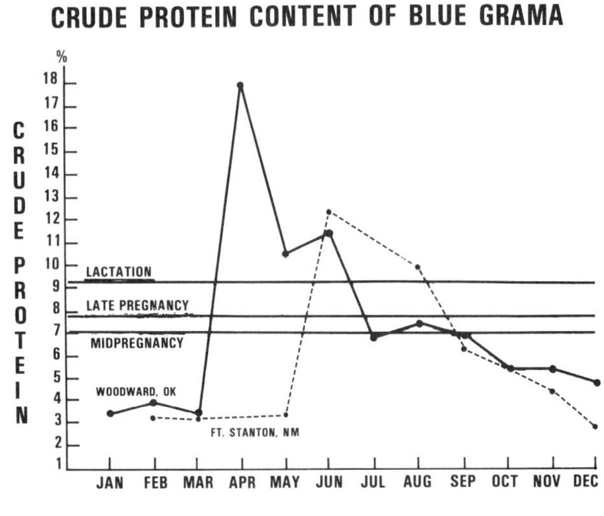

Fig. 5–20. Crude protein content (moisture-free basis) of blue grama (*Bouteloua gracilis*) grass. *Woodward, OK data from:* USDA Tech. Bull. 943, 1947, p.53, Table 12, by Savage, D. A. and V. G. Heller. *Ft. Stanton, NM data from:* New Mexico Ag. Exp. Sta. Bull. 662, 1978, p. 22, Table 1, by Pieper, R. D., *et al.* **Note:** The peaks in the protein content of the vegetation in the two areas reflect the periods of greatest rainfall. Thus, approximately 60% of the annual precipitation of the Woodward, OK area occurs in April, May, and June, whereas 60% of the annual precipitation of the Ft. Stanton, NM area occurs in July, August, and September.

Superimposed on the chart are the crude protein requirements in a moisture-free ration of a 1,100-lb cow during mid-pregnancy, late pregnancy, and lactation, respectively. Note that, in both areas, there is a protein deficiency except during periods of highest rainfall. (Chart provided by T. McCollum, Ph.D., Animal Science Department, Oklahoma State University, Stillwater)

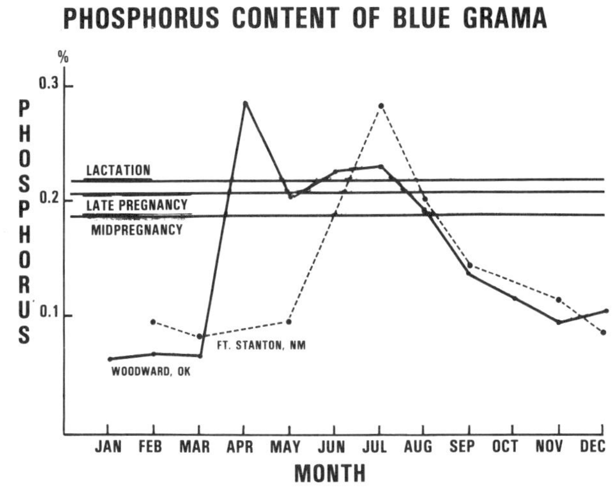

Fig. 5-21. Phosphorus content (moisture-free basis) of blue grama (*Bouteloua gracilis.*) *Woodward, OK data from:* USDA Tech. Bull. 943, 1947, p.52, Table 11, by Savage, D. A. and V. G. Heller. *Ft. Stanton, NM data from:* New Mexico Ag. Exp. Sta. Bull. 662, 1978, p. 28, Table 7, by Pieper, R. D., *et al.* **Note:** The peaks in the phosphorus content of the vegetation in the two areas reflect the periods of greatest rainfall. Thus, approximately 60% of the annual precipitation of the Woodward, OK area occurs in April, May, and June, whereas 60% of the annual precipitation of the Ft. Stanton, NM area occurs in July, August, and September.

Superimposed on the chart are the phosphorus requirements in a moisture-free ration of a 1,100-lb cow during mid-pregnancy, late pregnancy, and lactation, respectively. Note that, in both areas, there is a phosphorus deficiency except during periods of highest rainfall. (Chart provided by T. McCollum, Ph.D., Animal Science Department, Oklahoma State University, Stillwater)

• **Vitamin deficiencies**—Range plants are sometimes low in carotene. A severe deficiency of carotene (vitamin A) may result in low conception rate, a small calf crop, many calves weak or stillborn, with some calves born blind, more cows with retained afterbirth, low gains, greater susceptibility to calf scours, and more respiratory troubles.

Severe vitamin A deficiency in bulls may result in decreased sexual activity and lowered semen quality.

When grazing dry range longer than 4 to 6 months, it is recommended that a supplement containing 20,000 to 30,000 IU of vitamin A per pound be fed to brood cows and bulls.

Range cattle usually receive sufficient vitamin D from exposure to direct sunlight or from sun-cured forages.

RANGE LIVESTOCK SUPPLEMENTATION

Energy, protein, phosphorus, and vitamin A are the major nutrients limiting the performance of range livestock.

Energy supplementation may be advantageous when range forage is in short supply, as during a drought or heavy snowfall.

Protein is the major supplement cost of most ranches. Feeding protein supplements in the form of range cubes is common throughout much of the Great Plains areas where livestock are pastured on dry, dormant forage during winter months. Protein supplements can be provided to livestock on an every-other-day or every-third-day basis without affecting their performance. Salt or fat may be used to limit the consumption of cottonseed meal or other protein supplements. Although gains are enhanced on cattle fed protein supplements during late summer when protein levels in range forage drop below 6 to 7%, the extra gain realized is seldom of economic importance.

Phosphorus supplementation during periods of forage dormancy seems justified. Also, the routine inclusion of essential microminerals in salt blocks or mineral mixes is a good practice.

Vitamin A supplementation is recommended when livestock must be maintained for over 4 months without access to green grass or browse.

GRAZING PUBLICLY OWNED LANDS

The ownership of U.S. land is summarized in Table 5-3.

About one-third of U.S. public lands are in Alaska. Because of its remoteness and northern location, land development has been slow in this state. As a result, the Federal Government still owns almost 67% of all the lands in Alaska.

The other two-thirds of the public lands are located in the 48 contiguous states, but are not evenly distributed across the country. About 93% of these federal lands outside Alaska are in the 11 western states.

Today, in the 11 western public land states, the federal government owns and administers approximately 320 million acres on which grazing is allowed. At one time or another during the year, domestic cattle and sheep graze on about half of these public lands. More of the public lands are used for this purpose than for any other economic activity. In 1988, lands in the 11 western states administered by the Bureau of Land Management and the U.S. Forest Service provided grazing all or part of the year for an estimated 5,655,845 head of all classes of livestock. The number of livestock grazed on the publicly owned rangeland annually accounts for approximately 4 to 5% of the nation's total cattle numbers and slightly over 50% of sheep numbers.

TABLE TABLE 5-3
OWNERSHIP OF U.S. LAND (50 STATES)[1]

Ownership	Area		Percentage of Total
	(million acres)	*(million ha)*	(%)
Private ownership	1,329	*538*	58.7
Indian land	51	*20.6*	2.2
Public ownership	885	*358.3*	39.1
Federal	730	*295.5*	32.2
State and local governments	155	*62.7*	6.8

[1]*Statistical Abstract of the United States,* 1987, p. 182, Table 318.

AGENCIES ADMINISTERING PUBLIC LANDS

Because much of the grazing land that some ranchers rely upon to maintain their cattle and sheep enterprises is built up into operating units by leasing or by obtaining use permits from several federal and state agencies, private corporations, and individuals, it is imperative that the owner have a working knowledge of the most important of these agencies. Some range operators are placed in the position of using range rented from as many as six landlords—either private, state, and/or federal.

The bulk of federal land is administered by the following six agencies: the Bureau of Land Management, the U.S. Forest Service, the Bureau of Indian Affairs, the Department of Defense, the National Park Service, and the Bureau of Reclamation. The largest land area from the standpoint of grazing permits and utilization of grazing areas by animals is administered by the first two of these agencies; hence, each is discussed at this point, followed by pertinent information relative to Indian lands and state and local government-owned lands, and railroad-owned lands.

1. **Bureau of Land Management.** The Bureau of Land Management of the U.S. Department of the Interior administers more than 40% of all federal lands. More than one-third of the land it manages is in Alaska. The remainder is almost entirely in the 11 western states.

From the standpoint of the livestock producers, the most important function of the Bureau of Land Management is its administration of the grazing district established under the Taylor Grazing Act of 1934 and the unreserved public land situation outside of these districts which are subject to grazing lease under Section 15 of the Act. This federal act and its amendments authorize the withdrawal[4] of public domain from homestead entry and its organization into grazing districts administered by the Department of the Interior. Also, this legislation, as amended, allows the Bureau of Land Management to administer state and privately owned lands under a cooperative agreement.

In 1988, the Bureau of Land Management had 54 grazing districts, operating in the 11 western states and totaling 146 million acres of public lands. In these districts, 11,853 operators were granted privileges to graze 3,787,332 head of livestock for an average of about 5 months each year. These operators paid the United States, as grazing fees for this range use, a total of $12,416,598. In addition to this livestock use, in 1988, public lands supported an estimated 1.9 million big game animals, of which approximately 1.1 million were deer.

In addition to, and outside of, the grazing districts, in 1988 the Bureau of Land Management supervised 16 million acres of public domain in the western states, most of which was leased to 7,016 livestock producers for 604,758 head of livestock for about 5 months. These operators paid rentals in the amount of $2,039,033 for the use of these lands.

Each district is administered by a District Manager, who is a technically trained employee of the Bureau of Land Management. The District Manager is responsible to the state bureau office for the proper use, management, and welfare of the public land resources of the district. In turn, the state office is responsible to the Director's office in Washington, DC.

Grazing privileges are allocated to individual operators, associations, and corporations on the basis of (1) priority of use; (2) ownership or control of base property dependent on grazing district land for forage during certain seasons of the year, or control of permanent water needed to graze district land; (3) proximity of base property to public lands outside home ranch to the grazing district; and (4) adequate property to supply the feed needed along with grazing privileges, to maintain throughout the year the livestock permitted on public range.

[4]On May 28, 1954, a bill was signed by President Eisenhower lifting the 142 million acre limitation on public domain lands that can be included in Taylor Grazing Act districts.

All of these lands are subject to classification and disposal under Sections 7 and 14 of the Taylor Grazing Act, for any higher use or other appropriate purposes. Grazing privileges may, therefore, be cancelled whenever such lands are determined to be more suitable for other purposes.

A fee is charged for grazing privileges. In 1989, the basic fee was equivalent to $2.29 per animal unit month (AUM). An AUM is the equivalent of the grazing of a mature cow, 5 sheep, or 1 horse, for 1 month.

The Taylor Grazing Act has been responsible for many changes, not all of which have been popular. Some livestock producers complain about the loss of their ranges; others tell of increased costs; and there are those who resent government controls, and, above all, the confusions which results from dealing with several agencies. Without doubt, many of these criticisms are justified, and some errors in administration should be rectified; but those who would be fair are agreed that the ranges as a whole have improved under the supervision of the Bureau of Land Management and that further improvements are in the offing.

2. **USDA Forest Service.** Almost one-fourth of the federal lands are administered by the Forest Service. Over 100 million acres of the national forests and national grasslands are used for grazing under a system of permits issued to local farmers and ranchers by the Forest Service of the U.S. Department of Agriculture. In 1988, lands in the 11 western states administered by the USDA Forest Service provided grazing for all or part of the year for 1,868,513 head of all classes of livestock.

The Forest Service issues term grazing permits and annual permits. Among other things, the permit prescribes the boundaries of the range which they may use, the maximum number of animals allowed, the season in which grazing is permitted, and the expiration date of term permits.

Temporary permits may be waived back to the government when the permittees sell livestock or base property. Then, the purchaser of the permitted livestock or base property may apply for and be issued a permit if qualified.

The requisites in order to qualify for a term permit are:

a. U.S. citizenship.

b. Ownership. The ownership of both the livestock and commensurate ranch property.

A term grazing permit is not a property right. Rather, it is approved for the exclusive use and benefit of the person to whom it is issued. Permits may be revoked in whole or in part for a clearly established violation of the terms of the permit, the regulations upon which it is based, or the instructions of forest officers issued thereunder.

A ranger administers the grazing use on each National Forest Ranger District. Several districts (usually 3 to 6 or more) comprise a national forest. A forest supervisor administers the national forest. Several national forests, under the direction of a regional forester and staff, comprise a Forest Service region. The chief administers the Forest Service from Washington, DC, under the supervision of the Secretary of Agriculture.

Local farmers and ranchers act in an advisory capacity in reviewing allotment management plans and the use of range betterment funds.

Forest Service grazing fees are based on a formula which takes into account livestock prices over the past 10 years, the quality of forage on the allotment, and the cost of range operation. In 1990, average charges were $1.81 per animal unit month (AUM); or $1.81 for a mature cow or horse, or for 5 sheep, for a month.

Although shortcomings exist in the management of the national forests, it is generally agreed that these ranges have been vastly improved under the administration of the Forest Service. Some of them now approach the quality that existed in their virgin state. Perhaps the most heated arguments between livestock producers and the Forest Service arise over the relative importance attached to the multiple use of big game and other wildlife, recreation, etc.

3. **Bureau of Indian Affairs.** Most Indian lands, comprising 51 million acres, are really not public lands. Rather, these lands are held in trust for the benefit or use of the Indians and are merely administered by the Bureau of Indian Affairs of the Department of the Interior. Because over 80% of Indian lands are in the range area of the West, they are suited primarily to livestock. Thus, it is noteworthy that the sale of livestock and animal by-products regularly accounts for two-thirds of the total Indian agricultural income. Although the Indians themselves own most of the stock that graze these lands, animals owned by non-Indians utilize one-fourth of the Indian lands devoted to grazing. Provision for such use is handled under lease agreement jointly approved by the Indian owners and the Bureau of Indian Affairs.

Many of the Indian lands have suffered serious vegetative depletion, but a concerted effort is now being made to decrease livestock numbers in keeping with available feed supplies and to improve the quality of animals produced. However, overstocking continues to be a difficult problem on the Navajo, Hopi, and Papago Reservations.

4. **State and local government-owned lands.** A total of 134 million acres are owned by state and local governments. For the most part, the management of these areas is diverse and confused, each state and local government having established different regulations relative to the lands under its ownership. In general, however, such lands are operated on a stipulated lease arrangement. On many such areas, range depletion has been severe.

5. **Railroad-owned lands.** Recognizing that the main deterrent to rapid settlement and development of the West was lack of adequate transportation facilities, the Federal Government very early encouraged the construction and westward extension of the railroads by means of large grants of land. It was intended that the railroads should sell or otherwise utilize these lands in financing their costs of construction. These initial grants, totaling 94,355,739 acres, consisted of alternate sections extending in a checkerboard fashion for a distance of from 10 to 40 miles on each side of the right-of-way. Today, less than 20 million acres of these lands are held by railroads. Many of these holdings are leased to livestock producers; but because of the inconvenience, past abuses, or other reasons, some of these lands are considered worthless grazing. In general, railroad lease agreements do not restrict the number of stock to be grazed or the season during which the land may be so used.

PART III
MULTIPLE USE/CONSERVATION OF LAND

Fig. 5-22. Buffalo, once almost extinct, at home on the range. (Courtesy, USDA, Soil Conservation Service)

The multiple use of publicly-owned lands evolved in response to public pressure. The multiple use of privately-owned lands followed, in response to economic pressure—the need to increase net returns. Soil and water conservation evolved on both public and private lands in recognition that they are national resources that should be preserved for posterity.

The sections that follow pertain to the multiple use/conservation aspects of both public and private lands—to both introduced (seeded) pastures and native (range) pastures.

MULTIPLE-USE CONCEPT

Multiple-use of land is the management of all the various resources of lands, both public and private, so that they are utilized in combination. With federal lands, multiple use is based on their most profitable use, and management decisions are made privately by owners/managers. Important multiple uses include livestock grazing, mining, national heritage preservation, occupancy, recreation, water, wildlife, and wood/timber production.

The multiple use concept developed as a compromise relative to the use of public lands; it evolved as a result of attempting to placate individuals and groups who wish to have the land used for purposes which they consider desirable or to prevent others from using the land for purposes which they consider to be undesirable.

Fig. 5-23. Recreation on the range—trail rides are popular and a source of income for ranchers. (Courtesy, H. Dietz, USDA, Soil Conservation Service, Fort Worth, TX)

MULTISPECIES GRAZING

Grazing two or more species of livestock together or separately on the same land unit in a single growing season is known as multispecies grazing. Research indicates that multispecies grazing contributes to better and more uniform forage use and higher economic returns from livestock.

Multispecies grazing evolved in regions with diverse vegetation types and suitable climates. Grazing by a mix of domestic and wild animals can often result in more efficient use of forage and browse, more total animal gains, and a more vigorous plant community. While multispecies grazing is a common management practice on rangelands of the West, it is much less commonly practiced on the pasture lands of eastern United States.

Western rangelands are characterized by vast diversity in elevation, precipitation, temperature, and other climatic factors. These differences make for a multitude of range sites dispersed among several major vegetation or habitat types. It follows that great potential exists on these lands for multispecies grazing by livestock and wildlife to maintain forage production and species diversity.

Where multispecies grazing is practiced, cattle and sheep dominate. In the Southwest, goats are sometimes a component. Goats are without a peer in rough, unimproved areas and as browsers. Sheep prefer steeper terrain and eat more shrubs and forbs than cattle. Cattle stick to the more gentle slopes and prefer grasses. So, multispecies grazing can result in more complete and uniform utilization of multiplant species pastures and greater animal production. However, predators and labor problems have caused decreased sheep and goat numbers. In turn, this has resulted in lower income on many ranges.

In the past, wildlife has generally been incidental. Now, and in the future, economic pressures dictate that wildlife will be an integral part of multiple land use.

WILDLIFE

Wild animals and birds are becoming more valuable to today's landowner. Higher livestock production costs and demand for outdoor recreation have prompted practical landowners to seek means of increasing income by providing game for hunters. In some areas, wildlife income exceeds livestock income. This has caused landowners to include wildlife in farm and ranch planning.

There is a close association between kinds of plants and animals present in the habitat. Also, livestock numbers and grazing patterns can be manipulated to enhance wildlife habitat. Through proper habitat management, the farmer and rancher can maintain healthy, abundant wildlife populations. To accomplish this, land managers must place wildlife high on their priority list and consider wildlife in overall farm/ranch planning.

- **Kinds of wildlife**—Many kinds of wild animals and birds live on pastures and ranges. Identifying the kinds is necessary because management will vary for different species. Deer, for example, need browse, forbs, and grasses for feed, and timber and brushy areas for cover. Quail feed on weed seeds, nuts, and seeds of certain grasses and shrubs; and they prefer a mixture of wooded and open areas with small plots of low shrubs and vines for cover. Normally, management will involve meeting the needs of several different kinds of animals and birds.

• **Numbers of wild game**—Nationwide, in 1984 there were an estimated 10 to 11 million deer, 700,000 pronghorn antelope, and 500,000 elk, plus other species.[5]

In addition to the big game animals, there are many species of small game animals. Also, there are numerous species of game birds, including quail, partridge, and pheasants.

• **Wildlife management**—Wildlife can exist in harmony with livestock operations provided (1) wildlife needs and species are inventoried and included in the management plan, and (2) the following management aspects prevail:

1. **Grazing system.** For proper use of vegetation, it is important that the grazing system allow livestock and wildlife to harvest the forage without overgrazing. This protects the quality of the forages, provides wildlife cover, and prevents erosion. Grazing systems where livestock are concentrated and rotated between pastures reduce some of the competition with wildlife.
2. **Revegetation.** Some grasslands do not have adequate plants to meet the needs of livestock, wildlife, and erosion control. Such areas may be in need of reseeding. When reseeding, consideration should be given to including plants that have special value for wildlife. Many native forbs and shrubs have been selected and released and are now available for this purpose.
3. **Water.** Water is as important to wildlife as it is to livestock. Reliable and well distributed supplies of water should be provided and maintained.
4. **Brush control.** Controlling brush can improve grasslands for wildlife and livestock. But it must be done properly in harmony with other conservation practices. If poorly planned and not followed with good grassland management, it can harm the habitat for wildlife. One method of brush control is prescribed burning. If this method is used, a burning plan should be developed to meet the objectives of the owner/manager. Patterned brush control, or leaving strips or mottes of brush in pastures, increases the edge effect and enhanced wildlife habitat for many species.

SOIL EROSION CONTROL

Soil erosion control is any management plan to reduce soil and water losses.

Soil erosion is a natural occurrence. However, it may be increased by activities that disturb the natural balance of the pasture or range ecosystem. Poor grazing management is a major cause of erosion.

A raindrop that hits bare soil has a different effect from one that falls on a plant or litter. A racing raindrop smashes against bare soil with great force, splashing water and soil particles and packing the surface soil together; it seals pores, with the result that little water goes into the soil and runoff occurs. By contrast, when a raindrop hits a plant or litter, its force is broken and it trickles into the soil. Grasses are very effective in catching and holding moisture.

[5]Oldfield, J. E., Chairman, *et al., Forages,* Council for Agricultural Science and Technology, Ames, IA, 1986, p. 30.

Fig. 5-24. An Iowa farmer planting corn between terraces in the stubble of the previous year's crop, using a 16-row no-till planter. Conservation tillage and terraces help protect the soil from erosion. (Courtesy, USDA, Soil Conservation Service)

When plant cover is reduced by poor management and the distance between plants allows wind to reach the soil, wind erosion may result.

The amount of plant cover on the soil surface at the time of a rain or wind storm is the primary factor in preventing erosion. Both the bulk of cover and the distribution over the surface are important in reducing erosion.

The primary methods of controlling erosion on grasslands include brush control, deferred grazing, reseeding, and mechanical land treatments. Among the latter are contour furrowing, pitting, small dams, and diversions.

Fig. 5-25. Range furrowing in low rainfall area to stop water runoff and increase water penetration. (Courtesy, USDA, Soil Conservation Service)

The combination of practices used for erosion control will gradually result in better production of grass, improved condition of the range, and a better water supply for domestic animals and wild game. Also, and most important, it will lessen the sedimentation due to soil erosion, which (1) reduces channel capacity and reservoir storage, and (2) results in increasing flooding and reduced water supply.

WATER

Fig. 5–26. Energy-conserving watering facility on famed Lasater Ranch, Mattheson, Colorado. (Courtesy, T. Lasater, owner)

Water is often a limiting factor in pasture productivity, affecting forage production and/or drinking water for grazing animals.

The water cycle is the never-ending movement of water from clouds to soil, through plants, and back to clouds again. The cycle begins when precipitation strikes the land and ends when the water leaves the land either through runoff or evaporation. During the intervening time, a livestock producer should store as much water as possible, in the soil and in reservoirs. The shortage of water over much of the West is particularly important. In addition to limiting livestock production, lack of water may limit stream flow for fish, cultivated crops, and industries.

There are various types of stock water developments. These include *natural* water supplies such as lakes, ponds, streams, springs, and seeps, and *made* developments such as wells, reservoirs, dugouts, sand tanks, and catchment basins. A combination of two or more types of water development is often more advantageous than one type only.

Oats and alsike clover pasture. (Courtesy, USDA, Soil Conservation Service)

Haymaking has gone modern! Backbreaking pitchforks loading (left). Pickup baler with ejector—a bale loader (right). (Photo of bale loader courtesy of John Deere, Moline, IL)

SECTION

6

HAY AND CROP RESIDUES[1]

Contents *Page*

[1]The author gratefully acknowledges the helpful suggestions of the following eminent authorities who reviewed this chapter: J. E. Baylor, Ph.D., Professor of Agronomy, The Pennsylvania State University, State College; R. A. Forsberg, Ph.D., *et al.*, Department of Agronomy, University of Wisconsin-Madison; and S. C. Fransen, Ph.D., Forage Agronomist, Western Washington Research and Extension Center, Washington State University, Puyallup.

PART I—HAY

Fig. 6-1. Hay meadow on the Phillips Ranch, near Ridgway, Colorado, showing baled hay being removed for storage as winter feed. (Courtesy, Soil Conservation Service, Washington, DC)

Hay is forage harvested during the growing period and preserved by drying for subsequent use. Hays are made from legumes, grasses, and cereal crops. It is the most important harvested forage fed to livestock, and it ranks third among all livestock feeds, being exceeded only by pasture and corn. Hay is primarily a cattle, sheep, and horse feed, although alfalfa (especially ground, dehydrated alfalfa) may be included in swine and poultry rations. Average-quality hay runs 25 to 35% crude fiber and 45 to 55% TDN on an as-fed basis, whereas such concentrates as corn and wheat contain approximately 2 to 3% fiber and 80% TDN.

The object of haymaking is to (1) harvest the crop at the optimum stage of maturity which will provide the maximum yield of nutrients per acre without damage to the next crop, and (2) cure the crop properly by lowering the water content of the green herbage from 65 to 85% to 20% or less.

Drying, or making hay, is the most common method of preserving forage for storage, primarily because it is relatively easy to handle. It can be stored or transported long, chopped, pelleted, cubed, or packaged into various types and sizes of bales. Modern equipment and chemicals hasten drying time; and automated systems facilitate handling.

The great capacity and specialized functions of the rumen allow cattle and sheep to use hay, and other forages, in large amounts. Bacteria and protozoa in the rumen break down and make available to the host animal part of the nutrients in cellulose or fibrous material.

In addition to the nutrients that it contains, and to its value in providing feed throughout the year, hay has other values. Dry feed is essential for the proper functioning of the digestive tract; it acts as a stimulant in moving the feed through the intestines, and it maintains the proper conditions in the rumen for the microbial action which plays such a vital role in the digestion of the fibrous portions of feeds. Hay is often used as a supplement to "washy" pastures and succulent silages. Also, it speeds along the development of the rumen function of the young ruminant, lessens the incidence of displaced abomasum in cattle, and prevents a lowering of the fat content of the milk of lactating cows (unless it is finely ground). Also, and most important, good-quality hay is a hedge against high-concentrate prices, for when the price of such feeds increases disproportionately, increased amounts of hay may be fed and concentrates may be decreased, with a higher net return to the producer.

Despite its several advantages, hay has some shortcomings. It varies in nutrient content and palatability more than any other feed, because of differences in the (1) crops from which it is made, (2) stage of cutting, (3) handling, and (4) weather damage during curing. Not even ruminants can consume enough hay alone to meet the demands of high production; for example, when fed hay alone, dairy cows will produce only 50 to 70% as much milk as they would when fed a ration consisting of 50% concentrates. Also, fiber is poorly digested by monogastric animals, with the result that hay serves primarily as a source of minerals and vitamins for swine and poultry.

An estimated 80% of all hay is fed on the farms or ranches on which it is produced, rather than being purchased. It is important, therefore, that producers know how to produce good hay, as well as how to feed it, for most of them determine their own destiny from the standpoint of quality. For this reason, this chapter covers hay from production to feeding.

MAGNITUDE AND IMPORTANCE OF HAY

The importance of the nation's hay crop is attested to by the fact that the total area devoted to hay in the United States exceeds 60 million acres, the total production averages about 150 million tons, and the annual crop is worth approximately $10 billion—it is worth more than any other crop except corn and soybeans. On an air-dry tonnage basis, about 3 times as much hay is produced as silage.

Despite the importance of hay, no other feed crop suffers a higher loss of nutrients from the time it is cut to the time it is fed. During the curing process, the quality and feeding value of hay decreases rapidly by rain, sun bleaching, raking, handling when too dry, and storing with too much moisture. Studies by the U.S. Department of Agriculture revealed that the following losses accrued in field-cured, second-cut alfalfa hay from the time of cutting to the time of feeding: leaves, 35%; dry matter, 20%; and proteins, 29%. Moreover, the longer hay remains in the field until it is dry enough to store, the greater the nutrient losses (Fig. 6–2). These losses have been estimated to have a feeding value of more than a billion dollars annually.

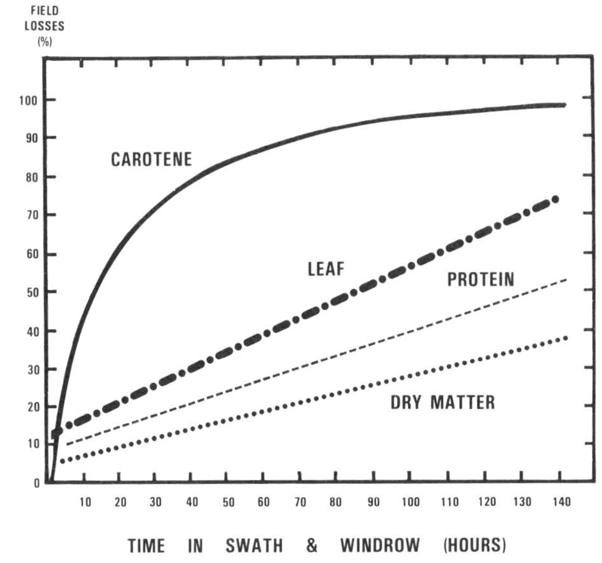

Fig. 6–2. Losses in sun-curing alfalfa hay as related to time in the field to reduce moisture to a safe level for storage. (Adapted by the author from USDA data.)

HAY AS AN ENERGY SOURCE

Fig. 6–3. Hay is a more important source of energy for dairy cows than for any other class of farm animal. (Courtesy, Union Pacific Railroad Company, Omaha, NE)

Hay is an important source of energy for cattle, sheep, and horses. Table 6–1 shows the percentage of total energy (TDN) intake provided to these species by hay and other kinds of feeds.

TABLE 6–1
PERCENTAGE OF ENERGY SUPPLIED BY HAY AND OTHER KINDS OF FEEDS[1]

Animal	Concentrates	Hay	Other Harvested Forages	Pasture	All Forage	Total
	(%)	(%)	(%)	(%)	(%)	(%)
Lactating cows	37.9	23.1	19.4	19.6	62.1	100
Other dairy cows	19.4	29.0	5.9	45.7	80.6	100
Finishing beef cattle	69.8	16.3	8.7	5.2	30.2	100
Other beef cattle	8.7	15.5	4.1	71.7	91.3	100
Sheep and goats	10.4	4.7	3.1	81.8	89.6	100
Horses and mules	20.6	18.3	10.2	50.9	79.4	100

[1]Based on USDA data. From paper entitled, "Hay Production, Preservation and Quality," by J. E. Baylor, The Pennsylvania State University, *Beef Cattle Science Handbook*, Vol. 13, p. 199, published by Agriservices Foundation, edited by M. E. Ensminger.

As shown in Table 6–1, hay is a more important source of energy for dairy cows than for any other class of farm animal. But it is also an important feed source for beef cattle and horses. It is noteworthy, too, that about one-half the total hay tonnage produced in North America is fed to dairy cattle, while beef cattle consume almost one-third of all hay produced. As increasing quantities of concentrates go to feed the world's hungry people, it is expected that livestock producers will depend even more on hay to meet a larger percentage of the total feed needs of ruminants.

COMPARATIVE VALUE OF HAY

In ruminant rations, hay is primarily a source of energy, but the legumes also serve as a source of protein. For swine and poultry, ground hay (especially alfalfa) is fed primarily as a source of minerals and vitamins.

Table 6–2 shows the dry matter (DM), crude protein (CP), and total digestible nutrients (TDN) in 100 lb of dry matter from corn grain, corn silage, and three different types of alfalfa hay (mature, midbloom, and early bloom). Note, too, that this table compares these crops on a per acre basis.

TABLE 6–2
COMPARATIVE ECONOMICS OF THE NUTRIENTS IN CORN GRAIN, CORN SILAGE, AND ALFALFA HAY OF THREE QUALITIES[1]

Item	Corn Grain	Corn Silage	Alfalfa Hay Mature	Alfalfa Hay Mid-Bloom	Alfalfa Hay Early Bloom
Analyses, DM basis, %					
Dry matter (DM)	88.0	33.0	90.0	90.0	90.0
Crude protein (CP)	10.1	8.1	12.9	17.0	18.0
Total Digestible Nutrients (TDN)	90.0	70.0	50.0	58.0	60.0
Value of 100 lb DM, $					
CP value[2]	1.72	1.38	2.19	2.89	3.06
TDN value[3]	5.04	3.92	2.80	3.25	3.36
Total value	6.76	5.30	4.99	6.13	6.42
Total value/acre,[2 3] $					
16 tons silage or 100 bu grain	333.00	560.00	–	–	–
21 tons silage or 150 bu grain	500.00	735.00	–	–	–
5 tons hay	–	–	449.00	552.00	578.00
8 tons hay	–	–	719.00	883.00	924.00

[1]Adapted by the author from: *Haymaker's Handbook*, by J. E. Baylor, Professor of Agronomy, The Pennsylvania State University, and M. A. Balas, New Holland, Inc., published by Ford New Holland, Inc., New Holland, PA, 1987, p. 140, Table 17.1.

[2]44% soybean meal used as a standard protein source, priced at $150 per ton or $.17 per pound of crude protein.

[3]Corn grain used as a standard TDN source, priced at $2.50 per bushel or $.056 per pound of TDN.

Of course, the economic comparisons in Table 6–2 are valid only at the stated prices of soybean meal and corn. However, in most practical feeding situations the comparisons are meaningful. It is noteworthy that, in terms of the economic value of the energy and protein provided by the different feeds listed in Table 6–2, early bloom alfalfa hay had a value nearly 95% (6.42 ÷ 6.76 × 100) that of corn grain. Note, too, that in terms of energy produced per acre, corn silage leads all other feeds.

HAY AS A GRAIN REPLACEMENT

In the future, livestock producers will increasingly rely upon the ability of ruminants to convert coarse forage, grass, and by-product feeds, along with a minimum of concentrate, into food for human consumption, thereby competing less for humanly edible grains.

The University of Arizona found that 60% alfalfa in a grower calf ration was converted as well as one containing only 15% roughage with the remainder consisting of high-priced concentrate or grain. The study involved steer calves averaging 400 lb at weaning, and a grow-out period of 84 days. Actual gains were 2.84 lb per head daily for the calves fed the 60% alfalfa ration vs 2.34 lb for those fed the ration containing 15% roughage. Feed conversion was practically the same—5.63 to 1 for the high roughage, and 5.60 lb of feed per pound gain for the control animals.

The U.S. Department of Agriculture conducted a study of all-forage rations for finishing cattle, the results of which are given in Table 6–3.

TABLE 6–3
FEEDLOT PERFORMANCE AND CARCASS EVALUATION OF STEERS FED ALL-FORAGE VS ALL-CONCENTRATE RATIONS[1]

Item		All-Forage Ration[2]	All-Concentrate Ration[2]
Average daily feed intake	(lb)	23.3	16.0
	(kg)	*10.59*	*7.26*
Average daily feed intake in % of body weight		3.23	2.15
Average daily gain	(lb)	2.3	2.8
	(kg)	*1.05*	*1.27*
Feed-gain ratio		10.06	5.71
Average carcass grade		Low Choice	Medium Choice
Dressing percentage	(%)	55.4	59.9
Marbling score		Abundant	Abundant
Rib eye area	(sq in.)	11.0	10.6
	(sq cm)	*71.1*	*68.5*
Fat over rib eye	(in.)	.37	.67
	(mm)	*9.4*	*17.0*
Taste panel evaluation[3]		7.6	7.2

[1]Oltjen, R. R., T. S. Rumsey, and P. A. Putnam, "All-Forage Diets for Fattening Beef Cattle," *Journal of Animal Science*, Vol. 32, No. 2, 1971, pp. 327–333.

[2]Corn grain provided 90% of the all-concentrate ration; pelleted alfalfa provided 98% of the all-forage ration.

[3]Overall desirability rated on a scale of 1 to 9, with 9 being the most desirable.

As a result of the experiment summarized in Table 6–3, the U.S. Department of Agriculture researchers concluded that (1) beef cattle of an acceptable quality were produced on a pelleted, all-forage ration; (2) steers on an all-forage ration had to be fed a month longer than those on the all-concentrate ration; (3) the all-forage-fed steers consumed about 95% as much metabolizable energy and were about 86% as efficient converters of it to body weight gains as were the all-concentrate steers; (4) the forage-fed steers had only 55% as much fat over the rib eye as did the all-concentrate-fed steers; and (5) there was a 4.5% difference in dressing percentage in favor of the animals receiving all-concentrate ration. Based on this study, the following conclusion may be drawn: Since cattle fed high-roughage rations normally have lower dressing percentages, forages must be cheaper than grain in order for the feeder to obtain the same net return; this situation usually exists relative to pastures, but it doesn't always apply to dry forages.

The Georgia Station compared (1) a conventional high-energy ration, composed of 72.8% corn, 20% peanut hulls, and 7.2% protein supplement; and (2) an all-forage ration consisting of Bermudagrass pellets (in drylot) and small grain pastures (in season), for finishing calves and yearlings.[2] The results of this study are reported in Table 6-4.

TABLE 6-4
FEEDLOT PERFORMANCE AND CARCASS CHARACTERISTICS OF CALVES AND YEARLINGS FINISHED ON A HIGH-ENERGY RATION AND AN ALL-FORAGE RATION

Item	Calves				Yearlings			
	High-Energy		All-Forage		High-Energy		All-Forage	
	(lb)	*(kg)*	(lb)	*(kg)*	(lb)	*(kg)*	(lb)	*(kg)*
Average initial weight	488	*222*	492	*224*	670	*305*	672	*306*
Average final weight	1,042	*474*	1,053	*479*	1,165	*530*	1,147	*521*
Average daily gain	2.84	*1.29*	2.33	*1.06*	3.07	*1.40*	2.31	*1.05*

Yearling steers fed the high-energy ration gained 3.07 lb daily vs 2.31 lb for the steers fed the all-forage ration. Average daily gains of steer calves and yearlings were similar when fed the all-forage ration, but average gain for yearlings was 8% faster than for calves when fed the high-energy ration. Steers fed the high-energy ration dressed out about 7.5% more carcass, and had more marbling, higher yield grades, and more fat covering over the rib eye than steers fed the all-forage ration. Under the conditions of these trials, finishing steers of both age groups were more profitable when an all-forage ration was fed.

KINDS OF HAY

Although there are favorite hays, a great variety of legumes, grasses, and cereal crops can be, and is, successfully used for hay. In terms of total tonnage produced annually, alfalfa (or lucerne), the "Queen of the Forages," accounts for approximately 57% of the U.S. hay crop. Many different kinds of hay make up the other 43% of the nation's hay supply; among them, other perennial legumes, cool season grasses, warm season grasses, cereal hays, summer annuals, and annual legumes. (See Table 6-5, next page.)

The kind of forage grown should be determined by soil type, soil drainage, soil pH, topography, climatic conditions, preferred use, and the animals to which it will be fed. Also, more and more farmers are coming to appreciate the flexibility afforded by growing varieties of grasses and legumes that may be used three ways: for pasture, for hay, or for silage. With such an arrangement, surplus pasture may be converted into hay, or, if the weather is not favorable for haymaking, the crop can be ensiled.

Generally speaking, legumes should be used as hay crops wherever they are adapted, either alone or in combination with grass(es). There is one possible exception to this recommendation—where horses are involved, sometimes a good-quality grass hay may be preferable.

Whenever feasible, it is recommended that a legume be grown for hay, for the reasons that, in comparison with grasses, legumes are (1) higher in protein, vitamins, and minerals; (2) higher yielding; and (3) nitrogen-fixing when inoculated, because the bacteria (rhizobia) on their roots take free atmospheric nitrogen from the air. However, a mixture of grasses and legumes is often preferred for reasons of palatability, ease in curing, erosion control, and lessening bloat.

Table 6-5 summarizes the adaptation and use of a number of species grown for hay. Note that soil drainage and pH are important. In addition to the adaptation and use of various species, consideration should be given to (1) variety, because some varieties may be better suited than others to a specific soil and management; and (2) quality of seed.

In addition to the plant species listed in Table 6-5 as suitable for hay, it is noteworthy that the grasses and legumes listed in this book in Section 5, Chart 5-1, are suitable for, and may be used, three ways—for hay, for silage, or for pasture; the three-way usage is achieved simply by varying the stage of harvesting.

[2]Utley, P. R., R. E. Hellwig, and W. C. McCormick, "Finishing Beef Cattle for Slaughter on All-forage Diets," *Journal of Animal Science*, Vol. 40, No. 6, 1975, p. 1034.

TABLE 6-5
ADAPTATION AND USE OF SOME COMMON FORAGE SPECIES GROWN FOR HAY[1]

Species	Estimated Life of Stand	Approx. Yield Pure Stand[2]	Soil Adaptation	Preferred Use	Preferred Associated Species for Mixtures
	(yrs)	(tons/acre)			
Legumes:					
Alfalfa	3-5	4-7	Deep, well drained, pH 6.5 or above.	Hay, silage, pasture.	Tall, cool season grasses and perennial ryegrass.
Red clover	1-2	3½-5	Medium to well drained, pH 6.0 or above.	Hay, silage, pasture.	Timothy, bluegrass, tall fescue.
Birdsfoot trefoil	3-5	2½-4	Somewhat poorly to well drained, pH 5.5 or above.	Hay, silage, pasture.	Timothy, Kentucky bluegrass, perennial ryegrass, reed canarygrass.
Alsike clover	1-2	1-2	Tolerates wet, acid soils, pH 6.0 or above.	Pasture, silage, hay.	Red clover, timothy.
Cool season grasses:					
Perennial ryegrass	3-4	2-4	Moist to well drained, pH 6.0 or above.	Pasture, silage, hay.	Alone, or with a mix of either grass or legume.
Orchardgrass	3+	3-5	Medium to well drained, pH 6.0 or above.	Pasture, silage, hay.	Alone or with alfalfa or ladino clover.
Smooth bromegrass	3+	3-4	Fertile, well drained, pH 6.0 or above.	Hay, silage, pasture.	Alone, or with alfalfa.
Timothy	2-3	3-4	Medium to well drained, pH 6.0 or above.	Hay, silage, pasture.	Alone, or with red clover.
Reed canarygrass	3+	3½-6	Widely adapted from wet to droughty, pH 5.5 or above.	Pasture, silage, hay.	Alone, or with a legume.
Tall fescue	3+	3-5	Widely adapted shallow to deep soil, pH 5.5 or above.	Pasture, silage, hay.	Alone, or with lespedeza.
Warm season grasses:					
Switchgrass	4+	5-6	Moderately deep, from somewhat dry to poorly drained, and tolerant to low fertility.	Pasture, hay.	Alone.
Big bluestem	4+	4-5	Moderately deep, from somewhat dry to poorly drained, and tolerant to low fertility.	Pasture, hay.	Alone.
Bermudagrass	4+	5-8	Moderate to well drained and fairly heavy soil, pH 5.5 or above.	Hay, silage, pasture.	Alone, or with crimson or arrowleaf clover.
Cereal hays:					
Barley	Annual	1-3	Well drained, pH 7.0 or above.	Grain, hay, silage.	Alone.
Oats	Annual	1-3	Well drained, pH 5.5 or above.	Grain, hay, pasture, silage.	Alone, or with field peas.
Rye	Annual	1-3	Sandy, acid, infertile, pH 5.5 or above.	Grain, pasture.	Alone, or with vetch or crimson clover.
Triticale	Annual	1-3	Droughty and poor soil, pH 5.5 or above.	Grain, pasture, hay.	Alone.
Wheat	Annual	1-3	Fertile, well drained, pH 6.0 or above.	Grain, pasture, hay.	Alone, or with vetch, crimson clover or Austrian peas.
Summer annuals:					
Millet	Annual	1-3	Poor soil, pH 7.0 or above.	Hay, silage.	Alone.
Sorghum hybrids	Annual	2-4	Fertile, well drained, drought resistant, pH 7.0 or above.	Grain, silage, green chop, pasture, hay.	Alone.
Sudangrass	Annual	2-4	Fertile, well drained, drought resistant, pH 7.0 or above.	Pasture, silage, green chop, hay.	Alone, or with soybeans or cowpeas.
Annual legumes:					
Cowpeas	Annual	2-4	Poorer soils, pH 6.0 or above.	Pasture, green manure, hay, seed.	Alone, or with corn or sorghum.
Soybeans	Annual	2-5	Fertile soil, pH 6.0 or above.	Seed, hay.	Alone, or with millet, sorghum or Sudangrass.
Vetch	Annual	2-2½	Poor soil, pH 6.5 or above.	Hay, pasture, soil improvement.	Alone, or with small grain.

[1]Adapted by the author from: *Haymaker's Handbook*, by J. E. Baylor, Professor of Agronomy, The Pennsylvania State University, and M. A. Balas, New Holland, Inc., published by Ford New Holland, Inc., New Holland, PA, 1987, pp. 17-18, Table 3.2.

[2]These are short tons (2,000 lb). To convert to metric tons/acre, multiply by 0.907; then, to obtain metric tons/hectare, multiply by 2.47.

HAY QUALITY

Hay quality is the degree of excellence, or the productive worth, that hay possesses. It refers to the nutritive value of hay. For hay to be of superior quality, it must be high in four factors: (1) nutrients, (2) palatability (intake), (3) digestibility, and (4) efficiency of utilization.

The most accurate method of determining hay quality involves live animal experiments on the farm or ranch where the forage is to be fed. However, this is often too costly, slow, and impractical. Therefore, forage value is predicted by visual inspection, chemical analysis, and/or new methods such as near infrared analysis.

Experiments and experiences show that, in addition to the low-nutrient content that characterizes poor-quality hay, a more serious loss may follow from feeding it. Studies show that part of the poor results obtained from feeding low-quality hay can be attributed to its failure to support maximum microflora in the rumen, with the result that the digestibility of the crude fiber suffers. Hand in hand with the decline in microflora activity, forage consumption goes down. Of course, if animals won't eat feed, it won't do them any good.

VISUAL INSPECTION

Although not as reliable as chemical analysis, most hay is still bought and sold on the basis of visual appraisal.

The factors to look for in high-quality hay are:

1. **Species of plants.** Determine what plants are present and the proportion of each. Hay with a high percentage of legume is usually higher in feed value than pure grass hay.
2. **State of maturity when cut.** Plants should not be in full bloom, nor should they have formed seeds. Early cut hay assures the maximum content of protein, minerals, and vitamins, and the highest digestibility.
3. **Percentage of leaves present.** Leaves are the part of the plant of highest quality; hence, a high proportion of leaves relative to stems is indicative of high quality.
4. **Green color.** A bright green color indicates (a) minimum of bleaching and leaching losses of carotene and other nutrients, and (b) palatability.
5. **Aroma and fragrance.** High-quality hay has a pleasing, fragrant aroma. Moldy smells are undesirable.
6. **Stemminess.** Large stiff, woody stems make for low acceptability and quality. High-quality hay is fine stemmed and pliable.
7. **Foreign material.** High-quality hay is free from such foreign materials as weeds, stubble, sticks, dirt, etc.
8. **Condition.** Hay that has been cured and stored properly does not contain excess moisture, is not in layers or chunks due to excess moisture or heating, is not moldy, and is not dry and brittle.

CHEMICAL ANALYSIS

Visual inspection of hay is needed for (1) weed detection and color, (2) predicting palatability, and (3) detecting the effects of mold, rain damage, and brittleness. So, chemical analyses should supplement, but not replace, visual inspection. Also, it is recognized that any method of determining hay quality by means of chemical analyses is of value only if it is related to feeding value.

Most livestock and hay producers are aware that hay harvested at an early stage of maturity is high in protein and low in fiber (cellulose), although hay yields at immature stages are low. But the magnitude of the variation is usually greater than suspected. Thus, both chemical composition and yields must be considered in a practical management system. The highest yield of protein and of most of the other important chemical constituents is obtained at near the $\frac{1}{10}$ bloom stage of growth. Although the yield of hay may continue to increase between $\frac{1}{10}$ bloom and full bloom, it is due largely to an increase in the yield of cellulose (Fig. 6–4).

As shown in Fig. 6–4, the yield of protein is greatest when alfalfa is harvested at the first flower ($\frac{1}{10}$ bloom) stage of maturity.

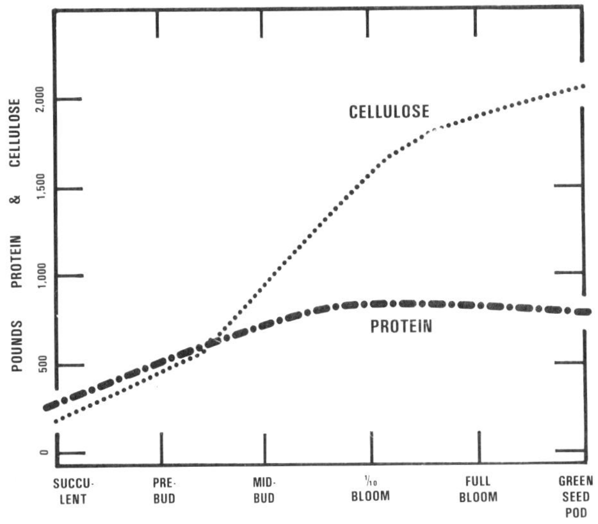

Fig. 6–4. Yield of crude protein and cellulose per acre from first cutting alfalfa. (*From:* Rohweder, D. A. and D. Smith, *Establishing and Managing Alfalfa,* Pub. A 1751, University of Wisconsin, 1982, p. 12.)

Research has generally shown a good relationship between the chemical composition of hay and its feeding value. As a result, a growing number of states now have laboratories where, at a nominal charge, a quick determination can be made of the chemical composition of hay. As hay matures, protein decreases (pounds of protein per acre decreased from 935 lb in early cut hay to 605 lb in late cut hay, according to a Cornell study) and fiber increases. Likewise, weathering lowers the protein and raises the fiber content, since soluble nutrients are washed out by rain and leaves are lost during harvest. It is also noteworthy that palatability is negatively correlated with crude fiber levels—the higher the fiber content, the lower the palatability; this is important, for if the animals won't eat it, they can't produce. Cornell investigators found that cows ate 2 lb more of the early-cut hay per day than of late-cut hay.

NDF, ADF, AND NIRS ANALYSES

Since 1865, fibrous materials traditionally have been analyzed by the *Weende proximate analysis* method. Although this method is still widely used, it is often supplemented with additional analyses. In a series of reports beginning in 1963,[3] Peter J. Van Soest proposed the *detergent analysis system,* better to evaluate the feeding value of fibrous materials; by using detergents, he separated the sample into two fibrous fractions: (1) a neutral detergent fibrous fraction (NDF), and (2) an acid detergent fibrous fraction (ADF). Further, he reported that, in comparison with traditional proximate analysis, NDF provided a better estimate of dry matter intake (consumption) by animals, and ADF provided a better estimate of the *in vivo* (inside the animal) dry matter digestibility. Van Soest partitioned forages into the following two fibrous fractions:

1. **Neutral detergent fiber (NDF),** which is the cell wall material or plant structure in feed, is comprised of hemicellulose, cellulose, lignin, lignified N, and insoluble ash. This constituent is insoluble in neutral detergent and is only partially available to animals. The lower the NDF percentage, the more the animal will eat; it is inversely related to voluntary intake (consumption). Thus, a low percentage of NDF is desirable.

The NDF content is also positively correlated with eating time and rumination, and this may be related to the rate of particle size reduction. Additionally, NDF is related to the proper function and health of the rumen; roughage value (the amount, source, and physical form of dietary NDF) is associated with chewing time, saliva flow (buffering and pH in the rumen), rumen fermentation patterns, milk fat test, and total energy output.

2. **Acid detergent fiber (ADF),** which is the highly indigestible plant material in a forage, is comprised of cellulose, lignin, and insoluble ash. This constituent is insoluble in acid detergent. ADF differs from crude fiber in that it contains silica. Silica and lignin in plants are associated with low *in vivo* (in animal) digestibility. The lower the ADF, the more feed an animal can digest. Thus, a low ADF percentage is desirable.

But chemical determinations are slow! The laboratory determination of crude protein, neutral detergent fiber, acid detergent fiber, *in vitro* (in a test tube or other artificial environment) dry matter digestibility, mineral, and vitamin analyses may take up to 2 weeks. This prompted the search for a more rapid, yet reliable, method of assessing the nutritive value of hay. In 1976, Norris, *et al.,* indicated (*Journal of Animal Science,* 47:747–759) that a relatively new procedure, known as *near infrared reflectance spectroscopy,* which had been applied successfully to grain quality evaluation, could be used for predicting forage quality, also. Additional research confirmed the initial findings of Norris and developed data bases and procedures for quickly analyzing hays.

The near infrared reflectance spectroscopy (NIRS) is a nonconsumptive instrumental method for fast, accurate, and precise evaluation of the chemical composition and associated feeding value attributes of forages and other feedstuffs.

The near infrared reflectance measures hay quality by comparing the energy reflected back from a hay sample with computerized standards established by conventional laboratory analysis of a large number of reference samples.

The NIRS method of analysis has four main advantages: speed, simplicity of sample preparation, multiplicity of analysis with one operation, and nonconsumption of the sample (it can be analyzed again by the same or another procedure). With the NIRS method of analysis, it is possible to take a sample from a truckload of hay and provide, in less than 3 minutes, an analysis for crude protein, NDF, ADF, dry matter, lignin, and *in vitro* dry matter digestibility.

MAKING QUALITY HAY

Fig. 6–5. Making quality hay on a Colorado ranch, showing hay raking in the foreground and hay baling in the background. (Courtesy, Soil Conservation Service, Washington, DC)

The object of haymaking is to (1) harvest the crop at the optimum stage of maturity which will provide the maximum yield of nutrients per acre without damage to the next crop; and (2) cure properly, which involves lowering the water content of the green herbage from 65 to 85% moisture to 20% or less.

Hay quality begins with the soil and ends with the manger, with many intermediate factors affecting it along the line. Once forage is cut, opportunities to increase nutrient content are over; from that point on, quality can only be preserved.

There is no one best haymaking method or kind of equipment. These must necessarily vary with the size of the operation, the kind of hay, the climate of the area, the individual farm or ranch conditions and buildings, and the available labor and machinery and their cost. Yet, the principles of good haymaking and the objectives sought are the same everywhere.

About 80% of all harvested forage is now baled. Only 10% is stored as loose hay and cubes, and the remaining 10% is stored as hay crop silage.

[3]Goldring, H. K. and P. J. Van Soest, *Forage Fiber Analyses,* Agriculture Handbook No. 379, ARS, USDA, 1970, p. 20 on which 12 papers are listed.

HARVESTING AT PROPER STAGE

Whether the crop is a grass or a legume, or a combination, the stage of maturity of the plants at the time of harvest affects digestibility, yield, and feeding value. Young, immature plants are high in protein and low in fiber or lignin. As hay crops mature, feeding value goes down and fiber content increases. Digestibility of the forage (TDN) declines about 0.5% each day cutting is delayed beyond the early bloom stage; and the intake of forage decreases during this same period at more than 0.5% each day. Thus, in total, the feeding value of forage drops more than 1% for each day's delay after early bloom.

Table 6–6 shows the primary changes in chemical constituents of alfalfa (a legume) and timothy (a grass) as affected by maturation. Note that, for both species, the DE, TDN, CP, lysine, Ca, and P decrease with maturity.

TABLE 6–6
EFFECT OF STAGE OF MATURITY AT HARVEST OF ALFALFA AND TIMOTHY ON COMPOSITION OF HAY[1]

Forage Crop	Stage of Maturity	DE[2]	TDN[3]	CP[4]	Lysine	Ca	P
		(Mcal/kg)	(%)	(%)	(%)	(%)	(%)
Alfalfa	Prebloom	2.78	63	19.4	1.10	2.10	0.34
	First flower[5]	2.42	56	17.9	0.94	1.75	0.28
	Midbloom	2.29	52	16.0	0.90	1.50	0.25
	Full bloom	2.16	49	15.0	0.64	1.29	0.25
Timothy	Prehead	2.20	50	11.5	—	0.50	0.25
	Head	1.98	45	9.0	—	0.41	0.19

[1]Adapted by the author from: Rohweder, D. A. and R. Antoniewicz, *Alfalfa, the high quality hay for horses,* produced by Certified Alfalfa Seed Council, Inc., Woodland, CA, p. 6, Table 1.

[2]Digestible energy. Divide by 2.2 to obtain Mcal/lb.

[3]Total digestible nutrients. Improved harvest techniques can increase values 10%.

[4]Crude protein.

[5]First flower to 1/10 bloom.

Stage of maturity also affects the vitamin content of hay. Carotene (precursor of vitamin A) and the B vitamins decrease as plants mature. Vitamin D content is the one exception—it increases as the forage is sun-cured.

Everything considered, there is a loss of about 1% in nutrient value for each day that hay harvest is delayed beyond the late vegetative stage of growth.

Table 6–7 gives guidelines relative to the proper forage-harvesting stage for maximum protein and minimum fiber.

TABLE 6–7
HAY CUTTING GUIDE

Kind of Hay	When to Cut
Alfalfa	Bud stage for first cutting; 1/10 bloom for second and later cuttings.
Alsike clover	Early bloom to 1/2 bloom stage.
Bermuda	When 16–18 in. tall, before lodging.
Birdsfoot trefoil	First flower to full bloom.
Bromegrass	Heads emerging.
Cowpeas	When pods are 1/2 to fully matured.
Crested wheatgrass	When the plants begin to head.
Crimson clover	From early bloom to 1/2 bloom.
Fescue	Boot to early head stage.
Grass-legume mixtures	When the legume is at the proper stage.
Johnsongrass, millet, Sudangrass, sorghum hybrids	40 in. height or early boot stage, whichever comes first.
Ladino clover	Few blooms to full bloom.
Lespedeza, annual	Early blossom.
Orchardgrass	Boot to early head stage.
Red clover	Late bud.
Sericea	When 12–15 in. high.
Small grains (oats, barley, wheat)	Boot stage to early dough stage.
Soybeans	Mid-to-full bloom and before bottom leaves begin to fall.
Sweet clover	Bud to very early flowering stage.
Timothy	Boot to early head stage.

CUTTING AND FIELD CURING HAY

Fig. 6-6. Hay mower-conditioner. This equipment reduces drying time and field losses. (Courtesy, Ford New Holland, New Holland, PA)

Proper cutting and field curing of hay embraces all the steps from cutting to ready-for-packaging or storing. In modern hay making, it includes the following steps:

• **Cutting/curing in the swath and windrow/raking or cocking**—The common steps, methods, and equipment used in cutting and field curing hay are as follows:

1. **Cutting and curing in the swath or windrow.** Cutting, followed by curing in the swath or windrow, is the first step in haymaking, regardless of the subsequent method or type of equipment employed. Any one of several types of mowers may be used, for all of them are designed to get the hay down. The most important thing is that the hay be cut at the proper stage of maturity.

2. **Raking.** After the hay has wilted sufficiently in the swath, but while it is still tough and the leaves will not shatter, it should be windrowed. For this assignment, the side-delivery rake is preferred to the dump rake.

Where windrowed hay is rained on, wait until the top half dries out, and then turn it upside down with the side-delivery rake (the use of the tedder for rewindrowing is not recommended because of excessive shattering).

3. **Cocking.** Formerly, well-made cocks, often adorned by hay caps, were considered a necessary part of good haymaking. However, this practice has greatly decreased, due primarily to higher labor costs and the advent of modern haymaking machinery. Today, the cocking of hay is confined almost entirely to use (a) in hot, arid regions where the leaves shatter if the hay is left in the swath or windrow for any appreciable length of time, and (b) as an emergency measure in order to protect hay when a storm is imminent.

• **Reducing moisture content/shattering/bleaching and fermenting**—Proper curing ensures that (1) the hay can be stored safely without heating excessively or becoming moldy; and (2) the maximum leafiness, green color, aroma, nutrient value, and palatability shall be retained. This calls for reducing the moisture content of the freshly cut forge, and minimizing shattering, bleaching, and fermenting losses. Details follow:

1. **Moisture content.** Freshly cut forage contains 75 to 80% moisture, whereas the maximum moisture content for safe hay storage is as follows:

For loose hay—25% moisture.
For baled hay—20 to 22% moisture (the lower figure for larger bales).
For chopped hay—18 to 20% moisture.
For cubes—16 to 17% moisture.

Hay of a higher moisture content than indicated should not be stored because (a) its value may be greatly lowered due to mold or to nutrient losses accompanying fermentations, and (b) of the ever-present danger of spontaneous combustion and a costly fire.

2. **Shattering losses.** In field curing hay, losses from leaf shattering range from 2 to 5% for grass hay and 3 to 39% for legume hays, with as much as 15 to 20% for legume hays field cured under the most favorable conditions.

3. **Bleaching and fermenting losses.** In general, the carotene or provitamin A content of freshly cured hay is proportional to the greenness. With severe bleaching, more than 90% of the vitamin A potency may be destroyed.

Even under the best of conditions, there is an unavoidable loss through fermentation, especially losses in sugars, starch, and carotene. With good weather and proper curing methods, however, these losses will not be excessive.

• **Reducing rain damage**—The leaching losses from rain are less severe soon after mowing, but increase in severity as curing progresses. Also, repeated showers are more damaging than one heavy rain. Experimental studies have revealed that damaging rains may lower the feeding value of hay by one-fourth to one-third, or even more with severe exposure.

Losses from weather damage may be reduced (1) by using haymaking equipment that reduces the field drying time, (2) by understanding and using existing weather aides, and (3) by using proven chemical conditioning and preserving agents.

• **Considering chemical conditioning and preserving agents**—Chemical hay drying agents and preservatives are giving haymakers a big assist in lessening hay making losses and improving hay quality. The big advantage of these products is that they speed up the haymaking process and reduce exposure to weather damage.

1. **Chemical conditioners.** Chemical drying agents, which are sprayed on the crop at mowing time, break down the waxy cutin layer on the wall of the stem and allow moisture to escape, thereby promoting faster drying time, with the drying rate of the stems approaching that of the leaves. Several chemicals can be, and are, used for conditioning, including potassium carbonate, sodium carbonate, and sodium silicate. Also, methyl esters of fats, vegetable oils, or animal fats have been mixed with potassium carbonate in an attempt to increase the effectiveness of chemical conditioning.

CAUTION: Since the crop is standing when desiccants are applied and won't be cut until a few milliseconds later, the EPA requires that some desiccants must be labeled as *pesticides.* So, before using a chemical conditioner, the haymaker should check with local regulatory authorities for their interpretation of this point.

2. **Preserving agents.** Under normal conditions, for safe baling a moisture content of 20% or less is a must. Studies in many states have shown that, if properly treated with an adequate amount of the right preservative, alfalfa hay can be baled at 25 to 30% moisture, thereby speeding harvesting and lessening losses significantly. Preservatives act as fungicides and inhibit the growth and reproduction of the microorganisms that cause heating and molding in wet hay. A brief description of each of the common preservatives follows:

a. **Organic acids.** Propionic acid is the organic acid of choice. It is sometimes mixed with acetic acid, inorganic acids, formaldehyde, water, flavoring ingredients, and/or antioxidants. But, to be most effective, organic acid formulations should have at least 60% propionic acid.

b. **Anhydrous ammonia.** When properly applied, anhydrous ammonia will stop bacteria and mold growth; and when applied to poor quality hay, it has the added advantage of increasing protein and digestibility. However, as a preservative it's not as effective as propionic acid. Also, unless large round bales are covered and/or contain less than 28% moisture, too much of the ammonia escapes. Excessive amounts of ammonia may cause animal disorders.

Where high-quality alfalfa hay is involved, which is already high in protein and digestibility, it's doubtful that the added expense of using ammonia can be justified.

c. **Bacterial inoculants.** Claims are made that most bacterial inoculants on the market will produce lactic acid, which acts as a fungicide and inhibits mold growth. More experimental work is needed, substantiating the effectiveness of bacterial inoculants as hay preservatives.

HAYMAKING SYSTEMS

In haymaking, the term *system* refers to a team of processes and machines that does the work from field through feeding, saves crop nutrients, reduces manpower requirements, and eliminates drudgery. When each step is mechanized, it must be matched; otherwise, workers and machines end up waiting.

In recent years, automation has had great impact on haymaking. Some haymaking systems are completely mechanized from field to feeding.

There is no one best haymaking system for all conditions. Nevertheless, all good systems are fast, make handling easy, save labor and nutrients, and increase profits. Baling is the most popular hay-handling system in North America.

LONG, LOOSE HAY

The acreage harvested as long, loose hay has declined sharply in recent years, especially in the humid areas, because (1) of high labor cost, and (2) long hay is too bulky for mechanized feeding. Nevertheless, long, loose hay is still popular in many western areas where specialized handling equipment is used. Moreover, some of the newer systems of handling and self-feeding loose hay show promise.

The two common methods of handling long hay are:

1. **Loading with hay-loader directly from windrows.** In this method, cured hay is loaded on a truck or wagon directly from the windrow by means of a hay-loader. Usually the hay is then transported to a barn or stack where it is unloaded by fork or sling and moved away by hand. Sometimes it is chopped into the barn or other storage area.

2. **Hauling cured hay from windrows or cocks with buck rakes, sweep rakes, or sled.** In the West, much of the hay is cured in windrows or cocks and then transported by buck rakes, sweep rakes, or sleds to field stacks where it is stacked by hay stackers or other large mechanical devices. Then after going through a sweat in the stack, the hay is fed out as loose hay, or, if intended for market, it is baled.

CHOPPED HAY

Chopped dry hay fits into some feeding systems, particularly in the West.

For safe storage, the moisture of chopped hay should not exceed 18 to 20%.

Two common methods of chopping freshly cured hay follow:

1. **Field chopping cured hay directly from the windrow.** In this method, a field chopper gathers the cured hay from the windrow, chops it, and blows it into a truck or trailer. The chopped hay is then blown into the barn, stack, or other storage area.

Fig. 6-7. Field chopping cured hay from the windrow. (Courtesy, Gehl Company, West Bend, IN)

2. **Chopping into the barn or other storage area.** In this method, the cured hay is generally hauled from the windrow or cock to the barn or other storage area where it is chopped by a hay chopper or silage cutter and blown directly into the storage area. This method is slower and requires more labor than where field-cured hay is chopped directly from the windrow, but less expensive equipment is necessary.

PACKAGED HAY

Great strides have been made in hay packaging in recent years, characterized by the advent of round bales, large rectangular bales, loaflike stacks, cubes, and pellets.

BALES

The following choices of bales are available:

1. **Rectangular bales.** Conventional, small, rectangular packages (often called *square bales*), weighing from 60 to 140 lb, were produced by the first baling machines; and they are still popular on farms and ranches that produce hay for their own use.

Today, hay is also being packaged in large rectangular bales, weighing from 1,000 to 2,000 lb, designed for custom operators or hay growers with large volumes of hay or straw.

Fig. 6-8. Baler making large rectangular bales which weigh up to 2,000 lb. Commercial hay producers like these big rectangular bales because they are easy to transport and store. (Courtesy, Ford New Holland, New Holland, PA)

2. **Large round bales.** Many makes and models of large round balers are on the market. All of them can be classified by the method of rolling the hay into a bale. One method is to pick up the hay from the windrow and roll it in a chamber, between a series of belts or chains. The other method is to roll the windrow on the ground, similar to rolling up a carpet. Both methods produce a rounded shape which gives weather protection. Large round bales range in weight from 850 to 2,000 lb, depending on the make of equipment.

Fig. 6-9. Round baler in operation. (Courtesy, Ford New Holland, New Holland, PA)

If not handled properly, big round bales can be dangerous. Three types of bale handling hazards have surfaced: (a) a downhill roll; (b) tractor mounted front-end loaders carrying such a heavy, bulky load have a tendency to upset if not properly counterbalanced; and (c) a bale can roll out of the loader bucket back down the loader arm onto the operator when a loader handling a round bale is raised to full height.

STACKS

These are loaf-shaped (one system makes a circular stack), mechanically pressed haystacks. Long, loose hay is blown into a wagon and pressed down by a hydraulically operated canopy roof. Stacks range in size from 7 to 10 ft wide, 8 to 22 ft long, and 8 to 11 ft high, and weigh from 1 to 6 tons.

Fig. 6-10. Loaflike stacks of hay. (Courtesy, Hesston Corp., Hesston, KS)

The stack system saves labor and permits nearly the same latitudes as loose hay, with the efficiencies of mechanization.

Limitations of the system are (1) Investment costs are high, (2) heavy rain or snow may seep down through the stacks if they are not formed properly, and (3) they are not efficiently transported over long distances. Nevertheless, the principles involved are good; hence, some of the problems will likely be overcome with more experience.

CUBES (WAFERS)

Field cubers are machines that move across hayfields, pick up windrows of forage, and produce dense, high-quality forage cubes or wafers. Stationary cubers are used to produce similar cubes from loose haystacks or bales.

The benefits derived from field cubing of hay are great. It (1) simplifies haymaking, (2) lessens transportation and storage space—cubed or wafered forages weigh between 45 and 55 lb per cu ft (baled hay density is 8 to 15 lb), (3) reduces labor,

Fig. 6-11. Hay cuber in operation, making cubes 1¼ in. square and 2 to 3 in. long. (Courtesy, Deere & Company, Moline, IL)

(4) makes automatic hay feeding possible, (5) decreases nutrient losses, (6) eliminates dust, and (7) makes for increased feed comsumption, gains, and feed efficiency. Also, with cubing, the spread between high- and low-quality roughage is narrowed; that is, within reason the poorer the quality of the roughage, the greater the advantage from cubing or wafering. The latter is so because such preparation assures complete consumption of the roughage.

Hay cubes are of special interest to dairy producers because they have the advantages of pellets, without their disadvantages. Like finely ground forage that is pelleted, cube-feeding can be readily automated; and, in comparison with long hay, there is less transportation and storage cost. Besides, cubes will not lower the fat content of the milk as much as pellets when appreciable quantities of them are fed.

PELLETS

Pelleted forages are finely ground, then condensed. The **advantages** of pellets are:

1. Pelleted feeds are less bulky than any other hay package (pelleted roughage requires one-fifth to one-third as much space as is required by the same roughage in loose or chopped form), and are easier to store and handle—thus lessening transportation, building, and labor costs. For these reasons, it is particularly advantageous to use pelleted feeds where storage space is limited and feed must be transported considerable distances, conditions which frequently characterize small enterprises.
2. Pelleting prevents animals from selectively refusing ingredients likely to be high in certain dietary essentials; each bite is a balanced feed.
3. Pelleting practically eliminates wastage. Since animals frequently waste up to 20% of long hay, less pelleted feed is required. Wastage of conventional feed is highest where low-quality hay is fed and/or feed containers are poorly designed.
4. Pelleting eliminates dustiness and lessens heaves.

The biggest deterrent to increased pelleting at the present time is the difficulty of processing chopped forage coarse enough so that it will not cause digestive disturbances. A minimum of a ¼-in. chop is recommended.

ARTIFICIAL DRYING

The use of forced air, either heated or unheated, for final drying is the most dependable way in which to preserve quality in hay. The application of heat provides faster drying, saves more leaves, and reduces losses of nutrients. Yet, artificial drying is on the decline in the United States because of the added cost in heating the air and the increased labor.

Hay-drying equipment permits handling while hay is green and tough enough to withstand mechanized processes without excessive leaf loss, and it minimizes weather damage. The most common methods of artificial drying are (1) mow curing, (2) artificial dehydrators, and (3) wagon driers.

• **Mow curing**—Mow curing, or drying, refers to the practice of curing partially dried hay—either long, chopped, or baled—in barn mows equipped with ventilation systems through which either unheated or heated air is forced.

Because of the relatively high cost involved in mow curing hay, in areas where poor haymaking weather generally prevails and field curing is hazardous, farmers may well consider the desirability of making silage.

• **Artificial dehydrators**—Artificial dehydrating refers to a process in which forage is taken from the field as soon as it is cut (or in some instances after wilting), put through a hay chopper or silage cutter, and dried in large driers of various types, following which it is finely ground. For the most part, this method of curing is limited to large commercial operations which process early cut alfalfa (or its leaves) and other legume and/or grass crops chiefly as a supplement for swine and poultry. Occasionally, artificially dehydrated hay is produced for other classes of animals, especially in those areas which rarely have good haymaking weather.

The producer is well justified in paying a premium price for high-quality dehydrated forage for use in swine and poultry rations. Except in special circumstances or as a partial grain replacement, however, it is seldom economical to feed artificially dehydrated hay to cattle or sheep. For ruminants, it is generally more practical to make silage or to resort to mow drying in those areas commonly having poor haying weather. This is especially true where the forage is to be fed on the farm on which it is produced.

• **Wagon dryers**—Wagon dryers were developed to reduce the high labor requirements of batch or platform drying. Essentially, they are batch dryers on wheels.

Some wagons are covered; they are known as *covered wagons* because of the outward appearance of the ballooned cover on the wagons during drying. The cover consists of a durable, lightweight material that won't leak water or air. Heat is conveyed under the cover. Other wagons are open and adapted to drying in a shed. In both types, the drying wagons are loaded in the field directly behind the baler, taken to the drying system where they are connected to a main air duct, and the hay is dried. After drying, they are taken to the storage place and unloaded. This procedure eliminates the unloading and reloading of wagons required in batch or platform drying.

Drying costs money, both for equipment and heat. Of course, the cost will vary according to (1) the initial moisture content of the forage, (2) the cost of electricity, gas, or fuel oil, and (3) how much the temperature of the outside air must be raised to bring it to at least 140°F. In any case, it will pay to let the sun do as much of the drying as practical without risking weather damage or shattering.

Although good-quality hay can be produced by wagon drying, it requires more labor and handling than most other systems of haymaking.

STORING HAY

Fig. 6–12. Hay stored in a pole-type hay shed. (Courtesy, Union Pacific Railroad, Co., Omaha, NE)

Good hay should never be poorly stored. Naturally, the type of storage will vary from area to area. In the more arid sections where little rainfall comes during the fall and early winter, a good stack of loose or baled hay may provide entirely satisfactory storage. On the other hand, in high-rainfall areas, more expensive waterproof storage should be provided. At and between these two extremes, hay may be and is successfully stored in many different ways in different sections of the country.

In the West, a considerable amount of hay is chopped (either at the time of gathering from the windrow or adjacent to the stack) and stack stored. Most such stacks are round, and are built by sliding a snow fence toward the top as the stack is built. Generally these stacks are rounded off at the top and left uncovered. The advantages claimed for stack storage of chopped hay are (1) minimum labor in haymaking, (2) minimum stack storage space and spoilage, and (3) ease of feeding.

Where different kinds and qualities of hay are produced or purchased, each kind and each quality should be stored in such manner that it will be accessible when needed. Otherwise, it may not be convenient to provide for variety in feeding and to feed some of the low-quality along with some of the high-quality hay.

• **Round bale losses/storage**—Prior to 1970, most hay was packaged in rectangular bales, and in humid areas it was stored inside. With the advent of large round bales weighing 850 to 2,000 lb, most of them were left uncovered in the field or in a fence row.

The nutritional losses from bales stored outside and unprotected depend on the storage area (it should be high and dry), the amount of rainfall during the storage period, the hay type and condition when baled, the bale shape and density, and the method of feeding; and the monetary value of the losses depends on the price of hay and the nutritional loss. Legume hays do not form as tight a thatch as grass hays; consequently, they are not as weather-resistant.

Table 6–8 shows the range of dry matter losses that can be expected in large round bales stored outside and unprotected. The lower value of each range represents well-formed bales located in areas with low rainfall (less than 25 in. per year) and low relative humidity. The higher values are for areas with high rainfall (greater than 40 in. per year) and high relative humidity.

TABLE 6–8
DRY MATTER LOSS IN LARGE ROUND BALES STORED OUTSIDE AND UNPROTECTED

Forage	Storage Period	
	Up To 9 Months	12 To 18 Months
	(% loss)	(% loss)
Alfalfa	6–24	16–50
Grass hay	4– 8	10–14
Sorghum X Sudan	6–12	10–24
Cereal hay	10–24	20–40

Not all the deterioration that occurs is the result of rain falling on the bale. Moisture movement at the bottom of the bale, where it contacts the ground, can also cause considerable loss. Dry matter losses can be reduced by as much as 10% if the bales are stored on high ground on a well-drained site; set on racks or pallets, fence posts, railroad ties, or a 3-in. base of rock; and spaced 12 to 28 in. between bales.

Large round bales stored outside and covered with plastic or canvas bonnets (or caps), sleeves, or bale bags sustain much less loss than unprotected bales.

Also, hay losses vary with the method of feeding. When animals have free access to hay, they trample and stand on it, with the result that losses may be as high as 40 to 50%. By using a barrier (a feeding rack, panels, feeding wagons, or gates), losses can be cut to 5 to 10%.

Fig. 6-13. Round bales can be fed with a minimum of waste if proper equipment is used such as this round bale feeder for sheep. (Courtesy, Shalom Valley Sheep Equipment, Inc., Gary, SD)

ADDITIVES FOR HAY

Farmers in many countries of the world have traditionally added about 20 lb of salt per ton of new hay at the time of stacking or putting it into the hay mow in the belief that the salt would prevent the hay from molding and heating. Carefully controlled experiments have failed to substantiate claims that salt will prevent excess heating or sweating; nor has it prevented spontaneous combustion of hay. However, when salt is used in moderate amounts, it may improve the color, aroma, and palatability of poor-quality hay. It is recognized, too, that much higher levels of salt—quantities sufficiently high to harm animals—may prevent mold.

Over the years a number of products, both liquids and powders, said to preserve hay have become commercially available. While claims have been made that there will be no heating or molding when these materials are used, the results have been highly variable.

Recent studies of several state experiment stations have shown that propionic acid or a combination of propionic and other organic acids can be used successfully to preserve baled or stacked hay stored at 35% or less moisture. Anhydrous ammonia and ammonium isobutyrate have also been found to be effective in preventing heating and preserving the quality of high-moisture hay.

SPONTANEOUS COMBUSTION

Wet hay ferments and generates heat. Sometimes this results in spontaneous combustion and fire, usually about a month to 6 weeks after storing. Here are the facts:

1. **Symptoms of heating.** The warning signals are hay that feels hot to the hands, strong burning odor, and visible vapor.

2. **Temperature of hot hay.** Hot spots may be located by probing the hay with a steel rod. Then the temperature of the hot spots may be tested with a thermometer (a dairy thermometer or other type) attached to a wire and dropped down a pipe. If the hay is over 140°F, it should be checked periodically during the day. If the hay is 160°F, it should be checked hourly. If the hay is 180°F, there are apt to be fire pockets, and it should be removed from a barn.

3. **Cooling hay.** Hay that is heating may be cooled by discharging through pipes into the hot areas either dry ice or liquid carbon dioxide.

4. **Removing hot hay.** When a fire is imminent and hot hay must be removed, it is important to have plenty of help on hand including the fire department. Then the hay should be removed cautiously and without wetting unless necessary.

5. **Precautions.** Never walk on hay that is heating—place planks over it, and do not breathe hot and noxious fumes.

BUYING AND SELLING HAY

Historically, most hay has been fed on the farms where it was produced. But this practice is changing. Today, about 29 million tons, or about 25% of the U.S. production, with a cash value of over $2 billion, is sold off the farm. The Northwestern states of Washington, Idaho, and Oregon, along with California and Texas, are the leading states in total value of hay sold. In California alone, the annual value of hay sold exceeded $425 million in the late 1980s. In four states—California, Arizona, New Mexico, and Washington— some 40% or more of the hay produced is sold. In California, nearly 70% of the hay produced is sold.

New hay markets are developing. As dairy producers, beef cow-calf producers, cattle feeders, and sheep feeders become more specialized, they prefer to grow animals and rely on other specialists to grow hay. Also, more high-quality hay is needed for expanding horse numbers. Mushroom growers are becoming an important market for hay, too; they require a high-energy hay, although it can be moldy.

Additionally, considerable hay is exported, primarily to Japan, Korea, and Taiwan. In order to save space and cut down transportation costs, this generally requires a special hay package—pellets, wafers, or high-density bales.

HOW HAY IS SOLD

New methods of marketing hay have made selling easier, but visual inspection is still the most common method of assessing quality and price when either selling or buying. However, the development of Near Infrared Reflectance Spectroscopy (NIRS) as a rapid, accurate, and precise method of measuring hay quality is having a major impact on the marketing of hay.

Fig. 6-14. Marketing hay is big business. (Courtesy, Ford New Holland, New Holland, PA)

The traditional sources of hay sold in the United States are as follows:

- **Hay dealers or brokers**—Hay dealers or brokers are important suppliers of hay, as evidenced by the continued growth of the National Hay Association. In California, a high percentage of the hay is marketed through this channel. For the nation as a whole, dealers and brokers market 5 to 10% of the hay produced. They purchase hay from growers and sell it to consumers.

- **Neighbor to neighbor**—This is the oldest market channel for hay, and it is still quite common. This is a disorganized market without any particular pricing structure. Hay is purchased on visual inspection.

- **Associations and cooperatives**—These organizations vary in size, from very small to very large. The San Joaquin Hay Producers Association in California is one of the largest. Such associations normally purchase hay for cash and store it or move it to the consumer. This is considered a high-risk operation, because there are no futures markets that enable hedging protection.

- **Contract**—This is an agreement between a hay grower and a livestock producer to supply hay of a specified quality at a prior agreed-upon price. It assures both the buyer and the seller an orderly market. Such eventualities as weather damaged hay should be covered in the contract.

- **Auctions**—This is the most rapidly growing method of marketing hay. Originally, auctions simply brought together producers with loads of hay, an auctioneer, and an assembly of prospective buyers. Hay was sold by visual inspection, and the price determined by supply and demand. This approach is still widely used in many areas. Today, *Quality-Tested Hay Auctions* are becoming popular in some states, including Wisconsin and Minnesota.

- **Quality-tested hay auctions**—In Wisconsin and Minnesota, two major hay producing and consuming states, the number of hay auctions has increased markedly in recent years. This is attributed to the introduction of *Near Infrared Reflectance Spectroscopy (NIRS)* equipment, usually in mobile units, as a rapid method of measuring hay quality prior to sale, plus the grading and selling of hay according to test results. These special auctions are known as *Quality-Tested Hay Auctions.*

Fig. 6-15. Automatic bale wagon stacking a load. This type of equipment facilitates hay auctions. (Courtesy, Ford New Holland, New Holland, PA)

The hay grading system used in Wisconsin and Minnesota is adapted from standards proposed by the American Forage and Grassland Council (AFGC) (see Table 6-9).

TABLE 6-9
FORAGE QUALITY STANDARDS FOR LEGUMES, GRASSES, AND LEGUME-GRASS MIXTURES[1]

Quality Standard	CP	ADF[2]	NDF[2]	DDM[3]	DMI[4]	RFV Index[5]
	(% of DM)					
Prime	>19	<31	<40	>65	>3.0	>151
1	17–19	31–35	40–46	62–65	3.0–2.6	151–125
2	14–16	36–40	47–53	58–61	2.5–2.3	124–103
3	11–13	41–42	54–60	56–57	2.2–2.0	102–87
4	8–10	43–45	61–65	53–55	1.9–1.8	86–75
5	<8	>45	>65	<53	<1.8	<75

[1]Standard assigned by Hay Marketing Task Force of AFGC.

[2]ADF = acid detergent fiber/NDF = neutral detergent fiber.

[3]Digestible dry matter (DDM,%) = 88.9 – (.779 × ADF%).

[4]Dry matter intake (DMI, % of body weight) = 120 ÷ forage NDF (% of DM).

[5]Relative feed value (RFV) calculated from DDM × DMI ÷ 1.29. Reference RFV of 100 = 41% ADF and 53% NDF.

Brief descriptions of the various factors included in the standards are as follows:

- **Acid Detergent Fiber (ADF)**—Both animal and laboratory trials indicate that ADF is highly related to the digestibility of a forage. Factors which increase ADF content, such as increasing maturity, weathering, rain damage, and weeds, decrease digestibility.

• **Neutral Detergent Fiber (NDF)**—Studies indicate that NDF is highly correlated with dry matter intake of the forage.

• **Digestible Dry Matter (DDM)**—The accepted equation for predicting DDM of legumes, grasses, and legume-grass mixtures from ADF is DDM % = 88.9 − (.779 × ADF %).

• **Dry Matter Intake (DMI)**—The amount of forage or feed DM an animal will consume is affected by how fast forages are digested and pass through the intestinal tract. The fiber fraction which appears to be most clearly related to the DMI of forages is Neutral Detergent Fiber (NDF). However, the exact NDF level in rations necessary to achieve optimum performance is uncertain. Wisconsin research indicates maximum feed intake in alfalfa-based dairy rations occurs when NDF is 1.2 lb per 100 lb of body weight.

$$\text{DMI (\% of body weight)} = \frac{120}{\text{Forage NDF (\% of DM)}}$$

• **Relative Feed Value (RFV)**—Relative feed value is an index which combines important nutritional factors (potential intake and digestibility) into one number for a quick, easy and effective method of evaluating feeding value or quality. The formula for calculating RFV is the estimated digestibility and potential intake of a forage calculated from ADF and NDF factions, respectively.

The calculation of RFV is made by multiplying DDM by DMI, then dividing by 1.29. The number derived from the RFV calculation has no units and is used only as an index for evaluating quality of hay or haylage made from legumes, grass, or legume-grass mixtures. The RFV concept should be used to evaluate quality only for those forages listed above.

$$\text{RFV} = \frac{\text{DDM} \times \text{DMI}}{1.29}$$

The RFV does not include crude protein (CP) because CP is influenced by factors unrelated to those affecting RFV. The CP should be considered, however, in pricing forages; and CP values are included in Table 6–9 to indicate the range of CP for forages of different qualities.

• **Other marketing approaches**—In several states, such as Oklahoma and Indiana, special computer assisted marketing systems, designed to help growers find buyers and to help buyers locate hay, have been developed. There is also a privately owned National Hay Exchange Corporation, which, by means of a computerized system, allows people to trade hay much easier, and in much less time, by matching buyer needs with seller offerings.

• **Improved hay marketing needed**—At today's feed prices, hay (especially alfalfa) is a profitable crop to grow, both for direct utilization through livestock and as a cash crop, *provided* it is of high quality and priced at its true value for feed. However, there is general agreement that if hay is to take its rightful place in the marketplace as a major source of protein and energy, many of the traditional selling arts must be replaced by developments of modern science.

The introduction of Near Infrared Reflectance Analysis and the development of standards and grades based on feeding value have alleviated some of the excuses for not marketing hay on the basis of analysis. Also, sampling procedures have been improved. So, systems of pricing hay consistent with the quality of the product sold are close at hand.

But other developments are necessary if hay is to become a leading cash crop, including:

1. **More markets.** There is need for more markets—such as auctions, dealers, and associations—to provide a ready market outlet for hay growers, and to attract both ample supplies of hay and buyers.
2. **Price protection for dealers while they have possession of the hay.** There is need for something similar to the futures market that will permit hedging and protection.
3. **More hay grower storage facilities.** There is need for additional grower storage so as to spread hay marketing over a longer period and avoid harvest market gluts that depress prices.
4. **Better hay market information.** There is need for hay growers and hay buyers to be better informed relative to supply, demand, and going prices.

FREEDOM FROM TOXIC RESIDUES

With the emphasis on residues in foods, it is important that hay be free from those residues which are prohibited. If meat or milk (or products derived from them) are found to have residues, the blame cannot be shifted to the hay grower by the livestock producer, unless there is a clear-cut case of fraudulent representation. The best assurance of freedom from such residue rests with the integrity of the hay growers, or those who represent them in selling their hay.

It is important that livestock producers and hay growers be well informed relative to (1) the chemicals which are banned, and (2) the conditions of application for those chemicals which are still permitted. In this way, disastrous financial effects from confiscation of market animals or products can be averted. So, *read the label.* Use proper pesticides and observe minimum days from last application to harvest. Wear protective rubber gloves, respirators, and coveralls when applying pesticides, especially those that are highly toxic. Wash before eating.

HAY SHRINKAGE

Hay buyers should figure hay shrinkage closely. Here's why: If a ton of hay containing 90% dry matter is bought for $75, 1,800 lb of dry matter have been purchased at this price. However, if the $75 per ton hay contains 80% dry matter, only 1,600 lb of dry matter have been purchased for this same price. Purchase of the high-moisture, 80% dry matter hay has resulted in a loss of 200 lb of hay, or 1/9 of the dry matter, worth $8.33 (1/9 × $75). Thus, if the 90% dry matter hay is worth $75 per ton, then the 80% dry matter hay is worth only $66.67 ($75 – $8.33) per ton. If 1,000 tons of hay are involved, that's a loss of $8,330 (1,000 × $8.33).

In addition to moisture losses, newly harvested hay may be expected to lose about 5% weight from going through the sweat.

HAY FEEDING FUNDAMENTALS

Fig. 6–16. Winter feed—*hay*. (Courtesy, Ford New Holland, New Holland, PA)

Feeding is the end of the line for hay. No matter how carefully it has been grown, harvested, and stored, all that has gone before can be dissipated if it is improperly fed—unless hay feeding fundamentals are observed.

Monogastric animals, including swine and poultry, must eat a large percentage of grains and other concentrates and depend almost entirely on digestive enzymes to break down these compounds. But ruminants, with their four stomach compartments and the help of microorganisms, can subsist largely, or entirely, on bulky, high-fiber forages which, because of their low energy per unit weight of dry matter, must be consumed in large quantities to supply their nutrient needs. The horse, because of its greatly enlarged cecum and large intestine, can utilize quantities of hay intermediate between simple-stomached and ruminant animals.

The economics of the situation—the relative price of forage and grain—call for greater emphasis on forage accompanied by less grain feeding. With greater quantities of forage incorporated in rations, it is expected that performance—the production of meat and milk—will decrease. However, maximum net returns, rather than just maximum production, will be the primary objective.

Increasingly, forage testing will be used in two ways: (1) to purchase hay on a quality basis, and (2) to balance rations more precisely.

HAY PREPARATION

Hay is fed as long hay or in processed form. The common methods of processing are chopping, grinding, cubing, and pelleting.

Fig. 6–17. Hay prepared in cubes, which (1) facilitate automation in both haymaking and feeding, and (2) do not affect milk fat materially. (Courtesy, Union Pacific Railroad Company, Omaha, NE)

Considerable hay is chopped in the West, for two reasons: (1) It facilitates handling, and (2) it lessens refusal and waste. Low-quality and coarse forages usually benefit more from chopping than high-quality forages.

Hay is usually finely ground when it is incorporated in mixed swine and poultry rations. Fine grinding is not desirable for ruminants; it results in reduced rumen acetate production and lower milk fat percentage.

Both cubing and pelleting (1) make automatic hay feeding feasible, (2) decrease nutrient losses, and (3) eliminate dust. Also, they narrow the spread between high- and low-quality forage; that is, the poorer the quality of the forage, the greater the advantage from cubing or pelleting. This is so because such preparation assures complete consumption. Also, cubing or pelleting, especially the latter, usually speeds up the passage of forage through the digestive system.

On the average, cattle fed high-roughage (above 80% roughage) or all-roughage rations will eat about 1/3 more pellets

pellets than long or chopped hay (due to increased density and more rapid passage through the digestive tract), make about ½ to ¾ lb faster daily gains, and require 200 to 250 lb less feed per 100 lb of gain. Also, it is recognized that the utilization of low-quality roughages is improved most by pelleting.

Cubes offer most of the advantages of pelleted forages, with few of the disadvantages. They alleviate fine grinding, and they facilitate automation in both haymaking and feeding, and they lower milk fat percentage only slightly, if at all.

Complete pelleted rations—in which the hay and grain are combined, then pelleted—are finding an increasing market for horses, and perhaps swine. Among the virtues ascribed to all-pelleted rations are (1) They prevent selective eating—if properly formulated, each mouthful is a balanced diet; (2) they alleviate waste; (3) they eliminate dust (thereby lessening heaves in horses); (4) they lessen labor and equipment; and (5) they lessen storage.

HAY FEEDING SYSTEMS

Hay may either be (1) self-fed, or (2) limit-fed.

Most hay is self-fed. With a manger full of hay in front of them, hay consumption is limited only by the capacity of animals—by the amount that they can hold. That's the reason that ruminants eat more cubes and pellets than long hay of equal quality.

Limited feeding of hay is accomplished either (1) by hand-feeding the hay and the concentrate allowances, or (2) by using a complete, mixed ration.

The vast majority of large cattle and sheep feedlots feed complete rations, in which the quantity of hay is limited. Also, an increasing number of large commercial dairies are switching to complete rations. Most experiments and experiences have not shown any difference between mixed rations and the feeding of roughage and concentrates separately insofar as rate and efficiency of gain are concerned. However, a mixed ration has the following **advantages**:

1. It makes for greater efficiency in feeding and lessens sorting at the feed bunk.

2. When the roughage is relatively unpalatable, a mixed ration forces consumption.

3. When concentrate consumption is to be limited, mixing with the roughage is desirable.

FEEDING HAY PACKAGES

Large round bales, large rectangular bales, and stacks are other alternatives to conventional, 60- to 140-lb, rectangular bales. There is little doubt that in many livestock operations large hay packages can greatly decrease labor and result in similar animal performance. However, special attention needs to be paid to methods of feeding these big hay packages; otherwise, waste can easily wipe out any saving in labor.

Under an in-field storage system, the bales are dropped where they are made, and remain there until needed for fall or winter grazing. Little or no labor is involved in making hay except for mowing, windrowing, and baling; and there is no manure to haul. Cattle graze the grass growth which occurs subsequent to baling and consume the bales in the field. There is little or no labor in feeding. The cattle go to the feed, rather than necessitating that the feed be taken to them.

The following methods are used in grazing round bales or small stacks, along with regrowth in the field:

1. **Continuous access to all bales.** In this system, the cattle are given continuous access to all the bales in a field.

2. **Strip grazing bales and stacks.** In this system, the cattle are given access only to those bales or stacks which are to be consumed within a given period of time. This is accomplished by using an electric fence and cross-stripping the field containing the bales or stacks. Such a strip-grazing program will increase the number of cattle days by at least 35%.

Strip grazing will work for both bales and small haystacks. With stacks, it is especially important to limit-feed or strip-graze in order to avoid excess wastage.

3. **Other feeding methods.** Other systems require more investment than in-field grazing of bales or small stacks, but the additional numbers of cattle carried per acre may justify the increased cost. Among such systems are the following:

a. **Hay packages (large round bales or small stacks) placed in rows and grazed with an electric fence.** Usually, one side of an existing hayfield is used for such a storage and feeding area.

b. **Portable feeding gates (fences).** Portable feeding gates (or fences), usually made of metal, may be placed around large bales or stacks.

c. **Feeding wagons.** There are several designs and sizes of self-feeding wagons, both commercial and home-made. They give much flexibility; wagons can be easily taken to the area where the round bales are stored, then pulled to various feeding locations.

d. **Three-sided feeder.** Bales may be placed on concrete and enclosed in a three-sided feeder.

PROPORTION OF HAY TO CONCENTRATE

Cattle, sheep, and horses will eat 2 to 3 lb of hay per 100 lb of body weight if fed hay alone. Also, it is noteworthy that the higher the quality of the hay, the more of it they will eat, with the result that the grain requirement will be lessened.

The economics of the situation—the comparative price and quality of hay and concentrate—along with the management practices, will determine the proportion of hay to concentrate.

Thus, during the period of low grain prices in relation to forage prices—from about 1950 to 1970—it was desirable to feed finishing cattle high-energy rations and to maximize gains. But the grain-fed cattle binge ended with the world grain shortages and high-priced grains of the early 1970s. In the years ahead, with grain becoming more scarce and higher in price than forages, comparatively speaking, more forage and less grain will be fed to finishing cattle, and net returns will be more important than high rate of gain. Cattle and sheep will increasingly be *roughage burners.* Livestock producers will rely upon the ability of the ruminant to convert coarse forage, grass, and by-product feeds, along with a minimum of concentrate, into palatable and nutritious food for human consumption, thereby competing less for humanly edible grains. Increasingly, the steer and the lamb of tomorrow will be produced on a maximum of milk and grass and a minimum of grain. More and more U.S. cereal grains will be used for human food, just as has been true, historically, in much of the rest of the world.

Ruminants can make the transition to more roughage with ease. For them, it is merely a return to nature, for they evolved as consumers of forage.

The best buy (hay vs grain) may be determined by calculating the cost per pound of TDN and of protein in the hay and grain being compared. Then, the proportion of hay to concentrate can be varied accordingly. If hay is the best nutrient buy, feed more hay and less grain. On the other hand, if grain is the best buy, feed more grain and less hay.

DIFFERENT QUALITIES OF HAY MAY BE USED

The type of ration which will be least costly and result in satisfactory performance will differ according to species, level of performance, reproductive status, age, etc. For example, the nutritive needs of a dry, pregnant beef cow are much lower than those of a high-producing dairy cow. Thus, a low-quality hay may be quite satisfactory for wintering a beef cow without calf at side, whereas high-producing, lactating cows should always have high-quality hay. Also, high-quality hay is important for swine and poultry. Where forage is incorporated in monogastric rations, high-quality dehydrated alfalfa is most commonly used. High-quality hay is also essential for horses.

HAY WASTE AND REFUSAL

In a recent Texas study involving 10 different hay feeding racks, the cows at the best conventional feeder still wasted 14% of their hay. Dairy producers have commonly accepted 10% refusal as normal.

High-priced feeds and smaller margins are causing livestock producers to scrutinize hay losses, and to do something about them. Chopping hay and/or adding molasses will lessen wastage. But feeding high-quality hay is the best way in which to lessen waste and refusal.

SUPPLEMENTING THE HAY RATION

Hay is generally lower in energy and higher in fiber than most grains and concentrates. Legume hays have a high-calcium content, but they vary in available phosphorus. If sun-cured properly, they are high in carotene and vitamin D, along with many of the B vitamins. A supplement should supply the nutrients that are most likely to be lacking in the hay; thus, supplements for alfalfa hay should be high in energy and phosphorus and low in fiber. Also, carotene or vitamin A should be provided if the hay has been bleached or turned brown. Salt is lacking in hays and other natural feedstuffs and should be provided as a supplement, along with trace minerals that are deficient in the local area.

STRETCHING THE HAY SUPPLY

When hay is scarce and high in price, the supply of it for ruminants and horses may be stretched. As the amount of hay fed is reduced, it must be replaced with other feeds so that the total ration is still balanced and fulfills all the nutrient requirements.

Most grains contain 75 to 80% TDN, while most medium- to good-quality hays contain 45 to 50% TDN. Hence, as a general rule of thumb, about 5 lb of grain equal 8 lb of hay, provided they are of comparable quality. Thus, it follows that if corn can be bought for $100.00 per ton, hay should be bought at $62.50 per ton, or less.

If the price of hay is less than five-eighths the price of grain, relatively more of it should be fed; whereas, if the price of hay is higher than this, relatively more grain will make for cheaper production.

When hay is scarce and high in price, the hay supply for ruminants and horses can be stretched as follows:

1. Feed only ½ to ⅔ the normal ration of hay, but be on the alert for digestive disturbances.

2. Replace 1 lb of hay with 3 lb of silage.

3. Replace 1 lb of hay with 4 lb of green chop.

4. Replace each 2 lb of hay deleted with 1 lb of grain.

5. Make the maximum use of such feeds as cottonseed hulls, corncobs, straw, and grass aftermath in the ration for (a) all but 5% of the alfalfa (or other legume) hay of grower rations; and (b) all of the "hottest" finishing ration, adding such supplementary proteins, minerals, and vitamins as necessary to balance the ration.

6. Get finishing cattle and lambs, and animals being fitted for show or sale, on high-concentrate rations as expeditiously as possible.

7. Provide such supplementary proteins, minerals, and vitamins as necessary.

PART II—CROP RESIDUES

CROP RESIDUES

Fig. 6-18. Cows winter grazing cornstalks—the residue remaining in the field after harvesting the corn grain. (Courtesy, Ron Baker, C & B Livestock, Inc., Hermiston, OR)

Most of the by-product feeds result from some sort of industrial manufacturing; for example, the by-products from milling grains. Another excellent source of feed is crop residues—parts of plants that are normally left in the field following harvest of the primary crops.

As production costs increase, livestock producers become more interested in the enormous potential for animal production through feeding crop residues.

KIND AND QUANTITY OF RESIDUE PRODUCED

The quantity of crop residues produced may be estimated by multiplying the annual grain production by a grain weight: residue weight ratio. Normally, grain-producing plants produce as much (or more) weight of vegetative material as of grain. The crops, ratios used for conversion from grain to residue, and estimated annual production of residues are given in Table 6-10.

TABLE 6-10
ESTIMATED SUPPLY OF CROP RESIDUES[1]

Crop Source	U.S. Grain Production	World Grain Production	Canadian Grain Production	Ratio Residue/Grain[2]	U.S. Residue	Residue % of Total
	(mil. metric tons)	(mil. metric tons)	(mil. metric tons)		(mil. metric tons)	(%)
Barley	13.3	182.0	14.6	2.0	26.6	6.53
Corn	209.6	476.6	5.9	1.0	209.6	51.48
Cottonseed	4.79	30.63	—	3.0	14.37	3.53
Flax	0.21	2.36	0.90	3.0	0.63	0.15
Oats	5.6	47.5	3.3	1.0	5.6	1.37
Peanuts	1.87	19.99	—	1.5	2.8	0.69
Rice, rough	6.0	466.9	—	1.0	6.0	1.47
Rye	0.5	31.0	0.6	1.0	0.5	0.12
Sorghum	23.8	64.3	—	1.0	23.8	5.84
Soybeans	57.11	97.03	1.01	1.0	57.11	14.03
Sugarbeets[3]	22.9	285.7	0.94	0.14	3.21	0.79
Wheat	56.9	529.7	31.4	1.0	56.9	13.98
Total	402.58	2,233.71	58.65		407.12	

[1]U.S., World, and Canadian grain production from: *World Agricultural Production,* USDA, Foreign Agricultural Service, Circular Series WAP 5-88, May 1988, except for sugarbeets.

[2]Ratio residue/grain from: *Underutilized Resources as Animal Feedstuffs,* National Research Council, National Academy Press, 1983, p. 180, Table 48.

[3]Sugarbeets production from: *1986 FAO Production Yearbook,* Vol. 40, p. 163.

As shown in Table 6–10, corn, wheat, and soybeans account for nearly 80% of the total residues. It follows that the major U.S. corn-, wheat-, and soybean-producing states produce the most residues.

Corn is the most widely produced grain crop in the United States. It usually produces an amount of residue equal to the quantity of grain produced. So, as shown in Table 6–10, the production of 209 million tons of corn in 1986 resulted in 209 million tons of corn residue, which was over ½ the total available crop residue that year.

Wheat, which produces much less tonnage of grain per acre than corn, accounts for about 14% of the residue. About 57 million tons of wheat straw were produced in 1986 (see Table 6–10).

Soybean residue usually provides another 14% of the crop residue (57 million metric tons in 1986) and grain sorghum 6% (23.8 million metric tons in 1986). Other crops account for the remaining residue (see Table 6–10).

More than 400 million metric tons of straws, stalks, and stubble are available in the United States each year (see Table 6–10) and another 60 million tons in Canada. Worldwide, more than 2.2 billion tons of crop residues are produced annually.

In addition to being used as livestock feeds, crop residues may be, and are, used for bedding, soil improvement, and as a substitute for fossil fuels.

ECONOMY OF CROP RESIDUES

Evaluating the nutrient content of crop residues, along with collecting, storing, treating, transporting, and feeding them, is much more difficult than determining the quantities available. Also, when evaluating crop residues for animal feeding, three major questions must be answered: (1) are they available in sufficient quantity to make their use as a feedstuff worthwhile, (2) do they have high enough nutrient content to justify feeding them to livestock, and (3) are they cost competitive? These and other important *field to feed* aspects of crop residues are discussed in the sections that follow.

HARVESTING

Because of differences in plant structure, grain harvesting methods, and moisture content, harvesting crop residues may not be easy. Straws from cereal grains are easily collected in dry state behind the combine. Corn and sorghum stovers often are too wet for dry storage, but they can be stored as silage. Soybean residue is difficult to harvest if allowed to drop on the ground behind the combine. Soil contamination during harvest may be a problem with all residues. Some of the residues, such as cottonseed hulls, rice milling by-products, and sugarcane bagasse, which are processed at central locations, have the advantage of being collected and available for treatment.

NUTRIENT VALUE OF CROP RESIDUES

Almost all crop residues are harvested after the plants reach physiological maturity; so, they are high in cell walls and lignin and low in protein and digestible dry matter. They are a feed energy source, but suitable only for ruminants—beef and dairy cattle, sheep, and goats.

Most of the energy in crop residues is in the lignocellulose complex (lignin, cellulose, and hemicellulose). The amount of the total lignocellulose and of its constituents varies widely with forage species and stage of maturity. Generally, the higher the lignin content the lower the digestibility of the cellulose material. The chemical and physical binding of cellulose and hemicellulose in the cell wall of plants is important; so, the amount of lignin *per se* is not always a good indicator of digestibility. Nutritionally, the lignocellulose complex consists of these fractions: (1) lignin, which is unavailable as an energy source, (2) a digestible energy source that can be utilized by rumen microorganisms, and (3) a fraction which is very resistant to bacterial action in the rumen, but which becomes an energy source after special treatment. The third fraction is of major interest because of the potential additional energy which can be made available in the cellulosic crop residues. In many of the crop residues, this third fraction is large enough that considerable research has been done on treatment to "unlock" its energy.

In addition to lack of available energy, crop residues may be deficient in protein, phosphorus and possibly other minerals, and vitamin A. Also, they are usually bulky and lacking in palatability. So, it is important to provide proper supplementation based on the performance expected of animals grazing or being fed crop residues.

TREATING CROP RESIDUES TO INCREASE DIGESTIBILITY

Crop residues are inefficiently utilized by animals because of the high content and poor digestibility of the fibrous fraction. This poor digestibility is related to the extent of lignification of the cell wall component of these low-quality forages. Although crop residues provide a satisfactory ration for dry gestating animals, they do not provide sufficient energy for either young or lactating ruminants—they simply cannot hold enough of these low-quality roughages to provide adequate energy. This prompts interest in increasing the digestibility of these crop residues.

Calcium hydroxide and ammonia treatments offer the greatest long-term potential for increasing the digestibility of crop residues. Also, several other treatments are being studied.

The potential of crop residue treatments becomes apparent when it is realized that straw, for example, is only 30 to 40% digestible before treatment. When pressure heated with water, it becomes 50 to 60% digestible; and digestibility increases to 70 to 80% when sodium hydroxide is added prior to cooking. By treating corn husklage and milo residue, workers at the Nebraska Station were able to increase the energy value of these residues to 90% that of corn silage.

Lowering the cost of treating crop residues to increase digestibility is the primary area which must be researched before these procedures can be applied to practical operations.

FEEDING SYSTEMS FOR CROP RESIDUES

Generally speaking, crop residues may be grazed, processed as dry feed, or made into silage. The important thing to remember is that their relatively low value, in comparison with grains, necessitates low-cost harvesting, storing, and feeding. Also, they must be fed to the right class of animals, and they must be properly supplemented. The use of low quality crop residues is restricted primarily to ruminants, such as wintering beef cows.

SOME CROP RESIDUES

A brief summary relative to several crop residues follows.

• **Corn residues**—Of all crop residues, the residue of corn is produced in greatest abundance and offers the greatest potential for expansion in ruminant numbers.

In 1986, 69,189,000 acres of corn, yielding 119.3 bu per acre, were harvested for grain in this country. For the most part, over and above the grain, about 2¾ tons of dry matter per acre were left to rot in the field. That's 190 million tons of potential feed wasted, enough to winter 96 million dry pregnant cows consuming an average of 33 lb of corn refuse per head daily during a 4–month period. Also, there are many other crop residues, which, if properly ultilized, could increase the 96 million figure given above. Since goats and sheep require even less feed per day, the potential for increasing animal numbers is even greater!

Although corn refuse offers tremendous potential as a feed for ruminants, there are difficulties in harvesting and storing it. But science and technology have teamed up and are working ceaselessly away at solving these problems.

Broadly speaking, three alternate methods of salvaging corn refuse are being used: (1) grazing, (2) harvesting and dry feeding, and (3) ensiling; with different ways of accomplishing each. The choice of the method should be determined primarily by cost, the proportion of refuse utilized, and how well it meshes with other farm enterprises—for example, in some cases the need for fall plowing will necessitate removal of the material from the land and eliminate grazing as an alternative.

Methods of utilizing corn refuse follow:

1. **Grazing.** This refers to turning the animals directly into the stalk field—the traditional way of utilizing cornstalks. Letting the animals do their own harvesting is the simplest and least expensive method devised for utilizing a crop. However, there is considerable wastage, and it is not possible to prolong the winter feeding period. In an open fall and winter, 2 acres of cornstalks will carry a pregnant cow for 100 to 120 days.

2. **Stalklage.** Stalklage refers to all the residue remaining after harvesting corn with a combine or picker. It may either be stored dry, or ensiled.

a. **Dry stalklage.** Stalklage is more difficult to collect than husklage, and more expensive, since it involves more equipment and another trip across the field. A number of different machines for harvesting stalklage are being used; among them, forage harvesters, balers, stackers with flail pickups, and choppers and stackers. By operating the machine a few inches above the ground to prevent excess soil pickup, a yield of 1 to 3 tons of residue per acre may be obtained, with the moisture content ranging from 20 to 55%, depending on the time of harvest. Stacked or baled cornstalks should be at the low end of this moisture range (20 to 35%) to reduce heating and spoilage.

Cows like dry stover. Self-feeders around a stack make feeding convenient. Leftover material may be used as bedding.

Fig. 6–19. The feeding of crop residues is becoming an important facet of livestock feeding. Here cornstalks are being harvested for feed. (Courtesy, Gehl Company, West Bend, WI)

b. **Stalklage ensilage (stover silage).** Stalklage may also be ensiled, producing a product known as corn stover silage or cornstalk silage. When this is done, the use of a forage harvester equipped with a screen or a recutter-blower at the silo is necessary in order to chop the material finely. Fine chopping will ensure good packing and improve consumption by avoiding selectivity. Where corn stover silage is to be made, the residue should be harvested as soon as possible after the grain is taken off, before it loses any moisture. At that time, the grain moisture will generally be under 30% and the refuse will have about twice the moisture content of the shelled grain. In an airtight silo, 40 to 45% moisture will suffice. In an unsealed or bunker silo, the moisture content should be 48 to 55% for proper lactic acid formation. Water may be added at the silo if necessary. As a precaution, some authorities recommend the addition of 56 lb of corn meal (or other finely ground grain) per ton of corn stover silage, as a means of providing carbohydrates from which acids will form and act as a preservative.

The biggest deterrent to harvesting stalklage, in either dry or ensiled form, is the cost—primarily for the equipment. Rather than own such expensive equipment, which is only used for a short period, custom harvesting of stalklage is likely cheapest for most operators. Although custom harvesting may be cheaper, the timeliness of harvest is important for quality and palatability. If too much time elapses before the stalklage is harvested, husks and leaves may be blown away, thereby lowering quality.

3. **Husklage (shucklage).** *Husklage is the forage discharged from the rear of a combine when harvesting corn.* It consists of the husks, cobs, and any grain carried over the combine, collected in a wagon or straw buncher pulled behind the combine. This operation minimizes labor and does not slow the grain harvest, because the husklage piles can be dumped at the end of the field for supplemental feeding or later pickup by a front-end loader and moved to another location for stacking or ensiling. The moisture content of this material will usually run between 30 and 40%, and the yields will be between 1 and 1½ tons per acre.

The greatest difficulty encountered in feeding husklage dumps at the end of the field is waste. Depending on weather conditions, as much as 50% of the material may be wasted. But wastage of husklage dumps can be materially lessened by controlling access to them.

Stacking of husklage has been satisfactory for some producers.

Ensiling husklage, along with recutting and adding water, results in increased cow consumption and less rejection of cobs.

• **Cottonseed hulls**—Cottonseed hulls are one of the most important roughages in the South, especially for cattle. On an as-fed basis, they supply about 42% TDN, which is about as much as is furnished by late-cut grass hay or by oat straw. They are low in protein (3.8%)—and practically none of it is digestible—low in calcium (0.13%), very low in phosphorus (0.09%), and lacking in carotene. To correct these deficiencies when fed to dry pregnant cows, hulls should be supplemented with a daily allowance of either (1) 6 lb of a good-quality legume hay, or (2) 2 lb of a 30 to 40% protein supplement, along with free access to a complete mineral, high in phosphorus unless a phosphorus-rich supplement such as cottonseed meal is fed. If no legume is fed, vitamin A should be fed or injected.

Cottonseed hulls can be fed without further processing—there is no chopping; and they are well liked by cattle, even when fed as the only roughage. In trials with lactating dairy cows, workers at the University of Arizona substituted pelleted or nonpelleted cottonseed hulls for 10, 30, and 50% alfalfa hay cubes in a ration containing 50% alfalfa hay cubes and 50% high-energy concentrate, with no differences due to ration in total milk production or percent fat.

Pelleted hulls are now on the market. In comparison with regular hulls, they are more digestible, require less transportation and storage space—because of their high density—and are easier to handle.

• **Pineapple cannery by-product, or pineapple bran**—When pineapples are canned, the outer portions and the core are discarded as wastes. These cannery by-products can be fed fresh, dried, or ensiled. When dried, it is commonly called *pineapple bran.* The dried product is generally supplemented with about 9% molasses to increase the energy content and enhance palatability.

• **Sorghum residues**—Ruminants will make good use of sorghum stover as a winter feed. It can be grazed or harvested and stored either as dry feed or silage. The sorghum plant stays green late in the fall; hence, good sorghum stover silage can be made without additional water. In comparison with corn residue, sorghum residue (1) is less palatable (if given a choice, cows will select corn refuse in preference to sorghum refuse); (2) comprises a lower percentage of the total plant dry matter than corn (40% of the total plant dry matter of sorghum is residue compared with 40 to 50% for corn); and (3) is lower yielding.

After harvesting, sorghum will send up new shoots if moisture permits. The prussic acid (hydrocyanic acid) content of these shoots may be harmful to grazing animals; hence, animal caretakers should be aware of this possible poisoning. These shoots can be grazed safely 4 to 6 days after a killing frost.

• **Soybean residues**—The stems and pods of soybean refuse available for feeding yield approximately ¼ ton per acre, with a ratio of stems to pods of about 2:1. The digestibility of stems is low—25 to 35%—due to their high-lignin content (18 to 20% for the stalk portion). The digestibility of pods is much higher, ranging from 58 to 63%. Soybean refuse should be used only as a filler or a high-quality feed stretcher. Cows consuming soybean refuse as a primary energy source have been shown to lose excessive weight.

• **Straws and tailings**—These are the chaff and grain behind the combine. In the days of binders and threshing machines, straw stacks, used extensively for winter cattle feed, were commonplace. With the advent of combines, much of the straw was left in the field. During periods of scarce and high-priced hay, straw is frequently used as either a *hay-stretcher* or *hay-replacer.*

Of the common cereal straws, oat straw is the most palatable and nutritious. Barley straw ranks second, and wheat straw is third.

Straw is a bulky feed, and it must be properly supplemented. It is low in protein (on an as-fed basis, wheat straw averages about 3.2% crude protein), low in phosphorus, and low in vitamin A.

In addition to the cereal straws, other low-cost roughages available in certain sections of the United States are lentil straw, field pea straw, bean straw, clover straw, and bluegrass straw.

• **Sweet corn stover**—This is the sweet corn plant without the ears. It may be ensiled, or allowed to cure on the stalk and harvested and fed dry or grazed in the field. If it is ensiled, it should be harvested soon after the ears are removed. Nutritionally, it is comparable to field corn stover, or about half as valuable as well-eared field corn.

Making grass silage. (Courtesy, D. A. Miller, Department of Agronomy, University of Illinois at Urbana-Champaign)

SECTION

7

SILAGE/ HAYLAGE/ HIGH-MOISTURE GRAIN[1]

Contents *Page*

[1]The author gratefully acknowledges the helpful suggestions of the following authority who reviewed this section and who provided several of the excellent pictures: Dr. R. L. Vetter, Director of Research, Harvestore Systems, A.O. Smith Harvestore Products, Inc., DeKalb, IL.

Fig. 7-1. Much silage is fed to dairy cattle. This shows Guernsey cows and silos at Hoard's Dairy Farm, Ft. Atkinson, WI. (Courtesy, American Guernsey Cattle Club, Reynoldsburg, OH)

Silage may be defined as fermented forage plants. It is a very old method of preserving feed. Columbus found that the Indians used pits or trenches in which to store their grain, and, centuries earlier in the Old World, silos were used as a means of preserving both grain and green forage. The first tower silo built in the United States is said to have been erected by F. Morris in Maryland in 1876.

Silage making is one of the 3 common methods of utilizing crops, the other 2 methods being pasturing and haying. Pasturing is the least expensive of the 3 methods, but it is seasonal in nature. In the spring and early summer, forage plants generally grow faster than they can be utilized by normal grazing, and become dormant in cold weather.

The surplus forage produced during the growing season may be preserved for feeding during the winter months and other periods of pasture scarcity by hay making which, next to grazing, is the most efficient method during dry weather. But weather conditions in many parts of the world are not always conducive to hay making. Ensiling, when preceded by a certain amount of moisture reduction in the forage, requires a shorter period of good weather than hay making. As a result, the forage can be harvested at close to optimum maturity, with a greater proportion of crop nutrients preserved for feeding. However, the costs of harvesting and storing silage are usually higher than for hay making.

Silage is used primarily as a beef and dairy feed, for which it serves as a good source of fiber and an inexpensive source of protein. Corn silage is also an important supplier of dietary energy for beef and dairy cattle. Silage is also a good sheep feed. Sometimes it is fed to brood sows. Very little silage is fed to horses.

The importance of silage in this country is attested to by the fact that about 120,000 tons are made annually. It is especially important in all beef and dairy regions of the United States that have humid climates and cold winters. A wider variety of silo types continues to appear; and the use of preservatives has grown enormously since 1980.

ENSILING PROCESS

The ensiling process refers to the changes which take place when forage or feed with sufficient moisture is allowed to ferment in a silo in the absence of air. The basic strategy of silage preservation is to exclude oxygen and to acidify the forage through bacterial fermentation. An understanding of the ensiling process is necessary for the production of high-quality silage.

The ensiling process requires 2 to 3 weeks, in which time the following aerobic (with air) and anaerobic (without air) phases occur:

1. **Aerobic phase.** The living plant cells of the forage continue to respire, consuming the oxygen of the entrapped air, producing carbon dioxide and water, and releasing energy or heat. Simultaneously, aerobic yeasts, molds, and bacteria, thrive and multiply.

In good ensiling conditions and with proper management, the aerobic phase is very short, and silage temperatures seldom rise above 100°F. However, slow filling, inadequate packing of the forage, or leaky sealing around the silage, lengthens the aerobic phase, increases losses, and causes excessive heating.

2. **Anaerobic (fermentation) phase.** When the available oxygen of the entrapped air has been consumed, anaerobic

bacteria multiply at a prodigious rate. Simultaneously, the molds and the yeasts die. Certain plant enzymes continue to function.

The combined anaerobic activity produces the following changes:

a. The nonstructural carbohydrates, especially sugars, are converted to lactic acid (the acid in sour milk), some acetic acid (the acid in vinegar), a small amount of other acids, alcohol, and carbon dioxide.

b. The plant proteins are broken down into peptides, ammonia, amino acids, amines, and amides. These non-protein nitrogen compounds are utilized less effectively by the animal than true protein.

c. When the acidity become high enough, the bacteria die and the silage stabilizes. This occurs when the silage pH reaches a low of 3.5 to 4.5 for corn and cereals, or 4.0 to 5.0 for grasses and legumes.

ADVANTAGES AND DISADVANTAGES OF SILAGE

Some of the **advantages** of silage are:

1. It retains a higher proportion of the nutrients of plants than can be accomplished by hay making, even if the weather is satisfactory for the latter, chiefly because shatter and bleaching losses are held to a minimum. Thus, grass or legume silage preserves 85% or more of the feed value of the crop, whereas hay making under the best of conditions will preserve only 80%, and under poor conditions only 50 to 60%.

2. It makes possible the production of the maximum quantity of feed per acre of land and increases the efficiency of production.

3. It is feasible to produce a high-quality feed during times of inclement weather when it would be impossible to dry the forage to hay moisture levels.

4. It is the most economical form in which the whole stalk of corn or sorghum can be processed and stored.

5. It requires less storage space per pound of dry matter than baled hay. A cubic foot of silage contains 2 to 3 times as much dry weight of feed as a cubic foot of long hay stored in the mow.

6. It practically eliminates the danger of loss by fire if stored within the recommended moisture range.

7. It is the most satisfactory and economical way in which to preserve a number of by-product feeds.

8. It improves the timeliness of harvest so as to maximize the quality of forage.

9. It is one of the best methods of controlling the European corn borer since the removal of cornstalks is required in making silage.

10. It helps to control weeds, which are often spread through hay or fodder.

11. It is a better source of protein and of certain vitamins (especially carotene, and perhaps some of the unknown factors), than dried forage.

12. It is a highly palatable feed that supports high intake and sustained production.

13. It may be completely mechanized as a feeding system, thereby reducing labor and time.

14. It offers advantages over pasture, including (a) no fencing required, (b) approximately one-third more forage from the same acreage, (c) harvesting at optimum maturity, (d) more uniform quality, (e) little or no bloat, and (f) closer observation of animals that are confined to a lot or corral.

Some of the **disadvantages** of silage are:

1. It requires a silo or storage structure and other special equipment, for best results. In comparison with the simpler method of *field* curing and storing hay, this is likely to mean higher costs, an important consideration for a small operator.

2. It possesses considerably less vitamin D than sun-cured hay.

3. It necessitates that 2 to 3 times as much tonnage be handled as when the same forage is dried for hay, due to the high water content.

4. It incurs an added expenditure when preservatives are necessary.

5. It may result in acidosis, milkfat depression, and other metabolic disorders, due to the short chop lengths required for good packing in the silo, if proper feeding management practices are not followed.

6. It may increase the need for more expensive dietary supplements because the degraded protein in silage may not be utilized effectively by the ruminant.

7. It may spoil during feeding, resulting in molds which may affect animal performance and contain toxins.

8. It may result in poor preservation if the moisture content is not right. With high-moisture content (above 65%), seepage occurs and a butyric fermentation may take place that results in an unpalatable and toxic feed. With low-moisture content (below 45%), excessive heating of the silage may occur, making the plant proteins indigestible and possibly leading to a silo fire.

9. It requires better management the year-round for stored silage than for stored hay.

SILOS

Silage may be stored in almost any kind of container. The main requisites of a good silo, regardless of kind, are:

1. That its size be in keeping with the number and kind of animals to be fed daily, the length of the feeding period, and the amount of forage available for ensiling. Directions on how to determine the size of silo to build are given in this book in Section 9, Buildings and Equipment.

2. That it exclude air from the stored material, including entrance of air around the doors of tower silos.

3. That the sidewalls be straight and smooth in order to prevent the formation of air around the doors of tower silos.

4. That it be of proper dimension, thus making for better packing and less surface area to total mass exposed.

5. That it be properly reinforced. This point is especially important where direct cut grass silage is made, because it exerts from ½ to 2½ times as much pressure on the walls as does corn silage. Thus, tower silos which were originally built for corn or sorghum silage but which are to be filled with wet grass silage should be either (a) reinforced with extra bands placed around the lower part to strengthen the walls if an inspection reveals that the existing strength is not adequate, or (b) not filled to more than half capacity.

6. That adequate provision be made for the escape of surplus juices, either by a drain or by a gravel bottom, along with positive containment to avoid surface pollution.

7. That it be conveniently located and accessible in all kinds of weather, from the standpoint of both filling and feeding.

Silos may be classified according to the five basic methods used for processing forages. Each method is associated with the shape and material of the structure, which also influences the efficiency of preserving the silage. The different shaped structures are also adapted to different methods of filling and unloading. Within each classification, there are many variations of each type depending upon the manufacturer.

The kind of silo decided upon and the choice of construction material should be determined primarily by the cost and by the suitability to the particular needs of the farm.

Silos may be classified as follows:

1. Conventional upright (tower) silos
   a. Concrete stave
   b. Galvanized steel
   c. Monolithic concrete (poured in place)
   d. Tile block
   e. Brick
   f. Wood stave
2. Gastight (oxygen-limiting) silos
   a. Glass-lined structures
   b. Concrete stave
   c. Galvanized steel
   d. Monolithic concrete
3. Pit silos
4. Horizontal silos
   a. Trench silos (below ground level)
   b. Bunker (above ground level)
5. Temporary silos
   a. Enclosed stacks
   b. Open stacks
   c. Modified trench-stack silos
   d. Long plastic or polyethylene bag silos
   e. Round bale bagged or wrapped silage

Some pertinent information relative to each main kind of silo is given in the discussion which follows, but it is not within the scope of this book to give detailed silo plans and specifications. The latter may be obtained from local authorities, from silo manufacturers, or by writing to the state agricultural college.

• **Conventional upright (tower) silos**—The upright or tower silo is a cylinder built above ground. Its round shape withstands pressure well and is adapted to good packing. Mechanized distributions and unloaders make filling and unloading relatively fast.

• **Gastight (oxygen-limiting) silos**—These silos resemble conventional tower silos, but are more expensive because of their construction. Practically all outside air is kept out of the oxygen-limiting silo, and carbon dioxide formed during fermentation is kept in.

• **Pit silos**—The pit silo is shaped like the tower silo, but inverted into the ground. It resembles a well or cistern. The walls of a pit silo may or may not be lined.

CAUTION: Before entering a pit silo, it is recommended that a lighted cigarette lighter, candle, or lantern be lowered into the silo. If the flame goes out, assume that the pit is dangerous to enter and replenish it with fresh air before entering.

• **Horizontal silos**—There are two types of horizontal silos, trench silos and bunker silos (or horizontal surface silos), both of which may be adapted to self-feeding.

1. **Trench silos.** The trench silo is a horizontal, trench-like structure that can be built quickly and at low cost. It is most popular in areas where the weather is not too severe and where there is good drainage. The walls of a trench silo may or may not be lined, but for making good silage they should always be smooth. There may or may not be a poured concrete floor. A trench silo should be wider at the top than at the bottom, and the bottom should slope away from one end so that excess juices will drain off.

Trench silos have the **advantages** of: (1) low initial cost; (2) low cost of filling machinery; (3) relative freedom from freezing; and (4) ease of construction. The chief **disadvantages** of trench silos in comparison with tower silos are the (1) larger area to seal, (2) higher spoilage losses, and (3) inconvenience in feeding during inclement weather. Because of shallowness, the forage should be packed very thoroughly in a trench silo by driving a wheel tractor back and forth over it. When filling is completed, the top should be carefully sealed with polyethylene, plastic, or other materials.

2. **Bunker silos.** Above ground horizontal silos are usually constructed with concrete floors and side walls of wood, concrete, or other materials. Bunker silos were originally intended for animals to self-feed directly, but skid loaders or front-end loaders on tractors are now the common methods of unloading.

• **Temporary silos**—Several kinds of above ground temporary silos are used. Generally, this kind of storage is used to meet emergencies, to supplement permanent silos, or to ensile such by-product feeds as cannery refuse, pea vines, and beet tops or pulp. Above ground temporary silos are low in cost, can be erected on short notice, require no special foundation, and can be set up on almost any level site convenient for filling and feeding.

Temporary silos can be classed as belonging to one of the following four kinds:

1. **Enclosed stacks.** These are built entirely above ground, without trenches or holes. They are upright, are generally circular, and are enclosed by snow or picket fences, poles, wooden staves, heavy woven wire, or other materials. Most of them are lined with tar paper, plastic, or tough fiber-reinforced paper made especially for the purpose.

2. **Open stacks.** These are similar to enclosed stack silos, except that no supports or walls are used.

3. **Modified trench-stack silos.** This silo, which is intermediate between a trench and a stack silo, is adapted to areas where the ground water level is high. It is constructed by excavating a shallow trench 12 to 18 in. deep, by piling the excavated earth on either side of the trench to support the silage and to keep out surface water, by packing silage thoroughly in and over the trench to a height of 10 to 15 ft, and by covering the stack with any one of the materials recommended for covering the trench silo (see trench silo).

4. **Plastic silos.** Plastic films are now available for use as temporary silos. Also, they are used (a) as covers for trench, bunker, and tower silos, and (b) as silo liners. If not punctured, plastic is airtight. Several types of plastic silos can be evacuated with vacuum pumps following filling, thereby using air pressure to compress the forage.

Among the special advantages attributed to plastic silos are (a) economy—the greater the size, the more economical; (b) adaptation to small quantities of silage and for out-of-the-way places; (c) the ability to store and use high and low quality forages separately; and (d) reduction of spoilage, provided sealing is complete and puncturing is prevented. However, good sealing and non-puncturing are difficult to accomplish. Present plastic materials are very susceptible to puncture by sharp objects (fingernails, plant stems, implements, animals, etc.). Hence, a minimum of 10% loss should be expected.

Fig. 7–2 Temporary silos—long plastic tubes. Note that cattle are self-feeding from the silo on the left. (Courtesy, Dr. R. L. Vetter, Harvestore Systems, DeKalb, IL)

SILAGE STORAGE LOSSES

Tight structures, good distribution and packing, and the proper use of plastic covers minimizes silage storage losses. Losses within type of silo storage vary widely based primarily on length of time and season of feedout; this is especially critical for silages continuously exposed to air. Silage losses also vary widely between kinds of silos, as shown in Table 7–1. Losses in trench and open stack silos are also influenced by depth; less surface is exposed in deeper silos.

Losses in the silo are of four types: (1) surface or top spoilage, (2) seepage, (3) gaseous, and (4) heating (browning reaction and spontaneous combustion).

TABLE 7–1
ESTIMATED (1) AVERAGE, AND (2) RANGE OF SILAGE STORAGE LOSSES

Type of Silo	Percent Of Loss	
	Average	Range
	(%)	(%)
Gastight upright	5	1–10
Conventional upright	10	5–15
Horizontal (trench)	15	10–20
Open stack	25	15–30

Surface or top spoilage losses of 20% or more may occur in stack silos and in any uncovered bunk, trench, or pit silo. These losses can be reduced by the use of suitable protection, such as a plastic cover.

Seepage losses can be high in high-moisture silage stored in upright silos. The higher the silo, the greater the pressure and the higher the losses through seepage. The seepage carries soluble feed nutrients with it. Horizontal silos have less seepage loss than upright (tower) silos because of lower vertical pressure. Seepage losses can be reduced by wilting forages to less than 65% moisture before ensiling.

Gaseous losses are unavoidable so long as the plant material respires and there is subsequent fermentation. However, these losses can be minimized by avoiding entry of air into the silo, by having the pH decline rapidly, and by encouraging favorable fermentations.

Lowering the moisture without excluding the air may lead to heat damage, known as the browning reaction or Maillard reaction.

Spontaneous ignitions sometimes occur in low-moisture silage (haylage). For such losses to occur, there must be a build-up of temperature to the combustion point in the silo mass, combined with a low transfer of heat. These fires are very difficult, and usually impossible, to extinguish. The addition of water may build up pressure and lead to an explosion. Most silo fires should be allowed to burn.

KINDS OF SILAGE

A great variety of crops can be and are made into silage. Generally speaking, crops that are palatable and nutritious to animals as pasture, as green chop, or as dry forage also make palatable and nutritious silage. Likewise, crops that are unpalatable and unnutritious as pasture, as green chop, or as dry forage make unpalatable and unnutritious silage. However, the palatability and nutrient value of some low quality forages and crop residues can be improved by nutrient additives and fermentation.

Most silage in the United States is made from either corn or sorghum, with corn silage far in the lead—over 15 times as much corn silage as sorghum silage is made. In 1987, 80.6 million tons of corn silage and 5.4 million tons of sorghum silage were produced in the United States. At the present time, it is estimated that 75% of the nation's silage is made from corn and sorghum and 25% from grasses, legumes, and other feeds. In addition to the kinds of silage already mentioned, silage is made from sunflowers, the small grains, sugar beet tops, crop residues, wastes from food processing (sweet corn, green beans, green peas), root crops, and various vegetable residues.

Many of the leading silage crops are listed in Table 7-2, which also includes an estimate of the expected and potential production from these crops, along with the normal stage of maturity at harvest.

TABLE 7-2
LEADING SILAGE CROPS[1]

Crop	Expected Production[2]			Potential Production[3]			Stage For Ensiling
	Dry Matter Yield	Digestible Protein	TDN	Dry Matter Yield	Digestible Protein	TDN	
	(tons/A)	(lb/A)	(lb/A)	(tons/A)	(lb/A)	(lb/A)	
Corn	6.0	600	8,100	10.5	1,100	14,300	Hard dough or early glaze
Grain sorghum	5.5	550	7,200	9.0	900	11,700	Soft to medium dough
Forage sorghum	6.0	600	6,900	9.0	900	10,400	Soft to medium dough
Sorghum-Sudangrass hybrids	4.5	540	5,000	8.0	960	8,800	Early bloom
Sudangrass	3.5	420	3,900	6.0	720	6,600	Early bloom
Alfalfa	5.0	1,500	6,500	9.0	2,700	11,700	Bud stage, first cut; early bloom, other cuts
Legume-grass mixtures	4.5	900	5,400	8.0	1,600	9,600	Legumes, early bloom
Small grains	4.0	560	4,800	7.0	980	8,400	Boot to early head

[1]From paper entitled "Silage Production, Preservation, and Quality," by John E. Baylor, The Pennsylvania State University, *Dairy Science Handbook*, Vol. 9, 1976, p. 224, published by Agriservices Foundation, edited by M. E. Ensminger. To convert to metric: tons/A = 0.907 metric ton/0.4 hectare; lb/A = 0.454 kg/0.4 hectare.

[2]Expected production—Figures presented are better than average yields for the country. They are yields which might be expected for average or better farmers who are following recommended practices.

[3]Potential production—Figures presented are production yields attainable where good soils, excellent growing conditions, and top management are present. These silage yields equivalents have been produced on farms and experiment stations prior to this time.

The average composition of the silages produced from several of these crops is given in Table 7-3.

TABLE 7-3
COMPOSITION OF VARIOUS SILAGES

Type Of Silage	Analyses On A Dry Matter Basis			
	Crude Protein	TDN	Ca	P
	(%)	(%)	(%)	(%)
Corn	8.3	68.0	0.31	0.27
Grain sorghum	7.9	55.0	0.34	0.19
Forage sorghum	9.2	57.9	0.30	0.24
Oats	10.0	57.0	0.47	0.33
Alfalfa	17.4	59.0	1.75	0.27

CORN AND SORGHUM SILAGE

For the United States as a whole, corn ranks first in importance as a silage crop. Generally more total digestible nutrients can be obtained from an acre of corn as silage—which will yield from 5 to 25 tons of forage per acre, with an average of about 14.5 tons—than can be obtained from an acre of any other crop. Also, corn ensiles easily without the aid of a preservative, and keeps almost indefinitely in a good silo, is highly palatable, is well adapted to mechanized feeding, and may be fed with little waste.

There are four kinds of corn silage; namely:

1. **The whole corn plant.** When at the peak of its nutritive value and right for ensiling, the whole corn plant contains 1½ times the nutrients of the ripened grain that the plant would have yielded. Also, in corn silage made from the whole crop more than 90% of the nutrients produced are saved.

2. **Ear corn silage.** The ensiled ears contain up to 68% of the nutrients of the entire corn plant.

Fig. 7-3. Harvesting corn silage. (Courtesy, Gehl Company, West Bend, WI)

3. **Corn stover silage.** The forage remaining after harvesting a grain crop. This accounts for about one-third of the total nutritive value of the crop.

4. **Shelled-corn silage.** This consists of the kernels only. At 70% dry matter (30% moisture), shelled-corn silage contains 61 to 66% of the nutrients in the whole crop (also see "High-Moisture Grain" later in this section).

The sorghums are more dependable and higher yielding than corn in certain areas, particularly in unirrigated, and relatively dry areas, of western and southwestern United States. Sorghum for silage is harvested with the same equipment as is used for corn silage. It should not be harvested for silage until the heads are soft to medium dough stage. Harvesting at this stage provides the highest yields of total feed material, enhances preservation, and makes silage that has good palatability.

Fig. 7-4. Harvesting grain sorghum for silage on a southeast Kansas dairy farm. (Courtesy, A. O. Smith Harvestore Products, Arlington Heights, IL)

On a dry-matter basis, corn silage contains an average of 8.3% crude protein, 68.0% total digestible nutrients, 0.31% Ca, and 0.27% P. Grain sorghum silage contains less protein and TDN than corn silage. Grass/legume silages contain more protein and less TDN than corn silage. The carotene content of corn silage is variable, but on the low side.

CORN AND SORGHUM RESIDUE SILAGE

Corn and sorghum residues—the forages that remain after harvesting a grain crop of corn or sorghum—may be used as cattle feed three ways: (1) grazed, (2) harvested (stacked or baled) and fed dry, or (3) ensiled and fed as silage.

The biggest deterrent to harvesting stalklage, in either dry or ensiled form, is the cost—primarily for equipment. Rather than own such expensive equipment, which is used for a short period only, custom harvesting of stalklage is likely cheaper for most operators.

Husklage—the forage discharged from the rear of a corn combine, and consisting of the husks, cobs, and any grain carried through the combine—may also be ensiled. Ensiling husklage, along with recutting and adding water, results in increased cow consumption and less rejection of cobs.

Like corn, sorghum stover may either be grazed or harvested and stored either as dry feed or silage. Because the sorghum plant stays green late in the fall, good sorghum stover silage can be made without additional water.

GRASS/LEGUME (HAY CROP) SILAGE

Grass/legume (hay crop) silage refers to silage made from any of the green crops which might otherwise be grazed or dried and made into hay. This includes grasses (such as timothy or fescues), legumes (such as alfalfa or clovers), grass-legume mixtures, and cereal grains (such as oats).

Chart 5-1, pasture table, (see Section 5) is equally applicable for the production of grass silage, because, in practical operations, any adapted grasses and/or legumes may be used three ways: for grazing, for hay, or for silage.

Grass/legume silage can be produced in areas where the climate is too cool and the growing season too short for corn or sorghum silage.

Although grass and legume crops have been ensiled in Europe for hundreds of years, the practice did not become widely used in the United States until the 1930s. At that time, interest in hay crops for silage increased as a result of farmers (1) becoming aware of the field losses that occur in hay making, (2) being provided with the information necessary to make high-quality silages from grasses and legumes, and (3) having access to field choppers, which facilitated making silage from hay crops.

The following are the most important **advantages** of grass/legume silage:

1. It minimizes field, harvest, and storage losses of grass/legume forages. (See Fig. 7–5.)

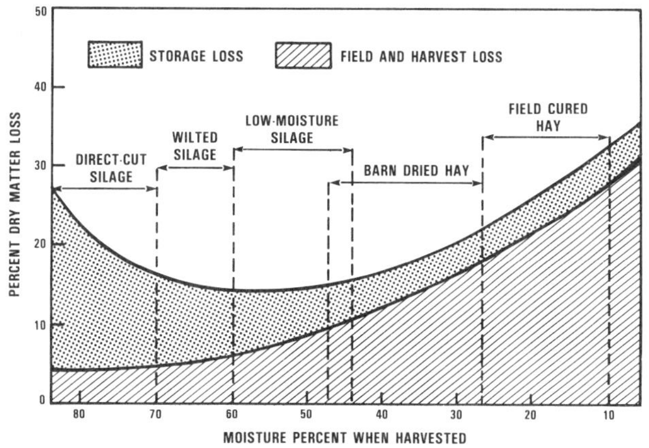

Fig. 7–5. Estimated total field, harvest, and storage loss when grass/legume forages are harvested at different moisture levels by alternative harvesting methods. (Courtesy, C. R. Hoglund, Professor Emeritus, Michigan State University, East Lansing)

2. It minimizes the dependence on favorable weather to harvest the crop.
3. It can be harvested with modern, efficient machinery and stored in large-volume structures.
4. It requires less supplemental feed than corn silage.
5. It can be handled and fed by mechanized methods, thereby reducing the labor requirements.
6. It kills weed seeds as a result of the fermentation.

Although grass/legume silage has important advantages, it also has the following **disadvantages**:

1. The initial investment costs for machinery, storage units, and feeding facilities are very high.
2. Silage-making machinery and storage and feeding facilities are highly specialized with the result that they have limited use for other purposes.
3. Inadequate fermentation may occur under certain conditions, resulting in poor-quality feed.
4. Storage and feeding losses may be high under poor management.
5. Like all silages, grass/legume silage is heavy, bulky, and costly to transport, thus its off-farm market value is limited.
6. Existing upright silos may not be in sufficiently good condition to store grass/legume silage.
7. Once grass/legume silage is removed from the silo, it must be fed within 12 to 24 hours to alleviate spoilage. Except for an oxygen-limiting silo, once feeding begins during warm weather, it is necessary to feed a minimum of 3 to 4 in. off the exposed surface daily to prevent spoilage in the silo.

Grass/legume silages are of three kinds based on moisture level:

1. Direct-cut silage, 70% moisture or above.
2. Wilted silage, 60 to 70% moisture.
3. Low-moisture silage (or haylage), 40 to 60% moisture.

• **Direct-cut silage**—*Direct-cut grass/legume silage is forage that is harvested and stored without field-drying, usually containing more than 70% moisture.* Although direct-cut ensiling is the standard practice with mature corn and sorghum, it is not recommended for grass/legume crops because of (1) the difficulty in getting good preservation due to the high moisture content, and (2) the increased nutrient losses due to seepage.

When making direct-cut silage, the following directions should be observed: (1) harvest at the proper stage of maturity; (2) avoid cutting when the forage is wet with dew or rain; (3) provide drainage for excess juice; (4) add an additive or preservative (coarsely ground cereal grain will absorb some of the excess moisture in addition to providing sugars for fermentation); (5) distribute evenly in the silo and pack thoroughly; and (6) cover with plastic or suitable material.

• **Wilted silage**—Today, a high percentage of grass/legume forage is dried to some degree prior to ensiling. Wilting gets rid of some water in the field, so less weight is handled. Also, in comparison with direct-cut silage, odor and seepage problems are reduced and additives or preservatives are usually not needed.

Fig. 7–6. Alfalfa forage wilted to 65% moisture being field chopped for ensiling. (Courtesy, Ford New Holland, New Holland, PA)

Authorities generally agree on the following rules for making good wilted silage: (1) harvest at the proper stage of maturity; (2) allow the forage to wilt in the swath and/or windrow until the moisture reaches about 65%, which may take from 1 to 4 hours, or longer, depending on the weather, but which may be expedited by the use of a forage conditioner; (3) use a short cut (about ⅜ in.) on the forage harvester; (4) fill the silo rapidly and continuously; (5) distribute evenly and pack thoroughly; and (6) top off with 2 ft of unwilted material and cover with plastic or other suitable cover. No additive or preservative is needed with properly wilted silage, although it may be added if desired.

• **Low-moisture silage (haylage)**—Low-moisture grass/legume silage (or haylage), containing 40 to 60% moisture, is made with limited bacterial growth and fermentation. The term *oatlage* is sometimes used specifically to indicate low-moisture silage made from oats.

Fermentation is of minor concern in making low-moisture silage since little acid is produced and pH is not a useful criterion of quality. The most important factor is the establishment and maintenance of air-free conditions through fine chopping, rapid filling, and a good silo. Because of the difficulty of maintaining air-free conditions, stacks, bunkers, and trench silos are seldom used for storing low-moisture silage. Infiltration of air into the silage mass will result in growth of yeasts and molds and increase in temperature. Temperatures about 95°F for a few days will cause certain proteins to combine chemically with carbohydrates to form a product that is indigestible, termed *bound protein*. When more than 12% of the total protein is in the "bound" form, the silage has undergone excessive heating.

Properly made and stored low-moisture silage has a pleasant aroma and is a palatable, high-quality feed. Animals usually receive more dry matter and net feed value in low-moisture silage than in wilted silage made from the same cut. Low-moisture silage is increasing in popularity, especially as a dairy feed. It may be fed like wilted silage, with adjustment for difference in moisture content.

In addition to excluding air by providing an airtight silo and by thorough packing, the following directions should be observed to make the best quality low-moisture silage: (1) harvest at the proper stage of maturity; (2) wilt in the swath and/or windrow until the moisture reaches 40 to 60% (35 to 40% for a bunker silo), with the required time determined by the weather, but which may be expedited by the use of a forage conditioner; (3) chop short, a ¼ in. cut is best; (4) fill silo rapidly and continuously; (5) add an additive or preservative if desired; (6) distribute silage evenly in the silo; (7) apply a top seal of forage containing 65 to 70% moisture, level, and tramp to remove air; and (8) crown the center slightly and cover with a plastic silo cap.

COMBINING CROPS FOR SILAGE

Sometimes, in order to lower the moisture content, to alleviate the necessity of a preservative, and to assure better quality silage, forages of high sugar content are combined with forages of low sugar content. Thus, excellent silage can be made by mixing 1 ton of sorghum forage with each 3 tons of grass/legume silage material, or a ton of corn forage with each ton of grass/legume forage material (less sorghum forage is necessary than corn forage, because of the higher sugar content of the former).

At times such combination silage crops are even grown together; for example, corn and soybeans; millet or Sudangrass; and soybeans, oats, and peas.

A major difficulty in combining ensiling crops is that it is almost impossible to synchronize the stage of maturity of different crops so that they reach maximum yield and nutrient level at the same time.

RAIN-DAMAGED HAY SILAGE

Partly cured hay that has been rained upon, but is not moldy, may be salvaged as silage (although it will not be of high quality), provided it is finely chopped, distributed evenly, and packed in the silo thoroughly enough to squeeze out the air. It is recommended that it be placed in the bottom of the silo, and, preferably, that alternate loads of a green crop be mixed with it. Otherwise, satisfactory packing can be obtained by putting a few loads of greener-than-ordinary material on top of it.

FROSTED CROP SILAGE

Sometimes corn, sorghum, sunflowers, small grains, beans, and other crops, which may or may not have been intended for silage, are frosted before they reach the silage cutting stage. Corn that has been frosted before reaching maturity is commonly known as *soft corn*. Such frosted crops may be salvaged as silage. They should be cut at recommended moisture contents and ensiled according to directions. If they are too dry, water should be added.

Frosted crops, especially frosted sorghum, may be high in cyanide (HCN). (See *CAUTION* under the heading "Drought Stricken Crop Silage.")

DROUGHT STRICKEN CROP SILAGE

Sometimes corn or sorghum, or other crops, are drought stricken to the extent that little or no grain will be produced. Such crops may be harvested for silage and used as an energy source for ruminants. They should be cut and ensiled like any other silage crop. If they are too dry, water should be added.

Drought stricken crop silage may be used in the same manner as any other low-energy source. It is well-suited for wintering breeding beef cattle and stockers, for backgrounding finishing cattle—to approximately 850-lb weights, and for dry dairy cows.

CAUTION: Danger of cyanide toxicity is much greater from sorghum than from corn. Drought stricken plants can accumulate cyanogenetic glycoside which hydrolizes to form free cyanide (HCN). The danger is increased when crops are grown on heavily nitrogen-fertilized soils or if any of the following have occurred: frosting, wilting, trampling, or hail. Any combination of these conditions can lead to a dangerous build-up or release of cyanide.

OTHER SILAGE CROPS

In the Northwest and North Central states, where the weather is cool and the growing season is short, sunflowers are sometimes grown for silage. Although they yield and ensile well, sunflower silage is neither as palatable nor as nutritious as corn, sorghum, or grass silage. Pound for pound, sunflower silage is about 80 to 85% as valuable as corn silage.

Throughout the United States, a great array of by-product feeds are ensiled, especially in the less expensive and temporary types of silos. Among such by-products are grain chaff, pea and bean vines, beet tops and pulp, sunflower hulls and chaff, potatoes, cannery refuse, cull and surplus fruits and vegetables, pulp and trimming wastes from market vegetables and fruits, wet brewers' and distillers' grains, almond hulls, and poultry litter. Sometimes Russian-thistles and other weeds are ensiled.

When potatoes, which contain about 80% moisture, are ensiled for cattle, it is recommended either (1) that 20 to 25 lb of dry hay, straw, or chaff be run through the ensilage cutter with each 100 lb of potatoes, or (2) that 1 ton of corn or sorghum silage be chopped with each 500 lb of potatoes. Frozen and sprouted potatoes should not be ensiled. Potato silage intended for swine should be made from cooked or steamed potatoes ensiled alone in a shallow pit or silo. Potato processing wastes (cull potatoes, off-flavor french fries and chips, etc.) can be ensiled in the same manner as unprocessed potatoes.

Either of the methods recommended for ensiling potatoes for cattle is equally adapted for the preservation of other high-moisture crops, such as apples, beets, pears, tomatoes, cauliflower, broccoli, kale, and trimming wastes from market vegetables—provided the added forage is in proportion to their respective moisture contents.

Cabbage, rape, and turnips should not be ensiled, as they make unsatisfactory, watery, foul-smelling silage.

SILAGE ADDITIVES AND PRESERVATIVES

Silage additives are products that provide supplemental nutrients which enhance the feeding value of silage.

Silage preservatives are products that enhance the keeping qualities of silage.

High-quality silage can be made without the use of additives or preservatives if good material is started with and all proven good practices are followed. But there are times when the ensiled material is either too wet or too dry; does not contain sufficient fermentable carbohydrates; is deficient in certain nutrients; is lacking in palatability; and/or the proven good practices cannot be followed. Under such circumstances, silage additives or preservatives may reduce silage losses and/or improve the feeding value of the silage.

Additives or preservatives should not be used as a substitute for a good silo or for proper chopping, packing, and sealing. Normally, additives or preservatives are neither needed nor added to corn or sorghum silages. But additives or preservatives may be very helpful if a grass/legume forage with over 70% moisture is ensiled.

A number of materials are available to incorporate in silage, with claims made that they will enhance the nutrient value, the preservation of the nutrients, and/or the palatability of the silage. *The bottom line when using any silage additive or preservative is how much it improves animal performance and net profit; and not whether it merely makes silage look better and smell better.*

Thorough testing of these materials would necessitate that each of them be used at several levels, with many kinds of silage, with each forage at various moisture contents, and under different storage conditions. Obviously, it would be highly impractical, if not impossible, to carry out such an extensive study. However, there is sufficient understanding of the process of silage formation, the requirements for the preservation of silage nutrients, and the mode of action of the ingredients used in various additives to make sound decisions as to whether they might be economically worthwhile. Also, some experimental testing has been done with certain additives.

In order to be effective, an additive or preservative should serve one or more of the following purposes:

1. Add nutrients.
2. Provide fermentable carbohydrates.
3. Furnish additional acids to increase acid conditions.
4. Inhibit undesirable types of bacteria and molds.
5. Reduce the amount of oxygen present, directly or indirectly.
6. Reduce the moisture content of the silage.
7. Absorb some nutrients which might otherwise be lost in seepage.
8. Reduce the fiber content of the forage through enzymatic action.

Four types of additives or preservatives are used in silage making: (1) feed additives, (2) acids, (3) fermentation aids, and (4) preservatives.

FEED ADDITIVES

Feed additives may be used to provide a readily available source of carbohydrates for fermentation into lactic acid, to reduce the moisture content, to provide needed nutrients, and/or to enhance palatability.

Feedstuffs used as silage additives include:

1. Corn-and-cob meal, ground corn, barley, or oats; applied in amounts varying from 100 to 300 lb per ton, depending on the moisture content of the crop.
2. Beet pulp, citrus pulp, chopped corncobs, or chopped hay to reduce seepage losses from the silo if moisture is high.
3. Molasses, either liquid or dehydrated, at rates of 40 to 80 lb per ton of green forage.
4. Dried whey, a product of the dairy industry, applied at the rate of 30 to 300 lb per ton, as a source of fermentable carbohydrate, protein, and minerals.
5. Nonprotein nitrogen (NPN) products such as urea and anhydrous ammonia. The addition of 10 lb of urea per ton of ensiled corn material will make for an approximate increase of the crude protein from 8.3 to 12.3%, on a dry-matter basis.
6. Ground limestone (calcium carbonate) added at a level of 0.5 to 1.0% to corn silage.

ACIDS

Both inorganic acids (hydrochloric, sulfuric, and phosphoric) and organic acids (propionic, acetic, lactic, citric, and formic) may be used as additives. Mineral acids lower the pH immediately, while organic acids have a limited effect on lowering pH. Both mineral and organic acids limit microbial growth and help to stabilize silage.

The use of inorganic acids, such as hydrochloric, sulfuric and phosphoric, for forage preservation was pioneered by A. I. Virtanen, Finnish biochemist, in the 1920s. He discovered the AIV method, named from his initials, for preserving silage by acidification, for which he was awarded the Nobel Prize in Chemistry in 1945. His work was highly regarded in that area of the world because hay drying was difficult and dairying was, and still is, important.

In general, the use of inorganic acid preservatives is not considered as desirable as the use of molasses or grain, because (1) they produce a more sour and less palatable silage; (2) they may damage clothing, machinery, and/or masonry silo walls, due to their corrosiveness; and (3) they do not add to the nutrient value of the silage except by enhancing the preservation of carotene.

Organic acids are used in a manner similar to inorganic acids, but they are much less corrosive and not so difficult to handle, although precautions must be taken. They will enhance the preservation of forage without the loss of palatability. Also, they serve as mold inhibitors.

FERMENTATION AIDS

This group includes bacterial cultures, yeast cultures, and enzyme supplements. Controlled experiments support the claims made for some of these products, but not all of them. So, they should be purchased only from reputable sources that have valid research data to support the claims made for them.

PRESERVATIVES

This group includes antibiotics, salt, and sterilants. These products preserve silage by inhibiting microbial action or undesirable fermentations. All of them are of questionable value if air is properly excluded from the silage. If air is not excluded, they must be added at very high levels in order to be effective.

SILAGE ADDITIVE AND PRESERVATIVE RECOMMENDATIONS

When added to silage, the following materials will increase the amount of nutrients it contains:

1. Grain or grain by-products will increase total digestible nutrients and dry matter.
2. Molasses will increase the total digestible nutrients (TDN, or energy) and may improve fermentation in legumes and certain grasses.
3. Urea or other NPN products will increase the nitrogen (crude protein).
4. Limestone will increase the calcium content.

Most of the arguments center around the use of the non-nutrient silage additives. A review of the literature indicates varying degrees of success from the use of such products. Some reports show positive effects from them while others show no effect. The most important consideration is whether the improved quantity and quality of forage from the use of an additive or preservative will offset its cost and application.

At the outset, it should be recognized that no silage additive or preservative can rectify mistakes that were made earlier—prior to incorporating the product. Neither can feeding more grain compensate for poor-quality silage.

Additives or preservatives are not essential to good silage formation when conditions of moisture and storage are right. Yet, under special circumstances they can be recommended for use. For example, molasses, grain, or grain by-products might be a wise addition to silage when conditions do not allow for proper wilting prior to ensiling, or when an *all-in-one* silage is being made. Urea may be an appropriate addition to an *all-in-one* silage or where increasing the protein content of the silage will simplify its feeding. It is doubtful that there is any justification for adding limestone unless this is a convenient method of calcium supplementation. The economy of most nutritive additives of this type depends largely on how well their nutrients are retained in the silage and the use made of them in balancing the rations.

When forages are stored at the proper moisture content, and when air is properly excluded, nutrient losses are low and a good-quality silage forms. Additives such as lactic acid bacteria, mold inhibitors, antibiotics, salt, enzymes, yeast cultures, and mineral acids, can, therefore, do little if anything to improve the preservation of the silage or its feeding value. When high-moisture material is ensiled, grain is superior to any of these additives. When air is not properly excluded, none of these additives will correct the large fermentation and spoilage losses.

In short, there is no substitute for good management of forage crops for silage, with proper control of such factors as stage of maturity at harvest, harvesting methods, moisture content, fineness of chopping, distribution and packing, and exclusion of air.

In order to assess the value of a silage additive or preservative, it is recommended that the following criteria be applied:

1. Does the product lower the ensiling temperature?
2. Does the product increase aerobic stability?
3. Does the product increase dry matter and nutrient recovery from the silo?
4. Does the product improve feed value and animal performance, particularly when silage is a major ingredient of the ration?
5. Does the product make for sufficient benefits to offset costs and give a return on investment?

HARVESTING METHODS AND MACHINERY

There is no one best silage-making method or kind of equipment. These must necessarily vary with the kind of forage, the kind of silo, the size of operation, and the labor and machinery cost.

Three principal kinds of machines are used for harvesting silage; namely, field forage harvesters, row-crop binders, and stationary silo fillers.

Fig. 7–7. A self-propelled forage harvester harvesting corn for silage. (Courtesy, Ford New Holland, New Holland, PA)

Fig. 7–8. Chopped forage being transferred from a forage truck into a blower which elevates it through a pipe to the top of, and onto, the silo. (Courtesy, Ford New Holland, New Holland, PA)

HOW TO MAKE GOOD SILAGE

In addition to using a sound silo of proper size, those who make good silage generally harvest at the proper stage of maturity, cut to proper length, control the moisture content, add an additive or preservative when needed, fill rapidly, distribute forage uniformly in the silo, and seal or top-off the silo. Each of these factors will be discussed.

HARVEST AT PROPER STAGE OF MATURITY

Harvesting at the proper stage of maturity assures the maximum yield and nutrient content.

Fig. 7–9 shows the effect of stage of maturity of the corn plant on total dry matter accumulation.

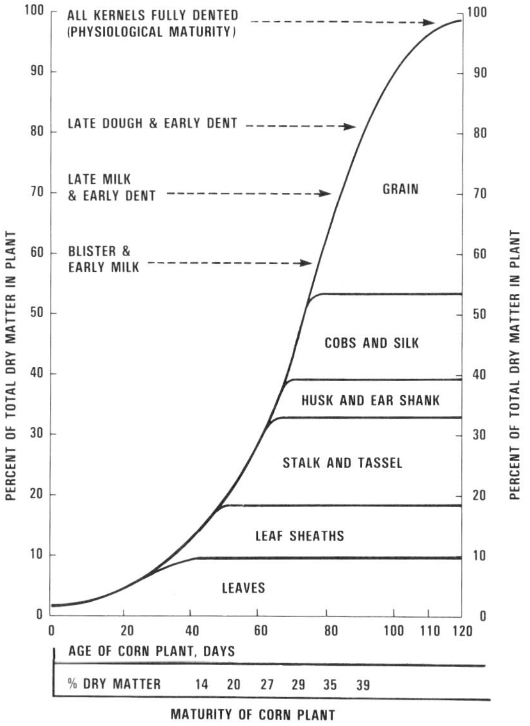

Fig. 7–9. Effect of maturity of corn plant on total dry matter accumulation.

The *black layer test* can be applied quickly and easily to determine when to harvest corn for maximum yield and nutrient quality (see Fig. 7–10).

Fig. 7–10. Black layer near the tip of the kernel indicates that the grain is physiologically mature and ready for the silo.

When the grain reaches physiological maturity, several layers of cells near the tip of the kernel turn black, forming the *black layer.* This layer can be detected by removing several kernels from the middle of the ear, then splitting them lengthwise or just cutting off the tip, and looking for the black layer near the tip. If the black layer is present, the grain is physiologically mature and ready for the silo.

At the black layer stage, the grains are usually dented and glazed, the lower 4 to 6 leaves of the corn plant are brown, and the plant contains 60 to 67% moisture. It can be cut 3 to 4 weeks past this stage with very little loss in dry matter or in feeding value.

Sorghum should be cut for silage when the seeds are hard.

Grass silage forages (grasses, legumes, and cereal crops) should be cut at the same stage at which they would make the best hay.

CUT TO PROPER LENGTH

The length of the cut sections affects the packing and, hence, the quality of the silage. Also, the proper length of cut varies with the crop and the moisture content. Thus, for corn and sorghum crops, forage harvesters should be set to make a theoretical cut of ¼ to ⅜ in. If the knives are sharp and set up to the cutter bar, this will result in about 15% of the particles being 1½ in. and over, 25% of the particles being ¾ to 1½ in., and 60% being ⅛ to ¾ in. in length. Such a combination of particle size is necessary for high-quality feed. Grass silages should be more finely chopped than corn or sorghum silage. Also, wilted and dry forage and forage with hollow stems should be chopped more finely than forage of high-moisture content, thus permitting more thorough packing and eliminating most air pockets.

CONTROL THE MOISTURE CONTENT

Moisture content is one of the most important factors in determining quality of silage. Experimental work and practical experience have indicated that 60 to 67% is the best moisture content for most crops to be ensiled. However, low-moisture silage of 40 to 60% moisture is now being preserved successfully in either oxygen-limiting silos, or tall conventional silos that are properly topped off with heavy, wet forage or sealed with a plastic cover.

HOW TO LOWER THE MOISTURE CONTENT

The moisture content of silage material may be lowered by any one or a combination of the following methods: by conditioning and/or wilting, by adding dry hay or straw, by combining with corn or sorghum silage, or by adding a dry additive/preservative of grain, dried molasses, or dried by-products of citrus or beets.

HOW TO INCREASE THE MOISTURE CONTENT

Drier material may be used for silage by cutting shorter and packing more thoroughly. If necessary, water should be added or the dry material should be mixed with very green, freshly cut material by alternating loads.

HOW TO DETERMINE THE MOISTURE CONTENT

Some methods of determining moisture content follow:

1. **The grab test (or squeeze method).** This test consists in taking a handful of the chopped forage and giving it a good hard squeeze for about 30 seconds. Then opening the hand slowly, noting the condition of the ball of forage in the hand, and referring to Fig. 7–11.

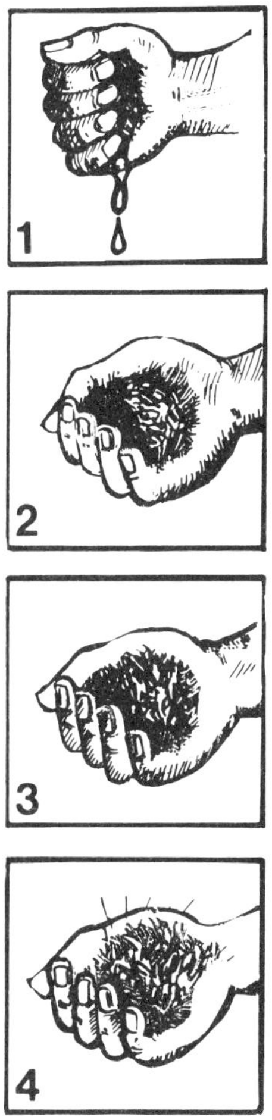

1. Juice runs freely or shows between the fingers. The crop contains 75 to 85% moisture and is too wet to make high-quality silage without treatment. Silages made from crops in this condition will lose large quantities of juice. When possible, wilt these crops. If they must be ensiled without wilting, use an effective chemical preservative (not all of them are effective) or 200 lb of ground grain per ton of crop.

2. The ball holds its shape—the hand is moist. The crop contains 68 to 75% moisture. Some juices will escape from tower silos. Additional wilting in the field is desirable. Where this is not done, use a chemical preservative or 150 lb of ground grain per ton of crop, or layer with wilted crops. Odors will be strong without some treatment.

3. The ball expands slowly—no dampness appears on the hand. The crop contains 60 to 67% moisture. This is the best condition for ensiling legumes without treatment.

4. The ball springs out in the opening hand. The crop contains less than 60% moisture. Only very young crops wilted in this condition can be safely ensiled. Others are likely to mold in the silo unless layered with wet crops or placed in gastight silos.

Fig. 7–11. The grab test.

2. **The twist method.** Before chopping, the forage should be so well wilted that the stems may be twisted without breaking, but the limp leaves should show no signs of dryness. This test cannot be used for such coarse crops as sweet clover.

3. **The oven-drying method.** If in doubt or until more experience is obtained, the moisture content of a sample of any kind of forage may be obtained in about an hour's time by drying it in an oven until the weight is constant. Determine the weight of water by subtracting the oven-dried weight from the weight of the green forage. Divide the weight of water by the weight of green forage, then multiply by 100 in order to determine the percent of moisture in the forage.

4. **Other methods.** The heated-oil method and certain patented devices (as forced air dryers) may be used for moisture determination. Recently, several electronic testers which give instantaneous moisture readings of forages have come on the market. Also, microwave ovens can be used for moisture determination, provided the forage is spread in a thin layer to prevent burning. The latter two methods are expensive, but fast.

FILL RAPIDLY

Once silo filling is started, it should be rapid, so as to avoid spoilage before the silo is filled and sealed. Generally speaking, a silo should be filled in two days or less.

DISTRIBUTE FORAGE UNIFORMLY IN THE SILO

In order to avoid the presence of air pockets and spoilage, it is essential that any kind of chopped forage be distributed uniformly in the silo and that it be packed well. Proper silage distribution is obtained by keeping the material nearly level or slightly higher at the center. Silage distributing equipment is available for keeping the material in an upright silo level. These devices are very helpful, especially in silos of 14–ft or larger diameters.

Where corn, sorghum, and sunflower forage is harvested at a green, immature stage and cut into short lengths, tramping in an upright silo will not be necessary; but uniform distribution is very important. The only filling precaution under these conditions is to see that the top is carefully leveled and well packed and covered whenever filling is completed.

Grass silage (especially when wilted), hollow-stemmed forages, and forages that have matured or dried beyond the best silage stage should always be trampled well, especially near the wall.

Packing in a trench silo should be obtained by use of a tractor.

SEAL OR TOP-OFF THE SILO

Sealing or topping-off is necessary in order to avoid excess spoilage, especially with grass silage, which tends to dry on the surface and to shrink away from the silo walls. This may be accomplished by carrying out one or more of the following procedures:

1. Leveling off the top and thoroughly trampling the last few feet, especially near the walls.

2. Topping-off the silo with 2 to 3 loads of wetter material.

3. Covering the top with plastic cut to fit the silo diameter and turned up against the silo wall a distance of 5 to 8 in.

FEEDING VALUE AND ECONOMY OF SILAGE

Fig. 7–12. Breeding ewes feeding on silage. (Courtesy, A. O. Smith Harvestore Products, DeKalb, IL)

A common rule of thumb is that 3 lb of 70% moisture grass silage or 2 lb of 40% haylage are equivalent to 1 lb of hay of similar kind and quality; a difference due primarily to the higher water content of silage or haylage. Suggested practical rations for different classes of livestock in which silage is incorporated, usually in combination with hay or some other dry forage, are given in this book in the chapters devoted to each species.

Many factors enter into any figures which propose to show the comparative economy of silage vs dry forages; among them, (1) the comparative yield of total digestible nutrients per acre, (2) the cost per ton for preserving and storing, (3) the relative nutrient and feeding value, (4) the distribution of labor, (5) the control of weeds, (6) the kind of hay-making weather, (7) the hazard of curing so much hay without it becoming too mature, (8) the price per ton, and (9) the machinery and efficiency of the methods used in each method.

CHARACTERISTICS OF GOOD SILAGE

In order to make good-quality silage, producers need to know what constitutes silage quality. They need to be acquainted with those recognizable characteristics of silage which indicate high palatability and nutrient content. These are:

1. **Odor.** It has a "clean," rather pleasing acid odor, in contrast to the foul or objectionable odor of poor silage.

2. **Taste.** The taste is pleasing, not bitter or sharp.

3. **Absence of mold and rot.** There is no visible mold, and it is not musty or slimy.

4. **Moisture and color.** It is uniform in moisture and color. Very high-moisture silage is likely to be dark colored, slimy textured, and have a disagreeable odor. Generally, green or brownish silage is good; tobacco brown, dark brown, caramelized, or charred silage indicate excessive heat; and black silage is rotten and should not be fed.

5. **Animal acceptance.** Animals like and thrive on good silage.

SILAGE POINTERS

Some additional pointers which may be of value to the farmer or rancher who is making or feeding silage follow.

- **Nutrient losses in seepage**—Seepage losses vary with the moisture content, depth of silage, distribution of the silage, and the amount of nutrients in the seepage. They may be as high as 14% of the dry matter stored.

The nutrient losses vary, but generally they are in proportion to the run-off. The nutrients lost in seepage from a 100-ton silo may equal the nutrients in ¾ ton or more of hay.

- **Exposure to air**—Spoilage begins the moment silage is exposed to the air. Therefore, once the silo is opened for use, feed should be removed daily. In the wintertime, a minimum of 1½ in. of silage should be removed daily from a tower silo; in the summertime, 3 in. Also, it should be realized that spoilage is likely to occur on the surface of the ensiled material if more than 2 days elapse between filling periods.

- **Removal of silage from silo**—In the past, the common method of feeding silage was by hand—with a fork. In the present era of bigness and automation, the removal and feeding of silage is being completely automated.

- **Moldy silage**—Moldy silage may be harmful. Any spoiled material that causes animals to go off feed, or that upsets the metabolic processes, should not be fed.

Some conditions cause certain molds to produce toxins. The toxins are called *mycotoxins* and the effects of the toxins on animals are called *mycotoxicoses.*

One way in which to determine the potential toxicity of moldy silage is to feed it to some less valuable animals for at least 2 weeks. Observe the animals daily for signs of toxicity—such as reduced gain and going off feed. If no toxic effects are noticed, it is probably safe to feed the suspect silage to other animals. If ill effects are noticed, switch them to other feed immediately and dispose of the suspect silage by spreading it on the land and plowing it under.

- **Silage for summer feeding**—Some farmers and ranchers, especially dairy producers, use silage effectively as a summer feed. This practice is especially desirable in areas where pastures dry up during the hot, dry months. It appears that more and more dairy producers will change to year-around silage feeding in confinement.

- **Effect of silage on milk odor and flavor**—Silage sometimes affects the flavor and odor of milk, especially when ensiled too wet. This effect may be somewhat more pronounced with some silages than with others. The dairy producer will do well, therefore, to feed all silages after, rather than before, milking.

- **Dangerous silage gases**—Two types of toxic gases may be formed when making silage: (1) carbon dioxide (CO_2), and/or (2) nitrogen dioxide (NO_2).

Carbon dioxide forms soon after filling begins and continues until fermentation stops. It is a colorless, suffocating gas, which is heavier than air and tends to collect in low places.

Under drought conditions, corn, sorghum, and other grass species may accumulate higher than normal levels of nitrates. When ensiled, nitrates are converted to nitrites, then nitrites are converted to nitrogen oxide by bacteria and plant cells. As the nitrogen oxide comes in contact with air, it is oxidized to form nitrogen dioxide, a reddish brown-colored gas, which is heavier than air. This gas is highly toxic to both humans and farm animals.

Precautions against hazards caused by silage gases include (1) operating the blower for a 15-minute period if it is still connected, (2) swinging a piece of canvas, a tree branch, or a burlap bag vigorously so as to agitate the air and dilute gases that may be present, or (3) taking proper life support equipment when entering an oxygen-limiting, or sealed, silo. Also, adequate provision for ventilation of the silo through the roof is essential.

A victim of silo gas should be moved into fresh air immediately, and artificial respiration should be applied. A physician should be called immediately.

HIGH-MOISTURE GRAIN

High-moisture grain refers to grain that is harvested at a moisture level of 22 to 40% and stored without drying.

Interest in storing and feeding high-moisture grain was prompted by the shift toward more field shelling of corn, instead of picking ear corn, because much shelled corn must be dried for safe storage whereas ear corn can be safely stored at moisture contents up to 24% without drying. Interest in high-moisture grain increased with high energy cost for drying. It takes approximately 1 gal of propane fuel to dry 4½ to 6 bu of corn, or 1 kilowatt hour of electricity to dry 10 to 12 bu of corn, with conventional high temperature drying to reduce the moisture content of wet grain 10 percentage units (i.e., to dry from 25% down to 15% moisture).

Perry and Beeson of Purdue University pioneered in the storage and feeding of high-moisture grain in the 1950s. They showed (1) that cattle utilize ensiled, fermented high-moisture corn 10% better than conventionally dried corn, and (2) that high-moisture grain is a low-energy, low-cost method of processing grain.

Fig. 7–13. High-moisture corn being stored in one of the cement-lined trench storages at the Farr Feeders, Inc., cattle feedlot northeast of Greeley, Colorado. The high-moisture corn is stored in large quantities (60,000 tons) to provide a year's continuous supply of feed. Note truck unloading high-moisture corn at bottom of trench and tractor packing it at top of the silo. (Courtesy, William D. Farr, Farr Feeders, Inc., Greeley, CO)

The **advantages** of high-moisture grain in comparison with dried grain are:

1. It alleviates the high energy cost for drying grain.
2. It lessens field losses at harvest time.
3. It permits harvesting earlier, at higher moisture content, and usually during more desirable weather. As a result, it releases land for fall plowing in the North and for fall seeding of a second crop in the South.
4. It makes it practical to use later maturing, higher yielding varieties of corn and sorghum in the northern areas which often have early frost.
5. It usually requires less investment in processing equipment.
6. It improves the feeding value of grain for beef and dairy cattle.

The **disadvantages** of high-moisture grain in comparison with dried grain are:

1. It requires a large inventory of high-moisture grain, which may increase capital requirements.
2. It limits market flexibility for the grain, since it must be fed to livestock.
3. It may result in higher storage losses than for dry grain if proper ensiling or acid-treatment is not followed.
4. It may freeze in the bunk in the winter, and flies may be a problem in the summer.

Today, considerable quantities of high-moisture ear corn, shelled corn, sorghum (milo), and small grains (wheat, barley, and oats), containing about 30% moisture, are stored and fed. Some high-moisture grains are planned for and intended. Others are the result of happenstances, such as crops planted late, early frost damage, or harvesting when wet.

HARVESTING

A number of field equipment combinations may be used to harvest high-moisture grains. Harvest cost plus product quality should be considered in selecting the most desirable combination.

High-moisture grain should be harvested when it reaches physiological maturity and the moisture is 22 to 40%; with ear corn, sorghum (milo), and small grains higher in moisture than corn at harvest time. At this stage, grains yield the maximum available nutrients per acre and preservation conditions are best. If grain is harvested when too wet and immature, dry matter yield per acre will be less. If the grain is too dry, mold will be high and fermentation will be low; thus, grain containing less than 22% moisture should be reconstituted by adding water.

STORAGE OF HIGH-MOISTURE GRAIN

There are three basic storage methods for high-moisture grains: (1) ensiling in sealed (airtight) storage, (2) ensiling in nonsealed storage, and (3) preservation with an organic acid or ammonia.

• **Sealed (airtight) storage**—High-moisture grain may be stored in an oxygen-limiting silo. It is not necessary to crack or grind grains before storing in this manner. Another advantage of sealed storage is that it can be unloaded from the bottom.

Sealed storage is the most popular method of storing high-moisture grains even though greater initial capital investment is required. This type of storage eliminates the 2 to 5% spoilage loss normally associated with unsealed storage.

Fig. 7-14. High-moisture grain (corn) coming straight from the combine and going directly into oxygen-limited storage. (Courtesy, A. O. Smith Harvestore Products, Arlington Heights, IL)

• **Unsealed storage**—Two types of unsealed high-moisture grain storage are used; (1) conventional upright silos made of concrete or steel, which are structurally adequate for storage of grass silage; or (2) horizontal silos. The grain should be ground into the storage unit, then firmly packed; otherwise, spoilage will result. With upright silos, about a 3 in. layer off the ground, high-moisture grain should be removed from the exposed surface each day during mild weather; with horizontal silos, 4 in. should be removed daily. Greater amounts should be removed from both types of storage units during warm weather to prevent spoilage.

• **Preservation with an organic acid or ammonia**—Storage involving the treatment of high-moisture grain with an organic acid or ammonia to inhibit mold or spoilage is favored by many farmers because it alleviates artificial drying or the necessity to store in an airtight silo. Several different organic acids may be used—propionic, acetic, isobutyric, formic, benzoic, or a combination of these acids—but the most commonly used acids are propionic or propionic-acetic acid mixtures, marketed under various trade names. Anhydrous ammonia and other gaseous mixtures are also effective.

The amount of acid required to treat high-moisture grain depends on moisture content of the grain, length of storage desired, and the temperature. The recommended rate of application of propionic acid to high-moisture grain to provide protection for 1 year is shown in Table 7-4. These rates are for corn, but they are suitable for other high-moisture grains.

TABLE 7-4
AMOUNT OF 100% PROPIONIC ACID REQUIRED FOR 1 YEAR OF STORAGE[1]

% Of Moisture Content Of Grain	% By Weight	Lb (*0.45 Kg*) Per Wet Bushel[2] (*35.2 l*)	Lb (*0.45 Kg*) Per Ton (*0.907 Metric Ton*)	Gal (*3.78 l*) Per Ton (*0.907 Metric Ton*)
16-18	0.50	0.28	10	1.3
20	0.75	0.42	15	1.8
25	1.00	0.56	20	2.4
30	1.25	0.70	25	3.0
35	1.50	0.84	30	3.6

[1]The amounts of acid listed are for long term storage (1 year). For storage periods of 6 months or less, the amount of acid used could be reduced by ½.

[2]56 lb (*25.5 kg*) of high-moisture corn.

FEEDING VALUE OF HIGH-MOISTURE GRAIN

The feeding value of high-moisture grain is equal or slightly superior to that of dry grain, with some variation according to class and productivity of livestock.

Beef cattle: Improvement is greater and more consistent with high-moisture ear corn than with high-moisture shelled corn. A summary of 14 experiments showed that high-moisture ear corn increased gains by 3% and feed efficiency by 10% over dried ear corn. Studies have also shown 3 to 5% improvement in the value of the dry matter in high-moisture shelled corn for cattle. Rate of gain has been similar for cattle fed dry or high-moisture shelled corn.

High-moisture storage improves the feeding value of sorghum (milo) more than it does corn for cattle. High-moisture harvested milo increases daily gains slightly (0 to 2%) and improves feed efficiency from 6 to 10% over dry sorghum grain for beef cattle.

High-moisture storage does not improve the feeding value of wheat for cattle.

Grain is stored whole in gastight silos, but it should be rolled or ground when removed for feeding. Grain should be ground or rolled when it is stored in horizontal or conventional upright silos in order for it to pack tightly and exclude air.

Dairy cows: Research studies have shown that the feeding value of properly ensiled or acid-treated high-moisture corn is equal to that of dry corn for lactating dairy cows; the milk yield and feed intake were similar. However, when high-moisture shelled corn supplies more than 50% of the total ration dry matter, depressed milk fat percentage may occur, with inadequate fiber given as the probable cause.

Some form of processing (*e.g.*, rolling or grinding) of high-moisture corn improves utilization.Whole kernels appearing in feces indicate incomplete digestion. Processed corn is higher in digestible, metabolizable, and net energy for dairy cows than rations containing whole shelled corn.

Swine: When compared on an equal dry matter basis and in mixed rations, rate of gain and feed efficiency are essentially the same for hogs fed high-moisture or dry grains. Free-choice feeding of high-moisture grain may be used successfully for hogs weighing more than 60 lb provided proper intake of the protein supplement relative to grain intake is assured. But free-choice feeding is not recommended for pigs weighing under 60 lb.

There is no advantage to grinding or cracking high-moisture corn for growing-finishing pigs other than for mixing purposes. However, grinding or cracking increases the feeding value of high-moisture sorghum (milo), barley, and wheat.

RECONSTITUTED GRAIN

Reconstituted grain is mature grain to which water has been added to raise the moisture content to 25 to 30% for storage.

Reconstituted sorghum (milo) stored whole in an airtight silo, then processed at feeding time and fed to finishing cattle, will improve rate of gain slightly and feed efficiency by 12 to 15%. However, if sorghum is ground before reconstituting, there is little improvement in feed value over dry milo for cattle. Reconstituted whole corn fed to finishing cattle will not improve rate of gain, but will improve feed efficiency by about 4.5%.

BUYING AND SELLING HIGH-MOISTURE GRAIN

Most high-moisture grain is grown by livestock producers for livestock feed, or grown as a cash crop and sold at harvest to livestock producers. High-moisture grain is bought/sold at a lower figure than dry grain because of the water content. Also a greater quantity of it must be fed because of the water content. In most cases, the buying and selling transactions of high-moisture grains are based on an 87% dry matter and 13% moisture basis.

Conventional upright (tower) silos—concrete stave silos. (Courtesy, Portland Cement Assn., Chicago, IL)

Gastight (oxygen-limiting) silos—glass lined structures. (Courtesy, USDA)

Temporary silos—round bale wrapped in plastic to produce "bale silage." (Courtesy, Ford New Holland, New Holland, PA)

Roundup! (Courtesy, USDA)

SECTION

8

MANAGEMENT

Contents	Page

Fig. 8–1. Branding time at Tequesquite Ranch, Albert, New Mexico. With this size crew, 700 calves are branded in 1 day. (Photo by H. D. Dolcater, Amarillo, TX. Courtesy, Tequesquite Ranch, Alberta, NM)

Management is the art of caring for, handling, or controlling. In a livestock operation, it gives point and purpose to everything else. It can make or break a livestock outfit.

Pertinent facts relative to, along with methods of accomplishing, some important livestock management practices are covered in this section. Management practices of importance to all classes of livestock are presented in the first part of the section, followed by specific class of livestock practices.

MARKING OR IDENTIFYING ANIMALS

The method of marking or identifying animals will vary according to the class of animals and the objectives sought. Thus, some methods of marking are well adapted to one class of animals but not to another; horn brands, for example, may be used on horned cattle only. Also, on the western range, marking or branding is primarily a method of establishing ownership and/or age; whereas in the small herd, particularly in the purebred herd, it is a means of ascertaining ancestry or pedigree.

MARKING OR IDENTIFYING BEEF AND DAIRY CATTLE

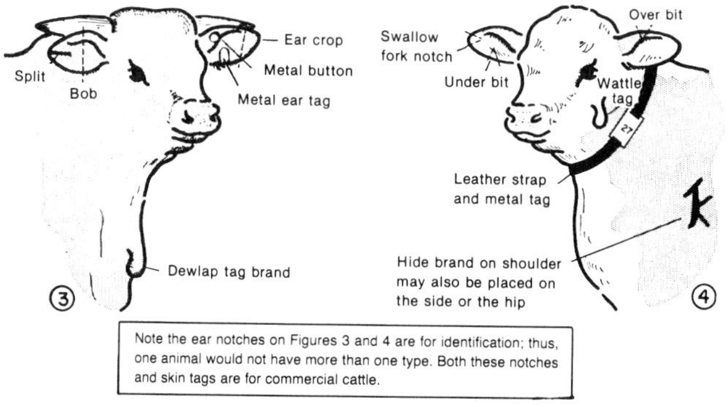

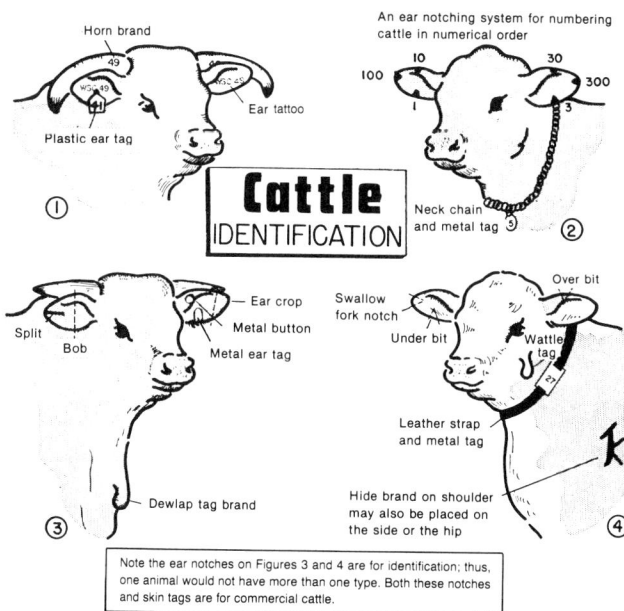

Fig. 8–2. Composite drawings showing a number of methods of cattle identification. It is unlikely that any individual animal will carry more than 1 or 2 of these methods of identification. (Drawing by R. F. Johnson)

The method employed to mark cattle should be determined primarily by the objective sought. Fig. 8–2 shows a number of methods of marking or identifying cattle. A description of each of these and other methods follows:

• **Hide brands**—When properly applied, hide brands are permanent. Throughout the range country, the hide brand is recognized as the rancher's trademark. Most of the western states require that each brand be recorded, as to both type and location, in order to avoid duplication. When stock are run close to a state boundary, the same brand may be recorded in two states.

In addition to the regular brand, many ranchers identify the age of the females by adding the last number of the year (usually at a different location). Thus, heifer calves born in 1991 might be identified by the number 1. At the end of 10 years, the numbers are used over again, for there is seldom any difficulty in determining ages where there is a 10-year spread. In those states where brands are recorded, these added numbers or brands must also be approved by and recorded with the registrar of brands.

Hide brands have the disadvantage of being unsightly, and hot iron brands lower the market value of the hide. For these reasons, they are not recommended except when necessary for identification purposes. Even then, it is desirable that their size be as small as possible, consistent with serving the primary objective of the brand. Pertinent facts about, and methods of applying, hide brands, along with characteristics of good brands, follow.

1. **Four methods of applying brands are:**

a. **Hot iron.** To date, this has been the preferred method. Ranchers heat irons to a temperature that will burn sufficiently deep to make the scab peel, but which will not leave deep scar tissue. The proper temperature of the hot iron is indicated by a yellowish color. Branding is accomplished by placing the heated branding iron firmly against the body area which it is desired to mark and by not allowing it to slip for the few seconds when the hide is burned. The branding iron should be kept free from dirt and adhering hair at all times.

Where electricity is available, the electric iron may be used; it keeps an even temperature, and if properly used, makes a clear, uniform brand.

b. **Freeze branding.** This method, developed by the U.S. Department of Agriculture, at Washington State University, makes use of a superchilled (by dry ice or liquid nitrogen) copper branding *iron* which is applied to the closely clipped surface for about 20 seconds, thereby depigmenting the hair follicles, following which the hair grows out white. When properly done, this method is painless, permanent, and there is no hide damage.

On white cattle, deliberate overbranding (30 seconds or more) will produce a bald brand suitable for identification after clipping.

Fig. 8–3. Freeze brand No. 851 on an Angus bull. (Courtesy, M. Jorgensen, Jorgensen Ranches, Ideal, SD)

c. **Branding fluids.** Branding fluids, which are less widely used in making hide brands, consist of caustic material applied by means of a cold iron. Best results are secured if the area is first clipped. The chemical method of producing hide brands is slower; the results are generally less satisfactory, particularly if the operator is inexperienced with the method; and the resulting brand is less permanent.

d. **Laser branding,** which is permanent, but which needs further development and experimental study.

2. **Characteristics of a good brand.** A good brand is one that is easily read, that is of simple design and yet cannot be easily changed or tampered with, that has no welds or thick points in the iron, and that interferes with the circulation as little as possible. Thick points mean deeper burning and slower healing; whereas small enclosed areas, such as a small O, will slough out entirely.

• **Plastic ear tags**—Plastic ear tags, which can be applied at any age, are a popular method of animal identification. Their **advantages:** they are economical; they can be read at a distance; they are flexible—for example, the individual animal number can be placed on the bottom of the tag, and the sire or dam numbers can be placed on the top. The major **disadvantages** are: plastics tend to become hard and brittle in cold weather; loss of tags, prenumbered tags with block-type numbers are difficult to read if tags get soiled; therefore, most herd owners prefer to purchase blank tags and then number them with the largest possible numbers.

Because of the frequency of loss, plastic ear tags should be duplicated by having a tag in each ear, or used in combination with a permanent means of identification.

Plastic ear tags come in a variety of styles, sizes, and colors. Basically, there are 3 styles: 1-piece plastic, 2-piece plastic, and over-top-of-ear or on-top-of-ear tags. The latter have the advantage of being located in the toughest part of the ear; also, the on-top-of-ear tags may be numbered on both sides for ease of reading. The larger plastic tags are more popular because they are economical, easy to install, easy to read from a distance, stay pliable in cold weather, and stay in the ear longer than metal tags. Yellow tags with black numerals are the most readable; red and white tags tend to fade or discolor.

Plastic tags can be purchased prenumbered or blank.

• **Metal ear tags and buttons**—Metal ear tags are difficult to read at a distance, and, due to their sharpness, cut through the ears rather easily. Metal buttons stay put, but they cannot be read at a distance. Both metal tags and metal buttons frequently rub and scratch the skin, thereby making openings for screwworm infestation. Like plastic tags, metal tags can be purchased numbered or blank.

• **Tattoos**—Tattooing is a permanent method of identification, required by most purebred beef cattle registry associations. It consists of marking by piercing the skin with instruments equipped with needle points which form letters or numbers. This operation is followed by rubbing indelible ink or paste into the freshly pierced area. After healing, the tattoo is permanent. It is well to disinfect the tattooing instruments carefully between each operation in order to alleviate the hazard of spreading warts to the pierced area, for warts make it impossible to read the tattoo.

Animals may be tattooed at any age, but it is most convenient to tattoo baby calves.

The **advantages** of a tattoo are: it is permanent, and it does not disfigure the animal.

The major **disadvantage:** cattle must be confined in order to read tattoo numbers. Even then, tattoos are difficult to decipher on dark-skinned animals. For this reason, most producers apply a brand or ear tag, in addition to the tattoo, so that the animal can be easily identified from a distance.

• **Earmarks**—Earmarks are permanent and easily recognized, but they're unsightly. They may be administered with either a sharp knife or a regular ear notcher. Sometimes polled animals are individually identified through ear notches. In such instances a definite value is assigned to each area location. When earmarks are used in commercial operations, however, they are uniform and recorded for any given ranch. Some of the more common earmarks are *crops, swallow forks, bobs, over-bits, under-bits,* and *splits.*

• **Neck chains or straps**—Neck chains or straps are the most frequently used means of identifying polled cattle. Occasionally, chains or straps may be lost, but this is not particularly serous if the caretaker is on the alert and immediately replaces each one that is lost, without allowing several losses to accumulate before taking action. In rare instances, an animal will hang itself by the chain.

Neck chains or straps must be adjusted, for young animals grow, or animals change in condition.

• **Horn brands**—Horn branding for individual identification is commonly used among breeding or sale animals of the horned breeds. Usually horn brands are made by heating small copper numbers with a blow torch or charcoal burner. On mature animals, this method of branding works fairly well, but it cannot be used on young animals while horns are still growing, unless it is repeated at intervals.

• **Other identification**—Other identification marks used on the range include: (1) *buds* formed by making a strip incision through the nose; (2) *wattles* made by cutting down a strip of skin on the jaw bone; and (3) *dewlaps* formed by cutting a strip of skin on the brisket.

The U.S. Department of Agriculture requires that most cattle 2 years of age or older be backtagged or eartagged to identify the animals to their herd of origin before they are shipped across state lines.

Various electronic devices are in different stages of research and development; among them—

1. **Radio transmitter in the second stomach.** The animal swallows a small radio transmitter enclosed in a ¾-in. × 2½-in. plastic capsule, which lodges in the second stomach. From there, it transmits a coded number when signaled by a receiving unit to do so. The transmitter can be retrieved at slaughter and reused.
2. **Implant under the skin.** Basically, the transponder is a silicone-coated minicircuit which is implanted under the animal's skin (for example, behind the poll) and powered by a microwave beam from a portable receiver unit. In addition to the animal's identification number, this device may include such information as a birth date, original owner, state of origin, year of implant, and temperature of the animal. The method holds great promise as a means of combatting cattle theft.

MARKING OR IDENTIFYING SHEEP AND GOATS

Sheep operators often find it necessary to mark sheep for one or more of the following reasons:

1. Identification of western sheep on ranges, especially public land, where the brands of different owners may get mixed.
2. Identification with a "buck brand" at breeding time.
3. Identification of ewes and lambs at lambing time.

The need is for a brand which will satisfactorily (1) serve for identification purposes, and (2) scour out.

The common methods of marking or identifying sheep are:

• **Branding fluid**—There are on the market commercial branding fluids which possess the following desirable features: (a) They will remain on the sheep for a year, and (b) they can be removed from the wool by normal scouring methods.

• **Earmarks**—Identification for sheep can be provided by earmarks, made with either a sharp knife or a regular ear notcher. Such marks are permanent and easily recognized, but unattractive.

Where individual identity is desired, as is necessary in a purebred flock, a definite value is assigned to each area location. With a commercial band, however, the same mark is administered to all animals.

• **Metal or plastic ear tags**—Most purebred sheep are provided with a metal or plastic ear tag; and sometimes two. Where two tags are used, generally one of them is the individual or flock number assigned by the owner, whereas the other is the individual number assigned by the breed registry association.

Metal or plastic ear tags are easily attached but easily lost. Also, they frequently rub or scratch the skin, thus, making an opening for screwworm infestation.

• **Ear tattoos**—These are administered in the same manner as described for cattle.

MARKING OR IDENTIFYING SWINE

Purebred breeders find it necessary to employ a system of marking so that they may determine the parentage of the individuals for purpose of registration and herd records. Even in the commercial herd, a system of identification is necessary if the gilts are to be selected from the larger and more efficient litters.

Plastic ear tags, branding, and tattooing are also used for identifying swine.

• **Ear notching system**—The common method of marking or identifying swine consists of ear notching the litters. Pigs are generally marked at the same time that the needle teeth are removed.

The ear notches are usually made with a special V-notcher. Most of the breed associations are in position to recommend a satisfactory marking system, and many require the Universal Ear Notching System shown in Fig. 8–4. This is the most common notching system though there are others.

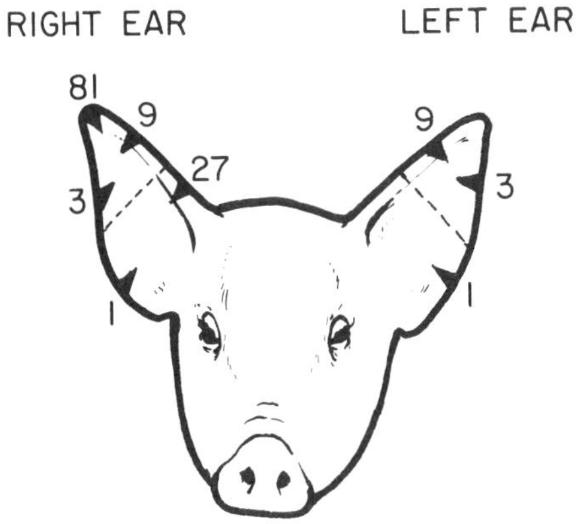

Fig. 8–4. Universal Ear Notching System (also known as the 1-3-9-27 system) used, or recommended, by most swine registry associations. Right ear is used for litter mark, and left ear is used for individual pig number. Up to 161 litters can be marked with this system.

• **Electronic I.D. implants**—Several different types of electronic devices for identifying individual animals are available. They are especially suited to swine, because they usually destroy or lose other identification equipment. The electronic equipment involves a transponder which is interrogated by an external electronic device to identify animals. In addition to identification of individual animals, electronic implants can be coupled with software that will permit linking dates from birth to buyers. Also, stolen animals can be traced to rightful buyers.

MARKING OR IDENTIFYING HORSES

The correct identification of Thoroughbred horses racing at the various tracks of the country is important. Thus, one of the duties of the steward, through the horse identification assistant, is that of assuring that each starter in a race is actually the horse named in the entries. This is necessary because only a relatively small percentage of the more prominent racehorses are fondly recognized by sight by the public, the vast majority of racehorses being known only by names and past performances.

To the end that the identity of each horse shall be guaranteed, The Jockey Club requires that all horses running at member tracks of the Thoroughbred Racing Association be lip tattooed.

The three common methods of marking or identifying horses—lip tattoo, freeze marking, and photographs of chestnuts—are herewith presented.

• **Lip tattoo**—The Thoroughbred Racing Protective Bureau (TRPB), Inc., utilizes this method to guarantee to the public the identity of each and every horse running at the tracks of their members. The system consists of tattoo branding, with forgery-proof dyes, The Jockey Club serial number (the registry number) under the upper lip of the horse, with a prefix letter added to denote the age of the horse (see Fig. 8–5). The process is both

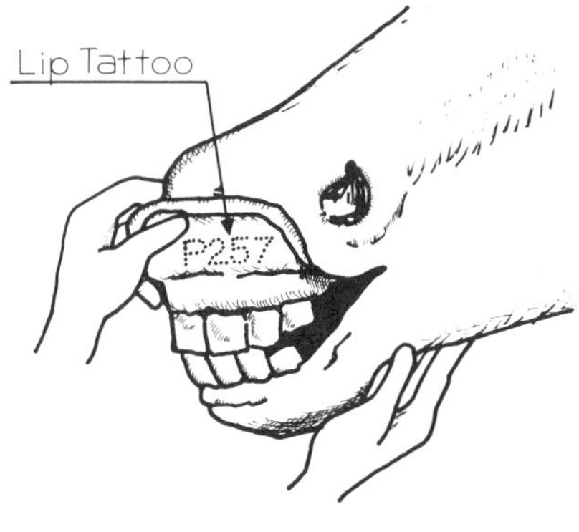

Fig. 8–5. A drawing showing the lip tattoo under the upper lip of a horse. The prefix denotes the age of the horse, and the numbers denote the Jockey Club registry number.

simple and painless. It is applied by expert crews of the TRPB to two-year-olds as they come to each TRA track. Details relative to lip tattooing follow:

1. After the digits are placed in the head of the tattoo gun, place the gun head with the digits in a dish of antiseptic (such as Zephiran chloride, available at any drug store).

2. Roll and hold upper lip back with fingers; do not place anything back of the lip.

3. Wipe upper lip clean with cotton saturated with rubbing alcohol.

4. Shake gun to dry off excess antiseptic.

5. Apply tattoo gun, making sure gun and digits are square with lip. Hold gun rigidly and with sufficient pressure to withstand recoil action of gun.

6. Apply ink and rub into perforations with thumb. Use more ink if bleeding persists. Leave any excess ink on lip.

Initially, the use of the horse lip tattoo was limited to identification of Thoroughbred racehorses running on member tracks of the Thoroughbred Racing Association, an exclusiveness which the Jockey Club maintained by not making available to others either the tattoo equipment or the formulation of the ink. However, both lip tattoo equipment and ink are presently available through some of the livestock supply houses.

- **Freeze marking (cold branding)**—This method of identifying horses, known as freeze marking (cold branding), was developed at Washington State University. Called the Angle System, it is derived from the ancient Arabic numeral system. It utilizes the basic principle that straight lines are easy to make with crude instruments. It offers simplicity, preciseness, universal application, and good visual communication. Also, it lends itself to a computerized data retrieval system. Freeze marking is now used to identify horses in the United States, and in several other countries.

Called freeze marking to escape painful association with the term *branding,* the technique utilizes heavy copper stamps, or marking rods, chilled in either liquid nitrogen (at −300°F) or dry ice and 95% alcohol. The area to be marked is shaved and scrubbed with 95% alcohol wetting solution to aid in conducting the intense cold and to withdraw body heat.

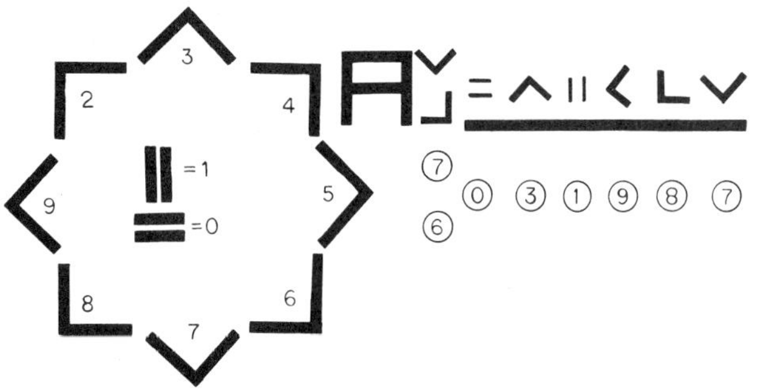

Fig. 8-6. Left is a series of right angles developed at Washington State University. The number that each angle represents is written inside the angle.

Right is a freeze brand as it would actually appear under the mane of a horse. The encircled numbers do not appear in the brand; they are used in this illustration and caption merely to enhance readability. The first symbol, the capital A, denotes that the horse is a purebred Arabian. The stacked symbols in the second position indicate the year of birth of the horse. The remaining symbols are the horse's registration number: 031987

Placing the copper stamp against the animal's body for 10 to 20 seconds destroys pigment producing cells (melanocytes) and produces a pigment-free skin area. Hairs growing back in this area will be white. Longer application times result in more balding, a condition necessary for producing legible marks on white or light colored animals.

A freeze mark that produces white hair causes only minimal changes in the hide and does not seriously impair leather properties. Freeze marks that produce baldness can cause some permanent scarring and hide damage. Severe freeze mark damage, however, is minimal compared with fire brand damage.

Freeze marking is more legible than fire branding. Marks are much more distinct, and last just as long. No open wound is produced, which eliminates disease and insect infestations, and freeze marking is relatively painless.

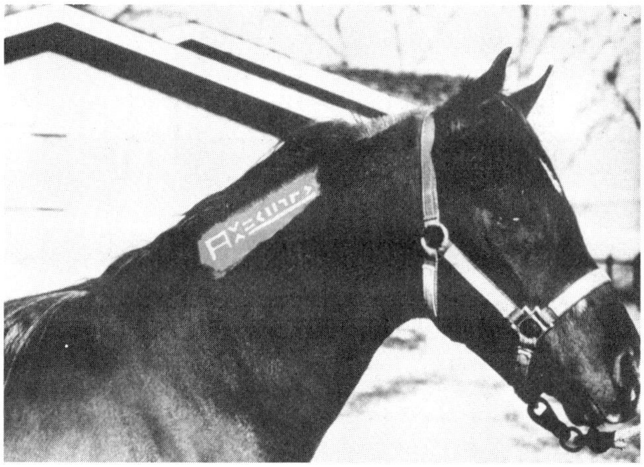

Fig. 8-7. Cold (freeze) branded horse. Note that the identity is placed on the neck under the mane, as is preferred by some exhibitors. (Courtesy, USDA)

Because some equestrians, particularly those who show horses, object to a visible mark, the mark is usually placed on the neck under the mane. It is applied approximately 2 in. below the eruption of the mane about midway between the poll and withers. An area approximately 12 in. × 7 in. is clipped close to the skin and washed with alcohol. The iron is then applied to the clipped area of the neck.

- **Photographs of chestnuts**—This consists of a life-size picture of the chestnuts (or night eyes) of each horse, together with pictures of the sides, front and rear of the animal, showing all markings. This corresponds to the human system of fingerprinting employed by the FBI and police departments throughout the world.

Studies have revealed (a) that no two chestnuts are exactly alike, and (b) that from the yearling stage on these chestnuts retain their distinctive size and shape. The chestnuts are photographed, and then classified according to (a) size, and (b) distinctive pattern.

In comparison with the lip tattoo system, "fingerprinting" horse chestnuts is more costly and necessitates more highly trained people to record and use it.

DEHORNING, CASTRATING, AND DOCKING FARM ANIMALS

Figs. 8–8 to 8–12 illustrate and Table 8–1, p. 314, lists the common methods of dehorning, castrating, and docking animals, and give the directions for accomplishing each. Many livestock producers routinely administer these management practices; others call upon a veterinarian for all or part of them. Perhaps the most important thing is that they be done at the proper time.

CATTLE DEHORNING EQUIPMENT

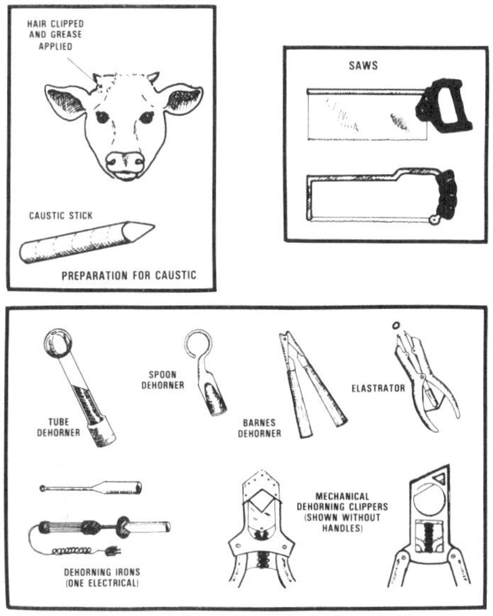

Fig. 8-8. Common instruments used for dehorning cattle.

CATTLE CASTRATING METHODS

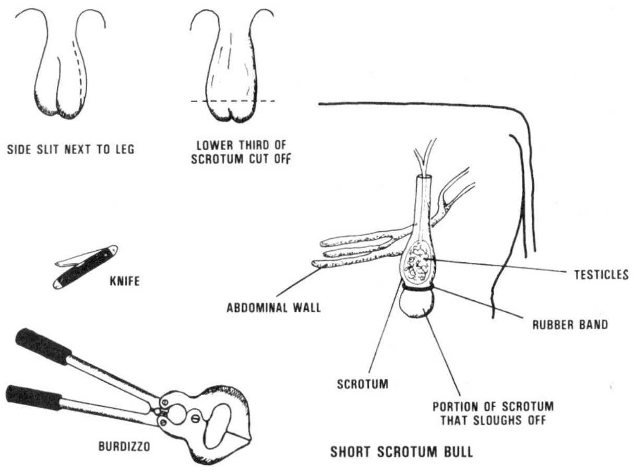

Fig. 8-9. Common methods of castrating cattle, and 2 common pieces of cattle castrating equipment (the knife and the Burdizzo). In using the knife, either the scrotum may be slit down the sides, or the lower third may be removed. With the Burdizzo, first the cord is worked to the side of the scrotum, and then the instrument is clamped on about 1½ to 2 in. above the testicle, where it is held for a few seconds. The short scrotum method of castrating is detailed in Table 8-1. (See Fig. 8-10 for an elastrator.)

LAMB DOCKING AND CASTRATING INSTRUMENTS

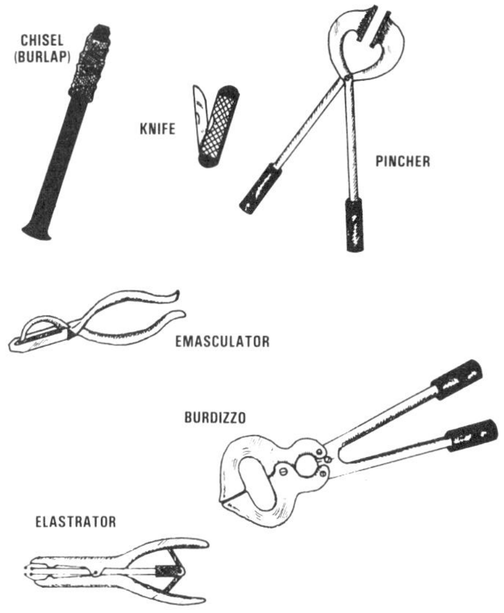

Fig. 8-10. Common instruments used for docking and castrating lambs.

LAMB DOCKING AND CASTRATING METHODS

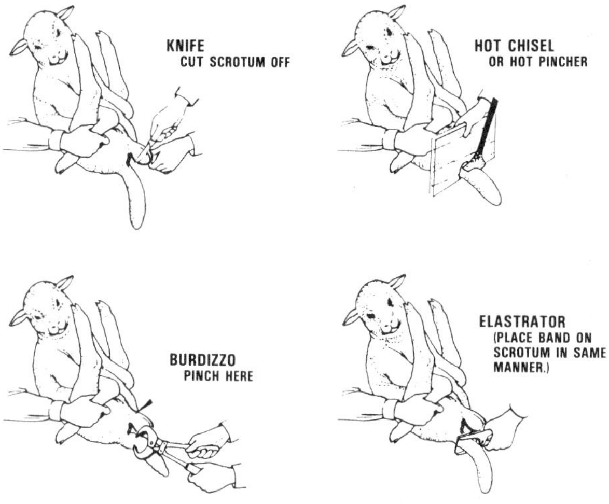

Fig. 8-11. Methods of docking and castrating lambs.

SWINE CASTRATING STEPS

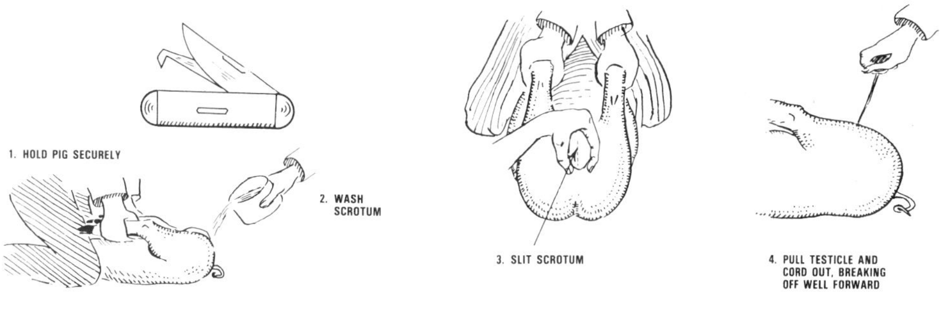

Fig. 8-12. Swine are castrated with a knife, by following four steps herewith illustrated.

TABLE 8-1
DEHORNING, CASTRATING, AND DOCKING METHODS

Class of Animal	Skill	Method	Directions	Remarks
Cattle, Beef & Dairy	**Dehorning**	Breeding polled cattle.	The use of polled bulls is the most humane method of securing cattle without horns. If a polled bull is *pure* polled (carrying in his blood no tendency to produce horns), practically all of his calves will be polled, even though their dams have horns.	This system of securing hornless calves saves labor and avoids pain and possible setback to the calves. Without doubt, the breeding of polled calves will increase in popularity. When dehorning other than by breeding polled cattle, it should be done at as young an age as possible, thus lessening weight loss, work, needed equipment, and bruising. About 2 wk are needed following dehorning for yearling animals to regain their original weight.
		Saws and clippers	Confine or restrain animal to be dehorned in a suitable chute, pinch gate, squeeze pen, or cattle stock. Calves may be handled by throwing, by snubbing them to a fence post, or by tying one side of the body against a strong fence or solid wall. Whatever the instrument used (saws or clippers), remove the horn with about ¼ to ½ in. of the skin around its base.	Clippers are satisfactory for removing the horns of younger cattle, but the hard, brittle horns of mature cattle can best be removed with a saw. With 2-yr-old or older cattle, the horns are usually tipped. Saws and various kinds of shears and clippers are the most widely used method of dehorning in the range country. Electrically operated saws are now available. Less insect trouble is encountered when dehorning is done in the early spring or late fall. If the operation is performed in fly season, apply a fly repellent to the wound.
		Dehorning spoon and dehorning tube	The steps and directions for using the dehorning tube are as follows: 1. Restrain the calf. 2. Select a tube of proper size to fit over the base of the horn and about ¼ in. of skin all the way around. 3. Place the cutting edge straight down over the horn and then push and twist until the skin has been cut through, making a cut ⅛ to ⅜ in. deep. 4. Hold the tube at about a 45° angle and rapidly turn and shove the cutting edge until the button comes off.	The dehorning spoon (or gouge) is a small instrument with which the horns of young calves can be gouged out. The dehorning tube is a newer instrument than the spoon, and is faster, less tiresome to use, and more certain to avoid regrowth. Either instrument can be used on calves up to 60 days of age. Dehorning tubes come in 4 sizes, varying from ¾ to 1⅛ in. in diameter. Practice cleanliness, and disinfect instruments (except hot irons) between animals in order to lessen infection and disease.
		Hot iron	The hot-iron method of dehorning consists of the application of a specially designed hot iron to the horns of young calves. Where electricity is available, the electric hot iron may be used; it keeps an even temperature, without getting too hot or too cold.	The hot-iron system of dehorning is bloodless and may be used at any time of the year, but it can be used on young calves only.
		Chemical (dehorning pastes or sticks)	Use caustic potash (potassium hydroxide) or caustic soda (sodium hydroxide) in either paste, stick, or lacquer base form. (The accompanying use of Vaseline is not necessary with the lacquer base.) Apply when the calf is 3 to 10 days old. Clip or shear the hair from around the buttons, and surround the area with a ring of heavy grease or Vaseline to protect the eyes against the chemical. Rub the chemical over the button until blood appears, protecting the hands while doing so.	The use of chemicals should be limited to small herds kept under supervision. Following the application of a dehorning paste or stick, keep calves away from their dams for a few hours and out of rain for 24 hr.
	Castrating	Slitting scrotum down the sides.	Pull one testicle down at a time and hold it firmly to the outside so that the skin of the scrotum is tight over the testicle. With a sharp knife, make an incision on the outside of the scrotum next to the leg. It is important that the incision extend well down to the end of the scrotum to allow for proper drainage and that it extend through both the scrotum and the membrane. If desired, the membrane need not be slit; simply remove it along with the testicle. Removal of all, or a substantial portion, of the membrane alleviates the possibility of it collecting blood and forming a clot. Remove the testicles by pulling them out. In older cattle, excessive bleeding may be prevented through severing the partially withdrawn cord by scraping with a knife or by clamping with an emasculator.	Castration is best done when calves are 4 to 12 wk of age. Young animals are usually thrown to be castrated, whereas animals of 8 mo of age or older may be more easily operated on in a standing position. It is best to perform the operation in the early spring or late fall to avoid infestation from flies; otherwise, a repellent should be applied to the wound. Keep cutting tools and hands clean and sterilized. Where screwworms exist, a fly repellent should be applied to the wound and the animals should be kept under close observation until the wound has healed over; or a bloodless method of castration (such as the Burdizzo) should be used. Currently, there is a trend toward castrating at slightly older ages than formerly, primarily (1) to take advantage of the higher gains and greater efficiency of bulls, and (2) to lessen the hazard of urinary calculi.

(Continued)

TABLE 8–1 *(Continued)*

Class of Animal	Skill	Method	Directions	Remarks
Cattle, Beef & Dairy *(Continued)*	**Castrating** *(Continued)*	Removal of lower end of scrotum.	Remove approximately the lower third of the scrotum, exposing the testicles from below. Slit the membrane covering each testicle. If desired, the membrane need not be slit; simply remove it along with the testicle. Remove the testicles by pulling them out. Cord may be pulled in calves up to 3 to 4 mo of age; emasculators may be used on older calves and bulls.	Except in small calves, post castration infection has been associated with castration by removal of the lower end of the scrotum, due to the bottom of the scrotum curing up and stopping drainage. The lateral sides of remaining scrotum may be slit up to the body to help drainage.
		Short scrotum bulls.	The scrotum is shortened by bringing it through a distended rubber band with an elastrator when the calf is 1 to 3 mo of age. Before the band is released, the testicles are moved near the abdominal wall. The scrotum below the rubber band sloughs off after 3 to 4 wk. Shortening the scrotum requires considerably less time than castration—it can be done in 15 to 30 seconds; and there is no weight loss. As a result of the shortened scrotum, the testicles lie close to the abdominal wall and the animal is sterile. The testicles then develop to about half the weight of those from fertile bulls of the same age and weight.	This patented method of rendering intact males infertile, known as "More-Lean Beef," was developed by the New Mexico Station. A short scrotum animal is really a pseudocryptorchid. The short scrotum treatment does not change either the temperament or the urge of animals; hence, there will be riding if sexes are not kept separate. The rate and efficiency of gains of short scrotum bulls and intact bulls are about the same, and the carcasses of short scrotum bulls are leaner than the carcasses of steers.
		Burdizzo pincers.	Throw the animal. Work the cord to the side of the scrotum, and then clamp the Burdizzo on about 1¾ to 2 in. above the testicle, where it is held for a few seconds. Then repeat this operation on the same cord at a location about ¼ in. removed from the first one. Repeat the same procedure on the other testicle.	Burdizzo pincers (named after their inventor, Dr. Burdizzo, and manufactured in Italy) make a "bloodless castration." In using the Burdizzo, it is important that the cord not slip out, that only one cord be clamped at a time, and that there is no interference with the circulation of the blood through the central portion of the scrotum. This method of castration is satisfactory if done properly and by an experienced operator.
		Elastrator rings.	The elastrator works best on young calves under 2 mo of age. Hold the calf in either a sitting or lying position. Press both testicles through the ring and to the lower end of the scrotum, and then release the rubber ring.	The elastrator (developed in New Zealand) is an instrument for use in stretching a specially made rubber ring, which may be placed over the scrotum to castrate young animals.
		"Russian method."	Make incision and remove spermatozoa-producing tissue of the testicle, but leave intact the sheathing layer that produces testosterone—a growth hormone.	This method of castration, which originated in the U.S.S.R., is now being evaluated in the U.S. The Russians claim that animals castrated in this manner gain like bulls, but have carcasses like steers.
		Injectable chemical castration (CHEM-CAST).	A patented, injectable chemical solution, sold under the brand name *Chem-Cast,* was approved by FDA in 1983 and is now available through veterinarians. When used according to label directions, it painlessly destroys the testicles and spermatic cords of bull calves weighing up to 150 lb. There is no weight loss, going off-feed, high risk infection, hemorrhaging, or pain; and it is nearly 100% effective. The procedure consists of injecting the prescribed dosage of Chem-Cast (for calves weighing up to 100 lb, the dosage is 1 ml per testicle; for calves weighing 101 to 150 lb, 1.5 ml per testicle) from the top into the middle third of each testicle via a small hypodermic needle.	About 24 hours following the injection, there is a swelling of the testicle, testicular vessels, and the spermatic cords. The swelling disappears within 2 weeks and the testicles begin to be reabsorbed and to be reduced in size. At the end of 60 days, only small, hard, nodules are left in the cod; and, in time, these disappear, also. Any cattle producer can use Chem-Cast after receiving a minimum of instruction and training.
Sheep	**Docking**	Hot iron (pincers or chisels)	Do not heat instruments beyond a very dull red color. Protect the lamb's buttocks by placing the tail in a slot in the end of a board or by putting it through a hole in a board. Sever the tail rather quickly, avoiding any more burning than necessary to prevent bleeding.	The use of hot instruments results in less loss of blood and less danger of infection than the use of a knife, but the wound heals more slowly.
		Knife or shears.	Press the skin toward the body before cutting, leaving loose skin above the cut which will close over the wound. With small flocks, some operators make a practice of tying a string or placing a rubber band around the tail prior to cutting, preventing a loss of blood in this manner. If this is done, the string or band should be removed 3 to 4 hr later. Sever the tail at the place desired.	All lambs should be docked when they are 7 to 14 days of age. Strong lambs may be docked and castrated at the same time, with castrating being done first. Where cutting instruments are being used, keep the hands clean and the instruments clean and disinfected. The lamb is usually held with its back to the assistant, who grasps the hind and front legs of the same side in each hand. Sever the tail about an inch from the body as measured on the underside of the tail.

(Continued)

TABLE 8-1 *(Continued)*

Class of Animal	Skill	Method	Directions	Remarks
Sheep *(Continued)*	**Docking** *(Continued)*	Burdizzo pincers.	Close the jaws over the tail at the point at which it is to be severed. Cut the tail off inside the closed jaws, after the Burdizzo jaws have been closed.	Burdizzo pincers may be used for castrating as well as for docking. This is a bloodless method.
		Emasculator.	Close the emasculator over the tail at the point at which it is to be severed.	The emasculator crushes a part of the tissue while cutting, thus lessening the loss of blood.
		Elastrator.	Draw the tail through the ring, and then release the rubber ring at the point where it is desired to sever it. Where there are scouring or unsanitary conditions, it may be advisable to cut off the tail of the lamb after the ring has been applied.	The elastrator is an instrument for use in stretching a specially made rubber ring, which may be placed around the tail to dock young lambs. The elastrator is a bloodless method of docking. The rubber band cuts off the blood supply, and atrophy follows.
	Castrating	Knife.	Grasp the tip of the scrotum, and hold it tight while cutting off the lower third. Draw out the exposed testicles together with the surrounding membranes with either the hands or teeth.	All male lambs not to be left as rams should be castrated when they are 7 to 14 days of age. Strong lambs may be castrated and docked at the same time, with castrating being done first. Where cutting instruments are used, keep the hands clean and the instruments clean and disinfected. The lamb is usually held with his back to the assistant, who grasps the hind and front legs of the same side in each hand.
		Short scrotum.	The technique is the same as for bulls; hence, see "short scrotum bulls."	The best time to shorten the scrotum is when lambs are 1 to 3 mo of age.
		Elastrator.	Press both testicles through the ring and to the lower end of the scrotum and then release the rubber ring.	The elastrator is an instrument for use in stretching a specially made rubber ring, which may be placed around the scrotum to castrate young lambs. The elastrator is a bloodless method of castrating.
		Burdizzo.	Work the cord to the side of the scrotum, and then clamp the Burdizzo on above the testicle, where it is held for a few seconds. Repeat the same procedure on the other testicle. Complete atrophy of the testicles follows in about 6 wk.	This is a bloodless method in which the testicles are made functionless through destroying their channels of nourishment.
Swine	**Castrating**	Knife.	Restrain or hold the animal in a manner in keeping with its age and size and the number of helpers available. A young pig is generally either (1) suspended by its hind legs with the back toward the helper (the helper also clamps his/her knees against the pig's ribs, near the shoulders,) or (2) held on its back on the top of a table (this requires either (a) a castration crate, or rack, or (b) two helpers—one grasping the front legs and the other the rear legs). Large boars are usually snared around the upper jaw and behind the tusks, with the free end of the snare tied to a post; then further restraint is applied either by tying all four legs or by hoisting the hind legs, with the animal castrated in either a standing or lying position. Wash the hands thoroughly with soap and water and rinse with a good disinfectant. If the scrotum is dirty, wash it with soapy water, using a coarse fiber brush. After washing, disinfect the area. Also, disinfect the knife before and between operations. With a sharp knife, slit the scrotum on each side (the one-incision method—with the cut made directly between the testicles—is satisfactory, and is preferred by some) as each testicle is pressed outward. Extend both cuts well down to allow for proper drainage, and cut deep enough to extend through the scrotum and membrane. (If desired, the membrane need not be slit on young boars; simply remove it along with the testicle. However, it is desirable to slit the membrane on old boars, rather than break down fibrous attachments of membrane to scrotum.) Pull the cord out (with tension directed backward; otherwise, the inguinal ring may be torn and evisceration produced)—or break the cord off well forward. Use the emasculator on old boars. In fly season, apply an insect repellent to the wound.	Male pigs not intended for breeding purposes should be castrated while they are still suckling their dams, and far enough in advance of weaning to allow healing before being separated from the dam (the incision usually heals in 2 to 3 wk). The operation should not be done at the same time that pigs are vaccinated. Generally this means that castration should be done within the first 4 wk (some castrate 5-day-old pigs); pigs that are weaned at 4 wk of age or earlier should not be castrated within 1 wk of the time of weaning. In preparation for the operation, pigs should be kept off feed a short time. Boars that are no longer useful in a breeding program may be castrated to remove the boar odor before marketing, which operation is known as stagging. By the time the castration wound has healed (in 3 to 4 wk), the odor usually disappears enough to allow the stag to be marketed. Pigs with undescended testicles or ruptures (scrotal hernias) should be operated on by a veterinarian.

SPAYING HEIFERS

In females, the operation corresponding to castration is known as spaying. Under most conditions, desexing of the heifers is not recommended because: (1) the operation is more complicated and difficult, requiring a very experienced person, (2) spaying is attended with more danger than castration; (3) it eliminates the heifers for possible replacement purposes or sale as breeding stock; and (4) experiments and practical operations with spayed heifers have generally shown that the selling price obtained is not sufficiently higher to compensate for the lower and less efficient gains plus the attendant risk of the operation. On the other hand, spaying does prevent the possibility of heifers becoming pregnant, and eliminates the necessity of separating heifers from bulls or steers.

CASTRATING HORSES

Regardless of age or time, castration of colts is best performed by an experienced veterinarian. A colt may be castrated when only a few days old, but most owners prefer to delay the operation until the animal is about one year of age. Although there is less real danger to the animal and much less setback with early altering, the practice results in imperfect development of the fore parts. On the other hand, leaving the colt entire for a time will result in more muscular, bold features and better carriage of the fore parts. Therefore, weather and management conditions permitting, the time of altering should be determined by the development of the individual. Thus, underdeveloped colts may be left entire six months or even a year longer than overdeveloped ones. Breeders of Thoroughbred horses usually prefer to have the horses first race as an entire.

There is less danger of infection if colts are castrated in the spring of the year soon after they are turned out to clean pasture. Naturally, this should be done sufficiently early to avoid hot weather and fly time.

WEANING

Weaning is the stopping of young animals from suckling their mothers.

Weaning age varies according to species, but normally is about as follows:

Species	Normal Weaning Age
Calves	6–8 months
Lambs	5 months
Pigs	2–7 weeks
Foals	4–6 months

Currently, there is much interest in early weaning—in weaning animals earlier than the normal ages indicated above. Without doubt, this practice will increase.

If the young are consuming considerable feed at the time of the separation (perhaps by creep feeding), weaning will result in very little disturbance or setback.

The separation should be complete and final, preferably with no opportunity for the young to see or hear their mothers again. In no case should the dam be returned to her offspring once the separation has been made. Such practice will only prolong the weaning process and give rise to digestive disorders in the young.

The feed of the dam should be decreased a few days prior to the separation, and should be more bulky for a few days after the removal of the young and until the udder has dried up.

When drying up lactating females, spoiled udders can be alleviated by adhering to the following procedure:

1. Decrease the ration, and do not feed milk-stimulating feeds just prior to, during, or immediately after weaning—until the udder has dried up.
2. Let "back pressure" in the udder build up. Examine the udders of cows or of mares at intervals, but do not milk them out. If the bag fills up and gets tight, rub an oil preparation on it, *but do not milk it out.* At the end of five to seven days, when the bag is soft and flabby, what little secretion remains (perhaps not more than half a cup) may be milked out if so desired.

BEDDING ANIMALS

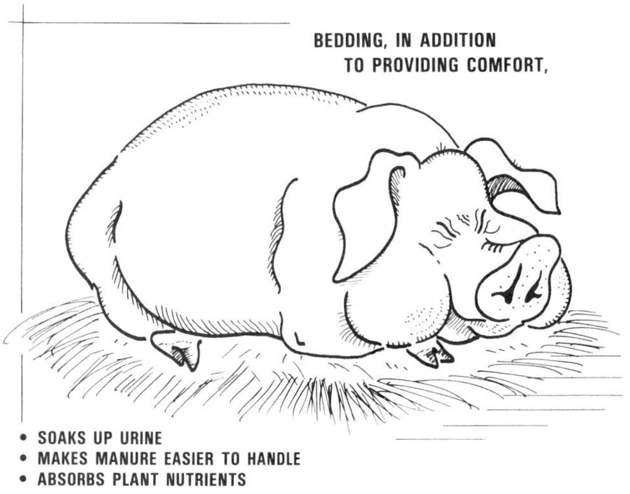
BEDDING, IN ADDITION
TO PROVIDING COMFORT,
• SOAKS UP URINE
• MAKES MANURE EASIER TO HANDLE
• ABSORBS PLANT NUTRIENTS

Fig. 8-13. Bedding is used primarily for the purpose of keeping animals clean and comfortable. (Drawing by R. F. Johnson)

Bedding or litter is used primarily for the purposes of keeping animals clean and comfortable. But bedding has the following added values from the standpoint of the manure.

1. It soaks up the urine which contains about one-half the total plant food of manure.
2. It makes manure easier to handle.
3. It absorbs plant nutrients, fixing both ammonia and potash in relatively insoluble forms that protect them against losses by leaching. This characteristic of bedding is especially important in peat moss, but of little significance with sawdust and shavings.

• **Kind and amount of bedding**—The kind of bedding material selected should be determined primarily by (1) availability and price, (2) absorptive capacity, (3) cleanness (this excludes dirt or dust which might cause odors or stain livestock), (4) ease of handling, (5) ease of cleanup and disposal, (6) non-irritability from dust or components causing allergies, (7) texture or size (for example, material that will not get into the wool of sheep), and (8) fertility value or plant nutrient content. In addition, a desirable bedding should not be excessively coarse, and should remain well in place and not be readily kicked aside.

Table 8–2 lists some common bedding materials and gives the average water absorptive capacity of each. In addition to these bedding materials, many other products can be and are successfully used for this purpose, including leaves of many kinds, tobacco stalks, buckwheat hulls, processed manure (made by separating solid fibers from the liquid and water soluble material in animal wastes), and shredded paper.

TABLE 8–2
WATER ABSORPTION OF BEDDING MATERIALS

Material	Lb of Water Absorbed/Cwt of Air-Dry Bedding
Barley straw	210
Cocoa shells	270
Corn stover (shredded)	250
Corncobs (crushed or ground)	210
Cottonseed hulls	250
Flax straw	260
Hay (mature, chopped)	300
Leaves (broadleaf)	200
(pine needles)	100
Oat hulls	200
Oat straw (long)	280
(chopped)	375
Peanut hulls	250
Peat moss	1,000
Rye straw	210
Sand	25
Sawdust (top-quality pine)	250
(run-of-the-mill hardwood)	150
Sugar cane bagasse	220
Tree bark (dry, fine)	250
(from tanneries)	400
Vermiculite[1]	350
Wheat straw (long)	220
(chopped)	295
Wood chips (top-quality pine)	300
(run-of-the-mill hardwood)	150
Wood shavings (top-quality pine)	200
(run-of-the-mill hardwood)	150

[1]This is a micralike mineral mined chiefly in South Carolina and Montana.

The availability and price per ton of various bedding materials may vary from area to area, and from year to year. Thus, in the New England states shavings and sawdust are available, and straws are more plentiful in the central and western states.

Table 8–2 shows that bedding materials differ considerably in their relative capacities to absorb liquid.

The minimum desirable amount of bedding to use is the amount necessary to absorb completely the liquids in manure. Some helpful guides to the end that this may be accomplished follow:

1. Per 24-hours' confinement, the minimum daily bedding requirements, based on uncut wheat or oats straw, of different kinds of livestock are as follows: cow, 9 lb; steers, 7 to 10 lb; sheep, 1 lb.; hogs, ½ to 1 lb; and horses, 10 to 15 lb. With other bedding materials these quantities will vary according to their respective absorptive capacities (see Table 8–2). Also, more than these minimum quantities of bedding may be desirable where cleanliness and comfort for the animal are important. Comfortable animals lie down more and utilize a higher proportion of the energy of the feed for reproductive purposes (cattle and sheep require 9% less energy when lying down than when standing).

2. Under average conditions, about 500 lb of bedding are used for each ton of excrement.

• **Reducing bedding needs**—In most areas, bedding materials are becoming scarcer and higher in price, primarily because (1) geneticists are breeding plants with shorter straws and stalks, (2) of more competitive and remunerative uses for some of the materials, and (3) the current trend toward more confinement rearing of livestock requires more bedding.

Caretakers may reduce needs and costs as follows:

1. Collect liquid excrement separately.
2. Chop bedding.
3. Use deep-bedding system.
4. Ventilate quarters properly.
5. Feed and water away from sleeping quarters.
6. Provide exercise area.
7. Mound cattle lots.
8. Consider slotted floors.
9. Consider use of rubber mats.

BEEF CATTLE MANAGEMENT PRACTICES

Fig. 8–14. Beef cattle management gives point and purpose to all that has gone before. (Courtesy, *Livestock Breeder Journal*, Macon, GA)

Successful cattle raisers practice good management. Some beef cattle management practices that are not covered elsewhere in this book follow.

SEMEN AND FERTILITY EVALUATION

Semen quality is based on evaluating the ejaculate for (1) density of concentration, (2) rate of movement, (3) motility, and (4) morphology. It is advisable to have a semen or fertility evaluation made before buying a bull, especially when purchasing for a single sire herd.

MANAGING BEEF COWS

Reproduction—the production of calves—is the first and most important requisite of the cow-calf system, for if cows fail to produce the cow-calf producer will soon be out of business.

- **Breeding and calving season**—The season at which the cows are bred depends primarily on the facilities at hand, taking into consideration the feed supply, pasture, equipment, labor, and weather conditions; whether the cattle are being produced for ordinary commercial or for purebred purposes; and whether they are strictly beef or dual-purpose cattle.

- **Advantages of spring calves**—The production of spring calves has the following advantages:

1. The cows are bred during the most natural breeding season—at a time when they are on pasture, gaining in flesh, and more likely to conceive. The calving percentage is usually higher, therefore, with a system of spring calving.
2. The calves will be in shape to sell directly from the cows in the fall, at which time there is a good demand for feeder calves.
3. If the calves are to be sold as yearlings, one winter is saved; or if they are to be sold at weaning time, no wintering is required.
4. Because of greater utilization of cheap roughage, dry cows may be wintered more cheaply.
5. Less labor and attention is required in caring for the calves the first winter.
6. Spring calves require less grain and utilize the maximum amount of pasture and forage.

- **Advantages of fall calves**—The production of fall calves has the following advantages:

1. The cows are in better condition at calving time.
2. The cows give more milk for a longer period.
3. The calves make better use of the grass during their first summer.
4. The calves escape flies, screwworms, and heat while they are small.
5. Upon being weaned the following spring, the calves can be placed directly on pasture instead of in a drylot; or, if the desire is to sell, they usually find a ready market ahead of the influx of fall feeder calves from the range areas.
6. When the intention is to sell market milk from dual-purpose cows, fall calves are usually best. The greater flow of milk is obtained during the period of highest prices.

- **Care of cows at calving**—The gestation period of a cow is about 283 days, but it may vary a few days in either direction. The careful and observant caretaker will make definite preparations in ample time. It is especially important that first-calf heifers be watched at calving time, for frequently they will need some assistance. Older cows that habitually have trouble in parturition may well be culled from the herd.

- **Signs of approaching parturition**—Usually, the first sign of approaching parturition is a distended udder, which may be observed some weeks before calving time. near the end of the gestation period, the content of the udder changes from a watery secretion to a thick, milky colostrum. As parturition approaches, there generally will be a marked shrinkage or falling away of the muscular part of the region of the tailhead and pinbones, together with a noticeable enlargement and swelling of the vulva. The immediate indication that the cow is about to calve are extreme nervousness and uneasiness, separation from the rest of the herd, and muscular exertion and distress.

- **Preparing for calving**—At the time the signs of approaching parturition seem to indicate that the calf may be expected within a short time, arrangements for the place of calving should be completed.

During the seasons of the year when the weather is warm, the most natural and ideal place for calving is a clean, open pasture away from other livestock. Hogs should not be allowed in the same place with the cow, for they are likely to injure or kill the young calf. They have even been known to injure the cow.

Under pasture conditions, there is decidedly less danger of either infection or mechanical injury to the cow and calf. In commercial range operations, it is common practice to ride the range more frequently at calving time. A better procedure consists in having a small pasture adjoining headquarters into which heavy springing cows are placed a few days before calving. With the added convenience of such an arrangement, the animals can be given more careful attention.

During inclement weather, the cow should be placed in a roomy (10 or 12 ft square), well-lighted, well-ventilated, comfortable box stall or maternity pen which should first be carefully cleaned, disinfected, and bedded for the occasion.

- **Normal presentation**—Labor pains in a mild form usually start some hours before actual parturition. After a time, the water bag appears on the outside, usually increasing in size until it ruptures from the weight of its own contents. This is followed closely by the appearance of the amniotic bladder (the second water bag), with the fetus. With the rupture of the second water bag, the straining becomes more violent, and presentation soon follows. Most commonly in presentation, the front feet come first followed by the nose which is resting on them, then the shoulders, the middle, the hips, and then the hind legs and feet.

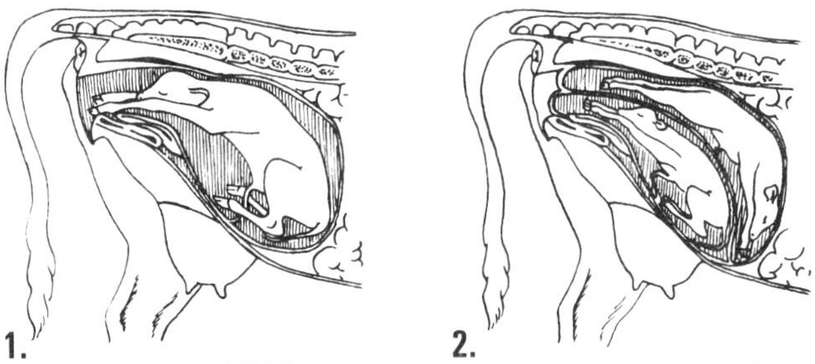

1. NORMAL PRESENTATION

2. NORMAL TWIN POSITIONS

Fig. 8–15

1. Normal single presentation; the back of the fetus is directly toward that of the mother, the forelegs are extended toward the vulva, and the head rests between the forelegs. If it is necessary to render assistance, apply ropes above the ankle joints and pull alternately downward on each leg as the cow strains.

2. Normal twin positions. If delivery does not proceed normally, this is a case for a veterinarian.

With posterior presentation (hind feet first), there is likely to be difficulty in calving. Moreover, there is considerably more danger of having the calf suffocate through the rupture of the umbilical cord and strangulation.

• **Rendering assistance**—If presentation is normal and within an hour or two after the onset of signs of calving, no assistance will be necessary. On the other hand, if the cow has labored for sometime with little progress or is laboring rather infrequently, it is usually time to give assistance. Such aid will usually consist of fastening small ropes around the pasterns and pulling the young outward and downward as the cow strains. This should be done by an experienced caretaker or a competent veterinarian. It is always well to be reminded that rough, careless, or unsanitary methods at such a time may do more harm than good.

• **The newborn calf**—If parturition has been normal, the cow can usually take care of the newborn calf, and it is best not to interfere. However, in unusual cases, it may be necessary to wipe the mucus from the nostrils to permit breathing; or, more rarely yet, artificial respiration methods must be applied to some calves. This may be done by blowing into the mouth, working the ribs, rubbing the body rather vigorously, and permitting the calf to fall gently. The cow should be permitted to lick the calf dry.

With calves born in sanitary quarters or out on clean pastures, there is little likelihood of infection. To lessen the danger of such infection, the navel cord of the newborn calf should be treated at once with a 10% solution of tincture of iodine.

A vigorous calf will attempt to rise in about 15 minutes and usually will be nursing in half an hour to an hour. The weaker the calf, the longer the time before it will be able to be up and nursing. Sometimes it may be necessary to assist the calf by holding it up to the cow's udder.

The colostrum (the milk yielded by the mother for a short period following the birth of the young) is most important for the well-being of the newborn calf. Aside from the difference in chemical composition, compared with later milk, the colostrum contains antibodies which temporarily protect the calf against certain infections, especially those of the digestive tract.

Usually it is best to keep the cow and calf in a small pasture for a few days. After this, they may be turned back with the main herd. Nothing is better for the cow at calving time than plenty of grass, and both the cow and calf will be helped by an abundance of fresh air and sunshine. The cow may deliberately hide the calf for the first few days, and the job may be so thoroughly done as to require considerable cleverness on the part of the caretaker to find it.

• **The afterbirth**—Under normal conditions, the fetal membranes (placenta or afterbirth) are expelled from 3 to 6 hours after parturition. Should they remain as long as 24 hours after calving, competent assistance should be given by an experienced caretaker or a licensed veterinarian. The operation of removing a retained afterbirth requires skill and experience; and, if improperly done, the cow may be made a nonbreeder. Furthermore, before doing this, the fingernails should be trimmed closely; the hands and arms should be thoroughly washed with soap and warm water, disinfected, and then lubricated with petroleum jelly or linseed oil. In no case should a weight be tied to the placenta in an attempt to force removal.

As soon as the afterbirth is ejected, it should be removed and burned or buried in lime, thus preventing the development of bacteria and foul odors. This step is less necessary in the open range, where animals traverse over a wide area.

(Also, see Section 4, under the heading "Feeding Brood Cows.")

HOOF TRIMMING

Regular inspection and care of the feet of all beef cattle is important. It will reduce the incidence of foot rot, lameness, and other foot troubles.

If hoofs are not properly trimmed, permanent damage may result in the form of crooked legs and weak pasterns, and the productive life of the animal may be lessened.

Young animals should not be neglected. Corrective trimming can be accomplished when animals are young and growing rapidly; for example, the hoofs can be trimmed to correct excessive growth and to rectify pigeon-toed and splay-footed (toe-in or toe-out) conditions.

Herd bulls should receive special attention. The hoofs should be checked and trimmed 1 to 2 months before the breeding season. A well-planned foot trimming program can result in improved reproductive performance.

If the hoofs are trimmed properly, the animal stands squarely and walks properly, with each leg directly under the weight it supports.

The reason for and the technique of trimming hoofs are the same whether an animal is in a commercial or purebred herd, or being fitted for show.

GENTLING CATTLE

Easy does it! Tom Lasater, famed cattleman and founder of the Beefmaster breed, uses the following method to gentle cattle; and it works.

He places up to 12 animals in a relatively small pen where he can walk among and stay fairly close to them. Then, he (1) carries a small bucket half-full of pellets, which he shakes to attract the animals' attention; (2) holds 1 pellet between his thumb and forefinger, and feeds each animal that will take the pellet from him, one by one; (3) repeats this procedure daily, or more often, until the animals get sufficiently gentle that they can be hand-fed and back-rubbed; and (4) *never, never* touches the animals on the forehead or horns, as this will teach them to butt.

The Lasater gentling procedure substitutes the application of animal behavior and human patience for rough handling.

MANAGING FEEDLOT CATTLE

There are many facets of cattle management. Only those that are unique to cattle feedlots will be discussed in the sections that follow.

- **Bird control**—Nationwide, starlings constitute the major feedlot bird problem. Other feed-consuming bird species commonly identified in feedlots are: brewer blackbirds, redwing blackbirds, and cowbirds.

Some large commercial feedlots estimate their starling population at 100,000 per lot. Some western feedlot operators compute the cost for overwintering each 1,000 starlings at $100.

In addition to feed consumption, birds contaminate feed and spread diseases—to both animals and humans. The starling has been incriminated in the spread of coccidiosis among animals, transmissible gastric enteritis (TGE) in swine, and histoplasmosis in humans.

Recordings of distressed bird calls, carbide cannons, and harassment achieve only partial control. Many chemicals and baits have been tested, and a few have been found to be effective. However, some states do not allow the use of chemicals in bird control. Therefore, before using any chemical, the cattle feeder should check with the appropriate federal, state, and local departments of health.

(Also, see later presentations in this chapter headed "Control Pests; Lessen Waste" and "English Sparrows, Starlings, and Pigeons.")

- **Feedlot pollution control**—Pollution control is a most critical factor in site selection and operation of a cattle feedlot. Remoteness from urban development is recommended because of dust and odor. Also, before constructing a cattle feedlot, the owner should become familiar with both state and federal regulations. The state regulations can be secured from the state water board. They differ from state to state, but most states require a catch basin (detention pond) sufficient to contain the runoff from a storm of the magnitude of the largest rainfall during a 48-hour period of the most recent 10 years. A feedlot may minimize runoff by locating near the top of the slope, and if necessary, by using diversion embankments to divert runoff from other areas.

Cattle feedlots located near centers of populations are having an increasing number of complaints lodged against them because of manure, dust, and odor. Lawsuits, based on the nuisance law, are being filed against them.

(Also, see Section 1, under the heading "Pollution of the Environment.")

DAIRY CATTLE MANAGEMENT PRACTICES

Fig. 8-16. Dairy cattle management is the art of caring for, and handling, dairy cattle. (Courtesy, Union Pacific Railroad Company, Omaha, NE)

Successful dairy producers pay close attention to the details of management. Some dairy cattle management practices that are not covered elsewhere in this book follow.

MANAGING DRY AND LACTATING COWS

The management of dry and lactating cows materially affects the efficiency of production and the quantity and quality of the milk produced.

(Also, see Section 4, under headings "Rations for Lactating Cows" and "Feeding Dry Cows.")

MANAGED MILKING

The physiology of the discharge of milk is a delicate process, and it requires the close cooperation of the milker and the cow if it is to be successful. A managed milking program is made up of the following coordinated steps:

1. **Preparing the equipment.** Prior to milking, the equipment to be used in the milking process should be assembled and sanitized. Also, it should be checked and adjusted if necessary.

2. **Preparing the cow.** Under natural conditions, the cow is primed or stimulated by the suckling of the calf. This process can be simulated by washing the cow's teats and udder with warm water (120 to 130°F), then massaging and drying them with a paper towel. Following this process, remove 2 or 3 streams of milk from each quarter into a strip cup (never strip milk onto the floor) and examine for visible evidence of mastitis. Also, this (a) washes out any debris adhering to the end of the teat, and (b) enhances the let down effect.

About 45 seconds after the priming stimulus, the udder becomes full and firm (especially in early lactation), and milk occasionally will leak from the teats. This is evidence that the cow has let down her milk and is ready for the next step.

3. **Attaching the teat cup and beginning.** About 1 minute after washing the udder, and not more than 1½ minutes, the teat cups should be attached and milking should begin. Most cows will milk out in 3 to 6 minutes, depending upon the amount of milk and the characteristics of the cow. Also, and most important, each quarter should be milked individually, because some quarters milk out faster than others.

4. **Stripping by machine.** When it is apparent that the cow is about milked out, she should be machine stripped. This consists of pulling down on the teat cup with one hand, and massaging the udder downward with the other. This process should not take over about 20 seconds.

5. **Removing the teat cup.** Both incomplete and overmilking should be avoided. The greatest cause of machine injury is leaving the teat cups on too long. Incomplete milking usually results because one or more quarters are more difficult to milk than the others.

As soon as the udder is empty, and before the teat cups crawl up, they should be removed, properly and gently. Then, tip the teats with a fresh disinfectant solution (100 ppm idophor or chlorine, or other sanitizing agent). This will remove the milk from the end of the teats and prevent the invasion of bacteria into the udder. Also, it will avoid attracting flies.

As soon as the teat cups have been removed from the udder of the cow, they should be cleaned. First, dip them in clean, cold water to remove milk inside the liners, then put them in a clean, warm, approved sanitizing solution. Change the solution after each five to seven cows.

6. **Cleaning up equipment.** After milking the last cow, all milking equipment should be thoroughly cleaned and put away.

7. **Milking time.** The actual milking time per cow will range from 3 to 6 minutes, with an average time of 3½ minutes for cows in midlactation. But additional time must be allowed for let-down, adjustments, and interval between cows.

The number of machines one milker can manage successfully depends upon the type of barn, the type of milking equipment, the ability of the milker, and the jobs that the milker performs other than milking.

One person should handle no more (preferably less) than the following number of units:

Type Milker	Units per Milker
Stanchion barns:	
Bucket	2
Pipeline	3
Milking parlor:	
Walk-through	3
Side-Opening	3
Herringbone	4

Within a 3-inch-line elevated parlor, one milker will average 18 to 25 cows per hour. With a 4-inch-line parlor, one milker will average 25 to 30 cows per hour. However, additional time must be allowed to bring the cows in from the outside, setting up, cleaning up, as well as milking problem cows.

8. **Milking order.** Cows that have mastitis or a history of chronic mastitis are a source of infection to noninfected cows. Hence, it is well to milk "clean" cows first. A desirable milking order in stanchion barns is:

a. First calf heifers that have been free of mastitis.
b. Older cows that have been free of mastitis.
c. Cows that have a previous history of mastitis, but which no longer show symptoms.
d. Cows with quarters producing abnormal milk.

FREQUENCY OF MILKING

Most cows are milked 2X per day. Increasing milking frequency to 3X per day increases milk production by 10 to 25% and milking 4X per day will stimulate milk yield another 5 to 15%. Whether these increases in milk production are worth the extra expense in labor, feed, utilities, and milking supplies depends upon economic conditions on each particular dairy farm. Also, with increasing genetic capability of cows, more frequent milking (3X and 4X) appears to improve udder health and lower the incidence of mastitis.

FEEDS AFFECTING MILK FLAVOR

Consumers want milk to taste like milk—not like silage, grass, or weeds.

Although feeds are not the only cause of milk flavors, they are major contributors. Feed flavors enter the milk through the digestive system, respiratory system, and by direct absorption. Research indicates that most feed flavors are detectable in the milk 20 minutes after the feed is consumed, and that they are usually most pronounced at the end of two hours.

Feed flavors that enter the milk through the respiratory system can usually be detected much sooner than those entering through the digestive system. For example, if a cow breathes air reeking with silage odors, those flavors can be detected in the milk almost immediately. Flavors that are directly absorbed by milk are less common, but they appear if the milk is left exposed for a long enough period.

The following control measures are recommended to alleviate feed flavors:

1. Avoid sudden change to fresh, lush pasture.
2. Control and avoid undesirable weeds.
3. Avoid silage flavor.

PESTICIDE RESIDUES

Pesticides are chemicals that are used to kill pests—insects, weeds, and rodents. These products are very necessary for food and milk production. Our abundant supply of wholesome foods would not have been available without their use. Yet, it is important that they be properly used, and that certain precautions be taken. The following points are pertinent to their proper use.

- **How pesticides contaminate milk**—They are absorbed by animal fat. Since milk contains fat, it is one channel through which the animal eliminates pesticides from its body.

- **Length of time that a contaminated cow may give contaminated milk**—Cases are known where residues have been detected in milk for four to eight months after discontinuing the feeding of contaminated feeds.

- **Ways that milk may become contaminated**—Milk becomes contaminated (a) by spraying animals with nonrecommended pesticides, (b) by using these materials in back rubbers and vaporizers, (c) by feeding forages and concentrates which have been contaminated with these materials, (d) by allowing cows to drink pesticide-contaminated water, and (e) by using milk utensils that have become contaminated through their use for chores other than handling milk or in milk production. Hence, milk contamination can be prevented by avoiding any of these avenues of contamination.

- **The meaning of the word *tolerance* as applied to a chemical residue**—A given tolerance is that amount of chemical residue, usually expressed in ppm (parts per million) set by the FDA (Food and Drug Administration), that remains on or in a commodity at harvest and which is at least 100 times less than that amount of the chemical known to be toxic to experimental animals.

A zero tolerance, as applied to chemical residue, means that no amount of the pesticide chemical may remain on or in the raw agricultural commodity when it is offered for shipment. The tolerance level for pesticides in milk is 1.25 ppm on a fat basis. This means that there must be less than 1.25 parts of the pesticide (DDT, for example) to 1 million parts of milk fat figured on a weight basis.

COW TESTING PROGRAMS

Individual cow records are a must in any progressive dairy production program. Milk producers use records as a guide for feeding, for location, and culling out the least profitable cows, and for maintaining a permanent, detailed record of each cow. Records necessitate that each cow be individually identified, and that there be milk and butterfat production records.

The various testing programs sponsored by federal and state extension services are discussed in Section 3 of this book; hence, the reader is referred thereto.

MANAGING DAIRY CALVES

The recommended practices in managing dairy calves follow, in summary form:

- Dehorn anytime after 10 days of age, using an electric dehorner or caustic potash.
- Remove extra teats before heifers are six months old; cut them off with clean scissors and disinfect with iodine.
- Check for scours.

(Also, see Section 4, under the heading "Feeding Dairy Calves.")

MANAGING REPLACEMENT HEIFERS

Good management of replacement heifers embraces the following principles and practices:

- Separate bulls and heifers before six months of age; do not have over three month's difference in age of animals within a given group.
- Treat for worms when the need is demonstrated.
- Breed heifers at 15 to 18 months of age, but also consider weight and size.
- Accustom bred heifers to milking barn procedure beginning about one month prior to calving.

(Also, see Section 4, under the heading "Feeding Dairy Heifer Replacements.")

HOOF TRIMMING

Hoof troubles make cows reticent to walk for feed and water and result in a drop in milk production. Hence, regular inspection and care of the feet of all dairy animals is important, especially for older and larger cows. It will reduce the incidence of footrot, lameness, and other foot troubles.

Cows in confinement tend to grow long toes and build up excessive tissue on the soles of the feet. As a result, more weight is carried by the heels and the hocks, and the pasterns are subjected to extra stress. If these conditions are not corrected by proper hoof trimming, permanent damage may result in the form of crooked legs and weak pasterns, and the productive life of the cow will be shortened.

If the hoofs are trimmed properly, the animal stands squarely and walks properly, with each leg directly under the weight it supports.

The reason for and the technique of trimming hoofs are the same whether a cow is in a commercial production string or being fitted for show.

SHEEP MANAGEMENT PRACTICES

Fig. 8–17. Sheep management is the art of caring for, and handling, sheep.

Several sheep management practices that are not covered elsewhere in this book follow.

CONFINEMENT VS PASTURE

Most breeding sheep utilize pasture in season. However, some ewe-lamb producers are drylotting all or part of the year, primarily for parasite and predator control. So, now there are two alternatives—confinement vs pasture, and the producer may choose between the two. (Also, see Section 4, under the heading "Feeding Breeding Ewes," "•Drylot (confinement) feeding.")

MANAGING THE RAM

If possible, the ram should be secured considerably in advance of the breeding season, thereby providing an opportunity to become acclimated before being placed in service. In case of show or sale rams, it may also be advisable to remove some of their surplus flesh.

Stud rams are usually kept separate from the ewes except during the breeding season. Their quarters need not be elaborate or expensive. Usually, a dry shelter that will provide protection during times of inclement weather is all that is necessary. Plenty of exercise should be provided at all times.

Rams may subsist largely on pasture and dry roughage. If the pasture has been scanty prior to the breeding season, the rams may be conditioned by feeding a little grain, usually not more than 1 lb daily. Rams are usually fed some grain when being fitted for show or sale, but it must be remembered that excess fat may actually be harmful from a breeding standpoint.

(Also, see Section 4, under the heading "Feeding Rams.")

• **Preparing the ram for mating**—As the weather is usually rather warm at the time of the breeding season, shearing the ram just prior to this will make him more active. This is especially true of old show or sale rams. Where rams are not sheared completely, they should at least have the wool clipped from the neck and from the belly in the regions of the penis, for this will result in copulation with greater ease. It is also important to see the the hoofs of the ram are properly trimmed prior to the breeding season.

• **Marking the ram**—When several rams are turned in with a large band of ewes, it is impossible to detect individual rams that may be failing to settle ewes. Moreover, it is quite likely that a different ram will serve the ewe should there be a recurrence of heat, or perhaps more than one ram may serve the ewe at the time of estrus. When only one ram is being used on a small flock, however, it is important to know whether the ewes are getting with lamb. Then, too, with a purebred flock individual breeding records are rather important.

A breeding record can best be kept by using a marking harness (breeding harness), containing a crayon (different colored crayons are available), on the ram, or by smearing the breast of the ram and the area between his forelegs every day or two with a thick paste. Then, as the ram serves the ewe, a mark will be left on her rump. Paint or tar should never be used for this purpose.

The color of the crayon or paste should be changed every 16 days (the approximate estrus cycle of the ewe) so that one can determine whether ewes that have been bred are returning in heat. For example, during the first 16-day interval, the thick paste used on the ram might well be a mixture of ordinary lubricating oil and yellow ochre; for the second 16-day interval, it might be lubricating oil and venetian red; and for the third 16-day interval (if there is still some question about some of the ewes having settled), it might be a paste made of using lubricating oil and lamb black (thus proceeding from light to dark colors).

Naturally, if a good percentage of the ewes are found coming in heat for a second time, the ram should be regarded with suspicion, and perhaps another ram should be obtained. The sterility in some instances may be temporary because of high condition and lack of exercise.

(Also, see Section 3, under the heading "Heat Detection Methods and Devices," "•Heat detection of sheep.")

MANAGING EWES

Managing ewes at breeding time and at lambing time is extremely important, because it determines the size lamb crop.

• **Trimming and tagging ewes for mating**[1]—Tagging is the practice of cutting the dung locks off sheep from around the udders and docks. Usually this is done immediately prior to lambing and shearing, and again prior to breeding season.

---

[1]In Australia, this practice is known as *crutching*.

Usually the wool around the eyes of animals subject to wool blindness is clipped at the same time.

Fig. 8-18. A properly tagged (crutched) ewe. The wool has been shorn from around the udder, flank, dock, and eyes. Tagging is a valuable aid to successful lambing, nursing, and breeding. (Courtesy, California Wool Growers Association, Sacramento, CA)

• **Managing ewes at lambing**—There is an appalling lamb death loss of 24.6% from birth to weaning. Most of these losses occur in the first few days of life.

As lambing time approaches, unsheared ewes should be tagged. This consists of shearing the wool from around the udder, flank, and dock. The ewe should be placed where she has plenty of room, away from any jamming or crowding. The grain allowance should be materially reduced, but the roughage allowance may be continued if it is certain that it is of good quality and palatable. Careless feeding at this stage is likely to result in milk fever following parturition. At this time, the wool around the udder should be clipped short in order to allow the lamb to find the teats readily. If breeding records have not been kept, the signs of approaching parturition must be relied upon. A nervous, uneasy disposition; a sinking in front of the hips; and fullness of udder are such indications.

1. **The lambing pen.** Just before lambing, or immediately thereafter, the ewe should be placed in a lambing pen. These pens are usually 4 ft square and are made by placing together two hinged hurdles, which are then set against the walls of the sheep barn. Use of the lambing pen prevents other sheep from trampling on the newborn lamb; eliminates the possibility of the lamb wandering away and becoming chilled; and, through keeping the dam and offspring together, lessens the danger of disowned lambs.

2. **Normal presentation.** Normal presentation of the lamb consists of having the forelegs extended with the head lying between them, although some lambs are delivered hind legs first. Even though the lambs are born in clean quarters, tincture of iodine should be applied to the navel soon after birth.

3. **Taking the lamb.** If the ewe has labored for some time with little progress or is laboring rather infrequently, it is usually time to give assistance. If the lamb is not in the proper position, such assistance consists of inserting the hand and arm in the vulva and turning the lamb so that the forefeet and head are in position to be delivered first. Delivery may then be helped by pulling the young outward and downward as the ewe strains. Before doing this, however, the fingernails should be trimmed closely and the hands and arms should be thoroughly washed with soap and warm water, disinfected, and then lubricated with Vaseline or linseed oil.

4. **Chilled and weak lambs.** Lambs arriving during cold weather may become chilled before they have dried. One of the most effective methods of reviving a chilled lamb is to immerse the body, except for the head, in water that is as warm as one's elbow can bear. The lamb should be kept in this for a few minutes and then removed and rubbed vigorously with towels. It then should be wrapped in an old blanket, a sheepskin, or other heavy material and should be given some warm milk as soon as possible. Another convenient and effective method of drying and warming a chilled lamb consists of putting it into a box containing a light bulb or electric heater.

After breathing has started and the navel cord has been painted with iodine, an attempt should be made to get the lamb to nurse. Quite often even a very weak lamb will nurse the ewe if it is held to the teat. If it refuses to nurse in this manner, some of the colostrum of the ewe should be milked into a sterilized bottle, and the lamb should be fed a few teaspoonsful each hour by means of the bottle and nipple, until it gains strength.

If the ewe has no milk, an attempt should be made to obtain milk from another ewe that has just lambed, and perhaps in a few hours the normal flow of milk will start.

5. **Disowned lambs.** When lambing pens are used, the number of disowned lambs is kept to a minimum. For the most part, disowning of lambs is due to improper feeding during pregnancy or because of a poor milk supply, an inflamed udder, or a maternal instinct that is not sufficiently developed, as is often true in ewes with their first lambs.

For the first few days, a ewe seems to recognize her young by scent or sense of smell. When difficulty is encountered in getting a ewe to own her own lamb or when it is desired to transfer or "graft" a lamb (as may be necessary with the loss of a ewe or when there are twins on an old ewe), deception in the sense of smell is an effective approach. One of the most common practices is to milk some of the ewe's milk on the rump of the lamb and then to smear some of it on the nose of the ewe. Many good caretakers take some of the mucus from the mouth and nose of the newborn lamb and smear it over the nose of the ewe. If these methods fail and the ewe persists in fighting the lamb away, blindfold her so that she cannot see the lamb. As a last resort, and when all other methods have failed, tie a dog in an adjoining pen. Sometimes the latter method will cause latent maternal instincts to rise to a surprising degree.

Occasionally, a ewe will fail to own one of a pair of twin lambs. When this condition exists, about all that can be done is that the attendant be patient in training the disowned lamb to nurse at the same time as its mate. Both lambs are usually kept from the ewe and turned with her at intervals.

6. **The orphan lamb.** A lamb may be orphaned through the death of its mother or because of the inability of the mother to suckle it. The most satisfactory arrangement for the orphan is to provide a foster mother. The good attendant will try to have every ewe raise a lamb. There may be a ewe that has just lost her lamb or a strong, healthy ewe with just one lamb. When a lamb dies at birth and it is desired to transfer or "graft" another lamb on the ewe, two procedures are common. Sometimes a ewe will accept another lamb provided that the lamb to be adopted is first rubbed with the body of the dead lamb that it is to replace. Though a bit more bothersome, a more effective approach consists of removing the skin from the dead lamb and tying it over the lamb to be adopted. After 2 to 3 days, the skin may be removed gradually, a piece at a time. The latter method is commonly used in the range bands of the West.

When it is impossible to transfer an orphan lamb to another ewe, it may be raised either on cow's milk or on milk replacer. Of course, the problem will be simplified if the lamb has received some colostrum (the first milk) from its mother or from another ewe.

(Also, see Section 4, under the heading "Feeding Growing-Finishing Lambs," "•Feeding orphan ("bummer") lambs.")

7. **Feed and water after lambing.** Following parturition, the ewe is in a feverish condition and should be handled carefully. She may be watered immediately after lambing, and at frequent intervals thereafter, but she should never be allowed to gorge. It is also a good plan to take the chill off the water before giving it to her. In general, feeds of a bulky and laxative nature should be provided during the first few days. A mixture of equal parts of oats and wheat bran may be fed in very limited quantities, with all the hay that can be consumed. Heavy grain feeding at this time may cause udder trouble in the ewe and digestive disturbances in the lamb. The feed may be gradually increased until the ewe is on full feed in about a week.

8. **Examination of the udder.** During the first two days following lambing, the udder should be examined night and morning. Sometimes a lamb will nurse one side only. If all the milk is not being taken by the lamb, the udder should be milked out and the ration lessened accordingly. If the udder becomes swollen and feverish, it should be milked out, bathed with warm water, and then dried. Following this, it should be painted with tincture of iodine. This treatment should be repeated once or twice daily, as necessary. Lambs should not be allowed to suckle their mothers when the udder is in such a condition. It is also a good plan to isolate the affected ewes from the rest of the flock.

(Also, see Section 4, under the heading "Feeding Breeding Ewes.")

FOOT TRIMMING

Most ewes keep their feet worn off sufficiently. However, some need to have their feet trimmed to avoid lameness. The rams' feet should always be trimmed prior to the breeding season.

SHEARING

Generally, shearing is an annual event, though some flocks and bands in the South and Southwest are shorn twice each year. The shearing season is almost entirely determined by climate, starting early in the spring in the South and ending in June or July in the North. Where animals are turned to pasture immediately after shearing and no shelter is available—the usual situation on the western range—cold rains or stormy weather at this time may result in excessive losses from pneumonia.

In addition to the problem of inclement weather in determining the time of shearing, consideration must also be given to the lambing season. Generally, the wool is removed before lambing, but there are good reasons for delaying under certain conditions. For example, in the Northwest where the production of grassfat lambs is the primary objective and warm weather comes rather late, few shed-lambed ewes are shorn before lambing.

When there is danger of either inclement weather or blistering from the sun, some operators take the added precaution of insisting that the sheep be shorn with thick combs or by hand shears, thus leaving a small amount of wool for protection. Most shearing is done by unionized professional shearers who travel from band to band, charging on a per head basis. Expert machine shearers will shear up to 200 head per day.

In addition to the experience of the shearer, the speed of shearing is affected by the size of the sheep, the degree of wrinkles, and the density and condition of the fleece. On the whole, the professional shearers do excellent work, but occasionally they may have to be cautioned against too many ugly cuts or unnecessary rough handling.

PREDATOR CONTROL

In the western United States, it is estimated that the annual loss of sheep caused by coyotes ranges from 4 to 8% of the lambs and from 1.5 to 2.5% of the ewes, and that the total annual loss of sheep and goats to predators is about $60 million.[2]

Predators may also cause damage to other animal species, especially the young.

Although predation is considered a problem for the entire sheep industry, its magnitude is highly variable among producers. Geography is a key factor; the greatest losses are in the West. Sheep producers in the West are most troubled by coyotes, but bears and cougars also inflict serious damage in some areas. Sheep producers in the Midwest and East frequently contend with sheep-killing dogs, although coyotes have spread across the continent.

But, not all predators, coyotes included, kill sheep or other livestock. Also, it is recognized that coyotes and other predators are an integral part of most wildlife communities, and that their predation on rodents and rabbits, and their feeding on carrion, may benefit agriculture in some areas. Thus, the challenge of sheep producers is to have an effective program of preventing sheep losses from predators without unnecessarily negatively impacting the nation's natural resources.

Today, predation management is largely the responsibility of each sheep producer. Many producers enlist the assistance of the USDA's Animal Damage Control (ADC) programs, state animal damage control programs, and/or private trappers. The federal ADC program, which is widely used in western United States, has legislative responsibility for control of damage to agriculture by wildlife species which are owned by the public. Control efforts are made at the request of, and in cooperation with, public or governmental entities. Because of public attitudes about wildlife and information about predator biology, control actions are designed to alleviate the damage, not necessarily to kill the offending animal, although this is often essential for stopping predation by coyotes. In some states, the county extension agent and the animal damage control specialist provide sheep producers with advice and self-help training relative to predation control methods.

[2]Estimates made by F. R. Henderson, Department of Animal Science and Industry, Kansas State University, Manhattan. The estimated total loss of $60 million is for 1984. Prof. Henderson also provided authoritative information for this presentation on "Predator Control."

GOAT MANAGEMENT

Fig. 8-19. Young kids. (Courtesy, Babson Bros. Co., Oak Brook, IL)

The principles and practices of goat and sheep management are very similar; hence, repetition is not necessary.

SWINE MANAGEMENT PRACTICES

Fig. 8-20. Swine management is the art of caring for, and handling swine. (Courtesy, Charles Pfizer & Co., Inc.)

Some management practices unique to swine, and not covered elsewhere in this book, follow.

CONFINEMENT VS PASTURE

In recent years, confinement rearing, in which breeding animals are confined part of the time, and growing-finishing pigs are confined from birth to market, has increased. By 1990, more than 60% of the market hogs of the United States were raised from farrow to finish in some type of confinement system—ranging all the way from simple shelters to environmentally controlled pig palaces. The main reasons given for going to confinement are savings in labor and land. The main problems encountered are high investment in buildings, more disease troubles, rations become more critical, manure disposal and odor control problems are greater, and sow fertility is lowered.

In season, most producers utilize pastures for breeding animals. The vast majority of the nation's gestating sows and herd boars are kept in movable houses, preferably on clean pasture on land that has been plowed since hogs were last on the area. In any event, pastures for sows and litters are not obsolete; rather, there now exists two alternatives for the breeding herd—confinement vs pasture—and the wise manager will choose between them.

GROUPING HOGS

Grouping, along with separating hogs by sexes, ages, and size, is important. The following practices are generally advocated by successful producers:

1. **Gilts to be retained for breeding herd.** They should be separated from market hogs at four to five months of age.
2. **Pregnant gilts and sows.** They should be kept separate during the gestation period, unless they are self-fed a bulky ration.
3. **Boars of different ages.** Junior and mature boars should not be run together. Boars of the same age or size can be run together during the off-breeding season.
4. **Adjusting size of litter.** Where possible, the size of litters should be adjusted to the number of functioning teats and the nursing ability of the sow. Transferring pigs from sow to sow should be done as early as possible; three to four days after farrowing is usually the maximum length of time that this can be done, unless the odor of the pigs is masked, when it may be possible to transfer at a later date.
5. **Running sows and litters together.** Pigs should be about 2 weeks old before placing sows and litters together, although small groups may be put together as early as 1 week. The age difference between such litters should not be more than 1 week in a central farrowing house or 2 weeks on pasture. Not more than 4 sows and litters should be grouped together in a central farrowing house; and not more than 6 on pasture.
6. **Creep feeding.** A maximum of 40 pigs per creep may be allowed.
7. **Early weaning.** In early weaning, not over 10 pigs should be placed together up to 3 weeks of age; 20 may be placed together at 3 to 4 weeks of age; and 25 at 5 weeks of age.
8. **Pigs of different weights.** Growing-finishing pigs of varying weights should not be run together. It is recommended that the range in weight should not exceed 20% above or below the average.

MANAGING THE BOAR

Proper care and management of the herd boar is most essential for successful swine production. Too frequently the boar is looked upon as a necessary evil and is neglected. Under such conditions, he is usually confined to a small, filthy pen—a typical pigsty—exercise is discouraged; and the feeding practices are anything but intelligent.

Outdoor exercise throughout the year is one of the first essentials in keeping the boar in a thrifty condition and virile. This may be accomplished by providing a well-fenced pasture. Even then, the caretaker may find it necessary to walk old boars or boars that are being fitted for the shows. In addition to the valuable exercise that is obtained in the pasture lot, green succulent pasture furnishes valuable nutrients for the herd boar. The amount of feed provided should be such as to keep the boar in a thrifty, vigorous condition at all times. He should be neither overfat nor in a thin, run-down condition. The concentrate allowance should be varied with age, development, and temperament of the individual; breeding demands; roughage consumed, etc. Feeding the boar is more fully covered in Section 4 of this book; hence, the reader is referred thereto.

A satisfactory but inexpensive shelter should be provided for the boar, and he should be allowed to run in and out at choice.

Boars of the same age or size can be run together during the off-breeding season, but boars of different ages should not be kept together.

(Also, see Section 4, under the heading "Feeding Boars.")

MANAGING BREEDING SWINE

Management, more than any other factor, determines how well swine reproduce and survive, and how nearly they perform to their genetic potential.

- **Normal breeding and farrowing seasons**—The season in which the sows are bred and the question of raising 1 or 2 litters a year vs multiple farrowing (scheduling breeding so that the litters arrive throughout the year, rather than once or twice per year as is the case in the conventional 1- or 2-litter systems), depend primarily on the facilities at hand. The location of the producer (particularly the weather conditions in the area), availability and price of feeds, condition and growth of the sows, equipment for handling pigs during the winter months, available labor, and the type of production (purebred or commercial) should be taken into consideration. No positive advice can be given, therefore, for any and all conditions. Sows will breed any time of the year; but, as in other farm animals, the conception rate is much higher during those seasons when temperature is moderate and the nutritive conditions are good. For the country as a whole, spring pigs are preferred, as is shown by the size of the spring pig crop in comparison with the fall pig crop.

No one expects the seasonal pattern of hog production to be completely eliminated, but, because of the several recognized advantages of multiple farrowing to both the processor and the producer, it is likely that it will increase sufficiently to make for a lessening of some of the market gluts of the past.

- **Breeding practices**—The following breeding practices are recommended:

1. **Breeding following early weaning.** When weaning under two weeks of age, breed sows on the second heat period after weaning. When weaning at three weeks or older, it is satisfactory to breed sows on the first heat period following weaning.

2. **First service of boars.** Whenever practical, it is recommended that boars be allowed to serve females outside the breeding herd (some market hogs) prior to serving those in the breeding herd.

- **Managing sows at farrowing**—It has been conservatively estimated that from 30 to 35% of the pigs farrowed never reach weaning age, and an additional loss of 5 to 10% occurs after weaning. This means that only 60% of the pig crop reaches market age.

The careful and observant caretaker realizes the importance of having everything in readiness for farrowing time. If the pregnant sows have been so fed and managed as to give birth to a crop of strong, vigorous pigs, the next problem is that of saving the pigs at farrowing time. Good management will give a powerful assist to this end.

1. **Signs of approaching parturition.** The immediate indications that the sow is about to farrow are extreme nervousness and uneasiness, an enlarged vulva, and a possible mucous discharge. She usually makes a nest for her young, and milk is present in the teats.

2. **Preparation for farrowing.** About three to four days prior to farrowing, the sow should be isolated from the rest of the herd. It is important, however, that moderate exercise be continued while the animal is in the farrowing quarters.

3. **Sanitary measures.** Before being moved into the farrowing quarters, the sow should be thoroughly scrubbed with soap and warm water, especially in the region of the sides, udder, and undersurface of the body. This removes adhering parasite eggs (especially the eggs of the common round worm) and disease germs.

The house should be thoroughly cleaned to reduce possible infection. This may be done by scrubbing the walls and floors with boiling-hot lye water made by using one can of lye to 15 gal of water. If the farrowing house has dirt floors, the top 2 or 3 in. of soil should be replaced by an equal quantity of clean clay soil. The sow should then be placed in her new quarters.

4. **The quarters.** Hogs are sensitive to extremes of heat and cold and require more protection than any other class of farm animals. This is especially true at the time of parturition. It is recommended that the farrowing house temperature be maintained at 60 to 70°F, and that it not go below 40°F or above 85°F. Along with this temperature, there should be adequate ventilation at all times. In cold areas and during the winter months, use heat lamps or pig brooders when the farrowing house temperature falls below 65°F.

The main requirements for satisfactory housing are that the quarters be dry, sanitary, and well ventilated and that they provide good protection from the heat, cold, and winds.

5. **Farrowing crates (stalls) or pens/guard rails.** Adequate facilities for farrowing are important because one-third or more of all death losses before weaning result from overlaying or crushing. Mechanical devices such as guard rails and farrowing stalls or crates are valuable aids in reducing these losses.

Fig. 8-21. Sow in farrowing stall, with pigs in adjacent brooder area warmed by heat lamp. (Courtesy, Iowa State University, Ames)

A variety of farrowing stalls or crates of different designs and size are available. Some are equipped with self-feeders and waterers. The important thing is that they include a brooder or creep area about 24 in. wide for baby pigs. Where crates are equipped with waterers, the waterers should be designed and located so that the pen area will not get wet.

Some producers keep sows in crates continuously from farrowing to weaning, but generally the sow and litter are moved to other quarters about 2 weeks after farrowing. By this time, normal pigs are quite active and crushing by the sow is unlikely.

A guard rail around the farrowing pen is an effective means of preventing sows from crushing their pigs. The rail should be raised 8 to 10 in. from the floor and should be 8 to 12 in. from the walls. It may be constructed of two-by-fours, two-by-sixes, or strong poles or steel pipe.

6. **Heat lamps and brooders.** Newborn pigs are quite comfortable at temperatures of 85 to 95°F, shiver when standing alone at 70°F, and are quite cold at 60°F. Recent estimates

indicate that swine producers can save an average of 1½ pigs per litter by using supplementary heat to keep baby pigs warm.

Brooders of various designs are used. Where sows are farrowed in conventional pens, a triangular-shaped brooder is usually secured in one corner of the pen.

In cold climates, heat may be supplied by infrared-heat lamps, by electric hovers, or by gas-fired hard pyrex glass (red filter type) that resists breakage when splashed with water. Use heavy porcelain sockets, a metal hood, and suspend the lamb by a chain. Never suspend the heat lamp by the electrical cord. A 250-watt lamp should be 24 in. above the floor of the brooder area.

Any type of heating device can cause a fire, so heaters should be installed carefully.

7. **Bedding.** The farrowing quarters should be lightly bedded with clean, fresh material. Any good absorbent that is not too long and coarse is satisfactory. Wheat, barley, rye, or oat straw; short or chopped hay; ground corn cobs; peanut hulls; cottonseed hulls; shredded corn fodder; and shavings are most commonly used.

8. **The attendant.** The attendant should be on the job, especially during time of inclement weather. It may be necessary to free the newborn pigs from the enveloping membrane and to help them reach the mother's teat. In cold weather the young should be dried off and other precautions taken to avoid chilling.

If the sow has labored for some time with little progress or is laboring rather infrequently, assistance should be given. This usually consists of inserting the hand and arm in the vulva and gently correcting the condition preventing delivery. Before doing this, the fingernails should be trimmed closely, and the hands and arm should be thoroughly washed with soap and warm water, disinfected, and then lubricated with petroleum jelly or linseed oil.

As soon as the afterbirth is expelled, it should be removed from the pen and burned or buried in lime. This prevents the sow from eating the afterbirth and prevents the development of bacteria and foul odors. Many good swine producers are convinced that eating the afterbirth encourages the development of the pig-eating vice. Dead pigs should be removed for the same reason.

It is also well to work over the bedding; remove wet, stained, or soiled bedding and provide clean, fresh material.

9. **Chilled and weak pigs.** One of the most effective methods of reviving a chilled pig is to immerse the body, except the head, in water as warm as one's elbow can bear. The pigs should be kept in this for a few minutes, then removed and rubbed vigorously with towels.

10. **Orphan pigs.** Pigs may be orphaned either through sickness or death of their mother. In either event, the most satisfactory arrangement for the orphans it to provide a foster mother. When it is impossible to transfer the pig to another sow, they may be raised on cow's milk or milk replacer. The problem will be simplified if the pigs have received a small amount of colostrum (the first milk) from their mother.

If cow's milk is used, it is preferable that it be from a low-testing cow. Do not add cream or sugar; however, skim milk powder at the rate of a tablespoonful to a pint of fluid milk may be added, if available. Milk replacer should be mixed according to the directions found on the container. The first 2 or 3 days the orphans should be fed regularly every 2 hours, and the milk should be at 100°F. Thereafter, the intervals may be spaced farther apart. All utensils (pan feeding or a bottle and nipple may be used) should be clean and sterilized.

Orphan pigs should be started on a prestarter or starter ration when they are one week old. Also, a source of iron should be provided (in keeping with instructions given in Section 4).

(Also, see Section 4, under the heading "Feeding Orphan Pigs.")

11. **Artificial heat.** During times of inclement weather, artificial heat usually must be provided, especially for pigs farrowed in northern United States. Most large central hog houses are equipped with a heating unit for use in winter farrowing, designed to maintain the temperature at 60 to 70°F.

Individual houses may be insulated by banking with straw and other insulating materials. Then a lantern or oil burner may be suspended from the top of the house. It must be remembered, however, that there is considerable fire hazard with this practice. The electric pig brooder is a much safer heating unit for either the central or the movable hog house. The principles involved are identical to those of the electric chick brooder.

(Also, see Section 4, under the heading "Feeding Brood Sows.")

SWINE SKILLS

Among the essential swine skills which the swine producer must perform are the following:

1. **Clipping the boar's tusks.** The common procedure in preparation for removing the tusks consists of drawing a strong rope over the upper jaw and tying the other end securely to a post or other object. As the animal pulls back and the mouth opens, the tusks may be cut with a bolt clipper.

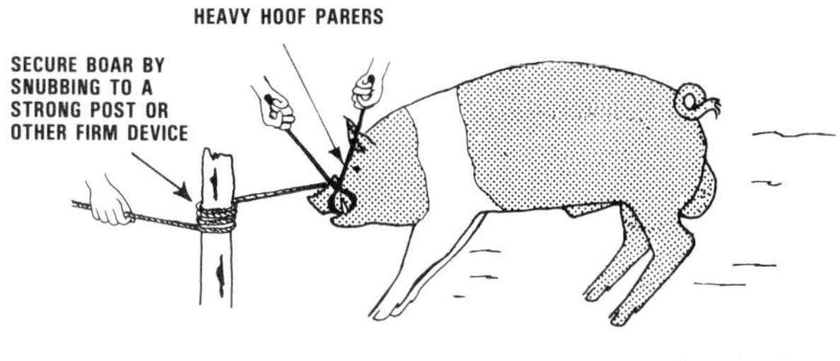

Fig. 8–22. Removing the boar's tusks.

2. **Removing the needle teeth.** Newborn pigs have eight small, tusklike teeth (so-called needle or black teeth), two on each side of both the upper and lower jaw, which are usually clipped by means of pliers or special forceps, thereby alleviating the likely possibility of the pigs inflicting injury upon the sow's udder or on each other. In removing the teeth, care should be taken to avoid injury to the jaw or gums, for injuries may provide an opening for germs; for this reason only the tips of needle teeth should be clipped.

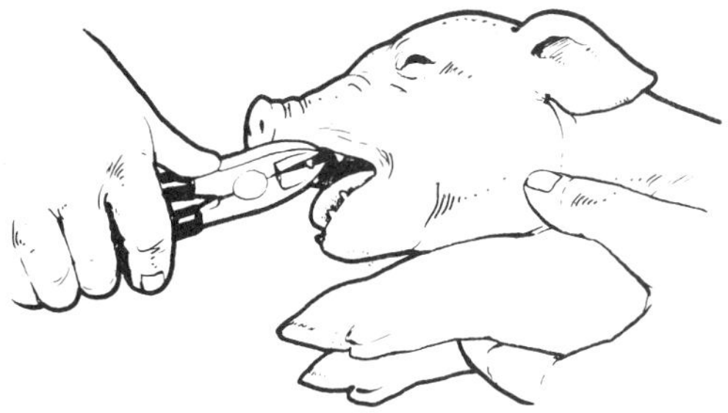

Fig. 8-23. Eight needle teeth—two on each side of both the upper and lower jaw—are removed soon after birth.

3. **Docking tails.** Trimming or *docking* baby pigs' tails is practical for farrow-to-finish operations, and it is mandatory in some graded feeder pig sales. Tail docking seems to be the best method of preventing, or at least reducing, tail biting.

The tail should be clipped to about .75 to 1.0 in. from the bone of the tail. Either sterilized wire cutters or an electric cauterizing blade can be used. Most times the procedure is performed when needle teeth are clipped and/or iron injections are given. A wound protectant spray or dip may be applied to the tail stump.

4. **Ringing.** When rooting starts, the herd should be *ringed;* and this applies to all hogs past weaning age. Older animals can be restrained by a rope or instrument placed around the snout, whereas young pigs can be held.

Many types of rings can be and are used, but the fishhook type is most common. Rings (generally 1 to 3 rings) are usually placed in the snout, just back of the cartilage but away from the bone; although some producers prefer to use a ring that is placed through the septum (the partition of the nose). Others cut the cartilage on top of the snout, but this causes a rather severe setback and should be practiced with caution.

HORSE MANAGEMENT PRACTICES

Fig. 8-24. Horse management is the art of caring for, and handling, horses. (Courtesy, American Quarter Horse Assn., Amarillo, TX)

Horse management practices vary between areas and individual managers. In general, however, the principles of good management are the same everywhere, and they apply whether one horse or many horses are involved. Some horse management practices that are not covered elsewhere in this book follow.

NORMAL BREEDING AND FOALING SEASONS

The most natural breeding season for the mare is in the spring of the year. Usually mares are gaining flesh at this time; the heat period is more evident, and they are more likely to conceive. Furthermore, the spring-born foal may be dropped on pasture—with less danger of infection and with an abundance of exercise, fresh air, and sunshine to aid in its development; and there will be good, green, succulent pasture for the mare. Such conditions are ideal.

Also, it must be remembered that the exhibitor will want to give consideration to having the foals dropped at such time that they may be exhibited to the best advantage. The same applies to the person who desires to sell well-developed yearlings of the light horse breeds or to race two-year-olds. It is noteworthy, however, that the percentage of barren mares that conceive at an early breeding is markedly lower than is obtained later in the season. Nevertheless, some mares do conceive early in the year, and even a small percentage is advantageous to some breeders.

MANAGING THE STALLION

Although certain general recommendations can be made, it should be remembered that each stallion should be studied as an individual, and his care, feeding, exercise, and handling should be varied accordingly.

- **Quarters for the stallion**—The most convenient arrangement for the stallion is a roomy box stall which opens directly into a 2- or 3-acre pasture paddock, preferably separated from the other horses by a double fence. A paddock fence made of heavy lumber is safest. The stall door opening into such a paddock may be left open except during extremely cold weather; this will give the stallion plenty of fresh air, sunshine, and additional exercise.

(Also, see Section 9, under the heading "Horse Buildings and Equipment.")

- **Feeding the stallion**—The feed and water requirements of the stallion are adequately discussed in Section 4. In addition to this, it may be well to reemphasize that, in season, clean lush pastures produced on fertile soils are excellent for the stallion. Grass is the horse's most natural feed, and it is a rich source of vitamins that are so necessary for vigor and reproduction. Perhaps the ideal arrangement in providing pasture for the stallion is to give him access to a well-sodded paddock.

- **Exercise for the stallion**—Most stallioners feel that regular, daily exercise for the stallion is important. Certainly, it is one of the best ways in which to keep a horse in a thrifty, natural condition.

Stallions of the light horse breeds are generally exercised under saddle or hitched to a cart. Thus, Standardbred stallions are usually jogged 3 to 5 miles daily while drawing a cart. Thoroughbred stallions and saddle stock stallions of all other breeds are best exercised under saddle for from 30 minutes to 1 hour daily, especially during the breeding season. Exercise should not be hurried or hard; the walk and the trot are the best gaits to use for this purpose. After the stallion is exercised, he should be rubbed down and cooled off before he is put up, especially if he is hot. Better yet, the ride should be so regulated at the end that the horse will be brought in cool, in which case he can be brushed off and turned into his corral.

Frequently, in light horses, bad feet exclude exercise on roads, and faulty tendons exclude exercise under the saddle. Under such conditions, one may have to depend upon (1) exercise taken voluntarily by the stallion in a large paddock, (2) longeing or exercising on a 30- to 40-ft rope, or (3) leading.

Longeing should be limited to a walk and a trot; and, if possible, the stallion should be worked on both hands; that is, made to circle both to the right and to the left. It is also best that this type of exercise be administered within an enclosure. Two precautions in longeing are: (1) do not longe a horse when the footing is slippery, and (2) do not pull the animal in such manner as to make him pivot sharply with the hazard of breaking a leg.

Leading is a satisfactory form of exercise for some stallions if it is not practical to ride them. In leading, a bridle should always be used—never a halter—and one should keep away from other horses and be careful that the horse being ridden is not a kicker.

Where several stallions are exercised, a properly installed mechanical exerciser driven by an electric motor may be used as a means of lessening labor. It is similar to the merry-go-round type of equipment used to exercise dairy bulls.

The objection to relying upon paddock exercise alone is that the exercise cannot be regulated, especially during inclement weather. Some animals may take too much exercise and others too little. Moreover, merely running in the paddock will seldom, if ever, properly condition a stallion. Nevertheless, a 2- to 3-acre grassy paddock should always be provided, even for horses that are regularly exercised. Stallions that are worked should be turned out at night and on idle days.

• **Grooming the stallion**—Proper grooming of the stallion is necessary, not only to make the horse more attractive in appearance, but to assist exercise in maintaining the best of health and condition. Grooming serves to keep the functions of the skin active. It should be thorough, with special care taken to keep all parts of the body clean and free from any foulness, but not so rough nor so severe as to cause irritation either of the skin or the temper.

MANAGING MARES AT FOALING

The period of parturition is one of the most critical stages in the life of the mare. Through carelessness or ignorance, all of the advantages gained in selecting genetically desirable and healthy parent stock and in providing the very best of environmental and nutritional conditions through gestation can be quickly dissipated at this time.

• **Work and exercise**—Saddle or light-harness mares should be exercised moderately in the accustomed manner. If they are not used, other gentle exercise, such as leading, should be provided. This is especially important if they have not been accustomed to being on pasture and if it is desirable to avoid any abrupt changes in feeding at this time.

• **Signs of approaching parturition**—Usually the first sign of approaching parturition is a distended udder, which may be observed 2 to 6 weeks before foaling time. About 7 to 10 days before the arrival, there will generally be a marked shrinkage or falling away of the muscular parts of the top of the buttocks near the tailhead and a falling of the abdomen. Although the udder may have filled out previously, the teats seldom fill out to the ends more than 4 to 6 days before foaling; and the wax on the ends of the nipples generally is not present until within 2 to 4 days before parturition. About this time the vulva becomes full and loose. As foaling time draws nearer, milk will drop from the teats; and the mare will show restlessness, break into a sweat, urinate frequently, lie down and get up, etc. It should be remembered, however, that there are times when all signs fail and a foal may be dropped when least expected. Therefore, it is well to be prepared as much as 30 days in advance of the expected foaling time.

• **Preparation for foaling**—When signs of approaching parturition seem to indicate that the foal may be expected within a week or 10 days, arrangements for the place of foaling should be completed. Thus, the mare will become accustomed to the new surroundings before the time arrives.

During the spring, summer, and fall months when the weather is warm, the most natural and ideal place for foaling is a clean, open pasture away from other livestock. Under these conditions, there is decidedly less danger of either infection or mechanical injury to the mare and foal. Of course, in following this practice, it is important that the ground be dry and warm. Small paddocks or lots that are unclean and foul with droppings are unsatisfactory and may cause such infectious troubles as navel ill.

During inclement weather, the mare should be placed in a roomy, well-lighted, well-ventilated, comfortable, quiet box stall which should first be carefully cleaned, disinfected, and bedded for the occasion. It is best that the mare be stabled therein at night a week or 10 days before foaling so that she may become accustomed to the new surroundings. The foaling stall should be at least 12 ft square and free from any low mangers, hay racks, or other obstructions that might cause injury to either the mare or the foal. After the foaling stall has been thoroughly cleaned, it should be disinfected to reduce possible infection. This may be done by scrubbing with boiling hot lye water, made by using 8 oz of lye to 20 gal of water (one-half this strength of solution should be used in scrubbing mangers and grain boxes). The floors should then be sprinkled with air-slaked lime. Plenty of clean, fresh bedding should be provided at all times.

A foaling stall somewhat away from other horses and with smooth, well-packed clay floor is to be preferred. The clay floor may be slightly more difficult to keep smooth and sanitary than concrete or other such surface materials, but there is less danger to the mare and the newborn foal from slipping and falling; and it is decidedly better for the hoofs.

• **Feed at foaling time**—Shortly before foaling, it is usually best to decrease the grain allowance slightly and to make more liberal use of light and laxative feeds, especially wheat bran. If there are any signs of constipation, a wet bran mash should be provided.

(Also, see Section 4, under the heading "Feeding Broodmares.")

• **The attendant**—A good rule for the attendant is to *be near but not in sight.* The presence of an attendant may prevent possible injury to the mare and foal; and, when necessary, the attendant may aid the mare or call a veterinarian.

• **Parturition**—The first actual indication of foaling is the rupture of the outer fetal membrane, followed by the escape of a large amount of fluid. This is commonly referred to as the rupture of the *water bag.* The inner membrane surrounding the foal appears next, and labor then becomes more marked.

With normal presentation, a mare foals rapidly, usually not taking more than 15 to 30 minutes. Usually, when the labor pains are at their height, the mare will be down, and it is in this position that the foal is generally born, while the mare is lying on her side with all legs stretched out.

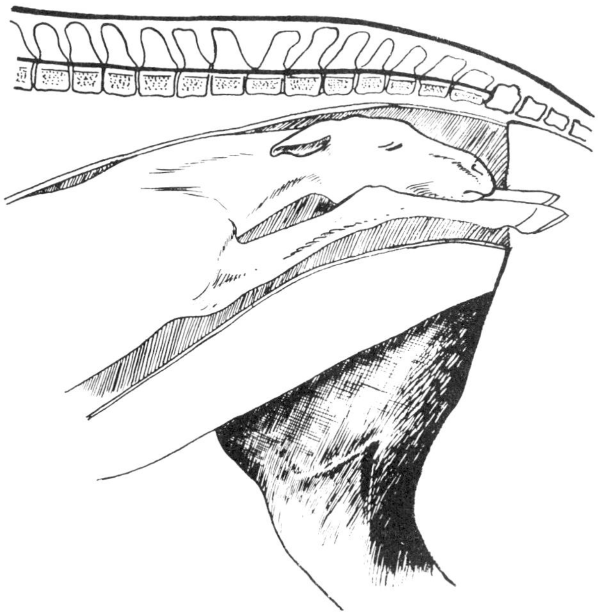

Fig. 8-25. Normal presentation. The back of the fetus is toward the back of the mother, the forelegs are extended toward the vulva with the heels down and the nose rests between the forelegs.

In normal presentation, the front feet, with the heels down, come first, followed by the nose, which is resting on them, then the shoulders, the middle (with the back up), the hips, and then the hind legs and feet. If the presentation is other than normal, a veterinarian should be summoned at once, for there is great danger that the foal will smother if birth is delayed. If the feet are presented bottoms up, it is a good indication that they are the hind ones, and there is likely to be difficulty.

If after reasonable time and effort have been expended a mare appears to be making no progress in parturition, it is advisable that an examination be made and assistance be rendered before the animal has completely exhausted her strength in futile efforts at expulsion. In rendering any such assistance, the following cardinal features should exist: (1) cleanliness; (2) quietness; (3) gentleness; (4) perseverance; and (5) knowledge, skills, and experience.

When parturition is unduly delayed or retarded, the fetus often dies from twists or knots in the umbilical cord, or from remaining too long in the passage. In either case, there may be stoppage of fetal circulation or lack of oxygen for the fetus, or both.

If foaling has been normal, the attendant should enter the stable to make certain that the foal is breathing and that the membrane has been removed from its mouth and nostrils. If the foal fails to breath immediately, artificial respiration should be applied. This may be done by blowing into the mouth of the foal, working the ribs, rubbing the body vigorously and permitting the foal to fall around. Then after the navel has been treated, the mare and foal should be left to lie and rest quietly together as long as possible so that they may gain strength.

• **The afterbirth**—If the afterbirth is not expelled soon after the mare gets up following foaling, it should either be tied up in a knot or tied to the tail of the mare. This should be done so that the foal or mare will not step on it, thereby increasing the danger of inflammation of the uterus and foal founder in the mare. Usually the afterbirth will be expelled within 1 to 6 hours after foaling. If it is retained for a longer period, or if lameness is evident, the mare should be blanketed, and an experienced veterinarian should be called. Retained afterbirth often causes laminitis, which is recognized by lameness in the mare. This is usually treated by feeding easily digested feed for a period of 36 hours and by applying cold applications to the mare's feet until the condition is relieved.

To prevent development of bacteria and foul odors, the afterbirth should be removed from the stall and burned or buried in lime as soon as possible.

• **Cleaning the stall**—Once the foal and mare are up, the stall should be cleaned. Wet, stained, or soiled bedding should be removed. The floor should be sprinkled with lime; and clean, fresh bedding should be provided. Such sanitary measures will be of great help in preventing the most common type of joint ill.

If the weather is extremely cold and the mare hot and sweaty, she should be rubbed down, dried, and blanketed soon after getting on her feet.

• **Feed and water after foaling**—Following foaling, the mare usually is somewhat hot and feverish. She should be given small quantities of lukewarm water at intervals, but she should never be allowed to gorge. It is also well to feed lightly and with laxative feeds for the first few days. The very first feed might well be a wet bran mash with a few oats or a little oat meal soaked in warm water. About one-half the usual amount should be fed. Usually, for the first week, no better grain ration can be provided than bran and oats. The quantity of feed given should be governed by the milk flow, the demands of the foal, and the appetite and condition of the mare. Usually the mare can be back on full feed within a week or 10 days after foaling.

(Also, see Section 4, under the heading "Feeding Broodmares.")

• **Observation**—The good caretaker will be ever alert to discover difficulties before it is too late. If the mare has much temperature (normal for horses is about 101°F), something is wrong and the veterinarian should be called. As a precautionary measure, many good caretakers take the mare's temperature a day or two after foaling. Any discharge from the vulva should be regarded with suspicion.

MANAGING THE NEWBORN FOAL

Immediately after the foal has arrived and breathing has started, it should be thoroughly rubbed and dried with warm towels. Then it should be placed in one corner of the stall on clean, fresh, straw. Usually the mare will be less restless if this corner is in the direction of the hear. The eyes of the newborn foal should be protected from bright light.

• **Navel cord**—To reduce the danger of navel infection (which causes a disease known as joint ill or navel ill) the navel cord of the newborn foal should be treated at once with a solution of tincture of iodine (or Metaphen or Merthiolate may be used). This may be done by placing the end of the cord in a wide mouthed bottle nearly full of tincture of iodine while pressing the bottle firmly against the abdomen. This is best done with the foal lying down. The cord should then be dusted with a good antiseptic powder. Dusting with the powder should be continued daily until the stump dries up and drops off and the scar heals, usually in three to four days. If an antiseptic powder is not available, air-slaked lime may be used. Any foreign matter that accumulates on the navel should be pressed out, and a disinfectant should be applied.

If left alone, the navel cord of the newborn foal usually breaks within 2 to 4 in. from the belly. Under such conditions, no cutting is necessary. However, if it does not break, it should be severed about 2 in. from the belly with clean, dull shears or it may be scraped with a knife. Never cut diagonally across. A torn or broken blood vessel will bleed very little, whereas one that is cut directly across may bleed excessively. If severing of the cord is resorted to, it should be immediately treated with iodine.

• **The colostrum**—The colostrum is the milk that is secreted by the dam for the first few days following parturition.

The strong, healthy foal will usually be up on its feet and ready to nurse within 30 minutes to 2 hours after birth. Occasionally, however, a big awkward foal will need a little assistance and guidance during its first time to nurse. The stubborn foal should be coaxed to the mare's teats (forcing is useless). This may be done by backing the mare up on additional bedding in one corner of the stall and coaxing the foal with a bottle and nipple. The attendant may hold the bottle while standing on the opposite side of the mare from the foal. The very weak foal should be given the mare's first milk even if it must be drawn in a bottle and fed by nipple for a time or two. Sometimes these weak individuals will nurse the mare if steadied by the attendant.

Aside from the difference in chemical composition, the colostrum (the milk yielded by the mother for a short period following the birth of the young) seems to have the following functions:

1. It contains antibodies that temporarily protect the foal against certain infections, especially those of the digestive tract.

2. It serves as a natural purgative, removing fecal matter that has accumulated in the digestive tract.

(Also, see Section 4, under the heading "Special Feeds," "•Colostrum," and "•Milk Replacers.")

• **Bowel movement of the foal**—The regulation of the bowel movement in the foal is very important. Two common abnormalities are constipation and diarrhea or scours.

Impaction in the bowels of the excrement accumulated during the development prior to birth—material called meconium—may prove fatal if not handled promptly. Usually a good feed of colostrum will cause elimination, but not always—especially when foals are from stall-fed mares.

Bowel movement of the foal should be observed within 4 to 12 hours after birth. If by this time there has been no discharge and the foal seems rather sluggish and fails to nurse, it should be given an enema. This may be made by using 1 to 2 qt of water at blood heat, to which a little glycerin has been added; or warm, soapy water is quite satisfactory. The solution may be injected with a baby syringe (one having a 3-in. nipple) or a tube and can. This treatment may be repeated as often as necessary until the normal yellow feces appear.

Diarrhea or scours in foals may be associated with infectious diseases or may be caused by unclean surroundings. Any of the following conditions may bring on diarrhea: contaminated udder or teats; nonremoval of fecal matter from the digestive tract; fretfulness or temperature above normal in the mare; an excess of feed affecting the quality of the mare's milk; cold, damp bed; or continued exposure to cold rains. As treatment is not always successful, the best practice is to avoid the undesirable conditions.

Some foals scour during the foal heat of the mare, which occurs between the seventh and ninth day following foaling.

Diarrhea is caused by an irritant in the digestive tract that should be removed if recovery is to be expected. Only in exceptional cases should an astringent be given with the idea of checking the diarrhea; and such treatment should be prescribed by the veterinarian.

If the foal is scouring, the ration of the mare should be reduced, and a part of the milk should be taken away by milking her at intervals.

(Also, see Section 4, under the heading "Feeding Foals.")

WEANING THE FOAL

Weaning of the foal is more a matter of preparation than of absolute separation from the dam. The simplicity with which it is accomplished depends very largely upon the thoroughness of the preparation.

• **Age of weaning**—Foals are usually weaned at 4 to 6 months of age, depending on conditions.

• **Separation of the mare and foal**—When all preliminary precautions and preparations for weaning have been made, the separation should be accomplished. This should be complete and final with no opportunity for the foal to see, hear, or smell its dam again.

After weanlings have remained in the stable for a day or two and have quieted down, they should be turned out to pasture. Where a group of weanlings is involved, undue running and possible injury hazard may be minimized in this transition by the following procedure: First turn two or three of the least valuable animals out and let them tire themselves out, and then turn the rest of the weanlings out and they will do very little running.

With a great number of weanlings, it is advisable to separate the sexes, and even to place some of the more timid ones to themselves. In all cases, it is best not to run weanlings with older horses.

• **Drying up the mare**—The following procedure for drying up the mare is recommended:

1. Rub an oil preparation on the bag. Take the mare from the foal and place her on less lush pasture or grass hay.
2. Examine the udder and rub oil on it at intervals, but do not milk it out for 5 to 7 days. It will fill up and get tight, *but do not milk it out.* At the end of 5 to 7 days, when the bag is soft and flabby, milk out what little secretions remain (probably not more than half a cup).

CARE OF THE FEET

The important points in the care of a horse's feet are to keep them clean, prevent them from drying out, trim them so they retain proper shape and length, and shoe them correctly when shoes are needed.

Each day, the feet of horses that are shod, stabled, or used should be cleaned and inspected for loose shoes and thrush. Thrush is a disease of the foot, caused by a necrotic fungus, and characterized by a pungent odor. It causes deterioration of tissues in the cleft of the frog or in the junction between the frog and bars. This disease produces lameness and, if not treated, can be serious.

• **Proper stance; correcting common faults**—Before trimming the feet or shoeing a horse, it is important to know what constitutes both proper and faulty conformation (see Section 11).

Fig. 8–26 shows the proper posture of the hoof and incorrect postures caused by hoofs grown too long in either toe or heel. The slope is considered normal when the toe of the hoof and the pastern have the same direction. This angle should always be kept in mind and changed only as a corrective measure. If it should become necessary to correct uneven wear of the hoof, the correction should be made gradually over a period of several trimmings.

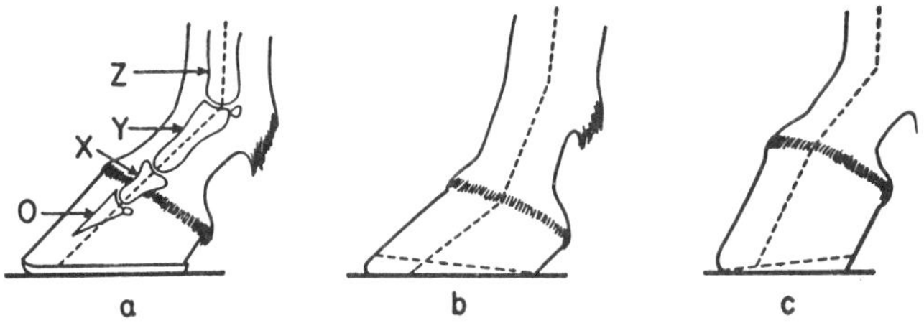

Fig. 8–26. (a) Properly trimmed hoof with normal foot axis: O—coffin bone; X—short pastern bone; Y—long pastern bone; Z—cannon bone. (b) Toe too long, which breaks the foot axis backward. Horizontal dotted line shows how hoof should be trimmed to restore normal posture. (c) Heel too long, which breaks the foot axis forward. Horizontal dotted line shows how trimming will restore the correct posture.

Prior to the trimming of the feet, the horse should be inspected while standing squarely on a level area—preferably a hard surface. Then it should be seen in action, both at the walk and the trot.

The hoofs should be trimmed every month or six weeks, whether the animal is shod or not. If shoes are left on too long, the hoofs grow out of proportion. This may throw the horse off balance and place extra stress upon the tendons. Hence, the hoofs should always be kept at the proper length and the correct posture. They should be trimmed near the level of the sole; otherwise, they will split off if the horse remains unshod. The frog should be trimmed carefully, with only ragged edges removed that allow the filth to accumulate in the crevices, and the sole should be trimmed sparingly, if at all. The wall of the hoof should never be rasped.

• **Care of the foal's feet**—Foals may become unsound of limb when the wear and tear is not equally distributed due to an unshapely hoof. On the other hand, faulty limbs may be helped or even corrected by regular and persistent trimming. Such practice also tends to educate the foal and to make shoeing easier at maturity. If the foal is run on pasture, trimming of the feet may be necessary long before weaning time. A good practice is to check the feet regularly every month or 6 weeks and to trim a small amount each time if trimming is needed, rather than to trim too much at any one time. Tendons should not receive undue strain by careless trimming of the feet. Usually, only the outer rim should be trimmed, though sometimes it is necessary to cut down the heel or frog or to shorten the toes.

The necessary trimming may be done with the rasp, the farrier's knife, and nippers (using the rasp for the most part).

Before the feet are trimmed, the foal should be inspected first while standing squarely on a hard surface. Then it should be seen in action, both at the walk and the trot.

• **Treatment of dry hoofs**—When hoofs become dry and brittle, they sometimes split and cause lameness. The frogs lose their elasticity and are no longer effective shock absorbers. If the dryness is prolonged, the frogs shrink and the heels contract.

Dry hoofs usually can be prevented by keeping the ground wet around the watering tank, attaching wet burlap sacks around the hoofs, or applying a hoof dressing. Several satisfactory commercial hoof dressings are on the market. Also, a good homemade product may be made as follows:

6 parts fish oil (cod liver oil)
1 part pine tar oil
1 part Creolin
2 parts glycerin

The above mixture should be stirred well before using and applied daily. If fish oil is not available, raw linseed oil may be substituted.

GROOMING

Proper grooming is necessary to keep a horse attractive and help maintain good health and condition. Grooming cleans the hair, helps keep the skin functioning naturally, lessens skin diseases and parasites, and improves the condition and fitness of the muscles.

Grooming should be rapid and thorough but not so severe that it makes the horse nervous or irritates the skin. Horses that are kept in stables or small corrals should be groomed thoroughly at least once a day. When horses are worked or exercised, they should be groomed both before and after the work or exercise.

Wet or sweating animals should be handled as follows:

1. Remove the tack as fast as possible, wipe it off, and put it away.

2. Remove excess water from the horse with a sweat scraper and then rub the coat with a grooming or drying cloth to dry it partially.

3. Cover the horse with a blanket and walk it until it is cool.

4. Allow the horse to drink two or three swallows of water every few minutes while it is cooling and drying.

To assure that the horse is groomed thoroughly and that no body parts are missed, follow a definite order of grooming. This may vary according to individual preference.

Grooming may be checked by rubbing the fingertips against the natural lay of the hair. If the coat and skin are not clean, the fingers will get dirty and gray lines will show on the coat where the fingers passed. The cleanliness of the ears, face, eyes, nostrils, lips, sheath, and dock can be determined by inspection.

TRANSPORTING HORSES

Fig. 8-27. A horse being boarded on a cargo liner in Kennedy Airport, New York, for transportation to Hollywood Park. (Courtesy, United Airlines)

Horses are transported via trailer, van, truck, rail, boat, and plane. Today, transportation by motor (trailer, van, or truck) is most common because of the distinct advantage of door-to-door movement. Regardless of the method, however, the objectives are the same: to move them safely, with the maximum of comfort, and as economically as possible. To this end, selection of the equipment is the first requisite. But equipment alone, no matter how good, will not suffice.

The trip must be preceded by proper preparation including the conditioning of horses; and horses must receive proper care, including smooth movement, *en route*.

In summary form, the requisites of good transportation, with special emphasis on motor transportation follow:

1. Provide good footing.
2. Drive carefully.
3. Make nurse stops.
4. Provide proper ventilation.
5. Teach horses to load early in life.
6. Provide health certificate and statement of ownership.
7. Schedule properly.
8. Have the horses relaxed.
9. Clean and disinfect public conveyance.
10. Have a competent caretaker accompany horses.
11. Use shanks except on stallions.
12. Feed lightly.
13. Water liberally.
14. Pad the stalls.

15. Take along tools and supplies.
16. Check shoes, blankets, and bandages.
17. Be calm when loading and unloading.
18. Control insects.

Trailers, vans, and trucks, have the very great advantage of being able to load from in front of one stable and unload in front of another.

The trailer is usually a 1- or 2-horse unit, which is drawn behind a car or truck. Generally speaking, this method of transportation is best adapted to short distances—less than 500 miles. Horses are trailered to shows, races, endurance rides, breeding establishments, to new owners, from one work area to another on the range.

The van or vanlike trailer is a common and satisfactory method of transportation where three to eight horses are involved. There is hardly any limit to the kinds of vans, ranging from rather simple to very palatial pieces of equipment.

Rail and boat shipment are seldom used anymore. Where valuable horses are involved, they have given way to the greater speed and flexibility of plane transportation.

STABLE MANAGEMENT

The following stable management practices are recommended;

1. Remove the top layer of clay floors yearly; replace with fresh clay, and level and tamp. Also, keep the stable floor higher than the surrounding area, thereby making for dryness.
2. Keep stalls well lighted.
3. Use properly constructed hayracks to lessen waste and contamination of hay, with the possible exception of maternity stalls.
4. Scrub concentrate containers at such intervals as necessary, and after feeding a wet bran mash.
5. Work over bedding daily, removing excrement and wet, stained, or soiled material, and provide fresh bedding.
6. Practice rigid stable sanitation to prevent fecal contamination of feed and water.
7. Lead foals when taking them from the stall to the paddock and back, as a way in which to further their training.
8. Restrict the ration when horses are idle, and provide either a wet bran mash the evening before an idle day or turn idle horses to pasture.
9. Provide proper ventilation at all times—by means of open doors, windows that open inwardly from the top, or stall partitions slatted at the top.
10. Keep stables in repair at all times, so as to lessen injury hazards.

BLEEDERS

Stress triggers different syndromes in different species and in different individuals. When subjected to heavy exercise, such as in racing, horses are prone to a condition known as *bleeders.*

It has been estimated that 70 to 80% of Thoroughbred racehorses are bleeders. In a recent major U.S. race, it was reliably reported that 5 horses in the field of 11 ran with the aid of Lasix, a diuretic used to combat bleeding. So, bleeders are a serious problem!

Bleeders are horses afflicted by blood flowing from their nostrils or bronchial tubes, but originating in the lungs, following strenuous exercise. But the problem is not confined to Thoroughbreds, nor is it limited to racehorses. It also afflicts Quarter Horses and Standardbreds, and it afflicts horses when they are subjected to maximum stress by strenuous physical activity of whatever kind; for example, in endurance rides. But it does not affect racing dogs or humans who race competitively. The species dissimilarity is attributed to the difference in the horse's anatomy. The horse has a sloping diaphragm and is primarily an abdominal breather, inhaling by movement of the diaphragm. This type of breathing appears to create stress in the equine lung.

Bleeding appears to affect the performance of racehorses. But the cause and cure remain elusive. The condition is exercise-induced; and there appears to be a higher frequency of bleeders among older horses. When bleeding is excessive, the most common treatment is to discontinue training and racing temporarily or permanently.

Research studies indicate that bleeders are a stress/environmental interaction of equines induced by heavy exercise, but the findings do not signal any restriction in racing horses; because galloping is a very natural behavior in horses, inherited from their wild ancestors who escaped the attacks of predators by flight.

MAKING AND KEEPING THE SOIL PRODUCTIVE[3]

Fig. 8-28. Spreading barnyard manure. (Courtesy, Gehl Company, West Bend, WI)

[3]This entire section, beginning with "Making and Keeping the Soil Productive" and extending through "Treat Saline and Alkaline Soils" benefitted from the authoritative review of the following: Dr. D. L. McCune, Managing Director, International Fertilizer Development Center, Muscle Shoals, AL; Mr. G. D. Myers, President, The Fertilizer Institute, Washington, DC; and Dr. H. M. Reisenauer, Professor, Land, Air, and Water Resources, University of California at Davis)

Making and keeping the soil productive is the very foundation of a successful agriculture, of national prosperity. It has been well said that good soils, good farms and ranches, and good living go hand in hand.

BARNYARD MANURE

Manure is the refuse from stables, barnyards, and feedlots, including both animal excreta and straw or other bedding (litter). In some other countries, the term *manure* is used more broadly and includes both "animal manure" and "chemical fertilizers," for which the term *fertilizer* is commonly used in the United States.

Animals provide manure for the fields, a fact which was often forgotten during the era when chemical fertilizers were relatively abundant and cheap. One ton of average manure contains 10 lb of nitrogen (N), 5 lb of phosphoric acid (P_2O_5), and 10 lb of potassium (K_2O). At current prices (per pound: N = 25¢, P_2O_5 = 25¢, and K_2O = 17¢), it's worth $5.45 per ton.

In the decades to come, the world will need to feed a growing and hungry population, while at the same time protecting the planet's increasingly fragile environment. This necessitates a balancing act—it calls for maximizing our production while minimizing the effect of this production on the environment. To meet this challenge, a growing number of American farmers are returning to organic farming; they are using more manure—the unwanted barnyard centerpiece of the past, and they are discovering that they can be just as good reapers of the land and far better stewards of the soil.

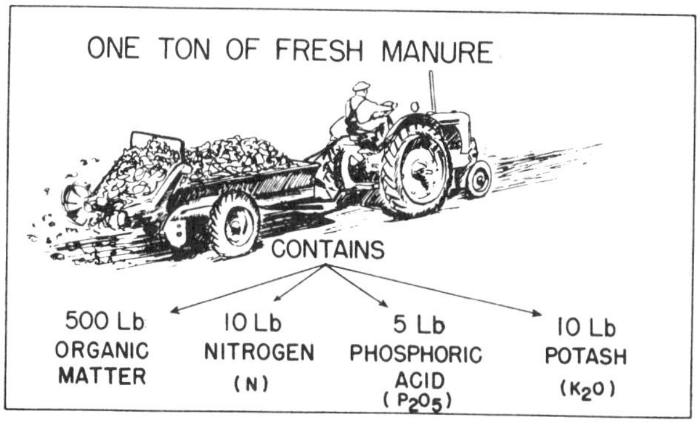

Fig. 8–29. The contents of one ton of average fresh manure.

AMOUNT, COMPOSITION, AND VALUE OF MANURE

The quantity, composition, and value of excrement produced vary according to species, weight, kind, and amount of feed, and kind and amount of bedding. The author's computations are on a fresh manure (exclusive of bedding) basis. Table 8–3 presents data by species per 1,000 lb liveweight, whereas Table 8–4 gives yearly tonnage and value.

The data in Table 8–3 and Fig. 8–30 are based on animals confined in stalls the year around. Actually, the manure

TABLE 8–3
QUANTITY, COMPOSITION, AND VALUE OF FRESH MANURE (FREE OF BEDDING) EXCRETED BY 1,000 POUNDS LIVEWEIGHT OF VARIOUS KINDS OF FARM ANIMALS

(1) Animal	(2) Tons Excreted/ Year/1,000 Lb Liveweight[1]	Composition and Value of Manure on a Tonnage Basis[2] (3) Excrement	(4) Lb/Ton[3]	(5) Water	(6) N	(7) P_2O_5[4]	(8) K_2O[4]	(9) Value/ Ton[5]
			(lb)	(%)	(lb)	(lb)	(lb)	($)
Cow (beef or dairy)	12	Liquid	600	79	11.2	4.6	12.0	5.99
		Solid	1,400					
		Total	2,000					
Steer (finishing cattle)	8.5	Liquid	600	80	14.0	9.2	10.8	7.64
		Solid	1,400					
		Total	2,000					
Sheep	6	Liquid	660	65	28.0	9.6	24.0	13.48
		Solid	1,340					
		Total	2,000					
Swine	16	Liquid	800	75	10.0	6.4	9.1	5.65
		Solid	1,200					
		Total	2,000					
Horse	8	Liquid	400	60	13.8	4.6	14.4	7.05
		Solid	1,600					
		Total	2,000					

[1]*Manure Is Worth Money—It Deserves Good Care,* University of Illinois Circ. 595, 1953, p.4.

[2]Columns 5, 6, 7, and 8 from *Farm Manures,* University of Kentucky Circ. 593, 1964, p.5, Table 2.

[3]From *Reference Material for 1951 Saddle and Sirloin Essay Contest,* compiled by M. E. Ensminger, p. 43; data from *Fertilizers and Crop Production* by Van Slyke, published by Orange Judd Publishing Co.

[4]P_2O_5 can be converted to phosphorus (P) by dividing the figure given above by 2.29, and K_2O can be converted to potassium (K) by dividing by 1.2.

[5]Calculated on the assumption that nitrogen (N) retails at 25¢, P_2O_5 at 25¢, and K_2O at 17¢ per pound in commercial fertilizers.

recovered and available to spread where desired is considerably less than indicated because (1) animals are kept on pasture and along road and lanes much of the year, where the manure is dropped, (2) of losses in weight which often run as high as 60% when manure is exposed to the weather for a considerable time, and (3) of losses in nitrogen as NH_3 volatilizes with drying. Almost one-fourth of the total nitrogen of cow manure may be lost in 12 hours of drying at high temperature.

Fig. 8-30. On the average, each class of stall-confined animals produces per year 1,000 lb weight the tonnages shown above. (Drawing by R. F. Johnson)

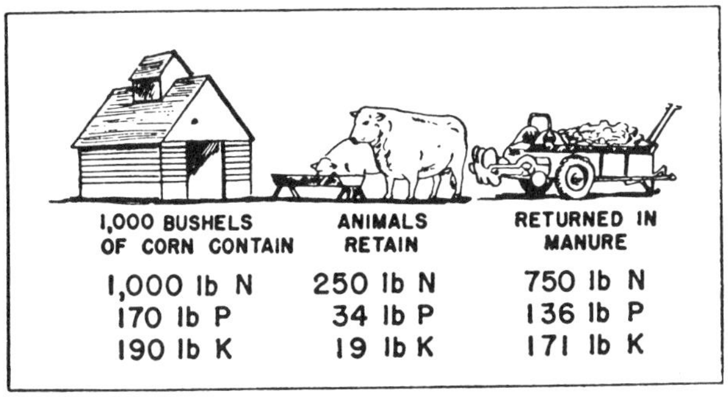

Fig. 8-31. Animals retain about 20% of the nutrients in feed; the rest is excreted in manure.

As shown in Fig. 8-31, about 75% of the nitrogen, 80% of the phosphorus, and 85% of the potassium contained in animal feeds are returned as manure. In addition, about 40% of the organic matter in feeds is excreted as manure. As a rule of thumb, it is commonly estimated that 80% of the total nutrients in feeds are excreted by animals as manure. A ton of fresh barnyard manure has approximately the composition shown in Fig. 8-29.

Naturally, it follows that the manure from well-fed animals is higher in nutrients and worth more than that from poorly fed ones. For example, steer manure produced by fattening cattle liberally fed on nutritious concentrates is more valuable than that produced from cattle wintered on hay.

The urine makes up 20% of the total weight of the excrement of horses, and 40% of that of hogs; these figures represent the 2 extremes in farm animals. Yet the urine, or liquid manure, contains nearly 50% of the nitrogen, 6% of the phosphorus, and 60% of the potassium of average manure; roughly ½ of the total plant food of manure (see Fig. 8-32). Also, it is noteworthy that the nutrients in liquid manure are more readily available to plants than the nutrients in the solid excrement. These are the reasons it is important to conserve urine.

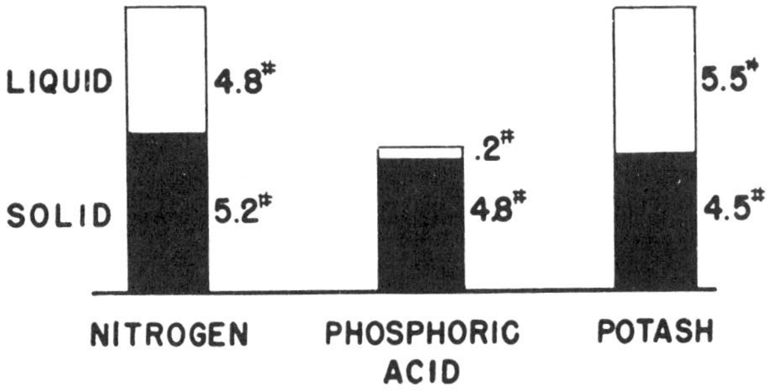

Fig. 8-32. Distribution of plant nutrients between liquid and solid portions of the average farm manure. As noted, the urine contains about half the fertility value of manure.

The actual monetary value of manure can and should be based on (1) increased crop yields, and (2) equivalent cost of a like amount of commercial fertilizer. Numerous experiments and practical observations have shown the measurable monetary value of manure in increased crop yield. Table 8-3 (footnote 5), gives the equivalent cost of a like amount of commercial fertilizer. Of course, the cost of application of manure vs the cost of application of fertilizer must be considered.

Currently, we are producing manure (exclusive of bedding) at the rate of 1.25 billion tons annually (see Table 8-4, next page). That is sufficient manure to add nearly ⅔ ton each year to every acre of the total land area (1.9 billion acres) of the United States.

Based on equivalent fertilizer prices (see Table 8-3, right-hand column) and livestock numbers (Table 8-4), the yearly manure crop is worth $6.8 billion.

Of course, the value of manure cannot be measured alone in terms of increased crop yields and equivalent costs of a like amount of commercial fertilizer. It has additional value for the organic matter which it contains, which almost all soils need, and which farmers and ranchers cannot buy in a sack or tank.

Also, it is noteworthy that, due to the slower availability of its nitrogen and to its contribution to the soil humus, manure produces rather lasting benefits, which may continue for many years. Approximately ½ of the plant nutrients in manure are available to and effective upon the crops in the immediate cycle of the rotation to which the application is made. Of the unused remainder, about ½, in turn, is taken up by the crop in the second cycle of the rotation; ½ of the remainder in the third cycle, etc. Likewise, the continuous use of manure through several rounds of a rotation builds up a backlog which brings additional benefits, and a measurable climb in yield levels.

TABLE 8-4
TONNAGE AND VALUE OF MANURE (EXCLUSIVE OF BEDDING) EXCRETED BY U.S. LIVESTOCK[1]

Class of Livestock	Number of Animals on Farms[2]	Average Liveweight	Tons Manure Excreted/Year/ 1,000 Lb Liveweight[3]	Total Manure Production	Total Value of Manure[4]
		(lb)	(tons)	(tons)	($)
Cattle (beef and dairy; including steers)	102,468,000	900	11	1,014,433,200	5,528,660,940
Sheep	10,328,000	100	6	33,772,560	86,011,584
Swine	50,960,000	200	16	888,742,400	947,448,320
Horses	8,519,000	1,000	8	68,152,000	371,428,400
				1,251,854,000	6,822,604,300

[1]In these computations, no provision was made for animals that died or were slaughtered during the year. Rather, it was assumed that their places were taken by younger animals, and that the population of each species was stable throughout the year.

[2]From USDA, *Agricultural Statistics 1987,* except horse numbers from American Horse Council, Inc., Washington, DC. The cattle and sheep figures are for Jan. 1, 1987; the swine and layer figures are for Dec. 1, 1986; the broiler and turkey figures are number raised during the year; the horse numbers are for 1986. All figures are assumed averages throughout the year.

[3]*Manure Is Worth Money—It Deserves Good Care,* University of Illinois Circ. 595, 1953, p. 4.

[4]Computed on the basis of the value per ton given in the right-hand column of Table 8-3.

Farmers and ranchers sometimes fail to recognize the value of this barnyard crop because (1) it is produced whether or not it is wanted, and (2) it is available without cost. Most of all, no one is selling it. Whoever heard of a traveling manure salesman?

MANURE AS A FERTILIZER

Fig. 8-33. "Honey wagon" spreading liquid manure on cropland in Idaho. (Courtesy, USDA, Soil Conservation Service)

With today's heavy animal concentration in one location, the question is being asked: How much manure can be applied to the land without depressing crop yields, making for salt problems in the soil, making for nitrate problems in feed, contributing excess nitrate to groundwater or surface streams, or violating state regulations?

Based on earlier studies in mid-western United States, before the rise of commercial fertilizers, it would appear that one can apply from 5 to 20 tons of manure per acre, year after year, with benefit.

Heavier applications can be made, but probably should not be repeated every year. With rates higher than 20 tons per annum, there may be excess salt and nitrate buildup. Excess nitrate from manure can pollute streams or groundwater and result in toxic levels of nitrate in crops. Without doubt the maximum rate at which manure can be applied to the land will vary widely according to soil type, rainfall, and temperature.

State regulations differ in limiting the rate of manure application. Missouri draws the line at 30 tons per acre on pasture, and 40 tons per acre on cropland. Indiana limits manure application according to the amount of nitrogen applied, with the maximum limit set at 225 lb per acre per year. Nebraska requires only one-half acre of land for liquid manure disposal per acre of feedlot, which appears to be the least acreage for manure disposal required by any state.

When a farmer has sufficient land, manure should be used at rates which supply only the nutrients needed by the crop rather than the maximum possible amounts suggested for pollution control.

MANURE AS A FEED

Recycling manure as a livestock feed is the most promising of the nonfertilizer uses. Various processing methods are being employed; and some manure is being fed without processing. More and more feedlot manure will be either (1) incorporated in a grower ration, or (2) fed to breeding herds during periods when pasture supplementation is beneficial, with the residues distributed over grazing areas where they would have fertilizing value.

• **Poultry waste**—Nearly 100 million tons of poultry wastes (from layers, broilers, and turkeys) are produced annually. Because poultry production is highly intensive, with many birds in a small area, waste disposal is a major problem. Most cage-layer operations produce manure free of litter as the primary form of waste. Broiler operations, generally produce litter.

Poultry litter is the most collectable and the most nutritious of all animal wastes. It follows that many experiments have been conducted with it, involving feeding trials with different species. The results of numerous experiments are summarized in Table 8–5. The mean values for waste-fed animals reported therein were obtained by averaging all levels of feeding poultry wastes in the respective categories, though some of the levels were excessive. As shown in Table 8–5, the performance of animals fed wastes was generally slightly lower than that of the controls that were fed traditional feed ingredients. But, on a dry-matter basis, animal wastes generally make for least cost rations and higher net returns.

Also, dried poultry litter has been fed successfully to dry and lactating cows, to growing and breeding sheep, to growing swine, and to broilers.

TABLE 8–5
PERFORMANCE OF ANIMALS FED RATIONS CONTAINING POULTRY WASTES[1]

Species of Experimental Animal Used	Kind of Poultry Waste Studied	Performance of Experimental Animals: Criteria	Control Group	Waste-Fed Group
Cattle	Dehydrated layer waste	Daily gain ... lb	2.35 (*1.07 kg*)	2.31 (*1.05 kg*)
		Daily feed dry-matter intake .. lb	15.82 (*7.19 kg*)	15.44 (*7.02 kg*)
		Feed/gain ratio	7.81	7.72
Lactating cows	Dehydrated layer waste	Milk yield ... lb/day	41.8 (*19.0 kg*)	38.94 (*17.7 kg*)
		Milk fat ... %	3.51	3.63
		Milk total solids ... %	12.04	12.01
Sheep	Dehydrated layer waste	Daily gain ... lb	0.42 (*0.19 kg*)	0.40 (*0.18 kg*)
		Feed/gain ratio	5.52	6.66
Swine	Dehydrated layer waste	Daily gain ... lb	1.32 (*0.60 kg*)	1.14 (*0.52 kg*)
		Feed/gain ratio	4.12	4.82
Growing chicks	Dehydrated layer waste	Daily gain ... grams	16.1	15.7
		Feed/gain ratio	2.36	2.60
Laying hens	Dehydrated layer waste	Egg production ... % lay	71.9	72.8
		Feed/dozen eggs ... lb	4.18 (*1.90 kg*)	4.18 (*1.90 kg*)
Cattle	Poultry litter[2]	Daily gain ... lb	2.2 (*1.0 kg*)	1.91 (*0.87 kg*)
		Feed/gain ratio	10.18	11.58

[1]Adapted by the author from *Unidentified Resources as Animal Feedstuffs,* NRC, National Academy Press, Washington, DC, 1983, pp. 132–144, Tables 35–41.

[2]Also, dried poultry litter has been fed successfully to dry and lactating dairy cows, to growing and breeding sheep, to growing swine, and to broilers.

MANURE AS A NONFEED ENERGY SOURCE

Manure can also serve as a source of nonfeed energy, which, of course, is not new. The pioneers burned dried bison dung, which they dubbed *buffalo chips,* to heat their sod shanties. In this century, methane from manure has been used for power in European farm hamlets when natural gas was hard to get. While the costs of constructing plants to produce energy from manure on a large-scale basis may be high, some energy specialists feel that a prolonged fuel shortage will make such plants economical. India now has many anaerobic digestion plants in operation.

Methane, of course, is usable like natural gas. There is nothing new or mysterious about this process. Sanitary engineers have long known that a family of bacteria produces methane when they ferment organic material under strictly anaerobic conditions. (Our grandparents called it *swamp gas:* their city cousins called it *sewer gas.*) However, it should be added that, due to capital and technical resources needed, for some time to come, the production of methane by anaerobic digestion will likely be limited. If all animal manure were converted to energy, it has been estimated that it could produce energy equal to 10% of the petroleum requirements or 12½% of our natural gas requirements.

COMMERCIAL FERTILIZERS

Valuable as it is, the average livestock farm or ranch simply doesn't produce enough manure to maintain the fertility of its soil. This is so because (1) animals take out about one-fifth of the feed nutrients,[4] (2) even with the most approved methods of handling manure, there are certain additional losses before it gets into the soil, (3) barnyard manure is low in phosphorus

[4]The sale of a 1,000-lb steer removes fertility equivalent to 150 lb of sodium nitrate (24 lb of N), 100 lb of superphosphate (20 lb P_2O_5), and 50 lb of limestone (47.5 lb $CaCo_3$). (Little potash is removed by animals; most of it is voided in the feces.) In the sale of 10,000 lb of milk—the annual production of a good cow—nitrogen and phosphoric acid are removed in amounts equivalent to that found in 300 lb of ammonium sulfate and 200 lb of superphosphate.

(but the availability coefficient of the manure is high compared to that of nitrogen), (4) it is not always profitable or good business to feed to livestock all of every crop produced, and (5) it is seldom practical to use enough purchased feeds to make up for all the plant food deficiencies inevitable in the use of homegrown feeds only, due to the forces indicated in points 1 and 2 above. In brief, few farmers buy enough feed or save enough manure to maintain the original fertility of the soil.

Thus, the addition of commercial fertilizers, green manure crops, and crop residues (such as straw, chaff, and stalks), is necessary if the fertility of the soil is to be maintained. All farmers should know a few important fertilizer facts so that they can use these products more profitably.

FERTILIZER GRADES

The grade of fertilizer is the nutrient percentage of the product by weight.

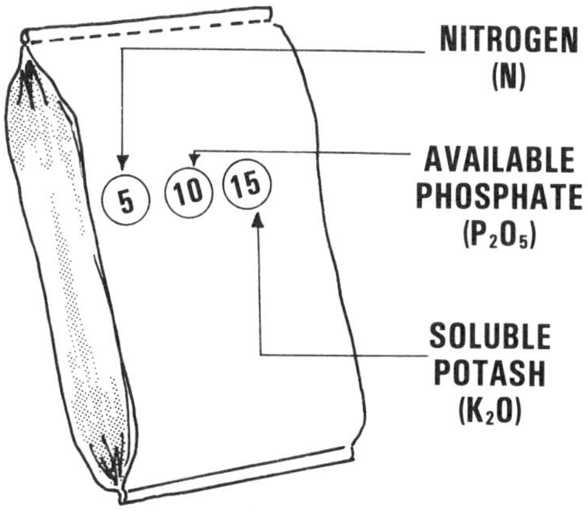

Fig. 8–34. The grade, or analysis, of a fertilizer is always designated by three numbers, which are always listed from left to right.

Grade is expressed in a set of three numbers, always read from left to right: (1) percentage of total nitrogen (N), (2) available phosphate (P_2O_5), and (3) soluble potash (K_2O). Thus, a 5–10–15 grade fertilizer contains 5% N, 10% P_2O_5, and 15% K_2O. The remaining 70% of the product consists of other elements, such as calcium, chlorine, and oxygen. If a nutrient is missing, it is represented by a zero, such as 45–0–0 for urea, 0–44–0 for triple superphosphate, 0–0–60 for potassium chloride (muriate), and 18–46–0 for diammonium phosphate.

HOW TO DETERMINE SOIL DEFICIENCIES

It is recognized that general information relative to the most common plant nutrient deficiencies of soil types or large areas is insufficient, for each farm and even each field presents an individual problem. Thus, in order that fertilizers and lime or gypsum may be used most effectively and profitably, farmers are in need of dependable methods for testing soils, and of other methods of determining plant food deficiencies. Some methods now employed to determine plant food deficiencies in soil are:

1. **Soil tests.** Laboratory soil tests are widely used to determine the supply of available nutrients. Essentially, soil testing is a chemical (or biological) procedure conducted in a laboratory under controlled conditions. For a reliable test, a representative sample must first be taken, for, regardless of how carefully the test itself is conducted, it is no better than the sample which was tested (for instructions on sampling, see the heading, "How to Take Soil Samples").

2. **Plant hunger signs (deficiency symptoms).** Any livestock producer can recognize when pigs or other animals are hungry. Plants, too, have ways of showing hunger—of showing deficiency symptoms when extreme shortages exist of certain essential elements. Also, characteristic plant symptoms indicate the presence of saline and alkali soils. But some experience is necessary for such recognition. Also, the visible hunger signs in plant may be revealed too late for effective treatment of current crops. The information obtained, however, may be used in preparing for the following season.

3. **Plant analysis.** The fertilizer needs of crops can also be determined through systematic sampling and chemical analysis of plants.

4. **Biological tests.** Most of the early experiments designed to determine fertilizer needs of soils made use of a biological method, and these methods are still used. The field plot is the oldest, as well as the best known and most used. Greenhouse pot and other biological tests are time-consuming and expensive; as a result, their use is limited primarily to research.

5. **Portable soil-testing kits.** Many commercial soil-testing kits are now on the market, ranging in price from a few dollars for a simple pH unit to $100 and over for more elaborate apparatus for testing most of the essential elements. Most of these kits have merit and will give desirable results if used on certain kinds of soil according to specific direction. However, all soil-testing kits have the following inherent weaknesses which must be recognized before they can be used successfully:

a. The difficulty of keeping the glassware clean.

b. The problem of contamination and deterioration of the reagents.

c. The hazard of properly interpreting the results in terms of lime and fertilizer needs.

HOW TO TAKE SOIL SAMPLES

If soil tests for lime and fertilizer needs are to be of value, the samples must be taken properly and the results interpreted correctly. Although the specific instructions relative to how to take such samples will vary somewhat with the laboratory running the test, the following general principles apply (also, see Fig. 8–35):

1. **Number of samples.** Where fields are uniform (from the standpoint of soils, topography, crop growth, and past treatment) one composite topsoil (plow depth) sample from every 10 acres is usually ample. Where fields are not uniform, however, more samples should be taken.

2. **Surface litter.** Scrape away surface litter before sampling.

3. **Sampling equipment.** A spade, trowel, or soil auger may be used in taking the samples.

4. **Sampling procedure.** For each sample, take a uniform, but thin, slice down from the surface to the usual plow depth. With soils that are in pastures or lawns, the sample should be limited to the upper 2 to 4 in. Take about one sample from each acre; then thoroughly mix these samples in a bucket or other container and take one composite sample therefrom.

5. **Dry sample.** Spread each composite sample out on a clean sheet of paper and let dry at room temperature.

6. **Pack.** Remove any stones or roots and pack about a pint of each sample in a soil sample carton, or other suitable container, and forward to the laboratory making the test. Wrap securely and address clearly.

7. **Added information.** Knowledge of past cropping and fertilizer history is important when making a fertilizer recommendation. An "information sheet" is usually provided by the chemical testing laboratory for use in submitting this information. Fill this out, place it in an envelope, stamp it, and secure it to the outside of the soil sample package.

8. **Further information.** Further information on soil sampling and soil testing can be obtained from the county extension agent, Soil Conservation Service, or fertilizer dealer.

9. **Follow instructions.** After receiving your soil test results and recommendations, study them carefully and lime and fertilize in keeping therewith.

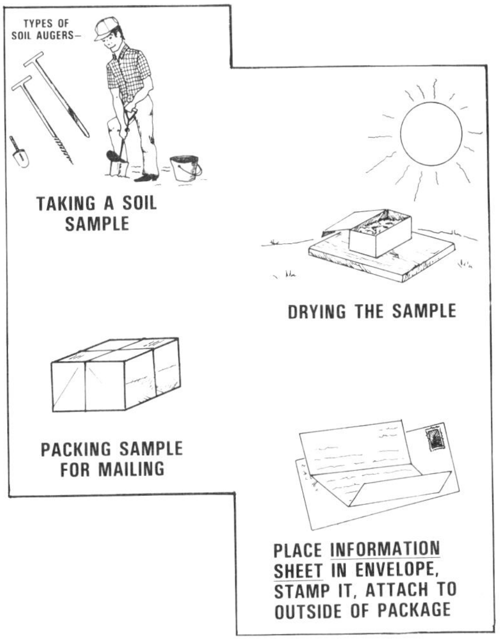

Fig. 8-35. Although some variations in instructions exist between laboratories, the soil sampler is generally admonished to pay particular attention to the four steps illustrated above. (Drawing by R. F. Johnson)

KIND AND AMOUNT OF COMMERCIAL FERTILIZER TO APPLY

It is impossible to give accurate directions relative to the kind and amount of fertilizer to apply to all soils because, in addition to the hundreds of different soil conditions, consideration should be given to (1) the previous crops, (2) the current crop rotation, (3) the farming system, (4) the price of crops, (5) the price of fertilizers, and (6) the amount and quality of manure and crop residue available.

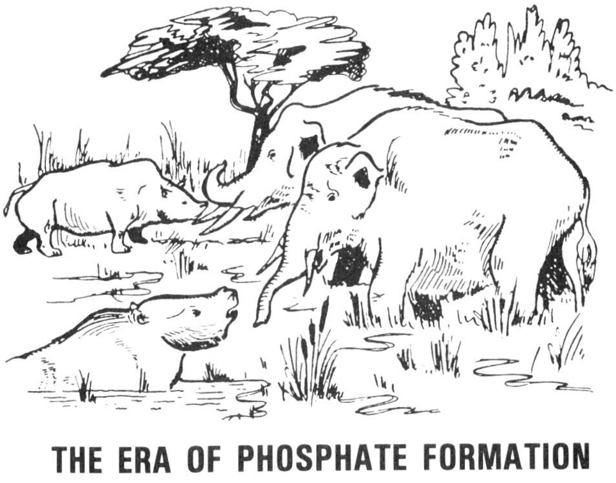

Fig. 8-36. Primeval giants, such as rhinoceros (left) and mastodons (right), roamed Florida during the phosphate forming era. (Drawing by R. F. Johnson)

On some soils and under some conditions, "complete fertilizers"—those furnishing nitrogen, phosphate, and potash—will pay best; on others, a two-element fertilizer may be best; and, under still other conditions a fertilizer containing only one plant nutrient may be best.

The amount of fertilizer to apply should be determined primarily by (1) the level of available nutrients as determined by soil test, (2) crop requirement for the nutrient, and (3) the amount of barnyard manure available. Fig. 8-37 (next page) shows that crop "appetites" do vary.[5] Also, it is generally recognized that the grain farmer will need to use about twice as much commercial fertilizer per acre as the livestock farmer who fertilizes with manure every three or four years.

Before purchasing and using fertilizers, the farmer is admonished to confer with the local county agent, vocational agriculture instructor, or fertilizer dealer.

[5]The amount of fertilizer to apply cannot be based entirely on how much plant food it is estimated that the crop will take out, primarily because crops do not use all plant foods added in fertilizers with the same efficiency. For example, under average conditions, crops probably take up to ½ of the nitrogen added and ⅔ of the potash, but only 10 to 20% of the phosphoric acid. Actually, one should add such amounts of fertilizer as will produce maximum economical yields under the specific crop and environmental conditions.

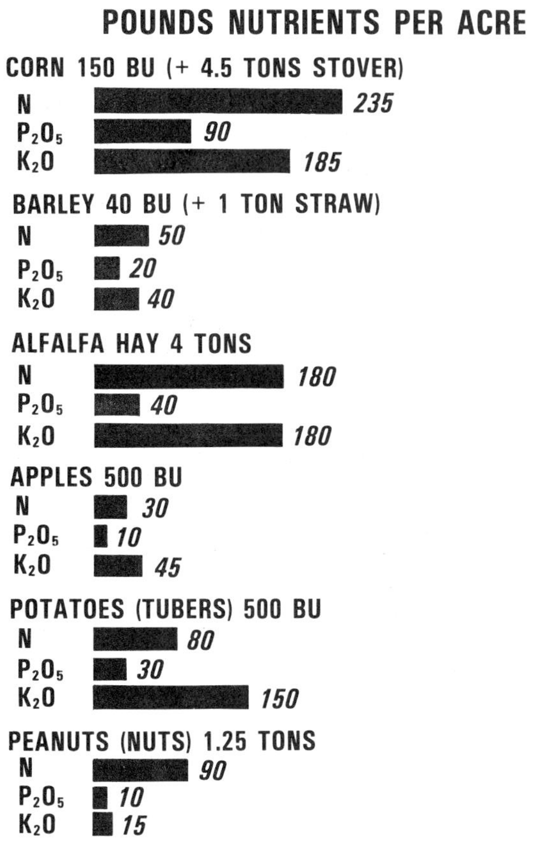

Fig. 8–37. Crop appetites vary. The above bar diagram shows the pounds of nutrients per acre required to produce six important crops. Legumes, such as peanuts and alfalfa, normally get the greater part of their nitrogen from the air. (Bar diagrams based on data from *The Fertilizer Handbook*, published by the Fertilizer Institute, Washington, DC)

HOW TO BUY COMMERCIAL FERTILIZERS

Each of the chief plant nutrients—nitrogen, phosphorus,[6] and potassium—has about the same value per pound in any of its respective forms. Usually the fertilizer with the highest composition is the cheapest, but this is not always the case. In addition to cost of nutrients, consideration should be given to the following:

1. **Additional nutrients.** Some fertilizers, such as ammonium sulfate and single superphosphate, contain additional nutrients (in this case sulfur) which may be of vital importance to crop production. In these cases, an additional value can be attached to these fertilizers, over and above their respective nitrogen and phosphorous content.
2. **Cost of application.** Where both nitrogen and sulfur are needed, it may cost the farmer less to purchase and apply ammonium sulfate in one operation than to apply a nitrogen fertilizer as anhydrous ammonia, and then a sulfur compound in a second operation.
3. **Service and reputation.** In addition to cost as such, consideration should be given to the reputation of the dealer and the services rendered. Often the services rendered are of greater value than the fertilizer.
4. **Cost of transportation, storage, and labor in applying.** In addition to the actual cost of the material, consideration must be given to the cost of transportation, storage, and labor used in applying the fertilizer. These costs may be difficult to evaluate; but if the actual cost of the nutrients from one source is the same as another, farmers will gradually learn to take the one requiring the least labor. The higher-analysis products require less labor in handling, and fewer stops in applying the material.

[6]This refers to available P_2O_5, for the method of computation is not necessarily accurate with reference to total P_2O_5.

APPLYING FERTILIZERS

It is important that commercial fertilizers be applied at the proper time. For most annual crops, this means just prior to, or at the time of, seeding; but supplemental applications may be made as either side-dressings or as liquid applications. Pastures and haylands may be fertilized almost any season of the year, but generally greater responses are obtained from early spring applications. Supplemental applications to pasture and hay crops may be made by top-dressing or through the sprinkler system.

There is no one best way in which to apply a fertilizer. Several methods are used, and there are advantages to each. The method of application should be determined by the crops, soil, climate, time and rate of application, and kinds of fertilizer and equipment available. Regardless of the method selected, the aim should be to get the fertilizer in the soil where it will do the most good.

Some of the common methods of applying fertilizer are:

1. **Broadcasting.** Consists in spreading fertilizer uniformly over the land by means of a fertilizer distributor.
2. **Side-dressing.** Consists in putting fertilizer along each side of the row after crops are up and growing, using a special fertilizer drill or attachment.
3. **Drilling with seed.** Consists in applying the fertilizer in the row when the seed is drilled or planted.
4. **Banding along the row.** Consists in distributing the fertilizer in bands one or more inches wide on either or both sides of the row while planting, usually at or below the seed level.
5. **Injection (deep drilling).** Consists in placing fertilizer in bands at desired depths. Anhydrous ammonia must be placed 4 to 6 in. deep.
6. **Plowsole of deep furrow.** Consists of placing fertilizer in the bottom of each furrow, a practice which is helpful to plants when the surface of the soil becomes dry during the growing season.
7. **Foliar application.** Consists in spraying dilute solutions of salts of certain metals (copper, manganese, iron, or zinc) when deficiencies of these trace elements show up.
8. **Liquid fertilizer distribution.** Consists in applying fertilizers in solutions either through attachments on planters and cultivators or through injection into irrigation water (either in ditches or in sprinklers) and distribution with the water. Also, liquid fertilizers are sometimes applied at the time of transplanting crops.

LIME ACID SOILS

Lime may be defined as any compound of calcium or of calcium and magnesium capable of neutralizing soil acidity.

Some soils originally had, and still have, plenty of lime in them. Others never did have enough of this constituent to produce successfully the best legumes. Still others once had enough lime, but subsequently it has been leached out and removed by crops. As much as 400 lb of lime per acre may be leached away annually with drainage water, and every crop takes out lime.

It has been estimated that 70% of the tillable land in the eastern half of the United States is acid or "sour" as a result of losses by leaching and crop removal of such basic elements as calcium, magnesium, and potassium. In the arid or semiarid regions of the West, many soils are alkaline or *sweet.*

Liming is the first step in the improvement of most acid soils, and, in addition, it serves as an economical source of calcium (and sometimes magnesium).

FUNCTIONS OF LIME

In addition to serving as a fertilizer material, especially in humid soils, lime is a soil amendment which functions in acid soil as follows:

1. It corrects soil acidity.
2. It supplies calcium for plants and animals, and sometimes magnesium (if the lime is dolomitic).
3. It stimulates soil microbe (bacteria) activity.
4. It speeds the decay of organic matter and the liberation of nitrogen and other plant foods.
5. It increases availability of molybdenum.
6. It improves crop yields.
7. It reduces the toxicity of aluminum, manganese, protons, and iron.
8. It increases the efficiency of manures and fertilizers.
9. It decreases soil erosion and improves soil structure as a result of increased vigor and density of plants.

THE pH VALUE OF THE SOIL

Soil pH is an excellent indicator of general soil conditions; hence, it should always be determined. However, pH values should be augmented by a lime-requirement determination, as described in the section on "How to Determine Lime Needs."

The pH value of a soil is a convenient method of expressing the degree of acidity or alkalinity. The pH scale is divided into 14 divisions, or pH units numbered from 0 to 14. A soil with a pH of 7.0 is neutral (see Fig. 8–38). Soils with pH values below 7.0 are acid or sour, while those above 7.0 are alkaline or sweet. A pH of 5.0 is 10 times more acid than a pH of 6.0, and a pH of 4.0 is 10 times more acid than a pH of 5.0. Thus, a soil having a pH value of 4.0 is 100 times more acid than one having a pH of 6.0. The pH of the majority of soils lies within the range of 4.0 to 8.0. Most soils in the humid region are acid or sour as a result of losses by leaching and crop removal, whereas most soils in arid or semiarid regions are alkaline or *sweet.*

It is recognized that the determination of pH does not give a direct measure of the amount of exchangeable acidity, but only the "free" or ionized acidity. Therefore, recommendations for the amount of lime had best be based on measurements of soil pH and lime requirement, along with information concerning the reactivity of the liming material to be used.

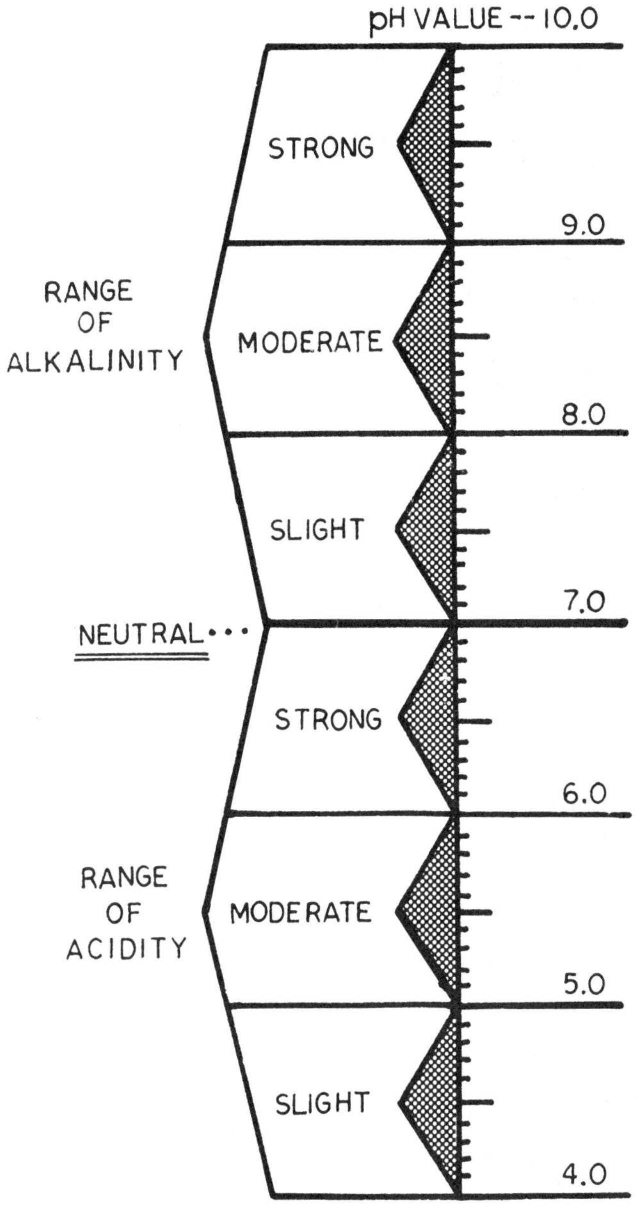

Fig. 8–38. pH scale for soil reaction.

Soils in the pH range of 6.0 to 8.0 are apt to be more trouble-free than those higher or lower. Values of pH 5.0 or under may be indicative (1) of deficiency or unavailability of such elements as calcium, magnesium, phosphorus, molybdenum, and boron, or (2) of toxic amounts of zinc, manganese, aluminum, nickel, and other elements because of increased solubility.

Values above pH 8.5 indicate (1) the presence of sodium carbonate and/or high exchangeable sodium, (2) the need to treat with gypsum, sulfur, or other acidic material, (3) the need to leach out acidic materials, and (4) the need to leach out excess salts if present.

• **The preferred pH range for crops**—Plants differ widely in their response to added lime. Certain plants will grow well in acid soils, while others will not. In considering the liming program, therefore, the type of crop to be grown is of importance (see Fig. 8–39).

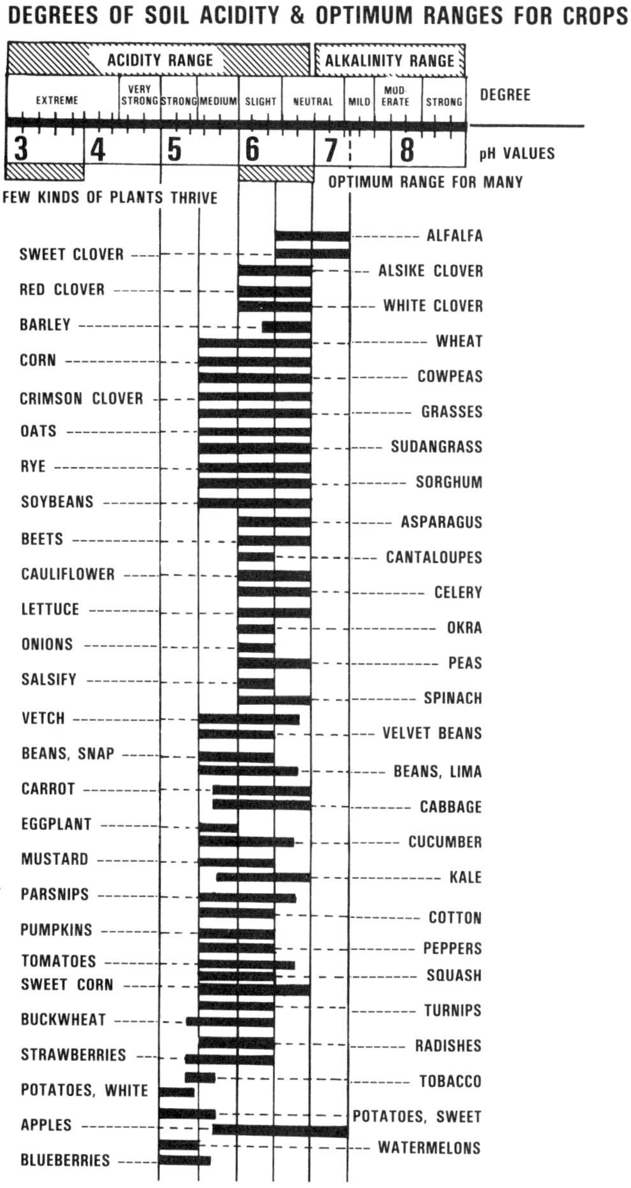

Fig. 8–39. Degrees of soil acidity and optimum ranges for crops. (From: *The Fertilizer Handbook*, published by the Fertilizer Institute, Washington, DC)

HOW TO DETERMINE LIME NEEDS

Different methods have been developed by soil testing laboratories to determine the amount of lime needed to bring the pH of an acid soil to a desirable range. All such tests presently used take into consideration the soil texture and organic matter content. Thus, it may require much more lime to raise the pH of a clay loam soil from 5.9 to 6.8 than will be required to raise the pH of a sand from 4.1 to 6.8. The difference in lime requirements is due primarily to the difference in the exchange capacity and the amount of reserve acidity.

Also, rates of lime applications and time between applications will depend on the crops being grown, amount of rainfall, and amount of leaching. A pH of 6.5 is recommended for temperate crops, but liming of grassland to pH 6 is preferred, since grasses are more tolerant of acidity and the possibility of inducing deficiencies of Mn, Cu, and Zn is minimized. Tropical species are more acid-tolerant and, furthermore, liming to pH 6–6.5 may reduce P availability in soils high in exchangeable aluminium and destabilize the structure of kaolinitic clay soils.

With a chemical soil analysis available, along with background information of the major soil series and types in the area they serve, together with their previous responses to lime, the state laboratory is in the best position to make liming recommendations for the specific farm, crops, and conditions. In the absence of a laboratory service, a soil test kit in the hands of an experienced person will give reasonably satisfactory results.

There may be considerable differences in soil acidity, even within a given field—especially on fields that are not level. In some fields, the tops of the ridges may be alkaline, the slopes acid, and the bottoms alkaline.

Soil samples for testing lime should be taken as directed in the earlier section entitled, "How to Take Soil Samples."

KIND AND AMOUNT OF LIME TO USE

Nature has provided an abundant supply of liming materials in most parts of the country. In general, the particular product selected should be determined by the following:

1. The comparative cost based on the neutralizing value (NV) of one ton of each material.
2. The convenience and cost of handling.
3. The relative speed of action in the soil, as determined by firmness, hardness, etc. The rate of reaction of a liming material with the soil is essentially a function of the size of particle; the finer the lime, the more rapid the changes brought about by its application.

The fineness of lime is measured by screen tests. A standard mesh screen is one in which the number of openings per linear inch corresponds to its screen number. Thus, a 10-mesh screen has 10 openings per linear inch, or 100 openings per square inch. Although the fineness of limestone upon which recommendations are based varies greatly among states, a common guarantee for lime is—

85% through a 15-mesh sieve.
30% through a 100-mesh sieve.

4. The magnesium content of the limestone where this element is needed. When the soil contains less than 50 lb to the acre of available magnesium, dolomitic limes should be used.

The amount of lime to apply per acre should be determined by (1) the degree of soil acidity, (2) the kind of soil, (3) the kind of lime, (4) the frequency of application, and (5) the kind of crops to be grown.

Under most general farming conditions, the application of lime should be regulated so as to maintain the soil reaction at a pH of about 6.5 or in the neutral range, as determined by a pH test. Generally, this will require an initial application of 2 to 5 tons of ground limestone (or equivalent lime) and subsequent applications of about 1 ton every 5 to 10 years thereafter. Smaller quantities should be applied to sandy soils at any one time, but the applications should be made more frequently.

A rule of thumb commonly followed is that it requires about 1 ton of ground limestone (or equivalent lime) per acre to change the soil reaction by one pH unit (say, from 5.5 to 6.5) in a strongly acid sandy loam; loams require approximately 2 tons; clay loams 3 tons; and muck soils 5 tons. If soils are fairly high in organic matter, as they are in the northern states, it will take more lime to accomplish the same change in pH. Conversely, in southern soils, which are lower in organic matter, less lime will be needed to accomplish the same pH change.

APPLYING LIME

Two facts are of paramount importance in determining when to apply lime: (1) It takes some time for lime to dissolve enough really to become effective, and (2) some crops, such as alfalfa and sweet clover, respond more to lime than other crops. Therefore, with a rotation such as CORN-CORN-OATS-CLOVER, it would be best to apply the lime to the crop preceding the seeding of the legume; between the first and second year corn in this case.

Ground limestone can be applied at any time of the year, and without hazards or injury to crops (except potatoes, where lime may make conditions more favorable for scab). Perhaps the only precaution is that it should not be applied when the ground is soft and there is danger of trucks puddling the soil. Burned and hydrated lime should not be applied to growing crops due to their burning effect on plants.

Lime is commonly applied on rough plowed land or after a single harrowing, and incorporated into soils through subsequent cultivation. Also, it is frequently applied as a top-dressing on pastures, lawns, and other grassed areas.

• **How to apply lime**—Since damage can be done by getting too much lime in spots, even spreading is important. Also, lime should be mixed with the surface soil.

Generally, lime is spread by the dealer who (1) has a truck with a specially built V-shaped bed and a centrifugal type of spreader, and (2) charges very little more for spreading the lime than for dumping it in one place on the field.

If the farmer spreads lime, it should be done with a regular lime spreader or an endgate seeder. Also, it can be spread with manure by putting lime in the gutters behind dairy cows or by putting lime on the loaded manure spreader.

TREAT SALINE AND ALKALINE SOILS

Saline and alkaline conditions lower the productivity and value of large areas of agricultural land in the United States—an estimated one-fourth of our irrigated land and less extensive acreages of nonirrigated cropland and pasturelands.

Saline and alkaline soils are soils that have been harmed by soluble salts, consisting mainly of sodium, calcium, magnesium, chloride, and sulfate, and secondarily of potassium, bicarbonate, carbonate, nitrate, and boron.

Saline soils contain excessive amounts of soluble salts only. Alkaline soils contain excessive absorbed sodium. Because leaching may have occurred previously, alkaline soils do not always contain excess soluble salt.

Salt-affected soils, which occur mostly in arid or semiarid regions, are problem soils that require special remedial measures and management practices.

Salt comes from the water and to a lesser degree from fertilizers. Colorado River water contains about 1¼ tons of soluble salts per acre foot. Salts accumulate in soils unless sufficient water is moved through the profile to carry dissolved salts deep below crop root zones.

• **Handling saline soils**—Excess salinity can injure plants. But some crops are extremely salt tolerant, notably cotton, sugar beets, and barley. Vegetables are relatively sensitive to salinity.

There are no magical chemicals that can tie up salts; however, there are ways to increase water penetration. Some soils may be improved by profile modification, such as chiseling or slip plowing. Organic amendments worked into the soil are helpful. If animal manures are used, the soil-manure mixture should be preirrigated before planting. This will allow for some leaching of the salts contained in manure as well as release of ammonia.

• **Handling alkaline soils**—Gypsum is widely used in reclaiming alkaline soils in western United States. Practically all cropland in California, Arizona, Nevada, and eastern Oregon is alkaline, rather than acidic, on the pH scale. Approximately one million tons of gypsum are incorporated in California soils each year. Gypsum ($CaSo_41 \cdot 2H_2o$) is the common name for calcium sulfate, a mineral used in the fertilizer industry as a source of calcium and sulfur. Another common name is landplaster.

Other amendments (in addition to gypsum) used as correctives for alkaline soils are sulfur, sulfuric acid, and ferric sulfate. Soil correctives for alkaline soils either contain calcium or make the calcium in the soil available for correcting alkaline conditions.

Leaching alkaline soils with irrigation water is an important step in their reclamation. Adequate drainage is needed to remove the excess soluble salts which result from the application of the soil amendment.

Calcium is extremely important in the reclamation of alkaline soils. Where there is an excess of sodium attached to clay particles, it tends to disperse or deflocculate the soil. This makes the particles pack together in such a way that water either permeates very slowly or cannot get through at all.

When gypsum is applied, it supplies calcium. When sulfur containing materials are used, they render the natural calcium in the soils more soluble. In turn, the calcium replaces the excess of the absorbed sodium.

The replacement of the sodium on the soil particles by calcium results in an aggregation, or grouping, of the particles so that there are more large pore spaces in the soil. This improved soil structure permits more rapid water and air penetration and improves soil tilth.

The quantities of soil amendments to be used in the reclamation of alkaline soils is dependent upon a number of factors, the most important of which is the natural calcium in the soil and its relation to the exchangeable sodium present. Other factors include buffer capacity, organic matter content, and the physical characteristics of the soil.

A PROGRAM OF SOIL MANAGEMENT

The following program of soil management is recommended for the livestock farm or ranch:

1. Feed livestock for income.
2. Handle the manure so as to return the maximum fertility value of the land.
3. Test soil and apply fertilizer, lime, or gypsum as needed.
4. Grow inoculated legumes to furnish nitrogen, humus, and good livestock feed.
5. Rotate crops to ensure good soil structure and keep down weeds.
6. Keep hilly and rolling ground in pasture as much as possible.
7. Utilize all crop residues by returning them to the soil.
8. Prevent environmental pollution.
9. Control soil erosion.

WHERE TO GO FOR SOIL FERTILITY HELP

Any one or all of the following sources may be called upon for counsel and advice on problems of soil fertility:

1. Soil testing laboratories.
2. County extension agent and the district conservationist.
3. The state agricultural college.
4. Successful neighboring farmers who have similar conditions.
5. Professional farm managers and consultants.
6. Local fertilizer dealers.
7. Fertilizer trade associations.

CONTROL PESTS; LESSEN WASTE

Fig. 8-40. The challenge: (1) control pests, but (2) protect the ecological food chain and human health.

Waste of food supplies will increasingly nag the consciences and pocketbooks of all people—producers and consumers alike.

Pests cause an estimated 30% annual loss in the worldwide potential production of crops, livestock, and forests.[7] Every part of our food, feed, and fiber supply—including marine life, wild and domestic animals, field crops, horticultural crops, and wild plants—is vulnerable to pest attack. Obviously, if these losses could be prevented, or reduced, world food supplies would be increased.

Remember that a worldwide annual loss of 30% potential food productivity occurs despite the use of advanced farming technology and mechanized agriculture. Remember, too, that in many of the developing countries losses greatly exceed this figure.

Pests of many kinds attack plants during all stages of their growth, and they attack food and food products after harvest—in storage, during transportation to market, in warehouses, in elevators, in ships, in supermarkets, and in homes after purchase. Here are a few notable pest losses:

- **Plant diseases and insects**—Disease organisms kill plants, cause rotting and blemishing of food products, and reduce crop yields and quality.

Insects devour growing crops, lower yields and quality, and attack grains and other food products in storage and during transport. Also, insects harbor and transmit diseases to plants, animals, and man.

More than 160 bacteria, 250 viruses, and 8,000 fungi are known to cause plant diseases. In the United States alone, approximately 10,000 species of insects are destructive enough to be called *enemies;* and about four-fifths of them are injurious to crops.

---

[7]Ennis, W. B., Jr., W. M. Dowler, and W. Klassen, "Crop Production to Increase Food Supplies," *Science,* Vol. 188, No. 4188, pp. 593–598. The authors are staff scientists on the National Program Staff, Agricultural Research Service, USDA, Beltsville, MD.

• **Weeds and brush**—Weeds and brush reduce yields and quality by competing with crops or forage species for water, nutrients, and light. Also, they may poison livestock, interfere with harvesting, and slow the flow of water for irrigation and drainage. The extent of this competition and damage depends on the growth as well as their density and distribution.

Under rangeland conditions, brush is the primary problem. It has been estimated that of the 321 million acres of grazing lands in the United States, over 230 million acres are infested with brush and could benefit from control measures.

In the United States, 2,000 species of weeds and brush cost farmers and ranchers an estimated $13 billion annually.

• **Rats**—Each year rats consume feed, damage additional feed, destroy property, and spread disease, for a total cost of $28 per rat. Thus, with an estimated U.S. rat population of 100 million, this means that the nation's yearly keep on rats totals about $2.8 billion.

Although they are not as damaging as rats, mice, gophers, and other rodents should also be controlled—and for the same reasons.

• **Birds**—Birds that are in livestock feeding areas consume and contaminate much feed and spread many diseases to both animals and humans. Hence, they should be controlled. In a study of a 12,000-head cattle feedlot in California, University of California researchers found that the 10,000 to 20,000 birds that came to "dinner" ate an average of about 350 lb of feed each day, for a total of 57,750 lb during the 5½-month winter season. Iowa cattle feeders figure that starlings add $3 to $4 to the cost of each steer marketed. Some western feedlot operators estimate that starling nuisance and feed costs add 2¢ to the cost of each pound of gain.

Fig. 8–41. Birds (blackbirds, starlings, and cowbirds) in tree roost, Lincoln, Nebraska. (Photo by R. J. Johnson, University of Nebraska, Lincoln. Courtesy, R. M. Timm, Superintendent and Extension Wildlife Specialist, University of California-Davis, Hopland Field Station, Hopland, CA)

In addition to the direct losses caused by pests, there are hidden, or indirect, losses: losses in efficiency, and losses in the input of energy involved in crop production—wasted energy. And losses in suffering!

By applying the science and technology that we already have, food and fiber losses can be reduced substantially—perhaps by as much as 30 to 50%. The net result would be an increase of 10 to 15% in the world food supply, with no new land required. In no other way can the hungry gap be filled so quickly and at so little cost.

WEED AND BRUSH CONTROL[8]

Weed may be defined as a plant (1) growing where it is not wanted and interfering with the desired land use, or (2) with a negative value within the framework of current land use.

Weeds are classified as follows:

1. **Summer annuals.** Seeds produced during the summer and plant dies before winter. Seeds germinate in the spring. Examples are: pigweed, lamb's quarter, and crabgrass.
2. **Winter annuals.** Produce seeds in spring and early summer and plant dies. Seeds germinate in the fall and plants live through the winter. Examples are: chickweed, peppergrass, and annual bluegrass.
3. **Biennials.** Seeds germinate in spring and produce leafy plants which lie dormant during the winter. The following season they develop seed stalks which produce flowers and seed. Examples are: wild carrot and bull thistle.
4. **Perennials.** Plants which live longer than two years and produce successive crops of seeds. This group includes woody plants and plants which are killed to the ground each winter and produce new tops the following year. The latter are called herbaceous perennials. Examples are quackgrass, Johnsongrass, Bermudagrass, bindweed, and dandelion. Woody plants are categorized as one of four growth types: (1) upright, single-stemmed trees, (2) bushes or trees with a running or creeping growth habitat, (3) multistemmed bushes, and (4) those plants which grow as vines or canes (see Fig. 8–42, next page).

Noxious woody and weedy plants occur frequently on western U.S. rangelands. One cause is preferential grazing by domestic livestock and wildlife. Palatable species of plants are grazed and weakened enabling ungrazed plants to gain advantage. Overgrazed and cultivated areas left to revert to native vegetation without seeding are prime areas for noxious plants to become established. Excessive erosion, trampling and bedding of animals also results in denuded areas for noxious plants to become established. In some areas these plants may be a natural part of the environment.

[8]The author gratefully acknowledges the help of the following authorities who reviewed and/or provided source material for this section: Dr. D. C. Cress, Extension Pesticide Coordinator, and Dr. D. Peterson, Extension Specialist, Weed Science, Department of Agronomy, Kansas State University, Manhattan; Dr. G. Crosby, Agricultural Chemical Division, FMC Corp., Princeton, NJ; Dr. K. C. McDaniel, Professor, Animal and Range Sciences Department, College of Agriculture and Home Economics, New Mexico State University, Las Cruces; Dr. S. D. Miller, Professor of Weed Science, University of Wyoming, Laramie; Dr. S. R. Muench, Manager, Product Development, Monsanto, St. Louis, MO; Dr. D. G. Shilling, Assistant Professor, Institute of Food and Agricultural Sciences, University of Florida, Gainesville; Dr. T. G. Welch, Range Science Department, Texas A&M, College Station; and Dr. R. L. Zimdahl, Department of Plant Pathology and Weed Science, Colorado State University, Fort Collins.

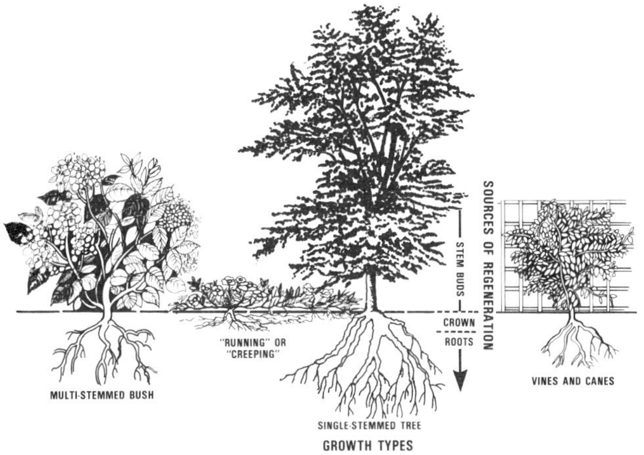

Fig. 8–42. Woody plants occur in four basic growth forms which vary in regrowth potential, source of asexual regeneration, and response to conventional control methods.

WEED AND BRUSH CONTROL OBJECTIVES

The primary objective of weed and brush control is to reduce coverage of undesirable plants at an economically feasible cost. Reducing the coverage of undesirable plants will increase levels of light, water, and nutrients for forage production.

With poisonous plants, the offset of livestock losses may determine the feasibility of a control program rather than increased forage production.

Also, livestock management is facilitated (made easier) where brush has been controlled.

WEED AND BRUSH CONTROL METHODS

The discussion that follows pertains primarily to weed and brush control in forages—the animal feed from native pastures, rotational pastures, rangeland, and hay.

• **Weed control**—Various methods are effective and should be used in the control of weeds, including—

1. Prevention by the use of clean, certified seed that is free of weed seed.

2. Cultivation.

3. Mechanical means—mowing and cutting.

4. Roguing, or individual weeding, if the plants are scattered and not found in large numbers.

5. Pasture improvement, especially through management and renovation.

6. Chemical control.

7. Biological control.

• **Brush control**—Brush is generally controlled by one, or a combination of, the following:

1. **Mechanical means.** These methods involve any treatment that physically damages the plant or removes it from the ground. Mechanical methods may be grouped as follows:

Fig. 8–43. Mechanical brush control. (Courtesy, Dr. K. C. McDaniel, Animal and Range Science Department, New Mexico State University, Las Cruces)

a. **Hand methods.** These methods are the oldest and often the most effective. They are applied to plants individually and are slow and costly. The common hand methods are:

(1) **Grubbing.** This consists of digging out the plant with the root system.

(2) **Cutting.** Cutting with an ax or other tool is effective on some species that do not produce basal or root sprouts.

(3) **Girdling.** This refers to cutting the cambium (under the bark) layer in a ring around the tree trunk, thereby stopping movement of sap and nutrients.

b. **Power methods.** These methods include a wide range of equipment from power saws to tractors. The cost of labor is often reduced, but the high cost of equipment makes many mechanical methods very expensive.

2. **Burning.** This is the oldest brush control method. Burning is an effective but sometimes hazardous treatment, which must be used with utmost care. Weeds and browse are injured more by fire than grass due to the location of the growing points. The growth point on grass is near the base of the plant whereas

in weeds and brush the growth point is on the extremities. The soil should be cool and moist with the grass plants dormant for controlling noxious plants while preserving forage plants. Before burning, construction of fire lanes and providing adequate fire control equipment are necessary to prevent wildfires. A permit is generally required in order to burn. Also, air pollution must be considered.

3. **Proper range management.** Proper range management, including reseeding when necessary, assures healthy, vigorous forage and prevents many brush problems.

Fig. 8-44. Range drill used for seeding range grasses in the Great Plains area. (Courtesy, H. Dietz, USDA, Soil Conservation, Fort Worth, TX)

4. **Chemicals.** Herbicides are the most widely used agricultural pesticides in the United States. In 1990, there were 121 chemicals registered as herbicides in the United States, with 242 major registered trade names and hundreds of trademarked products with varying active ingredients, formulations, and concentrations. A list of the herbicides, along with a description of each of them, is beyond the scope of this book.

Chemical treatment has been an important method of weed and brush control. However, concern about residues and other aspects of the environment has resulted in reduced use of herbicides. Careful planning and judicious application of approved chemicals is vital to their continued use.

Fig. 8-45. Chemical brush control. (Courtesy, Dr. K. C. McDaniel, Animal and Range Sciences Department, New Mexico State University, Las Cruces)

The successful use of any chemical weed control program depends primarily upon two factors:

a. Choosing the right herbicide, keeping in mind that more than one chemical may be effective.

b. Applying the chemical in keeping with the directions of the container label, and in compliance with the herbicide regulations of the area.

5. **Biological controls.**—*Biological control involves the use of one living organism to eat, damage, or kill another one.* Interest in the use of biological control methods is increasing. The classic example of biological control of an undesirable species was the practical elimination of prickly pear in the Moonie River Valley, Queensland, Australia, between 1927 and 1932, by a moth introduced from Argentina. Also, the eating of sprouts by big game and domestic animals is a form of biological control. Cattle and elk normally consume 80% grass and 20% browse and weeds. Consumption by sheep consists of about 60% grass and 40% browse and weeds. Antelope eat large amounts of weeds, while goats and deer consume more browse and less grass. However, not all weeds and brush are palatable to domestic and wild animals.

The use of insects and other organisms for biological control is being studied; and the search is on in various parts of the world, including the United States, for other biological control agents.

(Also, see Section 5, Pasture and Range Forages, under the heading of "Brush and Weed Control.")

RODENT AND BIRD CONTROL[9]

Without exception, U.S. livestock establishments are plagued by one or more rodents, and/or with birds. Among such pests are rats, mice, English sparrows, starlings, and pigeons, all of which were introduced into this country. They should be controlled and if possible, eliminated because they (1) spread diseases and parasites, (2) damage feeds and buildings, and (3) decrease profits.

Rodents and birds can be effectively and efficiently controlled or eliminated by (1) eliminating the food that maintains

[9]The author gratefully acknowledges the helpful suggestions of the following authorities who reviewed and/or provided source material for this section on "Rodent and Bird Control": Dr. D. C. Cress, Extension Pesticide Coordinator, and Mr. F. R. Henderson, Extension Specialist, Animal Damage Control, Department of Animal Science and Industry, Kansas State University, Manhattan; Dr. H. F. DeLuca, Steenbock Research Professor, Department of Biochemistry, University of Wisconsin-Madison; Dr. G. A. Green, Public Information Manager, Monsanto, St. Louis, MO; Dr. W. B. Jackson, Professor, Department of Biological Sciences, Bowling Green State University, Bowling Green, OH; and Mr. R. E. Marsh, Specialist in Vertebrate Ecology, Agricultural Experiment Station, University of California-Davis.

these pests, and (2) following the directions given in this section or by engaging the services of professionally trained pest control operators or government agents.

CONTROLLING AND ELIMINATING RATS

Rats have plagued people since the beginning of recorded history. They are referred to as the deadliest and most destructive of all animal enemies of humans. Other pertinent facts relative to them are:

1. There are an estimated 100,000,000 rats in the United States.
2. Each year, on the average, every rat (a) consumes and damages feed worth $10, and (b) contaminates additional feed, destroys property, and spreads diseases costing another $18. Thus, the yearly keep of each rat amounts to over $28 or 2.8 billion for all U.S. rats.

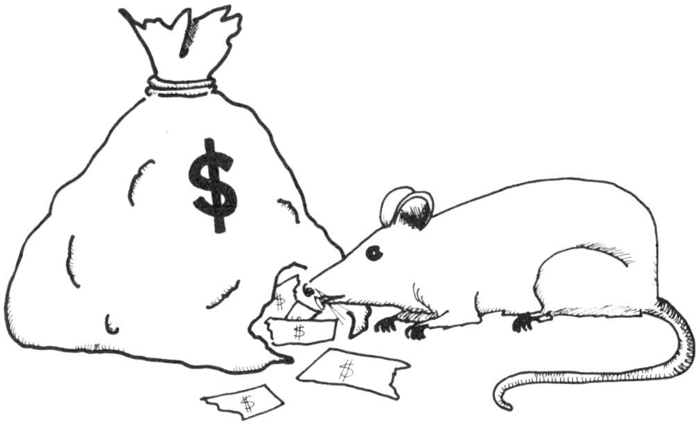

Fig. 8-46. Rats are costly! Annually, it costs more than $28 to keep a rat or nearly $2.8 billion for all U.S. rats.

3. The United Nations Food and Agriculture Organization estimates that rats eat or contaminate 42.5 million tons of the world's grains each year—enough to feed 200 million people.
4. Rats spread many serious human and animal diseases—including typhus, bubonic plague, infectious jaundice, rat bite fever, tularemia, and food poisoning of people, and trichinosis, pseudorabies, and various kinds of fleas, lice, mites, and several internal parasites of animals.
5. Rats start fires by gnawing through the insulation of wires, kill baby chicks, and weaken building and dike/dam foundations through burrowing under them.
6. Rats become capable of producing young when they are 90 to 120 days of age, produce annually 6 to 10 litters averaging 8 each, and under natural conditions, live to 1 year of age.

- **Anticoagulants**—Warfarin, pindone (Pival), diphacinone (Ramik), and chlorophacinone (RoZol) represent first-generation anticoagulants, whereas, bromadiolone (Maki, Contrac) and brodifacoum (Havoc, Talon) represent the newer second-generation anticoagulants. The anticoagulant of choice should be used according to the directions on the container.

Fig. 8-47. Some common types of homemade bait stations. But these must be used with caution, because they may not adequately prevent accidental poisoning of pets or livestock.

- **Other types of rodenticides**—While 95% of all rat and mouse control is conducted with anticoagulants, several other types of rodenticides are also available. Zinc phosphide is an acute (rapid-acting) rodenticide used since the 40s and may be the bait of choice where rats are numerous and a rapid kill is a must. It presents a greater hazard to children, domestic animals, and other nontarget species, especially free-roaming chickens, ducks, or geese.

Bromethalin (Assault, Vengeance) and cholecalciferol (Quintox, Rampage) are two of the newer single-dose rodenticides that have totally different modes of action than anticoagulants. With these two latter rodenticides, death generally occurs 2 to 4 days following a lethal ingestion. For use, follow label directions carefully.

- **Cleaning up premises and rat-proofing buildings**—The old adage that "an ounce of prevention is worth a pound of cure" still applies, despite the amazing results secured from the use of anticoagulants. Premises should be cleaned up and buildings should be rat-proofed.

CONTROLLING AND ELIMINATING OTHER RODENTS AND BIRDS

Although they may not be as damaging as rats, it is nevertheless, important that other rodents and birds be controlled—and for the same reasons.

- **House mice**—Although house mice are smaller than rats, their greater numbers and widespread activities make them at least equally dangerous insofar as damage to property and food contamination are concerned. It is important, therefore, that every effort be made to control them.

Mice may be controlled by using anticoagulants or other rodenticides in a similar manner to that indicated for rats.

Fig. 8-48. Open ends of corrugated metal siding permitting house mice to enter. (Photo by R. M. Timm. Courtesy, R. M. Timm, Superintendent and Extension Wildlife Specialist, University of California-Davis, Hopland Field Station, Hopland, CA)

• **Pocket gophers**—Evidence of the presence of pocket gophers is indicated by (1) many kidney shaped mounds thrown up in hay or pasture areas, and (2) young fruit and shade trees eaten off below the ground during the winter.

Pocket gophers can be controlled either by poisoning or trapping. Where large and heavily infested areas are involved, poisoning is fastest and cheapest—especially if a gopher-bait applicator is used. This tractor-mounted unit makes an artificial burrow and meters poisoned grain into it. Four to six acres can be treated per hour. For a demonstration of the mechanical burrow-builder, contact your county extension agent or nearest representative of the USDA, Animal and Plant Health Inspection Service (APHIS), Animal Damage Control Station.

For poisoning smaller areas or rough ground—areas not suited to use the mechanical applicator—the following procedure is recommended:

1. Select and purchase the registered pocket gopher grain-type bait available for use in your area. Strychnine is generally the rodenticide of choice as it gives the best results for the cost. Zinc phosphide and some of the anticoagulant rodenticides are also registered for gopher control; however, these are mostly used by home gardeners.

2. Locate the main runway (10 to 16 in. back from the mound on the side where the horseshoelike depression is found) by probing with a pipe, trowel, sharpened broomstick, or shovel handle (see Fig. 8-49). Then apply bait according to the label directions, followed by closing the burrow.

3. Drag a harrow down the mounds.

4. If new mounds appear, administer a second poisoning.

In trapping, the following procedure is recommended:

1. Locate the runway as indicated for poisoning.

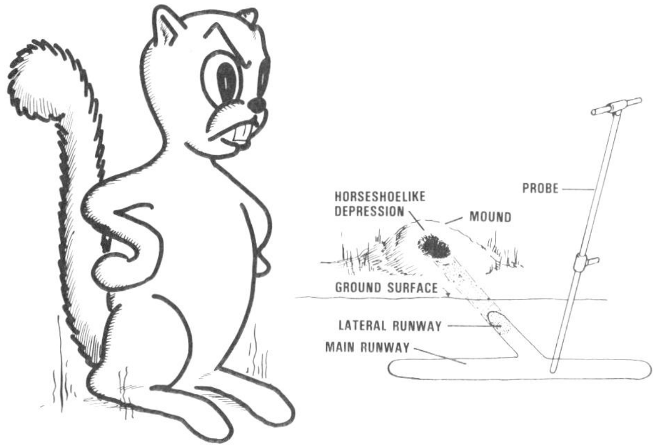

Fig. 8-49. Right way to use a probe to locate the main runway.

2. Open the lateral to the main runway with a garden trowel, or similar instrument, and insert commercially-made pocket gopher spring traps in the main runway (see Fig. 8-50).

3. Tie trap with a piece of light wire attached to a stake.

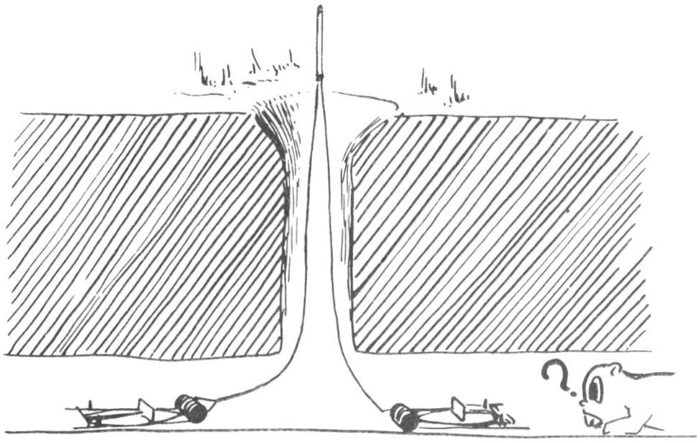

Fig. 8-50. Right way to trap pocket gophers.

• **Ground squirrels**—In certain areas, the population of ground squirrels is so high that they destroy as much as 25% of the growing and ripening grain. Also, ground squirrels may seriously compete with livestock for forage. Although these rodents may be trapped or shot, such methods are neither effective nor practical when there are many squirrels. With dense populations, the following control measures are recommended:

1. Zinc phosphide grain baits are currently the most commonly used for ground squirrel and prairie dog control. Several types of zinc phosphide baits are commercially available, some on whole or rolled grain and others are pelleted. For best results, the rodents should first be prebaited with clean grain of the same type used in the bait. This will accustom the rodents to feeding on grain and enhance their consumption of the poisoned bait when applied about a week later. Follow the directions on the bait container; to do otherwise may be illegal.

2. Anticoagulant rodenticides such as chlorophacinone (Rozol) and diphacinone (Ramik Green) are also available in some states for the control of ground squirrels. Some require that they be placed in bait stations to prevent access to nontarget species. Again, be sure to follow label directions.

3. Gassing ground squirrels in their burrows is particularly effective in the spring and often can be conducted before the new litters are produced. Smoke cartridges (gas bombs) are the easiest to use. Aluminum phosphide (Phostoxin) gas tablets or pellets, which produce a very toxic phosphine gas, can also be effective. Because of its toxic nature, the user must be more aware of the potential hazards to the applicator. Follow label directions carefully, especially with aluminum phosphide.

Poisoning and gassing are about equally effective in controlling ground squirrels, but the former is less time-consuming. Further, it may be desirable to use both methods, especially where difficulty is encountered in getting squirrels to eat poisoned grain.

• **English sparrows, starlings, and pigeons**—Most bird contamination around the farm is caused by English sparrows, starlings, and pigeons.

The following methods of control are effective:

1. Keep most such birds out of buildings by placing ½-in. hardware cloth over all windows and other openings, particularly around eaves.

2. Kill the birds by shooting. For best results, (a) induce the birds to feed at certain places, (b) scatter the grain in long, narrow lanes, along which shooting may be directed when the birds flock to feed, and (c) use No. 10 shot.

3. Trap the birds. In many localities shooting is impossible. In such places, unwanted bird populations may be controlled by the persistent use of traps. Satisfactory traps of many kinds and styles may be homemade at low cost by anyone with a moderate amount of skill. For specifications and information of the locations and operation of various traps for different birds, the farmer and rancher should contact the county agent, vocational agriculture instructor, or write to the state college of agriculture.

4. Poison baits prepared with strychnine, which were commonly available in the past, can no longer be used for bird control. Starlicide (DRC-1339) for starling control is available for use in specific situations.

5. A chemical frightening agent, Avitrol, is available in bait form for the control of pigeons, sparrows, and starlings. The affected birds display erratic behavior and emit distressing cries, which, in turn, frighten away the flock. This frightening agent, when used properly, is often very effective for a relatively long period in freeing certain buildings of birds. It may prove lethal to birds that ingest sufficient quantities.

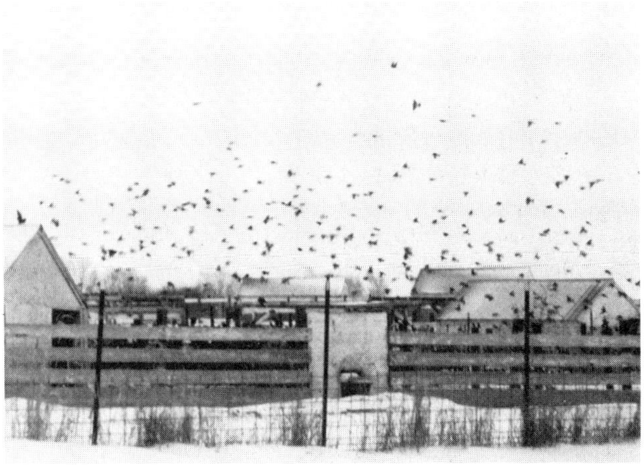

Fig. 8-51. Birds flocking over a Midwestern livestock farm. Photo by R. J. Johnson, University of Nebraska. (Courtesy, R. M. Timm, Superintendent and Extension Wildlife Specialist, University of California-Davis, Hopland Field Station, Hopland, CA)

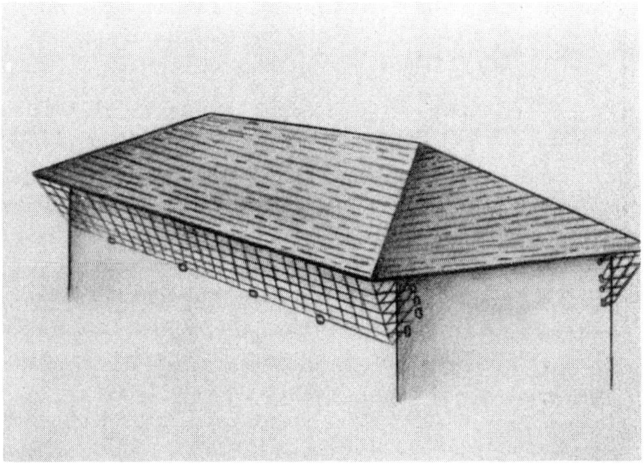

Fig. 8-52. Bird netting installed to exclude birds from under-rafter area of building. Photo by R. J. Johnson, University of Nebraska. (Courtesy, R. M. Timm, Superintendent and Extension Wildlife Specialist, University of California, Davis-Hopland Field Station, Hopland, CA)

FEEDING THE HUNGRY/ PROTECTING THE ENVIRONMENT

Fig. 8-53. "Don't do it in the stream." (Courtesy, USDA, Soil Conservation Service)

The challenge to the year 2000 and beyond will be (1) feeding the hungry, and (2) protecting the environment. These are equally important and challenging goals.

Conservative projections indicate that the world will have an increase of over 1 billion people during the decade of the 90s, bringing the world's population to 6.2 billion by the year 2000. Further, it is projected that we shall have a population of nearly 11 billion by the year 2050. Moreover, most of the increase will take place in that part of the world where 75% of the population can barely feed themselves, where nearly 500 million people are severely malnourished, and where 15 million children die each year from starvation.

To meet the food/environment need calls for a balancing act between producing food and preserving the environment. It calls for farmers and ranchers, worldwide, meeting this goal with (1) fewer crop protection chemicals and chemical fertilizers, which currently account for as much as 40 to 50% of the world's agricultural production; (2) a shortage of land, and millions of acres made unproductive because of erosion; scarcity of water due to droughts, heavy irrigation, and lowering of the water table; and (4) the inevitable time lag between biotechnology, discovery, and application.

To feed the hungry and protect the environment to the year 2000 and beyond, technology must play a major role. Above all, we cannot return to the "good old days" of wormy apples, unpasteurized milk, threshing grain by a separator powered by a smoke-belching steam engine, and fresh fruit and vegetables only in the summertime.

Feeding the hungry, protecting the environment, and applying technology in the decades to come calls for a nation of farmers and ranchers with an integrated approach involving traditional big and modern operations, along with knowledge and application of each of the following, which are fully covered in Section 1, Animal Behavior and Environment:

Animal Environment, see pp. 14 to 18.
Pollution of the Environment, see pp. 19 to 23.
Sustainable Agriculture, see p. 23.
Food Safety and Diet/Health Concerns, see p. 24.

Additionally, *organic farming*, which is covered in the section that follows, will increase in importance.

ORGANIC FARMING

The U.S. Department of Agriculture's definition of organic farming follows:

> *Organic farming is a production system which avoids or largely excludes the use of synthetically compounded fertilizers, pesticides, growth regulators, and livestock feed additives. To the maximum extent feasible, organic farming systems rely upon crop rotations, crop residues, animal manures, legumes, green legumes, off-farm organic wastes, mechanical cultivation, mineral-bearing rocks, and aspects of biological pest control to maintain soil productivity and tilth, to supply plant nutrients, and to control insects, weeds, and other pests.*[10]

Often, the meanings for *organic, natural,* and *health,* which people imply or conjure up in their minds, are misleading, harmful, and tend to polarize people. Frequently, it is proclaimed that such products are safer and more nutritious than conventionally grown and marketed foods. The growing interest of consumers in the safety and nutritional quality of the American diet is a welcome development. Regrettably, however, much

[10]*Report and Recommendations on Organic Farming,* USDA, 1980.

of this interest has been colored by alarmists who state or imply that the American food supply is unsafe or somehow inadequate to meet our nutritional needs.

Fig. 8-54. Organic farming—a hog, fish, grain—rotation in China. Grain is fed to hogs to produce manure, to feed more fish for human food; then the pond residue is used to fertilize more grain, to produce more hogs, to feed more fish. (Photo by A. H. Ensminger)

The FDA has taken no position on the use of the terms *organic, natural,* and *health* in food labeling, because the terms are often used loosely and interchangeably. The Federal Trade Commission (FTC) in its proposed Food Advertising Rule would prohibit use of the words *organic* and *natural* in food advertising because of concern about the ability of consumers to understand the terms in the conflicting and confusing ways they are used.

• **Feed/food quality**—Undeniably, many benefits accrue from organic farming, not the least of which is the valuable exercise that the practicing advocates get from growing their own organically produced vegetables. However, no laboratory test or animal feeding trial can distinguish between crops fertilized with inorganic fertilizers and those fertilized with organic fertilizers. Moreover, experiments designed to compare the levels of different essential nutrients in crops produced with organic fertilizers against those produced with comparable amounts of nutrients supplied as inorganic materials have shown little difference, with the advantage in favor of the inorganic as often as the organic. If the few experiments in which the plants produced under the two systems have been fed to test animals, the small differences noted in animal growth have not consistently favored either the organic or inorganic sources of the nutrients.

The above results are as one would expect, based on the function of plants in the food chain—to convert inorganic compounds to organic compounds. If organic materials containing the nutrients are incorporated in the soil, the microorganisms in the soil must first break down the organic matter into inorganic forms. Inorganic ions of the essential nutrients are then taken up by the plant roots and manufactured into new organic materials. In the plants, and in the bodies of animals, these essential nutrient elements have the same effect, regardless of whether they were added to the soil in the form of organic or inorganic fertilizers.

• **Environmental control**—An important reason for adding organic materials to agricultural and garden soils is that this practice can be used to recycle the organic material without damaging the environment. On the other hand, too much organic matter can be added to the soil. Problems from excessive application of organic materials are generally confined to fields in close proximity to large cattle feedlots or poultry operations. Under such circumstances, nitrate toxicity and grass tetany have been serious problems where pastures have received excessive applications of manure.

• **Feed/food quantity**—The use of chemical fertilizers has been partly responsible for the abundance of food available. If all farmers were to adopt the organic methods, there would be a decline in productivity as shown in Table 8-6.

TABLE 8-6
ESTIMATED NATIONAL AVERAGE CROP YIELDS UNDER CONVENTIONAL AND ORGANIC FARMING[1]

Crop	Bushels per Acre	
	Conventional	Organic
Corn	98	49
Wheat	43	20
Soybeans	40	20
Other grains	57	17

[1] *Organic and Conventional Farming Compared,* Council for Agricultural Science and Technology (CAST), Report No. 84, October 1980, p. 24, Table 6.

In a 1980 report issued by the Council for Agricultural Science and Technology (CAST), it was estimated that if organic farming were widely adopted the cost of food would increase because the total production from the land under cultivation would decrease. Those now practicing organic methods realize their yields are less, but they may receive higher prices for their products because they sell to a specialty market.

• **Some inorganic fertilizers and mineral supplements are essential**—It is noteworthy that some nutritional problems can be corrected only by inorganic fertilizers or mineral supplements. For example, iodine deficiency in animals or people cannot be corrected by organic fertilizers, unless, of course, they contain kelp (seaweed) or marine by-products, good sources of iodine. The cobalt deficiency that plagued the cattle of the colonists in New Hampshire did not respond to organic fertilizers that contained little cobalt.

• **Summary**—Undoubtedly, world food production of the future will make use of a combination of both organic and inorganic fertilizers, with the nature and proportions of the combination for different farms and for different countries dependent on their access to fossil fuels, the availability and price of fertilizers, their soils, their food production requirements, their environmental control problems, and many other factors. Regardless of the combination of organic and inorganic fertilizers used, feed and food plants of adequate nutritional quality can be produced.

Good buildings and equipment are an asset to the farm or ranch. (Courtesy, Ford New Holland, New Holland, PA)

SECTION

9 BUILDINGS AND EQUIPMENT

Contents *Page*

Fig. 9-1. Dairy farm in New York. (Courtesy, The American Jersey Cattle Club, Columbus, OH)

Properly designed, constructed, and arranged buildings and equipment are an asset to the farm or ranch. They increase animal production, make for labor and feed efficiency, conserve crops and manure, provide comfort for caretakers and beasts, and add to the beauty of the farm landscape. In serving these purposes, it is not necessary that buildings and equipment be either elaborate or expensive.

Fences should be considered. Repairing fencing or adding new fencing is expensive. Also, it should be kept in mind that the life of the fence depends to a great extent on the life of the posts, and the stability of the corner posts.

Increasingly, the U.S. livestock industry has moved to more and more confinement—to production in limited quarters, with or without shelter, but with no pasture. The reasons: (1) it lessens land cost, (2) it minimizes management time, (3) it minimizes labor, (4) it maximizes control of animals, (5) it maximizes genetic potential of animals, and (6) it maximizes feed conversion. These motivating forces will be accentuated in the future, with the result that confinement production will continue to increase. The trend and acceptance is toward more closed, insulated, mechanically ventilated, environmentally controlled structures—especially for poultry, swine, and dairy enterprises. But the shift to confinement structures and high-density production operations has introduced new problems and accentuated old ones, especially in the areas of animal behavior, manure management, optimum environment, and flexibility.

Family happiness is important to the success of any livestock venture. Thus, the home should come in for its share of consideration, also. If it is not completely modern so far as plumbing, heating, and lighting are concerned, it will need to be made so as soon as possible.

In this section, special attention is given to space requirements, environmental control, types of floors, handling manure, and labor saving buildings and equipment. The effect of changed animal environment on animal behavior is accorded special and rather complete treatment in Section 1.

Because of the variations in climatic conditions, sizes and types of enterprises, and systems of management, no attempt will be made herein to present detailed building and equipment plans and specifications. Rather, it is proposed merely to convey suggestions regarding some of the desirable features of buildings and equipment in use for each class of farm animals in various parts of the country. For detailed plans and specifications for a particular locality, the farmer/rancher should (1) study successful buildings and equipment on neighboring farms/ranches; (2) consult the local county agricultural agent; and/or (3) write to the state college of agriculture.

LOCATION OF FARM OR RANCH HEADQUARTERS

The headquarters or farmstead is the base of all farm or ranch operation. It is important, therefore, that it be located and planned correctly for efficient, profitable, and pleasant operation. So, when planning an entirely new farm or ranch headquarters, the choice of the location for the buildings is the first consideration. Likewise, when appraising the desirability of an existing headquarters, the same factors should be considered. These factors are:

1. Accessibility to fields.
2. Central location.
3. Drainage.
4. Electricity.
5. Erosion control.
6. Mail route.
7. Roads.
8. Schools and churches.
9. Size and shape of area.
10. Soil type.
11. Sun exposure and wind protection.
12. Telephone.
13. Topography.
14. Vegetation and windbreaks.
15. View.
16. Water supply.

The consideration of the above points in locating, planning, and improving the farm or ranch headquarters can add materially to the convenience, comfort, and pleasure of the operator, and to the profitable operation of the enterprise.

FARMSTEAD ARRANGEMENT

The arrangement—which means the location and orientation of individual buildings within the site—should make for ease of use, economy of labor and cost in operation, and attractiveness. In general, for conservation of space and time, the barn and other service buildings should be located around a central court and should be so arranged that most of them can be seen from the house.

In planning a new farm or ranch headquarters or in altering an old one, buildings, fences, lots, trees, etc., should be added according to an established master plan; for, once constructed, buildings are difficult and expensive to move.

In arriving at the best arrangement, the farmstead cannot and should not be modeled after one popular pattern. Instead, consideration should be given to the following pertinent points:

1. **The house location comes first.** As the farm or ranch house is the headquarters or office of the farm business as well as a home, its location is of greatest importance in farmstead arrangement.

2. **Orientation.** Fortunately, the farm or ranch headquarters need not be oriented with the compass.

3. **Direction of wind.** The house should be located on the windward side of the headquarters, with special consideration given to summer winds.

4. **Efficiency.** The buildings should be located so as to require a minimum of walking when doing the chores.

5. **Corrals and lots.** The buildings and their adjacent corrals and lots should be arranged so that the buildings are accessible without walking through feedlots and corrals.

6. **Fire protection.** Farm buildings should be far enough apart so that fire will not spread easily from one building to another. In general, this means at least 100 ft apart in the case of large buildings. In acquiring added fire protection through spacing buildings farther apart, one should avoid extreme distances that will mean inefficiency in operation; fire insurance is probably cheaper than labor.

7. **Appearance.** Careful attention to the headquarters arrangement can add to the attractiveness of the entire unit.

8. **Gates and lanes.** The adoption of larger machinery has necessitated wider gates and lanes than have been commonplace in the past.

9. **Expansion.** Provision should be made for easy expansion of the farmstead. Many times buildings can be expanded in size by extending their length, provided no other buildings or utilities interfere.

LAYOUT OF LIVESTOCK OPERATIONS

Prior to starting construction, farmers/ranchers may avoid much subsequent difficulty and expense by first doing some paper and pencil planning. They should first decide on the specific kind, or kinds, of livestock and the size of the enterprise. Then, they should sketch out the buildings and equipment required to meet these needs in the most efficient and economical manner. In particular, the preliminary layout of livestock operations should include the following:

1. **The management system.** The management system will greatly affect the kind, size, and amount of buildings and equipment. For example, swine producers must first decide whether they will (a) farrow out pigs, or buy feeders; (b) farrow twice a year, or multiple farrow—if they are going to maintain a breeding herd; (c) rely on multipurpose buildings, or have special buildings (for example, a special house for farrowing and a special house for finishing): (d) buy commercial, ready-mixed feeds, or use a maximum of homegrown feeds; or (e) follow confinement rearing, or pasture rearing—or a combination of the two.

2. **Plans for the flow of all materials.** A detailed plan should be developed for the flow of all materials, with primary consideration given to maximum automation and minimum labor and expense. These plans should include provision for (a) delivering the proper feed to the animals at the desired time and place, (b) providing a sanitary water supply, (c) delivering and distributing bedding, (d) removing manure, and (e) marketing animals. All these considerations, and more, enter into the handling of materials to, within, and from the buildings.

The above information, constituting the layout of operations, should first be put on paper in sketch form, by the producer. From this, an architect and/or engineer can design, or recommend for purchase, buildings and equipment which most effectively and economically meet the production requirements of the specific enterprise.

REQUISITES OF LIVESTOCK BUILDINGS

When planning to build, the owner should first ask, then answer, this searching question: "Why is this structure to be built?" Livestock buildings should not be monuments. Instead, they should be production tools; they should contribute to the farm or ranch operations, and they should not only pay for themselves but they should pay satisfactory returns on the investment.

Each farm or ranch is different, and the type and size of buildings will vary accordingly. Among the factors determining the type and size of buildings are: (1) kind and fertility of the soil, (2) available markets; (3) size of farm, (4) tenant or owner operation, (5) kind and amount of livestock and crops to be grown, (6) personal preference, (7) climatic conditions of the region, and (8) storage requirements. Thus, the specific requisites of animal buildings will vary according to the needs of the region, state, community, and individual farm or ranch.

TABLE
SPACE REQUIREMENTS OF BUILDINGS

	Barn or Shed		Shades		Feedlots[1]		Hay or Silage	
Class, Age, and Size of Animal	Floor Area per Animal	Height of Ceiling[3]	Shade per Animal	Shade Height	Area If Ordinary Dirt Lot	Area If Paved Lot	Length per Animal[4]	Width If Feeds from 1 Side
	(sq ft)	(ft)	(sq ft)	(ft)	(sq ft)	(sq ft)	(in.)	(in.)
Cows, 2 years or over	40–50	8½–10	30–40	10–12	300[5]	50–100	24–30	30
Yearling finishing cattle	Solid floor: 30–40 Slotted floor: 20–25	"	25–35	"	125–200[5]	30–50	20	"
Calves, 350 to 500 lb	20–30	"	15–25	"	100–175[5]	20–50	18	"
Cows in maternity stall	100–120	"	35–40	"	1–2 acre pasture paddock	—	30	—
Herd bulls	100–150	"	35–45	"	"	—	"	30

[1]Allow slope of ¼ to ½ in./ft in paved lots, and ½ in. or more in dirt lots (depending on soil and climate conditions).

[2]Feed bunks should be about 8 in. deep for calves and 12 in. for older cattle.

[3]Minimum ceiling height of 9 ft necessary where a power-operated manure loader is to be used.

[4]With liberal grain or other concentrate feeding, half the recommended space given herein. With bunker or self-feeder silos, allow 6 in./animal.

[5]More space is desirable under some soil and climatic conditions.

Before starting construction, there should be a complete plan. This step should precede arranging financing and starting site preparation and construction; and it should be observed whether you (a) do the construction yourself, (b) have a contractor build it, (c) buy a site-erected manufactured building, or (d) buy a prefabricated building. Much of the needed planning information may be obtained from this book. Additional planning information, along with working drawings for building construction, are available from manufacturers and dealers. Also, standard plans prepared by your university or the U.S. Department of Agriculture may be purchased from your extension agricultural engineer or from the following:

Midwest Plan Service
Iowa State University
Ames, IA 50011

Certain general requirements of animal buildings should always be considered. Once buildings are constructed, there is a practical limit to the changes that can be made in remodeling. Consequently, it is most important that very careful consideration be given to the following requisites:

1. Adapted to present and future needs.
2. Adequate space for animals.
3. Adequate space for feed and bedding storage.
4. Attractiveness.
5. Convenient.
6. Direct sunlight.
7. Dryness.
8. Durability.
9. Easily cleaned.
10. Environmental control.
11. Flexible design.
12. Insulated properly.
13. Keep proper humidity.
14. Minimum fire risk.
15. Modify extreme temperatures.
16. Multiple use.
17. Protect animal health.
18. Protect newborn animals.
19. Provide for manure disposal.
20. Provide protection from the elements.
21. Reasonable construction and maintenance costs.
22. Reduce labor.
23. Rodent control.
24. Safety.
25. Sanitary.
26. Surrounded by suitable corrals or lots.
27. Utility value.
28. Water source.
29. Well lighted.
30. Well ventilated.

SPACE REQUIREMENTS OF BUILDINGS AND EQUIPMENT

One of the first and frequently one of the most difficult problems confronting the farmer/rancher who wishes to construct a building or item of equipment is that of arriving at the proper size or dimensions. Tables 9–1 to 9–7, pages 360 to 366, contain some conservative average figures which, it is hoped, will prove helpful. In general, less space than indicated may jeopardize the health and well-being of the animals; whereas more space may make the buildings and equipment more expensive than necessary.

Where of comparable age and size and where handled similarly, the space requirements of dairy cattle are like those given for beef cattle in Table 9–1. But since dairy cattle are kept for milk whereas beef cattle are kept for meat, there are differences in their management and space requirements. Thus, the special space requirements of dairy cattle—for dry and lactating cows, dairy calves, and replacement heifers—are given later in this section under the heading "Dairy Cattle Buildings and Equipment."

The space requirements of most swine are given in Table 9–3 (next page). The space requirements and other requirements of early weaned pigs are given in Table 9–17, p. 383.

9–1
AND EQUIPMENT FOR BEEF CATTLE

Manger, or Rack			Feed Bunk or Trough for Hand-Feeding Grain[2]				Self-Feeder	Water	
Width If Feeds from 2 Sides	Width If Attached Side of Barn	Height at Throat	Length per Animal	Width If Feeds from 1 Side	Width If Feeds from 2 Sides	Height at Throat	Trough Length If Feeder Is Kept Filled	Water per Animal per Day	Water Trough
	(in.)	(in.)	(in.)	(in.)	(in.)	(in.)	(in.)	(gal)	
48–60	30	24	24–30	18–30	48	24	6–12 per animal	12	**A**llow 1 linear ft of open water tank space for each 10 cattle; or one automatic watering bowl for each 25 cattle. **A** satisfactory water temperature range in winter is 40–45°F; in summer, 60–80°F
"	"	20	18–24	18	"	22	6–9	10	
"	"	20	18	"	"	18	6–8	8	
—	—	26	30	—	—	30	9–12	15	
36–40	30	"	"	30	36–40	"	"	"	

Remarks:
Animals with horns require about 1 linear ft more manger or trough space per animal than the figures given in this table. Movable hayracks or feed bunks are usually 12 to 16 ft in length.
Provide a paved area of at least 10 ft around waterers, feed bunks, and roughage racks.
For specifications on slotted floors see "Confinement Feeding: Slotted Floors," later in this section.
Re: Water per animal per day. A minimum of 20 gal/day is needed for continuous flow to keep the water clean, and to keep it from freezing in the winter months.

TABLE
SPACE REQUIREMENTS OF

Class, Age, and Size of Animal	Barn or Shed: Floor Area per Animal	Barn or Shed: Height of Ceiling	Barn or Shed: Window Space (not including open sheds)	Feedlot: Area If Ordinary Dirt Lot	Feedlot: Area If Paved Lot	Shades: Area per Animal	Shades: Height
	(sq ft)	(ft)	(sq ft)	(sq ft)	(sq ft)	(sq ft)	(ft)
Dry ewes	16	8½–10	1 sq ft window space per 35 sq ft floor space	16–20	16	10–12	8–10
Ewes with lambs	20	"	"	30	20	14	8–10
Stud rams	20–30	"	"	30–60	25	15	8–10
Feeder lambs	6	"	"	25–30	16	6–8	8–10

[1]For self-feeding silage, allow 5 linear in. of manger or rack space for mature sheep and 4 in. for feeder lambs.

Remarks:

Wide barn doors are needed to prevent crowding and possible injury to pregnant ewes. Doors at least 8 ft wide are preferable.

For specifications of slotted floors, see section entitled, "Confinement of Sheep on Slotted Floors."

TABLE
SPACE REQUIREMENTS OF
(See Table 9–17 for

Age and Size of Animal	Swine Buildings: Inside Sleeping Space or Shelter per Animal[1]	Swine Buildings: Height of Ceiling[2]	Swine Buildings: Height of Pen Partition	Swine Buildings: Hog Door Height	Swine Buildings: Hog Door Width	Shades: Shade per Animal	Shades: Shade Height	Pasture or Feeding Floor: Good Pasture	Pasture or Feeding Floor: Paved Feeding Floor in Addition to Sleeping Space, When Confined, per Animal
	(sq ft)	(ft)	(in.)	(in.)	(in.)	(sq ft)	(ft)	(animals/acre)	(sq ft)
Sows before farrowing:									
Gilts	15–17	7–8	36	36	24	17	4–6	10–12	15–20
Mature sows	18–20	"	"	"	"	20	"	8–10	"
Sows with pigs:									
Gilts	48	"	"	"	"	50	"	6–8	48
Mature sows	64	"	"	"	"	60	"	6–8	64
Herd boars	15–20	"	48	"	"	15–20	"	¼ acre/boar	15–20
Growing-finishing swine:									
Weaning[8] to 75 lb	5–6[7, 8]	"	30	"	"	6	"	20 on full-feed; 10–15 limited feed	6–8[9]
75 lb to 125 lb	6–7[7, 8]	"	33	"	"	7	"	"	7–9[9]
125 lb to market	8–10[7, 8]	"	36	"	"	10	"	"	8–10[9]

[1]Space requirements are less with slotted floors (see section on slotted floors under "Buildings and Equipment for Swine").

[2]Ceiling heights in excess of 7 to 8 ft make for cold hog houses in the northern half of the United States.

[3]For example, a 6-ft feeder open on both sides has 12 linear ft of feeding space.

[4]The drinking water should not fall below 35° to 40°F during the winter.

[5]With creep provided for pigs in addition.

9–2
BUILDINGS AND EQUIPMENT FOR SHEEP

Hay or Silage Manger, or Rack (for hand-feeding)				Feed Trough (for grain or roots; hand-feeding)				Self-feeder (for concentrate or roughage)	Water
Length per Animal[1]	Width If Feeds from 1 Side	Width If Feeds from Both Sides	Height at Throat	Length per Animal	Width If Feeds from 1 Side	Width If Feeds from 2 Sides	Height at Throat	Trough Length If Feeder is Kept Filled	Water per Animal per Day
(in.)	(in.)	(in.)	(in.)	(in.)	(in.)	(in.)	(in.)	(in.)	(gal)
12	14–16	20–24	12–15	12	14–16	20–24	10–15	6 (when salt is used as a governor, 3 in. will suffice)	2
″	″	″	″	″	″	″	″	—	3
″	″	″	″	20–24	″	″	″	—	3
″	12–14	18–22	10–12	12	″	12	8–12	3 in. for conc. alone; 4 in. for complete ration, or hay, or silage.	1

Provide a paved area of at least 5 ft around waterers, feed bunks, roughage racks, and entrances to sheds.
Provide water space as follows: 1 linear foot of open tank per 10 head of sheep, or one automatic bowl per 15 head. Maintain water temperature above 35°F in winter and below 75°F in summer.

9–3
BUILDINGS AND EQUIPMENT FOR SWINE
Early Weaned Pigs)

Feeding Equipment: Self-feeder Space (animals per linear ft, or per hole)		Percent of Total Self-feeder Space Given to Protein Supplement		Watering Equipment[4]: Feed Trough Space per Animal for Hand-feeding	Water Trough Space per Animal for Hand-feeding	Automatic Watering Cups (two openings considered 2 cups)	Comments
Dry-Lot	Pasture	Dry-Lot	Pasture				
(no. animals/ linear ft)[3]	(no. animals/ linear ft)[3]	(% total feeder space)	(% total feeder space)	(linear ft/ animal)	(linear ft/ animal)		
3	3	15	10–15	1½	1½	1 cup/12 gilts	When alfalfa hay is fed in rack, allow 4 sows per linear ft.
2	4	″	″	2	2	1 cup/10 sows	
1[5]	1[5]	″	″	1½[5]	2	1 cup/4 sows	For the pig creep, provide a minimum of 1 ft of feeder space per 5 pigs, see that the edge of the feeder trough does not exceed 4 in. above the ground floor, and do not allow more than 40 pigs per creep.
1[5]	1[5]	″	″	1½[5]	2	1 cup/4 sows	
1	1	″	″	2	2	1 cup/2 boars	
4	4–5	25	20–25	¾	¾	1 cup/20 pigs	
3	3–4	20	15–20	1	1	″	When salt or mineral is fed free-choice provide 3 linear ft of mineral box space or 3 self-feeder holes per 100 pigs.
3	3–4	15	10–15	1¼	1¼	″	

[6]For early weaning (under 5 to 6 weeks) space requirements, see Table 9–17.

[7]Over and above the sleeping space given herein, pigs that are confined from weaning to market should be provided the feeding floor space recommended in the column headed "Paved feeding floor in addition to sleeping space, when confined, per animal."

[8]The larger are in the summertime.

[9]The larger area when fed from troughs; the smaller area is adequate where self-feeders are used.

TABLE 9-4
SPACE REQUIREMENTS OF BUILDINGS FOR HORSES

Kinds, Uses and Purposes	Recommended Plan	Box Stalls or Shed Areas: Size	Height of Ceiling	Height of Doors	Width of Doors	Tie Stalls (size)
Smaller Horse Establishments						
Horse barns for pleasure horses, ponies, and/or raising a few foals.	12 ft × 12 ft stalls in a row; combination tack-feed room for 1- and 2-stall units; separate tack and feed rooms for 3-stall units or more. Generally, not more than a month's supply of feed is stored at a time. Use of all-pelleted rations (hay and grain combined) lessens feed storage space requirements.	Horses: 12 ft × 12 ft Ponies: 10 ft × 10 ft	8 - 9 ft	8 ft	4 ft	5 ft wide; 10 - 12 ft long
Larger Horse Breeding Establishments: The following specially designed buildings may be provided for different purposes.						
Broodmare and Foaling Barn	A rectangular building, either (1) with a central aisle, and a row of stalls along each side, or (2) of the the *island* type, with 2 rows of stalls, back to back, surrounded by an alley or runway. Ample quarters for storage of hay, bedding, and grain. A record or office room, toilet facilities, hot water supply, veterinary supply room, and tack room are usually an integral part of a broodmare barn.	12 ft × 12 ft to 16 ft × 16 ft	9 ft	8 ft	4 ft	
Stallion Barn	Quarters for one or more stallions, with or without feed storage. A small tack and equipment room. Stallion paddocks, at least 300 ft on a side, adjacent or in close proximity.	14 ft × 14 ft	9 ft	8 ft	4 ft	
Barren Mare Barn	An open shed or rectangular building, with a combination rack and trough down the center or along the wall. Storage space for ample hay, grain, and bedding.	150 sq ft per animal	9 ft	8 ft	4 ft	
Weanling or Yearling Quarters	Open shed or stalls. The same type of building is adapted to both weanlings and yearlings; but different ages and sex groups should be kept separate. When stalls are used, 2 weanlings or 2 yearlings may be placed together.	10 ft × 10 ft	9 ft	8 ft	4 ft	
Breeding Shed	A large roofed enclosure with a high ceiling; should include laboratory for the veterinarian, hot water facilities, and stalls for preparing mares for breeding and holding foals.	24 ft × 24 ft	15 - 20 ft	8 ft	9 ft	
Isolation (quarantine) Quarters	Small barn, with feed and water facilities and adjacent paddock; for occupancy by new or sick animals.	12 ft × 12 ft	9 ft	8 ft	4 ft	
Riding Academies; Training and Boarding Stables	Either (1) stalls constructed back to back in the center of the barn, with an indoor ring around the outside; (2) stalls around the outside and a ring in the center; or (3) stalls on either side of a hallway or alleyway, and an outdoor ring.	12 ft × 12 ft	9 ft	8 ft	4 ft	5 ft wide; 10 - 12 ft long

[1]Even for ponies, a 12 × 12 ft stall is recommended since (1) it costs little more than a 10 × 10 ft, and (2) it affords more flexibility—it can be used for bigger horses when and if the occasion demands.

TABLE 9-5
SPECIFICATIONS OF FEED AND WATER EQUIPMENT FOR HORSES

Equipment for	Kind of Equipment	Materials and Design	Sizes for: Horses	Sizes for: Ponies	In Stall: Location	In Stall: Height	In Corral: Location	In Corral: Height	Remarks
Concentrates	Pail; tub	Metal, plastic, or rubber; usually with screweyes, hooks, or snaps for suspending.	16-20 qt	14-16 qt	Front of stall.	⅔ height of animal at withers; or 38-42 in. for horses, and 28-32 in. for ponies.	Along fence line.	Same height as stall.	For sanitary reasons, removable concentrate containers are preferable so that they can be taken out and easily and frequently cleaned—which is especially important after a wet bran mash has been fed.
	Box	Wood	Width 12-16 in. Length 24-30 in. Depth 8-10 in.	Width 10-12 in. Length 20-24 in. Depth 6- 8 in.	Front of stall.	⅔ height of animal at withers; or 38-42 in. for horses, and 28-32 in. for ponies.			If desired, a pie-shaped metal pan set in a wooden shelf can be mounted in a front corner of the stall and pivoted in such manner that it can be pulled outward for filling and cleaning, then returned into the stall and locked in place.

(Continued)

TABLE 9–5 *(Continued)*

Equipment for	Kind of Equipment	Materials and Design	Sizes for Horses	Sizes for Ponies	In Stall Location	In Stall Height	In Corral Location	In Corral Height	Remarks
Hay	Stall rack	Metal, fiber, or plastic	25-30 lb	10-15 lb	Corner of stall; in trailer or van.	Bottom of rack, same height as horse or pony at withers.			Hayracks (1) eliminate contaminated hay and lessen parasite infestation, and (2) lessen pawing and waste. Racks should open at bottom so that dirt, chaff, and trash may be removed or will fall out. For stallions and broodmares, always use high racks to avoid injury hazards.
	Manger	Wood	Width 30 in. Length 24-30 in.	Width 20 in. Length 20 in.	Front or corner of stall.	30-42 in. for horses; 20 - 24 in. for ponies.			
	Corral rack	Wood	Large enough to provide one day's supply of hay for intended number of horses.				In fence line if it feeds from one side only. On high ground if it feeds from both sides.	Top of rack may be 1 in. to 2 in. higher than height of horse at withers.	Corral hayracks that feed from both sides should be portable.
Mineral	Box	Wood			Corner of stall.	Same height as concentrate box	Fence corner	⅔ height of horse at withers.	If mineral container is stationed in the open—in a corral, or in a pasture—it should be protected from wind and rain. Mineral containers should have 2 compartments—1 for mineral mix, and the other for salt.
	Self-feeder	Metal or wood				Same height as concentrate box.	Fence corner		
Water	Stall, automatic	Metal; 1 cup or 2 cups			Front corner of stall.	24-30 in.			The daily water requirements are: Mature horse, 12 gal; foals to 2-year-olds, 6-8 gal; and ponies 6-8 gal. In colder areas, waterers should be heated, and equipped with thermostatic controls. A satisfactory water temperature range in the winter is 40-45° F; in summer, 60-80° F. Watering facilities should be designed so as to facilitate draining and cleaning. Also, they should be located proper distance from feed containers; otherwise, horses will (1) carry feed to the waterer, or (2) slobber water into the concentrate container. A 20 × 30 in. automatic waterer will accommodate about 25 horses, and a 2-cup waterer will serve 12 head. Automatic waterers should be checked daily.
	Corral, automatic				In fence corner	24-30 in.			
	Pail	Metal, plastic, or rubber			Front of stall.	⅔ height of horse at withers; or 38 - 42 in. for horses and 28 - 32 in. for ponies.			
	Tank	Concrete, steel					Set in fence so that there are no protruding corners; or painted white out in corral or pasture.	30-36 in.	One linear foot of tank space should be allowed for each 5 horses. Tanks should be equipped with a float valve, which should be protected.

RECOMMENDED MINIMUM WIDTH OF SERVICE PASSAGES

In general, the requirements for service passages are similar, regardless of the kind of animals. Accordingly, the suggestions contained in Table 9–6 are equally applicable to cattle, sheep, swine, and horse barns.

TABLE 9–6
RECOMMENDED MINIMUM WIDTHS FOR SERVICE PASSAGES

Kind of Passage	Use	Minimum Width
Feed alley	For feed cart	4 ft
Driveway	For wagon, spreader, or truck	9-12 ft
Doors and gate	Drive-through	8-9 ft

STORAGE SPACE REQUIREMENTS FOR FEED AND BEDDING

Table 9–7 gives the storage space requirements for feed and bedding. This information may be helpful to the individual operator who desires to compute the barn space required for a specific livestock enterprise. This table also provides a convenient means of estimating the amount of feed or bedding in storage.

TABLE 9–7
STORAGE SPACE REQUIREMENTS FOR FEED AND BEDDING[1]

Kind of Feed or Bedding	Pounds per Cubic Foot	Cubic Feet per Ton	Pounds per Bushel of Grain
Hay/Straw:[2]			
1. Loose			
Alfalfa	4.4 - 4.0	450 - 500	
Nonlegume	4.4 - 3.3	450 - 600	
Straw	3.0 - 2.0	670 - 1000	
2. Baled			
Alfalfa	10.0 - 6.0	200 - 330	
Nonlegume	8.0 - 6.0	250 - 330	
Straw	5.0 - 4.0	400 - 500	
3. Chopped			
Alfalfa	7.0 - 5.5	285 - 360	
Nonlegume	6.7 - 5.0	300 - 400	
Straw	8.0 - 5.7	250 - 350	
Corn:			
15½% moisture			
Shelled	44.8		56
Ear	28.0		70
Shelled, ground	38.0		48
Ear, ground	36.0		45
30% moisture			
Shelled	54.0		67.5
Ear, ground	35.8		89.6
Barley, 15%	38.4		48.0
ground	28.0		37.0
Flax, 11%	44.8		56.0
Oats, 16%	25.6		32.0
ground	18.0		23.0
Rye, 16%	44.8		56.0
ground	38.0		48.0
Sorghum, grain 15%	44.8		56.0
Soybeans, 14%	48.0		60.0
Wheat, 14%	48.0		60.0
ground	43.0		50.0

[1]*Housing and Equipment Handbook*, MWPS-6, Midwest Plan Service, Iowa State University, Ames.

[2]Many factors—other than kind of hay/straw, such as form (loose, baled, chopped), and period of settling—affect the density of hay/straw in a stack or in a barn, including (a) moisture content at haying time, and (b) texture and foreign material.

ENVIRONMENTALLY CONTROLLED BUILDINGS

Environment may be defined as all the conditions, circumstances, and influences surrounding and affecting the growth, development, and production of a living thing. In animals, this includes the air temperature, relative humidity, air velocity, wet bedding, dust, light, ammonia buildup, odors, and space requirements. Control or modification of these factors offers possibilities for improving animal performance. There is still much to be learned about environmental control, but the gap between awareness and application is becoming smaller.

The keepers of herds and flocks were little concerned with the effect of environment on animals as long as they roamed pastures and ranges. Space requirements, wet bedding, ammonia buildup, odors, and manure disposal were no problem. But the concentration of animals into smaller spaces changed all this. With the shift to confinement structures and high-density production operations, building design became more critical.

Environmentally controlled buildings are costly to construct, but they make for the ultimate in animal comfort, health, and efficiency of feed utilization. Also, they lend themselves to automation, which results in a savings in labor; and, because of minimizing space requirements, they effect a saving in land cost. Today, environmental control is rather common in poultry and swine housing, and it is on the increase with other classes of livestock.

VAPOR PRODUCTION OF ANIMALS

Animals give off moisture during normal respiration; and the higher the temperature, the greater the moisture. This moisture should be removed from buildings through the ventilation system. Most building designers govern the amount of winter ventilation by the need for moisture removal. Also, cognizance is taken of the fact that moisture removal in the winter is lower than in the summer; hence, less air is needed. However, lack of heat makes moisture removal more difficult in the wintertime.

Since ventilation also involves a transfer of heat, it is important to conserve heat in the building to maintain desired temperatures and reduce the need for supplemental heat. In a well-insulated building, mature animals may produce sufficient heat to provide a desirable balance between heat and moisture; but young animals will usually require supplemental heat. The major requirement for summer ventilation is temperature control, which requires moving more air than in the winter.

RECOMMENDED ENVIRONMENTAL CONDITIONS FOR ANIMALS

The comfort of animals (or people) is a function of temperature, humidity, and air movement. Likewise, the heat loss from animals is a function of these three items.

The prime function of the winter ventilation system is to control moisture, whereas the summer ventilation system is primarily for temperature control. If air in livestock barns is supplied at a rate sufficient to control moisture—that is, to keep the inside relative humidity in winter below 75%—then this will usually provide the needed fresh air, help suppress odors, and prevent an ammonia buildup.

Temperature, humidity, and ventilation recommendations for different classes of livestock are given in Table 9–8. This table will be helpful in obtaining a satisfactory environment in confinement livestock buildings, which require careful planning and design.

TABLE 9–8
RECOMMENDED ENVIRONMENTAL CONDITIONS FOR ANIMALS

Class of Animal	Temperature				Acceptable Humidity	Commonly Used Ventilation Rates[1]					Drinking Water			
	Comfort Zone		Optimum			Basis	Winter[2]		Summer		Winter		Summer	
	(°F)	*(°C)*	(°F)	*(°C)*	(%)		(cfm)	*(m³/min.)*	(cfm)	*(m³/min.)*	(°F)	*(°C)*	(°F)	*(°C)*
Beef cow	40–70	*5–21*	50–60	*10–15*	50–75	1,000 lb (or *454 kg*)	100	*2.8*	200	*5.7*	50	*10*	60–75	*15–24*
Steer, enclosed bldg. on slotted floor	40–70	*5–21*	50–60	*10–15*	50–75	1,000 lb (or *454 kg*)	100	*2.1–2.3*	200	*14.2*	50	*10*	60–75	*15–24*
Dairy cow	40–70	*5–21*	50–60	*10–15*	50–75	1,000 lb (or *454 kg*)	100	*2.8*	200	*5.7*	50	*10*	60–75	*15–24*
Dairy calves	50–75	*10–24*	65	*17*		per 100 lb (*45 kg*)	10		25					
Sheep:														
Ewe	45–75	*7–24*	55	*13*	50–75		20–25	*.6–.7*	40–50	*1.1–1.4*	40–45	*5–8*	60–75	*15–24*
Feeder lamb	40–70	*5–21*	50–60	*10–15*	50–75		15	*.3*	30	*.65*	40–45	*5–8*	60–75	*15–24*
Newborn lamb	75–80	*24–27*												
Swine:														
Sow, farrowing house	60–70	*15–20*	65	*17*	60–85	Sow and litter	80	*1.4*	210	*2.8*	50	*10*	60–75	*15–24*
Newborn pigs (brooder area)	80–90	*27–32*	85	*29*	60–85									
Growing-finishing hogs	60–65	*15–17*	60	*15*	60–85	125 lb (or *57 kg*)	15	*.7*	75	*2.1*	50	*10*	60–75	*15–24*
Horse	45–75	*7–24*	55	*13*	50–75	1,000 lb (or *454 kg*)	60	*1.7*	150	*4.5*	40–45	*5–8*	60–75	*15–24*
Newborn foal	75–80	*24–27*												
Poultry:														
Layers	50–75	*10–24*	55–70	*13–20*	50–75	per bird	2		5		50	*10*	60–75	*15–24*
Broilers	85–95	*21–27*	70	*24*	50–75	per lb body weight	½		1		50	*10*	60–75	*15–24*
Turkeys	95–100 (beginning poults)	*35–38*				per lb body weight	½		1		50	*10*	60–75	*15#24*

[1]Generally 2 different ventilating systems are provided: one for winter, and an additional one for summer. Hence, as shown in Table 9–10, the winter ventilating system in a beef cow barn should be designed to provide 100 cfm (cubic feet/minute) for each 1,000-lb cow. Then, the summer system should be designed to provide an added 100 cfm, thereby providing a total of 200 cfm for summer ventilation.

[2]In practice, in many buildings, added summer ventilation is provided by opening (1) barn doors and (2) high-up hinged walls.

[3]Provide approximately ¼ the winter rate continuously for moisture removal.

NATURALLY VENTILATED BUILDINGS

Farmers and ranchers continue to have interest in naturally ventilated buildings, as well as in environmentally controlled buildings, primarily because of their significantly lower construction and operating costs. Because no attempt is made to regulate temperature, the cost of heavy insulation, tight fitting doors and windows, and a mechanical ventilation system are averted. Of course, buildings which for management reasons must be maintained at temperatures above winter levels are not suited to natural ventilations (e.g., farrowing barns or dairy stanchion barns).

Naturally ventilated buildings can be successfully used for most livestock housing; among them (1) free-stall housing for dairy cattle; (2) loafing or bedded pack barns for dairy, beef, or sheep; (3) swine-finishing buildings; and (4) calf barns.

Naturally ventilated buildings are mainly a shell to protect animals from rain and snow, and to protect the building contents (grain, hay, etc.). Winter inside temperatures will often be within 3 to 10°F of outside temperatures. Thus, such buildings are often referred to as cold confinement livestock buildings.

A naturally ventilated building has a continuous opening at the high point (normally the ridge) of the building for air exhaust and continuous openings or inlets along the long sidewalls of the building for fresh air. The size of these openings is based on rules of thumb or experience. Air entering along the sidewalls (normally under the eaves) of the building is warmed by the heat from the animals in the building and picks up moisture as it rises toward the ridge. The continuous open ridge allows this warm, moist air to escape, thus completing the air exchange process.

During the warm weather, the building should serve mainly to keep rain out and act as a sunshade. Large continuous openings in the sidewalls allow summer breezes to blow through the building.

Typical naturally ventilated buildings can be divided into two types as follows:

1. **Open front.** These buildings have one long side completely open at least one-half the height of the sidewall. The open side faces away from the direction of prevailing winter winds, normally to the south or southeast.

2. **Enclosed.** These buildings have all sides closed but provide continuous eave openings and large doors or vent panels for summer conditions. Enclosed naturally ventilated buildings offer more protection from wind and precipitation than open front buildings.

ROOFS

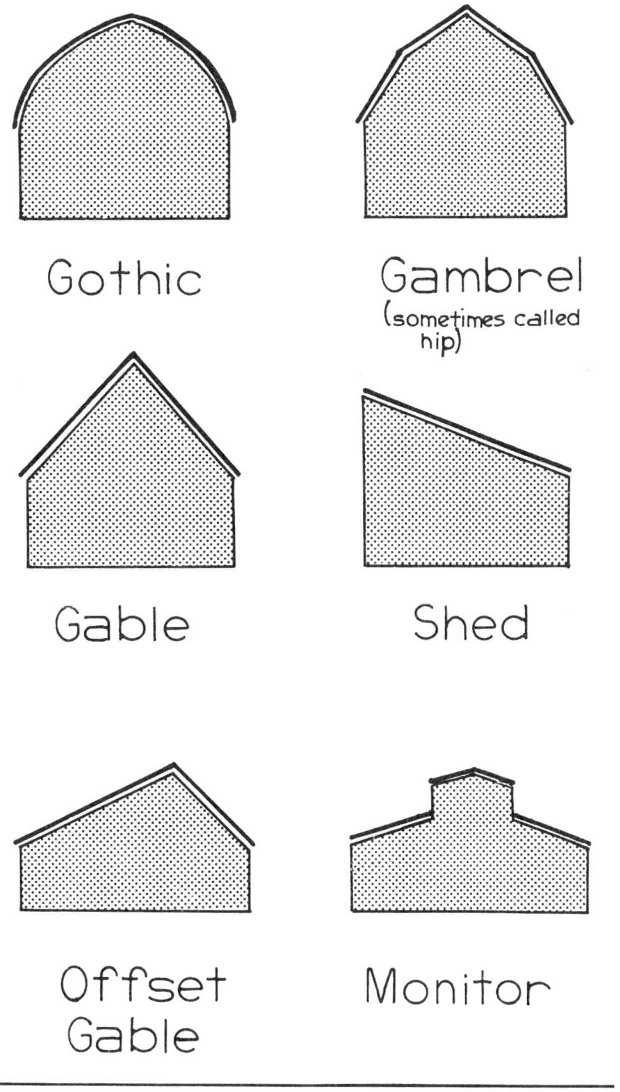

Fig. 9-2. Six roof shapes.

Six roof shapes are shown above. The three most widely used styles are shed, gable, and offset gable.

The shape, style, slope, and type of roof construction selected should be based upon the function to be served, economy of construction, strength, and appearance.

On permanent buildings, the roofing should last 15 to 20 years without replacement. Among the more durable roofings are: cedar shingles, cement-asbestos shingles, asphalt shingles, steel sheets, and aluminum.

FLOORS FOR STALLS OR STABLES

Most producers feel that a perfect flooring material has not yet been developed, as each of the existing types has certain disadvantages. Rough wooden floors furnish good traction for animals and are warm to lie upon; but they are absorbent and unsanitary. They also lack durability and often harbor rats and other rodents. Concrete floors are durable, imperious, easily drained, and sanitary; but they are rigid and without resilient qualities, are slippery when wet, and are cold to lie upon. Clay floors are noiseless and springy, and afford a firm natural footing unless wet; but they are difficult to keep clean and level.

• **Solid floor**—Solid floors for animals may be and are constructed of numerous materials—including clay, clay with a concrete border, plank, concrete, concrete with board surfacing, cork, brick, creosoted wooden blocks, cinders, or various combinations of these materials. Regardless of the type of flooring material, for a good dry bed there should be a combination of surface and subsurface drainage, together with a cover provided by a suitable absorbent litter.

• **Slotted floors**—*Slotted floors are floors with slots through which the feces and urine pass to a storage area below or nearby.* Such floors are not new; they have been used in Europe for over 200 years.

Fig. 9-3. Growing-finishing hogs on concrete slotted flooring in complete confinement. (Courtesy, American Landrace Association, Inc., Lebanon, IN)

More and more slotted floors are being used for swine and poultry in this country, and there is increased interest in using them for cattle and sheep.

The main **advantages** of slotted floors are (1) they facilitate automation and save labor; (2) they lessen or eliminate bedding; (3) they facilitate handling of manure; (4) they necessitate less space per animal; (5) they require less land; (6) they increase sanitation; (7) they lessen mud, dust, odor, and fly problems; and (8) they lessen pollution.

The chief **disadvantages** of slotted floors are (1) higher initial cost than convention solid floors, (2) less flexibility in the use of the building, (3) any spilled feed is lost through the slots (4) animals raised on slotted floors resist being driven over solid floor, and (5) environmental conditions become more critical.

Slats may be made of wood, concrete, or metal (steel or aluminum).

The design and spacing of concrete slats is important (see Table 9–9 for recommendations).

TABLE 9–9
CONCRETE SLAT DIMENSIONS[1, 2]

Class of Animal	Length	A	B	C	Space Slats	Floor Space
	(ft)	(in.)	(in.)	(in.)		
Cattle (beef and dairy): Designed for approx. 250 lb per linear foot live load	6	6	6	3	*Calves:* 1¼ in. apart.	For 1,000 lb: 20–25 sq ft.
	8	6	6	3	*Steers*, weaning to market: 1½ to 1¾ in.	
	10	6	7½	3		
Hogs and Sheep: Designed for approx. 100 lb per linear foot live load	4	4	3½	3	Space slats either a uniform ⅜ in. apart or ¾ to 1 in. apart.	*Dry ewe:* 6 sq ft.
	6	4	4	3	Spaces between ⅜ and ¾ in. are not recommended because pigs' legs may get caught. Space slats 1 in. apart behind the sow to improve cleaning.	*Ewe & lamb:* 8 sq ft.
	8	5	4½	4		*Feeder lamb:* 4 sq ft.
	10	5	5	4		*Sows:* 8 sq ft.
	12	5	5½	4		*Sow and litter:* 20–30 sq ft.
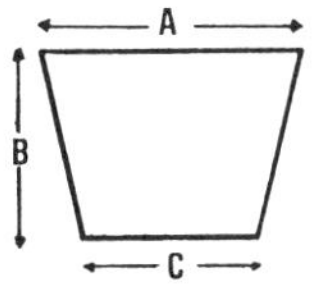 Concrete slat						*Growing-finishing:* 1. To 75 lb: 5 sq ft. 2. 75–125 lb: 6 sq ft. 3. 125 lb-market: 8 sq ft.

[1]For concrete slats, use a 7½ bag mix with ¾ in. maximum aggregate.

[2]The A, B, C dimensions given in columns 3, 4, and 5 correspond to the locations shown in the figure.

MANURE MANAGEMENT

Modern livestock buildings and equipment should be designed to handle the manure produced by the animals that they serve; and this should be done efficiently, with a minimum of labor and pollution, so as to retrieve the maximum value of the manure, and make for maximum animal sanitation and comfort.

MANURE PRODUCTION AND STORAGE

Table 9–10 shows the approximate daily manure production of each class of animal.

TABLE 9–10
APPROXIMATE DAILY MANURE PRODUCTION, WITHOUT BEDDING[1]

Animal	Cu Ft/Day Solids and Liquids[2]	Gallons/ Day[3]
1,000-lb cow	1½	11
1,000-lb steer	1	7½
10 head of sheep	½	4
10 head of hogs:		
50 lb	⅔	5
100 lb	1⅓	10
150 lb	2¼	17
200 lb	2¾	20½
250 lb	3½	26
1,000-lb horse	¾	5½

[1]Adapted by the author from *Michigan State University Circular Bull. 231.*

[2]There are about 34 cu ft in a ton of manure.

[3]One cu ft = 7½ gal.

Manure may be stored in a separate tank or it may be left to accumulate in a pit under slotted floors.

Storage capacity can be computed as follows:

Storage capacity = no. of animals × daily manure production × desired storage time in days + extra water.

Example: 80 cows (1,000 lb each) × 1½ cu ft × 120 days = 14,400 cu ft; 7½ gal × 14,400 cu ft = 108,000-gal capacity.

Extra water must often be added to liquify the wastes. Thus, if the manure is to be pumped, ⅕ to ⅗ of the storage volume may be needed for the extra water. For irrigation, there should be about 95% water and 5% manure. Water should be kept to a minimum if the manure is to be field spread with a tank wagon.

Generally, 3 to 6 months' storage capacity is desirable.

MANURE GASES

When stored inside a building, gases from liquid wastes create a hazard and undesirable odors. Most (95% or more) of the gas produced by manure decomposition is methane, ammonia, hydrogen sulfide, and carbon dioxide.

Animals and people can be killed (asphyxiated) because methane and carbon dioxide displace oxygen.

Most gas problems occur when manure is agitated or when ventilation fans fail.

No one should enter a storage tank, unless (1) the space over the wastes is first ventilated with a fan, (2) another person is standing by to give assistance if needed, and (3) wearing self-contained breathing equipment—the kind used for fire fighting or scuba diving.

It is important that maximum building ventilation be provided when agitating or pumping wastes from a pit. Also, an alarm system (loud bell) to warn of power failures in tightly enclosed buildings is important, because there can be a rapid buildup of gases when forced ventilation ceases.

MANURE HANDLING

Modern handling of manure involves maximum automation and a minimum loss of nutrients. Among the methods being used, with varying degrees of success, are slotted floors emptying or pumping into irrigation systems; storage vats; spreaders (including those designed to handle liquids alone or liquids and solids together); dehydration; power loaders; conveyers; industrial-type vacuums; lagoons; and oxidation ditches. Actually, there is no one best manure management system for all situations; rather, it is a matter of design and using that system which will be most practical for a particular set of conditions.

REQUISITES OF LIVESTOCK EQUIPMENT

Generally speaking, *livestock equipment refers to structures other than barns or shelters used in the care and management of animals. Much of this equipment is portable.*

The size and design of livestock equipment may differ; that is, not all hay racks or self-feeders, for example, are the same. Yet there are certain fundamentals of livestock equipment that are similar regardless of the kind of equipment, the design, or the size. These requisites are:

1. Accessible.
2. Conserve manure.
3. Dependable.
4. Durable.
5. Low annual cost and upkeep.
6. Movable.
7. Reduce labor.
8. Save feed.
9. Simple construction.
10. Utility value.

AUTOMATION

Automation is a coined word meaning the mechanical handling of materials. Farmers/ranchers automate to lessen labor and cut costs.

Modern equipment has practically eliminated the pitchfork, bucket, and basket. Such chores as feeding, watering, bedding, and barn cleaning have been, or are being, mechanized. Operators are using more self-loading trucks and trailers, self-feeders, feed bunk augers and belts, laborsaving grain- and forage-processing equipment (producing pellets, cubes, or wafers, etc.), automatic waterers, and manure disposal units. Automation of the livestock industry will increase.

SCALES

Scales are a valuable piece of equipment for the modern stock farm or ranch; for they make it possible to determine weights of animals on production-testing studies, to secure the accurate rate of gains of animals being finished, to sell animals on the farm or ranch on a weight basis, and to buy and sell feed on a weight basis. For greatest usefulness, scales should be so arranged that a pen may be set up quickly when weighing mature animals or may be removed when weighing feed.

A convenient place for farm scales is in the farm court, next to the corrals or feedlot. In this location, the scale is convenient for weighing livestock or loads of feed and supplies.

HOW TO DETERMINE THE SIZE BARN TO BUILD

The length and depth of the barn (its size) may be varied according to needs. The size barn to build for any given farm or ranch may be determined as follows:

1. Estimate the number of and kind of animals to be quartered and compute the total animal space requirements from Tables 9–1, 9–2, 9–3, and 9–4.
2. Compute the yearly feed requirements of the animals to be fed and quartered by referring to Section 4, Feeding.
3. Estimate the farm production of feeds and bedding to be stored in the barn. In most operations this should coincide closely to the total animal requirements (point No. 2), but there may be circumstances where the feed and bedding storage requirements are more or less than the animal feed requirements.
4. Estimate the total tonnage of feed and bedding to be stored by correlating the animal feed needs and the farm or ranch production (correlate the results of points 2 and 3). Then determine the total storage space requirements for feed and bedding from Table 9–7, p. 366.
5. Determine the size of barn to build from the total animal space requirements and the total yearly feed and bedding storage requirements (points 1 and 4).

HOW TO DETERMINE THE SIZE SILO TO BUILD

The size of silo to build should be determined by needs. With tower type and pit silos, this means (1) that the diameter should be determined by quantity of silage to be fed daily, and

(2) that the height (depth in a pit silo) should be determined by the length of the silage feeding period. Similar consideration should be accorded with trench silos.

SIZE OF TOWER SILO

If the diameter is too great, the silage will be exposed too long before it is fed; and, unless a quantity is thrown away each day, spoiled silage will be fed.

The minimum recommended rate of removal of silage varies with the temperature. In most sections of the United States, it is desirable that a minimum of 1½ in. of silage be removed from tower silos daily during the winter feeding period, with the quantity increased to a minimum of 3 in. when summer feeding is practiced. Of course, the total daily silage consumption on any given farm or ranch will be determined by (1) the class and size of animals, (2) the number of animals, and (3) the rate of silage feeding. Some suggestions on how much silage to feed cattle, sheep, and horses are found in the Feeding Guides given for each of the species in Section 4, Feeding.

Silo height should be determined primarily by the length of the intended feeding period. In general, however, the height should not be less than twice, nor more than three and one-half times the diameter. The greater the depth, the greater the unit capacity. Extreme height is to be avoided because (1) of the excessive power required to elevate the cut silage material, and (2) of the heavier construction material required. Also, it is noteworthy that, with silos of the larger diameters, more labor is required in carrying the silage to the silo door for removal.

Table 9-11 may be used as a guide in computing the proper diameter of tower silos for any given farm or ranch.

TABLE 9-11
MAXIMUM DIAMETER OF TOWER SILO TO BUILD IF SILAGE IS TO BE KEPT FRESH

Inches of Silage Removed Daily	Total Silage Removed Daily with an Inside Silo Diameter of: 10 Feet	12 Feet	14 Feet	16 Feet	18 Feet	20 Feet
	(lb)	(lb)	(lb)	(lb)	(lb)	(lb)
Summer: 3 in. daily will remove[1]	786	1,312	1,539	2,010	2,545	3,142
Winter: 1½ in. daily will remove[1]	393	656	770	1,005	1,272	1,571

[1]The pounds listed in each of the columns to the right are approximations based on an average constant weight of 40 lb of silage per cu ft.

Fig. 9-4 shows capacities of tower silos of different heights and diameters. It is based on well-eared corn silage harvested in the early dent stage, cut in ¼ in. lengths, well-tramped when filled, and with the silo refilled once after settling for a day.

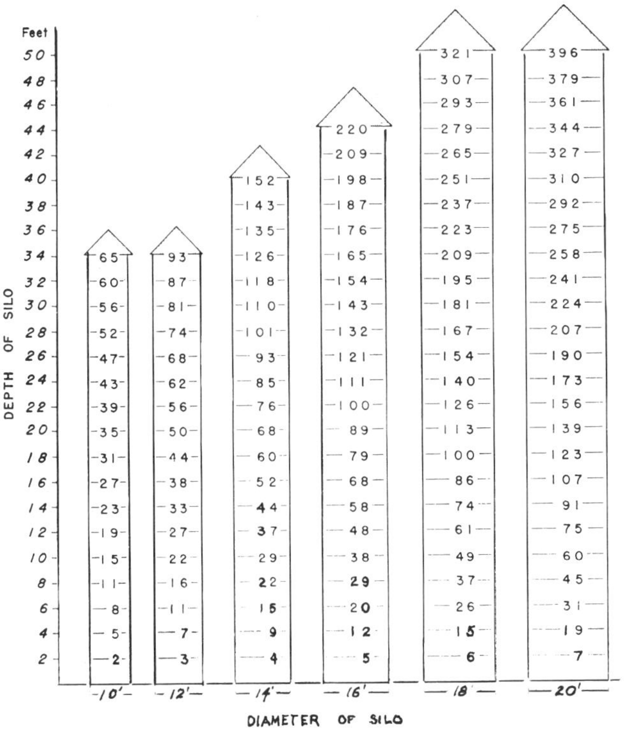

Fig. 9-4. Capacity in tons of settled corn silage in tower silos of varying sizes (based on data reported in USDA Circ. 603). (Drawing by R. F. Johnson)

Fig. 9-4 can be adapted for corn silage of different stages of maturity and grain content, and for other kinds of silage, by applying the rules of thumb given in Table 9-12.

TABLE 9-12
EFFECT OF KIND OF SILAGE ON WEIGHT

Kind of Silage	Changes to Be Made in the Number of Tons Shown in Fig. 9-4
1. For corn silage ensiled when less mature than usual	Add 5 to 10%
2. For corn ensiled when dry or overripe	Deduct 5 to 10%
3. For corn very rich in grain	Add 5 to 10%
4. For corn with very little grain	Deduct 5 to 10%
5. For sorghum silage	Use the same weights as used for corn silage of comparable grain and maturity.
6. For sunflower silage.	Add 5 to 10%
7. For grass silage.	Add 10 to 15%[1]

[1]For this reason, a stronger structure is necessary where grass silage is stored.

The following example will serve to illustrate how to determine the size tower silo to build:

> Over a period of years, a farmer plans to winter 34 head of 425-lb stocker calves on a ration of corn silage and protein supplement. There is a 240-day

wintering period. No increase in the herd is planned. What size tower silo should be built?

The answer is obtained as follows:

1. First, here are the silage requirements:

a. Section 4, Table 4–12, pp 160–161, indicates that 425 lb stocker calves on a high corn silage ration should receive up to 30 lb of silage per head per day.

b. 34 × 30 = 1,020 lb of silage required daily for the 30 calves.

c. 1,020 × 240 = 244,800 lb or 122.4 tons of silage required for the 240-day wintering period for the 34 calves.

2. Next, here is the size silo to build:

a. Table 9–11 shows that in order to remove 1,005 lb of silage daily (which is only slightly less than the 1,020 lb needed daily), with 1½ in. removed from the top of the silo each day, the diameter of the silo should not be greater than 16 ft.

b. Fig. 9–4 can now be used as a guide in determining both the proper height (or depth) and diameter of the silo. Fig. 9–4 shows that a silo 16 ft in diameter and 27 ft high will hold 127 tons of silage, which would allow for 4.6 tons spoilage in excess of the required 122.4 tons. However, the height of a silo should not be less than twice the diameter. It appears best, therefore, to plan on a 14-ft diameter silo. As noted in Fig. 9–4, 34 ft of settled silage in a 14-ft diameter silo will provide 126 tons of silage, which would allow for 3.6 tons spoilage in excess of the required 122.4 tons. To allow for settling, an additional 4 to 6 ft should be added to the height, thus making a 38- to 40-ft height.

c. The size silo to build to meet the needs outlined in this example, therefore, is one that is 14 ft in diameter and 38 to 40 ft high.

SIZE OF TRENCH SILO

As in an upright silo, the cross-sectional area of a trench silo should be determined by the quantity of silage to be fed daily. The length is determined by the number of days of the silage feeding period. The only difference is that generally greater allowance for spoilage is made in the case of a trench silo, though this factor varies rather widely.

Under most conditions, it is recommended that a minimum 4-in. slice be fed daily from the face (from the top to the bottom) of a trench silo during the winter months, with a somewhat thicker slice preferable during the summer months.

The dimensions, areas, and capacities given in Table 9–13 are based on the assumption that the silage weighs 35 lb per cubic foot,[1] which is an average figure for corn or sorghum silage. Thus, a trench silo 8 ft deep, 6 ft wide at the bottom, and 10 ft wide at the top has a cross-sectional area of 64 ft. This size silo will hold 747 lb of silage for each 4-in. slice, or 2,240 lb of silage for each 1-ft slice, or 112 tons in a trench 100 ft long.

For illustrative purposes, let us use the same example and silage requirements as were used in the section on "Size of Tower Silo," but this time determine the size of trench silo to build rather than the size of tower silo. Briefly, the requirements are for 1,020 lb of silage daily for a 240-day wintering period.

[1]Because the silage in trench silos is generally not so deep and well-packed as the silage in tower silos, an average figure of 35 lb per cubic foot is used herein for trench silos and 40 lb for upright silos. With all types of silos—including aboveground and belowground types—the weight of a cubic foot of silage varies with the kind and maturity of material, moisture content, length of cut, rate of filling, and depth of the silo. Corn silage harvested when about 74% of the grain has passed the milk stage and containing approximately 70% moisture is considered average silage. Volume for volume, sorghum silage weighs about the same as corn silage. Grass or grass-legume silage is 10 to 15% heavier than corn silage.

TABLE 9–13
DIMENSIONS, CROSS-SECTION AREA OF TRENCH SILO, AND WEIGHT OF SILAGE IN 4-INCH SLICE AND PER LINEAL FOOT[1]

Side Slope per Foot of Depth (inches)	Depth	Bottom Width	Top Width		Cross-Sectional Area	Weight of Silage 4-Inch Slice	Weight of Silage 1-Foot Slice
	(ft)	(ft)	(ft)	(in.)	(sq ft)	(lb)	(lb)
3	4	5	7	0	24	280	840
4	4	6	8	8	29	338	1,015
5	4	7	10	4	33	385	1,155
3	6	6	9	0	45	525	1,575
4	6	7	11	0	54	630	1,890
5	6	8	13	0	63	735	2,205
3	8	6	10	0	64	747	2,240
4	8	7	12	4	77	898	2,695
5	8	8	14	8	91	1,062	3,185
3	10	6	11	0	85	992	2,975
4	10	8	14	8	113	1,318	3,955
5	10	10	18	4	142	1,657	4,970

[1]*Silos, Types and Construction*, USDA, Farmers' Bulletin No. 1280, p. 55.

As noted in Table 9–13, one day's feed or 1,020 lb of silage (1,062 lb to be exact) can be obtained in each 4-in. slice of a trench silo 8 ft wide at the bottom, 14 ft 8 in. wide at the top, and 8 ft deep; or a 91 sq ft cross-sectional area. The cross-sectional area should not be larger than this if a 4-in. slice is to be removed daily in order to alleviate spoilage.

In order to obtain a 240-day feed supply, the filled trench must be 80 ft long (240 by ⅓—the ⅓ representing ⅓ ft or 4 in.).

The size trench silo to build to meet the specified needs, therefore, is one that is 8 ft wide at the bottom, 14 ft 8 in. wide at the top, 8 ft deep, and 80 ft long. In order to take care of spoilage and to provide a measure of safety, it is recommended that the actual length be from 85 to 90 ft.

About 8 ft for a trench silo is the economical depth from the standpoint of cost and feeding. Of course, in filling it is desirable to pile silage 3 ft higher over the center of the trench and round it off. This provides for settlement.

BEEF CATTLE BUILDINGS AND EQUIPMENT

Fig. 9–5. Automated cattle feeding, with the feed conveyed by auger. (Courtesy, Ralston Purina Co., St Louis, MO)

The economical production of beef cattle in most sections of the United States depends largely upon the investment in practical, durable, and convenient buildings and equipment, as well as upon the care, feeding, and management of the herd. As would be expected in a country so large and diverse as the United States, there are wide differences in the system of beef production. In a broad general way, a major difference in management exists between the farm herd method and the range cattle method. In addition, further management differences exist within each according to whether the enterprise is commercial or purebred, whether it is a cow-calf proposition or devoted to one of the many methods of growing stockers and feeders or cattle for finishing, or whether it is a combination of two or more of these systems of beef production. Climatic differences also vary, all the way from nearly year-round grazing in the deep South to a long winter-feeding period in the northern part of the United States. Then, too, the size of the herd may vary all the way from a few animals up to an operation involving many thousands of head. Finally, there is the matter of availability of materials and labor and individual preferences.

Beef cattle are not so sensitive to extremes in temperature—heat and cold—as are dairy cattle or swine. In fact, mature beef animals will withstand extremely cold weather if kept dry.

Beef cattle shelters are of two kinds, natural and artificial. The former includes hills and valleys, timber, and other natural windbreaks. The artificial shelters include those structures (solid fences, stacks, barns, and sheds) designed to protect cattle against the elements—heat, cold, wind, rains, and snows. It is with beef cattle barns and sheds that this discussion will deal.

In addition to protecting animals during severe cold and stormy weather and at winter calving time, these structures should (1) provide a reasonably dry bed for the animals, (2) simplify feeding and management, (3) provide storage for feed and bedding when necessary, and (4) protect young calves.

Although the discussion describes beef cattle barns and sheds, it must be recognized that on small farms, which have a limited number of beef cattle, the animals are usually housed in a general-purpose barn or shed or in extensions of other barns rather than in separate and specially designed beef cattle structures.

BEEF CATTLE BARNS

Barns are more substantial structures than sheds and provide more complete protection for stock in the colder areas. In addition to housing the animals, such structures usually provide adequate facilities for all of the roughage and bedding needed during the winter season and for a considerable proportion of the concentrates. Stalls, pens, and storerooms may also be included—additions which are especially important where a breeding herd is to be served. In general, beef cattle barns effect a saving of labor and time in feeding, and save feed.

The type of barn is determined by the kind of stock and the method of handling and management. Fig. 9–6 shows a pole barn.

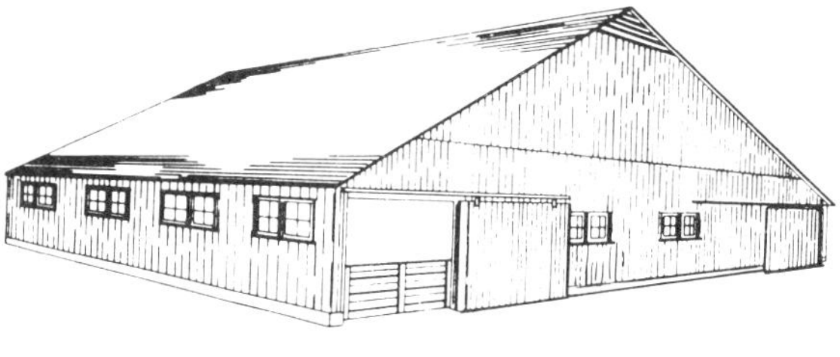

Fig. 9–6. Enclosed beef barn, pole type, construction. From *Handbook of Building Plans,* p. 137. (Courtesy, Midwest Plan Service, Iowa State University, Ames)

BEEF CATTLE SHEDS

Sheds are the most versatile and widely used beef cattle shelters throughout the United States. They are used for cattle in the feedlot, as range shelter for dry cows with calves, and for housing young stock. They usually open to the south or east, preferably opposite to the direction of the prevailing winds and toward the sun. They are enclosed on the ends and sides. Sometimes the front is partially closed, and in severe weather drop-doors may be used. The latter arrangement is especially desirable when the ceiling height is sufficient to accommodate a power manure loader. Fig. 9–7 shows a traditional beef cattle shed.

Fig. 9–7. An open beef cattle shed. Sheds are the most versatile and widely used beef cattle shelters throughout the United States. (Courtesy, Oklahoma State University, Stillwater)

Fig. 9–8 shows a large beef cattle headquarters with open sheds, feed storage, and outside feeding.

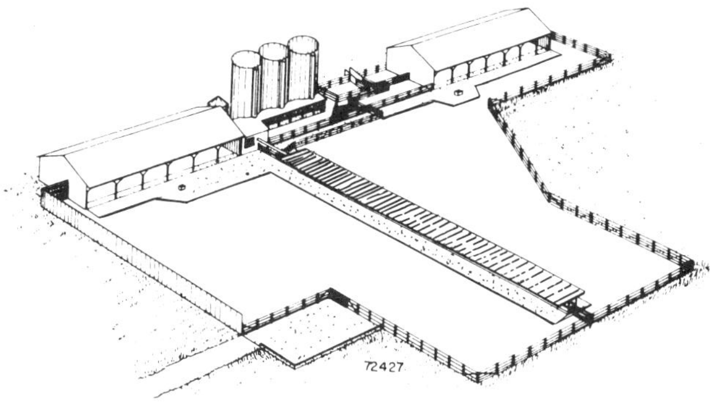

Fig. 9–8. Beef cattle layout, with open sheds and outside feeding. From *Handbook of Building Plans,* p. 133. (Courtesy, Midwest Plan Service, Iowa State University, Ames)

FACILITIES FOR FINISHING CATTLE

Cattle-feeding facilities and equipment are a manufacturing plant, wherein animate objects (cattle) convert feed into beef. Hence, they merit the same level of competence in planning and design as any other sophisticated manufacturing plant.

Some preliminary feedlot planning considerations follow:

1. Decide on the type of facilities: (a) feedlot (open pen), (b) cold confinement, or (c) warm confinement.

2. Decide on the number of cattle and the feed and storage requirements with provision for expansion.

3. Determine the justifiable investment in cattle-feeding facilities.

4. Select the facilities, equipment, and arrangement that best fit the management program you have chosen; for example, (a) fenceline bunks and a central feed-processing plant, (b) upright storage with distributors and bunks, or (c) self-feeders.

5. Design a system that is practical, laborsaving, environmentally suitable for economical gains of cattle, and attractive.

An open lot without shelter is the cheapest type of feedlot construction of any. Housing increases costs, and the more elaborate the housing, the greater the cost.

• **Open pen feedlot**—An open pen feedlot is, as indicated by the name, a lot in which the cattle are in the open—usually it is without shelter (except for such natural protection as may be afforded by trees, hills, or wind fences, or perhaps a roof over the feed bunks or shades).

• **Confinement feeding; slotted floors**—Currently, there is much interest in cattle confinement feeding and slotted floors. The main deterrent is cost; construction costs vary with type of structure and may range up to $300 per steer space.

Confinement cattle feeding refers to feeding in limited quarters, generally 20 to 25 sq ft per yearling animal, which is about 1/8 the space normally allotted to a yearling in an unsurfaced lot and 1/3 that of a paved lot. The confinement is usually under roof on slotted floors.

Slotted floors are floors with slots through which the feces and urine pass to a storage area immediately below or nearby.

Interest in confinement feeding and slotted floors was ushered in for the purposes of (1) automating and saving labor; (2) cutting down on bedding and facilitating manure handling; (3) lessening mud, dust, odor, and fly problems; (4) increasing gains and saving feed; (5) lessening land requirements; and (6) lessening pollution.

When planning confinement feeding facilities, a choice must be made between cold confinement and warm confinement, along with design requirements.

Cold confinement[2] refers to a more or less open shed for confining cattle; hence, winter temperatures therein are within a few degrees of outdoor temperatures. Open sheds should be faced away from the direction of the prevailing winds. Additionally, doors or other openings in the closed walls should be provided for summer ventilation.

Warm confinement[3] refers to a confinement building for cattle which is sufficiently insulated and ventilated to maintain inside winter conditions above 35°F in severe weather, and in the range of 50 to 60°F most of the time.

Based on information and experiences presently available, the following figures may be used as guides relative to design and space:

1. **Floor space.** Allot 15 to 30 sq ft per animal exclusive of the bunk and alley, with an average of 20 to 25 sq ft for a 1,000-lb animal
2. **Animals per pen.** 25 to 100 head per pen, with 25 to 30 being most common.
3. **Bunk.** Allow 6 to 18 in. of linear bunk space per animal, with the amount of feeding space determined by frequency of feeding and size of animal.
4. **Waterers.** Locate 1 waterer per 25 head at the back (opposite feed bunk) of each pen, preferably in the center.
5. **Slats.** Reinforced concrete, steel, or aluminum may be used. Most concrete slats are 5 to 6 in. wide across the top, 6 to 7 in. deep, tapered to 3 to 4 in. wide at the bottom, and placed so as to provide a slot width of 1½ to 1¾ in.
6. **Manure production and storage.** Manure production will vary with size of animal and *kind of feed,* but it will be approximately as follows:

Animal	Cu Ft/Day Solids and Liquid	% Water	Gallons/Day
1,000-lb steer	1–1½	80–90	7½–10¾

Here is how to determine how much manure will need to be stored:

Storage capacity = Number of animals × daily manure production × desired storage time (days) + extra water.

A rule of thumb is that when the pit occupies the entire area beneath the cattle, it will fill at a rate of 8 to 10 in. per month.

BEEF CATTLE EQUIPMENT

There is hardly any limit to the number of different items of beef cattle equipment, and the design of each. Suitable equipment saves feed and labor, conserves manure, and makes for increased production. Detailed plans and specifications for such equipment can usually be obtained through the local county agricultural agent, FFA instructor, lumber dealer, or through writing to the college of agriculture in the state.

[2]The terms *cold confinement* and *warm confinement* refer to winter conditions. Without mechanical cooling, both systems are warm during the summer months.

[3]Ibid.

DAIRY CATTLE BUILDINGS AND EQUIPMENT

Fig. 9-9. A dairy cattle farm in Rhode Island. (Courtesy, The American Jersey Cattle Club, Columbus, OH)

Modern dairy cattle buildings and equipment should be designed to facilitate (1) freedom and individual comfort of cows, (2) automation and laborsaving, (3) herd health and sanitation, (4) bedding conservation, and (5) manure disposal.

Economical and successful dairy production depends largely upon the investment in practical and convenient buildings and equipment, as well as upon the care, feeding, and management of the herd. As would be expected in a country so wide and diverse as the United States, there are wide differences in the systems and facilities of dairy production. A major difference exists between (1) the family-owned and operated farm herd, which usually relies on pastures in season and produces its own winter roughage; and (2) the large, commercial type, drylot operation. In addition, facilities differ according to whether the operation is devoted to producing milk or raising heifer replacements, or a combination of both. Further facility differences exist between a strictly commercial enterprise and a purebred herd. Climate also makes for differences; from the animal being out in the open much of the time in the South and in California, to the need for warmth and a long winter feeding period in the North. Finally, there are differences due to meeting the health regulations of the particular area and market, the availability of materials and labor, and individual preferences.

Two basic facility systems are used for lactating cows: (1) stall barns, and (2) loose housing. Thus, the starting point in designing lactating cow facilities is to select the system.

Many arguments are heard relative to the merits of each system. Experiments show that, with proper care and management, similar milk production can be achieved with either system. The stall barn allows the cows to be displayed to greater advantage, which is particularly important in the purebred herd; and, generally speaking, the cows are observed more frequently than in loose housing. On the other hand, loose housing requires less labor, saves bedding, results in less udder and leg injury, and usually costs less to construct because such items as expensive concrete work, stanchions, water cups, and ventilation are omitted.

Fig. 9-10 shows a layout for group housing. Fig 9-11 shows a free-stall arrangement for four different groups.

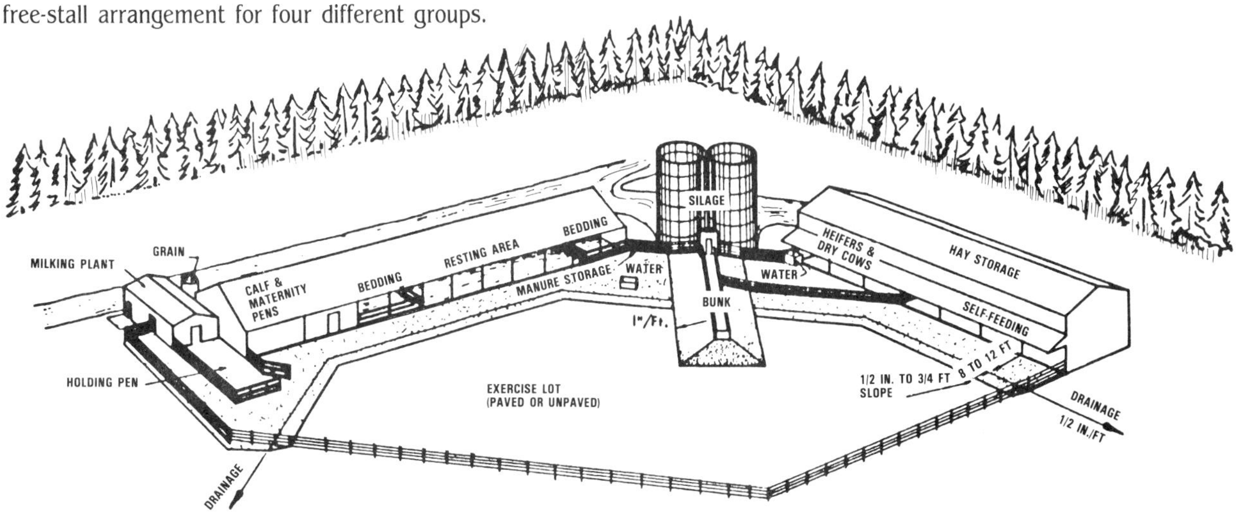

Fig. 9-10. Layout of group housing. From *Dairy Equipment Plans and Housing Needs.* (Courtesy, Midwest Plan Service, Iowa State University, Ames)

Fig. 9-11. *Right:* A 200 free-stall dairy barn (4 groups). Scrape manure to the center of the building and move it to storage. Space is included for a double-8 herringbone milking parlor. From *Dairy Housing and Equipment Handbook,* 1985, p. 4.8. (Courtesy, Midwest Plan Service, Iowa State University, Ames)

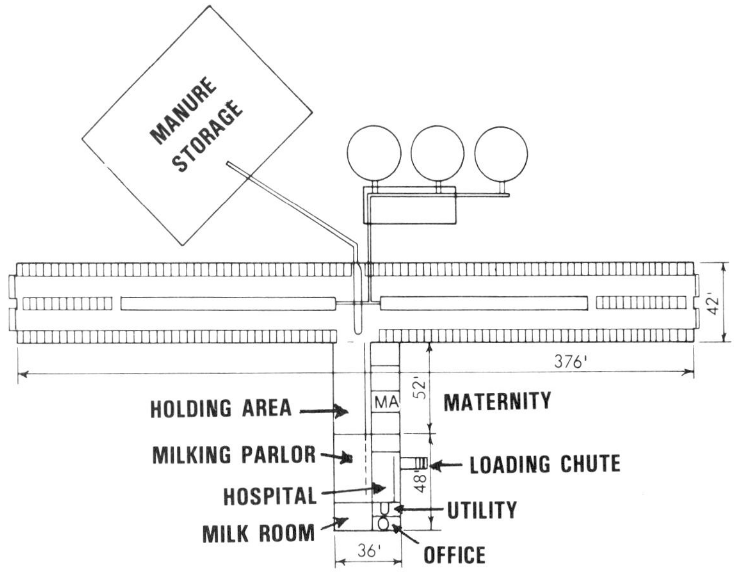

STALL BARNS

Fig. 9-12. Stanchioned cows in stall barn. Note automatic watering cups. (Courtesy, Holstein Friesian Assn. of America, Brattleboro, VT)

The stall barn consists of one or two rows of cows that are usually confined to stanchions (see Fig. 9–13), although tie or comfort-type stalls may be used. For the most part, concrete floors are used; and sometimes they are covered with rubber mats. The floor slopes into a gutter, which is usually 16 in. wide and 8 in. deep on the stall side and 6 in. deep on the alley side.

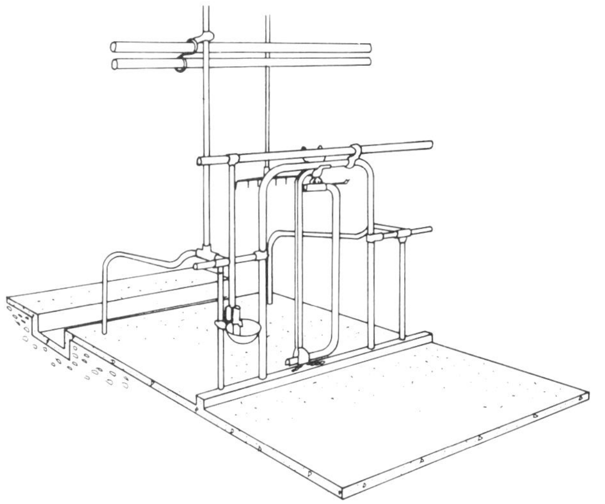

Fig. 9–13. Stall barn stanchion stall. From *Dairy Housing and Equipment Handbook*, p. 4. (Courtesy, Midwest Plan Service, Iowa State University, Ames)

Table 9–14 gives suggested stanchion-stall dimensions. Tie or comfort stalls should be 2 in. wider and 4 in. longer than the measurements given.

TABLE 9–14
SUGGESTED STANCHION-STALL DIMENSIONS

Size of Cow	Stall Width	Stall Length
Small cow (Jersey)	3′ 6″ to 3′ 10″	4′ 6″ to 4′ 10″
Medium cow (Guernsey, Milking Shorthorn, Red Poll)	4′ 0″ to 4′ 3″	5′ 0″ to 5′ 4″
Large cow (Holstein, Brown Swiss)	4′ 4″ to 4′ 6″	5′ 4″ to 5′ 8″

LOOSE HOUSING

Loose housing is that system in which the herd is handled on a group basis except at milking time.

Currently, there is considerable interest in this system, primarily because it requires less labor than the stall system.

The following functional areas are involved in most loose housing systems:

1. **Resting or loafing area.** There are two rather distinct types of resting or loafing arrangements:

   a. **Group housing.** In this system, the cows rest in a common, bedded area on a manure pack. About 60 to 70 sq ft of bedded area should be provided per cow. The bedding material should be deep to begin with; then each day the droppings should be removed and 10 to 12 lb of fresh bedding per cow added.

   The reported advantages of group housing over individual stall housing is that the cost per cow is less.

   b. **Freestall housing.** This is a modification of the group housing system. It consists of individual open stalls, which are bedded. For small breeds, use 4 × 7-ft stalls; for large breeds, use 4 × 7½-ft stalls.

   The reported advantage of freestall housing over grouping housing are: bedding costs may be reduced by as much as 75%; less labor is required to bed the cows; the cows are cleaner, with the result that less cow-washing time is required; and less space per cow is necessary.

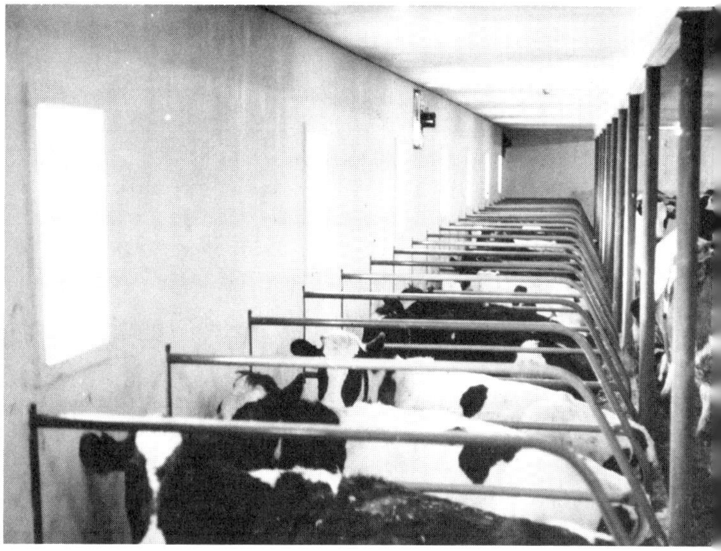

Fig. 9–14. Freestalls. (Courtesy, Babson Bros., Naperville, IL)

2. **Calf and maternity quarters.** The author recommends the use of individual calf stalls for the first 6 weeks, followed by the use of group pens thereafter. Generally, one individual pen is provided for each 10 cows. Calf pens should be dry and well ventilated.

   Also, maternity and isolation stalls should be provided. It is recommended that 100 to 120 sq ft be allowed per stall, with one such stall per 20 milk cows in the herd.

3. **Feeding areas.** Separate feeding and bedded areas are preferred. Also, more and more large dairies are emulating cattle fattening operations as a labor-saving device, by feeding complete rations in fenceline feeders, with the herd separated into two or more production groups.

4. **Exercise yard.** This is usually a paved area which serves as a place for exercise. Approximately 100 sq ft per cow should be allowed. It should be designed so that it can be cleaned daily by scraping.

5. **Holding areas.** This area is for the purpose of confining cows in preparation for milking. It should be paved, easy to clean daily, and funnelled to the parlor. About 15 to 20 sq ft per cow should be allowed.

6. **Milking parlors.** Loose housing systems lend themselves particularly well to the use of milking parlors. Although there is a wide range in choices in milking parlors, the three most common ones are:

a. **Tandem-type.** Where the cows stand single file in line and broadside to the operator's pit.

b. **Herringbone.** Where the cows stand in groups at an angle to the operator's pit—like herringbone (see Fig. 9–15).

c. **Rotary (or Carousel).** Where the parlor, which may incorporate either tandem or herringbone principle, rotates around the operator.

Fig. 9–15 shows diagrams of different milking parlor arrangements, whereas Table 9–15 shows the capacities of milking parlors.

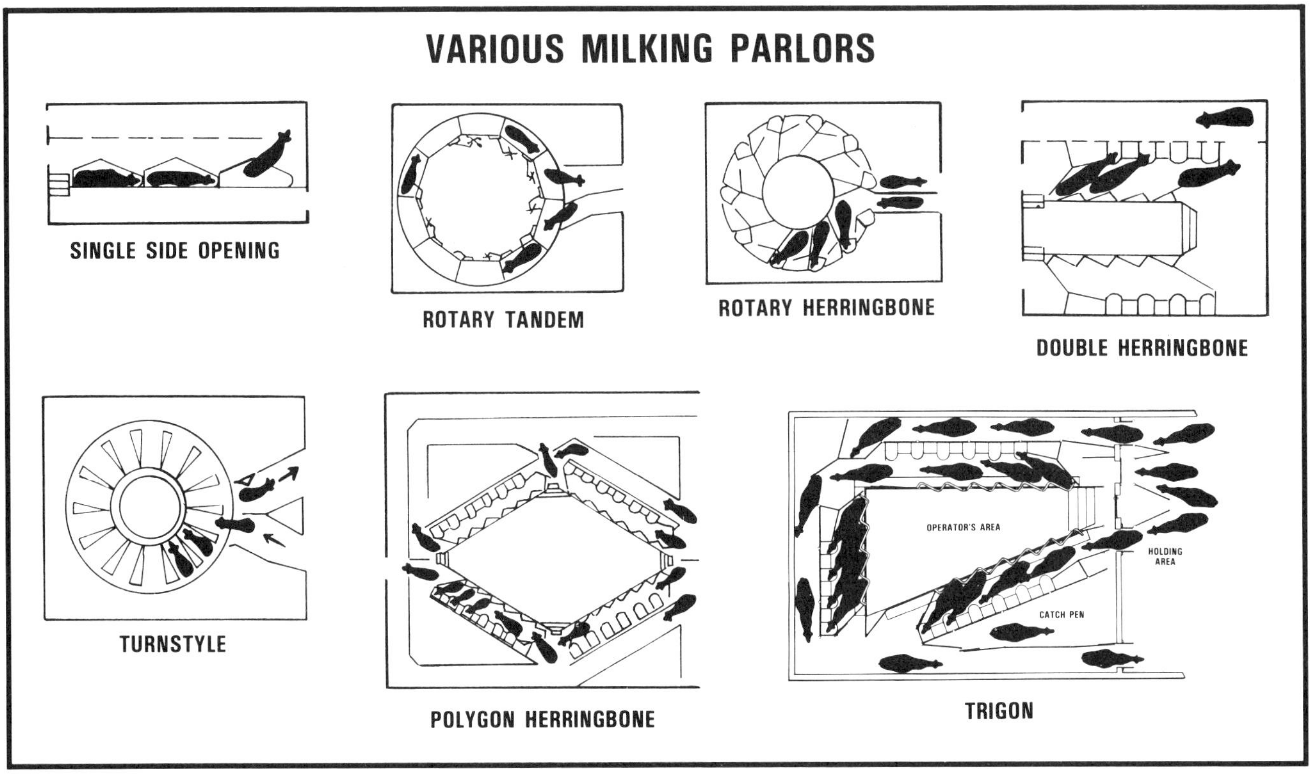

Fig. 9–15. Diagrams of various milking parlors.

TABLE 9–15
MILKING PARLOR CAPACITIES[1]

Parlor Size	Cows/Hr	Cows/Work Hr	Milk Lb/Work Hr	Parlor Size	Cows/Hr	Cows/Work Hr	Milk Lb/Work Hr
Double 3 HB	34	29	440	Double 2 SO	41	41	882
Double 4 HB	41	39	870	Double 3 SO	51	47	1,031
Double 6 HB	58	30	640	Rotary (turnstyle)	96	48	—
Double 8 HB	71	35	720	Polygon	134	67	1,566
Double 10 HB	92	44	1,040				

[1]From *Dairy Housing and Equipment Handbook*, p. 22, published by Midwest Plan Service, Iowa State University, Ames. Data from a field survey.
HB = herringbone parlor. SO = side opening parlor.

FACILITIES AND EQUIPMENT FOR DRY AND LACTATING COWS

Pertinent information relative to lots, housing, feed, and water facilities of dry and lactating cows follows:

- **Lot and Housing Facilities**—

1. Provide the following minimum lot space per head:

Kind of Lot	Sq Ft/Animal
All paved	100
Paved and dirt	150
Dirt	200

2. Slope paved lots ¼ to ½ in./ft, and slope dirt lots ½ in. or more per foot.

3. Pave (rough finish) a 15- to 20-ft area around waterers, feed bunks and racks, and entrances to sheds.

4. Open sheds may be used under a loose housing system. Free stall housing saves bedding, keeps the cows cleaner, and saves labor. For small breeds, use 7 × 4-ft stalls; for large breeds, 7½ × 4-ft stalls.

5. Keep the temperature in stall barns 40°F or more; and in milking parlors 50 to 60°F during the winter.

6. Provide for proper ventilation—changing of air—in cold climates and during the winter months, with care taken to avoid direct drafts and coldness. A fan system installed according to manufacturer's directions is best. For each 1,000 lb of animal weight in a well-insulated barn, the fan should be capable of removing a minimum of 100 cu ft per minute (cfm).

7. Bed all cows except in dry climates.

8. Provide for efficient manure disposal, selecting the system best adapted for the particular dairy and area.

• **Feed and Water Facilities**—

1. Provide 24 to 30 in. per head of manger space for roughage feeding.

2. Make bunks and roughage racks 24 to 30 in. wide when cattle are fed from one side; 36 in. wide when feeding from both sides.

3. Provide adequate water space: (a) 1 linear foot of open tank per 8 to 10 head, or (b) one automatic bowl per 15 head.

4. Keep water temperatures within the range of 35 to 80°F; warm it to 50°F in the winter.

MILKING EQUIPMENT

Basically, there are two types of milking equipment: (1) the bucket system, and (2) the pipeline system.

In the bucket system, the milk is received directly into a nearby vacuumized portable bucket which may be either of two types: (1) floor type, or (2) suspended type.

Milk may be handled by either of two systems: the can system, or the bulk system. Until 1939, when the bulk system was first introduced in California, all milk was handled in cans. Today, the trend is to more and more bulk tanks. Although the initial cost is greater than where cans are used, the greater returns over a period of time, justify the expense. Further, many dairies are facing the situation of being forced into going to bulk tanks if they are to retain a market outlet.

Generally speaking, the following advantages accrue to the use of bulk tanks, in comparison with cans: (1) a saving in labor, (2) less loss in milk, (3) alleviating 10-gallon cans, (4) higher butterfat tests (due to butterfat being left on lids of cans), (5) a saving in hauling costs, and (6) a premium paid by the plant.

FACILITIES AND EQUIPMENT FOR DAIRY CALVES

Pertinent information relative to facilities and equipment for dairy calves follows:

Fig. 9–16. Calf feeding facilities and individual pens at Dairy Research Unit, Ohio Agricultural and Development Center, Wooster.

• **Housing**—

1. House calves separately (in individual pens or tie stalls)—from birth until at least one week after milk or milk substitute is discontinued. Thereafter, they may be raised in groups, with (a) a maximum of 10 head per group (preferably 6 to 8 per group), and (b) a maximum age difference of 2 months between calves.

2. Provide a minimum of 24 sq ft pen space for individual calves; 2½ × 4-ft tie stalls; 30 sq ft for calves in groups without outside runs.

3. Solid partitions between individual pens reduce drafts and chilling. Front of pens should be wire or slatted.

4. Preferred pen temperature is within the range of 50 to 75°F.

• **Feed and Water**—

1. Feed boxes for individual calves should be 8 × 10 × 6 in. deep, and they should be removable so as to facilitate cleaning. Troughs for group feeding should be 10 in. wide and 6 in. deep, with 2 linear feet per calf; provide stanchions. Top of feed containers should be 20 in. from the floor, and feed containers should be located in corner of pen away from water.

2. Automatic drinking cups are preferred for both individual quarters and group pens (one cup for 5 to 8 calves). Where pails or tanks are used, keep them clean. Top of drinking cups for calves should be 20 in. from the floor.

3. Always locate water facilities at a corner away from the feed.

FACILITIES AND EQUIPMENT FOR REPLACEMENT HEIFERS

Pertinent information relative to facilities and equipment for replacement heifers follows:

Fig. 9–17. Inside view of open-to-the-south heifer barn showing clear-span trusses and tombstone design fenceline feeder; used for the bigger and older heifers that outgrow the first-stage hutches. (Courtesy, *Hoard's Dairyman*, Fort Atkinson, WI)

- **Lots and Housing**—

1. Provide the following lot space per head:

Kind of Lot	Sq Ft/Animal
All paved	50–75
Paved and dirt	75–100
Dirt	100–150

2. Slope paved (rough finish) lots ¼ to ½ in./ft, and dirt lots ½ in. or more per foot.

3. Pave (rough finish) a 15- to 20-ft area around waterers, feed bunks and racks, and entrances to sheds.

4. Provide an open shed; allow 20 to 30 sq ft per head for small cattle, and 30 to 40 sq ft per head for large cattle. Bed sheds as needed.

5. Provide artificial shade in hot climates if natural shade is not available. Allow 20 to 30 sq ft per animal, and build shade 8 to 10 ft high.

- **Feed and Water Facilities**—

1. Provide feed bunks that are 24 to 30 in. above the ground (to top of bunk; with height determined by the size of cattle); 8 to 12 in. deep (12 in. deep for silage); and 18 to 24 in. wide when feeding from one side, 36 in. wide when feeding from 2 sides.

2. Allow the following amount of feeder space per head:

	Grain	Roughage
Small cattle	12 in.	18 in.
Large cattle	18 in.	24 in.

3. Allow 1 linear foot of water tank space for each 10 animals and one automatic watering bowl for each 25 animals. Water temperature may range from 35 to 80°F; warming to 50°F in the winter is desirable.

MANURE MANAGEMENT ON THE DAIRY FARM

Planned manure management is an important part of a modern dairy production program. The collection, transport, storage, and land application of manure must be compatible with sanitary milk production, housing systems, and pollution control. Likewise, manure should be handled so as to retain its highest value as a fertilizer.

Prior to construction, any proposed waste management system should be approved by the appropriate regulatory, public health, and milk market officials.

Manure can be handled in either of two ways:

- **As a solid**, which runs 20 to 30% solids, and which refers to feces plus bedding, or feces after liquid separation.

- **As a liquid,** which may be up to 15% solids, and which refers to feces, urine, and sometimes dilution water.

Manure can be hauled (1) daily and spread on available land, usually as a solid; or (2) from storage, either as a solid or liquid, and spread on cropland at a convenient time.

The average dairy cattle production of manure is shown in Table 9–16.

TABLE 9–16
DAILY DAIRY CATTLE MANURE PRODUCTION, SOLIDS AND LIQUIDS[1]
(Manure at 87.3% water and 62 lb/cu ft density)

Animal Size	Total Manure Production			Nutrient Content		
				N	P	K
(lb)	(lb/day)	(cu ft/day)	(gal/day)	←	(lb/day)	→
150	12	0.19	1.5	0.06	0.010	0.04
250	20	0.33	2.4	0.10	0.020	0.07
500	41	0.66	5.0	0.20	0.036	0.14
1,000	82	1.32	9.9	0.41	0.073	0.27
1,400	115	1.85	13.9	0.57	0.102	0.38

[1]From *Dairy Housing and Equipment Handbook*, p. 37. (Courtesy, Midwest Plan Service, Iowa State University, Ames)

DAIRY CATTLE EQUIPMENT

There are numerous articles of dairy cattle equipment, and many designs of each.

Detailed plans and specifications for dairy cattle equipment can usually be obtained through the local county agricultural agent, FFA instructor, lumber dealer, or through writing to the college of agriculture in the state.

SHEEP BUILDINGS AND EQUIPMENT

Fig. 9-18. A laborsaving bank-sheep barn, built on a hillside. From the top of the hill, hay is trucked into the haymow, then gravity-fed to the sheep below. (Courtesy, Washington State University, Pullman)

Sheep do not require expensive or elaborate buildings and equipment, but this statement should not be construed to mean that the facilities for the sheep enterprise should not be carefully planned. On the contrary, it pays well to plan and construct sheep buildings and equipment that will promote sheep health and conserve feed and labor.

The shelter should be of such nature as to protect the flock from becoming soaked with rain or wet snow. Dry snow or bitter cold has no harmful effect, and up until lambing time, a shelter open to the south on well-drained ground may be entirely satisfactory.

Except for smaller space requirements, many sheep buildings and equipment closely resemble those used by beef cattle, and their function and requisites are similar.

SHEEP FINISHING FACILITIES AND EQUIPMENT

Drylot feeding refers to feeding under restricted conditions. This may either be (1) open-yard feeding, or (2) shelter or barn feeding.

• **Open-yard feeding**—Open-yard feeding is the common method of finishing lambs in the irrigated area of the West, though a few eastern lamb-feeding operations are in open yards. In this system, equipment costs are kept to a minimum—the facilities merely consisting of an enclosed and well-drained yard which may or may not have a natural or constructed windbreak, and the necessary feed bunks. Open-yard feeding is often used by large operators who feed thousands of lambs.

• **Shelter or barn feeding**—Because of inclement weather in the fall and early winter, many of the lamb-feeding operations in the central and eastern states are in drylots which afford shelter. In some instances, the lambs are kept under cover without an exercising lot. These barns may consist of anything from an open shed to more costly and elaborate structures, including slotted floors.

CONFINEMENT OF SHEEP ON SLOTTED FLOORS

The shelter above the slotted floor may range all the way from a mere shade over the top of the floor to a completely enclosed, environmentally controlled building; or it may be somewhere between those two extremes—for example, a shed open to one side.

• **Design requirements for slotted floors for sheep**—Unfortunately, there is limited experimental work on which to base design recommendations for slotted floors for sheep. Consequently, the recommendations given in this section are based on producer experiences and such research as is available. Some of them may, and likely will, change as further knowledge becomes available. Nevertheless, the recommendations that follow may serve as useful guides and stimulate further studies:

1. **Floor space.** Allow 8 to 10 sq ft per dry ewe, 10 to 12 sq ft for a ewe and a lamb, and 4 to 5 sq ft for a feeder lamb.

2. **Animals per pen.** Not more than 100 pregnant ewes, or 50 ewes and lambs, or 500 feeder lambs in each group.

3. **Bunk space.** Allow 12 in. per ewe. For feeder lambs, allow 6 in. per head for twice-a-day feeding, and 3 in. per head for self-feeding.

With slotted floors, hay should always be chopped or ground. Long hay may pile up on the slotted floors and prevent manure from dropping through, thereby providing a place for internal parasite development and causing the sheep to befoul themselves.

4. **Waterers.** Locate one automatic waterer at the back (opposite the feed bunk) of each pen. In cold areas, waterers should be equipped with electric heating. Provide one automatic waterer for 15 ewes or 20 feeder lambs. With an open tank, allow 1 ft per 10 ewes or 15 feeder lambs.

5. **Slats.** Slats are usually made from wood, concrete, or metal. Wood slats are less costly than concrete or metal, but they are also less durable; and, if they're green, they will warp

and expand. Slats should be strong enough to support a 200-lb concentrated load at the center.

Slat width and spacing (slot width) are governed by animal comfort and cleaning efficiency. Narrow slats and too wide spacing can cause injury to the feet. On the other hand, wide slats and narrow openings result in floors that are not completely self-cleaning.

Most sheep producers (including both ewe-lamb producers and sheep feeders) seem to favor ¾ to ⅞ in. spacing (slots) between 2- to 3-in. slats.

Access to the pit should be provided through the slotted floor if the manure is handled as a liquid and a manure pump is used. For this purpose, either a steel grid or removable slats may be used.

• **Manure production and storage**—Sheep manure may be handled either as a liquid or a solid. Because sheep feces are rather dry, in comparison with the feces of cattle and hogs, sheep manure lends itself to handling as a solid.[4] On the other hand, handling it as a liquid may, in some operations, offer certain advantages from the standpoint of automation and labor-saving. In either case, a storage area beneath the floor is required.

The quantity of manure produced varies according to the size of animal, kind and amount of feed, and amount of bedding; but it will be approximately as follows for each 1,000 lb of sheep:

Cu Ft/Day Solids and Liquid	Percent Water	Gallons/ Day
0.6	70	4.5

When sheep manure is handled as a liquid (rather than as a solid), extra water will need to be added to liquify the wastes. From ⅕ to ⅗ of the storage volume may be needed for extra water if manure is to be pumped. For irrigation, there should be about 95% water and 5% manure.

There are about 34 cu ft in a ton of manure.

Of course, the total manure storage capacity will depend on the frequency of cleaning. Here is how to determine how much manure will need to be stored:

Storage Capacity = no. of sheep × daily manure production × desired storage time (days) + extra water if handled as a liquid.

Pits range up to 10 ft deep. With finishing lambs, a 5-ft pit, cleaned every 100 days, will suffice.

When manure is to be handled as a liquid, storage tank dimensions and proportions should follow the recommendations of the manure agitator manufacturer. Access to the pit is usually provided from the slotted floor, via either a steel grid or removable slats.

[4]Fresh sheep manure is about 14% lower in moisture than cow manure.

Where manure is handled as a solid, the height of the storage area is determined by two factors: (1) frequency of cleaning, and (2) method of cleaning. The manure may be removed by means of a tractor-mounted loader or scraper. Less working height is required where the building is arranged so floor sections may be removed. But the removing of floor sections does require labor. So, consideration should be given to having greater floor height, with access to the pit via doors or removable panels opening from the ends or sides.

Regardless of whether the sheep manure is handled as a liquid or a solid, the area from the floor to the ground must be completely enclosed to prevent drafts.

SHEEP AND GOAT EQUIPMENT

Detailed plans and specifications for sheep and goat equipment can usually be obtained through the local county agricultural agent, FFA instructor, lumber dealer, or through writing to the college of agriculture in the state.

SWINE BUILDINGS AND EQUIPMENT

Fig. 9–19. Swine farrowing house. (Courtesy, Cedar Point Farms, Easton, MD)

Properly designed swine buildings and equipment should provide for quartering, feeding, and handling hogs in accordance with recommended production practices. The functions and requisites of swine buildings and equipment differ from those for other classes of livestock in that increased emphasis is placed on (1) temperature control—because hogs are so sensitive to extremes of heat and cold; and (2) ventilation, sanitation, and manure disposal—because swine are confined more (and confinement rearing is increasing) than other four-footed animals.

SPACE, TEMPERATURE, AND GROUPING OF EARLY WEANED PIGS

For early weaned pigs (pigs weaned under six weeks), warm, dry, well-ventilated, draft-free housing is essential, and supplemental heat (such as heat lamp) and special feeders and waterers are recommended; at its best, this entails an extra building—a nursery. Also, the conditions listed in Table 9-17 should prevail.

TABLE 9-17
GUIDELINES TO SUCCESSFUL EARLY WEANING[1]

Guideline	Age in Weeks				
	1	2	3	4	5
Minimum pig weight (lb)	5	9	12	15	21
Nursery temperature at pig level (°F)	85	85	83	81	79
Minimum floor space per pig (sq ft)[2]	3	3	3	3	3
Maximum number of pigs per linear foot of feeding space	5	5	5	5	5
Maximum number of pigs per nipple waterer[3]	8	8	8	8	8
Maximum number of pigs per group	10	10	10	15	25

[1]See Appendix for conversion of U.S. customary to metric.

[2]The figures given herein are for solid floors. On slotted floors, this may be lowered 2 sq ft per pig from 1 to 5 weeks of age.

[3]Where bowls are used instead of nipples, there should be one bowl for each 12 pigs.

SWINE BUILDINGS AND SYSTEMS

Fig. 9-20 illustrates swine movement through a sequence of stages of production.

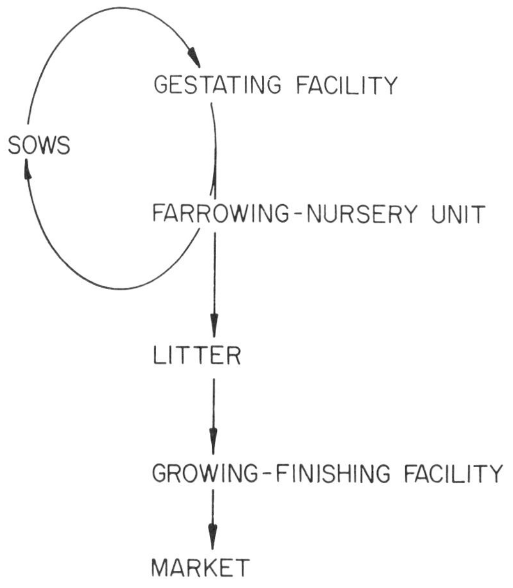

Fig. 9-20. A swine building system flow diagram. Each building is a part of the system.

As indicated, a swine building does not function alone; rather, it is one part of a system. A brief description of one-stage, two-stage, and three-stage production follows:

- **One-stage production**—Farrow-to-finish production, all in one pen of one house, is no longer common. Feeder pig production (which does not involve finishing by the breeder) is a one-stage system. Normally, about 4 litters per year are sold from the pens in which they are farrowed, at 40- to 60-lb weights.

- **Two-stage production**—Under this system, pigs are farrowed, nursed, weaned, and started in one pen to about 60 lb, or 12 weeks of age. Then they are moved to a finishing unit for next 12 weeks. Usually, 3 to 4 litters per year are raised this way.

- **Three-stage production**—The three-stage system is conducted as follows: (1) pigs farrowed in stalls of a farrowing-nursery unit, where they are held until weaning; (2) started or grown in pens with supplemental heat from about 25 to 100 lb; and (3) finished to market weight in still another unit.

SLOTTED FLOORS

Fig. 9-21. Young pigs on woven wire flooring. This type of flooring cleans easily and stays dry. (Courtesy, Delphi Products Co., Delphi, IN)

The main **advantages** of using slotted floors are:

1. **Less space is needed.** When using a slotted floor, only half as much floor space per hog is needed as when using a conventional solid floor. Thus, one can either (a) place twice as many hogs in a building with slotted floors as in a conventional building, or (b) construct a building of one-half the size.
2. **Bedding is eliminated.** Bedding is expensive; that is, when all costs are computed—initial purchase price, storage space, handling, spreading it in the pens, and removal of the straw with the manure.
3. **Manure handling is reduced.** Conventional solid floors require frequent removal of manure. With a slotted floor, manure may be handled every month, or every 6 months, depending on the type and size of the pit under the floor.

The chief **disadvantage** of slotted floors are:

1. There is a higher initial cost.
2. There is less flexibility in the use of the buildings.
3. Any spilled feed may be lost through the slots.
4. Pigs raised on slotted floors resist being driven over a solid floor.
5. Environmental conditions become more critical.

• **Slotted floor construction**—The main features that control the successful operation of a slotted floor system for swine are:

1. **The storage pit under the slats.** A concrete lined pit is located under the slat floors. Depth of the pit varies with the method of manure disposal. If the manure is to be drained to a lagoon, the pit usually ranges from 1½ to 2 ft deep. If 6 months' storage is desired, the pit should be 5 to 6 ft deep. The floor of the pit should slope slightly (about 1 in. per 25 ft) to the cleanout location.
2. **Slats.** Slats are usually made from concrete, steel, or wood. Wood slats are less costly than concrete or steel, but also less durable. Hardwood slats of oak, elm, hickory, or maple may last 2 to 5 years.

Slat width and spacing (slot width) are governed by size of hogs and cleaning efficiency. Narrow slats are usually more effective in farrowing and nursery units, but too wide spacing of narrow slats can cause injury to the feet and legs of finishing hogs. On the other hand, wide slats and narrow openings result in floors that are not completely self-cleaning.

The recommended slot widths are given in Table 9-18.

TABLE 9-18
RECOMMENDED SLOT WIDTH (SLOTTED FLOORS)

Age or Weight	Slot Width
Newborn pigs[1]	⅜ in. and 1 in.
25-40 lb	1 in.
40 lb to market, and farrowing	1 in.

[1]Cover openings with plywood, sheet metal, or mesh during farrowing. Use 1 in. slots behind sow; use ⅜ in. slots elsewhere.

Slats stay cleaner if they are run at right angles to the pig's major traffic pattern.

• **Partially slotted floors**—A solid section permits limited feeding on the floor to help keep pens clean, or permits underfloor heat in a farrowing or nursery area. Where a partially slotted floor is used, the ratio of slotted to solid floor should be about 3 to 1 or 4 to 1, the objective being to limit the space on the solid concrete floor so that there is just enough room for the pigs to lie down.

The following plans and practices are pertinent to the success of a partially slotted floor arrangement:

1. Place the feeder in one end of the pen, and the waterer in the other on slats.
2. Use a continuous feeder along one wall; the arrangement will alleviate dunging along the wall.
3. Avoid long pens; the distance from the feeder to the slots shouldn't be more than 12 ft. A hog will seldom move 12 ft before defecating.

Fig. 9-22. Partially slotted floor, showing automatic waterers in the slotted area. (Courtesy, Wyatt Manufacturing Company, Inc., Salina, KS)

HOG MANURE

Handling manure has become a major problem for producers who raise large numbers of hogs in confinement. Careful waste management is necessary in order to—

1. Maintain good swine health through sanitary facilities.
2. Avoid pollution of air and water.
3. Comply with local, state, and federal regulations.

On a 100-lb liveweight basis, the approximate daily production of manure is: ⅛ cu ft, 1.0 gal, or 7.5 lb. The average density of manure is 59 lb/cu ft.

Table 9-19 shows the approximate daily manure production.

TABLE 9-19
APPROXIMATE DAILY MANURE PRODUCTION, FREE OF BEDDING[1]

Class Weight	Waste Production			
	Liquids and Solids		Wet Solids Only	
(lb)	(cu ft)	(gal)	(cu ft)	(lb)
Pigs:				
40	.06	0.5	.04	2.4
100	.13	1.0	.10	5.9
150	.21	1.7	.15	8.8
210	.30	2.2	.20	12.0
Sows and boars:				
300	.43	3.0	.30	17.5
500	.71	5.0	.50	30.0
Sow and litter:	.55	4.0	.50	30.0

[1]From: *Swine Handbook Housing and Equipment*, p. 33, Table 6. (Courtesy, Midwest Plan Service, Iowa State University, Ames)

Generally, swine producers have a choice in the method of handling wastes. Manure from bedded areas and drained solid floors is usually handled as a solid. Wastes from unbedded floors and some lots are semisolid. Wastes under slotted floors are usually liquid.

PORTABLE HOUSES

Portable houses were originally designed to accommodate one sow and her litter. In recent years, however, the size of movable houses has been greatly increased. Consequently, double-unit portable houses are common, and some portable houses will accommodate as many as six sows.

Fig. 9–23. The portable double-unit Sunshine Hog House. Note folding doors in front.

SWINE EQUIPMENT

Detailed plans and specifications for swine equipment can usually be obtained through the local county agricultural agent, FFA instructor, lumber dealer, or through writing to the college of agriculture of the state.

HORSE BUILDINGS AND EQUIPMENT

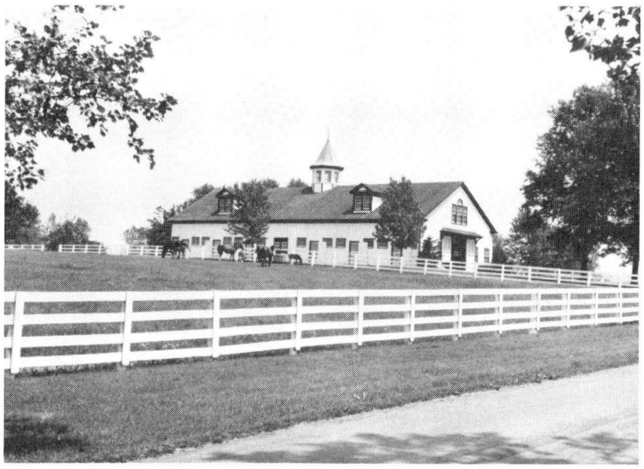

Fig. 9–24. Attractive barn on a Kentucky horse farm. Horse barns may be designed to serve different numbers of horses, ranging from one horse to many horses. (Courtesy, Kentucky Department of Travel Development, Frankfort)

Adequate and well-designed buildings and equipment make horse care easier and add to the personal satisfaction of the owner and caretaker.

Two of the most important considerations in planning horse facilities are health and safety—not only of the horses, but also of the people who either come in contact with the animals and facilities or who live nearby. Other important considerations include sound construction, laborsaving conveniences, and building style in harmony with the surroundings.

The same basic principles are involved when selecting and equipping a barn to stable only one or two animals as for a larger number.

It is recognized that the design of both buildings and equipment used for light horses is likely to be dominated by the fads and fancies of the owner. Figs. 9–25 to 9–28 suggest some designs of buildings and equipment for light horses.

HORSE BARNS

The horse barn, whether large or small, should be well planned, durable, and attractive. Basically, its purposes are to provide an environment that protects the horses from temperature extremes, keeps them dry and out of the wind, eliminates drafts, provides fresh air in both winter and summer, and protects them from injury.

Ample space should be provided for the well-being of the horses, and for the convenience, safety, and enjoyment of the people who care for and use them.

The needs for housing horses and storing materials vary according to the intended use of the building. Broadly speaking, horse barns are designed to serve (1) small horse establishments—the owner with one or a few head, (2) large horse breeding establishments, or (3) riding academies and training and boarding stables. A summary of the space requirements of buildings for horses is presented earlier in this section under the heading "Space Requirements of Buildings and Equipment," Table 9–4, page 364.

SMALL HORSE ESTABLISHMENTS

When 1 or 2 riding horses or ponies are kept, they are usually stabled close to the house, which makes for greater convenience in their care and use. In most cases, box stalls are built in a row and provision is made for limited feed, bedding, and tack storage; usually a combination feed and tack room for units with 1 to 2 stalls and separate feed and tack rooms with 3 or more stalls. Fig. 9–25 (next page) shows an attractive small barn for 2 horses; a barn which was designed by the author for the U.S. Department of Agriculture.

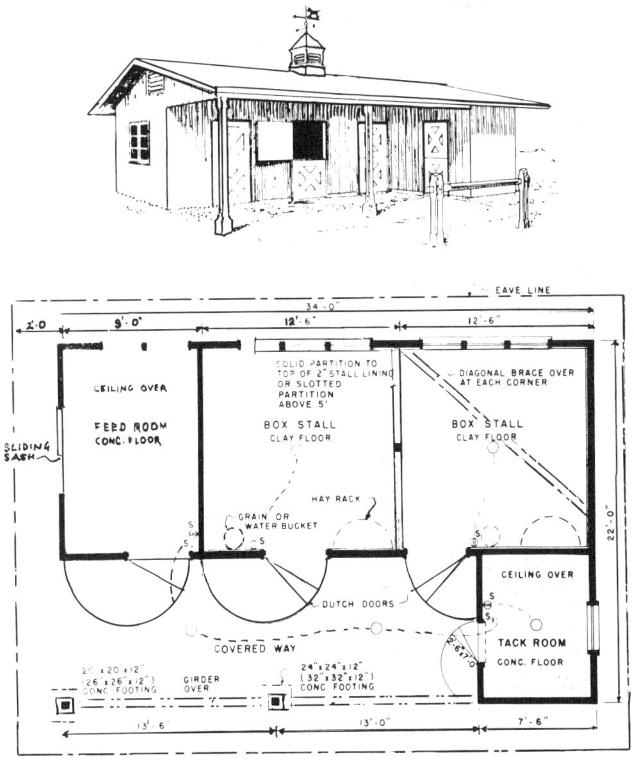

Fig. 9–25. Riding horse barn above and floor plan below. Barn has two box stalls, a feed room, and a tack room.

LARGE HORSE BREEDING ESTABLISHMENTS

With large horse breeding establishments, specially designed buildings are generally provided for different purposes; among them, the following: (1) broodmare and foaling barn, (2) barren mare barn, (3) stallion barn and paddock, (4) breeding shed, (5) weanling and yearling quarters, and (6) isolation (quarantine) quarters. Fig. 9–26 shows a 17-stall horse barn.

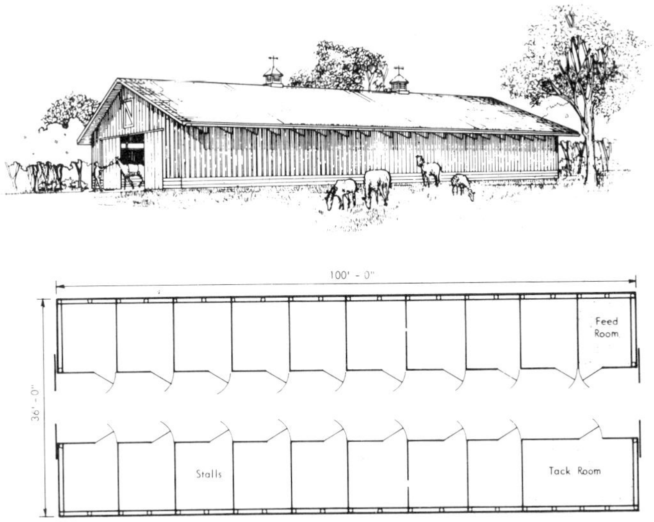

Fig. 9–26. A 17-stall horse barn; 36 ft × 100 ft in size, with a 20 ft × 20 ft tack room, a 10 ft × 12 ft feed room, and a 12 ft alley. From *Horse Handbook-Housing and Equipment*, 1971, p. 56. (Courtesy, Midwest Plan Service, Iowa State University, Ames)

RIDING, TRAINING, AND BOARDING STABLES

For this purpose, the quarters may consist of (1) stalls constructed back to back in the center of the barn with an indoor ring around the stalls, (2) stalls built around the sides of the barn with the ring in the center, or (3) stalls on either side of a hallway or alleyway and the ring outdoors.

STALLS

Stalls are of two general types; (1) box stall; and (2) tie, straight, standing, or slip stalls. As tie stalls differ primarily in the width of the area and their use is less common in breeding establishments, the discussion will be confined to loose or box stalls. The latter are preferred because they allow the horses more liberty, either when standing or lying down.

Box stalls range in size from 10 × 10 to 16 × 16 ft. Stalls 16 × 16 ft or larger are generally used for foaling mares. Tie stalls are usually 5 ft wide and 9 to 12 ft long. The length of the tie stall should be measured from the front of the manger or grain box to the rear of the stall partition. (See section headed, "Space Requirements of Buildings and Equipment" for stable specifications, Table 9–4, page 364.)

Adequate quarters for a horse should be (1) ample in size and height for the particular type of animal; (2) properly finished and without projections; (3) dry with good footing; (4) equipped with suitable doors; (5) provided with ample windows for proper lighting; (6) well ventilated; (7) cool in summer and warm in winter; (8) equipped with suitable mangers, grain containers, watering facilities, and mineral boxes; and (9) easy to keep clean.

STALL FLOOR

A raised clay floor covered with a good absorbent bedding, with proper drainage away from the building, is the most satisfactory flooring for horse stables. Clay floors are noiseless and springy, keep the hoofs moist, and afford firm natural footing unless wet; but they are difficult to keep clean and level. To lessen the latter problem, the top layer should be removed each year, replaced with fresh clay, and leveled. Also, a semicircular concrete apron extending into each stall at the doorway will prevent horses from digging a hole in a clay floor at this point. This arrangement is particularly desirable in barns for yearlings, as they are likely to fret around the door.

TACK ROOM

A tack room is an essential part of any barn. With one or two stall units, a combination tack and feed room is usually used, for practical reasons. On large establishments, the tack room is frequently the showplace of the stable. As such, the owner takes great pride in its equipment and arrangement. Also, depending upon the use of the horses, the tack room takes on an air and personality that represents the horses in the stalls.

Fig. 9–27 shows a formal tack room for a stable that features hunters, jumpers, and polo ponies, whereas Fig. 9–28 shows a tack room for parade and Palomino horses.

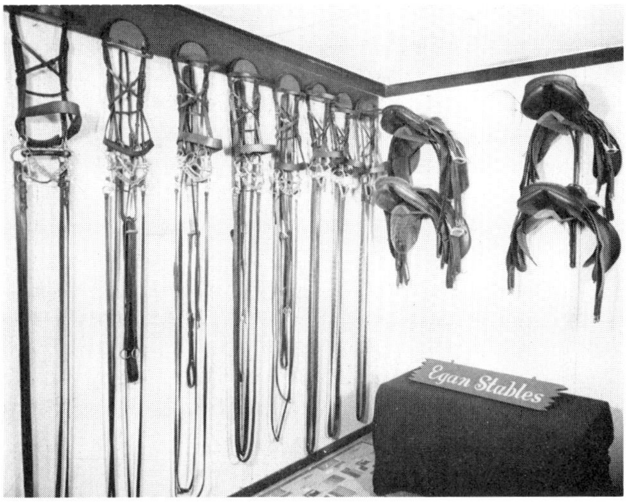

Fig. 9–27. The tack room of the Bob Egan stable, Pacific Palisades, California is rather typical for owners of hunters, jumpers, and polo ponies. The tack box holds coolers, rub rags, cleaning equipment, hammers, nails, and other gear necessary in taking the horses to the shows. (Photo by J. H. Williamson, Arcadia, CA)

Fig. 9–28. Tack room of Dwight Murphy's San Marcos Ranch, Santa Ynez Valley, California. Note the large natural color painting by Nicholas S. Ferfires, depicting the different costumes and saddles of riders and the beautiful silver decorated saddles used in parades on the noted Palominos of San Marcos Ranch. (Photo by J. H. Williamson, Arcadia, CA)

SHADES

A shade, either trees or artificial, should be provided for horses that are in the hot sun.

The most satisfactory constructed horse shades are (1) oriented with a north-south placement, (2) at least 12 to 15 ft in height (in addition to being cooler, high shades allow a mounted rider to pass under), and (3) open all around.

SHOW-RING

There are no standard specifications relative to size, type of construction, and maintenance of show-rings. Yet, all the better rings meet certain basics.

The National Horse Shows Association recommends the following ring sizes: indoor ring, 110 × 220 ft; outdoor ring, 120 × 240 ft. It is recognized, however, that many good show-rings are either smaller or larger than these dimensions.

In addition to ring size, consideration must be given to proper footing—to achieving resilience, yet firmness and freedom from dust. With an outdoor ring, establishing proper drainage and constructing a good track base are requisite to all-weather use. Drainage is usually secured by (1) locating the ring so that it is high, with the runoff away from it, and (2) installing a perforated steel pipe (with the perforations toward the bottom side), or drainage tile, underneath the track if necessary.

HORSE EQUIPMENT

Although the design of horse equipment is likely to be dominated by the fads and fancies of the owner, the basic needs are merely for simple but effective equipment with which to provide hay, concentrates, minerals, and water—without waste, and without hazard to the horse. Whenever possible, it is desirable that feed and water facilities be located so that they can be filled without necessitating that the caretaker enter the stall or corral, from the standpoint of both convenience and safety. In any event, it should not be necessary to walk behind horses in order to feed and water them.

Feed and water equipment may be built in or detached. Because specialty feed and water equipment is more sanitary, flexible, and suitable, many caretakers favor it over old-style wood mangers and concrete or steel tanks. Bulk-tank feed storage may be used to advantage on large horse establishments to eliminate sacks, lessen rodent and bird problems, and make it possible to obtain feed at lower prices by ordering large amounts.

The specifications of feed and water equipment for horses are given earlier in this section in Table 9–5, page 364. Detailed plans and specifications for horse equipment can usually be obtained through the local county agricultural agent, FFA instructor, lumber dealer, or through writing to the college of agriculture of the state.

CONCRETE STRUCTURES FOR THE FARM AND RANCH[5]

Progressive farmers and ranchers are making increased use of concrete. Although they usually either contract the job or hire experienced help where major structures are involved, most farmers and ranchers do their own concrete work on the smaller and simpler jobs.

---

[5]In the preparation of this section, invaluable assistance was given by the Portland Cement Association, Skokie, IL. Also, an authoritative review of the section was provided by S. H. Kosmatka, Assistant Manager, Construction Information Services, Portland Cement Assn., Skokie, IL.

SELECTING THE MIX FOR THE JOB

Fig. 9–29. Concrete water tank. (Courtesy, Portland Cement Assn., Skokie, IL)

Concrete mixes are often designated by volume proportion, such as 1:2¼:3; the figures from the left to right referring to the proportions or parts of cement, sand, and gravel—all measured by the same unit of volume. Such designation is incomplete, however, because it leaves unspecified the amount of one of the most important ingredients—water. In fact, the amount of water and the amount of cement determine the quality of the concrete. Aggregates such as sand and gravel are only fillers. The amounts of these to use varies, depending on the consistency of the mix desired. Therefore, in mixing concrete, it is essential to measure carefully the amount of cement and water. For most farm jobs, the recommended water-cement ratio is 6 gal of water to each sack of cement.

The sand used in the mix usually contains moisture, which means that less than 6 gal of water will actually be added to the mixer. The moisture condition of the sand may be evaluated by squeezing it in your hand. Dry sand falls apart; moist sand forms a ball; and wet sand glistens and leaves excess moisture on the hand. After the sand has been evaluated, refer to Table 9–20 to find the correct amount of water to use. Use this amount only! Other mixes for special uses are also shown in Table 9–20.

For half-sack batches, use half the amount of water shown in Table 9–20 with each half-sack of cement. Always maintain this proportion of water to cement and vary the aggregate amounts to get the desired mix workability. The durability of the concrete decreases if additional water is added.

TABLE 9–20
WATER-CEMENT GUIDE; AND SUGGESTED TRIAL MIXES

Type of Job	Total Amount of Water to Each Sack of Cement		Maximum Size Aggregate		Amount of Water to Use at the Mixer per Sack of Cement When Sand Is— Damp[1]		Wet[2] (average)		Very Wet[3]		Suggested Mixture for 1-sack Trial Batches[4] Cement Sacks		Aggregates Fine		Coarse	
	(gal)	(l)	(in.)	(cm)	(gal)	(l)	(gal)	(l)	(gal)	(l)	(cu ft)	(m³)	(cu ft)	(m³)	(cu ft)	(m³)
Concrete subject to severe wear, weather, or weak acid and alkali solutions; such as milk coolers, tanks, creamery floors, etc.	5	19	¾	1.9	4½	17	4	15	3½	13	1	.03	2	.06	2¼	.06
Most farm concrete jobs, such as floors, steps, walks, driveways, septic tanks, dairy and swine barn floors, silos, manure pits,	6	23	1	2.5	5½	21	5	19	4¼	16	1	.03	2¼	.06	3	.08
water tanks, basement walls, etc.	6	23	1½	3.8	5½	21	5	19	4¼	16	1	.03	2½	.07	3½	.10
Concrete in thick sections and not subject to freezing, such as thick footings, foundations, walls, engine bases, etc.	7	27	—	—	6¼	24	5½	21	4¾	18	1	.03	3	.08	4	.11

[1]Damp describes sand that will fall apart after being squeezed in the palm of the hand.

[2]Wet describes sand that will ball in the hand when squeezed but leaves no moisture on the palm.

[3]Very wet describes sand that has been subjected to a recent rain or recently pumped. Balling a sample in the hand will leave moisture on the palm and the sand-gravel glistens in the light.

[4]Mix proportions will vary slightly depending on gradation of aggregate.

MAKING QUALITY CONCRETE

High-quality concrete is made by (1) selecting suitable materials (cement, aggregate, and water); (2) thoroughly mixing them together in the right proportions; and (3) correctly placing, finishing, and curing the resulting mix. To this end, the following simple rules should be followed:

1. **Water.** The water used in mixing concrete should be clean enough to drink.

2. **Sand.** The sand should be clean, hard, and well graded.

3. **Gravel.** The gravel (or rock or slag) should be clean and hard and range in size below the maximum specified for the particular kind of work.

4. **Cement.** The cement should be free from hard lumps due to dampness (lumps which do not readily pulverize when squeezed in one's hand).

5. **Measuring.** Measure water and cement for each mix. It is especially important that no more water be used than is indicated in Table 9–20.

Water may be measured in a pail marked off in gallons and half gallons. Cement is measured by shovelfuls or by dividing the sack in the desired proportions.

6. **Mixing.** For machine-mixing, allow one or two minutes after all materials are in the mixer.

For hand-mixing, proceed as follows: (a) place the sand on a watertight mixing platform; (b) spread the cement evenly over the sand and turn the two materials with a shovel until they are thoroughly mixed, as evidenced by the uniform color; (c) spread the mixture out evenly, add the gravel, and again mix thoroughly; and (d) form a hollow in the materials, slowly add the measured quantity of water, and mix until every particle has been completely covered with cement paste. Hold rigidly to the predetermined amounts of water and cement; then vary the amounts of sand and gravel according to the consistency of the mix desired.

7. **Ready-mixed concrete.** When ordering ready-mixed concrete for average farm jobs, specify a mix based on Table 9–21.

TABLE 9–21
GUIDE FOR ORDERING READY-MIXED CONCRETE

When Ordering Concrete For					
Flat Work (using 1½-in. maximum size aggregate)			**Formed Work (using ¾-in. maximum size aggregate)**		
Garbage feeding floors, floors in dairy plants.	Paved barnyards, floors for farm buildings, sidewalks.	Footings, concrete improvements in mild climates.	Mangers for silage feeding, manure pits.	Reinforced concrete walls, beams, tanks, foundations.	Concrete improvements in mild climates.
Specify cement content: minimum number of sacks per cu yd of concrete.					
7	6	5	7¾	6½	5½
Water content (includes water contained in aggregates): maximum gal per sack of cement.					
5	6	7	5	6	7
Get medium-consistency concrete (3-in. slump).					

8. **Workable mix.** Table 9–20 gives the proportions of water to cement to use. Aggregates are used until the mix is workable (one which is smooth and plastic, and will place and finish well). For the first trial batch, use the proportions of sand to gravel shown in Table 9–20. If this batch is too stony, use more sand and less gravel in the next batch. If it is oversanded, reduce the amount of sand and increase the amount of gravel until the desired consistency is obtained. Remember; always adjust the mix by varying the amounts of aggregates.

9. **Forms.** Forms, constructed to conform to the desired shape, should be rigid, tight, and well-braced. Then they should be oiled so that they can be easily removed.

10. **Placing.** Freshly mixed concrete should be placed in 6- to 12-in. layers into the forms and then tamped, spaded, or vibrated enough to eliminate air pockets.

11. **Construction joints.** Where there is some delay in completing a concrete job (even from one day's run to another), (a) roughen the surface with a stiff broom before it hardens, and (b) wet the surface and cover it with a layer of cement mortar ½ in. thick before adding concrete.

12. **Cold weather.** During cold weather, the water, sand, and gravel should be heated before mixing, and the new concrete should be protected from freezing for at least three days. Do not deposit concrete on frozen ground or in forms containing frost or ice.

13. **Finishing.** Newly placed concrete should first be leveled off with a strikeboard or wood float and then allowed to stiffen before finishing in final form. For a gritty, nonskid floor (such as is best for barn floors and driveways), finish with a wood float; for a still rougher floor (such as is best for concrete corrals), finish with a broom. For a smooth, dense surface (as is best for feed mangers, poultry house floors, and basement floors), finish with a steel trowel.

14. **Curing.** For proper curing, concrete needs moisture and should be protected from drying out for at least 7 days. Depending on the type of structure, cover with canvas, burlap, earth, or straw, and keep wet for the required time.

15. **Reinforcing.** Where an important structure is involved, the advice of a competent architect or engineer should be sought in the design and installation of reinforcing. It is best to use regular reinforcing rods or mesh. Never use scrap iron or rusty fence wire for reinforcement.

16. **Dry floors.** For a dry concrete floor (such as is desired in hog houses, grain storage buildings, etc.), (a) select a well-drained site; (b) provide a 6- to 12-in. well tamped fill of gravel or crushed rock; (c) place a moisture barrier off 55-lb asphalt roll roofing, tough waterproof paper, or 6 mil (or heavier) plastic film over it, lapping the points 6 in.; (d) place 3 in. of sand over the plastic; and (e) place the concrete slab on top.

FENCES FOR LIVESTOCK[6]

Good fences (1) maintain farm boundaries, (2) make livestock operations possible, (3) reduce losses to both animals and crops, (4) increase land values, (5) promote better relationships between neighbors, (6) lessen accidents from animals getting on roads, and (7) add to the attractiveness and distinctiveness of the premises.

FENCE SPECIFICATIONS

Except for corrals and horses, woven and/or barbed wire fences are used for most classes of livestock. Fence construction specifications for different classes of livestock, and different materials, are given in Tables 9-22, 9-23, and 9-24.

[6]In the preparation of this section, the author had the benefit of the authoritative review and helpful suggestions of specialists at Keystone Steel & Wire Co., Peoria, IL.

TABLE 9-22
WOVEN WIRE FENCE

Kind of Stock	Recommended Woven Wire Height	Recommended Weight of Stay Wire	Recommended Mesh or Spacing Between Stays	Recommended Number of Strands of Barbed Wire to Add to Woven Wire[1]	Comments
	(in.)	(gauge)	(in.)		
Cattle	47, 48, or 55	9 or 11	12	1 strand 2 in. to 3 in. above top of woven wire, with points 4 in. or 5 in. apart, to prevent animals from breaking down woven wire.	Also satisfactory for all farm animals, except young pigs. Fences for cattle feedlots should be constructed of wood, cable, pipe, or other strong material, and should be 60 in. high.
Sheep	32	11 or 12½	12	2 strands on top.	Sheep fences should total 39 in. in height. Twelve-in. mesh is best for sheep as they will not get their heads caught if they attempt to reach through. With a heavy concentration of feeder lambs, use wooden fence 39 in. high.
Swine	26, 32, or 39	9 or 11	6	1 strand on bottom.	Barbed wire on bottom prevents rooting under.
Horses (also see Table 9-31, Horse Fence Chart)	55 or 58	9 or 11	12	1 strand on top; with points 4 in. or 5 in. apart.	Also satisfactory for all farm animals except young pigs. Cyclone, wood pole, or other durable and attractive materials are usually used around the headquarters.
All farm animals ..	26 or	9 or 11	6	3 strands on top; 1 strand on bottom.	
	32	9 or 11	6	2 strands on top; 1 strand on bottom.	

[1]The American Society of Agricultural Engineers' standard for barbed wire calls for 4 in. spacing with 2-point wire and 5 in. spacing with 4-point wire.

TABLE 9–23
BARBED WIRE FENCE CHART[1]

Kind of Stock	Recommended Number of Points	Recommended Spacing Between Points	Recommended Weight of Strands	Recommended No. of Lines of Barbed Wire to Install	Comments
		(in.)	(gauge)		
Cattle or horses; in farm pastures	2 or 4	4 or 5	12½	4–5	Two point barbs are 4 in. apart; 4-point are 5 in. apart.
Cattle or horses; on the range	2 or 4	4 or 5	12½	2 or 4	Not all animals will be restrained by 2 or 3 strands.
Sheep	Barbed wire is not considered suitable for sheep because it tears the fleece.				
Swine	2 or 4	4 or 5	12½	6	A 6-strand barbed wire fence for swine may cost more to build and maintain than a woven wire fence.

[1]The American Society of Agricultural Engineers' standard for barbed wire calls for 4 in. spacing with 2-point wire and 5-in. spacing with 4-point wire.

TABLE 9–24
HORSE FENCE CHART

	Material Specifications		Construction Details			
Material	Post	Line Fence or Rails	Fence Height	Number and Spacing Rails or Mesh of Wire	Distance Between Post on Centers	Comments
Board	7½ ft, 4–8 in. diameter 8½ ft, 4–8 in. diameter	1 in. × 6 in., or 1 in. × 8 in. 1 in. × 6 in., or 1 in. × 8 in.	60 in. 72 in.	4 boards 5 boards	8 ft 8 ft	Horses that are closely confined in small paddocks or show-rings require 2 in. × 6 in. or 2 in. × 8 in. boards.
Poles	7½ ft, 4–8 in. diameter 8½ ft, 4–8 in. diameter	4–6 in. diameter 4–6 in. diameter	60 in. 72 in.	4 poles 5 poles	8 ft 8 ft	
Polyvinyl chloride (PVC)	Round or square post available.	2 in. × 6 in. planks, or rails	50–60 in.	4 or 5 planks or rails	8 ft	The major disadvantage of this type of fence is the initial cost, which is twice that of a board fence.
Rubber-Nylon	7–8 ft line post, 6 in. diameter Corner post, 8 in. diameter	Long strips maintained under pressure	50–60 in.	Strips 2 in. × 4 in. wide	8 ft	This type of fencing comes in strips 2 in. to 4 in. wide. Anchor post must be well braced.
Steel or aluminum rail	7½ ft 7½ ft 8½ ft	10 ft or 20 ft rail 10 ft or 20 ft rail 10 ft or 20 ft rail	60 in. 60 in. 72 in.	3 rails; 20 in. centers 4 rails; 15 in. centers 4 rails; 18 in. centers	10 ft 10 ft 10 ft	Because of the strength of most metal rails, fewer rails and posts are necessary than where wood is used.
Wire mesh	7–8 ft line post, 6 in. diameter Corner post, 8 in. diameter		50–60 in.	Diamond-shaped or 2 in. × 4 in. rectangles	8 ft	Place pipe, pole, or 6 in. board at the top. Also, some fence builders recommend that a pole be added at the bottom, and that a diagonal brace be placed between posts for reinforcement.
Woven wire	7½ ft, 4–8 in. diameter	9 or 11 gauge stay wire	55–58 in.	12 in. mesh	12 ft	Woven wire is satisfactory for larger areas where the concentration of animals is not too great. But it is not recommended for corrals, paddocks, or small pastures. Use 1 or 2 strands of barbed wire (with points 3 in. to 4 in. apart) on top.
	7–8 ft heavy channel steel or wood posts	11 or 12½ gauge	60 in. or 72 in.	2 in. mesh, 4 in. vertical	10 to 15 ft	The 2 in. × 4 in. mesh prevents a hoof from going through and alleviates leg damage.

WIRE FENCES

Fig. 9-30. Woven wire cattle fences with one strand of barbed wire on top. (Courtesy, Keystone Steel & Wire Co., Peoria, IL)

The discussion which follows will be limited primarily to wire fencing, although it is recognized that such materials as rails, poles, boards, stone, and hedge have a place and are used under certain circumstances. Also, where there is a heavy concentration of animals, such as in corrals and in feed yards, there is need for a more rigid type of fencing material than wire. Moreover, certain fencing materials have more artistic appeal than others.

SELECTING WIRE

The kind of wire to purchase should be determined primarily by the class of animals to be confined. Table 9–22, 9–23, and 9–24 are suggested guides.

Twenty-six- or thirty-two-inch woven wire, with or without barbed wire, is frequently used as movable fence for hogs, especially when hogging down crops.

The following additional points are pertinent in the selection of wire:

1. **Styles of woven wire.** The standard styles of woven wire fences are designated by numbers as 958, 1155, 849, 1047, 741, 939, 832, and 726.

The first one or two digits represent the number of line (horizontal) wires; the last two, the height in inches; i.e., 1155 has 11 horizontal wires and is 55 in. in height. Each style can be obtained in either (a) 12-in. spacing of stays (or mesh), or (b) 6-in. spacing of stays. Also, a special 2-in. mesh is available for horses.

2. **Mesh.** Generally, a close-spaced fence with stay or vertical wires 6 in. apart (6-in. mesh) will give better service than a wide-spaced (12-in. mesh) fence. However, some fence manufacturers believe that a 12-in. spacing with a No. 9 wire is superior to a 6-in. spacing with No. 11 filler wire (about the same amount of material is involved in each case).

3. **Weight of wire.** A fence made of heavier weight wires will usually last longer and prove cheaper than one made of light wires. Heavier or larger size wire is designated by a smaller gauge number. Thus, No. 9 gauge wire is heavier and larger than No. 11 gauge. Woven wire fencing comes in Nos. 9, 11, 12½, and 16 gauges—which refers to the gauge of the wires other than the top and bottom wires. Heavy barbed wire is 12½ gauge. But there is a lighter, high tensile barbed wire which comes in 15½ to 16½ gauge.

Heavier or larger wire than normal should be used in those areas subject to (a) salty air from the ocean, (b) smoke from industries of close proximity, which give off chemical fumes into the atmosphere, (c) rapid temperature changes, or (d) overflow or flood.

Likewise, heavier wire than normal should be used in fencing (a) small areas, (b) where a dense concentration of animals is involved, and (c) where animals have already learned to get out.

4. **Styles of barbed wire.** Styles of barbed wire differ in the shape and number of the points of the barb, and the spacing of the barbs on the line wires. The two-point barbs are commonly spaced 4 in. apart while four-point barbs are generally spaced 5 in. apart. Since any style is satisfactory, selection is a matter of personal preference.

5. **Standard size rolls or spools.** Woven wire comes in 20 and 40 rod rolls; barbed wire in 80 rod spools.

6. **Wire coating.** The kind and amount of coating on wire definitely affects its lasting qualities. Galvanized coating is most commonly used to protect wire from corrosion. Coatings are specified as Class I, Class II, and Class III. The higher the class number, the greater the coating thickness and performance.

SELECTING POSTS

Three kinds of material are commonly used for fence posts: wood, metal, and concrete. The selection of the particular kind of posts should be determined by (1) the availability and cost of each, (2) the length of service desired (posts should last as long as the fencing material attached to it, or the maintenance cost may be too high), (3) the kind and amount of livestock to be confined, and (4) the cost of installation.

SUSPENSION FENCES

A suspension fence consists of 3 to 5 barbed wires stretched tightly and held up by line posts 80 to 120 ft apart. Twisted wire stays are placed about 16 ft apart. Suspension fences are in use in southwestern United States and in Australia. They cost about one-half as much as conventional woven-wire or barbed wire fences.

A suspension fence sways in the breeze, with the result that cattle are spooked by the movement. When they attempt to rub on it, the wires "give" and they back away. Suspension fences require less expensive upkeep than conventional fences.

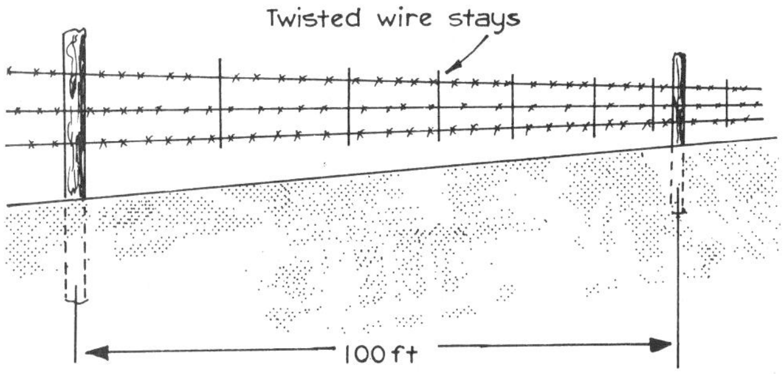

Fig. 9–31. Suspension fences save money—and work. The line posts are spaced 80 to 120 ft apart, and the twisted wire stays are about 16 ft apart.

If the bulls fight through it, the wire springs back into position after they run over or under it.

Suspension fences should work as well with sheep as with cattle.

Suspension fences are best adapted to even terrain. Whenever slope direction changes, a support post should be set. The fence needs either wooden or steel line posts, at 80- to 120-ft intervals, and it must be securely braced at the corners and every one-quarter mile. Wire is stretched taut so that there is not more than 3 in. of sag between the line posts.

The wire should be attached to wooden posts with L-shaped, deformed, shanked staples, long U-shaped staples or a piece of 18 to 20 gauge metal strip ½- by 1½ in. placed over the wire and held with a six penny nail on each end. Regular wire clips should be used on steel posts. The wires must move freely under the holders. Don't let the wire stays touch the ground.

A mile of four-wire fence with stays 16-ft apart takes 16 rolls of wire, 53 line posts, corner and stretch posts, 300 spiral wire stays, metal strips, nails, brace wire, and about 108 work-hours of labor.

ELECTRIC FENCES

Where a temporary enclosure is desired or where existing fences need bolstering from roguish or breachy animals, it may be desirable to install an electric fence, which can be done at minimum cost.

The following points are pertinent in the construction of an electric fence:

1. **Safety.** If an electric fence is to be installed and used, (a) necessary safety precautions against accidents to both persons and animals should be taken, and (b) the farmer or rancher should first check into the state regulations relative to the installation and use of electric fences. *Remember that an electric fence can be dangerous.* Fence controllers should be purchased from a reliable manufacturer; homemade controllers are dangerous and should not be used.

2. **Charger.** The charger should be safe and effective (purchase one made by a reputable manufacturer). There are four types of chargers: (a) *the battery charger,* which uses a 6-volt hot shot battery; (b) *the inductive discharge system,* in which the current is fed to an interrupter device called a circuit breaker or chopper which energizes a current limiting transformer; (c) *the capacitor discharge system,* in which the power line is rectified to direct current and the current is stowed in the capacitor; and (d) *the continuous current type,* in which a transformer regulates the flow of current from the powerline to the fence.

3. **Wire height.** As a rule of thumb, the correct wire height for an electric fence is about three-fourths the height of the animal; with two wires provided for sheep and swine. Following are average fence heights above the ground for different animals:

Cattle—30 to 40 in.
Calves—12 to 18 in.

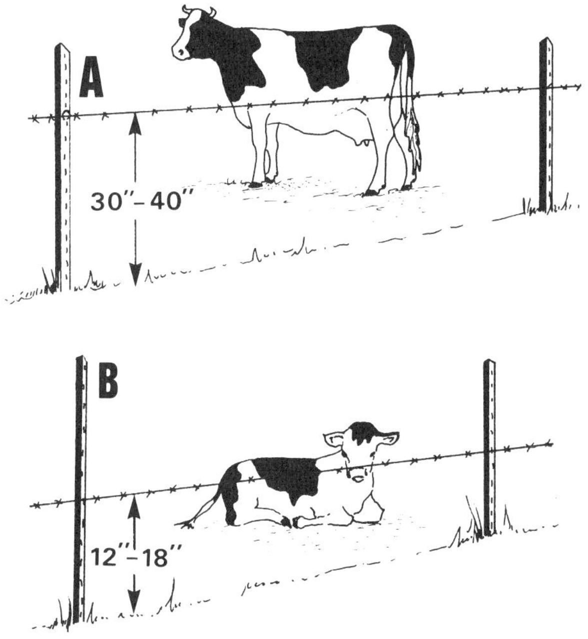

Fig. 9–32. Recommended height for electric fences for (A) cattle and (B) calves.

Sheep, two wires—One wire 8 to 10 in. and the other 16 to 18 in.

Swine, two wires—One wire 6 to 8 in. and the other 14 to 16 in.

Horses—30 to 40 in.
Mixed livestock, three wires—8, 12, and 32 in.

Note well: Traditional electric fences are more satisfactory for cattle and calves than for sheep, swine, or horses. An electric net fence is recommended for sheep.

4. **Posts.** Either plastic or steel posts may be used for electric fencing. Corner posts should be as firmly set and well braced as required for any nonelectric fence so as to stand the pull necessary to stretch the wire tight. Line posts (a) need only be heavy enough to support the wire and withstand the elements, and (b) may be spaced 40 to 50 ft apart for horses and cattle, and 25 to 40 ft apart for sheep and swine.

5. **Wire.** New 4 point 12½ gauge barbed wire is preferred, because the barbs will penetrate the hair of animals and touch the skin, but smooth wire can be used satisfactorily. Rusty wire should not be used, because rust is an insulator.

6. **Insulators.** Wire should be fastened to the posts by insulators and should not come into direct contact with posts, weeds, or the ground, unless plastic or fiberglass posts are used (plastic and fiberglass do not require insulators). Inexpensive solid glass, porcelain, or plastic insulators should be used, rather than old rubber or necks of bottles.

7. **Grounding.** One lead from the controller should be grounded to a pipe driven into the moist earth. *An electric fence should never be grounded to a water pipe, because it could carry lightning directly to connecting buildings.* A lightning arrestor should be installed on the ground wire.

- **Electric net fencing**—This is a movable electric net fence, which is particularly useful in the rotation of pastures for sheep and goats. It comes in 150-ft rolls, with removable push-in type fiber posts about every 15 ft. Heights range from 22 to 42 in. Each roll weighs about 10 lb, depending on the height of the fence. It is relatively dog proof.

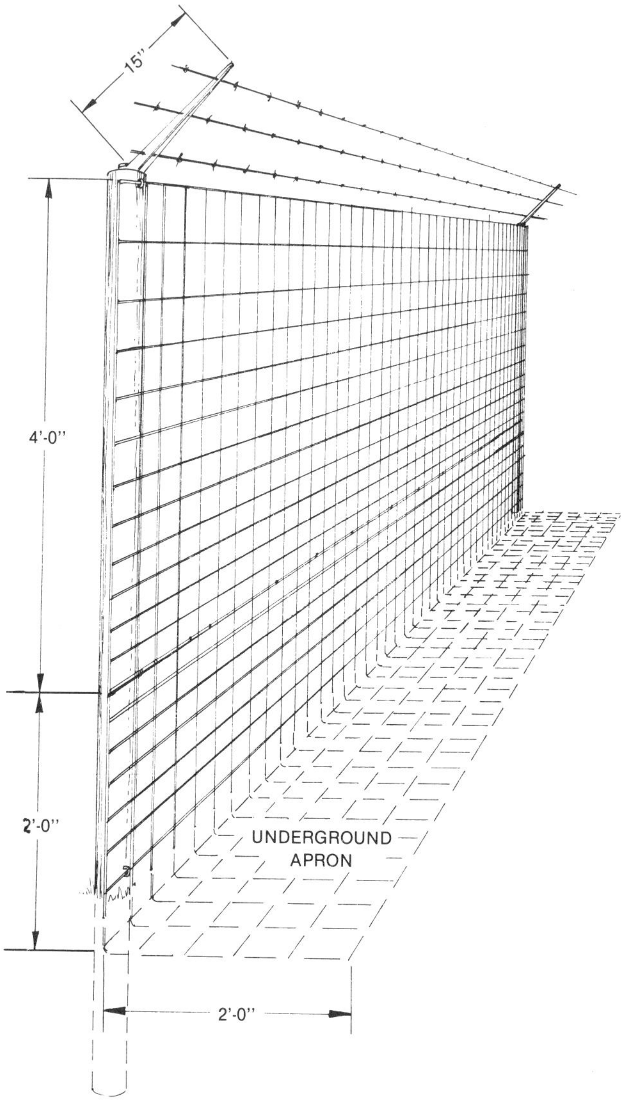

Fig. 9-33. A design for a dogproof or coyoteproof fence which prevents digging under or jumping over. Stays on metal posts should be added between wood posts 30 ft or more apart. The two rolls of woven wire can be joined with hog rings.

DOG-PROOF FENCES FOR SHEEP

Sheep corrals should be fenced with dogproof fencing. Although such fencing is difficult to construct and expensive, it will pay dividends in the protection afforded.

A satisfactory dogproof fence should be a minimum of 6 ft in height (preferably 7 ft). Two rolls of woven wire are used as shown in Fig. 9-33. To keep dogs from digging under the fence, the wide spaced horizontal wires are buried to form an apron. To prevent dogs from jumping the fence, three strands of barbed wire run across the top of an arm as shown in Fig. 9-33.

WOOD FENCES

Wood fencing, either boards or poles, is the traditional fence around the farm or ranch headquarters, especially horse farms. In larger pasture areas, 1 in. × 6 in. or 1 in. × 8 in. boards will suffice, but where animals are confined to small areas, 2 in. × 6 in. or 2 in. × 8 in. boards should be used in order to resist the extra pressure of crowding.

Board fencing protected with electric wire on top, and another electric wire on the corral or pasture side, to keep horses away from it, will last a long time.

The disadvantage of board fencing are the high initial cost, maintenance expense, and susceptibility to rotting, breaking, splitting, and chewing.

For a rustic appearance, rail fences are popular. These are prefabricated and cut from durable wood (such as cedar) which resists rot up to 30 years without treatment. Rail fences have a high initial cost and are eventually susceptible to rot, but are more resistant to splitting and breaking than board fences because rails are thicker than boards.

Fig. 9–34. A strong pole fence. Fences for valuable horses should always be constructed of metal, poles, or 2-in. lumber, and there should not be any projections that might cause injury. Barbed wire fence is always hazardous to horses. (Courtesy, *Sunset Magazine*, Menlo Park, CA)

WIRE MESH FENCES

Wire mesh fencing may be used for horses. The best wire mesh for horses is diamond shaped, in the form of 2 in. × 4 in. rectangles, or chain-linked; fencing with larger mesh may result in a hoof being caught in the wire. A mesh fence will stay tight a very long time provided (1) it is properly braced and stretched, and (2) it is topped by a board, pole, or pipe to prevent horses from leaning over it and mashing it down. Some fence specialists also recommend that a pole be added at the bottom of the fence, and that a diagonal brace be placed between posts.

STEEL PIPE FENCES

Originally, steel pipe fencing was made from surplus or discarded drilling rig pipe from the oil industry. It still is, but some manufacturers now produce steel pipe exclusively for horse fences. The pipe rails are fastened to metal posts set in concrete. Pipe fence is durable and lasts a very long time. Because of its strength and durability, pipe makes a good fencing for stallion paddocks or for boundaries along busy roads. It can inflict injury, however, if a horse runs into it.

RUBBER-NYLON FENCES

This fencing material is made from belting, cut in strips 2 in. to 4 in. wide. Because the strips are maintained under tension, it is important that the anchor posts be well braced. When properly constructed, rubber-nylon fence is durable, safe, low-maintenance, and gives the appearance of a black board fence when viewed from a distance. It is relatively inexpensive. If a horse chews on rubber-nylon fence, and consumes some of the indigestible fibers, it may result in impaction and colic. So, when this type of fence is used for a wood chewer, it is recommended that an electric wire be added on top as a deterrent to nibbling.

POLYMER PLASTIC OR POLYVINYL CHLORIDE (PVC) FENCES

Originally this material was developed for plumbing pipe. It was first used for fencing in the early 1980s. Subsequently, it has been improved and adapted for fencing, for which it is quite popular.

Fences made of this plastic or polymer (1) resemble traditional white board fences; (2) are durable, safe, and low-maintenance; and (3) cannot be chewed by horses. This material is available both as round rails and as 2 in. × 6 in. planks; with round or square posts to match, which are set in concrete. The major disadvantage of this type of fence is the initial cost, which is twice that of a board fence.

Fig. 9–35. PVC fence. (Courtesy, Saratoga Fence Co., Ballston Spa, NY)

PAINT FOR THE FARM AND RANCH[7]

The chief purposes of painting farm buildings and equipment are: (1) to preserve them from the effects of the weather, and (2) to add to their attractiveness. In addition, interior painting, such as is done in most homes, makes walls and ceilings more sanitary and dark rooms lighter.

ESTIMATING QUANTITY OF PAINT NEEDED

The chief factors determining the quantities of paint needed for a given building are: (1) the square feet of surface area to be painted, and (2) the kind of coating material, the character of the surface to be covered, and the number of coats to be applied.

Once the square feet of surface area to be painted has been determined, the number of gallons of paint required for the job can be obtained by referring to Table 9–25, Guide for Estimating Quantities of Paint Needed. Simply divide the total square feet to be painted by the number of square feet a gallon of paint will cover. Thus, in the case of a barn having a total area of 2,448 sq ft and assuming a smooth—previously painted or primed wood surface, the number of gallons of paint is:

2,448 ÷ 400 = 6.1 gal (roughly 6 gal) paint required for 1 coat

12.2 gal (roughly 12 gal) body paint required for 2 coats

12.2 × 10% = 1 gal (approximately) of trim paint required

TABLE 9–25
GUIDE FOR ESTIMATING QUANTITIES OF PAINT NEEDED

Coating Materials	Character of Surface to Be Painted	Square Area Covered by One Gallon: 1 Coat (sq ft)	1 Coat (sq m)	2 Coats (sq ft)	2 Coats (sq m)
Asphalt-asbestos roof cement	Smooth	75	7.0		
Asphalt roof paint	Smooth	175	16.3		
	Rough	125	11.6		
Latex wall paint (interior)	Smooth (prime plaster with solvent based wall primer)	400	37.2		
	Rough—previously painted stipple or textured walls	150–200	13.9–18.6		
Solvent based (interior)	Smooth (prime dry wall with latex primer)	350	32.5		
Latex flat or gloss (exterior)	Smooth—previously painted or primed wood	400	37.2	200	18.6
	Wood shakes	200	18.6	100	9.3
	Smooth sound masonry	300	27.9	125–175	11.6–16.3
	Rough masonry (stucco, brick)	175–250	16.3–23.2	100–150	9.3–13.9
	Filled aggregate block	250–350	23.2–32.5	125–200	11.6–18.6
Solvent based gloss or flat (exterior)	Smooth—previously painted or primed wood	350–400	32.5–37.2		
Solvent based stain	Smooth wood	500	46.5	300	27.9
	Rough siding or shakes	125–175	11.6–16.3	75–100	7.0–9.3
Latex based stain (exterior)	Smooth wood	500	46.5	300	27.9
	Rough siding or shakes	125–175	11.6–16.3	75–100	7.0–9.3
Shellac (interior)	Smooth wood	600	55.7	300	27.9
Spar varnish (exterior)	Smooth wood	500	46.5	275	25.5
Varnish (interior)	Smooth wood	550	51.1	275	25.5
Whitewash	Wood	250	23.2		
	Brick	200	18.6		
	Plaster	300	27.9		

Of course, the figures given in Table 9–25 are conservative estimates for painting under average conditions. As noted in Table 9–25, the quantity of paint is affected by (1) the kind of coating materials, (2) the character of surface to be painted, and (3) the number of coats of paint to be applied. Additional factors affecting the quantity of paint are: (4) the consistency of the paint, (5) the thickness of the applied paint film, (6) the temperature, (7) the manner of application, and (8) the experience of the painter.

[7]In the preparation of this section, the author had the benefit of the authoritative review and helpful suggestions of the following: M. G. Brodie, A. Ogurchak, R. S. Taub, paint specialists of The Sherwin-Williams, Co., Cleveland, OH; and T. A. Melody, paint specialist, The Glidden Co., San Francisco, CA.

PAINT POINTERS

Observation of the following pointers will ensure more durability and satisfaction in farm and ranch paint jobs:

1. **Kind and quality of paint.** The first requisite in planning to paint is to select the right kind of paint. A specific kind of paint, enamel, or varnish is essential for each type of surface to be painted (see Table 9–25).

The second requisite is that the paint shall be of high quality, for the labor cost is the same regardless of the quality, and the cost of paint represents only about one-fourth the total cost of a painting job.

2. **Choice of color.** Although choice of color should remain a matter of personal preference, farm folks should be aware:

a. that red, orange, and yellow are warm colors, which attract attention;

b. that cream, buff, peach, and light tan are also warm colors, but they blend into the surroundings and are pleasing;

c. that green, blue, and violet are cool;

d. that light colors give an impression of greater spaciousness and dark shades tend to shrink areas;

e. that farm buildings can be made to belong to the landscape through the selection of the proper shades of both body paint and trim;

Fig. 9–36. Farm buildings can be made to belong to the landscape through the selection of the proper shades of both body paint and trim.

f. that light colors reflect more illumination than dark ones, an important consideration in interior painting;

g. that the color of the finished job will appear somewhat deeper than that of the color chip (chart); and

h. that deep colors absorb solar heat which can cause marginal substrates to expand excessively, ultimately leading to coating failure. The heat absorption by deep colors can also cause the dwelling interior temperature to be uncomfortably warm.

3. **Preparation of surface.** Proper preparation of the surface prior to painting consists in doing the following:

a. Remove all dirt, dust, or other similar substances by wiping with a cloth or brushing with a stiff brush.

b. Remove all cracked, peeling, loose paint, and chalk. Treat mildewed surfaces with a solution made by mixing approximately 1 quart of household bleach to 3 quarts of water, with the concentration of the bleach varying according to the severity of the mildew. Scrub solution on with a stiff bristly brush. Allow to remain on the surface 30 minutes. Rinse thoroughly; use rubber gloves, goggles, and protective clothing when applying.

c. Smooth rough places with abrasive paper.

d. Touch up all knots, sappy streaks, and bare spots with a good exterior undercoater.

e. Putty, with a putty of a color to match that of the finish, all nail holes and cracks. Where a two- or three-coat job is involved, it is best to putty after the first coat.

f. If grease is present on metals that are to be painted, first clean with paint thinner and scrape and brush free of all dirt, rust, and loose paint.

For masonry, proper preparation of the surface consists of doing the following:

a. **When repainting masonry.** Clean off mildew and any excess chalk; remove loose and peeling paint; spot prime bare spots with a masonry conditioner; and apply latex based topcoats. Porous or sand-blasted stucco or masonry should first be treated with masonry conditioner prior to topcoating with latex paints.

b. **New masonry.** Apply two coats of latex if in sound condition. Observe the following:

(1) Fill aggregate block with a block filler prior to painting.

(2) Either remove form-saving oils or other surface treatments from concrete, or allow to weather two years before repainting.

(3) Allow glazed shingles to weather two years before painting.

(4) Allow new brick to weather at least one year before painting. Also, make certain that the mortar joints are sound.

All undereave and protected areas should be washed with detergent and water, rinsed thoroughly, and allowed to dry before painting. This will help prevent peeling caused by invisible salt deposits that are not removed by normal rain washing. Sand lightly any remaining glossy areas and remove dust prior to painting.

4. **Some paints can be poisonous.** Paints containing heavy metals (lead, mercury, selenium, antimony, or arsenic) are poisonous to humans, animals, poultry, and marine life. Care must be taken that children and animals are not allowed access to freshly painted surfaces or discarded paint containers, and that water supplies and gardens are kept free from contamination.

For safety, it is recommended that lead-free (zinc oxide and titanium base) paint or whitewash be used on the interior of stables and on corral fences for calves.

When painting inside buildings, the windows should be open in order to obtain air circulation and to keep inhalation of vapor to a minimum.

CAUTION: In 1971, Congress declared lead-based paint a health hazard. But, as a result of long prior use, it remains on the walls and woods of buildings across the nation, posing a lasting toxic threat to both people and animals. Lead poisoning has been associated with lead in drinking water and other products, and with inhaling dust tainted by lead-based paint, or ingesting lead-containing paint chips.

5. **Paint is combustible.** Because paints, varnishes, and oils are combustible, (a) they should be stored at safe distances from heat and open flames, and (b) paint-soaked rags should be disposed of promptly in an airtight container.

THE FARM POND[8]

Fig. 9–37. The farm pond. (Courtesy, H. Dietz, USDA, Soil Conservation Service, Fort Worth, TX)

It is estimated that about 50,000 farm ponds are developed annually in the United States; half of them by financial assistance provided by federal and state governments (the usual assistance to qualifying farmers amounts to 50% of the total cost of the pond). It is further estimated that there are approximately one million ponds on American farms. When properly planned and built, ponds are generally successful in regions where a tight subsoil prevails but unsuccessful in those areas where the subsoil is too pervious. If a pond is essential in an area where the subsoil is open, a bentonite clay lining, a geotextile mat, or a combination of both; or a heavy polyethylene lining may be used to prevent seepage from the submerged area of the pond. This will add expense, but it will allow a pond to operate successfully. If used for livestock, the bentonite or bentonite/geotextile liner is preferred since livestock can puncture a polyethylene liner.

[8]In the preparation of this section, the author benefitted from the authoritative review and helpful suggestions of the following: Dr. J. A. Ferguson, Professor, Department of Biological and Agricultural Engineering, University of Arkansas, Fayetteville; Dr. H. Luttrell, Professor and Head, Department of Agricultural Engineering, Institute of Agriculture, University of Tennessee, Knoxville; R. K. Matthes, Agricultural and Biological Engineering, Mississippi State University, Mississippi State; H. Neibling, Assistant Professor, Agricultural Engineering Department, University of Missouri-Columbia; and D. L. Reddell, Professor and Head, Department of Agricultural Engineering, Texas A&M University, College Station.

A well-planned pond contributes to greater farm earnings and may pay for itself in a short time if less expensive sources of water cannot be developed. In addition, it may provide a number of incidental benefits and pleasures for the farm family. Among the multiple uses that may be made of farm ponds are the following:

1. Water supply for livestock.
2. Irrigation of crops.
3. Fire protection.
4. Food and recreation. A pond may be stocked with fish, and it will attract birds and wildlife. Also, it may provide the family with a suitable swimming and skating area.
5. Control of gullies and flooding of farm land by impounding peak runoff water.
6. Domestic use. If special care is taken to protect the water from contaminants, ponds can be used for watering gardens and flushing toilets; and, with chemical purification or boiling (all surface water is considered contaminated according to public health standards), pond water can be used for household purposes.

HOW TO BUILD A POND

There are four major types of ponds in common use: (1) dug-out ponds fed by groundwater, (2) ponds fed by surface runoff, (3) ponds fed by springs or creeks, and (4) off-stream storage ponds. The discussion that follows pertains primarily to the latter three types, which normally require a small earth dam to impound water.

With a little know-how and technical assistance, satisfactory farm ponds can be built by using a farm tractor, scraper, and plow. However, the use of heavy equipment, which can usually be rented or contracted, is faster and often less expensive. Only a minimum of materials need be purchased.

The ensuing discussion will suggest the general procedure to follow in building a dam and the resulting reservoir in most localities.

PRELIMINARY CONSIDERATIONS

The first step is to secure the counsel and advice of the Soil Conservation Service or other local authorities. Specifically, reliable information should be secured on the following points: (1) the suitability of the soil, for holding water and dam construction; (2) the water requirements; (3) the adequacy of the water supply; (4) the quality of the water; (5) the existing state laws pertaining to impounded water; and (6) the probable costs and benefits. These, and similar considerations, will indicate whether the proposal is feasible and sound. It is also wise to see what experience the neighbors have had relative to the above and other points. If the decision is then reached to build the pond, these same authorities can usually provide suggestions relative to (1) design details, (2) construction, (3) periodic inspection during construction, and (4) operation and maintenance.

DESIGNING

Sound design of even a small earth-fill dam is usually so involved that competent engineering assistance is necessary. This can usually be obtained from agricultural engineers in the Soil Conservation Service or in other local or private organizations. These specialists will usually make surveys and plans and evolve construction blueprints similar to Figs. 9–38, 9–39, and 9–40 prior to starting construction.

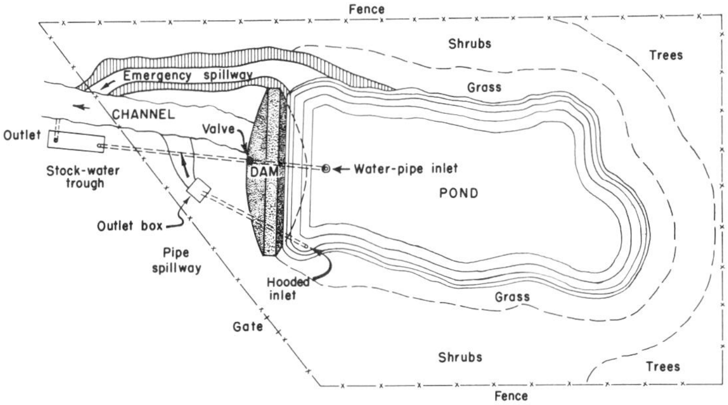

Fig. 9–38. *Left:* Farm pond layout.

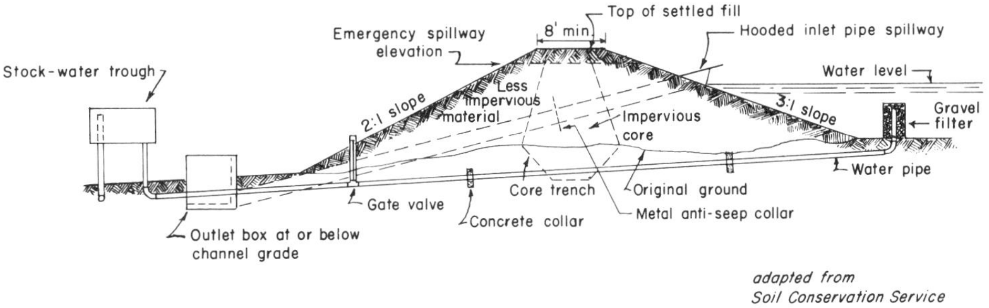

Fig. 9–39. *Right:* Cross section of fill showing certain fill dimensions, core trench, hooded inlet pipe spillway, emergency spillway elevation, and water pipe system.

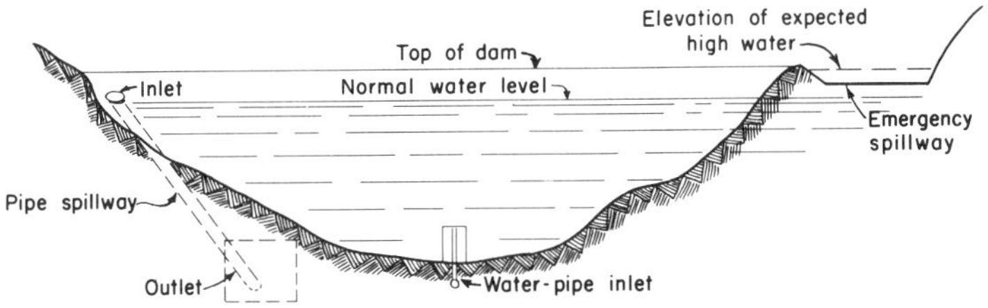

Fig. 9–40. *Left:* Profile through axis of dam, showing the relative elevations of the emergency spillway, pipe, spillway, and top of dam.

SURVEY OF THE AREA

A topographic survey of the watershed and the pond site should first be made by a trained technician. A general survey of the surface and subsurface soil conditions of the area is also necessary and can be noted on the topographic map, together with important features such as vegetative cover, farm unit boundaries, and building sites and roads. The subsurface soil survey should be taken to a depth of at least 10 ft below the bottom of the deepest point of the proposed pond. From these surveys, plans showing the detailed dimensions of the dam and other factors may be made.

LOCATION

The proper pond site is extremely important. Here are the most important considerations relative thereto:

1. **Convenience.** First and foremost, the pond should be located where it will conveniently serve the purposes for which it is intended.

2. **Size of watershed.** The drainage area should be large enough to keep water in the pond during dry periods, but not so large as to flood the pond and cause heavy spillway flow during heavy rains.

3. **Cover of watershed.** Whenever possible, the pond should be located to receive the drainage from a watershed that is covered with permanent grass or trees. If a pond must be constructed in a tilled or partially tilled watershed, precautions should be taken to minimize the loss of pond storage by siltation, such as—

a. Establishing a grass strip with a minimum width of 30 to 50 ft surrounding the perimeter of the pond to intercept much of the sediment from overland flow runoff. Research by Neibling at the University of Missouri shows that strips with relatively dense vegetation as narrow as 4 to 8 ft are effective in removing over 90% of sediment in runoff if flow is uniformly distributed across the slope and is not concentrated into small streams. Additional width will help remove sediment on flat slopes with dense vegetation.

b. Constructing grassed waterways in the major drainage ways, which will both reduce erosion by concentrated flow and remove sediment from the flowing water.

c. Constructing a smaller sediment pond immediately upstream of the main pond to preserve storage in the main pond. Runoff enters the sediment pond, slows, and allows sediment to settle to the bottom. Relatively clean water then flows into the main pond.

If the pond is to be used primarily as a water supply, it should not be located in a deep gully, as it will likely fill rapidly with silt unless an adequate program of gully stabilization is adopted.

4. **Depth of water.** In warm climates the depth of water at the deepest part should be at least 6 to 8 ft; in cold climates, depths of 10 to 12 ft are needed. Depths of 12 to 15 ft are needed in the arid and semiarid parts of the country.

5. **Type of soil material.** The soil material should be suitable; as a foundation for the dam, as fill material for the earth embankment, and as basin material to prevent excessive seepage from the bottom of the pond. Clay or sand clay subsoils are usually best for holding water.

6. **Spillway possibilities.** The most economical spillway is a wide, flat channel, built at either end of the dam in undisturbed soil.

7. **Cost.** Everything else being equal, the most desirable location from the standpoint of minimum construction cost is the one that requires the smallest dam, such as between two hills.

SIZE

A pond should have sufficient storage capacity to meet the intended needs. The size pond to build on a given farm should be determined by four factors; namely, (1) the volume of water needed, based on average daily requirements; (2) the annual rainfall; (3) the kind of vegetation and terrain; and (4) the evaporation and seepage, which vary according to geographic location but which may increase the design capacity by 50 to 100%.

If there is not sufficient volume of water available to fill a large enough reservoir to meet all the intended needs, the size pond should then be limited to the volume of water actually available.

For general planning purposes, the quantity of water, in cubic feet, that can be stored in a given pond may be estimated from the following formula:

Volume (cubic feet) = $\frac{4}{10}$ the greatest depth (ft) × the surface area of the pond (sq ft).

FENCE ANIMALS OUT

Animals should always be fenced away from the pond and dam site in order to ensure clean, clear water and to safeguard them against drinking water contaminated with their own feces. The fence should keep animals away from the upstream sides of the pond and away from the dam.

INSTALL A SPRINGBOARD OR DRIFTING RAFT IF SWIMMING IS DESIRED

If swimming is desired, a support for a springboard may be anchored to the floor of the pond at the time of construction; otherwise, a drifting raft may be anchored in the center of the pond.

STOCK WITH FISH IF DESIRED

Clear water ponds stocked with fish provide both recreation and food. A well-managed pond will yield upwards of 150 lb of fish per acre each year. In addition, a good supply of fish will lessen the mosquito menace.

Information on fertilizing the pond and on the proper kind of fish for stocking can be obtained from state and federal fish hatcheries, from the state agricultural college, or from the U.S. Department of Agriculture.

SAFETY MEASURES

Regardless of the intended use, it is very likely that a pond will be used for swimming, boating, or skating. So, safety measures should be taken. All trees, stumps, brush, rubbish, wire, and discarded machinery should be removed from the ponded area that is likely to be used for swimming. Also, all dropoffs and holes should be eliminated in the swimming area.

The swimming area should be marked with a float line, and warning signs should be installed at all danger points. Also, lifesaving equipment, such as ring buoys, ropes, planks, and long poles, should be placed on the shore near the swimming area. If ice skating is likely, place a long plank or ladder near the skating area for rescue operations.

SCS technicians and county agents can provide plans for a Red Cross approved lifesaving station.

Administering TLC—tender loving care.

SECTION

10

ANIMAL HEALTH/ DISEASE PREVENTION/ PARASITE CONTROL[1, 2]

By
Dr. Robert F. Behlow, D.V.M.,
Professor and Extension Veterinarian, North Carolina State University, Raleigh, North Carolina
and
Dr. M. E. Ensminger, Ph.D.,
Distinguished Professor, University of Wisconsin-River Falls;
Adjunct Professor, California State University-Fresno; and Collaborator, U.S. Department of Agriculture

Contents *Page*

[1]The material presented in this section is based on factual information believed to be accurate, but it is not guaranteed. Where the instructions and precautions herein are in disagreement with those of competent local authorities or reputable manufacturers, always follow the latter two.

[2]The use of trade names of wormers and insecticides in this section does not imply endorsement; rather, it is recognition that livestock producers and those who counsel with them are generally more familiar with the trade names than the generic names. Also, cognizance is taken that, from time to time, FDA (1) bans the use of some old drugs, and (2) approves the use of new drugs.

No claim is made that all wormers and insecticides are listed in this section; so, no criticism is implied of products not named.

Contents — Page

Contents — Page

THEN...AND...NOW!

Fig. 10-1. The cow doctor. How it used to be done. (Courtesy, Bettmann Archive, New York, NY)

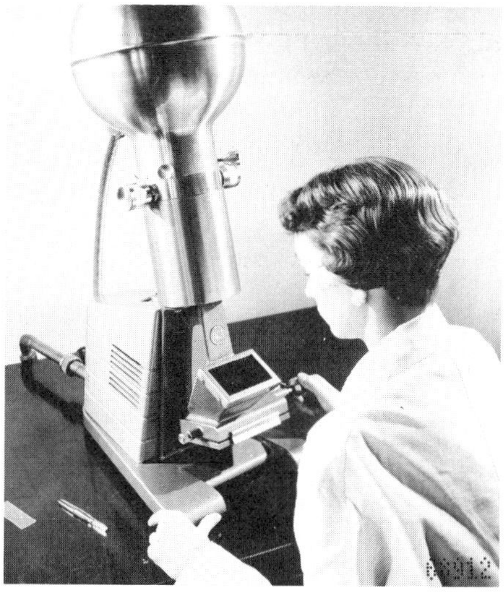

Fig. 10-2. The search goes on. An electron microscope. (Courtesy, Radio Corporation of America, Camden, NY)

Each year, American farmers and ranchers, from one end of the country to another, are robbed of their profits in livestock production—robbed by diseases and parasites. Inevitably, these losses are passed along to consumers, where they add substantially to the cost of food and fiber. It is conservatively estimated that annual U.S. losses from diseases, parasites, and pests of livestock and poultry are equivalent to about 15% of the cash receipts from marketings of livestock and livestock products.[3] Thus, with cash receipts of $76.218 billion in 1987 losses for that year on an equivalent basis of 15% would total $11.4 billion ($76.218 × 15% = $11.4 billion).

Studies, including extensive surveys made by the author, reveal that American farmers and ranchers suffer the following appalling losses:

1. Twelve percent of the cows that are bred never calve.
2. Calf losses from birth to weaning run 6%.
3. Ten percent of all calves (beef and dairy combined) are afflicted with calf scours, and 18% of all dairy calves so afflicted die.
4. Diseases cost cow-calf operators $26.95 per cow in 1990.
5. Cattle feedlot losses on calves run about 2.0%; and on yearlings, about 1.4%.
6. About 1.5 million head of cattle die in feedlots each year, at an estimated loss of more than $750 million.
7. Sterility and delayed breeding in dairy cattle make for an estimated yearly loss of $60 per cow, or a national total of $650 million.
8. Dairy herds average 10% breeding difficulty at any one time and 1.85 services per conception.
9. Retained placenta occurs in about 10% of the parturitions of dairy cattle.
10. Nearly 40% of all dairy cows have some form of mastitis, which according to the National Mastitis Council causes a yearly loss of $225 per afflicted cow.
11. Five percent of the ewes bred never lamb.
12. Twenty percent of the lambs born die between birth and weaning.
13. About 2.5% of lambs on finishing rations die.
14. Fifteen percent of the sows bred never farrow.
15. Twenty-five percent of the pigs born die between birth and weaning.
16. Horse owners and caretakers are spending millions on needless concoctions.
17. One-half of all pregnant mares either abort or produce weak foals.
18. We produce only a 50% foal crop, which means that 2 mares are kept a whole year to produce 1 foal.
19. Six percent of the foals born die between birth and weaning.

Although deaths of animals take a tremendous toll, even greater economic losses—hidden losses—result from failure to reproduce living young, and from losses due to retarded growth and poor feed efficiency, carcass condemnations and decreases in meat quality, and in added labor and drug costs.

Also, considerable cost is involved in keeping out diseases, such as foot-and-mouth disease, that do not exist in this country. Quarantine of a diseased area may cause depreciation in land values or even restrict whole agricultural programs. Additionally, and most important, some 200 different types of infectious and parasitic diseases can be transmitted from animals to human beings; among them, such dreaded diseases as brucellosis (undulant fever), leptospirosis, anthrax, Q fever, rabies, trichinosis, tuberculosis, and tularemia. Thus, rigid meat and milk inspection is necessary for the protection of human health. This is added expense which the producer, processor, and consumer must share.

Despite all these disturbing factors, it is satisfying to know that the United States is regarded as the safest country in the world for a flourishing livestock industry. In order to ensure further progress, however, thousands of workers—including scientists with the U.S. Department of Agriculture, colleges, pharmaceutical houses, practicing veterinarians, and others—are constantly striving to make this country even healthier for both people and animals.

It is not intended that this book shall serve as a source of home remedies. Rather, enlightened livestock producers will institute programs designed to assure herd health, disease prevention, and parasite control. When animal disease troubles are encountered, they will not attempt to diagnose or treat, but will call upon their local veterinarian in exactly the same manner as they call upon the family doctor when human ill health is encountered. But well-enlightened producers will (1) be in a better position to institute programs designed to assure herd health, (2) more readily recognize any serious outbreak of disease and promptly call their veterinarian, (3) prevent unnecessary suffering of sick animals, (4) be better qualified to assist the veterinarian in administering treatment, and (5) be more competent in carrying out a program designed to bring the disease under control with a minimum spread of the infection.

SIGNS OF GOOD HEALTH

In order that caretakers may know when animal disease strikes, they must first know the signs of good health, any departure from which constitutes a warning of trouble. Some of the signs of good health are:

1. **Contentment.** Healthy animals appear contented; the cow will stretch on rising, the sheep will stand or lie quietly, the pig will curl its tail, and the horse will look completely unworried when resting.

---

[3]The author arrived at the 15% by two methods:

1. **Method No. 1:** A report prepared for the Council of Deans, Association of American Veterinary Medical Colleges, based on information available on disease losses as of Feb. 1, 1981, and including cattle (beef and dairy), sheep, swine, poultry, horses, and fish, showed animal disease losses of $10 billion per year. In 1980, cash receipts from marketing of total livestock and products amounted to $67.991 billion. (Source: *Agricultural Statistics 1988*, USDA, p. 409, Table 581). So, a $10 billion disease/parasite loss was equivalent to 14.7% of the cash receipts from marketings of livestock and livestock products that year.

2. **Method No. 2:** *Losses in Agriculture*, Ag. Handbook No. 291, ARS, USDA, 1965, pp. 73–82, Tables 26–32, reported estimated annual U.S. losses from the more important diseases, parasites, and pests of livestock and poultry at $2.8 billion for the period 1951–60. During this same period (1951–60), cash receipts from marketings of total livestock and products averaged $17.76 billion. (Source: *Agricultural Statistics 1962*, USDA, p. 567, Table 688). So, a $2.8 billion disease/parasite loss was equivalent to 15.8% of the average annual cash receipts for the 10-year period 1951–60.

2. **Alertness.** Healthy animals are alert and bright-eyed and will prick their ears at the slightest provocation.

3. **Eating with relish, and cudding by ruminants.** In healthy animals, the appetite is good and the feed is attacked with relish (as indicated by eagerness to get to the trough, wagging the tail, etc.) In cattle and sheep, cudding is a sure sign of good health, and is one of the first things to disappear in sickness.

4. **Sleek coat and pliable, elastic skin.** A sleek, oily coat and a pliable and elastic skin characterize healthy animals. When the hair coat loses its luster and the skin becomes dry, scurfy, and hidebound, there is usually trouble.

5. **Bright eyes and pink eye membranes.** In healthy animals, the eyes are bright and the membranes—which can be seen when the lower lid is pulled down—are whitish pink in color and they are moist.

6. **Normal feces and urine.** The consistency of the feces varies with the diet; for example, when animals are first turned on lush grass they will be loose. Also, the consistency and dryness of the feces vary between species, but they should be firm and not dry. And there should not be large quantities of undigested feed. The urine should be clear. Both the feces and urine should be passed without effort, and should be free from blood, mucus, or pus.

7. **Normal temperature, pulse rate, and breathing rate.** Table 10–1 gives the normal temperature, pulse rate, and breathing rate of farm animals. In general, any marked and persistent deviations from these normals may be looked upon as a sign of animal ill health.

TABLE 10–1
NORMAL TEMPERATURE, PULSE RATE, AND BREATHING RATE OF FARM ANIMALS

Animal	Normal Rectal Temperature		Normal Pulse Rate	Normal Respiration Rate
	Average	Range		
	(degrees F)	(degrees F)	(rate/min.)	(rate/min.)
Cattle	101.5	100.4–102.8	60–70	10–30
Sheep	102.3	100.9–103.8	70–80	12–20
Goats	103.8	101.7–105.3	70–80	12–20
Swine	102.6	102.0–103.6	60–80	8–13
Horses	100.5	99.0–100.8	32–44	8–16

Every caretaker should have an animal thermometer, which is heavier and more rugged than the ordinary human thermometer. The temperature is measured by inserting the thermometer full length in the rectum, where it should be left 2 to 3 minutes. Prior to inserting the thermometer, a long string should be tied to the end.

In general, infectious diseases are ushered in with a rise in body temperature, but it must be remembered that body temperature is affected by stable or outside temperature, exercise, excitement, age, feed, etc. It is lower in cold weather, in older animals, and at night.

The pulse rate indicates the rapidity of the heart action. The pulse of different farm animals is taken at the following body areas: cattle, either on the outside of the jaw just above its lower border, on the soft place immediately above the inner dewclaw, or just above the hock joint; sheep and swine, on the inside of the thigh where the femoral artery comes in close proximity with the skin; and horse, either at the margin of the jaw where the artery winds around from the inner side, at the inside of the elbow, or under the tail. It should be remembered that the younger, smaller, and the more nervous the animals, the higher the pulse rate. Also, the pulse rate increases with exercise, excitement, digestion, and high outside temperature.

The breathing rate can be determined by placing the hand on the flank, by observing the rise and fall of the flanks, or, in the winter, by watching the breath condensate in coming from the nostrils. Rapid breathing due to recent exercise, excitement, hot weather, or stuffy buildings should not be confused with disease. Respiration is accelerated in pain and in febrile conditions.

SIGNS OF ILL HEALTH

Most sicknesses are ushered in by one or more departures from the signs of good health. They're foretold by signs of poor health, by indicators that tell expert caretakers that all is not well—that tell them that their animals will go off feed tomorrow, and that prompt them to do something about it today.

Among the signs of animal ill health are: lack of appetite—the animal does not eat or graze normally; listlessness; droopy ears; sunken eyes; humped-up appearance; abnormal dung—either very hard or watery dung suggests an upset in the water balance or some intestinal disturbance following infection; abnormal urine—repeated attempts to urinate without success or off-colored urine should be cause for suspicion; abnormal discharges from the nose, mouth, and eyes, or a swelling under the jaw; unusual posture—such as standing with the head down or extreme nervousness; persistent rubbing or licking; dull hair coat and dry, scurfy, hidebound skin; pale, red, or purple mucous membranes lining the eyes and gums; reluctance to move or unusual movements; higher than normal temperature; labored breathing—increased rate and depth; altered social behavior such as leaving the herd and going off alone; and sudden drop in production—weight gains, milk, wool, or work.

NUTRITION AND DISEASE/ PARASITE INTERACTION

In general, well-nourished animals are more resistant to bacterial, viral, and parasitic diseases than those poorly nourished. This is attributed to better body tissue integrity, more antibody production, more immunity to disease, greater detoxifying ability, increased blood regeneration, and other factors. Also, it is recognized that proper nutrition, which heads the list of what constitutes *good nursing*, is essential for fast recovery from diseases and the ravages of parasites.

It is estimated that, annually, diseases and parasites in the United States (1) decrease animal productivity by 15 to 20%, and (2) are involved in 85% of the veterinary cases. In the developing countries, diseases and parasites take an even greater toll—they decrease animal productivity by 30 to 40%.

A nutrition program can be fully effective only if the animals are healthy. The converse is equally true. Also, the higher the productivity level, the higher the nutritional and health requirements.

Some noteworthy interactions of nutrition and diseases/ parasites follow:

1. Some nutrients (for example, vitamin A) are important in keeping the epithelial tissue (the skin and mucous membranes), the body's first line of defense, in a healthy condition.

2. Protein, certain B-complex vitamins, and some trace minerals and other nutrients are essential for the production of antibodies and phagocytes, which serve as defenders against infectious agents that enter the body through one of the openings or through the skin.

3. Adequate nutrition is essential for an animal to respond properly to a vaccination. Although many factors affect the effectiveness of a vaccination, studies are showing that the good nutritional health of the animal is one of the most important requisites for a vaccination to induce an immune response.

4. Certain diseases increase the need for various nutrients by animals. This may be due to reduced appetite resulting in inadequate nutrient intake, vomiting or diarrhea resulting in a loss of nutrients from the intestinal tract, fever, decreased absorption or utilization of nutrients, or other causes.

5. Certain minerals, vitamins, and proteins (amino acids) are required at higher than normal levels in order to produce maximum immune response and resistance to diseases. Thus, the decline of immunoglobulins in the milk, followed by early weaning of calves, pigs, and lambs necessitate superior nutrition in order to avoid outbreaks of diarrhea (scours).

6. Many extra nutrients are needed for the repair and restoration of tissues, red blood cells, vital organs, and other parts of the body destroyed by diseases and parasites.

7. Diseases and parasites that cause diarrhea (scours) or vomiting decrease intestinal absorption of nutrients and cause electrolyte loss and dehydration; hence, extra nutrients and electrolytes may be beneficial in treating these conditions.

8. Diseases and parasites that reduce appetite and decrease total feed intake increase the need for higher levels of the nutrients in the feed consumed as a means of ensuring that the total daily nutrient needs will be met.

9. Parasites that cause severe damage to the digestive tract result in impairment of absorption of a number of essential nutrients. Thus, the protozoan parasites, particularly the coccidia, produce profound effects on the digestive physiology of animals.

10. Ketosis increases the need for niacin by dairy cows.

11. Certain nutrients are required at higher than normal levels during stress produced by such things as uncomfortable environmental conditions, vaccinations, crowding, loud noises, debeaking birds, and castrating and dehorning animals. Thus, it is noteworthy that shipping fever (bovine respiratory diseases complex), which causes estimated losses of $500 million annually, is most frequently associated with animals in which resistance has been lowered due to change in weather and feed, overcrowding, hard driving, lack of rest, and improper shelter that accompany shipping.

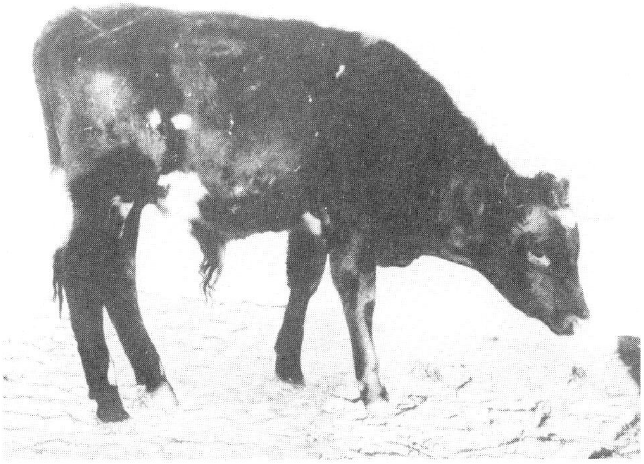

Fig. 10–3. An animal with bovine respiratory disease complex. The disease is most frequently associated with animals whose resistance has been lowered due to travel; hence, the name *shipping fever.* (Courtesy, USDA)

12. Well-fed animals have fewer parasites. For example, there is experimental evidence, substantiated by practical observation, that milk and its by-products are helpful in holding in check some of the internal parasites of swine. Milk-fed pigs make more rapid gains and have fewer parasites in the digestive tract than pigs not receiving milk.

13. Infectious diseases and parasites increase feed costs and lower feed efficiency.

The above list clearly shows that good nutrition is the first requisite of a disease and parasite prevention and control program. It also points up (1) that further intensive studies on these interrelationships hold great potential for discoveries of new ways to improve resistance to diseases and parasites, and (2) that there is a continuing need for cooperative effort between nutritionists and veterinarians.

ANIMAL DISEASES

Disease is defined as any departure from the state of health. Beyond a doubt, the most serious menace threatening the livestock industry is animal ill health. There are many degrees of ill health, but by far the largest loss is a result of the diseases that are due to a common factor transmitted from animal to animal. These disorders are classed as infectious, contagious, and parasitic diseases and are considered theoretically controllable. Today, with the modern rapid transportation facilities and the dense livestock population centers, the opportunity for animals to become infected are greatly increased compared with a generation ago.

CAUSES OF DISEASES

Any agent that may bring about an abnormal condition of any or all tissues of the body is a disease-producing entity. Among the chief causes may be listed infectious agents, including chemicals, poisons of various types, faulty nutrition (see Section 4), and injuries. In addition to the actual causative agents, any of the following conditions may predispose disease; overwork, exposure to cold, and long shipments—especially in cold weather.

Diseases are classified according to the following bases:

1. **Infectiousness.** As either (a) infectious—one caused by the presence in or on an animal body of a foreign living organism that creates disturbances and leads to symptoms; or (b) noninfectious.

2. **Communicability or contagiousness.** As (a) communicable; or (b) noncommunicable.

Fig. 10–4. Germ-free animals aid research workers in fighting diseases. This lamb, being fed by an Ohio Station (Wooster) veterinary researcher, is being reared under sterile conditions for studies aimed at pinning down specific disease-causing agents and finding means of disease prevention or cure. The lamb was taken from its mother by modified hysterectomy and immediately moved to the sterile isolation unit, thus avoiding any contamination by normally present organisms. The germ-free technique has helped the Ohio scientists find causes and control techniques for several important livestock diseases. The technique has been used successively with lambs, pigs, turkeys, and calves. (Courtesy, Agricultural Research and Development Center, Ohio State University)

3. **Manner of occurrence.** As (a) sporadic—one which occurs in isolated cases or outbreaks, like glanders or swamp fever in horses; (b) epizootic—one which appears suddenly and affects many animals over a large area at the same time, like influenza; or (c) enzootic—one which affects certain animals of a given area year after year, like goiter.

4. **Anatomic.** As (a) respiratory; (b) nervous; (c) urogenital, etc.

5. **Course and duration.** As (a) acute—one that runs a rapid course of a few days; (b) subacute—one which runs a slower course and lasts 2 or 3 weeks; or (c) chronic—one which lasts from 4 weeks to an indefinite period.

6. **Prognosis.** As (a) curable; (b) incurable; (c) malignant; or (d) benign.

7. **Origin.** As (a) inherited; (b) acquired; (c) prenatal; or (d) postnatal.

8. **Kinds or types of organisms that produce them.** As bacteria, molds, protozoa, viruses, etc. (see section that follows).

INFECTIOUS TYPES OF ORGANISMS

The principal types of infectious organisms that cause disease may be grouped as follows:

1. **Bacteria.** Are one of the smallest and simplest known forms of plant life. They are microscopic, possess just one cell, vary in shape, multiply by transverse fission, and possess no chlorophyll.
2. **Chalamydia** *(C. psittaci* and C. trachomatis). Are bacteria which lack some important mechanisms for production of metabolic energy and thus lead an intercellular existence.
3. **Flukes** *(trematodes)*. Are soft, flat, leaf-shaped parasitic worms.
4. **Insect larvae.** Are the immature, wingless form of insects.
5. **Moldlike bacteria.** Are somewhat higher in the evolutionary scale than ordinary bacteria. Some have referred to them as higher bacteria.
6. **Molds** *(fungi)*. Are fungi distinguished by the formation of mycelium (a network of filaments or threads), or by spore masses.
7. **Mycoplasmas** *(PPLO)*. Are microscopic organisms intermediate between viruses and bacteria.
8. **Protozoa.** Are the simplest and most primitive form of animal life; they consist of one single cell.
9. **Rickettsiae.** Appear to be intermediate between the bacteria and the viruses.
10. **Roundworms** *(nematodes)*. Are the unsegmented worms, usually cylindrical and elongated in shape and with tapered ends.
11. **Tapeworms** *(cestodes)*. Have bodies made of flattened segments joined together to make a chain. Each segment contains a set of male and female reproductive organs.
12. **Viruses.** May be defined as disease-producing agents that (a) are so small that they cannot be seen through an ordinary microscope (they can be seen by using an electron microscope), (b) are capable of passing through the pores of special filters which retain ordinary bacteria, and (c) propagate only in living tissue. They are generally classified according to the tissue they invade, although this is a very arbitrary method, as some viruses invade many tissues.
13. **Yeastlike fungi.** Are characterized by budding, yeastlike cells.

PARASITES

Broadly speaking, parasites are organisms living in, on, or at the expense of another living organism.

Animals are attacked by a wide variety of internal and external parasites. They include fungi, protozoa (or unicellular animals), arthropods (or insects, ticks, and related forms), and helminths (or worms).

Many diseases cannot be spread unless carried by insects. They are among the most ancient afflictions of people and they have played their part in shaping human history. Malaria has influenced the rise and fall of civilizations, epidemics of plague and yellow fever have decimated populations in both the old and new worlds, and outbreaks of louse-borne typhus have often determined the outcome of military campaigns.

Any animal that serves as a residence for a parasite is referred to as a host. In order to complete their life-span (cycle), some parasites require only one host while others need more.

While in residence, parasites usually seriously affect the host, but there are notable exceptions. Among the ways in which parasites may do harm are (1) absorbing food, (2) sucking blood or lymph, (3) feeding on the tissue of the host, (4) obstructing passages, (5) causing nodules or growths, (6) causing irritation, and (7) transmitting diseases. These may result in death of the affected animal; or they may cause large financial loss through stunted growth, lowered production, general unthriftiness, and emaciation.

Fig. 10-5. Calf with *bottle jaw.* Generally, this condition is indicative of a heavy infection of stomach worms. (Courtesy, School of Veterinary Medicine, Auburn University, Auburn, AL)

The prevention and control of parasites is one of the quickest, cheapest, and most dependable methods for increasing production with no extra animals, no additional feed, and little more labor. This is important, for, after all, the farmer or rancher bears the brunt of this reduced production, wasted feed, and damaged hides. It is hoped that the discussion that follows may be helpful in (1) preventing the propagation of parasites, and (2) causing the destruction of parasites through the use of the most effective anthelmintic or insecticide.

From time to time, new insecticides and vermifuges are approved and the old ones are banned or dropped. Where parasitism is encountered, therefore, it is suggested that the livestock producer obtain from local authorities the current recommendation relative to the choice and concentration of the insecticide and vermifuge to use. This information can be obtained from the county agent, extension entomologist, agricultural consultant, or local veterinarian.

TOXIC PLANTS AND CHEMICALS/DRUGS

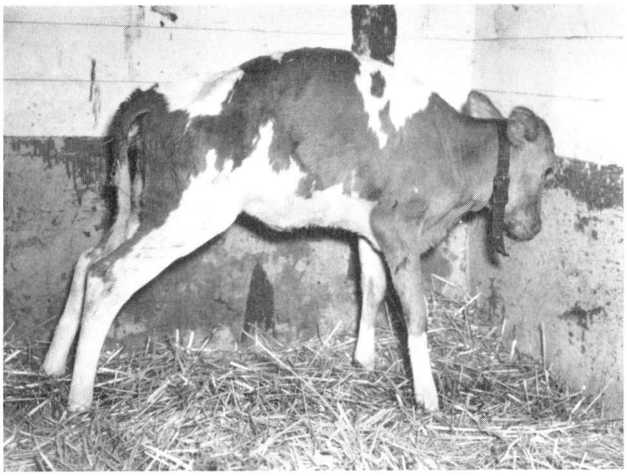

Fig. 10-6. Lead poisoning. Calf shows evidence of gastroenteritis and incoordination of the hind legs. (Courtesy, College of Veterinary Medicine, University of Illinois, Urbana)

Toxic substances are chemical substances that may present an unreasonable risk of injury to health or to the environment. Many poisons are called toxins. Among such toxic substances that may cause disease in animals are certain poisonous plants, chemicals, and drugs.

MODES OF SPREADING DISEASES

Infectious diseases may be spread from one animal to another in a variety of ways; among them, the following:

1. **Direct contact** with diseased animals, in which the infected host actually touches the susceptible animal and transmits the disease. Venereal diseases are spread in this manner.

2. **Indirect contact,** such as (a) by susceptible animals touching infected animals' excretions or secretions, like the placenta, or aborted fetuses; or (b) by susceptible animals breathing infected droplets exhaled from the nose and mouth of infected animals.

3. **Contaminated facilities and equipment,** including vehicles used to transport animals, contaminated feed, water, cattle chutes, syringes, poultry crates, clothes used to wash cows' udders, etc. Gastrointestinal infections, such as salmonellosis, and bovine mastitis are transmitted in this manner.

4. **Vectors,** which include insects, mites, ticks, and snails. In some cases, transmission of the agent by vector is purely mechanical; for example, a biting fly serves as a "flying needle." In other cases, a stage of development, or part of the life cycle, of the infectious agent in the vector is actually necessary before it may be passed on to a new host.

5. **Carrion feeders** (flesh eaters, such as dogs, foxes, or birds) may carry bits of infected carcasses to clean farms.

The rapidity of spread of different infections, their geographical and seasonal distributions, and the relative ease of their prevention and control may all depend, at least in part, upon their means of transmission.

MODES OF PATHOGEN BODY ENTRY

The infection of a tissue and the production of a disease by a living agent is not always easily accomplished. The agent must first gain entrance to the animal by one of the body openings or through the skin. It then usually multiplies and attacks the tissues. To accomplish this, it must be sufficiently powerful (virulent) to overcome the defenses of the animal's body. The defenses of the animal's body vary and may be weak or entirely lacking, especially under conditions of a low nutritional plane and poor management practices.

Pathogens commonly gain entrance into the body through one or more of the following channels:

1. Respiratory tract.
2. Digestive tract.
3. Genital tract, especially during mating or parturition.
4. Wounds.
5. Mucous membranes of the eye, e.g., pinkeye and leptospirosis (the latter may be acquired when the urine of an infected animal gets into the eye).
6. Teat canal, especially in lactating females.
7. Navel cord in the newborn.
8. Contaminated syringes or surgical instruments.
9. Insect bites.

ANIMAL DISEASES TRANSMISSIBLE TO HUMANS

The progress of people, from cave to condominium, has been greatly influenced by animals. From the remote day of their domestication forward, the most advanced civilizations of the day have been the keepers of herds and flocks. Although progress walks Indian file behind animals, certain diseases follow. Many of these diseases are transmitted through meat, milk, or eggs; others are transmitted through close contact with the animals, contact with its excreta, or contact with its products—such as hides, wool, or hair; still others are carried from animals to people by insect vectors.

This group of shared diseases is known as *zoonoses.* It includes African sleeping sickness, anthrax, brucellosis, leptospirosis, Q fever, rabies, trichinosis, tuberculosis, and tularemia. It is important that we realize that many zoonotic diseases cannot be prevented or controlled in people except through their control in animals.

HUMAN DISEASES TRANSMISSIBLE TO ANIMALS

Disease transmission is a two-way street! Some human diseases are transmitted from people to animals, thence back to people. Pertinent information relative to a few of these diseases follows:

1. **Viral infections.** The viruses that cause influenza (Type A) in humans appears to be closely related to those isolated from swine, horses, avians (chickens, ducks, and turkeys), and possibly cattle and sheep. This prompts much speculation relative to the role of animals in the epidemiology of human influenza, particularly as a source of new major antigenic variants. It is now generally accepted that the swine strain is a prototype of the influenza strain responsible for the widespread and severe human epidemic of 1917–18, and that swine got this strain from people. This gives rise to the following unanswered question: Does a human influenza strain have the potential to establish itself in animals, then at some later date be reintroduced into humans, when the latter's antibody status permits?

Chimpanzees are susceptible to the virus that causes the common cold and are capable of transmitting the infection back to people. Also, certain of the great apes may acquire chicken pox from children and serve as a source of infection to other children.

People may transmit the virus that causes mumps to dogs and hamsters, and possibly cats. Present evidence also indicates a close relationship between the virus of measles and canine distemper; thus, the measles virus may be used to immunize dogs against canine distemper.

2. **Bacterial infection.** *Streptococcus pyogens,* the streptococcus that causes scarlet fever in humans may be transmitted from people to cows, then be shed in milk and produce milk borne epidemics of the disease. However, staphylococci of people are usually less virulent in cows than in humans.

3. **Protozoan infections.** Several animal parasites, including amoebic dysentery, may be passed from people to animals (cats, dogs, monkeys, and rats), then back to people again.

DISEASES FOR WHICH ANIMALS SERVE AS PASSIVE CARRIERS

Animals may host the spores of several pathogenic organisms in their intestinal tracts, without exhibiting any symptoms of disease. Among such diseases are the following:

1. **Tetanus.** *Clostridium tetani,* the tetanus organism, is commonly found in the intestine of herbivorous animals (cattle and horses) and to a lesser extent in humans. Manure, and soil that has been fertilized with manure, are prime sources of these spores. It follows that people working around horses and stables are more likely to be carriers than those engaged in other occupations.

In people and farm animals (especially horses, sheep, and goats), tetanus often follows a deep puncture wound, such as may be inflicted by a nail, a firecracker, or a gunshot. The deep puncture provides both an entry (broken skin) and anaerobic conditions for growth of the organism. After such an injury, tetanus antitoxin will give a very effective passive immunity. However, active immunizations, using tetanus toxoid, is the preferred preventive measure.

2. **Gas gangrene.** Like tetanus, gas gangrene infection in people results from the introduction of spores through the broken skin. The clostridia organisms, especially *C. perfringens,* are frequently found in the intestinal tract of most farm animals and people. Gas gangrene is more prevalent in combat zones where people are more apt to be wounded and more likely to be in contact with soil contaminated with animal or human excreta.

3. **Botulism.** Botulism in people and animals is caused by the ingestion of food in which the organism *C. botulinum* has produced toxins. Among farm animals, horses are the most susceptible to the disease. They develop botulism from eating moldy or spoiled forages or grains in which botulism toxin has been produced. Poultry are also quite susceptible to the disease.

DIAGNOSING DISEASES

In addition to physical examination (the signs of good health vs the signs of disease), the veterinarian has several means of diagnosing disease; among them, the following:

1. Skin testing with antigen, e.g., tuberculosis in cattle. This test is considered sufficiently accurate to cause the slaughter of an apparently normal animal that reacts positively.
2. Agglutination tests, in which there is a characteristic clumping together of cells when serum from a positive animal is brought into contact with the specific antigen. This type of test is used in bloodtyping and in diagnosing such diseases as brucellosis, typhoid fever, and tularemia.
3. Microscopic examination of blood, used to identify the type of infectious organism and to make red and white blood cell counts.
4. Chemical tests on blood, urine, feces, milk, saliva, etc.
5. Skin scrapings, which are especially useful for identifying fungi or parasitic mite infections.
6. Microscopic and/or visual fecal examination, primarily for parasites.
7. Strip cup or other simple physical tests on milk.
8. Biopsy and/or autopsy.

DRUGS

Drugs, or medicinal agents, are substances of mineral, vegetable, or animal origin used in the relief of pain or for the cure of disease. Much superstition cloaks the reasons for the recommended use of many drugs that have been employed for centuries. An example of this is liverwort, which was heralded as a sure cure for liver disorders only because it was shaped like a liver. Unfortunately, there is no known cure-all for a large number of diseases or for the relief of a great number of different parasitisms.

Lacking the knowledge of limitations of drugs and the nature of disease, many farmers and ranchers have been sold worthless products. There is a flourishing business in various cure-alls that are sold under such names as *tonic, reconditioner, worm expeller, liver medicine, mineral mixture, mineral and vitamin mix, regulator,* and numerous others. It is poor practice to disregard the advice of reputable veterinarians and experimental workers and to rely on claims made by unscrupulous manufacturers of preparations of questionable or fraudulent nature. Most of these patent drugs are sold for fantastic prices, considering their actual cost, and most of their ingredients are never indicated. To avoid being swindled, purchase should be limited to preparations recommended by the local veterinarian. Fortunately, the Food and Drug Administration has been very vigilant and has been instrumental in the disappearance of many misbranded drugs and remedies from interstate channels.

GENERAL ANIMAL SANITATION AND DISEASE PREVENTION

In order to reduce the possibility of disease, one must adopt certain management practices relative to the environment of the animal. It has been said that domestication and increased animal numbers imply sort of a contract. Caretakers in fulfilling their obligations for services rendered, must protect their animals from the elements, parasites, and diseases, and furnish them sanitary quarters and suitable rations. Abuse leads to the reduction of profits—a case in which money and decency are on the same side of the ledger.

Animals require sanitary quarters. In the wild state, they had access to plenty of fresh air, clean feed, and plenty of room. They are naturally of clean habits and if given the choice will not voluntarily consume contaminated feed nor lie in filth.

VENTILATION

The need for ventilation is not as great for animals as it is for human beings, for most of the animal's life is spent out of doors where plenty of fresh air is available. Ventilation is significant only when animals are in crowded quarters.

Ventilation is the act of causing the movement of air through buildings with the objective of supplanting foul air with fresh air containing needed oxygen. Contrary to common opinion, when a feeling of discomfort is noticed, it is the result of oxygen starvation rather than carbon dioxide poisoning.

The amount of moisture in the air is important. When improper ventilation prevents proper evaporation, the moisture content of the air increases. If humidity rises too high, interfering with heat elimination, heat stroke may ensue. Moist air generally is a more favorable medium for the existence of microorganisms, thus lending itself well to the transmission of contagious diseases. When one animal is infected with a contagious disease and is closely housed with others, an epidemic will usually follow. The air may also pick up various noxious gases, such as ammonia from decomposing urine, which may cause irritation to the sensitive membranes of the mouth, eyes, nose, and respiratory tract.

Ventilation is measured in cubic feet per minute (cfm). The required ventilation differs according to species of animal, size of animal, and outside temperature. (Also, see Section 9.)

HOUSING

Although housing and close confinement predispose animals to more disease, it is often necessary. Housing must frequently be provided to facilitate handling, to combat the elements, or to furnish protection during illness or when young are arriving. Proper drainage and dryness, adequate space, and good lighting are some of the requirements for good housing. In addition, animal quarters must be of such construction as to facilitate proper cleaning, disinfection, and maintenance of sanitary conditions. This includes suitable floors, adequate waste disposal, and proper absorbent bedding. Further discussion of the requisites of livestock buildings is found in Section 9.

Fig. 10–7. Visitors can bring diseases, so precautions must be taken to ensure that a healthy herd stays that way. Plastic boots help ensure against diseases and parasite introduction. (Courtesy, DeKalb Poultry Research, Inc., DeKalb, IL)

ADEQUATE MANURE DISPOSAL

Situations that compel animals to live in close contact with their own excreta are most injurious to physical well-being. Urine, feces, exhalations, and nose and mouth discharges may often contain disease-producing agents, and furnish an ideal medium for the growth of microorganisms. Livestock producers are fully aware of the miraculous recovery many animals undergo when taken from small, unsanitary enclosures to good, clean pastures.

The importance of removing excrement frequently from the immediate surroundings (enclosures, barns, and loafing sheds) cannot be stressed too much. The method of disposal of solid and liquid manure is also very important. As this manure may contain a variety of parasites and eggs, proper disposal offers an excellent opportunity for breaking the life cycles of these parasites. On the other hand, if left in an accessible place for animals, manure can be a rich, never-ending source of disease and parasitism.

In order to ensure the killing of many harmful parasites, one may store manure (for two weeks to a month) so that the heat generated will cause their death. It should be stored in a covered concrete pit and located far enough away from the buildings to prevent contamination. These enclosures should be inaccessible to all animals. Spraying manure pits with a suitable insecticide will inhibit fly development. If the manure is believed to be free of specific infectious microorganisms (for example, tuberculosis, brucellosis, and blackleg), it may be spread daily on arable land containing no animals. Here the purifying elements—such as rain, sunshine, soil, and vegetable processes—will tend to render the manure sanitary. Food and water should always be protected from contamination by manure.

PASTURE ROTATION

Pasture rotation provides a very practical method of control of many diseases and parasites. Permanent pastures used by one species of animal may be regarded as highly dangerous for profitable endeavors. A method by which land areas for pasturage are systematically changed periodically to crop production is recommended.

As many parasites (including bacteria) are often specific for a certain host (for example, bots of horses affect no other animal), frequently pastures may be rotated between different species.

CARCASS DISPOSAL

In the disposal of carcasses, it is a safe rule to assume that all of them are a source of some infection, and subsequently, it is important to adopt proper sanitary precautions.

The most sanitary method of destroying a carcass is to burn it, preferably at the site of death in order to prevent the contamination of surrounding ground. A trench of sufficient size should be prepared, a fire built, and the animal placed on top so that it will be consumed in its entirety.

The most common method of large animal carcass disposal is by burial. So that this method will be effective, the carcass should be buried deep and covered with quicklime. The top of the carcass should be at least 4 ft below the surface of the ground and in soil from which there is no danger of contamination by drainage. Burial should not be near a flowing stream, for this will only serve to spread the disease downstream.

Near large centers of population, rendering plants will take carcasses, and they afford the easiest method of disposal.

When an animal dies, it is recommended that a veterinarian be called immediately to perform a postmortem examination. This is done in an attempt to study the abnormal conditions present and to determine the cause of death. It is never safe for one who is uninformed about specific disease lesions to open an animal carcass. Such practice may not only serve to spread a highly contagious disease, but may also expose the operator to a dangerous infection.

It is also unsafe to feed the carcass to other animals. Such procedure may cause the animal consuming it to become sick, or it may serve only to spread the disease.

DISINFECTANTS

Disinfectants are bactericidal or microbicidal agents that are free from infection (usually a chemical agent which destroys disease germs or other microorganisms, or inactive viruses). Related terms include the following:

1. *Antiseptics, which are chemicals which kill or prevent the growth of microorganisms.* They are formulated to minimize tissue irritation and damage.
2. *Germicides, which refers to the ability of an antiseptic or disinfectant to destroy pathogens.*
3. *Sanitizers, which refers to products used to clean.* Sanitizers may or may not have germicidal properties.

The high concentration of animals and continuous use of modern livestock buildings often results in a condition referred to as disease buildup. As disease-producing organisms—viruses, bacteria, fungi, and parasite eggs—accumulate in the environment, disease problems can become more severe and be transmitted to each succeeding group of animals raised on the same premises. Under these circumstances, cleaning and disinfection become extremely important in breaking the life cycle. Also, in the case of a disease outbreak, the premises must be disinfected.

Under ordinary conditions, proper cleaning of barns removes most of the microorganisms, along with the filth, thus eliminating the necessity of disinfection.

Effective disinfection depends on five things:

1. Thorough cleaning before application.
2. The phenol coefficient of the disinfectant, which indicates the killing strength of a disinfectant as compared to phenol (carbolic acid). It is determined by a standard laboratory test in which the typhoid fever germ often is used as the test organism.
3. The dilution at which the disinfectant is used.
4. The temperature; most disinfectants are much more effective if applied hot.
5. Thoroughness of application, and time of exposure.

In all cases, disinfection must be preceded by a very thorough cleaning, for organic matter serves to protect disease germs and otherwise interferes with the activity of the disinfecting agent.

Sunlight possesses disinfecting properties, but is variable and superficial in its action. Heat and some of the chemical disinfectants are more effective.

The application of heat by steam, by hot water, by burning, or by boiling is an effective method of disinfection. In many cases, however, it may not be practical to use heat.

A good disinfectant should (1) have the power to kill disease-producing organisms, (2) remain stable in the presence of organic matter (manure, hair, soil), (3) dissolve readily in water and remain in solution, (4) be nontoxic to animals and humans, (5) penetrate organic matter rapidly, (6) remove dirt and grease, and (7) be economical to use.

The number of available disinfectants is large because the ideal universally applicable disinfectant does not exist. Table 10–2 gives a summary of the limitations, usefulness, and strength of some of the common disinfectants.

When using a disinfectant, *always read and follow the manufacturer's directions.*

TABLE 10–2
DISINFECTANT GUIDE[1]

Kind of Disinfectant	Usefulness	Strength	Limitations and Comments
Alcohol (ethyl-ethanol, isopropyl, methanol)	Primarily as a skin disinfectant and for emergency purposes on instruments.	70% alcohol—the content usually found in rubbing alcohol.	They are too costly for general disinfection. They are ineffective against bacterial spores.
Boric acid[2]	As a wash for eyes, and other sensitive parts of the body.	1 oz in 1 pt water (about 6% solution).	It is a weak antiseptic. It may cause harm to the nervous system if absorbed into the body in large amounts. For this and other reasons, antibiotic solutions and saline solutions are replacing it.
Chlorines (sodium hypochlorate, chloramine-T)	They will kill all kinds of bacteria, fungi, and viruses, providing the concentration is sufficiently high. Used as a deodorant.	Generally used at about 200 ppm.	They are corrosive to metals and neutralized by organic materials. Not effective against TB organisms and spores.
Cresols (many commercial products available)	A generally reliable class of disinfectant. Effective against brucellosis, shipping fever, swine erysipelas, and tuberculosis. Cresols give good results in foot baths.	Cresol is usually used as a 2 to 4% solution (1 cup to 2 gal. of water makes a 4% solution).	Effective on organic material. Cannot be used where odor may be absorbed.

(Continued)

TABLE 10-2 *(Continued)*

Kind of Disinfectant	Usefulness	Strength	Limitations and Comments
Formaldehyde	Formaldehyde will kill anthrax spores, TB organisms, and animal viruses in a 1 to 2% solution. It is often used to disinfect buildings following a disease outbreak.	As a liquid disinfectant, it is usually used as a 1 to 2% solution. As a gaseous disinfectant (fumigant), use 1½ lb of potassium permanganate plus 3 pt of formaldehyde. Also, gas may be released by heating paraformaldehyde.	It has a disagreeable odor, destroys living tissue, and can be extremely poisonous. The bactericidal effectiveness of the gas is dependent upon having the proper relative humidity (above 75%) and temperature (above 86° F and preferably near 140° F, above 30° C and preferably near 60° C).
Heat (by steam, hot water, burning, or boiling)	In the burning of rubbish or articles of little value, and in disposing of infected body discharges. The steam "Jenny" is effective for disinfection *if properly employed,* particularly if used in conjunction with a phenolic germicide.	10 minutes' exposure to boiling water is usually sufficient.	Exposure to boiling water will destroy all ordinary disease germs, but sometimes fails to kill the spores of such diseases as anthrax and tetanus. Moist heat is preferred to dry heat, and steam under pressure is the most effective. Heat may be impractical or too expensive.
Iodine[2] (tincture)	Extensively used as skin disinfectant, for minor cuts and bruises.	Generally used as tincture of iodine, either 2% or 7%.	Never cover with a bandage. Clean skin before applying iodine. It is corrosive to metals.
Iodophors (tamed iodine)	Effective against all bacteria (both gram-negative and gram-positive), fungi, and most viruses.	Usually used as disinfectants at concentrations of 50 to 75 ppm titratable iodine, and as sanitizers at levels of 12.5 to 25 ppm. At 12.5 ppm titratable iodine, they can be used as an antiseptic in drinking water.	Iodophors are a combination of iodine and detergents. They are inhibited in their activity by organic matter. They are quite expensive. They should not be used near heat.
Lime (quicklime, burnt lime, calcium oxide)	As a deodorant when sprinkled on manure and animal discharges; or as a disinfectant when sprinkled on the floor or used as a newly made "milk of lime" or as a whitewash.	Use as a dust; as "milk of lime"; or as a whitewash, but *use fresh.*	Not effective against anthrax or tetanus spores. Wear goggles when adding water to quicklime.
Lye (sodium hydroxide, caustic soda)	On concrete floors; against microorganisms of brucellosis and the viruses of foot-and-mouth disease, hog cholera, and vesicular exanthema. In strong solution (5%), effective against anthrax.	Lye is usually used as either a 2% or a 5% solution. To prepare a 2% solution, add 1 can of lye to 5 gal of water. To prepare a 5% solution, add 1 can of lye to 2 gal of water. A 2% solution will destroy the organisms causing foot-and-mouth disease, but a 5% solution is necessary to destroy the spores of anthrax.	Damages fabrics, aluminum, and painted surfaces. Be careful, for it will burn the hands and face. Not effective against organisms of TB or Johne's disease. Lye solutions are most effective when used hot. It is relatively cheap. *Diluted vinegar can be used to neutralize lye.*
Lysol (the brand name of a product of cresol plus soap)	For disinfecting surgical instruments and instruments used in castrating and tatooing. Useful as a skin disinfectant before surgery, and for use on the hands before castrating.	0.5 to 2.0%	Has a disagreeable odor. Does not mix well with hard water. Less costly than phenol.
Phenols (carbolic acids): 1. Phenolics—coal tar derivatives. 2. Synthetic phenols.	They are ideal general-purpose disinfectants. Effective and inexpensive. They are very resistant to the inhibiting effects of organic residue; hence, they are suitable for barn disinfection, and foot and wheel dip-baths.	Both phenolics (coal tar) and synthetic phenols vary widely in efficacy from one compound to another. So, note and follow manufacturers' directions. Generally used in a 5% solution.	They are corrosive, and they are toxic to animals and humans. Ineffective on fungi and viruses. Effective against all bacteria including TB organisms.
Quarternary ammonium compounds (QAC)	Very water soluble, ultra-rapid kill rate, effective deodorizing properties, and moderately priced. Good detergent characteristics and harmless to skin.	Follow manufacturers' directions.	They can corrode metal. Not very potent in combating viruses. Adversely affected by organic matter; hence, they are of limited use for disinfecting livestock facilities. Not effective against TB organisms and spores. Not effective against anthrax and tetanus.
Sal soda	It may be used in place of lye against foot-and-mouth disease and vesicular exanthema.	10½% solution (13½ oz to 1 gal water).	
Sal soda and soda ash (or sodium carbonate)	They may be used in place of lye against foot-and-mouth disease and vesicular exanthema.	4% solution (1 lb to 3 gal water). Most effective in hot solution.	Commonly used as cleansing agents, but have disinfectant properties, especially when used as a hot solution.
Soap	Its power to kill germs is very limited. Greatest usefulness is in cleansing and dissolving coatings from various surfaces, including the skin, prior to application of a good disinfectant.	As commercially prepared.	Although indispensable to sanitizing surfaces, soaps should not be used as disinfectants. They are not regularly effective; staphylococci and organisms which cause diarrheal disease are resistant.

[1]For metric conversions, see Section 20, "Weights and Measures."

[2]Sometimes loosely classed as a disinfectant but actually an antiseptic and useful only on living tissue.

A GENERAL PROGRAM OF ANIMAL HEALTH, DISEASE PREVENTION, AND PARASITE CONTROL

Consciously or unconsciously, most livestock producers follow a program of animal health. But, unfortunately, few give sufficient thought to it. As a result, it is often a hit and miss proposition and little attention is given to disease prevention until it is too late—until some disease or parasite takes a heavy toll. At such times, one becomes very well aware that "an ounce of prevention is worth a pound of cure."

Because livestock diseases are costly, it behooves each producer to develop and follow a written-down program of animal health, disease prevention, and parasite control. The owner should take the lead in developing such a program; but the best wit, wisdom and judgment of others should be mobilized—including the local veterinarian, and perhaps the county agent, the vocational agriculture instructor, and/or successful neighbors. Also, it must be recognized that cumulative changes in this program will need to be made from time to time; in light of new problems, new biologics, etc.

Any general program of animal health, disease prevention, and parasite control should embrace certain management practices relative to the environment of the animal; among them, proper ventilation, housing, manure disposal, pasture rotation, and carcass disposal. Additionally, the following general program is applicable to any one or all classes of farm animals:

1. **Provide good housing and ventilation.** Although housing and close confinement predispose animals to disease, it is often necessary; to facilitate handling, to combat the elements, and/or to furnish protection when young are arriving.

Experimental studies indicate that a 1,000-lb cow breathes into the air approximately 10 lb of moisture per day. Thus, from a herd of 42 cows there would be given off 420 lb, or about 50 gal of water per day. This amount of moisture must be removed daily in order to keep the barn free from dampness.

When ventilation is poor and there is a difference of several degrees between inside and outside temperature, the moisture condensation forms frost. Thus, condensation on the walls or other surfaces of barns, along with wet animals, gives evidence of unsatisfactory moisture conditions. Such condensation is objectionable because it is harmful for animals to go from a moist, warm barn into the cold outside air and because the excess moisture causes the structure to decay or deteriorate. Thus, in no case should warmth of a building be obtained at the cost of poor ventilation.

Ventilation may be secured by various systems. The simplest method usually consists of one or more of the following: (a) open shed, (b) open doors, (c) windows that open inwardly from the top, or (d) building or stall partitions left slatted or open at the top. A more complete method of ventilating tight buildings consists of a system of intake and outtake flues operated on the basis of either gravity or forced ventilation. Whatever the system, proper ventilation is one in which the foul air is drawn off and harmful humidity conditions are eliminated without excessive heat loss or creation of drafts.

2. **Keep barn idle for one month or longer.** All barns should be emptied of animals for a minimum period of one month each year, thus permitting thorough cleaning, disinfecting, and drying out.

3. **Provide suitable feed containers.** Avoid feeding off the ground because of the hazard of spreading diseases and parasites.

4. **Control and exterminate rodents and birds.** Without exception, U.S. livestock establishments are plagued by one or more rodents and/or with birds. Among such pests are rats, mice, English sparrows, and pigeons. They should be controlled, and eliminated if possible, because they (a) spread diseases and parasites, (b) damage feeds and buildings, and (c) decrease profits. (For pointers on this subject, see discussion in Section 8 of this book.)

5. **Isolate new animals.** Strictly from a disease prevention standpoint, there is much merit in maintaining a closed herd and not bringing in new animals; but this is not always practical. When new animals must be added to the herd or flock, (a) secure a health certificate signed by the local veterinarian and (b) isolate them in separate quarters—in a separate barn, lot, or pasture—for a minimum period of three weeks and arrange for their feed and care by a separate caretaker.

Thoroughly clean and disinfect the isolation stall after each animal(s) is removed and before a new animal(s) is placed therein. Disinfect with a hot 3% lye solution, followed by the use of another recommended disinfectant such as Lysol (see Table 10–2 Disinfectant Guide).

6. **Restrict commercial stock trucks.** Do not permit commercial stock trucks to drive on the premises unless they have been thoroughly disinfected.

7. **Use caution in showing.** Despite all of their many virtues, there is a disease hazard in showing at livestock shows.

8. **Use disinfectants.** *A disinfectant is defined as any biological, physical, or chemical agent capable of exerting changes in the environment which makes it unfavorable for the continued survival of microorganisms.*

Under ordinary conditions, proper cleaning of barns removes most of the microorganisms, along with the filth, thus eliminating the necessity of disinfection. In case of a diseases outbreak, however, the premises must be disinfected. Also, it is desirable that an adequate foot disinfection program be maintained at all times for all visitors entering barns (see Table 10–2 for proper disinfectant).

9. **Call the veterinarian.** Effective animal health programs call for full cooperation between the producer and the

veterinarian, with the former calling upon the latter in exactly the same manner as well-informed people call upon the family doctor when human ill health is encountered.

SPECIFIC SPECIES PROGRAMS

In addition to the *General Health Program* presented above, each individual farm or ranch should, in cooperation with the local veterinarian and other advisors, develop a specific program adapted to the specific enterprise and specific species. Although the basics will be similar, there is need for a program tailored for each species—for beef cattle, for dairy cattle, for sheep/goats, for swine, and for horses. Among such specific species programs are (1) preconditioning beef cattle, and (2) specific pathogen-free (SPF) pigs.

PRECONDITIONING

Preconditioning is a way of preparing the calf to withstand the stress and rigors of leaving its mother, learning to eat new kinds of feed, and shipping from the farm or ranch to the feedlot. To the cow-calf producer, it is a program of management, nutrition, and immunization.

A study conducted by Washington State University revealed that preconditioning of calves increased profits by $10.56 per head on steers and $7.67 on heifers, after deducting all preconditioning costs. This explains why some feedlot operators are willing to pay $1.00 to $2.00 per cwt more for preconditioned calves, and why others are lowering the price if it is not done.

The term *preconditioning* is new, but the concept has long been recommended. Stated simply, preconditioning is the schedule of practices used in preparing feeder calves to withstand the stress of leaving their mothers, shipping, and adapting to feedlot conditions. It consists of administering generous amounts of TLC (tender, loving care) along with immunological practices and treatment for parasites.

Changed environment; excitement of sorting, loading, and shipping; long periods without feed and/or water; movement through one or more assembly points; change of feed; and exposure to disease—all add up to *fatigue, stress, shrink, and lowered disease resistance.*

Preconditioning is the answer. The steps used in preconditioning may, and should, vary somewhat among areas, farms, and ranches. The important thing is that the program be written down, adhered to rigidly, then certified by both the owner and the veterinarian. Producers should take the lead in developing such a program, but they should seek the counsel of their veterinarians and potential buyers.

Preconditioning is often confused with handling newly arrived feedlot cattle, and backgrounding. This is understandable because all three of them are important phases between weaning and finishing. Yet, each of them is a distinct and separate phase.

Opinions differ rather widely as to what constitutes properly preconditioned cattle. However, the following preconditioning program is presented with the hope that the beef producer will use it (1) as a yardstick with which to compare the existing program, or (2) as a guide so that the animal health team (the producer, the veterinarian, and other advisers) may develop a similar and specific program for their own enterprise.

1. **Handle quietly.** Calves should be handled quietly, with a minimum of excitement.

2. **Dehorn and castrate.** All calves that will eventually go into feedlots should be dehorned (although tipping of horns is acceptable), and they should be castrated unless they are to be fed out as bulls. There is far less stress if calves are dehorned and castrated well ahead of weaning—about 2 months of age is best.

3. **Wean.** Calves should be weaned 30 days ahead of shipment.

4. **Start on feed.** Adjust to feed bunks and water troughs and start on a ration similar to that which they will get in the feedlot. For the first 3 days following weaning, calves should have access to loose grass hay. Additionally, they should be started on a ration of about the following composition:

Crude protein, minimum %	12.0
Calcium, %	0.5
Phosphorus, %	0.3
Vitamin A, IU/lb	5,000
Net Energy for production, (NE_p), Mcal	38
Roughage-concentrate ratio, approx.	40:60

If weaning is totally impractical, calves should be started on a creep feed similar to the above ration.

This type of ration will be very similar to the starting ration that calves will receive when they arrive in the feedlot.

Use medicated feed only on the recommendation of your veterinarian.

5. **Vaccinate.** Vaccinate 2 weeks after weaning. If calves were vaccinated for blackleg, malignant edema, and leptospirosis before 3 months of age, revaccinate. Simultaneously, vaccinate for *red nose* (infectious bovine rhinotracheitis, or IBR), bovine virus diarrhea (BVD), and bovine respiratory disease complex. In some instances, clostridial toxoids for types C and D are needed. Follow your veterinarian's advice for vaccination procedures. If a direct sale to a feedlot is involved, the calves should be vaccinated in keeping with the regular program of the feedlot.

6. **Treat for parasites.** At the time of weaning, and prior to shipment, calves should be checked for both internal and external parasites, and treated as necessary. Usually this in-

volves (a) treating for grubs, through either spray, pour-on, or feed: (b) spraying for lice; and (c) checking for worm eggs, and worming if necessary.

7. **Reduce time from farm or ranch to feedlot.** Every effort should be made to reduce the total time between the moment calves leave the farm or ranch and when they arrive at the feedlot.

Where either truck or rail shipments are longer than 36 hours (the 28-hour law governing rail shipments may be extended to 36 hours upon written request of the owner), unload en route for the purpose of giving feed, water, and rest for a period of at least 5 consecutive hours before resuming transportation.

8. **Reduce stress and exposure to infection.** The stress and exposure to infection during the marketing and transportation periods should be reduced to a minimum.

9. **Provide preconditioning certificate.** It is extremely important that records be kept of all husbandry, nutritional, and medical histories, and that the seller of the feeder cattle should provide the person receiving them with a written record of all of them. This well help the feedlot operator fit the cattle to the existing program and minimize costly and unnecessary procedures.

SPECIFIC PATHOGEN-FREE (SPF) PIGS

Specific pathogen-free pigs are pigs that are free of disease at birth. Pathogen-free pigs are obtained from their dam 2 to 4 days prematurely by hysterectomy, caesarotomy, or hysterotomy. Also, pigs may be caught at natural birth in sterile canvas bags, in sterile basins, or on sterile canvas towels.

Hysterectomy means removal of the uterus. With this technique, the pigs are freed without transversing the birth canal, thereby eliminating any chance of the pigs becoming infected while passing through the birth canal. Although hysterectomy represented the ultimate in disease control, it had one major disadvantage. Many laboratories could not comply with inspection regulations; hence, they experienced difficulty in marketing the carcass of the sow. As a result, this forced commercial laboratories to obtain the pigs by Caesarean section (C-section), with methods developed to keep the newborn pigs separated from the contaminated environment of the sow.

This system, which is licensed, embraces the following provisions: (1) obtaining pigs by a surgical process (Caesarean section) 2 to 4 days before normal birth; (2) rearing pigs in individual isolation until 1 week old, and in groups of 8 to 12 until 4 weeks old; (3) rearing pigs in groups of 10 to 20 from 4 weeks old to maturity, on farms from which all other swine have been removed to which no new stock is introduced, and where the producer avoids contact with other swine; (4) resuming normal birth of SPF pigs on these clean farms; and (5) restocking other "clean" farms—farms that have no swine or only SPF swine, and on which the owner avoids contact with other swine.

Primary SPF herds are those originating from surgically derived stock and maintained in strict isolation. Any new blood lines added to the herd must also be obtained by surgical means. These primary herds are used to supply secondary multiplying herds, which in turn supply breeding stock to commercial swine producers.

The SPF method is drastic and costly. However, at the present time, it is the only means whereby atrophic rhinitis, mycoplasmal pneumonia (MP), and swine dysentery can be controlled and eradicated. In addition to these diseases SPF herds must be validated brucellosis-free, leptospirosis-free, and with no evidence of lice or mange.

The National Swine Repopulation Association, Lincoln, Nebraska, is responsible for supervising the SPF program and for issuing an Accreditation Certificate to those who qualify.

A diagrammatic outline of the SPF method is presented in Fig. 10–8.

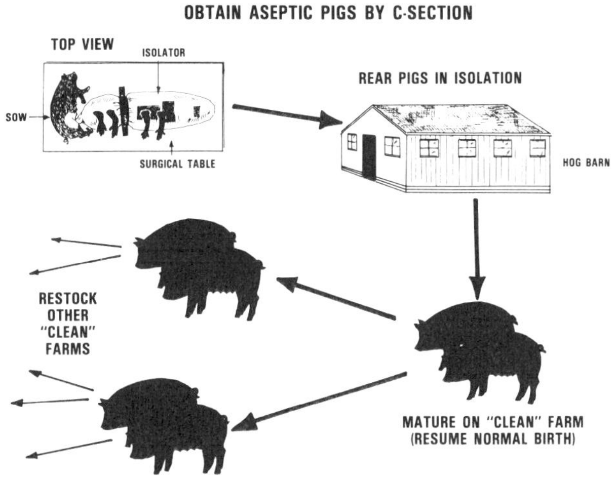

Fig. 10–8. Diagrammatic outline of swine repopulation method.

Table 10–3 is a summary of some of the common nonnutritional diseases and ailments affecting animals. (The nutritional diseases and ailments of animals are covered in Table 4–52 of Section 4.)

At the outset, the author recognized (1) that there are hundreds of animal diseases, (2) that the most important disease to livestock producers at any given time is the one with which they are concerned, and (3) that it is impossible to present a summary of all animal diseases in *Stockman's Handbook Digest*. So, the author arbitrarily selected the diseases which he felt would be most helpful to the readers of this book. For information relative to additional diseases, readers are referred to the latest edition of *The Merck Veterinary Manual* and other good books devoted to animal diseases.

SYMPTOMS & SIGNS OF DISEASES

Fig. 10-9. Blackleg. Heifer with blackleg, 6 hours before death. Note the lameness and the swelling over the neck and shoulder. (Courtesy, Veterinary Research Laboratory, Montana State University, Bozeman)

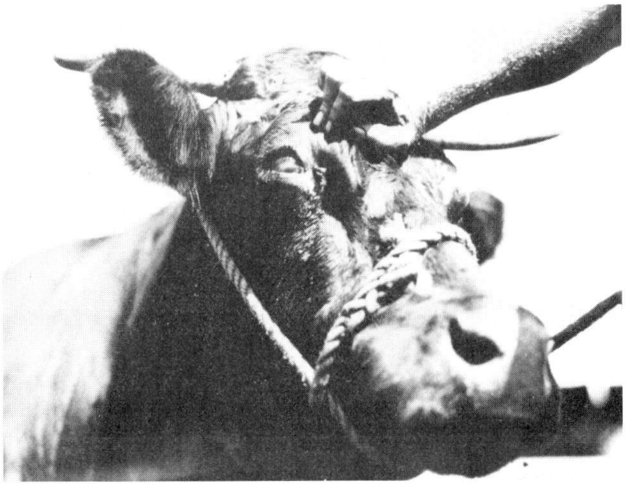

Fig. 10-10. Pinkeye. Note eye discharge and the cloudiness or milkiness of the cornea or covering of the eyeball. (Courtesy, Dept. of Veterinary Pathology and Hygiene, College of Veterinary Medicine, University of Illinois, Urbana)

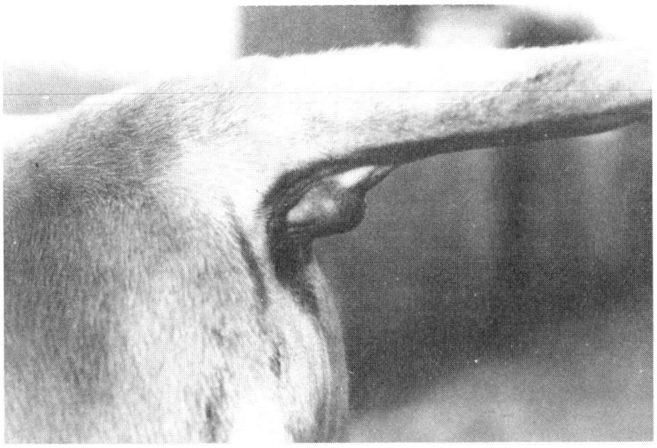

Fig. 10-11. Tuberculosis. A positive reaction (indicating the presence of tuberculosis) to the intradermic (into the true skin) tuberculin test in a cow. Reactors show a noticeable swelling, varying from the size of a pea to the size of a walnut, at the point of injection. The reading is made approximately 72 hours after injection. (Courtesy, Dept. of Veterinary Pathology and Hygiene, College of Veterinary Medicine, University of Illinois, Urbana)

Fig. 10-12. Sore mouth, showing lip lesions in a ewe. This is a highly contagious disease, caused by a filtrable virus. (Courtesy, USDA)

Fig. 10-13. Atrophic rhinitis. Atrophic rhinitis may cause twisted noses in some animals of infected herd. (Photo by J. C. Allen and Son, West Lafayette, IN)

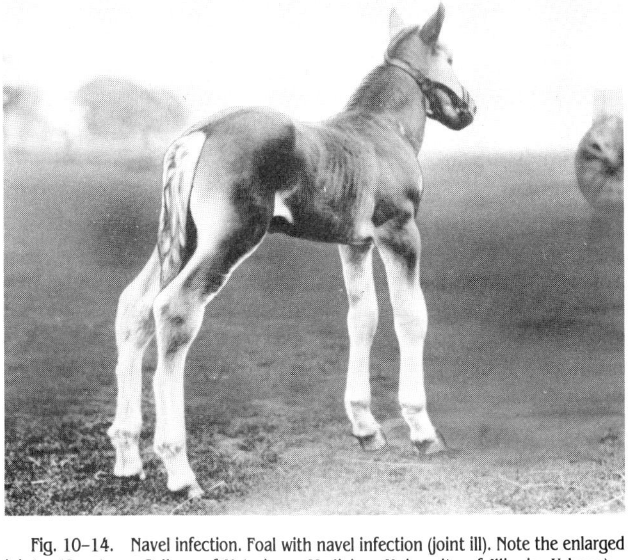

Fig. 10-14. Navel infection. Foal with navel infection (joint ill). Note the enlarged joints. (Courtesy, College of Veterinary Medicine, University of Illinois, Urbana)

ANIMAL DISEASES

TABLE 10-3
ANIMAL DISEASES AND HEALTH[1]

ACTINOMYCOSIS (ACTINOBACILLOSIS) (See LUMPY JAW)

ANAPLASMOSIS (See Table 10–7, Cattle Parasites and Their Control)

ANTHRAX (SPLENIC FEVER, CHARBON)

Anthrax is an acute, infectious disease.

Species Affected: All warm-blooded animals and humans.

Cause: *Bacillus anthracis*, a large, rod-shaped organism.

Symptoms and Signs (or age group most affected): History of sudden death. Sick animals are feverish, excitable, and later depressed. They carry head low and lag behind herd. Respiration is rapid. There are swellings over the body, especially around the neck region.

Milk secretion may turn bloody or cease entirely, and there may be a bloody discharge from all body openings.

Cattle are most susceptible. Most frequent in mature animals on summer pasture.

Distribution and Losses Caused By: General throughout the world in so-called anthrax districts.

Treatment: Early treatment with massive doses of penicillin may be effective.

Control and Eradication: Quarantine infected herds, and withhold all milk and other products from the market until the danger of disease transmission is past.

All carcasses and contaminated material should be burned completely or buried deeply and covered with quicklime, preferably on the spot.

Vaccinate all exposed but healthy animals, rotate pastures, and initiate a rigid sanitation program.

Spray affected and normal animals to avoid fly transmission of infection.

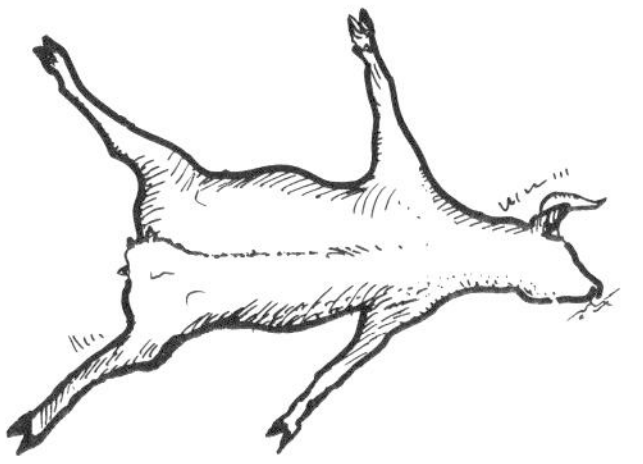

Fig. 10–15.

Prevention: In infected areas, vaccination should be repeated each year, usually in the spring; and there should be adequate fly control by spraying animals during the insect season.

The nonencapsulated Stern-strain vaccine is used almost exclusively for livestock immunization. Vaccination should be done 2 to 4 weeks prior to the season when outbreaks may be expected. Animals should not be vaccinated within 60 days of slaughter.

Prevention of anthrax in people depends on (1) eradication of the disease in animals; (2) elimination of industrial infections (tanneries, woolen mills, and factories utilizing animal hair); and (3) early diagnosis and prompt treatment of infected cases.

Remarks: The farmer or rancher should never open the carcass of a dead animal suspected of having died from anthrax; instead, the veterinarian should be summoned at the first sign of an outbreak.

Control measures should be carried out under the supervision of a veterinarian.

The bacillus that causes anthrax can survive for years in a spore stage, resisting all destructive agents.

ARTHRITIS (See NAVAL INFECTION)

ATROPHIC RHINITIS

Atrophic rhinitis is a transmissible disease of swine characterized by a twisting of the upper jaw.

Species Affected: Swine (apparently the disease in swine is not related to atrophic rhinitis in humans).

Cause: *Bordatella bronchiseptica* and other bacteria. Also, a calcium-phosphorus imbalance or a calcium deficiency in growing pigs has been shown to produce similar, if not identical, lesions.

Symptoms and Signs (or age group most affected): Presistent sneezing, which becomes more pronounced as the pigs grow older, is the first symptom. At 4 to 8 wks of age, the snout begins to show wrinkles, and the snout may bulge and thicken. At 8 to 16 wks of age, the snout and face may twist to one side. Affected pigs become rough all over, and make slow and inefficient gains. Nose bleeding is often seen. Actual death may be due to pneumonia.

Young pigs under 60 to 80 lb weight are most susceptible.

Distribution and Losses Caused By: The disease is quite widespread in the U.S. and has been reported in other countries.

It is estimated that 5 to 10% of the slaughtered swine coming from major swine-producing areas of the U.S. have turbinate atrophy.

Treatment: *B. bronchiseptica* is sensitive to the sulfonamide drugs. The 2 most commonly used ones are:

1. Sulfamethiazine medication in the feed at the level of 100 to 450 g of sulfamethiazine per ton of complete ration.
2. Sodium sulfathiazole administered in the drinking water, at the level of ⅓ to ½ g per gallon of water

Young animals should be treated for 5 weeks, whereas older animals need to be treated only 4 weeks.

Control and Eradication: The following control plans are effective:

1. Isolate bred females in separate lots and never allow contact with any other swine except their offspring until they are culled. Keep individual litters separate until a month after removal of the sow at weaning time. Then select and isolate new breeding stock from those litters which show no evidence of symptoms.
2. Allow the pigs to nurse the sow one or a few times (to obtain colostrum), and then remove and raise them as orphaned pigs; but never allow them to get near the head of their dam.
3. Obtain specific pathogen-free breeding stock.
4. Procure the offspring by hysterectomy and raise them in as nearly sterile an environment as possible.
5. Catch the pigs at birth on a sterile cloth, with subsequent removal to an isolated area.

When considering the 4th and 5th methods given above, caution must be given that hand-rearing baby pigs that have received no colostrum is beset by many difficulties.

Prevention: Select breeding stock from herds known to be clean, and isolate for a period of 30 days or until their litters have been weaned without showing symptoms.

Use clean farrowing quarters.

If feeder pigs are purchased, select animals above 60–80 lb in weight, as they are less susceptible.

Separate different age groups.

Vaccinate with a federally licensed and approved vaccine.

Remarks: Atrophic rhinitis is not the same as bull nose.

No simple test is available to check for carrier swine.

(Continued)

TABLE 10-3 *(Continued)*

BLACKLEG (BLACK QUARTER, QUARTER ILL, EMPHYSEMATOUS GANGRENE)

Blackleg is a very infectious, highly fatal disease.

Species Affected: Cattle, and less frequently sheep and goats.

Cause: *Clostridium chauvoei*, an anaerobic bacterium.

Symptoms and Signs (or age group most affected): *Cattle:* Lameness, and swellings over the neck, shoulder, flanks, thighs, and breast, which crackle under pressure. High fever, loss of appetite, and severe depression. Death usually occurs within 2 days after onset of symptoms. Most frequently seen in cattle ranging from 4 months to 2 years of age, but may occur in older animals

Sheep: Swellings are most frequently in the region of a recent wound; from shearing, castrating, docking, bruising from fighting, or from parturition. In sheep, all ages may be affected.

Distribution and Losses Caused By: Widespread, especially in the western range states. Infected territories are referred to as "hot areas."

Few recoveries.

Treatment: Parenteral and multiple local injections of penicillin will sometimes save an animal, provided given during the early stages of the disease. But a good immunization program is the key to preventing losses from blackleg.

Control and Eradication: Burn or bury carcasses.

Vaccinate all healthy animals.

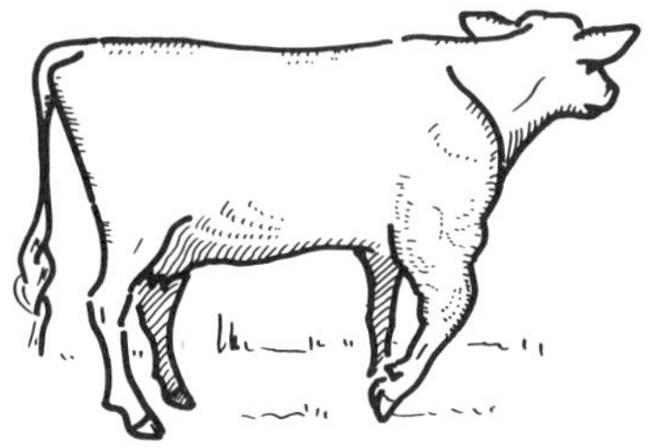

Fig. 10-16.

Eradication of blackleg from pastures is difficult, if not impossible.

When outbreaks are encountered, all susceptible cattle should be vaccinated and treated with penicillin to prevent new cases.

Prevention: Calves should be vaccinated twice, 2 weeks apart, between 2 and 6 months of age. In high-risk areas, revaccination may be necessary at 1 year and every 5 years thereafter.

In "hot areas," vaccinate sheep every 2 to 4 wks, before shearing, castrating, and docking.

Remarks: A bacterin containing *Cl chauvoei* and *Cl septicum*, is a safe and reliable immunizing agent for both cattle and sheep.

BLUE BAG (See MASTITIS)

BLUETONGUE

Bluetongue is an infectious but noncontagious virus disease.

Species Affected: Sheep. Goats. Cattle are sometimes mildly affected. Wild ruminants.

Cause: A virus, transmitted by insects of the *Culicoides* sp.

Symptoms and Signs (or age group most affected): A blue tongue, high temperature (104–107°F), depression and loss of appetite, rapid and extreme loss of weight, reddened mucous membrane of the mouth which turns purplish or blue in color, frothing of the saliva, formation of lip ulcers, offensive odor, discharge from the eyes and nose, weakness, appearance of a red band at the top of the animal's hoof, lameness (and in extreme cases the hoof may slough off), and loss of wool.

When the disease first appeared in the U.S., it was erroneously diagnosed as "sore mouth."

Distribution and Losses Caused By: Bluetongue has been known in South Africa since 1876. Also, it has appeared in Cyprus and Palestine. It was first definitely diagnosed in the U.S. in 1953, although it probably had been present in Texas since 1948 where it was known as "sore muzzle." In the U.S., found only in the Southwest.

In South Africa, mortality rates usually do not exceed 10%, but have run as high as 70%. So far, the disease has proved less virulent in the U.S. (from 5–30% of stricken sheep may die), and the mortality rate has been considerably lower. A severe economic loss results from the reduction in condition and wool.

Treatment: Methods of treatment tried to date have been of little value.

Good nursing will save some affected animals.

Secondary complications may be treated by the veterinarian with appropriate antibiotics and sulfonamides.

May also be necessary to treat or prevent secondary screwworm infections of the lesions in the Southwest.

Control and Eradication: Banning the inshipment of sheep from infected areas will help, but it must be remembered that gnats and other insects that transmit the disease are carried about by automobiles, trucks, and planes.

Prevention: Commercial vaccines are available.

Vaccinate all ewes and rams at shearing time each year, and all replacement lambs at 3½ mos. of age.

Do not vaccinate ewes during the first 2 mos. of pregnancy. Otherwise, a large number of "crazy lambs" may be born.

Night housing of sheep on high land beyond the flying range of the vectors also protects against the exposure of the sheep to the bites of virus-infected midges.

Remarks: It is not transmitted by contact, but it is transmitted by insects—such as sand flies.

Bluetongue becomes prevalent in summer and stops abruptly after the first hard frost.

Rams are more susceptible than ewes.

A sheep that has had bluetongue becomes immune for life against that strain, but not against other strains.

Suspected cases of bluetongue should be diagnosed by the local veterinarian and then reported to state or federal officials.

BOVINE RESPIRATORY DISEASE COMPLEX (BRDC, SHIPPING FEVER, HEMORRHAGIC SEPTICEMIA)

Bovine respiratory disease complex is a respiratory disease associated with *Pasteurella* spp in younger cattle.

Species Affected: Cattle.

Cause: Bovine respiratory disease complex is caused by multiple infection due to the interaction of viruses and bacteria, accentuated by environmental conditions creating physical tension or stress. Changes in weather and feed, overcrowding, hard driving, lack of rest, and improper shelter all help usher in the disease.

The three viruses which cause most bovine respiratory infections are: IBR (infectious bovine rhinotracheitis), BVD (bovine virus diarrhea), and PI3 (parainfluenza 3). Other viruses which may cause respiratory problems include adenovirus, syncytial virus, rhinovirus, and rotavirus. Unfortunately, two or more of these organisms may infect a herd at the same time. But viruses aren't the only agents of respiratory infection; bacteria can cause prob-

(Continued)

TABLE 10-3 *(Continued)*

BOVINE RESPIRATORY DISEASE COMPLEX *(Continued)*

lems, too, especially in cattle already weakened by infections. For example, infection by *Pasteurella multocida* and *Pasteurella haemolytica* is thought to be a major cause of shipping fever (hemorrhagic septicemia). Other bacteria which may infect weakened cattle include *Haemophilus somnus* and species of *Salmonella, Pseudomonas*, and *Leptospira.*

Symptoms and Signs (or age group most affected): The first sign of the disease (which may appear within 2 to 21 days after moving cattle) is a tired appearance and reduced appetite. The affected animal may show signs of depression, watery to slimelike nasal discharge, increased body temperature (rising to 105–107°F *[40.6–41.7°C]*), occasional soft or hacking cough, rapid breathing, loss of appetite, followed by loss of body weight and drop in milk production. In very acute forms, animals may die showing no symptoms. Death losses may be high in untreated cases.

Calves are more susceptible than older animals, but cattle of all ages are affected.

Distribution and Losses Caused By: It occurs widely throughout the world, especially among thin and poorly nourished young animals that are subjected to shipment by truck or rail during periods of inclement weather, though it may occur in animals in good condition.

Bovine respiratory disease complex (BRDC) causes estimated losses of $500 million annually.

Treatment: Antibiotics (e.g., oxytetracycline) and sulfa drugs (e.g., sulfamethazine) are effective treatments if given early in the course of the disease. Treatment after BRDC develops is often disappointing.

Control and Eradication: Bacterial infections can be controlled by bacterins (vaccines) which are composed of killed organisms. Two injections 14 or more days apart will provide adequate protection.

The cattle producer who plans to ship young stock should first confer with a veterinarian relative to the choice of and time to administer the vaccine(s).

Where calves have been subjected to great stress—weaning, long shipment, extensive handling, and/or exposure to severe weather conditions—it is recommended that they be handled as follows: given long grass or oat hay and a calf starter ration, plus access to plenty of clean, fresh water at all times. Also, the incidence of BRDC can be reduced by feeding a combination of an antibiotic (chlortetracycline) and a sulfonamide (sulfamethazine) for 30 days; at the end of the 30 days, the medicated feed should be slowly withdrawn so as to prevent bloat. Newly arrived cattle should also receive 50,000 IU of vitamin A per head daily and have free access to a good mineral mixture.

Prevention: As a preventive measure, one should eliminate as many as possible of the predisposing factors that lower the animal's vitality. Also, newly purchased animals should be isolated for 2 to 3 weeks before being placed in the herd.

Immunity against IBR, BVD, and PI3 can be achieved by administration of modified live or inactivated vaccines, in single or combination forms. The routes of administration of these vaccines are intramuscular (IBR, BVD, PI3) or intranasal (IBR, PI3 only). Both intramuscular and intranasal vaccines provide adequate immunity.

Remarks: The term *shipping fever* is losing favor because it is misleading and the term *hemorrhagic septicemia* is best reserved for the septicemic *Pasteurella* infections seen in cattle.

BOVINE VIRUS DIARRHEA (BVD, MUCOSAL DISEASE)

Bovine virus diarrhea is a disease of cattle caused by a virus.

Species Affected: Cattle.

Cause: As indicated by the name, the disease is caused by a virus.

Symptoms and Signs (or age group most affected): The incubation period is 7 to 9 days following exposure to the virus. The disease is characterized by a high temperature (104–107°F *[40–41.7°C]*), for 2 to 5 days, nasal discharge, rapid breathing, depression, and loss of appetite. Some animals make a prompt recovery. In other cases, signs persist, including nasal discharge and diarrhea (not all animals exhibit diarrhea). Sometimes blood flecks occur in the feces. Coughing, eye lesions, and lameness may affect 10% of the herd. In pregnant cows, abortions may appear 3 to 6 weeks after infection; and, in lactating cows, a marked loss in milk production occurs.

Distribution and Losses Caused By: The disease is widespread in the U.S. The greatest losses are in weight, condition, and feed. Mortality is low, rarely exceeding 5%.

Treatment: Antibiotics or sulfonamides effectively combat the secondary bacterial invaders that accompany the disease. Administration of balanced electrolytes and fluid is indicated to rehydrate animals with diarrhea.

Control and Eradication: Where virus diarrhea is a constant problem, cows and feedlot cattle should be vaccinated. Immunity against BVD can be achieved by the intramuscular administration of modified live or inactivated vaccines. But two **don'ts** should be observed: (1) **Don't** use the vaccine on pregnant cows because of possible abortions and birth defects; and (2) **don't** vaccinate calves under 6 months of age because it may be ineffective due to the temporary immunity from colostrum of immune dams. One vaccination should last a lifetime.

Prevention: The most effective preventive measures consist in avoiding contact with affected animals and in keeping away from contaminated feed and water. Also, all incoming animals should be isolated for at least 30 days. Once the disease makes its appearance, sick animals should be isolated and rigid sanitary measures should be initiated.

The disease may be prevented by use of a vaccine. Follow vaccine manufacturer's directions as to use and schedule.

Remarks: Bovine virus diarrhea (BVD) may also be involved in bovine respiratory disease complex (BRDC); so, see also the latter.

Bovine virus diarrhea is not a good name for this disease since not all animals exhibit diarrhea.

BRUCELLOSIS (BANG'S DISEASE, UNDULANT FEVER, MALTA FEVER)

Brucellosis is a hidden disease; one of the most serious and widespread affecting the livestock industry.

Species Affected: Cattle. Sheep (rarely). Goats. Swine. Humans.

The *suis, abortus*, and *melitensis* strains are seen in horses, and both the *suis* and melitensis strains are seen in cattle, but the incidence is less frequent than *abortus.*

Humans are susceptible to all 3 types.

Cause: *Brucella abortus. B. suis. B. melitensis.*

Symptoms and Signs (or age group most affected): The act of abortion is the most characteristic symptom (especially in cattle), although not all animals that abort are affected and not all affected animals abort.

In cattle, the typical symptoms are (1) abortion in the last third of pregnancy, (2) retained afterbirth, (3) several services per conception, and (4) uterine infections.

There are no marked symptoms in goats.

In swine, abortion and sterility are not so common as in cattle; infection may cause swollen joints and lameness, and swelling or atrophy of the testes, epididymus, and prostate in the male.

In people, the disease is characterized by chills, headache, severe night sweats, fever, and extreme weakness.

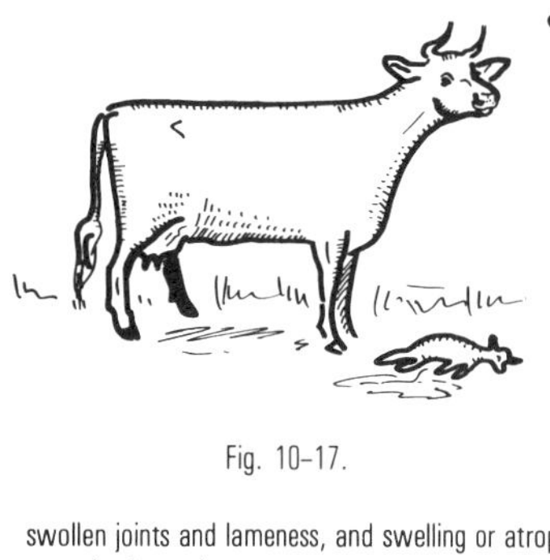

Fig. 10-17.

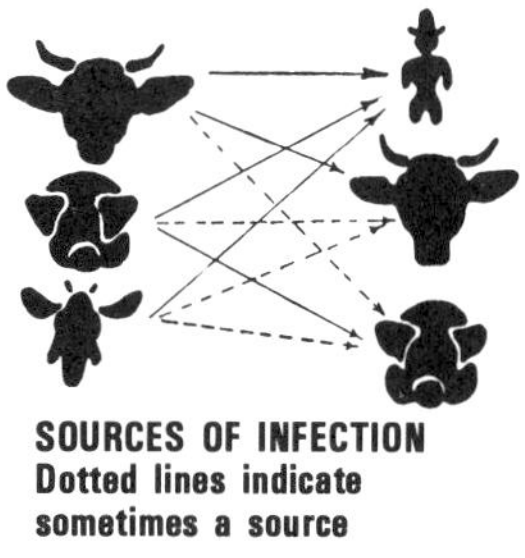

Fig. 10-18.

(Continued)

TABLE 10-3 *(Continued)*

BRUCELLOSIS *(Continued)*

Distribution and Losses Caused By: Worldwide. It is one of the most widespread and ravaging diseases in the world.

It is the most important animal-human disease, and there is great economic loss in fewer animal offspring, in breeding and parturition trouble, in lowered milk production, etc.

For the U.S. as a whole, fewer than ½ or 1% of all cattle tested (including both beef and dairy animals) react.

It is rather common in goats, but rare in sheep.

Treatment: Since there is no successful animal treatment, farmers and ranchers should not waste valuable time and money on so-called cures that are advocated by fraudulent operators.

In humans, the recommended treatment is the antibiotic Auremoycin administered for at least 3 weeks.

Control and Eradication: The nationwide cooperative federal-state brucellosis eradictaion program has been very successful in reducing the incidence of bovine brucellosis in the U.S. The program consists of blood testing and certifying brucellosis-free herds and areas. The certification progresses from an individual herd, thence to an area or county, thence to a state. In 1974, 30 states were certified as brucellosis free and 20 more had modified certified status.

Cattle: Two principles are involved: (1) finding infected animals and eliminating them from the herd, and (2) vaccination. In heavily infected herds where valuable animals are involved, the test-and-slaughter plan is not practical. In such herds, vaccination with Strain 19 at 2 to 10 months of age is recommended.

Currently, a new French vaccine, Strain H-38, is being tested in the U.S. It uses a killed brucellosis organism, whereas Strain 19 uses a live organism. Strain H-38 offers new hope for brucellosis eradication, but it is still in the experimental stage.

In lightly infected herds, blood testing and removal of reactors is recommended. If there is danger of exposure, calfhood vaccination should be used as a protective measure.

A federal-state cooperative plan for the control and eradication of brucellosis is in progress in the U.S. under this program. *Certified* herds are those that are free of the disease; *Modified Certified Areas* are areas that, as a result of complete testing, are considered nearly free of the disease; *Certified Brucellosis-free* are former Modified Certified Areas in which continued testing indicates that the disease has been completely eradicated.

Goats: Blood testing and the elimination of reactors is recommended. It is claimed that strains of *B. melitensis* used as a bacterin (killed vaccine) and as a vaccine induce a high degree of immunity in sheep and goats.

Swine: Several plans for eradication of brucellosis in swine are followed. With an infected commercial herd, it is recommended that the entire herd be sold for slaughter. With a valuable purebred herd, blood testing and slaughter is recommended, with the separation of the pigs from the sows if there are serveral reactors. Vaccination of swine has not been successful and is not recommended.

Prevention: Buy replacement animals that are free of the disease and that are from herds known to be free of the disese. Divert or fence off drainage from infected areas. Animals that are purchased or that are shown should be isolated for 30 days and tested before adding to the herd. Avoid visiting infected farms or premises, as the germs may be brought home on shoes or clothing. For the same reason feeds should not be bought from such farms, and one should be aware of used feed bags.

Vaccination of calves with Strain 19 is effective in preventing brucellosis. Dairy heifers should be vaccinated at 2 to 6 months of age and beef heifers at 2 to 10 months of age. (Check state regulations relative to age to vaccinate.)

Remarks: Brucellosis derives its name from a British Army surgeon, Sir David Bruce, who, in 1887, discovered the bacteria later named *Brucella melitensis*. In cattle, it is called Bang's disease after a Danish veterinarian, who isolated *B. abortus* in 1896.

Pasteurizing milk and cooking meat make these foods safe for human consumption.

There is ample evidence that boars transmit the disease. Bulls are less apt to do so.

The following tests are used for diagnosis of the disease in cattle:

1. *Agglutination test*, of which there are two common methods:

a. *The tube, or "slow" method*—in which a blood sample is taken from the jugular vein; the blood is allowed to clot and the serum to separate; and the serum is mixed in small test tubes with a suspension of specially selected strain of *B. abortus*. Complete agglutination in dilutions of 1:100 and higher are positive.

b. *The plate or rapid test*—This is a rapid agglutination test which is done on a glass slide or plate. The antigen consists of specially selected strains of *B. abortus* stained with gentian violet and brilliant green.

2. *Milk ring test*—This is a modification of the agglutination test which is done with milk. The test involves mixing the antigen with fresh milk. The test depends on the fact that clumps of agglutinated organisms are carried to the surface by rising fat globules. A positive test is indicated by a purple cream layer with white milk below. The milk ring test is a highly efficient and accurate screening test for locating infected dairy herds.

3. *Card test*—This test involves the use of a disposable card on which blood serum or plasma is mixed with buffered whole-cell suspension of *B. abortus* (antigen), which reacts (agglutinates) with antibodies in the blood serum of animals infected with brucellosis.

The agglutination test is also used for diagnosing swine brucellosis. A specially prepared *Brucella* antigen is added to swine blood serum at definite dilution rates. If brucellosis is present, clumping, or agglutination, occurs in the test sample.

CALF SCOURS (NEONATAL DIARRHEA)

Calf scours is an acute, contagious, and often rapidly fatal disease.

Species Affected: Cattle, especially young or newborn calves, under 2 days of age.

Cause: Enterotoxigenic K 99 + *Escherichia coli*, rotavirus, and corona-like virus are the common causative agents. The *E. coli* cause dairrhea in calves from 1 to 3 days of age, the rotavirus and cornavirus cause diarrhea in calves from 5 to 15 days of age.

Symptoms and Signs (or age group most affected): Calf scours can vary from a mild to a severe disease. In the mild form, the main symptom is softer than normal feces. The severely affected calf initially appears depressed and has a lack of appetite. Then begins a severe diarrhea which consists of yellowish, foul smelling, watery or foamy feces. These calves can have a rough hair coat, sunken eyes, and appear emaciated. In very acute cases, death can occur before diarrhea is observed; however, death usually occurs 2 to 3 days after the onset of diarrhea. Some degree of associated pneumonia occurs more frequently in stabled dairy calves than in beef calves on the range.

Distribution and Losses Caused By: Scours is the cause of more calf deaths than all other diseases combined. It is estimated that 10% of all calves in the U.S. are affected by the disease, and that 8% of beef calves and 18% of dairy calves so affected die.[2] It is further estimated that calf scours costs the cattle industry $200 million annually.[3]

Treatment: Treatment of severely affected diarrheic calves should include: discontinuing feeding milk for 24 to 48 hours; giving fluids orally and by infection to combat dehydratioin; administering gastrointestinal, protectants; and giving antibiotcis orally and by injection. The choice of the antibiotic should be made by the veterinarian, for in many areas the bacteria associated with calf diarrhea are resistant to many of the available drugs.

Control and Eradication: To keep the disease away from the herd, one must prevent primary infection of the newborn. This rests on strict sanitary measures and isolation. The disease can be introduced by adding calves or adult animals from another herd. Calf diarrhea frequently occurs when a newly assembled herd begins to calve.

Prevention: The most effective preventive measure of calf scours involves the following three practices: (1) reduce the degree of exposure of newborn animals to the infectious agent, (2) provide resistance with adequate colostrum and optimal husbandry, and (3) increase the resistance of the newborn by vaccination of the dam 2 to 6 weeks before parturition to stimulate antibodies which are then passed on to the newborn through the colostrum.

Remarks: Contributing causative factors include: insufficient colostrum, lack of antibodies in the colostrum in dams that have not been exposed to certain pathogens, stress due to weather, inferior milk-replacers, inadequate housing and hygiene, and poor management.

(Continued)

TABLE 10–3 *(Continued)*

CAPRINE ARTHRITIS ENCEPHALITIS (CAE)

Caprine arthritis is a chronic arthritis in adult dairy goats and encephalitis in young goats.

Species Affected: Goats.

Cause: CAE is caused by a retrovirus, a slow virus that produces disease after a long incubation period, persisting throughout an animal's life. The virus is transmitted from the doe to the kid(s) through the colostrum and milk. Does not showing any symptoms of CAE may carry the virus and transmit it.

Symptoms and Signs (or age group most affected): In kids, it causes paralysis, and in adults it causes arthritis. The late stages of CAE are similar to rheumatoid arthritis in humans.

In any herd, the expression of the disease ranges from 0 to 25% and in those animals that do show clinical symptoms, the rate of progression and the severity of the disease vary markedly. The expression in kids may vary from a barely noticeable unsteadiness of gait to a rapid fatal paralysis. Diligent nursing may maintain paralyzed goats for months, but the damage is permanent. In mature goats, the joints (front knees, hocks, and stifle joints) become swollen or disfigured, with a loss in body weight and a drop in production.

The severity of the arthritis form varies from intermittent lameness or stiffness for years to complete debilitation.

Distribution and Losses Caused By: Between 80 and 90% of the domestic goats in the U.S. are infected by the virus responsible for CAE.

Treatment: At present, there is no effective treatment available. Aspirin or phenylbutazone gives temporary relief from the arthritis pain. Good nutrition delays wasting, and avoidance of cold eases joint stiffness and pain.

Control and Eradication: The rate of infection of newborn kids can be reduced by more than 90% by removing them from infected does as they pass from the birth canal, providing them colostrum that has been heated to 132°F *(56°C)* for 1 hour, pasteurizing the milk, and raising them in isolation from infected goats. Also, the agar-gel immunodiffusion test can be used to monitor infection.

Prevention: Attempts to create a vaccine thus far have not been successful. Currently, the best method of prevention seems to be the establishment of CAE virus-free herds. This involves the routine testing of all animals in a herd with Agar Gel Immunodiffusion Test (AGID) for antibody against CAE. Animals testing positive are culled. However, if AGID-positive does are kept in the herd, then steps to establish a CAE-free herd begin with the kids. All births are observed and kids are removed from the does immediately at birth. Kids are then fed colostrum only from does identified as negative with the AGID test or heat-treated (135°F *[57.2°C]* for 1 hour) colostrum. After receiving colostrum, kids can be raised on pasteurized goat's milk or milk replacer. Another approach is to raise colostrum-deprived kids, but this requires know-how and experience.

CONTAGIOUS ECTHYMA (See SORE MOUTH)

DIARRHEA IN FOALS

Diarrhea in foals is caused by various factors.

Species Affected: Horses, in foals under 6 months of age.

Cause: Various factors may cause diarrhea, including the mare's first heat after foaling, dietary changes, parasites, and infectious agents such as bacteria or viruses. Most cases are mild and self-limiting, but the infectious diarrheas can be life-threatening and cause significant economic loss.

Recently, equine rotavirus has emerged as a significant cause of foal diarrhea. Rotavirus-induced diarrhea is generally seen in foals under 3 months of age and is of particular concern in the very young foal where the ensuing dehydration can be fatal.

Symptoms and Signs (or age group most affected): The symptoms and signs of foal diarrhea are depression, diarrhea, dehydration, and loss of appetite. In severe diarrhea, the foal may have fever and reddened mucous membranes.

Distribution and Losses Caused By: Diarrhea is one of the most common disorders seen in foals, occuring in 70 to 80% of foals under 6 months of age.

The most common diarrhea in foals occurs at about 7 to 12 days of age and coincides with the mare's first heat cycle. Usually, diarrhea is not serious, and the only treatment needed is to wash the feces off the buttocks and apply petroleum jelly to the area. The cause of foal heat diarrhea at the time of the mare's first heat is not known; it is conjectured that it may be due to changes in the mare's hormones at the time which affect the milk, or that it is caused by a change in the microorganisms in the foal's digestive tract. But the diarrhea is seldom serious, and it generally subsides a day or two after the end of the mare's first heat cycle.

Treatment: Diarrhea can most effectively be treated if discovered early. Treatment should be determined by the cause. If severe diarrhea persists for more than a day, fluids and electrolytes should be administered before the foal becomes too dehydrated. The veterinarian may also administer an antibiotic and/or a gut-soother such as Kaopectate.

Control and Eradication: Control and eradication require hygiene at foaling and in the foaling environment, adequate disinfection of the navel at birth, possible use of antibiotics, and adequate consumption of colostrum right after birth.

Prevention: Sanitation constitutes the best prevention. Foaling stalls should be thoroughly cleaned prior to each foaling, and the mare's udder should be washed. Staff hygiene should be strictly enforced. Foals should receive adequate colostrum. To date, no equine vaccine has been developed for the prevention of rotavirus.

DYSENTERY, SWINE
(BLOODY SCOURS, VIBRIONIC DYSENTERY, HEMORRHAGIC DYSENTERY)

Swine dysentery is an acute, infectious disease.

Species Affected: Swine.

Cause: Swine dysentery appears to be caused by the anaerobic bacterium *Treponema hyodysenteriae*, though it probably acts synergistically with other anaerobic bacteria normally present in the digestive track.

Symptoms and Signs (or age group most affected): The most characteristic symptom of swine dysentery is a profuse bloody diarrhea. Sometimes the feces are black instead of bloody and contain shreds of tissue.

Most affected animals go off feed, and there is a moderate rise in temperature. Some pigs die suddenly after a couple of days of illness, whereas others linger on for two weeks or longer. On autopsy or postmortem, the large intestine is found to be inflamed and bloody.

Distribution and Losses Caused By: Coast to coast, but it is most common in the Corn Belt, where the swine population is densest. Outbreaks of the disease are usually associated with animals that pass through central markets or public auctions.

The death rate may vary from less than 10% to more than 90% of the herd, with an average of about 25% unless treatment is effective.

(Continued)

TABLE 10–3 *(Continued)*

DYSENTERY, SWINE *(Continued)*

Treatment: Therapeutic use of antibacterial drugs is effective if treatment is started early. Water medication is preferred at first. Because drug resistant strains may be present, it is essential to choose a drug to which the organism is sensitive. Bacitracin, carbadox, linconycin, nitroimidazoles, tiamulin, and virginiamycin are commonly used therapeutic agents.

Good management and nursing will help. Milk seems to aid recovery. There is a tendency for relapses following treatment.

Control and Eradication: In case of an outbreak, sick animals should be removed from the healthy ones and a rigid program of sanitation initiated.

Prevention: Avoid public stockyards and auction rings, isolate newly acquired animals, and practice rigid sanitation.

Remarks: The disease may be eradicated from infected premises by carrying out a persistent and carefully planned program thaat includes treatment of carrier pigs with bactericidal drugs and thorough cleaning and disinfection of vacated facilities.

ENCEPHALOMYELITIS (See EQUINE ENCEPHALOMYELITIS)

ENTERIC COLIBACILLOSIS (BABY PIG SCOURS, COLIBACILLOSIS, SCOURS)

Enteric colibacillosis is an infectious disease of swine caused by *E. coli.*

Species Affected: Swine.

Cause: Diagnosis requires laboratory tests to be certain that the cause is a pathogenic strain of *E. coli* since other bacteria and viruses can cause diarrhea and not all strains of *E. coli* produce disease.

In pigs less than 7 days old, pathogenic (disease-causing) strains of *Escherichia coli* bacteria rapidly proliferate in the small intestine, producing toxins which produce diarrhea. The danger of diarrhea in baby pigs is the death of piglets through dehydration. This disease is often called baby pig scours, scours, or colibacillosis.

Symptoms and Signs (or age group most affected): In litters a few hours old to a few days old, the first signs may be dead pigs to a mild diarrhea with no indication of dehydration. Those pigs dying with no signs of diarrhea may do so because the loss of fluid into the intestine is so sudden and severe that dehydration results without diarrhea. In other animals, diarrhea may be so fluid and clear, and dribbling from the anus, that close inspection is required to detect it. Pigs which survive for a time become depressed and sluggish. Their eyes will appear sunken; the skin will be bluish gray with a parchmentlike texture; and the bony prominences become exaggerated. However, the piglets are thirsty and they will continue to nurse until they are too weak and depressed.

Distribution and Losses Caused By: Enteric colibacillosis occurs in all swine regions of the world. It is the most common infectious cause of mortality in unweaned pigs.

Baby pig scours is so acute that even with proper treatment death and setbacks in performance make it very costly to producers. It is far better to prevent this disease than to be fighting it.

Treatment: Therapy includes prompt treatment with anibacterial drugs, along with restoration of fluid and electrolyte balance. Antibacterial drug sensitivity testing is helpful to identify effective medication.

Control and Eradication: Prevention is far better than treatment.

The objective in the control of baby pig scours is to reduce the number of *E. coli* gaining entrance into the piglet below the level that it is able to control through its own defense mechanisms.

Quarantine is the method by which control is exercised over the introduction of different serotypes of *E. coli* and other infectious agents to those already existent in the herd.

Prevention: Prevention of baby pig scours can be viewed as a three-pronged approach:

1. A good sanitation program reduces the number of *E. coli* bacteria in the environment. In an environment where the ventilation is poor, the humidity high, and the facilities dirty and wet, large numbers of *E. coli* are present. Moreover, in herds with baby pig scours, affected pigs should be treated and liquid stools removed to limit the spread of the bacteria.
2. An overall program of good management—nutrition and herd health—ensures the delivery of vigourous piglets and satisfactory lactation. Moreover, proper management of the farrowing house to avoid stressful conditions for the piglets, primarily chilling, maintains the resistance of piglets to infections.
3. Immunity passed from the sow to the piglet in the colostrum and milk provides protection until piglets start producing their own antibodies at about 10 days of age. Vaccination of the sow late in gestation with the pathogenic strain of *E. coli* promotes the formation of colostral antibodies which provides the piglets with passive immunity to the *E. coli* until their immune system starts producing antibodies. The veterinarian and the producer need to work together for this portion of the approach to control baby pig scours.

Remarks: The most satisfying preventive procedures for this disease consist of sound management practices; i.e., reducing the infectious rate through hygiene; enhancing the immunity through management of the immune system; and maintaining an environment, especially an adequate temperature, that places few if any stresses on the piglet to reduce its innate ability to withstand the effects of the disease.

ENTEROTOXEMIA, TYPE D (OVEREATING DISEASE, PULPY KIDNEY DISEASE)

Enterotoxemia, Type D, is an acute disease of sheep.

Species Affected: Sheep.

Cause: *Clostridium perfringens*, Type D, an anaerobic bacterium.

Symptoms and Signs (or age group most affected): Loss of appetite, sluggishness, diarrhea, staggering blindly about, and convulsions.

Usually the course of the disease is acute, with affected animals dying within a few hours.

Affects sheep of all ages in a high state of nutrition—on a lush feed of grain, milk, or grass.

Distribution and Losses Caused By: Wherever lambs are finished out.

Overeating disease is responsible for the largest number of death losses in lambs in western feedlots, in unvaccinated feedlot lambs, minimum losses of 1% may be expected. In explosive outbreaks, losses may range from 10 to 40%.

Fig. 10–19.

(Continued)

TABLE 10-3 *(Continued)*

ENTEROTOXEMIA, TYPE D (Continued)

Treatment: Once the disease has developed, treatment is unsuccessful.

Great difficulty is encountered in getting affected lambs back on feed.

Control and Eradication: Control of explosive outbreaks late in the feeding period consists of the following:

1. Reduce the concentrate allowance by 50% for one week or longer.
2. Market all lambs carrying adequate condition for slaughter.
3. Vaccinate the remaining lambs with bacterin or toxoid and gradually return to full feed.
4. Consider the use of Type D antitoxin (which is expensive) to stop the losses. It will confer temporary immunity (2-3 weeks), following which a long-lasting immunity may be established by vaccinating with bacterin or toxoid.

Prevention: *Prevention by management:*

1. Make a gradual transition from range to feedlot.
2. Check lambs for parasites; if necessary worm with a recommended anthelmintic prior to vaccinating.
3. Have ample amount of feed available at all times.
4. Use antibiotic in feed (10 mg/lb feed) to reduce losses from enterotoxemia.

Prevention by vaccination:

1. Vaccinate lambs with either a bacterin or toxoid soon after their arrival in the feedlot, provided they are in good condition and not wet. Allow at least 10 days after vaccination for immunity to develop.
2. Under certain conditions, vaccinate with the bacterin or toxoid (a booster shot) is required 2 to 4 weeks following the first vaccination.
3. Young lamb losses during the first 6 weeks of life may be prevented by vaccinating the pregnant ewes. Ewes that have not been vaccinated previously should be vaccinated twice: 2 to 4 weeks apart, with the second vaccination being given 2 to 4 weeks prior to lambing. Thereafter, an annual booster shot should be given 2 to 4 weeks prior to lambing.

Remarks: This disease affects the biggest and fastest-growing lambs.

The feeling persists among some feeders that a higher than usual incidence of enterotoxemia occurs in feedlot lambs on high-silage rations.

ENZOOTIC ABORTION OF EWES (EAE)

Enzootic abortion of ewes is an infectious disease in sheep manifest by abortion.

Species Afflicted: Sheep.

Cause: EAE is caused by a strain of *Chlamydia psittaci* similar to, and possibly identical to, the one causing bovine abortion.

Symptoms and Signs (or age group most affected): Abortion occurs late in pregnancy, or lambs are full-term stillborn or weak and die soon after birth. Ewes may appear somewhat sick and depressed before and after they abort. Often the placental membranes are retained. The placental membranes and aborted fetuses are essential for proper laboratory diagnosis.

Distribution and Losses Caused By: It has been reported in Scotland, East Germany, West Germany, Hungary, Romania, Bulgaria, South Africa, and the United States. In the United States, EAE is more prevalent among sheep of the western states of the intermountain area. In some areas, EAE causes severe economic losses.

The usual percentage of abortions is small, probably 1 to 2% of the flock, but it has been reported to be as high as 30%.

Treatment: During an outbreak, antibiotics may help, especially when administered to infected newborn lambs and to ewes that carried dead fetuses for some time. Also, secondary bacterial infections resulting from retained placentas necessitate antibiotic treatment.

Control and Eradication: A positive laboratory diagnosis is essential to differentiate enzootic abortion from vibriosis and noninfectious causes of abortion.

Prevention: Some prevention is possible by segregation of aborting animals since the infection is primarily transmitted at lambing time. Also, a vaccine may be available that can be given in combination with the vaccine for vibriosis at the time of breeding.

Remarks: This disease appears to be very similar to vibriosis.

EQUINE ABORTION
(VIRUS ABORTION, EQUINE ARTERITIS, RHINOPNEUMONITIS, BACTERIAL ABORTION)

Equine abortion is a premature expulsion of the fetus.

Species Affected: Horses.

Cause: Causes of abortion in mares may be grouped into two types: (1) infectious agents, such as viruses, bacteria, and fungi; and (2) noninfectious abortions, such as twinning, hormonal deficiencies, congenital anomalies, and miscellaneous causes.

Fig. 10-20.

Symptoms and Signs (or age group most affected): Expulsion of the fetus at any period prior to the time that the foal can survive out of the uterus.

Rhinopneumonitis is a mild, usually nonfatal disease of the upper respiratory tract, commonly seen in young horses in the fall or winter. It is characterized by a cough, a nasal discharge, a loss of appetite, and a temperature of 102–105°F.

As the disease progresses, the temperature returns to normal, but the nasal discharge and cough may persist for several weeks. In older horses, the disease may be so mild as to go unnoticed. Most abortions due to this virus occur between the eighth and eleventh months of gestation, although they may occur as early as the fifth month. Sometimes the foal is born alive at term, but dies at 2 to 3 days of age due to infection by the virus.

The virus of *equine arteritis* may also cause abortion. It produces more obvious signs of illness than equine rhinopneumonitis, including discharges from the eyes and nose, fever (102–106°F), and filling (edema) of the limbs. A laboratory examination is necessary conclusively to establish the presence of the specific virus. Up to 50% of affected pregnant mares may abort.

Bacterial infection is a common cause of abortion in mares. Several species of bacteria have been incriminated. *Salmonella abortus equi*, which was formerly responsible for abortion storms, seems to have eradicated itself. However, the most common cause of bacterial abortion at the present time is organisms of the streptococci group. Other bacteria frequently cultured from aborted feti include *E. coli*, *Klebsiella*, and *Staphylococci*. They may cause abortion at any stage of pregnancy. But, generally *Streptococci* cause abortion during the first 5 months, whereas *E. coli* are more apt to cause abortion during the last half of pregnancy. Bacterial abortion is often characterized by retention of the placenta, as well as by metritis or inflammation of the uterus.

Fungi do not attack the fetus directly; rather they cause degeneration of the placenta so that the fetus has insufficient nourishment. For this reason, the aborted fetus is often small and only a fraction of the normal weight for its gestational age. If abortion does not occur, the foal may be carried to full term and be born in a reasonably vigorous, but undersized and undernourished state. Most mycotic abortions occur during the second half of pregnancy. There is no vaccine.

(Continued)

TABLE 10-3 *(Continued)*

EQUINE ABORTION *(Continued)*

Distribution and Losses Caused By: Incidence is highest in areas where the greatest number of foals are produced.

It is estimated that, for the U.S. as a whole, ⅓ of all pregnant mares either abort or produce weak, infected foals.

Treatment: Isolate mares that have aborted, and accord them good feed and care.

Control and Eradication: Quarantine animals that have aborted.

Burn or bury the bedding and fetus.

Disinfect contaminated premises.

Isolate newly introduced animals.

Prevention: Preventive measures embrace avoidance of all possible causes. It begins with mating only healthy mares to healthy stallions and with being scrupulously clean at the time of breeding.

New horses should always be isolated as a preventative measure, and aborting mares should be quarantined. Where abortions have occurred in the broodmare band, the special cause in the matter of feed, water, exposure to injuries, overwork, lack of exercise, and so forth may often be identified and removed.

Avoid constipation, diarrhea, indigestion, bloating, violent purgatives or other potent medicines—including administering cortisones in late pregnancy, painful operations, and slippery roads.

The following points are pertinent in controlling abortion in a band of broodmares:

1. Prevent rhinopneumonitis by choosing between the following vaccination program and schedule:
   a. Killed vacine given to (1) pregnant mares—5th, 7th, and 9th months of pregnancy, and (2) young animals—2 doses 4 to 6 weeks apart; followed by a booster vaccination at intervals of 6 months.
   b. Modified live virus given to all horses except pregnant mares—2 doses 4 to 8 weeks apart; followed by booster vaccination at intervals of 6 months.
2. Prevent equine arteritis in areas where the disease is a problem by administering the vaccine.
3. Control and prevent bacterial abortion by mating only healthy mares to healthy stallions and observing scrupulous cleanliness at the time of service and examination. Suture mares where necessary.
4. Keep broodmares healthy and in good flesh, and feed a ration that contains all the essential elements of nutrition.

Remarks: Sanitation and herd health are important factors in lessening the incidence of abortions, regardless of kind.

Consult the local veterinarian whenever abortion occurs.

Cattle abortion is not a factor in producing abortion in mares.

EQUINE ENCEPHALOMYELITIS (SLEEPING SICKNESS)

Equine encephalomyelitis is a virus, epizootic (epidemic) disease transmitted by insects.

Species Affected: Horses. Mules. Humans. Birds (chickens, pheasants, etc.). Wild rodents.

Cause: The disease is caused by several distinct viruses. The 3 most active types in the U.S. are Eastern equine encephalomyelitis, Western equine encephalomyelitis, and Venezuelan equine encephalomyelitis.

Symptoms and Signs (or age group most affected): In early stages, the animal walks aimlessly about, crashing into objects. Later it may appear sleepy, standing with a depressed head. Grinding of the teeth, inability to swallow, paralysis of the lips, and blindness may be noted. Paralysis may cause the animal to go down.

If affected animal does not recover, death occurs in 2 to 4 days.

Distribution and Losses Caused By: Since 1930, the Eastern and Western types of the disease have assumed alarming proportions in the U.S. Then, in 1971, Venezuelan equine encephalomyelitis first occurred in the U.S., when an outbreak was reported in Texas.

Generally speaking, mortality from the Western type does not exceed 50%, whereas that from the Eastern and Venezuelan types is 90% or higher.

Treatment: Treatment is not very effective, because of the rapid course of the disease. Since the Western type progresses more slowly and results in a lower mortality rate than the Eastern and Venezuelan types, it lends itself to more supportive treatment. Good nursing is perhaps the best and most important treatment. The maintenance of fluid and electrolyte balance is recommended. No specific therapeutic agent is known to influence the course of the disease.

Control and Eradication: Prompt disposal of all infected carcasses; destruction, if possible of insect breeding grounds; and discouragement of movement of animals from an epizootic area to a clean one.

Fig. 10-21.

Prevention: Prevention entails vaccination of all horses against the 3 strains as follows:

1. For Eastern, Western, and Venezuelan strains: trivalent vaccine with EEE, WEE, and VEE, given IM; initial vaccination given one month before mosquito season, and repeated within the year if the mosquito season is quite long; both injections repeated annually as boosters; given to all ages, including foals beginning at 2-3 mo. of age.
2. For Venezuelan strain: attenuated virus cell culture, given IM; one injection only (do not give to pregnant mares); one booster shot repeated annually; including foals beginning at 2-3 mo. of age.

Remarks: Birds and wild rodents are natural disease hosts for Western type.

Mosquitoes *(Culex tarsalis)* transmit the disease.

Human and horse infections are incidental, out of species infections.

Public health aspects of this disease are unrelated to horse infections.

EQUINE INFECTIOUS ANEMIA (EIA, SWAMP FEVER)

EIA is an infectious virus disease.

Species Affected: Horses. Mules.

Cause: Virus.

Symptoms and Signs (or age group most affected): Symptoms vary, but some of the following are usually seen; high and intermittent fever, depression, stiffness and weakness—especially in the hindquarters, anemia, jaundice, edema and swelling of the lower body and legs, unthriftiness, and loss of condition and weight—even though the appetite remains good. Most affected animals die within 2 to 4 weeks.

Distribution and Losses Caused By: It was first reported in France in 1843, and it has existed in different sections of the U.S. for at least 60 years.

The USDA reported that of the horses tested for equine infectious anemia in 1974, 9,089 or 2.56% tested positive.

Treatment: No successful treatment known.

Control and Eradication: After horse owners test and eliminate all infected animals from their herds, the Coggins Test can be used to protect their stock from reinfection, (1) by buying horses only after they have been tested and found free from the disease, (2) by not allowing untested horses to be stabled or pastured with their own, and (3)

(Continued)

TABLE 10-3 *(Continued)*

EQUINE INFECTIOUS ANEMIA *(Continued)*

by not taking their horses to any assembly point (show, sale, racetrack, trail ride, etc.) where prior testing is not required.

Prevention: Apply the Coggins Test for diagnosing EIA. Repeat the test to confirm all positive reactions. Positive reactors are identified with an "A" in a visible brand or lip tatoo, which stands for anemia. Animals so branded are quarantined and cannot be moved except for slaughter or approved research purposes.

In order to prevent bringing in the disease, in 1976 the USDA amended the import regulations to require that imported horses pass the Coggins Test to assure that they are free of equine infectious anemia.

Remarks: Infected horses may be virus carriers for years and represent a source of danger for susceptible horses.

Beginning in 1977, several states modified or repealed their Coggins testing regulations, in response to industry pressure on the following bases:

1. The Coggins Test is adequate and valuable to confirm a diagnosis of EIA, but it should not be used indiscriminately as a screening test on apparently healthy animals.
2. None of the healthy positive reactors have been shown to have transmitted the disease.
3. The positive horse may never have been sick.
4. The old gelding which grazes across the road likely never has and never will be tested.

EQUINE INFLUENZA

Equine influenza is an acute, highly contagious, respiratory disease.

Species Affected: Horses. Mules.

Cause: Influenza is caused by any one of a group of related viruses.

Symptoms and Signs (or age group most affected): All age horses are susceptible to these viruses; however, young horses, 1 to 3 years of age, are most susceptible. Older animals are more resistant due to previous exposure to these viruses.

Symptoms develop 2 to 10 days after exposure.

Onset is marked by rapidly rising temperature which may reach 106°F. The fever persists 2 to 10 days.

Other signs include loss of appetite, extreme weakness and depression, rapid breathing, a dry cough, and a watery discharge from the eyes and nostrils which is later followed by a white- to yellow-colored nasal discharge.

Distribution and Losses Caused By: Widespread throughout the world. It frequently appears where a number of horses are assembled, such as racetracks, sales, and shows.

Death rate is low, but economic loss is high. It interrrupts training, racing, and showing schedules, and it may force the withdrawal of animals from sales.

Treatment: Horses without complications require only rest and nursing, but antibiotics are indicated where fever persists more than 3 to 4 days or when nasal discharge or pulmonary involvement are evident.

Control and Eradication: Avoid transmission of the virus through contaminated feed, bedding, water, buckets, brooms, on the clothing and hands of attendants, and on transportation facilities.

Prevention: Where available, bivalent inactivated vaccines should be administered according to schedule. Racing and other equestrian authorities require all horses to have received a basic 2-injection immunization, followed by a first booster injection 6 months later, and further boosters annually. However, many authorities recommend that booster injections be given more frequently, viz., every 3 to 6 months.

FESCUE FOOT (FESCUE TOXICITY)

Fescue foot is a dry gangrene of the extremities occuring as a result of consuming toxic fescue.

Species Affected: Cattle, both beef and dairy. Sheep. Horses.

Cause: Fescue foot is associated with a fungus, which lives in the leaves, stems, and seed of the tall fescue plant and is not visible externally. The toxin is a vaso constrictor that affects the blood vessels, which explains the extreme lameness found in the winter months. Occasionally, the circulation is closed to a degree that causes the entire foot to slough off. Such animals walk on stumps of bone.

Symptoms and Signs (or age group most affected): There are variations in the severity of symptoms in animals on toxic pastures. Some animals show no apparent lameness, whereas others show varying degrees of sloughing (necrosis) on the ends of their tails.

Cattle grazing toxic pastures show a poor growth rate, increased temperatures, and increased pulse and respiratory rates. The only complaint that cattle producers make is the fact that the cattle are not doing as well as in previous years. In some herds, the weaning weights decline for 2 or 3 years before producers realize that they have a problem.

The symptoms in mares are: decrease or absence of milk production, prolonged gestation, abortion, and thickened placenta.

Distribution and Losses Caused By: Fescue foot occurs wherever tall fescue is grown.

Cattle that graze on such pastures do not perform well. In cold climates, they occasionally develop a crippling disease, known as fescue foot or fescue toxicity.

Most cases of fescue toxicity occur among animals that graze pure stands of fescue during late fall and winter; and most toxic stands of fescue pasture are several years old. Fescue toxicity is more prevalent in animals suffering from malnutrition or parasitism.

Treatment: No medication is effective for cattle with fescue foot. In severe cases where sloughing has occurred, the animals should be destroyed for humane reasons.

Cattle usually recover completely if they are removed from fescue pasture or fescue hay and are given other feed or pasture as soon as the first signs of the disease appear.

Control and Eradication: Control is based on using fungus-free fescue seed.

When fescue foot is a problem, the animals should be removed from the pasture.

Prevention: The seeding of fungus-free fescue is the best way to prevent fescue foot. Toxic pastures should be renovated and some legume should be seeded with the fescue. It requires good pasture management, along with fertilization, to maintain a good fescue-legume pasture.

FOOT ROT (FOUL FOOT)

Foot rot is an infectious disease of the foot that causes lameness.

Species Affected: Cattle. Sheep. Goats. Swine.

Cause: Foot rot is a contagious, infectious disease caused by the organism *Bacterioides nodosus* in conjunction with *Fusobacterium necrophorum*. Possibly other organisms such as *Spirocheata penortha* and *Corynebacterium pyogenes* may be involved. Since the disease can be spread only by infected animals, early diagnosis followed by proper treatment

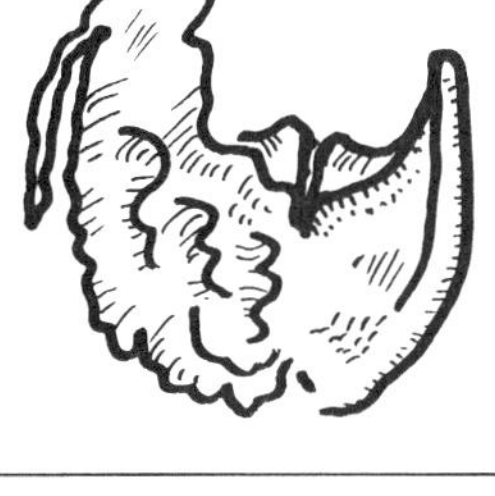

Fig. 10-22.

(Continued)

TABLE 10-3 *(Continued)*

FOOT ROT *(Continued)*

will prevent spread of the disease and will result in considerable saving in time and expense. Walking animals over contaminated areas where infected animals have been is the principal means of spreading the disease, though the incidence is influenced by the weather and temperature.

Symptoms and Signs (or age group most affected): Usually the first indication of the presence of foot rot is severe lameness in one or more animals, though not all lameness is caused by foot rot. In the early stages of the disease, the skin in the cleft between the toes and the soft tissues on the inside surface of the heels is slightly swollen, reddened, and moist. Within 3 or 4 days, this skin becomes grayish-yellow or dead-looking. If not checked at this stage, the disease spreads under the hoof, causing a separation of the hoof from the underlying soft tissue. A grayish-yellow pus oozes from the affected parts.

Foot rot is accompanied by a characteristic foul odor which attracts flies in warm weather. Frequently the affected feet become infested with maggots.

Sheep symptoms/signs: Foot rot in sheep rarely involves the coronary band or extends above the top of the hoof. Since the tissue from which the hoof is formed is not affected, the hoof continues to grow, resulting in a misshapen hoof.

A sheep may have foot rot in more than one foot at the same time. If both front feet are affected, the animal may kneel when grazing. In cases where more than two feet are involved, the sheep will spend most of the time lying down, and may even refuse to stand.

Distribution and Losses Caused By: Foot rot occurs in all seasons, but tends to be most prevalent during wet seasons and in muddy areas. It is the major cause of lameness of cattle and sheep.

Although foot rot is not highly fatal, it is of considerable economic importance. Affected animals become unthrifty and less productive; young animals become stunted; and males may be rendered useless at breeding time.

It is not uncommon for sheep producers to dispose of entire flocks because of this disease..

Treatment: *Cattle treatment:* Systemic and local treatment with antibiotics and sulfonamides is recommended. Other procedures that may speed recovery are cleaning the foot, applying a protective dressing, wiring the claws together, and removing the necrotic interdigital mass. Zinc methionine has been recommended for both treatment and prevention. Walking cattle through a 3% formalin foot bath, a 5% copper sulfate foot bath, or mixed powdered copper suflate and lime twice a day decreases the incidence of foot rot.

Sheep treatment: Place infected sheep in a clean, dry pen and treat as follows:

1. Examine every foot of every animal. Trim each foot showing infection, removing enough of the horn of the hoof thoroughly to expose all diseased tissue.

2. Walk all animals through a suitable disinfectant solution and move to clean ground. Visibly affected animals should be kept standing in the solutioin 5 to 10 minutes. The two most widely used disinfectants are (1) formaldehyde, 10%, and (2) copper sulfate, 20%.[4] Repeat foot bath at weekly intervals until foot rot disappears. Then continue at 2-week intervals for another 2 months. Two weeks after initial antiseptic treatment, examine feet of each animal a second time, to detect and trim infections overlooked the first time or developed subsequently.

3. After trimming, and treatment in a foot bath, place animals in a clean, dry pasture or lot. One that has not been used for 30 days would be considered clean. An animal with foot rot may spread infection up to 3 years, but contaminated land loses its ability to infect within 3 weeks.

Sulfonamide or antibiotic therapy may accompany trimming and foot baths with good results.

Swine treatment: Application of sulfonamide ointment, formalin solution, or copper sulfate. Also, large doses of penicillin have been reported to be effective.

Control and Eradication: Control of foot rot in cattle, sheep, and swine is best achieved by alleviating muddy or abrasive walking areas, regular trimming of hooves, sanitation, isolating infected animals, and use of a suitable disinfectant.

A foot rot vaccine for sheep has been used on an experimental basis with variable results. It has been beneficial in some flocks, but not in others.

Prevention: Preventioin of foot rot includes draining muddy pastures and segregating new animals. Foundation and replacement animals should be purchased from known clean sources. If animals are from a questionable or unknown source, pass through a public market, or are transported by a public conveyance, their hoofs should be trimmed on arrival, and then they should be walked through a foot bath and isolated for 1 month. Cross infections of foot rot between cattle and sheep do not occur, but cross infections between sheep and goats do occur.

Sheep prevention: The best preventive measure is a 10% zinc sulfate bath, made by mixing 8 lb of zinc sulfate/10 gal of water, and 1 to 2 in. deep. For best results, the foot bath should be placed between the pasture and the water supply.

Cattle prevention: Oral iodides have been beneficial as preventives in some cases.

Swine prevention: Swine housed in confinement, or on artificial floors, develop more foot lesions than animals on pasture. The ideal floor should provide adequate traction, be non-abrasive, and comfortable to lie upon, easily cleaned or self-cleaning, durable, and reasonable in cost.

GARGET (See MASTITIS)

HEMORRHAGIC SEPTICEMIA (See SHIPPING FEVER)

INFECTIOUS ATROPHIC RHINITIS (See ATROPHIC RHINITIS)

INFECTIOUS BOVINE RHINOTRACHEITIS (IBR, RED NOSE)

Infectious bovine rhinotracheitis is a disease of cattle caused by a virus.

Species Affected: Cattle.

Cause: Virus.

Symptoms and Signs (or age group most affected): Affected animals go off feed and lose weight; generally cough; may show pain in swallowing; usually slobber and show a nasal discharge; breathe rapidly, with difficulty, and in severe cases through the mouth; show severe inflammation of the nostrils, trachea, and windpipe; have a high fever, 104–107°F; and may remain sick for as long as a week. When the disease breaks out, 25 to 100% of the animals are affected. Death loss rarely exceeds 5%.

Although IBR is usually thought of as a respiratory disease, it may cause inflammation of the eyes and/or vagina. Also, it may cause abortion.

Distribution and Losses Caused By: Throughout the U.S.

The main economic losses are in poor growth, loss of weight, loss of milk production, abortions, loss of time, and cost of drugs.

Death losses rarely exceed 5%.

Treatment: No known treatment, but sulfonamides and antibiotics effectively combat the secondary bacterial invaders that accompany the disease.

Control and Eradication: Practice good sanitation and disease preventive measures; isolate sick animals.

Prevention: Infectious bovine rhinotracheitis can be prevented by the use of a vaccine, of which there are two types. The modified live virus vaccine provides lasting immunity, but it should not be used on pregnant cows or on calves under 6 months of age. Killed virus vaccines must be repeated.

Remarks: IBR was first found in a Colorado feedlot in 1950.

(Continued)

TABLE 10–3 *(Continued)*

INFECTIOUS EMBOLIC MENINGOENCEPHALITIS (THROMBOEMBOLIC MENINGOENCEPHALITIS)

Infectious embolic meningoencephalitis is a bacterial disease of cattle.

Species Affected: Cattle.

Cause: A hemophiluslike gram-negative bacterium, *Hemophilus somnus.* Further investigation is needed.

Symptoms and Signs (or age group most affected): Feedlot cattle, in fall and winter months. Affects both sexes, usually animals 1 to 2 years of age. Characterized by incoordination, coma, sometimes blindness, and always fever (near 107°F). Death usually follows in 2 to 4 days. Positive diagnosis can be made upon autopsy, by the inflamed areas of infection observed in the brain.

Distribution and Losses Caused By: Western region of the U.S.

Only 1 to 2 cases develop in a lot at a time, but 10% of cattle may be affected before the disease runs its course.

Treatment: Affected animals should be isolated and given systemic antibiotics for several days.

High levels of tetracyclines in the drinking water are recommended if many animals are afflicted.

Control and Eradication: Feedlot pens should be checked frequently to detect newly afflicted animals and to provide prompt treatment.

Most outbreaks will run their course in 2 to 3 weeks.

Prevention: Application of the usual sanitary measures may help prevent the disease.

A killed vaccine is available.

Remarks: Polioencephalomalacia, which also affects feedlot cattle and causes incoordination, may be confused with infectious embolic meningoencephalitis; but fever is rarely associated with polioencephalomalacia.

INFLUENZA (See SWINE INFLUENZA)

JOHNE'S DISEASE (CHRONIC BACTERIAL DYSENTERY, PARATUBERCULOSIS)

Johne's disease is a chronic, incurable, infectious disease.

Species Affected: Cattle. Sheep and goats, sometimes. Swine and horses, rarely.

Cause: *Mycobacterium paratuberculosis,* a bacterium.

Symptoms and Signs (or age group most affected): Loss of flesh, and intermittent diarrhea and constipation—with the former becoming more prevalent. Affected animals may retain a good appetite and normal temperature. The feces are watery but contain no blood and have a normal odor.

The disease is almost always fatal, but with the animal living from a month to 2 years.

Upon autopsy, the thickening of the infected part of the intestines, covered by slimy discharge, is all that is evident.

Researchers around the world have been working to develop an improved, and more rapid, diagnostic test for Johne's. This goal is near. Producers desiring information relative to the new test should check with their local veterinarian.

Distribution and Losses Caused By: Widespread; in practically every country where cattle are raised on a large scale. Apparently it is increasing in the U.S.

Treatment: No satisfactory treatment is known.

Control and Eradication: If infection strikes, have herd tested with "Johnin" at intervals of 3 to 6 mos., remove reactors, disinfect quarters, and isolate young stock from mature animals.

The Johnin test, as with many other tests, is not entirely accurate, as some affected animals fail to react to it.

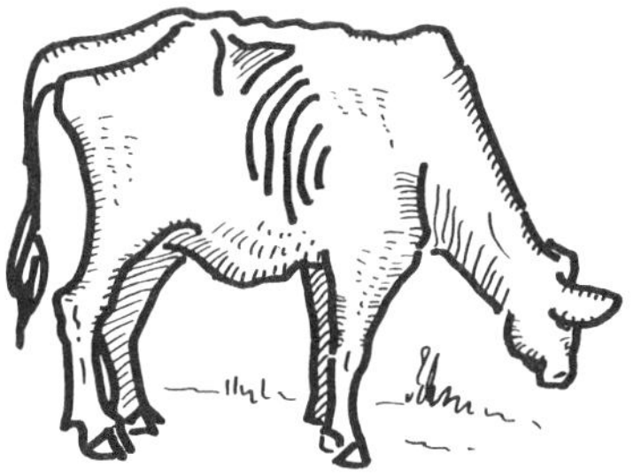

Fig. 10–23.

Difficult to eradicate from a herd.

Prevention: Effective prevention is accomplished by keeping the herd away from infected animals. Purchase new or replacement animals from reputable breeders and disease-free herds.

A new vaccine for Johne's has been approved by the USDA, but individual approval of each state is required for its use.

Remarks: This is a chronic incurable infectious disease. It resembles tuberculosis in many respects.

The disease seems to involve calfhood exposure, with no evidence of infection until 6 to 18 months later.

JOINT ILL (See NAVEL INFECTION)

KERATITIS (See PINKEYE)

LAMB DYSENTERY (CLOSTRIDIUM PERFRINGES TYPE B)
(Also see ENTEROTOXEMIA)

Lamb dysentery is a bacterial disease of sheep.

Species Affected: Sheep.

Cause: The disease is caused by the bacterium *Clostridium perfringes* Type B.

Symptoms and Signs (or age group most affected): The symptoms are variable. The disease may appear suddenly in lambs under 2 weeks of age. Some lambs die without showing signs of disease. Other lambs show up with yellow liquid feces, which become brown and bloody. Coma and death may follow.

Distribution and Losses Caused By: Occurs in sheep throughout the world.

Lambs born on the range are less susceptible than those born in a lambing shed.

The mortality is very high.

Treatment: There is no highly effective treatment, once lambs are affected. Various drugs such as sulfonamides, antibiotics, antidiarrheals, and anthelmintics may save a few.

Control and Eradication: Practice rigid sanitation; isolate diseased animals and move healthy animals to clean quarters or ground; keep lambs clean, warm, and dry.

Prevention: Vaccinate all ewes twice, 6 and 3 weeks before lambing. Must be type "B" vaccine or "BCD" combination.

(Continued)

TABLE 10–3 *(Continued)*

LEPTOSPIROSIS

Leptospirosis is an infectious disease of animals caused by various leptospiral organisms.

Species Affected: Cattle. Sheep. Goats. Swine. Horses. Dogs. Foxes. Rats and other rodents. Humans (the disease is transferable between species).

Cause: Several species of corkscrew-shaped organisms of the spirochete group.
Leptospira pomona primarily affects cattle and swine.
In the U.S., the disease in cattle is primarily due to *Leptospira pomona, hardjo*, and *grippotyphosa.*
Leptospira pomona is the most common serovar found in U.S. sheep, swine, and horses.

Symptoms and Signs (or age group most affected): In most herds, leptospirosis is a mild disease. However, the symptoms may vary from herd to herd. In general, the symptoms noted in cattle are (1) high fever (103–107°F), (2) poor appetite, (3) abortion anytime, (4) bloody urine, (5) anemia, and (6) ropy milk.
All ages of cattle, and both sexes, are affected (including steers).
Swine: Leptospirosis is usually characterized by abortion, pigs born dead or weak, and unthrifty market hogs.
Equine: Leptospirosis is characterized by fever, inappetence, mild depression, and occasionally jaundice.
Human: Leptosprisos is characterized by abrupt onset of fever with chills, headache, vomiting, and pains in the extremities, joints, and muscles.

Distribution and Losses Caused By: Leptospirosis was first observed in humans in 1915–1916, in dogs in 1931, and in cattle in 1934.
It was first reported in cattle and swine in the U.S. in 1944 and 1952, respectively, although it has been found in dogs in the U.S. since 1939. Bovine leptospirosis has been reported in Europe, Australia, and the U.S.
Surveys indicate this disease is widespread in the cattle population of the U.S. as well as in many other parts of the world.
Losses due to this disease amount to over $100 million annually in the U.S. Mortality is low in most outbreaks; however, in young calves, it may be high. The main losses are from poor growth in beef cattle and loss of milk production in dairy cows. If it were not for abortions, this disease would go undetected in many herds.

Treatment: Treatment of animals, which should be prescribed by a veterinarian, may include blood transfusions, administration of selected antibiotics, and good care.
Antibiotics give fairly good results if cases are treated promptly. It appears that selected antibiotics must be used to eliminate shedders.
High levels of the tetracycline drugs (400–500 g per ton) can be used in swine complete feeds for 14 days to help remove the carrier phase.
In human leptospirosis the M.D. should be consulted relative to treatment.

Control and Eradication: The disease is spread by infective urine; therefore, spread animals out over a large area; avoid congestion in a corral or barn.
Fence off waterholes or ponds of slow running streams.
Isolate sick animals or new additions to the herd.
Discard milk from diseased cows.
Clean and disinfect the barns; exterminate rodents.
Administer leptospirosis vaccine to all cows and sows on problem farms each year.
Keep different classes of livestock separated, because leptospirosis can be spread from one species to another.

Prevention: Vaccinate susceptible animals annually if disease is present in area.
Purchase clean animals, isolate for 30 days and retest.
Vaccination of people with a suspension of killed leptospires has been employed and reported successful in several countries.

Remarks: Carrier animals—animals that have had leptospirosis and survived—may spread the infection by shedding spirochetes in the urine. The infected urine may then either (1) be breathed as a mist in cow barns, or (2) contaminate feed and/or water and thus spread the infection. Breeding bulls can transmit this disease to cows.
It is known that recovered cattle can remain carriers for up to 3 months and swine can remain carriers up to 1 year.
Leptospirosis is mainly a warm-weather disease.
The spirochetes seldom survive for more than 30 days outside the animal. Stagnant water favors their survival.

LISTERIOSIS (CIRCLING DISEASE)

Listeriosis is an infectious disease caused by a bacterium.

Species Affected: Cattle. Sheep. Goats. Also reported in swine, horses, and humans.

Cause: *Listeria monocytogenes*, a bacterial infection.

Symptoms and Signs (or age group most affected): Depression, staggering, circling, and strange awkward movements. Cows may abort. Positive diagnosis can be made only by laboratory examination of the brain.

Distribution and Losses Caused By: Widespread.

Treatment: A satisfactory therapeutic agent has not been found.
Antibiotics and electrolytes give variable results.

Control and Eradication: Isolate animals with circling disease.
Move unaffected animals to clean premises.
Use caution in handling infected tissues as humans are susceptible to the disease.

Prevention: Results with bacterins have been inconclusive.
The following program may aid in preventing the disease:
1. Do not store silage in a silo that is in poor repair.
2. Do not feed silage from the top layer of an upright silo.

Fig. 10–24.

3. Never feed moldy or spoiled silage.
4. Provide clean, dry quarters during inclement weather.
5. Provide clean drinking water.
6. Control parasites.
7. Avoid stress.

Remarks: Incidence ranges from 1 to 7% in an infected herd of cattle.
The mortality rate of affected animals is extremely high.
Silage samples can be submitted to a diagnostic laboratory to determine if *Listeria* are present.

LOCKJAW (See TETANUS)

(Continued)

TABLE 10–3 *(Continued)*

LUMPY JAW AND WOODEN TONGUE

Lumpy jaw and wooden tongue are two infectious, chronic diseases.

Species Affected: Cattle. Swine. Horses. Humans.

Cause: *Actinomyces bovis* causes lumpy jaw.
Actinobacillus lignieresei causes wooden tongue.

Symptoms and Signs (or age group most affected): *Lumpy jaw* is usually confined to the bones of the lower jaw, although the upper jaw and nasal bones may be involved. Affected bones become enlarged, spongy, and filled with creamy pus. Inflamed cauliflower masses of tissue may spread out and appear on the surface, discharging a pus of foul odor. The teeth may become loosened.

Wooden tongue attacks chiefly the tissue in the throat area of cattle, but it is also seen in the tongue, stomachs, lungs, and lymph glands. Usually it first makes its appearance as a small swelling under the skin in the infected areas. The enlargements usually break open and discharge pus. If the tongue is involved, it will increase in size and hardness. There will be constant drooling and the animal will lose weight due to impaired eating.

Lumpy jaw and wooden tongue occur most frequently in young cattle during the period of changing teeth.

Distribution and Losses Caused By: Throughout the U.S.

Treatment: The veterinarian may, under certain conditions, (1) administer a water solution of an iodine salt of sodium or potassium, (2) prescribe an antibiotic, (3) resort to surgery, or (4) use X-ray therapy.

Treatment of lumpy jaw is not very satisfactory, but most cases of wooden tongue yield readily to treatment.

Sometimes treatment with organic iodine (EDDI) is effective. Add to the ration 250 to 500 mg/head/day for 2 to 3 weeks.

Superficial abscesses should be opened, drained, and swabbed with tincture of iodine.

Control and Eradication: Segregate and treat or eliminate infected animals, and, if practical, do not feed materials having sharp awns (foxtail, barley, rye, bearded wheat, etc.)

Fig. 10–25.

Prevention: Since the organisms causing lumpy jaw and wooden tongue gain entrance to the body through injuries or abrasions, the only prevention consists in not feeding such feeds as foxtail, barley, rye, and bearded wheat—unless there is no practical alternative.

Under some conditions, organic iodine appears to be effective in the prevention of lumpy jaw in cattle. For prevention, add to the ration 50 mg of ethylenediamine dihydriodide (EDDI)/head/daily.

Remarks: These two different and distinct microorganisms produce similar chronic diseases affecting mainly the head of cattle—hence the name "big head."

The same fungus that causes lumpy jaw occasionally attacks the udder of sows where it is characterized by many small abscesses filled with calcified granules. On rare occasions, the lumpy jaw organism has also been found in the fistulous withers of the horse in conjunction with *Brucella* organisms.

The condition may spread throughout the body, resulting in emaciation and a condemned carcass upon slaughter.

LYME DISEASE

Lyme disease is a disease of animals and humans of which the vector is a tick.

Species Affected: Horses. Cattle. Dogs. Rodents. Humans.

Cause: In 1981, the disease agent (a spirochete) was isolated from the tick and named *Borrelia bargdorferi.*

Symptoms and Signs (or age group most affected): The most common symptoms/signs in horses are lameness (arthritis), fever, muscle aches and pains, limb swelling, eye inflammation, encephalitis, hepatitis, nephritis, and abortion. A positive diagnosis of Lyme disease is based on clinical signs, opportunity for infection, antibody titers, and response to treatment.

Distribution and Losses Caused By: The disease is widely distributed over the U.S.

Treatment: Once diagnosed, Lyme disease is fairly simple to treat. Penicillin and tetracycline have been shown to be effective, particularly in the early stges of the disease; response to early treatment can be dramatic, often overnight. However, chronic cases may require treatment with antibiotics for a long time, up to 6 months.

Control and Eradication: Control of ticks appears to be essential for the control of Lyme disease.

Prevention: The best prevention is a diligent tick and fly control program. In the meantime, fervent research is in progress to develop a vaccine.

Remarks: Lyme disease was first recognized in people in Old Lyme, Connecticut in 1975; hence, the name. At that time, the outbreak was traced to the bite of a small deer tick, *Ixodes dammini.* (In California, the vector is the western black-legged tick, *Ixodes pacificus.*) The first case in equines was diagnosed in a pony, in Wisconsin, in 1985.

MALIGNANT EDEMA (GAS GANGRENE)

Malignant edema is an acute, general fatal toxemia of animals.

Species Affected: Cattle. Sheep. Goats. Swine. Horses.

Cause: Malignant edema is caused by *Clostridium septicum* and related bacteria.

Symptoms and Signs (or age group most affected): This is an acute infectious, but noncontagious disease characterized by gangrene and emphysema around a wound.

The affected animal goes off feed, breathes rapidly, and is profoundly depressed. A swelling forms around the wound. A gaseous and malodorous fluid exudes from the wound. In advanced stages of the disease, the animal is prostrated and often disoriented. There may or may not be a rise in temperature. Death occurs after a course of 12 to 48 hours. The mortality rate is high.

(Continued)

TABLE 10-3 *(Continued)*

MALIGNANT EDEMA *(Continued)*

Distribution and Losses Caused By: The incidence in a single herd may be high following castration, dehorning, or accidental wounds.

Treatment: In the early stages of the disease, treatment with massive doses of antibiotics may be effective.

Control and Eradication: Annual vaccination is recommended in high-risk areas.

Prevention: Since malignant edema is associated with contamination of wounds, the disease can be partially prevented by minimizing wounds and by castrating and dehorning under hygienic conditions.

Vaccination of young cattle with a vaccine containing *C. septicum* (for malignant edema) along with *C. chauvoe* (for blackleg) at the time of the blackleg vaccination(s) will give some protection against malignant edema. Also, antibiotics may be administered 4 to 5 days following surgery.

MASTITIS (GARGET, BLUE BAG in sheep)

Mastitis is an infectious inflammation or irritation in the udder which interferes with the normal flow of milk and/or its quality.

Species Affected: Cattle. Sheep (known as blue bag). Goats. Swine.

Cause: Mastitis may be either infectious or noninfectious. Infectious mastitis, resulting from the invasion of bacteria in the gland, may be from several different types of bacteria.

Over 95% of all cases are caused by the following species of streptococci and staphylococci: *Streptococcus agalactiae, Streptococcus dysgalactiae, Streptococcus uberis,* and *Staphylococcus aureus.* Infection with any of these organisms is usually chronic, with flare-ups occurring at regular intervals. No amount of drugs given to cows today can prevent another attack next month under the same conditions.

Noninfectious mastitis is the result of injury, chilling, bruising, or rough or improper milking.

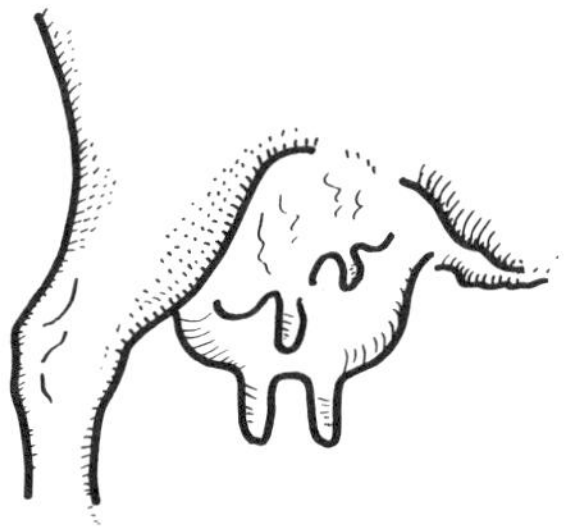

Fig. 10-26.

Symptoms and Signs (or age group most affected): In acute mastitis, the udder is hot, very hard, and tender. The animal will have an increase in temperature, refuse to eat, "lose its cud," have dull eyes and a rough coat. Because of the soreness of the udder, the animal stands in an awkward position, moves about and lies down with reluctance and difficulty. Milk will be reduced greatly, and may become lumpy or watery. Abscesses may appear on the udder. Death often occurs in untreated acute mastitis.

In chronic mastitis, the only symptom which may be noted is that the milk will be thick or lumpy.

Sheep: Mastitis of sheep, if untreated, usually results in chronic discharging teats or gangrene (blue bag) and death.

Affected ewes tend to spearate out from the flock or band. Generally only one side of the udder is affected. It is hot and painful; and, in early stages, secretion from teat is (1) thin and reddish due to hemorrhage, or (2) thick, creamy, and yellowish in color. Usually accompanied by rise in temperature, 104–106°F. Following gangrene (when udder becomes cold and turns dark in color), death may follow; or, with recovery, a portion of the udder may slough off.

Distribution and Losses Caused By: Worldwide.

Mastitis takes heavier toll from the dairy industry than any other single disease. Losses are chiefly in decreased milk production and poor quality milk.

A 1981 survey conducted by *Hoard's Dairyman* showed that 87% of the respondents encountered mastitis during the year.

The listing prepared by the associate deans and directors of veterinary research programs for the Council of Deans, Association of American Veterinary Medical Colleges, showed estimated mastitis losses of $368 million per year, based on losses as of February 1, 1981.

Mastitis causes a yearly loss of $225 per afflicted cow, according to the National Mastitis Council.

Treatment: Consult a veterinarian. Mastitis is usually treated by the intramammary injection of antibiotics, sulfa drugs, nitrofurans, or combinations of these drugs.

Acute cases should also be treated systemically by the veterinarian.

Local application of hot packs and udder massage will increase circulation in chronic cases; however, hot, acute, painful glands should be treated with cold applications.

Sheep: Antibiotics or sulfas are the treatments of choice. If treatment is started early enough, the udder may return to normal milk production. Surgical removal of gangrenous half of udder is not successful; thus, ewes not capable of producing milk should be marketed.

Control and Eradication: Although mastitis is usually apparent, it may be a "hidden" disease. Therefore, several different tests have been developed for detecting the presence of the causative microorganisms in lactating cows; among them, (1) *screening tests*, or *presumptive tests*, made either at the side of the cow or at the bulk tank, of which the California Mastitis Test (CMT) is the most widely used one; and (2) *specific laboratory tests* designed to detect the causative organism. A reasonable goal, based on using CMT test, is to have at least 75% of the bucket milk samples score negative or trace; less than 75% negative (–) and trace (T) bucket readings indicates a milking management problem. On an individual quarter basis, 90% of the samples scoring negative or trace indicates a well-managed herd.

Prevention: Milk all diseased cows last.

Use strip cup before milking. Promptly remove new cases from the milking line and segregate.

Wash udders with clean individual towels placed in chlorine solution (with a strength of 200 ppm); then wring the towel dry and wipe.

Milk properly, in a regular, rapid, and thorough manner.

Dip the ends of the teats in a chlorine solution after milking (chlorine with a strength of 200 ppm)

Before milking each cow, wash hands with soap and water, disinfect them with chlorine solution, and wipe them dry, preferably with paper towels.

Do not milk on the floor.

Do not permit wet hand milking.

Before sweeping the floor, sprinkle it with lime or superphosphate.

Use plenty of straw bedding for each cow.

Remarks: Although mastitis in sows and in ewes is somewhat different than that found in milk cows, the same general principles in preventive and control measures are applicable

The methods of treatment, control and eradication, and prevention of mastitis are the same for both milk cows and milk goats.

Most screening tests for inflammation measure the number of leukocytes and mammary cells in milk. Noninfected quarters usually contain less than 500,000 cells per ml of milk, whereas higher counts may signify inflammation.

(Continued)

TABLE 10–3 *(Continued)*

MASTITIS-METRITIS-AGALACTIA COMPLEX (MMA, AGALACTIA)

Mastitis-metritis-agalactia complex is a syndrome of complex etiology which causes lactation failure in the sow.

Species Affected: Swine.

Cause: The organism commonly thought to cause the MMA syndrome is *Escherichia coli*. Other organisms that have been associated with the syndrome include: *Actinobacillus, Actinomyces, Aerobacter, Citrobacter, Clostridia, Corynebacterium, Enterobacter, Klebsiella, Pseudomonas, Mycoplasma, Proteus, Streptococcus, Staphylococcus*, and *Chlamydia*. It is particularly noteworthy that *E. coli* is also the most common organism isolated as the cause of diarrhea in baby pigs during the first three weeks of life. However, the disease is complex, and proof of the cause is often difficult.

Symptoms and Signs (or age group most affected): This is a disease of sows and gilts characterized by an inflammation of the mammary glands (mastitis), and inflammation of the uterus (metritis), and/or a failure to secrete milk (agalactia). Often, however, there is not total lack of milk but rather a reduction in the normal amount of milk (hypogalactia). The death rate in sows is low, but the loss of baby pigs may be high.

The first signs of the disease usually appear within three days after farrowing, although symptoms usually can be seen before farrowing or before the pigs are weaned. A whitish or yellowish discharge of pus appears from the vagina of the infected animal and the temperature rises to 103–106°F *(39–41°C)* or higher. The sow goes off feed and stops milking, or produces less milk, and the pigs often develop diarrhea. Sometimes the problem is not recognized until the pigs starve to death.

Distribution and Losses Caused By: The disease is worldwide. A Missouri survey showed that 13% of the farrowing sows were afflicted with MMA.

MMA appears to be a more serious problem in "multiple" and "confinement" farrowing than where portable houses on pasture are used.

Treatment: The list of suggested treatments is long, confusing, and not always effective; among them, are:

1. Cross-fostering the piglets to a normal mother. This appears to be the most effective solution.
2. Antibiotics or nitrofurans to eliminate infections.
3. Oxytocin, a hormone, to cause milk let-down.
4. Relief of constipation by (a) using a warm, soapy enema, (b) feeding a ration high in molasses or wheat bran, or (c) providing morning and evening exercise.
5. Supplemental milk and glucose to keep the baby pigs alive.
6. Oral antibiotic or nitrofuran for the baby pigs if diarrhea occurs.

Control and Eradication: There is no confirmed evidence that vaccines or antibacterial agents have any beneficial prophylactic effect.

Prevention: Prevention revolves around sound management, good nutrition, superior sanitation, and proper swine husbandry. When sows are kept in confinement, it is especially important to keep the sows from becoming overfat and constipated, and to reduce stress, particularly near the time of farrowing. The first condition can be controlled by limiting the ration, and the addition of 6 to 10% molasses or extra bran to the ration will prevent constipation. Also, underfeeding may contribute to the problem.

MUCOSAL DISEASE (See BOVINE VIRUS DIARRHEA)

MYCOPLASMAL PNEUMONIA (See ENZOOTIC PNEUMONIA)

NAVEL INFECTION
(JOINT ILL, NAVEL ILL, ACTINOBACILLOSIS, ARTHRITIS in lambs)

Navel infection is an infectious disease of newborn animals.

Species Affected: Horses. Cattle. Sheep: in lambs it is commonly called arthritis. It affectcs newborn foals, calves, and lambs.

Cause: Several kinds of bacteria are involved.

Symptoms and Signs (or age group most affected): Loss of appetite, swelling, soreness and stiffness in the joints, and general listlessness. Umbilical swelling and discharge.

Two forms occur in lambs: (1) suppurative (pus-forming) caused by pus-producing bacteria and marked by swelling of and pus in the affected joints; and (2) nonsuppurative, caused by the swine erysipelas organism, in which most lambs recover within a month without treatment, but some develop a chronic lameness.

Distribution and Losses Caused By: Throughout the U.S.

About 50% of the infected foals die and many that survive have deformed joints. In calves and lambs, the mortality is not so high.

In suppurative arthritis in lambs, the mortality rate is high and chances of recovery poor.

Treatment: The veterinarian may give a blood transfusion from the dam to the offspring, or in other cases may administer a sulfa drug, an antibiotic, a serum, or a bacterin.

Control and Eradication: (See Prevention.)

Fig. 10–27.

Prevention: Practice sanitation and hygiene at mating and parturition. Feed iodized salt to pregnant mares in iodine-deficient areas. Soon after birth, treat the navel cord of newborn animals with iodine.

Remarks: Navel infection in calves and lambs occurs less frequently than in foals. Providing clean quarters for the newborn and painting the navel cord with iodine constitute the best preventive measures for all classes of livestock.

NECROTIC ENTERITIS (See ENTERITIS)

(Continued)

TABLE 10–3 *(Continued)*

PINKEYE (KERATITIS)

Pinkeye is an infectious eye ailment.

Species Affected: Cattle. Sheep. Goats.

Cause: The most common bacterial form of the disease is caused by *Moraxella bovis (Hemophilus bovis)*. Infection with this bacteria can be started with vitamin A deficiency, injuries, dust, insects, or strong sunlight.

Also, a virus form of pinkeye is caused by the "red nose" or infectious bovine rhinotracheitis (IBR) virus.

Symptoms and Signs (or age group most affected): Liberal flow of tears and tendency to keep the eyes closed; redness and swelling of the lining membrane of the eyelids and sometimes of the visible part of the eye; and there may be discharge of pus and ulcers of the cornea. If unchecked, blindness may follow.

In viral pinkeye, the eyeball itself is only slightly affected, IBR mainly affects the eyelids and the tissues surrounding the eyes. It causes a severe swelling of the lining of the lids.

Distribution and Losses Caused By: The disease is widespread throughout the U.S., especially among range and feedlot cattle. Pinkeye is encountered in nearly half of U.S. beef cattle herds and affects 3% of all beef cattle.

Deaths are rare.

One record showed that affected steers gained an average of 50 lb less during the grazing season than those not affected.

Treatment: The most common treatment for bacterial pinkeye is the application of antibiotics or sulfa drugs to the affected eye as ointments, powders, or sprays; preferably, with treatment made twice daily. Recovery is speeded up by keeping the infected animals in a dark barn. A commercially produced protective eye patch is now available. It completely covers the infected eye, holding the medication in place, protects the eye from insects and bright sunlight, and reduces the work and expense of handling and isolation. Held in place by a special adhesive, the eye patch drops off and decomposes after about 7 to 10 days.

Fig. 10–28.

Treatment of IBR conjunctivitis is seldom of value, although antibiotics sometimes help reduce the secondary bacterial infection.

Control and Eradication: Isolate affected animals, control insects, provide good nutrition—including adequate vitamin A, and control IBR conjunctivitis by a vaccination program.

Prevention: Prevention of bacterial pinkeye consists in the following: controlling face flies and other insects that feed around the eyes; good nutrition, including adequate vitamin A; and isolation of affected animals.

IBR conjunctivitis may be prevented by proper vaccination of animals prior to onset of the disease. The herd should not be vaccinated once the disease appears; nor should pregnant cows be vaccinated. Affected animals should be isolated.

Remarks: Pinkeye is highly infectious and may spread rapidly through a herd, producing a drop in production or occasional blindness in some animals.

PLEUROPNEUMONIA (HPP)

Pleuropneumonia is a severe and contagious respiratory disease, primarily of young pigs.

Species Affected: Swine, especially growing-finishing pigs.

Cause: Often a tentative diagnosis can be made based on the history, age of affected pigs, signs and symptoms, and lung lesions, but a definitive diagnosis may require the culture of the causative bacteria—*Haemophilus pleuropneumoniae*—from a lung lesion.

Symptoms and Signs (or age group most affected): The incubation period is short, sometimes 8 to 12 hours, and it commonly strikes pigs from 40 lb *(18 kg)* to market weight. Often the first indication of hemophilus is the sudden death of apparently healthy pigs, though this sudden death follows a stressful period; for example, moving, mixing or rapid weather change. When observed, healthy pigs may develop labored breathing and die within minutes of a stress. Bleeding from the nose may be observed in some pigs, but not all. Pigs less severely affected may demonstrate thumping, a fever of 104–107°F *(40.0–41.7°C)*, depression, and reluctance to move. Prominent lesions include hemorrhagic, edema-filled lungs.

Distribution and Losses Caused By: The distribution of hemophilus is worldwide, causing significant economic losses to the swine industries of such countries as Brazil, Canada, Denmark, East Germany, West Germany, Mexico, Switzerland, and, of course, the United States. Each country has noted a sudden rise in the occurrences during the past 10 years.

Since the causative organism is spread via aerosol droplets, morbidity reaches 100% while mortality is 20 to 40%, when immediate and effective treatment is instituted.

While pigs affected with hemophilus will often die, the greatest economic losses result from pigs surviving the disease, since the chronic lung lesions from the disease create "poor doers." Average daily gains decrease and feed per pound of gain increases.

Treatment: Treatment during an acute outbreak consists of maintaining high blood levels of antibiotics such as procaine penicillin and long-acting oxytetracycline (LA-200) by injection.

Following an outbreak, antibiotic sensitivity tests should be used to determine the most effective drug. To reduce death losses, both healthy and sick swine should be treated. The treatment of animals showing clinical signs has not demonstrated uniform success due to the development of lung damage.

Control and Eradication: Because survivors frequently remain carriers, control is difficult. All-in, all-out management and good ventilation are recommended. If possible, replacements should be from herds free of the organism.

Prevention: Presently, the best prevention seems to be prevention of its spread and introduction into a herd. The major source of infection is the carrier pig who has recovered from hemophilus. New animals should be isolated and tested to determine if they are carriers before being introduced into an uninfected herd. Vaccines are available but to date their effectiveness ranges from excellent to nil; hence, additional research is needed.

Remarks: Although pleuropneumonia is primarily a disease of young pigs, adults may suffer abortions or fatal infections.

PNEUMONIA

Pneumonia is an inflammation of the lungs.

Species Affected: All animals.

Cause: Causes are numerous, including (1) many microorganisms, (2) a number of different viruses, (3) inhalation of water or medicines given by untrained persons as a drench, and (4) changeable weather during the spring and fall.

(Continued)

TABLE 10–3 *(Continued)*

PNEUMONIA *(Continued)*

Symptoms and Signs (or age group most affected): Ushered in by a chill, followed by elevated temperature.

Quick, shallow respiration, discharge from the nostrils and perhaps eyes, and a cough. Legs wide apart, drop in milk production, loss of appetite, and constipation. Crackling noises with breathing, and gasping for breath.

Distribution and Losses Caused By: Nationwide.

If untreated, 50 to 75% of affected animals die.

Pneumonia causes ⅕ of all nonnutritional mortality in U.S. beef cattle.

Treatment: Place sick animals in quiet, clean quarters away from drafts, and give easily digested nutritious feeds. Sulfonamides and antibiotics are effective in the treatment of most acute pneumonias. But they are ineffective against virus pneumonia, except for keeping down secondary bacterial pathogens.

Control and Eradication: Segregate sick animals.

Pigs are subject to epizootic pneumonia, known as virus pneumonia, VPP, or infectious pneumonia. It is caused by a *Mycoplasma*.

Two methods have been used for eradication of virus pig pneumonia: (1) specific pathogen-free (SPF) pigs, and (2) farrowing old sows in isolation and ascertaining that the pigs are free of VPP before adding them to the herd.

Prevention: Provide good hygienic surroundings and practice good, sound husbandry.

Remarks: Changeable weather during the spring and fall and drafty, damp barns are conducive to pneumonia.

Always try to establish what organism is causing the condition. Many terminal bacterial pneumonias start from virus diseases.

PORCINE PARVOVIRUS (PPV)

Porcine parvovirus is a genetic disease which is triggered by stress.

Species Affected: Swine.

Cause: The disease is caused by a virus of the genus *Parvovirus*, though there is no indication that the porcine parvovirus is the same as the parvovirus affecting species such as cows, sheep, dogs, cats, rats, and mice. They are antigenically different.

Symptoms and Signs (or age group most affected): Usually, the only clinical sign is maternal reproductive failure such as (1) return to estrus, (2) failure to farrow despite not showing signs of heat, (3) farrowing few pigs per litter, or (4) farrowing a large proportion of mummified fetuses. The observable signs depend upon the stage of pregnancy when the sow or gilt is infected. Gilts are especially vulnerable.

Diagnosis is made on the basis of differentiating parvovirus from other causes of reproductive failure in swine, and, if available, mummified fetuses can be submitted for immunofluorescence testing.

Distribution and Losses Caused By: Parvovirus occurs throughout the world.

Losses caused by this disease are usually unexpected since the sow or gilt typically appears healthy during gestation. The incidence of parvovirus is high and it has resulted in large economic losses. It seems that porcine parvovirus is one of the most important infectious agents indicated in SMEDI.

Treatment: There is no treatment for porcine parvovirus.

Control and Eradication: Experience indicates that few herds can be expected to remain free of porcine parvovirus even if access is carefully controlled.

Prevention: Prevention consists of naturally infecting or vaccinating gilts with porcine parvovirus before they are bred. Commonly, gilts are exposed to sows or moved to a potentially contaminated area to promote a natural infection and immunity. The effectiveness of this method is, however, uncertain. A vaccination can ensure that gilts develop immunity before breeding. Both inactivated and modified live virus vaccines are available and effective.

PORCINE STRESS SYNDROME (PSS)

Porcine stress syndrome is a genetic disease which is triggered by stress.

Species Affected: Swine.

Cause: The disease is genetic, following inheritance patterns that are consistent with those expected with recessive inheritance. In other words, the gene or genes responsible for PSS must be present in the dam and the sire.

Symptoms and Signs (or age group most affected): Some pigs are unable to adapt successfully to stressful management practices, and they exhibit a nonpathological disorder that is best described as a syndrome rather than a single abnormal characteristic. It commonly occurs in rapidly growing, heavily muscled pigs, resulting from intensive genetic selection in systems of partial or total confinement. It is often manifested as sudden unexplained death losses. Furthermore, the syndrome is related to low-quality pork products or so-called pale, soft, exudative (PSE) pork, when the stress susceptible pigs survive until slaughter.

A series of signs develop in PSS pigs when they are subjected to some stressful management practices such as loading and transport to market, overcrowding and fighting, vaccination, or a sudden change in the weather. These signs occur in the following order:

1. Muscle and tail tremors.
2. Irregular and difficult breathing.
3. Alternating blanched and reddened areas of the skin—blotches.
4. Rapid rise in body temperature with signs of heat stress even in cool weather.
5. Total collapse, marked muscle rigidity, and hyperthermia.
6. Death in a shocklike state.

Distribution and Losses Caused By: Porcine stress syndrome exists throughout the world.

About 35% of the hog producers in the U.S. have encountered the condition and have experienced death losses from PSS.

While fewer than 1% of the swine in the U.S. actually die of PSS under normal marketing and management practices, the incidence of low-quality pork in ham and loin muscles runs about 8 to 10%.

Treatment: Both the treatment and the prevention of PSS require that those stress susceptible pigs be identified. This can be accomplished by three methods: (1) visual appraisal, (2) halothane screening, and (3) creatine phosphokinase (CPK) levels. Visual appraisal is the evaluation of extreme muscled, small statured pigs which display some of the initial signs of PSS when physically stressed. This requires a great deal of experience for accurate appraisal.

Control and Eradication: Halothane screening is involved, but reliable. It requires putting the animal to sleep with the inhalant anesthesia halothane and then observing pigs for evidence of muscle rigidity. Collected blood samples analyzed for the enzyme CPK are about 60 to 85% accurate. CPK is abnormally high in PSS pigs. Neither of these methods will detect carriers.

Prevention: The best prevention is through genetic selection. All possible within-herd replacements should be appraised for the possibility of being PSS pigs, and replacement sires should be those that can be confidently predicted as noncarriers.

If some pigs in the herd are identified as being PSS pigs, they should be managed carefully until they reach an acceptable market weight. Moreover, when these pigs are marketed they should continue to be handled carefully throughout marketing procedures on preferably dry, cool days after pigs have fasted for 12 to 24 hours. During some stressful management practices a tranquilizer may be advisable for these animals.

Remarks: Genetic selection is the only sensible approach to reducing the incidence of PSS in problem herds.

(Continued)

TABLE 10–3 *(Continued)*

POTOMAC HORSE FEVER

Potomac horse fever is an acute, often fatal, diarrhea, caused by microorganisms called *rickettsia.*

Species Affected: Horses.

Cause: The causative agent is named *Ehrlichia risticii* after its discoverer. *E. risticii* is a member of the class of microorganisms called *rickettsia*, which are between bacteria and viruses in size. To date, researchers have not been able to determine how the disease is transmitted, but biting insects are strongly suspected.

Symptoms and Signs (or age group most affected): Signs are: fever (102–108°F), depression, loss of appetite, colic, edema of the underline, and stocking of the limbs. These symptoms are usually followed within 48 hours by the onset of diarrhea which, in severe cases, is watery and explosive.

A high percentage of horses with Potomac fever develop founder, or laminitis.

Distribution and Losses Caused By: Potomac horse fever, which was first reported in Maryland in 1979, has spread over most of the United States.

The fatality rate is approximately 10%.

The fatality rate approaches 30%, with death occurring primarily from dehydration and shock. A new rapid diagnostic test is available, which is capable of giving a positive diagnosis in less than 5 minutes.

Treatment: The diagnostic test makes rapid treatment possible. Correct diagnosis is important since the symptoms of Potomac horse fever often mimic those of salmonellosis, yet the drug tetracycline, which is helpful against Potomac horse fever, can be deadly to horses with other diarrheal disorders.

Treatment consists of large volumes of intravenous fluids, tetracycline, and supportive treatment to control fever and reduce laminitis.

Control and Eradication: Control and eradication involve control of insects, vaccination, and isolation of sick animals.

Prevention: A successful vaccine is available, which is administered intramuscularly in two doses, repeated 21 days apart, followed by an annual booster shot. So, if Potomac horse fever is present in a state or area, all horses therein should be vaccinated against the disease.

Remarks: This acute, often fatal, diarrhea was first noticed in a number of horses in Montgomery County, Maryland, in the summer of 1979. At first the disorder was called *acute equine diarrhea syndrome*, but soon it became known as *Potomac horse fever*, after the region where it was first recognized.

PSEUDORABIES (PRV, AUJESZKY'S DISEASE, MAD ITCH)

Pseudorabies is a viral infection of which swine are the natural hosts and chief reservoirs.

Species Affected: Most species of domestic and wild animals. Swine are the natural hosts and chief reservoir of the disease.

Cause: Virus of the herpesvirus group.

Symptoms and Signs (or age group most affected): Baby pigs may die with few if any, clinical signs evidenced. However, death is usually preceded by fever which may exceed 105°F, dullness, loss of appetite, vomiting, weakness, incoordination, and convulsions. In pigs less than 3 weeks old, death losses frequently approach 100%. After 3 weeks of age, the signs and death losses decrease.

In adult pigs, the signs may be very mild and include fever, off feed, coughing, sneezing, vomiting, diarrhea, constipation, convulsions, itching, middle ear infections, and blindness.

Sows infected in middle pregnancy may abort mummified fetuses. Sows infected late in pregnancy often abort or give birth to weak, shaker, or stillborn pigs. Piglets infected prior to birth usually die within 2 days.

In animals other than swine, the disease is characterized by intense itching, self-mutilation, convulsions, and death.

Distribution and Losses Caused By: Widespread and of considerable economic importance in midwesern U.S.

The disease is widespread in Europe, where it causes heavy losses.

Treatment: Drugs and feed additives are not effective for control or treatment of pseudorabies.

Control and Eradication: The following control measures are recommended for the protection of herds:

1. Dispose of dead pigs properly (burning or burying).
2. Buy tested breeding stock and isolate them for 30 days.
3. If you raise breeding stock, do not buy feeder pigs.
4. Get feeder pigs from a farm that has not had the disease.
5. Keep visitors away from swine premises.
6. Keep stray dogs, cats, and wildlife off the premises.
7. Keep swine and cattle separate.
8. Isolate show stock for 30 days after the fair is over.

Prevention: Attenuated and inactivated vaccines that are effective in prevention of the disease are available commercially, but neither kind will prevent infection by field virus.

Breeding stock should be vaccinated twice per year prior to breeding. Pigs from unvaccinated sows may be vaccinated after 3 days of age. Pigs from immunized sows should be vaccinated when 3 to 8 weeks old.

Remarks: Pseudorabies is not related to rabies.

The disease was first recognized as an infectious disease of cattle and dogs in Hungary by Aujeszky in 1902.

The continued increase and severity of the disease may be due to rearing more pigs in large confinement units or by the cessation of cholera vaccination.

A serum test is the most practical herd test.

RAM EPIDIDYMITIS (REO)

Ram epididymitis is a disease of rams, caused by *Brucella ovis* which causes impaired fertility in rams.

Species Affected: Sheep, in rams.

Cause: The cause of the disease is the ram epididymitis organism (REO), or *Brucella ovis.*

Symptoms and Signs (or age group most affected): Ram epididymitis is an infection of the epididymis which affects the fertility of rams.

Distribution and Losses Caused By: The disease appears to be worldwide.

The disease is widely distributed in the U.S., with the highest prevalence in range bands. It is estimated that ram epididymitis occurs in 5 to 6% of the rams of the U.S.

This disease results in complete or partial infertility of the ram and a marked increase in dry ewes. It also causes the lambing season to be spread out over several months. In some rams, palpation of the scrotal contents reveals a lump in the epididymis.

Economic losses result from reduced fertility, shortened breeding life, reduced marketability, and lowered lambing rates. Additionally, an infection in the ewe causes placentitis, abortion, and in the lamb, perinatal mortality.

Treatment: Treatment of infected rams has not been very satisfactory.

Control and Eradication: The disease can be controlled by a rigid culling and vaccination program as follows:

1. Examine all rams before the breeding season and again at shearing. Market affected rams for slaughter only.
2. Vaccinate young rams at 9 to 12 months of age or 3 weeks before the breeding season with *B. abortus* strain 19 plus bacterin of *B. ovis.*
3. Vaccinate replacement rams before putting them with older rams.

Prevention: In some areas of the United States, where there have been no reports of epididymitis, a close annual inspection with rigid culling may be a preferred approach.

(Continued)

TABLE 10–3 *(Continued)*

RHINOPNEUMONITIS (See EQUINE ABORTION)

ROTAVIRAL DIARRHEA

Rotaviral diarrhea is a viral disease of baby pigs which causes diarrhea.

Species Affected: Swine, in baby pigs.

Rotaviruses have been isolated from several species of mammals; and the diarrheas associated with all of them have close similarities. Thus, it occurs in calves, lambs, pigs, foals, rabbits, and mice.

Cause: The diarrhea is caused by a group of viruses known as rotaviruses—a name derived from the wheellike appearance of the virus through the electron microscope. A rotavirus infection can be confused with transmissible gastroenteritis (TGE); a laboratory diagnosis is necessary to differentiate.

Symptoms and Signs (or age group most affected): The disease causes diarrhea in piglets 1 to 6 weeks old. This diarrhea is characterized by a white or yellow stool which is liquid at first but later becomes creamy and then pasty before returning to normal. Piglets also become depressed, fail to eat, and resist moving. The severity of the disease is influenced by (1) the age of the piglet, (2) stresses such as chilling, (3) concurrent infections with pathogenic *E. coli* or the TGE virus, and (4) inadequate intake of immune milk. In general, the older the pig, the less severe the outcome, but the severity increases with weaning at any age. In many pigs, however, the infection may occur with no clinical signs.

Distribution and Losses Caused By: It is recognized that the infection is widespread and common in pigs. The mortality rate in piglets varies from 0 to 50%.

Treatment: No antibotics or other drugs are effective against rotavirus; hence, they are of no value in treatment, unless there is concurrent bacterial infection. Research shows that most often *E. coli* are secondary invaders.

Providing diarrheic weaned pigs with warm, dry, draft-free environment, along with frequent, limited feeding will help prevent starvation, secondary diseases, and permanent stunting.

Control and Eradication: The best methods for the prevention and control of rotavirus in young pigs appears to be sure piglets get adequate colostrum and milk at an early age and to provide good sanitation and a comfortable (nonstressful) environment.

There is no vaccine available to control rotaviral infections in pigs; and most swine herds are infected with rotavirus. Piglets do, however, receive protection from the antibodies in the sow's colostrum.

Prevention: Good sanitation prevents the build-up of rotavirus and the chance of concurrent *E. coli* infection.

SALMONELLOSIS

Salmonellosis is a disease of all animals caused by many species of salmonella.

Species Affected: All animals and humans.

Cause: The disease is caused by many species of salmonellae and is characterized clinically by 1 or more of 3 major syndromes.

Stressors that may precipate the disease include deprivation of feed and water, long transportation, recent calving, and mixing and crowding in feedlots.

Symptoms and Signs (or age group most affected): Young calves, lambs, piglets and foals usually develop the septicemic form.

Adult cattle, sheep, and horses commonly develop acute enteritis.

Chronic enteritis may occur in young pigs and occassionally in cattle.

Septicemia (caused by the presence of microorganisms or their toxins in the circulating blood) is the usual syndrome in newborn calves, lambs, pigs, and foals. Illness is acute, depression is marked, fever is usual, and death occurs in 24–48 hours. Nervous signs may occur in calves and pigs. In pigs, dark red-to-purple discoloration may appear on the ears and abdomen, and pigs may also suffer from pneumonia.

Distribution and Losses Caused By: The disease occurs worldwide, and the incidence is increasing with intensification and confinement.

The case fatality may reach 100%.

Treatment: Broad spectrum antibiotics are used parenterally to treat the septicemia. A mixture of trimethoprim and sulfadiazine is effective for the treatment of salmonellosis in calves. Ampicillin also may be used for the treatment of septicemic salmonellosis in all species.

Control and Eradication: The principles of control include prevention of introduction and limitation of spread within a herd.

Control measures, when the disease is present, include: (1) isolation of sick animals since the sick animal contaminates the environment and infects other pigs, (2) scrupulous pen sanitation, (3) decreased stress of potential victims, and (4) restricted movement of animals and people from contaminated areas to clean areas.

Control measures when the disease is not evident include, primarily, keeping stresses to a minimum, cleanliness, and the use of disinfectants.

Prevention: Prevention of *Salmonella* infection is virtually impossible; rather, control is more realistic. The major sources of *Salmonella* introduction into a herd are (1) the introduction of new animals into a herd, and (2) contaminated feed. Quarantining new animals allows time for fecal shedding of *Salmonella* to stop. Wet feed and inadequately cooked feed support the growth of *Salmonella*.

SCOURS (INFECTIOUS DIARRHEA, LAMB DYSENTERY) (See ENTEROTOXEMIA)

SHEATH-ROT (PIZZLE-ROT, POSTHITIS, URINE SCALD)

Sheath-rot is a moderately contagious disease of sheep and goats, and ocassionally of cattle, characterized by lesions on the prepuce of males and sometimes on the vulval lips of females.

Species Affected: Sheep, especially in wethers and rams. Goats. Cattle (steers and bulls), sometimes.

Cause: Although the disease is associated with castration and protein-rich diets, it is caused by a bacterium indistinguishable from *Corynebacterium renale*, which acts on urea and initiates the ulceration.

Symptoms and Signs (or age group most affected): Sheath-rot is characterized by a spreading ulceratioin of the skin of the prepuce and may sometimes involve the penis. Ulcers are small at first, but they become wet during urination. Gradually, they enlarge, coalesce with others, and become blocked, and the sheath will become distended with foul-smelling urine and pus. Animals may be humped, may walk stiffly, and may lose condition.

Distribution and Losses Caused By: The highest incidence occurs in Merino and Angora wethers.

Producers consigning breeding rams to sales should be especially watchful for sheath-rot since this condition is cause for rejection at most sales.

Treatment: Treatment involves the application of (1) antiseptics such as 5 to 10% copper sulfate ointment, quaternary ammonium compounds, or aluminum silicone, or (2) ointment containing penicillin or bacitracin to the affected area. Before application, any scabs, urine, pus, and debris should be removed. Changing the diet to low-quality pasture or straw and implanting testosterone, the male hormone, may hasten healing.

Control and Eradication: There is no practical means of control of sheath-rot.

Prevention: For control, animals could be switched to a lower-quality ration, but this is not practical when these animals are those that the producers want to gain well.

(Continued)

TABLE 10-3 *(Continued)*

SHIPPING FEVER (HEMORRHAGIC SEPTICEMIA)

Shipping fever is one of a group of infectious diseases.

Species Affected: Cattle. Sheep, especially lambs.

Cause: *Pasteurella* generally occurs in younger animals following shipping; hence, the common name *shipping fever.*

Shipping fever is caused from multiple infection due to the interaction of viruses and bacteria, accentuated by environmental conditions creating physical tension or stress. Change in weather and feed, overcrowding, hard driving, lack of rest, and improper shelter all help usher in the disease.

Symptoms and Signs (or age group most affected): Develops rapidly and lasts for a week or less. Usually high temperature, discharge from the eyes and nose, a hacking cough, difficulty in breathing, and there may be a swelling in the region of the neck. Sometimes there is diarrhea. In very acute forms, animals may die without showing symptoms.

Young animals most susceptible, but animals of all ages affected.

Distribution and Losses Caused By: Worldwide, especially among thin and poorly nourished young animals that are subjected to shipment by rail or truck during bad weather.

It is a serious problem to both shippers and receivers of animals. Death losses may be high in untreated cases.

Treatment: Isolate affected animals.

Treatment should be handled by a veterinarian. In the early stages, the use of antibiotics and/or sulfonamides will control the accompanying bacterial infection.

Control and Eradication: Institute good feeding and management.

Prevention: *Cattle:* As a preventative measure, one should eliminate as many as possible of the predisposing factors that lower the animal's vitality. Also, newly purchased animals should be isolated for 2 to 3 weeks before being placed in the herd.

Fig. 10-29.

Several vaccines, both modified and inactivated, are available.

Vaccination should be done 3 to 4 weeks before exposure. Where cattle have been subjected to great stress—long shipment, extensive handling, and/or exposure to severe weather conditions—it is recommended that they be handled as follows: Adult cattle should be given long grass or oat hay, rolled oats, and/or wheat bran during the first week; newly weaned calves can be given a calf starter ration in addition; all animals should have access to plenty of clear fresh water at all times.

For the first 28 days after arrival, fortify the ration with 350 mg of Aureomycin plus 350 mg of sulfamethiazine per head per day.

Newly arrived cattle should also receive 50,000 IU of vitamin A per head daily and have free access to a good mineral mixture.

Sheep: Reducing stress and adding high levels of antibiotics or sulfas to the feed during susceptible periods may be of some value in reducing the incidence of pneumonia.

Remarks: The name hemorrhagic septicemia should be discarded because this name describes neither the disease nor its cause.

SMEDI (STILLBIRTH, MUMMIFICATION, EMBRYONIC DEATH, INFERTILITY)

SMEDI is caused by several viriuses and results in lowered reproduction.

Species Affected: Swine.

Cause: SMEDI can be caused by several viruses which affect the unborn pig and result in early embryonic death, mummification, or stillbirths. Viruses which have been suggested include enteroviruses, influenza virus, pseudorabies virus, parvovirus, and possibly other viruses.

Symptoms and Signs (or age group most affected): The signs are those from which the name is derived. Producers do not observe any illness in the sows but they notice lower conception rates, irregular heat cycles, small litters, fetal death and mummification, stillbirths, and sterility.

Treatment: As with many viral infections, there is no treatment, so control measures depend on prevention.

Control and Eradication: Since animals build immuminty to the viruses, those animals affected at one farrowing will usually carry normal litters the next breeding. If the problem is due to parvoviruses, a vaccine is available.

Prevention: The best approach to the prevention of SMEDI is management practices which ensure that gilts and new breeding stock entering the breeding herd are exposed to existing viruses at least 1 month before breeding. This establishes a common viral flora and provides uniform immunity. Exposure can be accomplished (1) if animals are reared in a single building, (2) by fecal contamination from recently weaned pigs, (3) by commingling sows and gilts, or (4) by fence line contact with new boars.

Remarks: SMEDI refers to a syndrome rather than a disease caused by a specific pathogen. The viruses involved are associated with stillbirth (S), mummified fetuses (M), embryonic death (ED), and infertility (I).

SORE MOUTH (CONTAGIOUS ECTHYMA)

Sore mouth is a highly contagious virus disease.

Species Affected: Sheep. Goats. Humans.

Cause: A specific virus, often complicated by the necrosis organism, *Spherophorus necrophorus.*

Symptoms and Signs (or age group most affected): Infected lambs and kids refuse to feed and appear depressed. Small vesicles appear on the lips, gums, and tongue, causing these parts to be red and swollen. Vesicles break and form sores that bleed easily and become encrusted with a scab. The sores may become infected or may spread to the teats, udder, and feet (just above the coronet) of the mother.

This is especially a disease of lambs and kids.

Fig. 10-30.

(Continued)

TABLE 10-3 *(Continued)*

SORE MOUTH *(Continued)*

Distribution and Losses Caused By: Throughout the U.S.
The mortality (from secondary causes) is low, but of economic importance in that it results in unthriftiness and loss of weight.

Treatment: Infected lips, mouths, and nostrils should be treated by applying an ointment containing a broad-spectrum antibiotic to the lesions.

Control and Eradication: If the disease is in serious proportion, immunization with a vaccine may be necessary in order to bring it under control.
Isolate, if practical, infected animals and prevent secondary invaders.

Prevention: General sanitation. A specific vaccine, similar in nature to smallpox vaccine, will also produce an immunity. The latter should be used either (1) at the time of docking and castrating, or (2) at least 10 days before shipping feeder lambs in areas in which the disease is known to be prevalent. But do not vaccinate noninfected sheep on noninfected ground because this will infect the premises.

Remarks: If young lambs or kids go off feed for an extended time, the udder of the dam may become caked, predisposing blue bag or mastitis.
Sore mouth is often complicated by necrosis organism and by screwworms.
Scabs from sore mouth lesions remain infective for several years and carry infection over from year to year, on the premises.

STRANGLES (DISTEMPER)

Strangles is a widespread contagious disease.

Species Affected: Horses. Mules.

Cause: *Streptococcus equi*, a bacterium.

Symptoms and Signs (or age group most affected): Depression, loss of appetite, high fever, and discharge of pus from the nose. By the 3rd or 4th day of the disease, the glands under the jaw start to enlarge, become sensitive, and eventually break open and discharge pus. A cough is present.
Any age but most common in young stock.

Distribution and Losses Caused By: Worldwide.
Death loss is very low.

Treatment: Good nursing is the most important treatment. This includes clean, fresh water, good feed, and shelter with uniform temperature away from drafts. The veterinarian may prescribe one of the sulfas and/or antibiotics, or both.
Early treatment is of the utmost importance in distemper.

Control and Eradication: Put affected animals in strict quarantine. Clean and disinfect contaminated quarters and premises.

Fig. 10–31.

Prevention: Vaccination is the best preventive for strangles.
Foals may be vaccinated at 3 months of age.
Vaccination lasts 6 to 12 months. A booster should be given every 6 to 12 months.

Remarks: Affected animals are usually immune for the remainder of life.

SWINE DYSENTERY (See DYSENTERY, SWINE)

SWINE INFLUENZA (HOG FLU)

Swine influenza is a severe and contagious respiratory disease, primarily of young pigs.

Species Affected: Swine.

Cause: Type A influenza virus, which is similar to the virus that caused the worldwide flu epidemic of 1918 in humans. Indeed, there is substantial evidence that flu was first introduced to swine at that time—that this is a classic example of a human disease that was transmitted to animals.

Symptoms and Signs (or age group most affected): Makes appearance suddenly. High fever, loss of appetite, a cough, and discharge from the eyes and nose are seen. Animals reluctant to move, but may sit up like dogs to facilitate breathing.

Distribution and Losses Caused By: Throughout the U.S.
Mortality seldom exceeds 2%. The principal loss from swine flu is the lingering debility, which results in the animals not making economic gains.

Treatment: Provide warm, dry, clean quarters and minimum rations.
Antibiotics and sulfonamides may be used on a herd basis to control various bacterial invaders.
Expectorants aid if respiration is difficult.

Control and Eradication: Correct any faulty feeding and management with the herd.
Avoid bringing in new animals.
Good husbandry and freedom from stress help to reduce losses.

Prevention: There are no licensed vaccines in the U.S. for the prevention of swine influenza.

Remarks: This disease is precipitated by sharp cold or windy days, particularly during the spring and fall.
There is serologic evidence of cross infection of humans and swine with the type A influenza virus.

TETANUS (LOCKJAW)

Tetanus is chiefly a wound infection disease.

Species Affected: Chiefly in horses (and other equines) and humans, but occurs in cattle, swine, sheep, goats.

Cause: A powerful toxin (more than 100 times as toxic as strychnine) liberated by the bacterium *Clostridium tetani*, an anaerobe.

(Continued)

TABLE 10–3 *(Continued)*

TETANUS *(Continued)*

Symptoms and Signs (or age group most affected): Usually associated with a wound. First sign of tetanus is a stiffness about the head. The animal often chews slowly and weakly and swallows awkwardly. Third or inner eyelid protrudes over forward surface of eyeball. With the slightest movement or noise, animal shows violent spasms. Usually remains standing until close to death.

All ages susceptible.

Distribution and Losses Caused By: Worldwide, but in the U.S. occurs most frequently in the South.

Death occurs in over 80% of the affected cases.

Treatment: Place the animal under the care of a veterinarian and keep it quiet. Good nursing is important.

Tetanus antitoxin may be helpful if administered at the time of the first symptoms.

Control and Eradication: All surgical procedures should be conducted with the best possible operative techniques.

Prevention: Immunity against tetanus can be obtained through inoculation with either toxoid or antitoxin. Toxoid is an injection of neutralized tetanus toxin to stimulate the

Fig. 10–32.

horse to build its own antibodies. Antitoxin is a concentrated serum with tetanus toxin antibodies taken from another horse and administered as a preventive measure following wounds, surgery, or foaling.

Active immunization is achieved through 2 injections of tetanus toxoid at 2- to 4-week intervals, followed by annual booster injections. If an immunized horse is wounded 2 months or more following such immunization, it is recommended that the veterinarian administer another toxoid injection at that time. If a horse not previously immunized is wounded, it is recommended that the veterinarian administer antitoxin, which will give passive protection for up to 2 weeks.

Remarks: On some "hot" premises, all surgery should be accompanied with tetanus antitoxin.

TUBERCULOSIS

Tuberculosis is a chronic infectious disease. (Fig. 10–33 shows a positive reaction to the TB test.)

Species Affected: All animals and humans.

Tuberculosis in sheep and goats is rare and of chronic character.

Cause: *Mycobacterium tuberculosis*, of which there are 3 kinds: (1) the human, (2) the bovine, and (3) the avian (bird) types.

Symptoms and Signs (or age group most affected): Animals usually get tuberculosis of the lungs and lymph nodes; although in poultry the liver, spleen, and intestines are chiefly affected. In cows, the udder sometimes becomes infected and swollen in chronic cases. Many times infected animals show no outward physical signs of the disease. There may be loss in weight, swelling of joints, and a chronic cough and labored breathing. Other seats of infection are genitals, central nervous system, and the digestive system.

Sheep: Manifest few symptoms; observed on post-mortem. Coughing is prominent in goats, but not in sheep.

Distribution and Losses Caused By: Worldwide.

The incidence of tuberculosis in the U.S. is steadily declining.

Treatment: In humans, tuberculosis can be arrested by hospitalization and complete rest, but in animals this method of treatment is neither effective nor practical. Also, no known medical treatment is effective with animals.

Control and Eradication: *Cattle:* Periodic testing and removal of reactors is the only effective method of control. The following is an effective control program for animals:

1. Disposing of tubercular swine, cattle, and poultry.
2. Applying strict sanitation.

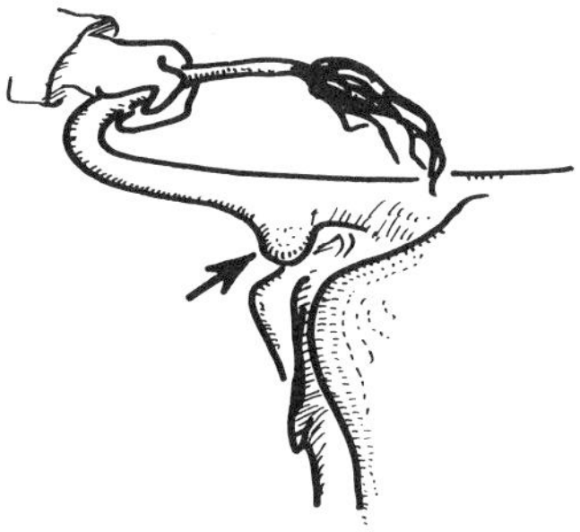

Fig. 10–33.

3. Rotating feedlots and pastures,

Sheep: Testing with avian tuberculin may be of assistance.

Prevention: Removal and supervised slaughter of reactor animals, and pasteurization of milk and creamery by-products.

Fig. 10–42 shows a positive reaction to the intradermic (into the skin) tuberculin test in a cow. This reaction indicates the presence of TB.

Avoid pasturing or housing cattle and swine with chickens.

Remarks: All states will accept for entry the following: (1) accredited herds which have been tested for TB within the past 12 months, or (2) cattle which have had individual negative tests within the past 30 days.

Cattle for export must be tested for tuberculosis and found free of the disease within 90 days of shipment.

VIBRIOSIS (VIBRIONIC ABORTION, VIBRIO FETUS)

Vibriosis is a bacterial disease of sheep and cattle which lowers reproduction.

Species Affected: Sheep. Cattle.

Cause: The bacterium *Vibrio fetus*.

Symptoms and Signs (or age group most affected): Many abortions among ewes, especially during the last 4 to 6 weeks of pregnancy, with lambs stillborn.

Infected cattle herds are characterized by (1) abortions in the middle third of pregnancy, (2) several services per conception, and (3) irrregular heat periods.

For diagnosis, laboratory methods must be used.

(Continued)

TABLE 10-3 *(Continued)*

VIBRIOSIS *(Continued)*

Distribution and Losses Caused By: Vibriosis occurs in sheep and cattle throughout the U.S. Reports indicate that the distribution is worldwide.

In affected flocks, abortions may be up to 80%. In general, ewes show little clinical evidence of infection.

Treatment: There is no treatment for sheep.

Cattle are treated by injecting drugs into the uterus and/or by allowing sexual rest.

Control and Eradication: Females that have aborted should be isolated and all aborted fetuses and membranes should be burned.

Control of vibriosis in sheep or cattle requires sound management, strict sanitation, early diagnosis, and proper vaccination.

Prevention: A vaccine is available. Repeat annually before breeding.

Avoid contact with diseased animals and contaminated feed, water, and materials.

Artificial insemination is a rapid and practical method of stopping infection from cow to cow.

Remarks: Ewes that abort one year may raise perfectly healthy lambs in succeeding years.

Some probably remain as carriers.

Transmission in sheep appears to be from consuming infected feed and water, whereas in cattle it is from an infected bull.

[1]The illustrations for this table were prepared by R. F. Johnson.

[2]*Better Beef Business*, April 1973.

[3]*Successful Farming*, May 1973.

[4]A 10% formaldehyde solution may be made by mixing 1 gal *(3.8 liters)* of 38% formaldehyde and 9 gal *(34 liters)* of water, and a 20% copper sulfate solution may be made by mixing 1.66 lb *(754 g)* of copper sulfate per gallon of water. Since copper sulfate is corrosive for most metals, it should be prepared in earthenware or wooden containers.

NUTRITION RELATED DISORDERS

Not all noncontagious diseases are nutritionally responsive. Some of them are due to physical factors. Others are caused by faulty management. Regardless of the nature or the cause, however, all of them have an impact on the nutritional well-being of the affected animal; hence, they are nutrition related. Several of the most important of these disorders are discussed in the sections that follow.

CHOKING

Occasionally, feeds become lodged in the esophagus of cattle or horses, causing them to choke.

Cattle may choke on such feeds as beets, potatoes, apples, hay cubes, or ears of corn. Afflicted animals drool saliva from the mouth, make frequent attempts at swallowing, bloat rapidly due to closure of the outlet for gas from the stomach, and switch the tail. If the obstruction is in the region of the neck, it can be felt from the outside. Treatment consists in an attempt to work the object into the stomach. A speculum should be applied to hold the mouth open, then the hand should be inserted into the animal's throat to grasp and remove the obstacle. (*CAUTION:* Beware of the hazard of a bitten hand or arm.) If this fails, a stomach tube or rubber hose lubricated with water or oil and pushed down the esophagus usually will free the object: but care must be taken not to damage the lining of the esophagus. Should there be marked bloating, the stomach may have to be punctured.

Horses choke most frequently from bolting (eating too rapidly) their grain, although they may choke on hay cubes, ears of corn, potatoes, and apples. Afflicted animals become excited, squeal, and thrust the head forward. Treatment consists in controlling the pain with sedatives, confining the animals, and allowing access to water but not feed. Passage of a stomach tube to the obstruction and repeated pumping and siphoning may relieve grain choke. As a last resort, the obstruction may be gently pushed into the stomach with a large stomach tube or rubber hose.

Minimizing choking in both cattle and horses consists in avoiding, to the extent practical, feeds that are most likely to cause choking. Such feeds as potatoes, apples, and roots are less apt to cause trouble if they are sliced or chopped. When feeding potatoes to cattle, choking can be materially lessened by forcing the animals to eat them from the ground level with their heads down. This can be accomplished by having a cable or pole arrangement at the top of their necks to keep their heads down. Choking can be lessened in gluttonous horses (horses that eat their feed too rapidly—that bolt their feed) by putting into the grain box several smooth stones about the size of a baseball, thereby slowing their eating.

DISPLACED ABOMASUM

This disorder is being diagnosed with increasing frequency in dairy cows.

Normally, the abomasum is located on the right side of the rumen of the cow and rests on the floor of the abdomen. It is attached at the front end to the omasum and at the back end to the small intestine. The mid-portion of the abomasum is not held rigidly in place by other structures; however, the weight of the material within it usually keeps it in place on the floor of the abdomen. Sometimes when the metabolism of the stomach is abnormal, the abomasum slides under the rumen, becomes partially filled with gas, and rises on the left side between the rumen and the body wall. When this happens, the afflicted cow goes off feed and exhibits many symptoms similar to ketosis. In fact, secondary ketosis may result from displaced abomasum because of decreased feed intake, thereby inducing a ketotic state. However, an experienced veterinarian can differentiate between displaced abomasum and primary ketosis.

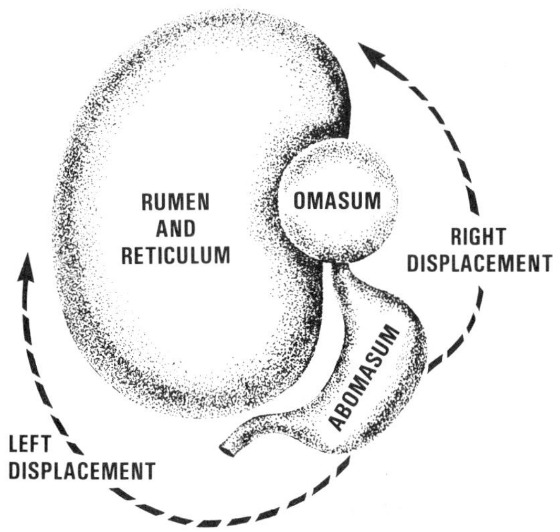

Fig. 10-34. Normally, the abomasum rests near the belly floor and to the right of the rumen. In most displaced abomasums, it shifts to the left.

Several treatments for displaced abomasum have been used with variable success; most commonly, the following two:

1. Rolling the cow onto her back and getting the abomasum back into normal position. Frequently, the problem recurs following this treatment.

2. Performing rather simple surgery, involving the veterinarian suturing the bottom of the rumen wall to the body wall, thereby preventing the abomasum from slipping between the two. When properly done, this usually provides a permanent cure. The operation can be done using local anesthesia. Frequently, cows so treated do not miss any milkings.

The cause of displaced abomasum is not known; however, the following theories have been advanced:

1. High-concentrate rations and inadequate bulk, as a result of which the rumen pulls away from the body wall.

2. Reduced ruminal contents for various reasons, such as cows going off feed, resulting in a gap between the rumen and the body floor.

When displaced abomasum is a problem, the incorporation of more roughage in the ration is recommended. A rule of thumb is that at least half of the dry matter should be in the form of roughage-type feeds, such as hay, silage, or pasture.

HARDWARE DISEASE

The term *hardware disease* (traumatic gastritis) is used to describe the condition that results from swallowing foreign materials, usually metal (nails, wire, screws, and pins). Cattle are involved more than other classes of animals; however, cases have been reported in horses and goats. In most cases, the metal is found in the reticulum (second stomach). However, the material may puncture the reticulum lining and pass into the body cavity. Some objects may pierce the diaphragm and enter the thoracic cavity, thence work their way to the heart or lungs, where they may cause serious damage or death.

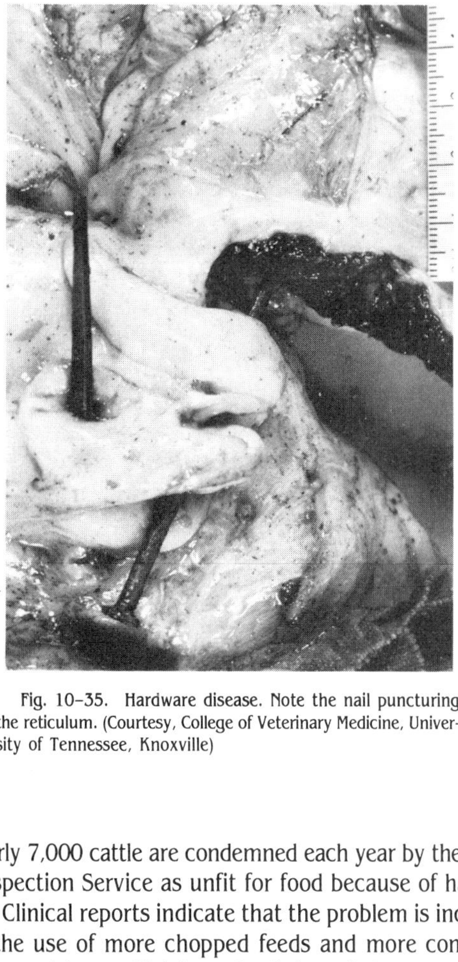

Fig. 10-35. Hardware disease. Note the nail puncturing the reticulum. (Courtesy, College of Veterinary Medicine, University of Tennessee, Knoxville)

Nearly 7,000 cattle are condemned each year by the Federal Meat Inspection Service as unfit for food because of hardware disease. Clinical reports indicate that the problem is increasing due to the use of more chopped feeds and more contamination. Sharp objects will injure the lining of the stomach and cause infection and inflammation, a condition known as traumatic gastritis.

Hardware disease is a problem in cattle because of their eating habits and stomach arrangement. The usual source of metals is in the feed. The animals eat rapidly and are not able to sort foreign objects from their feed.

The most common symptoms of hardware disease are loss of appetite and digestive disturbance; slow and stiff movement and arched back; elbows that bow outward; decreased rumen movement and chewing; tendency to stand with the front feet elevated so as to lessen the pressure of the viscera on the inflamed area; rise in body temperature; and swelling under the jaw, at the brisket, and at the hock joints. Bulls may be reluctant to mate.

Prevention consists in avoiding foreign objects getting into the feed through good management. Also, it is recommended that strong magnets be installed, in keeping with the manufacturer's directions, (1) at the outlets of mechanical silo unloaders, and (2) in feed processing equipment.

Magnets may also be permanently placed in the cow's second stomach, for the purpose of holding objects that have not penetrated the stomach wall. However, the only sure cure for traumatic gastritis is veterinary surgery. Surgery may be successful only if performed before the condition has progressed to the point that damage has been done to the heart or other organs.

IMMUNOGLOBULIN DEFICIENCY

Immunoglobulin, found in colostrum, is the means by which most newborn farm mammals acquire passive immunity against pathogenic diseases and infections.

Deficiencies in immunoglobulins may result (1) when the young fail to nurse properly following birth, (2) when the mother is exhausted or sick after giving birth, (3) when the mother's colostrum contains low levels of specific antibodies, (4) when the caretaker fails to give colostrum to newborn, and/or (5) when other feeds are consumed prior to colostrum. Also, immunoglobulin deficiencies may exist in milk from first-calf heifers and in milk from cows that haven't dried up or have had less than a 30-day dry period. Older cows have been exposed to a wider range of diseases than first-calf heifers and therefore produce more immunoglobulins against them. So, if livability problems are encountered, colostrum from older cows should be fed to calves from heifers or from cows that haven't dried up or have had too short a dry period.

The highest absorption of intact immunoglobulins directly into the lymphatic system (thoracic duct) occurs immediately after birth (within 15 to 30 minutes) and during the first 6 hours after birth, but by 24 hours most of the absorption capacity is lost. Hence, for maximum protection against infection, newborn mammals should be fed colostrum very early in life.

Where the immunoglobulin phenomenon exists, as in the calf, the intestinal villi of the newborn are able to absorb the globulins by pinocytosis (engulfing). This enables those species which do not normally obtain adequate immune protection through placental transfer to acquire instant immunity by ingesting colostrum high in immunoglobulins. Aside from this unique situation, protein must be digested.

IMPACTION IN HORSES

Impaction is a form of colic caused by obstruction of the cecum or colon by fibrous feeds. It is usually caused by feeding horses large amounts of straw, cornstalks, or other coarse, high-fiber feeds, along with lack of water intake. Distention of the large colon with gas causes acute abdominal pain. The usual signs described by colic are seen. Medical treatments, which should be administered by a veterinarian, meet with variable success. Mineral oil and magnesium sulfate are popular laxatives. Dioctyl calcium sulfosuccinate may be used to penetrate the impacted mass. Antiferments and oral antibiotics such as neomycin are helpful in preventing gas formation. Surgery should be the last resort.

POISONOUS PLANTS[4]

Poisonous plants have been known to people since time immemorial. Biblical literature alludes to the poisonous properties of certain plants, and history records that hemlock (a poison made from the plant from which it takes its name) was administered by the Greeks to Socrates and other state prisoners.

Fig. 10–36. Lupine toxicity. Calf showing abnormalities due to ingestion by dam of lupines during 40th to 70th days of gestation. (Courtesy, The American Institute of Nutrition, Bethesda, MD)

No section of the United States is entirely free of poisonous plants, for there are hundreds of them. But the heaviest livestock losses from them occur on the western ranges because (1) there has been less cultivation and destruction of poisonous plants in range areas, and (2) the frequent overgrazing on some of the western ranges has resulted in the elimination of some of the more nutritious and desirable plants, and these have been replaced by increased numbers of the less desirable and poisonous species. It is estimated that poisonous plants account for 8 to 10% of all range animal losses each year; and even more in some areas.

DIAGNOSIS OF PLANT POISONING

The diagnosis of plant poisoning in animals is not an easy or precise procedure. Any case of sudden illness or death with no apparent cause is commonly considered to be a poisoning. This may not always be correct. When large numbers of animals are suddenly affected, however, a suspicion of poisoning is justified until it has been proven otherwise.

[4]In the preparation of this section, the author had the benefit of the authoritative review and suggestions of the Director of the Poisonous Plant Research Laboratory, USDA, Logan, UT

Symptoms or signs induced by eating poisonous plants may include (1) sudden death; (2) transitory illness; (3) general body weakness; (4) disturbance of the central nervous, vascular, and endocrine systems; (5) photosensitization; (6) frequent urination; (7) diarrhea; (8) bloating; (9) chronic debilitation and death; (10) embryonic death; (11) fetal death; (12) abortion; (13) extensive liver necrosis and/or cirrhosis; (14) edema and/or abdominal dropsy; (15) tumor growths in tissues; (16) congenital deformities; (17) metabolic deficiencies; and (18) physical injury.

No general set of symptoms and signs per se irrefutably provides all the information necessary to make a diagnosis of plant poisonings. Nevertheless, a careful description of the toxic signs coupled with information pertaining to available plants provides a meaningful basis for a tentative diagnosis. Additional information essential to a poisonous plant diagnosis includes (1) type of feed, site grazed, and availability of water; (2) identification and relative abundance of all poisonous plants available to animals; (3) amount and stage of growth of the various poisonous plants being grazed; (4) the toxicity and palatability of the plants in relation to their stage of growth; (5) time from eating the plants until onset of toxic signs; (6) species, age, and sex of animals affected; (7) clinical signs of toxic reactions; (8) chemical analysis of plants; and (9) a careful evaluation of all the information relative to the etiology of the disease.

WHY ANIMALS EAT POISONOUS PLANTS

A frequently asked question is: Why do animals eat poisonous plants? The answer is not simple, but among the reasons are the following: (1) total lack of sufficient palatable forage—the animals are hungry; (2) decrease in palatability and nutrients of mature, weathered range grasses, with the result that poisonous plants become more appealing, comparatively speaking; (3) insufficient spring grass; (4) rain, melting snow, and heavy dew may enhance the palatability of some poisonous plants; and (5) going without water too long, which results in a reduction in feed intake, then, after watering, they develop a ravenous appetite and eat anything in sight—including less palatable poisonous plants.

Poisonous plants vary in palatability—between species, and within species, and at different stages of growth. For example, poisonous hemlock is never palatable and is eaten only as a last resort—when palatable forage is not available. Locoweed and black nightshade are eaten at any stage of growth or when mixed with hay. Others, such as lupines, horsebrush, and death camas may be eaten only at certain stages of growth. Still others, such as milk vetch, larkspur, and halogeton, are highly palatable to livestock at any and all times, with the result that if they're present, animals will seek them out and there will be losses. Then, too, certain plants are poisonous to cattle but not to sheep (and vice versa), as shown in Table 10-4.

TABLE 10-4
TYPE OF RANGE ANIMAL SUSCEPTIBLE TO POISONOUS PLANTS AT DEFINITE SEASONS

Poisonous to Cattle	Time of Year	Poisonous to Sheep	Time of Year	Poisonous to Cattle & Sheep	Time of Year
Low larkspur	Spring	Death camas	Spring	Broomweed	Spring and summer
Oak	Spring	Greasewood	Fall	Chokecherry	Spring
Tall larkspur	Early summer & early fall	Horsebrush	Spring	Copperweed	Summer
Timber milk vetch	Spring	Rubberweed	Summer	Desert parsley	Spring
Water hemlock	Spring	Sneezeweed	Summer	Halogeton	All year
				Loco	Spring
				Lupine	Summer and fall
				Milkweeds	Summer
				Veratrum	Summer

PREVENTING LOSSES FROM POISONOUS PLANTS

With poisonous plants, the emphasis should be on prevention of losses rather than on treatment, no matter how successful the latter. The following are effective preventative measures:

1. **Follow good pasture or range management** in order to improve the quality of the pasture or range. Plant poisoning is nature's sign of a "sick" pasture or range, usually resulting from misuse. When a sufficient supply of desirable forage is available, poisonous plants may not be eaten, for they are usually less palatable. On the other hand, when overgrazing reduces the available supply of the more palatable and safe vegetation, animals may, through sheer hunger, consume toxic plants.

2. **Know the poisonous plants common to the area.** This can usually be accomplished through (a) studying drawings, photographs, and/or descriptions; (b) checking with local authorities; or (c) sending two or three fresh whole plants (if possible, include the roots, stems, leaves, and flowers) to the state agricultural college—first wrapping the plants in several thicknesses of moist paper.

By knowing the poisonous plants common to the area, it will be possible—

a. To avoid areas heavily infested with poisonous plants which, due to animal concentration and overgrazing, usually include waterholes, salt grounds, bed grounds, and trails.

b. To control and eradicate the poisonous plants effectively, by mechanical or chemical means (as recommended by local authorities) or by fencing off.

c. To recognize more surely and readily the particular kind of plant poisoning when it strikes, for time is important.

d. To know what first aid, if any, to apply, especially when death is imminent or where a veterinarian is not readily available.

e. To graze with a class of livestock not harmed by the particular plant or plants, where this is possible. Many plants seriously poisonous to one kind of livestock are not poisonous to another, at least under practical conditions.

f. To shift the grazing season to a time when the plant is not dangerous, where this is possible. That is, some plants are poisonous at certain seasons of the year, but comparatively harmless at other seasons.

g. To avoid cutting poison-infested meadows for hay when it is known that the dried cured plant is poisonous. Some plants are poisonous in either green or dry form, whereas others are harmless when dry. When poisonous plants (or seeds) become mixed with hay (or grain), it is difficult for animals to separate the safe from the toxic material.

3. **Know the symptoms that generally indicate plant poisoning,** thus making for early action.

4. **Avoid turning to pasture in early spring.** Nature has ordained most poisonous plants as early growers—earlier than the desirable forage. For this reason, as well as from the standpoint of desirable pasture management, animals should not be turned to pasture in the early spring before the usual forage has become plentiful.

5. **Provide supplemental feed during droughts, after plants become mature, and after early frost.** Otherwise, hungry animals may eat poisonous plants in an effort to survive.

6. **Avoid turning very hungry animals where there are poisonous plants,** especially those that have been in corrals for branding, etc; that have been recently shipped or trailed long distances; or that have been wintered on dry forage. First feed the animals to satisfy their hunger or allow a fill on an area known to be free from poisonous plants.

7. **Avoid driving animals too fast when trailing.** On long drives, either allow them to graze along the way or stop frequently and provide supplemental feed.

8. **Remove promptly all animals from infested areas when plant poisoning strikes.** Hopefully, this will check further losses.

9. **Treat promptly, preferably by a veterinarian.**

TREATMENT OF PLANT-POISONED ANIMALS

Unfortunately, plant-poisoned animals are not generally discovered in sufficient time to prevent loss. Thus, prevention is decidedly superior to treatment.

When trouble is encountered, the owner or caretaker should *promptly* call a veterinarian. In the meantime, the animal should be (1) placed where adequate care and treatment can be given, (2) protected from excessive heat and cold, and (3) allowed to eat only feeds known to be safe.

The veterinarian may determine the kind of poisonous plant involved (1) by observing the symptoms, and/or (2) by finding out exactly what poisonous plant was eaten through looking over the pasture and/or hay and identifying leaves or other plant parts found in the animal's digestive tract at the time of autopsy.

It is to be emphasized, however, that many poisoned animals that would have recovered had they been left undisturbed, have been killed by attempts to administer home remedies by well-meaning but untrained persons.

COMMON POISONOUS PLANTS

The list of poisonous plants is very extensive; so, it is impossible to present a summary of all of them in *Stockman's Handbook Digest*. Nevertheless, both the livestock producer and the veterinarian should have a working knowledge of the principal poisonous species in the area in which they operate. Usually, this information can be obtained from a bulletin(s) published by the Agricultural Extension Service and available from the County Agent.

AGRICULTURAL CHEMICALS AND DRUGS

In the everyday pursuit of modern agriculture, more and more chemicals and drugs are being used. Hand in hand with this development, there has been increased public concern over the use of the products, for fear of poisoning human food.

Chemicals and drugs must be used with discretion, especially those designed to kill some living organism. But sometimes choices must be made; for example, between malaria-carrying mosquitoes and some fish, or between hordes of locusts and grasshoppers and the crops they devour. This merely underscores the need for (1) careful testing through properly designed experiments of all products prior to use, (2) conforming with federal and state laws, and (3) accurate labeling and use of products.

The vast majority of agricultural chemicals and drugs have been properly used. Of course, it shouldn't be too surprising that a few have been improperly used when it is realized that there are approximately 300,000 trade name products on the market.

When chemical poisoning or drug misuse happens, it can be both devastating and perplexing. Usually, the causative agent can be diagnosed after an investigation of the environment and the feed. However, few poisons can be diagnosed with certainty by clinical signs alone. When trouble is encountered, the producer should promptly call a veterinarian if animals are involved, or a medical doctor if people are involved.

DIAGNOSING AND TREATING LIVESTOCK POISONING

It is often difficult to make a definite diagnosis of an animal poisoning. Clinical signs are not usually specific, and all signs are not always seen in every poisoned animal. However, in a herd of poisoned animals, every sign or toxic effect will likely be seen in some animal. The recommended procedure for making a diagnosis of the cause of poisoning follows:

1. Check on the accessibility of a poisonous substance. A highly toxic substance may or may not be hazardous to livestock, depending on whether the animals could conceivably come into contact with it.
2. Study the clinical signs. This may be difficult, especially with possible combinations of toxins or infectious agents.
3. Use a few test animals in a feeding trial.
4. Make a pathological examination of the animal's internal organs and tissues.
5. Chemically analyze the feed, water, and animal tissues for the presence of suspected toxin. It is necessary to have enough information so that certain poisons or groups of poisons can be suspected, because the analytical methods are quite specific and certain tissues are required.
6. Use a specific antidote (where available) for the suspected poison. If it alleviates the clinical signs, it gives evidence of the cause.

The principles of treatment are directed toward accomplishing the following:

1. Preventing injury and controlling convulsions with a sedative, usually a barbiturate.
2. Relieving pain by use of chemical analgesics.
3. Removing or neutralizing the poison by—
   a. Washing of any surface poison.
   b. Using gastric lavage with activated charcoal for absorbing toxins in the stomach.
   c. Using cathartics to help fecal elimination of absorbed toxins.
   d. Using diuretics to help urinary elimination of absorbed toxins.
   e. Performing a rumenotomy for physical removal of unabsorbed toxins.
   f. Using a specific antidote, if available.
4. Maintaining the vital signs of respiratory, circulatory, and renal functions by physical or chemical resuscitation, fluid therapy, etc.
5. Observing the animal for further treatment needs, because the toxin may continue to be absorbed from the skin, gut, or respiratory system of the animal.

NATIONAL ANIMAL POISON CONTROL CENTER

Established in 1978 and maintained at the University of Illinois, Urbana-Champaign, the National Animal Poison Control Center hotline number is 217/333-3611. Recognizing that accidents don't wait for business hours, the Center is open 24 hours a day, every day of the week. The toxicology group is staffed to answer questions about known or suspected cases of poisoning or chemical contaminations involving any species of animal. It is not intended to replace local veterinarians or state toxicology laboratories, but to complement them.

The toxicologists at the center constantly update their files on chemicals, feed additives, human and veterinary drugs, pesticides, environmental contaminants, and plant and mold toxins. Their comprehensive file of information contains comparative species toxicity data, product ingredients, and recommended therapeutic and decontamination measures. The goal is a computer database containing 200,000 entries to facilitate quick and accurate responses to all types of poisoning/contamination incidents and inquiries.

Many times a proper treatment regime can be recommended over the telephone. When telephone consultation is inadequate or the problem is of major proportion, a team of veterinary specialists can arrive at the scene of a toxic or contamination problem within a short time.

The cost of an investigation varies according to distance traveled, personnel time, and laboratory services required. Where consultation over the telephone is adequate, there is no charge to the veterinarian or producer.

POTENTIAL POISONS

A poison is a substance which in sufficient quantities and/or over a period of time kills or harms living things. Toxic substances are chemical substances that may present an unreasonable risk of injury to health or to the environment. Many poisons are called toxins. The study of poisons is called *toxicology.* The discussion that follows and Table 10-6 pertain primarily to feed-related poisons that may be eaten by animals. For most of these, there is both a safe level and a poisonous level; and the severity of the side effect depends upon (1) the amount taken, (2) the period of time over which the substance is taken (certain poisons are cumulative), and (3) the age and physical condition of the animal. This lends credence to the toxicological adage: "Only the dose makes the poison."

There are more than 4 million chemical compounds of which more than 60,000 are commercially produced; and about 1,000 new ones are introduced each year. Some of these make their way into feed and water. With the growth and use of chemicals, feed supplies are subject to contamination from or treatment with chemicals in the course of growing, fertilizing, harvesting, processing, and storing. In addition to manufactured chemical poisons, there are many naturally occurring poisonous substances.

Farmers know that unless they follow state and federal regulations they risk having their products condemned and seized, or refused by food processors. Nevertheless, the economics dictate that new products be used as soon as they prove useful and are approved. On the other hand, food faddists may feel that they are being poisoned; wildlife conservationists may be concerned over possible damage to songbirds and other animals; beekeepers become unhappy if insecticides kill honeybees; and public health agencies are concerned about

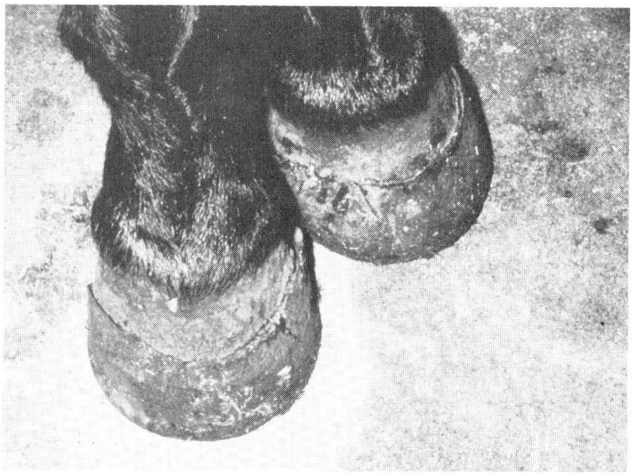

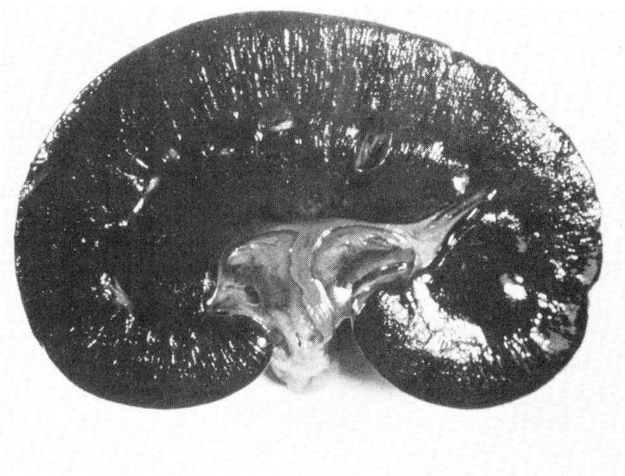

Fig. 10-37. For some substances, such as selenium and copper, a little is good but more may be poisonous. *Left:* Selenium toxicity resulting in severe hoof damage of a horse—note the horizontal cracks. (Courtesy, Colorado State University, Ft. Collins) *Right:* Copper toxicity evidenced by gunmetal-colored kidneys. (Courtesy, University of Tennessee, Knoxville)

contamination of soil, water, and food supplies. Thus, great care should be exercised in handling chemicals and drugs; the labels on the containers should be read and heeded carefully, and partly used packages and empty containers should not be left where animals have access to them.

When poisoning happens, it can be both devastating and perplexing. No part of veterinary diagnostics is as difficult and complex as toxicology. First, what compound is being tested for out of the many thousand known? Second, detecting trace levels, such as parts per billion, of pesticides and other chemicals in feed and water by low level residue analysis can be as difficult as understanding them.

Table 10-5 lists the most common chemical poisons and presents pertinent facts pertaining to each.

TABLE 10-5
POTENTIALLY POISONOUS ELEMENTS

ERGOT (a parasitic fungus)

Source: It replaces the seed in the heads of grasses and cereal grains, in which it appears as a purplish-black, hard banana-shaped dense mass from ¼ to ¾ in. long.
Most common in rye, wild rye, bromegrass, and dallisgrass.

Species Affected: Cattle. Sheep. Horses. Humans.

Symptoms and Signs: Acute ergot poisoning, caused by large quantities eaten at one time, may produce paralysis of the limbs and tongue, disturbance of the gastrointestinal tract and abortion.
Chronic poisoning produces gangrene of the extremities, with subsequent sloughing off of hooves, ears, and tail.
Delirium, spasms, and paralysis may occur before death.

Distribution and Losses Caused By: Ergot is found throughout the world.

Prevention: Never feed heavily ergot-infested hay or grain.

Treatment: If noticed in time, stricken animals may recover if put on good feed.

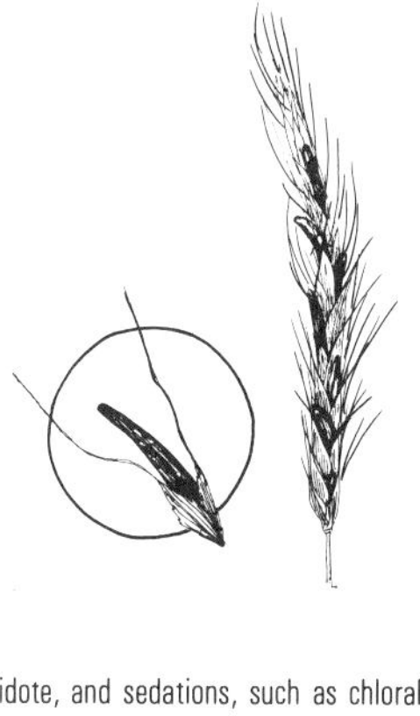

Fig. 10-38.

Tannin used as a drench is an antidote, and sedations, such as chloral hydrate, may be given to nervous animals.
Control of ergotism consists of an immediate change to an ergot-free diet.

(Continued)

TABLE 10–5 *(Continued)*

FLUORINE (FLUOROSIS) *(F)*

Source: Ingesting excessive quantities of fluorine through either the feed or water.

Species Affected: All farm animals, poultry, and humans.

Symptoms and Signs: Abnormal teeth (especially mottled enamel) and bones, stiffness of joints, loss of appetite, emaciation, reduction of milk flow, diarrhea, and salt hunger.

Distribution and Losses Caused By: The water in parts of Arkansas, California, South Carolina, and Texas has been reported to contain excess fluorine. Occasionally, throughout the U.S. high-fluorine phosphates are used in mineral mixtures.

Prevention: Avoid the use of feeds, water, or mineral supplements containing excessive fluorine.

Treatment: Any damage may be permanent, but animals which have not developed severe symptoms may be helped to some extent, if the source of excess fluorine is eliminated.

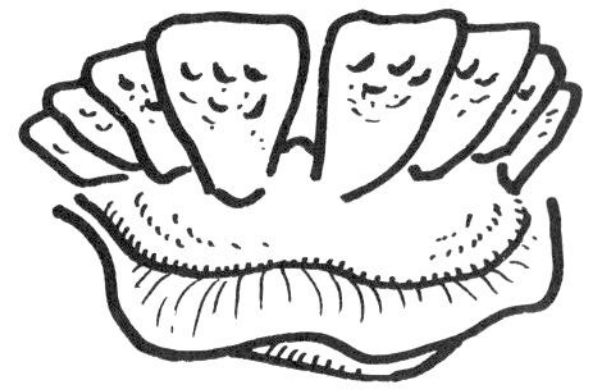

Fig. 10–39.

Remarks: Fluorine is a cumulative poison. 100 ppm (0.01%) fluorine of the total dry ration is the borderline in toxicity for cattle, sheep, and pigs. At levels of 25 to 100 ppm, some mottling of the teeth may occur over periods of 3 to 5 years. In breeding animals, whose usefulness exceed 3 to 5 years, the permissible level is 50 ppm of the total dry ration.

Not more than 65 to 100 ppm fluorine should be present in dry matter of rations when rock phosphate is fed.

LEAD *(Pb)*

Source: Lead is discharged into the air from auto exhaust fumes and other sources.

Lead pollution of feed and food crops as a result of lead being deposited on the leaves and other edible portiions of the plant by direct fallout.

Inhaling airborn lead.

lead may get into feed or food and water from contact with lead pipes, utensils, or discharged storage batteries.

Species Affected: All farm animals; but cattle and sheep are especially susceptible.

Symptoms and Signs: Symptoms develop rapidly in young animals, but slowly in mature animals.

Feces may become very dark gray and be tinged with blood.

Salivation, champing of the jaws, frenzy, blindness, convulsions, coma, and death.

Mature animals usually have diarrhea and show incoordination, especially in the hind limbs and prostration.

Distribution and Losses Caused By: Rather extensive, because of the wide use of lead preparations in agriculture.

Prevention: Avoid sources of lead.

Treatment: If damage to tissue has been extensive, treatment is of little value; in any event it should be handled by a veterinarian.

The best chemical antidote is protein (milk, eggs, blood serum).

Remarks: Lead poisoning can be diagnosed positively by analyzing the blood tissue for lead content.

It is a cumulative poison.

MERCURY *(Hg)*

Source: Mercury is discharged into air and water from industrial operations and is used in herbicide and fungicide treatments.

Consumption of seed grains treated with fungicides that contain mercury, for the control of fungus diseases of oats, wheat, barley, and flax.

Mercury poisoning has occurred where mercury from industrial plants has been discharged into water and then accumulated in fish and shellfish.

Species Affected: All farm animals, but especially cattle and hogs.

Symptoms and Signs: Gastrointestinal, renal and nervous disturbances; but impossible, on basis of symptoms, to differentiate mercury from other poisons.

Case history of animals consuming mercury-treated grains should be considered strong circumstantial evidence.

Distribution and Losses Caused By: When, through ignorance or negligence, mercury-treated grain is fed to animals.

Prevention: Do not feed livestock seed grains treated with a mercury-containing fungicide.

Surplus of treated grain should be burned and the ash buried deep in the ground.

Treatment: Treatment is not too satisfactory.

The best antidote is protein (milk, egg, blood serum).

Remarks: Ultimate diagnosis depends upon demonstrating the presence of mercury in the tissues, especially in the kidneys and liver.

Food and Drug Administration prohibits use of mercury-treated grain for feed or food.

Mercury is a cumulative poison.

MYCOTOXINS (toxin-producing molds; e.g., *Aspergillus flavus, Penicillium cyclopium, P. islandicum,* and *P. palitans)*

Source: Aflatoxin (most studied of the group) associated with peanuts, brazil nuts, silage, corn and most other cereals, hay, and grasses. The mold can produce toxic compounds on virtually any food (even synthetic) that will support growth.

While aflatoxin appears to cause most of the problem, it is not the only mycotoxin to be feared. Other mycotoxins are being studied.

Species Affected: Turkeys. Ducklings. Pheasants. Trout. Cattle. Swine. Humans.

In all species, the young are far more susceptible than mature animals.

Generally, ruminants appear to tolerate higher levels of mycotoxins and longer periods of intake than simple-stomached animals.

(Continued)

TABLE 10-5 *(Continued)*

MYCOTOXINS *(Continued)*

Symptoms and Signs: *The toxic symptoms from continued (long-term) intake of aflatoxin are:*

Animal	Aflatoxin Level[1] (ppb)	Symptoms
Beef cattle:		
450 lb	700	Liver damage.
	1,000	Reduced growth and feed efficiency.
Dairy cattle:		
Calves (milk fed)	200	Fatal.
Lactating cows	20	Drop in milk yield. Aflatoxin secreted in milk.
Sheep	1,750	Reduced fertility.
Swine:		
50 lb	280	Reduced growth and feed efficiency.
80 lb	450	Liver damage.
	615 810	Reduced growth and feed efficiency.
Breeding herd	450–1,500	Abortions, dead pigs at birth.

Distribution and Losses Caused By: Widely distributed throughout the world.

In addition to the effect of mycotoxins on the animal's health, milk and eggs are contaminated by the residues or mycotoxins, or their metabolic products.

Prevention: The prime cause of aflatoxin is moisture; hence, proper harvesting, drying, and storage are important factors in lessening contamination and toxin production.

Propionic and acetic acids will inhibit mold growth; hence, their use in preserving high moisture grains is encouraged.

Treatment: Remove the source of the mold.

Animals suffering from molds frequently respond to vitamin B injections.

Iron therapy may be helpful, since hemorrhaging is a frequent problem.

Remarks: Certain molds produce toxins, or mycotoxins.

Aflatoxin has been clearly shown to be a carcinogen (tumor producing).

Ultraviolet irradiation and anhydrous ammonia under pressure will reduce the toxicity of aflatoxins and, if continued long enough, will deactivate them entirely.

Not all toxins are harmful. For example, zearalenol is being commercially produced as a growth promotant hormone for cattle.

NITRATE-NITRITE POISONING (OAT HAY POISONING, CORNSTALK POISONING)

Source: Consuming high-nitrate feeds—feeds with a high-nitrate content due to nitrate fertilization, drought, etc.

Eating nitrate or nitrite fertilizer, or drinking pond water containing same.

Species Affected: Cattle, sheep, and horses; especially cattle.

Symptoms and Signs: Accelerated respiration and pulse rate; diarrhea; frequent urination; loss of appetite; general weakness, trembling, and a staggering gait; frothing from the mouth; lowered milk production; abortion; blue color of the mucuous membrane, muzzle, and udder due to lack of oxygen; and death in 4½ to 9 hr after eating lethal doses of nitrate.

Distribution and Losses Caused By: Excessive nitrate content of feeds is an increasingly important cause of poisoning in farm animals, due primarily to more and more high nitrogen fertilization.

Prevention: Regard any amount of nitrate nitrogen over 0.5% of the total ration (moisture-free basis) as a potential source of trouble.

When in doubt, have the feed analyzed.

Nitrate poisoning may be lowered by (1) feeding high levels of carbohydrates or energy feeds (grain or molasses) and vitamin A, (2) feeding limited amounts of high-nitrate forage, (3) alternating or mixing high-and low-nitrate forages, and (4) ensiling forages high in nitrates, since fermentation reduces some of the nitrates to gas.

Treatment: A 4% solution of methylene blue (in a 5% glucose or a 1.8% sodium sulfate solution) administered by a veterinarian intravenously at the rate of 100 cc/1,000 lb liveweight.

Yeast culture fed as follows: ¼ lb daily for cattle, and ⅒ lb per day for lambs.

Remarks: Nitrate does not appear to cause the actual toxicity. During digestion, the nitrate is reduced to nitrite, a far more toxic form (10 to 15 times more toxic than nitrates). In cows and sheep, this conversion takes place in the rumen (paunch); in horses in the cecum.

SELENIUM POISONING (ALKALI DISEASE) *(Se)*

Source: Consumption of plants grown on soils containing selenium.

Species Affected: All farm animals and humans.

Symptoms and Signs: Loss of hair from the mane and tail in horses, from the tail in cattle, and a general loss of hair in swine. In severe cases, the hoofs slough off, lameness occurs, food consumption decreases, and death may occur by starvation.

Distribution and Losses Caused By: In certain regions of western U.S.—especially certain areas in South Dakota, Montana, Wyoming, Nebraska, Kansas, and perhaps areas in other states in the Great Plains and Rocky Mountains. Also, in Canada.

Prevention: Abandon areas where soils contain selenium, because crops produced on such soils constitute a menace to both animals and humans.

Fig. 10-40.

Treatment: Although arsenic has been shown to counteract the effects of selenium toxicity, there appears to be no practical method of treating other than removal of animals from affected areas.

Remarks: The dietary requirements for selenium are 0.1–0.3 ppm and the maximum tolerable levels for all species has been established as 2 ppm.

[1]FDA regulations do not permit grain or feed containing more than 20 parts per billion of aflatoxin to be fed to animals.

ANIMAL PARASITES[5]

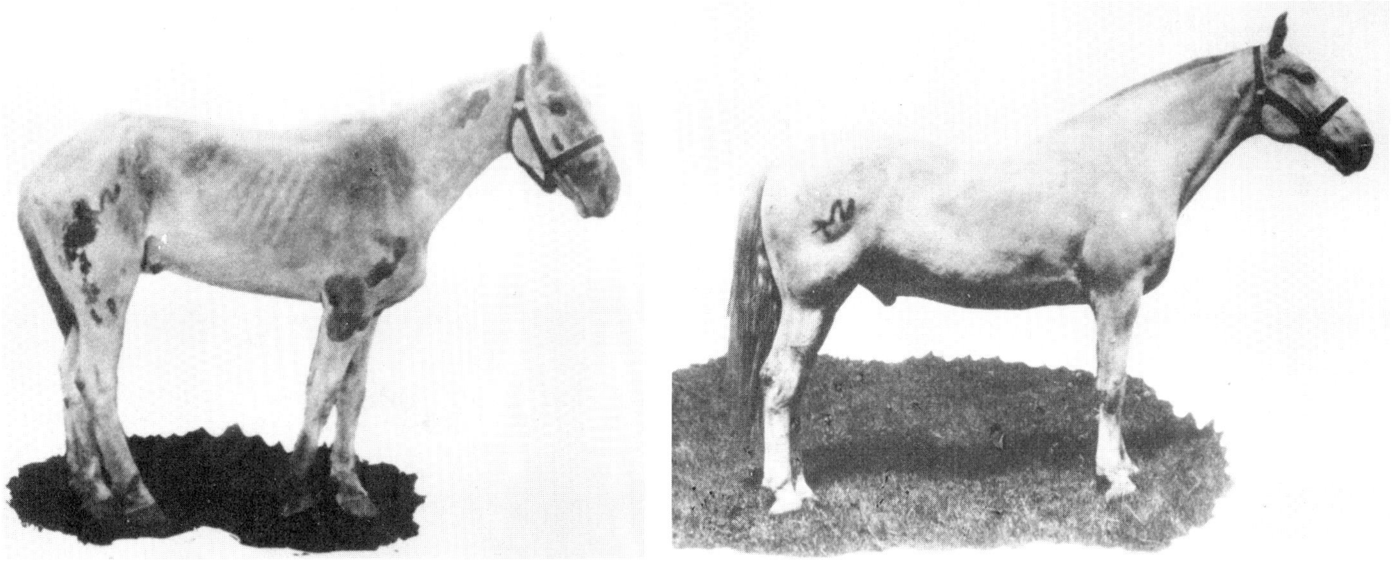

Fig. 10–41. Same horse before (left) and after (right) treatment for internal parasites. Parasites retard the foal's development and lower the efficiency of mature horses. Also, feed is always too costly to give to parasites. (Courtesy, College of Veterinary Medicine, University of Illinois, Urbana)

Animals are attacked by a wide variety of internal and external parasites, the prevention and control of which is one of the quickest, cheapest, and most dependable methods of increasing production with no extra animals, no additional feed, and little more labor. This is important, for, after all, the farmer or rancher bears the brunt of this reduced meat, milk, and wool production, wasted feed, and damaged hides. It is hoped that the discussion that follows may be helpful in (1) preventing the propagation of parasites, and (2) causing the destruction of parasites through the use of the most effective wormer or insecticide.

[5]This section was authoritatively reviewed by the following: J. J. Arends, Extension Entomologist, North Carolina State University, Raleigh; W. J. Gojmerac, Extension Entomologist, University of Wisconsin-Madison; M. C. Marquardt, Biology Department, Colorado State University, Ft. Collins; and R. E. Williams, Professor of Entomology, Purdue University, West Lafayette, IN.

The use of trade names of wormers and insecticides in this section does not imply endorsement, nor is any criticism implied of similar products not named; rather, it is recognition of the fact that farmers and ranchers, and those who counsel with them, are generally more familiar with the trade names than the generic names.

Note well: Cognizance is taken that, from time to time, FDA (1) bans the use of some old drugs, and (2) approves the use of new drugs.

Also, no claim is made that all wormers and insecticides are listed in this section; so, no criticism is implied of products that are not listed.

CHOICE OF WORMER

Knowing what internal parasites are present within an animal is the first requisite to the choice of the proper drug, or anthelmintic. Since no one drug is appropriate or economical for all conditions, the next requisite is to select the right one; the one which, when used according to directions, will be most effective and produce a minimum of side effects on the animal treated. So, coupled with knowledge of the kind of parasites present, an individual assessment of each animal is necessary. Among the factors to consider are age, pregnancy, other illnesses and medications, and the method by which the drug is to be administered. Some drugs characteristically put animals off performance for several days after treatment, whereas others often have less tendency to do so. Some drugs are unnecessarily harsh or expensive for the problem at hand, whereas a safe inexpensive alternative would be equally suitable.

Each livestock establishment should, in cooperation with the local veterinarian and/or other advisor, evolve with a parasite control program and schedule. It is recommended that several different wormers be used, and that they should be rotated. Also, a schedule of treatments should be prepared, based on knowledge of the life cycles of the various parasites.

Table 10–6 lists the common chemical compounds for the control of internal parasites, by species. Although this is a valuable guide, it is recognized that wormers are constantly being improved, and that new ones are becoming available. So, the producer should consult the local veterinarian relative to the choice of drug to use on the animals at the time.

TABLE 10-6
RECOMMENDED COMPOUNDS FOR CONTROL OF INTERNAL PARASITES, BY ANIMAL SPECIES[1]

Wormer	Trade Name—Manufacturer	Cattle: Coccidiosis	Cattle: Gastrointestinal Nematode Worm	Cattle: Lungworm	Sheep: Brown Stomach Worm	Sheep: Coccidiosis	Sheep: Cooperias	Sheep: Hookworm	Sheep: Lungworm	Sheep: Nodular Worm	Sheep: Stomach Worm	Sheep: Trichostrongyles	Sheep: Whipworm	Swine: Ascarids	Swine: Coccidiosis	Swine: Kidney Worm	Swine: Lungworm	Swine: Nodular Worm	Swine: Stomach Worm	Swine: Threadworm	Swine: Whipworm	Horses: Ascarids	Horses: Bots	Horses: Pinworm	Horses: Stomach Worm	Horses: Strongyles, Large	Horses: Strongyles, Small	Horses: Tapeworm	Horses: Threadworm	Wormer
Albendazole	Valbazen		x	x	x		x	x	x	x	x	x																		Albendazole
Amprolium[2]	Amprol (Merck)	x				x									x															Amprolium[2]
Cambendazole	Camvet																					x		x					x	Cambendazole
Carbon disulfide																						x	x		x					Carbon disulfide
Coumaphos	Co-Ral Baymix (Bayvet Corp.)		x		x		x	x		x		x	x																	Coumaphos
Dichlorvos	Atgard Equigard Equigel													x				x	x		x	x	x	x		x	x			Dichlorvos
Dithiazanine iodide and piperazine citrate	Dizan (Elanco)																					x		x			x			Dithiazanine iodide and piperazine citrate
Fenbendazole	Safe Guard (Hoescht)		x	x										x		x	x	x			x									Fenbendazole
Haloxon	Luxon Loxon		x		x		x	x		x	x	x																		Haloxon
Hygromycin B	Hygromix (Elanco)													x				x			x									Hygromycin B
Ivermectin	Ivomec (Merck)		x	x	x					x	x	x		x		x	x	x		x			x			x	x			Ivermectin
Lead arsenate	Bi-forma (Texas Pheno Co.)																											x		Lead arsenate
Levamisole	Tramisol		x	x	x		x	x	x	x	x	x	x	x		x	x	x	x	x	x	x			x	x	x			Levamisole
Levamisole-piperazine	Am. Cyanamid)																							x	x				x	Levamisole-piperazine
Mebendazole	Telmin (Pitman-Moore)																					x		x		x	x			Mebendazole
Morantel			x																											Morantel
Oxyfendazole			x	x																										Oxyfendazole
Parbendazole	(Helmatac)		x		x		x	x		x																				Parbendazole
Phenothiazine	(Du Pont)				x		x	x		x	x															x	x			Phenothiazine
Phenothiazine, low level	Pheno-Sweet (Farnam)																									x	x			Phenothiazine, low level
Phenothiazine-piperazine	Pheno-Pip (Haver-Lockhart)																					x				x	x			Phenothiazine-piperazine
Phenothiazine-trichlorfon	Equiverm (Texas Pheno Co.)																					x	x	x		x	x			Phenothiazine-trichlorfon
Piperazine	Wonder Wormer (Farnam)													x				x				x					x			Piperazine
Piperazine-carbon disulfide	Parvex (Upjohn)																	x				x					x			Piperazine-carbon disulfide
Piperazine-carbon disulfide-phenothiazine	Parvex-Pheno (Upjohn)																					x				x	x			Piperazine-carbon disulfide-phenothiazine
Pyrantel tartrate	Banmith (Pfixer)													x				x				x		x		x	x			Pyrantel tartrate
Tetramisole	Tetramizole Bayer 9051								x																					Tetramisole
Thiabendazole	Thibenzole (Merck) Equizole (Merck)		x		x		x	x		x	x	x						x		x		x		x		x	x		x	Thiabendazole
Thiabendazole-piperazine	Equizole-A (Ft. Dodge)																					x		x		x	x		x	Thiabendazole piperazine
Thiabendazole trichlorfon	Equivet-14 (Farnam)																					x	x	x		x	x		x	Thiabendazole trichlorfon
Trichlorfon	Anthon Bot-X (Farnam) Dyrex (Ft. Dodge)												x									x	x	x						Trichlorfon
Trichlorfon-phenothiazine-piperazine	Dyrex T.F. (Ft. Dodge)																					x	x	x		x	x			Trichlorfon-phenothiazine-piperazine

[1]The products listed have 90% efficacy or more. This list is not complete. Inclusion of trade manes does not imply endorsement.

[2]In the U.S., it is permitted in feed only as an aid in the control of coccidiosis of chickens and turkeys.

APPLICATION OF INSECTICIDES

The availability of an insecticide and the type of application(s) for which it was formulated are of prime importance, but the treatment of a herd or flock is dictated pretty much by the animal species, number of animals, available handling facilities, time or season, the target pest, management practices, and cost. The common methods of insecticide application are (1) spraying, (2) dipping, (3) back rubbers, (4) pour-on, (5) feed additives, (6) injectables, (7) insecticide impregnated ear tags, and (8) boluses.

USE INSECTICIDES SAFELY

Certain basic precautions must be observed when insecticides are to be used because, used improperly, they can be injurious to people, domestic animals, wildlife, and beneficial insects. Follow the directions and heed all the precautions on the labels.

- **Selecting insecticides**—Always select the formulation and insecticide labeled for the purpose for which it is to be used.

- **Applying**—Use only amounts recommended. Apply at the correct time to avoid unlawful residues in meat. Avoid treating animals younger than specified on the label. Avoid retreating more often than label restrictions. Avoid drift on nearby crops, pastures, livestock, or other nontarget areas. Avoid prolonged contact with all insecticides. Do not eat, drink, or smoke until all operations have ceased and hands and face are thoroughly washed. Change and launder clothing after each day's work.

- **Withdrawal**—After treating animals with pesticides, observe the prescribed number of days interval between the last treatment and slaughter. Refer to the product labels for this information.

PARASITES AND THEIR CONTROL

A summary, by species, of the common animal parasites and their control is given in the following tables:

Table 10–7, Cattle Parasites and Their Control, p. 450–455.
Table 10–8, Sheep and Goat Parasites and Their Control, p. 456–458.
Table 10–9, Swine Parasites and Their Control, p. 459–461
Table 10–10, Horse Parasites and Their Control, p. 462–465.

Few dosages for the control of internal parasites are given; instead, users are admonished to *follow the directions on the label.* Also, few insecticides are suggested for the control of external parasites because of (1) the diversity of environments and management practices under which they occur, (2) the varying restrictions on the use of insecticides from area to area, and (3) the fact that registered users of insecticides change from time to time. Information about what is available and registered for use in a specific area can be obtained from the county agent, extension entomologist, or agricultural consultant.

The insecticide recommendations given for horses are for animals not used for human food. Where horses are to be slaughtered for human food, the tolerance levels and withdrawal periods given on the manufacturer's label should be followed with care.

TABLE 10–7
CATTLE PARASITES AND THEIR CONTROL

ANAPLASMOSIS

Anaplasmosis is an infectious disease caused by a minute parasite, *Anaplasma marginale*, which invades the red blood cells. See Fig. 10–42, which shows greatly enlarged red blood cells from an animal with anaplasmosis. The black dots near the margins of the cells are *Anaplasma marginale.*

Species Affected: Cattle, especially adults.

Symptoms and Signs of Affected Animals (or Damage Inflicted): Calves usually have the mild type, either not manifesting any symptoms or simply becoming "dumpy" for a few days and then apparently recovering, though their blood remains the permanent abode of the parasite.

Mature animals develop severe anemia and usually show a rapid, pounding heart action, labored and difficult breathing, dry muzzle, marked depression, tremors of the muscles, loss of appetite, and marked reduction in milk flow. Also, the eyes and other mucous membranes and the skin may become yellow, there may be depraved appetite, and sick animals may show brain symptoms and an inclination to fight. Urine is normal color. Blood is thin and watery. In severe cases, death may follow in a few days.

Distribution and Losses Caused By: Throughout the world, especially in warm climates.

From 25 to 60% of infected animals may die.

The mortality rate may vary from 2.5% to 50–60%.

Treatment: Chlortetracycline (Aureomycin) or tetracycline (Terramycin) is effective as an early treatment, and as a means of eliminating carriers when used at high levels.

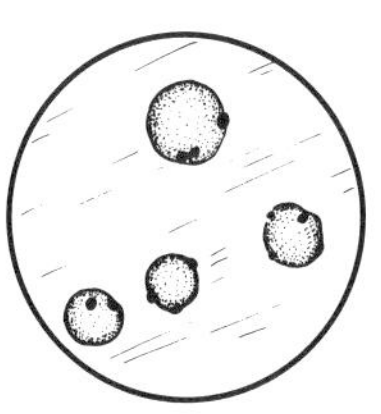

Fig. 10–42.

Good care and nursing and keeping animals quiet will help.

Blood transfusions are effective.

Prevention and Control: Immunize cattle by administering 2 doses of anaplasmosis vaccine at least 4 weeks apart. Consult with your veterinarian relative to vaccination or elimination of carrier programs.

Spray cattle and buildings to ward off biting insects that spread the disease.

Sterilize surgical instruments, dehorners, and needles.

Market animals that have recovered from the disease, except in endemic areas.

Remarks: All animals that recover remain permanent carriers of the parasite.

Anaplasmosis is transmitted by ticks, horse flies, mosquitoes, and probably other biting insects, and by such mechanical agencies as needles, and surgical and dehorning instruments.

Usually a summer disease.

(Continued)

TABLE 10–7 *(Continued)*

BOVINE TRICHOMONIASIS

Bovine trichomoniasis is a protozoan venereal disease of cattle caused by *Trichomonas foetus* (see Fig. 10–43), which are one-celled, microscopic in size, and capable of movement.

Species Affected: Cattle.

Symptoms and Signs of Affected Animals (or Damage Inflicted): Infected bulls do not usually exhibit visible symptoms, but they are the source of spread from cow to cow. Infected cows are characterized by (1) abortions in the first third of pregnancy, (2) uterine infections, (3) irregular heat periods, and (4) several services per conception.
Diagnosis can be confirmed microscopically.

Distribution and Losses Caused By: Throughout the U.S.
Economic loss in beef cattle is primarily due to the low percentage calf crop in infected herds. In dairy cattle, delayed return to milk production lowers profits.

Treatment: Infected cows should be rested for a minimum of 3 months and then test-bred by artificial insemination to a known clean bull.

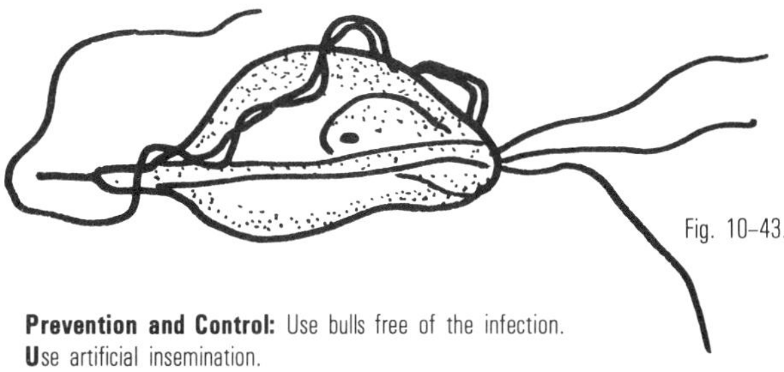

Fig. 10–43.

Prevention and Control: Use bulls free of the infection.
Use artificial insemination.
If practical, sell infected animals for slaughter or allow 90 days of sexual rest.
Exercise great precaution in introducing new animals in the herd, in breeding outside cows, and in taking cows outside the herd for breeding purposes.

Remarks: The infected bull is the source of the infection.

CATTLE TICK FEVER (TEXAS FEVER, SPLENIC FEVER)

Cattle tick fever is an infectious protozoan disease of cattle caused by a one-celled protozoa called *Babesia bigemina*, which depends upon the tick for transmission and survival.

Species Affected: Cattle, especially adults.

Symptoms and Signs of Affected Animals (or Damage Inflicted): High temperature, rapid breathing, enlarged spleen, engorged liver, pale and yellow membranes, and red to black urine.

Distribution and Losses Caused By: Confined to the Gulf Coast area.
Death occurs in about 10% of the chronic and 90% of the acute cases. Infected young animals are stunted, mature animals are emaciated, and the milk flow of infected dairy animals is greatly reduced.
Death losses are much higher in countries where anaplasmosis also occurs in cattle.

Treatment: Successful treatment of sick animals depends upon early recognition of the disease and prompt treatment. Agents traditionally used include trypan blue, trypaflavine, and quinuronium sulfate.

Prevention and Control: Avoid contact with the cattle fever tick the only natural agent by which cattle tick fever is transmitted.

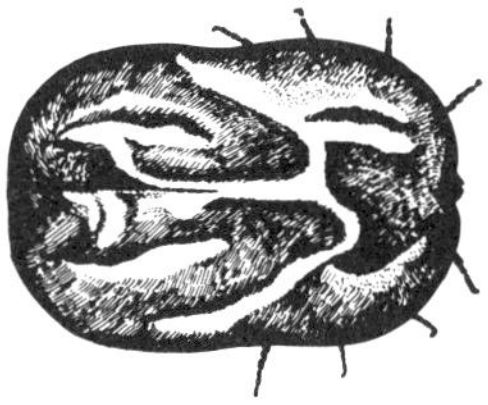

Fig. 10–44.

Treat animals at regular intervals with a suitable insecticide.
If practical, render pastures tick-free by excluding host animals (cattle, horses, and mules) for 8 to 10 mos., thus starving the ticks.

Remarks: The protozoa invade the red blood cells of cattle.
The parasite is transmitted to cattle by ticks and is carried over in ticks by egg transmission.
Although immune, recovered animals are permanent carriers of the disease.
In infected areas, native cattle are either immune or only slightly affected.
With the exception of a few possible areas along the Gulf Coast, the cattle fever tick has been eradicated—thus controlling tick fever.

COCCIDIOSIS

Coccidiosis is a parasitic disease caused by protozoan organisms known as coccidia.

Species Affected: Cattle. Sheep. Goats. Pet stock. Poultry.
Each class of animals harbors its own species of coccidia; thus, there is no cross infection between animals.

Symptoms and Signs of Affected Animals (or Damage Inflicted): Diarrhea and bloody feces, and pronounced unthriftiness and weakness.

Distribution and Losses Caused By: Worldwide.
There is lowered gain and production in infected animals, along with some death losses. Most severe in calves.

Treatment: Amprolium (Amprol) and a decoquinate effective. They are approved for use in beef and dairy calves, chickens, and turkeys.

Prevention and Control: Avoid feed and water contaminated with the protozoa that causes the disease.
Segregate affected animals.
Remove and properly dispose of manure and contaminated bedding daily.
Drain low, wet areas.
Keep animals in a sunny, dry place.

Remarks: In the oocyst stage, the parasite may resist freezing and certain disinfectants and may remain viable outside the body for months, but it is readily destroyed by direct sunlight or complete drying.
Most cattle outbreaks are among dairy calves and feeder calves.

FLIES

Several species of flies attack or annoy animals. The most common ones follow:

1. **BITING MIDGE (ceratopogonid no-see-um,** *Culicoides variipennis*). This fly is common in the U.S. It is a small ($<$ 0.5 mm) fly that is usually brownish, and has a hump-backed appearance similar to the black fly. Wings are often patterned. They are blood-feeders which often leave a bloody lesion at the site of the bite.

Species Affected: All classes of livestock and humans.

Symptoms and Signs of Affected Animals (or Damage Inflicted): Biting midge flies cause production losses, restless animals, and large, weeping, crusting lesions on

(Continued)

TABLE 10–7 *(Continued)*

FLIES *(Continued)*

BITING MIDGE *(Continued)*

the skin where large numbers of flies have fed.

Distribution and Losses Caused By: Worldwide, in both warm and temperate climates.

Production losses result from blood-feeding of the flies. Biting midges also transmit the virus that causes blue tongue in cattle, sheep, and some wild ruminants.

Preadult stages are found most often in boggy areas near bodies of water—streams or ponds. The larvae prefer conditions where there is a large amount of organic matter from the feces and urine of farm animals.

Treatment: Treat areas where larvae are found with insecticide.

Prevention and Control: Stabilize banks of bodies of water to reduce the habitat of the larvae.

Spray farm animals with insecticide.

Remarks: Blue tongue disease in cattle can be prevented by the administration of a vaccine.

2. **BLACK FLY** (buffalo gnat, no-see-um). These are small (< 0.5 mm), dark colored flies, with a hump-backed appearance. They are blood-suckers that leave a small, bloody lesion at the site of feeding.

Species Affected: All farm animals and humans.

Symptoms and Signs of Affected Animals (or Damage Inflicted): Restlessness; production losses; and swollen, weeping lesions on the skin.

Distribution and Losses Caused By: Worldwide, in all climates.

Losses result from irritation of animals, blood loss, and transmission of some animal diseases in the tropics.

Treatment: No satisfactory treatment.

Prevention and Control: Black flies develop in well-aerated, running water. Flies often emerge in hoards in the spring, and attack both animals and humans.

Prevention consists of treatment of streams with insecticides.

Remarks: Sometimes black flies occur in such large numbers that cattle are smothered by the onslaught.

Irrigation structure and downstream from dams are often sites of larval development.

3. **FACE FLY** *(Musca autumnalis)*. The face fly was first found in this country in New York in 1953. It is a close relative of and similar in appearance to the house fly.

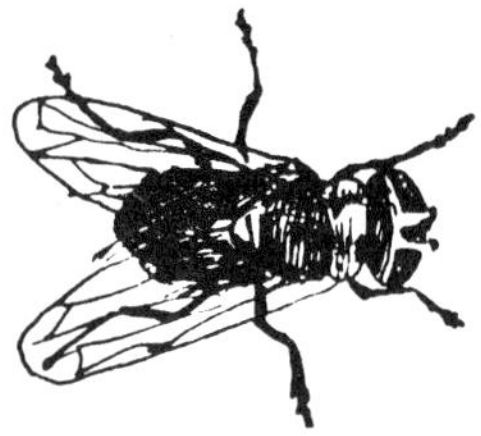

Fig. 10–45.

Species Affected: The face fly is primarily a pest of cattle, although it also attacks horses and sheep.

Symptoms and Signs of Affected Animals (or Damage Inflicted): The face fly does not bite, but its habit of clustering around the eyes, mouth, and nostrils is extremely annoying to animals, interfering with their vision and breathing, and preventing normal grazing. Large populations force animals to leave pastures and seek relief in wooded areas and shelters.

Distribution and Losses Caused By: Throughout continental U.S. except Arizona, Florida, New Mexico, and Texas.

The face fly causes direct irritation to eye tissue and aids in the transmission of pinkeye.

Treatment: Insecticide impregnated ear tags, or insecticide boluses.

The following insecticides are effective for face fly control:

1. Crotoxyphos plus.
2. Dichlorvos (Cio Vap).
3. Dichlorvos (Vapona).
4. Fenoalerate (Ectrin).
5. Flucythrinate (Guardian).
6. Permethrin.

Prevention and Control: Prevention consists in scattering or removing fresh cow manure.

Remarks: When cattle enter a barn or darkened area, the fly leaves the animal's face and rests on fence posts, gates, sides of barns, etc. The adult fly hibernates in attics and other protected places during the winter.

The face fly lays its eggs in fresh manure,where the larvae develop.

Insecticides should be applied in keeping with the directions of the manufacturer.

4. **HORN FLY** *(Haematobia irritans)*. This fly, which is about one-half the size of an ordinary house fly, is one of the most numerous and worst annoyances of cattle.

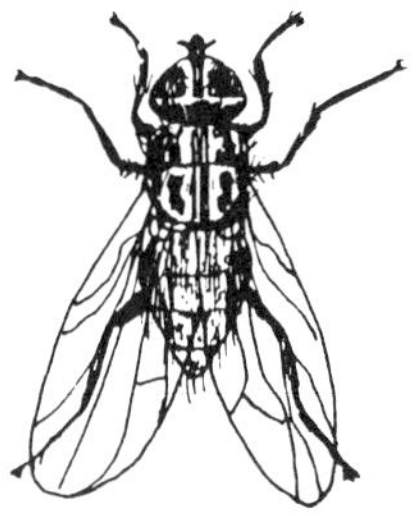

Fig. 10–46.

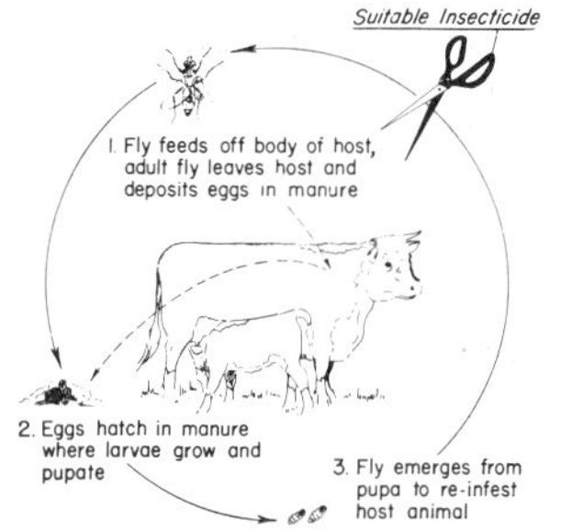

Fig. 10–47.

Species Affected: Cattle. Horses.

Symptoms and Signs of Affected Animals (or Damage Inflicted): Tormented cattle often refuse to graze during the day and seek protection by hiding in dark buildings, brush, or tall grass.

Heavily infested cattle may also have rough, sore skin, and suffer an inevitable loss in condition.

Distribution and Losses Caused By: Throughout the U.S.

Losses inflicted by horn flies include lowered gains, and lowered milk production.

Treatment: The horn fly can be effectively controlled.

For beef animals, the following insecticides are effective:

1. Chlorpyrifos.
2. Coumaphos (CoRal).
3. Dioxathion (Delnan).
4. Malathion.
5. Methoxychlor.
6. Permethrin.
7. Rabon.

For both beef and dairy animals, the following are among the many insecticides that are approved for horn fly control:

1. Coumaphos (CoRal).
2. Crotoxyphos (Ciodrin).
3. Dichlorvos (Vapona).
4. Methoxychlor.
5. Pyrethrins.
6. Rabon.

For both beef and dairy animals, the following types of ear tags may be used for horn fly and face fly control, but avoid wastage and beware of fly resistance:

1. Cyfluthrin (Cutter Gold).
2. Diazinon (Optimizer).
3. Lambdacythalothrin (Saber).
4. Pirimiphos-methyl (Tomahawk).

(Continued)

TABLE 10–7 *(Continued)*

FLIES *(Continued)*

HORN FLY *(Continued)*

Prevention and Control: In small pastures and where it is practical, spread fresh droppings with a spring-tooth harrow in order to hasten their drying.

Remarks: The horn fly is often found resting at the base of the horn, hence the name.
Each horn fly insecticide should be used in keeping with the directions of the manufacturer. Some of them may be used in self-applicating devices such as ear tags and back rubbers.

5. **HORSE FLY AND DEER FLY.** Horse flies and deer flies are biting flies that attack cattle. The two most troublesome genera are *Tabanus* (horse flies) and *Chrysops* (deer flies).

Species Affected: Cattle. Horses. Also, less common on other animals.

Symptoms and Signs of Affected Animals (or Damage Inflicted): The bite from the slashing mouthparts of these insects is very painful, and animals try to dislodge the fly with their tail or tongue or by stamping their feet. Heavily attacked animals stop grazing and tend to bunch together or seek shelter. Severe outbreaks can seriously affect weight gain.

Distribution and Losses Caused By: Tabanids are found in all parts of the U.S., and large numbers may be expected wherever there are extended areas of permanently wet, undeveloped land and a mild climate. Generally, horse flies are more of a problem to livestock than deer flies, but deer flies are often extremely annoying in the coastal areas of the South and the mountain areas of the West.
In many areas, tabanids are the principal source of *Anaplasma* transmission between cattle.

Treatment: For both beef and dairy animals, the following insecticides are effective for horse flies, deer flies, stable flies, and mosquitoes:
1. Crotoxyphos plus.
2. Dichlorvos (Cio Vap).
3. Fenvalerate (Ectrin).
4. Permethrin.

Prevention and Control: If possible, avoid pasturing cattle near swampy wooded areas when these flies are numerous. Also, sheltering animals is often beneficial since tabanids do not ordinarily enter enclosures.

Remarks: Horse flies are also implicated in disease transmission because their habit of feeding on one animal and immediately attacking another can result in the direct mechanical transfer of pathogenic organisms that live in blood.

6. **HOUSE FLY** *(Musca domestica).* House flies are nonbiting flies that are common around barns and lots.

Species Affected: Cattle. Sheep. Swine. Horses. Poultry.

Symptoms and Signs of Affected Animals (or Damage Inflicted): Although house flies are nonbiting, they cause serious economic losses through annoyance of livestock and by disease transmission. Also, they create public health problems.

Distribution and Losses Caused By: House flies become numerous both inside and outside barns and farm buildings. Perhaps they are the most abundant insect pest of feedlots and confinement livestock housing. House flies are annoying to livestock and people, and they can spread human and animal diseases.

Treatment: Several insecticides in fogs, mists, surface sprays, or baits may be used. The following are approved:
Fog, mist, or surface spray:
1. Dichlorvos (Vapona).
2. Naled (Dibrom).
Fog or mist:
1. Pyrethrins.
Surface sprays (with animals removed):
1. Dimethorate (Cygon).
2. Fenvalerate (Ectrin).
3. Permethrin.
4. Rabon.

Prevention and Control: Insecticides alone will not control house flies. Adequate sanitary measures, including proper disposition or handling of manure, are necessary to eliminate fly breeding areas. Spread manure thinly in fields so fly eggs and larvae will be killed by drying and heat.

Remarks: House flies breed in manure, garbage, and decaying vegetable matter. The eggs hatch after an incubation period of 12 to 36 hours. The larvae feed on the organic medium and grow to full size in 6 to 11 days.

7. **STABLE FLY** *(Stomoxys calcitrans).* The stable fly, which is about the size of a house fly, is usually found in the vicinity of animals. See Fig. 10–49 for the life history and eating habits of the stable fly.

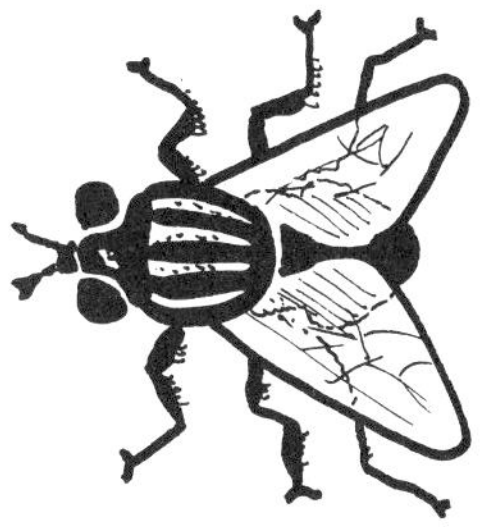

Fig. 10–48.

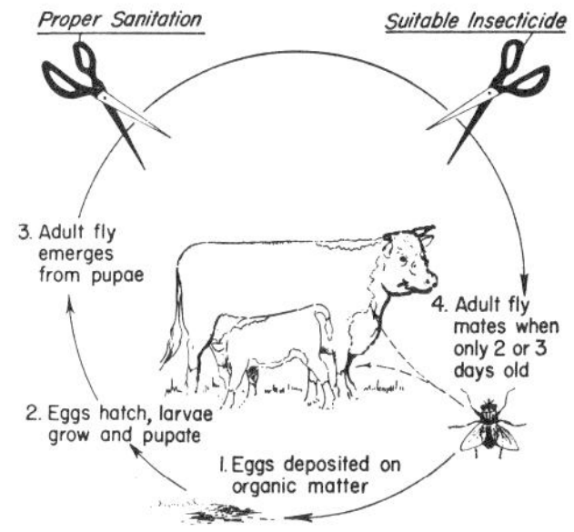

Fig. 10–49.

Species Affected: All farm animals and humans.
The recommended insecticides for both beef and dairy animals are the same as approved for horse flies, deer flies, and mosquitoes. (See Horse flies and Deer flies).

Symptoms and Signs of Affected Animals (or Damage Inflicted): Fly-fighting and restlessness.
In seeking natural protection, animals frequently restort to mudholes, brush, etc.

Distribution and Losses Caused By: All temperate regions of the world. Losses include:
1. Decreased gains in beef cattle.
2. Lowered milk production in dairy cattle.
3. Possible transmission of certain diseases and parasites.
4. Gains and/or milk production may be lowered by as much as 50% in seasons when the number of flies becomes large.

Treatment: With stable flies, insecticides should be used as a supplement to good sanitation, rather than as the principal method of control, because alone they may not do a satisfactory job. Residual sprays are the most effective method of treatment. They should be applied inside and outside barns and other farm structures where stable flies rest. The spray may be applied by spray gun or by fogging or misting devices. Care should be taken to prevent contamination of feed and drinking water, and animals should be removed from the area during the spraying. Application of insecticides to the cattle may afford only temporary relief. The preferred method of application is spraying since the insecticide should be applied to the legs and lower body of the animal.

Prevention and Control: Control of stable flies by direct application of insecticides to cattle is usually not satisfactory. They are best controlled by sanitation and by application of insecticides to the resting surfaces. Sanitation is the most effective method of controlling stable flies in such areas as feedlots and barnyards because it breaks the cycle by removing the breeding sites. Barnyards and feedlots should be well drained; manure and decaying organic matter should be removed from inside and outside buildings and disposed of weekly or more often, if possible, by spreading it out to dry (this kills developing larvae). If manure cannot be spread, it should be placed in compact piles where the surface will dry quickly and become unattractive to females.

Remarks: As a blood-feeding fly, the stable fly is implicated in carrying disease, and high populations cause reduced weight gains in cattle.

(Continued)

TABLE 10–7 *(Continued)*

GASTROINTESTINAL WORMS

Gastrointestinal worms—including many species—may be found in the stomach, small intestine, and colon.

Species Affected: Cattle. Sheep. Goats. Swine. Horses. Poultry. Dogs. Cats.

Symptoms and Signs of Affected Animals (or Damage Inflicted): The symptoms are not specific, but infested animals generally show loss of weight, anemia, and/or diarrhea.

Distribution and Losses Caused By: One or more species are found in most areas throughout the U.S.

The losses are in terms of lowered feed efficiency caused by disturbed digestion, lowered meat and milk production, and some death losses.

Treatment: Therapeutic doses before breeding, after calving, and when grazing is at its height of one of the wormers listed in Table 10–6, Recommended Compounds for Control of Internal Parasites, by Species. Use drug of choice according to manufacturer's directions.

Cattle producers and their veterinarians can (1) rotate drugs to prevent parasitic resistance, and (2) select the method of administration easiest to follow (although varying according to drug, they may be given as a drench, bolus, feed or mineral mix, paste, or injection).

Prevention and Control: Rotate pastures.

Segregate calves from mature animals.

Avoid overstocking or overgrazing of pastures since the infective larvae are mainly on the bottom inch of grass.

Cross-graze with cattle and horses.

Keep feeders and waterers sanitary.

Remarks: In areas of constant exposure, routine treatment is recommended. Consult with your veterinarian relative to a program.

GRUBS (WARBLES, HEEL FLY)

Cattle grubs are the maggot stage of honeybeelike insects known as heel flies, warble flies, or gad flies. In the U.S. there are 2 species of grubs with similar habits; namely, the common cattle grub and the northern cattle grub.

Fig. 10–50.

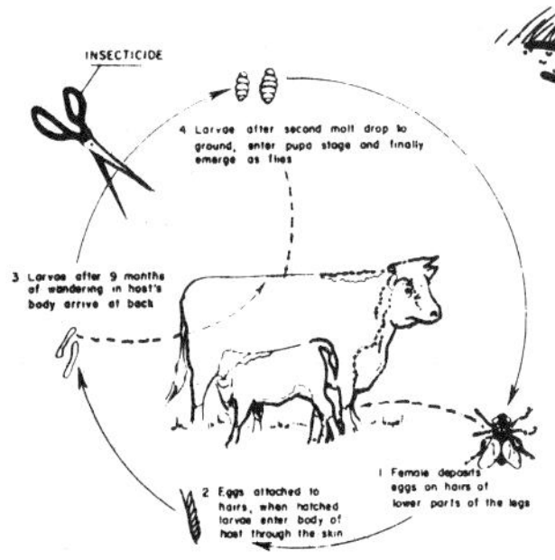

Fig. 10–51.

Species Affected: Cattle.

Symptoms and Signs of Affected Animals (or Damage Inflicted): Attack of heel fly, in spring or early summer, causes cattle to run madly with their tails high over their backs in an attempt to escape.

Grub (larva) in the back, usually from December to May, causes a conspicuous swelling.

Distribution and Losses Caused By: Throughout the U.S.

The following kinds of losses are incurred:

1. Decreased gains or milk production, mechanical injury, and even death.
2. Carcass damage.
3. Shock to animals.
4. Hides are downgraded.

Treatment: Applying a systemic insecticide to cattle as soon as possible after the activity of the heel fly ceases since these insecticides kill the young larvae in the animal's body.

When the grubs are near the back or located in the back, treatments are less effective, and the possible side effects are more likely. Side effects may also occur when there is concentration of grubs in the gullet or spinal cord of treated cattle. A single treatment with a systemic insecticide should give excellent control of cattle grubs. For the correct timing, each owner is advised to check with the local county agent or consultant. Systemics may be administered as sprays, dips, injectables, or as feed additives. Never use more than one systemic insecticide at a time, and always use a systemic in keeping with the manufacturer's directions.

For beef and nonlactating dairy animals, the following insecticides are effective:

1. Coumaphos (Co Ral).
2. Famphur (Warbex).
3. Fenthion (Tigavon).
4. Ivermectin (Ivomec).
5. Prolate (GX–118).
6. Trichlorfon (Neguvon).

Prevention and Control: Effective and complete eradication necessitates area campaigns; farm by farm, county by county, and state by state.

Remarks: The cattle grub or heel fly is probably the most destructive insect attacking beef and dairy animals.

Follow the label instructions carefully, including the minimum intervals to slaughter and freshening (dairy).

LICE

Lice are small, flattened, wingless insect parasites of which there are several species, most of which are specific for a particular class of animal.

Species Affected: Cattle (with other species for other classes of animals).

Symptoms and Signs of Affected Animals (or Damage Inflicted): Intense irritation, restlessness, and loss of condition. There may be severe itching and the animal may be seen scratching, rubbing, and gnawing at the skin. The hair may be rough, thin, and lack luster; and scabs may be evident. Lice are apt to be most plentiful around the root of the tail, on the inside of the thighs, over the ankle region and along the neck and shoulders. One type sucks blood and may cause the animal to become anemic.

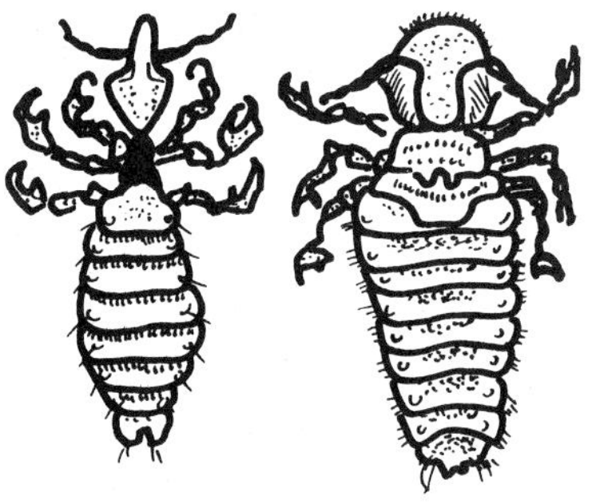

Fig. 10–52.

(Continued)

TABLE 10-7 *(Continued)*

LICE *(Continued)*

Distribution and Losses Caused By: Widespread.
Lice retard growth, lower milk production, and produce unthriftiness.

Treatment: For effective control, all members of the herd must be treated simultaneously at intervals, and this is especially necessary during the autumn months about the time they are placed in winter quarters. Cattle should be inspected for lice periodically throughout the winter and spring and retreated when necessary.
For beef animals, the following insecticides are effective for lice countrol.
1. Chlorpyrifos.
2. Coumaphos (Co Ral).
3. Dioxathion (Delnav).
4. Dursban.
5. Famphur (Warbex).
6. Fenvalerate (Ectrin).
7. Ivermectin (Ivomec).
8. Lindane.
9. Malathion.
10. Rabon.
11. RaVap.
12. Trichlorfon (Neguvon).

For both beef and dairy animals, the following insecticides are approved for lice control:
1. Amatraz (Taktic).
2. Cio Vap.
3. Coumaphos (Co Ral).
4. Crotoxyphos (Ciodrin).
5. Permethrin.
6. Rabon.

Prevention and Control: Because of the close contact of cattle during the winter months, it is practically impossible to keep them from becoming infested with lice.

Remarks: Lice show up most commonly in winter and on ill-nourished and neglected animals.
Self-applicating devices can aid in louse control.

MITES

Mites are very small parasites that produce mange (scabies, scab, itch).

Species Affected: Cattle.
Each class of animals has its own species or subspecies of mange mites.

Symptoms and Signs of Affected Animals (or Damage Inflicted): Marked irritation, itching, and scratching, crusting over of the skin, accompanied by formation of thick, tough, wrinkled skin.

Distribution and Losses Caused By: Widespread.
Mites retard growth, lower milk production and gains, and produce unthriftiness. Also, the skin is made less valuable for leather.

Treatment: Mites can be controlled by spraying or dipping infested animals with suitable insecticidal solutions, and by quarantine of affected herds.
The following insecticides may be used in keeping with the manufacturer's label:
1. Coumaphos (Co Ral).
2. Crotoxyphos.
3. Invermectin (Ivomec).
4. Lime-sulphur.
5. Phosmet.
6. Toxaphene

Only lime-sulfur is registered for use on lacatating cows.

Prevention and Control: Avoid contact with diseased animals or infested premises.
Scabies is a reportable disease in the U.S. So in the case of an outbreak, contact the local veterinarian or livestock sanitary official.
Control by spraying or dipping infested animals with suitable insecticides, and quarantine affected herds.

Remarks: There are 2 chief forms of mange: sarcoptic mange (caused by burrowing mites), and psoroptic mange (caused by mites that bite the skin and suck blood but do not burrow.
The disease appears to spread most rapidly during the winter months and among young and poorly nourished animals.

RINGWORM

Ringworm is a contagious disease of the outer layer of skin caused by certain microscopic molds or fungi.

Species Affected: All animals and humans.

Symptoms and Signs of Affected Animals (or Damage Inflicted): Round, scaly areas almost devoid of hair appear mainly in the vicinity of the eyes, ears, side of the neck, or the root of the tail.
Mild itching usually accompanies the disease.

Distribution and Losses Caused By: Throughout the U.S. It is unslightly and affected animals may experience considerable discomfort, but actual economic losses are not too great.

Treatment: Clip the hair from the affected areas, remove scabs with a brush and a mild soap. Paint affected areas with tincture of iodine or salicylic acid and alcohol (1 part in 10) every 3 days until cleared up.
Certain proprietary remedies available only from verterinarians have proved very effective in treatment.

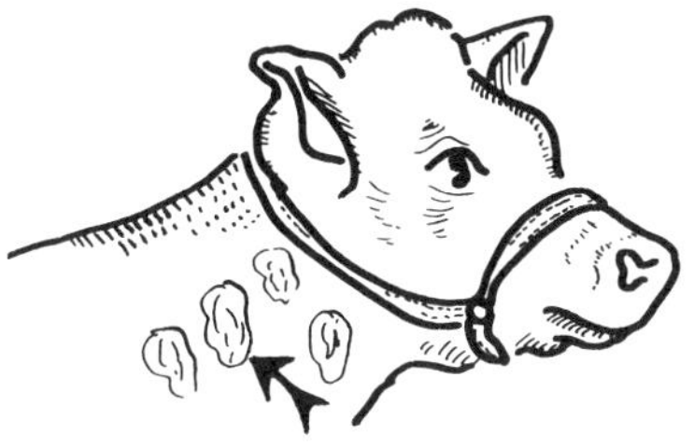

Fig. 10-53.

Prevention and Control: Isolate affected animals.
Disinfect everything that has been in contact with infested animals, including curry combs and brushes.
Practice strict sanitation.

Remarks: Though ringworm may appear among animals on pasture, it is far more prevalent as a stable disease.
It is usually a winter disease, with recovery the following summer after the animals are turned out to pasture.

WARBLES (See GRUBS)

TABLE 10-8
SHEEP AND GOAT PARASITES AND THEIR CONTROL

BANKRUPT WORMS (see TRICHOSTRONGYLES)

COCCIDIOSIS

Coccidiosis is a parasitic disease caused by protozoan organisms known as coccidia (see Fig. 10-54 for the life history and habits of coccidia).

Species Affected: Sheep. Cattle. Goats. Pet stock. Poultry.
Each class of animals harbors its own species of coccidia; thus, there is no cross infection between animals.

Symptoms and Signs of Affected Animals (or Damage Inflicted): Diarrhea and bloody feces, and pronounced unthriftiness and weakness.

Distribution and Losses Caused By: Worldwide.
There are lowered gains and production, and frequently high mortality in feedlot lambs.

Treatment: Lasalocid is approved for prevention of coccidiosis in sheep maintained in confinement.
Decoquinate is approved for the prevention of coccidiosis in young goats.
Both the above products should be used in keeping with the manufacturer's instructions on the label.
In the U.S., amprolium is permitted in feed only as an aid in the control of coccidiosis of chickens and turkeys.

Prevention and Control: *Feedlot lambs:* In feedlot lambs, where the disease is most prevalent, good management and natural resistance are important. To these ends, move feeders into feedlots with a minimum of stress and shrink, allow for plenty of space, keep water and feed troughs free from fecal pellets, maintain dry lots and bedding, start animals on grain feed gradually, and segregate affected animals if practical. The same principles and practices apply to other sheep, when trouble is encountered.
Also, when practical, drain or fence low, wet areas; and keep animals in a sunny, dry place.

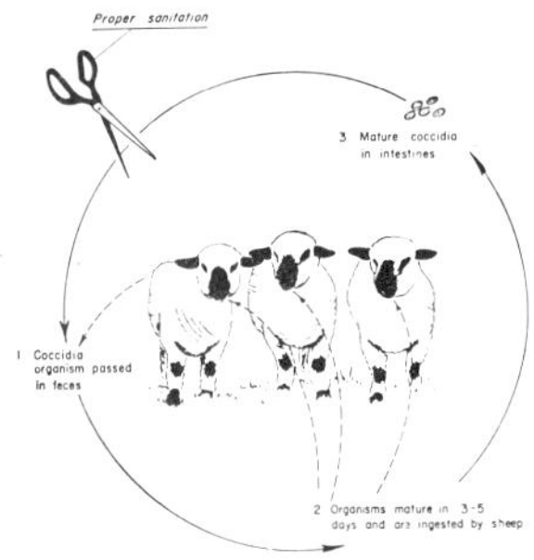

Fig. 10-54.

Remarks: In the oocyst stage, the parasite resists many disinfectants and may remain viable outside the body for months, but it is readily destroyed by direct sunlight or complete drying.
Some sheep producers protect feeder lambs from coccidiosis by using a feed mixture containing sulfa drugs (such as sulfaguanidine) which check the growth of parasites. Medicated feed may also be fed to lambs when they have to be kept closely confined with ewes.

FLIES

The common species of flies are fully discussed in Table 10-7, Cattle Parasites and Their Control; hence, the reader is referred thereto.

GID TAPEWORM
(Coenurus cerebralis)

Gid tapeworm is the larval form of one of the 4 species of bladder worm or tapeworm, found in dogs and related carnivores, which also affect sheep and goats (see Fig. 10-55)

Species Affected: Sheep. Goats. Pet stock.

Symptoms and Signs of Affected Animals (or Damage Inflicted): Disease known as coenurosis, the symptoms of which are defects in vision and disturbances in movements. Affected animals may stumble, run into objects, walk with the head high or in circles and there may be at least a partial paralysis of the hindquarters.

Distribution and Losses Caused By: Spotted and rare over the U.S.

Treatment: No known treatment. Surgery is successful in some cases.

Prevention and Control: Elimination of stray dogs.
Examination, and proper worm treatment when necessary, of all dogs that may come in contact with sheep and goats.
Proper disposal of all carcasses of infested animals.

Fig. 10-55.

Remarks: In afflicted sheep and goats, cysts containing the larvae *(Coenurus cerebralis)* of the tapeworm eggs voided by dogs or other carnivorous animals are found on the brain and spinal cord.

GRUB-IN-THE-HEAD (See SHEEP BOTS)

HEAD BOTS (NOSE BOTS) (See SHEEP BOTS)

KEDS (See SHEEP KEDS)

(Continued)

TABLE 10–8 *(Continued)*

LICE

Lice are small, flattened, wingless insect parasites of which there are 2 groups: sucking lice and biting lice. Biting lice, *Damalinia (Bovicola) ovis*, are most common but least harmful. The sucking species are (1) body lice *(Linognahus ovillus* and *L. africanus)*, and (2) the foot louse *(L. pedalis)*. Sucking lice are larger than biting lice. Body lice are found anywhere on the body where wool is dense, while food lice are usually found on legs below knees and hock.

Species Affected: Sheep, with other species for other classes of animals.

Symptoms and Signs of Affected Animals (or Damage Inflicted): Intense irritation, restlessness, and loss of condition. There may be severe itching and the animal may be seen scratching, rubbing, and gnawing at the skin. The wool may be matted and lack luster; and scabs may be evident.

Distribution and Losses Caused By: Lice are not as common on sheep as on other domestic animals, but they do occur occasionally.
Lice retard growth, lower wool production, and produce unthriftiness.

Fig. 10–56.

Treatment: For the control of lice, apply one of the following insecticides in keeping with the manufacturer's label:
1. Coumaphos (Co Ral).
2. Fenvalerate (Ectrin).
3. Lindane.
4. Malathion.
5. Methoxychlor.

Prevention and Control: For effective control, all sheep should be treated simultaneously.

Remarks: Lice show up most commonly in winter and on ill-nourished and neglected animals.
Spray at high pressure (up to 400 psi).

SHEEP BOTS (NASAL BOTFLY)
(Oestrus ovis)

Sheep bots, commonly called grub-in-the-head, is due to a beelike fly about the size of the common horse fly. See Fig. 10–57 for the life history and habits of the sheep nasal fly.

Species Affected: Sheep.

Symptoms and Signs of Affected Animals (or Damage Inflicted): When the flies attempt to deposit their larvae around the nostrils of sheep, the animals cease to feed, become restless, press their noses against other sheep, and/or seek shelter.
Grub infestation results in a snotty nose, and there may be difficulty breathing and frequent sneezing.

Distribution and Losses Caused By: Worldwide.
Although death losses are rare, there is loss in condition, both at the time the fly attacks and while the larvae are in the nasal passage.

Treatment: No insecticide is registered and approved for sheep nose bot control.
Infestation can be prevented by painting, weekly, the nostrils of sheep with pine tar during the season when the adult bot fly is active.

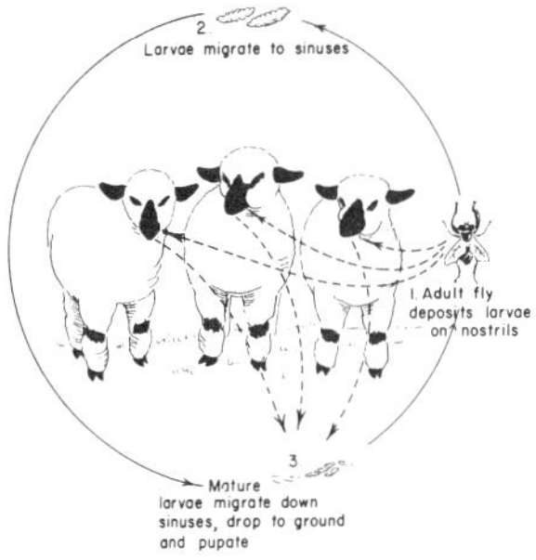

Fig. 10–57.

Prevention and Control: The following measures, during the sheep nasal fly season, may lessen, but not eliminate, sheep nasal fly:
With a small flock and where practical, keep sheep in a darkened barn during the day.
Apply pine-tar oil about the nostils of the sheep every few days, thus repelling some of the files.

SHEEP KEDS (SHEEP TICKS)
(Melophagus ovinus

Sheep keds are hairy, bloodsucking flies without wings, which range up to ¼ in. in length.

Species Affected: Sheep. Goats.

Symptoms and Signs of Affected Animals (or Damage Inflicted): Marked reduction in condition, anemia, biting, and scratching, and loss of and damage to wool.

Distribution and Losses Caused By: Throughout the U.S., but especially prevalent in the northern states.
Losses include retarded growth of young animals, loss in condition of mature animals, and damage to the fleece.
Fine wool breeds of sheep and Angora goats are not seriously affected.

Treatment: Treatment after shearing as soon as the shear cuts heal (including unshorn lambs) with any one of the following insecticides, used according to the manufacturer's label:
1. Coumaphos (Co Ral).
2. Fenvalerate (Ectrin).
3. Malathion.

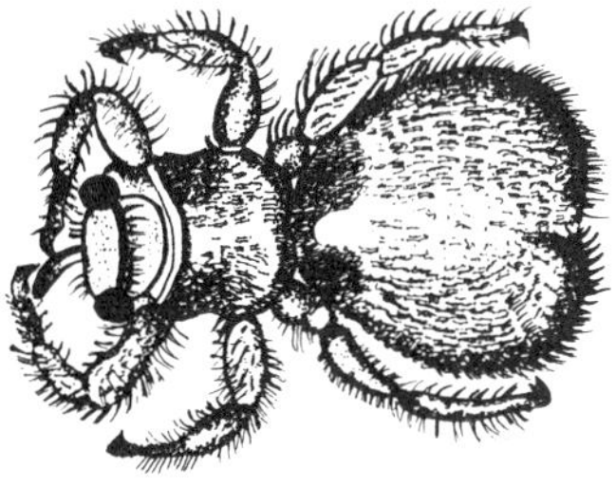
Fig. 10–58.

Prevention and Control: Prevention and control involves spraying or dipping all sheep as soon as the cuts heal up following shearing.

Remarks: Poorly housed and poorly fed animals are most likely to suffer from sheep ticks.
Method of application of insecticide depends on product used; it may involve dipping, spraying, pour-ons, or dusting.

(Continued)

TABLE 10-8 *(Continued)*

SMALL STOMACH and INTESTINAL WORMS (See TRICHOSTRONGYLES)

STOMACH WORMS (TWISTED STOMACH WORM, COMMON STOMACH WORM)
Haemonchus contortus)

Stomach worms are the most destructive parasite of sheep and goats. Worms are ¾ to 1½ in. long, about the size of a horse hair in diameter, and the live females are striped like a barber pole (see Fig. 10–59 for the life history and habits of the stomach worm).

Species Affected: Sheep. Goats. Cattle, although cattle are usually not seriously affected.

Symptoms and Signs of Affected Animals (or Damage Inflicted): No specific symptoms, because (1) sheep are seldom infested with one kind of parasite only, and (2) identical symptoms may be exhibited in cases of infestation by other parasites. Sheep heavily infested with parasites become unthrifty, listless, thin, and weak. The membranes of the eyes, nose, and mouth become pale, and there may be diarrhea, loss of wool, and a watery swelling under the lower jaw and along the abdomen.
Lambs and kids are more seriously affected than older animals.

Distribution and Losses Caused By: Mortality from stomach worms is high in sheep; however, in cattle, mortality is low but economic losses may be large.
The incidence is greatest in warm, wet areas of the U.S.

Treatment: See Table 10–6, Recommended Compounds for Control of Internal Parasites, by Animal Species. Currently, the wormers of choice are: haloxon, levamisole, phenothiazine, and thiabendazole. Always use drugs according to manufacturer's directions.

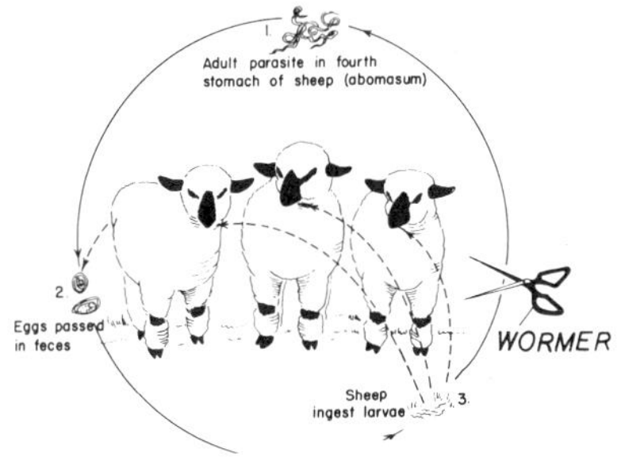

Fig. 10–59.

Prevention and Control: Rotate pastures, changing the flock to clean, fresh pastures at about 2-week intervals. But, in the event of an outbreak of stomach worms, treat the sheep before turning them on to new pastures.
Horses and hogs, which are not harmed by the stomach worm, may be rotated with sheep and goats.
Sheep and goats can be kept free from stomach worms by confining then to a drylot (without grass or weeds) or to a barn.

Remarks: See manufacturer's label for proper drug withdrawal prior to slaughter for food.

THIN-NECKED BLADDER WORM
(Cysticercus tenuicollis)

Thin-necked bladder worm is the larval form of one of the 4 species of bladder worm or tapeworm, found in dogs and related carnivores, which also affect sheep and goats (see Fig. 10–60).

Species Affected: Sheep. Goats.

Symptoms and Signs of Affected Animals (or Damage Inflicted): Usually there are no external symptoms, and since light infestations are the rule, no attention is called to the parasite.
Thin-necked bladder worms burrow into the liver and/or thin membranes of the abdominal cavity of sheep and goats, producing tissue damage.

Distribution and Losses Caused By: Worldwide.

Treatment: No known treatment.

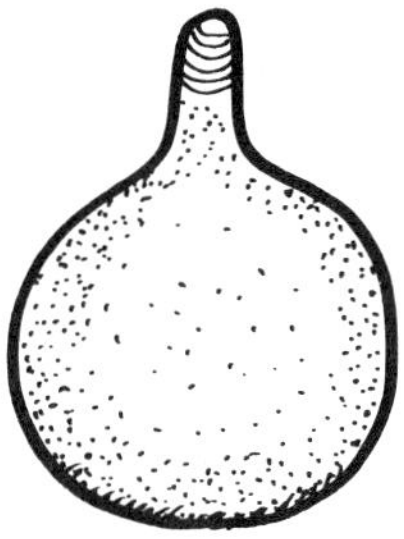

Fig. 10–60.

Prevention and Control: Elimination of stray dogs.
Examination and proper worm treatment, when necessary, of all dogs that must come in contact with sheep and goats.
Proper disposal of all carcasses of infested animals.

TRICHOSTRONGYLES (SMALL STOMACH and INTESTINAL WORMS)
(Trichostrongylus axei, T. colubriformis, T. vitrinus, T. capricola)

Trichostrongyles are small hairlike worms less than ⅓ in. in length.

Species Affected: Sheep. Goats. Cattle.

Symptoms and Signs of Affected Animals (or Damage Inflicted): Severe unthriftiness, diarrhea, and anemia.

Distribution and Losses Caused By: Death losses are not unusual, particularly when animals are on scant rations.

Treatment: See Table 10–6, Recommended Compounds for Control of Internal Parasites, by Animal Species. Use drug of choice according to manufacturer's directions.

Prevention and Control: The recommended measures for prevention and control of trichostrongyles are the same as those of the stomach worm; so, see the latter.

Remarks: See manufacturer's label for proper drug withdrawal prior to slaughter for food.

TWISTED STOMACH WORM (See STOMACH WORM)

TABLE 10-9
SWINE PARASITES AND THEIR CONTROL

ASCARIDS (LARGE INTESTINAL ROUNDWORM)
(Ascaris lumbricoides)

Ascarids are yellowish or pinkish worms, 8 to 15 in. long, almost the size of a lead pencil (see Fig. 10-61 for the life history and habits of ascarids).

Species Affected: Swine.

Symptoms and Signs of Affected Animals (or Damage Inflicted): Young pigs become unthrifty and stunted, and there is usually coughing, "thumpy" breathing, and there may be a yellow color to the mucous membrane due to blockage of the bile ducts.

Principal damage is produced by migrating larvae which produce liver damage and lung lesions resulting in verminous pneumonia.

Distribution and Losses Caused By: Worldwide.

Losses include stunted growth, uneconomical gains, and sometimes death in young animals.

Treatment: See Table 10-6, Recommended Compounds for Control of Internal Parasites, by Animal Species. Use drug of choice according to manufacturer's directions.

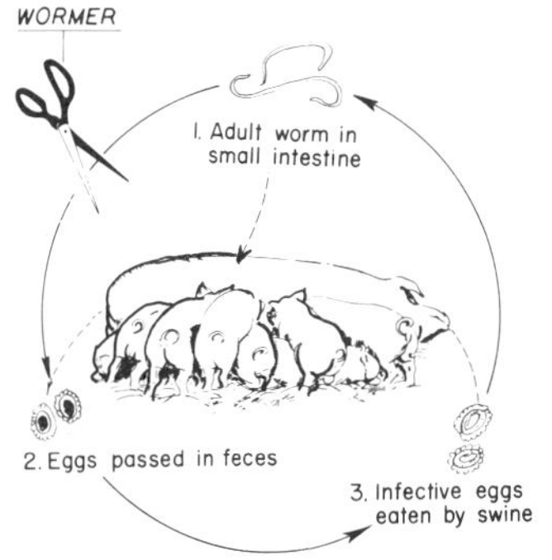

Fig. 10-61.

Prevention and Control: Prevention consists in keeping the young pigs away from infection.

Remarks: See manufacturer's label for proper drug withdrawal prior to slaughter for food.

COCCIDIOSIS

Coccidiosis is a parasitic disease caused by protozoan organisms known as coccidia (see Fig. 10-62 for the life history and haibts of coccidia).

Species Affected: Swine. Cattle. Sheep. Goats. Pet stock. Poultry.

Each class of animals harbors its own species of coccidia; thus, there is no cross infection between animals.

Symptoms and Signs of Affected Animals (or Damage Inflicted): Diarrhea and bloody feces, and pronounced unthriftiness and weakness. Piglets may appear weak, dehydrated, and undersized; and they may die.

Distribution and Losses Caused By: Worldwide.

Death losses are rare, but there is lowered gain and production in infected animals.

Treatment: Good nursing will help. Sulfamerazine, sulfamethazine, or sulfaquinoxaline.

Amprolium (Amprol) is reported to be effective. But, in the U.S.,it is permitted in feed only as an aid in the control of coccidiosis of chickens and turkeys.

Prevention and Control: Avoid feed and water contaminated with the protozoa that causes the disease.

Segregate affected animals.

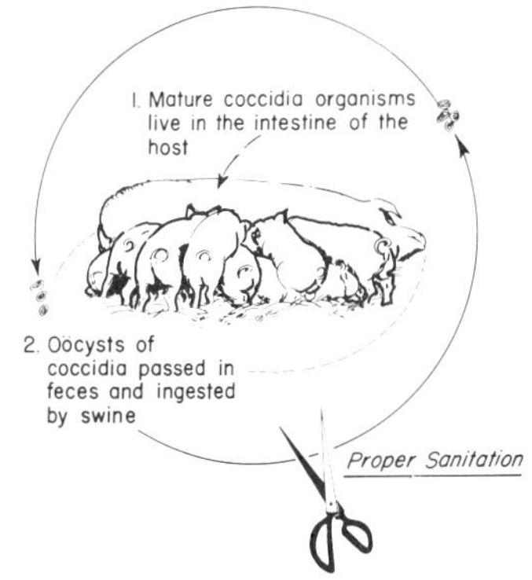

Fig. 10-62.

Remove and properly dispose of manure and contaminated bedding daily.

Drain low, wet areas.

Keep animals in a sunny, dry place.

Remarks: In the oocyst stage, the parasite resists low temperatures and disinfectants and may remain viable outside the body for months, but it is readily destroyed by sunlight or complete drying.

FLIES

The common species of flies are fully discussed in Table 10-7, Cattle Parasites and Their Control; hence, the reader is referred thereto.

Treatment: For house fly control, the following insecticides are recommended:

1. Femvalerate (Ectrin).
2. Rabon oral larvacide.
3. Also, see Cattle flies.

LARGE INTESTINAL ROUNDWORM (See ASCARIDS)

LICE

Lice are small, flattened, wingless insect parasites of which there are several species, most of which are specific for a particular class of animals. Only one species is found on swine.

Species Affected: Swine, with other species for other classes of animals.

Symptoms and Signs of Affected Animals (or Damage Inflicted): Intense irritation, restlessness, and loss of condition. There may be severe itching and the animal may be seen scratching and rubbing. The hair may be rough, thin, and lack luster; and scabs may be evident. Lice are apt to be most plentiful around the root of the tail, on the inside of the thighs, and around the neck and ears.

(Continued)

TABLE 10-9 *(Continued)*

LICE *(Continued)*

Distribution and Losses Caused By: Widespread.
Lice retard growth, lower milk production, and produce unthriftiness.

Treatment: For lice control, the presently available and effective insecticides include:
1. Coumaphos (Co Ral).
2. Crotoxyphos (Ciodrin).
3. Crotoxyphos + dichlorvos (Ciovap).
4. Fenthion (Liguvon).
5. Fenvalerate (Ectrin).
6. Ivermectin.
7. Malathion.
8. Permethrin.
9. Prolate.
10. Rabon.

Prevention and Control: Because of the close contact of swine during the winter months, it is practically impossible to keep them from becoming infected with lice.

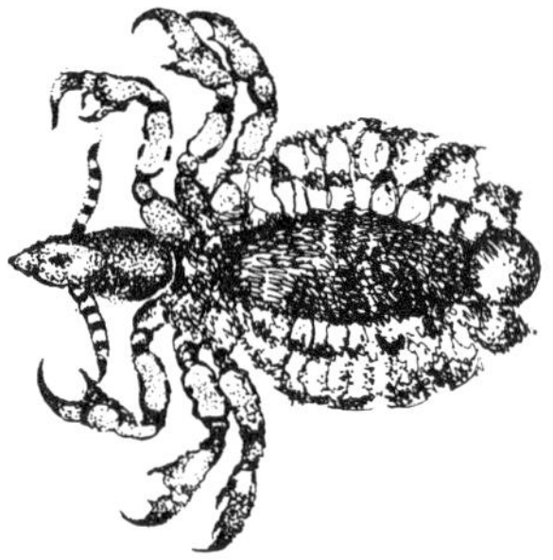

Fig. 10-63

For effective control, all swine should be treated simultaneously at intervals, especially in the fall about the time they are placed in winter quarters.

Remarks: Lice show up most commonly in winter and on ill-nourished and neglected animals.
Always use insecticides according to manufacturer's directions.

MANGE (See MITES)

MITES

Mites are very small parasites that produce mange (scabies, scab, itch).

Species Affected: Swine.
Each class of animals has its own species or subspecies of mange mites.

Symptoms and Signs of Affected Animals (or Damage Inflicted): Marked irritation, itching, and scratching. Crusting over of the skin, accompanied by formation of thick, tough, wrinkled skin.

Distribution and Losses Caused By: Widespread.
Mites retard growth, lower gains, and produce unthriftiness.

Treatment: The following insecticides are approved for the control of mange mites:
1. Chlordane.
2. Ivermectrin (Ivomec).
3. Lindane.
4. Malathion.

Prevention and Control: Avoid contact with diseased animals or infested premises.
Control by spraying or dipping infested animals with suitable insecticides, and quarantine affected herds.

Remarks: The disease appears to spread most rapidly during the winter months and among young and poorly nourished animals.
Follow container label for mixing directions, application, and safety precautions.

NODULAR WORM

Nodular worms consist of four species occuring in swine, all of which are slender, whitish to grayish in color, and ⅓ to ½ in. in length (see Fig. 10-64 for the life history and habits of the nodular worm).

Species Affected: Swine.

Symptoms and Signs of Affected Animals (or Damage Inflicted): No specific symptoms. Weakness, anemia, emaciation, diarrhea, and general unthriftiness occur.

Distribution and Losses Caused By: Widely distributed over U.S., but damage is heaviest in southeastern states. In addition to the usual lack of thrift, the intestines of severely infested animals are not suited for either sausage casings or food (chitterlings).

Treatment: See Table 10-6, Recommended Compounds for Control of Internal Parasites, by Animal Species. Use drug of choice according to manufacturer's directions.

Prevention and Control: A strict program of swine sanitation, accompanied by pasture rotation, constitutes a successful and practical preventive measure.

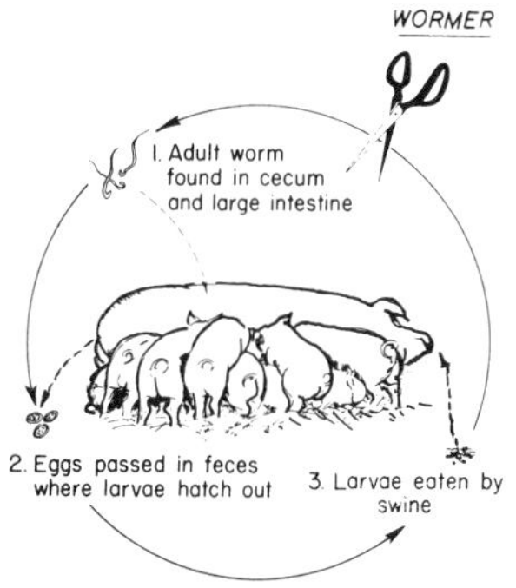
WORMER
1. Adult worm found in cecum and large intestine
2. Eggs passed in feces where larvae hatch out
3. Larvae eaten by swine

Fig. 10-64.

Remarks: Dichlorvos and levamisole are broad-spectrum wormers; hence, they control ascarids and whipworms, in addition to nodular worms.

(Continued)

TABLE 10–9 *(Continued)*

THORN-HEADED WORMS
(Macracanthorhynchus hirudinaceus)

Thorn-headed worms are white to bluish worms, cylindrical to flat, up to the size of a lead pencil, with rows of hooks which it uses for attachment purposes (see Fig. 10–65 for the life history and habits of the thorn-headed worm).

Species Affected: Swine.

Symptoms and Signs of Affected Animals (or Damage Inflicted): No specific symptoms, although swine infested with thorn-headed worms exhibit the general unthriftiness commonly associated with parasites. Digestive disturbance accompanies severe cases.

Distribution and Losses Caused By: Common in southern U.S.
Losses include slow growth, inefficient feed utilization, death losses, and damaged intestines that are unfit for sausage casings.

Treatment: No known drug treatment is entirely satisfactory for removing thorn-headed worms.

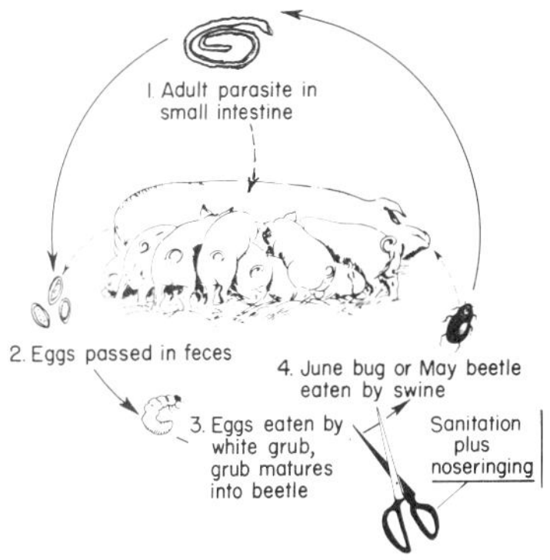

Fig. 10–65.

Prevention and Control: Keep pigs from feeding in areas where they might obtain the white grub of the June bug, the intermediate host.
Sanitation, clean ground, and nose-ringing are effective preventive measures.

TRICHINELLA (TRICHINOSIS)
(Trichinella spiralis)

Trichinella is a parasitic disease of humans contracted largely by consuming infested pork, eaten raw, or imperfectly cooked (see Fig. 10–66 for the life history and habits of trichina).

Species Affected: Swine. Humans.

Symptoms and Signs of Affected Animals (or Damage Inflicted): No specific symptoms in hogs, even when the parasite is present in the muscle tissue, its usual abode.

Distribution and Losses Caused By: Old studies (conducted prior to current garbage-cooking laws) showed (1) less than 1% of pork from grain-fed hogs infected with trichinosis, and (2) 5 to 6% infection of pork in hogs fed uncooked garbage.

Treatment: There is no practical treatment for infected hogs. Infected humans should be under care of an M.D.

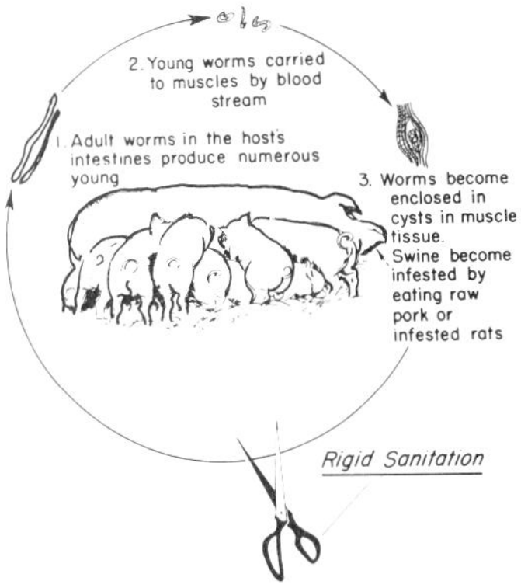

Fig. 10–66.

Prevention and Control: Prevention of trichinosis in humans may be obtained by:
1. Thoroughly cooking all pork at a temperature of 137°F before it is consumed; or
2. Freezing pork for a continuous period of not less than 20 days at a temperature not higher than 5°F.

Trichinosis in swine may be lessened by:
1. Destruction of all rats on the farm;
2. Proper carcass disposal of hogs and other animals that die on the farm; and
3. Cooking all garbage and offal from slaughterhouses.

Remarks: In humans, the disease is usually accompanied by a fever, digestive disturbances, swelling of infected muscles, and severe muscular pain (in the breathing muscles as well as others).
The presence of trichina can be detected by a microscopic examination of pork, but such a method is regarded as impractical in meat inspection procedure.
The FA and ELISA tests are the most reliable of currently available blood tests to detect infections in animals.

TABLE 10–10
HORSE PARASITES AND THEIR CONTROL

ASCARIDS (WHITE WORM, LARGE ROUNDWORM)

(Parascaris equorum)

Ascarids female varies from 6 to 14 in. in length and the male from 5 to 13 in. When full grown, both are about the diameter of a lead pencil. Fig. 10–67 shows the life history and habits of the ascarid, *Parascaris equorum.*

Species Affected: Horses. Mules. Zebras.

Symptoms and Signs of Affected Animals (or Damage Inflicted): The injury produced by ascarids covers a wide range from light infections producing moderate effects to heavy infections which may be the essential cause of death. Death is usually due to a ruptured intestine. Serious lung damage caused by migrating ascarid larvae may result in pneumonia. More common, and probably more important, is a retarded or impaired growth and development manifested by potbellies, rough hair coats, and digestive disturbances.

Especially affect foals and young animals, but are rarely important in horses over 5 years of age; older animals develop acquired immunity from earlier infections.

Distribution and Losses Caused By: Throughout the U.S.

Presence of ascarids results in:

Loss of feed through feeding worms.

Lowered work efficiency (including performance on the track and in the show-ring).

Retarded growth of young animals.

Lowered breeding efficiency.

Death in severe infestations.

Treatment: See Table 10–6, Recommended Compounds for Control of Internal Parasites, by Animal Species. Use drug of choice according to manufacturer's directions.

In addition to selecting the particular drug(s) for ascarid control, the caretaker should set up a definite treatment schedule, then follow it. The advice of the veterinarian should be sought on both points. Also, to preclude the possibility that worms may become resistant to a drug that is used continuously, the veterinarian may recommend a rotation of drugs.

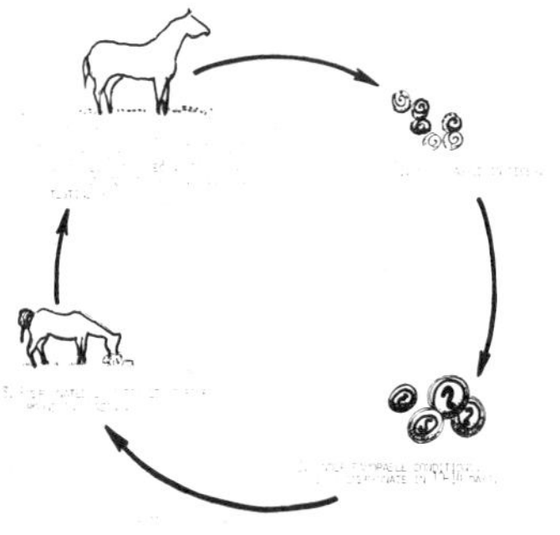

Fig. 10–67.

Prevention and Control: Keep foaling barn and paddocks clean.

Store manure in pit 2 to 3 weeks.

Provide clean feed and water.

Place young foals on clean pasture.

Worm all mares 4 to 6 weeks before foaling.

Worm foals and yearlings on a regular basis.

Remarks: Foals usually first acquire ascarid infection from contaminated stalls and paddocks.

Foals should be treated early in life, before the ascarids have a chance to mature and become large enough to block the intestine.

As a precaution, mares should not be treated closer than 30 days before foaling or within 14 days after foaling.

BOTS

Bots consist of three species of horse bot flies which are pests of horses in the U.S.: the common horse bot or nit fly *(Gastrophilus intestinalis)*, the throat bot or chin fly *(G. nasalis)*, and the nose bot or nose fly *(G. hemorrhoidalis).*

Species Affected: Horses. Mules. Asses. Zebras.

Symptoms and Signs of Affected Animals (or Damage Inflicted): Animals attacked by the bot fly may toss their heads in the air, strike the ground with their front feet, and rub their noses on each other or on any convenient object.

Infected animals may show frequent digestive upsets and even colic, lowered vitality and emaciation, and reduced work output.

Distribution and Losses Caused By: Worldwide.

Presence of bots results in:

Loss of feed through feeding worms.

Itching and loss of tail hair due to rubbing.

Lowered work efficiency.

Retarded growth of young animals.

Lowered breeding efficiency.

Death in severe infestations.

Treatment: See Table 10–6, Recommended Compounds for Control of Internal Parasites, by Animal Species. Use drug of choice according to manufacturer's directions.

In the late fall or early winter at least one month after the first killing frost, administer one of the recommended drugs.

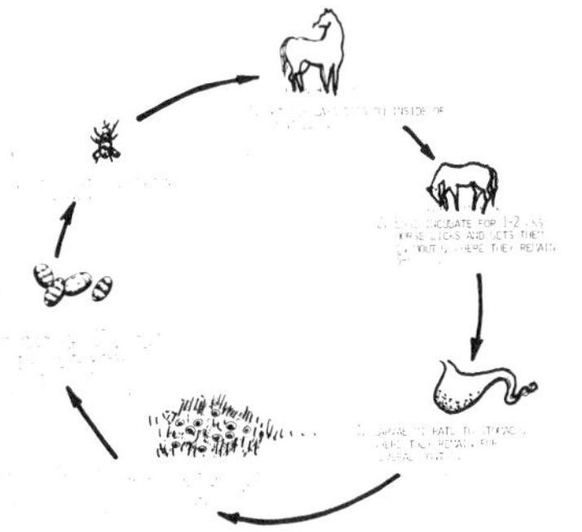

Fig. 10–68.

Prevention and Control: Frequent grooming, washing, and clipping.

Prevention of reinfestation is best assured through community campaigns in which all horses within the area are treated.

Fly nets and nose covers offer some relief from the attacks of adult bot flies.

Thirty days prior to worming, the eggs of the bot fly which may be clinging to the body should be destroyed by (1) clipping the hair, and/or (2) washing with warm water at 120°F. The insides of the knees and the fetlocks especially should be treated in this manner.

Remarks: As a precaution, mares should not be treated closer than 30 days before foaling or within 14 days after foaling.

(Continued)

TABLE 10-10 *(Continued)*

FLIES and MOSQUITOES

Flies and **mosquitoes** are usually classed as follows:
1. *Biting flies and mosquitoes*—this includes horse flies, deer flies, stable flies, horn flies, black flies, biting midges, and mosquitoes.
2. *Nonbiting flies*—include the face fly and house fly.

Species Affected: Horses.

Symptoms and Signs of Affected Animals (or Damage Inflicted): They lower the vitality of horses, mar the hair coat and skin, produce a general unthrifty condition, lower performance, and make for hazards when riding or using horses. Also, they may temporarily or permanently impair the development of foals and young stock.

Distribution and Losses Caused By: Wherever there are horses. Flies and mosquitoes are probably the most important insect pests of horses.

Treatment: For materials and control recommendations for horse flies, deer flies, and mosquitoes, see Table 10-7, Cattle Parasites and Their Control.

For control of horn flies, face flies, house flies, and stable flies, the following materials and controls are recommended, with the admonition to follow label instructions:
1. Coumaphos (Co Ral).
2. Dioxathion (Delnan).
3. Dioxathion + dichlorvos.
4. Ectrin strips (Farnum), use on halter or brow-band.
5. Fenvalerate (Ectrin).
6. Permethrin (Ectiban).
7. Permethrin + piperonyl butoxide.
8. Rabon oral larvacide.
9. Vapona + pyrethrin + piperonyl butoxide.

Also, the following compounds may be used in automatic spray systems in keeping with the label directions:
1. Natural pyrethrins + piperonyl butoxide.
2. Permethrin.
3. Resmethrin.
4. Vapona.

Fig. 10-69.

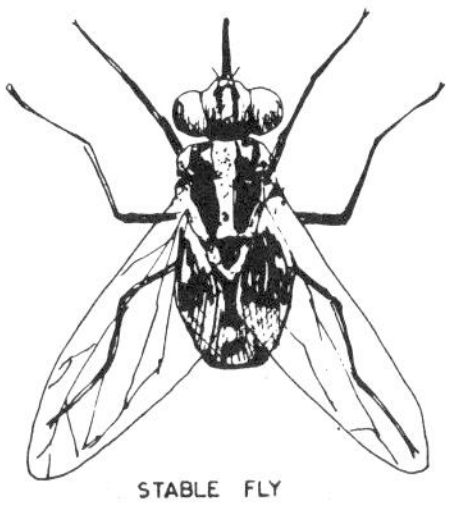

Fig. 10-70.

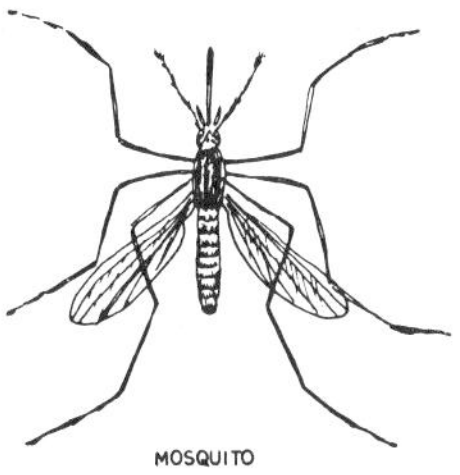

Fig. 10-71.

Prevention and Control: *Biting flies and mosquitoes:* Sanitation—the destruction of the breeding areas of the nests—is the key to the control of biting flies and mosquitoes. Do not allow manure or other breeding areas to accumulate. Spread manure in fields (to dry) every day or two. Control horse flies, deer flies, and mosquitoes by filling low spots in corrals or paddocks and draining all water-holding areas.

Nonbiting flies: Sanitation is the most efficient method of reducing populations of house flies. Sanitation may be additionally important if the horses are located near an urban area, in order to avoid complaints from neighbors. Residual sprays will eliminate many house flies. Also, house flies are attracted to baits (insecticides mixed with sugar or other attractive material), which are effective house fly killers.

Remarks: Flies and mosquitoes can be the vector (carrier) of serious diseases.

LICE

Lice are small, flattened, wingless insect parasites of which there are several species, most of which are specific for a particular class of animals.

Species Affected: Horses. Mules, with other species for other classes of animals.

Symptoms and Signs of Affected Animals (or Damage Inflicted): Intense irritation, restlessness, and loss of condition. There may be severe itching and the animal may be seen scratching, rubbing, and gnawing at the skin. The hair may be rough, thin, and lack luster; and scabs may be evident. Lice are apt to be most plentiful around the root of the tail, on the inside of the thighs, over the fetlock region, and along the neck and shoulders.

Distribution and Losses Caused By: Widespread. Lice retard growth, lower work efficiency, and produce unthriftiness.

Treatment: Period application, according to the directions on the label, of one of the following insecticides:
1. Coumaphos (Co Ral).
2. Crotoxyphos (Ciodrin).
3. Fenvalerate (Ectrin).
4. Permethrin.

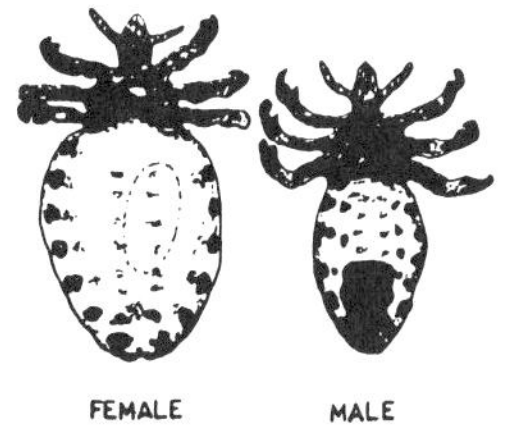

Fig. 10-72.

Prevention and Control: Because of the close contact of horses during the winter months, it is practically impossible to keep them from becoming infested with lice.

For effective control, all horses should be treated simultaneously at intervals, especially in the fall about the time they are placed in winter quarters.

Remarks: Lice show up most commonly in winter and on ill-nourished and neglected animals.

Although rarely dipped, horses may be so treated for lice, using any one of the mixtures and procedures recommended for dipping cattle.

(Continued)

TABLE 10–10 *(Continued)*

PINWORMS (RECTAL WORMS)

Two species of **pinworms** are frequently found in horses. *Oxyuris equi* are whitish worms with long, slender tails, whereas *Probstmyria vivipara* are so small as to be scarcely visible to the naked eye (see Fig. 10–73 for the life history and habits of the pinworm).

Species Affected: Horses. Also occurs in humans, but with different species of worms.

Symptoms and Signs of Affected Animals (or Damage Inflicted): Irritation of the anus and tail rubbing. Heavy infections may also cause digestive disturbances and produce anemia.

The large pinworm is most damaging to the horse.

Distribution and Losses Caused By: Throughout the U.S.

Treatment: See Table 10–6, Recommended Compounds for Control of Internal Parasites, by Animal Species. Use drug of choice according to manufacturer's directions.

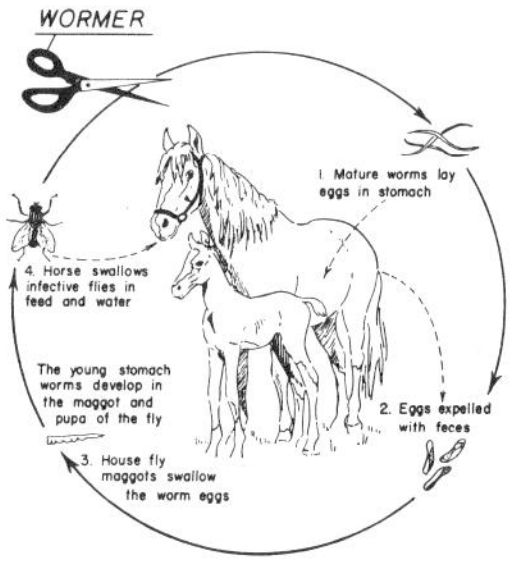

Fig. 10–73.

Prevention and Control: Sanitation and keeping animals separated from their own excrement.

STOMACH WORMS

Stomach worms are a group (3 kinds are important in horses) of parasitic worms that produce an inflammation of the stomach (see Fig. 10–74 for the life history and habits of the stomach worm).

Species Affected: Horses.

Trichostrongylus axei is also a common parasite of cattle, sheep, and a number of other animals.

Symptoms and Signs of Affected Animals (or Damage Inflicted): Loss of condition.

Severe gastritis.

Summer sores, a skin disease.

Distribution and Losses Caused By: Throughout the U.S.

Wasted feed and lowered efficiency are the chief losses.

Treatment: See Table 10–6, Recommended Compounds for Control of Internal Parasites, by Animal Species. Use drug of choice according to manufacturer's directions.

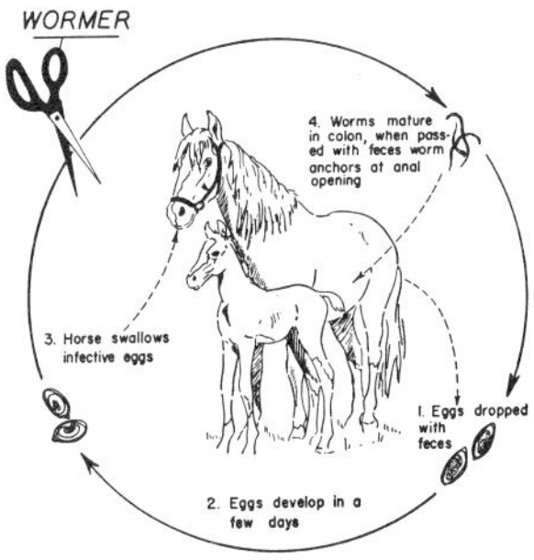

Fig. 10–74.

Prevention and Control: House flies and stable flies are the vectors. Hence, the control of flies is the best method of preventing and controlling stomach worms.

Remarks: Infection of *T. axei*. Also occurs in ruminants.

Dilute formaldehyde and astringents are commonly used in the treatment of summer sores.

(Continued)

TABLE 10–10 *(Continued)*

STRONGYLES

Strongyles consist of about 60 species; 3 are large (up to 2 in. in length) and the rest small (some scarcely visible to the naked eye). The large strongyles are variously referred to as bloodworms *(Strongylus vulgaris)*, palisade worms, sclerostomes and red worms (see Fig. 10–75 for the life history and habits of *S. vulgaris*).

Species Affected: Horses. Mules.

Symptoms and Signs of Affected Animals (or Damage Inflicted): Lack of appetite, anemia, progressive emaciation, a rough hair coat, sunken eyes, digestive disturbances including colic, a tucked-up appearance, and sometimes posterior paralysis and death. Collectively these symptoms indicate the disease known as strongyles.

Harmful effects greatest with younger animals.

One species of large strongyles *(S. vulgaris)* may permanently damage an intestinal blood vessel wall, resulting in death at any age.

Also, they may cause severe colic, which may terminate in death.

Distribution and Losses Caused By: Throughout the U.S., wherever horses and mules are pastured.

Presence of strongyles results in:

Loss of feed through feeding worms.

Lowered work efficiency (including on the track and in the show-ring).

Retarded growth of young animals.

Lowered breeding efficiency.

Death in severe infestations.

Treatment: See Table 10–7, Recommended Compounds for Control of Internal Parasites, by Animal Species. Use drug of choice according to manufacturer's directions.

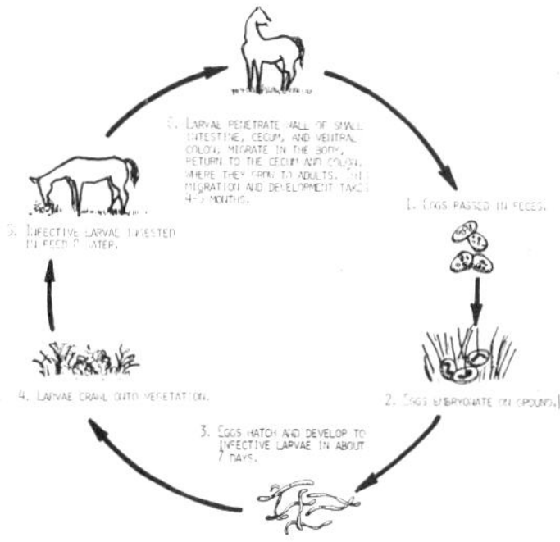

Fig. 10–75.

Prevention and Control: First and foremost, it is important that there be a barrier between the horse and its excrement. It is recommended that manure be picked up from horse pastures twice each week. It may be removed with a scoop shovel or with a vacuum pump powered by a tractor or small engine. Also, avoid overgrazing.

Approved drugs should be rotated on an annual basis; that is, use one product for about a year, then rotate. Also, it is very important that rotation be between drug classes.

Remarks: Strongyles are not transmissible to ruminants or swine.

Heavily infected animals may have one or more of the 3 species of large strongyles along with 10 or 12 species of small strongyles.

TICKS

Tick control may vary, depending on the tick species.

Species Affected: All farm animals and humans.

Symptoms and Signs of Affected Animals (or Damage Inflicted): Ticks reduce the vitality of horses through constant irritation and loss of blood.

Massive infestations may cause anemia, loss of weight, and even death. "Head heaviness" is often associated with massive infestations of ear ticks. Other losses may result from the simple presence of the ticks on the animal, a factor called *tick worry.*

Distribution and Losses Caused By: Ticks are particularly prevalent on horses in the southern and western parts of the U.S.

Treatment: The following insecticides may be used for tick control:

1. Amatraz.
2. Coumaphos (Co Ral).
3. Crotoxyphos (Ciodrin).
4. Fenvalerate (Ectrin).
5. Permethrin.
6. Rabon.

Prevention and Control: Because most species of ticks, except for the ear tick and tropical horse tick, attach to the external surfaces of horses, an application of the recommended insecticide by spray or wipe-on will give effective control. Ear ticks and tropical horse ticks should be treated by applying the chemical into the ears of the horses. Since horses are often confined to rather small areas, treatments of the premises may also help control heavy infestations of ticks.

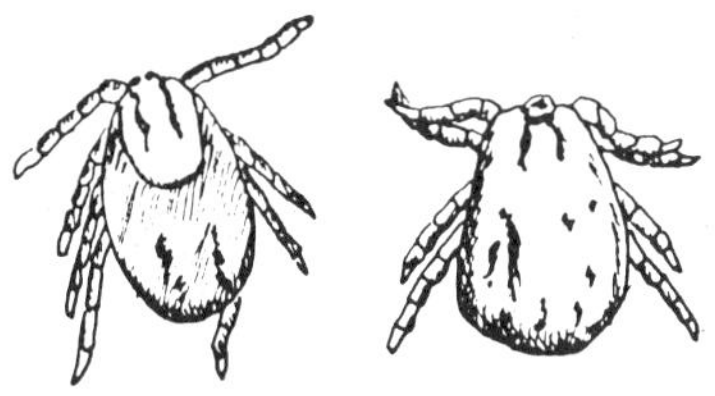

FEMALE MALE

WINTER TICKS

Fig. 10–76.

Remarks: Ticks are important to those taking care of horses because they may transmit diseases such as equine piroplasmosis (carried by *Anocentor nitens*) or cattle fever (carried by the *Boophilus* species). Also, most of the ticks mentioned may be vectors of anaplasmosis, and several species can cause tick paralysis in hosts.

ANIMAL PARASITES

Fig. 10–77. Face flies on cow. (Courtesy, R. E. Williams, Professor of Entomology, Purdue University, West Lafayette, IN)

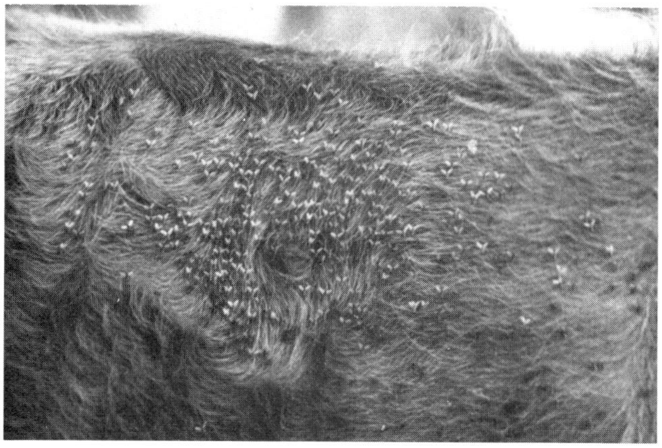

Fig. 10–78. Horn flies on cow. (Courtesy, R. E. Williams, Professor of Entomology, Purdue University, West Lafayette, IN)

Fig. 10–79. Heel flies (cattle grubs). Heifers running away from heel flies, the maggot stage of which is the cattle grub. Though the fly does not bite or sting, when it lays its eggs on the lower leg, it usually terrifies the animal, causing it to run with tail hoisted, seeking relief. (Courtesy, Livestock Conservation, Inc.)

Fig. 10–80. Grubs in back of cow. (Courtesy, R. E. Williams, Professor of Entomology, Purdue University, West Lafayette, IN)

Fig. 10–81. Ringworm. Heifer with ringworm. The fungi causing these raised circular areas may be transmitted to man. (Courtesy, Dept of Pathology and Hygiene, College of Veterinary Medicine, University of Illinois, Urbana)

Fig. 10-82. Gid tapeworm. Sheep infected with gid tapeworms. Giddy sheep show defects in vision and disturbances in movements. (Courtesy, USDA)

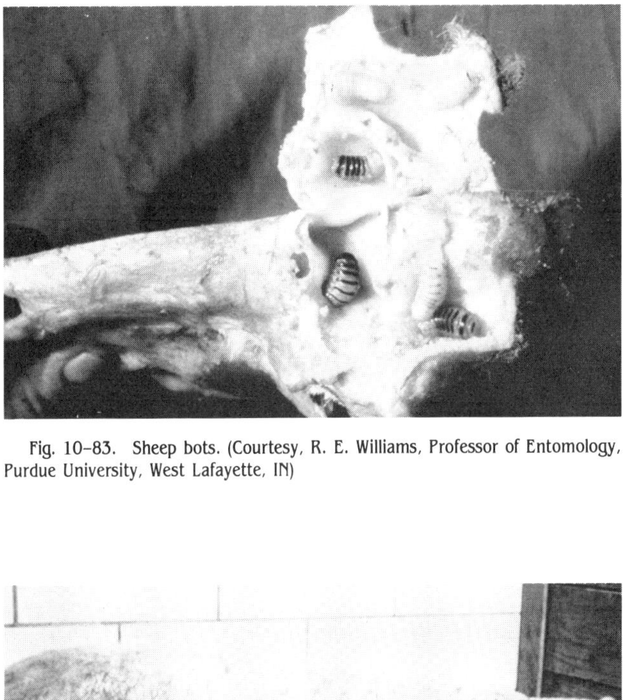

Fig. 10-83. Sheep bots. (Courtesy, R. E. Williams, Professor of Entomology, Purdue University, West Lafayette, IN)

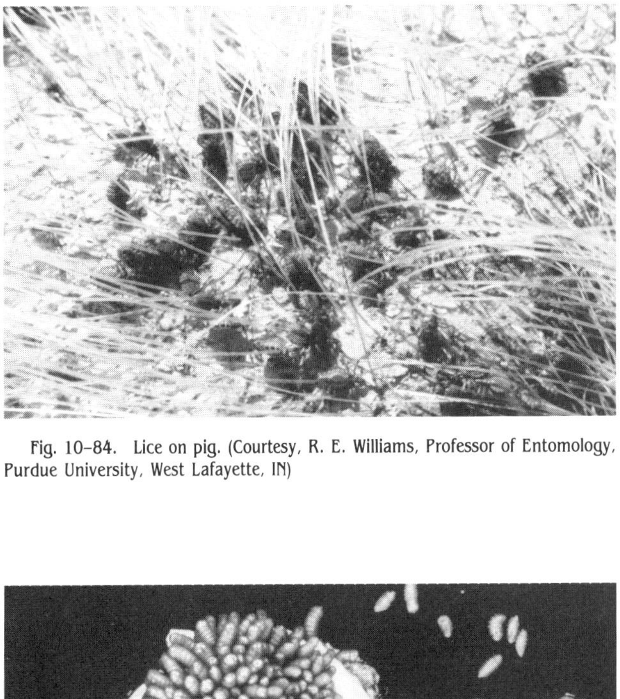

Fig. 10-84. Lice on pig. (Courtesy, R. E. Williams, Professor of Entomology, Purdue University, West Lafayette, IN)

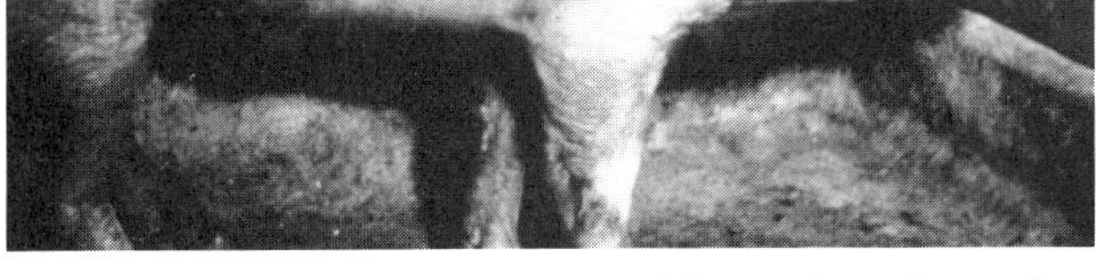

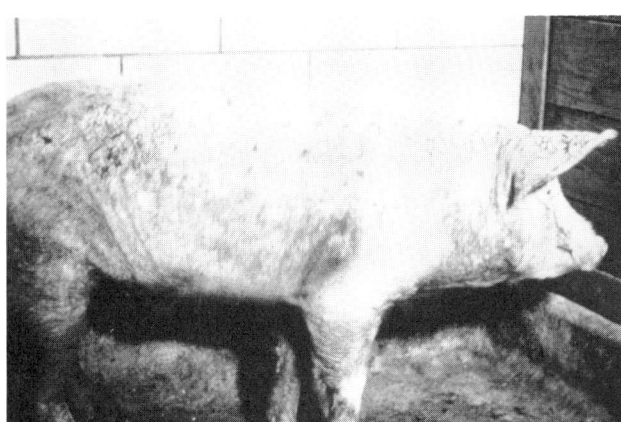

Fig. 10-85. Mange on hog. (Courtesy, R. E. Williams, Professor of Entomology, Purdue University, West Lafayette, IN)

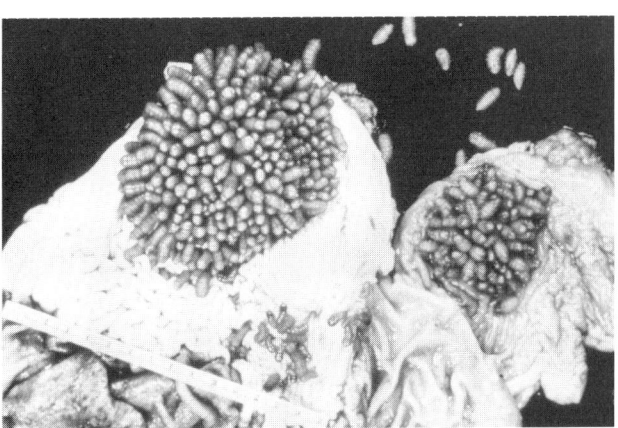

Fig. 10-86. Horse bots. (Courtesy, R. E. Williams, Professor of Entomology, Purdue University, West Lafayette, IN)

REGULATIONS RELATIVE TO DISEASE CONTROL

Certain animal diseases are so devastating that no individual farmer or rancher could long protect privately owned herds and flock against their invasion. Moreover, where human health is involved, the problem is much too important to be entrusted to individual action. In the United States, therefore, certain regulatory activities in animal disease control are under the supervision of various federal and state organizations. Federally, this responsibility is entrusted to the following agency:

Veterinary Service
Animal and Plant Health Inspection Service
U.S. Department of Agriculture
Federal Center Building
Hyattsville, MD 20782

FOOD AND DRUG ADMINISTRATION (FDA)

In 1906, the U.S. Congress enacted the Pure Food and Drug Law. Concurrently, the Federal Meat Inspection Act was passed. Both laws became effective in 1907. The U.S. Food and Drug Administration (FDA) was established as a separate unit of the U.S. Department of Agriculture in 1927. Then, in 1940, the FDA was transferred from the USDA to the Federal Security Agency, presently, the U.S. Department of Health and Human Services.

The FDA is charged with the responsibility of safeguarding American consumers against injury, unsanitary food, and fraud. It inspects and analyzes samples and conducts independent research on such things as toxicity (using laboratory animals), disappearance curves for pesticides, and long-range effects of drugs.

U.S. DEPARTMENT OF AGRICULTURE (USDA)

The following four divisions of the U.S. Department of Agriculture have primary responsibilities in the area of animal and human health.

1. **The Animal and Plant Health Inspection Service.** This division is charged with maintaining the wholesomeness and safety of meats processed in packing plants that ship meat and meat products, including poultry and poultry products, interstate. Veterinarians and other trained personnel make the inspections. Its purpose it to protect consumers against infected meats and fraudulent and unsanitary preparation of meat products. The inspection first consists of an examination of the live animals so that any unfit beast may be removed and disposed of properly. Secondly, the carcasses and internal organs are inspected for any abnormalities of animals carrying infectious diseases. Centers of infection sources may be located, thus assisting the livestock owners in the vicinity. The records of meat inspection also serve a useful purpose to the research scientist.

Fig. 10-87. Animal health inspection of meat animals consists of two examinations: (a) live animals, and (b) carcasses and internal organs. This shows live hogs being inspected prior to slaughter. (Courtesy, USDA)

2. **The Labeling and Registration Section.** This section in the USDA has responsibility for the proper labeling and safe use of pesticides. Manufacturers of pesticides must present new products with their proposed labels for approval before they are authorized to sell them. The label must indicate, as a minimum, the following: the name of the product; the active and inactive ingredients, together with percentages of each, in the formulation; the pest(s) controlled; directions for use—including the method and rate of application; any restrictions to be observed in application and handling; and an antidote—if known.

It is the responsibility of FDA, however, to set legal tolerances for pesticides on or in raw agricultural products. Also, it sets the safe interval between last application of the insecticide and the time of harvest of the crop or the slaughter of the animal.

Thus, through cooperative supervision of the USDA and the FDA, both the pesticide user and the consumer of the product are safeguarded.

3. **The Veterinary Services Division (VSD).** This division of the USDA is responsible for programs to control and eradicate

(if possible), certain diseases of livestock, e.g., brucellosis, tuberculosis, scabies, and hog cholera. It does the following things: conducts nationwide federal-state cooperative programs for the control and eradication of animal diseases; suppresses spread of disease through control of interstate and international movements of livestock; keeps informed of the overall disease situation nationally and internationally; administers laws to ensure human treatment of livestock and certain laboratory animals; collects and disseminates information on morbidity and mortality; and provides training for USDA employees and others in related government agencies.

4. **Stockyard Inspection.** With the advent of large public markets, public stockyards inspection was initiated. This is an addition to the regular inspection performed on animals by meat inspectors prior to slaughter. Among the principal diseases for which inspections are made are: anthrax, scabies of cattle and sheep, tick or splenetic fever, hog cholera, and erysipelas of swine.

Not only are the incoming shipments of livestock inspected, but a reinspection is made of outgoing shipments. Tests for tuberculosis and brucellosis are accomplished, and dipping for scabies is performed before shipments are allowed to return to farms and ranches.

U.S. PUBLIC HEALTH SERVICES (USPHS)

This section of the Department of Health and Human Services is concerned with the prevention and treatment of disease. It works in the areas of vector control, pollution control, and control of communicable diseases of people. A part of this important complex is the National Institute of Health (NIH), which was formed in 1930, and which is composed of the following nine sister institutes: the National Cancer Institute, the National Heart Institute, the National Institute of Allergy and Infectious Diseases, the National Institute of Arthritis and Metabolic Diseases, the National Institute of Dental Research, the National Institute of Mental Health, the National Institute of Neurological Diseases and Blindness (including multiple sclerosis, epilepsy, cerebral palsy, and blindness), the National Institute of Child Health and Human Development, and the National Institute of General Medical Science. In addition to its own research program, the USPHS provides grants for health-related research at many universities and research institutes in the United States.

STATE VETERINARIANS, SANITARY COMMISSIONS, AND BOARDS

Most states have state veterinarians, or comparable officials, who direct the livestock sanitary and regulatory programs within their respective states. Livestock producers may secure the regulations applicable to the state in which they reside by writing their state department of agriculture.

QUARANTINE[6]

Many highly infectious diseases are prevented by quarantine from (1) gaining a foothold in this country, or (2) spreading. By quarantine is meant (1) segregation and confinement of one or more animals in the smallest possible area to prevent any direct or indirect contact with animals not so restrained; or (2) regulating movement of animals at points of entry.

When an infectious disease outbreak occurs, drastic quarantine must be imposed to restrict movement out of an area or within areas. The type of quarantine varies from one involving a mere physical examination and movement under proper certification to the complete prohibition against the movement of animals, produce, vehicles, and even human beings.

FOREIGN DISEASE PROTECTION

Fig. 10–88. Cow with foot-and-mouth disease. This dreaded disease is capable of crippling the entire U.S. cattle industry. So, drastic measures are taken to prevent the introduction of the disease into the U.S. (Courtesy, USDA)

Distance no longer provides a buffer against the invasion of foreign diseases. More than 90% of animals imported into the United States arrive by air. An airplane can outpace the development of clinical signs of diseases in an animal that has been exposed to infection just prior to shipment. This prompts great concern for epizootic diseases capable of crippling or destroying entire livestock populations. Such diseases still exist in Europe, Asia, Africa, and Latin America; among them,

[6]This portion from "Quarantine" to "State Indemnity Payments" was authoritatively reviewed by, and helpful suggestions were received from, Dr. M. A. Mixson, Chief Staff Veterinarian, Emergency Programs, Veterinary Services, Animal and Plant Health Inspection Service, U.S. Department of Agriculture, Washington, DC)

are such dreaded diseases as rinderpest, contagious bovine pleuropneumonia, foot-and-mouth disease, hog cholera, Africa horse sickness, Africa swine fever, exotic Newcastle disease, African trypanosomiasis, East Coast fever, and piroplasmosis.

Until 1875, the importation of livestock into the United States was free and easy. But, that year the United States prohibited the importation of cattle and hides from Spain, where foot-and-mouth disease was rampant. By 1880, European countries were refusing to buy cattle or beef from the United States for fear of getting contagious pleuropneumonia. Then, in 1884, Congress established the Bureau of Animal Industry in the U.S. Department of Agriculture (USDA) and gave the Secretary of Agriculture authority to enforce quarantine laws.

Today, there are stations at several entry points, where inspectors of the USDA's Animal and Plant Health Inspection Service (APHIS) inspect all animals and poultry to be imported to the United States. If no communicable diseases are found, the animals may be quarantined for a period of time, during which time they are treated for external parasites and subjected to various tests—e.g., horses are tested for glanders and equine infectious anemia, and cattle are tested for brucellosis. At the end of the quarantine period, if no communicable diseases are found, they are released to the purchaser. APHIS is also charged with the responsibility of safeguarding against diseases introduced by the importation of zoo animals into this country. Wild animals brought into this country must undergo an extensive quarantine period abroad, followed by a further quarantine period at the animal quarantine station at Newburg, New York. Moreover, they are allowed to go only to certain approved zoos, where the zoo animals are isolated from domestic livestock and where proper measures are taken to dispose of waste to prevent the spread of diseases.

FEDERAL QUARANTINE CENTER

A Federal Quarantine Center was authorized in Public Law 91–239, signed by the President on May 6, 1970; and a 16.1-acre site for the Center was selected at Fleming Key, near Key West, Florida.

The quarantine center is designed to hold some 400 head of cattle, or other species in equivalent numbers, at one time, for a 5-month quarantine period. This maximum security station enables American livestock producers to import breeding animals from all parts of the world, while at the same time safeguarding our domestic herds and flocks from such diseases as foot-and-mouth disease, rinderpest, piroplasmosis, and others.

INDEMNITY PAYMENTS

Where certain animal diseases are involved, the livestock producer can obtain financial assistance in eradication programs through Federal and State sources.

Note well: Both Federal and State indemnity payments are subject to change. So, for current regulations, the livestock producer should contact the local veterinarian or State Department of Agriculture.

FEDERAL INDEMNITY PAYMENTS

Information relative to indemnities paid to owners by the federal government for animals disposed of as a result of outbreaks of certain diseases is given in Chapter 1, Subchapter B, Title 9 of the Code of Federal Regulations, a summary of which follows:

- **Brucellosis and tuberculosis**—The indemnity payments to owners by the federal government where brucellosis and tuberculosis are involved change from time to time. But the pertinent regulations that existed when this book was written follow:

1. **Brucellosis.**

a. **Affected cattle.** Owners of cattle destroyed which are affected with brucellosis may be paid an indemnity by the USDA not to exceed $50 for any grade animal or $250 for any registered or nonregistered dairy animal, except in Alaska, Hawaii, Puerto Rico, and the Virgin Islands, where no payment for any animal destroyed shall exceed $250 for any registered cattle or bison or $150 for any nonregistered cattle or bison, except that, for nonregistered dairy cattle, the indemnity shall not exceed $250.

b. **Herd depopulation.** The deputy administrator may authorize the payment of federal indemnity to owners whose cattle are destroyed because of brucellosis not to exceed $50 for any nonregistered animal or $250 for any registered or nonregistered dairy animal which (1) has been found to be exposed; and (2) is a part of a known infected herd, the destruction of which will, in the opinion of the deputy administrator, contribute to the brucellosis eradication program.

c. **Exposed female calves.** The deputy administrator may authorize the payment of federal indemnity to owners for exposed female calves destroyed because of brucellosis not to exceed $50 per head. Indemnity payments shall be made only for exposed female calves and only when the deputy administrator determines the destruction of such calves will contribute to the brucellosis eradication program.

2. **Tuberculosis.**

a. **Affected cattle.** The Department may pay owners an indemnity for cattle and bison affected with tuberculosis not to exceed $750 for each animal, but any joint state-federal indemnity payment, plus salvage, must not exceed the appraised value of each animal.

Also, the following regulations apply specifically to tuberculosis: The deputy administrator may authorize the payment of indemnity to owners of cattle which are destroyed because of tuberculosis not to exceed $450 for any animal which is a part of a known infected herd when it has been determined by the deputy administrator that the destruction of all the exposed cattle and bison in the herd will contribute to the tuberculosis eradication program; but the joint state-federal indemnity payments, plus salvage, must not exceed the appraised value of each animal.

b. **Appraisals of cattle destroyed because of tuberculosis.** Cattle to be destroyed because of tuberculosis shall be appraised by an independent, professional appraiser at veterinary services' expense, except that the veterinarian in charge may waive the requirement for an independent professional appraiser for reasons which are considered satisfactory. Due consideration shall be given to their breeding value as well as to their dairy or meat value. Where purebreds are involved, the owner shall either (1) see that the animals are accompanied by their registration papers at time of appraisal; or (2) be granted a reasonable time, by the veterinarian in charge, in which to present papers. Veterinary services may decline to accept any appraisal that appears to be unreasonable or out of proportion to the market value of cattle of like quality.

3. **Marking (or identifying) and slaughtering brucellosis and tuberculosis reactors.** Prior to marketing, the cattle must be marked and identified as follows:

a. Brucellosis reactor cattle must be branded with a 2- to 3-in. high letter "B" on the left jaw, and tagged with a metal federal or state reactor tag in the left ear; provided, however, that in lieu of branding and tagging, reactors and exposed cattle and bison in herds scheduled for herd depopulation may be identified by USDA-approved backtags and either accompanied to slaughter by a veterinary services or state representative, or moved directly to slaughter in vehicles closed with official seals.

b. Tuberculosis reactor cattle must be branded with a 2- to 3-in. high letter "T" on the left jaw, and tagged with a metal federal or state reactor tag in the left ear.

c. The cattle on which indemnity payments are made must be slaughtered within 15 days after the appraisal is made, unless an extension of time is granted by APHIS.

• **Foot-and-mouth disease, pleuropneumonia, rinderpest, and other contagious or infectious animal diseases which constitute an emergency and threaten the livestock industry of the country.** Under Title 9, Part 53, of the U.S. Code of Federal Regulations, the Secretary of Agriculture of the U.S. Department of Agriculture may declare a national emergency due to the existence of foot-and-mouth disease, rinderpest, contagious pleuropneumonia, or any other communicable disease of livestock and poultry which threatens the livestock industry of the country. Upon agreement with state authorities to enforce quarantine restrictions and orders and to participate equally in the payment of indemnities:

1. The federal government will pay to the owner 50% (and, in the case of exotic Newcastle disease, up to 100%) of the expense of purchase, destruction, and disposition of animals and materials required to be destroyed because of being contaminated or exposed to such diseases. The appraisal of animals shall be based on the fair market value and shall be determined by the meat, egg production, dairy, or breeding value of such animals. (In the case of hog cholera, under Part 56, of the U.S. Code of Federal Regulations, the federal government can pay up to 100% of the appraised value up to a maximum of $360 per head for purebred, inbred, hybrid, and breeding swine, and a maximum of $180 per head for all other swine.)

2. The federal government will pay to the owner 50% (and, in the case of exotic Newcastle disease or lethal avian influenza, up to 100%) of the expenses of purchase, destruction, and disposition of animals and materials required to be destroyed because of being contaminated by or exposed to such disease.

3. The appraised value of animals and materials must be established by either (a) veterinary services (VS) and a state representative jointly, or (b) a VS representative alone, provided the state authorities approve.

4. Animals affected by or exposed to disease shall be destroyed promptly after appraisal and disposed of by burial or burning, unless otherwise specifically authorized by veterinary services of the Animal and Plant Health Inspection Service of the U.S. Department of Agriculture.

In order to reduce the cost of eradicating emergency disease to the livestock producer and to the state and federal governments, it is essential that suspicious cases be promptly reported. If such a disease is suspected, a report should promptly be made to your practicing veterinarian and to state and federal animal health officials.

STATE INDEMNITY PAYMENTS

It is suggested that livestock producers secure the regulations applicable to the state in which they reside by writing to their state department of agriculture.

A heifer lunges from the squeeze chute after treatment by the veterinarian. (Courtesy, USDA)

SECTION 11

SELECTING AND JUDGING LIVESTOCK[1]

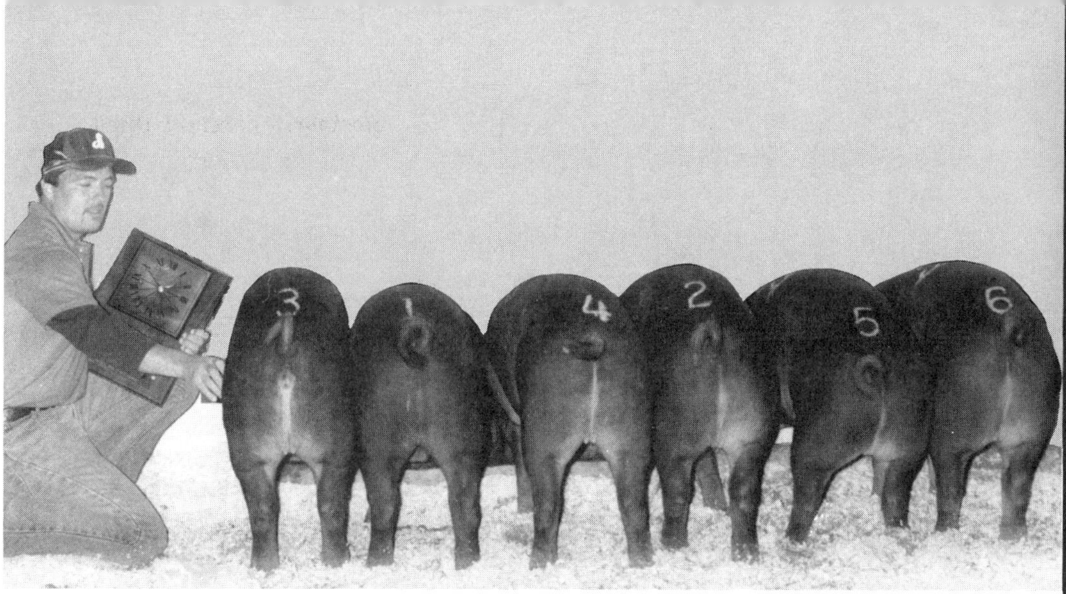

National barrow show champion truckload of barrows exhibited by Waldo Farms, De Witt, IL.

Contents *Page*

The great livestock shows throughout the land have exerted a powerful influence in molding animal types. At the same time, producers of meat animals are ever aware of market demands as influenced by consumer preferences, milk producers recognize that the main function of dairy cattle is the production of milk, and producers of light horses are cognizant of the importance of performance. It is realized, however, that only a comparatively few animals on farms and ranches are subjected annually to the scrutiny of experienced show-ring judges or market specialists. Rather, the vast majority of purebred animals and practically all commercial animals are evaluated by practical livestock producers—people who conduct their own buying and selling operations. In general, these operators are intensely practical; no animal meets with their favor unless it has utility value. Such producers have no interest in the so-called breed fancy points, and they may not be able to express fluently their reasons for selecting certain animals while culling others. But successful producers become quite deft in their evaluations. They are generally good judges of livestock.

BASES OF SELECTION

There are four bases of selection; namely, (1) selection based on type or individuality, (2) selection based on pedigree, (3) selection based on show-ring winnings, and (4) selection based on production testing. Only the first one—commonly called judging—will be enlarged upon in the sections which follow.

[1]Most of the line drawings in this section were made by Prof. R. F. Johnson, Dept. of Animal Sciences and Industry, California Polytechnic State University, San Luis Obispo.

PARADE OF CHAMPIONS

Fig. 11-1. Reserve Grand Champion Hereford female at the Western Nugget National, Reno, Exhibited by Bill Bennett, B&B Cattle Co., Connell, WA. Owner exhibitor Bennett is holding the heifer. (Courtesy, Bill Bennett)

Fig. 11-2. National Grand Champion Jersey cow. (Courtesy, The American Jersey Cattle Club, Reynoldsburg, OH)

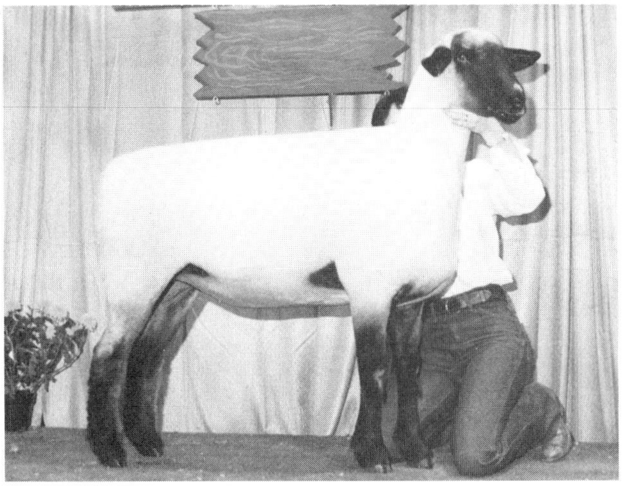

Fig. 11-3. Champion Hampshire ewe. (Courtesy, American Hampshire Sheep Assn., Ashland, MO)

Fig. 11-4. Champion Chester White boar. (Courtesy, Chester White Swine Record Assn., Peoria, IL)

Fig. 11-5. Champion Arabian in "Pleasure Driving Class." (Courtesy, International Arabian Horse Assn., Denver, CO)

QUALIFICATIONS OF A GOOD JUDGE

The essential qualifications which a good judge of any class of stock must possess are:

1. **Knowledge of the parts of an animal.** This consists of mastering the language that describes and locates the different parts of an animal.

2. **A clearly defined ideal or standard of perfection.** The successful livestock judge must know what to look for.

3. **Keen observation and sound judgment.** The good judge possesses the ability to observe both good conformation and defects, and to weigh and evaluate the relative importance of the various good and bad features.

4. **Honesty and courage.** The good judge of any class of livestock must possess honesty and courage, whether it be in making a show-ring placing or in conducting a breeding and marketing program. For example, it often requires considerable courage to place a class of animals without regard to (a) placings in previous shows, (b) ownership, and (c) public applause. It may take even greater courage and honesty with oneself to discard from the herd a costly animal whose progeny has failed to measure up.

5. **Logical procedure in evaluating.** There is always great danger of beginners making too close an inspection; they often get "so close to the trees that they fail to see the forest." Good judging procedure consists of the following three separate steps: (a) observing at a distance (20 to 30 ft) and securing a panoramic view where several animals are involved, (b) using close inspection (and handling cattle and sheep), and (c) moving the animal in order to observe action. Also, it is important that a logical method be used in viewing an animal from all directions, as for example (a) side view, (b) rear view, and (c) front view; thus avoiding overlooking anything and making it easier to retain the observations that are made.

6. **Tact.** In discussing either (a) a show-ring class, or (b) animals on a farm or ranch, it is important that the judge be tactful. Owners are likely to resent any remarks which indicate that their animals are inferior.

DO'S AND DON'TS FOR CONTEST JUDGES

Members of 4-H Clubs, FFA students, college judging classes, and other prospective livestock judges should first become thoroughly familiar with the six qualifications of a good judge as outlined in the section above. Next, they should observe the following do's and don'ts:

1. **Do's:**

a. Make certain how the class is numbered, and keep the numbers straight.

b. Get a clear picture of the class and of each individual animal in mind, so that they will be remembered when giving reasons.

c. Keep in a position of vantage where the class can be seen at all times; usually this means some distance away rather than too close.

d. Make placings on the basis of the big things.

e. Make certain that the card is filled out completely and correctly, and that the correct numbers are kept in mind.

f. If permissible, make concise notes that will assist in recalling each individual in the class; record such things as distinctive color markings, outstanding faults, etc.

g. When giving reasons, use good poise and look the judge in the eye.

h. Talk reasons clearly, and with conviction and confidence.

i. Give reasons in logical sequence; give the major reason first.

j. Use breeding terms in a breeding class and market terms in a market class; and use terms appropriate to the class of animals (for example, round for beef cattle, mammary system for dairy cows, leg for sheep, ham for hogs, and croup for horses.

k. Use comparative and descriptive terms in giving reasons. Avoid such vague terms as good, better, and best.

l. Concede or grant good points and faults, regardless of the placing of the animal.

2. **Don'ts:**

a. Don't act on hunches; if the first placing is arrived at after due consideration and in a logical manner, stick to it.

b. Don't place animals on the basis of small relatively unimportant characters.

c. Don't destroy self-confidence and self-respect by discussing the class with others before giving reasons.

d. Don't pay attention to what you overhear others say about a class; be an independent judge.

e. Don't give wordy and meaningless reasons.

f. Don't bluff; if you don't know the answer to a question, say so.

HOW TO USE THE JUDGING GUIDES, TABLES 11-1 THROUGH 11-5

Tables 11-1 through 11-5 (pp. 477-478, 481-482, 484-486, 488-489, and 492-493) are handy guides for judging beef cattle, dairy cattle, sheep, swine, and horses. It is suggested that the beginner use these as follows:

1. Examine animals in the order indicated; namely, (a) side view, (b) rear view, (c) front view, and (d) handling (etc., etc.).

2. Study the points listed under "ideal type," and know "the common faults."

3. Rank or place the animals according to their consistent rating on all points, especially the most important ones, or if you prefer, use the scorecard method which follows.

SCORECARD JUDGING

A scorecard is a listing of the different parts of an animal, with a numerical value assigned to each part according to its relative importance. It is a standard of excellence. The use of the scorecard involves studying each part, then assigning a score to each.

Different methods of scoring individual animals have evolved. All of them are based on visual appraisal. This point bears emphasis because livestock producers and students often get the erroneous impression that, just because some visual scoring system (scoring systems based on visual appearance, in contrast to actual weights, measurements, etc.) is recommended for or used in conjunction with a production testing program, it must be more accurate than all other scoring systems. This isn't true. All are visual methods, and the score resulting from the use of any of them is no better than the person making it. Some method of selecting all animals by score, preferably on a systematic and written down basis, is the important thing.

The following new and modern scorecards are presented in the sections that follow:

Fig. 11-16, Breeding Beef Cattle Scorecard, p. 479.

Fig. 11-17, Market Steer Scorecard, p. 480.[2]

Fig. 11-28, Dairy Cow Scorecard, p. 483.[3]

Fig. 11-39, Breeding Sheep Scorecard, p. 486.[2]

Fig. 11-40, Market Lamb Scorecard, p. 487.[3]

Fig. 11-51, Breeding Swine Scorecard, p. 490.[2]

Fig. 11-52, Market Barrow Scorecard, p. 491.[3]

Fig. 11-64, Horse Scorecard, p. 494.

[2]Adapted by the author from *Animal Science and Industry Laboratory Manual*, by Able, B. V., D. M. Allen, R. H. Hines, and M. McKee, Kansas State University, Kendall/Hunt Publishing Company, Dubuque, IA.

[3]Adapted by the author from unified scorecard prepared by The Purebred Dairy Cattle Association, and approved by the American Dairy Science Association.

As noted, the scorecard gives each of several traits a value, which total 100 for a perfect score.

A scorecard is a valuable teaching aid. It systemizes judging and avoids any part of the animal being overlooked. However, a scorecard has the following limitations: (1) It is not adapted to evaluating a great number of animals, or to comparative or show-ring judging, because of the time involved in using it; (2) a nearly worthless animal may score quite high—for example, an animal that is so structurally unsound that it can hardly walk may have a rather high total score; (3) it evaluates each part of an animal, rather than the system—the skeletal system, the muscle system, etc.; (4) it is based almost entirely on consumer needs (for example, on the end product—meat, in meat animal(s); and (5) it accords precious little consideration as to whether, or how, an animal can be changed better to conform to the needs and desires of producers.

JUDGING BEEF CATTLE

The parts of a beef animal are shown in Fig. 11-6, whereas Table 11-1 is a judging guide for beef cattle. Also, two beef cattle scorecards follow; one for breeding beef cattle, the other for steers (see Figs. 11-16 and 11-17, pp. 479 and 480).

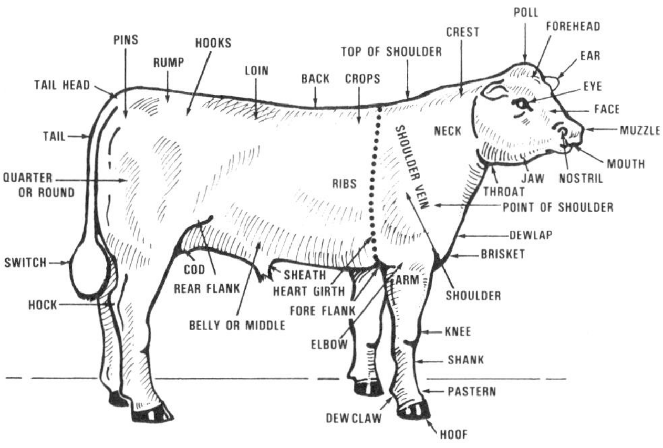

Fig. 11-6. Parts of a steer. The first step in preparation for judging beef cattle consists in mastering the language that describes and locates the different parts of an animal.

Beef type and size have changed through the years. Today, ideal beef animals are growthy and well muscled—they're meat type. Such animals gain rapidly and possess carcass excellence. They are longer, less deep bodied, trimmer through the middle, more upstanding, and less smooth than the type previously in vogue. Meat-type animals must show muscling when viewed from any angle, plus necessary minimum outside finish when ready for slaughter. Muscling is indicated by bulges and creases; hence, great smoothness is no longer a virtue.

TABLE 11-1
JUDGING GUIDE FOR BEEF CATTLE

Procedure for Examining, and What to Look For	Ideal Type	Common Faults
Side view: 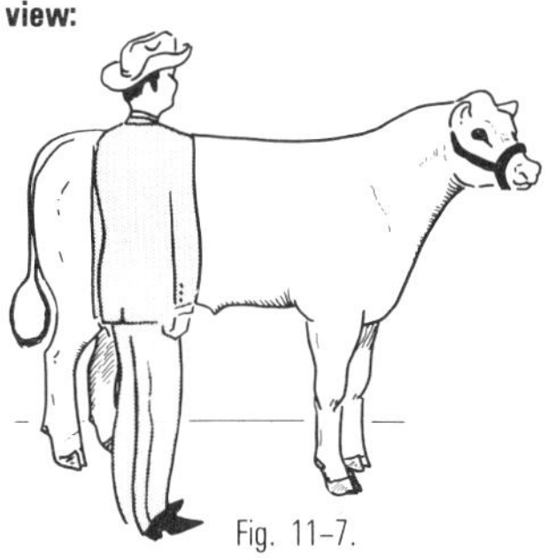 Fig. 11-7.	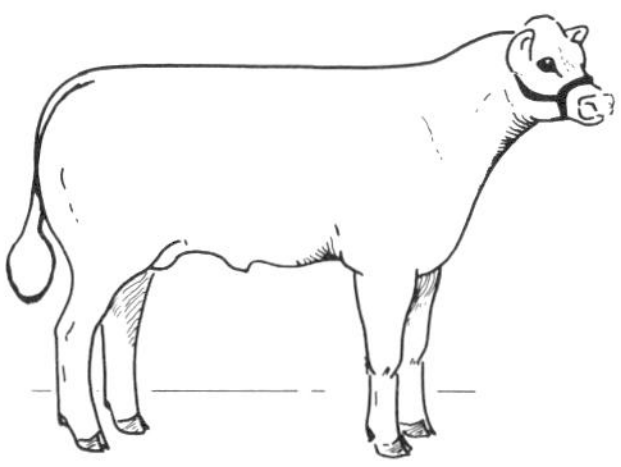 Fig. 11-8.	 Fig. 11-9.
1. **Reproductive efficiency** (not applicable to market animals)—Femininity in females, and masculinity in males; giving evidence of reproductive efficiency—which is the first requisite of beef cattle kept for breeding purposes.	1. **V**ery feminine females; long bodied, lean, smooth muscled; refined, feminine head; lean shoulder and hindquarters; and a good, functional udder. **V**ery masculine bull. Muscles well developed and clearly defined; and well developed testicles of equal size.	1. **F**emales with steery appearance—coarse front; protruding brisket; bristly hair on neck and top of shoulders; rounded hindquarters; and fat deposits over body. **B**ull lacking masculinity; underdeveloped crest; muscles lacking development and not clearly defined; testicles small, imbalanced, or with one carried high; scrotum that is twisted or filled with fat.
2. **Muscling**—Bulginess of muscles in those areas least affected by fatness—the arm, forearm, gaskin, and stifle; movement and bulge of muscles as animal walks.	2. **H**eavily muscled in arm, forearm, gaskin, and stifle; muscles that move and bulge as the animal walks.	2. **L**acking muscling. Light arm, forearm, gaskin, and stifle. Little movement or bulging of muscles as animal walks.
3. **Size**—Indicated by height to top of shoulders and length from nose to tailhead. Young breeding animals and steers that are long, tall, and not excessively fat—indications that they will continue to grow.	3. **A**dequate size, as evidenced by height to top of shoulders and length from nose to tailhead. Growthy young animals are long, tall, and not excessively fat.	3. **U**ndersized; small and dumpy. Bulls showing signs of early sexual maturity; they are not likely to make continued rapid growth and reach large mature size.
4. **Freedom from waste**—Trimness in both breeding and slaughter cattle.	4. **V**ery trim and free from waste; not too fat; and with no loose hide on the throat, dewlap, brisket, fore flank, navel or sheath, or twist.	4. **L**oose hide that is filled, or will fill, with fat. **E**xcessively fat breeding cattle, accompanied by lowered reproduction. **E**xcessively fat slaughter cattle, with reduced carcass value.
*5. **Structural soundness**—Straight legs, that are properly set; feet that are large and properly shaped; and hocks and knee joints that are clean and free from puffiness.	5. **L**egs that are straight, true, and squarely set; feet that are large, wide, and deep at the heel, with toes of equal size and shape that point straight ahead; and hocks and knee joints that are correctly set and clean.	5. **S**ickle-hocked, post-legged, back at the knees (calf-kneed), over at the knees (buck-kneed), or puffiness or swelling of knee or hock joints.
*6. **Breed type**—Characteristics true to breed, as distinguished by color and markings, shape of head, presence or absence of horns (and shape of horns if present), set of ears, body shape, and size. **C**ommercial cattle producers can disregard this trait. Likewise, it is unimportant in steers.	6. **T**rue-to-breed characteristics; in color and markings; body shape, and size.	6. **B**reed characteristics associated with undesirable traits.
Rear view: 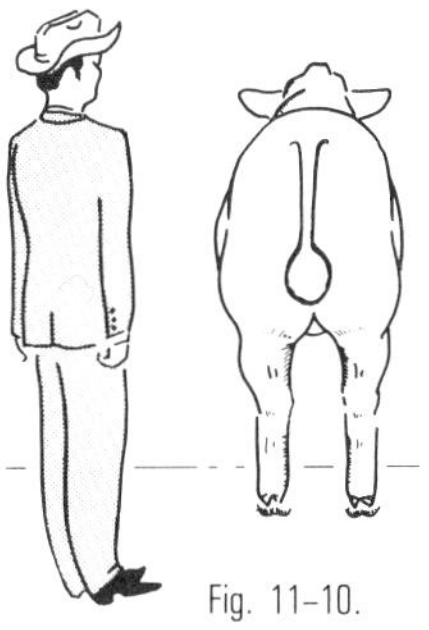 Fig. 11-10.	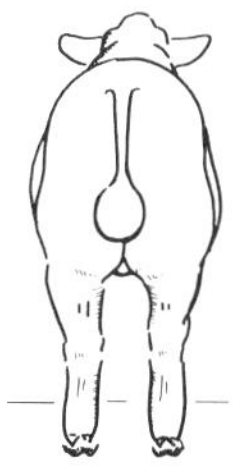 Fig. 11-11.	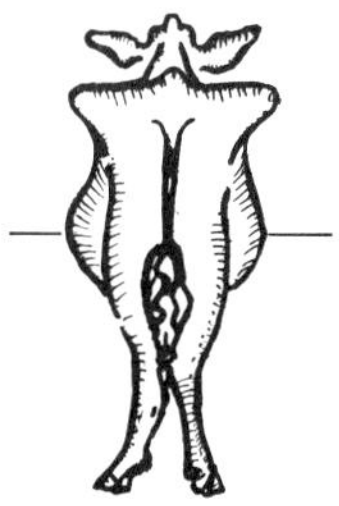 Fig. 11-12.
1. **C**urve of loin and round; crease in thigh; groove down topline, and bulging of loin eye on each side of backbone.	1. **W**ell curved loin and round; marked crease in thigh; well-defined groove down topline, with loin eye bulging on each side of backbone.	1. **F**lat loin and round; little or no crease in the thigh or groove down topline.

(Continued)

TABLE 11-1 *(Continued)*

Procedure for Examining, and What to Look For	**Ideal Type**	**Common Faults**
Rear view: *(Continued)*		
*2. **S**traightness and set to hind legs. Size and shape of hind feet. Cleanness of hock.	2. **H**ind legs straight and squarely set. Large feet that are wide and deep at the heels, with toes of equal size that point straight ahead; hock correctly set and clean.	2. **C**ow-hocked; puffiness and selling of hock joints. Small feet.
Front view:		
 Fig. 11-13.	 Fig. 11-14.	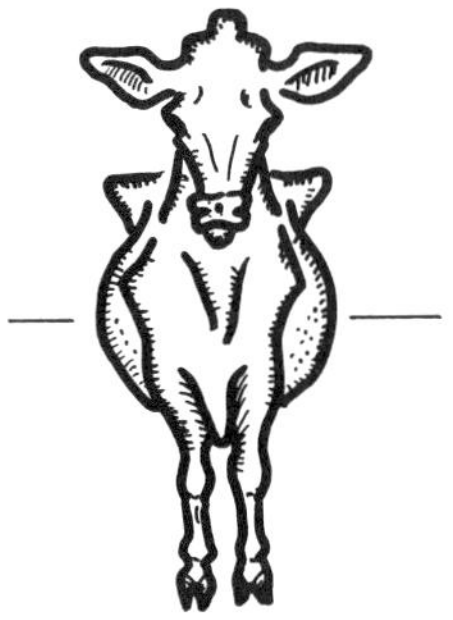 Fig. 11-15.
*1. **S**hapeliness of head.	1. **A** shapely head, true to breed type—as indicated by presence or absence of horns (shape of horns if present), and set of ears.	1. **A** plain head.
*2. **S**ex character.	2. **F**emales show femininity about the head and front end—they have lean cheecks, jaw, neck, brisket, and shoulders. **B**ulls show masculinity; they are on the "look," with head up and ears cocked, and they have a well-developed crest.	2. **C**ows lacking femininity; bulls lacking masculinity.
3. **B**risket.	3. **A** neat, trim brisket.	3. **H**eavy and wasty in the brisket.
4. **W**idth of chest.	4. **A** wide chest.	4. **A** narrow chest.
*5. **S**et to front legs; size and shape of front feet.	5. **C**orrectly set front legs, with front feet of proper size and shape.	5. **C**rooked front legs; puffiness and swelling of the knee joints; curled toes.
Handling:		
1. Muscling.	1. **H**eavily muscled, meaty.	1. **L**ight muscled.
2. **Q**uality of hide and mellowness.	2. **A** loose, pliable, mellow hide.	2. **C**oarse hided and hard.
3. **F**inish.	3. **D**esirable finish.	3. **L**acking finish; or overdone, soft and flabby.

*Not as important in market steers.

BREEDING BEEF CATTLE SCORECARD

	Perfect Score	ANIMAL No. 1	No. 2	No. 3	No. 4	Etc.
REPRODUCTIVE EFFICIENCY:	20					
Highly fertile female—Feminine—long body, lean, smooth muscled; refined, feminine head; lean cheek, jaw, neck, brisket, shoulder, and hindquarters; and a good functional udder (or promise of udder development in a heifer).						
Avoid lowly fertile female—Steery appearance—coarse, heavy front, masculine rather than feminine; protruding brisket; bristly hair on neck and top of shoulders; rounded hindquarters; and fat deposits on the face, brisket, shoulders, hips, rump, pins, below the vulva, and in front of the udder.						
Highly fertile bull—Masculine—"he's on the look," with head up and ears cocked; well-developed crest; muscles well developed and clearly defined, especially in the regions of the neck, loin, and thigh; and well-developed genitalia, with testicles of equal size and well defined, and a proper neck to the scrotum.						
Avoid lowly fertile bull—Lacking masculinity—ears not alert; undeveloped crest; muscles lacking development and not clearly defined; testicles small, unbalanced, or with one carried high; scrotum that is twisted or filled with fat.						
MUSCLING:	20					
Well muscled—Bulging in those areas least affected by fatness—the arm, forearm, gaskin, and stifle muscles move and bulge as animal walks. Look for curved loin and round; crease in thigh; well-defined groove down topline, with loin eye bulging on each side of backbone. Look for calves with long, smooth muscling, indicating continued growth. Since muscling is a masculine trait, it is more important in bulls and steers than in heifers.						
Avoid coarse shoulders in breeding cattle, because it is usually associated with calving problems.						
SIZE:	15					
Adequate size—As indicated by height to top of shoulders and length from nose to tailhead. Young breeding animals and steers should be long, tall, and not excessively fat—indications that they will continue to grow.						
Avoid bulls showing signs of early sexual maturity; they are not likely to make continued rapid growth and reach large mature size.						
FREEDOM FROM WASTE:	15					
Freedom from waste; trimness—In both breeding and slaughter cattle. Excessively fat breeding cattle usually have lowered reproduction. Excessively fat slaughter cattle have reduced carcass value.						
Avoid loose hide that is filled, or will fill, with fat. Look for loose hide on the throat, dewlap, brisket, fore flank, navel or sheath, and twist. Look for fat over back ribs, point of shoulder, and along backbone; since no muscle should be found at these places, if you feel something, it is fat.						
STRUCTURAL SOUNDNESS:	15					
Structurally sound—Legs straight, true, and squarely set; feet large, wide, and deep at the heel, with toes of equal size and shape that point straight ahead; hock and knee joints correctly set and clean.						
Avoid sickle-hocked, post-legged, back at the knees (calf-kneed), over at the knees (buck-kneed), or puffiness or swelling of knee or hock joints.						
BREED TYPE:	15					
Characteristics true to breed—Breed distinguished by color and markings, shape of head, presence or absence of horns (and shape of horns if present), set of ears, body shape, and size.						
Avoid breed characteristics associated with undesirable traits. Commercial cow-calf producers can disregard this trait. Likewise, it is unimportant in steers.						
TOTAL	100					

Fig. 11-16. Breeding Beef Cattle Scorecard.

MARKET STEER SCORECARD

	Perfect Score	ANIMAL No. 1	No. 2	No. 3	No. 4	Etc.
CONFORMATION:	60					
General appearance—(10 points)						
Muscular, thick, legs set wide apart, stylish, well balanced, large framed, adequate size for age. (10)						
Hindquarters—(29 points)						
Loin—meaty, thick, full deep loin edge. (10)						
Rump—long, level, full, and square. (7)						
Quarter—long, deep, thick, bulging, meaty. (10)						
Legs—correct, set wide apart. (2)						
Forequarters—(16 points)						
Back—thick, muscular, strong. (7)						
Ribs—bold spring, deep forerib. (3)						
Shoulders—smooth, muscular. (3)						
Crops—full. (0.5)						
Neck—clean, balanced. (0.5)						
Brisket—trim, neat dewlap. (1)						
Legs—correct, set wide apart. (1)						
Middle—(5 points)						
Stretchy, trim, straight underline. (5)						
Finish—(35 points)	35					
Desirable finish—over the back, loins, ribs, rump, and shoulder. (30)						
Trim in flanks, cod, and along underline. (5)						
Quality—(5 points)	5					
Trimness of head, hide, fineness of hair, ample bone.						
TOTAL	100					

Fig. 11-17. Market Steer Scorecard.

JUDGING DAIRY CATTLE

The main function of the dairy cow is to produce milk. Her appearance is not always indicative of her productive ability. However, a cow's appearance does tell us about her potential and her wearing ability.

The parts of a dairy cow are shown in Fig. 11-18, whereas Table 11-2 is a judging guide for dairy cattle. Also, a scorecard for dairy cattle follows as Fig. 11-28 (see page 483).

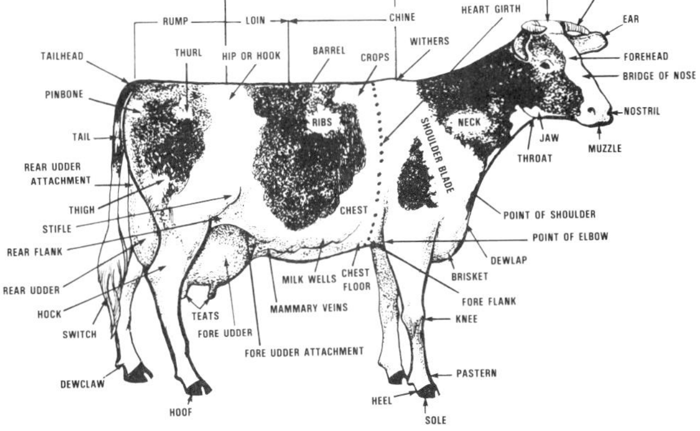

Fig. 11-18. Parts of a dairy cow.

TABLE 11–2
JUDGING GUIDE FOR DAIRY CATTLE

Procedure for Examining, and What to Look For	Ideal Type	Common Faults
Side view:		
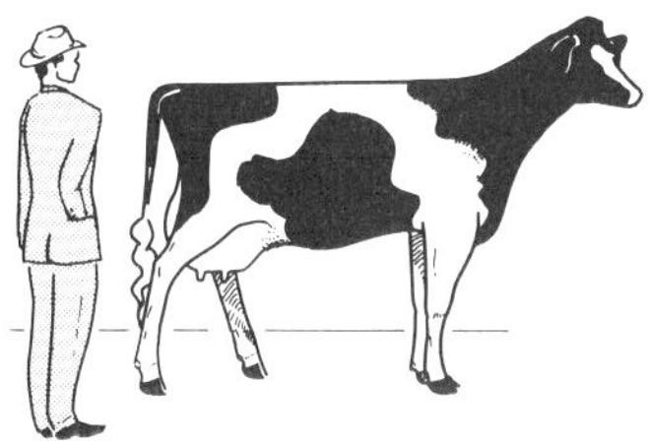 Fig. 11–19.	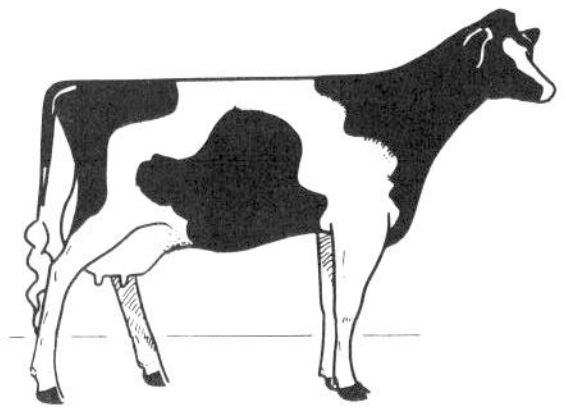 Fig. 11–20.	 Fig. 11–21.
1. **General appearance**—Style and attractiveness, size, relative length and depth throughout, straightness of back and levelness of rump, attachment and shape of udder, and quality as denoted by fineness of hair, smoothness of joints, fineness of bone, and absence of coarseness throughout. At the walk, observe the style and carriage, straightness of the legs, and the blending of the parts.	1. Great style and beauty.	1. Plain; lacking in attractiveness and in symmetry and balance.
a. **Breed characteristics**—Breed type and size.	a. True to the breed as evidenced by the conformation of the head and color markings. Adequate size for the breed.	a. Lacking in breed character. Undersized.
b. **Head**—Relative proportion of body; clean cutness.	b. Head in proportion to body, moderate length, clean-cut, and alert.	b. Head either too big or too little for the body; plain headed.
c. **Shoulder blades**—Smoothness.	c. Shoulder blades blend smoothly into the body.	c. Loose or out at the shoulders.
d. **Back**—Straightness and strength.	d. Back straight from withers to tailhead; strong.	d. Weak, sway backed; high tailhead.
e. **Rump**—Length, width, and levelness.	e. A long and nearly level rump, from the hip bones to the pin bones; high thurls; tailhead set level with the back bone and free of coarseness.	e. Short, sloping rump; low thurls; high tailhead.
f. **Legs and feet**—Set to the legs, strength of pasterns, cleanness of joints, and size of feet.	f. Proper set to the legs; hind legs nearly perpendicular from hock to pastern as viewed from the side; front legs straight as viewed from the side; standing well on the pasterns; clean joints.	f. Sickle hocks (hind legs set too far forward); back at the knees (calf-kneed); over at the knees (buck-kneed); down on the pastern; puffy around the joints.
2. **Dairy character**—Evidence of milking ability; angularity; openness, without weakness; not coarse.	2. A triangular body; loose, without being weak; and showing quality.	2. A round, thick neck; open shoulders; heavy, wasty brisket.
a. **Neck**—Length, leanness, smoothness, cleanness about the throat, dewlap, and brisket.	a. A long lean neck, which blends smoothly into the shoulders; clean-cut throat, dewlap, and brisket.	a. A short, thick neck; open shoulders; heavy, wasty brisket.
b. **Withers, ribs, flanks, thighs**—Shape of withers; width between ribs, and shape and length of ribs; shape of thigh.	b. Withers sharp and free of excessive flesh; ribs wide apart, and rib bones wide, flat, and long; thighs rather thin and flat.	b. Excessive flesh over the withers; close ribbed, and rib bones narrow, round, and short; and full, fleshy thighs.
3. **Body capacity**—Relatively large in proportion to size of animal; ample capacity, strength and vigor.	3. Great capacity, primarily achieved through length of body rather than extreme depth.	3. Lacking capacity; short bodied; and lacking depth.
a. **Barrel**—Support of barrel; length and depth.	a. Strongly supported; long, with adequate depth; well sprung ribs; greater depth of barrel toward rear.	a. Lacking capacity; short bodied; and lacking depth in the middle.
b. **Heart girth**—Spring of forerib; fullness of crops; width of chest floor.	b. Good spring of forerib; full crops; wide chest floor.	b. Lacking spring of forerib; slack in the crops; and narrow chest floor.
4. **Mammary system**—Attachment; balance; capacity; texture.	4. Udder strongly attached, well balanced, capacious, and fine textured.	4. Udder broken away from the body; different size halves and quarters; small; coarse.
a. **Udder**—Shape, attachment, texture.	a. Symmetrical; moderately long, wide, and deep; strongly attached; noticeable but not deep division between right and left halves; division between front and rear quarters not marked, and right and left quarters evenly developed; soft, pliable, and collapsed after milking.	a. Quarters not evenly developed; deeply cut between quarters or halves; excessive depth of udder; meaty textured; weakly attached.
b. **Fore udder**—Length, width, depth, and attachment.	b. Moderate length, uniform width from front to rear, and strongly attached.	b. Fore udder not extended well forward, lacking width, and weakly attached.
c. **Rear udder**—Height, width, shape, and attachment.	c. The rear udder should extend high up, and be wide, slightly rounded, of fairly uniform width from top to floor, and strongly attached.	c. Not attached high, and narrow, flat, not uniform in width from top to floor, and weakly attached.
d. **Teats**—Size and placement.	d. Teats of convenient size and squarely placed.	d. Teats too large or too small, and poorly placed.
e. **Mammary veins**—Size, length, crookedness, branching.	e. The milk veins are large, long, crooked, and branching; the milk wells are large and numerous; and the veins on the udder are large, crooked, and numerous.	e. Small milk veins and wells.

(Continued)

TABLE 11–2 *(Continued)*

Procedure for Examining, and What to Look For	Ideal Type	Common Faults
Rear view:  Fig. 11–22.	 Fig. 11–23.	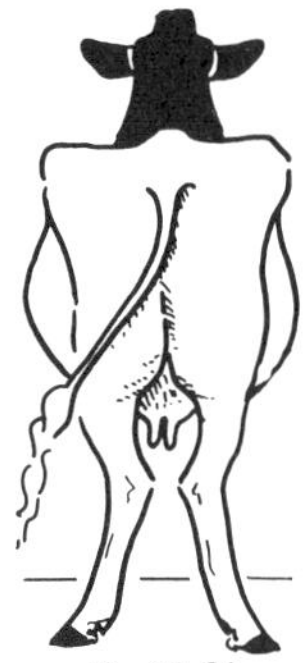 Fig. 11–24.
1. **W**idth over back.	1. **V**ertebra well defined.	1. **V**ertebra not well defined; meaty.
2. **W**idth over loins and rump.	2. **W**ide over the loins and rump.	2. **N**arrow over the loins and rump.
3. **W**idth between hip and pin bones.	3. **W**ide between the hip and pin bones.	3. **N**arrow between the hip and pin bones.
4. **H**eight and width of thurls. **T**he width between hips and pins and the levelness of rump are believed to be associated with the size and shape of the udder.	4. **T**hurls high and wide apart.	4. **L**ow thurls.
5. **S**traightness of hind legs.	5. **H**ind legs straight as viewed from the rear.	5. **C**ow hocked; puffiness and swelling of hock joints.
6. **S**ize and shape of hind feet.	6. **F**eet short, compact, and well rounded with deep heel and level sole.	6. **S**mall feet; toes not of equal size.
7. **R**ear udder attachment.	7. **R**ear udder that extends high and is wide and strongly attached.	7. **R**ear udder not extended high up; weakly attached.
Front view: 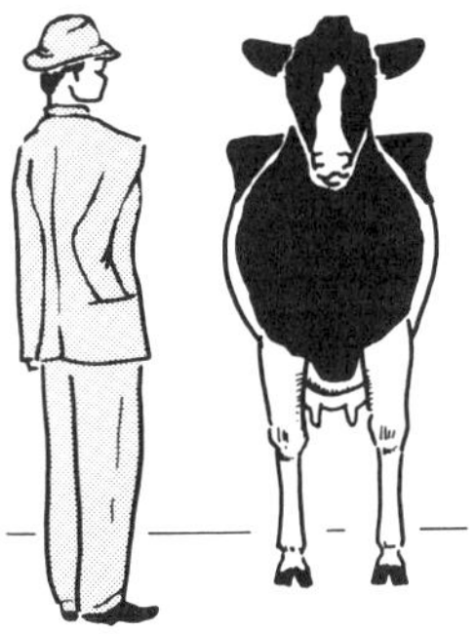 Fig. 11–25.	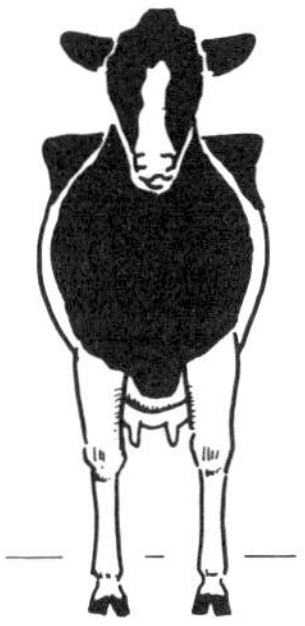 Fig. 11–26.	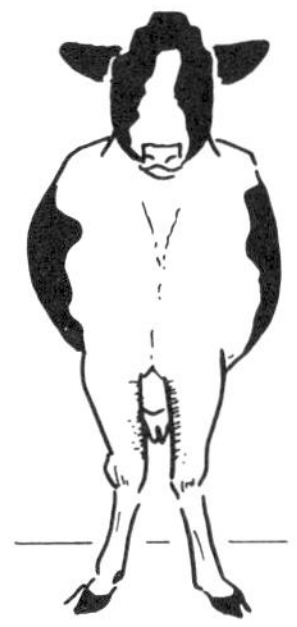 Fig. 11–27.
1. **S**hapeliness of head.	1. **A** shapely head, with a broad muzzle, large nostrils, and a strong jaw.	1. **A** plain head.
2. **S**ex character.	2. **C**ows show femininity about the head and front end; bulls show masculinity and have a well-developed crest.	2. **C**ows lacking femininity; bulls lacking masculinity.
3. **Neck**—Length, leanness; blending of neck and shoulders; throat, dewlap, and brisket.	3. **N**eck long, lean, and blended smoothly into the shoulders, with clean-cut throat, dewlap, and brisket.	3. **S**hort, thick neck; neck not blending smoothly into shoulders; leathery and wasty about the throat, dewlap, and brisket.
4. **Chest**—Width of floor.	4. **W**ide chest floor, indicating constitution and chest capacity.	4. **A** narrow chest.
5. **Front legs and feet**—Set to front legs; size and shape of front feet.	5. **F**orelegs medium in length, straight, wide apart, and squarely placed.	5. **C**rooked front legs; puffiness and swelling at knee joints; curled toes.

DAIRY COW SCORECARD

	Perfect Score	ANIMAL No. 1	No. 2	No. 3	No. 4	Etc.
GENERAL APPEARANCE:	20					
Attractive individuality with femininity, vigor, stretch, scale, harmonious blending of all parts and impressive style and carriage. All parts of a cow should be considered in evaluating a cow's general appearance. (10)						
Breed characteristics—Breed type and size.						
Head—Clean-cut, proportionate to body; broad muzzle with large, open nostrils; strong jaws; large, bright eyes; forehead, broad and moderately dished; bridge of nose straight; ears medium size and alertly carried.						
Shoulder blades—Set smoothly and tightly against the body.						
Back—Straight and strong; loin, broad and nearly level. (10)						
Rump—Long, wide, and nearly level from hook bones to pin bones; clean-cut and free from patchiness; thurls, high and wide apart; tailhead, set level with back line and free from coarseness; tail, slender						
Legs and feet—Bone flat and strong, pasterns short and strong, hocks cleanly moulded. Feet, short and compact and well rounded with deep heel and level sole. Forelegs medium in length, straight, wide apart, and squarely placed. Hind legs, nearly perpendicular from hock to pastern, from the side view, and straight from the rear view. (10)						
DAIRY CHARACTER:	20					
Evidence of milking ability, angularity, and general openness, without weakness; freedom from coarseness, giving due regard to period of lactation.						
Neck—Long, lean, and blending smoothly into the shoulders, clean-cut throat, dewlap, and brisket.						
Withers—Sharp.						
Ribs—Wide apart, rib bones wide, flat, and long.						
Flanks—Deep and refined. (20)						
Thighs—Incurving to flat, and wide apart from the rear view, providing ample room for the udder and its rear attachment.						
Skin—Loose, and pliable.						
BODY CAPACITY:	20					
Relatively large in proportion to size of animal, providing ample capacity, strength, and vigor.						
Barrel—Strongly supported, long and deep; ribs highly and widely sprung; depth and width of barrel tending to increase toward rear. (10)						
Heart girth—Large and deep, with well-sprung foreribs blending into the shoulders; full crops; full at elbows; wide chest floor. (10)						
MAMMARY SYSTEM:	30					
A strongly attached, well balanced, capacious udder of fine texture indicating heavy production and long period of usefulness.						
Udder—Symmetrical, moderately long, wide, and deep, strongly attached, showing moderate cleavage between halves, no quartering on sides; soft, pliable, and well collapsed after milking; quarters evenly balanced. (10)						
Fore udder—Moderate length, uniform width from front to rear and strongly attached. (6)						
Rear udder—High, wide, slightly rounded, fairly uniform width from top to floor, and strongly attached. (7)						
Teats—Uniform size, of medium length and diameter, cylindrical, squarely placed under each quarter, plumb, and well spaced from side and rear views. (5)						
Mammary veins—Large, long, tortuous, branching "Because of the natural undeveloped mammary system in heifer calves and yearlings, less emphasis is placed on mammary system and more on general appearance, dairy character, and body capacity. A slight to serious discrimination applies to overdeveloped, fatty udders in heifer calves and yearlings." (2)						
TOTAL	100					

Fig. 11-28. Dairy Cow Scorecard.

JUDGING SHEEP

Basically, the description of body parts and the desirable meat cuts of sheep and beef cattle, are similar. Sheep simply come in a smaller package. However, sheep are covered with wool, which makes it important that they be handled in order to be sure of their conformation. Also, the fleece should be considered in breeding classes.

The parts of a sheep are shown in Fig. 11–29, whereas Table 11–3 is a judging guide for sheep. Also, two scorecards for sheep follow; one for breeding sheep, and the other for market lambs (see Figs. 11–39 and 11–40, pp. 486 and 487).

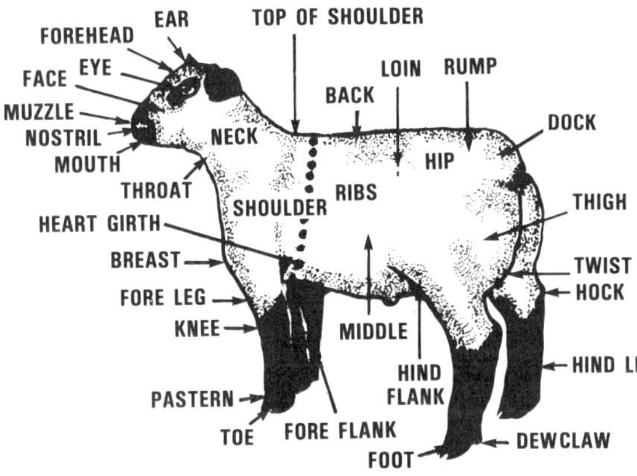

Fig. 11–29. Parts of a sheep. The first step in preparation for judging sheep consists in mastering the language that describes and locates the different parts of an animal.

TABLE 11–3
JUDGING GUIDE FOR SHEEP

Procedure for Examining, and What to Look For	Ideal Type	Common Faults
Side view:		
 Fig. 11–30.	 Fig. 11–31.	 Fig. 11–32.
1. **S**ize, as indicated by height to top of shoulder and length from nose to dock.	1. **A**dequate size, for breed and age.	1. **U**ndersized; too small.
2. **B**alance and symmetry.	2. **B**alanced and symmetrical.	2. **L**acking in balance and symmetry.
3. **S**tretch, with variation according to breed and age.	3. **B**reeds vary; some are more stretchy than others. **Y**oung animals should be growthy—they should be long, tall, and not excessively fat.	3. **D**umpy lambs—which are not likely to make continued rapid growth or reach large mature size.
4. **S**trength of back.	4. **A** strong, straight back.	4. **A** weak back.
5. **L**evelness of rump.	5. **A** level rump.	5. **A** sloping rump.
6. **T**rimness of underline.	6. **A** trim underline.	6. **H**igh in the flanks.
*7. **S**traightness of legs and strength of pasterns.	7. **S**traight, true, and squarely set legs, and strong pasterns.	7. **C**rooked legs, and weak pasterns.
8. **S**ize of bone.	8. **A**mple bone, with quality (wethers rather fine boned).	8. **C**oarse boned, lacking quality (or breeding animals too fine boned).
9. **S**tyle.	9. **P**lenty of style; a pleasing, alert appearance.	9. **L**acking in style.
*10. **B**reed type (markings, shape of head, and fleece characteristics true to the breed).	10. **S**howing plenty of breed type.	10. **L**acking breed type.
11. **F**reedom from wrinkles.	11. **S**mooth bodied.	11. **W**rinkles along the neck, especially in fine-wool breeds.

(Continued)

TABLE 11-3 *(Continued)*

Procedure for Examining, and What to Look For	Ideal Type	Common Faults
Rear view:		
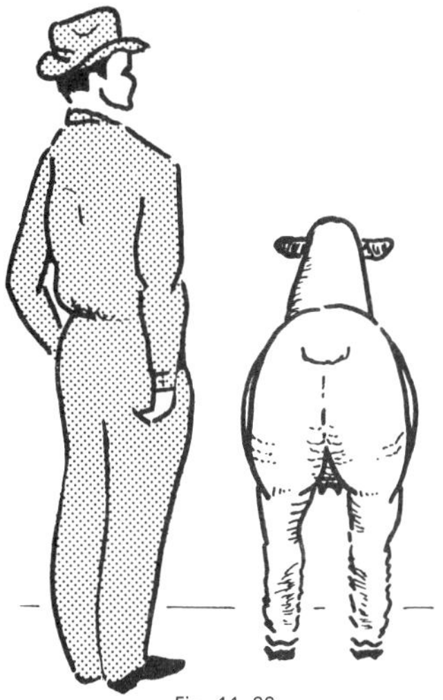 Fig. 11-33.	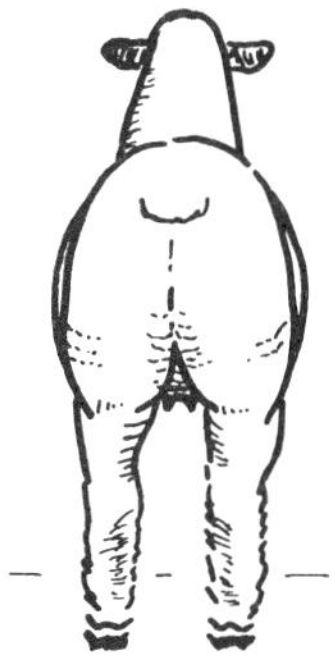 Fig. 11-34.	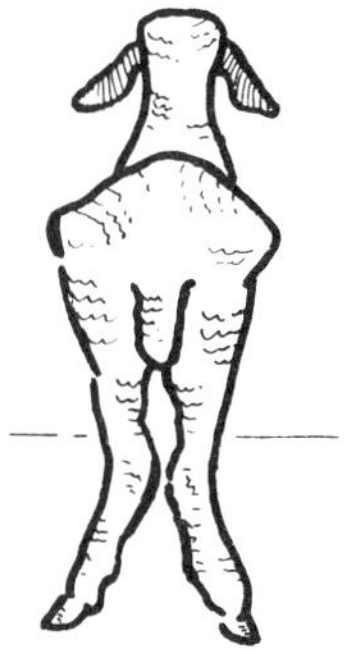 Fig. 11-35.
1. Uniformity of width from front to rear.	1. Uniformly wide from front to rear.	1. Narrow bodied.
2. Curve over back, loin, and round.	2. Well-covered or rounded-over back, loin, and rump.	2. Narrow over the back and loin.
3. Trimness of middle.	3. Trim in the middle.	3. Paunchy.
4. Depth and plumpness of leg.	4. Deep, plump leg.	4. A light leg.
*5. Set of hind legs.	5. Legs set wide apart.	5. Legs too close together; cow hocked; or bow legged.
Front view:		
 Fig. 11-36.	 Fig. 11-37.	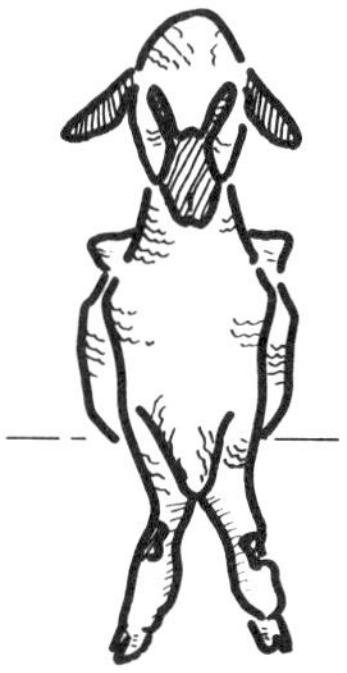 Fig. 11-38.
*1. Shapeliness of head.	1. Head shape true to breed.	1. A plain head.
*2. Sex character.	2. Ewes show femininity; rams masculinity.	2. Ewes lacking femininity; rams lacking masculinity.
3. Brisket.	3. A neat, trim brisket.	3. Heavy in the brisket; and "apron" or folds of skin.
4. Width of chest.	4. A wide chest.	4. A narrow chest.
*5. Set of front legs.	5. Correctly set front legs.	5. Crooked front legs.

(Continued)

TABLE 11-3 *(Continued)*

Procedure for Examining, and What to Look For	Ideal Type	Common Faults
Handling and fleece:		
1. **M**uscling.	1. **W**ell muscled.	1. **L**acking muscling.
2. **S**moothness over the shoulders.	2. **S**moothly laid in shoulders.	2. **O**pen in the shoulders.
3. **F**ullness of fore rib.	3. **W**ell-sprung fore rib.	3. **P**inched in the heart girth.
4. **R**ounding of back, loin, and rump.	4. **R**ounded over the back, loin, and rump.	4. **N**arrow over the back and loin; and not carrying width out to the dock.
5. **D**epth and plumpness of leg.	5. **A** deep, plump leg.	5. **A** light leg.
†6. **F**inish.	6. **D**esirable finish.	6. **L**acking finish.
*7. **Q**uality of fleece.	7. **A** long, dense, clean fleece, with uniformly good quality throughout.	7. **A** coarse, open fleece, lacking quality and uniformity; kempy black fibers.

*Not as important in market wethers as in breeding animals.
†Not important in breeding animals.

BREEDING SHEEP SCORECARD

	Perfect Score	ANIMAL No. 1	No. 2	No. 3	No. 4	Etc.
CONFORMATION:	73					
General appearance—(25 points)						
Size and scale—big for age, roomy, heavy bone. (15)						
Type—straight lined, balanced, deep ribbed, long, stylish. (10)						
Hindquarters—(26 points)						
Leg—muscluar, plump, thick, deep. (9)						
Rump—long, level, full, square dock. (7)						
Loin—wide, strong, meaty. (9)						
Twist—deep, full. (1)						
Forequarters—(22 points)						
Back—wide, straight, strong. (8)						
Ribs—bold spring, deep ribbed. (6)						
Shoulders—muscular, smooth. (4)						
Chest—deep, wide chest floor. (3)						
Neck—short, thick. (1)						
BREEDING QUALITIES:	27					
Head—clean-cut, bright eyes, feminine or masculine, proper color of face, free from wool blindness. (5)						
Underpinning—strong pasterns, legs correctly and squarely placed, rugged bone. (15)						
Fleece—dense, uniform crimp, long staple, pink skin, fineness according to standard of the breed, free from black fiber. (7)						
TOTAL	100					

Fig. 11-39. Breeding Sheep Scorecard.

MARKET LAMB SCORECARD

		Perfect Score	ANIMAL No. 1	No. 2	No. 3	No. 4	Etc.
CONFORMATION:		55					
General appearance—(10 points)							
Straight top and underline, muscular, thick, legs set wide apart, stylish, well balanced, adequate size for age.							
Hindquarters—(26 points)							
Legs—straight, set wide apart.	(2)						
Twist—clean, muscular.	(1)						
Leg—meaty, plump, long, deep, thick.	(8)						
Rump—long, level, thickly muscled.	(6)						
Loin—meaty, thick, deep loin edge, straight.	(9)						
Forequarters—(13 points)							
Back—thick, straight.	(6)						
Ribs—bold spring, deep forerib.	(3)						
Shoulders—muscular, smooth.	(2)						
Neck—short, thick.	(1)						
Breast—wide, deep chest floor, trim.	(0.5)						
Legs—straight, set wide apart.	(0.5)						
Middle—(6 points)							
Middle—trim, free from wastiness.							
FINISH:		40					
Uniformly covered with the correct amount of finish over back, ribs, loin, rump.	(30)						
Covering over shoulder, dock.	(5)						
Trim in flanks, cod, etc.	(5)						
QUALITY:		5					
Smooth pelt, head trim and refined, ample bone.							
TOTAL		100					

Fig. 11-40. Market Lamb Scorecard.

JUDGING SWINE

The desired meat-type hog yields a high percentage of lean meat with a minimum amount of fat. It is heavily muscled, correct in finish, firm, well-balanced, and has adequate length of side.

The parts of a hog are shown in Fig. 11-41, whereas Table 11-4 is a judging guide for swine. Also, two swine scorecards follow; one for breeding swine, and the other for market barrows (see Figs. 11-51 and 11-52, pp. 490 and 491).

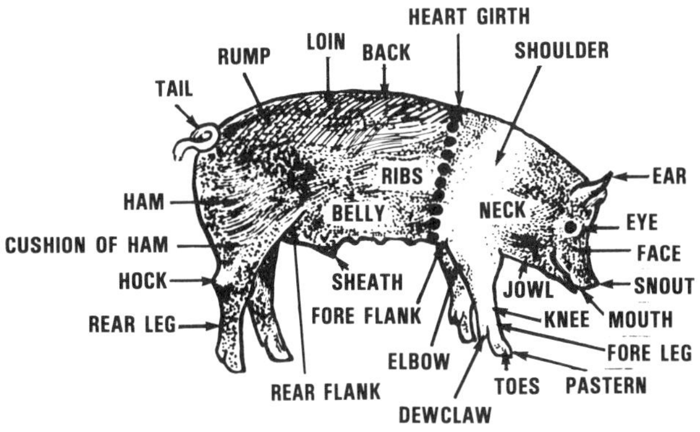

Fig. 11-41. Parts of a hog. The first step in preparation for judging hogs consists in mastering the language that describes and locates the different parts of the animal.

TABLE 11-4
JUDGING GUIDE FOR SWINE

Procedure for Examining, and What to Look For	Ideal Type	Common Faults
Side view:		
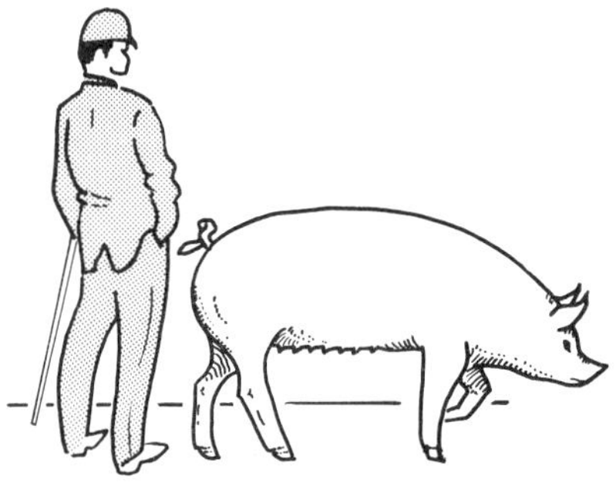 Fig. 11-42.	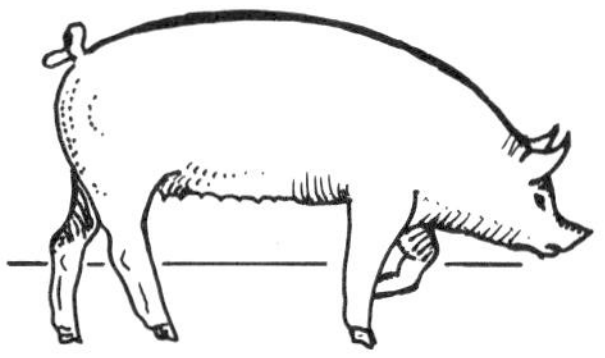 Fig. 11-43.	 Fig. 11-44.
1. **S**ize.	1. **A**dequate size.	1. **U**ndersized; too small.
2. **B**alance and symmetry.	2. **B**alanced and symmetrical, with all parts in the right proportion.	2. **L**acking in balance and symmetry, such as long necked, and slack framed.
3. **L**ength of side from a point in the center of the ham to the forepart of the shoulder. (This corresponds to measuring from the front of the aitch bone to the front of the rib in the carcass.	3. **A** long side. Meat-type barrows should be a minimum of 29.5 in. long at 230- to 240-lb weight.	3. **A** short side.
4. **D**epth of body.	4. **M**oderate depth of body, and moderately deep in the fore and rear flanks.	4 **A** shallow body; high in the flanks.
5. **T**opline.	5. **A** moderate and evenly arched back; high tail setting.	5. **L**ow back of the shoulders; steep in the rump; low tail setting.
6. **U**nderline; *teats.	6. **U**nderline is firm, trim, and free from wrinkles. Breeding animals should have at least 12 evenly spaced, well-developed teats.	6. **W**asty middle. Blind or inverted nipples.
7. **T**rimness of jowl and length of neck.	7. **A** trim jowl and medium length neck.	7. **H**eavy jowl and short, thick neck.
8. **S**houlders; smoothness, wrinkles.	8. **M**uscular and free from wrinkles.	8. **R**ough, open shoulders; wrinkles.
9. **S**ize of bone.	9. **A**mple bone with quality (barrows rather fine boned).	9. **C**oarse bone, lacking quality (or breeding animals too fine boned).
10. **S**traightness and placement of legs; *strength of pasterns.	10. **L**egs straight, true, and squarely set on the corners; pasterns strong.	10. **C**rooked legs; weak pasterns.
†11. **F**inish.	11. **D**esirable finish.	11. **T**oo fat and lardy.
12. **S**tyle.	12. **P**lenty of style.	12. **L**acking in style.
*13. **B**reed type true to breed (as shown by color, head, face, and ears).	13. **S**howing plenty of breed type.	13. **L**acking breed type.

(Continued)

TABLE 11–4 *(Continued)*

Procedure for Examining, and What to Look For	Ideal Type	Common Faults
Rear view:		
 Fig. 11–45.	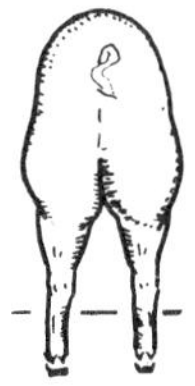 Fig. 11–46.	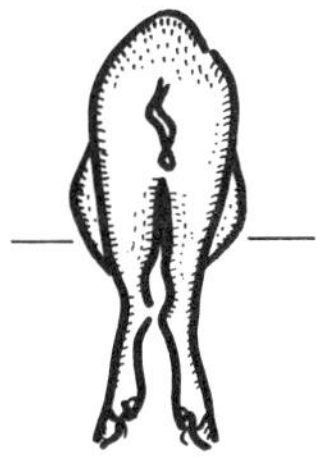 Fig. 11–47.
1. Width over the top.	1. A gradual rounding over the top which indicates meatiness.	1. Either "fish backed" or square topped (the latter indicates excess fatness).
2. Shoulder muscling.	2. Shoulder slightly bulging with the space between the shoulder blades well filled in with muscle.	2. Heavy and/or rough in the shoulders
3. Width and fullness of loin.	3. Wide, strong loin.	3. Narrow and pinched over the loin.
4. Rump.	4. Long, with a gradual slope toward a high set tail. Slightly rounded from side to side over the top, with no sign of excessive fatness.	4. A steep rump and low tail setting, which cut down on the size of the ham.
5. Plumpness, fullness, trimness, and firmness of ham; length of shank.	5. Deep, thick, slightly bulging, and firm ham; meated well down to hocks.	5. A light ham; long shank.
*6. Set of hind legs.	6. Legs set well apart.	6. Crooked hind legs.
Front view:		
 Fig. 11–48.	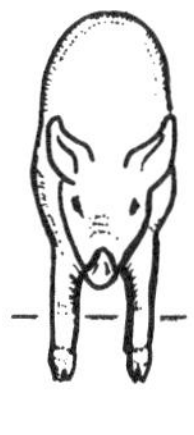 Fig. 11–49.	 Fig. 11–50.
*1. Shapeliness and trimness of head.	1. Head of medium length, wide between the eyes, and trim. The face, ears, and nose of breeding animals true to breed characteristics.	1. A plain head.
*2. Sex character.	2. Females show femininity; males show masculinity.	2. Females lacking femininity; males lacking masculinity.
*3. Set of front legs.	3. Correctly set front legs.	3. Crooked front legs.

*Not as important in market barrows as in breeding animals.

†Not important in breeding animals.

BREEDING SWINE SCORECARD

	Perfect Score	ANIMAL No. 1	No. 2	No. 3	No. 4	Etc.
GENERAL APPEARANCE:	25					
Type—heavy muscled, lean, trim, firm, smooth, long bodied, ham and rump should be wider than rest of body, moderately deep forerib, uniformly arched top, well balanced and stylish with a high degree of development in the valuable region of the hindquarters, same standards as for the market barrow. (12)						
Size—ample size, scale, and ruggedness for age. (13)						
CONFORMATION:	45					
Hindquarters—(26 points)						
Hind legs—set wide apart, out on the corners giving an indication of abundant muscling. (3)						
Ham—wide, deep, long, full, firm, meaty, deep in the seam. (9)						
Rump—long, wide, uniformly turned, high tail setting, meaty. (6)						
Loin—muscular turn, long, lean. (8)						
Forequarters—(14 points)						
Back—muscular turn, long, lean, uniformly arched, full spring of rib. (7)						
Shoulders—smooth, muscluar, free from fatty creases and wrinkles, no evidence of fat deposits at the elbow. (4.5)						
Head—clean-cut, trim, firm jowl. (2)						
Neck—short. (0.5)						
Middle—(5 points)						
Deep, roomy middle but not loose or wasty, belly trim and firm, deep ribbed.						
BREEDING QUALITIES:	30					
Underpinning—(13 points)						
Legs—straight as viewed from side, front, and rear, squarely set under corners of body, strong, straight toes. (8)						
Pasterns—strong, short, straight, but not buckled over. (1)						
Action—free, easy, unhindered walk, not stiff or "peggy." (4)						
Mammary system—(12 points)						
Six (6) sound nipples on a side, nipples prominent and evenly spaced; no evidence of inverted or blind teats.						
Breed character—(5 points)						
Head—varies with the breed, wide between the eyes, sows feminine and boars masculine. (3)						
Ears—relatively small and refined, not large and coarse thereby hindering vision; breeds having erect ears should show no tendency for drooping; breeds having drooping ears should show no tendency for being erect. (2)						
TOTAL	100					

Fig. 11-51. Breeding Swine Scorecard.

MARKET BARROW SCORECARD

		Perfect Score	ANIMAL No. 1	No. 2	No. 3	No. 4	Etc.
CONFORMATION:		60					
General appearance—(12 points)							
Heavy muscled, lean, trim, firm, long bodied, ham and rump should be wider than the rest of the body, uniformly arched top, well balanced and stylish with a high degree of development in the valuable region of the ham and loin, adequate size for age.							
Hindquarters—(28 points)							
Hind legs—set wide apart, out on the corners, giving an indication of abundant inner muscling, straight.	(2)						
Ham—wide, deep, long, full, firm, meaty, deep in the seam.	(10)						
Rump—long, wide, uniformly turned, high tail setting, meaty.	(7)						
Loin—muscular turn, lean, long, uniformly arched.	(9)						
Forequarters—(14 points)							
Back—muscular turn, long, lean, uniformly arched, uniform width, full spring of rib.	(7)						
Shoulders—smooth, muscular.	(5)						
Neck—short.	(0.5)						
Head—clean-cut, refined, trim firm jowl.	(1.5)						
Middle—(6 points)							
Side—long, moderately deep forerib, smooth, free from wrinkles.	(3)						
Underline (belly)—trim, firm, no evidence of looseness or wastiness.	(3)						
FINISH:		36					
Ham—firm and free from wrinkles at the base, firm in the crotch.	(5)						
Rump—no evidence of a counter sunk tail setting.	(5)						
Back and loin—lean, meaty turn, evidence of abundant muscling accompanied by minimum amount of backfat.	(8)						
Shoulders—firm and smooth, free from fatty creases and wrinkles, no evidence of fat deposit at the elbow.	(8)						
Jowl—trim and firm.	(5)						
Belly—trim and firm.	(5)						
QUALITY:		4					
Smooth throughout, not creased or wrinkled.	(2)						
Bone—ample substance of bone, definitely not fine but not overly coarse either.	(2)						
TOTAL		100					

Fig. 11-52. Market Barrow Scorecard.

JUDGING HORSES

Horses are judged on the basis of body conformation and performance. A horse must conform to the specific type that is needed for the function it is to perform; and, additionally, it should conform to the characteristics of the breed that it represents.

The parts of a horse are shown in Fig. 11-53, whereas Table 11-5 is a judging guide for light horses. Although light horses only are covered in this section, the same methods and principles apply to judging draft horses and mules. Also, a horse scorecard follows (see Fig. 11-64, p. 494).

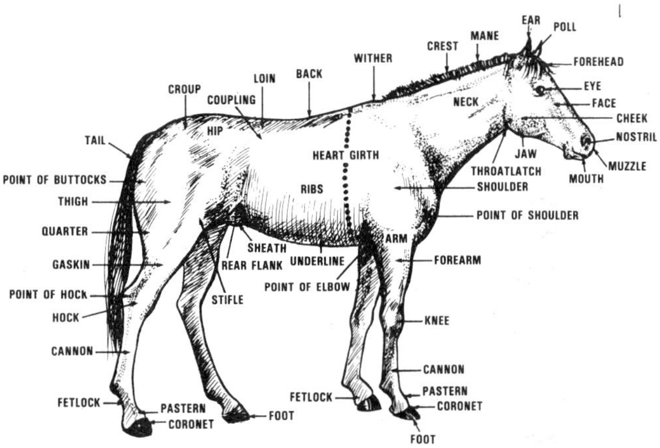

Fig. 11-53. Parts of a horse. The first step in preparation for judging horses consists in mastering the language that describes and locates the different parts of the animal.

TABLE 11-5
JUDGING GUIDE FOR HORSES[1]

Procedure for Examining, and What to Look For	Ideal Type	Common Faults
Side view:		
 Fig. 11-54.	 Fig. 11-55.	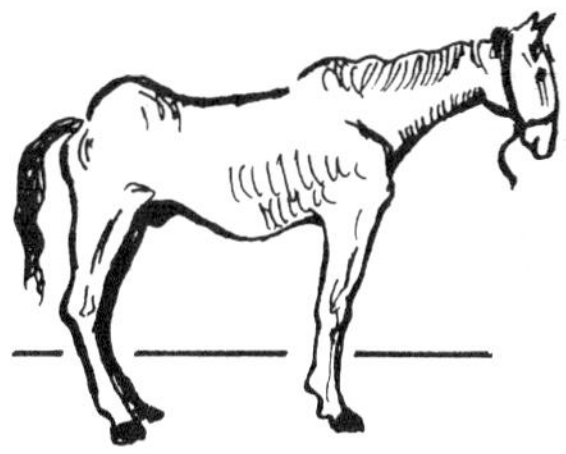 Fig. 11-56.
1. **S**tyle and beauty.	1. **H**igh carriage of head, active ears, alert disposition and beauty of conformation.	1. **L**acking style and beauty
2. **B**alance and symmetry.	2. **A**ll parts well developed and nicely blended together.	2. **L**acking in balance and symmetry.
3. **N**eck.	3. **F**airly long neck; carried high; clean-cut about the throat latch; with head well set on.	3. **L**acking style and beauty.
4. **S**houlders.	4. **S**loping shoulders (about a 45° angle).	4. **S**traight in the shoulders.
5. **T**opline.	5. **A** short, strong back and loin, with a long, nicely turned and heavily muscled croup, and a high, well-set tail; withers clearly defined and of the same height as the high point of croup.	5. **S**way backed; steep croup.
6. **C**oupling.	6. **A** short coupling as denoted by the last rib being close to the hip.	6. **L**ong in the coupling.
7. **M**iddle.	7. **A**mple middle due to long, well-sprung ribs.	7. **L**acking middle.
8. **R**ear flank.	8. **W**ell let down in the rear flank.	8. **H**igh cut rear flank or "wasp waisted."
9. **A**rm, forearm, and gaskin.	9. **W**ell-muscled arm, forearm, and gaskin.	9. **L**ight-muscled arm, forearm, and gaskin.
10. **L**egs, feet, and pasterns.	10. **S**traight, true, and squarely set legs; pasterns sloping about 45°, hoofs large, dense, and wide at the heels.	10. **C**rooked legs; straight pasterns; hoofs small, contracted at the heels, and shelly.
11. **Q**uality.	11. **P**lenty of quality, as denoted by clean, flat bone, well-defined joints and tendons, refined head and ears, and fine skin and hair.	11. **L**acking quality.
12. **B**reed type (size, color, shape of body and head, and action true to the breed represented).	12. **S**howing plenty of breed type.	12. **L**acking breed type.

(Continued)

TABLE 11–5
JUDGING GUIDE FOR HORSES[1]

Procedure for Examining, and What to Look For	Ideal Type	Common Faults
Rear view:		
 Fig. 11–57.	 Fig. 11–58.	 Fig. 11–59.
1. **W**idth of croup and through rear quarters.	1. **W**ide and muscular over the croup and through the rear quarters.	1. **L**acking width over the croup and muscling through the rear quarters.
2. **S**et to the hind legs.	2. **S**traight, true, and squarely set.	2. **C**rooked hind legs.
Front view:		
 Fig. 11–60.	 Fig. 11–61.	 Fig. 11–62.
1. **H**ead.	1. **H**ead well proportioned to rest of body, refined, clean-cut, with chiseled appearance; broad, full forehead with great width between the eyes; jaw broad and strongly muscled; ears medium sized, well carried, and attractive.	1. **P**lain headed; weak jaw.
2. **S**ex character.	2. **R**efinement and femininity in the brood mare; boldness and masculinity in the stallion.	2. **M**ares lacking femininity; stallions lacking masculinity.
3. **C**hest capacity.	3. **A** deep, wide chest.	3. **A** narrow chest.
4. **S**et to the front legs.	4. **S**traight, true, and squarely set.	4. **C**rooked front legs.
Soundness:		
1. **S**oundness, and freedom from defects in conformation that may predispose unsoundness.	1. **S**ound, and free from blemishes.	1. **U**nsound; blemished (wire cuts, capped hocks, etc.).
Action:[2]		
1. **A**t the trot.  Fig. 11–63.	1. **R**apid, straight, elastic trot, with the joints well flexed.	1. **W**inging, forging, and interfering.

[1]The illustrations for this table were prepared by R. F. Johnson.

[2]Three-gaited horses are shown at the walk, trot, and canter. Five-gaited horses must perform two additional gaits—the running walk and the rack. In judging, (1) every horse should be observed at each intended gait, and (2) trained horses should be examined when performing at the use for which they are intended.

HORSE SCORECARD

	Perfect Score	ANIMAL No. 1	No. 2	No. 3	No. 4	Etc.
BREED TYPE:	15					
Animals should possess the distinctive characteristics of breed represented, including— **Color, Height at maturity, Weight at maturity**						
FORM:	35					
Style and beauty—Attractive, good carriage, alert, refined, symmetrical, and all parts nicely blended together.						
Body—Nicely turned, long well-sprung ribs, heavily muscled.						
Back and loin—Short and strong, wide, well muscled, and short coupled.						
Croup—Long, level, wide, muscular, with a high-set tail.						
Rear quarters—Deep and muscular.						
Gaskin—Heavily muscled.						
Withers—Prominent, and of the same height as the high point of the croup.						
Shoulders—Deep, well laid in, and sloping about a 45° angle.						
Chest—Fairly wide, deep, and full.						
Arm and forearm—Well muscled.						
FEET AND LEGS:	15					
Legs—Correct position and set when viewed from front, side, and rear.						
Pasterns—Long, and sloping at about a 45° angle.						
Feet—In proportion to size of horse, good shape, wide and deep at heels, dense texture of hoof.						
Hocks—Deep, clean-cut, and well supported.						
Knees—Broad, tapered gradually into cannon.						
Cannons—Clean, flat, with tendons well defined.						
HEAD AND NECK:	10					
Alertly carried, showing style and character.						
Head—Well proportioned to rest of body, refined, clean-cut, with chiseled appearance; broad, full forehead with great width between the eyes; ears medium sized, well carried, and attractive; eyes large and prominent.						
Neck—Long, nicely arched, clean-cut about the throat-latch, with head well set on, gracefully carried.						
QUALITY:	10					
Clean, flat bone; well-defined and clean joints and tendons and fine skin and hair.						
ACTION:	10					
Walk—Easy, springy, prompt, balanced, a long step, with each foot carried forward in a straight line; feet lifted clear of the ground.						
Trot—Prompt, straight, elastic, balanced, with hocks carried closely, and high flection of knees and hocks.						
DISCRIMINATION:						
Any abnormality that affects the serviceability of the horse.						
DISQUALIFICATION:						
In keeping with breed registry or show regulations.						
TOTAL	100					

Fig. 11-64. Horse Scorecard.

DETERMINING THE AGE OF HORSES BY THE TEETH

The mature male horse has a total of 40 teeth,[4] whereas the young animal, whether male of female, has 24. These are as listed in Table 11-6.

TABLE 11-6
NUMBER AND TYPES OF HORSE TEETH

Number of Teeth of Mature Animal	Number of Teeth of Young Animal	Types of Teeth
24	12	**M**olars or grinders.
12	12	**I**ncisors or front teeth (the 2 central incisors are known as centrals or nippers; the next 2—one on each side of the nippers—are called intermediates or middles; and the last—or outer pair—the corners).
4	None	**T**ushes or pointed teeth. These are located between the incisors and the molars in the male.

As the tushes are usually not present in the mare, the mature female may be considered as having a total of 36[4] teeth rather than 40 as in the male.

The permanent incisor teeth of young horses five to seven years of age are elliptical or long from side to side; whereas when the young animal become older, these teeth become triangular, with the apex of the triangle pointed upward. As the animal advances still more in age, the teeth become more slanting. Instead of curving to approach a right angle with the jaws, they slant outward.

From 5 to 12 years of age the wearing surface of the cups is the most reliable indication of age. At fairly regular intervals, according to age, the cups disappear with wear.

After 12 years of age, even the most experienced horse specialist cannot accurately determine the age of an animal. It is known, however, that with more advanced age the teeth change from oval to triangular and that they project or slant forward more and more each year.

It must also be realized that the environment of the animal can very materially affect the wear on the teeth, often making it impossible to determine accurately the age of animals. For example, the teeth of horses raised in a dry, sandy area will show more than normal wear. Thus, the 5-year-old western horse may have a 6- or even 8-year-old mouth. The unnatural wear resulting in the teeth of cribbers or animals with parrot mouth or undershot jaw also make it difficult to estimate age.

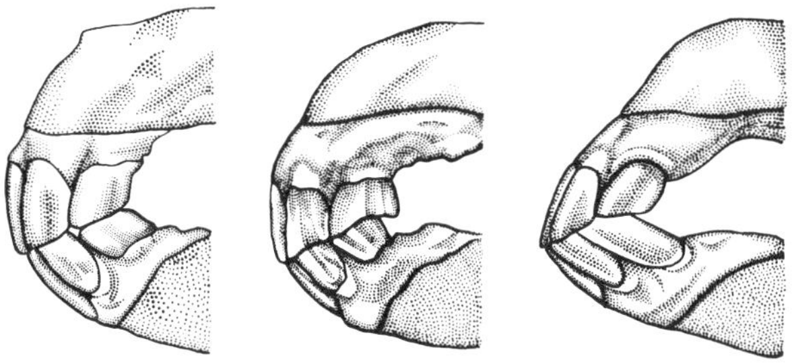

Fig. 11-65. Side view of 5-, 7-, and 20-year-old horse mouth. Note that as the horse advances in age the teeth change from nearly perpendicular to slanting sharply toward the front. (Drawing by R. F. Johnson)

(For details on how to determine the age of horses by the teeth, including drawings and descriptions of the teeth of horses of different ages, see *Horses and Horsemanship*, Chapter 5, a book by the same author and publisher as *Stockman's Handbook Digest.*

COMPARATIVE ANIMAL AGES

Most people use a direct ratio, or rule of thumb, in predicting the ages of animals. For example, a dog's age is commonly assumed to be one-seventh of that of a human; hence, a dog 7 years old would be equivalent in age to a human age 49. But since the oldest dog on record lived to the age of 34, according to the 1- to 7-year theory, a human could live to 34 × 7, or 238 years. This, of course, is more than twice the highest reliable recorded human age. The fallacy of the rule-of-thumb method (the 1-to-7 ratio in the case of dog to human) is that it fails to take into consideration *both* (1) the average age of maturity, and (2) the average age of death.

Mr. David P. Willoughby developed an ingenious, but relatively simple, "prediction formula," which takes into consideration both the time required to reach maturity and the maximum age recorded.[5]

[4]Quite commonly, a small, pointed tooth, known as a "wolf tooth," may appear in front of each first molar tooth in the upper jaw, thus increasing the total number of teeth to 42 in the male and 38 in the female. Less frequently, there are 2 more "wolf teeth" in the lower jaw, which increases the total number of teeth in the male and female to 44 and 40, respectively.

[5]Mr. Willoughby is Honorary Associate in Vertebrate Paleontology at the Los Angeles County Museum of Natural History. Table 11-11 reproduced herein is taken from *Natural History*, Vol. LXXVIII, No. 10, published by The American Museum of Natural History, New York, NY, December, 1969, pp. 56-59.

As shown in the first two columns of Table 11–7, he assumes (based on information available to him) for each species (1) the average age of maturity, and (2) the average age of death.

In Willoughby's formula, the average age of death is B, and the average age at maturity is A. Then the raes of the maximum potential age, C to B is C/B = 0.32 B/A + 0.6.

Using Willoughby's formula, and securing from Table 11–7 the average age of a horse at death (22 years) and the average age of a horse at maturity (5 years), the maximum predicted age of a horse is—

$$C/22 = [0.32(22/5) + 0.6], \text{ or}$$
$$C = 22 \times [0.32\ (22/5) + 0.6] = 44 \text{ years.}$$

By applying this formula to each species listed in Table 11–7, the maximum predicted ages shown in the right-hand column were obtained. Although recognizing that the Willoughby formula makes use of 2 assumed values (average age at maturity, and average age at death), in the opinion of the author of this book, it is a more accurate way to predict maximum age than the direct ratio method (1 to 7 dog to human ratio, for example.

TABLE 11–7
COMPARATIVE AGES IN YEARS, OF HUMANS AND SOME LOWER ANIMALS[1]

Species	Avg. Age at Maturity	Avg. Age at Death	Maximum Age	
			Reported	Predicted
(all males)	(years)	(years)	(years)	(years)
Human	21	70	115+	117
Bovine (cattle)	4	19	30+	40
Camel	8	31?	40+	57
Cat	1	10	35	38
Dog	1.5	11.5	34	35
Elephant, Asiatic	25	60	77+	82
Hog	1	8	13	25
Horse	5	22	62	44
Mouse	.182	2.2	6	9.8
Sheep	1.5	9	15+	22.7

[1]Adapted by the author. Also, the author added swine.

Collegiate contest in progress at the National Western Stock Show & Rodeo, Denver, Colorado. (Courtesy, National Western)

Fitting and showing dairy cattle. (Courtesy, California Polytechnic State University, San Luis Obispo)

SECTION

12

FITTING AND SHOWING LIVESTOCK[1]

Contents *Page*

There is no higher achievement than that of breeding and fitting a champion; an animal representing an ideal which has been produced through intelligent breeding and then fitted to the height of perfection. Also, it is realized that most of the advertising value of the show-ring accrues to those who exhibit winners, thus behooving the exhibitor to select, fit, and show animals to the best possible advantage.

The selection of the prospective show animals is the first and most important assignment. Some pointers relative to type or individuality are given in Section 11, Selecting and Judging Livestock.

The second requisite for successful showing is to fit the animals to the peak of perfection. Directions on feeding show animals and suggested fitting rations are contained in Section 4, Feeding.

[1]Grateful appreciation is expressed to the following person who contributed liberally of his time and talents in perfecting Section 12; as a reviewer, in providing source material, and in making the clever line drawings: Professor R. F. Johnson, Department of Animal Sciences and Industry, California Polytechnic State University, San Luis Obispo.

FITTING AND SHOWING TO WIN

Fig. 12-1. The judge is "toothing" a Brahma heifer to determine her age. (Courtesy, San Antonio Livestock Exposition, Inc., San Antonio, TX)

Fig. 12-2. Holstein heifers being shown by junior exhibitors at the San Antonio Livestock Exposition, Inc., San Antonio, TX. (Courtesy, San Antonio Livestock Exposition)

Fig. 12-3. Market lambs being shown by junior exhibitors at San Antonio Livestock Exposition, Inc., San Antonio, TX. (Courtesy, San Antonio Livestock Exposition)

Fig. 12-4. Angora goats being shown by junior exhibitors at San Antonio Livestock Exposition, Inc., San Antonio, TX. (Courtesy, San Antonio Livestock Exposition)

Fig. 12-5. Market barrow judging at the San Antonio Livestock Exposition, Inc., San Antonio, TX. (Courtesy, San Antonio Livestock Exposition)

Fig. 12-6. Shannon Dorstewitz showing a Half-Arabian in an English class. (Courtesy, International Arabian Horse Assn., Denver, CO)

Assuming that show animals have been carefully selected and properly fed, there yet remains the assignment of parading before the judge. In order to present a pleasing appearance in the show-ring, animals must be well trained, thoroughly groomed, and properly shown. Competition is keen, and often the winner will be selected by a very narrow margin. Close attention to details may, therefore, be a determining factor in the decisions.

Also, most successful exhibitors attempt (1) to select animals as old as possible within the respective age classifications, so that they may show to the best possible advantage, and (2) to fill as many show classes as possible, thus increasing their chances of winning. In order to meet these requisites, the exhibitor must be fully versed relative to the usual show classifications. The age and show classification of each species for each show may be obtained from the premium list.

FITTING AND SHOWING BEEF CATTLE[2]

Today's show trend favors large-framed, growthy, well-muscled, efficient, clean-cut, straight-lined, eye-appealing cattle

Pertinent information relative to fitting and showing beef cattle so as to accentuate these qualities follows.

TRAINING, GROOMING, AND SHOWING BEEF CATTLE

Table 12–1 is an illustrated guide for training, grooming, and showing beef cattle.

[2]The presentation on "Showing Beef Cattle," including Table 12–1, benefited from the authoritative review and helpful suggestions of the following persons: B. Bennett and M. Moore, BB Cattle Co., Connell, WA.

TABLE 12–1
TRAINING, GROOMING, AND SHOWING BEEF CATTLE[1]

GENTLING, HALTER BREAKING, AND POSING

Why: Such schooling makes it possible for the judge to see the animal at its best.

How: Proper training of the animal requires time, patience, firmness, and persistence.

First, the animal should be gentled. Easy does it! Here is how to gentle the animal with a minimum of stress. Place the animal in a relatively small pen where you can walk fairly close to it. Then, (1) carry a small bucket containing a few pellets of feed, which you should shake to attract the animal's attention; (2) hold one pellet between your thumb and forefinger, and feed the animal when it will take the pellet from you; and (3) repeat this procedure daily, or more often, until the animal gets sufficiently gentle that it can be hand fed and backrubbed. Never, never touch the animal on the forehead or horns, as this will teach it to butt.

Following the gentling procedure outlined above, the animal may be haltered. Halter breaking can most easily be accomplished at an early age, rather than waiting until the animal is stronger and has *a mind of its own.* Begin by tying the calf in the stall or corral with a rope halter (preferably, nylon ½ to ⅝ in. in diameter with a long lead). Leave it tied for a period of 1 to 2 hours, brushing and talking to it occasionally, so that it realizes that you aren't going to hurt it. When you tie the animal, be sure to leave just the right amount of slack, so that it is able to lie down without getting its legs caught in the rope. After the animal has learned to stand tied, which may take 3 to 4 days days, or much longer, start leading it around in the stall or corral. Later, lead the animal outside the stall or corral.

As an alternate method to tying the animal at the outset, some experienced fitters prefer to let the rope-haltered animal drag the long rope lead for 1 to 2 days, during which time it learns to accede to pressure when it steps on the rope repeatedly.

CAUTION: Never let an animal break away from you while you are training it; if you do, it will discover that it is master and you will have a problem animal.

Leading should be correctly done from the left side and with the halter strap or rope neatly coiled in the right hand. Beef exhibitors follow the custom of walking forward, glancing back over the right shoulder at frequent intervals. The rope halter is preferable when starting the training program, but it is very important that the animal become accustomed to being led with the leather show halter well in advance of the show. The latter precaution is important because the animal reacts differently when led with the show halter than when led with the rope halter.

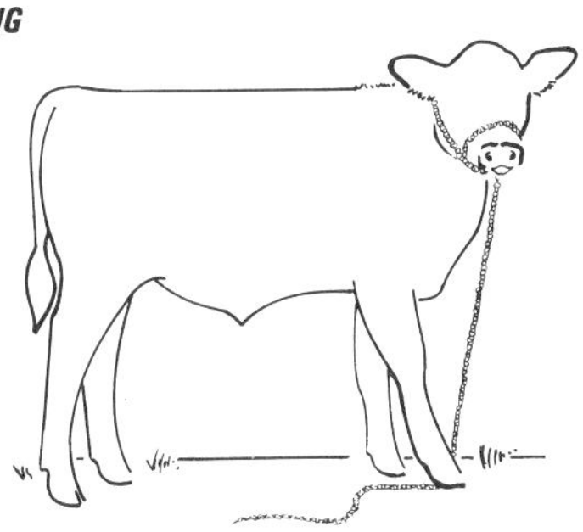

Fig. 12–7. Rope-haltered animal dragging the rope lead.

The next step is that of teaching the animal to stand or *pose* properly, so that the judge may have an opportunity to examine it carefully. For correct posing the animal must stand squarely on all 4 feet (preferably with the forefeet on slightly higher ground than the rear feet). The back should be held perfectly straight and the head held on a level with the top of the back. At first, this position may be quite strained and unnatural for the animal. For this reason, it should not be required to hold this position too long. Later, it should be possible to *pose* the animal for 15 to 20 minutes at a time. In *posing*, the exhibitor should hold the strap in the left hand and face toward the animal. A show stick is usually used in placing the hindfeet but the exhibitor's foot is best used in obtaining correct placement of the front feet.

TRIMMING AND CARING FOR THE FEET

Why: The feet should be trimmed regularly so that the animal will stand squarely and walk properly. Besides, long toes or unevenly worn hoofs are unsightly in appearance.

How: Where available, the animal should be secured in either (1) a tip-over type trimming table or chute, or (2) a tip-over hydraulic chute. Where tip-over equipment is not available, stocks may be used or the animal may be thrown. Use a foot nipper to cut the feet back, then use an electric sander to shape them. Some experienced fitters also use a chisel, farmer's knife, and rasp. Trimming should be done, or supervised, by a person with experience. In some areas, professional trimmers are available for hire.

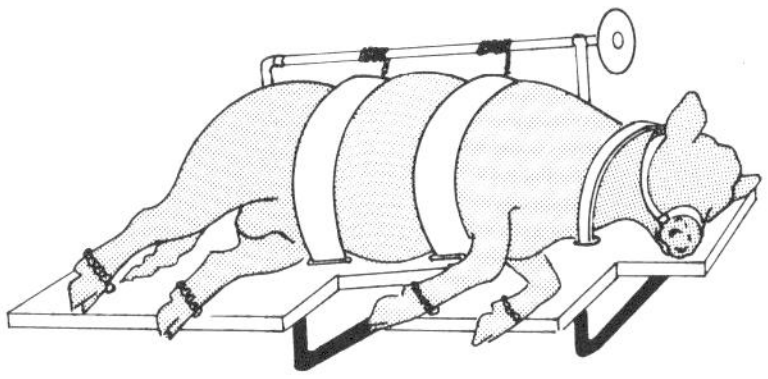

Fig. 12–8. Animal restrained on tilting hoof-trimming table.

(Continued)

TABLE 12–1 *(Continued)*

TRIMMING AND CARING FOR THE FEET *(Continued)*

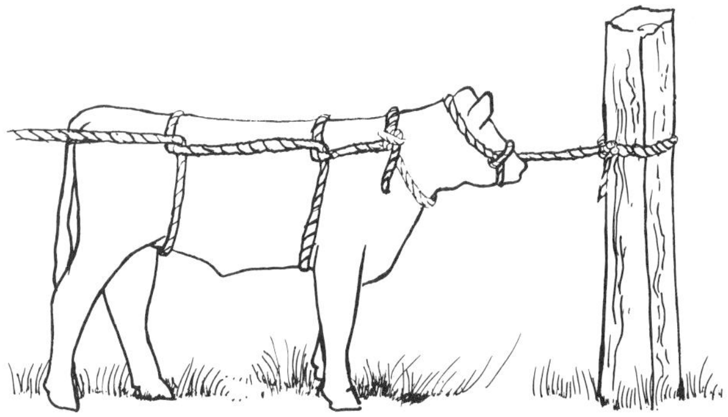

Fig. 12–9. A simple method of throwing an animal is illustrated here using a rope halter and a rope about 40 ft long. With the animal haltered, tie the halter shank or lead to a stout post, tree, etc. Then, with one end of the rope around the neck, tie a bowline (nonslip) knot. Next, circle the animal's body just behind the shoulder and a half hitch at the withers; continue the rope back to the loin and make a second half hitch and circle the rope around the body at the flanks. Make sure the second half hitch is just in front of the hooks. A strong backward pull on the rope will cause the animal to sink, and a shift in the pull to the side on which the animal is to fall will result in an easy, soft fall to the ground. Maintaining the strong pull on the rope will keep the animal lying on the ground, making it possible to do minor doctoring, foot trimming, etc.

The feet of some animals should be trimmed regularly as often as every 2 months. Too much trimming at any one time, however, may result in lameness. For this reason, it is not advisable to work on the feet immediately before the show.

• **Pointers on general trimming follow—**

1. Trim the inside toe and heel before trimming the outside toe and heel, because it generally grows faster and longer. Trim the toe before the heel to ensure that the animal will walk up on its toes.
2. Remove the outgrowth or rim of the sole around the edge of the toes and along the side of the foot with the nippers. Be careful to keep the foot level while trimming.
3. When the bottom of the foot is springy to the touch, the next cut will probably draw blood—and you have gone too far.
4. Shape the foot and all rough edges with a rasp. An electric sander generates too much heat and may seal the pores of the foot. However, a sander may be used for steers, because they are slaughtered at an early age.
5. The bottom of the foot, between the toes, should be hollowed out to allow mud, etc., to ooze up through the toes. This functions as a self-cleaning mechanism.
6. Make the side of the toes relatively straight on the inside by rasping between them.
7. Apply a disinfectant to any cracks or cuts in the foot, especially between the toes and along the hoof head, to aid healing.

• **Common leg problems and how to correct them follow—**

1. Bow-legged behind.
   a. Trim inside heel down.
   b. Trim rim off the inside claw out to the toe, but leave the toe long.
   c. Build up the outside heel.
   d. Trim the outside toe short.
2. Toe-out in front. Trim both the inside toe and heel short.
3. Pigeon-toed in front. Same as trimming for the bow-legged condition (see No. 1).
4. Cow-hocked behind. Same as trimming for the toe-out condition (see No. 2).

Note well: To correct leg problems, trimming should be performed regularly every 30 days.

• **Treating dry, brittle hoofs—**Quite often, when cattle are kept constantly in stables, the feet may become dry and brittle. This condition can usually be corrected by turning the animals out in a pasture paddock at night when there is dew on the grass. Packing the hoofs with wet clay, or applying neat's-foot oil will also be helpful in such cases. If the animal gets sore feet from standing in a filthy stable, the soreness should first be corrected. Following this, the feet should be washed and disinfected.

TRAINING AND POLISHING HORNS

Why: Horned bulls may be left horned. But the heifers of all breeds of beef cattle are generally dehorned soon after birth, using a dehorning paste.

On horned bulls, a well-curved set of horns will command the admiration of the judge, but poorly shaped horns will give the head a coarse, unattractive appearance.

How: There is a difference in the desired shape with different breeds. The horns of the Hereford should curve downward, whereas the horns of the Shorthorn should curve slightly forward and inward.

As soon as the horns are long enough (3 to 4 in.) and sufficiently strong to bear the weight, it is time to begin training. For this purpose ½-lb weights are usually used on small horns, ¾-lb weights for medium size horns, and 1-lb weights for bulls about 1 year old. Care should be taken to prevent making a sharp turn in the horn by using a weight that is too heavy or by allowing the weights to remain on for too long a time. If the horns yield too readily, it is best to remove the weights and give the horns a rest of from 10 days to a month, the length of time depending upon their condition. Then replace the weights until the desired effect is obtained. Weights should be removed when the horns become level with the top of the head, or not more than 1 in. below this level. Leaving the weights on longer tends to cause the horns to curve inward too much, thus causing a problem later. If the screw type of fastener is used, one should be careful not to force the screw into the horn so deeply that the depression cannot be removed. Horn weight losses may be reduced by tying a strong cord around the screws on the two weights; then if one weight is knocked off, it will not be lost. Horns may be pulled forward when they are 3 to 4 in. long by using a suitable spring or strap device for the purpose.

As an alternative to the use of horn weights for curving the horns, some professionals cut the horns of bulls at an angle about half way up on the horn, thereby causing them to turn down and curve nicely.

Extremely long horns may appear out of proportion and unsightly. In such cases, they can often be cut back as much as 2 to 3 in., provided not more than ½ in. is removed at any one time and at no more frequent intervals than a month or 6 weeks. As a rule, most of the black tip can be removed without harming the sensitive part.

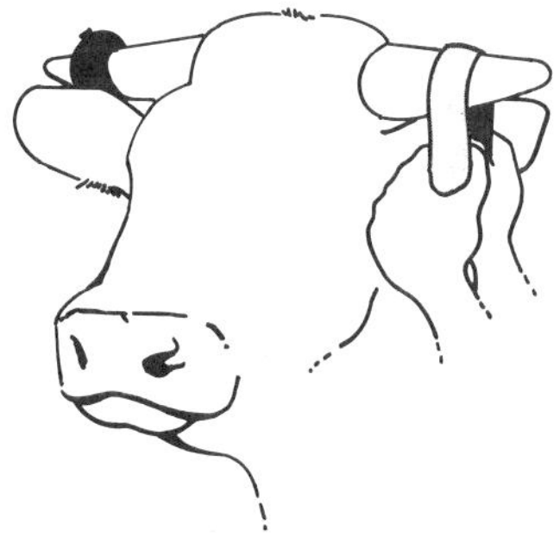

Fig. 12–10. Training and polishing the horns.

After the horns have been properly shaped, the next job is that of trimming and polishing. Usually it is best to smooth them down a week or two before the show. The rough surface may be smoothed with a sharp knife, a rasp, or a steel scraper; always scrape from the base toward the top. The final smoothing or finishing touches may then be given by using sandpaper, fine emery cloth, steel wool, or a flannel cloth and emery dust.

Horns are usually polished just before the show. An excellent polish that will not collect dust can be obtained as follows: Apply a paste which is made by mixing olive oil or sweet oil with pumice stone or tripoli. Polish by rubbing briskly with a flannel cloth. A quick and more simple polish can be obtained by the use of glycerine, linseed oil, or mineral oil. However, a polish obtained in this manner is rather temporary, and the oil will collect dust quickly.

(Continued)

TABLE 12-1 *(Continued)*

WASHING

Why: Frequent washing in the months preceding the show keeps the animal clean; stimulates a heavy growth of loose, fluffy hair; and keeps the skin smooth and mellow.

How: Beginning 4 to 6 weeks before the show, wash the animal regularly once a week. Before washing, dirt and debris may be removed by using a hot air blower designed for cattle, provided such equipment is available.

In preparation for washing, place a chain about the neck; never use a rope about the animal's neck when washing, for a wet rope cannot be easily loosened should the animal fall or otherwise get into trouble.

Most exhibitors use a high sudsing soap, such as Orvas or Castile, or a special livestock soap. However, if desired, an excellent preparation may be made as follows: mix 1 to 2 cups of good concentrated liquid coconut oil shampoo in 1½ gal of lukewarm water. With a bristle wash brush, thoroughly wet the dry animal with the soap solution. Then, with the hands and brush work the soap into a good lather, making sure that all parts of the body are well scrubbed and clean. Parts of the animal that are frequently neglected in washing are the head, tail, legs, brisket, and belly. Unless the animal is free of dandruff and the hide is thoroughly clean, double soaping is recommended.

Always wash the animal from head-to-tail and top-to-bottom. In washing the head, avoid getting soap or water in the eyes, ears, nostrils, and mouth. Cattle do not like to have their heads washed. A precaution commonly used in washing the head is to wash one side at a time while firmly holding the ear on the side being washed. Death of the animal may result from getting water into the lungs through the nostrils or mouth.

Following washing, the animal should be rinsed off very thoroughly in order to remove all traces of soap from the hair and skin, because soap left on the animal causes dandruff. For animals with light parts, a little bluing added to the last rinse water will improve results. Excess rinse water may be removed by using the back of a Scotch comb, a water scraper, or by brushing downward.

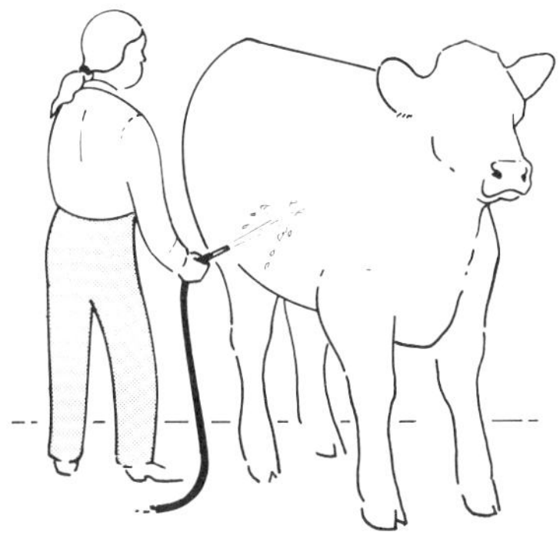

Fig. 12-11. Washing with soap and water.

Most experienced fitters discontinue the use of soap 1 to 3 days before the show so that the hair will be more manageable on show day. During this period, they keep the animal as clean as possible, spot wash or rinse with clean water when necessary, and use a blower to remove dust and dandruff.

After brushing the wet hair into place, a hot air blower may be used, if available. The excess water should be blown out in the same direction that the hair has been brushed. Brushing and blowing the hair in the same direction helps train the hair to stay in that position.

GROWING AND CARING FOR HAIR

Why: A good coat of hair can, in the hands of an experienced exhibitor, cover up a multitude of defects.

How: Proper blocking and hair care necessitates that there first be sufficient hair.

Growing hair, especially in hot weather, involves the application of the following practices, beginning several months in advance of the show:

1. Keeping the animal as cool as possible.
2. Stimulating hair follicle growth with plenty of brushing.

Hair growth can be stimulated by wetting (*not* washing with soap) an animal down every morning and evening, during the coolest part of the day. After completely soaking the animal, the bar of the Scotch comb should be used to remove excess water before combing and brushing.

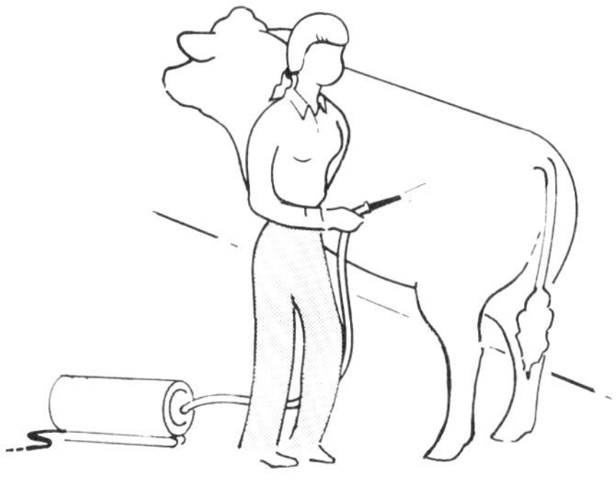

Fig. 12-13. Blowing the hair. This is done either (1) to dry the hair, or (2) to remove dust at the time of final grooming for the show.

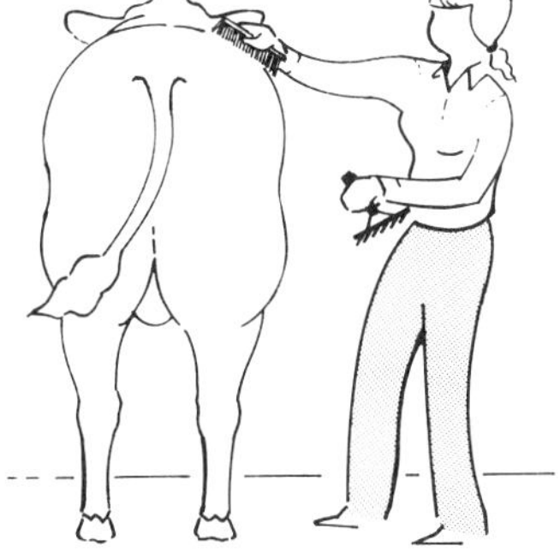

Fig. 12-12. Grooming with Scotch comb and brush.

While the hair is still damp, spend at least 20 minutes brushing and 10 minutes Scotch-combing. Never brush or comb down on long-haired animals the object is to get the hair to stand up and out so as to add extra thickness and dimension. If you have a blower, alternate the brush and comb routine with some blowing in the direction you're brushing. When the hair is dry, apply a mist of water or hairset mixture; then brush, Scotch comb, and blow some more. For the first 2 weeks, work the hair forward and slightly down; during the second 2 weeks, work the hair straight forward; during the third 2 weeks, work the hair forward and upward to a 45° angle.

Remember that the key to a well-trained haircoat is the three Bs—*brush, brush, brush*. You cannot overbrush. Brushing stimulates the hair follicles and causes extra hair growth.

After working the hair, tie the animal in the shade where there is some air circulation. Better yet, tie it in an open barn or shed with a fan blowing on it to increase air circulation.

If the caretaker does not have sufficient time to grow hair and block, as outlined above, a popular alternative is to show *slick*. This consists of brushing all the hair on the body straight down, except for the legs and tailhead; saddle soap is applied to the legs and tailhead and the hair is pulled up with a Scotch comb (see Fig. 12-14).

(Continued)

TABLE 12-1 *(Continued)*

CLIPPING AND BLOCKING

Why: Clipping and blocking are designed to accentuate the animal's most desirable qualities and to minimize its weaknesses. To this end, successful fitters study each animal; they observe it when walking, when standing relaxed, and when posed with the show stick. The fitter forms a picture of the animal as nature made it, and visualizes how it can be changed to approach the ideal. Following this preliminary study, the careful fitter plies the art, step by step, to emulate the ideal counterpart. Each bit of clipping and blocking is designed to assist in molding the final form. Each animal is treated as an individual, and the job of clipping and blocking is tailor-made for that individual.

How: Professional show people agree that no other assignment in fitting and showing cattle requires so much skill, patience, and time as does the art of proper clipping and blocking. The best way in which to master this art is to watch and emulate the experts.

Final clipping is best done about a week before the show, so that the clipped hair will lose its stubby, *fresh haircut* appearance.

In preparation for clipping and blocking, the animal may be tied to a stout fence. However, if many animals are involved, most showpeople build or buy a simple metal pipe blocking chute with a headgate.

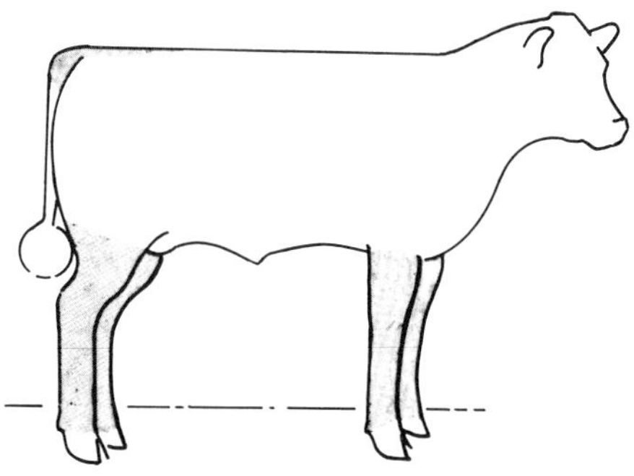

Fig. 12-14. For cattle with short hair, train and pull the hair up as shown in the shaded areas.

• **Clipping**—Standard straight-edge clippers are used to clip the head, brisket, underline, and tail. Clipping these areas makes the animal look neat and trim, and not wasty.

Custom decrees certain breed differences in clipping and grooming (in *haircuts* and *hair-dos*), and differences between steers and breeding animals. These will be pointed up in the discussion that follows:

1. **Head.** Head clipping styles vary, depending upon breed and sex. For cattle on which heads are normally clipped (show steers and Angus), use an electric clipper and clip in front of a line that starts directly back of the ear; however, the long hairs on the poll should be left so as (a) to give the poll more prominence, and (b) to give the head a longer appearance (see Fig. 12-15). On Herefords (horned and polled), heads are clipped from the jaw forward but leaving the hair on the front of the face longer. On Shorthorns, heads are clipped except for point of the poll, where a "top notch" is left. Other breeds vary; hence, professional advice should be sought relative to clipping them. Do not clip hair from the ears, the eyelashes, and around the nose.

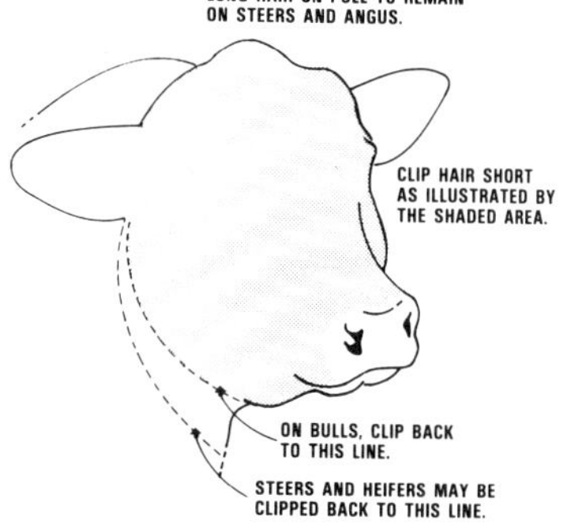
LONG HAIR ON POLL TO REMAIN ON STEERS AND ANGUS.
CLIP HAIR SHORT AS ILLUSTRATED BY THE SHADED AREA.
ON BULLS, CLIP BACK TO THIS LINE.
STEERS AND HEIFERS MAY BE CLIPPED BACK TO THIS LINE.

Fig. 12-15. Clip the head and face as shown (shaded areas), but do not clip the ears, the eyelashes, and around the nose.

2. **Brisket and underline.** Most beef animals need to have their briskets and underlines clipped, so that they appear trimmer. Clipping the belly also imparts the impression that the animal has more altitude. It is especially important that steers look trim, and not wasty. Also, the belly, brisket, shoulders, and front legs of breeding heifers may be clipped to emphasize femininity. If the heifer is shallow bodied, just clip the long hairs, and do not clip too close on the belly.

The brisket and underline should be clipped 3 to 4 weeks in advance of the show. Also, such clipped areas should gradually blend into the longer hair of the middle, neck, and shoulders, thereby avoiding any unartistic clipper line. This requires a steady hand. Also, the blending can be facilitated by use of a clipper guard (a plastic comblike attachment) or by use of thinning shears.

3. **Tail.** One of the main objects in clipping the tail is to show the fullness of the twist and the thickness or beefiness of the hindquarters. In order to do this to best advantage with each individual, good judgment should be exercised in determining the extreme points of clipping that will show these characteristics to advantage. A general guide is to clip the tail from the high point of the twist to the tailhead (see Fig. 12-16). At the tailhead, the hair should be gradually tapered off near the body so that the tail blends nicely with the rump. To avoid leaving ridges of long hair extending down the center of the tail, the clipper should be run across the tail after it has been run upward to the tailhead.

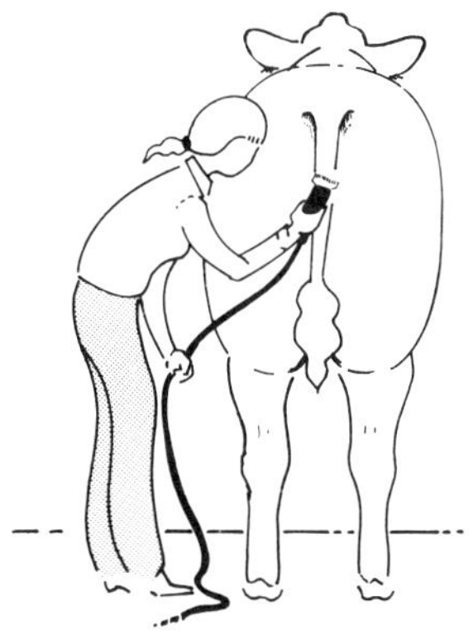

Fig. 12-16. Clipping the hair from the tail of an Angus. One of the main objects in clipping the tail is to show the thickness or beefiness of the hindquarters. Clipping should begin above the switch of the tail opposite the point where the twist is fullest; and extend to the tailhead where it gradually tapers off, giving a blended effect with the rump.

• **Blocking**—After the head, brisket, underline, and tail have been clipped, the animal should be blocked. Blocking is done to emphasize the animal's strong points—to make the animal appear at its best. Usually, a different type of clipper other than the standard straight-edge type is used for this job; the preferred clipper for use in blocking is known as a sheep-head clipper. However, when putting the finishing touches on an animal, a smaller, quieter set of animal clippers may be preferred.

Blocking the animal consists of clipping over the top and down the sides, and in blending in the points where hair has been closely clipped.

If the animal is to be blocked all over, care must be taken to insure that the hair is in the proper position before clipping. Sweeping the hair forward—from the rear quarters, along the sides, and to the shoulders—makes an animal look fatter.

Start at the animal's underside, and blend the underline into the hair on the side by gradually tapering the length of the hair until there is no longer a definite line. Shorten the hair through the middle, fore and rear flanks, and through the neck and shoulders to make the animal appear trimmer. Leave the hair on the rear quarters a little longer in order to make the animal appear very heavily muscled. In order to achieve the desired smooth look, professionals work both horizontally and vertically with the clippers, until they get to the topline of the animal.

(Continued)

TABLE 12-1 *(Continued)*

CLIPPING AND BLOCKING *(Continued)*

Work the topline hair forward with a Scotch comb, getting it to stand up. Clip the topline hair so that the topline is as straight and level as possible as viewed from the side, and is rounding and muscular as viewed from the rear. Also, the topline hair should be blended into the tailhead, shoulder, and crest.

Professionals also block the following:

1. **Front legs.** The hair from above the knees should be clipped so as to blend smoothly into the forearm. Likewise, the hair should taper from the knees down; and long and bunchy hair should be lightly trimmed.

2. **Hind legs.** Pull the hair on the back legs up with a hard rubber brush or Scotch comb; and lightly trim the long and bunchy hair. Also, the legs should be trimmed to give the appearance of proper set and heavy bone. Trim the area between the hock and twist, and also above the hock on the outside of the quarter so as to give it more bulge and expression of muscle. As viewed from behind, clip the top to give a round muscular appearance rather than a flat, square one.

CAUTION: Go slowly, and don't take off too much hair. Remember that you can always go back and take off more hair, but you cannot put back what you have already cut off. At intervals, step back and take a critical look at your work. Remember, too, that a good blocking job is an art, and that it takes plenty of time and patience.

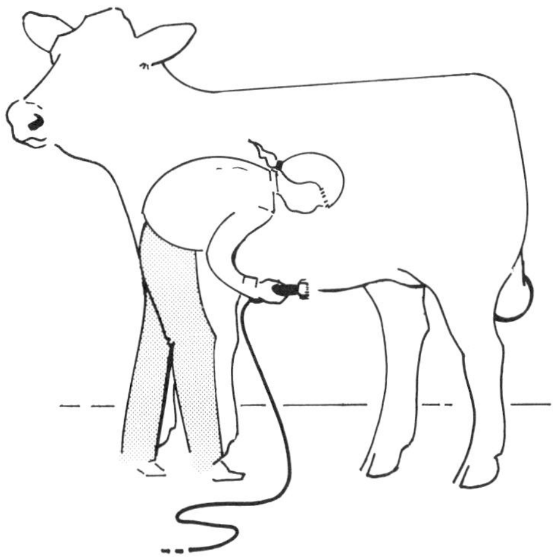

Fig. 12-17. Trimming the underline and blocking the animal along the sides and back is practiced by most fitters to help complement the good points and diminish the bad points of an animal.

SHOW DAY

Why: Show day gives point and purpose to all the hard work, long hours, and patience that has gone before. So, the final touches on show day may make the difference between a winner and a loser.

How: On show day, the exhibitor should be well rested and get up early. The show animal(s) should be rinsed (*not* soaped) early and dried (a blower-dryer hastens the drying). Feed and water at the regular time. If the animal has a heavy middle, limit the feed and water, or do not feed and water. If the animal is tight middled, feed and water well. After feeding, let the animal rest.

Keep in mind that the natural look is in vogue now.

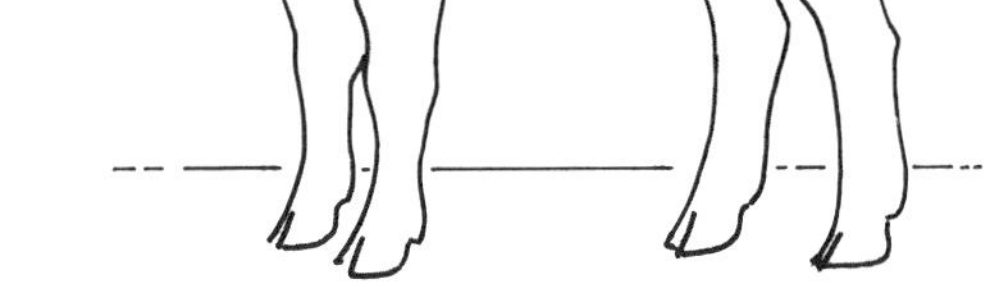

Fig. 12-18. The *natural look* of the well-groomed animal.

About 45 minutes before entering the ring, put the show halter on the animal; make sure that the nose strap is adjusted correctly. Blow or brush all dust and straw off the animal. Then proceed as follows:

1. **Brush the body.** The method of brushing the hair on the body depends on the length of the hair.

A long-haired animal should be groomed (brushed and Scotch-combed) forward and up (at about a 45° angle), following which the hair should be sprayed with a hair-setting product. (several of these products are available through livestock supply companies.) If the hair is unruly, spray lightly with adhesive. But work quickly with a Scotch comb, because adhesive sets fast.

If the animal has short hair, the body hair should be brushed down and smooth. Brush the hair up on the lower round; brush the hair on the shoulder and neck of heifers down to emphasize femininity; comb the hair on top of the tailhead up so that the tailhead looks square.

2. **Topknot the poll.** Pull the poll topknot up with your hand, and spray with adhesive.

3. **Prepare the tailhead.** The tailhead needs special treatment, also. Brush up the tailhead hair; spray with adhesive; bring up again with your hand and a Scotch comb; and trim with the clippers and scissors so as to level out the topline. The objective is a natural look.

4. **Bone the legs.** Many judges still try to select heavy boned cattle in breeding classes, despite the fact that studies reveal (a) that bone is of little importance in a selection program, (b) that bone varies little between animals of the same age, and (c) that bone cannot be determined by *eyeballing*, because what appears to be bone is really a combination of hair, hide, connective tissue, tendon, and bone. Nevertheless, the exhibitor is showing before the judge, not educating the judge. Hence, the exhibitor should, to the extent possible, please the judge. Since many judges insist on evaluating bone in breeding classes, and since no judge will place an animal down for appearing to have heavy bone, it is good business to accentuate the bone. This is done by *boning the legs*.

The legs are boned for two reasons: (a) to impart an illusion of bigger bone, and (b) to make the legs appear more correctly set. First, rub a bar of saddle soap (either clear

(Continued)

TABLE 12–1 *(Continued)*

SHOW DAY *(Continued)*

or colored according to the color of the animal) on the legs. Next, pull the hair on the legs upward, to the forearms in front and to the stifle behind; using a Scotch comb. Then spray with an adhesive, followed by running a Scotch comb through immediately. **Note well:** With heifers, only bone the hind legs. Brush the hair on the front legs straight down to give a feminine appearance.

5. **Rat the tail.** Rat the tail into a tight ball by starting at the top of the tail and ratting everything high and to the center. Trim off excess long strands and spray with adhesive. The ball should be done up so that it falls just below the quarter when the tail hangs naturally. Also, the ball may be shaped according to what will look best on the individual. On cattle that are long and flat quartered, the long, narrow ball is best. On cattle that are short quartered, the fuller ball will give more depth to the quarter when viewed from the side.

6. **Oil the coat lightly.** Apply a light coat of oil either by (a) using a rag, oiled with a spray bottle; or (b) spraying oil on the animal, then brushing with a soft bristled brush. Do not show an animal with too much oil. This will give the coat an extra sheen so that the animal will stand out in the show-ring. CAUTION: Do not get oil on any area with adhesive on it, because the oil will take the adhesive out.

7. **Paint the hoofs.** The hoofs may be spray painted; using black lacquer on black cattle and clear lacquer on the rest.

8. **Make a final inspection.** Before entering the ring, give the animal a final check. Use the clipper and scissors to remove unruly hairs. Are the ears brushed and cleaned out? Did you remember to oil the face? Is the tail up and in shape to stay? *Remember, a natural-looking animal is desired.*

SHOWING

Why: To win, think positive. When the class is called, enter the ring with confidence and a winning attitude. Be proud of the fitting that you have done.

How: Expertise in showing livestock cannot be achieved through reading any set of instructions. Each show and each ring will be found to present unusual circumstances. However, there are certain guiding principles which are always adhered to by the most successful cattle exhibitors. Some of these are:

1. Train the animal long before entering the ring.
2. Have the animal carefully groomed and ready for the parade before the judge.
3. Dress neatly for the occasion.
4. Enter the ring promptly and in clockwise direction when the class is called.
5. Lead the animal from the left side (walking near the left shoulder), with the halter strap in the right hand.
6. When asked to line up, go quickly but not brashly.
7. When stopped, pose the animal correctly, and so as to minimize faults. Take the strap in the left hand and set the animal up with a leg under each corner. Generally, it is best to set the hind feet before setting the front freet. Keep the animal's head up and the back straight. A firm pressure near the navel, applied with the show stick, will help keep the weak-backed animal straighter. Animals with high loins can be pinched down with your fingers to straighten their tops. Cow-hocked animals can be made to look straight by pulling on the hocks with your hands.
8. Stroke the animal along the back or under the belly while posing, calming it.
9. When the judge handles your animal, react properly. If you feel that your animal may be slightly overdone, or too soft, turn its head away from the judge—thereby imparting firmness to the touch. If you think your animal is too bare, turn its head toward the judge—thereby imparting softness. After the judge handles the animal, comb the hair up where the judge handled it.
10. Keep one eye on the judge and the other on the animal. Center your attention entirely on showing the animal. The animal may be under observation of the judge at a time when you least suspect it.
11. Let the animal stand *at ease* if you are in a big class and the judge is working at the other end of the ring. Calm it by scratching it with the show stick.
12. Never stand so that you block the judge's view; the judge is interested in seeing your animal—not you.
13. If you find that you are hemmed in and that the judge cannot see your animal, move to another location of vantage, unless, of course, the judge has asked you to hold your position.
14. Keep calm and collected. Remember that the nervous exhibitor creates an unfavorable impression.
15. Work in close partnership with the animal.
16. Be courteous and respect the rights of other exhibitors.
17. Do not enter into conversation with the judge. Speak to the judge only when you are asked a question; and never question the judge's placings.
18. Be a good sport. *Win without bragging and lose without squealing.*

Fig. 12–19. Showing.

[1]All the line drawings in this table were prepared by R. F. Johnson, Department of Animal Sciences and Industry, California Polytechnic State University, San Luis Obispo.

FITTING AND SHOWING DAIRY CATTLE[3]

Showing is an important part of the purebred dairy business. It provides an excellent means of publicizing individual herds, of evaluating and comparing breeding stock, and of observing and evaluating the type of a bull's daughters. Additionally, a number of breed programs are furthered through dairy cattle shows; and many members of 4-H Clubs and the National FFA acquire and build interest in dairy cattle through exhibiting animals and participating in judging contests at dairy cattle shows.

TRAINING, GROOMING, AND SHOWING DAIRY CATTLE

Table 12–2 is an illustrated guide for training, grooming, and showing dairy cattle.

[3]The presentation on "Showing Dairy Cattle," including Table 12–2, benefited from the authoritative review and helpful suggestions of the following persons: Professor L. Ferreira, Dairy Science Department, California Polytechnic State University, San Luis Obispo; and K. Melvold, Fresno, CA.

TABLE 12–2
TRAINING, GROOMING, AND SHOWING DAIRY CATTLE[1]

SELECTING AND FITTING

Why: Selection of the prospective show animal is the first and most important assignment. Unless the right kind of animal is selected, no amount of fitting and showing can make a champion.

The second requisite is proper fitting so as to enhance the attractiveness of the animal, without excessive fatness.

How: Select show animals and begin preparing at least 2 months ahead of the show. Try to select animals that will be "full aged" in the class. Heifers sired by plus-proven sires and from high index (type and production) dams cost more, but they will have a higher retail value. Separate show animals from the rest of the herd.

Animals with excess condition will be placed down in the show because of lack of dairy character. So, if the animal is carrying excess condition (coarse at the withers, patchy over the pinbones, throaty, or fat), place her on a low maintenance ration, If the animal is thin and in poor condition, feed extra grain daily. Some young animals will grow faster than others, so carefully observe their growth pattern. Feed plenty of hay since this will develop capacity and body depth. The grain mixture can be any homegrown grain (corn, oats, barley, etc.) plus minerals.

BRUSHING DAILY; BLANKETING

Why: Brush to encourage the hair to lie flat and appear smooth and sleek.
Blanket to keep the animal clean.

How: With a brush and a woolen cloth. Use a curry comb sparingly.
Use a thick blanket, either a commercial or a homemade one.

CLIPPING

Why: To accentuate quality and dairy character.
Clip the *tail* to accentuate the switch and make the tail appear slender.
Clip the *legs* to accentuate quality of bone and correct stance.
Clip the *udder* to bring out veining and show quality.
Clip the *head and neck* to impart a clean-cut appearance.
Encourage hair to grow on the topline and clip so as to straighten the topline. Give the vertebrae a sharp appearance.

How: Do not clip the entire animal (1) unless it has an extremely rough hair coat, has not lost the winter hair coat, has stained areas, or shows excessive sun bleaching; or (2) if the show is less than 2 months away.
Clip the animal on the tail, legs, udder, head and neck.
Start at the top of the switch and clip up the tail against the grain of the hair to within 4 to 5 in. of the tailhead. Clip the tailhead with the grain of the hair (referred to as blending) being careful not to call attention to defects in the rump region. If there is a high area, clip closely. If a low point exists, leave the hair.

Clip the inside of the rear legs from the hoof to the body against the hair. Clip the outside and back of the rear legs from the hoof to the hock closely against the grain of the hair. Take advantage of natural lines and attempt to correct the legs by carefully removing hair. The blood vein in the hock region makes an excellent point for blending.

Clip the entire udder, except leave hair on the fore-udder if the cow has a bulgy fore-udder. Do not clip the belly of calves or heifers since it makes the animals appear shallow bodied.

Extremely long, woolly hair can be removed by holding the clipper away from the body when clipping. The belly on cows should be clipped only enough to show milk veining to advantage.

Clip the head and neck closely against the grain of the hair.

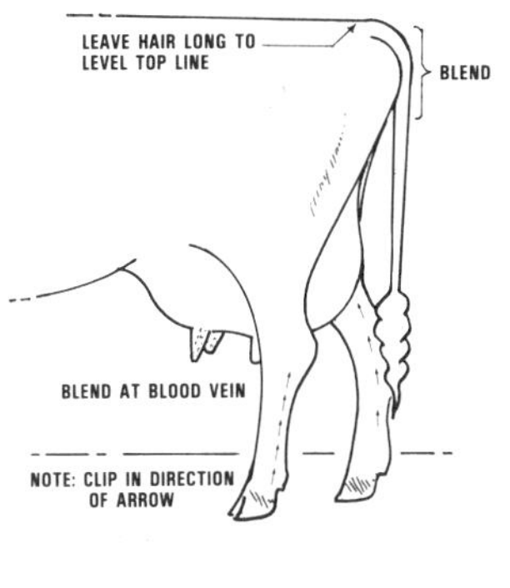

Fig. 12–20. How to clip the legs and tail.

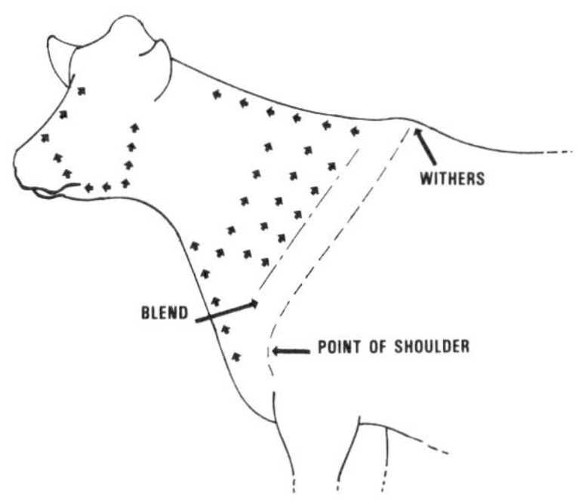

Fig. 12–21. How to clip the head and neck.

(Continued)

TABLE 12–2 *(Continued)*

CLIPPING *(Continued)*

Clip the area forward of a line formed by the point of the shoulders and the front of the withers. The natural crease formed by the neck and shoulders can be used to blend the long hair. The hair left on the withers should be clipped to create a sharp triangular appearance. Care should be exercised when clipping about the head. Hair should be removed from inside the ears. If the animal will not allow the clipper, use scissors to clip the long hair in the ears. Be careful with the scissors!

At intervals, lubricate the clipper by dipping the blades in a shallow widemouthed can of light oil or kerosene to remove dirt, dust, and minimize wear and dulling.

When a good job of clipping has been done, no lines are visible where blending has occurred.

TRIMMING HOOVES

Why: So that the animal will stand squarely and walk properly, and that the feet will be clean and attractive.

Long toes increase chances of foot rot, punctures from the sole, and broken or cracked toes. Each of these leads to lameness in the animal.

The bottom or sole of the foot should be trimmed so that the animal stands squarely on her feet with the wall (outside shell) supporting her weight.

How: The normal foot should appear as shown in Fig. 12–22, top; it should be well rounded (see A), have short toes (see B), and be deep at the heel (see C). The proper leg formation puts the leg directly under the weight it is to support (see D). Be familiar with the bone structure in the foot, and see that the weight of the animal is carried properly by that structure (see E).

The bottom drawing of Fig. 12–22 shows a foot that needs attention. The dark area shows the amount of growth that needs to be removed. When the toes are too long, it causes the animal to carry too much weight in an unfavorable position; too much weight is on the heel and not enough on the toes.

If hooves are too long, shorten them with a hoof trimmer, nipper, or a chisel and mallet. A hoof knife (tool with a U-shaped tip) will remove excess growth and is safer than other tools. The foot can be smoothed with a rasp. Wear leather gloves to protect your hands.

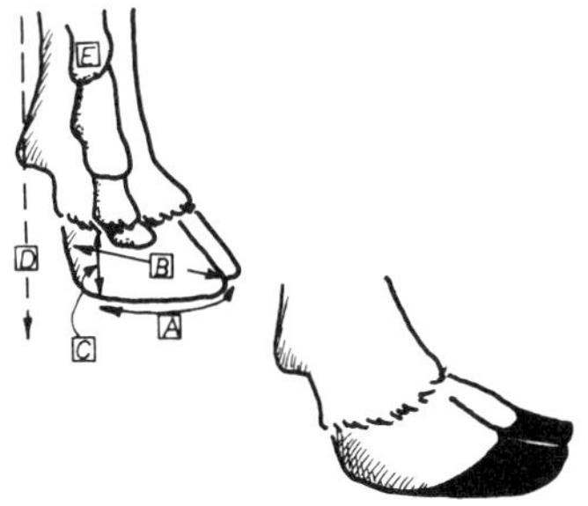

Fig. 12–22. *Top:* normal foot. *Bottom:* foot that needs attention.

The following methods can be used to work on the feet: Restrain the animal in a set of stocks or a restraining table, or throw it. Work on one foot at a time. Proceed with care to avoid injuring yourself or your animal. Final trimming of hooves should be done at least two weeks prior to show, so animal will walk normally on show day.

WASHING

Why: To make and keep the animal clean.

How: Use a mild soap (detergent can cause skin irritation) and scrub with a coarse brush. Avoid getting water in the ears and slowly accustom the animal to water (avoid cold water and high pressure hoses). Remove all soap with a good rinse and extra water with a squeegee. Blanket the animal. Avoid washing more than once a week since it removes the animal's natural oils and coarsens the hair. Wash the stained areas and switch frequently until clean. Check the poll and head area since dirt accumulates there quickly.

Remove wax and dirt from inside the ears with a clean cloth dampened with rubbing alcohol.

OILING

Why: To impart "bloom" to the animal. Oil is not needed unless the hair is unusually dry and lifeless.

How: Several good commercial oils are on the market. However, a good homemade oil can be prepared by mixing equal parts of glycerine, sweet oil, and rubbing alcohol.

If the animal is oiled, use it only on dark hair because it may yellow white hair.

TRAINING

Why: Because practice makes for perfection, and proper training is necessary if the animal is to be shown to advantage.

How: Practice well ahead of the show. The exhibitor and the animal must work as a team, each knowing what to expect of the other. Get the animal used to the halter that you will be using in the show. Train it to walk slowly one half step at a time in a clockwise direction. Always keep the animal between you and the judge. The lead strap should be on the left side of the animal loosely coiled in one hand (not wrapped around trapping your hand). The other hand should grasp the halter next to the head of your animal for control. You should be able to walk forward and backward with your animal under complete control and in view of the judge at all times. When you stop your animal, her feet should be positioned correctly as follows:

Calves and heifers: The front feet should be parallel or straight across from each other. The hind foot *nearest* the judge should be one half step BACK. Usually this is the right rear leg (the heifer appears longer and more stretchy this way).

Cows: Again the front feet are parallel, but the hind foot *nearest* the judge is one half step *FORWARD* (this is normally the right hind leg). This allows the judge to see the rear and fore udder attachments.

All movements and positioning of the animal and her feet should be done with halter commands (not with your feet or body pushing). Allow 2 to 4 ft between you and the animal ahead of you when circling. When you are called in by the judge, line up closely to the animal next to you (less chance of the judge placing another animal above you). Never stop your animal with her front feet in a hole or going downhill since it makes her look smaller.

AT THE SHOW

Why: In order to win.

How: At the show, you should know the answers to the following questions:

1. What time does the show begin, and when do you show?
2. Animal's birth date, days in milk (if lactating), calving date, and sire.
3. Are all health, registration, and entry papers in order and checked in?

The morning of the show, follow the same feeding, watering, bedding, and grooming routine established at home. Just before going into the ring, give her a final drink of water, but watch her sides (good spring of rib, but not rounded). If your animal dislikes the water, add a little molasses to cover up chlorine, mineral, and other tastes. You may do this at home to get your animal used to the molasses-tasting water.

Be prompt and ready to go into the show-ring. Watch the judge at all times and follow directions closely. Present your animal to best advantage, moving slowly, and keeping her between you and the judge. Once you are called in to line up, move smartly, but do not run. Don't cut in front of someone who is placed above you. Above all, exhibit good sportsmanship and a positive attitude. Be a modest winner and a gracious loser.

[1]The line drawings in this table were prepared by Professor R. F. Johnson, Department of Animal Sciences and Industry, California Polytechnic State University, San Luis Obispo.

FITTING AND SHOWING SHEEP[4]

Pertinent information relative to showing sheep is summarized in the two sections which follow.

TRAINING, GROOMING, AND SHOWING SHEEP

Table 12-3 is an illustrated procedure for training, grooming, and showing sheep.

[4]The presentation on "Showing Sheep," including Table 12-3, benefited from the authoritative review and helpful suggestions of the following persons: B. Lane, Extension Livestock Specialist, University of Missouri, Columbia; and Dr. D. Nelson, Department of Animal Science, California State University-Fresno)

TABLE 12-3
TRAINING, GROOMING, AND SHOWING SHEEP[1]

GENTLING AND POSING

Why: In order that the judge may see the animal at its best.

How: Hold the sheep by grasping the wool lightly under the chin with the left hand; standing or squatting to the left and front of the sheep.
Train the animal to stand with the legs squarely placed, the back straight, and the head held erect.

Fig. 12-23. Proper holding.

TRIMMING THE FEET

Why: To keep the sheep strong and straight on the pasterns.

How: Trim feet at approximately 2-month intervals or as required. Ideally, trim feet a few days to one week prior to the show. Hold sheep in a sitting position and trim hooves with a pair of sheep hoof trimmers or a sharp knife.

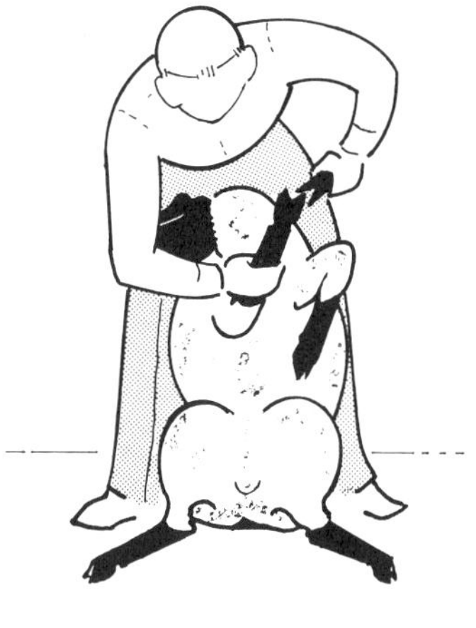

Fig. 12-24. Trimming position.

BLANKETING BREEDING SHEEP

Why: To keep the fleece clean, to "condition" it, and to keep it compact.

How: By using either (a) a commercially-made show blanket, or (2) a homemade blanket made from heavy cotton sacks. Hoods may also be helpful in keeping the head and neck clean.

Fig. 12-25. Blanketed sheep.

(Continued)

TABLE 12-3 *(Continued)*

DECIDING HOW AND WHAT TO TRIM

Why: In order to enhance the attractiveness of the individual through accentuating its strong points and covering its weaknesses.

How: Carefully study each individual animal and visualize how it could and should appear after trimming.

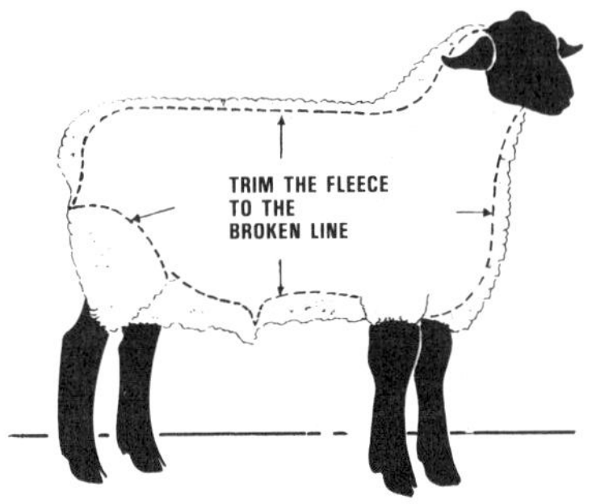
TRIM THE FLEECE TO THE BROKEN LINE

Fig. 12–26. How and what to trim.

SECURE ANIMAL ON TRIMMING STAND

Why: To hold the sheep in approximate show position.

How: The trimming stand or blocking table, which may be homemade or purchased, is a platform on which the animal is stood and which has a headstand with yoke at one end for securing the head in approximate show position.

Fig. 12–27. Sheep on trimming stand.

WASHING THE SHEEP FOR SHOW

Why: To prepare the animal for trimming or clipping. Only meat-type breeds are washed. Wool-type breeds are *not* washed.

How: Step by step, wash meat-type breeds as follows:

1. Prepare the pen or stall where sheep will be placed after washing. Use clean straw, and make sure that the walls of the pen are clean.
2. Trim hooves.
3. Shear the belly.
4. Place sheep on trimming stand. Never leave animal unattended while on the stand. Wash about 4–5 days prior to show if possible.
5. Run water, with hose, over entire body in this pre-rinse stage. Protect ears and eyes when around those areas.
6. Wash with a special livestock soap, such as ORVUS, or with a liquid dishwater soap that is white or clear in color. Use plenty of soap, lather up and wash entire sheep. If animal is especially dirty, it may be necessary to rinse and wash certain areas again. Usually however, one washing will be sufficient, provided it is done right.
7. Rinse soap out of fleece. This will take some time, but it must be done right. Use *moderate* water pressure. With high pressure, wool fibers will be more difficult to card out later.
8. Curry as much water out of fleece as possible following rinsing. A clean curry comb can be used for this purpose.
9. Card the fleece out, using a wool card. This will facilitate drying and make later trimming easier.
10. Blow-dry the animal, using a hot blow dryer, or place it in a clean pen to air dry.
11. Blanket and hood the sheep after it is relatively dry in order to keep it clean.

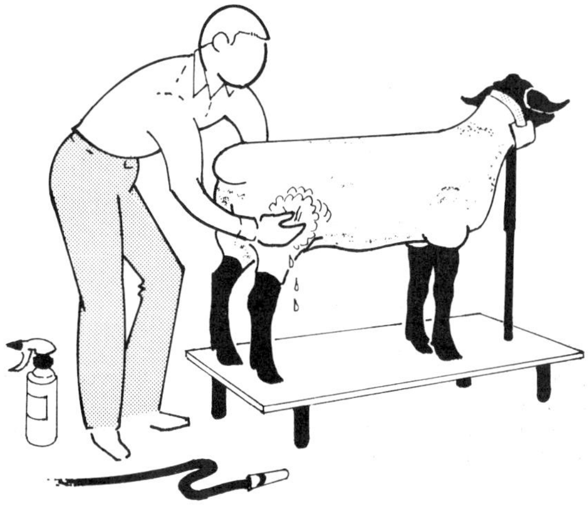

Fig. 12–28. Washing.

(Continued)

TABLE 12–3 *(Continued)*

TRIMMING THE SHEEP

Why: To impart the desired appearance and to result in a blending together of all body parts.

How:

• **Breeding sheep**—If done properly, a breeding sheep needs to be trimmed only once prior to a show. Of course, it may be necessary to go over the same body parts more than once to get a well-finished animal. Plan to finish trimming the day before the show if possible. Check individual show rules regarding fitting practices, shearing dates, and other rules, in order to be in compliance.

1. **Breeding Sheep (meat-type breeds).** Trimming procedures:
   a. Place the sheep on the stand.
   b. Brush off all dirt, straw, hay, and other foreign material.
   c. Spray the area to be trimmed with water and soap (should be slippery) solution.
   d. Follow a trimming sequence. There is no hard and fast rule relative to the sequence, or order, but experienced fitters generally develop a sequence. A common sequence is: (1) back, (2) sides and belly, (3) hindquarters, (4) neck and breast, and (5) head.
   e. Use suitable shears. Many types and sizes of shears are available. Commonly, 6½ in. rigged (those with a leather hand strap) shears are used.
   f. After carding the wool fibers out straight, trim them off smooth. Remember what the ideal sheep looks like so you can shape the animal correctly as well as get it smooth.
   g. After getting the animal shaped correctly, and the fleece trimmed smooth, use a packing card (straight tooth) and "pack" the fleece after respraying with water and soap solution. This will help "hold" the job in place.
   h. After packing, blanket and hood the sheep, and place it back in a clean pen.
2. **Breeding Sheep (wool breeds).** Trimming procedure:
   a. Do not wash.
   b. Have about a 90 to 120 day fleece to work with at show time.
   c. Shear the belly as indicated in the meat-type section.

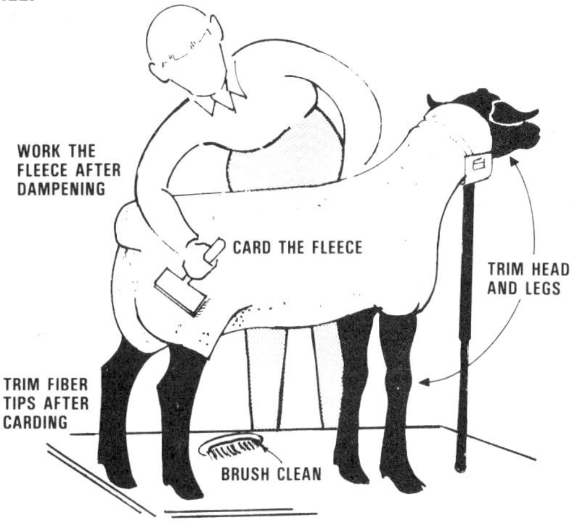

Fig. 12–29. Trimming.

   d. Use trimming sequence as shown in meat-type section.
   e. Trim hooves prior to trimming fleece.

• **Club Lamb(s) or Market Lamb(s)**—Trimming procedure:

1. Shear completely 45 to 50 days prior to the show. Be sure the lamb is sheared smoothly.
2. Wash about 2 days before the show.
3. After drying, use a pair of "shearmasters" with 20 tooth comb and reshear the animal.

GIVING PRESHOW TOUCHING UP

Why: To remove any marks or irregularities in the fleece caused by shipment or blanketing, and to impart freshness and bloom.

How: Dampen the fleece, pack with packing card (which will help smooth out the fitting job), touch up where needed with hand shears, wipe off face and legs of black face sheep with oil solution.

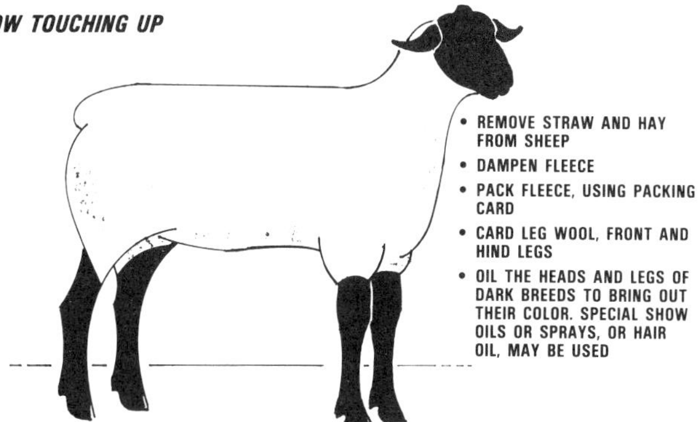

• REMOVE STRAW AND HAY FROM SHEEP
• DAMPEN FLEECE
• PACK FLEECE, USING PACKING CARD
• CARD LEG WOOL, FRONT AND HIND LEGS
• OIL THE HEADS AND LEGS OF DARK BREEDS TO BRING OUT THEIR COLOR. SPECIAL SHOW OILS OR SPRAYS, OR HAIR OIL, MAY BE USED

Fig. 12–30. Touching up.

SHOWING

Why: In order to win.

How: The guiding principles in showing sheep are:

1. "Touch-up" the sheep prior to showing.
2. Be on time for the class.
3. Dress appropriately and neatly for the occasion.
4. Be courteous and a good sport.
5. Follow the directions of the judge and/or the show officials.
6. Line up so as to give yourself and those around you plenty of room to work without interfering with their animals or them.
7. Pose the animal correctly; standing squarely on its feet and legs, head up in natural and alert position, and back straight.
8. Always keep the animal between the judge and yourself whenever possible.
9. Never go behind or step over the back of the sheep to get to the other side.
10. As long as you have total control of the animal, you may kneel (one or both knees), squat, or stand while showing your sheep.

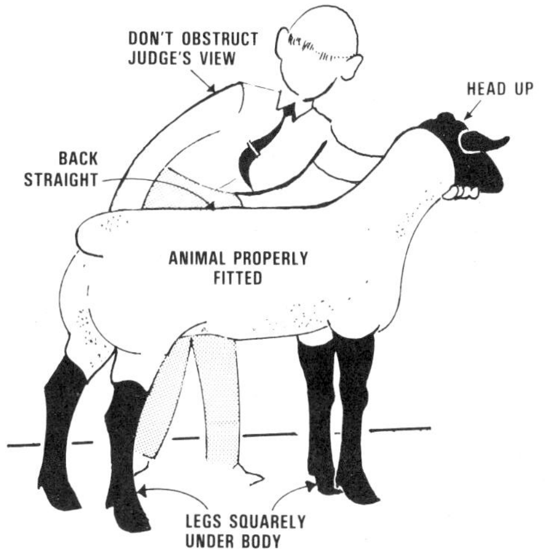

Fig. 12–31. Showing.

(Continued)

TABLE 12-3 *(Continued)*

SHOWING *(Continued)*

11. "Brace" the animal gently when the judge handles your sheep. This will make the animal feel more muscular and indicate to the judge that you are paying close attention to him/her as well. Bracing can be accomplished with slight pressure of your knee into the chest of the sheep or by bowing the neck slightly and pushing the animal back slightly. Either, if done gently, is acceptable.

12. When standing still, watch the judge 50% of the time and pay attention to your sheep 50% of the time.

13. When moving your sheep, break up your eye contact as follows:
   a. Watch where you are going ⅓ of the time.
   b. Watch what your animal looks like ⅓ of the time.
   c. Watch the judge ⅓ of the time.

14. Keep calm and collected.

15. When moving around your sheep and others, move slowly and easily. Do not make quick, fast movements which may startle your animal and others.

16. Pay close attention and follow the directions of the judge closely.

17. Be a gracious winner and an equally good loser.

[1]The line drawings in this table were prepared by Professor R. F. Johnson, Department of Animal Sciences and Industry, California Polytechnic State University, San Luis Obispo.

The procedure given in Table 12–3 is particularly applicable to mutton-type sheep. With animals of the long-wool and the fine-wool breeds, the following differences should be observed:

• **The long-wooled breeds (Lincoln, Cotswolds, and Leicesters)**—

1. There is a tendency to accentuate the length of the fleece through stubble shearing and allowing more than 12 months growth before showing.

2. The fleece is usually oiled or dressed, most commonly with wool fat thinned down by heating, and rubbed on with a brush or cloth.

3. Blankets are not used immediately after oiling or dressing, for they may cause sweating and fleece discoloration.

4. The fleece is neither carded nor blocked. Shears are used only for trimming and docking.

• **The fine-wool breeds (Rambouillet and Merino)**—

1. Sometimes the fleece is colored. One such product is made by mixing 1 tablespoonful of burnt amber and 2 tablespoons of lampblack to each 1½ pints of dressing. It should be noted, however, that the practices of artificial oiling and coloring are now being discouraged or even outlawed in many shows.

2. Blankets are always used sparingly on fine-wool sheep, because their presence may cause excess sweating and fleece discoloration.

FITTING AND SHOWING SWINE[5]

Pertinent information relative to showing swine is summarized in the two sections which follow.

TRAINING, GROOMING, AND SHOWING SWINE

Table 12–5 is an illustrated guide for training, grooming, and showing swine.

[5]The presentation on "Showing Swine," including Table 12-4, benefited from the authoritative review and helpful suggestions of the following persons: Dr. R. Anderson, Department of Animal Sciences and Industry, California Polytechnic State University, San Luis Obispo; and D. Spalding, Spalding Farms, Visalia, CA.

TABLE 12-4
TRAINING, GROOMING, AND SHOWING SWINE[1]

GENTLING AND POSING

Why: So that the judge may see the animal to best advantage.

How: The pig should be gentled by handling, without becoming a pet. It should be trained to respond to either the cane or the whip so that it can be guided by merely placing the cane or whip alongside the jowl. Always keep the animal on the move, slowly. Guide it with the whip or cane by tapping lightly on the jowl or rib. Never hit the animal on the back, legs, or ham.

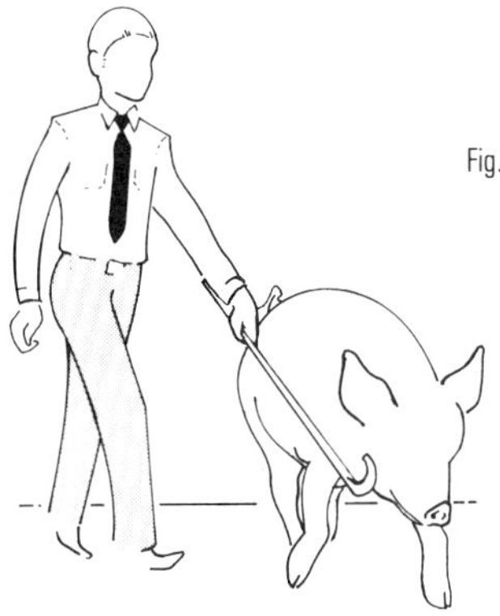

Fig. 12-32. Gentling and handling.

(Continued)

TABLE 12–4 *(Continued)*

TRIMMING THE FEET

Why: In order that the animal may stand squarely and walk properly and that the pasterns will appear straight and strong.

How: Exercise on sandy soil may alleviate the need for foot trimming. However, where trimming is necessary, the following procedure is recommended:

First, get the animal to lie on its side, which can usually be accomplished by merely stroking its belly. Then trim the toes to the proper length and square up the bottoms by using a small rasp and a knife. Also, shorten and dress down the dew claws.

The toes should be trimmed regularly, with some animals as often as every 6 weeks. Do not trim within 2 weeks prior to the show.

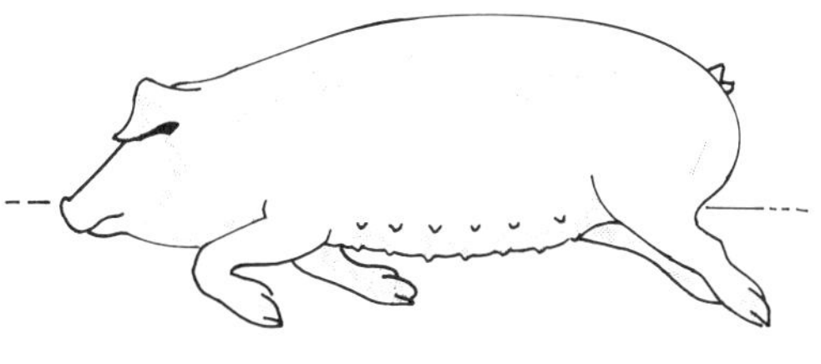

Fig. 12–33. Position for trimming feet.

WASHING

Why: To keep the animal clean, to make the skin smooth and mellow, and to assist in shedding the coat of older animals.

How: Daily, starting about 2 weeks prior to show, rinse hog with water, but no soap, and apply a balsam conditioner. Brush the conditioner into coat, then rinse. This softens the hair coat.

During the 2-week period prior to the show, use soap and water once or twice only, as the soap tends to dry out the hair and skin. Wet the skin and hair thoroughly with lukewarm water, rub the hair with soap until a suds is formed, and then work the suds into the hide with the hands and brush. Add a small amount of bluing to the water when washing white hogs. Following washing, rinse off all traces of soap.

Following each washing, place the animal in a clean pen on fresh bedding.

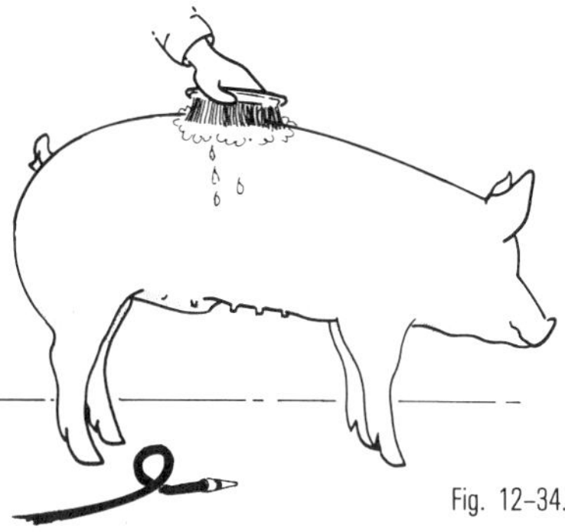

Fig. 12–34. Washing.

OILING COLORED HOGS

Why: In order to soften the skin and hair and to give the necessary bloom to the coat.

How: Just before entering the ring, apply with a spray bottle a light oil mixed with water or alcohol. Lightly mist the hair and coat and rub or brush in oil. Oil should not be used in hot weather; instead, just mist coat with water.

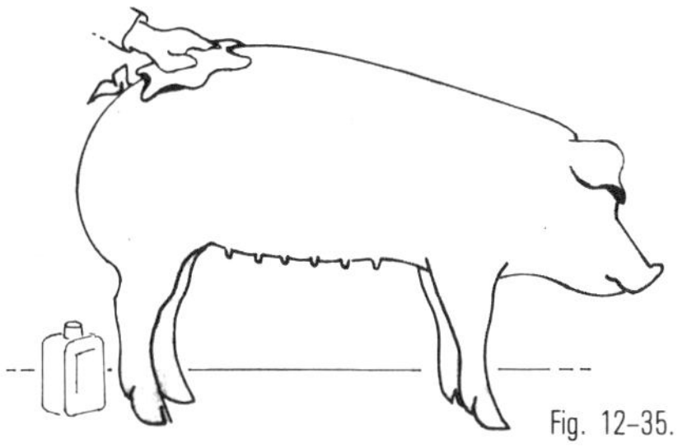

Fig. 12–35. Oiling.

POWDERING OR BLUING WHITE HOGS

Why: In order to make white hogs or white spots on dark hogs appear clean and attractive.

How: If powder is used, animal should first be thoroughly washed and allowed to dry. Before entering the ring, white hogs or white spots on dark hogs may be powdered lightly with talcum powder or corn starch.

If bluing is used, mix a small amount of bluing in the water when washing white hogs. This imparts a more natural look than powder.

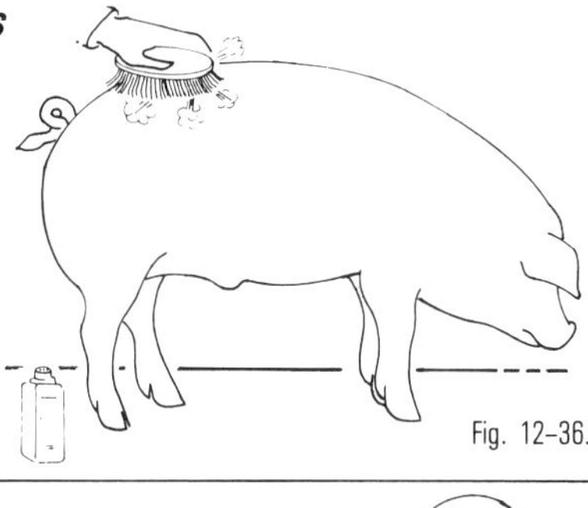

Fig. 12–36. Powdering.

SHOWING

Why: In order to win.

How: The guiding principles adhered to by most successful swine exhibitors are:

1. Train the animal long before entering the ring.
2. Have the animal carefully groomed and ready for the parade before the judge.
3. Dress neatly for the occasion.
4. Enter the ring promptly when the class is called.
5. Do not crowd the judge, but keep your animal in a position of vantage at all times.
6. Avoid being smothered by the mob. Keep in the open.
7. Keep one eye on the judge and the other on the pig. Center your attention entirely on showing the hog. The animal may be under the observation of the judge when you least suspect it.
8. Keep calm, confident, and collected. Remember that the nervous exhibitor creates an unfavorable impression.
9. Work in close partnership with the animal. Never lose your temper; and never strike your animal. Gentle prodding works best.
10. Be concerned about the well-being of your animal. For example, during hot weather, have a helper with a bucket of water near the ring for use in case your animal gets overheated.
11. Be courteous and respect the rights of other exhibitors.
12. Do not allow the hog to bite or fight other animals.
13. Be a good sport. Win without bragging and lose without squealing.

Fig. 12–37. Showing.

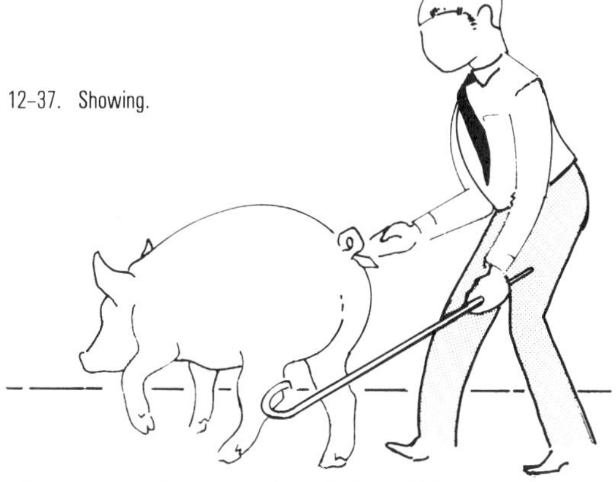

[1]The line drawings in this table were prepared by Professor R. F. Johnson, Department of Animal Sciences and Industry, California Polytechnic State University, San Luis Obispo.

FITTING AND SHOWING HORSES[6]

Horses are shown either (1) in hand (in breeding classes), wearing either a halter or bridle; or (2) in performance classes.

The performance classes for horses are so many and varied that it is not practical to describe them in this book. Instead, the reader is referred to the official Rule Book of the American Horse Shows Association and to the rules printed in the programs of the local horse shows.

[6]The presentation on "Showing Horses," including Table 12–5, benefited from the authoritative review and helpful suggestions of the following person: N. K. Dunn, Professor, Animal Science, Director of Equine Operations, California Polytechnic State University, Pomona.

TRAINING, GROOMING, AND SHOWING HORSES

Table 12–5 is an illustrated guide for training, grooming, and showing horses.

TABLE 12–5
TRAINING, GROOMING, AND SHOWING HORSES[1]

TRAINING

Why: To reach a high degree of proficiency for the intended use—pleasure riding, racing, jumping or whatnot.

How: There are many different ways in which to train a horse. Although the approach may differ, most trainers schedule the schooling of the horse as follows:

1. **Training the foal.** Put a halter on the foal when it is 10 to 14 days old. Teach it to lead and stand properly. This is an ongoing process.
2. **Training the yearling.** Teach the yearling the meaning of "whoa" and its name; to stand when asked; and to get used to the blanket and saddle.
3. **Training at 18 months old.** At 18 months of age, teach the young horse to drive, turn, stop, and back up—by using longlines; to flex its neck and set its head; to respond to the bosal, and to get used to leg pressure.
4. **Training the 2-year-old.** Train the 2-year-old to respond to the aids (legs, hands, reins, and voice); to back; and if a western horse, train it to pivot and make a sliding stop. *Note:* Some horses are not ridden until they are 3 years of age.

CLIPPING AND SHEARING

Why: It makes horses look sharp and feel sharp. Show-ring custom decrees certain breed differences in hair-cuts and hairdos.

How: When clipping and shearing, proceed as follows:

1. **Protect the ears** by placing a wad of cotton in them, to cut down on the noise from clippers and prevent hair from falling into the ears.
2. **Clip long hairs** by removing long hairs from about the head, the inside of the ears, on the jaw, and about the fetlocks.

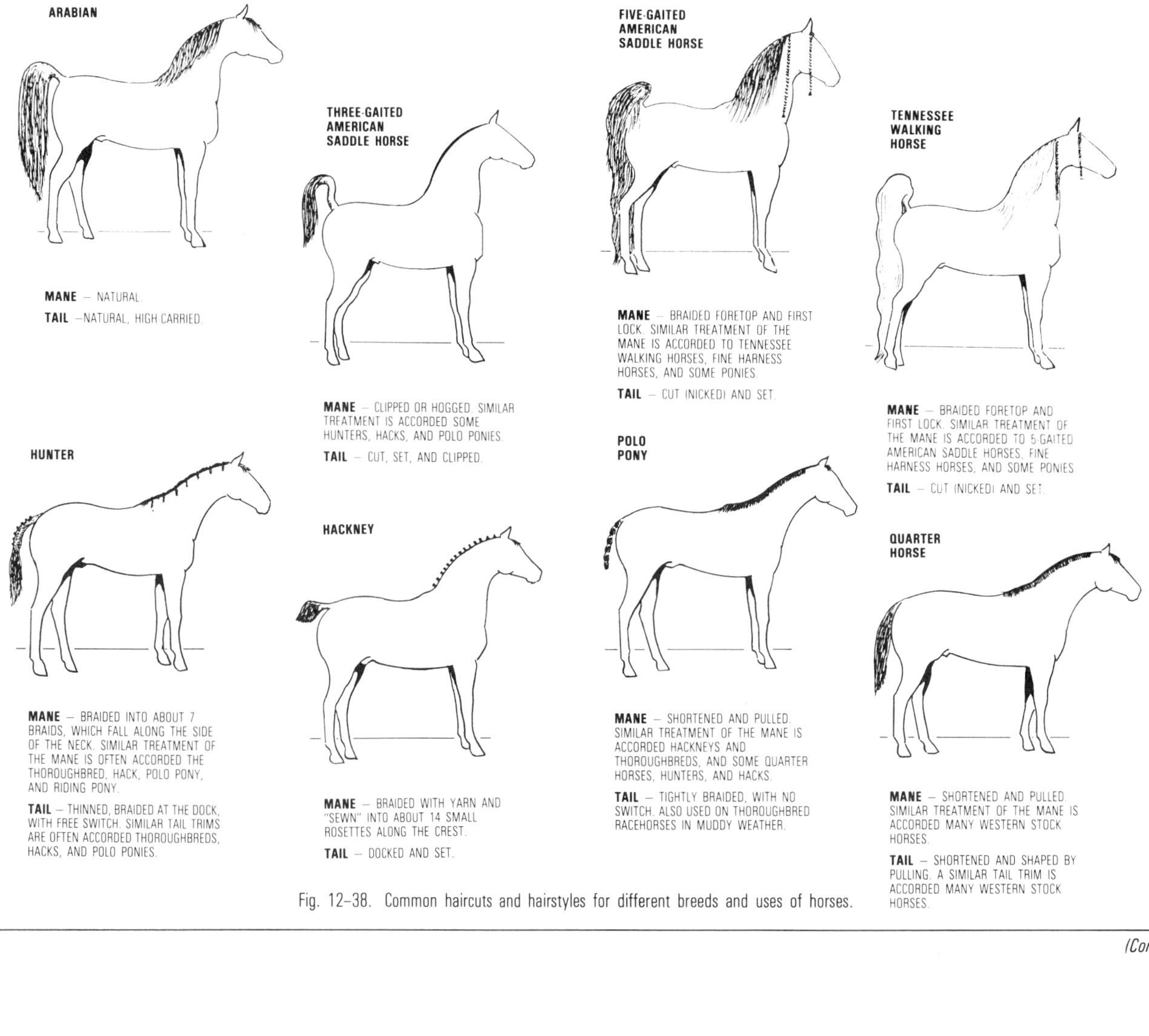

ARABIAN
MANE — NATURAL
TAIL —NATURAL, HIGH CARRIED
THREE-GAITED AMERICAN SADDLE HORSE
MANE — CLIPPED OR HOGGED. SIMILAR TREATMENT IS ACCORDED SOME HUNTERS, HACKS, AND POLO PONIES.
TAIL — CUT, SET, AND CLIPPED.
FIVE-GAITED AMERICAN SADDLE HORSE
MANE — BRAIDED FORETOP AND FIRST LOCK. SIMILAR TREATMENT OF THE MANE IS ACCORDED TO TENNESSEE WALKING HORSES, FINE HARNESS HORSES, AND SOME PONIES.
TAIL — CUT (NICKED) AND SET.
TENNESSEE WALKING HORSE
MANE — BRAIDED FORETOP AND FIRST LOCK. SIMILAR TREATMENT OF THE MANE IS ACCORDED TO 5-GAITED AMERICAN SADDLE HORSES, FINE HARNESS HORSES, AND SOME PONIES.
TAIL — CUT (NICKED) AND SET.
HUNTER
MANE — BRAIDED INTO ABOUT 7 BRAIDS, WHICH FALL ALONG THE SIDE OF THE NECK. SIMILAR TREATMENT OF THE MANE IS OFTEN ACCORDED THE THOROUGHBRED, HACK, POLO PONY, AND RIDING PONY.
TAIL — THINNED, BRAIDED AT THE DOCK, WITH FREE SWITCH. SIMILAR TAIL TRIMS ARE OFTEN ACCORDED THOROUGHBREDS, HACKS, AND POLO PONIES.
HACKNEY
MANE — BRAIDED WITH YARN AND "SEWN" INTO ABOUT 14 SMALL ROSETTES ALONG THE CREST.
TAIL — DOCKED AND SET.
POLO PONY
MANE — SHORTENED AND PULLED. SIMILAR TREATMENT OF THE MANE IS ACCORDED HACKNEYS AND THOROUGHBREDS, AND SOME QUARTER HORSES, HUNTERS, AND HACKS.
TAIL — TIGHTLY BRAIDED, WITH NO SWITCH. ALSO USED ON THOROUGHBRED RACEHORSES IN MUDDY WEATHER.
QUARTER HORSE
MANE — SHORTENED AND PULLED. SIMILAR TREATMENT OF THE MANE IS ACCORDED MANY WESTERN STOCK HORSES.
TAIL — SHORTENED AND SHAPED BY PULLING. A SIMILAR TAIL TRIM IS ACCORDED MANY WESTERN STOCK HORSES.

Fig. 12–38. Common haircuts and hairstyles for different breeds and uses of horses.

(Continued)

TABLE 12–5 *(Continued)*

GROOMING

Why: Proper grooming is necessary to (1) make and keep the horse attractive, and (2) maintain good health and condition.

Grooming cleans the hair, keeps the skin functioning naturally, lessens skin diseases and parasites, and improves the condition and fitness of the muscles.

How: To assure that the horse will be groomed thoroughly and that no body parts will be missed, follow a definite order. This may differ according to individual preference, but the following procedure is most common:

1. Clean out the feet.
2. Groom the body.
3. Brush the head, comb and brush the mane and tail.
4. Wipe with the grooming cloth.
5. Check the grooming.
6. Wash and disinfect grooming equipment.

WASHING (SHAMPOOING)

Why: Washing (1) cleans the animal—it removes the dirt, stains, and sweat that cannot be removed by grooming; (2) makes for a fine hair coat with a good sheen; and (3) keeps the skin smooth and mellow.

How: In preparation for shampooing, (1) groom the horse carefully, (2) secure the animal for washing either by having someone hold it by the shank or by tying, and (3) have shampoo concentrate, warm water, buckets, and sponges available. To assure that the horse will be washed thoroughly and that no body part will be missed, follow a definite order.

After shampooing, rinse the horse with warm water, using either a bucket and sponge or a hose (if the horse is used to the latter). Then complete the washing operation as follows:

1. Scrape with a "sweat scraper" held snugly against the hair to remove excess water, using long sweeping strokes, except do not scrape the head and legs.
2. Dry with a clean dry sponge or coarse towel, squeezing it out at intervals.
3. Blanket the horse and walk it until completely dry.
4. Apply a coat of dressing if desired.

COAT DRESSING

Why: To achieve the all-important "bloom" or eye appeal in show, parade, and sale animals.

A coat dressing will not take the place of the natural conditioning of the horse, which can be achieved only through proper feeding, health, grooming, and shampooing.

How: Proper grooming should always precede the use of coat dressing. Coat dressing is best applied by means of a heavy cloth (preferably terry cloth). Moisten the rag with the dressing and rub the coat vigorously in the direction of the natural lay of the hair; then brush to bring out the bloom.

Coat dressing should be used following washing, and for show, parade, or sale. It is best to apply a heavier application of coat dressing 12 to 24 hours ahead of the event, then go over the horse with a lightly dressed rag just prior to showing.

SHOWING

Why: Horse shows provide entertainment for spectators, and recreation, sport, and competition for the exhibitors. Also, the show-ring has been, and will continue to be, an important medium for getting horses and people together in one place and at one time to compare the quality of their horses and their breeding programs.

Also, horse shows stimulate improved breeding and training procedures.

How: The following practices are recommended for showing in hand, or at halter:

1. Train the horse early.
2. Groom the horse thoroughly.
3. Dress neatly for the show.
4. Enter the ring promptly and in tandem when the class is called. Line up at the location indicated by the show official or judge unless directed to continue around the ring in tandem.
5. Stand the horse squarely on all 4 feet with the forefeet on higher ground than the hind feet if possible. The standing position of the horse should vary according to the breed. For example, Arabians are not stretched, but American Saddlers are trained to stand with their front legs straight under them and their hind legs stretched behind them. Other breeds generally stand in a slightly stretched position, somewhat intermediate between these two examples. When standing and facing the horse, hold the lead strap or rope in the left hand 10 to 12 in. from the halter ring. Try to make the horse keep its head up.
6. Unless the judge directs otherwise, the horse should first be shown at the walk and then at the trot.
7. Keep the horse posed at all times; keep one eye on the judge and the other on the horse.
8. When the judge signals to change positions, the exhibitor should follow instructions.
9. Keep calm; a nervous exhibitor creates an unfavorable impression.

Fig. 12–39. Correct method of leading when showing "in hand."

10. Work in close partnership with the horse.
11. Be courteous and respect the rights of other exhibitors.
12. Do not stand between the judge and the horse.
13. Be a good sport; win without bragging and lose without complaining.

[1]The line drawings in this table were prepared by Professor R. F. Johnson, Department of Animal Sciences and Industry, California Polytechnic State University, San Luis Obispo.

SHIPPING TO THE FAIR

Today, most animals are shipped via either truck or air. Very few are transported via rail nowadays.

Regardless of the method of shipping, it is important that the following details receive consideration.

1. **Schedule properly.** Schedule the transportation so that animals will arrive within the limitations imposed by the show, and at least two to three days in advance of the date that they vie for awards.

2. **Clean and disinfect any public conveyance.** Before using, thoroughly clean and disinfect any type of public conveyance.

3. **Place feed and supplies on a deck.** If space is at a premium, place the feed supply, bedding, and show equipment on a deck or platform in the truck.

For cattle, a deck 5½ to 6 ft above the floor will allow for air circulation and placing younger cattle thereunder; for sheep and swine, a clearance of 5 ft is adequate.

4. **Bed properly.** For cattle and sheep and for hogs in cool weather, the floor should first be sanded so that the animals will not slip, and then covered with long, clean, bright straw.

During warm weather, properly wetted sand makes the best bedding for swine. In extremely hot weather, it may be necessary to air-condition the conveyance.

The floor of vehicles used for transporting horses should be covered with coconut matting made for the purpose, rubber mats, or sand covered with straw.

5. **Load animals properly.** The loading arrangement will vary between classes of animals. Herewith are the preferred arrangements:

a. **For cattle.** In transporting by truck, cattle are generally stood crosswise of the truck, with the largest animal near the cab and tied facing to one side. The direction of facing the remaining animals is alternated; the second animal is faced in the opposite direction from the first, and so on. In trucking, it is usually best to tie cattle fairly short and near enough together so that they will not lie down.

b. **For sheep.** Provide suitable and necessary partitions for separate penning of the sexes and for separating out the rams that are not accustomed to running with each other. Do not overcrowd; allow sufficient space for sheep to bed down in comfort.

c. **For swine.** Provide suitable and necessary partitions for separate penning of animals of different ages and sexes. Show hogs should not be crowded.

d. **For horses.** Most horses are transported via trailer or van. The trailer is usually a 1- or 2-horse unit, in which the horse(s) face toward the direction of travel. Vans are commonly used where 3 to 8 horses are transported, with the horses either (1) stabled abreast and facing the direction of travel, or (2) stabled crosswise, with the alternating animals facing in opposite directions.

6. **Take feed along.** When mixed feeds are used, as is usually the case in fitting rations, a supply adequate for the entire trip should be taken along. This will reduce the hazard of animals going off feed because of feed changes.

7. **Feed limited rations.** Limit show animals to half feed at the last feed before loading out and while in transit. A heavy *fill* is likely to result in digestive disturbances and overheating in warm weather.

8. **Handle animal quietly.** In transit, the animals should be handled quietly and should not be allowed to become hot or to be in a draft.

Western Pleasure Horse Class at the National Western Stock Show and Rodeo, Denver, Colorado. (Courtesy, National Western Stock Show)

Two Egyptian cattle raisers taking an ox to market, for bartering for crafts. Photograph from Bas Relief found in Sakara in the tomb of King Ephto Stoptep. (Courtesy, The Bettmann Archive, New York, NY)

SECTION

13

MARKETING LIVESTOCK AND MILK

Contents *Page*

Fig. 13-1. Marketing milk. (Courtesy, American Dairy Association, Rosemont, IL)

Marketing is the end of the line. From the producer's standpoint, it is that part which gives point and purpose, and profit or loss, to all that has gone before. Market receipts constitute the only source of reimbursement to producers for their work.

In the past, producers of meat animals or milk could be successful if they knew how to breed, feed, and manage their stock. Today, this is not enough; preconsidered, if not prearranged, markets are essential.

In Part I, discussion is limited to the marketing of four-footed meat animals—cattle, sheep, and hogs. Because of its distinct and different market channels and procedures, milk is treated in a separate section, Part II.

PART I—MARKETING LIVESTOCK

Fig. 13-2. The historic stone gate entrance to the Union Stock Yards, Chicago. (Courtesy, Chicago Historical Society, Chicago, IL)

Market day is the payday for producers—hence, it is the most important single day of operation for them.

Livestock marketing embraces those operations beginning with loading animals out on the farm, ranch, or feedlot and extending until they are sold to go into processing channels.

METHODS OF MARKETING LIVESTOCK

The producer of livestock is confronted with the perplexing problem of where and how to market animals. Usually there is a choice of market outlets, and the one selected often varies with different species of livestock and among sections of the country. The methods of marketing also differ between slaughter and feeder animals, and all of these differ from the marketing of purebreds.

In 1987, meat packers (722 of them) purchased their animals through the following channels:[1]

	Cattle	Calves	Sheep	Hogs
	(%)	(%)	(%)	(%)
Nonpublic markets	80.2	61.0	81.4	88.8
Auction markets	15.6	35.8	13.0	4.9
Terminal markets	4.2	3.2	5.5	6.3

Most U.S. animals are marketed through four channels—nonpublic markets, auctions, terminals, or carcass grade and weight basis.

- **Nonpublic markets**—*Nonpublic markets include all purchases from sources except from public markets—from all sources except from terminals and auctions.* The term *nonpublic markets* evolved to replace the term *direct markets* because some questions had been raised as to whether *directs* included *only* purchases direct from sellers to plants or to packer buyer stations.

Nonpublic markets do not involve a recognized market. The selling usually takes place at the farm, ranch, feedlot, or some other nonmarket buying station or collection yard.

The out-of-pocket cost to producers for nonpublic market selling is zero. Their only selling expense is their time, which, of course, is not a direct out-of-pocket cost.

- **Auction market method**—*Auction markets (also referred to as sales barns, livestock auction agencies, community sales, and community auctions) are trading centers where animals are sold by public bidding to the buyer who offers the highest price per hundredweight or per head.* Auctions may be owned by individuals, partnerships, corporations, or cooperative associations.

The cost to the producer of using the auction market is the combined cost of selling, yardage, feed, and service. Auction market charges are generally somewhat higher than terminal charges.

Fig. 13-3. Cattle in auction pens of Producers Livestock Auction Co., San Angelo, TX. (Courtesy, *Livestock Weekly*, San Angelo, TX)

- **Terminal or central market method**—*Terminal or central markets are livestock trading centers which generally have several commission firms and an independent stockyards company.* Formerly, terminal markets were synonymous with private treaty selling. Today, however, many terminal markets operate their own sale ring and all, or almost all, of their livestock are sold by auction.

Fig. 13-4. Penned cattle at a central market. (Courtesy, Oklahoma State University, Stillwater)

The direct out-of-pocket cost to the producer is the combined cost of yardage, sales, commissions, feed, and service. These costs vary somewhat from market to market. Commission fees are the largest terminal market costs. They usually account for one-half of the total charges, with yardage plus feed and bedding accounting for the other half.

- **Carcass grade and weight basis**—It is generally agreed that there is need for a system of marketing which favors payment for a high cutout value of primal cuts and a quality product. Selling on the basis of carcass grade and weight fulfills these needs.

For the United States (48 contiguous states) as a whole, the following percentages of total slaughter were purchased on a carcass grade and weight basis in 1987:

(%)	(%)	(%)	(%)
30.4	36.7	35.9	13.5

[1]*Packers and Stockyard's Statistical Report,* USDA, Packers and Stockyards Administration, P&SA Statistical Report No. 88-1, Jan. 1989, p.9.

PREPARING AND SHIPPING LIVESTOCK

Improper handling of livestock immediately prior to and during shipment may result in excess shrinkage; high death, bruises, and crippling losses; disappointing sales; and dissatisfied buyers. The following general considerations should be accorded in preparing livestock for shipment and in transporting them to market:

1. Select the best-suited method of transportation.
2. Feed and water properly prior to loading. Withhold grain feeding of all classes of livestock 12 hours before loading (omit one feeding). Cattle and sheep may be allowed free choice to dry, well-cured hay up to loading time, but they should not be allowed access to water within 2 to 3 hours of shipment.
3. Keep animals quiet. Remember that "easy does it."
4. Comply with the requirements for health certificates, permits, and brand inspections where interstate shipments are involved.
5. Comply with the federal 28-hour law[2] in rail shipments. This prohibits transporting livestock by rail for a longer period than 28 consecutive hours without unloading, feeding, watering, and resting 5 consecutive hours before resuming transportation. On request of the owner, the period can be extended to 36 hours.
6. Feed or graze cattle or sheep in transit if advantageous. This refers to a provision of railroads whereby livestock producers may be granted permission to graze or finish animals for a period of up to 12 months, at some intermediate stop between their point of origin and the market to which they will be consigned at the end of the finishing period.
7. Use partitions in the truck or car when necessary to separate species, sexes, or age groups.
8. Avoid shipping during extremes in weather, either when it is very cold or very hot.

PREVENT SHIPPING BRUISES, CRIPPLING, AND DEATH LOSSES

Losses from bruising, crippling, and death that occur during the marketing process represent a part of the cost of marketing livestock; and indirectly, the producer pays most of the bill.

The following precautions are suggested (in addition to those already covered under the main heading, "Preparing and Shipping Livestock") as a means of reducing livestock marketing losses from bruises, crippling, and death.

1. Dehorn cattle, preferably when young.
2. Remove projecting nails, splinters, and broken boards from feed containers and fences.
3. Keep feedlot free from old machinery, trash, and any obstacle that may bruise.
4. Remove protruding nails, bolts, or any sharp objects in truck or car.
5. Bed properly (see Table 13–2).
6. Use good loading chutes; not too steep.
7. With two or more decks, have upper deck(s) high enough to prevent back bruises on animals below.
8. Use partitions (a) in cars and trucks that are not fully loaded, to keep animals closer together; and (b) in very long trucks, to keep animals from crowding from one location to another.
9. In rail shipments, place "bull board" in position and secure before car door is closed on loaded cattle.
10. Drive trucks carefully. Slow down on sharp turns and avoid sudden stops.
11. Inspect load enroute to prevent trampling of animals that may be down. If an animal goes down, get it back on its feet immediately.
12. Back truck slowly and squarely against unloading dock.
13. Unload slowly. Do not drop animals from upper to lower deck; use cleated inclines.
14. Never lift sheep by the wool.

All of these precautions are simple to apply; yet all are violated every day of the year.

In a nationwide survey involving 775,000 hogs and 163,000 cattle, Livestock Conservation Institute found that 8.5% of all market hogs and 6.4% of all market cattle showed unmistakable and costly carcass bruises.

NUMBER OF ANIMALS IN A TRUCK

Overcrowding of market animals causes heavy losses. Sometimes a truck or a railroad car is overloaded in an attempt to effect a saving in hauling charges. More frequently, however, it is simply the result of not knowing space requirements.

Table 13–1 gives some indication as to the number of market animals that may be loaded in a truck.

TABLE 13–1
NUMBER OF ANIMALS FOR SAFE LOADING IN A TRUCK[1]

Length of Truck Floor	Kind and Weight of Animals		
	Cattle	Lambs	Hogs and Calves
(ft)	(1,000 lb) *(454 kg)*	(100 lb) *(45.4 kg)*	(225 lb) *(102 kg)*
8 *(2.4 m)*	4	20	16
12 *(3.7 m)*	7	31	24
15 *(4.6 m)*	9	40	30
20 *(6.1 m)*	12	54	40
24 *(7.3 m)*	15	65	48

[1]Recommendations of Livestock Conservation Institute, Madison, WI.

KIND OF BEDDING FOR ANIMALS IN TRANSIT

Among the several factors affecting livestock losses, perhaps none is more important than proper bedding and footing in transit. This applies to both truck and rail shipments.

[2]No such law applies to truck transportation of animals.

Recommended kinds and amounts of bedding and footing materials are give in Table 13–2.

TABLE 13–2
GUIDE RELATIVE TO BEDDING AND FOOTING MATERIAL WHEN TRANSPORTING LIVESTOCK[1, 2, 3]

Class of Livestock	Kind of Bedding for Moderate or Warm Weather; above 50°F	Kind of Bedding for Cool or Cold Weather; below 50°F
Cattle	Sand, 2 inches.	Sand; for calves use sand covered with straw.
Sheep and goats	Sand.	Sand covered with straw.
Swine	Sand, ½ in. to 2 in.[4]	Sand covered with straw.
Horses and mules	Sand.	Sand.

[1]Straw or other suitable bedding (covered over sand) should be used for protection and cushioning breeding stock that are loaded lightly enough to permit their lying down in the car or truck.

[2]Sand should be clean and medium-fine, and free from brick, stones, coarse gravel, dirt, or dust.

[3]Fine cinders may be used as footing for cattle, horses, and mules, but not for sheep or hogs. They are picked up by and damage the wool of sheep, and they damage hog casings.

[4]In hot weather, wet sand down before loading and while *en route*. Drench hogs when necessary, but never apply water to the backs of hot hogs—it may kill them.

SHRINKAGE IN MARKETING ANIMALS

The shrinkage (or drift) refers to the weight loss encountered from the time animals leave the farm or feedlot until they are weighed over the scales at the market. Thus, if a steer weighed 1,000 lb at the feedlot and had a market weight of 970 lb, the shrinkage would be 30 lb or 3.0%. Shrink is usually expressed in terms of percentage. Most of this weight loss is *excretory shrink*, or in the form of feces and urine, and the moisture of the expired air. On the other hand, there is some *tissue shrinkage*, which results from metabolic breaking-down changes.

Shrinkage is of importance because the carcass meat is the most valuable portion of the animal. For this reason, dressed yield is one of the most important factors taken into consideration by packers in buying livestock for slaughter.

The most important factor affecting shrinkage is the fill which refers to *the amount of feed and water consumed by animals upon their arrival at the market and prior to selling.* Normally, the fill of hogs consists of a feed or some grain (small grains are ground) common to the area, and water; whereas the fill of cattle and sheep consists of hay and water, although grain-fed animals may be given some concentrate. Naturally, the larger fill animals take, the smaller the shrinkage.

Because animals transported via motor truck may not have remained off feed too long, there is an increasing tendency not to feed and water them prior to selling on the market. This saving in feed and expense seems economically sound.

On the average, the following shrinkage is obtained on market animals:

Cattle .... from 3 to 6%
Sheep .... from 6 to 10%
Hogs .... from 1 to 2%

MARKET CLASSES AND GRADES OF LIVESTOCK

Broadly speaking, *the market class is the use to which animals are put*, whereas, *the market grade is a measure of how well the animal fulfills the requirements for the class.* More accurately, however, the market class is determined by all of those factors affecting the use and value of the animal, except the final grade. Thus, in cattle, the market class is determined by whether the animals are cattle or calves; by the general use to which the animals are put (slaughter, stocker and feeder cattle, milkers and springers, vealers, slaughter calves, and stocker and feeder calves); by the sex (steers, heifers, cows, bulls), by the age (yearlings, or 2-year-old or over steers), and by weight.

Grading livestock is the act of sorting, dividing, or designating animals of similar classes and grades. The grade is the final subdivision in the classification process. It indicates the relative degree of excellence of an animal or group of animals. When grading is properly and expertly done, each individual of a specific class and grade group is quite similar to other individuals in that group, regardless of whether the animals are in the same pen or in separate markets hundreds of miles removed from each other.

FACTORS DETERMINING MARKET GRADES

Market grades are determined by attributes associated with market preferences and valuation. But the relevant attributes differ widely among species—in numbers, in range of variability, and in ease of objective measurement. Among species grade differences are the following:

1. Eight grades are used to cover the range in quality of steer and heifer carcasses, in comparison with only three for ready-to-cook poultry.

2. In addition to quality grades, five separate yield grades or, cutability grades (i.e., yield of boneless, closely trimmed retail cuts) are used in conjunction with quality grades of both beef and lamb.

Until 1989, meat packers could choose to grade or not to grade beef. But, if they graded, they were *required* to grade for both yield and quality. In 1989, the law was changed, separating quality and yield grades of beef; and allowing packers to choose whether beef carcasses are graded for quality, for yield, or for both quality and yield. But packers could continue to choose to grade or not to grade.

Quality grades, such as Prime, Choice, and Select, gauge differences in the taste of beef, primarily by examining marbling and maturity of the meat. Yield grades, numbered 1 to 5 identify differences in the percentage of lean meat obtained from a carcass.

3. Pork grades incorporate yield and quality consideration without separate quality and yield designations.

Thus, drawing up standards or specifications for a system of grades is a complex undertaking. Some of the attributes upon which grades are based can be evaluated directly; others must be evaluated indirectly, through indicators. For example, the yield of lean cuts of pork is related to (or indicated by) backfat thickness, carcass weight, and carcass length.

MARKET CLASSES AND GRADES OF CATTLE

The generally accepted market classes and grades of live cattle are summarized in Table 13–3. The first five divisions and subdivisions include those factors that determine the class of the animal or the use to which it will be put. The grades indicate how well the cattle fulfill the requirements to which they are put.

TABLE 13–3
THE MARKET CLASSES AND QUALITY GRADES OF CATTLE

Cattle or Calves	Use Selection	Sex Classes	Age	Weight Divisions Wt. (Group)	(lb)	Commonly Used Quality Grades[1]
Cattle	Slaughter cattle[1]	Steers	Yearlings	Light Medium Heavy	750 down 750–950 950 up	Prime, Choice, Select, Standard, Utility, Cutter, Canner
			2-year-olds and over	Light Medium Heavy	1,100 down 1,100–1,300 1,300 up	Prime, Choice, Select, Standard, Commercial, Utility, Cutter, Canner
		Heifers	Yearlings	Light Medium Heavy	750 down 750–900 900 up	Prime, Choice, Select, Standard, Utility, Cutter, Canner
			2-year-olds and over	Light Medium Heavy	900 down 900–1,050 1,050 up	Prime, Choice, Select, Standard, Commercial, Utility, Cutter, Canner
		Cows	All ages	All weights		Choice, Select, Standard, Commercial, Utility, Cutter, Canner
		Bullocks	24 mo. & under	All weights		Prime, Choice, Select, Standard, Utility
		Bulls		All weights		None (yield graded only)
	Feeder cattle	Steers	Yearlings	Light Medium Heavy Mixed		Prime, Choice, Select, Standard, Utility, Inferior
			2-year-olds and over	Light Medium Heavy Mixed		Prime, Choice, Select, Standard, Commercial, Utility, Inferior
		Heifers	Yearlings	Light Medium Heavy Mixed		Prime, Choice, Select, Standard, Utility, Inferior
			2-year-olds and over	Light Medium Heavy Mixed		Prime, Choice, Select, Standard, Commercial, Utility, Inferior
		Cows	All ages	All weights		Choice, Select, Standard, Commercial, Utility, Inferior
		Bullocks	24 mo. & under	All weights		Prime, Choice, Select, Standard, Utility, Inferior
		Bulls	24 mo. & under	All weights		None
	Milkers & springers	Cows (milkers or springers)	All ages	All weights		None
Calves	Vealers	No sex class (Sex characteristics of no importance at this age)	Under 3 mo.	Light Medium Heavy	110 down 110–180 180 up	Prime, Choice, Select, Standard, Utility
	Slaughter calves	Steers Heifers Bulls	3 mo. to 1 year	Light Medium Heavy	200 down 200–300 300 up	Prime, Choice, Select, Standard, Utility
	Feeder calves	Steers Heifers Bulls	Usually 6 mo. to 1 year	Light Medium Heavy Mixed		Prime, Choice, Select, Standard, Utility, Inferior

[1]In addition to the quality grades, there are the following yield grades for all slaughter cattle, except bulls: Yield Grade 1, Yield Grade 2, Yield Grade 3, Yield Grade 4, and Yield Grade 5; with Yield Grade 1 representing the highest cutability, and Yield Grade 5 the lowest. Thus, slaughter cattle are graded for both quality and yield grade.

MARKET CLASSES AND GRADES OF SHEEP

The market classes and grades of sheep (see Table 13–4) follow closely the pattern for the classes and grades of cattle and swine. One notable difference is that a sizable number of sheep are sold as breeders. For the most part, this class is made up of mature western ewes that are sold to country buyers for the purpose of producing one or two more crops of lambs before again being returned to market. Usually such ewes can be acquired at a lower cost than ewe lambs. Another difference between sheep and other species is found in the fact that one feeder class, namely the shearers, is based on wool value as well as adaptability for further feeding.

TABLE 13–4
MARKET CLASSES AND QUALITY GRADES OF SHEEP

Sheep or Lambs	Use Selection	Sex Classes	Age	Weight Division	(lb)	(kg)	Commonly Used Grades
Sheep	Slaughter sheep	Ewes	Yearling	Light Medium Heavy	120 down 120–140 140 up	*54.5 down* *54.5–63.6* *63.6 up*	Prime, Choice, Good, Utility[1]
Sheep	Slaughter sheep	Ewes	Mature (2-year-old or older)	Light Medium Heavy	140 down 140–160 160 up	*63.6 down* *63.6–72.7* *72.7 up*	Choice, Good, Utility, Cull[1]
Sheep	Slaughter sheep	Wethers	Yearling	Light Medium Heavy	130 down 130–150 150 up	*59.0 down* *59.0–68.2* *68.2 up*	Prime, Choice, Good, Utility[1]
Sheep	Slaughter sheep	Wethers	Mature (2-year-old or older)	Light Medium Heavy	150 down 150–170 170 up	*68.2 down* *68.2–77.3* *77.3 up*	Choice, Good, Utility, Cull[1]
Sheep	Slaughter sheep	Rams	Yearling	All weights			Prime, Choice, Good, Utility[1]
Sheep	Slaughter sheep	Rams	Mature (2-year-old or older)	All weights			Choice, Good, Utility, Cull[1]
Sheep	Feeder sheep	Ewes and wethers	Yearlings	All weights			Fancy, Choice, Good, Medium, Cull
Sheep	Feeder sheep	Ewes	Mature (2-year-old or older)	All weights			Choice, Good, Medium, Cull
Sheep	Breeding sheep	Ewes (rams occasionally purchased as breeders, but not listed in market reports)	Yearlings, 2-, 3-, or 4-year-old or older	All weights			Fancy, Choice, Good, Medium, Cull
Lambs	Slaughter lambs	Ewes, wethers, and rams	Hothouse lambs	60 down			Prime, Choice, Good, Utility[1]
Lambs	Slaughter lambs	Ewes, wethers, and rams	Spring lambs	Light Medium Heavy	100 down 100–110 110 up	*45.4 down* *45.4–50.0* *50.0 up*	Prime, Choice, Good, Utility[1]
Lambs	Slaughter lambs	Ewes, wethers, and rams	Lambs	Light Medium Heavy	105 down 105–120 120 up	*47.7 down* *47.7–54.5* *54.5 up*	Prime, Choice, Good, Utility[1]
Lambs	Feeder lambs	Ewes and wethers	All ages	All weights			Fancy, Choice, Good, Medium, Cull
Lambs	Shearer lambs	Ewes and wethers	All ages	All weights			Choice, Good, Medium

[1]In addition to the above quality grades, there are five yield grades applicable to all lamb and mutton carcasses, denoted by numbers 1 through 5, with the Yield Grade 1 representing the highest degree of cutability. Thus, slaughter sheep and lambs may be graded for (1) quality alone, (2) yield grade alone, or (3) both quality and yield grades.

MARKET CLASSES AND GRADES OF HOGS

The market classes and grades of hogs are summarized in Table 13–5.

The market classes and grades of swine were developed in much the same manner as the classifications of cattle were developed and brought into use. They also serve much the same purpose. Swine classes and grades do differ from those used in cattle and sheep in that (1) there are no age divisions by years (e.g., cattle are classified as yearling and 2-year-old and over), (2) only a limited number of hogs are returned to the country as feeders for further growth or finishing, and (3) rarely are hogs of any kind purchased on the market for use as breeding animals. As in the classification of market cattle, the class of market hogs indicates the use to which the animals are best adapted, whereas the grade indicates the degree of perfection within the class.

TABLE 13–5
THE MARKET CLASSES AND GRADES OF HOGS

Hogs or Pigs	Use Selection	Sex Class	Weight Divisions (lb)		Weight Divisions (kg)		Commonly Used Grades
Hogs	Slaughter hogs	Barrows and Gilts (often called butcher hogs)	120–140	240–270	*55–64*	*109–123*	U.S. No. 1
			140–160	270–300	*64–73*	*123–136*	U.S. No. 2
			160–180	300–330	*73–82*	*136–150*	U.S. No. 3
			180–200	330–360	*82–91*	*150–163*	U.S. No. 4
			200–220	360–400	*91–100*	*163–182*	U.S. Utility
			220–240	400 lb up	*100–109*	*182 kg up*	
		Sows (or packing sows)	270–300	400–450	*123–136*	*182–204*	U.S. No. 1
			300–330	450–500	*136–150*	*204–227*	U.S. No. 2
			330–360	500–600	*150–163*	*227–272*	U.S. No. 3
			360–400	600 lb up	*163–182*	*272 kg up*	Medium, Cull
		Stags	All weights				Ungraded
		Boars	All weights				Ungraded
	Feeder hogs	Barrows and Gilts	120–140		*55–64*		U.S. No. 1, U.S. No. 2,
			140–160		*64–73*		U.S. No. 3, U.S. No. 4,
			160–180		*73–82*		U.S. Utility, Cull
Pigs	Slaughter pigs	Barrows, Gilts and Boars	Under 30		*Under 13.6*		Ungraded
			30–60		*13.6–27.2*		
		Barrows and Gilts	60–80		*27.2–36.3*		Ungraded
			80–100		*36.3–45.4*		
			100–120		*45.4–54.5*		
	Feeder pigs	Barrows and Gilts	80–100		*36.3–45.4*		U.S. No. 1, U.S. No. 2
			100–120		*45.4–54.5*		U.S. No. 3, U.S. No. 4,
							U.S. Utility, Cull

SOME LIVESTOCK MARKETING CONSIDERATIONS

Enlightened and shrewd marketing practices generally characterize the successful livestock enterprise. Among the considerations of importance in marketing live animals are those which follow.

CYCLICAL TRENDS IN MARKET LIVESTOCK

The price cycle as it applies to livestock may be defined as that period of time during which the price for a certain kind of livestock advances from a low point to a high point and then declines to a low point again. In reality, it is a change in animal numbers that represents the producer's response to prices. Until about 1970, the price cycle of the different classes of animals was about as follows: hogs, 4 years; sheep, 9 to 10 years; and cattle, 10 years. (See Fig. 13–5.)

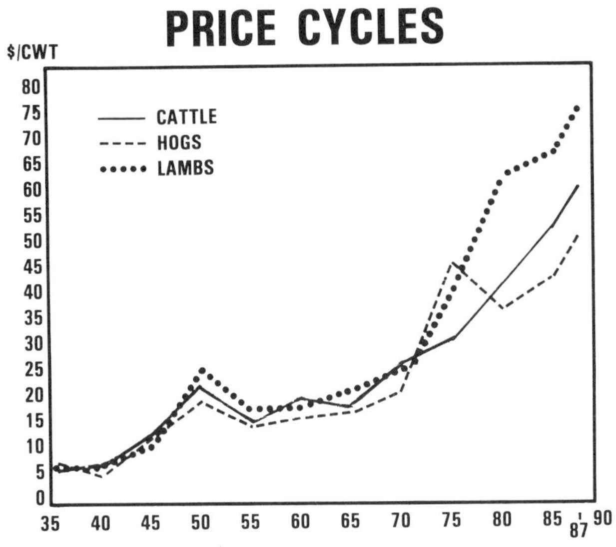

Fig. 13–5. Average price received by U.S. farmers for each class of livestock, 1935–87. Until about 1970, this showed that the price of animals was approximately as follows: hogs, 4 years; sheep, 9 to 10 years; and cattle, 10 years. But price cycles may be outmoded. (Source: Data for 1935–60 from *Agricultural Statistics 1962*, USDA, pp. 371, 387, 402. Data for 1965–70 from *Agricultural Statistics 1975*, USDA, pp. 306, 315. Data for 1975–1987 from *Agricultural Statistics 1988*, USDA, pp. 265, 276, 287.

The species cycles were a direct reflection of the rapidity with which the numbers of each class of farm animals may be shifted under practical conditions to meet consumer meat demands. Litter-bearing and early-producing swine could be increased in numbers much more rapidly than either sheep or cattle. When market hog prices were favorable, established swine enterprises were expanded, and new herds were founded, so that about every 4 years, on the average, the market was glutted and prices fell, only to rise again because too few hogs were being produced to take care of the demand for meats.

It is noteworthy that cattle cycles were formerly 15 years or longer, but that they were shortened to about 10 years, due to the earlier maturity of modern cattle and the marketing of cattle at younger ages. This is clearly shown in Fig. 13–6.

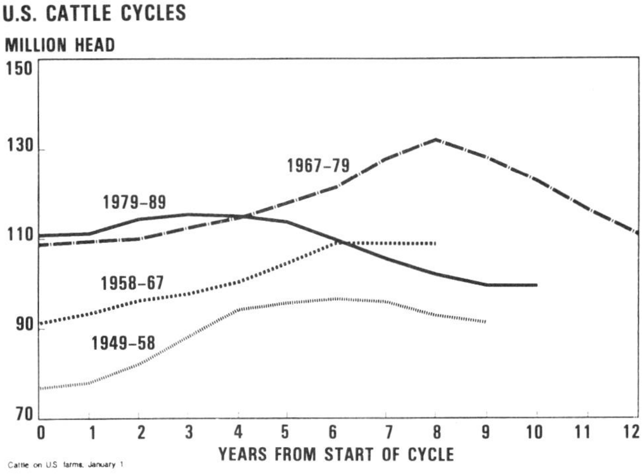

Fig. 13–6. U.S. cattle cycles have shortened from about 15 years to about 10 years. (Source: *1989 Agricultural Outlook*, USDA, Agricultural Handbook No. 684, p. 21, T33)

But big operations, confinement production, more year-round births of young, and less seasonal production, have lessened the impact of price cycles.

SEASONAL CHANGES IN MARKET LIVESTOCK

In recent years, seasonal patterns have not been as marked as they used to be. Year-round finishing of cattle and lambs in large commercial feedlots, year-round farrowing of sows in confinement, and control of estrus by light and hormones, have made for more uniform marketing throughout the year, and lessened seasonality in livestock marketing, in both receipts and prices. Thus, when arriving at livestock forecasts and marketing advice, proper reservation should be exercised in considering seasonal patterns.

Fig. 13-7 shows the seasonality of livestock receipts.

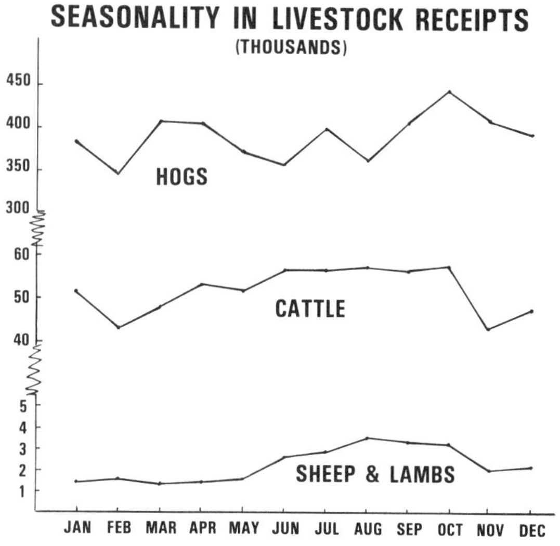

Fig. 13-7. Seasonality in livestock receipts. This shows the average monthly receipts for the combined years of 1977, 1980, 1985, and 1987, of (1) hogs, (2) cattle, and (3) sheep and lambs, in the interior Iowa and southern Minnesota area. (Source: Chart based on data from *Livestock and Meat Statistics*, 1984-85, USDA, ERC, Statistical Bulletin No. 784, p. 64, Table 59, 1989)

Fig. 13-7 shows the following seasonality in receipts of (1) hogs, (2) cattle, and (3) sheep and lambs:

1. The highest receipts of hogs is in March and October, and the lowest in May to August.
2. The highest receipts of cattle are from June to October—during the pasture season; and the lowest receipts are in November to March.
3. The highest receipts of sheep and lambs occur in June to October, when spring lambs are marketed; and the lowest receipts occur in January to May.

Fig. 13-8 shows the seasonality of lamb prices.

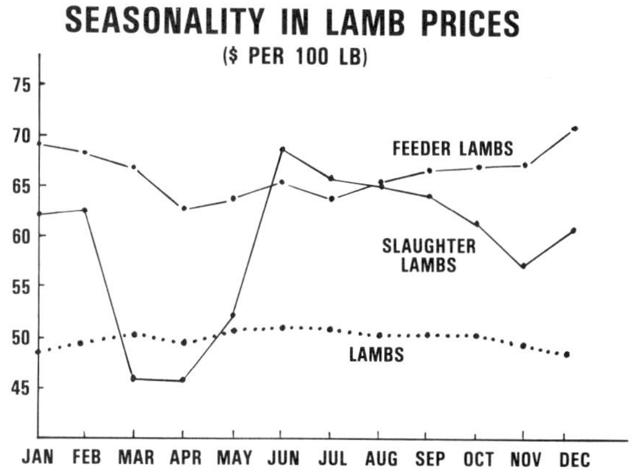

Fig. 13-8. Seasonality in lamb prices. This shows average monthly prices of (1) feeder lambs, (2) slaughter lambs, and (3) lambs in the United States. The feeder lamb prices are based on the South St. Paul market, for the combined years 1977, 1980, 1985, and 1987. The slaughter lamb prices are based on the Omaha market, for the combined years 1977, 1980, 1985, and 1987. The lamb prices are based on markets throughout the United States, for the combined years 1970, 1975, 1980, and 1985. (Source: USDA)

Fig. 13-8 shows the following seasonality in lamb prices:

1. Feeder lambs, which consist of young animals under one year of age that carry insufficient finish for slaughter, are highest from November to March—when few feeder lambs are available, and feeder lamb prices are lowest from April to August—when the bulk of the spring lamb crop is being marketed.
2. Slaughter lambs, which consist of young animals under one year of age that carry sufficient finish for immediate slaughter, show the greatest spread in seasonality between high and low prices of any species. The highest slaughter lamb prices are in May and June—when slaughter lambs are in short supply, and the lowest prices are in March and April—when many lambs finished in feedlots are marketed.
3. Lamb prices, which consist of all lambs under one year of age, and includes feeder lambs, slaughter lambs, and shearer lambs, shift very little throughout the year, but they are slightly higher in May to July, and slightly lower in December.

SHORT-TIME CHANGES

Day-to-day variations in livestock prices usually are caused by uneven distribution of receipts on a given market because of such factors as weather, interference with transportation strikes, uncertain or threatened federal policies and stock market fluctuations. Although such changes are not large, the shrewd producer and market specialist is quick to take advantage of fluctuations of financial interest.

JEWISH HOLIDAYS

The Jewish abstinence from eating meat during certain holidays exerts a minor influence on livestock prices. The important thing is that no kosher slaughtering is done on certain of these holidays. Likewise, producers should know that no animals are kosher slaughtered on Saturdays (the Jewish Sabbath). Normally, therefore, it is not good business to have too many kosher-type cattle, calves, or lambs on the eastern markets just prior to or during these holidays.

DOCKAGE

Dockage refers to deductions made in the liveweight of market animals because of excessive dressing losses, or because part of the product is of low quality. Some common dockages on livestock markets are:

1. **Piggy sows.** Usually docked 40 lb, but it may range from 0 to 50 lb, depending on the market.
2. **Stags (hogs).** Usually docked 70 lb, but it may range from 40 to 80 lb, depending on the market.
3. **Cattle with lumpy jaw.** Usually bought subject to the amount of wastage.

MEAT PROMOTION (CHECKOFF)

Effective meat promotion—which should be conceived in a broad sense and embrace research, educational, and sales approaches—necessitates full knowledge of the nutritive qualities of the product. To this end, we need to recognize that (1) meat contains 15 to 20% high-quality protein, on a fresh basis; (2) meat is a rich source of energy, the energy value being dependent largely upon the amount of fat it contains; (3) meat is a rich source of several vitamins, but it is especially good as a source of phosphorus and iron; (4) meat is one of the richest sources of the important B group of vitamins, especially thiamin, riboflavin, niacin, and vitamin B–12, and (5) meat is highly digestible, with about 97% of meat proteins and 95% of meat fats being digested. Thus, meat is one of the best foods with which to alleviate human malnutrition, a most important consideration in light of the estimation that 35% to 40% of the U.S. population is now failing to receive an adequate diet.

Also, it is noteworthy that the per capita consumption of red meat in five countries exceeds that of the United States: by rank, based on 1989 meat consumption, these are: (1) Germany, East 202 lb; (2) Uruguay, 194.5 lb; (3) Czechoslovakia, 188.8 lb; (4) New Zealand, 176 lb; and (5) Australia, 169.6 lb.

Thus, based on (1) its nutritive qualities and (2) per capita consumption in those five countries exceeding us, it would appear that there is a place and a need for increased meat promotion, thereby increasing meat consumption and price.

There will always be controversy as to how meat promotion funds should be raised and spent. There can be no doubt, however, that all segments of the meat industry would benefit by working together in a unified approach.

Several beef and pork checkoff/promotion programs are in operation. Each program is under different sponsorship; some nationwide, others statewide.

- **Beef promotion**—The 1985 Farm Bill contained the Beef Promotion and Research Act—enabling legislation which made it possible for the beef industry to establish a uniform national checkoff of $1.00 per head to fund promotion and research programs. So, beginning October 1, 1986, $1.00 per head was collected on all cattle sold or imported into the United States. This checkoff program generates approximately $60 million per year. A 113-member Cattlemen's Beef Promotion and Research Board—made up of representatives of the different states—has basic responsibility for the program, with 10 persons elected therefrom serving as the Beef Promotion and Operating Committee.

The Beef Promotion and Research Board has promoted effectively the slogan:

Beef, real food for real people.

The headquarters/address is:
Beef Promotion and Research Board
P.O. Box 3316
Englewood, CO 80155

- **Pork promotion**—The National Pork Producers Council (NPPC), organized in 1954, is today the largest commodity organization in the nation with an identified membership. The NPPC is the pork producer's voice and advocate to solve problems efficiently for the industry. Headquartered in Des Moines, Iowa, with an additional office in Washington, DC, the Council's purpose is to improve and increase the quality, production, distribution, and sales of pork and pork products.

The NPPC is funded by a voluntary 20¢ checkoff on market hogs and a 10¢ checkoff on feeder pigs. These funds are collected by NPPC which retains 60%. The remaining 40% is divided between the National Live Stock and Meat Board (20%) and member states (approximately 20%). The National Pork Board approved a $26.2 million checkoff-funded budget for 1990.

The National Pork Board has scored well with its promotion of:

Pork, the other white meat.

The headquarters/address is:
National Pork Producers Council
P.O. Box 10383
Des Moines, IA 50306

PARITY AND PARITY PRICES IN FARM ANIMALS

Parity may be defined as a yardstick for measuring the relationship between the prices farmers receive for the products they sell and the prices they pay for the things they buy, including interest, taxes, and farm wage rates. The term *parity* comes from the Latin word *paritas*, meaning equal.

The price of a farm commodity may be said to be at parity when a given unit has the same purchasing power that it had in the base period. Thus, if swine producers receive parity price per hundred weight for hogs, they should be able to take the money and buy as much with it as they could back in 1910 to 1914, the base period. If the price of things farmers buy for their production program doubles, the parity price of farm commodities also doubles, to keep in line or equal parity.

It is important that the farmers and ranchers be informed relative to parity prices because most farm price support legislation is based on some percentage of parity. To this end, the following facts are presented:

1. **The old parity formula.** The base period 1910 to 1914 was long used in figuring the parity price of most farm products. This period was selected originally because it was felt that price relationships between farm and nonfarm products were fair to both farmers and nonfarmers during this period.
2. **The new parity formula which became effective January 1, 1950**—In 1948 and again in 1949, Congress passed acts which revised and brought up to date the parity formula, called the *new formula.* Although over-all price relationships between what farmers get and what they pay are still based on the years 1910 to 1914, the parity prices of individual farm commodities are figured by a new formula that reflects recent price relationships. The latter is accomplished by using the latest 10-year averages. This is an improvement, because it helps reflect changes in price relationships that have taken place in farming since 1910 to 1914.

The new parity formula lowers parity prices for many crops and raises them for some livestock and livestock products.

PART II—MARKETING MILK[3]

Fig. 13-9. Satisfactory milk marketing necessitates one basic ingredient—quality milk, which begins on the dairy farm. (Courtesy, National Milk Producers Federation, Arlington, VA)

In our present system, the marketing of milk and dairy products is handled largely by specialists, usually under a multitude of regulations and controls. However, successful milk producers must understand milk markets and factors affecting them if they are to take full advantage of their opportunities.

The farm value of dairy products in 1986 was $17.8 billion. The total marketing bill—the cost of transporting, processing, and distributing dairy products, or the difference between consumer expenditures and farm value—that same year came to $34.2 billion. Upon being retailed, consumers spent $52 billion for these products (see Table 13-6).

TABLE 13-6
FARM VALUE, MARKETING BILL, AND CONSUMER EXPENDITURES FOR DAIRY PRODUCTS, 1986[1]

Item	Value
	(million $)
Farm value	17,800
Marketing bill (cost for transporting, processing, and distributing)	34,200
Consumer expenditures (retail cost)	52,000

[1]Source: *1988 Dairy Producer Highlights*, National Milk Producers Federation, Arlington, VA, pp. 6 and 30.

[3]"Part II—Marketing Milk" to the end of the section, benefitted from the authoritative review and helpful suggestions of the following person: Dr. R. F. Fallert, Leader, Dairy Research Section, USDA, Washington, DC.

Other noteworthy statistics are: In 1986, (1) dairy products accounted for 13.1% of the cash income of the nation's farmers, and (2) consumers spent about 12.0% of their food dollar for dairy products.

FARM PRODUCTION AND HANDLING OF MILK

Satisfactory milk marketing necessitates one basic ingredient—quality milk; and, ultimately, this means more income for the producer.

The difference between Grade A milk and the lower grades is considerable. But it goes beyond this; quality can mean increased consumer demand.

Buyers, consumers, and health departments all have a distinct interest in the quality of milk marketed.

Quality milk can be produced only when producers pay special attention to a number of factors; among them, herd health, the layout and structure of the barn and milk house, clean cows, care of the utensils, cooling and storage of milk, and transportation of milk to market.

In 1987, only 1.6% of the total milk production was used on farms, compared with 15% in 1950.

MARKET STAGES FOR MILK AND DAIRY PRODUCTS

Milk moves from the farm to the consumer in the following three stages:

1. Assembly and transportation from farms to processing plants.
2. Processing and packaging or manufacturing into various dairy products.
3. Distribution of packaged milk and manufactured milk products to consumers.

Also, producers market their milk as (1) Grade A milk, or (2) manufacturing milk (Grade B).

HOW MILK IS PRICED AND REGULATED

Chaotic conditions in milk marketing, resulting from the breakdown of private controls and the serious economic plight of farmers during the depression years of the early 1930s, brought requests from organized producers and distributors for government control. Out of this evolved two forms of government controls—those established by the federal government, and those established by the state governments; both were designed to bring more stability into the marketing of milk.

Today, federal and state agencies, directly or indirectly, affect the pricing of milk marketed by dairy farmers in the United States.

FEDERAL MILK MARKETING ORDERS

Milk marketing orders are designed to stabilize the marketing of fluid milk and to assist farmers in negotiating with distributors for the sale of their milk. Prices paid to farmers are controlled, but there is no direct control of retail prices.

Federal orders are not concerned with sanitary regulations. These are administered by state and local health authorities. In 1990, there were 41 different federal order markets, each with a market administrator and provision for setting minimum farm prices and regulating transactions between farmers and milk dealers in their area. About 80% of fluid grade milk, and about 71% of all milk, produced in the United States is marketed under Federal Orders.

STATE MILK CONTROL

Through State Orders, 17 states have authority to set minimum farm prices and/or retail prices at the wholesale and retail levels. In some states, milk control commissions determine not only what farmers are to be paid but what price the stores can charge customers.

COOPERATIVES

The practice of dealing separately with a large number of producers led to dissatisfaction in a number of cases. To rectify this situation, cooperatives were organized. These cooperative associations are of two general types:

1. Bargaining associations which do not handle any milk, but make all the business arrangements.
2. Associations which process and distribute milk or assemble it for fluid use.

About 75% of the total deliveries of milk to plants and dealers in the United States is handled by cooperatives. In 1986, the net volume of 354 U.S. dairy cooperatives was $14.8 billion.

OTHER REGULATORY PROGRAMS

Because of the essential nature of milk, plus the fact that it is easily contaminated and a favorable medium for bacterial growth, it is inevitable that numerous programs have evolved around it—federal, state, and local, some having been designed to control prices and assure a reasonably uniform flow of milk, and others for sanitary reasons.

- **Sanitary regulations**—The sanitation of milk and dairy products is assured by the enforcement of sanitary regulations by federal, state, and local authorities. There are more the 15,000 state, county, local, and municipal health and sanitation jurisdictions in the United States.

In 1923, the U.S. Public Health Service (USPHS) established an Office of Milk Investigations, and in 1924, the USPHS published its first Grade A pasteurized milk ordinance. Subsequently, this regulation has been revised several times.

Producers are issued permits allowing them to ship Grade A milk. The permit is revoked if either the bacteria count of raw milk exceeds 100,000 per milliliter or the cooling temperature exceeds 40°F in three of the last five samples.

The standard plate count of Grade A pasteurized milk may not exceed 20,000 per milliliter nor the coliform count 10 per milliliter in three of the last five samples or the processor's permit will be revoked.

The Food and Drug Administration (FDA) is charged with inspecting dairy products and processing plants for contamination and adulteration.

- **Standards and grades**—The U.S. Department of Agriculture has responsibility for the development of standards and grades for milk and dairy products. Milk is graded as Certified, Class I (Grade A), or Class II (Grade B).

The major dairy products for which the USDA has established grades, and the proportion graded are shown in Table 13–7.

TABLE 13–7
SELECTED DAIRY PRODUCTS GRADED BY USDA[1]

Product	Volume	Share of U.S. Production
	(million lb)	(%)
Butter	107.7	9.75
Cheese	5.3	0.1
Nonfat dried milk	16.2	1.53

[1]Source: USDA.

METHODS OF PRICING OR PAYING FOR FLUID MILK

Economists refer to the different system of paying for milk as *price plans.* These plans, which in actual practice generally involve two or more plans—for example, pricing based on (1) class, (2) grade, and (3) base-surplus—are:

1. **Flat price plan.** This was the common method up to World War I. The milk producer was paid a uniform price for all milk sold, regardless of quality or the use made of it.

2. **Use classification plan.** Most marketing orders established two use classes—Class I and Class II.

Class I milk generally includes milk used in fluid form such as whole fluid milk, or milk for creamed drinks which must be made from milk approved by local health authorities. Generally speaking, Class I prices are 10 to 15% higher than Class II prices.

Class II milk usually includes milk in excess of fluid needs, which is used to make manufactured dairy products—primarily butter, nonfat dry milk, and cheese.

On some markets, a further division is made, primarily for going into cottage cheese, with the result that there are three classes of milk—Class I, Class II, and Class III.

3. **Blend price.** When dealers buy according to classification prices, they may pay producers a blend price. The blend is an average of class prices weighted by volume of milk in each class, usually quoted at a specific point and for a specific test of milk.

4. **Quality grade plan.** Frequently, the terms Grade A and Grade B (Grade B is usually called *Manufacturing Grade Milk*) are encountered in milk marketing. Although there may be some local variations in their use, Grade A usually refers to milk produced under conditions which make it acceptable for fluid use in a given market. Grade B often refers to milk produced under conditions which do not make it acceptable for fluid milk use—it's manufacturing milk.

The production of Grade A milk relative to that of Grade B milk has been increasing in recent years (see Fig. 13–10). In 1987, U.S. farmers marketed 139,058 million lb of milk, of which 123,768 million lb, or 89%, was Grade A, and 15,290 million lb, or 11%, was Grade B.

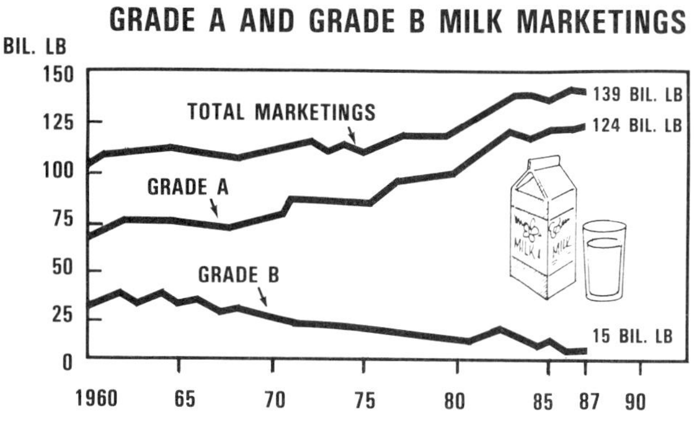

Fig. 13–10. Grade A and Grade B milk marketings, 1960 to 1987. (Source: USDA)

5. **Base surplus plan.** The base surplus plan (or base rating plan) is designed to encourage that a uniform supply of milk be available throughout the year. It compensates the producer who maintains a high fall production, when more milk is needed.

6. **Butterfat test price plan.** The butterfat test of milk affects the price. The common practice is to establish a price for 100 lb of milk of a specified butterfat test. Usually 3.5% butterfat is the basis for pricing, although several markets have established their base as high as 4.0% butterfat. Then, a price differential (per point or 0.1%) is set up for milk testing above or below this amount.

7. **Solids-not-fat price plan.** Today, the emphasis on the food value of milk is shifting from fat content to the other solids, especially protein. This is feasible because tests for solids-not-fat have been devised, and these are proving practical for field use. It is anticipated that this system of pricing milk will expand in the future.

On the average, whole milk contains about 2¼ lb of solids-not-fat for each pound of milk fat. Thus, milk testing 4% butterfat contains approximately 9 lb of solids-not-fat, to a total of 13 lb of solids per hundredweight.

8. **Component pricing plans.** Although various component pricing plans being formulated or adopted are not uniform, they continue to give price credit for butterfat in farm milk; in addition, they give credit for solids-not-fat, including protein. Some also involve end product pricing in which farm milk prices are based on the yield and market value of cheese and other dairy products that can be manufactured from the milk. Most also either (1) establish a maximum somatic cell count at which component premiums will be paid, or (2) pay a premium of 6 to 12¢ per hundredweight of milk for minimum somatic cell counts.

9. **Gallon or quart plan.** Occasionally, a producer supplies milk to a distributor on a per gallon or per quart basis. Since average milk weighs 2.15 lb to the quart and 8.6 lb to the gallon, 100 lb of milk would be equivalent to 46.5 qt or 11.6 gal. Thus, one can easily compute the possible returns from selling milk by different methods.

10. **Special milks.** Certain milks are sold under special labels. Among them are:

a. **Certified milk.** This is milk that is produced under special sanitary conditions prescribed by the American Association of Medical Commissioners. It is sold at a higher price than ordinary milk.

b. **Golden Guernsey milk.** Golden Guernsey milk is produced by owners of purebred Guernsey breeds who comply with the regulations of the American Guernsey Cattle Club. Such milk is sold under the trade name "Golden Guernsey," at a premium price.

c. **All-Jersey milk.** This is produced by registered Jersey herds whose owners comply with the regulations of The American Jersey Cattle Club. It is sold at a premium price under the trademark "All-Jersey."

THE PRICE SUPPORT PROGRAM

The price support program for milk comes under the Agricultural Act, authorized and directed by the Secretary of Agriculture. The Act has been amended several times.

In 1987, U.S. farmers received $12.03 per cwt for all milk sold to plants; with a price differentiation of $12.40 per cwt for fluid (Grade A) milk, and $11.10 per cwt for manufacturing (Grade B) milk.

In several of the postwar years, the government made substantial purchases of dairy products—through CCC and other purchase programs—to support prices at announced levels.

Meat and milk are good—and good for you. This shows kabobs and milk served in outdoor cooking. (Courtesy, United States Dairy Assn., Rosemont, IL)

SECTION

14

MEAT AND MILK

Contents *Page*

Perhaps most people consume meat and milk simply because they like them. They derive a rich enjoyment and satisfaction therefrom. For flavor, variety, and appetite appeal, they are unsurpassed.

But animal products are far more than just very tempting and delicious foods. From a nutrition standpoint, they contain certain essentials of an adequate diet. This is important, for how we live and how long we live are determined in large part by our diet.

It is estimated that the average American gets the percentages of food nutrients shown in Fig. 14–1 from animal products. Foods of animal origin (meat, milk, and their various by-products) are especially important in the American diet; they provide nearly all of the vitamin B-12, ⅔ of the total protein, about ⅓ of the total energy, about ⅘ of the calcium, ⅔ of the phosphorus, and significant amounts of the other minerals and vitamins needed in the human diet.

Although this section is devoted primarily to the final animal products—meat and milk—it must be remembered that the top grades of these important food constituents represent the culmination of years of progressive breeding, the best in nutrition, vigilant sanitation and disease prevention, superior care and management, and modern marketing, processing, and distribution. Thus, the efficient availability of the highest quality meat and milk is dependent upon the well-coordinated operation of the whole field of animal science. Much effort and years of progress have gone into the production of meat and milk.

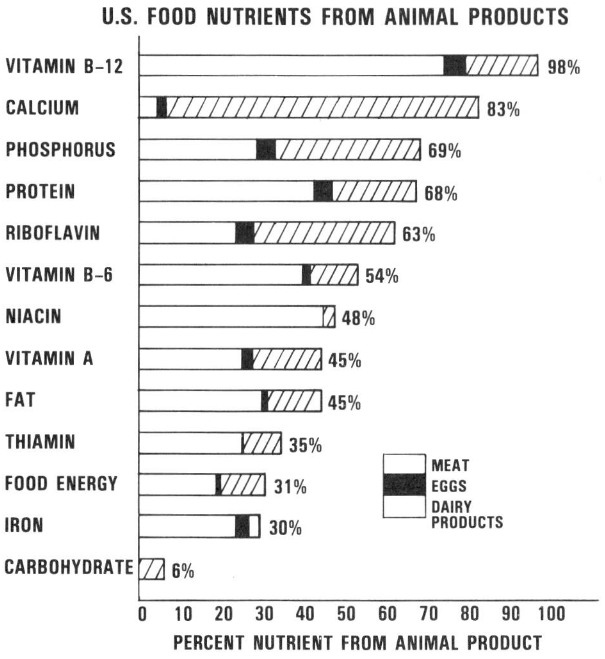

Fig. 14–1. Percentage of food nutrients contributed by animal products of the total nutrient supply in the United States in 1985. (Source: *Agricultural Statistics 1988*, USDA, p. 493, Table 679, data for 1985.)

PART I—MEAT

Although comprising only 5% of the world's population, the people of the United States consume 11.4% of the total world production of meat. The amount of meat consumed in this country varies from year to year (see Fig. 14–2). In 1989, the average per capita meat consumption, exclusive of lard, was 219.8 lb, with distribution of meat as shown in Table 14–1.

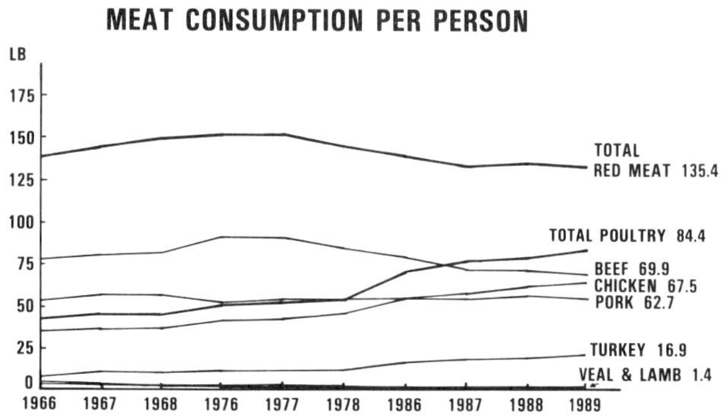

Fig. 14–2. U.S. per capita meat consumption on a retail basis, by kind of meat. Note that per capita beef consumption has gradually declined, while poultry consumption has increased dramatically—almost doubling since 1966. (Source: *National Food Review*, USDA, ERC, April-June, 1989, Vol. 12, Issue 2, p. 2, Table 1)

TABLE 14–1
UNITED STATES PER CAPITA CONSUMPTION OF MEAT, BY KINDS, RETAIL WEIGHT BASIS[1]

Type of Meat	1989 Annual Per Capita Consumption (retail weight basis)	
	(lb)	*(kg)*
Red Meat:		
Beef	69.9	*31.8*
Veal	1.4	*0.6*
Lamb	1.4	*0.6*
Pork	62.7	*28.5*
Total	135.4	*61.5*
Poultry:		
Chicken	67.5	*30.7*
Turkey	16.9	*7.7*
Total	84.4	*38.4*

[1]Data for 1989. Source: *National Food Review*, USDA, ERC, April-June 1989, Vol. 12, Issue 2, p. 2, Table 1.

Part I contains the latest information relative to the processing and preparation of meat for the table.

MEAT ON THE TABLE[1]

Fig. 14-3. Porterhouse steaks.

Fig. 14-4. Beef burgers.

Fig. 14-5. Bacon.

Fig. 14-6. Lamb chops.

[1]Courtesy, National Live Stock and Meat Board, Chicago, IL.

LIVESTOCK MARKETING AND MEAT SUPERVISION

With the growth of the far-flung livestock-marketing system and meat slaughtering, processing, and distribution, it soon became apparent that federal and state supervision was necessary to prevent unfair and discriminatory trade practices, to protect the public health, to ensure human methods of handling animals, and to establish fair and equitable rates and charges for all agencies operating on a given public market. The various services rendered, legislative acts, and state and federal organizations carrying out these functions will be discussed briefly.

FEDERAL MEAT INSPECTION

The federal government requires supervision of establishments which slaughter, pack, render, and prepare meats and meat products for interstate shipment and foreign export; it is the responsibility of the respective states to have and enforce legislation governing the slaughtering, packaging, and handling of meats shipped intrastate, but state standards cannot be lower than federal levels. The meat inspection laws do not apply to farm slaughtering for home consumption, although all states require inspection if the meat is sold.

The purposes of meat inspection are (1) to safeguard the public by eliminating diseased or otherwise unwholesome meat from the food supply, (2) to enforce the sanitary preparation of meat and meat products, (3) to guard against the use of harmful ingredients, (4) to prevent the use of false or misleading names or statements on labels. Personnel for carrying out the provisions of the act are of two types: Professionals or veterinary inspectors who are graduates of accredited veterinary colleges, and non-professional food inspectors who are required to pass a Civil Service examination. In brief, the inspections consist of the following two types:

1. **Antemortem** inspection is made in the pens or as the animals move from the scales after weighing. The inspection is performed to detect evidence of disease or any abnormal condition that would indicate a disease. Suspects are provided with a metal ear tag bearing the notation "U.S. Suspect No. . . . ," and are given special postmortem scrutiny. If in the antemortem examination there is definite and conclusive evidence that the animal is not fit for human consumption, it is *condemned*, and no further postmortem examination is necessary.

Fig. 14–7. Antemortem inspection of cattle being made by a federal veterinarian. Animals that are clearly diseased, emaciated, or otherwise unfit for human consumption are destroyed. Their carcasses may be used only in making inedible grease, fertilizer, or other nonfood products. Animals that appear slightly abnormal on foot are tagged "U.S. Suspect," and are given special postmortem scrutiny. (Courtesy, USDA)

2. **Postmortem** inspection is made at the time of slaughter and includes a careful examination of the carcass and the viscera (internal organs). All good carcasses are stamped "U.S. Inspected and Passed," whereas the inedible carcasses are stamped "U.S. Inspected and Condemned." The latter are sent to the rendering tanks, the products of which are not used for human food.

Fig. 14–8. Two veterinarians making postmortem inspection. (Courtesy, USDA)

In addition to the antemortem and postmortem inspections referred to, the government meat inspectors have the power to refuse the application of the mark of inspection to meat products produced in a plant that is not sanitary. All parts of the plant and its equipment must be maintained in a sanitary condition at all times. In addition, plant employees must wear clean, washable garments, and suitable lavatory facilities must be provided for hand washing.

Meat inspection regulations require the condemnation of all or affected portions of carcasses of animals with various disease conditions, including pneumonia, peritonitis, abscesses and pyemia, uremia, tetanus, rabies, anthrax, tuberculosis, various neoplasms (cancer), arthritic, actinobacillosis, and many others.

Most of the larger meat packers are under federal inspection; hence, they are allowed to ship interstate. Fig. 14–9 shows the proportion of the U.S. meat slaughter that was produced in (1) federally inspected plants, (2) nonfederally inspected plants, and (3) farm slaughter in 1987.

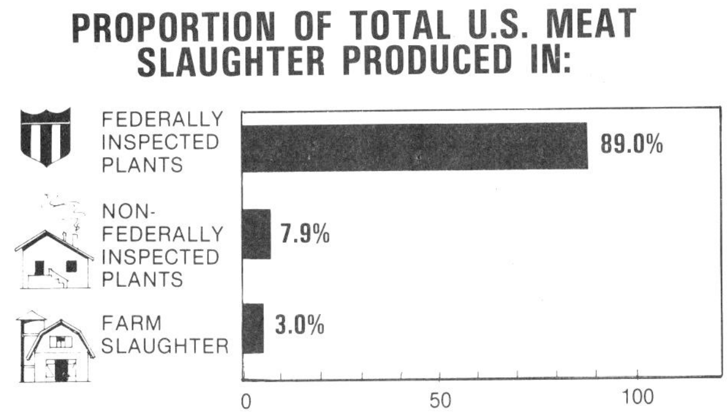

Fig. 14-9. Proportion of total U. S. meat slaughter produced in (1) federally inspected plants, (2) nonfederally inspected plants, and (3) farm slaughter. (Source: *Agricultural Statistics 1988*, USDA, p. 299, Table 442)

STATE MEAT INSPECTION

The Wholesome Meat Act gave the states the option of either conducting their own inspection service, or turning the responsibility over to the federal government. In most of these states, the service is administered by the State Department of Agriculture. Quite frequently, they simply apply the federal regulations.

HEALTH CONTROL OFFICIALS

At every large central market, two groups of health control officials are represented: The U.S. Department of Agriculture and regulatory officials of the state deparment of agriculture. These health control officials are vested with the responsibility of preventing the spread of animal diseases. The federal government has the jurisdiction over interstate shipments, and the laws governing such movement apply uniformly to all parts of the country. The several states have separate and different laws governing the inshipment of livestock from other states and the control of livestock diseases within the state.

BRAND INSPECTION

Brand inspection of western cattle is undertaken primarily to prevent the stealing (rustling) of animals on the range and their subsequent shipment and sale to innocent parties at the big markets. This service is also very useful in recovering stray animals that have become mixed in with other cattle and cattle that, through oversight, have been shipped to market by parties other than their real owners. Brand inspection is supported by, and is under the supervision of, the different stock growers' associations of the range states. In Montana, however, the state government manages the work. Brand inspection is taken care of by inspectors expert in the reading and deciphering of brands. They examine each individual in every shipment of branded cattle.

MEAT GRADING

Broadly speaking, there are three meat grading systems: (1) Federal Grading, (2) Packer Brand Names, and (3) Kosher Meats.

Fig. 14-10. Federal graders shown measuring rib eye of beef before applying the grade with an edible vegetable dye. (Courtesy, *Livestock Breeder Journal*, Macon, GA)

FEDERAL GRADES OF MEAT

At first, federal grading of meat was limited to beef, but it now includes mutton, lamb, calf, veal, and pork carcasses.

Government grading, unlike meat inspection, is not compulsory. Official graders are subjected to the call of anyone who wishes their services (packer, wholesaler, or retailer) with a charge per hour made. Government meat graders are appointed from a list of eligibles submitted by the Civil Service Commission. To qualify, a candidate must have had at least 6 years of suitable practical experience in wholesale meat marketing or grading. Beginners are also trained for a period of time under experienced graders.

Fig. 14–11 shows the proportion of the total U.S. meat

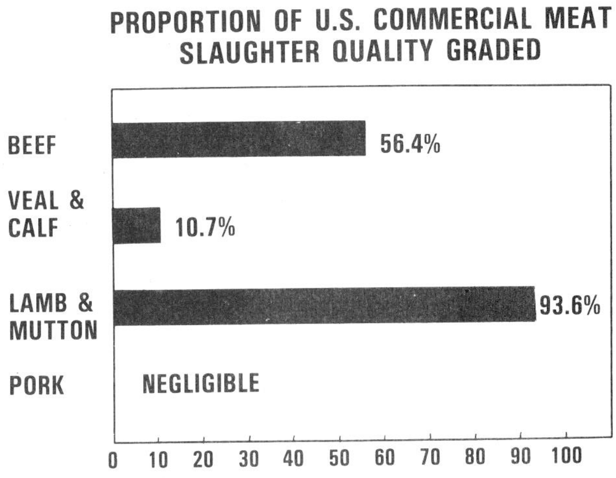

Fig. 14-11. Proportion of U.S. commercial meat production federally graded in 1988. (Source, USDA, Agricultural Marketing Service, Livestock and Seed Division, Meat Grading and Certification Branch)

slaughter (including federally inspected, nonfederally inspected, and farm slaughter) which was quality graded in 1988. It is noteworthy (1) that 93.6% of all lamb and mutton was quality graded, (2) that 56.4% of the beef was quality graded, and (3) that only negligible amounts of pork were quality graded. No farm-produced meat is federally graded.

The grade of meat may be defined as a measure of its degree of excellence based on quality, or eating characteristics of the meat, and the yield, or total proportion of primal cuts. Naturally, the attributes upon which the grades are based vary between species. Nevertheless, it is intended that the specifications for each grade shall be sufficiently definite to make for uniform grades throughout the country and from season to season, and that on-hook grades shall be correlated with on-foot grades.

Both producers and consumers should know the federal grades of meats and have a reasonably clear understanding of the specifications of each grade. From the standpoint of producers, this is important, for, after all, meat over the block is the ultimate objective. From the standpoint of consumers, especially the one who buys most of the meat, this is important, because (1) in these days of self-service, prepackaged meats there is less opportunity to secure the counsel and advice of the meat cutter when making purchases, and (2) the average consumer is not the best judge of the quality of the various kinds of meats on display in the meat counter.

Federally graded meats are so stamped (with edible vegetable dye) that the grade will appear on the retail cuts as well as on the carcass and wholesale cuts. These are summarized in Table 14–2.

TABLE 14–2
QUALITY GRADES OF MEATS, BY CLASSES[1]

Beef[2]	Veal	Mutton and Lamb	Pork
1. Prime[3]	1. Prime	1. Prime[4]	1. U.S. No. 1
2. Choice	2. Choice	2. Choice	2. U.S. No. 2
3. Select	3. Select	3. Select	3. U.S. No. 3
4. Standard	4. Standard	4. Utility	4. U.S. No. 4
5. Commercial	5. Utility	5. Cull[5]	5. U.S. Utility
6. Utility			
7. Cutter			
8. Canner			

[1]In rolling meat, the letters "U.S." precedes each federal grade name. This is important, as only the government-graded meat can be so marked. For convenience, however, the letters "U.S." are not used in this table or in the discussion which follows.

[2]In addition to the quality grades given herein, there are the following yield grades for beef and lamb (and mutton) carcasses: Yield Grade 1, Yield Grade 2, Yield Grade 3, Yield Grade 4, and Yield Grade 5.

[3]Cow beef is not eligible for the Prime grade.

[4]Limited to lamb and yearling carcasses.

[5]Limited to mutton carcasses.

In addition to the quality grades given in Table 14–2, there are the following yield grades for beef and lamb (mutton):

Grade #1
Grade #2
Grade #3
Grade #4
Grade #5

Quality grades are designed to gauge differences in taste (palatability), whereas yield grades identify the percentages of lean meat obtained from a carcass.

Until 1989, meat packers could choose to grade or not to grade beef. But, if they graded, they were *required* to grade for both yield and quality. In 1989, the law was changed, separating quality and yield grades of beef; and allowing packers to choose whether beef carcasses are graded for quality, for yield, or for both quality and yield. But packers could continue to choose between *to grade or not to grade.*

Because of the different grade designations and the smaller number of grades in comparison with beef, the following elucidation is given relative to the grades of pork: Pork is more uniform in quality than any other class of meat and, therefore, there is need for fewer grades. Also, the grades of barrow and gilt carcasses are based on two general considerations: (1) the quality-indicating characteristics of the lean, and (2) the expected combined yields of the four lean cuts (ham, loin, picnic shoulder, and Boston butt). Although the quality of the lean is best evaluated by a direct observation of its characteristics in a cut surface, such observations are impractical in grading carcasses. Thus, in carcasses, the quality of the lean is evaluated indirectly, on the basis of such quality-indicating characteristics as (1) firmness of fat and lean; (2) amount of feathering between the ribs; (3) color of lean; and (4) belly thickness, determined primarily by the thickness of the belly pocket. Research has shown that the actual average thickness of the backfat in relation to carcass length is a rather reliable guide to the yield of the four lean cuts. Therefore, in determining the grade of pork carcasses, the actual thickness of backfat and the carcass length are considered.

Some additional and pertinent facts relative to the federal grades of meat are:

1. **There is no differentiation between steer, heifer, and cow beef.** Federal grades make no distinction between steer, heifer, and cow beef.

2. **Bull and bullock beef are identified.** Bull and bullock beef are identified by class as *bull* and *bullock* beef, respectively.

3. **Lower grades seldom sold as retail cuts.** It is seldom that the lower grades—Cutter and Canner beef, Utility veal, Cull mutton and lamb, and Utility pork—are sold as retail cuts. The consumer, therefore, only needs to become familiar with the upper grades of each kind of meat.

As would be expected, in order to make the top grade in the respective classes, the carcass or cut must possess a very high degree of the attributes upon which grades are based. Fig. 14–12 gives the percentage distribution by grades of the total beef graded during 1988.

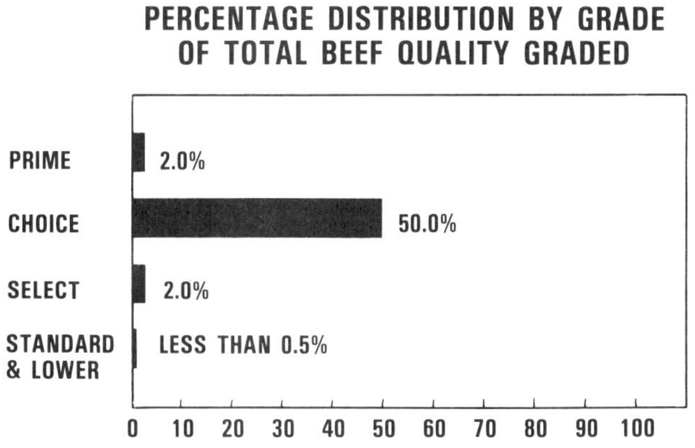

Fig. 14–12. Percentage distribution of beef by quality grades, in 1988. About 56.4% of the commercial beef slaughter was quality graded in 1988. (Source: USDA)

PACKER BRAND NAMES

Practically all packers identify their higher grades of meats with alluring private brands so that the consumer as well as the retailer can recognize the quality of a particular cut. Thus, the top brand names synonymous with the names of the big four early day meat packers and applied to their highest quality beef, pork, and lamb were: Armour's *Star*, Swift's *Premium*, Wilson's *Certified*, and Cudahy's *Puritan.*

A meat packer's reputation depends upon consistent standards of quality for all meats that carry the company brand names. The brand names are also effectively used in advertising campaigns.

KOSHER MEATS

Meat for the Jewish trade—known as kosher meat—is slaughtered and processed according to ancient Biblical laws, called *Kashruth*, dating back to the days of Moses, more than 3,000 years ago. The Hebrew religion holds that God issued these instructions directly to Moses, who, in turn, transmitted them to the Jewish people while they were wandering in the wilderness near Mount Sinai.

The Hebrew word *kosher* means *fit or proper,* and this is the guiding principle in the handling of meats for the Jewish trade. Also, only those classes of animals considered clean—those that both chew the cud and have cloven hooves—are used. Thus, cattle, sheep, and goats—but not hogs—are koshered. Poultry is also koshered.

Contrary to common belief, both forequarters and hindquarters of kosher slaughtered meat may be used by orthodox Jews. But because all the veins must be removed before the meat is delivered to the customer, the Jewish trade usually confines itself to the forequarters, from which the veins can be removed with a minimum of tearing of the flesh. In order to devein a hindquarter, it is necessary to cut it up into such small pieces that it is very unattractive and unsalable for anything but ground meat or stews. Because the forequarters do not contain such choice cuts as the hinds, the kosher trade attempts to secure the best possible fores; thus, there is a secondary reason the kosher trade demands well-finished beef; namely, the ban against the use of lard in cooking, and the need for beef fat therefor.

Kosher meat must be sold by the packer or the retailer within 72 hours after slaughter, or it must be washed (a treatment known as *begiss*, meaning to wash) and reinstated by a representative of the synagogue every subsequent 72 hours. At the expiration of 216 hours after the time of slaughter (after begissing 3 times), however, it is declared *trafeh,* meaning forbidden food, and is automatically rejected for kosher trade. It is then sold in the regular meat channels. Because of these regulations, kosher meat is moved out very soon after slaughter. Also, it is easier to devein the meat while it is still warm than after it has been chilled.

The Jewish law also provides that before kosher meat is cooked, it must be soaked in water for one-half hour. After soaking, the meat is placed on a perforated board in order to drain off the excess moisture. It is then sprinkled liberally with salt. One hour later, it is thoroughly washed. Such meat is then considered to remain kosher as long as it is fresh and wholesome.

BEEF AND VEAL

Beef exceeds pork in farm production but not in processing. Considerably less beef than pork is cured. The vast majority of it is either consumed fresh or frozen. With some minor variations, veal is slaughtered and processed in the same manner as beef.

DRESSING PERCENTAGE OF CATTLE AND CALVES

Cattle are not all beef, and beef is not all steak! It is important, therefore, that those who slaughter animals and those who purchase wholesale or retail cuts know the approximate (1) percentage yield of chilled carcass in relation to the weight of the animal on foot, and (2) yield of different retail cuts.

Fig. 14–13 illustrates these points. As noted, an average 1,050-lb steer will yield about a 650-lb carcass or 448.8 lb retail cuts, only 29.4 lb of which will be porterhouse and T-bone steaks.

A STEER IS NOT ALL STEAK!

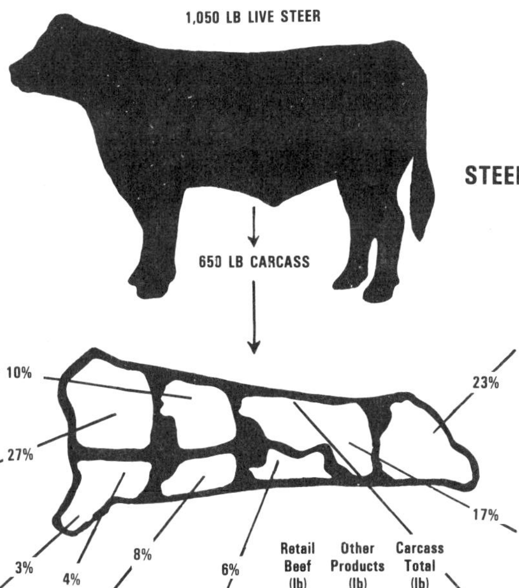

STEER CARCASS BREAKDOWN

	Retail Beef (lb)	Other Products (lb)	Carcass Total (lb)
RIB (62.0 lb)			
Rib roast	25.4		
Rib steak	13.0		
Short ribs	4.9		
Braising beef, ground beef	6.5		
Fat, bone		12.2	
Total	49.8	12.2	62.0
CHUCK (176.8 lb)			
Blade roast	63.6		
Arm pot roast	23.9		
Cross rib pot roast	22.1		
Stew, ground beef	34.4		
Fat, bone		32.8	
Total	144.0	32.8	176.8
FORESHANK (18.9 lb)			
Shank meat	9.0		
Bone		9.9	
Total	9.0	9.9	18.9
BRISKET (27.3 lb)			
Boneless brisket	11.0		
Fat, bone		16.3	
Total	11.0	16.3	27.3

	Retail Beef (lb)	Other Products (lb)	Carcass Total (lb)
PLATE (54.0 lb)			
Plate, stew, short ribs	43.2		
Fat, bone		10.8	
Total	43.2	10.8	54.0
FLANK (37.1 lb)			
Flank steak	3.7		
Ground beef	13.8		
Fat		19.6	
Total	17.5	19.6	37.1
TOTAL	**448.8**	**201.2**	**650.0**

	Retail Beef (lb)	Other Products (lb)	Carcass Total (lb)
ROUND (147.6 lb)			
Top round	22.5		
Bottom round	21.7		
Tip	14.0		
Rump	5.1		
Stewing beef, ground beef, etc.	26.3		
Fat, bone		58.0	
Total	89.6	58.0	147.6
LOIN (110.5 lb)			
Porterhouse steak	19.5		
T-bone steak	9.9		
Top loin steak	5.4		
Sirloin steak	43.2		
Ground beef	3.0		
Fat, bone		29.5	
Total	81.0	29.5	110.5
MISCELLANEOUS (15.8 lb)			
Kidney, hanging tender	3.7		
Fat, suet, shrink, and loss		12.1	
Total	3.7	12.1	15.8

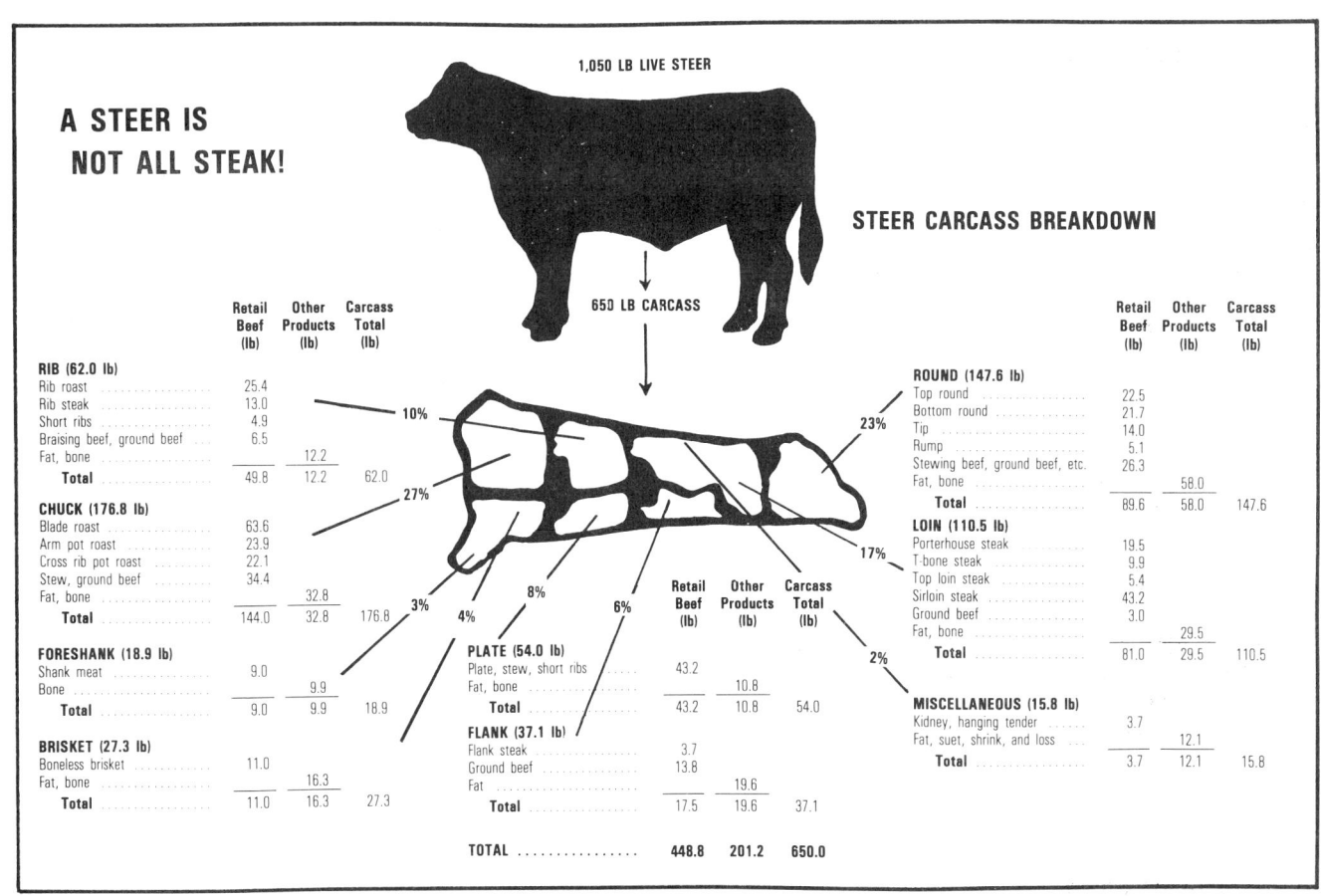

Fig. 14–13. Cattle are not all beef, and beef is not all steak! This shows the approximate (1) percentage yield of carcass in relation to the weight of the animal on foot, and (2) the yield of different retail cuts. Note that a 1,050-lb live steer produces approximately a 650-lb carcass, and ends up with only 448.9 lb of retail beef. Note, too, the small amount of steak. (Source: Adapted by the author from *Meat Facts*, published by the American Meat Institute, Washington, DC. Data derived from USDA and industry figures.)

Dressing percentage may be defined as the percentage yield of chilled carcass in relation to the weight of the animal on foot. For example, a steer which weighed 1,050 lb on foot and yielded a carcass weighing 650 lb may be said to have a dressing percentage of 61.9. The offal—so-called because formerly (with the exception of the hide, tallow, and tongue) the offal (waste) was thrown away—consists of the blood, head, shanks, tail, hide, viscera, and loose fat.

A high carcass yield is desirable because the carcass is much more valuable than the by-products. Although the packers have done a marvelous job in utilizing by-products, about 88% of the income from cattle and calves is derived from the sale of the carcass and only 12% from the by-products. Thus, the estimated dressing percentage of slaughter cattle is justifiably a major factor in determining the price or value of the live animal.

The chief factors determining the dressing percentage of cattle are (1) the amount of fill, (2) the finish or degree of fatness, (3) the general quality and refinement (refinement of head, bone, hide, etc), and (4) the size of udder. The better grades of steers have the highest dressing percentage, with thin Canner cows showing the lowest yield. Table 14-3 gives the dressing percentages that may be expected for different grades of cattle and calves.

TABLE 14-3
DRESSING PERCENTAGE OF CATTLE AND CALVES, BY GRADE[1]

Grade	Range	Average
Cattle		
Prime	62-67	64
Choice	59-65	62
Select	58-62	60
Standard	55-60	57
Commercial	54-62	57
Utility	49-57	53
Cutter	45-54	49
Canner	40-48	45
Calves and Vealers		
Prime	59-65	62
Choice	59-60	58
Select	52-57	55
Standard	47-54	51
Utility	40-48	46

[1]From USDA sources.

The average liveweights of cattle and calves dressed by commercial meat-packing plants, and their percentage yields in meats, for the year 1987 are given in Table 14-4. As shown, cattle average 60% and calves 61%.

TABLE 14-4
AVERAGE LIVEWEIGHT, CARCASS YIELD, AND DRESSING PERCENTAGES OF ALL CATTLE AND CALVES COMMERCIALLY SLAUGHTERED IN THE UNITED STATES IN 1987[1]

	Average Liveweight		Dressed Weight		Dressing Percentage
	(lb)	*(kg)*	(lb)	*(kg)*	(%)
Cattle	1,105	*502*	659	*300*	60
Calves	249	*113*	151	*69*	61

[1]Source: *Livestock and Meat Statistics, 1984-88*, Stat. Bull. No. 784, USDA, pp. 107, 132.

DISPOSITION OF THE BEEF CARCASS

Beef carcasses are disposed of in one of three ways: (1) block beef; (2) fabricated, boxed beef; or (3) processed meats.

1. **Block beef.** Block beef refers to beef that is suitable for sale over the block. Such beef is purchased by the retailer in sides, quarters, or wholesale cuts.

2. **Fabricated, boxed beef.** Fabricated beef is broken into subprimal cuts, vacuum sealed, boxed, and shipped to retailers.

Fig. 14-14. Boxed beef. After chilling, the carcass is fabricated, or broken into subprimal cuts, vacuum sealed, boxed, moved into storage, then trucked to retailers. (Courtesy, Iowa Beef Processors, Dakota City, NE)

3. **Processed meats.** Beef that is not suitable for sale as block or fabricated meat is boned out and disposed of as boneless cuts, is canned, is made into sausage, or is cured by drying and smoking. It is estimated that about one-fifth of all slaughter cattle are disposed of as processed meats.

BEEF CUTS AND HOW TO COOK THEM

Fig. 14-15. Retail meat market. (Courtesy, National Live Stock and Meat Board, Chicago, IL)

Fig. 14–16 shows the wholesale and retail cuts of beef and gives the recommended method(s) of cooking each.

In order to buy and/or process beef wisely, and to make the best use of each part of the carcass, the consumer should be familiar with the types of cuts and how each should be processed.

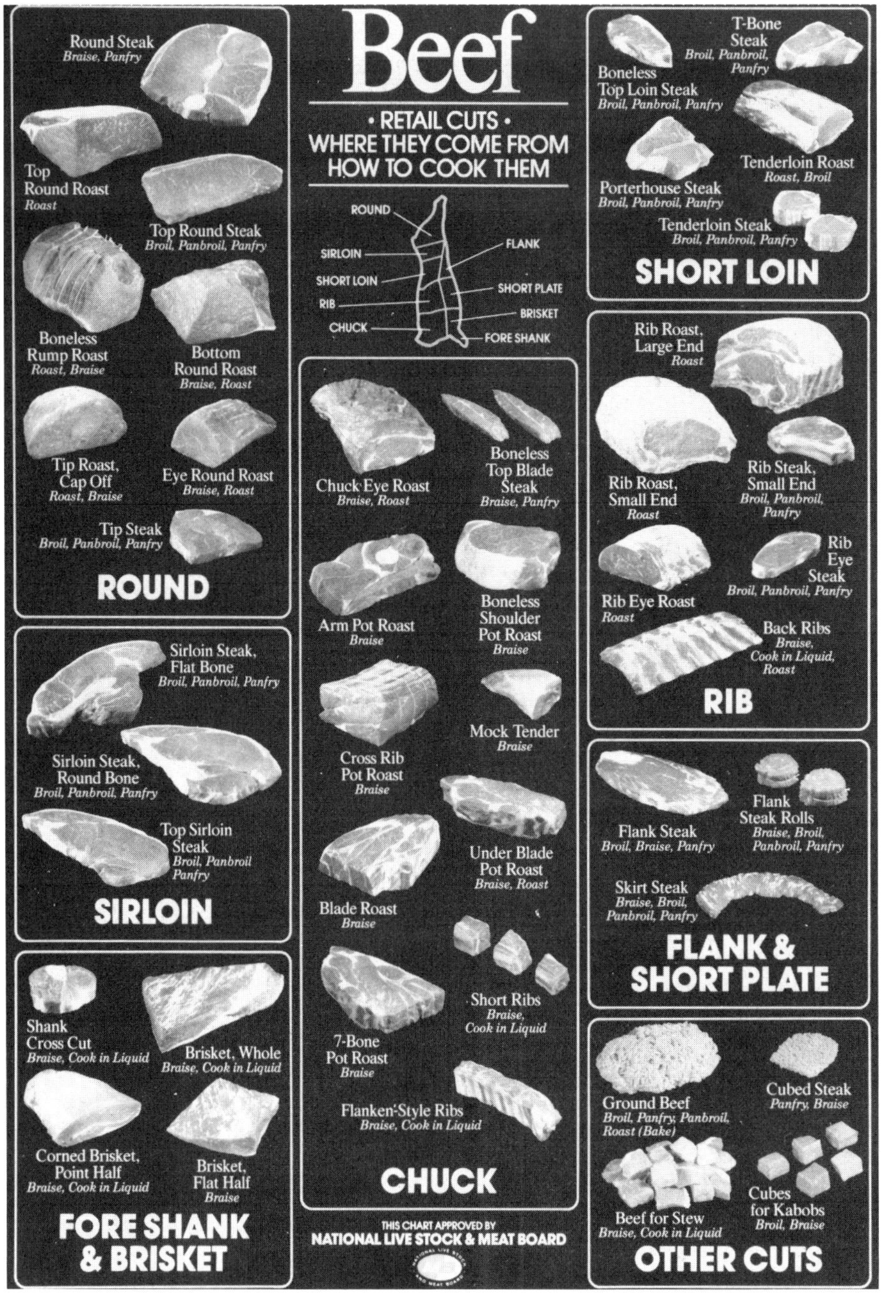

Fig. 14–16. Beef. The retail cuts of beef; where they come from; and how to cook them. (Courtesy, National Live Stock and Meat Board, Chicago, IL)

LAMB

Most of the sheep meat eaten in this country is lamb. Accordingly, the discussion which follows will be limited to lamb, and mutton will not be covered.

In general, the packer classes as lamb all carcasses in which the forefeet are removed at the break-joint or lamb-joint. This point—which can be severed on all lambs, most yearling wethers, and some yearling ewes (ewes mature earlier than wethers or rams)—is a temporary cartilage located just above the ankle. In lambs, the break-joint has four well-defined ridges that are smooth, moist, and red. In yearlings, the break-joint is more porous and dry. In mature sheep, the cartilage is knit or ossified and will no longer break, thus making it necessary to take the foot off at the ankle instead. This makes a round-joint (commonly called spool-joint). All carcasses possessing the round-joint are sold as mutton rather than lamb.

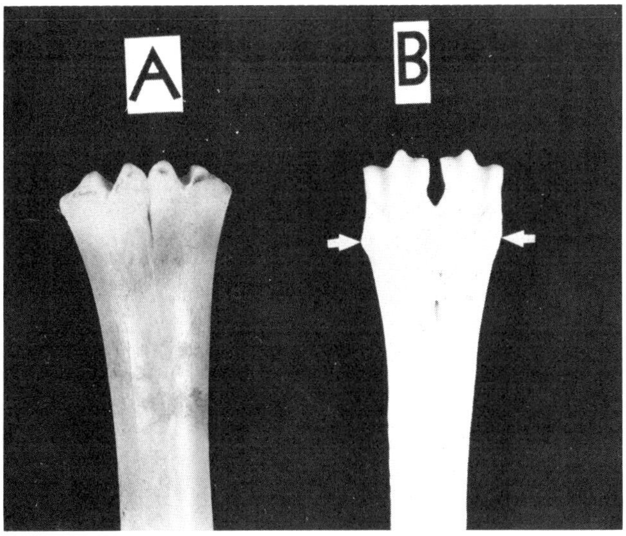

Fig. 14–17. The two joints of the foreleg of a sheep: A, the break-joint or lamb-joint, and B, the round-joint or spool-joint. Arrow indicates the location of the break-joint or ossification. All carcasses possessing the round-joint are sold as mutton rather than lamb. (Courtesy, Washington State University, Pullman)

Even though lamb constitutes a smaller proportion of the total U.S. meat supply than any other class of meat, these facts are noteworthy: (1) lamb, like other meats, is easily digested, and, therefore, is widely used in the diet of convalescents, (2) there is less religious prejudice against lamb and mutton than any other meat except fish, and (3) fewer lamb and mutton carcasses (percentagewise) are condemned by meat inspectors than any other class of livestock.

DRESSING PERCENTAGE OF LAMBS

Table 14–5 gives the dressing percentages that may be expected from the different grades of sheep and lambs. As would be expected, the highest dressing percentage is obtained when animals are slaughtered following shearing. Lambs of mutton breeds yield a somewhat higher percentage of carcass than so-called wool breeds. The offal and by-products from the slaughter of sheep and lambs consist of the blood, pelt, feet, head, and viscera.

The average liveweight of sheep and lambs dressed by federally inspected meat-packing plants and their percentage of yield in meat for the year 1987 was as shown in Table 14–6.

TABLE 14–5
DRESSING PERCENTAGE OF LAMBS AND SHEEP (MUTTON) BY GRADE[1]

Lambs (wooled)			Sheep (excludes yearlings)		
Grade	Range	Average	Grade	Range	Average
Prime	49–55	52	Choice	49–54	52
Choice	47–52	50	Good	47–52	49
Good	45–49	47	Utility	44–48	46
Utility	43–47	45	Cull	40–46	43
Cull	40–45	42			

[1]From USDA sources.

TABLE 14–6
AVERAGE LIVEWEIGHT, CARCASS YIELD, AND DRESSING PERCENTAGES OF ALL SHEEP AND LAMBS COMMERCIALLY SLAUGHTERED IN THE U.S. IN 1987[1]

	Average Liveweight		Average Dressed Weight		Average Dressing Percentage
	(lb)	*(kg)*	(lb)	*(kg)*	(%)
Sheep and lambs	120	*55*	60	*27*	50

[1]Source: *Livestock and Meat Statistics, 1984–88*, Stat. Bull. No. 784, USDA, pp. 119, 135, 136.

LAMB CUTS AND HOW TO COOK THEM

The two major wholesale cuts of lamb are the (1) hindsaddle, and (2) foresaddle. The division into hindsaddle and foresaddle is made between the twelfth and thirteenth rib, with one pair of ribs remaining in the hindsaddle. Each of these two cuts comprises about 50% of the carcass weight.

The hindsaddle is further divided into the leg and loin. The foresaddle is subdivided into the shoulder, rib, foreshank, and breast.

Fig. 14–18 shows the common retail cuts of lamb, where they come from, and how to cook them.

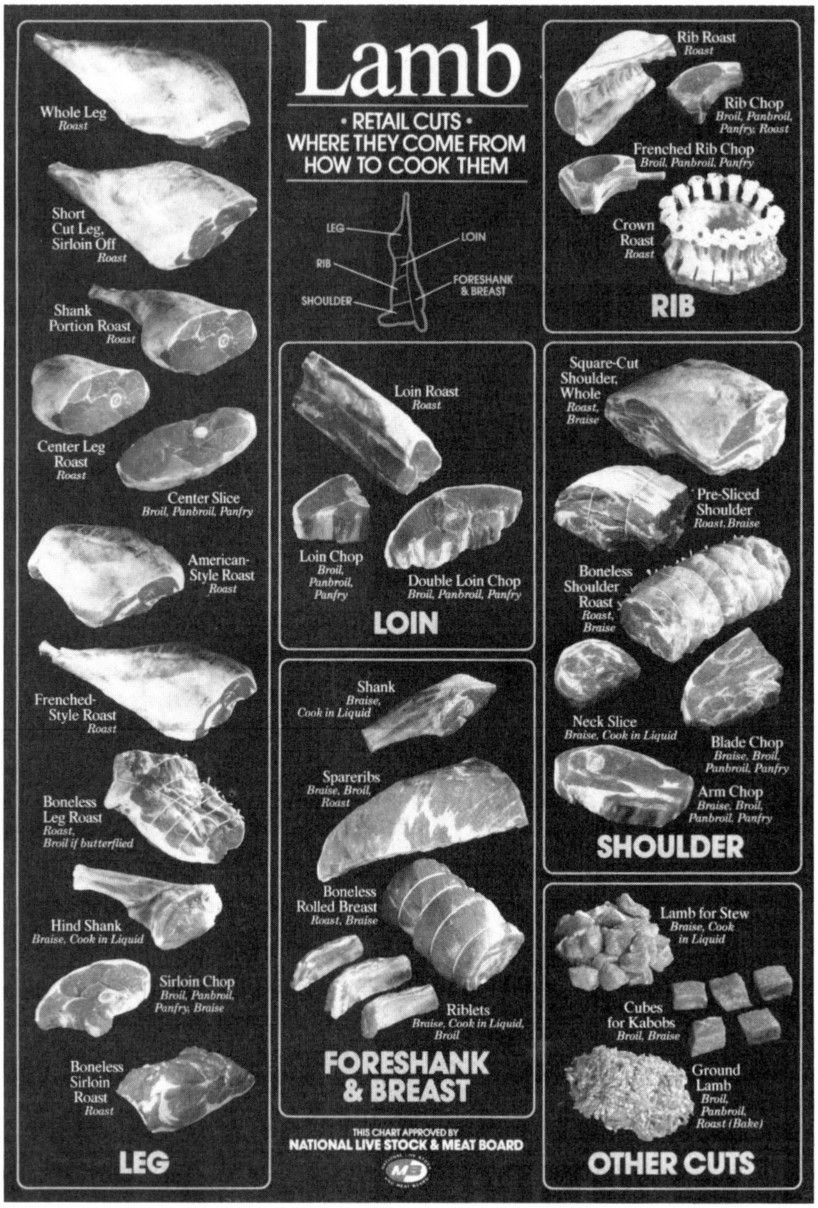

Lamb
• RETAIL CUTS •
WHERE THEY COME FROM
HOW TO COOK THEM
LEG
LOIN
RIB
FORESHANK & BREAST
SHOULDER
Whole Leg Roast
Short Cut Leg, Sirloin Off Roast
Shank Portion Roast Roast
Center Leg Roast Roast
Center Slice Broil, Panbroil, Panfry
American-Style Roast Roast
Frenched-Style Roast Roast
Boneless Leg Roast Roast, Broil if butterflied
Hind Shank Braise, Cook in Liquid
Sirloin Chop Broil, Panbroil, Panfry, Braise
Boneless Sirloin Roast Roast
LEG
Loin Roast Roast
Loin Chop Broil, Panbroil, Panfry
Double Loin Chop Broil, Panbroil, Panfry
LOIN
Shank Braise, Cook in Liquid
Spareribs Braise, Broil, Roast
Boneless Rolled Breast Roast, Braise
Riblets Braise, Cook in Liquid, Broil
FORESHANK & BREAST
THIS CHART APPROVED BY NATIONAL LIVE STOCK & MEAT BOARD
Rib Roast Roast
Rib Chop Broil, Panbroil, Panfry, Roast
Frenched Rib Chop Broil, Panbroil, Panfry
Crown Roast Roast
RIB
Square-Cut Shoulder, Whole Roast, Braise
Pre-Sliced Shoulder Roast, Braise
Boneless Shoulder Roast Roast, Braise
Neck Slice Braise, Cook in Liquid
Blade Chop Braise, Broil, Panbroil, Panfry
Arm Chop Braise, Broil, Panbroil, Panfry
SHOULDER
Lamb for Stew Braise, Cook in Liquid
Cubes for Kabobs Broil, Braise
Ground Lamb Broil, Panbroil, Roast (Bake)
OTHER CUTS

Fig. 14–18. Retail cuts of lamb—where they come from, and how to cook them. (Courtesy, National Live Stock & Meat Board, Chicago, IL)

PORK

Generally, pork is cheaper than beef or lamb, and the lower grades are more tender and palatable than the comparable lower grades of other meats. Also, more hogs are farm slaughtered than any other class of animals.

DRESSING PERCENTAGE OF HOGS

Pigs are not all pork chops! It is important, therefore, that those who slaughter hogs and the consumer know the approximate (1) percentage of yield of chilled carcass in relation to the weight of the animal on foot, and (2) yield of different retail cuts. Fig. 14–19 illustrates these points. As noted, on a liveweight basis only about 7.5% of a pig is center cut pork chops.

The average liveweight of hogs, dressed packer style by federally inspected meat-packing plants, and their percentage yield in meat for the years 1977 and 1987 were as shown in Table 14–7.

TABLE 14–7
AVERAGE LIVEWEIGHT, CARCASS YIELD, AND DRESSING PERCENTAGES OF ALL HOGS COMMERCIALLY SLAUGHTERED IN THE U.S.[1]

Hogs	Average Liveweight		Average Dressing Weight[2]		Dressing
	(lb)	*(kg)*	(lb)	*(kg)*	(%)
1977	237	*107*	170	*77*	71.7
1987	248	*113*	177	*80*	71.4

[1]*Livestock and Meat Statistics, 1984-88*, USDA, Stat. Bull. No. 784, pp. 107, 135, 136.
[2]Packer style.

PIGS ARE NOT ALL PORK CHOPS!

232 LB LIVE HOG

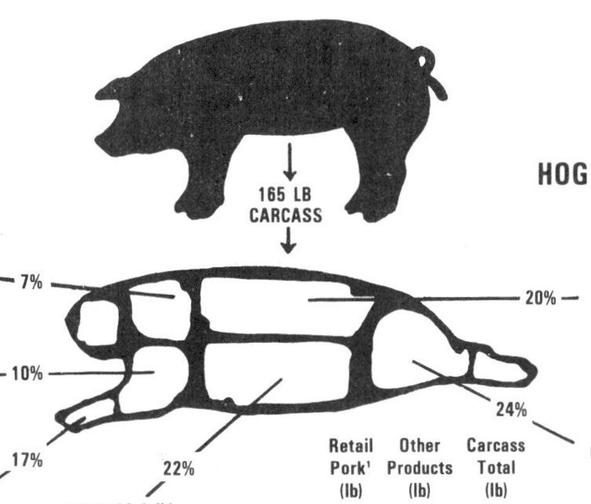

SHOULDER (28.0 lb)	Retail Pork[1] (lb)	Other Products (lb)	Carcass Total (lb)
Boston Butt			
Blade steaks	7.0		
Blade roast	4.2		
Trimmings	0.8		
Total	12.0		12.0
Picnic			
Arm roast	7.7		
Trimmings	3.7		
Skin, fat, bone		4.6	
Total	11.4	4.6	16.0
MISCELLANEOUS (28.3 lb)			
Jowls, feet, tails, neckbones, etc.	6.5		
Trimmings	3.9		
Fat, skin, bone		14.4	
Shrink and loss		3.5	
Total	10.4	17.9	28.3

SIDE (36.4 lb)	Retail Pork[1] (lb)	Other Products (lb)	Carcass Total (lb)
Cured bacon	18.6		
Spareribs	6.4		
Trimmings	9.4		
Fat		2.0	
Total	34.4	2.0	36.4
TOTAL	126.3	38.7	165.0

HOG CARCASS BREAKDOWN

	Retail Pork[1] (lb)	Other Products (lb)	Carcass Total (lb)
LOIN (32.9 lb)			
Blade roast	8.0		
Center chops	17.3		
Sirloin roast	5.6		
Fat		2.0	
Total	30.9	2.0	32.9
HAM (39.4 lb)			
Cured Ham	20.1		
Fresh Ham	4.1		
Trimmings	3.0		
Skin, fat, bone		12.2	
Total	27.2	12.2	39.4

[1]Retail cuts on semiboneless basis. Fully boneless would show lower retail weight.

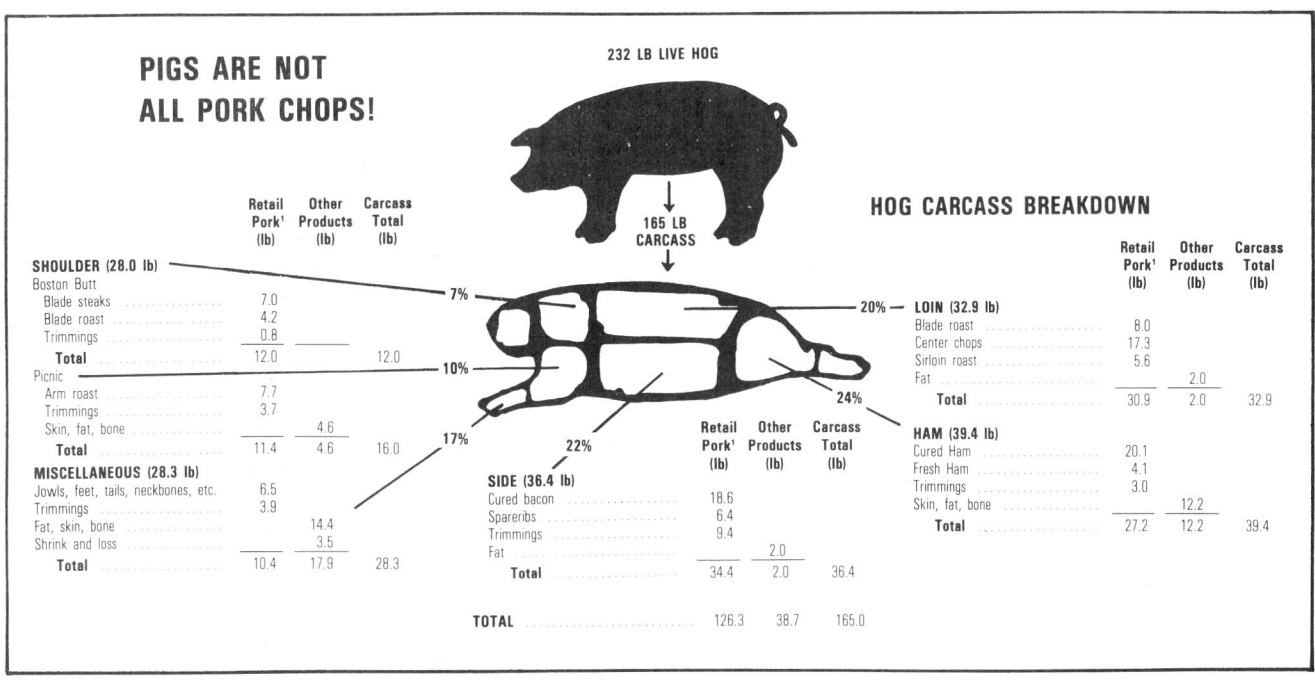

Fig. 14-19. Pigs are not all pork chops! This shows the approximate (1) percentage of yield of carcass in relation to the weight of the animal on foot, and (2) yield of different retail cuts. Note that a 232-lb live hog produces approximately a 165-lb carcass, and ends up with only 126.3 lb of retail pork. Note, too, the small amount of center chops, only 17.3 lb. (Source: Adapted by the author from *Meat Facts*, published by the American Meat Institute, Washington, DC. Data derived from USDA and industry sources.)

DISPOSITION OF THE PORK CARCASS

The handling of pork differs from that of beef and lamb in that only a relatively small percentage, about 30%, of the pork is sold fresh. The remaining 70% is either cured by various methods, rendered into lard, or manufactured into meat products, In general, loins, shoulders, and spareribs are most likely to be sold as fresh cuts. But it must be remembered that practically every pork cut may be cured, and, under certain conditions, is cured. Because pork is well adapted to curing, it has a decided advantage over beef and mutton, which are sold almost entirely in the fresh state. The hog market is stabilized to some extent by this factor.

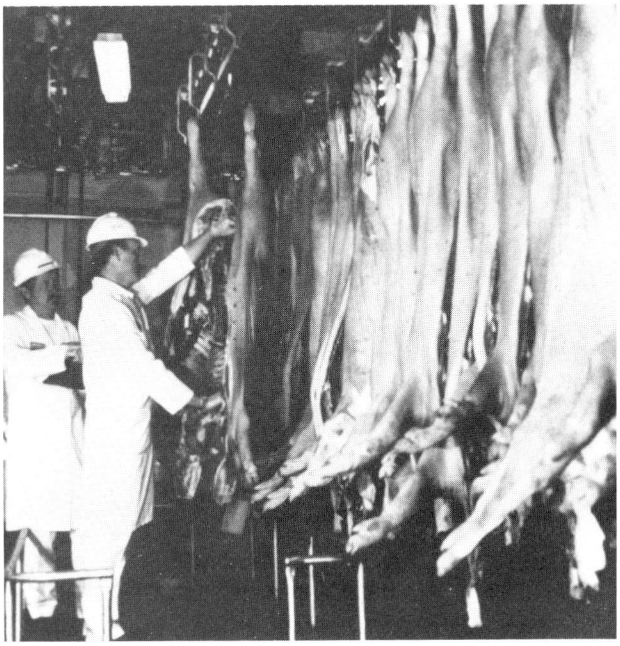

Fig. 14-20. Pork carcasses. (Courtesy, National Pork Producers Council, Des Moines, IA)

PORK CUTS AND HOW TO COOK THEM

Market hogs weighing from 240 lb will yield from 54% of their liveweight in the four primal cuts: the ham, loin, picnic shoulder, and Boston butt. However, because of the relatively higher value per pound of these cuts, they make up three-quarters of the value of the entire carcass.

Fig. 14–21 shows retail cuts of pork, where they come from, and how to cook them. Also, the wholesale cuts of pork are shown in the drawing immediately below the heading.

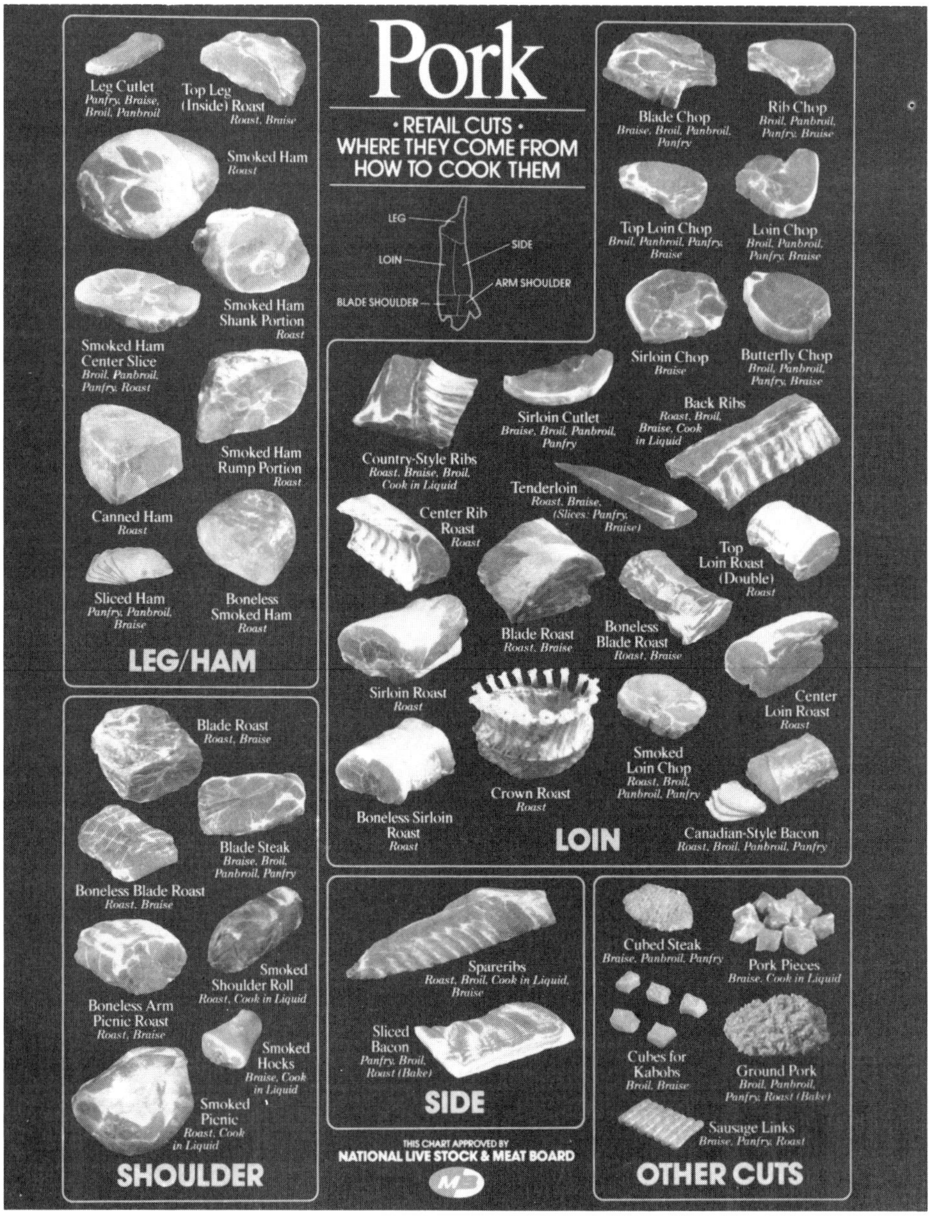

Fig. 14–21. PORK, retail cuts, where they come from, and how to cook them. Note, too, that the wholesale cuts are shown in the drawing immediately below the heading. (Courtesy, National Live Stock and Meat Board, Chicago, IL)

MEAT PRESERVATION

From time immemorial, one of humankind's major food problems has been that of preserving meats over a period of time and in a condition suitable as food. Fundamentally, meat preservation is a matter of controlling putrifactive bacterial action. Various methods of preserving meats have been practiced through the ages, the most common of which are (1) drying, (2) smoking, (3) salting, (4) freezing, (5) canning, and (6) making into sausages.

In this country, meat curing is largely confined to pork, primarily because of the keeping qualities and palatability of cured pork products. Considerable beef is dried or corned and some lamb and veal are cured, but none of these is of such great importance as cured pork.

No phase of meat preservation and merchandising has received greater interest in recent years than frozen meats.

Fig. 14-22. Indians drying and smoking venison.

FREEZING MEATS

Freezing is not a new method of meat preservation, for in arctic regions meats have been frozen since time immemorial. But special freezing methods are important recent developments. The rapid growth of freezer lockers, and the increased popularity of food freezers, have made this method of meat preservation available in homes throughout America.

When properly prepared and stored, frozen meats resemble fresh meats in appearance, flavor, appetite appeal, and food value; and they furnish a welcome diversion to the familiar stocks of canned, salted, and cellar stored food.

Among the reasons for the increased interest in frozen meats in recent years are the following:

1. Uniform supplies of meats can be available throughout the year by freezing. This is particularly advantageous for such products as lamb, which is highly seasonable.
2. The consumer is accepting frozen foods in greater quantities, especially frozen prepared foods, fruits, and vegetables. Hence, consumer confidence in frozen food has improved; and with it, frozen meat is becoming more acceptable.
3. Home freezers are in widespread use. With frozen food space available, there is a tendency to fill it up by purchasing greater quantities in frozen form.
4. Improved quality control is possible in frozen meat merchandising. Cutting and packaging can be done at scheduled times and under strict supervision, rather than as dictated by the need to refill the fresh meat display case.
5. Central fabricating (cutting) is facilitated by frozen meat merchandising, thereby eliminating or greatly reducing the need for in-store cutting.
6. Improved utilization of the entire carcass results from frozen meat merchandising. Cuts do not depreciate in retail display cases as is true in fresh meat merchandising; and shrinkage and spoilage are practically eliminated. As a result, the retailer is able to merchandise all of the cuts to the best advantage.
7. No case of botulism from frozen foods is on record, and 3 weeks of zero storage will kill any trichinae (the parasites that cause trichinosis) in pork.

SELECTING, PREPARING, FREEZING, THAWING, AND COOKING MEATS

Generally, the local freezer locker operator(s) has the expertise, and is available and willing, to provide the necessary help, advice, and/or literature relative to selecting, preparing, and freezing meats. Also, reliable literature and information is usually available from manufacturers of home food freezers and from the Home Demonstration Agent in the County Agricultural Extension Office. Information on how long meat can be frozen, and on how to thaw and cook frozen meats, follows.

• **How long can meat be frozen?**[2]—This is a controversial subject. It depends upon many factors, such as proper slaughtering, chilling, aging, wrapping, slicing, and the speed with which meat is frozen after wrapping. On the average, however, it appears that meat may be safely stored at 0°F for the periods shown in Table 14-8.

TABLE 14-8
STORAGE GUIDE FOR FROZEN MEATS

Product	Storage Period	Comments
	(month)	
Beef	6 to 12	
Fresh pork	4 to 6	
Lamb and veal	6 to 9	
Ground beef and lamb	3 to 4	
Ground pork	4 to 6	
Sausage	3	
Liver, heart, etc.	3 to 4	
Ready-cooked beef stew, without potatoes	6	**To prepare:** Remove from package. Heat one hour on low heat in covered saucepan, or in casserole for one hour in oven at 325°F.
Baked ham	3 to 4	**To prepare:** Thaw; then heat 15 minutes per pound in 350°F oven.
Ready-cooked meat loaf	6	**To prepare:** Heat for 1 to 1½ hours at 350°F or thaw at room temperature and serve cold.

• **How to thaw and cook frozen meats**—Frozen meat may be either (1) removed from frozen storage, unwrapped, and cooked from the frozen state, or (2) left wrapped and thawed before being cooked.

The following pointers are pertinent to this subject:

1. **Thawing time.** The following rules of thumb may be used in estimating the thawing time:

In a refrigerator	5 to 10 hours per pound.
Room temperature	2 to 3 hours per pound.
Before a fan	1 to 1½ hours per pound.

[2]An interesting historical sidelight on the length of time that meat may be preserved by freezing is found in the discovery of elephants, once hunted by primitive people, frozen in ice cliffs along the coast of Siberia. Perfectly preserved in their airtight tombs, these elephants have been kept in frozen storage for unnumbered thousands of years.

Unless meat is thawed slowly, there will be excessive loss of juices and the cooked product will be dry and tasteless.

2. **Increased roasting time from frozen state.** In roasting meat from the frozen state, add an additional 20 minutes per pound of meat. Better yet, if this procedure is followed, (a) remove the frozen meat from the freezer, (b) place it in the oven, and (c) when it is completely thawed, place a meat thermometer in the thickest part to determine desired doneness.

3. **Increased broiling time from frozen state.** In broiling steaks from the frozen state, add an additional 20 to 30 minutes.

4. **Cooking time.** The cooking time for thawed meat is the same as for fresh meat.

5. **Ground meats and heart.** Such meats are usually thawed completely before cooking.

METHODS OF CURING PORK ON THE FARM

Farm meat curing other than freezing is largely confined to pork, primarily because of the keeping qualities and palatability of cured pork products. Accordingly, the discussion in this section will be limited to pork only.

The secret of pork curing is to use good sound meat, the correct curing method and formula, clean containers, and to be fortunate enough to secure cool curing weather.

The primary meat curing ingredients and the functions of each are:

1. **Salt.** It preserves by inhibiting or retarding the bacteria that cause spoilage. In excess quantities, salt makes meat less palatable.

2. **Sugar.** Sugar is used to offset some of the harshness of salt. A combination of 7 lb of salt and 3 lb of white or brown sugar is a basic mixture.

3. **Saltpeter** (nitrate of potash). It is used to enhance the bright red color of the lean meat. Also, and most important, nitrite in meat ensures against the development of botulism.

4. **Commercial cures.** In addition to salt, or salt and sugar, commercial cures frequently contain spices and flavorings to impart characteristic flavor, appearance, and/or aroma.

Note well: There is considerable concern over nitrite-cured bacon, which produces low levels of nitrosamines when cooked at high temperatures, because nitrosamines have been found to be carcinogenic in rats. An expert panel appointed by the U.S. Secretary of Agriculture has recommended that nitrates be eliminated entirely from meat cures, that the initial use of nitrite be limited to 156 ppm, and that the residual nitrite be restricted according to the class of product from 50 ppm in sterile canned meats up to 125 ppm in cooked sausage types like frankfurters. The desire and intent is to lower the levels of nitrite so cooking does not produce nitrosamines, yet maintain enough nitrite to prevent botulism.

Table 14–9 details the common methods of curing pork on the farm.

TABLE 14–9
METHODS OF CURING PORK

Curing Method and Formula	Kinds of Cuts to Which Adapted	Curing Directions	Comments
Dry-salting: Salt only.	Backs, sides, and other cuts of pork; especially heavy fat backs and bacons.	Rub and sprinkle over the cuts 7 to 10 lb of salt per 100 lb of meat. Pile meat closely in layers and resalt at intervals. Let cure for 3 to 4 weeks. A curing temperature of about 40°F is preferable.	Dry-salting was formerly the most common method of preserving farm pork, but for the most part, it has now been replaced by the dry-cure and sweet-pickle methods.
Dry-cure (or dry-sugar cure): Mixture per 100 lb meat: 7 lb table salt 3 lb sugar (brown, or granulated sugar)	Ham, bacon, and shoulders.	**For bacon**, (1) rub on 1 oz of the dry-cure mix per pound of bacon (sprinkling any surplus of this amount on rib side), and (2) let cure 7 days per inch of thickness (thus a side of bacon 2 in. thick would be left in cure 14 days). **For hams**, (1) rub on 1 oz (if hams weigh over 20 lb use 1¼ to 1½ oz) of the dry-cure mix per pound of ham, applying in 3 rubbings at 3 to 5 day intervals, and (2) let cure 7 days per inch of thickness (as measured directly back of the aitch bone). **For shoulders**, (1) rub on 1 oz of the dry-cure mix per pound of shoulder, applying in 2 rubbings at 3 to 5 day intervals, and (2) let cure 7 days per inch of thickness Meats in cure should be placed on a clean shelf, on a table, or in a barrel with a drain at the bottom so the juice can drain off.	The dry-cure produces more palatable products than dry-salting. In comparison with the sweet-pickle cure, the dry-cure is more rapid, requires less equipment, results in a higher shrink, and gives a stronger cure. It is the preferred cure in warm areas. Curing should not be attempted where the temperature is over 50°F because spoilage is apt to take place before the cure can penetrate the ham or shoulder.

(Continued)

TABLE 14-9 *(Continued)*

Curing Method and Formula	Kinds of Cuts to Which Adapted	Curing Directions	Comments
Immersed in brine method (or sweet-pickle cure): Mixture per 100 lb meat: 7 lb table salt 3 lb sugar (brown or granulated sugar) 5 gal water (8 lb/gal, boiling water or unboiled, cold water—depending on its purity) As noted, the ingredients are the same in the dry-cure and in the sweet-pickle cure; the difference being that in the dry-cure the mixture is applied directly to the meat, while in the brine cure the mixture is dissolved in water and the meat submerged in it.	Hams, bacon, and shoulders.	Weigh out the ingredients of the formula (water may be measured), and stir until thoroughly mixed. Pack the meat skin side down into either a crock or a clean, well-soaked, odorless, hardwood barrel. Place the thicker and heavier cuts in the bottom of the container. Pour cool pickle or brine over the meat until the liquid just covers it. Four to five gal of pickle will cover 100 lb of meat. Place a weight, such as a lid, on top of the meat to hold it under the liquid, and weight it down with a clean stone or other suitable object. Let the meat cure in the pickle 11 days per inch of thickness. Thus the thinner and lighter cuts on the top of the container must be removed first. When large vats are used, overhauling (meaning the rehandling or repacking) of the meat is recommended in order to permit a more uniform distribution of the pickle, but this is not necessary with the small quantities cured on most farms. If the brine should spoil (as evidenced by cloudiness), remove all the cuts, wash them with cold water, and place them in a new brine with ⅔ of the original ingredients.	A combination of salt and water is called a brine or pickle; and with the addition of sugar, it is known as a sweet-pickle cure. In comparison with the dry-cure, the sweet-pickle cure results in a milder flavor and gives less shrinkage. For best results, the curing-room temperature should not rise above 40°F; temperatures above 50°F are too high for safe pickle curing. A salometer or salinometer is a ballasted, glass vacuum tube graduated in degrees, which is sometimes used for testing the strength or salinity of pickle. A test of about 75°F is preferred under most conditions for storage without refrigeration; 65°F where refrigerated storage is available.
Pumping, followed by either dry-cure or sweet-pickle cure: By means of a plunger-type syringe, sweet-pickle cure (see formula under "sweet-pickle cure") is injected into the ham or shoulder and then the ham or shoulder is subjected to either the dry-cure or sweet-pickle cure for the balance of the curing.	Hams.	Mix the sweet-pickle formula as directed under "sweet-pickle cure." Inject into the center of the ham an amount of sweet-pickle cure equivalent to 8 to 10% of the weight of the cut. Following pumping, complete the process by either the dry-cure or sweet-pickle cure methods, but *lessen the curing time by ⅓.*	Pumping hastens the curing process in the center of the cut and lessens spoilage. Meat-packers use a modified pumping method for hams, known as the artery cure. In this process, the sweet-pickle is injected into the femoral artery under 25 lb of pressure, and the subsequent dry-cure or sweet-pickle cure is limited to 2 weeks. The advantages of the artery cure are speed and uniform flavor.
Smoking: Hickory, oak, hard maple, and apple are favorite smoke woods. Hardwood sawdust is also satisfactory. Two methods of smoking are: 1. *For light mahogany smoke*, smoke 24 to 48 hours at a temperature of approximately 125 to 135°F. 2. *For meat that is stored for summer use*, smoke at a temperature of 80 to 100°F at intervals of approximately 5 to 10 days, over a period of several weeks.	Hams, bacon, and shoulders.	If the meat has been either dry-cured or sweet-pickle cured, prior to smoking, it should be (1) soaked from ½ to 3 hours in cold water (using the longer soaking period for the heavier cuts), (2) scrubbed with a clean, stiff brush, and (3) hung to dry overnight in the smokehouse. This removes the excess salt on the outside and alleviates the formation of salt streaks. Hang the cuts so that they will not touch each other, since this will cause streaking. After meat is smoked, many people like to season it heavily with black pepper.	Smoking is a common practice in the home-curing of pork, and most packer dry-cured pork cuts are smoked, as well as many items of sausage. Smoking adds flavor, makes for a more desirable appearance, and improves the keeping qualities of meats. The higher the temperature, the greater the absorption of smoke and the darker the color. After the smoked meat has cooled, it should be carefully wrapped with heavy parchment paper, put into muslin bags, and hung in a dry, dark, cool, well-ventilated place.

METHODS OF COOKING MEATS

Every grade and cut of meat can be made tender and palatable provided it is cooked by the proper method. Also, it is important that meat be cooked at a low temperature, usually between 300 and 350°F. At this temperature, it cooks slowly, and as a result it is juicier, shrinks less, and is better flavored than when cooked at high temperatures.

The method used in meat cookery depends on the nature of the cut to which it is applied. In general, the type of meat cookery may be summarized as follows:

1. **Dry heat cooking.** Dry-heat cooking is used in preparing the more tender cuts; those that contain little connective tissue. The common methods of cooking by dry heat are: (a) roasting, (b) broiling, and (c) panbroiling (see Fig. 14–23, next page).

2. **Moist-heat cooking.** Moist-heat cooking is generally used in preparing the less tender cuts, those containing more connective tissues that require moist heat to soften them and make them tender. In this type of cooking, the meat is surrounded by hot liquid or by steam. The common methods of

moist-heat cooking are: (a) braising, and (b) cooking in liquid. (See Fig. 14–23).

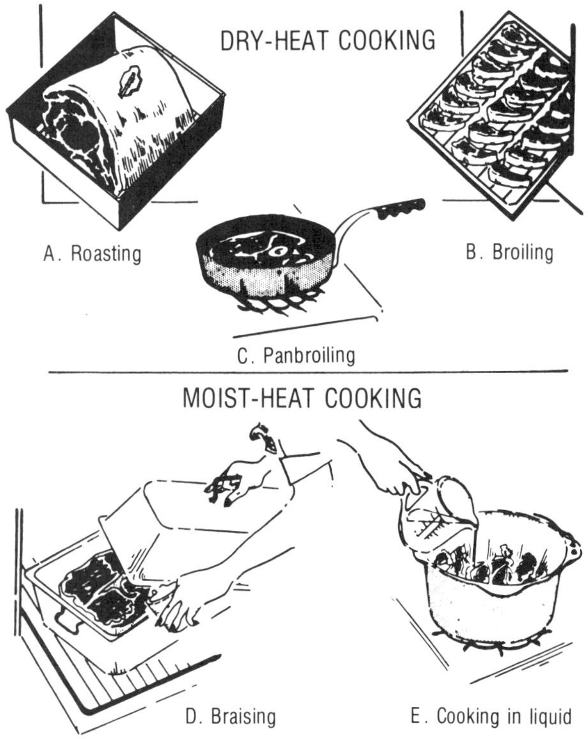

Fig. 14–23. Common methods of meat cookery. *Dry-heat cooking:* A, roasting; B, broiling; and C, panbroiling. *Moist-heat cooking:* D, braising; and E, cooking in liquid. (Drawing by R. F. Johnson)

For best results, a meat thermometer should be used to test the doneness of roasts (and also for thick steaks and chops). It takes the guess work out of meat cooking. Allowing a certain number of minutes to the pound is not always accurate; for example, rolled roasts take longer to cook than ones with bones.

The thermometer is inserted into the cut of meat so that the bulb reaches the center of the largest muscle, and so that it is not in contact with fat or bone. Naturally, frozen roasts need to be partially thawed before the thermometer is inserted, or a metal skewer or ice pick will have to be employed in order to make a hole in frozen meat.

As the oven heat penetrates, the temperature at the center of the meat gradually rises and is registered on the thermometer. The meat can be cooked as desired—rare, medium, or well-done, except for pork which should always be cooked well-done (160 to 170°F for cured pork; 185°F for fresh pork).

PACKINGHOUSE BY-PRODUCTS FROM SLAUGHTER

The meat or flesh of animals is the primary object of slaughtering. The numerous other products are obtained incidentally. Thus, all products other than the carcass meat are designated as by-products, even though many of them are wholesome and highly nutritious articles of the human diet. Yet it must be realized that, upon slaughter, cattle, sheep, and hogs produced in the United States yield an average of 40%, 50.5%, and 32%, respectively, of products other than carcass meat.[3] When meat packers buy a steer, lamb, or hog, they buy far more than the cuts of meat that will eventually be obtained from the carcass; that is, only about 50% of a meat animal is meat.

The complete utilization of by-products is one of the chief reasons large packers are able to compete so successfully with local butchers. Were it not for this conversion of waste material into salable form, the price of meat would be much higher then under existing conditions. In fact, under normal conditions, the wholesale value of the carcass is about the same as the cost of the animal on foot. The returns from the sale of by-products cover all operating costs and return a reasonable profit.

It is not intended that this book should describe all of the by-products obtained from animal slaughter. Rather Fig. 14–24 is presented in order to show some of the more important items, and a few select ones are discussed:

1. **Hides.** Hides are particularly important as a by-product of the cattle slaughter. In the mid 1960s, cattle hides represented 2.5 to 3.0% of the total on-foot value of steers. In the 1980s, cattle hides rose to 8.5% of the value of steers.

The leather from animal hides is used for shoes, harness and saddles, belting, traveling bags, razor straps, footballs,

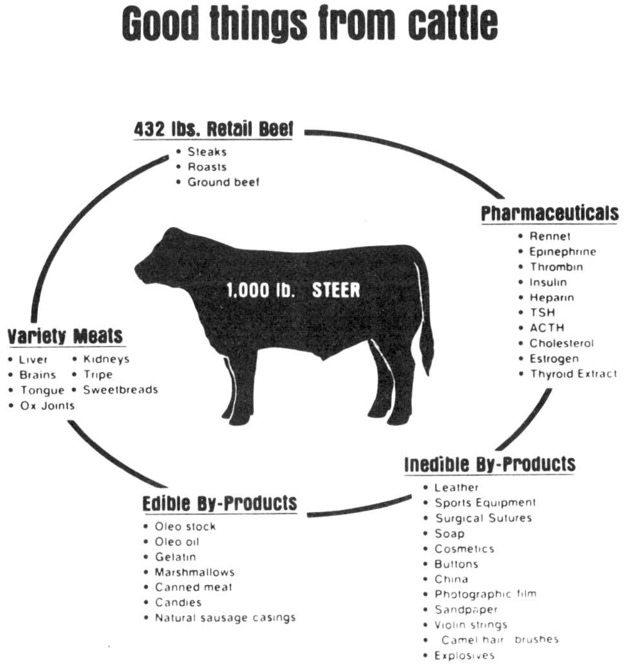

Fig. 14–24. Many good things come from cattle in addition to about 432 lb of steaks, roasts, and hamburger normally yielded by a 1,000-lb steer. Several of these products are shown in the above figure. (Courtesy, National Live Stock and Meat Board, Chicago, IL)

[3]USDA figures.

baseball mitts, "sheepskins" for diplomas, sweatbands for hats, gloves, and numerous other leather goods.

2. **Sheep pelts.** The sheep pelt is the most valuable by-product of sheep slaughtering. Sheep skins with short wool, ¾ in. or less in length, are usually tanned with the wool on and are used for coats, robes, felts, slippers, and other articles. Pelts with longer wool are sent to the pullery. Usually they are temporarily preserved by the addition of salt until the wool is removed. The pulling process consists of applying a depilatory solution (made of sodium sulfide, slaked lime, and water) to the skin side of the pelt and then pulling the wool loose from the skin after the chemical action has loosened the hold of the fibers.

3. **Fats.** Next to hides and pelts, the fats (not including lard) are the most valuable by-product derived from slaughtering. Products rendered from them are used in the manufacture of oleomargarine, soaps, animal feeds, lubricants, leather dressings, candles, fertilizers, etc.

Oleomargarine, which is one of the better known of the products in which rendered fat is incorporated, is usually a mixture of vegetable oils and select animal fat.[4] Oleo oil, one of the chief animal fats of this product, is obtained from beef, mutton, or lamb.

4. **Variety meats.** The heart, liver, brains, kidneys, tongue, cheek meat, tail, feet, sweetbreads (thymus and pancreatic glands), and tripe (pickled rumen of cattle and sheep) are sold over the counter as variety meats or fancy meats.

5. **Glands.** Various glands of the body are used in the manufacture of numerous pharmaceutical preparations.

MEAT IMPORT LAW (PUBLIC LAW 88–482)

The U.S. quotas and tariffs of live animals and meats are given in Table 14–10.

[4]Oleomargarine is of 2 kinds: (1) a mixture of 50 to 80% animal fat and 20 to 50% vegetable oil, churned with pasteurized skimmed milk; or (2) 100% vegetable oil, churned with pasteurized or skimmed milk. Oleomargarine was first perfected in 1869 by the Frenchman, Mege, who won a prize offered by Napoleon III for a palatable table fat which would be cheaper than butter, keep better, and be less subject to rancidity.

TABLE 14–10
U.S. QUOTAS AND TARIFFS OF LIVE ANIMALS AND MEATS[1]

Import Item	Quotas (no. head/year)[2]	Tariff (or duty) 1[3]	Special[4]	2[5]
Animals for breeding[6]	None	Free		Free
Cattle:				
Cattle weighing:				
under 200 lb	200,000	1¢/lb	Free (E, I)	2.5¢/lb
between 200 and 700 lb		1¢/lb	Free (E, I)	2.5¢/lb
Dairy cattle weighing:				
over 700 lb		Free		3¢/lb
Other cattle[7]	400,000	1¢/lb	Free (E, I)	3¢/lb
Beef and veal (fresh, chilled, or frozen)	(See footnote 8)	2¢/lb	Free (E, I)	6¢/lb
Sheep:				
Live sheep		Free		$3/head
Mutton		1.5¢/lb	Free (E, I)	5¢/lb
Lamb		0.5¢/lb	Free (E, I)	7¢/lb
Goats:				
Live goats		$1.50/head	Free (E, I)	$3/head
Goat meat		Free		5¢/lb
Swine:				
Live hogs		Free		2¢/lb
Pork		Free		2.5¢/lb
Horses:				
Valued under $150/head		Free		$30/head
Valued over $150/head		Free		20% *ad valorem*
Mules:				
Valued under $150/head		$15/head	Free (E, I)	$30/head
Valued over $150/head		10% *ad valorem*	Free (E)	20% *ad valorem*

[1]*Tariff Schedules of the United States Annotated (1987)*, USITC Publication 1317, United States International Trade Commission, Washington, DC.

[2]Includes Canada, Mexico, and all other countries.

[3]Products of Canada and all other countries not designated LDDC or 2.

[4]Products accorded special consideration.

[5]Products of communist countries.

[6]Must be purebreds of a recognized breed and registered in a recognized registry book.

[7]For not over 400,000 head entered in the 12-months period beginning April 1, in any year, of which not over 120,000 shall be entered in any quarter beginning April 1, Oct. 1, or Jan. 1.

[8]Legislation of Aug. 1964 establishes a basic limit of 725.4 million lb plus an added factor based on U.S. production.

The Secretary of Agriculture is required to estimate at the beginning of each quarter year the quantity of prospective imports. If the estimated quantity of prospective imports exceeds the trigger point, the President is required to invoke a quota on imports of these meats. In case quotas are imposed, the total import quota is allocated among the countries from whom the United States is importing on the basis of shares supplied by those countries during a representative period.

The law does contain provisions under which the President is empowered to suspend or increase quotas when it is deemed that it is in the best interest of the nation to do so because (1) of overriding economic or national security interest, (2) the supply of meat is inadequate to meet domestic demand at reasonable prices, or (3) international agreements have been entered into which will have the same effect as the Act.

It appears that voluntary agreements will be the chief means of controlling future imports. But the fact that the law exists may have a considerable psychological effect on negotiations.

PART II—MILK

Fig. 14–25. Milk and milk products. (Courtesy, American Dairy Assn., Rosemont, IL)

The first food that nature provides for all young mammals, including human infants, is milk. Long before recorded history, humans found that milk was good—and good for them. As a result, they augmented the secretions of women's mammary glands by domesticating milk-producing animals and selecting them for higher production for their own use. For the most part, this included the cow, whose importance in milk production is attested to by her well-earned designation as, "the foster mother of the human race."

HOW MILK IS USED

Today, dairy products are not consumed for nutritional value alone. Some of them have great appeal as health and/or dietary foods.

Fig. 14–26 shows change in U.S. per capita dairy product sales from 1977 to 1987. It is noteworthy that declines in per capita consumption were registered in the following products: ice milk, butter, buttermilk, flavored milk and drinks, evaporated and condensed whole milk, creamed cottage cheese, whole milk, and nonfat dry milk. It is noteworthy, too, that yogurt and low fat cottage cheese led the increases. Increase in yogurt reflects the rising interest in health foods, whereas increased consumption of low fat cottage cheese reflects weight watching.

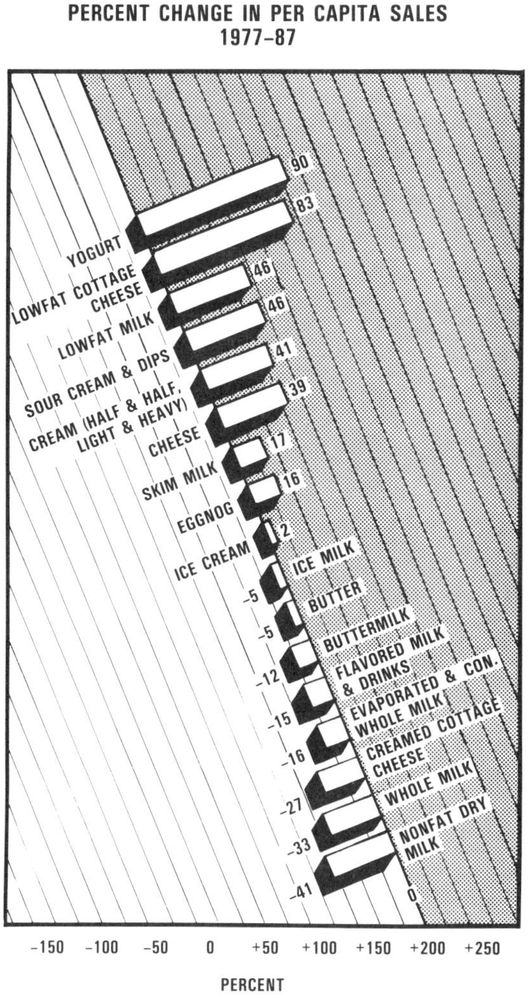

Fig. 14–26. Change in U.S. per capita consumption of milk and dairy products 1977 to 1987, based on sales. Bars to the left of the vertical line represent decreases; bars to the right represent increases. (Source: *1988 Milk Facts,* Milk Industry Foundation, Washington, DC, p. 16)

PRODUCE QUALITY MILK

Consumers and health departments all have a distinct interest in the quality of milk.

Quality milk can be produced only when the producer pays special attention to the following factors:

1. **Health of the herd.** The herd should be free from diseases that might be spread to human beings through the milk, and free from mastitis.

2. **Clean animals.** The milker should clean the flanks and udders of cows just prior to milking to prevent dirt from getting into the milk. Clean floors and bedding and a well-drained yard make the cleaning job easier.

3. **Clean equipment.** All milking equipment should be kept as clean and free from bacteria as possible.

4. **Cool and store milk properly.** Proper cooling and storage of milk on the dairy farm requires facilities which will cool the milk promptly from a temperature of about 90°F down to 40°F, and then hold it at that temperature until it is collected.

5. **Keep barn and milk house clean.** The milking barn should be clean and should have a concrete floor. Barn odors may be eliminated by having a building well ventilated.

6. **Control flies.** Fly control measures are important to milk producers. Flies add to the bacterial count of milk; cases are on record of flies carrying as many as 1,250,000 bacteria. They can carry typhoid, dysentery, and other contagious diseases.

7. **Control bacteria.** The bacteria count can be kept down by sanitation and by cooling the milk to 40°F as quickly as possible.

PHYSIOLOGICAL FACTORS AFFECTING AMOUNT AND COMPOSITION OF MILK

The variation in the butterfat composition of milk at the plant has puzzled dairy producers. And since the fat content of the milk has a bearing on the paycheck, it's an economical factor, too.

A number of physiological factors affect the amount and composition of milk:

1. **Breed and individual inheritance.** Variation in the ability of cows to produce total milk, fat, and solids-not-fat is an inherited characteristic. There is both a breed difference and an individual difference. In general, total milk production decreases and butterfat content increases by breeds in the following order: Holstein, Brown Swiss, Ayrshire, Guernsey, and Jersey.

With the Holstein breed, a range in butterfat from 2.6 to 6.0% has been reported; and within the Jersey breed, from 3.3 to 8.4%. Similar variation between breeds and individuals exist in total milk production.

2. **Stage of lactation.** The greatest variation in the composition of milk takes place immediately following parturition, within the first five days after freshening. The secretory product known as colostrum, found in the udder at the time of calving and produced for a short time thereafter, is not milk as such. It contains more globulins, vitamins A and D, iron, calcium, magnesium, chlorine, and phosphorus than does milk; but it contains less lactose and potassium than milk (see Table 14–11).

TABLE 14–11
COMPOSITION OF COLOSTRUM AND NORMAL MILK OBTAINED TWO WEEKS AFTER CALVING[1]

	Colostrum	Normal Milk
	(%)	(%)
Total solids	23.9	12.9
Minerals	1.1	0.7
Protein	14.0	3.1
Fat	6.7	4.0
Lactose	2.7	5.0

[1]Source: *Dairy Guide,* The Ohio State University, Columbus, OH.

Total milk production generally increases for the first month following freshening, then decreases gradually thereafter. Conversely, the butterfat test is usually higher toward the end of the lactation period than soon after freshening.

3. **Persistency.** This refers to the level at which milk production is maintained as lactation progresses. Generally speaking, following the peak lactation period, about a month after freshening, the total milk production each month is approximately 90% of that of the previous month.

4. **Estrus; pregnancy.** Milk and butterfat production may fluctuate, usually downward, on the day of or the day following the heat period. Pregnancy seems to have little effect on milk composition.

5. **Calving interval.** Research indicates that it is most profitable for cows to calve at 12-month intervals, rather than longer intervals. With an 8-week dry period, this means a lactation period of 10 months.

6. **First- and last-drawn milk.** The percentage of fat in last-drawn milk is higher than that in first-drawn milk.

7. **Age.** The age of a cow has a definite effect on production. Most cows reach maturity and maximum milk production at about 6 years of age, following which there is a decline in production. Records indicate that cows produce approximately 25% more milk at maturity than they do as 2-year-olds. Also, after passing their prime—after 6 years of age—butterfat gradually decreases with advancing age.

8. **Size.** Within a breed, large cows usually produce more milk than small cows. However, according to Brody of the Missouri station, for each 100 lb increase in body weight, production increases only 70% of the proportional increase in body size.

ENVIRONMENTAL FACTORS AFFECTING AMOUNT AND COMPOSITION OF MILK

All animals, including dairy animals, are the result of two forces—heredity and environment. Because of this, the maximum development of dairy cattle characteristics of economic importance—particularly total milk production—cannot be achieved unless there are optimum conditions of environment. Among the environmental factors affecting amount and composition of milk are the following:

1. **Feed.** If milk cows are not fed, or they do not eat, they will not produce. There are a number of ways in which feed may affect the quantity and/or composition of milk. Among them—

a. **Underfeeding.** The degree of milk reduction therefrom is related to the extent of underfeeding and the length of time it exists.

b. **Deficiency of nutrients.** A deficiency of any essential nutrient required by the cow will lower milk production and feed efficiency, rather than make for significant changes in the composition of milk.

c. **Some feed ingredients and rations influence feed consumption.** The fat percentage of milk is lowered by cod-liver oil and other fish oils, certain pasturages (especially lush spring pastures), and pearl millet. Also, fine grinding of forage, too small an amount of roughage, or heated starch will lower the butterfat content of milk. On the other hand, such feeds as whole cottonseed, soybeans, and coconut oil result in an increase in the fat content of milk.

The amounts of fat-soluble vitamins A, D, and E in milk are influenced by the amounts of these particular vitamins in the ration, and in the case of vitamin D, exposure to sunlight is a factor, also.

2. **Length of dry period.** A dry period of approximately 60 days is recommended following each lactation period. This is important because it permits the cow's body to store up reserves so as to meet the rigorous demand of the next lactation, and it permits proper involution and conditioning of the udder. A shorter dry period usually results in lower milk production.

3. **Conditioning at calving time.** Cows that are in a thin, run-down condition at calving time produce less milk than cows in good condition. Excessive condition will also lower milk production after freshening, but it should be added that this seldom happens in good producing dairy cows.

Cows in good flesh at calving time have been observed to start their lactation with 25% more milk production than those calving in poor condition. Generous feeding of thin cows following freshening may eliminate some of this difference, but it is questionable that thin, high-producing cows can ever consume enough to catch up.

4. **Frequency of milking.** Frequency of milking does result in more total milk produced; cows milked 3 times a day consistently produce more milk than those milked twice a day, and cows milked 4 times a day produce more milk than those milked 3 times daily. However, frequency of milking has no effect on butterfat percentage.

5. **Irregular feeding and milking.** Unequal intervals between milkings affects both quantity and composition of milk; more milk of slightly lower fat content is obtained following the longer intervals.

6. **Change of milkers.** High-producing dairy cows are under stress, with the result that they are usually very sensitive to any changes, including that of the caretaker.

7. **Environmental temperatures; season.** Butterfat percentage of milk varies with the season, being higher in the

fall and winter and lower in the spring and summer. It may vary up and down seasonally by an average of 0.3 to 0.5%. Solids-not-fat also show a seasonable variation, with the low point in the spring and summer.

Severe weather conditions usually decrease the amount of total milk produced and may influence the fat test either up or down. Temperatures about 85°F greatly affect cows and the situation is accentuated when high temperatures are accompanied by high humidity.

It is also noteworthy that cows calving in the fall months consistently produce more than those calving at other times of the year. Cows calving in the spring produce the least. This difference may be as much as 10 to 15%. This phenomenon may be due in part to the temperature; but more than likely available feeds, including spring pastures to which fall-calving cows respond so well, may be a factor.

8. **Day-to-day variation.** Research has shown that day-to-day butterfat tests vary from 0.1 to 2.0%.

9. **Disease.** Disease does affect milk secretion, in both total production and composition, with the degree of the effect determined by the kind and severity of disease. Mastitis will, for example, lower both the total production of milk and the composition thereof.

10. **Drugs.** Many types of drugs have been used in an effort to increase milk production and affect its composition. Most of them have no effect, so it is questionable that they can be used on a practical basis.

Oxytocin will, on a temporary basis, increase yields of both milk and fat.

11. **Prepartum milking.** Where prepartum milking is done, it is necessary to save (freeze) the early milk in order to have colostrum available for the newborn calf.

MASTITIS

Mastitis is an inflammatory reaction of udder tissue to bacterial, chemical, thermal, or mechanical injury. The term *mastitis* is from the Greek word *mastos*, for breast, and *itis* refers to inflammation of.

According to the National Mastitis Council, nearly 40% of all dairy cows have some form of mastitis, which causes a yearly loss of $225 per afflicted cow.

It has been said that producers themselves are responsible directly, or indirectly, for 90% of their mastitis troubles; however, most producers blame their milking machine. The three main routes through which mastitis comes are: (1) dirty, or poorly adjusted, milking equipment; (2) poor milking practices; and (3) injuries to cows because of their surroundings.

Although mastitis is usually apparent, it may be a hidden disease. Therefore, several different tests have been developed for detecting the presence of the causative microorganisms; among them (1) *screening tests*, or *presumptive tests*, made either at the side of the cow or at the bulk tank, of which the California Mastitis Tests (CMT) is the most widely used one, and (2) *specific laboratory tests* designed to detect the causative organism. A reasonable goal, based on using the CMT test, is to have at least 75% of the bucket milk samples score negative or trace; less than 75% negative (−) and trace (T) bucket readings indicates a milking management problem. On an individual quarter basis, 90% of the samples scoring negative or trace indicates a well-managed herd.

MILK FLAVOR

Most consumers base the quality of any product on its flavor; and milk is no exception. They want milk that tastes good. The objectionable flavors most often found in milk are:

1. **Feed and weed flavors.** Certain feeds and weeds may impart flavors to milk. For example, silage will flavor milk if it is fed ahead of milking rather than following milking. Also, many weeds will flavor milk if they are eaten by cows; among them, wild onions and skunk cabbage.

2. **Oxidized flavor.** This has been described as a cardboard flavor. Some causes of oxidized flavor are (a) metallic contamination from copper and iron, which may be alleviated by using stainless steel; (b) exposure to sunlight or just daylight; (c) foaming; and (d) dry lot feeding. Feeding vitamin E to the milking herd will reduce or eliminate oxidized flavors.

3. **Rancid flavor.** This flavor is caused by a breakdown of the butterfat which releases strong-flavored acids. This action is caused by the enzyme lipase, which is present in milk. The primary causes of rancid milk are (a) stripper cows (those well-advanced in lactation), (b) excessive agitation of milk, due to high lifts and sharp turns in pipeline milking; and (c) slow cooling with foaming.

4. **Barney.** This flavor(s) is caused by dirty stables, poor ventilation, unclean milking, and unclean cows—all of which can be alleviated.

5. **Salty.** This flavor, which masks the slightly sweet flavor of milk, is caused by mastitis, stripper cows, or certain individual cows. Milk from cows that have mastitis, or from strippers, should not be marketed.

6. **Malty.** Malty flavor is primarily due to high bacteria count. The remedy is to keep bacteria out of milk as much as possible, and to prevent growth of those that do get into it.

Clean and cold milk will practically eliminate malty flavor. Also, milk handlers should pick up all the milk and not leave any of it in the farm bulk tank.

7. **High-acid, sour milk.** This is due to a very high bacterial count. In these days of mechanical refrigeration, there is no excuse for sour milk; simply cool it as rapidly as possible from the 90°F temperature of the milk pail to 40°F.

8. **Unnatural or foreign.** This refers to flavors that come from medicinal agents and disinfectants. The control of such off-flavors consists in (a) handling medicines and disinfectants so that the flavor or odor from them will not get into the milk, and (b) using chemical sanitizers only in the concentrations indicated by the directions. Do not market milk from drug treated cows for at least 72 hours after last treatment, or longer if so prescribed on the drug label or by the veterinarian.

For good tasting milk, the producer should keep it clean, keep it cool, feed silage after milking (not before), use good quality feed, and not ship milk from problem cows.

DRUGS AND PESTICIDES

Many drugs used in the treatment of cattle diseases, along with many pesticides, are excreted in milk. Such milk should be discarded to prevent the drugs from entering the human food supply. The presence of antibiotics, sulfas, and pesticides in milk is illegal. Dairy producers should follow a residue avoidance program.

On October 1, 1990, the Food and Drug Administration (FDA) began a new program to monitor antibiotics and other drug residues in milk. The safety level set by FDA is: Sulfonamides, 10 parts per billion; tetracycline, 80 ppb; and for chlortetracycline and oxytetracycline, 30 ppb.

SANITARY REGULATIONS[5]

The sanitation of milk and dairy products is assured by the enforcement of sanitary regulations by federal, state, and local authorities.

There are more than 15,000 state, county, local, and municipal health and sanitation jurisdictions in the United States. Inspectors from these agencies regularly visit farms, plants, and stores, making sure dairy products keep their high quality. Unfortunately, from area to area, there are a bewildering number of different regulations, with the result that milk going to more than one city market is often subjected to duplication and confusion in inspection. Also, sanitary and health regulations have sometimes been used as barriers to keep milk out of a certain area for competitive reasons.

In 1923, the U.S. Public Health Service (USPHS) established an Office of Milk Investigations, and in 1924, the USPHS) published its first Grade A pasteurized milk ordinance. Subsequently, this regulation has been revised several times.

Producers are issued permits allowing them to ship Grade A milk. The permit is revoked if either the bacteria count of raw milk exceeds 100,000 per milliliter or the cooling temperature exceeds 40°F in three of the last five samples.

The standard plate count of Grade A pasteurized milk may not exceed 20,000 per milliliter nor the coliform count 10 per milliliter in three of the last five samples or the processor's permit will be revoked.

In addition to cleanliness and freedom from mastitis, temperature is important in processing quality milk. Bacteria cannot reproduce effectively below 40°F; so, dairy farmers should cool milk below 40°F as quickly as possible. By law, all fluid milk sold for human consumption must be pasteurized; so, at dairy processing plants, milk is pasteurized at either (1) 145°F for 30 minutes, or (2) 161°F for 15 seconds. Additionally, milk should be refrigerated while in the store or in the home.

The Food and Drug Administration (FDA) is charged with inspecting dairy products and processing plants for contamination and adulteration.

Presently, many cooperatives and some milk dealers pay a premium to dairy farmers for producing high quality milk. A variety of premiums and penalty bases are in use. The following standards for milk to qualify for the highest premiums are proposed as reasonable goals:

1. Standard plate count (SPC) or plate loop count (PLC), less than 100,000 per ml.
2. Preliminary incubation count (PIC), less than 20,000 per ml.
3. Somatic cell count (SCC), less than 200,000 per ml.
4. Antibiotic and chemicals, no detectable levels.
5. Temperature, 40°F or lower.
6. Odors and flavors, none objectionable.
7. Acid degree level, 1.0 or below.
8. Milk fat, 3.25% or above.
9. Protein, 3.2% or above.
10. Farm inspection score, 90 or above.

USES OF MILK

About 38% of the milk marketed by dairy farmers is consumed in fluid form. Fluid milk is retailed as pasteurized milk, homogenized milk, fortified milk (vitamin D), skimmed milk, flavored milk (whole milk with flavor added), or flavored milk drink (skimmed milk with flavor added).

Manufactured dairy products utilized about 62% of U.S. milk production. A few pertinent points relative to each of the manufactured products will be presented in the sections which follow.

---

[5]This part, from "Sanitary Regulations" to the end of this section, benefited from the authoritative review and helpful suggestions of the following person: R. F. Fallert, Leader, Dairy Research Section, USDA, Washington, DC.

MILK AND MILK PRODUCTS[6]

Fig. 14-27. Milk.

Fig. 14-28. Cheese and fruit.

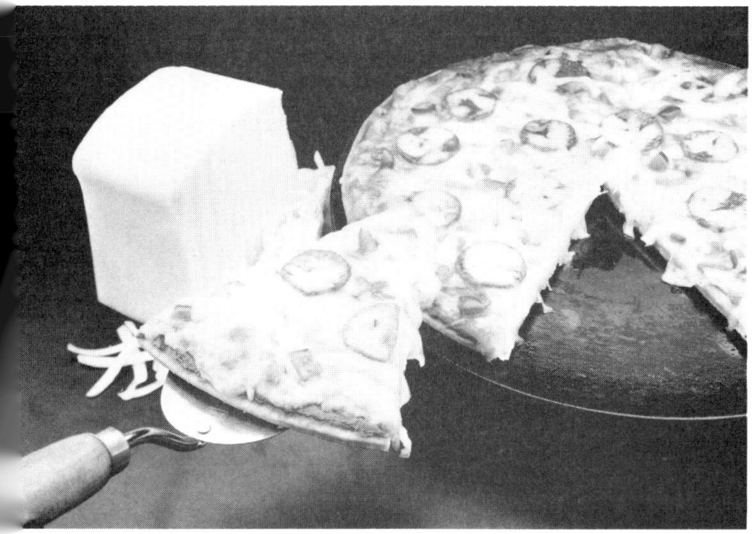

Fig. 14-29. Pizza.

Fig. 14-30. Dish of ice cream.

Fig. 14-31. Ice cream cones.

Fig. 14-32. Soft serve frozen yogurt.

[6]Courtesy, United Dairy Industry Assn., Rosemont, IL

• **Butter**—Butter is made from cream. As marketed, it consists of about 80% milk fat. The remainder is water, salt, and traces of other substances.

Fig. 14-33 shows the per capita consumption of butter and margarine, from 1910 to 1987. As noted, the per capita consumption of margarine surpassed butter in 1957. In 1987, the per capita consumption of butter was 4.7 lb, whereas the per capita consumption of margarine was 10.5 lb.

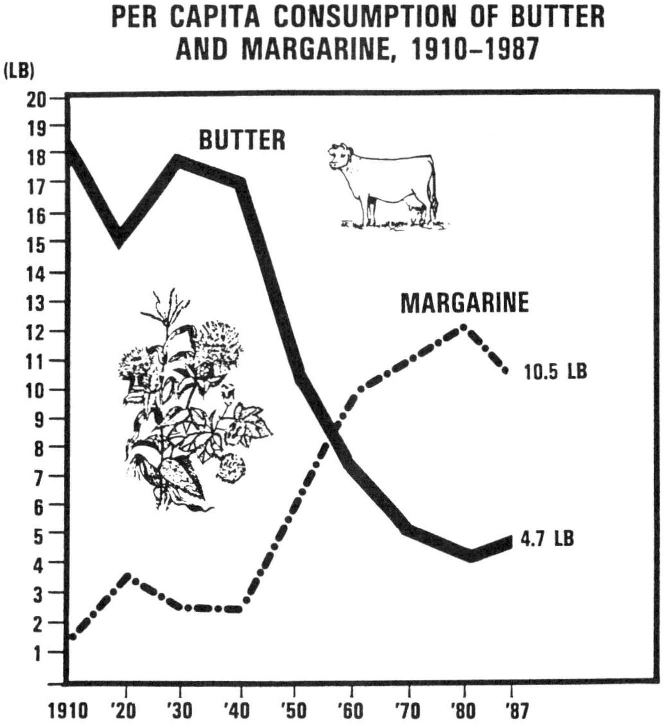

Fig. 14-33. The per capita consumption of butter and margarine from 1910 to 1987. (Source: USDA)

Wisconsin is the leading butter-producing state; California ranks second.

• **Casein**—Casein, which is the major protein of milk, is found only in milk. It is obtained by acid or rennet coagulation of defatted milk. Casein contains a minimum of 80% crude protein. It gives milk its white color.

Casein is used as an ingredient of coffee whitener and whipped toppings, in baked goods, and as the main source of protein in the manufacture of meat analogs and in the protein supplementation of some meat products.

In 1987, the United States imported 108,136 metric tons of casein, which far exceeded imports of dried milk (1,301 metric tons), and of butter (905 metric tons).

• **Cheese**—Cheese is made by (1) exposing milk to specific bacterial fermentation, or (2) treating with enzymes, or both methods, to coagulate some of the proteins.

Milk can be, and is, processed into many different varieties of cheese. Some are made from whole milk, others from milk that has had part of the fat removed, and still others from skimmed milk. American types of cheese (Cheddar, Colby, washed curd, stirred curd, Monterey, and Jack) make up 60% of the nation's cheese output. The most important variety produced from skimmed milk is cottage cheese. Other important types of cheese are Italian (mostly soft varieties), Swiss, Muenster, brick, blue, and processed cheese.

In 1987, 30% of all the milk used in manufactured dairy products was processed into cheese (exclusive of cottage cheese). The rising popularity of pizza in the United States account for much of the increase in cheese production and consumption in recent years.

The leading states in the production of cheese, excluding cottage cheese, by rank are: Wisconsin (with 35% of the total production), Minnesota, New York, California, Iowa, Pennsylvania, Missouri, Idaho, and South Dakota.

• **Condensed and evaporated milk**—The primary products within this category are evaporated milk and condensed milk placed in cans for consumer use, and condensed whole and skimmed milk shipped in bulk. Condensed and evaporated milk are manufactured by removing a major portion of the water from the whole milk in a machine called a vacuum pan. Condensed milk is further treated by the addition of large amounts of sugar.

Candy manufacturers, especially bakers and ice cream processors, are large users of condensed milk.

The production of condensed and evaporated milk is declining. In 1945, 10.8% of the milk marketed in the United States was used in condensed and evaporated milk, compared with only 1.5% in 1987.

• **Cream**—Cream is made by concentrating the fat portion of milk. Prior to the advent of the cream separator, this was accomplished by gravity separation. Today, it is done by passing milk through a cream separator. In commerce, whipping cream contains about 40% fat; coffee or table cream, 18 to 20%; and half-and-half, 12%.

• **Dried milk (whole milk, skimmed milk, and whey)**—Among the dried products produced from milk are nonfat dried milk (skimmed milk), for both human food and animal feed; dried whey, for both human food and animal feed; and dried whole milk.

In 1987, the following quantities of these dried products were produced in the United States: 1,067 million lb of dry whole milk, 1,059 million lb of nonfat dried milk, and 1,034 million lb of dried whey.

Dried milk products have many uses, principally as ingredients in other dairy and food products, although their use in the home has grown considerably in recent years. Despite its wide variety of uses, nonfat dried milk has been surplus much of the time.

• **Ice cream and similar frozen desserts**—Currently, 99% of all frozen desserts in the United States consist of ice cream, ice milk, sherbet, and mellorine (made with a vegetable fat base). Other frozen desserts include frozen custard, frozen malted milk, artificially sweetened ice cream and ice milk, and water ices.

Left: Majestic sheep from whence wool comes. *Right:* Majestic goats from whence mohair comes.

SECTION

15

WOOL AND MOHAIR

Contents *Page*

It is important that sheep and goat producers, and those who counsel with them, be familiar with the classes and grades of wool and mohair. With this information at hand, producers can more intelligently (1) place their production in line with market demands, and (2) market their products. To this end, a rather complete discussion follows relative to the classes and grades of wool and mohair.

WOOL

Fig. 15-1. Wool is the natural clothing of sheep. The Merino ram pictured is a member of the breed that produces the world's finest wool. (Courtesy, The Wool Bureau, Inc., New York, NY)

With all the perfection and modification in the fleece that has been wrought through centuries of domestication, breeding, selection, and improved environmental conditions, it must not be forgotten that wool is the natural hair as well as clothing of sheep. A covering of hair or feathers performs a thermoregulatory function for warm-blooded animals, the original intent being to protect the body from heat or cold. As wool fibers are poor conductors of heat, they serve to prevent any abnormal loss of heat from the body.

The unique characteristics and virtues of wool are:

1. It is porous and will absorb water more readily than any other textile fiber. It can absorb as much as 18% of its own weight in moisture without even feeling damp, and up to 50% of its weight without becoming saturated. This is an important health factor in clothing because body perspiration and outer dampness are prevented from clinging to the body in heat or cold, thus removing the chill line from the body.
2. It generates heat in itself.
3. It is a superior insulator, keeping the heat of the body from escaping and the cold air from entering. Because of this quality, wool is as effective as a protection from tropical heat and sun as it is against gale-driven storms of winter.
4. It is light.
5. It is very elastic; the average fiber will stretch 30% of its normal length and still spring back in shape. Because of this resilience, wool garments resist wrinkling, stretching, or sagging during wear. Wool can be bent 20,000 times without breaking (silk breaks after 1,800 bends, rayon after 75).
6. It transmits the health-giving ultraviolet rays.
7. It dyes well.
8. It is durable.
9. It is strong. Diameter for diameter, a wool fiber is stronger than steel.
10. It is almost nonflammable. It will stop burning almost as soon as it is taken away from a flame.
11. It can be felted or matted easily.

PRODUCTION AND HANDLING OF THE WOOL CLIP

Fig. 15-2. Sheep are unsurpassed in converting what would be wasted roughage into wool and meat. (Courtesy, USDA)

If U.S. sheep producers are to survive the inroads of imports and synthetic fibers, it is imperative that they market a higher quality product—one that does not require unnecessary processing expenditures in the textile mills.

Observance of the following wool production and handling practices will result in marketing a higher quality product:

1. Producing superior fleeces through (a) feeding properly, (b) protecting the on-the-back fleece from foreign material, and (c) tagging sheep at intervals.
2. Using a scourable branding material, where branding or identification is necessary.
3. Shearing in a clean place.
4. Using skilled shearers.
5. Packing properly.
6. Shipping in clean trucks or cars, and keeping the wool bags dry.

SHEARING

Fig. 15-3. Master shearer, Charles Swain, demonstrates shearing techniques to students in a shearing school. (Courtesy, American Sheep Industry Assn., Englewood, CO)

Up to and during the early part of the 1900s, most shearing was done by hand shears. Then clippers, similar to barber's clippers were developed; hand-powered at first, then electrically powered. A skilled shearer can clip 200 or more sheep per day using electric clippers. In the range states, the shearers travel in crews from ranch to ranch, staying on each ranch just long enough to shear the band.

High shearing cost and scarcity of sheep shearers have spurred interest in finding an easier and less costly way of removing the fleece from sheep. The following three innovative approaches are in various stages of experimentation and application: chemical shearing, laser beam shearing, and computerized shearing.

MARKETING WOOL

Fig. 15-4. Core sampling a wool bag with a 2 in. coring tool. This technique is used to obtain a sample for shrinkage determination. (Courtesy, C. J. Lupton, Texas A&M Agriculture Experiment Station, San Angelo, TX)

There are several differences between the marketing of animals on foot and the marketing of wool. In the first place, the average livestock producer is usually familiar with more than one of the several avenues through which live animals may be disposed; whereas, except for the larger wool growers, there is generally little knowledge concerning possible market outlets for wool.

Unlike most farm products, there are no open- or auction-markets for wool. Most wool is bought and sold by private treaty. Buyers may be representatives from woolen mills or brokers who represent the mills.

There are several channels through which growers may sell their clips. Producers in the major wool growing areas of the United States (Texas and the western states) usually have more options than those in the farm flock states. The predominant market channels are:

1. Private treaty.
2. Wool pools.
3. Sealed bid.
4. Consignment.
5. Direct selling.
6. Wool warehouses.

Fig. 15-5. Australian wool auctioneers. (Courtesy, C. J. Lupton, Texas A&M Agriculture Experiment Station, San Angelo, TX)

CLASSES OF WOOL

The wool trade recognizes two major classes of wool—apparel wool and carpet wool. As the names imply, most apparel wools are those suitable for manufacture into yarns and fabrics for human clothing, whereas most wools of the carpet class are used in making floor covering. In 1987, 91% of the wool consumed in the United States was apparel wool, and 9% was carpet wool.

Apparel wools are further classified according to use as (1) combing wool or staple wool—the long-fibered wools within the class, (2) French combing wool—the wools of intermediate length, and (3) clothing wool—the short-fibered wools. Although these three classes are based largely on length of fiber, other factors—such as supply and demand, fiber diameter, purity, condition, etc.—are important in determining the use made of wools. Thus, many wools used by the woolen industry are longer than some used in worsted manufacture; and a considerable amount of wool classed as clothing is used in the worsted industry. In general, however, the manufacturer can realize the greatest profit by utilizing apparel wools according to their best adaptation as indicated by the three classes. Further, carpet wool is not suited for use as apparel wool.

• **Combing or staple wool**—Combing or staple wools are usually referred to as the highest priced and best wool obtained from sheep. Both fineness and length are requisite. By and large, combing wools are used for making worsted fabrics. They take their name from the fact that one of the main processes in worsted manufacturing is the combing operation, which separates the long fibers from the short ones. The long fibers are used to make worsted cloths, and the short fibers (called noil) are used in the making of woolen cloths. In the former, the fibers are laid parallel to each other; whereas in the latter, the shorter wool fibers that are used in making wool cloths and felts are laid in every direction—in fact, the more mixing the better in woolens. These differences are of importance to the consumer. Among other things, they explain why worsted suits hold their press better than woolen suits.

In 1987, 53% of the mill consumption of apparel wool was processed on the worsted system and 47% on the woolen system.

• **French combing wool**—French combing wools are in-between the combing wools and the clothing wools in length. These wools are manufactured on the French or Heilman comb, which is designed to use shorter wools and still produce worsted fabrics.

• **Clothing wool**—Clothing wool is the name usually given to the shortest wool. This wool is too short to be manufactured on the worsted system, but it can be used successfully on the woolen system. The term *clothing wool,* however, does not mean that the wool is suitable only for fabrics to be made into clothing. This type of wool is also used to make felts.

• **Carpet wool**—Carpet wools, which are usually the coarsest wools, are low quality because they (1) contain mixtures of very coarse, hairy fibers and finer fibers, and (2) vary markedly in fiber length. The chief requisite of carpet wool is resilience, the quality that makes it resistant to matting down and to wear under the constant scuffing of passing feet. Most of this wool comes from long-wooled sheep and from sheep that show lack of breeding. Most carpet wools are imported. More than 90% of U.S. imports of carpet wool come from three countries: New Zealand, United Kingdom, and Argentina.

WOOL GRADING

Wool grading is based primarily on fiber diameter or fineness, but consideration is also given to length. Many manufacturers desire wool of a certain fineness only. This means that the wool must be separated at the warehouse and like fleeces must be piled by themselves. This process is called *wool grading,* and it is done by highly trained wool graders.

GRADES OF WOOL

The average diameter of fiber and the limits for the variation in diameter for the various grades are shown in Table 15–1. Maximum limits to the variation allowed for each grade are expressed by the statistical term—*standard deviation.* In application, if there is too much variation in fiber diameter, the wool is assigned to the next coarser grade. Wool can be separated roughly, after a little experience, into three broad market grades according to its diameter: (1) fine wool, (2) medium wool, and (3) coarse or braid wool. More accurately speaking, however, there are three distinct methods of grading wool according to diameter with several grades in each. The older method is called the blood system; the newer methods are the *numerical count system* and the *micron system.* A comparison of these three systems is contained in Table 15–1.

TABLE 15–1
COMPARATIVE WOOL GRADES AND CLASSES

		Standard Specifications[1]			Length Classes[2]					
			Micron System[3]							
Type of Wool	Old Blood Grade	Numerical Count Grade	Limit for Average Fiber Diameter	Variability Limit[4]	Combing Wool		French Combing Wool		Clothing Wool	
					over				under	
			(microns)[5]	(microns)[5]	(in.)	*(cm)*	(in.)	*(cm)*	(in.)	*(cm)*
Fine	Fine	Finer than 80s	Under 17.70	3.59	—	—	—	—	—	—
Fine	Fine	80s	17.70–19.14	4.09	2.75	*6.99*	1.25–2.75	*3.18–6.99*	1.25	*3.18*
Fine	Fine	70s	19.15–20.59	4.59	2.75	*6.99*	1.25–2.75	*3.18–6.99*	1.25	*3.18*
Fine	Fine	64s	20.60–22.04	5.19	2.75	*6.99*	1.25–2.75	*3.18–6.99*	1.25	*3.18*
Medium	½ blood	62s	22.05–23.49	5.89	3.0	*7.62*	1.5–3.0	*3.81–7.62*	1.5	*3.81*
Medium	½ blood	60s	23.50–24.94	6.49	3.0	*7.62*	1.5–3.0	*3.81–7.62*	1.5	*3.81*
Medium	⅜ blood	58s	24.95–26.39	7.09	3.25	*8.26*	2.0–3.25	*5.08–8.26*	2.0	*5.08*
Medium	⅜ blood	56s	26.40–27.84	7.59	3.25	*8.26*	2.0–3.25	*5.08–8.26*	2.0	*5.08*
Medium	¼ blood	54s	27.85–29.29	8.19	3.5	*8.89*	2.5–3.5	*6.35–8.89*	2.5	*6.35*
Medium	¼ blood	50s	29.30–30.99	8.69	3.5	*8.89*	2.5–3.5	*6.35–8.89*	2.5	*6.35*
Coarse	Low ¼	48s	31.00–32.69	9.09	4.0	*10.16*	—	—	4.0	*10.16*
Coarse	Low ¼	46s	32.70–34.39	9.59	4.0	*10.16*	—	—	4.0	*10.16*
Coarse	Common[6]	44s	34.40–36.19	10.09	5.0	*12.70*	—	—	5.0	*12.70*
Very coarse	Braid	40s	36.20–38.09	10.69	5.00	*12.70*	—	—	5.0	*12.70*
Very coarse	Braid	36s	38.10–40.20	11.19	5.00	*12.70*	—	—	—	—
Very coarse	Braid[6]	Coarser than 36s	Over 40.20	—	—	—	—	—	—	—

[1]Standards for grades of wool, as published by the USDA, August 20, 1965. *Federal Register* (7 CFR Part 31). These standards became effective January 1, 1966.

[2]There are no USDA official lengths for the different classes. The lengths given herein are in keeping with trade practices and were provided for use in this book by the Livestock Division Wool Laboratory, Standardization Branch, USDA, Denver, CO 80225.

[3]Beginning January, 1976, the unit designation terminology for wool prices changed to microns.

[4]Standard deviation maximum.

[5]A micron is 1/25,400 of an inch.

[6]Common and braid are not classified according to length because these wools are practically always of combing length. Carpet wool includes all those not suited to the three classes listed.

An experienced grader determines the grade of wool by the senses of sight and touch. However, for use in more objective grade determination and for arbitration purposes where there may be a dispute as to grade before final settlement, there is a scientific method of test prescribed. A copy of this method of test, which explains micro-projector equipment recommended and also sampling and testing procedures, may be obtained from the U.S. Department of Agriculture, Federal Center, Standardization Branch, Wool and Mohair Laboratory, Denver, Colorado. Testers and research workers may also use calipers or photographic or air-flow equipment for grade determination.

• **Blood system**—The blood system divides all wool, from finest to coarsest, into six market grades. These are (1) Fine, (2) ½ Blood, (3) ⅜ Blood, (4) ¼ Blood, (5) Low ¼ Blood, and (6) Common and Braid. Originally, these fractional Blood names denoted the amount of Merino blood in the sheep producing the wool. At the present time, these names indicate wool of a certain diameter only and have no connection whatsoever with the amount of Merino blood in the sheep. As a matter of fact, it is possible to have ⅜ Blood wool from a sheep with no Merino blood at all. The blood grades, therefore, are merely trade names identifying the different grades of wool, without relationship to the breeding of sheep and are rapidly being replaced with numerical count.

• **Numerical count system**—The numerical count system divides all wool into 14 grades, and each grade is designated by number. The numbers range from 80s for the finest wool down to 36 for the coarsest. This method gives more grades, and thus finer divisions can be made; and this is more satisfactory to the wool dealers and manufacturers. Table 15–1 shows the correlation between the two grade systems.

• **Worsted spinning count**—Theoretically, the numerical count system is based on the number of hanks of yarn (each hank representing 560 yards) that can be spun from 1 lb of such wool in the form of top. Wool of 50s quality, therefore, should spin 50 × 560 yards per pound of top, if spun to the maximum on the worsted system of manufacture. Unfortunately, this is not always true; the lower grades will not spin up to their number. Moreover, it is noteworthy that, in actual practice, wools are rarely spun to their maximum limit. Furthermore, spinning count is not determined by diameter alone; such factors as fiber length, moisture conditions, and the skill of the workers influence the count that may be spun. It may be concluded, therefore, that neither the blood system nor the numerical count system denotes accurately what it is supposed to indicate according to derivation of the respective term.

• **Micron system**—The micron system is a substantially more technical and accurate measurement of the wool fiber in a lot of wool. Sixteen grades are used, and are based on the average fiber thickness as measured by a micrometer. An 80s wool, for example, averages about 18 microns, which is less than half a 36s wool that average 39 microns. A micron is 1/25,400 of an inch. Wool too variable to fit within the limits of one grade is placed down a grade.

The micron system was largely developed at the USDA Denver Wool Laboratory. This system may eventually become the standard for describing wools in the United States. In January, 1976, the U.S. Economic Research Service began reporting the unit designation terminology for wool prices in microns.

Table 15–1 compares the old blood system, the numerical count system, and the micron system of wool grades.

WOOL SORTING

Sorting is the operation of taking an individual fleece, untying the twine, opening the fleece, and separating the fleece into the various grades that it possesses in the different body areas. This operation is usually done in the mill, but occasionally it is done in a warehouse. The reason for this is that a mill knows exactly what qualities of wool it wishes to put into a fabric. The object of sorting is to obtain large lots of wool that are very even and uniform in diameter, length, strength, and other characteristics. It is easy for an inexperienced person to distinguish a very fine wool from a very coarse wool, but it takes considerable training to be able to separate 2 consecutive grades, such as 56s from 58s. Sorting always is done on the grease wool. The dusting and scouring operation break up the fleece into small pieces, so that sorting of scoured wool is impracticable. Sorting is necessary on wool if a uniform worsted yarn with high spinning count is desired. If the wool is not to be spun to the maximum count, then only a superficial sorting is necessary. The thoroughness of the sorting varies according to the type of fabric into which the wool is to be made.

USES OF WOOL

About 91% of the wool consumed in the United States is apparel wool and 9% is carpet wool.

About 2% of apparel wool is used in batting and in the manufacture of pressed felt, mostly for hat bodies. The other 98% is consumed in the spinning of woolen and worsted yarn. About 15% of the woolen and worsted yarn is used in the production of knit goods, including sweaters, hosiery, underwear, gloves, and mittens. The remainder is used in the weaving of fabrics, and coatings and such nonapparel fabrics as blanketing, upholstery, draperies, and woven industrial felts.

The greater part of carpet wool is used in the manufacture of floor coverings, although small quantities are used in the manufacture of press cloth, knit and felt boots, and heavy fulled socks.

MADE OF WOOL AND MOHAIR

Fig. 15–6. A casual woolen pant suit, suitable for all occasions. (Courtesy, Pendelton Woolen Mills, Pendleton, OR)

Fig. 15–7. Wool, the versatile fiber, can be used successfully for draperies and upholstery fabrics, as shown in the headquarters of the International Wool Secretariat, New York, NY. (Courtesy, The Wool Bureau, New York, NY)

Fig. 15–8. Two fine mohair suits, designed in classic styles means that these suits will look fashionable and wear beautifully through many seasons. (Courtesy, Mohair Council of America, San Angelo, TX)

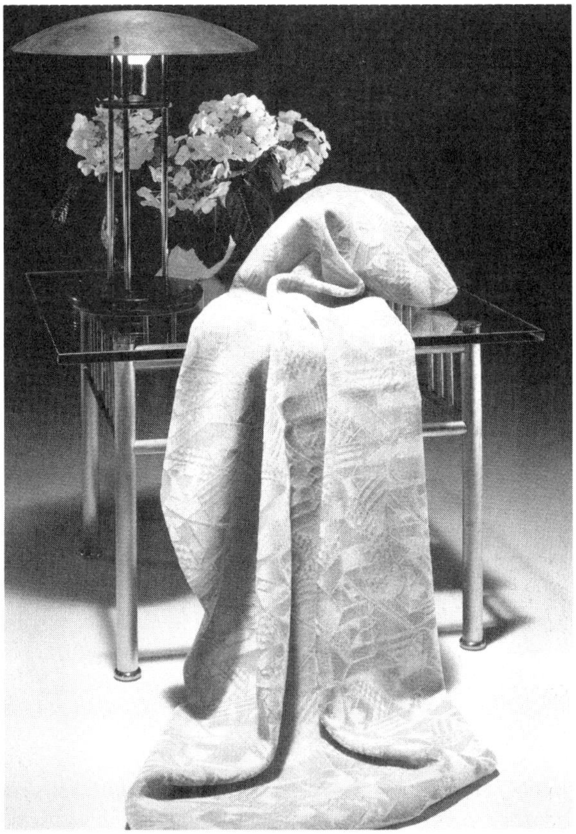

Fig. 15–9. An afghan made from mohair will, literally, wear forever. (Courtesy, Mohair Council of America, San Angelo, TX)

MOHAIR

Fig. 15-10. Angora goats, producers of mohair, in a rugged area. (Courtesy, Mohair Council of America, San Angelo, TX)

Mohair, known as the most versatile of fibers, is produced by the Angora goat, one of the oldest animals known. Yet, few citizens of the United States are more than casually aware of the Angora goat or its existence, despite the fact that these animals graze millions of acres of land and the hardwearing fabrics made from their lustrous coats are used and admired from coast to coast. Mohair possesses qualities all its own, found in no other animal fiber. It has less crimp and smoother surface scales than sheep's wool. These qualities add luster, softness, and dust resistance to the other fine qualities mohair shares with wool. Mohair has remarkable resistance to wrinkles, great strength, and unequalled affinity to brilliant, deep colors that resist time, the elements, and hard wear.

PRODUCTION, HANDLING, AND MARKETING MOHAIR

Fig. 15-11. Angora goats, producers of the diamond fiber. (Courtesy, Mohair Council of America, San Angelo, TX)

Although mohair is usually accorded more neglect than wool, the principles involved in the economical production and advantageous marketing of a high-quality product are the same with both fibers. For practical reasons, chiefly as a means of lessening fleece losses caused by shedding or brush, more goats are shorn twice per year than is the case with sheep. Also, in the Southwest, goats are shorn twice each year because of the warm weather. Except for this difference, and the fact that it is not recommended that the mohair fleece be tied at shearing time, the discussion already presented relative to the production, handling, and marketing of wool is equally applicable to mohair. The market channels and leading market centers for wool and mohair are identical.

It is unfortunate that a large amount of mohair produced in this country continues to be marketed by placing all grades in a single bag, with little attention given to sorting. On most shearing floors, altogether too little attention is given to keeping the fleece intact and rolling it together in order that an intelligent job of grading and sorting may be done later. So long as these careless production methods are followed, mohair will neither meet the highest requirements of the manufacturer nor command a top price for the grower.

Most of the mohair is exported through local buyers, representing English and European firms. Much of it is exported in original bags, then graded and processed abroad. However, some graded mohair is exported. Some mohair is graded and processed into *top* for domestic use.

CLASSES AND GRADES OF MOHAIR

The official grades of grease mohair and the specifications of each are given in Table 15-2.

TABLE 15-2
SPECIFICATIONS FOR THE OFFICAL GRADES OF GREASE MOHAIR

	Fiber Diameter		
	Limits for Average	Maximum Standard Deviation	Approximate Number of Fiber Measurements[1]
	(microns)	(microns)	
Finer than 40s	Under 23.01	7.2	1,000
40s	23.01-25.00	7.6	1,000
36s	25.01-27.00	8.0	1,200
32s	27.01-29.00	8.4	1,200
30s	29.01-31.00	8.8	1,400
28s	31.01-33.00	9.2	1,400
26s	33.01-35.00	9.6	1,600
24s	35.01-37.00	10.0	1,600
22s	37.01-39.00	10.5	1,800
20s	39.01-41.00	11.0	2,200
18s	41.01-43.00	11.5	2,200
Coarser than 18s	43.01 and over		2,600

[1]The number of fibers to measure for each test shall be the number needed to attain confidence limits of the mean within ± 0.40 micron at a probability of 95%. Measurement of the approximate number of fibers for the grades listed above may serve as a guide to meet the required confidence limits. The numbers indicated are based on mohair matchings.

Kid hair is finest and is especially sought by mills. The fleeces from adults—especially bucks and old wethers—are the coarsest; and that from yearlings is intermediate between the other classes. These classes can be recognized by the grower and should be packed separately at shearing time. In addition, those fleeces that are extremely coarse, weak, and shorter than 6 in. or those having an excess of kemp, burs, or other foreign matter should be kept separate from clean, strong fleeces of desirable length and fineness.

As with wool, the grades of mohair are based primarily on the presumed spinning count obtainable on the Bradford system (for the number of 560-yard hanks to the pound). In practice, fineness is associated with softness and is recognized by the experienced touch when handled between the thumb and fingers.

USES OF MOHAIR

Mohair is used for car upholstery, portieres, robes, rugs, braids, and artificial f urs; and there is considerable use of superior mohair in suit linings and for men's summer suitings. The long-fibered mohair is also in demand for use in manufacturing wigs and switches for theatrical purposes.

NATIONAL WOOL ACT

Through passage, and subsequent extension, of the National Wool Act (first passed in 1954), Congress recognizes wool as an essential and strategic commodity which is not produced in the United States in sufficient quantity to meet domestic needs.

The incentive payments are financed from the duties collected on the imports of wool. Also, the Act authorized an industry self-help program for the purpose of developing and conducting advertising and sales promotion programs for lamb and wool.

The incentive program for shorn wool and unshorn lambs follows:

1. **Payment for shorn wool.** Payments are made to wool producers after the marketing year is over, to bring the national average price received by all producers, for wool sold, up to the incentive price.

The rate of payment for the marketing year is announced after the year is over, and the average price received by all producers is known. The rate paid will be the percentage required to bring the national average price, received by all producers, up to the incentive price.

The percentage rate recognizes quality production and encourages producers to improve the quality and marketing of their wool. The higher the price the individual producer receives, the greater the payment.

2. **Payment for unshorn lambs.** The payment on unshorn lambs is at a comparable rate to the shorn wool payment and is designed to encourage the normal practice of marketing lambs without shearing the wool. The quantity of wool on unshorn lambs sold for slaughter averages around 5 lb per 100 lb of animal, liveweight.

In the 1988 marketing year, the incentive price per pound of wool was $1.78.

In order to secure the most benefit from this Act, the wool grower should (1) sell for the highest price possible, and (2) obtain complete sale records. For example, let us assume that the national average wool price is $1.38 per pound. To bring the national average price of $1.38 to the incentive level of $1.78, each producer's price would need to be increased by 29% (178 − 138 = 40; then 40 ÷ 138 = 29%). Therefore, if you sell 1,000 lb of wool for $1.38 per pound, you will get (1) $1,380 from the buyer, and (2) $400.20 (29% more) from the U.S. Department of Agriculture, making a total of $1,780.20.

But if you sell your wool for $1.85 per pound, instead of $1.38, the story is as follows: You will get (1) $1,850 from the buyer, and (2) $536.50 (29% of $1,850) from the U.S. Department of Agriculture, making a total return of $2,386.50. This shows how the returns of 1,000 lb of wool could be increased by $606.30 through careful marketing.

In 1989, the incentive price for mohair was $4.59 per pound, while the 1989 average market price was $1.58 per pound.

The incentive payments for wool and mohair have been both criticized and praised. The critics point out that the program has not resulted in an increase in wool and mohair as projected. The supporters of the program emphasize that it has helped (1) maintain viable sheep and Angora industries, and (2) improve the quality of wool and mohair produced. Producers of wool and mohair and leaders of grower organizations also suggest that, without the program, the decreases in production would likely have been much greater.

The National Wool Act also implemented the establishment of the American Sheep Producers Council (ASPC), the headquarters/address of which is:

American Sheep Producers Council, Inc.
200 Clayton Street
Denver, CO 80206

Fig. 15–12. Spinning, the way it used to be done. Recently, there has been a resurgence of hand processing. (Courtesy, American Sheep Producers Council, Wool Education Center, Denver, CO)

SECTION

16

LAW ON THE LIVESTOCK FARM

Athena, goddess of wisdom and agriculture, standing among the ruins of the Parthenon, on the Acropolis, which the Greeks built in her honor between 447 and 432 B.C. Athena was noted for her good judgment. The owl, emblem of wisdom and reason, was her symbol. (Photos by A. H. Ensminger; drawing of Athena by Margo Williams)

Contents — *Page*

The primary function of laws is protection. Also, they provide certain guides for dealing with others.

Observance of the principles herewith outlined will help avoid legal difficulties and losses. When the livestock producer is contemplating business deals involving questions about which there are uncertainties, however, a reliable attorney should be consulted. Observance of the latter point is important, for "ignorance of the law excuses no one." Also, it is generally recognized (1) that attorney fees are cheaper than court trial, and (2) that, when differences of opinion exist, it is usually less expensive and more satisfactory to compromise than to take the question to court.

FARM LEASES AND LEASING

Leases are the legal agreements by which landowners and farm operators do business. Since about 240,000 farmers rent land, a total of about 2 million people (landowners plus farm renters) face basic leasing problems.

A lease may be oral or written, long or short, adapted or not adapted; and, as a written instrument, it may be valid or invalid. In short, as a means of aiding the parties to it, a lease may be good, bad, or indifferent. Good written farm leases can lead to better understanding and closer cooperation between land owners and tenants.

CHOOSING THE RIGHT FARM OPERATOR

Most of the problems are solved when the right operator is selected. Although there are no exact formulae, most of the better tenants possess the following characteristics:

1. Honesty.
2. Willingness to cooperate in a successful farm management program.
3. Thorough knowledge of the proper care of all crops and livestock to be included in the farm business.
4. Ability, health, energy, and initiative to do good work in proper season.
5. Sufficient equipment and financial backing to operate the farm efficiently, unless these are furnished by the owner.
6. Favorable attitude toward adoption of new methods and practices, as rapidly as their merit is established.
7. Interest in soil conservation and sustainable agriculture.
8. Interest in preventing the introduction and spread of weeds.
9. Pride and interest in the farm and the community life.
10. Willingness to make minor repairs to buildings and farm.
11. Neatness about the farm and their person.
12. Spouses who are interested in farm life and the farm operation.

MINIMUM LEGAL ESSENTIALS OF A WRITTEN LEASE

To be legal, a written lease must meet the following requirements:

1. It must give the name and address of both parties,.
2. It must give the legal description of the property.
3. It must state the beginning and ending dates.
4. It must give the divisions of crops and designate the place of delivery.
5. It must state cash provisions, if any, and give the time and place of payment of same.
6. It must provide for possession.
7. It must be signed by both parties; preferably by both husband and wife.
8. It must be notarized if so required by state law.

PROVISIONS IN ADDITION TO MINIMUM LEGAL ESSENTIALS

A desirable farm lease contains more than the minimum legal essentials. Usually, it is to the mutual best interest of both the owner and the tenant that the following points be considered in written agreements:

1. Owner's right of entry to maintain improvements, and to plan and inspect farm operations.
2. Storage space for the rental share of crops.
3. Prohibiting subletting without owner's consent.
4. Yielding possession at the end of the lease period.
5. Restricting work of tenant off the farm.
6. Tenant's guarantee of good work.
7. Upkeep of fences, buildings, trees, and natural resources.
8. Keeping the premises in repair.
9. Interest on deferred rents.
10. Penalty in the case of failure or default of the tenant.
11. Owner's lien on crops and livestock.
12. Responsibility for costs in case of disputes.
13. Agreement that is binding on the heirs and assigns.
14. Agreement for renewals.
15. Some details of management. The crop rotation, and soil and livestock management plans should be worked out and made a part of the lease. They should not be so rigid as to be unworkable; nor should they be so general as to preclude the best use of land and improvements.
16. Reimbursement for capital improvements performed by operator, and provision for removal of portable improvements provided by the tenant.
17. Adjusting cash rental to a changed price level.
18. Control of weeds, hauling of manure, etc.
19. Kind of records that will be kept.
20. Special agreement not covered by normal practice; such as provisions for the use of legume crops, homegrown lumber, wood, gravel, and other items of special nature.
21. Information on the rights of third parties; hunting, use of water, rights of way, etc.
22. Provision for arbitration in case of disagreement.

VERBAL LEASES VS WRITTEN LEASES

Although verbal leases are legal in many states, most oral leases protect neither party and are often the cause of disagreement.

Written leases have the following advantages:

1. They serve as a basis of settlement in case of dispute.
2. They serve as a memorandum to prevent disputes, because both parties can refer to the written terms.
4. They give assurance that both parties will consider all phases of the contract.
5. They protect the heirs in case of death of a principal.
6. They serve as a history of the farm's operation.
7. They make the terms of rental definite.
8. They give an opportunity to provide for a reasonable notice period to terminate lease.

ONE-YEAR VS LONGTIME LEASES

Generally speaking, one-year leases are preferable to longtime leases for the following reasons:

1. They can be renewed annually when there is no disagreement.

2. They can be adjusted from year to year to care for changing conditions.

In one-year leases, the tenant can and should be protected by clauses agreeing to pay for the following improvements made by the tenant:

1. Soil improvements that extend beyond the year concerned.
2. Permanent grass or legume seedings, for which the tenant contributed expenses, but did not benefit.
3. Buildings, or major repairs to buildings, not normally the expenditure of the tenant.

TYPES OF FARM LEASES

The most common types of farm leases are:

1. **Cash lease.** This is a good type of lease for (a) the small farm or where the owner lives at a distance, and (b) a tenant who has adequate livestock, equipment, and working capital. It encourages livestock farming because all of the crop can easily be fed on the farm. Also, it is simple, with little chance for controversy.

There are two types of cash lease: (a) the type in which a fixed rent per acre is agreed upon when the lease is drawn, and (b) that type in which the rent is adjusted to prices of farm products which prevail during the lease year. Under the second plan, the owner bears part of the risk of price changes; however, it is difficult to keep cash rent in line with farm product prices. If product prices are used as a basis for rent changes, the products, markets, and dates should be specified.

The owner may prefer a cash lease because (a) the amount paid is definite, and (b) it requires less supervision by the owner. On the other hand, it may not always be desirable from the standpoint of the owner because (a) it generally makes for lower income, (b) it gives the owner less control of the farm, and (c) it is difficult to collect rent if crops fail.

The tenant may prefer a cash lease because (a) it will make for more profit if he/she is a successful manager, (b) it makes for more independence in the operation, and (c) it makes for more profit in the good years.

2. **Crop share lease.** In this type, crops are shared and cash is paid for pasture and/or hay land. This is the most used lease in grain areas.

Crop share leases are adapted to (a) areas where land is good and nearly all tillable, and (b) young tenants and those with less capital. However, they lend little encouragement to livestock farming.

The owner may prefer a crop share lease because (a) the rental more nearly equals the value of the land, and (b) there is more opportunity for supervision of the farm.

The tenant may prefer a crop share lease because (a) there is less risk, especially because of crop failures, (b) it requires less capital, and (c) the owner is more inclined to improve the farm and increase productivity.

3. **The livestock share lease.** This type of lease fits the tenant who wants to raise livestock, but cannot finance a program. It is especially suited where tenant and owner get along well and where the owner can make a good contribution in management.

In order for this type of lease to work best, the owner should live close to the farm, and either give it personal attention or arrange for adequate management help such as can be provided through a professional farm management service.

The owner may prefer a livestock share lease because (a) it encourages more livestock and more manure, (b) low-quality crops can be utilized more easily, (c) he/she retains an active interest in management, and (d) it generally makes for more profit.

The tenant may prefer a livestock share lease because (a) the risk is less since rent is based on net income on the farm, (b) it requires less tenant capital, (c) the landlord is more willing to make improvements, and (d) he/she can gain experience from the guidance of a successful owner.

4. **Manager operator or partnership.** This includes a number of leases in which the landowner normally furnishes all the capital, while the manager furnishes only the labor but shares in the gross returns, usually on a basis of a stipulated percentage—normally 35 to 40%. Among such leases, are the various parent-and-child business agreements.

With this type of lease, generally the owner supplies the equipment, livestock, capital, and general supervision; and the young person handles most of the work and the day-to-day management decisions. Among the basic requirements for this type of lease are separate living quarters for each of the families, a farm business large enough to support two families, ability to get along together, a good set of records, and revision of the agreement from time to time.

After a few years under this type of lease, usually the operator prefers, and is able, to switch to a regular crop share, livestock share, or cash lease.

Except for a parent-and-child agreement, this type of lease is normally not too satisfactory and should be avoided because (a) it is rather hazardous as the operator has only labor at stake, and (b) disputes arise over the system of accounting.

ADDITIONAL PERTINENT POINTS RELATIVE TO LEASES

Herewith are additional pertinent points relative to leases:

1. **One-year leases.** A tenant with a one-year lease should have it renewed at the end of the year; otherwise, the tenant may find (a) that the owner can require him/her to move on less notice than that established in the original lease, or (b) that the issues which the original lease covered are not now covered. Without renewal of the lease in writing, the lessee becomes a year-to-year tenant.

2. **Notice to vacate land.** Under a written agreement, a tenant is entitled only to the period of notice specified in the lease; unless the laws of the state stipulate otherwise. If the tenant does not have a written agreement, or if the written lease has expired and the lessee is holding the land as a year-to-year tenant, the period of notice will be variable according to state law.

It is in the best interest of both owner and tenant that the written lease provide for a period of notice, and that it contain an automatic renewal clause in the case of no notification.

3. **Owner's lien.** Most states provide for an owner's lien, which gives the owner of the land a claim for rent against the crops of the tenant. These vary somewhat from state to state.

4. **Owner's right to harvest crops.** Poor health, discouragement, poor crops, or better immediate prospects elsewhere sometimes lead a tenant to move before the term for which he/she has rented a farm has expired. Occasionally, a tenant even leaves before harvest without having made any arrangements to take care of growing crops. Under such circumstances, usually the state law (a) protects the owner's right to harvest crops, and (b) permits the tenant later to redeem crops matured and harvested by the owner provided the lessee first pays any rent due and other reasonable compensation to the lessor.

5. **Tenant's right to take removable fixtures.** Most state laws permit the tenant to take removable improvements that were either (a) brought with him/her, or (b) constructed at his/her own expense; usually these include hog houses, brooder houses, temporary fences, and cribs. Generally, a fixture is considered removable when the parties intend it to be and when it can be removed without undue injury to the land or other buildings. Actually, these matters should be fully covered in the lease or in a subsequent written agreement with the owner. The lease or agreement should either (a) allow the tenant to remove the structure(s), or (b) provide for payment of a fair market value when, and if, the tenant leaves.

In order to be entitled to this privilege, however, usually the law states that the tenant must meet the following three stipulations:

a. Tenant must not owe the owner back rent; otherwise, the owner may hold the improvements.

b. Tenant must have put the improvements on the land.

c. Tenant must remove them before the rental term expires.

LEASE FORMS AND ASSISTANCE

Several good lease forms are available. Usually, they can be obtained from county agents, state colleges of agriculture, or professional farm managers.

The person who has not had experience writing a lease form will need some help. Since a good lease must be both agriculturally and legally sound, it is well to call upon both an experienced agriculturist and an attorney.

OTHER LEGAL DOCUMENTS

In addition to leases, some other legal documents are: Bill of Sale, Chattel Mortgages, Uniform Commercial Code (U.C.C.), and Syndicated Animals (partnerships).

BILL OF SALE

This is a document given by a seller to a buyer, establishing the fact that legal title has passed hands to the new owner; and by the bill of sale the seller warrants that he/she is the owner or agent for the owner and has the lawful right to sell the property, that the property is free of encumbrance, and that he/she will defend the buyer's title. Such bill of sale should be received in the purchase of livestock and equipment. In many states, the title certificate replaces a bill of sale for all licensed vehicles such as trucks and automobiles. Also, in certain states the brand slip comprises the identification of animals with a bill of sale.

CHATTEL MORTGAGES

Livestock producers use short-term credit to finance livestock (especially feeder cattle and lambs), crops, and equipment and supplies. The chattel mortgage is the most common form of security for such loans. It allows the debtor (mortgagor) possession and use of the mortgaged chattels (personal property) and at the same time gives to the seller or lender (mortgagee) a security interest superior to the rights of purchasers or transferees of the mortgagor, regardless of the actual knowledge of such persons concerning encumbrances on the chattel. It enables farmers to obtain loans because it gives the lender a high type of security.

The following facts about chattel mortgages are important:

1. They must be recorded within the time and in the manner provided by state law; they must be prepared and signed in accordance with state law; and they become effective when not renewed or foreclosed within the time specified by law.
2. They must contain an accurate description of the property mortgaged.
3. They usually (in most states) attach to progeny born of mortgaged females.
4. They do not attach to crops until the seed is in the ground unless the state law provides otherwise.
5. They are good on feed, seed, or fertilizer if the same mortgage includes animals to which the feed is fed, or the crop on which the seed and fertilizer are used.
6. They prohibit the sale of mortgaged chattels without the consent of the mortgagee. If they are sold without consent, the buyer does not acquire title and the mortgagor is subject to penalty. The mortgagee then may recover the property from the purchaser or the purchaser's transferee, and may recover damages from the mortgagor. Also, many states stipulate that it is a crime to dispose wrongfully of mortgaged property.
7. They differ from state to state. Accordingly, the appropriate state law must be studied thoroughly by those whose business requires the use of chattel mortgages.

UNIFORM COMMERCIAL CODE (U.C.C.)

In recent years, most states have adopted the Uniform Commercial Code. It allows such security instruments as chattel mortgages and conditioned sale contracts to be replaced by a single security agreement that creates a security interest in personal property. The old instruments may be used, but their effect will be to create a security interest.

In those states having the Uniform Commercial Code, the form required by the particular state must be filed in compliance with the state law.

The Uniform Commercial Code can be used to integrate all the financial needs of the farmer in one transaction. The approval of future advances and provision for after-acquired personal property as collateral will give flexibility to farm production financing.

SYNDICATED ANIMALS (PARTNERSHIP)

Like livestock leases and parent-child farming plans, syndicated animals are a form of "farm partnership." Such animals may be owned by two or more partners. In a syndication, liability is mutually shared; that is, all parties are mutually or fully responsible for the liabilities of the syndicate. However, the syndicate is not liable for personal, as distinct from syndicate, debts. Syndicate agreements may be written in practically any form desired by the partners. For example, the services of a syndicated sire and the expenses for the sire's keep may be shared according to shares (investment) just as contributions to any other partnership in the form of capital, labor, and management may be shared in any proportion.

(Also, see Section 2, under the heading "Syndicated Horses.")

LEGAL ASPECTS OF FENCING

Most states have laws pertaining to boundary fences. In some states, however, the fencing regulations are left largely to the counties and townships. Although these laws vary greatly from state to state and are subject to frequent change, the following conditions usually prevail:

1. **Inside fences.** No stipulations are made relative to the materials used on inside fences (fences other than between boundaries).

2. **Boundary fences.** Usually state laws require every landowner to enclose his/her land with a fence tight enough and strong enough to turn livestock.

Some states deny the landowner any damages for trespass of livestock if the holding is not properly fenced; whereas other states permit collection of damage for trespass by animals even though the landowner suffering the damages does not have his/her land fenced.

It is a rather common point of law that the condition of the fence at the point where the stock passes over or through in trespassing determines whether it is a suitable fence. The fact that it is not high enough at some other point, or that someone left a gate open on the other side of the farm, has nothing to do with the case. Thus, the argument is settled solely by the condition of the fence at the place where animals went through, and not by its condition at any other place.

Although state laws vary rather widely, the predominant decisions of state courts on various situations involving livestock and fences are as follows:

a. When the livestock owner has good fences, is not aware that his/her animals habitually break out, has not been negligent, and makes an immediate attempt to get them back when they do break out, he/she is not liable for damage caused by them.

b. When animals break through an adjoining owner's part of a division of fence, and such fence is not good, the owner of the animals cannot be held liable for damages inflicted by their trespass.

c. The owner of animals may be held liable for damages inflicted by their trespass provided:

(1) The animals are known to be in the habit of breaking out, regardless of how good the fences may be.

(2) The fences are not good.

(3) The trespass of animals was caused through negligence, such as by leaving a gate open, or by stampeding animals until they break out.

3. **Misdemeanor.** In some states, if anyone willfully or negligently (a) leaves open or tears down a gate provided for the convenience of the public, (b) tears down a fence on another person's property, or (c) allows livestock to run at large, the act is classed as a misdemeanor. Upon conviction, such person is subject to fine or imprisonment, or both.

4. **Responsibility for division fences.** The responsibility of neighbors for the construction and maintenance of boundary or division fences is the subject of frequent controversy.

Where the partition fence extends north and south, custom decrees that the owner whose land is east of the fence must build the north half, and the owner whose land lies on the west side of the fence must build the south half. Where the partition fence extends east and west, the owner whose land lies north of the fence builds the west half and the owner whose land lies south of the fence builds the east half. A simple customary rule regarding the apportionment of a division fence is one which gives a responsibility to the landowner for that portion of the fence which is on his/her right when standing on the property and facing the fence. Where landowners agree to some other division, the agreement should be put in writing, acknowledged before a Notary Public, and recorded by the County Recorder.

5. **Railroad fences.** Railroad companies are generally required by state laws to construct and maintain fences along their rights-of-way, provided the land adjacent thereto is otherwise enclosed. Also, they are required to maintain at road crossings cattle guards sufficient to prevent stock from getting on the railroad. When such fences and guards are not constructed and maintained properly, the railroad company may be liable for all damages which may be done by its trains.

6. **Check state laws.** Before constructing a boundary fence, the farmer or rancher should first examine with care the state laws in order to make certain of compliance with the legal requirements.

Many states have legal fence statutes which settle the question as to what kind of division fence must be built.

ANIMALS ON HIGHWAYS

Sometimes farm animals get out on highways. If under these circumstances, a user of the highway runs into a loose animal and is injured and/or has his/her vehicle damaged, he/she frequently tries to collect damage from the owner of the animal. Although state, county, and/or township laws vary, and it is not possible to predict with accuracy what damages, if any, may be recovered in particular instances, the following general rules apply:

1. If a farmer or rancher is negligent in maintaining his/her fences and allows animals to get on the road, he/she can be held liable for damage resulting to persons using the highway.

2. If a farmer or rancher has good fences that are well maintained, but has one or more animals which are known to be in the habit of breaking out, he/she may be held liable for damages caused by such animals.

3. If animals get on to the highway, despite the facts that there are both good fences and no animals that are known habitually to get out, the owner may be held liable for any damage inflicted provided he/she knew that the animals were out and made no reasonable effort to get them in.

4. If the farmer or rancher is not negligent in any way, he/she may or may not be judged liable for the damage inflicted by the animals, depending on the state law and other circumstances.

5. If a farmer is driving animals along or across a highway, he/she is not likely to suffer liability for any damages unless it can be proved that there was negligence. Stock-crossing signs usually increase the caution exercised by motorists, but do not excuse a farmer from exercising due care.

6. In some states, laws provide that a farmer may, under the supervision of and with varying amounts of assistance from highway authorities, construct an underpass for animals and for general farm use.

TRESPASS BY ANIMALS

Livestock owners who do not use reasonable care in restraining animals may be held liable for damages caused by their trespassing. Among the kinds of damage for which the courts have held that the owner may be responsible for are:

1. The destruction of growing crops.
2. The transmitting of disease.
3. The breeding of females by trespassing sires.

The amount of damages in such cases is based on the difference in value to the owner between the actual progeny and the intended progeny. Damage may be considered where the female is a registered purebred and the culprit is a scrub.

Generally state laws stipulate that the owner of land on which animals are trespassing may do anything reasonable to terminate the trespass, including the following:

1. Drive them back to the place from whence they came.
2. Call the owner and ask him/her to get them.
3. Confine, feed, and care for the animals until the owner comes and takes them; collect costs for same.

HANDLING ESTRAYS (STRAYS)

An estray is usually defined as a domestic animal (not including dogs or cats) of unknown ownership running at large. In some states, poultry at large may be regarded as other straying animals and may be taken up to prevent damage.

Although there is no uniformity in the state laws governing the handling of estrays (strays), some of the more common provisions are:

1. That either (a) landowners, or (b) local authorities may confine such animals and care for them.

2. That following confinement of estrays, a reasonable attempt must be made to locate the owner. Some laws specify public posting and giving of notice in local papers.

3. That the taker-up is entitled to make reasonable use of estrays while they are in custody; for example, work a horse or milk a cow.

4. That upon coming for estrays, the owner must satisfy the claims of the taker-up for feed, housing, care, and other costs.

5. That if the owner does not claim his/her animals, they either (a) become the property of the taker-up, or (b) must be sold at public auction, with reimbursement made to the taker-up for expenses incurred and with the balance turned in as county funds.

AGISTERS

The term agister is taken from English law. *It refers to one who pastures, feeds, or cares for the livestock of another for hire.*

Most states possess lien laws in favor of agisters. Such laws generally provide that agisters shall have a claim against the animals for agreed or reasonable charges, and that such claim may be enforced by retention and sale of enough animals to satisfy it. Unless the law so states, it is usually assumed that the lien expires when possession of the animals terminates.

In order to be entitled to lien consideration, the agister must keep the premises properly fenced, take reasonable precautions against injury to animals, and provide suitable feed, water, and shelter. Also, in case of death or injury loss due to neglect, such as from feeding poisoned grain, the agister is liable if he/she knows or should have known of the circumstances.

Although interpretations have varied, generally the courts have established that the following rules shall prevail:

1. **That agisters must have the animals in their possession.** To be entitled to a statutory lien, the person claiming it must have the animals in his/her charge and under his/her control. Thus, commercial feed companies are not entitled to a lien on the basis that they supply feed on credit.

2. **That there must be a signed or implied agreement.** To be entitled to a lien, there must have been an agreement, either signed or implied, covering their pasture, feed, and/or care.

3. **That a security agreement shall take precedence.** A security agreement takes precedence over an agisters lien unless—

   a. The mortgagee consents to an agreement whereby persons other than the mortgagor shall feed and care for animals, or

   b. The mortgage is executed while animals are under agistment.

4. **That the agister must give the owner written notice of sale and publicize same.** After unsuccessfully requesting reasonable compensation and while still in possession of the

animals, the agister must give the owner written notice of the time and place of sale as required by the particular lien statute. Also, there must be due publication of notice. The animals may then be sold, with the agister retaining the amount which he/she claims and paying the balance, if any, to the owner.

SHEEP-KILLING DOGS

In many areas, one of the greatest causes for discouragement in sheep production is the problem of sheep-killing dogs. The problem is accentuated by the failure of dog owners to recognize that even the most lovable pet may roam the countryside at night, molesting and killing sheep and other domestic animals.

Unfortunately, from state to state, there is wide variation in laws pertaining to sheep-killing dogs. Few such laws are entirely satisfactory from the standpoint of the sheep owner, and most of them are not aggressively enforced. Most state laws provide for one or more of the following forms of legal protection against dogs:

1. **That dogs must be licensed.** This provision has two objectives; namely (a) to eliminate those dogs which the owner does not consider worthy of a license fee, and (b) to build up a county indemnity fund for payment to animal owners who suffer damage from dogs. Usually the maximum indemnity payment for various kinds of animals is stipulated by law, and claims must be presented through the township supervisor or other designated official.

2. **That it is a misdemeanor to allow a dog known to possess harmful tendencies to run at large.** Some state laws make it a misdemeanor to keep such a dog unless it is confined or chained. Also, these states usually make the owner an insurer for any and all damages inflicted by such a dog.

3. **That animal-molesting dogs may be killed.** Some states allow the owner of domestic animals the right to pursue and kill dogs not accompanied by their owners when they are discovered in the act of killing, wounding, or chasing domestic animals.

4. **That animal-molesting dogs may be poisoned.** Some state laws allow a sheep owner to put out poison for dogs on his/her premises, provided it is done with reasonable care and good intentions.

5. **That damages may be collected from the dog's owner.** In some states, the law provides that the owner of animals killed or injured by dogs has a right of action against the dog's owner for all damages caused by the dog.

Such laws as the above have done some good, but—regardless of the printed law—it is generally recognized that sheep-killing dogs make for much ill-feeling and that any damages collected seldom cover the actual losses. Under these circumstances, the best protection for a flock owner still consists of a dog-proof corral for lotting at night.

HORSE PROTECTION ACT (SORING HORSES)

Soring is the use of painful methods and devices to enhance a horse's gait in the show-ring. It evolved as a means of producing a fast, flashy gait in Tennessee Walkers, for the show-ring. Soring is the practice of using caustic liquid, commonly called "scooter juice," along with chains or shackles, to make a walking horse's front ankles sore. This process, combined with feet seven or more inches long, heavy shoes, and some drastic training, create the desired show-ring gait.

A true running walk is executed at a speed of 6 to 8 miles per hour and will not exceed 10 miles per hour. It is done with economy of effort to both the horse and the rider and is not very showy. In an effort to increase the speed to 15 to 18 miles per hour and obtain high action in front, yet keep the gait from being classed as a rack, horses are sometimes sored by means of blisters, chains and whatnot, so that they scoot their hind feet far under them in order to keep the weight off their sore front feet. The soreness, along with accompanying long toes and heavy shoes (secured by bands over the feet, in addition to nails), cause the horse to pick the front feet up very high as it leaps through the air. Actually the fast, artificial gait that results more nearly resembles a rack than a running walk.

The Horse Protection Act, making it illegal to show or exhibit sored horses or to conduct a horse show in which sored horses are allowed to participate, was passed by the U.S. Congress and signed into law in 1970. But soring persisted, finally culminating in the Horse Protection Act Amendment of 1976, with stricter provisions. The act as amended defines soring as any practice that causes a horse to suffer physical pain, distress, or lameness while walking, trotting, or otherwise moving.

The U.S. Department of Agriculture, Animal and Plant Health Inspection Service, is charged with enforcing the Act.

A first offender, violating the provisions of the Act is subject to a civil penalty up to $2,000. Criminal penalties carry a maximum fine of $3,000 or a year in jail, or both. Second offenders may be fined up to $5,000 or 2 years in jail, or both.

PERSONS INJURED BY ANIMALS

The owner of farm animals may be held liable for personal injuries caused by them under the following circumstances:

1. When he/she negligently allows or causes them to commit the injury.
2. When he/she is aware of the viscousness of such an animal, and when such an animal inflicts injury upon someone who was not acting negligently.

It is a common-law rule that "a dog is entitled to one bite," and that after that the owner may be liable for injury to others. However, some states have passed laws removing this protection and holding the owner liable for the first attack.

BRANDS AND BRAND INSPECTION

In the range country, brands are used as a means of determining the ownership of animals and of lessening theft. To meet these needs, each of the western states has laws governing the recording and inspection of brands and the transfer of branded animals. These laws generally contain the following provisions:

1. **Recorded brands.** Ranchers are required to register any brand they use, and, after its approval by the Registrar of Brands of the state agency in charge, to use that specific brand on their livestock—usually on cattle, sheep, horses, and mules: (For information relative to types of identification, the reader is referred to Section 8, Management, under the heading, "Marking or Identifying Animals.")
2. **Bill of sale.** When animals are sold, a bill of sale or other written evidence of transfer must be signed by the seller and given to the purchaser.
3. **Local brand inspector.** Local brand inspectors, usually under the supervision of the state department or commissioner of agriculture, inspect all animals leaving their district to determine if any are being sold by a person other than the rightful owner.
4. **Inspection of hides.** Frequently there is provision for an inspection of hides at slaughter houses as a further means of disclosing theft and wrongful sale.
5. **Slaughtering in remote places.** Usually the slaughtering of animals in remote places is prohibited, for purposes of lessening theft.
6. **Penalties.** Violations, especially theft and effacing or changing brands, are subject to severe penalties.

Livestock producers operating in those states which have brand laws should become thoroughly familiar with the provisions thereof, and should recognize that law enforcement against rustlers and thieves can only be as good as the existing brands and brand inspection program. In cases of suspected theft, the first question that the sheriff is prone to ask is "what brand did the lost animal have?" Unbranded range animals are an open invitation to thieves, and in the case of loss, make for a cold reception from law enforcement officials, for they can be of little help unless there is positive animal identification.

LIVESTOCK OPERATIONS WHICH MUST BE LICENSED

Although state laws vary, the following livestock and livestock operations are generally subject to license and regulation:

Auctioneers
Auction sale ring operations
Bull lessors
Commission merchants handling meat, livestock, and livestock products
Dealers of livestock
Dead animal disposal
Dogs
Feed dealers
Horseshoers
Meat dealers
Meat and produce peddlers
Pet dealers
Poultry dealers
Public carriers of livestock
Racetrack operators
Rendering plant operation
Sires for public service
Slaughter house operation
Stockyards (public) operation
Traders (itinerant) of horses and mules
Veterinarians
Weighing (public) of livestock

LIABILITY

Most farmers are in such financial position that they are vulnerable to damage suits. Moreover, the number of damage suits arising each year is increasing at an almost alarming rate, and astronomical damages are being claimed. Studies reveal that about 95% of the court cases involving injury result in damages being awarded.

Comprehensive personal liability insurance protects a farm operator who is sued for alleged damages suffered from an accident involving his/her property or family. The kinds of situations from which a claim might arise are quite broad, including suits for personal injuries caused by animals, equipment, or personal acts.

Both workers' compensation insurance and employer's liability insurance protect farmers against claims or court awards resulting from injury to hired help. Workers' compensation usually costs slightly more than straight employer's liability insurance, but it carries more benefits to the worker. An injured employee must prove negligence by his/her employer before the company will pay a claim under employer's liability insurance, whereas workers' compensation benefits are established by state law, and settlements are made by the insurance company without regard to who was negligent in causing the injury. Conditions governing participation in workers' compensation insurance vary among the states.

WORKERS' COMPENSATION[1]

Workers' compensation laws, now in full force in every one of the 50 states, cover on-the-job injuries and protect disabled workers regardless of whether their disabilities are temporary or permanent. Although broad differences exist among the individual states in their workers' compensation laws, principally in their benefit provisions, all statutes follow a definite pattern as to employment covered, benefits, insurance and the like.

Workers' compensation is a program designed to provide employees with assured payment for medical expenses or lost income due to injury on the job. Whenever an employment-related injury results in death, compensation benefits are generally paid to the worker's suriviving dependents.

[1]This report relative to "Workers' Compensation" was authoritatively prepared by S. R. Sutter, Personnel Management Advisor, University of California Cooperative Extension, Fresno County, Fresno, CA.

Generally all employment is covered by workers' compensation, although a few states provide exemptions for farm labor, or exempt farm employers of fewer than 10 full-time employees, for example. Farm employers in these states, however, may elect workers' compensation protection. Livestock producers in these states may wish to consider coverage as a financial protection strategy because under workers' compensation, the upper limits for settlement of lawsuits are set by state law.

This government-required employee benefit is costly for livestock producers, however. The 1990 basic premium rate in California, where workers' compensation is mandatory for agricultural employers, the basic rate for stock farms and feed yards is $25.03 per $100 of payroll. Costs vary among insurance companies due to dividends paid, surcharges and minimum premiums. Some companies, as a matter of policy, will not write workers' compensation in agricultural industries. Some states have a quasi-government provider of workers' compensation to assure availability of coverage for small businesses and high-risk industries.

For information, contact your area extension farm management or personnel management advisor and an insurance agent experienced in marketing workers' compensation and liability insurance.

SOCIAL SECURITY LAW[2]

The pertinent provisions of the present Social Security Law as it pertains to farmers and ranchers are:

1. **Who is covered.** The law covers the following:

a. **Agricultural workers.** If you are an agricultural worker, your employer must report your cash wages by January 31 of each year if he or she—

(1) spends $2,500 or more on agricultural labor in a year; or

(2) spends less than $2,500, but you were paid $150 or more in a given year.

If you commute to work daily from your home for a season picking fruit or vegetables by hand, your employer needs to report your wages if he or she paid you at least $150 in cash.

b. **Household workers.** If your are a household worker on a farm operated for profit, your employer must report your wages if he or she—

(1) spends more than $2,500 during a year on agricultural labor; or

(2) spends less than $2,500 during a year on agricultural labor, but your annual wages amount to $150 or more.

Agricultural workers admitted to the United States on a temporary basis from any foreign country are not eligible.

[2]In the preparation of this section, the author had the benefit of the review and suggestions of T. Butler, Associate Commissioner of Public Affairs, Department of Health & Human Services, Social Security Administration, Baltimore, MD.

2. **Amount paid in—**

a. Self employed farmers report their earnings and pay the social security self-employment tax at the time they file their annual tax return with the Internal Revenue Service.

They may report their actual net earnings or an amount under an optional method. If their gross income from farming is $2,400 or less, they may report their actual net (if $400 or more) or two-thirds of their gross; if their gross income is more than $2,400 and their net farm income is less than $1,600, they may report for social security (but not for income tax purposes) either their actual net or $1,600. If gross income is over $2,400 and the actual net is over $1,600, actual net earnings must be reported.

In 1990, a self-employed person paid 15.3% on net earnings up to $51,300. The taxable base will continue to increase as earning levels rise.

b. The self-employed farmer is also responsible for reporting the wage of his/her farm laborers and any domestic help he/she may have. In 1990, he/she should have deducted 7.65% of each employee's pay up to $51,300. He/she adds 7.65% of his/her own to this and pays the total (15.3%) to the Internal Revenue Service.

3. **What are the benefits?** Depending on creditable earnings, the benefits are approximately as follows:

a. For a retired farmer (65 in 1990), up to $975 per month; or up to 20% less if he/she chooses to take benefits between 62 and 65.

b. For a retired farmer and spouse (both 65 in 1990), up to $1,463 per month.

c. For a widow or widower, up to $975 per month.

d. For a farmer under 65 who is suffering from a severe disability which has lasted, or is expected to last, 12 calendar months, or to result in death, up to $1,149 per month, provided he/she has had at least 5 years of work under social security in the 10-year period just before the disability began. (A worker who becomes disabled before 31 needs fewer work credits—in some cases as little as 1½ years; a worker who become disabled at 43 or older needs more than 5 years of credit.) Also, his/her eligible dependents can get the same benefits as the dependents of a farmer retired at 65.

e. Besides monthly payments, an eligible survivor can receive a lump-sum death payment of $255 on the record of the deceased worker.

f. Medicare offers both hospital insurance and medical insurance under social security for most people 65 and over and for some disabled people under 65. Older people eligible for monthly social security benefits have hospital insurance automatically. Or those who are not eligible for monthly benefits can get it by paying a monthly premium ($175 in 1990). People who have been entitled to disability checks for 24 or more months, and insured workers and their dependents who need dialysis treatment or a kidney transplant because of permanent kidney failure, also have this protection.

People who are covered under hospital insurance have medical insurance automatically unless they state they do not want it. The premium for this coverage was $28.60 a month in 1990.

The Social Security Administration administers another program called Supplemental Security Income. It is financed from general revenues rather than from social security taxes. It pays monthly checks to people in financial need who are 65 or older, blind, or disabled. More information about this program is available at any social security office.

The number on his/her social security card is very important to the farm operator as well as to the hired farm worker. It identifies the individual's social security record and is key to future benefit payments. It is important, therefore, that a person's social security number is on the social security reports for both the self-employed farmer and the agricultural worker.

Those who expect to draw social security payments later should check with the Social Security Administration every three years, especially if they change jobs frequently, to make sure that their records are in order and that their correct earnings are credited to their individual social security records.

For a social security card—either a new card or a duplicate of one that has been lost—or for more information about retirement, survivors, and disability insurance, Medicare health insurance, or Supplemental Security Income, get in touch with the nearest social security office or call Social Security's toll-free number: 1-800-2345-SSA (1-800-234-5772).

GUIDES TO KEEPING OUT OF LEGAL DIFFICULTIES

Herewith are some guides to keeping out of legal difficulties:

1. **Use written contracts.** Use written contracts instead of verbal contracts whenever possible, because there is less opportunity for dispute later.

2. **Pay for an option.** An option or promise to leave an offer open should always be secured by a small payment; otherwise, the agreement may be revoked at pleasure.

3. **Require surrender of a note.** Upon paying a note, requires its surrender. Otherwise, it may be sold and you may be required to pay it again.

4. **Give adequate warning when lending a treacherous animal.** If a treacherous animal is lent to a neighbor, he/she should be warned of these traits; otherwise, the owner may be held liable for any harm or damage that the animal may inflict.

5. **Consider trees on boundary lines as joint property.** Trees standing on boundary lines are the property of both owners, and their disposal must be by mutual agreement. Also, one cannot legally claim fruit from a tree standing upon another person's property even though the branches extend over the boundary. However, fruit or nuts that fall on adjacent property may be eaten but not sold.

6. **Be aware of auto passenger responsibility.** If the owner of an auto offers a pedestrian (or hitchhiker) a ride, he/she may be liable for any injury to the passenger because of careless driving, defective equipment, or any action whereby an accident results.

7. **Pay money only to an authorized agent.** Never pay money to an agent unless you know he/she is authorized to make collections. When payment is made, be certain to secure a signed receipt.

8. **Pay by check.** Pay debts and bills by check; then there is written proof of payment.

9. **Secure adequate protection through insurance.** In these times of high court judgments, it is imperative that the livestock producer have adequate insurance protection. Without such protection, or without substantial wealth, he/she is at the mercy of the claims-conscious public. A judgment could put the livestock producer out of business unless there is adequate insurance to cover the judgment.

Most livestock farmers and ranchers strive to keep their fences in proper repair, their equipment in satisfactory order, their employees properly educated about the hazards of the occupation in which they are engaged; and to handle their entire operation in a safe and sane manner. Yet, accidents do happen, and, when that time comes, an insurance policy is the answer to the financial part of the problem.

Recently, the insurance industry developed liability policies designed specifically for the livestock farmer and rancher. These policies are blanket-type liability policies designed to cover the farmer's legal liability arising out of the operation of the farm or ranch. Some of the general provisions covered by such policies are:

a. Liability for bodily injury or property damage to employees or guests.

b. Medical aid where the policyholder is liable.

c. Property damage as a result of breachy animals.

d. Liability for accidents on highways and public roads caused by animals.

e. Bodily injury and property damage liability for personal acts of the livestock producer and his/her family.

Such policies, being tailored to the actual needs of farmers and ranchers, are quite flexible and can be written to suit each individual's particular needs.

The important thing is that the livestock producer should take advantage of adequate liability coverage, which can be obtained at little added cost, for the time is past when one can have a secure feeling with a $10,000 policy. The high cost of claim settlement and the increased amount of jury verdicts make it desirable that limits of liability be increased. There should be adequate coverage to assure the livestock producer protection when and if needed and to keep the business going.

10. **Have a will made.** Most important of all, the farmer or rancher should have a will that covers his/her property and disposes of it in keeping with his/her wishes. (See Section 2, Business Aspects, "Wills.")

Choosing the class of animal and the selection of the breed are important!

SECTION 17

BREEDS AND BREED REGISTRY ASSOCIATIONS

Contents *Page*

This section is a one-source summary of the breeds and breed registries of U.S. livestock. Tables 17–1 to 17–8 present pertinent information relative to a total of 170 breeds.

BREEDS AND BREED REGISTRIES

Choosing the particular class (or classes) of animal comes first—ahead of the choice of a breed(s). Usually a combination of several factors suggests the livestock enterprise or enterprises best adapted to the particular farm or ranch; among them, available feeds, market outlets for animals and animal products, and the labor requirements. Table 2–1, in Section 2 of this book, presents the labor requirements for each class of animal.

- **Selection of the breed or cross**—No one breed of each animal species excels all others in all points of production for all conditions. For the purebred breeder, the selection of a particular breed is most often a matter of personal preference, and usually the preferred breed will make for the greatest success. Where no strong breed preference exists, however, it is usually best to choose the breed (or cross) that is most popular in the community—if one breed predominates. Usually, the longtime popularity of a breed in a certain area is indicative of breed adaptation; for example, the adaptation of Brahma cattle to hot, insect-plagued areas, and the adaptation of Rambouillet sheep to herding and migrating on the ranges of the West.

Germ plasm choice for the commercial livestock producer is becoming increasingly difficult because of the large number of breeds and breed cross combinations now available in all species. This section was prepared to give an assist in the choice of breeds, along with pertinent information relative to breed registries. First, the following definitions and preliminary information is important to an understanding of breeds and breed registries.

• **Breeds**—*A breed is a group of animals related by descent from common ancestors and visibly similar in most characters.* A breed may come about as a result of planned matings, or, as has been more frequently the case, it may be pure happenstance. Once a breed has evolved, a breed association is usually organized.

A purebred animal may be defined as a member of a breed, the animals of which possess a common ancestry and distinctive characteristics, which is either registered or eligible for registry in the herd book of that breed.

Sometimes people construe the write-up of a breed of livestock in a book or in a U.S. Department of Agriculture bulletin as an official recognition of the breed. Nothing could be further from the truth, for no person or office has authority to approve a breed. The only legal basis for recognizing a breed is contained in the Tariff Act of 1930, which provides for the duty-free admission of purebred breeding stock provided they are registered in the country of origin. But this stipulation applies to imported animals only.

In this book, no *official* recognition of any breed is intended or implied. Rather, the author has tried earnestly and without favoritism to present the factual story of the breeds. In particular, such information relative to the new and/or less widely distributed breeds is needed, and often difficult to come by.

• **Breed registry associations**—A breed registry association consists of a group of breeders banded together for the purposes of: (1) recording the lineage of their animals, (2) protecting the purity of the breed, (3) encouraging further improvement of the breed, and (4) promoting the interest of the breed. The breed registry association(s) for each breed is given in Tables 17-1 to 17-8.

Tables 17-1 to 17-8, which follow, present in summary form, the U.S. breeds of livestock and their characteristics. For each breed, the presentation includes the following:

1. A picture of a representative of the breed.
2. The registry association(s).
3. The origin, color, and characteristics.

TABLE 17-1
BREEDS AND BREED REGISTRIES OF BEEF AND DUAL-PURPOSE CATTLE

AMERICAN BREED

Registry Association:
American Breed Assn.
P.O. Box 10679
Midwest City, OK 73140

Fig. 17-1. American Breed heifer.

Origin: On Art Jones' Cactus Ranch, Portales, NM. Today, the breed carries the following mix (blood): $\frac{1}{2}$ Brahman, $\frac{1}{4}$ Charolais, $\frac{1}{8}$ buffalo, $\frac{1}{16}$ Hereford, and $\frac{1}{16}$ Shorthorn.

Color: Defies description.

Characteristics: It was selected for (1) doing well on alkaline range grass, (2) ability to travel long distances for water, (3) high fertility, (4) ease of calving, (5) high percentage calf crop of small calves with high weaning weights, (6) high natural immunity to most diseases and parasites, and (7) carcass cutability and quality.

AMERIFAX

Registry Association:
Amerifax Cattle Assn.
P.O. Box 149
Hastings, NE 68900

Fig. 17-2. Amerifax cow.

Origin: In the U.S., in Kansas, Nebraska, South Dakota, and Wyoming. Today, the breed carries $\frac{5}{8}$ Angus and $\frac{3}{8}$ Beef Friesian breeding.

Color: Either solid black or red in color.

Characteristics: They are polled and moderate in size.

ANGUS

Registry Association:
American Angus Assn.
3201 Frederick Blvd.
St. Joseph, MO 64501

Fig. 17-3. Angus cow.

Origin: In the northeast of Scotland, in the counties of Aberdeen, Angus, Kincardine, and Forfar.

Color: They are solid black or red.

Characteristics: They are polled and moderate in size.

ANKINA

Registry Association:
Ankina Breeders
5803 Oakes Road
Clayton, OH 45315

Fig. 17-4. Ankina heifer.

Origin: In Clayton, Ohio. The breed registry was founded in 1975. Today, the breed carries $\frac{5}{8}$ Angus and $\frac{3}{8}$ Chianina breeding.

Color: Black or dark brown.

Characteristics: Polled and medium to large size.

(Continued)

TABLE 17–1 *(Continued)*

ANKOLE-WATUSI

Registry Association:
Ankole-Watusi International Registry
Box 319
Phippsburg, CO 80469

Fig. 17–5. Ankole-Watusi bull.

Origin: In Africa, in the districts of Uganda, Rwanda, and Kenya.

Color: A great array of solid colors. Also, some animals are spotted.

Characteristics: Ankole-Watusi cattle have the largest horns of any cattle in the world; the horns are outswept, 5 ft long, and 16 in. in diameter. They have a slight hump, a moderate dewlap, and moderate to fine bone.

BARZONA

Registry Association:
Barzona Breeders Assn.
P.O. Box 631
Prescott, AZ 86302

Fig. 17–6. Barzona bull.

Origin: In Arizona, on the Bard Ranches. The foundation animals were Africander, Hereford, Santa Gertrudis, and Angus.

Color: Red.

Characteristics: The cattle are horned, and have a long head and a straight profile. They are well adapted to arid ranges.

BEEFALO

Registry Association:
American Beefalo World Registry.
116 Executive Park
Louisville, KY 40207

Fig. 17–7. Beefalo heifer.

Origin: By D. C. Basolo, Tracy, CA, beginning in 1966. They are ⅜ buffalo (bison) and ⅝ domestic cattle.

Color: Variable.

Characteristics: The breed is promoted for its foraging ability, adaptability, hardiness, calving ease, low maintenance cost, and long life.

BEEF FRIESIAN

Registry Association:
Beef Friesian Society
118 Livestock Exchange Building
4701 Marion Street
Denver, CO 80216

Fig. 17–8. Beef Friesian bull.

Origin: In the U.S., from three approaches: (1) from purebred Beef Friesians brought from Europe, (2) from crossing Beef Friesian bulls on Holstein females, and (3) from Beef Friesian and Angus crosses.

Color: Black and white; Beef Friesian X Angus are usually black.

Characteristics: They may be either horned or polled. They make excellent rate and efficiency of gains, have little calving difficulty, and have good milking ability.

BEEFMASTER

Registry Association:
Beefmaster Breeders Universal
6800 Park Ten Blvd.
Suite 290 West
San Antonio, TX 78213

Foundation Beefmaster Assn.
Livestock Exchange Bldg., Suite 200
4701 Marion Street
Denver, CO 80216

National Beefmaster Assn.
P.O. Box 368
Canton, TX 75103

Fig. 17–9. Beefmaster bull.

Origin: On the Lasater Ranch, Falfurias, TX, in 1908. It is approximately ½ Brahman, and ¼ each Shorthorn and Hereford.

Color: Red is the dominant color, but color is variable and disregarded in selection.

Characteristics: The majority are horned, although a few are naturally polled. They are good milk producers under range conditions, with heavy weaning weights.

BELGIAN BLUE

Registry Association:
Canadian Belgian Beef Cattle Assn.
R.R. #2
Orangeville, Ont. L9W 2Y9 Canada

Fig. 17–10. Belgian Blue bull.

Origin: In Belgium, starting in the early 1950s, from a cross between native Belgian cattle and Shorthorns.

Color: Variable, mostly roan and red-and-white spotted.

Characteristics: Belgian Blues are noted for their docile temperament, adaptation to various environments, and double-muscling, high dressing percentage, and lean carcasses. The double-muscling has created some calving difficulty in the breed.

BELTED GALLOWAY

Fig. 17–11. Belted Galloway bull.

Origin: In Scotland, in the southwestern district of Galloway.

Color: Black with a brownish tinge, or dun; with a white belt completely encircling the body between the shoulders and hooks.

Characteristics: Polled, striking white belt, and heavy coat of hair.

BLONDE D'AQUITAINE

Registry Association:
Canadian Blonde d'Aquitaine Assn., 207
1606 Centre Street North
Calgary, Alta. T2E 2R9 Canada

Fig. 17–12. Blonde d'Aquitaine bull.

Origin: In southwestern France, when 3 French strains of similar background—Garonne, Quercy, and Pyrenee—were combined. Other infusions of Shorthorn, Charolais, and Limousin followed.

Color: Yellow, brown, fawn, or wheat colored.

Characteristics: The cattle are horned, long bodied, long rumped, relatively fine boned, have little calving difficulty, and are considered large in size.

(Continued)

TABLE 17–1 *(Continued)*

BRAFORD

Registry Association:
American Braford Assn.
1501 Wyandotte
Kansas City, MO 64101

International Braford Assn., Inc.
P.O. Box 2727
Fort Pierce, FL 34954

Fig. 17–13. Braford bull.

Origin: On the Adams Ranches, Fort Pierce, FL. It is approximately 5/8 Hereford and 3/8 Brahman breeding.

Color: Red or brindle, with white markings on the head and pigmentation around the eyes.

Characteristics: There are both horned and polled strains. Brafords are short haired, heat tolerant, fertile, and good milk producers. They have only a slight hump.

BRAHMAN

Registry Association:
American Brahman Breeders Assn.
1313 La Concha Lane
Houston, TX 77054

Fig. 17–14. Brahman bull.

Origin: In India, where there are 30 or more breeds or strains of *Indicus* cattle. The U.S. Brahman is an amalgamation of several Indian types, probably with a small infusion of European breeding.

Color: Gray or red are preferred, but there are brown, black, white, and spotted Brahman.

Characteristics: They have drooping ears, a long face, a prominent hump over the shoulders, and an abundance of loose, pendulous skin under the throat and along the dewlap; and a voice that resembles a grunt rather than a low.

BRALER

Registry Association:
American Bralers Assn.
Star Route, Box 47
Ganado, TX 77962

Fig. 17–15. Braler heifer.

Origin: In Texas, on the R-HOL-D Farms, Brenham, TX. Bralers carry 5/8 Salers and 3/8 Brahman breeding.

Color: Dark mahogany.

Characteristics: They are horned, have a long head and a straight profile, and are highly fertile and easy calvers.

BRANGUS

Registry Association:
International Brangus Breeders Assn., Inc.
5750 Epsilon
San Antonio, TX 78230

Fig. 17–16. Brangus bull.

Origin: In Oklahoma, on Frank Buttram's Clear Creek Ranch, Welch, OK. They carry 3/8 Brahman and 5/8 Angus breeding.

Color: Black.

Characteristics: They are polled. They show evidence of the Brahman influence by the slight crest over the neck and the smooth, sleek coat.

CHARBRAY

Registry Association:
American-International Charolais Assn.
P.O. Box 20247
Kansas City, MO 64195

Fig. 17–17. Charbray bull.

Origin: In Texas, in the Rio Grande Valley. They carry from 3/4 Charolais X 1/4 Brahman to 7/8 Charolais X 1/8 Brahman breeding.

Color: Light tan at birth, but usually change to a creamy white in a few weeks.

Characteristics: Charbray cattle show a slight hint of the Brahman dewlap. The breed is reputed to have the growth thrust of the Charolais and the heat-insect tolerance of the Brahman.

CHAROLAIS

Registry Association:
American-International Charolais Assn.
P.O. Box 20247
Kansas City, MO 64195

Fig. 17–18. Charolais bull.

Origin: In France, in the province of Charolais.

Color: Charolais are light tan at birth, but change to a cream white in a few weeks.

Characteristics: They are horned; have a pink skin and mucous membranes; and are noted for large size, growth thrust, and bred-in red meat.

CHIANINA

Registry Association:
American Chianina Assn.
P.O. Box 890
Platte City, MO 64079

Fig. 17–19. Chianina bull.

Origin: In Italy, in the Chianina Valley, Tuscany Province, central Italy.

Color: Chianina have porcelain white hair, black switch, and dark skin; calves are born tan colored which gradually turns to white at about 60 days of age.

Characteristics: They have horns, a narrow head with black pigmentation around the eyes, and a black tongue, nose, and palate. They are the largest breed of cattle in the world; mature bulls stand about 6 ft at the withers. The breed is also known for trimness of middle; fineness of head, horn, and bone; absence of excessive dewlap and brisket; and poor milking ability.

DEVON

Registry Association:
Devon Cattle Assn., Inc.
P.O. Box 61
The Plains, VA 22171

Fig. 17–20. Devon cow.

Origin: In England, in the countries of Devon and Somerset.

Color: Devons are a rich red, dark red is preferred; hence, the name *Ruby Reds*.

Characteristics: They have creamy white horns with black tips, but some are polled. The switch varies from whitish red to nearly white at tip; and the skin is orange-yellow with pigment especially noticeable around the eyes and muzzle.

(Continued)

TABLE 17–1 *(Continued)*

DEXTER

Registry Association:
American Dexter Cattle Assn.
Route 1, Box 378
Concordia, MT 64020

Fig. 17–21. Dexter cow.

Origin: In Ireland, in the southern and southwestern parts. The breed was named after their founder, a Mr. Dexter.

Color: Dexters are black or red.

Characteristics: They are horned, have a long head, are small and have short legs; some mature animals are under 40 in. high.

GALLOWAY

Registry Association:
American Galloway Breeders Assn.
Route 1, Box 106A
Athol, ID 83801

Galloway Cattle Society of America
Hennepin, IL 61327

Fig. 17–22. Galloway bull.

Origin: In Scotland, in the southwestern province of Galloway.

Color: They are black, dun, belted, white or red.

Characteristics: Galloways are polled, have thick wavy hair, calve easily, are hardy, and have the ability to rustle in cold weather.

GELBVIEH

Registry Association:
American Gelbvieh Assn.
5001 National Western Drive
Denver, CO 80216

Canadian Gelbvieh Assn.
Box 536, Marlborough P.O. Centre
Calgary, Alta. T2A 7L4 Canada

Fig. 17–23. Gelbvieh bull.

Origin: In Germany, in Bavaria. The Gelbvieh evolved as a result of amalgamating 4 breeds—Franconian, Glan-Donnersberg, Lahn, and Limpurg, around 1920.

Color: They are golden red to rust; and solid colored.

Characteristics: Gelbviehs are horned; large, long-bodied, well-muscled, fast-gaining, and have high-quality carcasses.

HAYS CONVERTER

Origin: In Canada, by the former Minister of Agriculture, Harry Hays, beginning in 1957. The foundation breeds were: Hereford, Brown Swiss, and Holstein.

Fig. 17–24. Hays Converter bull.

Color: The predominant color is black with a white face, white feet, and a white tail. About 30% are red with white faces.

Characteristics: The traits upon which Senator Hays built the breed include: (1) growth, (2) fertility, (3) minimum calving problems, (4) well-attached udders, (5) abundant milk, (6) sound feet and legs, and (7) pigmentation.

HEREFORD

Registry Association:
American Hereford Assn., The
P.O. Box 014059
Kansas City, MO 64101

International Hereford Organization
Box 1HO
Dayton, PA 16222

Fig. 17–25. Hereford bull.

Origin: In England, in the county of Hereford.

Color: Herefords are red with white markings; white face and white on the underline, flank, crest, switch, breast and below the knees and hocks.

Characteristics: They are horned; and they have a thick coat of hair.

INDU-BRAZIL

Registry Association:
International Zebu Breeders Assn.
783 N. Loop 337
New Braunfels, TX 78130

Fig. 17–26. Indu-Brazil bull.

Origin: In Brazil.

Color: They are light gray to silver gray; dun to red.

Characteristics: Indu-Brazil animals have metrical horns drawing upward and to the rear; a prominent forehead and long, drooping ears; a prominent hump over the shoulders; an abundance of loose, pendulous skin under the throat and along the dewlap; and a voice that resembles a grunt rather than a low.

LIMOUSIN

Registry Association:
North American Limousin Foundation
100 Livestock Exchange Building
P.O. Box 16767
Denver, CO 80216-0767

Fig. 17–27. Limousin cow and calf.

Origin: In southwestern France. The breed is named after the Limousin Mountains.

Color: Limousins are wheat to rust red.

Characteristics: Limousins are horned; long and relatively shallow bodied, with moderate to heavy muscling; and are noted for ease of calving with high carcass quality.

LINCOLN RED

Registry Association:
Canadian Lincoln Red Promotion Assn.
Box 447
Richmond Hill, Ont. L4C 4Y8 Canada

Fig. 17–28. Lincoln Red bull.

Origin: In England, in Lincolnshire—the rugged coast.

Color: Lincoln Reds are deep cherry red, with occasional white markings.

Characteristics: There are both horned and polled strains. The breed is characterized by a long body, light birth weights and ease of calving, excellent milk production, fast growth rate, and good fertility. In England, they are a dual-purpose breed, with separate strains for beef and dairy.

(Continued)

TABLE 17-1 *(Continued)*

MAINE-ANJOU

Registry Association:
American Maine-Anjou Assn.
528 Livestock Exchange Building
1600 Genesee Street
Kansas City, MO 64102

Maine-Anjou International
334 9th Avenue N.E.
Calgary, Alta. T2E 7A6 Canada

Fig. 17–29. Maine-Anjou bull.

Origin: In France, in the provinces of Maine and Anjou.

Color: Maine-Anjou are dark red with white underline, often with small white patches on the body. Also, dark roans are found. Most heads are either red, or the eyes are surrounded by red.

Characteristics: Maine-Anjou are the largest of the French breeds; they are considered a dual-purpose breed, with emphasis on beef. They are long, rather up-standing, have a long rump; and are noted for rapid growth.

MARCHIGIANA

Registry Association:
American Int'l Marchigiana Society
P.O. Box 198
Walton, KS 67151-0198

Fig. 17–30. Marchigiana bull.

Origin: In Italy, in the Marche region, near Rome.

Color: They are grayish white, although bulls may be darker; with dark skin pigmentation, a dark muzzle and switch, and dark below or around the eyes. Calves are born tan, but turn white at two months of age.

Characteristics: Marchigianas have horns that appear small in proportion to the size of the cattle; and they are noted for ability to do well under adverse conditions.

MURRAY GREY

Registry Association:
American Murray Grey Assn.
1222 N. 27th, Suite 208
Billings, MT 59101

Fig. 17–31. Murray Grey bull.

Origin: In Australia, by the Sutherlands of Murray Valley, Victoria; using a very light roan (almost white) Shorthorn cow and an Angus bull. Because of the further use of Angus bulls, the breed is predominantly Angus.

Color: It is silver gray in color, with dark skin pigmentation which lessens cancer eye.

Characteristics: The breed is polled, adapts well to sunny and colder areas, and is noted for small calves and ease of calving, good dispositions, and superior carcasses.

NORMANDE

Registry Association:
American Normande Assn.
P.O. Box 350
Kearney, MO 64060

Fig. 17–32. Normande bull.

Origin: In France, in the areas of Normandy, Brittany, and Maine.

Color: They are primarily dark red and white, with colored patches around the eyes, giving them a "bespectacled" appearance and resistance to cancer eye and pinkeye. Also, dark pigmentation on the udder prevents sunburn.

Characteristics: The Normande is known as a dual-purpose breed in France.

NORWEGIAN RED

Fig. 17–33. Norwegian red cow.

Origin: In Norway, where they are known as Norwegian Red-and-Whites.

Color: Rust red and white.

Characteristics: They are horned; and they are noted for abundant milk production, excellent feed conversion, and good carcasses.

PIEDMONTESE

Registry Association:
Piedmontese Assn. of the U.S.
Route 1
Cost, TX 78614

Canadian Piedmontese Assn.
Box 11
Admiral, Sask. S0N 0B0 Canada.

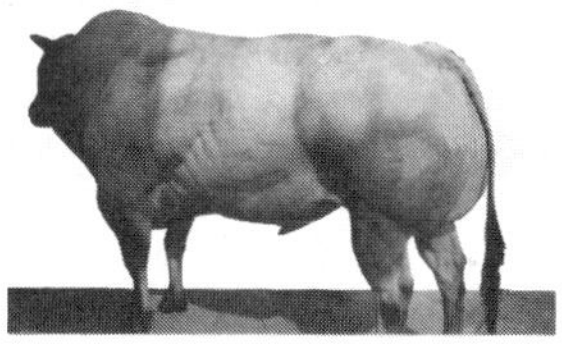

Fig. 17–34. Piedmontese bull.

Origin: In Italy, where they are the most popular breed.

Color: White or pale gray.

Characteristics: About 80% of Piedmontese are double muscled. Breeders report 9% higher dressing percentage and twice the steaks from double muscled cattle; thus, they command a premium price on the Italian market.

PINZGAUER

Registry Association:
American Pinzgauer Assn.
Route 1, Box 104E
Kelly, IA 50134-9801

Canadian Pinzgauer Assn.
233 Stockman's Centre
2116 – 27th Avenue, N.E.
Calgary, Alta. T2E 7A6 Canada

Fig. 17–35. Pinzgauer heifer.

Origin: In Austria, in the Pinz Valley.

Color: They have chestnut brown sides, with a white top-line and underline, and usually white feet; and with deep orange pigment around the eyes and on the udder.

Characteristics: They are horned, hardy, highly fertile, good foragers, and long lived.

POLLED HEREFORD

Registry Association:
American Polled Hereford Assn.
4700 E. 63rd Street
Kansas City, MO 64130

Fig. 17–36. Polled Hereford bull.

Origin: In Iowa, by Warren Gammon, from a mutation that appeared in the horned Hereford breed.

Color: They have identical color markings to the Hereford breed; they are red with white markings, white face and white on the underline, flanks, crest, switch, breast, and below the knees and hocks.

Characteristics: They are polled; and the white face is the distinctive trademark of the breed.

(Continued)

TABLE 17–1 *(Continued)*

RED ANGUS

Registry Association:
Red Angus Assn. of America
4201 I-35 North
Denton, TX 76201

Origin: In Scotland; evolved from black Angus as a result of a recessive red.

Color: Red.

Characteristics: Polled. Similar to black Angus except for the red color.

Fig. 17–37. Red Angus bull.

RED BRANGUS

Registry Association:
American Red Brangus Assn.
P.O. Box 1326
Austin, TX 78767

Origin: In Texas, on Paleface Ranch, Spicewood, TX, from Brahman X Angus cross.

Color: Red, with red pigmentation around the eyes.

Fig. 17–38. Red Brangus bull.

Characteristics: Smooth sleek coat, drooping ears; males have crest.

RED POLL

Registry Association:
American Red Poll Assn.
Box 35519
Louisville, KY 40232

Origin: In England, in the eastern middle coastal counties of Norfolk and Suffolk.

Color: Red, varying from light to dark red.

Characteristics: Polled. The Red Poll is a dual-purpose breed, adapted to both meat and milk production.

Fig. 17–39. Red Poll bull.

ROMAGNOLA

Origin: In Italy, in the lower Po Valley, by crossing the Podolic and native cattle.

Color: They are solid off-white to light gray; bulls have a characteristically darker color of hair about the shoulders, black color around the eyes, and a black switch.

Characteristics: The head appears to be short; horns are longer and sharper than in the other white breeds; tongue, muzzle, body orifices, tail, and hooves are black; in Italy, it is used for both milk and meat.

Fig. 17–40. Romagnola bull.

SALERS

Registry Association:
American Salers Assn.
5600 S. Quebec, Suite 220A
Englewood, CO 80111

Origin: In France, in the south-central mountainous area.

Color: Solid, deep cherry red, with a white switch and sometimes white spots under the belly.

Fig. 17–41. Salers cow.

Characteristics: Most Salers are horned, although there are polled strains. The breed is noted for rapid gains, hardiness, and adaptability.

SANTA GERTRUDIS

Registry Association:
Santa Gertrudis Breeders International
P.O. Box 1257
Kingsville, TX 78363

Origin: In Texas, on King Ranch; based on a Shorthorn X Brahman cross; named after the Spanish Land Grant, Santa Gertrudis, where it originated, now the headquarters for King Ranch.

Fig. 17–42. Santa Gertrudis cow.

Color: Red or cherry red.

Characteristics: Santa Gertrudis are generally horned, but there are polled strains; the hair is short, straight, and slick; the hide is loose, with the surface area increased by neck folds and sheath or navel flap, but neither should be excessive.

SCOTCH HIGHLAND

Registry Association:
American Scotch Highland Breeders' Assn.
P.O. Box 81
Remer, MN 56672

Origin: In Scotland.

Color: Variable, including red, yellow, silver, white, dun, black, or brindle.

Fig. 17–43. Scotch Highland cow.

Characteristics: Scotch Highland cattle have a short head; long, widespread horns; heavy foretop; long, shaggy hair; short legs; and they are hardy and can rustle in cold weather.

SENEPOL

Registry Association:
North American Senepol Assn.
P.O. Box 300168
Kansas City, MO 64130

Origin: On St. Croix, Virgin Islands, by combining Red Poll cattle from England and the N'Dama cattle of West Africa, a long-horned breed that is heat tolerant and disease resistant.

Fig. 17–44. Senepol bull.

Color: Cherry red.

Characteristics: Predominantly polled, highly heat and insect resistant, docile, fertile, very maternal, and good rustlers.

SHORTHORN/POLLED SHORTHORN

Registry Association:
American Shorthorn Assn.
8288 Hascall Street
Omaha, NE 68124

Canadian Shorthorn Assn.
Gummer Bldg., 5 Douglas Street
Guelph, Ont. H1H 2S8 Canada

Origin: In England, in the northeastern counties of Durham, Northumberland, York, and Lincoln. The name *Shorthorn* stems from the fact that, through breeding and selection, the early improvers of the breed shortened the horns of the native cattle.

Fig. 17–45. Shorthorn bull.

Color: Shorthorns may be red, white, or any combination of red and white.

Characteristics: They are rather short and refined, with in-curving horns.

(Continued)

TABLE 17-1 *(Continued)*

SIMBRAH

Registry Association:
American Simmental Assn.
1 Simmental Way
Bozeman, MT 59715

Fig. 17-46. Simbrah bull.

Origin: In the U.S., mostly in Texas where a few producers started crossing Simmental and Brahman in the late 1960s, but the first registration of Simbrah occurred in 1977.

Color: Red, straw, or gray with white intermixed.

Characteristics: Simbrahs may be polled, scurred, or horned. They are fertile, hardy, adaptable, and fast-gaining.

SIMMENTAL[1]

Registry Association:
American Simmental Assn.
1 Simmental Way
Bozeman, MT 59715

Fig. 17-47. Simmental bull.

Origin: In western Switzerland, in the Simme Valley, from which it derives its name.

Color: Generally red and white spotted, although some are nearly solid in color. The red varies from dark to a more common diluted, almost yellow, shade. The white face is dominant.

Characteristics: Most Simmentals are horned, although polled strains exist. It was developed as a dual-purpose breed; hence, it combines meat and milk to an unusually high degree, along with rapid growth.

SOUTH DEVON

Registry Association:
North American and International South Devon Assn.
P.O. Box 68
Lynville, IA 50153

Fig. 17-48. South Devon cow.

Origin: In England, in southern Devonshire, through infusion of Guernsey blood into the Devon breed.

Color: They are medium light red in color.

Characteristics: They are a horned, dual-purpose breed; the cows are heavy milkers.

SUSSEX

Registry Association:
In the U.S., the Sussex Cattle Assn. of America registers in the *English Herd Book.*

Fig. 17-49. Sussex cow.

Origin: In England.

Color: Deep mahogany red, with white switch, and ivory horns with dark tips.

Characteristics: Most Sussex cattle are polled; have flesh colored noses; and produce excellent meat carcasses with a high dressing percentage.

TARENTAISE

Registry Association:
American Tarentaise Assn.
P.O. Box 446
Reedpoint, MT 59069

Fig. 17-50. Tarentaise bull.

Origin: In the French Alps.

Color: Solid wheat-colored, ranging from a cherry to dark blond. Bulls tend to darken around the neck and shoulders with maturity, and often have a darker dorsal stripe.

Characteristics: The breed is characterized by black pigmentation of the muzzle and around the eyes. It is noted for ease of calving, hardiness, good fertility, good milking ability, and freedom from cancer eye and sunburned teats as a result of the pigmentation around the eyes and on the udder and teats. Tarentaise cattle are smaller than most of the exotics.

TEXAS LONGHORN

Registry Association:
Texas Longhorn Breeders Assn. of America
2315 N. Main Street, Suite 402
Fort Worth, TX 76106

Fig. 17-51. Texas Longhorn cow.

Origin: In the U.S., from cattle of Spanish extraction, beginning with cattle first brought to America by Columbus. By 1900, the Texas Longhorn was driven to near extinction by the European breeds—the Shorthorn, Hereford, and Angus. The Texas Longhorn Breeders Assn. was organized in 1964, at which time there were only about 1,500 head of genuine Texas Longhorn Cattle in existence. Since that time, the breed has made a strong comeback.

Color: The breed is noted for a great array of colors, in all degrees of richness, and in all possible combinations and patterns.

Characteristics: Texas Longhorns have large, spreading horns that curve upward; a long head, with small ears; long legs; fertility; calving ease; hardiness; good rustling ability; longevity; and adaptability.

WELSH BLACK

Registry Association:
United States Welsh Black Cattle Assn.
Route 1, Box 768
Shelburn, IN 47879

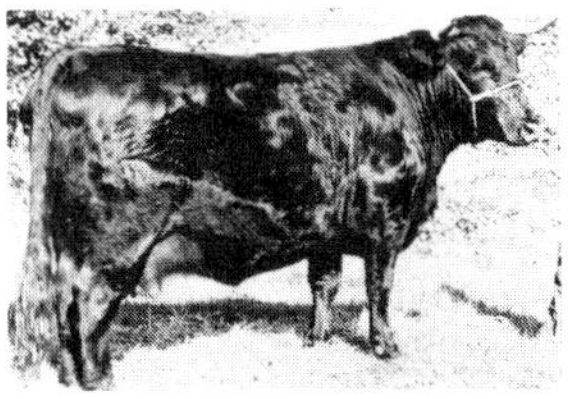

Fig. 17-52. Welsh Black cow.

Origin: In Wales as dual-purpose cattle.

Color: Black.

Characteristics: The breed is horned, although there is a polled strain. It is noted for high fertility, little calving difficulty, good milk production, adaptation to harsh conditions, longevity, and relative freedom from sunburned udders and cancer eye.

WHITE PARK

Registry Association:
White Park Cattle Assn.
419 N. Water Street
Madrid, IA 50156

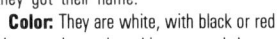

Fig. 17-53. White Park bull.

Origin: In England, where their ancestors were brought by the Romans, in 55 B.C. But they roamed free until domestication in the 18th century, at which time they were put in parklike enclosures from whence they got their name.

Color: They are white, with black or red pigmentation and markings around the eyes, ears, nose, feet, legs, teats, and anal area; they have a white head with black ears, eyes, and nose.

Characteristics: The breed is predominantly polled; and comparable in size to Angus, Hereford, and Shorthorn. It is noted for strong maternal instinct, ease of calving, excellent milking ability, and a docile disposition and being easy to handle.

[1]In Switzerland, the breed retained the name Simmental; in France it is known as Pie Rouge; in Germany, it is called Fleckvich; and in Austria, the name Austrovich is most common.

TABLE 17-2
BREEDS AND BREED REGISTRIES OF DAIRY CATTLE

AYRSHIRE

Registry Association:
Ayrshire Breeders' Assn.
2 Union Street
Brandon, VT 05733

Fig. 17-54. Ayrshire cow.

Origin: In Scotland, in the county of Ayrshire.

Color: Light to deep cherry red; mahogany; brown; or a combination of these colors with white; or white alone.

Characteristics: The horns are widespread and tend to curve upward and outward. Also, there is a polled strain. The udders are especially symmetrical and well attached to the body. The breed is noted for its style and animation, its good feet and legs, and grazing ability.

BROWN SWISS

Registry Association:
Brown Swiss Cattle Breeders' Assn., The
P.O. Box 1038
Beloit, WI 53511-1038

Fig. 17-55. Brown Swiss cow.

Origin: In the Alps of Switzerland.

Color: Solid brown, varying from very light to dark; white markings are objectionable. The nose and tongue are black, and there is a light colored band around the muzzle.

Characteristics: Brown Swiss have medium-length horns. They are strong and rugged, with some tendency toward the heavy muscling of the beef breeds; and they are calm and unexcitable.

DUTCH BELTED

Registry Association:
Dutch Belted Cattle Assn. of American, Inc.
P.O. Box 358
Venus, FL 33960

Fig. 17-56. Dutch Belted cow.

Origin: In the Netherlands.

Color: Black and white, with white belt extending entirely around the body from a little back of the shoulder to just in front of the hips.

Characteristics: Horned, with a long and somewhat dished head.

GUERNSEY

Registry Association:
American Guernsey Assn.
P.O. Box 666
Reynoldburg, OH 43068-0666

Fig. 17-57. Guernsey cow.

Origin: On the Island of Guernsey.

Color: Fawn with white markings clearly defined; preferably a clear muzzle.

GUERNSEY (Continued)

Characteristics: The horns incline forward, are refined and medium in length, and taper toward the tips. The head is of good length. Guernseys are noted for milk that is especially yellow in color, and for golden yellow skin pigmentation. The unhaired portions are light or pinkish in color. Birth weights are relatively light.

HOLSTEIN

Registry Association:
Holstein-Friesian Assn. of America
P.O. Box 808
Brattleboro, VT 05301

Fig. 17-58. Holstein cow.

Origin: In the Netherlands and northern Germany.

Color: Black and white, or red and white.

Characteristics: Clean-cut; broad muzzle; open nostrils; strong jaw, broad and slightly dished forehead; straight, bridged nose; and large and angular.

JERSEY

Registry Association:
American Jersey Cattle Club, The
6486 E. Main St.
Reynoldsburg, OH 43068-2362

Fig. 17-59. Jersey cow.

Origin: On the island of Jersey.

Color: Jerseys vary greatly in color, but usually are some shade of fawn, with or without white markings.

Characteristics: The forehead is broad and moderately dished, and the eyes are large and bright. They are especially known for their well-shaped udders, strong udder attachments, ease of calving, and for their very angular, but refined, appearance.

MILKING SHORTHORN (ILLAWARA)[1]

Registry Association:
American Milking Shorthorn Society
P.O. Box 449
Beloit, WI 53511

Fig. 17-60. Milking Shorthorn cow.

Origin: In England; the breed traces to a milking strain of Shorthorns developed by Thomas Bates.

Color: Red, white, or any combination of red and white.

Characteristics: Milking Shorthorns have fine horns that are rather short; the breed is very adaptable; and compete favorably in either milk or beef production.

[1]Beginning in 1969, semen of the Illawara, a Shorthorn developed in Australia especially for milk production, was imported. The American Milking Shorthorn Society must approve each importation of semen before the offspring can be registered in *The Herd Book*.

TABLE 17-3
BREEDS AND BREED REGISTRIES OF SHEEP

AMERICAN MERINO

Registry Association:
American & Delaine Merino Record Assn.
1193 Township Road 346
Nova, OH 44859

Origin: In Spain.

Color of face, ears, and legs: White, but reddish-brown spots may occasionally appear on the lips, ears, and pasterns.

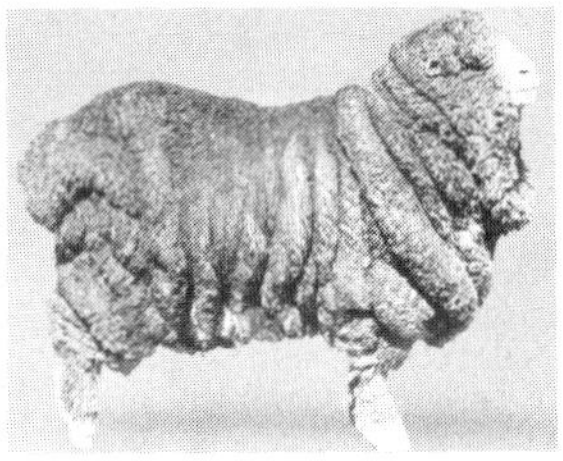
Fig. 17-61. American Merino ram (B-type).

Characteristics: Most rams have horns, but there are some polled strains. The more wrinkled American Merinos are known as the "A" and "B" types. The breed is noted for its strong flocking instinct. The ewes will breed out of season.

BOOROOLA MERINO

Origin: In Australia, at a government research station.

Color of face, ears, and legs: White, similar to Merino from which it is descended.

Fig. 17-62. Booroola Merino ewe.

Characteristics: Very prolific; ewes wean an average of 2.9 lambs. A single gene, known as the F gene, controls the Booroola's prolificacy; and the gene is transferable. In addition to prolificacy, Booroola Merino sheep produce high-quality wool.

CHEVIOT

Registry Association:
American Cheviot Sheep Society
R.R. 1, Box 100
Clarks Hill, IN 47930

American North Country Cheviot Sheep Assn.
833 Fall Creek Road
Longview, WA 98632

Fig. 17-63. Cheviot ram.

Origin: In Scotland, in the Cheviot Hills between Scotland and England.

Color of face, ears, and legs: They have a white face with a black nose; often black spots are found on the ears.

Characteristics: Both sexes are polled. Cheviots are stylish, alert, and active. The head and legs are free from wool.

CLUN FOREST

Origin: In the mountainous region of Clun Forest, England.

Color of face, ears, and legs: Dark brown.

Characteristics: The face is bare below the eyes. The ewes are prolific; easy lambers; and good milkers. Clun Forest are medium in size.

Fig. 17-64. Clun Forest ram.

COLUMBIA

Registry Association:
Columbia Sheep Breeders Assn. of America
P.O. Box 272
Upper Sandusky, OH 43351

Origin: In the U.S., in Wyoming and Idaho, by crossing Lincoln rams on Rambouillet ewes.

Color of face, ears, and legs: White.

Fig. 17-65. Columbia ram.

Characteristics: Columbias are polled, and open faced, with no tendency to wool blindness.

CORMO

Registry Association:
American Cormo Sheep Assn.
576 W. Funk Road
Lake Arthur, NM 88253

Origin: In Australia, on the island of Tasmania.

Color of face, ears, and legs: White.

Fig. 17-66. Cormo ram.

Characteristics: Cormo sheep are polled; open faced, with silky, translucent hair on their faces; and are noted for fertility (twinning); good mothering, and strong herding instincts.

CORRIEDALE

Registry Association:
American Corriedale Assn., Inc.
Box 29C
Seneca, IL 61360

Origin: In New Zealand.

Color of face, ears, and legs: White, with dark points. Black spots are sometimes present.

Fig. 17-67. Corridale ewe.

Characteristics: Both sexes are polled.

COTSWOLD

Registry Association:
American Cotswold Record Assn.
282 Meaderboro Road
Rochester, NH 03867

Origin: In the Cotswold hills of Gloucestershire, England.

Color of face, ears, and legs: White, although graying specks and bluish tinge are common.

Fig. 17-68. Cotswold ewe.

Characteristics: Both sexes are polled, although rams frequently have scurs. The fleece hangs in natural wavy ringlets or curls all over the body, with a tuft of wool on forehead. The Cotswold is second only to the Lincoln in size.

(Continued)

TABLE 17–3 *(Continued)*

DEBOUILLET

Registry Association:
Debouillet Sheep Breeders Assn.
300 S. Kentucky Avenue
Roswell, NM 88201

Origin: In the U.S., on the Amos Dee Jones Ranches of Roswell, NM.

Color of face, ears, and legs: White.

Characteristics: Rams may have horns, but there are polled strains; open face; comparatively smooth body; long staple

Fig. 17–69. Debouillet ram.

DELAINE MERINO

Registry Association:
American & Delaine Merino Record Assn.
1193 Township Road 346
Nova, OH 44859

Black Top & National Delaine Merino Sheep Assn.
290 Beach Street
Muse, PA 15350

Texas Delaine Sheep Assn.
Route 1
Burnet, TX 78611

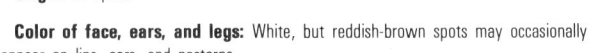

Fig. 17–70. Delaine Merino—"C" type Merino.

Origin: In Spain.

Color of face, ears, and legs: White, but reddish-brown spots may occasionally appear on lips, ears, and pasterns.

Characteristics: Most rams have horns, but there are some polled strains. Females should be free from horns or scurs. Delaine Merinos have a comparatively smooth body (of "C" type); possess strong flocking instinct; and the ewes will breed out of season.

DORSET

Registry Association:
Continental Dorset Club, Inc.
P.O. Box 506
Hudson, IA 50643

Origin: In England, in the southern counties of Dorset and Somerset. Polled Dorsets were developed at North Carolina State University.

Color of face, ears, and legs: White.

Fig. 17–71. Dorset ewe and twin lambs.

Characteristics: The face, ears, and legs are practically free from wool. There are both polled and horned strains. The ewes will breed out of season and are good milkers.

FINNSHEEP

Registry Association:
Finnsheep Breeders Assn., Inc.
P.O. Box 512
Zionsville, IN 46077-0512

Origin: In Finland.

Color of face, ears, and legs: White.

Characteristics: The head is free of wool; both sexes are usually hornless, but a few rams have light horns. Finnsheep are very prolific, averaging 2.41 lambs/birth.

Fig. 17–72. Finnsheep ewe.

HAMPSHIRE

Registry Association:
American Hampshire Sheep Assn.
Box 345
Ashland, MO 65010

Origin: In England, in the south-central county of Hampshire.

Color of face, ears, and legs: Rich, deep brown approaching black in color.

Characteristics: Both sexes are hornless, although rams sometimes have scurs. Hampshires are noted for large size and early maturity.

Fig. 17–73. Hampshire ram.

KARAKUL

Origin: In Asia, in the region of Bokhara, U.S.S.R.

Color of face, ears, and legs: They are black or brown.

Characteristics: The rams have horns, but the ewes are hornless. Karakuls have drooping ears and a fat tail. The lamb pelts are suitable for fur production.

Fig. 17–74. Karakul ram.

KATAHDIN

Registry Association:
Katahdin Registry
Piel Farm
P.O. Box 89
Abbot, ME 04406

Origin: On the Piel Farm, Abbot, ME; from Virgin Island sheep, with infusion of some British breeds.

Color of face, ears, and legs: Most Katahdins are white faced, but they can be any color or pattern.

Fig. 17–75. Katahdin ram.

Characteristics: Polled animals are preferred. This breed is a woolless meat-type sheep; they do not require shearing.

LEICESTER (BORDER/ENGLISH LEICESTER)

Registry Association:
American Border Leicester Assn.
7594 S.R. 534
West Farmington, OH 44491

Origin: In England, in the central county of Leicester.

Color of face, ears, and legs: White, but some animals may have a bluish tinge or black spots.

Fig. 17–76. Border Leicester ewe.

Characteristics: Both sexes are polled. There are two strains of Leicester sheep—the Border Leicesters, and the English Leicester. The Border Leicester is distinguished from the English Leicester by being more open faced and bare legged, and by having a shorter fleece.

(Continued)

TABLE 17–3 *(Continued)*

LINCOLN

Fig. 17–77. Lincoln ram.

Registry Association:
National Lincoln Sheep Breeders' Assn.
R.R. 6, Box 24
Decatur, IL 62521

Origin: In England, along the eastern coast and bordering the North Sea, in Lincolnshire.

Color of face, ears, and legs: White.

Characteristics: Both sexes are polled. The Lincoln is the largest of all breeds of sheep; and produces the heaviest fleece of any mutton breed.

MONTADALE

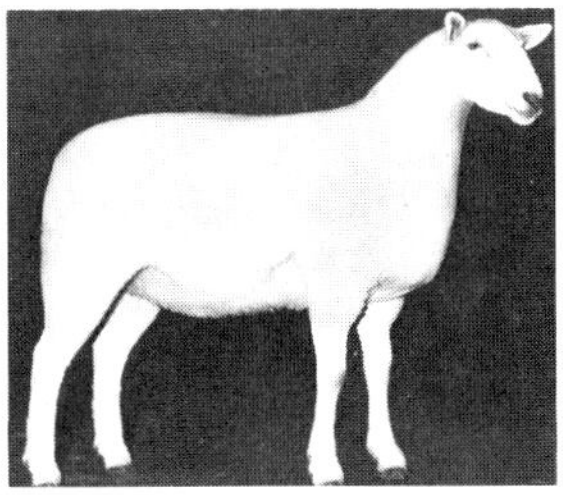

Fig. 17–78. Montadale ewe.

Registry Association:
Montadale Sheep Breeders' Assn., Inc.
P.O. Box 44300
Indianapolis, IN 46244

Origin: In the U.S. by E. H. Mattingly, St. Louis, MO, from a Columbia X Cheviot cross.

Color of face, ears, and legs: White.

Characteristics: Both sexes are polled; and the head is free from wool. Montadales have black hoofs and nose, and black spots in the ears.

NAVAJO

Fig. 17–79. Navajo ewe.

Origin: In the U.S., developed by the Navajo Indians from the Churro—a Spanish breed.

Color of face, ears, and legs: White, colored, or spotted, but generally a single color.

Characteristics: Rams are horned (some are multihorned), but ewes are usually polled. Ewes frequently breed out of season. Navajo sheep produce a very coarse wool which is easy to spin into blankets and rugs.

NORTH COUNTRY CHEVIOT

Fig. 17–80. North Country Cheviot ram.

Registry Association:
American North Country Cheviot Sheep Assn.
833 Fall Creek Road
Longview, WA 98632

Origin: In Scotland; from the old Long Hill sheep, but with infusion of Merino, Ryeland, and Southdown blood in formative period.

Color of face, ears, and legs: White.

Characteristics: The rams are sometimes horned. North Country Cheviots have a straight, slightly Roman nose, and produce wool that grades 50s and 56s.

OXFORD

Fig. 17–81. Oxford ewe lamb.

Registry Association:
American Oxford Down Record Assn.
Route 4
Ottawa, IL 61350

Origin: In England, in the south-central county of Oxford. The Oxford stems from a Hampshire X Cotswold cross made in the 1830s.

Color of face, ears, and legs: Variable, from gray to brown.

Characteristics: Both sexes are polled. Oxfords have a topknot of wool. They are the largest of the down breeds.

PANAMA

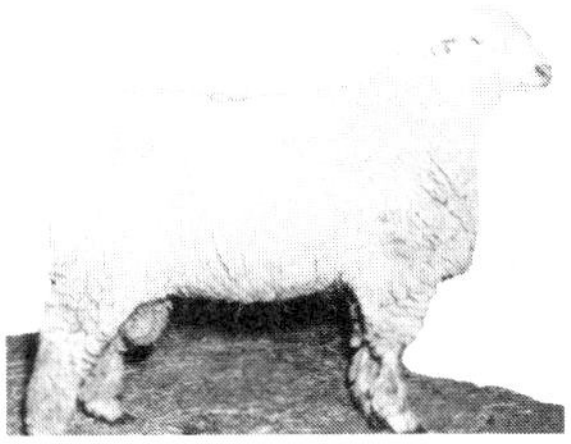

Fig. 17–82. Panama ram lamb.

Registry Association:
American Panama Registry Assn.
HC 85, Box 297
Grandview, ID 83624

Origin: The breed was developed by Laidlaw and Brockie of Muldoon, ID, from Rambouillet rams and Lincoln ewes.

Color of face, ears, and legs: White.

Characteristics: Both sexes are polled.

POLYPAY

Fig. 17–83. Polypay ewe and twin lambs.

Registry Association:
American Polypay Sheep Assn.
1934 E. Rua Bronco
Sandy, UT 84092

Origin: At the U.S. Sheep Experiment Station, Dubois, ID, using (Dorset X Targhee) and (Finnsheep X Rambouillet), then crossing the offspring.

Color of face, ears, and legs: White.

Characteristics: Open faced; ewes mature early, lamb with ease, and are prolific.

RAMBOUILLET

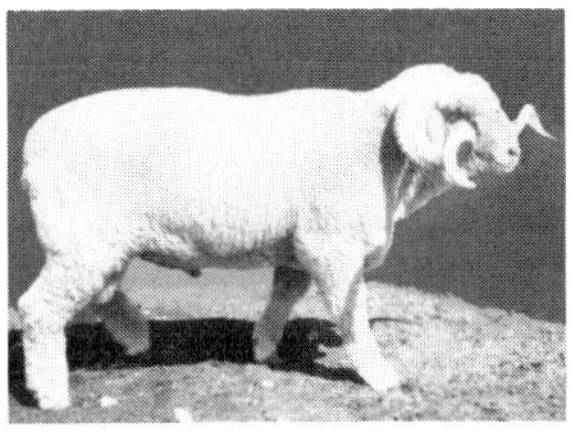

Fig. 17–84. Rambouillet ram.

Registry Association:
American Rambouillet Sheep Breeders' Assn., The
2709 Sherwood Way
San Angelo, TX 76901

Origin: In France, from Spanish Merino parent stock imported from Spain.

Color of face, ears, and legs: Cream to white.

Characteristics: Most rams have horns, but there are some polled strains; ewes are hornless. Rambouillets are the largest of the fine-wool breeds. They are noted for strong flocking instinct, and for ewes that will breed out of season.

(Continued)

TABLE 17-3 *(Continued)*

ROMANOV

Origin: In the Soviet Union, in the Volga Valley.

Characteristics: They are noted for a high lambing rate, ranging from 184–320%; early maturity; and the ability to breed out of season. In the U.S.S.R., Romanov sheep are raised for furs and pelts.

Fig. 17-85. Romanov ewe and triplet lambs.

ROMNEY

Registry Association:
American Romney Breeders Assn.
4375 N.E. Weslinn Drive
Corvallis, OR 97333

Origin: In England, in the Romney Marsh region of the county of Kent.

Color of face, ears, and legs: White.

Fig. 17-86. Romney ram lamb.

Characteristics: Both sexes are polled. They are open faced. In comparison with other long-wool breeds, the Romney is shorter legged, more rugged, and its fleece is shorter, finer, and less open.

SCOTCH HIGHLAND

Origin: In Scotland, in the highland country.

Color of face, ears, and legs: Black or mottled.

Characteristics: Both sexes have horns. Scotch Highland sheep are noted for their striking, stylish appearance. Their fleece consists of a long coarse outer coat and a finer inner coat.

Fig. 17-87. Scotch Highland ram.

SHROPSHIRE

Registry Association:
American Shropshire Registry Assn., Inc.
6508 West "R" Avenue
Schoolcraft, MI 49087

Origin: In England, in the central-western counties of Shropshire and Stafford.

Color of face, ears, and legs: The face is dark, but a gray nose is not objectionable.

Fig. 17-88. Shropshire ram.

Characteristics: Both sexes are polled, although rams frequently have scurs. Shropshires have a covering of dense wool well over the poll.

SOUTHDOWN

Registry Association:
American Southdown Breeders' Assn.
Route 4, Box 148
Bellefonte, PA 16283

Origin: In England, in the southeastern county of Sussex.

Color of face, ears, and legs: A light or mouse brown color is preferred.

Characteristics: Both sexes are polled, although rams sometimes have scurs. The breed is noted for superior conformation and quality of carcass.

Fig. 17-89. Southdown ram.

SUFFOLK

Registry Association:
American Suffolk Sheep Society
P.O. Box 256
Newton, UT 84327

National Suffolk Sheep Assn.
P.O. Box 324
Columbia, MO 63205

Origin: In England, in the southeastern counties of Suffolk, Essex, and Norfolk.

Color of face, ears, and legs: Suffolks have a very black head, ears, and legs.

Fig. 17-90. Suffolk lamb.

Characteristics: Both sexes are polled, although rams frequently have scurs; and the head and ears are entirely free from wool.

TARGHEE

Registry Association:
U.S. Targhee Sheep Assn.
P.O. Box 34
Jordan, MT 59337

Origin: The breed was developed by the USDA at the Dubois Experimental Station, Dubois, ID, using Rambouillet rams X Lincolnshire-Rambouillet-Corriedale ewes.

Color of face, ears, and legs: White.

Characteristics: Both sexes are polled. Targhees are open faced; moderately low set; and have a long productive life.

Fig. 17-91. Targhee ram.

TUNIS

Registry Association:
National Tunis Sheep Registry
R.D. 1
Wayland, NY 14572

Origin: In North Africa, in the province of Tunis.

Color of face, ears, and legs: Reddish brown to bright tan.

Fig. 17-92. Tunis ram.

Characteristics: Both sexes are polled. Tunis sheep have long drooping ears, and the head is free from wool. It was originally a fat-tailed breed, but breeders have selected away from this trait. Tunis will mate almost any season of the year; and the ewes are good milkers.

TABLE 17-4
BREEDS AND BREED REGISTRIES OF GOATS

ANGORA AND PIGMY GOATS

ANGORA

Registry Association:
American Angora and Goat Breeders' Assn.
P.O. Box 195
Rocksprings, TX 78880

Fig. 17-93. Angora buck.

Origin: In Turkey, in the province of Angora, from which it takes its name.

Color: Angoras are always pure white, although a black one appears occasionally. Red kids will shed their hair and produce white mohair later, but it is recommended that these animals be culled out and sent to slaughter.

Characteristics: Both sexes are usually horned, but polled individuals occur. The outer coat consists of long locks or strands of mohair. Angoras have rather thin, long, and pendulous ears.

PYGMY

Registry Association:
National Pygmy Goat Assn.
5621 W. Michigan
Tucson, AZ 85746

Fig. 17-94. Pygmy kid.

Origin: In the western part of Africa, particularly in the Cameroons. The first Pygmy goats were brought to zoos in the U.S. during the 1950s, from Sweden and Germany. In 1961, Dr. James Metcalfe established a colony at the Oregon Medical School, in Portland.

Color: The preferred colors range from white to gray and black in a predominantly grizzled (agouti) pattern. The muzzle, forehead, eyes, and ears are accented in lighter tones. The hoofs and cannons (socks) are black, as are the crown and dorsal stripe.

Characteristics: The Pygmy goat is small, cobby, and compact. The head and legs are short. At maturity, they measure only 16 to 21 in. at the withers. They are hardy, alert, animated, good-natured, gregarious, and docile. They are precocious breeders.

DAIRY (MILK) GOATS

ALPINE

Registry Association:
Alpines International
Route 1, Box 2065
Ft. Pierce, FL 33451

Fig. 17-95. Alpine doe.

Origin: In Switzerland, but developed by France (French Alpine); by the English (British Alpine); and by Mary Edna Rock in the U.S. (Rock Alpine). Swiss Alpine are now known as Oberhasli.

Color: There are a number of color patterns.

Characteristics: Some have horns at birth and are disbudded, others are hornless; the hair is medium to short; the ears are erect; and the nose is straight.

LaMANCHA

Fig. 17-96. LaMancha doe.

Origin: In the U.S., from a short-eared Spanish breed crossed on leading purebred breeds.

Color: Any color or combination of colors.

Characteristics: They virtually lack ears. LaManchas possess excellent dairy temperament, and are sturdy, productive, and adaptable.

NUBIAN

Registry Association:
International Nubian Breeders' Assn.
P.O. Box 130
Crewell, OR 97426

Fig. 17-97. Nubian doe.

Origin: The Nubian in the U.S. is of mixed origin. It evolved out of crossing Indian Jumna Pari and Egyptian Zariby types on British dairy goats.

Color: A variety of colors and patterns is accepted.

Characteristics: Some Nubians are born with horns and disbudded, others are hornless. They have long, drooping ears, a Roman nose, and a prominent forehead. The does are beardless. It is a relatively large breed. Nubians are noted for high milk and butterfat production.

OBERHASLI

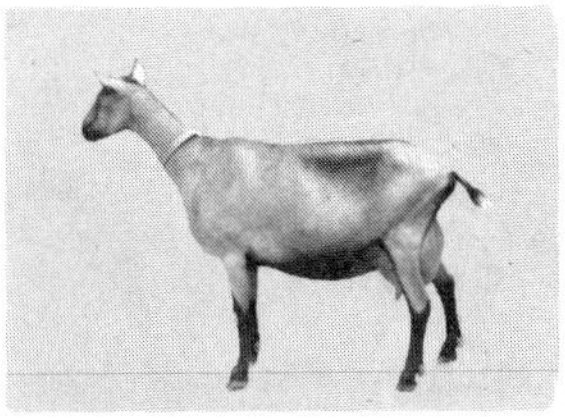

Fig. 17-98. Oberhasli doe.

Origin: In Switzerland.

Color: Chamois, which is a bay—ranging from light to deep red, with the latter preferred; with black markings and a few white hairs.

Characteristics: Oberhaslis are medium sized and alert, with a straight face.

SAANEN

Fig. 17-99. Saanen doe.

Origin: In Switzerland, in the Saanen Valley.

Color: Pure white or creamy white; the cream color may vary from light to dark fawn.

Characteristics: Hornless animals are preferred. Saanens are straight or dished faced, have erect ears, and are medium to large in size.

SABLE

Fig. 17-100. Sable doe.

Origin: In Switzerland, from Saanen parents.

Color: Dark—or sable-colored. Color results from Saanen being heterozygous for white; thus, 25% of the offspring from heterozygous Saanens will be sable color. Sable X sable always produces sable.

Characteristics: Same as Saanen except for color. This means that hornless animals are preferred, that they have a straight or dished face and erect ears, and that they are of medium size.

TOGGENBURG

Fig. 17-101. Toggenburg doe.

Origin: In Switzerland, in the Toggenburg Valley. But they originally came from the Alps.

Color: Light fawn to dark chocolate, with two white stripes on the face and white on the legs below the knees.

Characteristics: Toggenburgs are hornless or disbudded, have erect ears that are carried forward, have a straight or dished face, are of medium size and alert.

TABLE 17-5
BREEDS AND BREED REGISTRIES OF SWINE

BERKSHIRE

Registry Association:
American Berkshire Assn.
1769 U.S. 52 North, Box 2436
West Lafayette, IN 47906

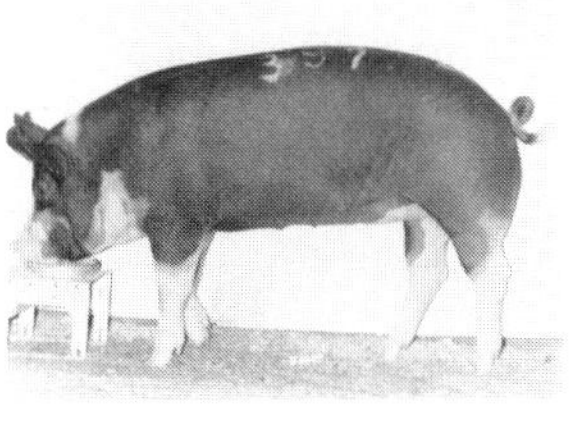
Fig. 17–102. Berkshire gilt.

Origin: In England, in the south central counties of Berkshire and Wiltshire.

Color: Black with 6 white points, 4 white feet, some white on face, and a white switch on the tail.

Characteristics: Medium short nose and erect ears; striking style and carriage.

CHESTER WHITE

Registry Association:
Chester White Swine Record Assn.
1803 W. Detweiller Drive
Peoria, IL 61615

Fig. 17–103. Chester White barrow.

Origin: In the U.S., chiefly in Chester and Delaware Counties of Pennsylvania.

Color: White.

Characteristics: Chester Whites mature early; adapt well to conditions; are very popular in the northern areas; and the sows are prolific and exceptional mothers.

DUROC

Registry Association:
United Duroc Swine Registry
1803 W. Detweiller Drive
Peoria, IL 61615

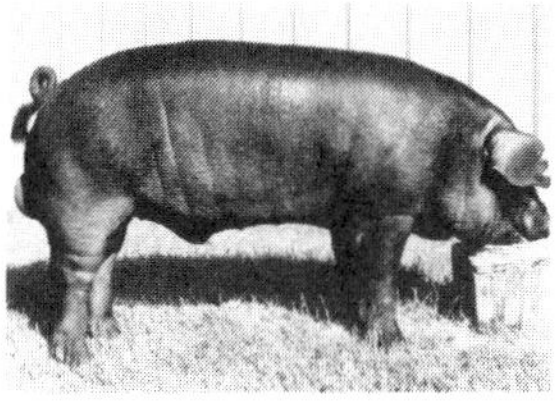
Fig. 17–104. Duroc boar.

Origin: In the U.S., chiefly in New York and New Jersey.

Color: Red, varying from light to dark.

Characteristics: Medium-sized ear, tipping forward; good feeding capacity; prolific; and hardy.

HAMPSHIRE

Registry Association:
Hampshire Swine Registry
6784 N. Frostwood Parkway
Peoria, IL 61615

Fig. 17–105. Hampshire gilt.

Origin: In the U.S., in Boone County, Kentucky.

Color: Black with white belt around the shoulders, body, and front legs.

Characteristics: The face is longer and straighter than most breeds; and the ears are carried erect. It is a medium sized breed.

HEREFORD

Registry Association:
National Hereford Hog Record Assn.
Route 1, Box 37
Flandreau, SD 57028

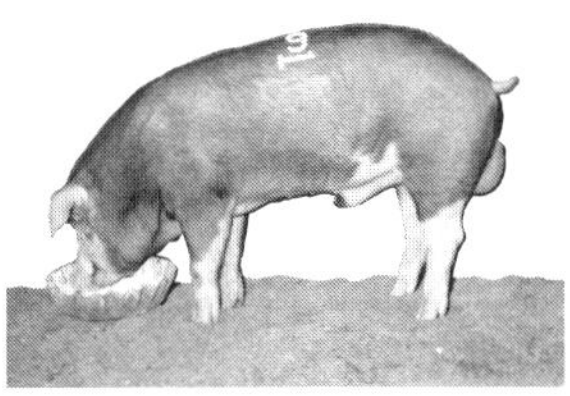
Fig. 17–106. Hereford boar.

Origin: In the U.S., by R. U. Webber of LaPlata, MO.

Color: Red body, with white face, legs, and switch.

Characteristics: The color and markings, which are similar to Hereford cattle, are distinctive. The breed is smaller than the other U.S. swine breeds.

LACOMBE

Origin: In Canada, at the Lacombe Experimental Farm, Alberta, using 55% Landrace, 23% Berkshire, and 22% Chester White.

Fig. 17–107. Lacombe gilt.

Color: White.

Characteristics: Medium sized flop ears of medium length; slightly dished face. Of the three parent breeds, it resembles the Landrace most closely.

LANDRACE (AMERICAN)

Registry Association:
American Landrace Assn., Inc.
P.O. Box 2340
West Lafayette, IN 47906

Fig. 17–108. Landrace boar.

Origin: In Denmark.

Color: White; black spots or freckles are acceptable.

Characteristics: Very long side; level top and well-defined underline; deep flanks; trim jowl, straight snout, and medium lop ears; prolific; and efficient feed utilization.

POLAND CHINA

Registry Association:
Poland China Record Assn.
P.O. Box 2537
West Lafayette, IN 47906

Fig. 17–109. Poland China boar.

Origin: In the U.S., in southwestern Ohio in the fertile area known as the Miami Valley, particularly in Warren and Butler Counties.

Color: Black with 6 white points—feet, nose, and tip of tail.

Characteristics: It is a large breed. In addition to the distinctive color, it has drooping ears.

(Continued)

TABLE 17–5 *(Continued)*

SPOTTED

Registry Association:
National Spotted Swiss Record, Inc.
P.O. Box 2807
West Lafayette, IN 47906

Origin: In the U.S., in the north central part, principally in Indiana.

Color: Spotted black and white, about 50% each color.

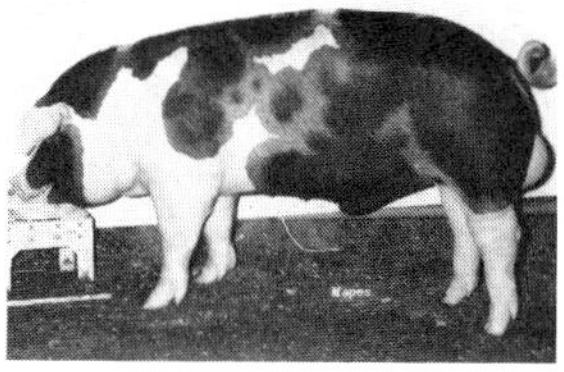

Fig. 17–110. Spotted boar.

Characteristics: It resembles the Poland China, but the animals have more white and the breed has been subjected to less type change than the Poland China.

TAMWORTH

Registry Association:
Tamworth Swine Assn.
2656 Horner Road
Winchester, OH 45697

Origin: In England, in the central counties of Stafford, Leicester, Warwick, and Northhampton.

Color: Red, varying from light to dark.

Fig. 17–111. Tamworth boar.

TAMWORTH *(Continued)*

Characteristics: Tamworths are the most extreme bacon type of any breed. They are wide between the ears, the snout is moderately long and straight, the jowl is neat, and the ears are medium-sized and carried erect. They produce a high proportion of lean meat.

YORKSHIRE

Registry Association:
American Yorkshire Club, Inc.
P.O. Box 2417
West Lafayette, IN 47906

Origin: In England (where it is known as *Large White*), in Yorkshire and neighboring counties, where it is a popular bacon breed.

Fig. 17–112. Yorkshire gilt.

Color: White.

Characteristics: The face is slightly dished; the ears erect; the sows are good mothers, farrowing and raising large litters, and they are great milkers. Yorkshires produce fine quality carcasses.

Racing for feed. (Courtesy, National Pork Producers Council, Des Moines, IA)

TABLE 17-6
BREEDS AND BREED REGISTRIES OF LIGHT HORSES AND PONIES

AKHAL-TEKE

Registry Association:
Akhal-Teke Assn. of America, Inc., The Shenandoah Farm
Route 5, Box 110
Staunton, VA 24401

Origin: In the U.S.S.R., in southern Turkmania, by the *Teke* tribe.

Color: Prevailing color is gold; either golden dun, golden bay, or golden chestnut, although other colors do occur.

Fig. 17-113. Akhal-Teke stallion.

Characteristics: The Akhal-Teke is a desert-bred horse, with a light, elegant build; long, tapering face; wide nostrils; large eyes; mobile ears; long, straight neck, short sparse mane; sloping shoulders; prominent withers; long, lean, narrow body; long legs; dense hooves; and gliding, elastic action. Akhal-Tekes are 15 to 15-2 hands high, and weigh 900 to 1,000 lb. It is a popular breed for use in all competitive equine sports.

AMERICAN BASHKIR CURLY

Registry Association:
American Bashkir Curly Registry
P.O. Box 453
Ely, NV 89301

Origin: In the U.S.S.R., in Bashkiria, on the eastern slopes of the Ural Mountains.

Color: All colors are accepted.

Fig. 17-114. American Bashkir Curly filly.

Characteristics: A curly coat, with corkscrew mane and wavy tail; medium size and chunky, somewhat resembling the early-day Morgan; small nostrils; gentle disposition; heavy milkers; and natural fox-trot gait. They are used for pleasure horses, for utility purposes—including light draft work, for family trail horses, and for children's mounts.

AMERICAN CREME

Registry Association:
International American Albino Assn.
Box 194
Naper, NE 68755

Origin: In the U.S., mostly in Oregon and Washington.

Color: Ivory, white, or shades of cream; eyes blue or dark.

Fig. 17-115. American Creme gelding.

Characteristics: Both horses and ponies accepted for registry—animals above 14-2 hands are classed as horses, those below 14-2 are classed as ponies. American Creme horses are used for pleasure and stock horses, exhibitions, parades, and flag-bearing.

AMERICAN GOTLAND HORSE

Registry Association:
American Gotland Horse Assn.
R.R. 2, Box 181
Elkland, MO 65644

Origin: In Sweden, on the Baltic island of Gotland.

Color: Bay, brown, dun, chestnut, palomino, roan, and some leopard and blanket markings.

Fig. 17-116. American Gotland mare.

Characteristics: They average 51 in. high, with a range of 11 to 14 hands. Gotlands are all-purpose horses, suitable for children and medium-sized adults.

AMERICAN MUSTANG

Registry Association:
American Mustang Assn., Inc.
P.O. Box 338
Yucaipa, CA 92399

Origin: Along the Barbary Coast of North Africa; eventually they were taken to Spain; thence brought to the U.S. by the conquistadors.

Color: Any color accepted.

Fig. 17-117. American Mustang stallion.

Characteristics: They are hardy and versatile—they are used for pleasure riding, show, trail riding, endurance trials, stock horses, and jumping.

AMERICAN SADDLEBRED

Registry Association:
American Saddlebred Horse Assn., Inc.
4093 Iron Works Pike
Lexington, KY 40511

Origin: In the U.S., in Fayette County, KY.

Color: They may be bay, brown, chestnut, gray, roan, black, or golden, but gaudy white markings are frowned upon.

Fig. 17-118. American Saddlebred stallion.

Characteristics: Ability to furnish an easy ride with great style and animation; long, graceful neck; proud action, used for 3- and 5-gaited saddle horses, fine harness, pleasure, and stock horses.

AMERICAN WALKING PONY

Registry Association:
American Walking Pony Assn.
Rt. 27, Box 605
Upper River Road
Macon, GA 31211

Origin: In the U.S., near Macon, GA, by crossing Welsh Pony X Tennessee Walking Horse.

Color: No color stipulation; the colors of both parent breeds occur.

Fig. 17-119. American Walking Pony.

Characteristics: American Walking Ponies range in height from 13 to 14-2 hands, perform the running walk gait, and are used for pleasure, children, and small adults.

AMERICAN WHITE

Registry Association:
International American Albino Assn.
Box 194
Naper, NE 68755

Origin: In the U.S. on White Horse Ranch, Naper, NE.

Color: Snow-white hair; pink skin; and light blue, dark blue, brown, or hazel eyes.

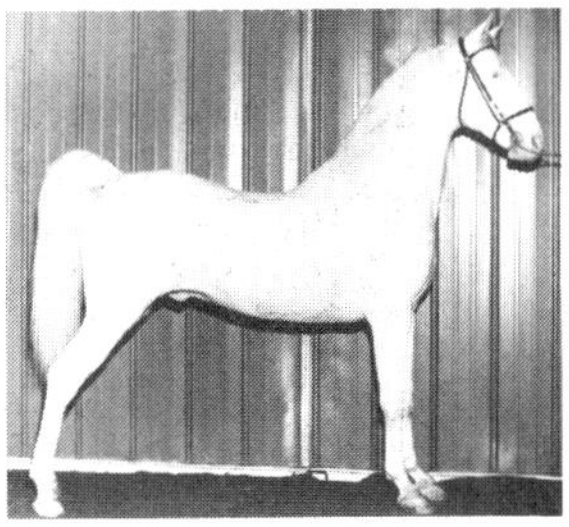

Fig. 17-120. American White stallion.

Characteristics: Animals above 14-2 hands are classed as horses, those below as ponies. Used for pleasure horses, exhibition purposes, and parade and flag-bearer horses.

(Continued)

TABLE 17–6 *(Continued)*

ANDALUSIAN

Registry Association:
American Andalusian Horse Assn.
P.O. Box 68
Tumacacori, AZ 85640

International Andalusian Horse Assn.
256 S. Robertson, No. 9378
Beverly Hills, CA 90211

Fig. 17–121. Andalusian stallion.

Origin: In Spain, from desert-bred Barbs crossed on the light, agile horses of southern Spain.

Color: Whites, grays, and bays are most common. Also, there are a few blacks, roans, and chestnuts.

Characteristics: They are used for bullfighting, parade, dressage, jumping, and pleasure riding.

APPALOOSA

Registry Association:
Appaloosa Horse Club, Inc.
P.O. Box 8403
Moscow, ID 83843

Fig. 17–122. Appaloosa stallion.

Origin: In the U.S., in Oregon, Washington, and Idaho; from animals originating in Fergana, Central Asia.

Color: Variable, but usually white over the loin and hips, with dark round or egg-shaped spots thereon.

Characteristics: The eye is encircled by white; the skin is mottled; and the hoofs are striped vertically black and white. Appaloosas are used for stock horses, pleasure horses, parade horses, and racing.

ARABIAN

Registry Association:
Arabian Horse Registry of America, Inc.
12000 Zuni Street
Westminister, CO 80234

International Arabian Horse Registry
of North America
P.O. Box 325
Delphi Falls, NY 13051

Fig. 17–123. Arabian mare.

Origin: In Arabia.

Color: Bay, gray, chestnut, with an occasional white or black. White marks on the head and legs are common. The skin is always dark.

Characteristics: Beautiful head; short coupling; docile; great endurance; airy way of going; used for saddle horses, show purposes, pleasure horses, stock horses, and racing.

BUCKSKIN

Registry Association:
International Buckskin Horse Assn., Inc.
P.O. Box 357
St. John, IN 46373

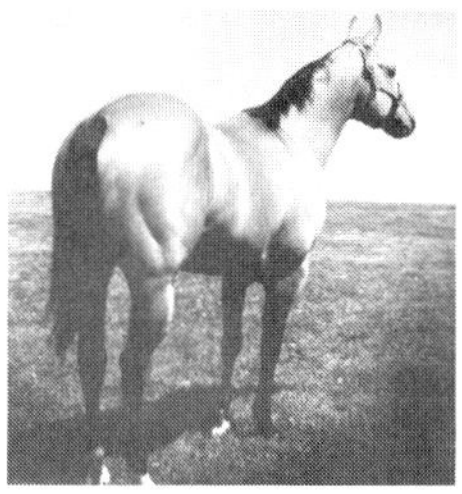

Fig. 17–124. Buckskin stallion.

Origin: Originated in the United States.

Color: Buckskin, red dun, grulla (mouse dun).

Characteristics: Used for stock horses, pleasure horses, and show purposes.

CHICKASAW

Registry Association:
Chickasaw Horse Assn., Inc., The
P.O. Box 8
Love Valley, NC 28677

National Chickasaw Horse Assn.
Route 2
Clarinda, IA 51232

Fig. 17–125. Chickasaw gelding.

Origin: In the U.S., by the Chickasaw Indians of Tennessee, North Carolina, and Oklahoma.

Color: Bay, black, chestnut, gray, roan, sorrel, or palomino.

Characteristics: Short head and ears; short back; short neck; square, stocky hips; low-set tail; wide chest; and great width between the eyes. Preferably 53 to 59 in. high. Used as cow ponies.

CONNEMARA PONY

Registry Association:
American Connemara Pony Society
R.D. 1
Hoshiekon Farm
Goshen, CT 06756

Fig. 17–126. Connemara Pony stallion.

Origin: In Ireland, along the west coast.

Color: Gray, black, bay, brown, dun, cream, with occasional roans and chestnuts.

Characteristics: Connemaras range in height from under 14–2 hands (pony) to over 14–2 hands (small horse). They are used as jumpers; for showing under saddle and in harness; and for both adults and children.

GALICENO

Registry Association:
Galiceno Horse Breeders Assn., Inc.
111 E. Elm Street.
Tyler, TX 75701

Fig. 17–127. Galiceno mare.

Origin: In Galiceno, a province in northwestern Spain.

Color: Solid colors prevail. Bay, black, chestnut, dun, gray, brown, and palomino are most common.

Characteristics: They are relatively small in size; at maturity, they stand 12 to 13–2 hands and weigh 625 to 700 lb. Galicenos are used as riding horses.

HACKNEY

Registry Association:
American Hackney Horse Society
P.O. Box 174
Pittsfield, IL 62363

Fig. 17–128. Hackney stallion.

Origin: In England; on the eastern coast in Norfolk and adjoining counties.

Color: Chestnut, bay, and brown are most common; roans and blacks are seen; white marks are common and desired.

Characteristics: Hackney horses are noted for high natural action. They are used for heavy harness or carriage horses and ponies; and for crossbreeding to produce hunters and jumpers.

(Continued)

TABLE 17–6 *(Continued)*

HAFLINGER

Registry Association:
Haflinger Assn. of America
14570 Gratiot Road
Hemlock, MI 48626

Haflinger Registry of North America
14640 State Route 83
Coshocton, OH 32812

Fig. 17–129. Haflinger yearling.

Origin: The Haflinger originated near the village of Hafling, which was part of Austria to the end of World War I, now a part of Italy.

Color: Chestnut, ranging from honey blond to chocolate, with white or flaxen mane and tail.

Characteristics: Haflingers are relatively small; noted for their beauty, strength, vitality, intelligence, and good disposition; and are used for either work or pleasure.

HANOVERIAN

Registry Association:
American Hanoverian Society, The
Office 2-E, 831 Bay Avenue
Capitola, CA 95010

Fig. 17–130. Hanoverian stallion.

Origin: In Germany, in the Hanover area; it was developed as a superior horse for military use, with emphasis on size, intelligence, and temperament.

Color: Variable.

Characteristics: Hanoverians are big and powerful; many stand 16½ hands or better. They combine nobility, size, and strength in a unique way. They are used for riding, driving, hunting, jumping, dressage, and utility purposes.

HUNGARIAN HORSE

Registry Association:
Hungarian Horse Assn.
P.O. Box 98
Anselmo, NE 68813

Fig. 17–131. Hungarian Horse mare.

Origin: In Hungary.

Color: All colors accepted, either solid or broken.

Characteristics: Hungarian Horses possess a unique combination of style and beauty with ruggedness. They are used for stock, cutting, pleasure, trail, hunting, and jumping.

LIPIZZAN

Registry Association:
United States Lipizzan Registry
12479 Duncan Plains Road N.W.
Johnstown, OH 43031

Fig. 17–132. Lipizzan mare.

Origin: In Lippiza, Yugoslavia.

Color: Most mature animals are white, but foals are born dark brown or gray, then turn white at 4 to 6 years of age.

Characteristics: Lipizzans have an elastic walk, with considerable knee action. They are used for dressage (for which purpose they are without peer), harness horses, pleasure horses, hunters, jumpers, and parade horses.

MINIATURE HORSE

Registry Association:
American Miniature Horse Assn., The
P.O. Box 129
Burleson, TX 76028

A.S.P.C./A.M.H.R.
P.O. Box 3415
Peoria, IL 61614

Fig. 17–133. Miniature Horse stallion.

Origin: In England and northern Europe, where they are used in the mines, and as pets by the titled families of Europe.

Color: All colors are accepted.

Characteristics: Miniature Horses cannot exceed 34 in. at the withers. There are two types: (1) the more refined Arabian type, and (2) the heavier Quarter Horse type. They are used as pets and for driving purposes—pulling various small vehicles.

MISSOURI FOX TROTTING HORSE

Registry Association:
Missouri Fox Trotting Horse
Breed Assn., Inc.
P.O. Box 1027
Ava, MO 65608

Fig. 17–134. Missouri Fox Trotting mare.

Origin: In the Ozark Hills of Missouri and Arkansas.

Color: Sorrels predominate, but any color is accepted.

Characteristics: The distinctive breed trait is the fox trot gait. They are used as pleasure horses, stock horses, and for trail riding.

MORAB

Registry Association:
North American Morab Horse
Assn., Inc.
W3174 Faro Springs Road
Hilbert, WI 54129

Fig. 17–135. Morab stallion.

Origin: Although Morgan X Arabian crosses were made in the early 1800s, the name *Morab* was coined in the 1920s when William Randolph Hearst crossed Arabian stallions on Morgan mares to produce horses for his San Simeon ranch in California.

Color: Bay, black, brown, buckskin, chestnut, dun, gray, grulla, palomino, and roan.

Characteristics: Morabs possess the ruggedness of the Morgan and the refinement of the Arabian. they are used for show purposes, pleasure riding, endurance rides, and ranch work.

MORGAN

Registry Association:
American Morgan Horse Assn.,
Inc., The
P.O. Box 960
3 Bostwick Road
Shelburne, VT 05482

Fig. 17–136. Morgan stallion.

Origin: In the U.S., in New England.

Color: Bay, brown, black, and chestnut, sometimes with extensive white markings.

Characteristics: They are easy keepers, possess good endurance, and are docile. Morgans are used for saddle horses, stock horses, and driving.

(Continued)

TABLE 17-6 *(Continued)*

NATIONAL APPALOOSA PONY

Registry Association:
National Appaloosa Pony, Inc.
Box 206
Gaston, IN 47342

Fig. 17-137. National Appaloosa Pony gelding.

Origin: In the U.S., near Rochester, IN.

Color: They are vari-colored, but leopard, blanket-type, snowflake, and roans are most popular. The skin, nose and area around eyes are mottled. White sclera encircles the eyes.

Characteristics: They stand under 14-2 hands. Their primary uses are: show ponies, trail riding, jumping, and racing.

NATIONAL SHOW HORSE

Registry Association:
National Show Horse Registry
Plainview Triad North, Suite 237
10401 Linn Station Road
Louisville, KY 40223

Fig. 17-138. National Show Horse foal.

Origin: In the U.S., from an Arabian X Saddlebred cross.

Color: All color commons to Arabians and Saddlebreds occur and are accepted.

Characteristics: National Show Horses combine the beauty, refinement, and stamina of the Arabian, and the size and high-stepping action of the Saddlebred. The ideal animal has a long neck, pretty head, relatively short back, with long legs, and a high, animated trot. National Show Horses are shown in the following classes: English pleasure, 3-gaited, 5-gaited, pleasure driving, fine harness, equitation, and country pleasure.

NORWEGIAN FJORD

Registry Association:
Norwegian Fjord Assn. of
North America
24570 W. Chardon Road
Grayslake, IL 60030

Fig. 17-139. Norwegian Fjord stallion.

Origin: In Norway over 2,000 years ago.

Color: Dun, ranging from light to dark, with a dorsal strip and dark bars on the legs, hooves, eyes, and ear tips.

Characteristics: The body is compact and muscular. Norwegian Fjords are used for draft, carriage, pleasure, jumping, dressage, and trail riding.

PAINT HORSE

Registry Association:
American Paint Horse Assn.
P.O. Box 18519
Ft. Worth, TX 76118

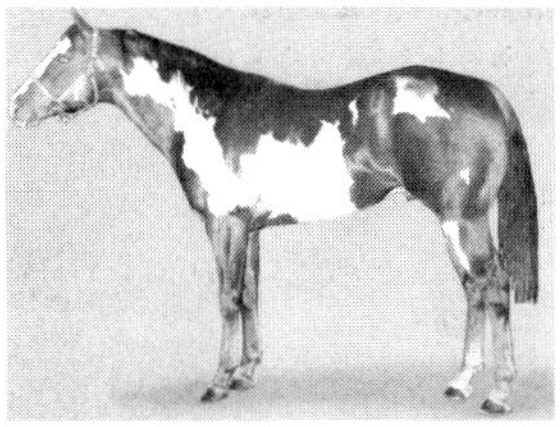

Fig. 17-140. Paint Horse stallion.

Origin: In the U.S.

Color: White, plus any other color; must be a recognizable paint. Paint Horses are distinguished by two color patterns—overo or tobiano.

Characteristics: No discrimination is made against glass, blue, or light-colored eyes. Paint Horses are used for stock horses, pleasure horses, show purposes, and racing.

PALOMINO

Registry Association:
Palomino Horse Breeders of America, Inc.
15253 E. Skelly Drive
Tulsa, OK 74116-2620

Palomino Horse Assn., Inc., The
P.O. Box 324
Jefferson City, MO 65101

Fig. 17-141. Palomino stallion.

Origin: In the U.S., from animals of Spanish extraction.

Color: Golden (the color of a newly minted gold coin, or 3 shades lighter or darker), with a light-colored mane and tail (white, silver, or ivory, with not more than 15% dark or chestnut hair in either). White markings on face or below the knees are acceptable. The skin is dark.

Characteristics: Palominos are used for stock horses, parade horses, pleasure horses, saddle horses, and fine harness horses.

PASO FINO

Registry Association:
Paso Fino Horse Assn., Inc.
P.O. Box 600
Bowling Green, FL 33834

Fig. 17-142. Paso Fino stallion.

Origin: In the Caribbean area; mostly in Puerto Rico, Columbia, Dominican Republic, and Cuba.

Color: Any color is accepted, although solid colors are preferred. Bay, chestnut, and black with white markings are most common. Occasionally palominos and pintos appear.

Characteristics: The paso fino gait, which distinguishes the breed, may be described as a broken pace. The legs on the same side move together, but the hind foot strikes the ground a fraction of a second before the front foot, producing a 4-beat gait. Paso Finos are used for pleasure, cutting, and parade horses; and for endurance riding and drill team work.

PERUVIAN PASO

Registry Association:
American Assn. of Owners & Breeders
of Peruvian Paso Horses
221 N. Alameda Avenue
Burbank, CA 91502

Peruvian Paso Horse Registry
of North America
1038 4th Street, Suite 4
Santa Rosa, CA 95404

Fig. 17-143. Peruvian Paso stallion.

Origin: In Peru.

Color: Any color is accepted, although solid colors are preferred.

Characteristics: The breed is characterized by a natural 4-beat, lateral gait know as the *piso.* Peruvian Pasos do only 2 acceptable gaits—the paso llano and sobreandando. They are used for pleasure horses, parade horses, and endurance horses.

(Continued)

TABLE 17–6 *(Continued)*

PINTO HORSE

Registry Association:
Pinto Horse Assn. of America, Inc.
1900 Samuels Avenue
Ft. Worth, TX 76102

Fig. 17–144. Pinto Horse stallion.

Origin: In the U.S., from horses brought in by the Spanish conquistadors.

Color: Preferably half color or colors and half white, with many spots well placed. The two distinct pattern markings are: overo and tobiano.

Characteristics: Glass eyes are not discounted. Pinto Horses are used for any light horse purposes, but especially for show, parade, novice, pleasure, and stock horses.

PONY OF THE AMERICAS

Registry Association:
Pony of the Americas Club, Inc.
5240 Elmwood Avenue
Indianapolis, IN 46203

Fig. 17–145. Pony of the Americas stallion.

Origin: In the U.S., near Mason City, IA.

Color: Similar to Appaloosas; white over the loin and hips, with dark, round or egg-shaped spots.

Characteristics: Pony of the Americas is a happy medium of Arabian and Quarter Horse in miniature, ranging in height from 46 to 54 in., with the appaloosa color. They are used mostly by children as a western type pony.

QUARTER HORSE

Registry Association:
American Quarter Horse Assn.
P.O. Box 200
Amarillo, TX 79168

Fig. 17–146. Quarter Horse mare and foal.

Origin: In the U.S.

Color: Chestnut, sorrel, bay, and dun are most common, although they may be palomino, black, brown, roan, or copper colored.

Characteristics: Quarter Horses are well-muscled and powerfully built. They have small, alert ears, and well-muscled cheeks and jaws. They are used for stock horses, racing, and pleasure horses.

QUARTER PONY

Registry Association:
National Quarter Pony Assn.
5131 Country Road, #5, Rt. 1
Marengo, OH 43334

Origin: In the U.S.

Color: Quarter Ponies have the same colors as Quarter Horses, from which they descend.

Characteristics: They have all the same characteristics as Quarter Horses, only they are smaller. They are used for children and juniors as stock, racing, and pleasure horses.

SHETLAND PONY

Registry Association:
American Shetland Pony Club
P.O. Box 3415
Peoria, IL 61614

Fig. 17–147. Shetland Pony mare.

Origin: On the Shetland Islands.

Color: All colors, either solid or broken.

Characteristics: Shetlands are small in size, with two types: (1) the classic type, which is short and chunky; and (2) the modern type, which is fine boned, long necked, and high going. They are used as children's mounts, harness horses, roadsters, and racing.

SPANISH-BARB

Registry Association:
Spanish-Barb Breeders Assn.
2888 Bluff Street, Box 465
Boulder, CO 80301

Fig. 17–148. Spanish-Barb stallion.

Origin: In Africa, from whence they migrated to Spain, Cuba, and Mexico, before coming to southwestern U.S., and to Florida a few years later.

Color: All colors are represented, but dun, grulla, sorrel, and roan are most common. Most animals are solid colored, but a dorsal stripe and zebra markings occur in all duns and grullas and in some sorrels.

Characteristics: Spanish-Barbs are small horses (standing from 13–3 to 14–3 hands), with short coupling, deep bodies, good action, and without extreme muscling. They are used for cow ponies, western riding, English riding, and pack horses.

SPANISH MUSTANG

Registry Association:
Spanish Mustang Registry, Inc., The
8328 Stevenson Avenue
Sacramento, CA 95828

Fig. 17–149. Spanish Mustang stallion.

Origin: In the U.S., from Indian-owned horses of Barb and Andalusian ancestry brought to America by the Spanish.

Color: The whole gamut of colors occurs, including all the solid colors and all the broken colors.

Characteristics: Spanish Mustangs stand 13–2 to 15–2 hands. They are used for cow ponies, western riding, English riding, and pack horses.

SPOTTED SADDLE HORSE

Registry Association:
National Spotted Saddle Horse Assn., Inc.
P.O. Box 898
Murfreesboro, TN 37130

Fig. 17–150. Spotted Saddle Horse stallion.

Origin: In the U.S., primarily in middle Tennessee.

Color: Spotted.

Characteristics: For registry, animals must be spotted and must be saddle horses. Spotted Saddle Horses are rugged; have intelligent eyes and fox-type ears; are surefooted; and have good dispositions. They are used for pleasure riding, jumping, bird hunting, and coon hunting.

(Continued)

TABLE 17-6 *(Continued)*

STANDARDBRED (RIDDEN STANDARDBRED)

Registry Association:
United States Trotting Assn.
750 Michigan Avenue
Columbus, OH 43215

Origin: In the U.S.

Color: Bay, brown, chestnut, and black are most common; but grays, roans, and duns are found.

Characteristics: Standardbreds are smaller, less leggy, and possess more substance and ruggedness than the Thoroughbred. They are used for harness racing, either trotting or pacing; and as harness horses in horse shows.

The Ridden Standardbred consists of purebred Standardbreds, half-bred Standardbreds, and partial-bred Standardbreds. It has a separate registry (Ridden Standardbred Assn., 1578 Fleet Road, Troy, OH 49373) which promoted Standardbreds for uses other than sulky racing.

Fig. 17-151. Standardbred stallion.

TENNESSEE WALKING HORSE

Registry Association:
Tennessee Walking Horse Breeders' and Exhibitors Assn.
P.O. Box 286
Lewisburg, TN 37091

Origin: In the U.S., in the Middle Basin of Tennessee.

Color: Sorrel, chestnut, black, roan, white, bay, brown, gray, or golden. White markings on the face and legs are common.

Characteristics: The Tennessee Walking Horse is uniquely adapted to the running walk gait. It is used for plantation walking horses, pleasure walking horses, and show horses.

Fig. 17-152. Tennessee Walking Horse stallion.

THORCHERON

Registry Association:
Thorcheron Hunter Assn.
3749 S. 4th Street
Kalamazoo, MI 49009

Origin: In the U.S., in Michigan, by mating Thoroughbred mares to Percheron stallions.

Color: All the colors common to Thoroughbreds and Percherons, the two parent breeds.

Characteristics: Thorcherons are hunter/jumper type, ranging in size from 16 to 17 hands. They are short-coupled and good movers, with the refinement of Thoroughbreds and the muscle and good nature of Percherons. They resemble the big Irish hunters; and are used mostly for hunting and jumping.

Fig. 17-153. Thorcheron mare.

THOROUGHBRED

Registry Association:
Jockey Club, The
380 Madison Avenue
New York, NY 10017

Origin: In England.

Color: Bay, brown, chestnut, and black, less frequently roan, and gray. White markings on the face and legs are common.

Characteristics: The breed traits are: refined conformation; and long, straight, and well-muscled legs. Thoroughbreds are used for racing, stock horses, saddle horses, polo mounts, and hunters.

Fig. 17-154. Thoroughbred stallion

TRAKEHNER

Registry Association:
American Trakehner Assn., Inc.
1520 West Church Street
Newark, OH 43055

Origin: In Trakehnen, East Prussia, by blending Prussian horses, Thoroughbreds, and Arabians.

Color: Variable, with all the colors of the 3 parent breeds found.

Characteristics: Trakehners are about the size of the Thoroughbred, but more rugged; and they possess the elegance of the Arabian.

Fig. 17-155. Trakehner stallion.

TROTTINGBRED

Registry Association:
International Trotting & Pacing Assn., Inc.
575 Broadway
Hanover, PA 17331

Origin: In the U.S.

Color: All the colors common to Standardbreds and the pony breed that is used.

Characteristics: They are miniature Standardbreds, with a dash of pony breeding. They must not be over 13-1 hands. Trottingbreds are used for harness racing, which is a do-it-yourself and family oriented type of racing.

WELARA PONY

Registry Association:
American Welara Pony Society
P.O. Box 401
Yucca Valley, CA 92284

Origin: In the U.S., from Welsh X Arabian crosses.

Color: All the colors common to the two parent breeds; any color except Appaloosa is acceptable.

Characteristics: The Welara resembles a miniature coach horse. It is noted for its beauty. The breed is used for fine harness, English pleasure, halter, hunter, and native costume classes, as well as trail riding.

WELSH PONY AND COB/WELSH COB

Registry Association:
Welsh Cob Society of America
Grazing Field Farm
Head of the Bay Road
Buzzard Bay, MA 02532

Welsh Pony and Cob Society of America
P.O. Box 2977
Winchester, VA 22601

Origin: In Wales, Great Britain.

Color: Any color except piebald or skewbald. Gaudy white markings are not popular.

Characteristics: It is intermediate in size between the Shetland Pony and the light horse breeds. The breed registry maintains various divisions based on height. Welsh serve as mounts for children and small adults. They are used for racing, roadsters, trail riding, parade horses, stock cutting, and hunting.

Fig. 17-156. Welsh Pony stallion.

TABLE 17-7
BREEDS AND BREED REGISTRIES OF DRAFT HORSES

AMERICAN CREAM HORSE

Registry Association:
American Cream Horse Assn.
Route 1, Box 88
Hubbard IA 50122

Fig. 17-157. American Cream Horse.

Origin: In U.S.A.

Color: Medium cream, with white mane and tail, pink skin, and amber colored eyes.

Characteristics: Creams are a medium heavy draft type. They are good all-purpose horses for the average farmer. They have excellent dispositions.

BELGIAN

Registry Association:
Belgian Draft Horse Corporation
of America
P.O. Box 335
Wabash, IN 46992

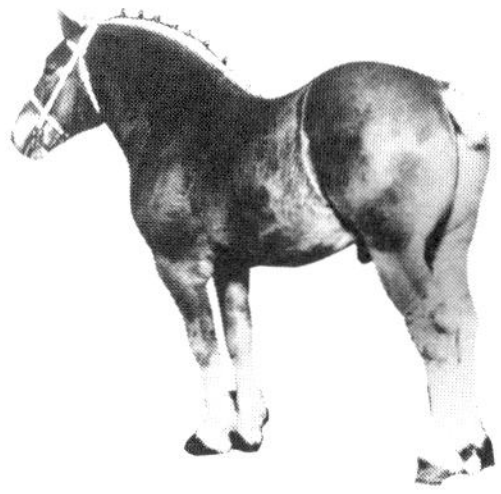

Fig. 17-158. Belgian stallion.

Origin: In Belgium.

Color: Bay, chestnut, and roan are most common, but browns, grays, and blacks are occasionally seen. Many Belgians have a flaxen mane and tail and a white-blazed face.

Characteristics: Belgians are the lowest set and most massive of all draft breeds. They are used for farm work horses and exhibition purposes.

CLYDESDALE

Registry Association:
Clydesdale Breeders Assn. of
the United States
Route 3
Waverly, IA 50677

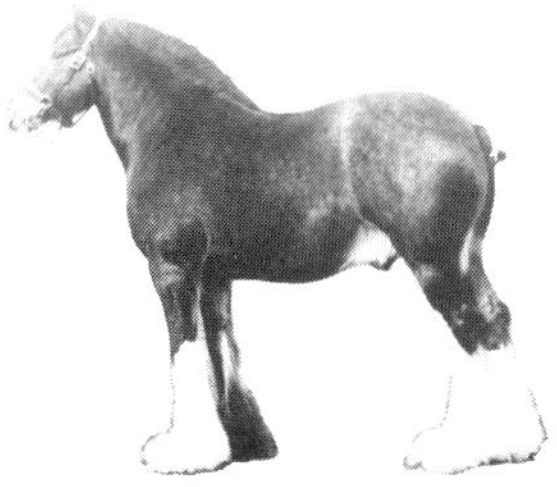

Fig. 17-159. Clydesdale stallion.

Origin: In Scotland, along the River Clyde.

Color: Bay or brown with white markings are most common, but blacks, grays, chestnuts, and roans are occasionally seen.

Characteristics: Clydesdales are distinguished by superior style and action, and feather or hair on the legs. They are used for farm work and exhibition purposes.

PERCHERON

Registry Association:
Percheron Horse Assn. of America
P.O. Box 141
Fredericktown, OH 43819

Fig. 17-160. Percheron stallion.

Origin: In France, in the Northwestern district of La Perche.

Color: Mostly black or gray, but bays, browns, chestnuts, and roans are seen.

Characteristics: In comparison with the other draft breeds, Percherons are noted for handsome, clean-cut heads. They are used for farm work and exhibition purposes.

SHIRE

Registry Association:
American Shire Horse Assn.
Route 1, Box 10
Adel, IA 50003-9702

Fig. 17-161. Shire stallion.

Origin: In England, primarily in the east central counties of Lincolnshire and Cambridgeshire.

Color: Common colors are bay, brown, and black, with white markings; although grays, chestnuts, and roans are occasionally seen.

Characteristics: Shires are taller than the other draft breeds, and they have feather or hair on the legs. They are used for farm work and exhibition purposes.

SUFFOLK

Registry Association:
American Suffolk Horse Assn., Inc.
Route 1, Box 212
Ledbetter, TX 78946

Fig. 17-162. Suffolk stallion.

Origin: In England, in the eastern county of Suffolk.

Color: Chestnut only.

Characteristics: Suffolks are the smallest of the draft breeds; also, they have short legs and are chunky. They are used for farm work and exhibition purposes.

TABLE 17-8
BREEDS AND BREED REGISTRIES OF JACKS, DONKEYS, AND MULES

MAMMOTH JACK (OR AMERICAN STANDARD JACK)

Registry Association:
Standard Jack and Jennet Registry of America
P.O. Box 1155
Pulaski, TN 38478-1155

Fig. 17-163. Mammoth Jack.

Origin: In southern Europe, mostly in the area bordering the Mediterranean.

Color: Variable, but mostly black with white nose, gray, or sorrel red.

Characteristics: The Mammoth Jack is the largest of the asses. It is used for the production of mules.

LARGE STANDARD DONKEY (SPANISH DONKEY)

Registry Association:
American Donkey and Mule Society, Inc.
2901 N. Elm
Denton, TX 76201

Fig. 17-164. Large Standard Donkey.

Origin: In southern Europe.

Color: Variable, but mostly black with white nose, gray, or sorrel red.

Characteristics: Males stand from 48 to 56 in. high and females from 48 to 54 in. They are used for the production of mules.

STANDARD DONKEY (BURRO)

Registry Association:
American Donkey and Mule Society, Inc.
2901 N. Elm
Denton, TX 76201

Fig. 17-165. Standard Donkey.

Origin: Developed throughout the world.

Color: Variable, but many are mouse colored.

Characteristics: They range in height from 36 to 48 in. at the withers. They are used for driving, riding, working, and as pets.

AMERICAN SPOTTED ASS

Registry Association:
American Donkey and Mule Society, Inc.
2901 N. Elm
Denton, TX 76201

Fig. 17-166. American Spotted Ass.

Origin: In most countries throughout the world.

Color: Spotted.

Characteristics: They are a blending of all breeds and sizes. Used for pets and show.

MINIATURE DONKEY

Registry Association:
Miniature Donkey Registry of the United States, Inc.
2901 N. Elm
Denton, TX 76201

Fig. 17-167. Miniature Donkey.

Origin: In Sicily and Sardinia.

Color: They have a dorsal strip, forming a cross with stripe over the withers and down the shoulders.

Characteristics: They must be under 36 in. at the withers. They are used as pets.

MULE (HINNY)

Registry Association:
American Mule Registry
2901 N. Elm
Denton, TX 76201

Fig. 17-168. Mule.

Origin: The mule is a hybrid, a cross between a horse and the ass.

Color: Variable, but sorrel is the preferred color.

Characteristics: The mule resembles his sire, the jack, more than his mother, the mare. But the desired conformation is identical to that of a horse for similar use, with more stress placed upon size, and set and quality of ear. Mules are used for driving, riding, packing, working, and showing.

SECTION

18

WHERE TO GO FOR HELP

Ask the Farm Advisor (County Agent)! This shows Aaron O. Nelson, Farm Advisor, Fresno, California, discussing USDA beef grades with ranchers at the annual meeting of the Fresno-Kings County Cattlemen's Assn.

Contents *Page*

Large manufacturing companies can and do hire their own specialists. For them, it is good business, for they know that they can sell more of their products if they are good than if they are poor.

Small farmers and ranchers, on the other hand, cannot afford to hire their own specialists, despite the fact that modern farming is fully as complex as manufacturing. This is so chiefly because of the smaller size of farms. There are, for example, about 2.0 million U.S. farms and ranches, whereas only four manufacturers make over nine-tenths of the nation's automobiles.

Under these circumstances, the farmer needs to know where to go for help—the help of specialists provided by various government agencies (federal, state, and county), and by commercial companies and trade associations. This section is presented for this purpose.

U.S. LAND-GRANT UNIVERSITIES AND CANADIAN PROVINCIAL UNIVERSITIES

U.S. producers can obtain a list of available bulletins and circulars, and other information by writing to (1) their state agricultural college (land-grant institution), and (2) the U.S. Superintendent of Documents, Washington, DC; or by going to the local county extension office (farm advisor) of the county in which they reside. Canadian producers may write to the Department of Agriculture of their province or to their provincial university. A list of U.S. land-grant institutions and Canadian provincial universities follows in Table 18–1.

TABLE 18–1
U.S. LAND-GRANT INSTITUTIONS AND CANADIAN PROVINCIAL UNIVERSITIES

State	Address
Alabama	School of Agriculture, Auburn University, Auburn, AL 36830
Alaska	Department of Agriculture, University of Alaska, Fairbanks, AK 99701
Arizona	College of Agriculture, The University of Arizona, Tucson, AZ 85721
Arkansas	Division of Agriculture, University of Arkansas, Fayetteville, AR 72701
California	College of Agricultural and Environmental Sciences, University of California, Davis, CA 95616
Colorado	College of Agricultural Sciences, Colorodao State University, Fort Collins, CO 80521
Connecticut	College of Agriculture and Natural Resources, University of Connecticut, Storrs, CT 06268
Delaware	College of Agricultural Sciences, University of Delaware, Newark, DE 19711
Florida	College of Agriculture, University of Florida, Gainesville, FL 32611
Georgia	College of Agriculture, University of Georgia, Athens, GA 30602
Hawaii	College of Tropical Agriculture, University of Hawaii, Honolulu, HI 96822
Idaho	College of Agriculture, University of Idaho, Moscow, ID 83843
Illinois	College of Agriculture, University of Illinois, Urbana-Champaign, IL 61801
Indiana	School of Agriculture, Purdue University, West Lafayette, IN 47907
Iowa	College of Agriculture, Iowa State University, Ames, IA 50010
Kansas	College of Agriculture, Kansas State University, Manhattan, KS 66506
Kentucky	College of Agriculture, University of Kentucky, Lexington, KY 40506
Louisiana	College of Agriculture, Louisiana State University and A&M College, University Station, Baton Rouge, LA 70803
Maine	College of Life Sciences and Agriculture, University of Maine, Orono, ME 04473
Maryland	College of Agriculture, University of Maryland, College Park, MD 20742
Massachusetts	College of Food and Natural Resources, University of Massachusetts, Amherst, MA 01002
Michigan	College of Agriculture and Natural Resources, Michigan State University, East Lansing, MI 48823
Minnesota	College of Agriculture, University of Minnesota, St. Paul, MN 55101
Mississippi	College of Agriculture, Mississippi State University, Mississippi State, MS 39762
Missouri	College of Agriculture, University of Missouri, Columbia, MO 65201
Montana	College of Agriculture, Montana State University, Bozeman, MT 59715
Nebraska	College of Agriculture, University of Nebraska, Lincoln, NE 68503
Nevada	The Max C. Fleischmann College of Agriculture, University of Nevada, Reno, NV 89507
New Hampshire	College of Life Sciences and Agriculture, University of New Hampshire, Durham, NH 03824
New Jersey	College of Agriculture and Environmental Science, Rutgers University, New Brunswick, NJ 08903
New Mexico	College of Agriculture and Home Economics, New Mexico State University, Las Cruces, NM 88003
New York	New York State College of Agriculture, Cornell University, Ithaca, NY 14850
North Carolina	School of Agriculture, North Carolina State University, Raleigh, NC 27607
North Dakota	College of Agriculture, North Dakota State University, State University Station, Fargo, ND 58102
Ohio	College of Agriculture and Home Economics, The Ohio State University, Columbus, OH 43210
Oklahoma	College of Agriculture and Applied Science, Oklahoma State University, Stillwater, OK 74074
Oregon	School of Agriculture, Oregon State University, Corvallis, OR 97331
Pennsylvania	College of Agriculture, The Pennsylvania State University, University Park, PA 16802
Puerto Rico	College of Agricultural Sciences, University of Puerto Rico, Mayaguez, PR 00708

(Continued)

TABLE 18–1 *(Continued)*

State	Address
Rhode Island	College of Resource Development, University of Rhode Island, Kingston, RI 02881
South Carolina	College of Agricultural Sciences, Clemson University, Clemson, SC 29631
South Dakota	College of Agriculture and Biological Sciences, South Dakota State University, Brookings, SD 57006
Tennessee	College of Agriculture, University of Tennessee, P.O. Box 1071, Knoxville, TN 37901
Texas	College of Agriculture, Texas A&M University, College Station, TX 77843
Utah	College of Agriculture, Utah State University, Logan, UT 84321
Vermont	College of Agriculture, University of Vermont, Burlington, VT 05401
Virginia	College of Agriculture, Virginia Polytechnic Institute and State University, Blacksburg, VA 24061
Washington	College of Agriculture, Washington State University, Pullman, WA 99163
West Virginia	College of Agriculture and Forestry, West Virginia University, Morgantown, WV 26506
Wisconsin	College of Agricultural and Life Sciences, University of Wisconsin, Madison, WI 53706
Wyoming	College of Agriculture, University of Wyoming, University Station, P.O. Box 3354, Laramie, WY 82070
Canada	**Address**
Alberta	University of Alberta, Edmonton, Alberta T6H 3K6
British Columbia	University of British Columbia, Vancouver, British Columbia V6T 1W5
Manitoba	University of Manitoba, Winnipeg, Manitoba R3T 2N2
New Brunswick	University of New Brunswick, Fredericton, New Brunswick E3B 4Z7
Ontario	University of Guelph, Guelph, Ontario N1G 2W1
Quebec	Faculty d'Agriculture, L'Universite Laval, Quebec City, Quebec G1K 7D4; and Macdonald College of McGill University, Ste. Anne de Bellevue, Quebec H9X 1C0
Saskatchewan	University of Saskatchewan, Saskatoon, Saskatchewan S7N 0W0

ORGANIZATIONS AND AGENCIES SERVING THE LIVESTOCK INDUSTRY

Listed herein are various organizations and agencies whose purpose is to assist farmers, ranchers, and related industries. Some of these are nationwide and are affiliated with separate regional or state organizations of similar kind and interest, and, in turn, some of the latter are organized into district or county groups.

AGRICULTURAL COUNCIL OF AMERICA

1250 "I" Street, N.W.
Washington, DC 20005-3922

AGRICULTURAL EDUCATION INSTRUCTOR

Those who teach FFA in high schools are known as *Agricultural Education Instructors* (changed from Vocational Agricultural Instructors in 1989). Their students are members of the FFA. Like the local county extension agent, the local agricultural education instructors are usually well acquainted with the agricultural problems and practices in their respective areas.

AGRICULTURAL STABILIZATION AND CONSERVATION

The U.S. Department of Agriculture maintains in each county (usually at the county seat) an office of the Agricultural Stabilization and Conservation Committee (ASC), which is directed by a committee of local farmers. They process the subsidy payments for the USDA. Also, for those who qualify, they may provide cost-sharing to farmers and ranchers (1) to improve the vegetative cover on their pasture land by means of artificial reseeding or by control of competitive shrubs; (2) to improve water penetration on soil by means of furrowing, chiseling, ripping, scarifying, or listing; or (3) to get better distribution of grazing by constructing wells, developing springs or seeps, installing pipelines for livestock water, or constructing permanent-type fences.

Cost sharing by the ASC usually amounts to their paying approximately 50% of the cost involved in performing the practice and providing the necessary engineering and other technical assistance required.

AGRISERVICES FOUNDATION

648 West Sierra Avenue
P.O. Box 429
Clovis, CA 93613

AMERICAN FARM BUREAU FEDERATION

225 W. Touhy Avenue
Park Ridge, IL 60068-4202

AMERICAN FEED INDUSTRY ASSOCIATION (AFIA)

1701 N. Ft. Myer Drive
Arlington, VA 22209

AMERICAN HORSE COUNCIL, INC. (AHC)

1700 "K" Street, N.W., Suite 300
Washington, DC 20006-3805

AMERICAN MEAT INSTITUTE (AMI)

P.O. Box 3556
Washington, DC 20007

AMERICAN SHEEP INDUSTRY ASSOCIATION

6911 S. Yosemite Street
Englewood, CA 80112-1414

AMERICAN SOCIETY OF AGRICULTURAL CONSULTANTS

8301 Greensboro Drive, Suite 260
McLean, VA 22102

AMERICAN SOCIETY OF ANIMAL SCIENCE

425 Illinois Building
113 N. Neil Street
Champaign, IL 61820

AMERICAN VETERINARY MEDICAL ASSOCIATION

930 North Meacham Road
Schaumburg, IL 60196-0001

CANADIAN DEPARTMENT OF AGRICULTURE

Sir John Carling Building
930 Carling Avenue
Ottawa, Ont. K1A 0C5 Canada

Canadian farmers and ranchers can receive assistance and publications by writing to the Canadian department of Agriculture at the address given above.

CANADIAN SOCIETY OF ANIMAL SCIENCE

151 Slater Street, Suite 907
Ottawa, Ont. K1P 5H4 Canada

COUNTY AGRICULTURAL AGENT

The county agricultural agent (or farm advisor), usually located in the County Court House, is qualified to give the latest findings of the state agricultural college, and the USDA, and to help in applying these findings to the individual farm or ranch.

DAIRY HERD IMPROVEMENT ASSOCIATION (DHIA)

This program, first adopted in 1926, is the most complete of all dairy production and record plans. More than 3 million cows in the United States on production test are on this program. Both registered and grade cows can be enrolled.

State and local Dairy Herd Improvement Associations (DHIA) conduct the program among dairy producers working through the Cooperative Extension Service in cooperation with the Federal Extension Service and the Animal Science Research Service of the USDA.

DAIRY HERD IMPROVEMENT REGISTRY (DHIR)

Each breed registry association has a program for testing registered cows under the Unified Rules for Official Testing as adopted by the Purebred Dairy Cattle Association and the American Dairy Science Association. This program is conducted cooperatively by the breed association and the Division of Dairy Herd Improvement Investigations of the U.S. Department of Agriculture (the division which, in cooperation with the states, is responsible for DHIR records). Records recognized as official by both groups are included in one program.

Under DHIR, milk testing is conducted once each month, with the tester obtaining a 24-hour milk weight and butterfat test. Also, the tester secures data on each cow that has freshened, including: feed consumption and quality, labor, price of milk, etc. All this data is sent to a central laboratory for analysis, with many states cooperating on a regional basis in the Electronic Data Processing Method (EDPM). The machine processed records are then returned to the producer, giving current information on milk yield, income over feed costs, milk produced per cow, and other pertinent information to help in making culling and managerial decisions.

In 1966–67, two breed registry production-testing programs—(a) Advanced Registry (AR), and (b) Herd Improvement Registry (HIR)—were discontinued, and the Dairy Herd Improvement Registry (DHIR) became the official milk-recording program of all the breeds.

DEPARTMENT OF VETERANS AFFAIRS

810 Vermont Avenue, N.W.
Washington, DC 20420

FARMERS' EDUCATIONAL AND COOPERATIVE UNION OF AMERICA

10065 E. Harvard Ave.
P.O. Box 2251
Denver, CO 80231-5964

FERTILIZER INSTITUTE

501 2nd St., N.E.
Washington, DC 20002-4916

4-H CLUBS

This is an organization of farm boys and girls, 9 through 19 years of age, under the sponsorship of the agricultural extension service of the land grant colleges. This organization is for the purpose of developing knowledge, leadership, and ability through agriculture, home economics, and community service projects. For further information about 4-H clubs, contact the local county agent.

INTERNATIONAL SILO ASSOCIATION, INC.

410 North Michigan Avenue
Chicago, IL 60611

LIVESTOCK CONSERVATION INSTITUTE (LCI)

6414 Copps Avenue, Suite 204
Madison, WI 53716

LIVESTOCK MARKETING ASSOCIATION

7509 Tiffany Springs Pkwy.
Kansas City, MO 64190

MIDWEST PLAN SERVICE

122 Davidson Hall
Iowa State University
Ames, IA 50011-3080

NATIONAL ACADEMY OF SCIENCES

National Research Council
2101 Constitution Avenue, N.W.
Washington, DC 20418

NATIONAL ASSOCIATION OF ANIMAL BREEDERS (NAAB)

P.O. Box 1033
Columbia, MO 65201

NATIONAL CATTLEMEN'S ASSOCIATION (NCA)

5420 S. Quebec Street
Englewood, CO 80111-1904

NATIONAL COUNCIL OF FARMER COOPERATIVES

50 "F" Street, N.W.
Washington, DC 20001-1530

NATIONAL CUTTING HORSE ASSOCIATION

4704 Highway 377 South
Fort Worth, TX 76116

NATIONAL FARMERS' ORGANIZATION (NFO)

720 Davis Avenue
Corning, IA 50841-0501

NATIONAL FFA ORGANIZATION (FFA)

5632 Mt. Vernon Memorial Hwy.
Alexandria, VA 22309-1502

This is an organization of farm boys and girls enrolled in agricultural education. Those who teach high school FFA are known as Agricultural Education Instructors (formerly Vocational Agricultural Instructors).

For further information about the FFA, contact the local agricultural education instructor or write to the National FFA Organization.

NATIONAL GRANGE

1616 "H" Street, N.W.
Washington, DC 20006-4999

NATIONAL LIVE STOCK & MEAT BOARD

444 N. Michigan Avenue
Chicago, IL 60611-3978

NATIONAL LIVESTOCK EXCHANGE

c/o Coburn, Endsley & Sedgwick
2577 Kentucky Ave.
Indianapolis, IN 46241

NATIONAL LIVE STOCK PRODUCERS ASSOCIATION

4851 Independence Street #200
Wheat Ridge, CO 80033

NATIONAL SWINE REPOPULATION ASSOCIATION

SPF Accreditation
P.O. Box 4405
Lincoln, NE 68504

SOIL CONSERVATION SERVICE (SCS)

The Soil Conservation Service is part of the U.S. Department of Agriculture. Technicians are specialists in soil and water management. They are qualified to determine contour lines and to make farm plans. In organized Soil Conservation Districts, facilities are available for classifying land according to its capabilities for use and resistance to erosion damage. From this information, a complete conservation program can be worked out for each farm or ranch, to be applied in large or small steps as desired.

Fig. 18–1. Soil survey mapping by USDA-SCS Soil Scientist Craig Busskhol. In organized Soil Conservation Districts, facilities are available for classifying land according to its capabilities for use and resistance to erosion damage. (Courtesy, USDA—Soil Conservation Service)

STATE DEPARTMENTS OF AGRICULTURE

Most states have a State Department of Agriculture, which renders various valuable services, including livestock sanitary and regulatory work. Usually the headquarters offices are at the respective state capitols.

U.S. DEPARTMENT OF AGRICULTURE (USDA)

Washington, DC 20250

The U.S. Department of Agriculture has a competent staff of specialists serving all the various branches of the nation's agriculture. Also, it operates a Research Center, with headquarters at Beltsville, Maryland, and numerous other laboratories and offices throughout the country.

A list of U.S. Department of Agriculture publications can be obtained by directing a request to the Office of Information, U.S. Department of Agriculture, Washington, DC 20250. Many of these publications are also available at the offices of county agricultural agents or through U.S. senators and representatives.

UNITED STATES ANIMAL HEALTH ASSOCIATION

P.O. Box 28176
Richmond, VA 23228

The primary responsibility of the United States Animal Health Association is to establish uniform methods and rules for the control of brucellosis.

WESTERN STATES MEAT ASSOCIATION (WSMA)

P.O. Box 12944
Oakland, CA 94604

POISON INFORMATION CENTERS

With the large number of chemical sprays, dusts, and gases now on the market for use in agriculture, accidents may arise because of operators being careless in their use. Also, there is always the hazard that a child may eat or drink something that may be harmful. Centers have been established in various parts of the country where doctors can obtain prompt and up-to-date information on treatment of such cases, if desired.

Local medical doctors have information relative to the Poison Information Centers of their area, along with some of the names of their directors, telephone numbers, and street numbers. When calling any of these centers, one should ask for the "Poison Information Center." If this information cannot be obtained locally, call the U.S. Public Health Service at Atlanta, Georgia; or Wenatchee, Washington.

Also, the *National Poison Control Center* is located at the University of Illinois, Urbana-Champaign. It is open 24 hours a day, every day of the week. The *hot line* number is: 217/333-3611. The toxicology group is staffed to answer questions about known or suspected cases of poisoning or chemical contaminations involving any species of animal. It is not intended to replace local veterinarians or state toxicology laboratories, but to complement them. Where consultation over the telephone is adequate, there is no charge to the veterinarian or producer. Where telephone consultation is inadequate or the problem is of major proportion, a team of veterinary specialists can arrive at the scene of a toxic or contamination problem within a short time. The cost of a personal visitation varies according to the distance traveled, personnel time, and laboratory services required.

SECTION 19

FEED COMPOSITION TABLES

Contents *Page*

In addition to the discussion that follows pertaining to composition of feeds, the reader is referred to Section 4, Feeding, of this book, under the headings: "Evaluating Feedstuffs" (pp. 142–144), "Feeding Standards" (p. 146), and "Balanced Rations" (pp. 147–154).

Both nutritionists and livestock producers should have access to accurate and up-to-date compositions of feedstuffs in order to formulate rations for maximum production and net returns. The ultimate goal of feedstuff analysis, and the reason for feed composition tables, is to be able to predict the productive responses of animals when they are fed rations of a given composition. At the outset, it was decided to utilize, to the extent available, the feed compositions which, for many years, were compiled by Lorin Harris, Utah State University, now carried forward by the USDA, National Agricultural Library, Feed Composition Bank. These data were augmented by the author with feed compositions from the National Academy of Sciences, NRC, and from experimental reports, industries, and other reliable sources.

To facilitate quick and easy use, the feeds in Section 19 are classified and separated into the following subtables:

1. **Table 19–1A, Energy Feeds.** This includes feeds which are high in energy and low in fiber (under 18%), and which generally contain less than 20% protein.
2. **Table 19–1B, Protein Feeds.** This includes feeds that contain more than 20% protein or protein equivalent.
3. **Table 19–1C, Dry Forages.** This includes feeds which are bulky, low in weight per unit volume, and relatively low in energy, and which in the dry state contain more than 18% crude fiber.
4. **Table 19–1D, Silages and Haylages.** Silage (ensilage) is fermentable high-moisture forage stored under anaerobic conditions in a silo, consisting of either green crops or crops to which moisture has been added, chopped when stored and containing 65 to 70% moisture.

   Haylages are low-moisture silages, made from grasses and/or legumes that are wilted to 40 to 55% moisture content before ensiling.
5. **Table 19–1E, Pasture and Range Plants.** This includes grass, browse, and other plants that are harvested by grazing animals.
6. **Table 19–1F, Mineral Supplements.** This includes rich natural and synthetic sources of inorganic elements needed to perform certain essential body functions.

7. **Table 19–1G, Vitamin Supplements.** This includes rich synthetic or natural feed sources of one or more complex compounds, called vitamins, that are required by animals in minute amounts for normal growth, production, reproduction, and/or health.

8. **Table 19–2, Amino Acids.** This gives the known amino acid composition of certain feeds.

9. **Table 19–3, Apparent Ileal Digestibility of Crude Protein and Essential Amino Acids in Feedstuffs for swine, and Digestible Amino Acid Recommendations for Swine Feed Formulation.** The protein requirement of the pig is primarily a requirement for essential amino acids. Currently, apparent ileal digestibility data give the best practical assessment of the availability of amino acids in various feed ingredients.

Some feeds fit the criteria of more than one of the above classes. For example, whole soybeans are used as both an energy feed and a protein feed; hence, they are listed in both Table 19–1A and Table 19–1B.

In Tables 19–1A to 19–1E, and 19–1G, covering 6 of the respective feedstuff classifications indicated above, values for each feed are presented in tabular form on 4 pages (2 double-page spreads), with one page devoted to each of the following categories:

Left-hand, proximate analysis
Right-hand, energy
Left-hand, minerals
Right-hand, vitamins

FEED NAMES

Ideally, a feed name should conjure up the same meaning to all those who use it, and it should provide helpful information. This was the guiding philosophy of the author when choosing the names given in the Feed Composition Tables. Genus and species—latin names—are also included. To facilitate worldwide usage, the International Feed Number of each feed is given. To the extent possible, consideration was also given to source (or parent material), variety or kind, stage of maturity, processing, part eaten, and grade.

Where feeds are known by more than one name, cross-referencing was used.

MOISTURE CONTENT OF FEEDS

It is necessary to know the moisture content of feeds in ration formulation and buying. Usually, the composition of a feed is expressed according to one or more of the following bases:

1. **As-fed; A-F (wet, fresh).** This refers to feed as normally fed to animals. As-fed may range from near 0% to 100% dry matter.

2. **Air-dry (approximately 90% dry matter).** This refers to feed that is dried by means of natural air movement, usually in the open. It may either be an actual or an assumed dry matter content; the latter is approximately 90%. Most feeds are fed in an air-dry state.

3. **Moisture-free; M-F (oven-dry, 100% dry matter).** This refers to a sample of feed that has been dried in an oven at 221°F until all the moisture has been removed.

CAROTENE

Where carotene has been converted to vitamin A, the conversion rate of the rat has been used as the standard value, with 1 mg of beta-carotene equal to 1,667 IU of vitamin A.

PERTINENT INFORMATION ABOUT DATA

The information which follows is pertinent to the feed composition tables presented in this section.

• **Variations in composition**—Feeds vary in their composition. Thus, actual analysis of a feedstuff should be obtained and used whenever possible, especially where a large lot of feed from one source is involved. Many times, however, it is either impossible to determine actual compositions or there is insufficient time to obtain such analysis. Under such circumstances, tabulated data may be the only information available.

• **Feed compositions change**—Feed compositions change over a period of time, primarily due to (1) the introduction of new varieties, and (2) modifications in the manufacturing process from which by-products evolve.

• **Biological value**—The response of animals when fed a feed is termed the biological value, which is a function of its chemical composition and the ability of the animal to derive useful nutrient value from the feed. The latter relates to the digestibility, or availability, of the nutrients in the feed. Biological tests of feeds are more laborious and costly than chemical analysis, but they are much more accurate in predicting the response of animals to a feed.

• **Where information is not available**—Where information is not available or reasonable estimates could not be made, no values are shown. Hopefully, such information will become available in the future.

• **Calculated on a dry matter (DM) basis**—All data were calculated on a 100% dry matter basis (moisture-free), then converted to an as-fed basis by multiplying the decimal equivalent of the DM content times the compositional value shown in the table.

• **Fiber**—Four values relating to dietary fiber are given in the feed composition tables—crude fiber, neutral detergent fiber (NDF), acid detergent fiber (ADF), and lignin.

Crude fiber, methods for the determination of which were developed more than 100 years ago, is declining as a measure of low digestible material in the more fibrous feeds. The newer method of forage analysis, developed by Van Soest and associates of the U.S. Department of Agriculture, separates feed dry matter into two fractions: a neutral detergent fibrous fraction; and an acid detergent fibrous fraction. Also the amount of lignin in the ADF may be determined.

1. **Crude fiber (CF).** This fraction is an indicator of the relative indigestibility and bulkiness of the sample. It is the residue that remains after boiling a feed in a weak acid, and then in a weak alkali, in an attempt to imitate the process that occurs in the digestive tract. This procedure is based on the supposition that carbohydrates which are readily dissolved also will be readily digested by animals, and that those not soluble under such conditions are not readily digested. Unfortunately, the treatment dissolves much of the lignin, a nondigestible component. Hence, crude fiber is only an approximation of the indigestible material in feedstuffs. Nevertheless, it is a rough indicator of the energy value of feeds. Also, the crude fiber value is needed for the computation of TDN.

2. **Neutral detergent fiber (NDF).** This is the fraction of the feed which is not soluble in neutral detergent. It consists of plant cell walls, including lignin, cellulose, and hemicellulose. NDF is closely related to feed intake because it contains all the fiber components that occupy space in the rumen and are slowly digested. The lower the NDF, the more forage the animal will eat; hence, a low percentage of NDF is desirable.

3. **Acid detergent fiber (ADF).** This is the fraction of the feed which is not soluble in acid detergent. It consists of cellulose (digestible) and lignin (indigestible). ADF is an indicator of forage digestibility because it contains a high proportion of lignin which is the indigestible fiber fraction. The lower the ADF, the more feed an animal can digest; hence, a low percentage of ADF is desirable.

4. **Lignin.** This fraction is essentially indigestible by all animals and is the substance that limits the availability of cellulose carbohydrates in the plant cell wall to rumen bacteria.

The acid detergent fiber procedure is used as a preparatory step in determining the lignin content of a forage sample. Hemicellulose is solubilized during this procedure, while the lignocellulose fraction of the feed remains insoluble. Cellulose is then separated from lignin by the addition of sulfuric acid. Only lignin and acid-insoluble ash remain upon completion of this step. This residue is then ashed, and the difference of the weights before and after ashing yields the amount of lignin present in the feed.

• **Nitrogen-free extract**—The nitrogen-free extract was calculated with mean data as: mean nitrogen-free extract (%) = 100 − % ash − % crude fiber − % ether extract − % protein.

• **Protein values**—Both crude protein and digestible protein values are given. Crude protein is determined by finding the nitrogen content and multiplying the result by 6.25. The nitrogen content of proteins averages about 16% (100 ÷ 16 = 6.25).

• **Ruminant values**—The ruminant values represent a pooling of cattle, sheep, and goat data.

• **Energy**—Many of the energy values given in the feed composition tables were derived from complex formulas developed by L. E. Harris and other animal scientists.

The following four measures of energy are shown:

1. **Total digestible nutrients (TDN).** This value is given because there are more of them, and because it has been the standard method of expressing the energy value of feeds for many years. However, the following disadvantages are inherent in the TDN system: (a) Only digestive losses are considered—it does not take into account other important losses, such as those in the urine, gases, and increased heat production; (b) there is a poor relationship between crude fiber and NFE digestibility in certain feeds; and (c) it overestimates roughages in relation to concentrates when animals are fed for high rates of production, due to the higher heat loss per pound of TDN in high-fiber feeds.

2. **Digestible energy (DE).** Digestible energy is that portion of the gross energy in a feed that is not excreted in the feces. It is roughly comparable to TDN.

3. **Metabolizable energy (ME).** Metabolizable energy represents that portion of the gross energy that is not lost in the feces, urine, and gas (mainly methane). It does not take into account the energy lost as heat, commonly called heat increment. As a result, it overevaluates roughages compared with concentrates, as do TDN and DE.

4. **Net energy (NE).** Net energy represents the energy fraction in a feed that is left after the fecal, urinary, gas, and heat losses are deducted from the GE. Because of its greater accuracy, net energy is being used increasingly in ration formulations, especially in computerized formulations for large operations. However, net energy is difficult and expensive to determine.

Two systems of net energy evaluation are presently used: (a) net energy for maintenance (NE_m) and net energy for gain (NE_g), and (b) net energy for lactation (NE_{lc}).

Note to dairy producers and nutritionists: In *Nutrient Requirements of Dairy Cattle,* Sixth Revised Edition, 1988, Table 7–1, Composition of Feeds Commonly Used in Dairy Cattle Diets on a 100% Dry Matter Basis, the NRC committee assumed an average decrease of 4% per unit of dry matter intake above maintenance in calculating NE_{lc} values for feed ingredients, or an average discount of 8% based on their assumption that lactating cows are fed at 3X maintenance.

• **Minerals**—The level of minerals in forages is largely determined by the mineral content of the soil on which the feeds are grown. Calcium, phosphorus, iodine, and selenium are well-known examples of soil nutrient—plant nutrient relationships.

• **Vitamins**—Generally speaking, it is unwise to rely on harvested feeds as a source of carotene (vitamin A value), unless the forage being fed is fresh (pasture or green chop) or of a good green color and not over a year old.

The author is very grateful to Lorin E. Harris, Ph.D., and Clyde R. Richards, Ph.D., Utah State University, Logan, for their interest and invaluable assistance in preparing the Feed Composition Tables for this book. Also, in the preparation of Table 19–1F, Composition of Mineral Supplements, the authors acknowledge with appreciation the review and added values provided by International Minerals and Chemical Corporation, Northbrook, Illinois; and R. F. Klay, Ph.D., Research Department, Moorman Mfg. Co., Quincy, Illinois.

TABLE 19–1A ENERGY FEEDS, COMPOSITION OF FEEDS, DATA EXPRESSED **AS-FED**

Entry Number	Feed Name Description	International Feed Number	Proximate Analysis: Dry Matter	Ash	Crude Fiber	Neutral Det. Fib. (NDF)	Acid Det. Fib. (ADF)	Lignin	Ether Extract (Fat)	N-Free Extract	Crude Protein	Digestible Protein: Ruminant	Swine	Horse
			%	%	%	%	%	%	%	%	%	%	%	%
1	**ALMOND,** HULLS	4-00-359	90	6.1	13.5	28.8	25.2	–	2.9	63.8	4.1	0.0	2.0	2.4
	ANIMAL													
2	FAT, HYDROLYZED	5-00-376	99	–	–	–	–	–	98.4	–	–	–	–	–
3	TALLOW	4-08-127	97	0.1	–	–	–	–	96.8	–	1.5	–1.6	–0.4	0.3
4	**ANIMAL–POULTRY,** FAT	4-00-409	99	–	–	–	–	–	99.1	–	–	–	–	–
	BARLEY													
5	GRAIN, ALL ANALYSES	4-00-549	88	2.4	5.0	16.8	10.7	1.5	1.7	67.7	11.7	8.8	9.6	9.6
6	GRAIN, PACIFIC COAST	4-07-939	89	2.5	6.5	–	–	–	2.0	68.2	9.5	7.1	6.9	6.7
7	MALT SPROUTS, DEHY	4-00-545	93	5.6	14.2	–	–	–	1.4	48.8	22.9	19.2	21.5	17.2
	BEAN													
8	SEEDS, KIDNEY	5-00-600	89	3.6	4.1	–	–	–	1.2	57.8	21.8	14.6	20.6	16.5
9	SEEDS, NAVY	5-00-623	89	4.0	4.4	–	–	–	1.4	56.7	22.9	20.2	21.6	17.6
10	SEEDS, PINTO	5-00-624	90	4.3	4.0	–	–	–	1.3	57.9	22.7	17.2	21.4	17.3
11	**BEAN, LIMA,** SEEDS	5-00-613	90	4.1	4.5	–	–	–	1.3	58.9	20.7	16.7	19.4	15.1
12	**BEAN, MUNG,** SEEDS	5-08-185	90	3.8	3.9	–	–	–	1.3	57.1	23.9	19.6	22.6	18.7
13	**BEAN, TEPARY,** SEEDS	5-08-349	90	4.2	3.4	–	–	–	1.4	59.3	22.2	18.0	20.9	16.7
	BEET, SUGAR													
14	MOLASSES, MORE THAN 48% INVERT SUGAR, MORE THAN 79.5 DEGREES BRIX	4-00-668	78	8.9	–	–	–	–	0.2	62.2	6.6	3.6	4.5	4.5
15	PULP, DEHY	4-00-669	91	4.8	18.2	53.6	26.3	4.5	0.5	58.4	8.8	4.3	3.6	6.2
16	PULP WITH MOLASSES, DEHY	4-00-672	92	5.7	15.2	–	24.5	2.4	0.6	61.1	9.3	6.1	2.3	6.5
17	**BREWERS GRAINS,** DEHY	5-02-141	92	3.6	13.0	38.7	23.9	4.6	6.6	41.6	27.3	20.1	21.8	21.0
18	**BUCKWHEAT, COMMON,** GRAIN	4-00-994	88	2.1	10.6	–	14.9	–	2.4	61.5	11.1	8.0	8.5	7.2
19	**CARROT,** ROOTS, FRESH	4-01-145	11	1.0	1.1	1.0	0.9	–	0.2	8.1	1.2	0.8	0.8	0.8
20	**CASSAVA,** TUBERS, FRESH	4-01-150	32	1.3	1.5	–	–	–	0.3	28.2	1.2	–0.1	0.5	0.6
	CITRUS													
21	PULP WITHOUT FINES, DEHY (DRIED CITRUS PULP)	4-01-237	91	6.0	11.6	20.9	20.0	–	3.4	63.9	6.1	2.7	3.8	4.0
22	SYRUP (MOLASSES)	4-01-241	67	5.1	–	–	–	–	0.2	55.7	5.8	2.0	3.9	3.9
	CORN, DENT YELLOW													
23	GRAIN, ALL ANALYSES	4-02-935	88	1.3	2.3	–	3.8	–	3.6	71.0	9.9	8.1	7.3	7.0
24	GRAIN, GRADE 2, 54 lb/bu *(69.5 kg/hl)*	4-02-931	87	1.2	2.1	–	–	–	4.0	71.3	8.9	6.8	7.0	6.2
25	GRAIN, HIGH MOISTURE	4-20-770	77	1.2	2.1	17.5	3.8	1.3	3.3	61.9	8.1	4.9	5.9	5.7
26	DISTILLERS GRAINS, DEHY	5-02-842	93	2.2	11.5	40.0	15.8	–	8.9	43.1	27.8	20.1	19.4	22.7
27	DISTILLERS SOLUBLES, DEHY	5-28-237	93	7.2	4.6	21.4	6.5	–	8.6	45.2	27.4	21.5	20.0	22.3
28	EARS, GROUND (CORN AND COB MEAL)	4-28-238	87	1.7	8.2	24.4	9.6	–	3.2	65.7	7.8	4.7	5.6	5.4
29	GLUTEN, MEAL, 60% PROTEIN	5-28-242	90	1.7	1.8	12.6	4.5	–	2.1	23.7	60.8	54.5	–1.3	–0.4
30	GLUTEN FEED	5-28-243	90	6.6	8.7	40.5	10.8	–	2.1	49.4	23.0	19.8	18.2	17.7
31	GRITS (HOMINY GRITS)	4-03-011	90	2.8	4.8	49.5	11.7	–	6.5	65.8	10.3	7.3	7.6	7.3
32	OIL	4-16-450	99	–	–	–	–	–	99.0	–	–	–	–	–
33	**CORN, OPAQUE 2, HIGH LYSINE,** GRAIN	4-11-445	90	1.6	3.0	–	–	–	4.4	71.0	10.1	6.3	7.4	7.2
34	**CORN, SWEET,** CANNERY RESIDUE, FRESH	2-02-975	77	3.5	17.0	–	22.3	–	1.9	47.6	6.8	3.9	4.1	–
35	**COTTON,** SEEDS	5-13-749	91	4.7	19.4	40.0	30.9	–	20.4	24.5	21.8	17.7	20.5	16.2
	DISTILLERS PRODUCTS (ALSO SEE CORN; RYE; SORGHUM; WHEAT)													
36	GRAINS, DEHY	5-02-144	93	1.5	12.8	–	–	–	7.4	43.5	27.3	18.9	26.0	22.3
37	SOLUBLES, DEHY	5-02-147	92	6.2	3.4	–	–	–	8.9	44.8	28.8	24.0	27.4	24.0
	FATS AND OILS													
38	FAT, ANIMAL, HYDROLYZED	4-00-376	99	–	–	–	–	–	98.4	–	–	–	–	–
39	FAT, ANIMAL-POULTRY	4-00-409	99	–	–	–	–	–	99.1	–	–	–	–	–
40	**FLAX, COMMON,** SEEDS	5-02-042	96	5.0	6.3	–	–	–	35.9	30.7	17.8	13.9	16.5	11.2
41	**GARBAGE, HOTEL AND RESTAURANT,** BOILED, WET	4-07-865	23	1.3	0.7	–	–	–	5.4	11.9	3.6	2.5	2.8	2.6
42	**HOMINY GRITS (CORN, DENT YELLOW, GRITS)**	4-03-011	90	2.8	4.8	–	–	–	6.5	65.8	10.3	7.3	7.6	7.3
43	**KAFIR SORGHUM,** GRAIN	4-04-428	89	1.5	2.0	–	1.8	–	2.8	72.0	10.8	6.6	8.3	7.7
44	**KELP (SEAWEED),** WHOLE, DEHY	1-08-073	91	35.0	6.5	–	–	–	0.5	42.4	6.5	2.9	2.2	3.4
45	**LARD (FAT),** SWINE	4-04-790	99	–	–	–	–	–	99.3	–	–	–	–	–
46	**MANGEL, BEET,** ROOTS, FRESH	4-00-637	11	1.1	0.8	–	–	–	0.1	7.7	1.3	0.9	1.0	0.9
47	**MANURE, CATTLE,** WITHOUT BEDDING, DEHY	1-01-190	93	17.9	30.7	–	43.8	–	2.5	30.1	12.0	7.7	6.9	7.6
48	**MILLET,** GRAIN	4-03-098	90	2.7	5.8	–	–	–	4.0	65.2	12.1	6.8	8.8	8.7
49	**MILLET, FOXTAIL,** GRAIN	4-03-102	89	3.4	7.4	–	–	–	4.1	63.0	11.4	7.5	8.6	8.2
50	**MILO SORGHUM,** GRAIN	4-04-444	89	1.6	2.2	20.6	4.7	–	2.8	71.9	10.1	6.8	7.2	7.1
	MOLASSES AND SYRUP													
51	BEET, SUGAR, MOLASSES, MORE THAN 48% INVERT SUGAR, MORE THAN 79.5 DEGREES BRIX	4-00-668	78	8.9	–	–	–	–	0.2	62.2	6.6	3.6	4.5	4.5
52	CITRUS, SYRUP (CITRUS MOLASSES)	4-01-241	67	5.1	–	–	–	–	0.2	55.7	5.8	2.0	3.9	3.9
53	SUGAR CANE, MOLASSES, DEHY	4-04-695	94	12.5	6.3	–	–	–	0.9	65.0	9.7	5.3	7.0	6.8
54	SUGAR CANE, MOLASSES, MORE THAN 46% INVERT SUGAR, MORE THAN 79.5 DEGREES BRIX (BLACKSTRAP)	4-04-696	74	9.8	0.4	–	0.3	0.2	0.2	59.7	4.3	0.6	1.3	2.7
55	WOOD, MOLASSES	4-05-502	62	4.1	0.5	–	1.2	–	0.3	56.6	0.6	–1.3	–0.6	–0.1
	OATS													
56	GRAIN, ALL ANALYSES	4-03-309	89	3.1	10.7	26.4	14.2	2.7	4.7	58.9	11.9	9.2	9.7	9.1
57	CEREAL BY-PRODUCT (FEEDING OAT MEAL; OAT MIDDLINGS)	4-03-303	91	2.3	3.6	–	–	–	6.4	63.7	14.8	10.9	11.6	10.8
58	GROATS	4-03-331	90	2.1	2.5	–	–	–	6.2	63.0	15.8	11.1	13.7	11.6
59	**PEA,** SPLIT SEED BY-PRODUCT (PEA FEED; PEA MEAL)	1-08-478	90	3.5	23.7	–	–	–	1.4	43.7	17.7	14.5	12.1	12.0
60	**PEA, FIELD,** SEEDS	5-08-481	91	2.9	5.9	–	–	–	1.3	57.8	23.2	19.9	21.8	17.7
61	**PEARL MILLET,** GRAIN	4-03-118	90	2.2	3.7	–	–	–	4.3	67.2	13.0	8.9	7.9	9.4
62	**PINEAPPLE,** CANNERY RESIDUE, DEHY (PINEAPPLE BRAN)	4-03-722	87	3.0	18.2	63.5	32.2	–	1.3	60.5	4.0	0.7	2.0	2.4
	POTATO													
63	TUBERS, FRESH	4-03-787	24	1.1	0.6	–	–	–	0.1	19.5	2.2	1.4	0.7	1.5
64	TUBERS, BOILED	4-03-784	24	1.3	0.7	–	–	–	0.1	19.6	2.2	0.3	1.5	1.5
	RICE													
65	GRAIN, GROUND (GROUND ROUGH RICE; GROUND PADDY RICE)	4-03-938	89	5.3	8.6	–	–	–	1.6	65.9	7.5	4.0	5.1	5.1
66	BRAN WITH GERMS (RICE BRAN)	4-03-928	91	11.3	11.9	28.0	25.7	3.6	13.5	41.0	13.0	8.6	9.5	9.4
67	GROATS, POLISHED (RICE, POLISHED)	4-03-942	89	0.5	0.4	14.2	0.9	–	0.5	80.3	7.0	3.6	5.9	4.7
68	GROATS (RICE, BROWN)	4-03-936	88	1.0	0.8	–	–	–	1.8	77.2	7.4	3.2	5.1	5.1
69	POLISHINGS	4-03-943	90	7.6	3.2	–	3.6	–	12.6	54.9	12.0	8.6	10.1	8.6

Entry Number	TDN Ruminant	TDN Swine	TDN Horse	Digestible Energy Ruminant		Digestible Energy Swine		Digestible Energy Horse		Metabolizable Energy Ruminant		Metabolizable Energy Swine		Metabolizable Energy Poultry MEn		Metabolizable Energy Horse		Net Energy Ruminant NEm		Net Energy Ruminant NEg		Net Energy Lactating Cows NElc	
	%	%	%	Mcal lb	Mcal kg	kcal lb	kcal kg	Mcal lb	Mcal kg	Mcal lb	Mcal kg	kcal lb	kcal kg	kcal lb	kcal kg	Mcal lb	Mcal kg	Mcal lb	Mcal kg	Mcal lb	Mcal kg	Mcal lb	Mcal kg
1	66	73	45	1.34	2.95	1457	3213	0.86	1.89	1.15	2.54	1375	3032	–	–	0.70	1.55	0.77	1.70	0.51	1.12	0.66	1.46
2	223	209	–	4.46	9.84	4144	9135	–	–	4.31	9.49	3834	8452	3685	8125	–	–	2.95	6.49	2.23	4.91	2.43	5.35
3	203	–	–	4.07	8.97	–	–	–	–	3.91	8.62	–	–	3304	7285	–	–	2.60	5.73	1.95	4.29	2.21	4.87
4	188	196	–	3.57	7.87	3750	8267	–	–	3.40	7.51	3617	7973	3482	7677	–	–	2.65	5.85	1.99	4.38	2.03	4.48
5	75	70	73	1.55	3.42	1396	3078	1.31	2.88	1.17	2.57	1331	2934	1180	2602	1.07	2.36	0.78	1.73	0.52	1.16	0.82	1.81
6	75	70	66	1.51	3.32	1405	3097	1.20	2.64	1.34	2.96	1324	2918	1171	2582	0.98	2.16	0.81	1.77	0.54	1.20	0.76	1.67
7	66	61	–	1.31	2.89	1219	2688	–	–	1.13	2.50	1046	2307	722	1592	–	–	0.69	1.52	0.43	0.95	0.67	1.48
8	73	79	–	1.46	3.22	1584	3493	–	–	1.29	2.85	1388	3059	–	–	–	–	–	–	–	–	–	–
9	76	78	–	1.53	3.37	1737	3830	–	–	1.36	3.00	1531	3376	1052	2320	–	–	0.83	1.83	0.56	1.24	0.78	1.72
10	74	78	–	1.47	3.25	1569	3460	–	–	1.30	2.87	1372	3025	–	–	–	–	0.81	1.79	0.54	1.20	0.77	1.69
11	78	78	–	1.55	3.42	1556	3430	–	–	1.39	3.05	1360	2999	–	–	–	–	–	–	–	–	–	–
12	79	81	–	1.53	3.37	1612	3554	–	–	1.36	3.00	1412	3113	–	–	–	–	0.82	1.80	0.55	1.21	0.77	1.70
13	69	80	–	1.37	3.02	1606	3542	–	–	1.20	2.65	1406	3100	–	–	–	–	–	–	–	–	–	–
14	61	57	–	1.20	2.64	1130	2491	–	–	1.04	2.29	1061	2338	875	1929	–	–	0.70	1.54	0.47	1.04	0.65	1.43
15	67	67	59	1.31	2.89	1305	2878	1.09	2.40	1.10	2.43	1225	2700	294	648	0.89	1.97	0.73	1.60	0.47	1.03	0.70	1.55
16	69	70	62	1.39	3.05	1390	3064	1.13	2.50	1.21	2.67	1307	2881	299	660	0.93	2.05	0.73	1.61	0.47	1.04	0.71	1.55
17	65	66	48	1.25	2.76	1045	2303	–	–	1.01	2.22	1038	2288	1047	2308	–	–	0.64	1.41	0.39	0.86	0.67	1.48
18	63	69	62	1.26	2.77	1377	3036	1.13	2.50	1.09	2.40	1297	2859	1200	2645	0.93	2.05	0.65	1.43	0.41	0.90	0.63	1.40
19	10	10	8	0.19	0.43	208	458	0.14	0.31	0.17	0.38	195	430	208	458	0.12	0.26	0.10	0.23	0.07	0.16	0.10	0.22
20	26	26	24	0.51	1.13	511	1127	0.43	0.95	0.45	0.99	482	1063	–	–	0.35	0.78	0.28	0.61	0.18	0.41	0.26	0.58
21	75	46	48	1.41	3.10	1386	3055	–	–	1.12	2.46	1097	2420	606	1336	–	–	0.74	1.63	0.48	1.06	0.77	1.71
22	51	54	–	1.01	2.22	1084	2390	–	–	0.86	1.89	1022	2253	–	–	–	–	0.57	1.26	0.38	0.83	0.52	1.15
23	80	79	64	1.63	3.59	1514	3338	1.17	2.57	1.26	2.78	1472	3246	1523	3359	0.96	2.11	0.86	1.90	0.59	1.31	0.82	1.81
24	80	81	61	1.57	3.47	1586	3498	1.11	2.45	1.45	3.19	1526	3365	1567	3456	0.91	2.01	0.89	1.96	0.61	1.35	0.84	1.85
25	71	68	54	1.41	3.11	1354	2984	0.99	2.18	1.27	2.80	1282	2826	–	–	0.81	1.79	0.79	1.75	0.55	1.22	0.74	1.63
26	81	65	–	1.54	3.41	1246	2746	–	–	1.28	2.82	1196	2636	894	1970	–	–	0.87	1.92	0.59	1.30	0.84	1.86
27	80	78	–	1.55	3.42	1474	3250	–	–	1.34	2.96	1419	3128	1324	2919	–	–	0.92	2.03	0.63	1.40	0.86	1.89
28	72	69	56	1.45	3.20	1410	3109	1.03	2.28	1.26	2.78	1260	2779	1238	2730	0.85	1.87	0.86	1.90	0.60	1.31	0.76	1.68
29	–	–	–	–	–	–	–	–	–	–	–	–	–	3370	7431	–	–	–	–	–	–	–	–
30	75	75	–	1.44	3.17	1375	3031	–	–	1.21	2.67	1037	2287	784	1729	–	–	0.82	1.80	0.55	1.21	0.78	1.72
31	84	82	49	1.69	3.73	1620	3571	0.91	2.01	1.29	2.85	1533	3381	1313	2894	0.75	1.65	0.88	1.95	0.61	1.34	0.91	2.00
32	203	208	–	3.57	7.86	3421	7543	–	–	3.40	7.50	3438	7579	3917	8635	–	–	2.65	5.84	1.99	4.38	2.20	4.86
33	80	79	61	1.53	3.38	1643	3623	1.12	2.47	1.37	3.01	1557	3432	1527	3367	0.92	2.03	0.81	1.78	0.54	1.20	0.76	1.68
34	54	–	42	1.05	2.31	–	–	0.79	1.74	0.90	1.99	–	–	–	–	0.65	1.43	0.57	1.26	0.36	0.79	0.56	1.23
35	87	71	–	1.57	3.46	1430	3152	–	–	1.39	3.07	1243	2739	–	–	–	–	0.97	2.13	0.67	1.49	0.91	2.01
36	78	82	–	1.61	3.54	1637	3609	–	–	1.43	3.16	1433	3160	1316	2901	–	–	0.93	2.06	0.65	1.43	0.87	1.92
37	68	75	31	1.39	3.07	1493	3291	0.62	1.37	1.22	2.69	1409	3107	–	–	0.51	1.12	0.77	1.70	0.51	1.12	0.74	1.63
38	223	209	–	4.46	9.84	4144	9135	–	–	4.31	9.49	3834	8452	3685	8125	–	–	2.95	6.49	2.23	4.91	2.43	5.35
39	188	196	–	3.57	7.87	3750	8267	–	–	3.40	7.51	3617	7973	3482	7677	–	–	2.65	5.85	1.99	4.38	2.03	4.48
40	–	–	–	–	–	–	–	–	–	–	–	–	–	–	–	–	–	–	–	–	–	–	–
41	20	25	–	0.42	0.92	501	1105	–	–	0.38	0.83	478	1053	–	–	–	–	0.24	0.53	0.17	0.37	0.22	0.49
42	84	82	49	1.69	3.73	1620	3571	0.91	2.01	1.29	2.85	1533	3381	1313	2894	0.75	1.65	0.88	1.95	0.61	1.34	0.91	2.00
43	75	81	70	1.47	3.24	1521	3354	1.26	2.77	1.30	2.87	1437	3169	1529	3372	1.03	2.27	0.79	1.74	0.53	1.16	0.75	1.64
44	29	–	–	0.58	1.27	–	–	–	–	0.40	0.87	–	–	–	–	–	–	–	–	–	–	–	–
45	–	–	–	–	–	3669	8089	–	–	–	–	3537	7798	3936	8677	–	–	–	–	–	–	–	–
46	9	9	8	0.18	0.39	181	399	0.14	0.32	0.15	0.34	171	377	–	–	0.12	0.26	0.09	0.21	0.06	0.14	0.09	0.20
47	44	–	25	0.88	1.94	1281	2825	0.53	1.17	0.70	1.54	1222	2695	–	–	0.43	0.96	0.36	0.79	0.13	0.29	0.43	0.96
48	61	66	61	1.54	3.39	1315	2900	1.11	2.46	1.17	2.58	1224	2697	1441	3178	0.91	2.01	0.86	1.90	0.59	1.29	0.81	1.79
49	76	72	57	1.45	3.19	1436	3167	1.04	2.30	1.28	2.82	1355	2988	–	–	0.86	1.89	0.75	1.65	0.49	1.09	0.71	1.58
50	76	77	68	1.41	3.12	1520	3350	1.23	2.71	1.13	2.49	1474	3250	1467	3234	1.01	2.23	0.75	1.66	0.50	1.09	0.74	1.62
51	61	57	–	1.20	2.64	1130	2491	–	–	1.04	2.29	1061	2338	875	1929	–	–	0.70	1.54	0.47	1.04	0.65	1.43
52	51	54	–	1.01	2.22	1084	2390	–	–	0.86	1.89	1022	2253	–	–	–	–	0.57	1.26	0.38	0.83	0.52	1.15
53	66	70	61	1.36	2.99	1208	2663	1.12	2.48	1.18	2.60	1127	2485	1227	2706	0.92	2.03	0.70	1.55	0.44	0.97	0.68	1.51
54	60	56	–	1.22	2.68	1135	2502	–	–	1.12	2.46	995	2194	870	1918	–	–	0.77	1.70	0.53	1.18	0.64	1.41
55	52	51	–	1.03	2.27	1033	2278	–	–	0.91	2.02	976	2151	–	–	–	–	–	–	–	–	–	–
56	69	65	65	1.36	3.00	1278	2818	1.19	2.63	1.19	2.62	1026	2263	1150	2536	0.98	2.15	0.80	1.77	0.54	1.19	0.70	1.55
57	86	77	57	1.72	3.79	1641	3618	1.04	2.30	1.55	3.42	1554	3426	1432	3158	0.86	1.89	0.95	2.10	0.67	1.47	0.89	1.95
58	87	84	59	1.72	3.80	1410	3108	1.08	2.39	1.46	3.21	1328	2928	1475	3251	0.89	1.96	1.02	2.24	0.72	1.58	0.88	1.94
59	78	55	62	1.34	2.96	–	–	1.13	2.49	1.17	2.59	–	–	–	–	0.93	2.04	0.57	1.26	0.33	0.74	0.58	1.28
60	76	82	–	1.51	3.34	1638	3611	–	–	1.34	2.96	1435	3163	1108	2442	–	–	0.80	1.77	0.54	1.18	0.76	1.68
61	65	65	64	1.36	3.01	1351	2977	1.16	2.57	0.99	2.19	1269	2799	1155	2546	0.96	2.11	0.63	1.39	0.39	0.85	0.67	1.48
62	64	61	50	1.27	2.81	1223	2696	0.93	2.04	1.11	2.45	1147	2530	–	–	0.76	1.67	0.70	1.55	0.46	1.00	0.67	1.49
63	19	20	19	0.38	0.84	398	878	0.35	0.77	0.34	0.74	377	830	–	–	0.29	0.63	0.20	0.45	0.14	0.30	0.19	0.43
64	17	21	19	0.33	0.73	416	918	0.35	0.77	0.29	0.63	394	869	–	–	0.29	0.63	0.16	0.35	0.10	0.21	0.16	0.35
65	68	72	59	1.35	2.98	1492	3290	1.09	2.39	1.18	2.61	1409	3107	1210	2668	0.89	1.96	0.71	1.57	0.46	1.01	0.68	1.51
66	64	69	–	1.10	2.42	1474	3250	–	–	0.99	2.18	1347	2971	920	2028	–	–	0.62	1.38	0.38	0.84	0.53	1.18
67	78	86	–	1.59	3.51	1697	3741	–	–	1.44	3.17	1658	3656	1399	3085	–	–	0.90	1.98	0.62	1.37	0.85	1.87
68	78	–	–	1.56	3.43	1666	3674	–	–	1.39	3.07	1642	3619	–	–	–	–	0.83	1.82	0.56	1.23	0.78	1.71
69	81	88	–	1.56	3.45	1684	3713	–	–	1.43	3.16	1555	3428	1367	3015	–	–	0.88	1.93	0.60	1.32	0.83	1.82

(Continued)

TABLE 19–1A ENERGY FEEDS, MINERAL AND VITAMIN COMPOSITION OF FEEDS, DATA EXPRESSED **AS-FED**

			Macro Minerals							Micro Minerals						
Entry Number	Feed Name Description	Dry Matter	Calcium (Ca)	Phos-phorus (P)	Sodium (Na)	Chlo-rine (Cl)	Mag-nesium (Mg)	Potas-sium (K)	Sulfur (S)	Cobalt (Co)	Copper (Cu)	Iodine (I)	Iron (Fe)	Man-ganese (Mn)	Sele-nium (Se)	Zinc (Zn)
		%	%	%	%	%	%	%	%	ppm or mg/kg	ppm or mg/kg	ppm or mg/kg	%	ppm or mg/kg	ppm or mg/kg	ppm or mg/kg
1	**ALMOND,** HULLS	90	0.19	0.09	–	–	–	0.48	0.10	–	–	–	–	–	–	–
	ANIMAL															
2	FAT, HYDROLYZED	99	–	–	–	–	–	–	–	–	–	–	–	–	–	–
3	TALLOW	97	–	–	–	–	–	–	–	–	–	–	–	–	–	–
4	**ANIMAL-POULTRY,** FAT	99	–	–	–	–	–	0.23	–	–	–	–	–	–	–	–
	BARLEY															
5	GRAIN, ALL ANALYSES	88	0.05	0.34	0.03	0.12	0.13	0.46	0.15	0.171	7.6	0.044	0.008	16.0	0.158	39.3
6	GRAIN, PACIFIC COAST	89	0.05	0.34	0.02	0.15	0.12	0.51	0.14	0.087	8.1	–	0.009	16.0	0.101	15.2
7	MALT SPROUTS, DEHY	93	0.18	0.63	0.88	0.36	0.17	0.25	0.79	–	5.9	–	0.018	29.4	0.416	56.4
	BEAN															
8	SEEDS, KIDNEY	89	0.11	0.43	0.01	–	0.10	0.93	–	0.508	6.1	–	0.007	17.0	0.355	24.3
9	SEEDS, NAVY	89	0.17	0.54	0.04	0.06	0.13	1.31	0.23	–	9.9	–	0.010	21.1	–	–
10	SEEDS, PINTO	90	0.18	0.48	0.01	–	0.24	2.59	–	0.551	13.0	–	0.014	17.3	–	40.0
11	**BEAN, LIMA,** SEEDS	90	0.08	0.38	0.02	0.03	0.18	1.61	0.20	–	8.2	–	0.009	16.1	–	–
12	**BEAN, MUNG,** SEEDS	90	0.13	0.35	0.01	–	–	1.04	–	–	–	–	0.009	–	–	–
13	**BEAN, TEPARY,** SEEDS	90	–	–	–	–	–	–	–	–	–	–	–	–	–	–
	BEET, SUGAR															
14	MOLASSES, MORE THAN 48% INVERT SUGAR, MORE THAN 79.5 DEGREES BRIX	78	0.12	0.03	1.16	1.28	0.23	4.73	0.46	0.362	16.8	–	0.007	4.5	–	14.0
15	PULP, DEHY	91	0.63	0.09	0.19	0.04	0.26	0.18	0.20	0.074	12.5	–	0.027	34.2	–	0.7
16	PULP WITH MOLASSES, DEHY	92	0.56	0.09	0.48	–	0.15	1.63	0.39	0.209	14.7	–	0.017	18.4	–	5.1
17	**BREWERS GRAINS,** DEHY	92	0.30	0.51	0.21	0.15	0.15	0.09	0.30	0.076	21.7	0.066	0.024	37.2	–	27.3
18	**BUCKWHEAT, COMMON,** GRAIN	88	0.10	0.33	0.05	0.04	0.10	0.45	0.14	0.049	9.5	–	0.005	33.7	–	8.8
19	**CARROT,** ROOTS, FRESH	11	0.05	0.04	0.06	0.06	0.02	0.32	0.02	–	1.2	–	0.002	3.6	–	–
20	**CASSAVA,** TUBERS, FRESH	32	0.05	0.05	–	–	–	0.33	–	–	–	–	–	–	–	–
	CITRUS															
21	PULP WITHOUT FINES, DEHY (DRIED CITRUS PULP)	91	1.69	0.12	0.07	–	0.15	0.71	0.17	0.169	5.0	–	0.033	6.6	–	13.7
22	SYRUP (MOLASSES)	67	1.18	0.09	0.28	0.07	0.14	0.09	0.16	0.109	72.8	–	0.035	40.9	–	92.4
	CORN, DENT YELLOW															
23	GRAIN, ALL ANALYSES	88	0.05	0.28	0.01	0.05	0.11	0.33	0.11	0.378	3.5	–	0.004	5.7	0.127	19.4
24	GRAIN, GRADE 2, 54 lb/bu *(69.5 kg/hl)*	87	0.02	0.29	0.02	0.04	0.11	0.31	0.12	0.029	3.8	–	0.003	5.3	–	13.7
25	GRAIN, HIGH MOISTURE	77	0.01	0.25	0.01	0.04	0.11	0.28	0.11	–	2.2	–	0.003	5.3	–	25.4
26	DISTILLERS GRAINS, DEHY	93	0.09	0.39	0.09	0.07	0.07	0.16	0.43	0.076	38.9	0.048	0.020	19.3	0.352	41.7
27	DISTILLERS SOLUBLES, DEHY	93	0.30	1.30	0.23	0.26	0.60	1.70	0.37	0.167	77.9	0.079	0.052	72.0	0.371	88.0
28	EARS, GROUND (CORN AND COB MEAL)	87	0.06	0.24	0.02	0.04	0.12	0.46	0.14	0.273	6.8	0.023	0.008	19.9	0.074	12.1
29	GLUTEN, MEAL, 60% PROTEIN	90	0.07	0.45	0.05	0.09	0.08	0.18	0.65	0.045	26.1	0.018	0.023	6.3	0.829	30.6
30	GLUTEN FEED	90	0.32	0.74	0.12	0.22	0.33	0.57	0.21	0.087	47.1	0.066	0.043	23.1	0.272	64.6
31	GRITS (HOMINY GRITS)	90	0.05	0.51	0.08	0.05	0.24	0.59	0.03	0.055	13.6	–	0.007	14.5	–	–
32	OIL	99	–	–	–	–	–	–	–	–	–	–	–	–	–	–
33	**CORN, OPAQUE 2, HIGH LYSINE,** GRAIN	90	0.03	0.20	–	–	0.13	0.35	0.10	–	–	–	–	–	–	–
34	**CORN, SWEET,** CANNERY RESIDUE, FRESH	77	0.25	0.54	0.02	–	0.18	0.88	0.10	–	5.4	–	0.016	–	–	–
35	**COTTON,** SEEDS	91	0.14	0.69	0.03	–	0.32	1.11	0.24	–	49.0	–	0.014	11.1	–	–
	DISTILLERS PRODUCTS (ALSO SEE CORN; RYE; SORGHUM; WHEAT)															
36	GRAINS, DEHY	93	0.12	0.54	0.05	0.05	0.09	0.20	0.46	0.092	47.9	–	0.027	35.0	–	–
37	SOLUBLES, DEHY	92	0.24	1.35	0.45	–	0.53	1.97	–	0.196	71.6	–	0.031	64.1	–	138.0
	FATS AND OILS															
38	FAT, ANIMAL, HYDROLYZED	99	–	–	–	–	–	–	–	–	–	–	–	–	–	–
39	FAT, ANIMAL-POULTRY	99	–	–	–	–	–	0.23	–	–	–	–	–	–	–	–
40	**FLAX, COMMON,** SEEDS	96	0.28	0.55	–	–	–	–	–	–	–	–	–	–	–	–
41	**GARBAGE, HOTEL AND RESTAURANT,** BOILED, WET	23	0.10	0.06	–	–	0.01	–	–	–	5.0	–	0.010	5.0	–	–
42	**HOMINY GRITS (CORN, DENT YELLOW, GRITS)**	90	0.05	0.51	0.08	0.05	0.24	0.59	0.03	0.055	13.6	–	0.007	14.5	–	–
43	**KAFIR SORGHUM,** GRAIN	89	0.03	0.31	0.05	0.10	0.15	0.34	0.16	0.387	7.0	–	0.007	15.8	0.797	13.5
44	**KELP (SEAWEED),** WHOLE, DEHY	91	2.47	0.28	–	–	0.85	–	–	–	–	–	–	–	–	–
45	**LARD (FAT),** SWINE	99	–	–	–	–	–	–	–	–	–	–	–	–	–	–
46	**MANGEL, BEET,** ROOTS, FRESH	11	0.02	0.02	0.07	0.16	0.02	0.25	0.02	–	0.6	–	0.002	–	–	–
47	**MANURE, CATTLE,** WITHOUT BEDDING, DEHY	93	1.35	1.08	–	–	–	0.47	–	–	–	–	–	–	–	–
48	**MILLET,** GRAIN	90	0.05	0.29	0.04	0.14	0.16	0.43	0.13	0.044	21.8	–	0.007	29.9	–	13.9
49	**MILLET, FOXTAIL,** GRAIN	89	–	0.41	–	–	–	0.31	–	–	–	–	0.010	–	–	–
50	**MILO SORGHUM,** GRAIN	89	0.04	0.30	0.04	0.08	0.13	0.31	0.11	0.471	4.3	0.061	0.005	15.8	0.201	16.9
	MOLASSES AND SYRUP															
51	BEET, SUGAR, MOLASSES, MORE THAN 48% INVERT SUGAR, MORE THAN 79.5 DEGREES BRIX	78	0.12	0.03	1.16	1.28	0.23	4.73	0.46	0.362	16.8	–	0.007	4.5	–	14.0
52	CITRUS, SYRUP (CITRUS MOLASSES)	67	1.18	0.09	0.28	0.07	0.14	0.09	0.16	0.109	72.8	–	0.035	40.9	–	92.4
53	SUGAR CANE, MOLASSES, DEHY	94	1.04	0.42	0.19	–	0.44	3.40	0.43	1.145	74.9	–	0.024	54.1	–	31.2
54	SUGAR CANE, MOLASSES, MORE THAN 46% INVERT SUGAR, MORE THAN 79.5 DEGREES BRIX (BLACKSTRAP)	74	0.74	0.08	0.16	2.26	0.31	2.98	0.35	1.180	48.9	1.564	0.020	43.7	–	15.6
55	WOOD, MOLASSES	62	1.17	0.05	0.03	0.12	0.07	0.04	0.03	–	–	–	–	12.6	–	–
	OATS															
56	GRAIN, ALL ANALYSES	89	0.08	0.34	0.05	0.09	0.14	0.40	0.21	0.056	6.0	0.112	0.007	35.8	0.215	34.9
57	CEREAL BY-PRODUCT (FEEDING OAT MEAL; OAT MIDDLINGS)	91	0.07	0.44	0.09	0.05	0.14	0.50	0.22	0.046	5.2	–	0.039	43.8	–	139.5
58	GROATS	90	0.08	0.43	0.05	0.08	0.11	0.35	0.20	–	6.0	0.108	0.008	27.8	–	0.0
59	**PEA,** SPLIT SEED BY-PRODUCT (PEA FEED; PEA MEAL)	90	–	–	–	–	–	–	–	–	–	–	–	–	–	–
60	**PEA, FIELD,** SEEDS	91	0.16	0.38	0.00	–	0.14	1.36	–	1.700	11.7	–	0.020	21.2	0.393	46.8
61	**PEARL MILLET,** GRAIN	90	0.05	0.31	0.04	0.14	0.16	0.43	0.13	0.045	22.1	–	0.006	31.0	–	13.3
62	**PINEAPPLE,** CANNERY RESIDUE, DEHY (PINEAPPLE BRAN)	87	0.20	0.11	–	–	–	–	–	–	–	–	0.049	–	–	–
	POTATO															
63	TUBERS, FRESH	24	0.01	0.06	0.02	0.07	0.03	0.51	0.02	–	6.7	–	0.002	9.8	–	–
64	TUBERS, BOILED	24	0.01	0.05	–	–	–	–	–	–	–	–	–	–	–	–
	RICE															
65	GRAIN, GROUND (GROUND ROUGH RICE; GROUND PADDY RICE)	89	0.07	0.32	0.06	0.07	0.13	0.47	0.05	–	–	–	–	18.0	–	15.0
66	BRAN WITH GERMS (RICE BRAN)	91	0.07	1.44	0.03	0.07	0.85	1.69	0.18	1.383	11.0	–	0.019	337.6	–	37.4
67	GROATS, POLISHED (RICE, POLISHED)	89	0.02	0.11	0.01	0.04	0.09	0.23	0.08	0.846	5.4	–	0.002	29.6	–	13.7
68	GROATS (RICE, BROWN)	88	0.03	0.20	0.02	0.07	0.08	0.30	0.04	0.727	3.7	–	0.003	20.3	–	14.3
69	POLISHINGS	90	0.05	1.34	0.04	0.11	0.60	1.28	0.17	3.890	8.0	–	0.009	126.8	–	63.2

ENERGY FEEDS

Entry Number	Fat-Soluble Vitamins: A (1 mg Carotene = 1667 IU Vit A)	Carotene (Provitamin A)	D	E	K	Water-Soluble Vitamins: B–12	Biotin	Choline	Folacin (Folic Acid)	Niacin	Pantothenic Acid (B–3)	(Pyridoxine) B–6	Riboflavin (B–2)	Thiamin (B–1)
	IU/g	ppm or mg/kg	IU/kg	ppm or mg/kg	ppm or mg/kg	ppb or mcg/kg	ppm or mg/kg	ppm or mg/kg	ppm or mg/kg	ppm or mg/kg	ppm or mg/kg	ppm or mg/kg	ppm or mg/kg	ppm or mg/kg
1	–	–	–	–	–	–	–	–	–	–	–	–	–	–
2	–	–	–	–	–	–	–	–	–	–	–	–	–	–
3	–	–	–	–	–	–	–	–	–	–	–	–	–	–
4	–	–	–	7.9	–	–	–	–	–	–	–	–	–	–
5	3.4	2.0	–	23.2	0.22	–	0.15	1036	0.57	76	7.9	5.80	1.6	4.5
6	–	–	–	26.2	–	–	0.15	976	0.50	47	7.1	2.89	1.5	4.2
7	–	–	–	3.7	–	–	4.09	1591	0.20	55	9.0	8.62	2.8	8.3
8	–	–	–	–	–	–	–	–	–	24	–	–	1.8	5.7
9	–	–	–	1.0	–	–	0.11	1017	1.29	24	2.4	0.30	1.7	6.3
10	–	–	–	–	–	–	–	–	–	22	2.2	–	3.1	8.6
11	–	–	–	–	–	–	–	–	3.31	20	8.4	–	1.7	4.6
12	–	–	–	–	–	–	–	–	–	25	–	–	2.1	3.9
13	–	–	–	–	–	–	–	–	–	–	–	–	–	–
14	–	–	–	4.0	–	–	–	827	–	41	4.5	–	2.3	–
15	0.4	0.2	1	–	–	–	–	820	–	17	1.4	–	0.7	0.4
16	0.4	0.2	–	–	–	–	–	814	–	16	1.5	–	0.7	–
17	0.8	0.5	–	26.7	–	3.6	0.44	1651	0.22	44	8.2	1.03	1.5	0.6
18	–	–	–	–	–	–	–	439	–	18	11.5	–	4.7	3.7
19	129.9	77.9	–	6.9	–	–	0.01	–	0.14	7	3.5	1.39	0.6	0.7
20	–	–	–	–	–	–	–	–	–	–	–	–	–	–
21	0.4	0.2	–	–	–	–	–	789	–	22	14.0	–	2.1	1.5
22	–	–	–	–	–	–	–	–	–	27	17.2	–	6.2	–
23	9.5	5.7	–	20.9	0.22	–	0.07	504	0.31	23	5.1	6.16	1.1	3.7
24	2.9	1.7	–	21.6	–	–	0.06	569	0.35	24	3.9	6.88	1.3	3.5
25	–	–	–	–	–	–	–	–	–	–	–	–	–	–
26	5.2	3.1	–	–	–	0.3	0.41	1113	1.00	38	11.3	4.22	5.0	1.8
27	1.1	0.7	–	45.9	–	4.2	1.49	4751	1.34	124	23.3	9.41	15.1	6.8
28	5.3	3.2	–	17.5	–	–	0.03	357	0.24	17	4.2	5.97	0.9	2.9
29	–	–	–	14.6	–	–	–	–	–	–	–	6.39	–	–
30	9.8	5.9	–	12.1	–	–	0.33	1514	0.27	70	13.6	13.93	2.2	2.0
31	15.4	9.2	–	–	–	–	0.13	1154	0.31	47	8.2	10.95	2.1	8.1
32	–	–	–	–	–	–	–	–	–	–	–	–	–	–
33	7.8	4.7	–	–	–	–	–	518	–	19	4.7	–	1.1	–
34	17.3	10.4	–	–	–	–	–	–	–	–	–	–	–	–
35	–	–	–	–	–	–	–	–	–	–	–	–	–	–
36	13.0	7.8	–	30.5	–	–	–	2645	–	47	11.9	6.00	6.6	2.5
37	1.9	1.1	–	–	–	2.9	2.84	4992	–	143	25.3	8.66	11.3	6.9
38	–	–	–	–	–	–	–	–	–	–	–	–	–	–
39	–	–	–	7.9	–	–	–	–	–	–	–	–	–	–
40	–	–	–	–	–	–	–	–	–	–	–	–	–	–
41	–	–	–	–	–	–	–	–	–	–	–	–	–	–
42	15.4	9.2	–	–	–	–	0.13	1154	0.31	47	8.2	10.95	2.1	8.1
43	0.6	0.4	–	–	–	–	0.24	439	0.20	38	12.0	6.68	1.2	3.8
44	–	–	–	–	–	–	–	–	–	–	–	–	–	–
45	–	–	–	22.8	–	–	–	–	–	–	–	–	–	–
46	0.2	0.1	–	–	–	–	–	–	0.17	3	1.0	0.43	0.4	0.3
47	–	–	–	–	–	–	–	–	–	–	–	–	–	–
48	–	–	–	–	–	–	–	739	0.22	48	9.0	–	1.5	6.6
49	–	–	–	–	–	–	–	–	–	33	–	–	1.1	3.8
50	0.4	0.2	0	12.1	0.22	–	0.23	638	0.21	37	11.0	4.69	1.1	4.1
51	–	–	–	4.0	–	–	–	827	–	41	4.5	–	2.3	–
52	–	–	–	–	–	–	–	–	–	27	17.2	–	6.2	–
53	–	–	–	5.2	–	–	–	–	–	–	–	–	–	–
54	–	–	–	5.4	–	–	0.69	764	0.11	36	37.4	4.21	2.8	0.9
55	–	–	–	–	–	–	–	–	–	–	–	–	–	–
56	0.2	0.1	–	14.9	–	–	0.27	967	0.39	14	9.9	2.53	1.4	6.0
57	–	–	–	23.7	–	–	0.22	1157	0.46	24	17.6	–	1.7	7.0
58	–	–	–	14.8	–	–	–	1132	0.51	10	13.8	1.00	1.2	6.5
59	–	–	–	–	–	–	–	–	–	–	–	–	–	–
60	–	–	–	–	–	–	0.19	654	0.36	34	7.4	1.01	1.4	4.1
61	4.3	2.6	–	–	–	–	–	790	–	52	8.8	–	1.8	7.1
62	78.4	47.0	–	–	–	–	–	–	–	–	–	–	–	–
63	–	–	–	–	–	–	–	–	–	17	–	–	0.5	1.2
64	–	–	–	–	–	–	–	–	–	–	–	–	–	–
65	–	–	–	14.0	–	–	–	926	0.25	40	7.1	–	0.7	–
66	–	–	–	60.4	–	–	0.43	1230	2.20	299	22.8	13.24	2.6	22.4
67	–	–	–	3.5	–	–	–	901	0.15	15	3.5	0.39	0.6	0.7
68	–	–	–	10.3	–	–	0.09	–	0.19	43	10.7	7.00	0.6	2.9
69	–	–	–	90.2	–	–	0.62	1248	–	506	46.4	27.89	1.8	20.0

(Continued)

TABLE 19–1A ENERGY FEEDS, COMPOSITION OF FEEDS, DATA EXPRESSED AS-FED—*(Continued)*

Entry Number	Feed Name Description	International Feed Number	Proximate Analysis									Digestible Protein		
			Dry Matter	Ash	Crude Fiber	Neutral Det. Fib. (NDF)	Acid Det. Fib. (ADF)	Lignin	Ether Extract (Fat)	N-Free Extract	Crude Protein	Ruminant	Swine	Horse
			%	%	%	%	%	%	%	%	%	%	%	%
	RYE													
70	GRAIN, ALL ANALYSES	4-04-047	87	1.6	2.2	–	–	–	1.5	70.0	12.0	8.4	9.1	8.7
71	DISTILLERS GRAINS, DEHY	5-04-023	92	2.3	12.3	–	–	–	6.0	48.3	23.0	13.8	21.7	17.5
72	DISTILLERS GRAINS WITH SOLUBLES, DEHY	5-04-024	90	6.4	8.1	–	–	–	4.1	44.7	27.2	22.6	25.9	22.3
73	**SAFFLOWER,** SEEDS	4-07-958	93	3.0	23.6	–	37.2	–	30.8	20.9	14.9	7.2	11.7	10.9
74	**SCREENINGS, GRAIN, CEREAL,** ALL ANALYSES (ALSO SEE BARLEY; WHEAT)	4-02-156	90	5.4	12.0	–	–	–	3.7	56.7	12.1	8.7	7.8	8.7
	SORGHUM													
75	GRAIN, ALL ANALYSES	4-04-383	90	1.8	2.6	16.2	8.1	1.2	2.7	71.6	11.5	7.1	8.7	8.2
76	KAFIR, GRAIN	4-04-428	89	1.5	2.0	–	1.8	–	2.8	72.0	10.8	6.6	8.3	7.7
77	MILO, GRAIN	4-04-444	89	1.6	2.2	20.6	4.7	–	2.8	71.9	10.1	6.8	7.2	7.1
78	SUDANGRASS, GRAIN	4-08-520	92	12.0	25.4	–	–	–	2.4	38.4	14.2	9.9	11.1	10.3
	SOYBEAN													
79	SEEDS	5-04-610	92	5.1	5.4	–	10.1	–	17.2	25.9	38.4	34.5	31.5	34.9
80	OIL	4-07-983	100	0.3	–	–	–	–	95.0	7.3	1.4	–1.8	–0.6	0.2
	SUGARCANE													
81	MOLASSES, DEHY	4-04-695	94	12.5	6.3	–	–	–	0.9	65.0	9.7	5.3	7.0	6.8
82	MOLASSES, MORE THAN 46% INVERT SUGAR, MORE THAN 79.5 DEGREES BRIX	4-04-696	74	9.8	0.4	–	0.3	0.2	0.2	59.7	4.3	0.6	1.3	2.7
83	**SUNFLOWER,** SEEDS	5-08-530	94	3.7	22.7	–	–	–	32.3	14.4	20.9	16.7	19.6	14.9
	SWEET POTATO													
84	TUBERS, DEHY, MEAL	4-08-536	89	3.8	3.7	–	–	–	1.0	74.5	6.4	0.9	4.9	4.3
85	TUBERS, FRESH	4-04-788	33	1.1	1.4	–	2.6	–	0.4	28.5	1.7	0.5	0.9	1.0
86	**TRITICALE,** GRAIN	4-20-362	89	1.8	3.0	11.9	–	–	1.5	67.3	15.4	11.1	12.2	11.3
87	**TURNIP,** ROOTS, FRESH	4-05-067	9	0.8	1.1	4.0	3.1	–	0.2	5.9	1.2	0.9	0.4	0.9
	WHEAT													
88	GRAIN, ALL ANALYSES	4-05-211	89	1.8	2.6	–	7.1	–	1.8	69.7	13.1	10.5	11.1	9.5
89	BRAN	4-05-190	89	5.9	10.0	40.9	12.0	2.6	4.0	53.6	15.5	12.0	11.8	13.1
90	MIDDLINGS, LESS THAN 9.5% FIBER	4-05-205	89	4.7	7.7	32.9	8.9	–	4.3	55.7	16.4	12.4	13.9	12.1
91	MILL RUN, LESS THAN 9.5% FIBER	4-05-206	90	5.1	8.2	–	9.9	–	4.1	57.4	15.1	11.4	12.0	11.0
92	RED DOG, LESS THAN 4% FIBER	4-05-203	88	2.4	2.9	–	–	–	3.4	64.0	15.6	12.9	13.6	11.5
93	SHORTS, LESS THAN 7% FIBER	4-05-201	88	4.4	6.4	–	–	–	4.6	56.5	16.5	12.9	13.2	12.2
94	**WOOD,** MOLASSES	4-05-502	62	4.1	0.5	–	1.2	–	0.3	56.6	0.6	–1.3	–0.6	–0.1

TABLE 19–1B PROTEIN FEEDS, COMPOSITION OF FEEDS, DATA EXPRESSED AS-FED

Entry Number	Feed Name Description	International Feed Number	Proximate Analysis									Digestible Protein		
			Dry Matter	Ash	Crude Fiber	Neutral Det. Fib. (NDF)	Acid Det. Fib. (ADF)	Lignin	Ether Extract (Fat)	N-Free Extract	Crude Protein	Ruminant	Swine	Horse
			%	%	%	%	%	%	%	%	%	%	%	%
1	**AMMONIUM, POLYPHOSPHATE SOLUTION**	6-08-042	60	–	–	–	–	–	–	–	54.8	–	–	–
2	**BARLEY,** MALT SPROUTS, DEHY	5-00-545	93	5.6	14.2	43.7	16.7	–	1.4	48.8	22.9	19.2	21.5	17.2
	BEAN													
3	SEEDS, KIDNEY	5-00-600	89	3.6	4.1	–	–	–	1.2	57.8	21.8	14.6	20.6	16.5
4	SEEDS, NAVY	5-00-623	89	4.0	4.4	–	–	–	1.4	56.7	22.9	20.2	21.6	17.6
5	SEEDS, PINTO	5-00-624	90	4.3	4.0	–	–	–	1.3	57.9	22.7	17.2	21.4	17.3
6	**BEAN, LIMA,** SEEDS	5-00-613	90	4.1	4.5	–	–	–	1.3	58.9	20.7	16.7	19.4	15.1
7	**BEAN, MUNG,** SEEDS	5-08-185	90	3.8	3.9	–	–	–	1.3	57.1	23.9	19.6	22.6	18.7
8	**BEAN, TEPARY,** SEEDS	5-08-349	90	4.2	3.4	–	–	–	1.4	59.3	22.2	18.0	20.9	16.7
9	**BLOOD,** MEAL	5-00-380	91	5.3	1.0	–	–	–	1.3	3.1	80.5	57.2	60.1	82.5
	BREWERS (SEE SPECIFIC GRAINS)													
10	GRAINS, DEHY	5-02-141	92	3.6	13.0	38.7	23.9	4.6	6.6	41.6	27.3	20.1	21.8	21.0
11	GRAINS, WET	5-02-142	22	0.9	3.1	9.2	5.0	1.1	1.5	10.7	5.8	4.2	5.5	4.5
	BUTTERMILK, CATTLE													
12	CONDENSED	5-01-159	29	3.6	0.1	–	–	–	2.4	12.4	10.8	9.2	10.0	9.5
13	DEHY	5-01-160	92	9.1	0.3	–	–	–	5.2	45.9	31.7	28.7	29.5	27.3
14	**CASEIN,** ACID PRECIPITATED, DEHY	5-01-162	91	2.2	0.2	0	0	–	0.6	3.6	84.0	81.5	82.2	86.5
15	**CHEESE, CATTLE,** RIND	5-01-163	85	7.7	0.2	–	–	–	19.8	11.7	45.5	39.4	44.1	43.6
	CHICKEN													
16	BY-PRODUCT, FRESH (VISCERA WITH FEET, WITH HEADS)	5-07-951	39	4.9	–	–	0.5	–	14.7	3.0	17.6	15.1	17.0	16.2
17	BY-PRODUCT, WITHOUT FEET, FRESH (VISCERA WITHOUT FEET, WITH HEADS)	5-07-952	34	1.1	0.2	–	0.7	–	16.1	3.0	13.8	11.7	13.2	12.4
	CITRUS													
18	MOLASSES, AMMONIATED	5-01-240	61	4.7	–	–	–	–	2.2	32.5	21.4	18.0	20.4	18.5
19	PULP AMMONIATED, DEHY	4-01-238	87	4.6	13.2	–	–	–	5.6	51.9	12.1	8.2	9.3	8.7
20	**COCONUT,** KERNELS WITH COATS, MEAL MECH EXTD (COPRA MEAL)	5-01-572	92	6.4	12.1	–	18.3	–	6.8	45.0	21.2	17.6	15.5	15.4
	CORN													
21	DISTILLERS GRAINS, DEHY	5-02-842	93	2.2	11.5	–	–	–	8.9	43.1	27.8	20.1	19.4	22.7
22	DISTILLERS SOLUBLES, DEHY	5-02-844	93	7.2	4.6	–	–	–	8.6	45.2	27.4	21.5	20.0	22.3
23	GERM MEAL, WET MILLED, SOLV EXTD	5-02-898	92	3.9	12.2	–	–	–	1.5	53.5	20.7	16.6	19.4	14.9
24	GLUTEN FEED	5-02-903	90	6.6	8.7	–	–	–	2.1	49.4	23.0	19.8	18.2	17.7
25	GLUTEN MEAL	5-02-900	91	3.1	4.5	33.7	8.2	–	2.2	38.4	43.2	36.3	41.7	40.3
	COTTON													
26	SEEDS WITHOUT LINT	5-13-749	91	4.7	19.4	–	–	–	20.4	24.5	21.8	17.7	20.5	16.2

ENERGY FEEDS

Entry Number	TDN Ruminant	TDN Swine	TDN Horse	DE Ruminant		DE Swine		DE Horse		ME Ruminant		ME Swine		Poultry ME_n		ME Horse		Ruminant NE_m		Ruminant NE_g		Lactating Cows NE_{lc}	
	%	%	%	Mcal		kcal		Mcal		Mcal		kcal		kcal		Mcal		Mcal		Mcal		Mcal	
				lb	*kg*	lb	*kg*	lb	*kg*	lb	*kg*	lb	*kg*	lb	*kg*	lb	*kg*	lb	*kg*	lb	*kg*	lb	*kg*
70	73	75	75	1.42	*3.12*	1474	*3249*	1.35	*2.96*	1.18	*2.60*	1319	*2909*	1202	*2651*	1.10	*2.43*	0.80	*1.75*	0.54	*1.18*	0.74	*1.63*
71	54	79	–	1.08	*2.38*	1586	*3497*	–	–	0.90	*1.99*	1387	*3057*	–	–	–	–	0.45	*1.00*	0.22	*0.49*	0.50	*1.09*
72	73	68	–	1.46	*3.21*	1354	*2985*	–	–	1.29	*2.83*	1173	*2586*	–	–	–	–	0.78	*1.72*	0.52	*1.14*	0.74	*1.64*
73	83	–	–	1.07	*2.36*	–	–	–	–	1.19	*2.62*	–	–	–	–	–	–	0.93	*2.06*	0.64	*1.42*	0.87	*1.92*
74	62	55	53	1.23	*2.71*	1493	*3291*	0.98	*2.16*	1.06	*2.34*	1409	*3107*	831	*1833*	0.80	*1.77*	0.63	*1.39*	0.39	*0.85*	0.62	*1.37*
75	67	–	71	1.34	*2.96*	1570	*3462*	1.28	*2.82*	1.17	*2.58*	1426	*3143*	–	–	1.05	*2.31*	0.53	*1.18*	0.30	*0.66*	0.55	*1.21*
76	75	81	70	1.47	*3.24*	1521	*3354*	1.26	*2.77*	1.30	*2.87*	1437	*3169*	1529	*3372*	1.03	*2.27*	0.79	*1.74*	0.53	*1.16*	0.75	*1.64*
77	76	77	68	1.41	*3.12*	1520	*3350*	1.23	*2.71*	1.13	*2.49*	1474	*3250*	1467	*3234*	1.01	*2.23*	0.75	*1.66*	0.50	*1.09*	0.74	*1.62*
78	47	–	–	0.94	*2.08*	–	–	–	–	0.76	*1.68*	–	–	–	–	–	–	–	–	–	–	–	–
79	84	93	–	1.69	*3.72*	1820	*4012*	–	–	1.52	*3.34*	1605	*3539*	1534	*3382*	–	–	0.93	*2.04*	0.64	*1.41*	0.86	*1.90*
80	193	–	–	3.86	*8.51*	3412	*7523*	–	–	3.70	*8.15*	3287	*7247*	4050	*8929*	–	–	2.40	*5.30*	1.79	*3.94*	2.09	*4.61*
81	66	70	61	1.36	*2.99*	1208	*2663*	1.12	*2.48*	1.18	*2.60*	1127	*2485*	1227	*2706*	0.92	*2.03*	0.70	*1.55*	0.44	*0.97*	0.68	*1.51*
82	60	56	–	1.22	*2.68*	1135	*2502*	–	–	1.12	*2.46*	995	*2194*	870	*1918*	–	–	0.77	*1.70*	0.53	*1.18*	0.64	*1.41*
83	78	84	–	1.56	*3.44*	1685	*3715*	–	–	1.38	*3.05*	1476	*3254*	–	–	–	–	0.89	*1.95*	0.60	*1.33*	0.83	*1.83*
84	72	64	68	1.42	*3.12*	1280	*2823*	1.23	*2.72*	1.25	*2.75*	1202	*2651*	–	–	1.01	*2.23*	0.76	*1.67*	0.50	*1.11*	0.72	*1.60*
85	27	27	25	0.52	*1.16*	530	*1169*	0.45	*0.99*	0.46	*1.02*	501	*1104*	–	–	0.37	*0.81*	0.28	*0.63*	0.19	*0.42*	0.27	*0.60*
86	75	–	79	1.44	*3.17*	1453	*3203*	1.42	*3.13*	1.27	*2.80*	1420	*3130*	1420	*3130*	1.16	*2.57*	0.75	*1.65*	0.49	*1.09*	0.71	*1.57*
87	8	7	6	0.16	*0.34*	146	*322*	0.11	*0.24*	0.14	*0.31*	138	*304*	–	–	0.09	*0.20*	0.09	*0.19*	0.06	*0.13*	0.08	*0.18*
88	77	79	76	1.54	*3.40*	1544	*3404*	1.36	*2.99*	1.28	*2.82*	1485	*3274*	1402	*3092*	1.11	*2.45*	0.88	*1.93*	0.60	*1.33*	0.81	*1.79*
89	63	57	44	1.26	*2.78*	1119	*2466*	0.84	*1.84*	1.09	*2.40*	1027	*2265*	556	*1225*	0.69	*1.51*	0.67	*1.48*	0.42	*0.93*	0.64	*1.41*
90	74	68	59	1.39	*3.07*	1321	*2912*	1.08	*2.39*	1.16	*2.57*	1225	*2702*	940	*2072*	0.89	*1.96*	0.78	*1.72*	0.52	*1.15*	0.79	*1.73*
91	71	72	58	1.46	*3.21*	1438	*3170*	1.07	*2.35*	1.14	*2.52*	1254	*2765*	803	*1771*	0.88	*1.93*	0.76	*1.68*	0.50	*1.11*	0.76	*1.68*
92	77	72	69	1.47	*3.23*	1429	*3149*	1.26	*2.77*	1.39	*3.06*	1305	*2876*	1164	*2566*	1.03	*2.27*	0.82	*1.80*	0.55	*1.22*	0.77	*1.70*
93	76	71	59	1.47	*3.24*	1413	*3116*	1.08	*2.39*	1.25	*2.75*	1327	*2926*	1001	*2206*	0.89	*1.96*	0.85	*1.88*	0.58	*1.28*	0.79	*1.75*
94	52	51	–	1.03	*2.27*	1033	*2278*	–	–	0.91	*2.02*	976	*2151*	–	–	–	–	–	–	–	–	–	–

PROTEIN FEEDS

Entry Number	TDN Ruminant	TDN Swine	TDN Horse	DE Ruminant		DE Swine		DE Horse		ME Ruminant		ME Swine		Poultry ME_n		ME Horse		Ruminant NE_m		Ruminant NE_g		Lactating Cows NE_{lc}	
	%	%	%	Mcal		kcal		Mcal		Mcal		kcal		kcal		Mcal		Mcal		Mcal		Mcal	
				lb	*kg*	lb	*kg*	lb	*kg*	lb	*kg*	lb	*kg*	lb	*kg*	lb	*kg*	lb	*kg*	lb	*kg*	lb	*kg*
1	–	–	–	–	–	–	–	–	–	–	–	–	–	–	–	–	–	–	–	–	–	–	–
2	66	61	–	1.31	*2.89*	1219	*2688*	–	–	1.13	*2.50*	1046	*2307*	722	*1592*	–	–	0.69	*1.52*	0.43	*0.95*	0.67	*1.48*
3	73	79	–	1.46	*3.22*	1584	*3493*	–	–	1.29	*2.85*	1388	*3059*	–	–	–	–	–	–	–	–	–	–
4	76	78	–	1.53	*3.37*	1737	*3830*	–	–	1.36	*3.00*	1531	*3376*	1052	*2320*	–	–	0.83	*1.83*	0.56	*1.24*	0.78	*1.72*
5	74	78	–	1.47	*3.25*	1569	*3460*	–	–	1.30	*2.87*	1372	*3025*	–	–	–	–	0.81	*1.79*	0.54	*1.20*	0.77	*1.69*
6	78	78	–	1.55	*3.42*	1556	*3430*	–	–	1.39	*3.05*	1360	*2999*	–	–	–	–	–	–	–	–	–	–
7	79	81	–	1.53	*3.37*	1612	*3554*	–	–	1.36	*3.00*	1412	*3113*	–	–	–	–	0.82	*1.80*	0.55	*1.21*	0.77	*1.70*
8	69	80	–	1.37	*3.02*	1606	*3542*	–	–	1.20	*2.65*	1406	*3100*	–	–	–	–	–	–	–	–	–	–
9	61	61	–	1.20	*2.65*	1220	*2690*	–	–	0.99	*2.19*	1012	*2231*	1282	*2826*	–	–	0.63	*1.38*	0.38	*0.84*	0.60	*1.33*
10	65	66	48	1.25	*2.76*	1045	*2303*	–	–	1.01	*2.22*	1038	*2288*	1047	*2308*	–	–	0.64	*1.41*	0.39	*0.86*	0.67	*1.48*
11	15	18	–	0.32	*0.70*	351	*774*	–	–	0.27	*0.60*	305	*673*	–	–	–	–	0.18	*0.40*	0.12	*0.26*	0.17	*0.38*
12	26	22	–	0.52	*1.14*	442	*974*	–	–	0.46	*1.02*	383	*844*	–	–	–	–	0.29	*0.63*	0.20	*0.43*	0.27	*0.59*
13	82	77	–	1.56	*3.44*	1578	*3479*	–	–	1.29	*2.84*	1381	*3045*	1248	*2752*	–	–	0.88	*1.94*	0.60	*1.32*	0.85	*1.87*
14	81	80	–	1.56	*3.44*	1590	*3506*	–	–	1.22	*2.68*	1391	*3068*	1867	*4116*	–	–	0.82	*1.81*	0.55	*1.22*	0.78	*1.72*
15	78	80	–	1.69	*3.73*	1609	*3547*	–	–	1.54	*3.39*	1371	*3023*	–	–	–	–	1.06	*2.34*	0.76	*1.68*	0.98	*2.15*
16	–	–	–	–	–	–	–	–	–	–	–	–	–	–	–	–	–	–	–	–	–	–	–
17	–	–	–	–	–	–	–	–	–	–	–	–	–	–	–	–	–	–	–	–	–	–	–
18	46	–	–	0.93	*2.05*	–	–	–	–	0.81	*1.80*	–	–	–	–	–	–	–	–	–	–	–	–
19	76	65	42	1.40	*3.09*	1293	*2850*	0.80	*1.77*	1.24	*2.73*	1216	*2680*	–	–	0.66	*1.45*	0.69	*1.53*	0.45	*0.99*	0.67	*1.47*
20	69	73	–	1.37	*3.01*	1460	*3218*	–	–	1.20	*2.64*	1386	*3055*	692	*1525*	–	–	0.74	*1.63*	0.48	*1.06*	0.70	*1.55*
21	81	65	–	1.54	*3.41*	1246	*2746*	–	–	1.28	*2.82*	1196	*2636*	894	*1970*	–	–	0.87	*1.92*	0.59	*1.30*	0.84	*1.86*
22	80	78	–	1.55	*3.42*	1474	*3250*	–	–	1.34	*2.96*	1419	*3128*	1324	*2919*	–	–	0.92	*2.03*	0.63	*1.40*	0.86	*1.89*
23	68	69	–	1.41	*3.11*	1568	*3456*	–	–	1.24	*2.73*	1372	*3025*	772	*1702*	–	–	0.73	*1.61*	0.47	*1.04*	0.71	*1.56*
24	75	75	–	1.44	*3.17*	1375	*3031*	–	–	1.21	*2.67*	1037	*2287*	784	*1729*	–	–	0.82	*1.80*	0.55	*1.21*	0.78	*1.72*
25	78	80	–	1.50	*3.31*	1637	*3609*	–	–	1.24	*2.74*	1437	*3168*	1367	*3015*	–	–	0.84	*1.85*	0.57	*1.25*	0.80	*1.75*
26	87	71	–	1.57	*3.46*	1430	*3152*	–	–	1.39	*3.07*	1243	*2739*	–	–	–	–	0.97	*2.13*	0.67	*1.49*	0.91	*2.01*

(Continued)

TABLE 19–1A ENERGY FEEDS, MINERAL AND VITAMIN COMPOSITION OF FEEDS, DATA EXPRESSED AS-FED—*(Continued)*

			Macro Minerals							Micro Minerals						
Entry Number	Feed Name Description	Dry Matter	Calcium (Ca)	Phos-phorus (P)	Sodium (Na)	Chlo-rine (Cl)	Mag-nesium (Mg)	Potas-sium (K)	Sulfur (S)	Cobalt (Co)	Copper (Cu)	Iodine (I)	Iron (Fe)	Man-ganese (Mn)	Sele-nium (Se)	Zinc (Zn)
		%	%	%	%	%	%	%	%	ppm or mg/kg	ppm or mg/kg	ppm or mg/kg	%	ppm or mg/kg	ppm or mg/kg	ppm or mg/kg
	RYE															
70	GRAIN, ALL ANALYSES	87	0.06	0.31	0.02	0.03	0.12	0.46	0.15	–	7.5	–	0.007	72.0	–	28.1
71	DILTILLERS GRAINS, DEHY	92	0.15	0.48	0.17	0.05	0.17	0.07	0.44	–	–	–	–	18.4	–	–
72	DISTILLERS GRAINS WITH SOLUBLES, DEHY	90	–	–	–	–	–	–	–	–	–	–	–	–	–	–
73	**SAFFLOWER,** SEEDS	93	0.24	0.57	0.06	–	0.34	0.74	0.06	–	10.0	–	0.032	1.1	–	30.0
74	**SCREENINGS, GRAIN, CEREAL,** ALL ANALYSES (ALSO SEE BARLEY; WHEAT)	90	0.33	0.35	0.40	–	0.12	0.30	–	–	–	–	–	44.4	–	–
	SORGHUM															
75	GRAIN, ALL ANALYSES	90	0.05	0.32	0.03	0.08	0.14	0.35	0.15	0.275	9.7	–	0.007	9.8	–	42.4
76	KAFIR, GRAIN	89	0.03	0.31	0.05	0.10	0.15	0.34	0.16	0.387	7.0	–	0.007	15.8	0.797	13.5
77	MILO, GRAIN	89	0.04	0.30	0.04	0.08	0.13	0.31	0.11	0.471	4.3	0.061	0.005	15.8	0.201	16.9
78	SUDANGRASS, GRAIN	92	–	–	–	–	–	–	–	–	–	–	–	–	–	–
	SOYBEAN															
79	SEEDS	92	0.25	0.60	0.00	0.03	0.27	1.66	0.22	–	18.2	–	0.009	36.4	0.111	56.9
80	OIL	100	–	–	–	–	–	–	–	–	–	–	–	–	–	–
	SUGARCANE															
81	MOLASSES, DEHY	94	1.04	0.42	0.19	–	0.44	3.40	0.43	1.145	74.9	–	0.024	54.1	–	31.2
82	MOLASSES, MORE THAN 46% INVERT SUGAR, MORE THAN 79.5 DEGREES BRIX	74	0.74	0.08	0.16	2.26	0.31	2.98	0.35	1.180	48.9	1.564	0.020	43.7	–	15.6
83	**SUNFLOWER,** SEEDS	94	0.16	0.67	0.02	–	0.37	0.68	0.28	–	23.5	–	0.006	21.9	–	68.6
	SWEET POTATO															
84	TUBERS, DEHY, MEAL	89	0.12	0.15	–	–	–	–	–	–	–	–	–	–	–	–
85	TUBERS, FRESH	33	0.03	0.05	0.02	0.02	0.05	0.35	0.04	–	1.4	–	0.002	3.7	–	–
86	**TRITICALE,** GRAIN	89	0.04	0.30	0.01	–	0.23	0.51	–	0.078	8.3	–	0.005	42.5	–	31.2
87	**TURNIP,** ROOTS, FRESH	9	0.06	0.03	0.01	0.06	0.02	0.26	0.04	–	2.0	–	0.002	3.9	–	2.7
	WHEAT															
88	GRAIN, ALL ANALYSES	89	0.05	0.35	0.06	0.08	0.14	0.41	0.18	0.442	5.8	0.090	0.006	41.5	0.256	31.4
89	BRAN	89	0.13	1.16	0.06	0.05	0.58	1.23	0.22	0.075	11.0	0.066	0.015	114.9	0.641	94.6
90	MIDDLINGS, LESS THAN 9.5% FIBER	89	0.13	0.89	0.01	0.04	0.34	0.98	0.17	0.502	15.9	0.109	0.009	114.0	0.736	96.9
91	MILL RUN, LESS THAN 9.5% FIBER	90	0.10	1.02	–	–	0.48	1.20	0.30	0.209	18.5	–	0.010	104.1	–	–
92	RED DOG, LESS THAN 4% FIBER	88	0.06	0.51	0.01	0.14	0.18	0.52	0.24	0.117	6.3	–	0.005	52.1	0.324	65.0
93	SHORTS, LESS THAN 7% FIBER	88	0.09	0.80	0.03	0.05	0.27	0.93	0.21	0.105	11.5	–	0.008	114.1	0.476	102.4
94	**WOOD,** MOLASSES	62	1.17	0.05	0.03	0.12	0.07	0.04	0.03	–	–	–	–	12.6	–	–

TABLE 19–1B PROTEIN FEEDS, MINERAL AND VITAMIN COMPOSITION OF FEEDS, DATA EXPRESSED AS-FED

			Macro Minerals							Micro Minerals						
Entry Number	Feed Name Description	Dry Matter	Calcium (Ca)	Phos-phorus (P)	Sodium (Na)	Chlo-rine (Cl)	Mag-nesium (Mg)	Potas-sium (K)	Sulfur (S)	Cobalt (Co)	Copper (Cu)	Iodine (I)	Iron (Fe)	Man-ganese (Mn)	Sele-nium (Se)	Zinc (Zn)
		%	%	%	%	%	%	%	%	ppm or mg/kg	ppm or mg/kg	ppm or mg/kg	%	ppm or mg/kg	ppm or mg/kg	ppm or mg/kg
1	**AMMONIUM, POLYPHOSPHATE SOLUTION**	60	0.10	13.44	–	–	–	–	0.50	–	–	–	0.505	–	–	–
2	**BARLEY,** MALT SPROUTS, DEHY	93	0.18	0.63	0.88	0.36	0.17	0.25	0.79	–	5.9	–	0.018	29.4	0.416	56.4
	BEAN															
3	SEEDS, KIDNEY	89	0.11	0.43	0.01	–	0.10	0.93	–	0.508	6.1	–	0.007	17.0	0.355	24.3
4	SEEDS, NAVY	89	0.17	0.54	0.04	0.06	0.13	1.31	0.23	–	9.9	–	0.010	21.1	–	–
5	SEEDS, PINTO	90	0.18	0.48	0.01	–	0.24	2.59	–	0.551	13.0	–	0.014	17.3	–	40.0
6	**BEAN, LIMA,** SEEDS	90	0.08	0.38	0.02	0.03	0.18	1.61	0.20	–	8.2	–	0.009	16.1	–	–
7	**BEAN, MUNG,** SEEDS	90	0.13	0.35	0.01	–	–	1.04	–	–	–	–	0.009	–	–	–
8	**BEAN, TEPARY,** SEEDS	90	–	–	–	–	–	–	–	–	–	–	–	–	–	–
9	**BLOOD,** MEAL	91	0.29	0.25	0.32	0.30	0.22	0.09	0.34	0.088	12.6	–	0.372	5.3	0.731	4.4
	BREWERS (ALSO SEE CORN; RYE; SORGHUM; WHEAT)															
10	GRAINS, DEHY	92	0.30	0.51	0.21	0.15	0.15	0.09	0.30	0.076	21.7	0.066	0.024	37.2	–	27.3
11	GRAINS, WET	22	0.06	0.12	0.06	0.03	0.03	0.02	0.07	0.022	4.9	–	0.006	9.0	–	23.2
	BUTTERMILK, CATTLE															
12	CONDENSED	29	0.44	0.26	0.31	0.12	0.19	0.23	0.03	–	–	–	–	–	–	–
13	DEHY	92	1.32	0.94	0.83	0.44	0.48	0.83	0.08	–	1.0	–	0.001	3.5	–	40.2
14	**CASEIN,** ACID PRECIPITATED, DEHY	91	0.61	0.82	0.01	–	0.01	0.01	–	–	4.1	–	0.002	3.5	–	31.8
15	**CHEESE, CATTLE,** RIND	85	0.98	0.56	0.81	0.60	0.02	0.28	–	–	–	–	–	–	–	–
	CHICKEN															
16	BY-PRODUCT, FRESH (VISCERA WITH FEET, WITH HEADS)	39	–	–	–	–	–	–	–	–	–	–	–	–	–	–
17	BY-PRODUCT, WITHOUT FEET, FRESH (VISCERA WITHOUT FEET, WITH HEADS)	34	0.34	0.24	–	–	–	–	–	–	–	–	–	–	–	–
	CITRUS															
18	MOLASSES, AMMONIATED	61	0.76	0.16	–	–	0.08	–	–	–	–	–	–	–	–	–
19	PULP, AMMONIATED, DEHY	87	1.66	0.12	–	–	0.07	–	–	–	–	–	–	–	–	–
20	**COCONUT,** KERNELS WITH COATS, MEAL MECH EXTD (COPRA MEAL)	92	0.19	0.60	0.04	–	0.30	1.65	0.34	0.127	16.7	–	0.068	70.1	–	48.5
	CORN															
21	DISTILLERS' GRAINS, DEHY	93	0.09	0.39	0.09	0.07	0.07	0.16	0.43	0.076	38.9	0.048	0.020	19.3	0.352	41.7
22	DISTILLERS' SOLUBLES, DEHY	93	0.30	1.30	0.23	0.26	0.60	1.70	0.37	0.167	77.9	0.079	0.052	72.0	0.371	88.0
23	GERM MEAL, WET MILLED, SOLV EXTD	92	0.04	0.51	0.04	0.04	0.16	0.35	0.31	–	4.5	–	0.034	3.8	0.340	104.8
24	GLUTEN FEED	90	0.32	0.74	0.12	0.22	0.33	0.57	0.21	0.087	47.1	0.066	0.043	23.1	0.272	64.6
25	GLUTEN MEAL	91	0.15	0.46	0.09	0.06	0.06	0.03	0.20	0.077	27.7	–	0.039	7.7	1.015	173.7
	COTTCN															
26	SEEDS WITHOUT LINT	91	0.14	0.69	0.03	–	0.32	1.11	0.24	–	49.0	–	0.014	11.1	–	–

ENERGY FEEDS

Entry Number	Fat-Soluble Vitamins					Water-Soluble Vitamins								
	A (1 mg Carotene = 1667 IU Vit A)	Carotene (Provitamin A)	D	E	K	B–12	Biotin	Choline	Folacin (Folic Acid)	Niacin	Pantothenic Acid (B–3)	(Pyridoxine) B–6	Riboflavin (B–2)	Thiamin (B–1)
	IU/g	ppm or mg/kg	IU/kg	ppm or mg/kg	ppm or mg/kg	ppb or mcg/kg	ppm or mg/kg	ppm or mg/kg	ppm or mg/kg	ppm or mg/kg	ppm or mg/kg	ppm or mg/kg	ppm or mg/kg	ppm or mg/kg
70	0.1	0.1	–	14.5	–	–	0.06	419	0.62	14	7.5	–	1.7	4.1
71	–	–	–	–	–	–	–	–	–	17	5.2	–	3.3	1.3
72	–	–	–	–	–	–	–	–	–	63	17.4	–	8.2	3.1
73	–	–	–	–	–	–	–	–	–	–	–	–	–	–
74	–	–	–	–	–	–	–	1044	1.06	10	12.8	–	1.8	–
75	2.0	1.2	–	–	–	–	0.26	686	0.22	47	10.2	5.41	1.2	4.5
76	0.6	0.4	–	–	–	–	0.24	439	0.20	38	12.0	6.68	1.2	3.8
77	0.4	0.2	0	12.1	0.22	–	0.23	638	0.21	37	11.0	4.69	1.1	4.1
78	–	–	–	–	–	–	–	–	–	–	–	–	–	–
79	1.5	0.9	–	33.7	–	–	0.38	2931	–	23	16.0	11.04	2.9	11.3
80	–	–	–	–	–	–	–	–	–	–	–	–	–	–
81	–	–	–	5.2	–	–	–	–	–	–	–	–	–	–
82	–	–	–	5.4	–	–	0.69	764	0.11	36	37.4	4.21	2.8	0.9
83	–	–	–	–	–	–	–	–	–	–	–	–	3.3	0.4
84	117.3	70.4	–	–	–	–	–	–	–	–	–	–	–	–
85	236.8	142.1	–	–	–	–	–	–	–	7	–	–	0.7	1.1
86	–	–	–	–	–	–	–	457	–	–	–	–	0.4	–
87	–	–	–	–	–	–	–	92	0.26	7	1.7	–	0.6	0.7
88	–	–	–	15.5	–	0.9	0.10	918	0.43	59	11.3	3.74	1.3	4.3
89	4.4	2.6	–	14.3	–	–	0.38	1232	1.77	197	28.0	10.34	3.6	8.4
90	5.1	3.1	–	23.8	–	–	0.24	1246	1.24	95	17.8	9.14	2.0	14.2
91	–	–	–	31.9	–	–	0.31	1005	1.08	116	13.7	11.09	2.1	15.2
92	–	–	–	37.4	–	–	0.11	1453	0.82	46	13.3	5.40	2.2	21.8
93	5.1	3.1	–	36.0	–	–	–	1697	1.51	105	21.9	–	4.1	19.5
94	–	–	–	–	–	–	–	–	–	–	–	–	–	–

PROTEIN FEEDS

Entry Number	Fat-Soluble Vitamins					Water-Soluble Vitamins								
	A (1 mg Carotene = 1667 IU Vit A)	Carotene (Provitamin A)	D	E	K	B–12	Biotin	Choline	Folacin (Folic Acid)	Niacin	Pantothenic Acid (B–3)	(Pyridoxine) B–6	Riboflavin (B–2)	Thiamin (B–1)
	IU/g	ppm or mg/kg	IU/kg	ppm or mg/kg	ppm or mg/kg	ppb or mcg/kg	ppm or mg/kg	ppm or mg/kg	ppm or mg/kg	ppm or mg/kg	ppm or mg/kg	ppm or mg/kg	ppm or mg/kg	ppm or mg/kg
1	–	–	–	–	–	–	–	–	–	–	–	–	–	–
2	–	–	–	3.7	–	–	4.09	1591	0.20	55	9.0	8.62	2.8	8.3
3	–	–	–	–	–	–	–	–	–	24	–	–	1.8	5.7
4	–	–	–	1.0	–	–	0.11	1017	1.29	24	2.4	0.30	1.7	6.3
5	–	–	–	–	–	–	–	–	–	22	2.2	–	3.1	8.6
6	–	–	–	–	–	–	–	–	3.31	20	8.4	–	1.7	4.6
7	–	–	–	–	–	–	–	–	–	25	–	–	2.1	3.9
8	–	–	–	–	–	–	–	–	–	–	–	–	–	–
9	–	–	–	–	–	44.3	0.09	780	0.10	31	2.3	4.41	2.0	0.3
10	0.8	0.5	–	26.7	–	3.6	0.44	1651	0.22	44	8.2	1.03	1.5	0.6
11	–	–	–	5.5	–	–	–	–	–	–	–	–	–	–
12	25.6	15.4	–	–	–	–	–	–	–	–	–	–	12.6	–
13	–	–	–	6.3	–	19.6	0.29	1746	0.39	9	37.0	2.47	30.6	3.4
14	–	–	–	–	–	–	0.04	208	0.47	1	2.7	0.42	1.5	0.4
15	–	–	–	–	–	–	–	–	–	–	–	–	–	–
16	–	–	–	–	–	–	–	–	–	–	–	–	–	–
17	–	–	–	–	–	–	–	–	–	–	–	–	–	–
18	–	–	–	–	–	–	–	–	–	–	–	–	–	–
19	–	–	–	–	–	–	–	–	–	–	–	–	–	–
20	–	–	–	–	–	–	–	1089	0.30	24	6.5	4.36	3.5	–
21	5.2	3.1	–	–	–	0.3	0.41	1113	1.00	38	11.3	4.22	5.0	1.8
22	1.1	0.7	–	45.9	–	4.2	1.49	4751	1.34	124	23.3	9.41	15.1	6.8
23	3.4	2.0	–	85.8	–	–	0.22	1586	0.20	39	4.2	–	3.8	4.5
24	9.8	5.9	–	12.1	–	–	0.33	1514	0.27	70	13.6	13.93	2.2	2.0
25	27.3	16.3	–	29.3	–	–	0.19	360	0.30	50	10.0	7.98	1.5	0.2
26	–	–	–	–	–	–	–	–	–	–	–	–	–	–

(Continued)

TABLE 19–1B PROTEIN FEEDS, COMPOSITION OF FEEDS, DATA EXPRESSED **AS-FED**—*(Continued)*

			Proximate Analysis									Digestible Protein		
Entry Number	Feed Name Description	International Feed Number	Dry Matter	Ash	Crude Fiber	Neutral Det. Fib. (NDF)	Acid Det. Fib. (ADF)	Lignin	Ether Extract (Fat)	N-Free Extract	Crude Protein	Ruminant	Swine	Horse
			%	%	%	%	%	%	%	%	%	%	%	%
	COTTON (Continued)													
27	SEEDS, MEAL MECH EXTD, 41% PROTEIN	5-01-617	93	6.1	11.9	25.9	18.5	5.6	4.7	28.9	41.0	35.1	33.8	37.7
28	SEEDS, MEAL SOLV EXTD, 41% PROTEIN	5-01-621	91	6.5	12.1	23.6	18.4	5.5	1.5	29.6	41.2	31.3	35.1	38.2
29	SEEDS, MEAL SOLV EXTD, 46% PROTEIN	5-26-100	92	7.2	8.9	25.8	19.3	–	1.6	26.8	47.6	41.1	46.1	45.2
30	SEEDS, MEAL SOLV EXTD, 48% PROTEIN	5-26-101	90	7.4	7.0	–	–	–	1.9	24.7	49.3	42.7	47.8	47.4
31	SEEDS WITHOUT HULLS, MEAL, PREPRESSED, SOLV EXTD, 50% PROTEIN	5-07-874	93	6.6	8.2	–	–	–	1.3	26.8	50.3	40.7	48.7	48.2
32	**COTTON, GLANDLESS,** SEEDS, MEAL SOLV EXTD	5-08-979	95	7.3	2.5	–	–	–	1.9	23.8	59.8	52.1	58.2	58.7
33	**COWPEA, COMMON,** SEEDS	5-01-661	89	3.5	5.0	–	–	–	1.5	55.5	23.8	19.6	22.4	18.7
34	**DIAMMONIUM PHOSPHATE,** $(NH_4)_2HPO_4$	6-00-370	98	35.5	–	–	–	–	–	–	112.9	–	–	–
	DISTILLERS PRODUCTS (ALSO SEE CORN; RYE; SORGHUM; WHEAT)													
35	GRAINS, DEHY	5-02-144	93	1.5	12.8	–	–	–	7.4	43.5	27.3	18.9	26.0	22.3
36	SOLUBLES, DEHY	5-02-147	92	6.2	3.4	–	–	–	8.9	44.8	28.8	24.0	27.4	24.0
37	**FEATHERS, POULTRY,** MEAL, HYDROLYZED	5-03-795	93	3.2	1.4	–	6.1	–	5.1	–	83.8	74.1	61.2	86.1
	FISH													
38	LIVER, MEAL MECH EXTD	5-01-968	93	6.1	1.2	–	–	–	17.3	5.4	62.8	54.9	61.2	62.3
39	SOLUBLES, CONDENSED	5-01-969	50	10.1	0.5	–	–	–	6.1	2.2	31.5	28.0	29.3	30.9
40	SOLUBLES, DEHY	5-01-971	93	12.7	2.0	–	–	–	9.0	8.7	60.4	52.7	58.8	59.7
41	**FISH, ANCHOVY,** MEAL MECH EXTD	5-01-985	92	14.7	1.0	–	–	–	4.1	6.7	65.4	57.3	55.0	65.4
42	**FISH, MENHADEN,** MEAL MECH EXTD	5-02-009	92	19.1	0.9	–	–	–	9.6	0.8	61.2	49.6	49.6	60.7
43	**FISH, SARDINE,** MEAL MECH EXTD	5-02-015	93	15.8	1.0	–	–	–	5.0	6.1	65.2	53.5	63.6	65.1
44	**FLAX, COMMON,** SEEDS, MEAL SOLV EXTD, 35% PROTEIN (LINSEED MEAL)	5-26-090	90	5.8	8.9	–	–	–	1.7	38.2	35.7	30.3	34.3	32.0
45	**LENTIL, COMMON,** SEEDS	5-02-506	88	2.6	3.4	–	–	–	1.0	57.1	24.4	19.3	23.1	19.4
46	**LINSEED,** SEEDS, MEAL SOLV EXTD, 35% PROTEIN (LINSEED MEAL)	5-26-090	90	5.8	8.9	–	–	–	1.7	38.2	35.7	30.3	34.3	32.0
47	**LIVER,** MEAL	5-00-389	93	6.3	1.4	–	–	–	15.7	3.2	66.1	57.9	64.4	66.1
	MANURE, POULTRY													
48	WITH LITTER, DEHY	5-05-587	86	17.8	15.0	–	–	8.0	2.6	25.7	24.6	14.3	23.3	19.8
49	WITHOUT LITTER, DEHY	5-14-015	90	29.2	12.0	36.3	14.4	2.1	2.0	21.6	25.4	21.0	24.1	20.4
	MEAT													
50	MEAL RENDERED	5-00-385	94	28.1	2.7	–	–	–	9.1	3.3	50.7	43.8	49.1	48.5
51	WITH BLOOD, MEAL RENDERED (TANKAGE)	5-00-386	92	21.1	1.8	–	–	–	8.7	0.1	60.5	52.8	58.9	59.8
52	WITH BLOOD, WITH BONE, MEAL RENDERED (TANKAGE)	5-00-387	93	28.2	2.2	–	–	–	12.7	3.1	46.6	40.2	45.1	44.1
53	WITH BONE, MEAL RENDERED	5-00-388	93	28.0	2.4	–	–	–	10.0	2.6	50.4	45.8	43.3	48.3
	MILK													
54	FRESH (CATTLE)	5-01-168	12	0.8	–	–	–	–	3.6	4.7	3.3	3.2	3.2	2.6
55	SKIMMED, FRESH (CATTLE)	5-01-170	10	0.7	–	–	–	–	0.1	5.8	3.0	2.5	3.0	2.5
56	SKIMMED, DEHY (CATTLE)	5-01-175	94	8.0	0.2	0.0	–	–	1.1	51.6	33.3	30.0	31.8	28.9
57	FRESH (GOAT)	5-02-128	13	0.8	–	–	–	–	4.2	4.8	3.4	2.8	3.2	2.6
58	**MOLASSES AND SYRUP,** SUGARCANE, MOLASSES, AMMONIATED	5-04-702	65	5.9	–	–	–	–	–	–	26.3	15.5	25.3	23.7
59	**MONOAMMONIUM PHOSPHATE,** $NH_4H_2PO_4$	6-09-338	98	53.0	–	–	–	–	–	–	69.4	–	–	–
	PEA													
60	SEEDS	5-03-600	89	2.9	5.5	–	–	–	1.1	56.7	23.2	19.0	21.9	18.0
61	SPLIT SEED BY-PRODUCT (PEA FEED; PEA MEAL)	1-08-478	90	3.5	23.7	–	–	–	1.4	43.7	17.7	14.5	12.1	12.0
62	**PEA, FIELD,** SEEDS	5-08-481	91	2.9	5.9	–	–	–	1.3	57.8	23.2	19.9	21.8	17.7
63	**PEANUT,** SEEDS WITHOUT HULLS, MEAL SOLV EXTD (PEANUT MEAL)	5-03-650	93	5.8	7.7	–	–	–	2.2	27.9	49.0	42.3	47.4	46.7
	POULTRY													
64	BY-PRODUCT, MEAL RENDERED	5-03-798	94	14.8	2.2	–	–	–	13.1	2.6	61.2	53.4	53.3	60.5
65	FEATHERS, MEAL, HYDROLYZED	5-03-795	93	3.2	1.4	–	6.1	–	5.1	–	83.8	74.1	61.2	86.1
66	**RAPE (CANOLA),** SEEDS, MEAL SOLV EXTD, 34% PROTEIN	5-26-092	90	7.0	13.0	–	–	–	2.5	33.5	34.0	–	–	–
67	**RICE,** HULLS, AMMONIATED	1-05-698	92	–	44.7	–	–	–	0.9	16.9	10.4	6.3	5.6	6.4
	RYE													
68	DISTILLERS GRAINS, DEHY	5-04-023	92	2.3	12.3	–	–	–	6.0	48.3	23.0	13.8	21.7	17.5
69	DISTILLERS GRAINS WITH SOLUBLES, DEHY	5-04-024	90	6.4	8.1	–	–	–	4.1	44.7	27.2	22.6	25.9	22.3
	SAFFLOWER													
70	SEEDS, MEAL SOLV EXTD, 20% PROTEIN	5-26-095	92	4.6	32.2	–	39.6	–	1.1	32.7	21.6	17.4	20.2	15.8
71	SEEDS WITHOUT HULLS, MEAL SOLV EXTD, 42% PROTEIN	5-26-094	92	6.5	14.6	–	19.2	–	1.3	26.3	42.7	36.7	41.3	39.8
72	**SESAME,** SEEDS, MEAL SOLV EXTD, 44% PROTEIN	5-26-096	92	13.1	6.8	–	–	–	1.4	25.8	45.0	39.4	44.4	42.3
73	**SHRIMP,** CANNERY RESIDUE, MEAL (SHRIMP MEAL)	5-04-226	90	26.7	14.1	–	16.6	–	3.9	6.7	38.7	33.0	37.3	35.4
	SORGHUM													
74	DISTILLERS GRAINS, DEHY	5-04-374	94	4.3	12.1	–	–	–	8.3	38.3	30.8	24.7	29.4	26.1
75	DISTILLERS GRAINS WITH SOLUBLES, DEHY	5-04-375	95	4.2	10.1	–	–	–	9.4	38.0	33.1	27.8	31.7	28.6
76	GLUTEN MEAL	5-04-388	90	1.7	4.9	–	–	–	4.4	34.9	44.4	38.2	43.0	41.8
	SOYBEAN													
77	SEEDS	5-04-610	92	5.1	5.4	–	9.2	–	17.2	25.9	38.4	34.5	31.5	34.9
78	SEEDS, MEAL SOLV EXTD, 44% PROTEIN	5-20-637	89	6.4	6.2	12.5	8.9	–	1.5	30.6	44.4	37.8	40.4	41.9
79	SEEDS WITHOUT HULLS, MEAL SOLV EXTD, 49% PROTEIN	5-20-638	90	6.1	3.7	6.6	6.2	–	1.2	29.8	49.0	42.4	44.6	47.1
80	**SUNFLOWER, COMMON,** SEEDS WITHOUT HULLS, MEAL SOLV EXTD, 44% PROTEIN	5-26-098	93	7.7	11.0	–	–	–	2.9	24.6	46.8	42.1	–	–
81	**UREA,** 45% NITROGEN, 281% PROTEIN EQUIVALENT	5-05-070	99	–	–	0	0	–	–	–	281.7	–	–	–
82	**UREA, CONDITIONER ADDED,** 42% NITROGEN, 262% PROTEIN EQUIVALENT	5-20-705	100	–	–	–	–	–	–	–	261.9	–	–	–
	WHEAT													
83	DISTILLERS GRAINS, DEHY	5-05-193	93	3.0	11.8	–	–	–	6.7	40.2	31.6	26.5	30.2	27.1
84	DISTILLERS SOLUBLES, DEHY	5-05-195	94	6.8	3.4	–	–	–	2.2	50.5	31.1	–	–	–
85	GERM MEAL	5-05-218	88	4.3	3.1	–	4.4	–	8.5	48.1	24.4	22.9	23.1	19.4
86	GLUTEN	5-05-221	91	0.9	0.4	–	–	–	0.8	9.8	79.0	71.1	77.3	80.9
	WHEY, CATTLE													
87	FRESH	4-08-134	7	0.7	–	0	0	–	0.3	5.1	0.9	0.6	0.7	0.7
88	DEHY	4-01-182	93	8.8	0.2	0.3	0.2	–	0.8	70.2	13.3	8.9	13.1	9.6
89	**YEAST, BREWERS,** DEHY	7-05-527	93	6.5	3.0	–	3.7	–	0.9	38.8	43.8	39.0	–	–
90	**YEAST, IRRADIATED,** DEHY	7-05-529	94	6.2	6.2	–	–	–	1.1	32.4	48.1	–	–	–
91	**YEAST, TORULA,** DEHY	7-05-534	93	8.0	2.5	–	3.7	–	1.6	31.5	49.6	45.1	40.6	–

Entry Number	TDN			Digestible Energy						Metabolizable Energy								Net Energy					
	Ruminant	Swine	Horse	Ruminant		Swine		Horse		Ruminant		Swine		Poultry ME_n		Horse		Ruminant NE_m		Ruminant NE_g		Lactating Cows NE_{lc}	
	%	%	%	Mcal		kcal		Mcal		Mcal		kcal		kcal		Mcal		Mcal		Mcal		Mcal	
				lb	kg	lb	kg	lb	kg	lb	kg	lb	kg	lb	kg	lb	kg	lb	kg	lb	kg	lb	kg
27	72	69	–	1.49	3.29	1305	2878	–	–	1.12	2.47	1197	2638	1025	2261	–	–	0.72	1.59	0.64	1.41	0.76	1.67
28	68	61	–	1.48	3.27	1209	2666	–	–	1.17	2.57	1069	2356	889	1960	–	–	0.74	1.64	0.63	1.40	0.73	1.61
29	71	64	–	1.42	3.13	1282	2826	–	–	1.25	2.74	1105	2436	950	2095	–	–	0.77	1.70	0.51	1.12	0.74	1.63
30	73	65	–	1.44	3.17	1292	2847	–	–	1.27	2.79	1116	2459	983	2166	–	–	0.80	1.77	0.54	1.18	0.76	1.67
31	70	–	–	1.43	3.15	1295	2854	–	–	1.25	2.76	1185	2613	971	2141	–	–	0.76	1.67	0.49	1.08	0.73	1.60
32	81	78	–	1.62	3.58	1560	3438	–	–	1.45	3.19	1359	2996	948	2089	–	–	0.92	2.03	0.63	1.39	0.86	1.90
33	76	75	–	1.48	3.27	1500	3307	–	–	1.32	2.90	1309	2886	–	–	–	–	0.80	1.76	0.53	1.18	0.75	1.66
34	–	–	–	–	–	–	–	–	–	–	–	–	–	–	–	–	–	–	–	–	–	–	–
35	76	83	–	1.52	3.36	1669	3680	–	–	1.35	2.97	1463	3225	1138	2509	–	–	0.87	1.92	0.59	1.31	0.82	1.80
36	78	82	–	1.61	3.54	1637	3609	–	–	1.43	3.16	1433	3160	1316	2901	–	–	0.93	2.06	0.65	1.43	0.87	1.92
37	67	62	–	1.21	2.66	1238	2729	–	–	0.84	1.84	1004	2213	1104	2434	–	–	0.48	1.05	0.24	0.53	0.65	1.43
38	97	96	–	1.94	4.28	1914	4219	–	–	1.77	3.90	1688	3722	–	–	–	–	1.12	2.48	0.80	1.76	1.03	2.28
39	41	44	–	0.85	1.87	866	1909	–	–	0.78	1.73	736	1623	755	1665	–	–	0.54	1.20	0.38	0.84	0.44	0.97
40	77	66	–	1.50	3.30	1467	3234	–	–	1.21	2.66	1278	2818	1322	2915	–	–	0.81	1.78	0.54	1.19	0.77	1.70
41	72	69	–	1.45	3.19	1370	3020	–	–	1.27	2.81	1124	2478	1245	2745	–	–	0.79	1.73	0.52	1.14	0.75	1.65
42	67	61	–	1.33	2.94	1578	3479	–	–	1.16	2.55	1194	2633	1292	2848	–	–	0.75	1.65	0.49	1.08	0.72	1.58
43	70	67	–	1.40	3.09	1327	2925	–	–	1.23	2.71	1148	2531	1313	2896	–	–	0.76	1.67	0.49	1.08	0.73	1.60
44	70	67	–	1.41	3.10	1336	2945	–	–	1.24	2.73	1156	2549	–	–	–	–	0.75	1.65	0.49	1.08	0.72	1.58
45	74	84	–	1.48	3.26	1675	3692	–	–	1.31	2.89	1471	3244	–	–	–	–	0.81	1.79	0.55	1.21	0.77	1.69
46	70	67	–	1.41	3.10	1336	2945	–	–	1.24	2.73	1156	2549	–	–	–	–	0.75	1.65	0.49	1.08	0.72	1.58
47	89	93	–	1.79	3.94	1867	4116	–	–	1.62	3.57	1645	3627	1306	2878	–	–	1.00	2.21	0.70	1.55	0.93	2.04
48	47	9	–	0.99	2.19	175	386	–	–	0.83	1.83	88	194	–	–	–	–	0.52	1.16	0.30	0.66	0.54	1.18
49	46	–	–	0.92	2.02	–	–	–	–	0.74	1.63	–	–	–	–	–	–	0.54	1.18	0.30	0.66	0.55	1.22
50	67	64	–	1.20	2.64	936	2064	–	–	0.89	1.95	1009	2225	947	2088	–	–	0.52	1.15	0.28	0.62	0.69	1.53
51	67	67	–	1.34	2.96	1112	2451	–	–	1.17	2.58	951	2096	1212	2673	–	–	0.74	1.62	0.48	1.05	0.71	1.56
52	63	68	–	1.44	3.17	1382	3047	–	–	1.26	2.79	1199	2644	1188	2618	–	–	0.90	1.99	0.62	1.36	0.85	1.86
53	66	68	–	1.32	2.91	1028	2267	–	–	1.14	2.51	981	2162	946	2086	–	–	0.71	1.57	0.45	1.00	0.69	1.52
54	16	15	–	0.32	0.71	309	681	–	–	0.30	0.65	275	606	–	–	–	–	0.19	0.42	0.14	0.30	0.18	0.40
55	9	9	–	0.18	0.39	188	415	–	–	0.16	0.35	166	365	–	–	–	–	0.10	0.22	0.07	0.16	0.10	0.21
56	80	86	–	1.43	3.15	1758	3876	–	–	1.07	2.37	1630	3593	1152	2539	–	–	0.69	1.52	0.43	0.95	0.83	1.82
57	17	–	–	0.34	0.75	–	–	–	–	0.32	0.70	–	–	–	–	–	–	–	–	–	–	–	–
58	47	–	–	0.95	2.08	–	–	–	–	0.82	1.81	–	–	–	–	–	–	–	–	–	–	–	–
59	–	–	–	–	–	–	–	–	–	–	–	–	–	–	–	–	–	–	–	–	–	–	–
60	77	–	–	1.56	3.44	1483	3268	–	–	1.39	3.07	1293	2850	959	2115	–	–	0.87	1.91	0.59	1.31	0.81	1.79
61	78	55	62	1.34	2.96	–	–	1.13	2.49	1.17	2.59	–	–	–	–	0.93	2.04	0.57	1.26	0.33	0.74	0.58	1.28
62	76	82	–	1.51	3.34	1638	3611	–	–	1.34	2.96	1435	3163	1108	2442	–	–	0.80	1.77	0.54	1.18	0.76	1.68
63	73	74	–	1.43	3.15	1296	2857	–	–	1.29	2.84	1264	2787	1229	2709	–	–	0.78	1.72	0.51	1.13	0.74	1.63
64	74	76	–	1.44	3.17	1406	3101	–	–	1.21	2.66	1301	2869	1300	2865	–	–	0.81	1.78	0.53	1.18	0.76	1.68
65	67	62	–	1.21	2.66	1238	2729	–	–	0.84	1.84	1004	2213	1104	2434	–	–	0.48	1.05	0.24	0.53	0.65	1.43
66	–	–	–	–	–	–	–	–	–	–	–	–	–	–	–	–	–	–	–	–	–	–	–
67	–	–	–	–	–	–	–	–	–	–	–	–	–	–	–	–	–	–	–	–	–	–	–
68	54	79	–	1.08	2.38	1586	3497	–	–	0.90	1.99	1387	3057	–	–	–	–	0.45	1.00	0.22	0.49	0.50	1.09
69	73	68	–	1.46	3.21	1354	2985	–	–	1.29	2.83	1173	2586	–	–	–	–	0.78	1.72	0.52	1.14	0.74	1.64
70	46	–	–	0.87	1.92	1089	2401	–	–	0.86	1.90	929	2047	623	1374	–	–	0.51	1.12	0.27	0.59	0.40	0.89
71	66	56	–	1.15	2.53	1508	3325	–	–	0.97	2.15	1317	2904	885	1951	–	–	0.58	1.27	0.33	0.74	0.56	1.24
72	69	81	–	1.38	3.04	1610	3549	–	–	1.13	2.48	1389	3063	1178	2598	–	–	0.71	1.56	0.46	1.01	0.79	1.75
73	41	–	–	0.82	1.82	–	–	–	–	0.65	1.43	–	–	871	1920	–	–	0.29	0.65	0.08	0.17	0.38	0.84
74	78	77	–	1.56	3.45	1535	3385	–	–	1.39	3.06	1338	2949	–	–	–	–	0.85	1.88	0.57	1.27	0.80	1.77
75	81	82	–	1.63	3.58	1649	3635	–	–	1.45	3.19	1442	3178	–	–	–	–	0.88	1.94	0.60	1.32	0.83	1.83
76	80	91	–	1.60	3.52	1581	3486	–	–	1.43	3.15	1386	3056	1237	2727	–	–	0.85	1.88	0.58	1.28	0.80	1.76
77	84	93	–	1.69	3.72	1820	4012	–	–	1.52	3.34	1605	3539	1534	3382	–	–	0.93	2.04	0.64	1.41	0.86	1.90
78	76	75	–	1.45	3.19	1565	3450	–	–	1.17	2.59	1430	3153	1005	2216	–	–	0.79	1.74	0.53	1.16	0.75	1.66
79	78	77	–	1.51	3.32	1591	3508	–	–	1.10	2.42	1433	3160	1124	2478	–	–	0.72	1.59	0.47	1.03	0.79	1.75
80	65	69	–	1.22	2.70	1366	3012	–	–	1.01	2.23	1173	2585	919	2026	–	–	0.59	1.30	0.34	0.74	0.63	1.40
81	–	–	–	–	–	–	–	–	–	–	–	–	–	–	–	–	–	–	–	–	–	–	–
82	–	–	–	–	–	–	–	–	–	–	–	–	–	–	–	–	–	–	–	–	–	–	–
83	78	80	–	1.52	3.35	1600	3527	–	–	1.35	2.96	1398	3081	–	–	–	–	0.81	1.79	0.54	1.19	0.77	1.70
84	–	–	–	–	–	–	–	–	–	–	–	–	–	–	–	–	–	–	–	–	–	–	–
85	83	80	–	1.66	3.66	1727	3807	–	–	1.50	3.30	1522	3357	1223	2696	–	–	0.96	2.11	0.67	1.48	0.89	1.95
86	84	–	–	1.69	3.73	2002	4413	–	–	1.52	3.35	1771	3905	–	–	–	–	0.96	2.11	0.67	1.47	0.89	1.96
87	7	–	–	0.13	0.29	–	–	–	–	0.12	0.26	–	–	–	–	–	–	–	–	–	–	–	–
88	76	77	–	1.51	3.33	1444	3183	–	–	1.28	2.83	1411	3110	880	1939	–	–	0.87	1.92	0.59	1.30	0.78	1.71
89	73	70	–	1.46	3.21	–	–	–	–	1.28	2.83	1299	2865	928	2047	–	–	0.81	1.79	0.54	1.19	0.77	1.69
90	72	–	–	1.43	3.16	–	–	–	–	1.26	2.77	–	–	–	–	–	–	–	–	–	–	–	–
91	72	64	–	1.49	3.29	1287	2837	–	–	1.27	2.81	1096	2416	840	1851	–	–	0.82	1.81	0.55	1.21	0.78	1.71

PROTEIN FEEDS

TABLE 19–1B PROTEIN FEEDS, MINERAL AND VITAMIN COMPOSITION OF FEEDS, DATA EXPRESSED **AS-FED**—*(Continued)*

			Macro Minerals							Micro Minerals						
Entry Number	Feed Name Description	Dry Matter	Calcium (Ca)	Phosphorus (P)	Sodium (Na)	Chlorine (Cl)	Magnesium (Mg)	Potassium (K)	Sulfur (S)	Cobalt (Co)	Copper (Cu)	Iodine (I)	Iron (Fe)	Manganese (Mn)	Selenium (Se)	Zinc (Zn)
		%	%	%	%	%	%	%	%	ppm or mg/kg	ppm or mg/kg	ppm or mg/kg	%	ppm or mg/kg	ppm or mg/kg	ppm or mg/kg
	COTTON (Continued)															
27	SEEDS, MEAL MECH EXTD, 41% PROTEIN	93	0.19	1.07	0.04	0.04	0.53	1.33	0.40	0.626	18.5	–	0.018	22.3	–	61.8
28	SEEDS, MEAL SOLV EXTD, 41% PROTEIN	91	0.17	1.11	0.04	0.04	0.54	1.37	0.25	0.483	19.5	–	0.019	20.6	–	60.7
29	SEEDS, MEAL SOLV EXTD, 46% PROTEIN	92	–	–	–	–	–	–	–	–	–	–	–	–	–	–
30	SEEDS, MEAL SOLV EXTD, 48% PROTEIN	90	0.20	1.20	–	–	–	–	–	–	–	–	–	–	–	–
31	SEEDS WITHOUT HULLS, MEAL, PREPRESSED, SOLV EXTD, 50% PROTEIN	93	0.18	1.16	0.05	0.05	0.46	1.45	0.52	0.042	14.5	–	0.012	23.0	–	73.8
32	**COTTON, GLANDLESS,** SEEDS, MEAL SOLV EXTD	95	–	–	–	–	–	–	–	–	–	–	–	–	–	–
33	**COWPEA, COMMON,** SEEDS	89	0.09	0.44	0.04	0.04	0.26	1.16	0.25	–	4.4	–	0.021	40.2	–	–
34	**DIAMMONIUM PHOSPHATE,** $(NH_4)_2HPO_4$	98	0.50	20.09	0.04	–	0.45	–	2.47	–	80.7	–	1.514	504.3	–	302.6
	DISTILLERS PRODUCTS (ALSO SEE CORN; RYE; SORGHUM; WHEAT)															
35	GRAINS, DEHY	93	0.12	0.54	0.05	0.05	0.09	0.20	0.46	0.092	47.9	–	0.027	35.0	–	–
36	SOLUBLES, DEHY	92	0.24	1.35	0.45	–	0.53	1.97	–	0.196	71.6	–	0.031	64.1	–	138.0
37	**FEATHERS, POULTRY,** MEAL, HYDROLYZED	93	0.30	0.62	0.63	0.28	0.18	0.27	1.50	0.116	7.3	0.044	0.023	11.9	0.913	71.9
	FISH															
38	LIVER, MEAL MECH EXTD	93	–	–	–	–	–	–	–	–	–	–	–	–	–	–
39	SOLUBLES, CONDENSED	50	0.16	0.57	2.45	2.93	0.03	1.64	0.12	0.069	46.6	1.111	0.028	13.2	–	43.2
40	SOLUBLES, DEHY	93	0.40	1.27	1.70	–	0.30	2.50	0.45	–	20.0	–	0.095	50.4	2.692	76.7
41	**FISH, ANCHOVY,** MEAL MECH EXTD	92	3.74	2.48	0.88	1.00	0.25	0.72	0.78	0.173	9.1	3.137	0.022	11.0	1.355	105.0
42	**FISH, MENHADEN,** MEAL MECH EXTD	92	5.19	2.88	0.41	0.55	0.15	0.70	0.56	0.153	10.3	1.091	0.055	37.0	2.147	144.2
43	**FISH, SARDINE,** MEAL MECH EXTD	93	4.61	2.68	0.18	0.41	0.10	0.32	–	0.183	20.2	–	0.030	23.2	1.772	–
44	**FLAX, COMMON,** SEEDS, MEAL SOLV EXTD, 35% PROTEIN (LINSEED MEAL)	90	0.40	0.82	0.14	–	0.60	1.37	0.39	–	–	–	–	–	–	–
45	**LENTIL, COMMON,** SEEDS	88	0.08	0.38	0.03	–	–	0.79	–	–	–	–	0.007	–	–	–
46	**LINSEED,** SEEDS, MEAL SOLV EXTD, 35% PROTEIN (LINSEED MEAL)	90	0.40	0.82	0.14	–	0.60	1.37	0.39	–	–	–	–	–	–	–
47	**LIVER,** MEAL	93	0.56	1.26	–	–	0.10	–	–	0.135	89.4	–	0.064	8.8	–	61.8
	MANURE, POULTRY															
48	WITH LITTER, DEHY	86	2.67	1.69	0.41	–	0.43	1.32	–	–	283.3	–	0.046	281.1	0.559	359.4
49	WITHOUT LITTER, DEHY	90	8.07	2.22	0.61	0.86	0.56	1.99	0.16	–	24.6	–	–	–	–	366.4
	MEAT															
50	MEAL RENDERED	94	8.61	4.58	1.05	1.11	0.25	0.55	0.46	2.250	9.6	–	0.050	11.8	0.505	74.3
51	WITH BLOOD, MEAL RENDERED (TANKAGE)	92	5.87	3.09	1.67	1.73	0.36	0.55	0.70	0.153	38.8	–	0.211	19.2	–	–
52	WITH BLOOD, WITH BONE, MEAL RENDERED (TANKAGE)	93	11.16	5.41	–	–	–	–	0.26	–	–	–	–	–	0.261	–
53	WITH BONE, MEAL RENDERED	93	10.00	4.94	0.72	0.75	1.02	1.33	0.25	0.181	1.5	1.317	0.066	13.3	0.263	94.3
	MILK															
54	FRESH (CATTLE)	12	0.12	0.09	0.05	0.11	0.01	0.14	0.04	0.001	0.1	–	0.002	–	–	2.3
55	SKIMMED, FRESH (CATTLE)	10	0.13	0.10	0.04	0.05	0.01	0.12	0.03	0.011	1.1	–	0.001	0.2	–	4.9
56	SKIMMED, DEHY (CATTLE)	94	1.28	1.02	0.51	0.90	0.12	1.60	0.32	0.113	11.7	–	0.001	2.1	0.124	38.5
57	FRESH (GOAT)	13	0.13	0.11	0.04	0.18	0.03	0.19	0.00	–	0.3	–	0.001	–	–	–
58	**MOLASSES AND SYRUP,** SUGARCANE, MOLASSES, AMMONIATED	65	0.79	0.13	–	–	–	–	–	–	–	–	–	–	–	–
59	**MONOAMMONIUM PHOSPHATE,** $NH_4H_2PO_4$	98	0.38	24.42	0.08	–	0.46	0.14	0.82	–	85.7	–	0.991	461.7	–	639.6
	PEA															
60	SEEDS	89	0.12	0.41	0.04	0.05	0.12	0.95	–	–	–	–	0.007	2.9	–	23.0
61	SPLIT SEED BY-PRODUCT (PEA FEED; PEA MEAL)	90	–	–	–	–	–	–	–	–	–	–	–	–	–	–
62	**PEA, FIELD,** SEEDS	91	0.16	0.38	0.00	–	0.14	1.36	–	1.700	11.7	–	0.020	21.2	0.393	46.8
63	**PEANUT,** SEEDS WITHOUT HULLS, MEAL SOLV EXTD (PEANUT MEAL)	93	0.36	0.61	0.03	0.03	0.27	1.16	0.31	–	–	–	–	–	–	–
	POULTRY															
64	BY-PRODUCT, MEAL RENDERED	94	3.97	2.06	0.78	0.54	0.14	0.51	0.53	4.926	19.9	3.101	0.064	16.5	0.920	193.5
65	FEATHERS, MEAL, HYDROLYZED	93	0.30	0.62	0.63	0.28	0.18	0.27	1.50	0.116	7.3	0.044	0.023	11.9	0.913	71.9
66	**RAPE,** SEEDS, MEAL SOLV EXTD, 34% PROTEIN	90	–	–	–	–	–	–	–	–	–	–	–	–	–	–
67	**RICE,** HULLS, AMMONIATED	92	0.15	0.19	–	–	–	–	–	–	–	–	–	–	–	–
	RYE															
68	DISTILLERS' GRAINS, DEHY	92	0.15	0.48	0.17	0.05	0.17	0.07	0.44	–	–	–	–	18.4	–	–
69	DISTILLERS' GRAINS WITH SOLUBLES, DEHY	90	–	–	–	–	–	–	–	–	–	–	–	–	–	–
	SAFFLOWER															
70	SEEDS, MEAL SOLV EXTD, 20% PROTEIN	92	0.31	0.61	–	–	0.32	0.74	0.20	–	9.6	–	0.043	17.7	–	39.6
71	SEEDS WITHOUT HULLS, MEAL SOLV EXTD, 42% PROTEIN	92	0.38	1.08	–	–	1.18	1.18	0.34	1.832	80.6	–	0.091	36.6	–	168.5
72	**SESAME,** SEEDS, MEAL SOLV EXTD, 44% PROTEIN	92	2.01	1.28	–	–	–	–	–	–	–	–	–	47.5	–	–
73	**SHRIMP,** CANNERY RESIDUE, MEAL (SHRIMP MEAL)	90	10.40	1.85	1.57	1.04	0.54	0.83	–	–	–	–	0.011	29.8	–	28.4
	SORGHUM															
74	DISTILLERS' GRAINS, DEHY	94	0.15	0.69	0.05	–	0.18	0.36	0.17	–	–	–	0.005	–	–	–
75	DISTILLERS' GRAINS WITH SOLUBLES, DEHY	95	0.17	0.92	–	–	–	–	–	–	–	–	–	104.5	–	–
76	GLUTEN MEAL	90	0.03	0.27	–	–	0.16	0.48	–	–	–	–	–	15.6	–	–
	SOYBEAN															
77	SEEDS	92	0.25	0.60	0.00	0.03	0.27	1.66	0.22	–	18.2	–	0.009	36.4	0.111	56.9
78	SEEDS, MEAL SOLV EXTD, 44% PROTEIN	89	0.35	0.64	0.03	–	0.27	1.98	0.41	1.381	19.9	–	0.017	31.6	0.486	50.5
79	SEEDS WITHOUT HULLS, MEAL SOLV EXTD, 49% PROTEIN	90	0.25	0.63	0.00	0.07	0.37	1.79	0.41	2.693	13.5	0.152	0.010	49.5	–	51.1
80	**SUNFLOWER, COMMON,** SEEDS WITHOUT HULLS, MEAL SOLV EXTD, 44% PROTEIN	93	–	–	–	–	–	–	–	–	–	–	–	–	–	–
81	**UREA,** 45% NITROGEN, 281% PROTEIN EQUIVALENT	99	–	–	–	–	0.00	–	–	–	6.9	–	0.018	–	–	6.9
82	**UREA, CONDITIONER ADDED,** 42% NITROGEN, 262% PROTEIN EQUIVALENT	100	–	–	–	–	–	–	–	–	–	–	–	–	–	–
	WHEAT															
83	DISTILLERS' GRAINS, DEHY	93	0.11	0.58	–	–	–	–	–	–	–	–	–	15.0	–	–
84	DISTILLERS' SOLUBLES, DEHY	94	–	–	–	–	–	–	–	–	–	–	–	–	–	–
85	GERM MEAL	88	0.06	0.95	0.02	0.06	0.25	0.94	0.27	0.120	9.2	–	0.006	132.5	0.463	119.4
86	GLUTEN	90	0.06	0.23	0.06	–	0.04	0.02	0.95	0.049	11.6	0.058	0.006	18.1	3.753	38.5
	WHEY, CATTLE															
87	FRESH	7	0.06	0.05	–	–	–	0.19	–	–	–	–	0.003	0.2	–	–
88	DEHY	93	0.86	0.76	0.62	0.07	0.13	1.11	1.04	0.111	46.5	–	0.017	5.9	–	3.2
89	**YEAST, BREWERS,** DEHY	93	0.14	1.36	0.07	0.07	0.24	1.69	0.43	0.506	38.4	0.358	0.009	6.7	0.911	39.0
90	**YEAST, IRRADIATED,** DEHY	94	0.78	1.42	–	–	–	2.14	–	–	–	–	–	–	–	–
91	**YEAST, TORULA,** DEHY	93	0.55	1.61	0.01	0.02	0.14	1.92	0.55	0.031	11.9	2.502	0.011	9.3	–	99.5

Entry Number	Fat-Soluble Vitamins: A (1 mg Carotene = 1667 IU Vit A)	Carotene (Provitamin A)	D	E	K	Water-Soluble Vitamins: B-12	Biotin	Choline	Folacin (Folic Acid)	Niacin	Pantothenic Acid (B-3)	(Pyridoxine) B-6	Riboflavin (B-2)	Thiamin (B-1)
	IU/g	ppm or mg/kg	IU/kg	ppm or mg/kg	ppm or mg/kg	ppb or mcg/kg	ppm or mg/kg	ppm or mg/kg	ppm or mg/kg	ppm or mg/kg	ppm or mg/kg	ppm or mg/kg	ppm or mg/kg	ppm or mg/kg
27	0.4	0.2	–	32.3	–	–	0.91	2753	2.45	35	10.2	5.00	5.2	7.1
28	–	–	–	14.6	–	–	0.55	2780	2.55	41	13.7	5.41	4.7	7.3
29	–	–	–	–	–	–	–	–	–	–	–	–	–	–
30	–	–	–	–	–	–	–	3316	–	51	15.5	–	6.0	–
31	–	–	–	11.3	–	–	0.44	2962	0.93	45	14.3	6.29	4.9	8.2
32	–	–	–	–	–	–	–	–	–	–	–	–	–	–
33	0.4	0.2	–	–	–	–	–	–	–	24	15.5	–	2.3	9.3
34	–	–	–	–	–	–	–	–	–	–	–	–	4.0	–
35	13.0	7.8	–	30.5	–	–	–	2645	–	47	11.9	6.00	6.6	2.5
36	1.9	1.1	–	–	–	2.9	2.84	4992	–	143	25.3	8.66	11.3	6.9
37	–	–	–	–	–	80.4	0.04	894	0.22	21	8.9	4.39	2.0	0.1
38	–	–	–	–	–	–	–	–	–	–	–	–	–	–
39	2.2	1.3	–	–	–	506.6	0.14	3370	0.22	176	35.7	12.20	12.9	5.5
40	–	–	–	6.1	–	485.9	0.40	5525	0.57	256	50.4	19.71	13.5	7.4
41	–	–	–	3.7	–	214.5	0.20	3700	0.16	81	10.0	4.71	7.3	0.5
42	–	–	–	6.8	–	122.0	0.18	3112	0.15	55	8.6	3.80	4.8	0.6
43	–	–	–	–	–	238.0	0.10	3277	–	75	11.0	–	5.4	0.3
44	–	–	–	5.9	–	–	–	1216	2.85	30	–	9.93	2.9	9.4
45	–	–	–	–	–	–	–	–	–	20	–	–	2.2	3.7
46	–	–	–	5.9	–	–	–	1216	2.85	30	–	9.93	2.9	9.4
47	–	–	–	–	–	501.3	0.02	11370	5.56	205	29.2	–	36.2	0.2
48	–	–	–	–	–	–	–	–	–	–	–	–	–	–
49	–	–	–	–	–	21.1	–	–	–	19	–	–	11.7	–
50	–	–	–	0.9	–	75.2	0.12	1980	0.39	56	6.0	4.23	5.2	0.2
51	–	–	–	–	–	89.4	–	2203	1.54	38	3.2	–	2.2	0.3
52	–	–	–	0.8	–	104.4	0.07	2067	0.57	58	4.8	–	5.0	0.2
53	–	–	–	0.9	–	118.4	0.10	2049	0.37	51	5.5	5.86	4.7	0.2
54	–	–	–	–	–	–	–	904	–	1	8.4	–	1.7	0.3
55	–	–	–	–	–	–	–	–	–	1	3.5	–	2.0	0.4
56	–	–	0	9.1	–	50.9	0.33	1394	0.62	11	36.4	4.10	19.1	3.7
57	–	–	–	–	–	–	–	–	–	–	–	–	–	–
58	–	–	–	–	–	–	–	–	–	–	–	–	–	–
59	–	–	–	–	–	–	–	–	–	–	–	–	–	–
60	1.2	0.7	–	3.0	–	–	0.18	547	0.22	31	27.8	1.97	1.8	4.6
61	–	–	–	–	–	–	–	–	–	–	–	–	–	–
62	–	–	–	–	–	–	0.19	654	0.36	34	7.4	1.01	1.4	4.1
63	–	–	–	2.9	–	–	–	1896	–	178	36.8	5.95	5.3	–
64	–	–	–	2.2	–	304.1	0.09	6052	0.51	54	12.4	4.43	10.6	0.2
65	–	–	–	–	–	80.4	0.04	894	0.22	21	8.9	4.39	2.0	0.1
66	–	–	–	–	–	–	–	–	–	–	–	–	–	–
67	–	–	–	–	–	–	–	–	–	–	–	–	–	–
68	–	–	–	–	–	–	–	–	–	17	5.2	–	3.3	1.3
69	–	–	–	–	–	–	–	–	–	63	17.4	–	8.2	3.1
70	–	–	–	0.9	–	–	–	1541	–	12	36.2	474.43	2.2	–
71	–	–	–	0.6	–	–	1.56	3156	1.47	21	38.2	10.71	2.3	4.2
72	–	–	–	–	–	–	–	1517	–	–	6.3	–	3.7	–
73	–	–	–	–	–	–	–	5497	–	–	–	–	3.9	–
74	–	–	–	–	–	–	0.31	805	–	–	–	–	–	–
75	–	–	–	–	–	–	–	844	–	61	12.3	–	4.2	1.3
76	–	–	–	–	–	–	–	680	–	37	9.3	–	1.5	–
77	1.5	0.9	–	33.7	–	–	0.38	2931	–	23	16.0	11.04	2.9	11.3
78	–	–	–	3.0	0.22	2.0	0.36	2706	0.69	26	13.8	5.90	3.0	6.6
79	–	–	0	3.3	–	2.0	0.38	2772	0.59	24	14.1	5.59	2.9	3.5
80	–	–	–	–	–	–	–	–	–	–	–	–	–	–
81	–	–	–	–	–	–	–	–	–	–	–	–	–	–
82	–	–	–	–	–	–	–	–	–	–	–	–	–	–
83	1.8	1.1	–	–	–	–	–	–	–	56	8.2	–	3.7	2.0
84	–	–	–	–	–	–	–	–	–	–	–	–	–	–
85	–	–	–	141.2	–	–	0.22	3062	2.12	68	18.6	9.97	6.0	23.1
86	–	–	0	34.1	–	73.1	0.00	577	0.74	74	5.8	2.26	0.7	0.9
87	–	–	–	–	–	–	–	–	–	1	5.3	–	1.4	0.3
88	–	–	–	0.2	–	18.9	0.35	1790	0.85	11	46.2	3.21	27.4	4.0
89	–	–	–	2.1	–	1.1	1.04	3847	9.69	443	81.5	36.67	34.1	85.2
90	–	–	–	–	–	–	–	–	–	–	–	–	18.5	–
91	–	–	–	–	–	4.0	1.19	2981	25.66	512	107.5	34.48	47.7	6.8

TABLE 19–1C DRY FORAGES, COMPOSITION OF FEEDS, DATA EXPRESSED AS-FED

Entry Number	Feed Name Description	International Feed Number	Proximate Analysis: Dry Matter	Ash	Crude Fiber	Neutral Det. Fib. (NDF)	Acid Det. Fib. (ADF)	Lignin	Ether Extract (Fat)	N-Free Extract	Crude Protein	Digestible Protein: Ruminant	Swine	Horse
			%	%	%	%	%	%	%	%	%	%	%	%
	ALFALFA (LUCERNE)													
1	HAY, SUN-CURED, ALL ANALYSES	1-00-078	90	8.6	28.2	35.4	30.9	8.9	1.7	35.9	16.0	11.2	7.5	11.9
2	HAY, PREBLOOM, SUN-CURED	1-00-054	90	8.3	20.7	38.3	29.6	5.5	4.1	36.4	20.2	15.3	14.3	14.0
3	HAY, EARLY BLOOM, SUN-CURED	1-00-059	91	8.4	25.8	36.8	29.0	5.8	2.6	35.8	17.9	13.3	12.3	12.2
4	HAY, MIDBLOOM, SUN-CURED	1-00-063	91	7.8	25.5	43.2	33.4	6.7	3.3	37.4	17.1	12.0	11.5	11.6
5	HAY, FULL BLOOM, SUN-CURED	1-00-068	91	7.1	27.3	45.0	35.2	6.9	3.1	37.9	15.5	11.3	10.1	10.3
6	HAY, MATURE, SUN-CURED	1-00-071	91	6.7	29.3	50.1	40.1	11.3	2.9	37.0	15.2	11.3	9.8	10.1
7	HAY, RAINED ON, SUN-CURED	1-00-130	89	6.6	33.5	–	–	–	0.8	30.4	17.9	12.9	12.2	12.2
8	MEAL, DEHY, 17% PROTEIN	1-00-023	92	9.7	24.0	41.3	31.5	9.7	2.8	37.8	17.4	12.6	11.7	11.8
9	MEAL, DEHY, 20% PROTEIN	1-00-024	92	10.2	20.8	38.6	28.5	–	3.3	37.1	20.2	15.1	14.4	14.0
10	MEAL, DEHY, 22% PROTEIN	1-07-851	93	10.2	18.3	36.3	26.0	–	4.1	38.1	22.2	16.4	16.4	15.5
11	**ALFALFA-BROMEGRASS, SMOOTH,** HAY, SUN-CURED	1-00-255	91	6.1	31.0	–	–	–	2.1	37.7	14.1	9.6	8.9	9.2
12	**ALFALFA-GRASS,** HAY, SUN-CURED	1-08-331	91	6.7	30.3	–	36.6	–	2.1	37.8	14.5	9.9	9.2	9.5
13	**ALFALFA-ORCHARDGRASS,** HAY, SUN-CURED	1-00-322	89	7.3	28.7	–	–	–	1.9	36.8	14.8	10.2	9.5	9.8
14	**ALMOND,** HULLS	4-00-359	90	6.1	13.5	28.8	25.2	–	2.9	63.8	4.1	0.0	2.0	2.4
15	**APPLE,** POMACE, DEHY	4-00-423	89	3.0	16.2	–	23.1	–	4.3	61.2	4.4	–0.4	2.3	2.7
16	**BAGASSE (SUGARCANE),** PULP, DEHY	1-04-686	91	2.9	42.3	78.8	54.5	12.8	0.7	43.8	1.4	–1.5	–2.3	–0.5
	BARLEY													
17	HAY, SUN-CURED	1-00-495	88	6.6	23.6	–	–	–	1.9	48.5	7.8	4.3	3.4	4.4
18	STRAW	1-00-498	91	6.7	37.9	77.5	51.1	6.9	1.7	41.1	4.0	0.6	0.0	1.4
	BEAN													
19	HAY, SUN-CURED	1-00-583	90	6.3	24.1	–	–	–	1.7	44.4	13.9	9.4	8.7	9.1
20	STRAW	1-00-585	89	8.7	36.8	–	50.0	–	1.3	35.2	7.1	3.5	2.8	3.9
21	**BERMUDAGRASS,** HAY, SUN-CURED	1-00-703	91	8.0	28.4	–	–	–	1.8	43.7	9.2	5.5	4.5	5.4
22	**BERMUDAGRASS, COASTAL,** HAY, SUN-CURED	1-00-716	91	6.3	27.0	69.2	34.6	–	2.0	43.9	11.7	7.5	6.7	7.4
23	**BIRDSFOOT TREFOIL,** HAY, SUN-CURED	1-05-044	91	6.7	29.3	–	–	–	1.9	38.9	13.9	9.6	8.7	9.1
24	**BLUEGRASS, CANADA,** HAY, SUN-CURED	1-00-762	92	6.5	27.6	–	–	–	2.4	45.8	9.5	4.1	4.8	5.7
25	**BLUEGRASS, KENTUCKY,** HAY, SUN-CURED	1-00-776	89	5.9	26.8	–	–	–	3.0	44.3	9.1	5.3	4.6	5.4
26	**BLUESTEM,** HAY, SUN-CURED	1-00-819	90	6.3	30.8	–	–	–	2.2	45.6	4.9	1.5	0.8	2.1
27	**BROMEGRASS,** HAY, SUN-CURED	1-00-890	91	7.1	29.8	–	31.7	4.3	1.9	43.3	8.7	4.8	4.1	5.1
28	**BUFFALOGRASS,** HAY, SUN-CURED	1-01-003	90	11.9	24.9	–	–	–	1.5	45.1	6.9	3.7	2.5	3.7
29	**BUR-CLOVER, TOOTHED,** HAY, SUN-CURED	1-01-030	90	8.7	22.5	–	–	–	2.5	38.9	17.0	12.2	11.5	11.5
30	**CANADA BLUEGRASS,** HAY, SUN-CURED	1-00-762	92	6.5	27.6	–	–	–	2.4	45.8	9.5	4.1	4.8	5.7
31	**CANARYGRASS, REED,** HAY, SUN-CURED	1-01-104	89	7.3	30.2	62.9	32.7	–	2.7	40.0	9.1	5.8	4.6	5.5
32	**CEREALS,** IMMATURE, DEHY	1-26-069	92	14.4	16.0	–	–	–	4.8	32.5	24.4	19.0	–	–
33	**CLOVER, ALSIKE,** HAY, SUN-CURED	1-01-313	88	7.6	26.2	–	–	–	2.4	39.1	12.4	8.3	7.5	8.0
34	**CLOVER, ALYCE,** HAY, SUN-CURED	1-00-361	90	5.7	36.2	–	–	–	1.6	35.3	10.9	6.7	6.1	6.8
35	**CLOVER, CRIMSON,** HAY, SUN-CURED	1-01-328	88	7.8	28.1	–	–	–	2.0	35.2	14.7	10.2	9.5	9.8
36	**CLOVER, LADINO,** HAY, SUN-CURED	1-01-378	89	8.4	18.5	32.1	28.5	5.9	2.4	39.9	20.0	15.4	14.1	13.8
37	**CLOVER, RED,** HAY, SUN-CURED, ALL ANALYSES	1-01-415	88	6.7	27.1	49.5	36.2	8.8	2.5	39.2	13.0	7.7	8.0	8.4
38	**CLOVER, RED-GRASS,** HAY, FULL BLOOM, SUN-CURED	1-01-532	89	5.9	29.9	–	–	–	1.6	37.4	14.0	9.6	8.9	9.2
39	**CLOVER, SUBTERRANEAN,** HAY, SUN-CURED	1-20-278	90	10.0	9.1	–	–	–	3.3	40.1	27.5	21.7	–	20.5
40	**CLOVER, SWEET, YELLOW,** HAY, SUN-CURED, ALL ANALYSES	1-04-754	89	7.6	28.9	–	–	–	1.9	36.4	13.7	9.5	8.6	9.0
41	**CLOVER, WHITE,** HAY, SUN-CURED	1-01-464	90	9.0	21.9	–	–	–	2.4	40.1	16.9	12.4	11.4	11.5
42	**CLOVER-TIMOTHY,** HAY, FULL BLOOM, SUN-CURED	1-01-484	90	5.7	32.2	–	–	–	3.1	39.6	9.5	5.5	4.8	5.7
	CORN													
43	FODDER WITH EARS, WITH HUSKS, SUN-CURED	1-02-775	90	4.9	32.6	–	–	–	7.4	29.5	15.6	4.0	3.6	11.0
44	STOVER WITHOUT EARS, WITHOUT HUSKS, SUN-CURED	1-02-776	85	6.1	29.3	57.0	33.2	–	1.1	43.2	5.4	2.3	1.5	2.7
45	HUSKS, SUN-CURED	1-02-785	89	3.2	30.0	–	–	–	0.8	51.2	3.3	0.7	–0.5	0.9
46	COBS, GROUND	1-02-782	90	1.6	32.2	80.1	31.5	–	0.6	52.7	2.8	–0.4	–1.0	0.4
47	**CORN, SWEET,** FODDER WITH EARS, WITH HUSKS, SUN-CURED	1-08-407	88	9.0	26.4	–	–	–	1.8	41.3	9.2	5.9	4.7	5.5
48	**COTTON,** HULLS	1-01-599	90	2.6	43.2	81.0	65.7	21.0	1.5	39.3	3.8	–0.4	–0.2	1.3
49	**COWPEA, COMMON,** HAY, SUN-CURED	1-01-645	90	10.5	24.4	–	–	–	2.6	35.1	17.7	12.2	12.0	12.0
50	**CRESTED WHEATGRASS,** HAY, SUN-CURED	1-05-418	92	6.4	30.8	–	33.2	5.1	2.1	42.2	10.3	6.4	5.5	6.3
51	**CRIMSON CLOVER,** HAY, SUN-CURED	1-01-328	88	7.8	28.1	–	–	–	2.0	35.2	14.7	10.2	9.5	9.8
52	**DALLISGRASS,** HAY, SUN-CURED	1-01-737	91	7.9	30.8	–	–	–	1.9	41.1	9.2	5.3	4.6	5.5
53	**FESCUE, TALL (ALTA),** HAY, SUN-CURED	1-05-684	89	5.9	32.6	61.7	35.6	–	2.0	41.4	7.2	3.6	2.9	3.9
54	**FLAX, COMMON,** STRAW	1-02-038	93	5.5	48.1	–	–	–	2.2	32.4	5.0	2.2	0.8	2.2
55	**FOXTAIL, MEADOW,** HAY, SUN-CURED	1-02-072	90	8.4	25.5	–	–	–	2.0	41.1	12.8	8.5	7.7	8.2
56	**GRAMA,** HAY, SUN-CURED	1-02-162	89	8.4	29.1	–	–	–	1.5	44.5	5.6	2.2	1.5	2.7
57	**GRASS,** HAY, SUN-CURED, ALL ANALYSES	1-02-250	89	7.3	29.7	–	34.4	–	2.3	41.2	8.9	5.2	4.4	5.7
58	**GRASS-LEGUME,** HAY, SUN-CURED	1-02-301	89	5.6	31.6	–	33.9	–	2.3	39.5	10.3	6.0	5.6	6.3
59	**JOHNSONGRASS SORGHUM,** HAY, SUN-CURED	1-04-407	91	7.7	30.4	–	–	–	2.0	43.7	6.7	3.0	2.4	3.6
	KAFIR SORGHUM—SEE SORGHUM													
60	**KELP (SEAWEED),** WHOLE, DEHY	1-08-073	91	35.0	6.5	–	–	–	0.5	42.4	6.5	2.9	2.2	3.4
61	**KENTUCKY BLUEGRASS,** HAY, SUN-CURED	1-00-776	89	5.9	26.8	–	–	–	3.0	44.3	9.1	5.3	4.6	5.4
62	**LESPEDEZA, COMMON,** HAY, SUN-CURED, ALL ANALYSES	1-08-591	89	4.7	28.4	–	–	–	2.5	39.4	13.8	5.9	8.7	9.1
63	**LESPEDEZA, SERICEA (CHINESE LESPEDEZA),** HAY, SUN-CURED, ALL ANALYSES	1-02-607	90	4.8	29.9	–	–	–	2.0	42.9	10.7	3.5	5.9	6.6
64	**MEADOW FESCUE,** HAY, SUN-CURED	1-01-912	88	7.9	28.0	74.0	43.8	–	2.4	41.0	8.2	4.6	3.9	5.3
65	**MEADOW FOXTAIL,** HAY, SUN-CURED	1-02-072	90	8.4	25.5	–	–	–	2.0	41.1	12.8	8.5	7.7	8.2
66	**MILLET, FOXTAIL,** HAY, SUN-CURED	1-03-099	87	7.5	25.7	–	–	–	2.5	43.7	7.5	3.7	3.2	4.2
67	**MILLET, JAPANESE (BARNYARD GRASS),** HAY, SUN-CURED	1-03-105	87	8.5	26.3	–	–	–	1.7	41.8	8.8	5.4	4.4	5.3
	MILO SORGHUM—SEE SORGHUM													
68	**NEEDLEGRASS,** HAY, SUN-CURED	1-03-202	88	6.2	30.0	–	–	–	2.2	41.9	7.8	4.1	3.4	4.4
69	**OATGRASS, TALL,** HAY, SUN-CURED	1-03-259	88	6.7	29.6	–	–	–	2.0	42.5	6.7	3.1	2.5	3.6
	OATS													
70	HAY, SUN-CURED, ALL ANALYSES	1-03-280	91	7.2	29.1	–	34.8	–	2.2	43.6	8.6	4.5	4.1	5.0
71	STRAW	1-03-283	92	7.2	37.2	65.7	43.1	7.0	2.0	41.6	4.1	0.8	0.0	2.4
72	**OATS-PEA,** HAY, SUN-CURED	1-03-398	88	7.4	26.9	–	–	–	3.0	38.7	11.7	8.3	6.9	7.4
73	**ORCHARDGRASS,** HAY, SUN-CURED	1-03-438	89	6.5	31.0	64.1	36.0	–	2.8	39.7	9.4	5.7	4.8	5.6
74	**PEA,** HAY, SUN-CURED	1-03-572	87	7.4	23.7	–	–	–	2.6	41.7	11.8	7.7	7.0	7.6
75	**PEA, FIELD,** HAY (VINES WITHOUT SEEDS, WITH PODS), SUN-CURED	1-03-607	88	6.8	23.4	–	–	–	2.3	42.5	12.7	8.5	7.8	8.2

Entry Number	TDN			Digestible Energy						Metabolizable Energy								Net Energy					
	Ruminant	Swine	Horse	Ruminant		Swine		Horse		Ruminant		Swine		Poultry ME_n		Horse		Ruminant NE_m		Ruminant NE_g		Lactating Cows NE_{lc}	
	%	%	%	Mcal		kcal		Mcal		Mcal		kcal		kcal		Mcal		Mcal		Mcal		Mcal	
				lb	*kg*	lb	*kg*	lb	*kg*	lb	*kg*	lb	*kg*	lb	*kg*	lb	*kg*	lb	*kg*	lb	*kg*	lb	*kg*
1	51	32	48	1.03	*2.28*	–	–	0.90	*1.98*	0.90	*1.99*	–	–	–	–	0.74	*1.62*	0.48	*1.06*	0.25	*0.55*	0.49	*1.07*
2	54	64	48	1.25	*2.75*	–	–	0.90	*1.98*	1.02	*2.24*	–	–	–	–	0.74	*1.63*	0.57	*1.26*	0.34	*0.74*	0.58	*1.27*
3	52	50	48	1.15	*2.54*	–	–	0.90	*1.99*	0.96	*2.12*	–	–	–	–	0.74	*1.63*	0.60	*1.33*	0.36	*0.80*	0.55	*1.21*
4	52	54	46	1.12	*2.46*	–	–	0.86	*1.90*	0.94	*2.07*	–	–	–	–	0.71	*1.56*	0.52	*1.14*	0.28	*0.62*	0.51	*1.13*
5	51	50	44	1.09	*2.39*	–	–	0.83	*1.83*	0.88	*1.95*	–	–	–	–	0.68	*1.50*	0.49	*1.09*	0.26	*0.57*	0.52	*1.15*
6	55	47	43	1.22	*2.68*	–	–	0.82	*1.81*	0.93	*2.06*	–	–	–	–	0.67	*1.48*	0.47	*1.04*	0.24	*0.53*	0.51	*1.12*
7	47	–	50	1.00	*2.20*	–	–	0.94	*2.07*	0.83	*1.83*	–	–	–	–	0.77	*1.70*	0.44	*0.96*	0.21	*0.46*	0.48	*1.05*
8	55	44	47	1.12	*2.47*	643	*1418*	0.89	*1.96*	0.96	*2.12*	601	*1326*	682	*1504*	0.73	*1.61*	0.60	*1.32*	0.36	*0.79*	0.57	*1.25*
9	57	48	50	1.07	*2.36*	943	*2079*	0.94	*2.08*	0.93	*2.06*	872	*1923*	737	*1625*	0.77	*1.70*	0.58	*1.28*	0.34	*0.75*	0.59	*1.30*
10	60	49	52	1.20	*2.65*	991	*2186*	0.98	*2.15*	1.02	*2.26*	841	*1855*	768	*1692*	0.80	*1.76*	0.63	*1.39*	0.38	*0.84*	0.63	*1.38*
11	58	41	45	1.11	*2.44*	–	–	0.85	*1.88*	0.93	*2.06*	–	–	–	–	0.70	*1.54*	0.59	*1.30*	0.35	*0.76*	0.59	*1.31*
12	51	42	45	1.01	*2.23*	–	–	0.86	*1.89*	0.83	*1.84*	–	–	–	–	0.70	*1.55*	0.50	*1.10*	0.26	*0.58*	0.53	*1.16*
13	52	42	47	1.03	*2.27*	–	–	0.88	*1.93*	0.86	*1.89*	–	–	–	–	0.72	*1.58*	0.49	*1.08*	0.26	*0.57*	0.52	*1.14*
14	66	73	45	1.34	*2.95*	1457	*3213*	0.86	*1.89*	1.15	*2.54*	1375	*3032*	–	–	0.70	*1.55*	0.77	*1.70*	0.51	*1.12*	0.66	*1.46*
15	60	69	40	1.21	*2.66*	1380	*3043*	0.77	*1.71*	1.04	*2.28*	1300	*2867*	–	–	0.64	*1.40*	0.66	*1.46*	0.42	*0.91*	0.64	*1.42*
16	43	–	–	0.88	*1.93*	–	–	–	–	0.69	*1.53*	–	–	–	–	–	–	0.37	*0.82*	0.15	*0.32*	0.40	*0.88*
17	50	40	42	0.99	*2.17*	–	–	0.81	*1.77*	0.79	*1.73*	–	–	–	–	0.66	*1.46*	0.45	*0.99*	0.22	*0.49*	0.50	*1.10*
18	43	–	28	0.84	*1.86*	–	–	0.57	*1.26*	0.64	*1.40*	–	–	–	–	0.47	*1.03*	0.29	*0.64*	0.07	*0.15*	0.42	*0.91*
19	59	48	52	1.13	*2.48*	–	–	0.97	*2.13*	0.95	*2.10*	–	–	–	–	0.79	*1.74*	0.52	*1.15*	0.29	*0.64*	0.54	*1.20*
20	43	–	30	0.87	*1.91*	–	–	0.60	*1.32*	0.69	*1.53*	–	–	–	–	0.49	*1.08*	0.36	*0.79*	0.14	*0.30*	0.42	*0.93*
21	43	35	40	0.88	*1.93*	–	–	0.77	*1.70*	0.64	*1.41*	–	–	–	–	0.63	*1.40*	0.29	*0.65*	0.07	*0.16*	0.40	*0.88*
22	49	43	46	1.04	*2.30*	–	–	0.86	*1.90*	0.66	*1.44*	–	–	–	–	0.71	*1.56*	0.31	*0.68*	0.09	*0.20*	0.53	*1.16*
23	54	41	46	0.91	*2.01*	–	–	0.87	*1.91*	0.83	*1.84*	–	–	–	–	0.71	*1.56*	0.55	*1.22*	0.32	*0.69*	0.57	*1.25*
24	57	42	41	1.14	*2.52*	–	–	0.79	*1.75*	0.97	*2.13*	–	–	–	–	0.65	*1.43*	0.61	*1.35*	0.37	*0.81*	0.61	*1.35*
25	54	44	37	1.06	*2.34*	–	–	0.73	*1.60*	0.89	*1.96*	–	–	–	–	0.60	*1.31*	0.51	*1.12*	0.28	*0.62*	0.53	*1.17*
26	41	31	33	0.88	*1.95*	–	–	0.65	*1.43*	0.71	*1.56*	–	–	–	–	0.53	*1.17*	0.44	*0.96*	0.21	*0.47*	0.48	*1.06*
27	48	35	39	1.12	*2.48*	–	–	0.75	*1.65*	0.95	*2.09*	–	–	–	–	0.61	*1.35*	0.50	*1.11*	0.27	*0.60*	0.50	*1.11*
28	48	30	35	0.93	*2.05*	–	–	0.69	*1.53*	0.76	*1.66*	–	–	–	–	0.57	*1.25*	0.40	*0.88*	0.18	*0.39*	0.45	*1.00*
29	54	52	49	1.08	*2.39*	–	–	0.92	*2.03*	0.91	*2.01*	–	–	–	–	0.75	*1.66*	0.54	*1.18*	0.30	*0.66*	0.55	*1.21*
30	57	42	41	1.14	*2.52*	–	–	0.79	*1.75*	0.97	*2.13*	–	–	–	–	0.65	*1.43*	0.61	*1.35*	0.37	*0.81*	0.61	*1.35*
31	44	36	43	0.93	*2.06*	–	–	0.82	*1.80*	0.76	*1.68*	–	–	–	–	0.67	*1.47*	0.46	*1.02*	0.24	*0.52*	0.50	*1.10*
32	56	–	–	1.13	*2.48*	–	–	–	–	0.92	*2.02*	–	–	–	–	–	–	–	–	–	–	–	–
33	51	42	41	1.01	*2.24*	–	–	0.78	*1.72*	0.85	*1.86*	–	–	–	–	0.64	*1.41*	0.49	*1.08*	0.26	*0.58*	0.51	*1.13*
34	44	–	38	0.89	*1.96*	–	–	0.74	*1.63*	0.71	*1.57*	–	–	–	–	0.61	*1.34*	0.30	*0.66*	0.08	*0.18*	0.38	*0.85*
35	50	41	43	1.00	*2.21*	–	–	0.82	*1.81*	0.84	*1.84*	–	–	–	–	0.67	*1.49*	0.54	*1.19*	0.31	*0.68*	0.55	*1.21*
36	58	60	57	1.16	*2.55*	–	–	1.05	*2.31*	0.99	*2.18*	–	–	–	–	0.86	*1.89*	0.58	*1.27*	0.34	*0.75*	0.58	*1.28*
37	52	44	42	1.21	*2.67*	–	–	0.81	*1.78*	0.95	*2.10*	–	–	–	–	0.66	*1.46*	0.50	*1.11*	0.28	*0.61*	0.53	*1.16*
38	51	39	47	1.02	*2.25*	–	–	0.88	*1.93*	0.85	*1.87*	–	–	–	–	0.72	*1.59*	0.48	*1.06*	0.25	*0.56*	0.51	*1.12*
39	–	–	–	–	–	–	–	–	–	–	–	–	–	–	–	–	–	–	–	–	–	–	–
40	49	39	43	0.98	*2.16*	–	–	0.81	*1.79*	0.81	*1.78*	–	–	–	–	0.67	*1.47*	0.49	*1.07*	0.26	*0.57*	0.51	*1.13*
41	57	52	50	1.11	*2.45*	–	–	0.94	*2.07*	0.94	*2.07*	–	–	–	–	0.77	*1.70*	0.53	*1.18*	0.30	*0.66*	0.55	*1.22*
42	51	39	34	1.01	*2.23*	–	–	0.67	*1.47*	0.84	*1.85*	–	–	–	–	0.55	*1.21*	0.49	*1.08*	0.26	*0.57*	0.52	*1.14*
43	45	45	44	0.91	*2.00*	–	–	0.84	*1.85*	0.74	*1.64*	–	–	–	–	0.69	*1.52*	0.44	*0.97*	0.12	*0.26*	0.46	*1.01*
44	51	26	36	1.01	*2.23*	–	–	0.70	*1.54*	0.85	*1.87*	–	–	–	–	0.57	*1.26*	0.51	*1.12*	0.29	*0.63*	0.52	*1.15*
45	55	29	41	1.11	*2.44*	–	–	0.79	*1.74*	0.94	*2.06*	–	–	–	–	0.65	*1.43*	0.69	*1.51*	0.44	*0.97*	0.66	*1.46*
46	44	28	28	0.91	*2.00*	–	–	0.57	*1.25*	0.74	*1.62*	–	–	–	–	0.46	*1.02*	0.39	*0.87*	0.17	*0.37*	0.44	*0.96*
47	56	34	38	1.03	*2.28*	–	–	0.73	*1.61*	0.86	*1.90*	–	–	–	–	0.60	*1.32*	0.43	*0.95*	0.21	*0.46*	0.47	*1.04*
48	42	–	29	0.86	*1.90*	–	–	0.59	*1.30*	0.60	*1.31*	–	–	–	–	0.48	*1.07*	0.25	*0.55*	0.03	*0.08*	0.39	*0.86*
49	54	48	46	1.14	*2.52*	–	–	0.86	*1.90*	0.92	*2.02*	–	–	–	–	0.71	*1.56*	0.58	*1.28*	0.34	*0.75*	0.56	*1.24*
50	49	38	41	0.99	*2.19*	–	–	0.78	*1.73*	0.82	*1.80*	–	–	–	–	0.64	*1.42*	0.48	*1.06*	0.25	*0.55*	0.52	*1.14*
51	50	41	43	1.00	*2.21*	–	–	0.82	*1.81*	0.84	*1.84*	–	–	–	–	0.67	*1.49*	0.54	*1.19*	0.31	*0.68*	0.55	*1.21*
52	49	33	37	0.97	*2.14*	–	–	0.73	*1.60*	0.80	*1.76*	–	–	–	–	0.60	*1.32*	0.44	*0.98*	0.22	*0.47*	0.49	*1.07*
53	48	31	41	0.95	*2.10*	–	–	0.78	*1.71*	0.78	*1.72*	–	–	–	–	0.64	*1.40*	0.44	*0.96*	0.21	*0.47*	0.48	*1.06*
54	38	–	–	0.81	*1.78*	–	–	–	–	0.66	*1.46*	–	–	–	–	–	–	0.37	*0.81*	0.14	*0.30*	0.44	*0.96*
55	57	42	44	1.08	*2.37*	–	–	0.84	*1.86*	0.90	*1.99*	–	–	–	–	0.69	*1.52*	0.48	*1.07*	0.25	*0.56*	0.51	*1.13*
56	39	27	35	0.84	*1.86*	–	–	0.68	*1.49*	0.67	*1.47*	–	–	–	–	0.55	*1.22*	0.40	*0.89*	0.18	*0.40*	0.46	*1.00*
57	51	35	40	1.08	*2.38*	–	–	0.76	*1.68*	0.91	*2.00*	–	–	–	–	0.63	*1.38*	0.53	*1.17*	0.30	*0.66*	0.52	*1.15*
58	52	37	38	1.04	*2.30*	–	–	0.74	*1.64*	0.87	*1.92*	–	–	–	–	0.61	*1.34*	0.47	*1.04*	0.24	*0.54*	0.50	*1.11*
59	51	31	35	1.01	*2.23*	–	–	0.68	*1.51*	0.84	*1.85*	–	–	–	–	0.56	*1.24*	0.48	*1.06*	0.25	*0.56*	0.51	*1.13*
60	29	–	–	0.58	*1.27*	–	–	–	–	0.40	*0.87*	–	–	–	–	–	–	–	–	–	–	–	–
61	54	44	37	1.06	*2.34*	–	–	0.73	*1.60*	0.89	*1.96*	–	–	–	–	0.60	*1.31*	0.51	*1.12*	0.28	*0.62*	0.53	*1.17*
62	44	46	45	0.98	*2.15*	–	–	0.86	*1.89*	0.80	*1.77*	–	–	–	–	0.70	*1.55*	0.53	*1.16*	0.29	*0.65*	0.54	*1.19*
63	42	40	44	0.93	*2.06*	–	–	0.83	*1.84*	0.76	*1.67*	–	–	–	–	0.68	*1.51*	0.50	*1.09*	0.26	*0.58*	0.52	*1.15*
64	53	35	37	1.06	*2.35*	–	–	0.72	*1.58*	0.90	*1.98*	–	–	–	–	0.59	*1.30*	0.55	*1.20*	0.32	*0.70*	0.56	*1.22*
65	57	42	44	1.08	*2.37*	–	–	0.84	*1.86*	0.90	*1.99*	–	–	–	–	0.69	*1.52*	0.48	*1.07*	0.25	*0.56*	0.51	*1.13*
66	51	38	35	1.00	*2.19*	–	–	0.68	*1.49*	0.83	*1.83*	–	–	–	–	0.56	*1.22*	0.46	*1.01*	0.24	*0.52*	0.49	*1.08*
67	47	34	38	0.94	*2.07*	–	–	0.73	*1.62*	0.77	*1.70*	–	–	–	–	0.60	*1.32*	0.43	*0.95*	0.21	*0.46*	0.47	*1.04*
68	42	34	35	0.90	*1.98*	–	–	0.69	*1.52*	0.73	*1.60*	–	–	–	–	0.57	*1.25*	0.45	*0.99*	0.23	*0.50*	0.49	*1.07*
69	46	32	34	0.93	*2.04*	–	–	0.67	*1.48*	0.76	*1.67*	–	–	–	–	0.55	*1.21*	0.43	*0.95*	0.21	*0.47*	0.47	*1.04*
70	52	36	38	1.04	*2.29*	–	–	0.74	*1.62*	0.93	*2.04*	–	–	–	–	0.60	*1.33*	0.57	*1.26*	0.33	*0.73*	0.55	*1.22*
71	46	–	44	1.15	*2.53*	–	–	0.84	*1.85*	0.78	*1.72*	–	–	–	–	0.69	*1.52*	0.43	*0.94*	0.20	*0.43*	0.48	*1.06*
72	52	43	37	1.03	*2.26*	–	–	0.72	*1.58*	0.86	*1.89*	–	–	–	–	0.59	*1.30*	0.49	*1.08*	0.27	*0.59*	0.52	*1.14*
73	51	37	40	1.23	*2.71*	–	–	0.77	*1.69*	0.96	*2.11*	–	–	–	–	0.63	*1.38*	0.50	*1.09*	0.27	*0.59*	0.52	*1.15*
74	51	45	42	1.02	*2.24*	–	–	0.79	*1.75*	0.85	*1.87*	–	–	–	–	0.65	*1.43*	0.50	*1.10*	0.27	*0.60*	0.52	*1.15*
75	52	47	46	1.04	*2.29*	–	–	0.86	*1.89*	0.87	*1.91*	–	–	–	–	0.70	*1.55*	0.51	*1.12*	0.28	*0.62*	0.53	*1.17*

(Continued)

DRY FORAGES

TABLE 19–1C DRY FORAGES, MINERAL AND VITAMIN COMPOSITION OF FEEDS, DATA EXPRESSED **AS-FED**

			Macro Minerals							Micro Minerals						
Entry Number	Feed Name Description	Dry Matter	Calcium (Ca)	Phos-phorus (P)	Sodium (Na)	Chlo-rine (Cl)	Mag-nesium (Mg)	Potas-sium (K)	Sulfur (S)	Cobalt (Co)	Copper (Cu)	Iodine (I)	Iron (Fe)	Man-ganese (Mn)	Sele-nium (Se)	Zinc (Zn)
		%	%	%	%	%	%	%	%	ppm or mg/kg	ppm or mg/kg	ppm or mg/kg	%	ppm or mg/kg	ppm or mg/kg	ppm or mg/kg
	ALFALFA (LUCERNE)															
1	HAY, SUN-CURED, ALL ANALYSES	90	1.28	0.24	0.07	0.33	0.30	1.85	0.25	0.250	10.0	–	0.019	41.4	–	21.9
2	HAY, PREBLOOM, SUN-CURED	90	1.34	0.30	0.10	0.31	0.19	2.25	0.48	0.256	10.2	–	0.021	42.2	–	33.5
3	HAY, EARLY BLOOM, SUN-CURED	91	1.48	0.20	0.14	0.34	0.31	2.32	0.27	0.264	11.4	–	0.021	32.8	0.497	27.3
4	HAY, MIDBLOOM, SUN-CURED	91	1.27	0.22	0.11	0.34	0.32	1.42	0.26	0.359	16.1	–	0.021	55.1	–	28.1
5	HAY, FULL BLOOM, SUN-CURED	91	1.08	0.22	0.06	–	0.25	1.42	0.27	0.210	9.0	–	0.015	38.5	–	23.7
6	HAY, MATURE, SUN-CURED	91	1.07	0.19	0.07	–	0.20	1.88	0.23	0.370	12.5	–	0.015	35.1	–	20.1
7	HAY, RAINED ON, SUN-CURED	89	2.04	0.21	0.05	–	0.24	2.16	–	–	2.5	–	0.026	22.1	–	23.8
8	MEAL, DEHY, 17% PROTEIN	92	1.40	0.23	0.10	0.47	0.29	2.38	0.23	0.302	8.6	0.148	0.041	31.0	0.335	19.3
9	MEAL, DEHY, 20% PROTEIN	92	1.59	0.28	0.11	0.47	0.33	2.41	0.50	0.259	12.2	0.135	0.036	45.2	0.285	21.8
10	MEAL, DEHY, 22% PROTEIN	93	1.69	0.30	0.12	0.52	0.31	2.40	0.30	0.311	9.8	0.166	0.036	36.4	0.534	19.5
11	**ALFALFA-BROMEGRASS, SMOOTH,** HAY, SUN-CURED	91	1.02	0.25	0.21	0.43	0.52	1.77	0.21	0.079	15.5	–	0.012	36.2	–	–
12	**ALFALFA-GRASS,** HAY, SUN-CURED	91	1.33	0.23	0.11	–	0.28	2.31	0.22	–	11.4	–	0.020	60.7	–	20.1
13	**ALFALFA-ORCHARDGRASS,** HAY, SUN-CURED	89	–	–	–	–	–	–	–	–	–	–	–	–	–	–
14	**ALMOND,** HULLS	90	0.19	0.09	–	–	–	0.48	0.10	–	–	–	–	–	–	–
15	**APPLE,** POMACE, DEHY	89	0.11	0.10	0.12	–	0.06	0.43	0.02	–	–	–	0.027	7.2	–	–
16	**BAGASSE (SUGARCANE),** PULP, DEHY	91	0.47	0.26	0.04	–	0.08	0.34	0.09	–	–	–	0.019	–	–	–
	BARLEY															
17	HAY, SUN-CURED	88	0.21	0.25	0.12	–	0.14	1.30	0.15	0.059	3.9	–	0.027	34.8	–	–
18	STRAW	91	0.27	0.07	0.13	0.61	0.21	2.16	0.16	0.061	4.9	–	0.019	15.1	–	6.8
	BEAN															
19	HAY, SUN-CURED	90	–	–	–	–	–	–	–	–	–	–	–	–	–	–
20	STRAW	89	1.65	0.13	–	–	0.12	1.02	–	–	–	–	–	–	–	–
21	**BERMUDAGRASS,** HAY, SUN-CURED	91	0.43	0.16	0.07	–	0.16	1.40	0.19	0.111	24.3	0.105	0.027	99.4	–	53.0
22	**BERMUDAGRASS, COASTAL,** HAY, SUN-CURED	91	0.38	0.17	–	–	0.16	1.46	0.19	–	–	–	0.028	–	–	18.2
23	**BIRDSFOOT TREFOIL,** HAY, SUN-CURED	91	1.54	0.21	0.06	–	0.46	1.74	0.23	0.100	8.4	–	0.021	26.0	–	69.9
24	**BLUEGRASS, CANADA,** HAY, SUN-CURED	92	0.28	0.24	0.10	–	0.30	1.73	0.12	–	–	–	0.028	84.9	–	–
25	**BLUEGRASS, KENTUCKY,** HAY, SUN-CURED	89	0.40	0.27	0.10	0.55	0.19	1.66	0.12	–	8.8	–	0.025	76.2	–	–
26	**BLUESTEM,** HAY, SUN-CURED	90	–	–	0.01	0.04	–	–	–	–	–	–	–	–	–	–
27	**BROMEGRASS,** HAY, SUN-CURED	91	0.32	0.15	0.03	–	0.09	1.49	0.18	–	–	–	0.019	–	–	–
28	**BUFFALOGRASS,** HAY, SUN-CURED	90	0.56	0.12	–	–	–	1.38	–	–	–	–	–	–	–	–
29	**BUR-CLOVER, TOOTHED,** HAY, SUN-CURED	90	–	–	–	–	–	–	–	–	–	–	–	–	–	–
30	**CANADA BLUEGRASS,** HAY, SUN-CURED	92	0.28	0.24	0.10	–	0.30	1.73	0.12	–	–	–	0.028	84.9	–	–
31	**CANARYGRASS, REED,** HAY, SUN-CURED	89	0.32	0.21	0.01	–	0.19	2.60	–	–	10.6	–	0.014	82.5	–	–
32	**CEREALS,** IMMATURE, DEHY	92	0.65	0.46	–	–	–	–	–	–	–	–	–	–	–	–
33	**CLOVER, ALSIKE,** HAY, SUN-CURED	88	1.14	0.22	0.40	0.68	0.40	1.95	0.17	–	5.3	–	0.023	60.5	–	–
34	**CLOVER, ALYCE,** HAY, SUN-CURED	90	–	–	–	–	–	–	–	–	–	–	–	–	–	–
35	**CLOVER, CRIMSON,** HAY, SUN-CURED	88	1.23	0.19	0.34	0.55	0.25	2.10	0.25	–	–	0.059	0.062	183.3	–	–
36	**CLOVER, LADINO,** HAY, SUN-CURED	89	1.30	0.30	0.12	0.27	0.42	2.17	0.19	0.144	8.4	0.268	0.042	109.7	–	15.2
37	**CLOVER, RED,** HAY, SUN-CURED, ALL ANALYSES	88	1.22	0.22	0.16	0.28	0.34	1.60	0.15	0.138	18.8	0.217	0.022	95.2	–	32.5
38	**CLOVER, RED-GRASS,** HAY, FULL BLOOM, SUN-CURED	89	–	–	–	–	–	–	–	0.137	–	–	–	–	–	–
39	**CLOVER, SUBTERRANEAN,** HAY, SUN-CURED	90	–	–	–	–	–	–	–	–	–	–	–	–	–	–
40	**CLOVER, SWEET, YELLOW,** HAY, SUN-CURED, ALL ANALYSES	89	1.44	0.24	0.08	0.33	0.39	1.35	0.42	–	8.8	–	0.015	95.4	–	–
41	**CLOVER, WHITE,** HAY, SUN-CURED	90	1.71	0.29	–	–	–	–	–	–	–	–	–	–	–	–
42	**CLOVER-TIMOTHY,** HAY, FULL BLOOM, SUN-CURED	90	–	–	–	–	–	–	–	–	–	–	–	–	–	–
	CORN															
43	FODDER WITH EARS, WITH HUSKS, SUN-CURED	90	0.45	0.23	0.03	0.17	0.26	0.84	0.13	–	6.9	–	0.009	61.4	–	–
44	STOVER WITHOUT EARS, WITHOUT HUSKS, SUN-CURED	85	0.49	0.08	0.06	–	0.34	1.24	0.15	–	4.3	–	0.018	115.9	–	–
45	HUSKS, SUN-CURED	89	0.16	0.12	–	–	–	0.57	–	–	–	–	–	–	–	–
46	COBS, GROUND	90	0.11	0.04	–	–	0.06	0.78	0.42	0.117	6.6	–	0.021	5.6	–	–
47	**CORN, SWEET,** FODDER WITH EARS, WITH HUSKS, SUN-CURED	88	–	0.17	–	–	–	0.98	–	–	–	–	–	–	–	–
48	**COTTON,** HULLS	90	0.13	0.09	0.02	0.02	0.13	0.78	0.08	0.018	12.0	–	0.012	107.8	–	19.8
49	**COWPEA, COMMON,** HAY, SUN-CURED	90	1.26	0.31	0.24	0.15	0.41	2.04	0.32	0.064	–	–	0.055	438.4	–	–
50	**CRESTED WHEATGRASS,** HAY, SUN-CURED	92	0.24	0.14	–	–	–	–	–	0.219	–	–	–	–	–	–
51	**CRIMSON CLOVER,** HAY, SUN-CURED	88	1.23	0.19	0.34	0.55	0.25	2.10	0.25	–	–	0.059	0.062	183.3	–	–
52	**DALLISGRASS,** HAY, SUN-CURED	91	0.46	0.18	–	–	0.67	–	–	–	–	–	0.011	–	–	–
53	**FESCUE, TALL (ALTA),** HAY, SUN-CURED	89	0.35	0.21	0.05	–	0.20	2.12	–	–	–	–	–	–	–	–
54	**FLAX, COMMON,** STRAW	93	0.51	0.05	–	0.25	0.29	1.62	0.25	–	–	–	–	7.6	–	–
55	**FOXTAIL, MEADOW,** HAY, SUN-CURED	90	0.95	0.18	0.01	–	0.12	1.57	–	–	–	–	–	–	–	–
56	**GRAMA,** HAY, SUN-CURED	89	0.34	0.18	–	–	–	–	–	–	–	–	–	–	–	–
57	**GRASS,** HAY, SUN-CURED, ALL ANALYSES	89	0.44	0.18	0.01	–	0.20	1.38	0.19	0.133	6.5	–	0.013	75.4	–	15.1
58	**GRASS-LEGUME,** HAY, SUN-CURED	89	0.66	0.19	0.06	–	0.23	1.62	0.13	0.118	5.0	–	0.014	42.5	–	23.2
59	**JOHNSONGRASS SORGHUM,** HAY, SUN-CURED	91	0.80	0.27	0.01	–	0.31	1.22	0.09	–	–	–	0.054	–	–	–
	KAFIR SORGHUM—SEE SORGHUM															
60	**KELP (SEAWEED),** WHOLE, DEHY	91	2.47	0.28	–	–	0.85	–	–	–	–	–	–	–	–	–
61	**KENTUCKY BLUEGRASS,** HAY, SUN-CURED	89	0.40	0.27	0.10	0.55	0.19	1.66	0.12	–	8.8	–	0.025	76.2	–	–
62	**LESPEDEZA, COMMON,** HAY, SUN-CURED, ALL ANALYSES	89	0.78	0.25	–	–	0.20	1.21	0.21	–	7.1	–	0.019	99.6	–	21.3
63	**LESPEDEZA, SERICEA (CHINESE LESPEDEZA),** HAY, SUN-CURED, ALL ANALYSES	90	0.93	0.22	–	–	0.20	0.99	–	–	–	–	0.027	91.1	–	–
64	**MEADOW FESCUE,** HAY, SUN-CURED	88	0.33	0.25	–	–	0.44	1.61	–	0.119	–	–	–	21.4	–	–
65	**MEADOW FOXTAIL,** HAY, SUN-CURED	90	0.95	0.18	0.01	–	0.12	1.57	–	–	–	–	–	–	–	–
66	**MILLET, FOXTAIL,** HAY, SUN-CURED	87	0.29	0.16	0.09	0.11	0.20	1.69	0.14	–	–	–	–	120.1	–	–
67	**MILLET, JAPANESE (BARNYARD GRASS),** HAY, SUN-CURED	87	0.20	–	–	–	–	2.11	–	–	–	–	–	–	–	–
	MILO SORGHUM—SEE SORGHUM															
68	**NEEDLEGRASS,** HAY, SUN-CURED	88	–	–	–	–	–	–	–	–	–	–	–	–	–	–
69	**OATGRASS, TALL,** HAY, SUN-CURED	88	0.30	0.27	0.01	–	0.17	1.74	–	–	–	–	–	32.2	–	–
	OATS															
70	HAY, SUN-CURED, ALL ANALYSES	91	0.29	0.23	0.17	0.47	0.26	1.35	0.21	0.067	4.4	–	0.037	89.6	–	40.8
71	STRAW	92	0.22	0.06	0.39	0.72	0.16	2.35	0.21	–	9.5	–	0.016	29.0	–	5.5
72	**OATS-PEA,** HAY, SUN-CURED	88	0.71	0.22	–	–	–	1.02	–	–	–	–	–	–	–	–
73	**ORCHARDGRASS,** HAY, SUN-CURED	89	0.34	0.23	0.01	0.37	0.16	2.68	0.23	0.339	12.9	–	0.014	162.7	–	32.0
74	**PEA,** HAY, SUN-CURED	87	–	–	–	–	–	–	–	–	–	–	–	–	–	–
75	**PEA, FIELD,** HAY (VINES WITHOUT SEEDS, WITH PODS), SUN-CURED	88	–	–	–	–	–	–	–	–	–	–	–	–	–	–

DRY FORAGES

DRY FORAGES

Entry Number	Fat-Soluble Vitamins: A (1 mg Carotene = 1667 IU Vit A)	Carotene (Provitamin A)	D	E	K	Water-Soluble Vitamins: B–12	Biotin	Choline	Folacin (Folic Acid)	Niacin	Pantothenic Acid (B–3)	(Pyridoxine) B–6	Riboflavin (B–2)	Thiamin (B–1)
	IU/g	ppm or mg/kg	IU/kg	ppm or mg/kg	ppm or mg/kg	ppb or mcg/kg	ppm or mg/kg	ppm or mg/kg	ppm or mg/kg	ppm or mg/kg	ppm or mg/kg	ppm or mg/kg	ppm or mg/kg	ppm or mg/kg
1	45.0	27.0	2	55.9	–	–	0.18	892	3.07	43	18.1	–	9.5	3.1
2	300.2	180.1	–	–	–	–	–	–	–	–	–	–	–	–
3	210.9	126.5	2	23.5	–	–	–	–	–	–	–	–	–	–
4	50.5	30.3	1	–	–	–	–	–	–	–	–	–	9.6	–
5	98.5	59.1	–	–	–	–	–	–	–	–	–	–	–	–
6	17.6	10.6	1	–	–	–	–	–	–	–	–	–	–	–
7	–	–	1	–	–	–	–	–	–	–	–	–	–	–
8	200.3	120.2	–	105.7	8.24	–	0.33	1369	4.37	37	29.7	7.18	12.9	3.4
9	265.4	159.2	–	143.3	14.19	–	0.35	1417	2.96	48	35.5	8.72	15.2	5.4
10	391.4	234.8	–	221.3	11.65	–	0.33	1605	5.15	50	39.0	8.28	17.6	5.9
11	39.4	23.7	–	–	–	–	–	–	–	25	21.4	–	6.1	–
12	28.9	17.3	–	–	–	–	–	–	–	–	–	–	–	–
13	–	–	–	–	–	–	–	–	–	–	–	–	–	–
14	–	–	–	–	–	–	–	–	–	–	–	–	–	–
15	–	–	–	–	–	–	–	–	–	–	–	–	–	–
16	–	–	–	–	–	–	–	–	–	–	–	–	–	–
17	77.4	46.4	1	–	–	–	–	–	–	–	–	–	–	–
18	3.5	2.1	1	–	–	–	–	–	–	–	–	–	–	–
19	–	–	–	–	–	–	–	–	–	–	–	–	–	–
20	–	–	–	–	–	–	–	–	–	–	–	–	–	–
21	87.5	52.5	–	–	–	–	–	–	–	–	–	–	–	–
22	123.7	74.2	–	–	–	–	–	–	–	–	–	–	–	–
23	217.8	130.6	1	–	–	–	–	–	–	–	–	–	14.6	6.2
24	378.4	227.0	–	–	–	–	–	–	–	–	–	–	–	–
25	–	–	–	–	–	–	–	–	–	–	–	–	9.9	–
26	62.4	37.4	–	–	–	–	–	–	–	–	–	–	–	–
27	50.3	30.2	–	–	–	–	–	–	–	–	–	–	–	–
28	–	–	–	–	–	–	–	–	–	–	–	–	–	–
29	–	–	–	–	–	–	–	–	–	–	–	–	–	–
30	378.4	227.0	–	–	–	–	–	–	–	–	–	–	–	–
31	28.2	16.9	–	–	–	–	–	–	–	–	–	–	8.5	3.6
32	–	–	–	–	–	–	–	–	–	–	–	–	–	–
33	272.1	163.2	–	–	–	–	–	–	–	–	–	–	15.1	4.2
34	–	–	–	–	–	–	–	–	–	–	–	–	–	–
35	32.9	19.8	–	–	–	–	–	–	–	–	–	–	–	–
36	239.5	143.7	–	–	–	–	–	–	–	10	1.0	–	15.2	3.7
37	40.5	24.3	–	–	–	–	0.09	–	–	38	9.9	–	15.7	2.0
38	16.6	10.0	–	–	–	–	–	–	–	–	–	–	–	–
39	–	–	–	–	–	–	–	–	–	–	–	–	–	–
40	145.8	87.4	2	–	–	–	–	–	–	–	–	–	–	–
41	92.1	55.3	–	115.5	–	–	–	–	–	–	–	–	–	–
42	–	–	–	–	–	–	–	–	–	–	–	–	–	–
43	6.6	4.0	1	–	–	–	–	–	–	–	–	–	–	–
44	6.3	3.8	1	–	–	–	–	–	–	–	–	–	–	–
45	–	–	–	–	–	–	–	–	–	–	–	–	–	–
46	1.0	0.6	–	–	–	–	–	–	–	7	3.8	–	1.0	0.9
47	–	–	–	–	–	–	–	–	–	–	–	–	–	–
48	–	–	–	–	–	–	–	–	–	–	–	–	3.7	–
49	52.7	31.6	–	–	–	–	–	–	–	–	–	–	–	–
50	34.2	20.5	–	–	–	–	–	–	–	–	–	–	–	–
51	32.9	19.8	–	–	–	–	–	–	–	–	–	–	–	–
52	–	–	–	–	–	–	–	–	–	–	–	–	–	–
53	30.8	18.5	–	–	–	–	–	–	–	–	–	–	–	–
54	–	–	–	–	–	–	–	–	–	–	–	–	–	–
55	–	–	–	–	–	–	–	–	–	–	–	–	–	–
56	–	–	–	–	–	–	–	–	–	–	–	–	–	–
57	40.8	24.5	–	–	–	–	–	–	–	–	–	–	–	–
58	18.0	10.8	2	–	–	–	–	–	–	–	–	–	–	–
59	58.8	35.3	–	–	–	–	–	–	–	–	–	–	–	–
60	–	–	–	–	–	–	–	–	–	–	–	–	–	–
61	–	–	–	–	–	–	–	–	–	–	–	–	9.9	–
62	73.9	44.3	–	–	–	–	–	–	–	–	–	–	8.7	–
63	59.3	35.6	–	–	–	–	–	–	–	–	–	–	8.7	–
64	105.8	63.4	–	118.6	–	–	–	–	–	–	–	–	–	–
65	–	–	–	–	–	–	–	–	–	–	–	–	–	–
66	86.9	52.1	–	–	–	–	–	–	–	–	–	–	–	–
67	–	–	–	–	–	–	–	–	–	–	–	–	–	–
68	–	–	–	–	–	–	–	–	–	–	–	–	–	–
69	–	–	–	–	–	–	–	–	–	–	–	–	–	–
70	45.0	27.0	1	–	–	–	–	–	–	–	–	–	–	–
71	5.8	3.5	1	–	–	–	–	–	–	–	–	–	–	–
72	–	–	–	–	–	–	–	–	–	–	–	–	–	–
73	28.9	17.3	–	170.7	–	–	–	–	–	–	–	–	6.1	2.6
74	–	–	–	–	–	–	–	–	–	–	–	–	–	–
75	–	–	–	–	–	–	–	–	–	–	–	–	–	–

(Continued)

TABLE 19–1C DRY FORAGES, COMPOSITION OF FEEDS, DATA EXPRESSED **AS-FED**—*(Continued)*

Entry Number	Feed Name Description	International Feed Number	Proximate Analysis									Digestible Protein		
			Dry Matter	Ash	Crude Fiber	Neutral Det. Fib. (NDF)	Acid Det. Fib. (ADF)	Lignin	Ether Extract (Fat)	N-Free Extract	Crude Protein	Ruminant	Swine	Horse
			%	%	%	%	%	%	%	%	%	%	%	%
76	**PEANUT,** HAY, SUN-CURED	1-03-619	91	8.2	30.3	–	37.2	–	3.3	39.1	9.9	5.7	5.2	6.0
77	**PEARL MILLET,** HAY, SUN-CURED	1-03-112	87	8.9	32.2	–	–	–	1.8	37.2	7.3	4.6	3.1	4.1
78	**PEAVINE,** HAY, SUN-CURED	1-03-666	91	6.8	24.9	–	–	–	2.6	36.6	19.8	15.4	13.9	13.6
79	**PRAIRIE GRASS, MIDWEST (PRAIRIE HAY),** HAY, SUN-CURED	1-03-191	91	7.2	30.7	–	–	–	2.1	45.2	5.8	2.9	1.6	2.9
80	**REDTOP,** HAY, SUN-CURED	1-03-885	92	6.0	28.4	–	–	–	2.8	47.4	7.4	3.7	3.0	4.1
81	**REED CANARYGRASS,** HAY, SUN-CURED	1-01-104	89	7.3	30.2	62.9	32.7	–	2.7	40.0	9.1	5.8	4.6	5.5
	RICE													
82	HULLS	1-08-075	92	19.0	38.9	71.9	62.3	9.6	1.0	30.3	3.0	0.1	–0.9	0.6
83	STRAW	1-03-925	91	15.4	31.9	64.4	50.1	4.4	1.3	38.2	3.9	0.7	–0.1	1.4
84	**RYE,** STRAW	1-04-007	91	3.8	38.3	–	–	–	1.4	44.7	2.8	–0.3	–1.1	0.5
85	**RYEGRASS,** HAY, SUN-CURED	1-04-057	88	7.1	25.3	56.3	37.0	–	1.8	46.2	7.5	3.8	3.2	4.2
86	**SAWDUST,** WOOD	1-07-714	90	0.7	71.5	–	–	–	0.8	16.7	0.3	–2.5	–3.2	–1.4
87	**SEAWEED (KELP),** WHOLE, DEHY	1-08-073	91	35.0	6.5	–	–	–	0.5	42.4	6.5	2.9	2.2	3.4
	SORGHUM													
88	FODDER WITH HEADS, SUN-CURED	1-07-960	90	8.9	25.6	–	–	–	2.0	47.4	6.2	2.4	2.0	3.2
89	STOVER WITHOUT HEADS, SUN-CURED	1-04-302	92	8.9	29.9	–	39.9	–	1.6	46.8	4.4	0.7	0.4	1.8
90	KAFIR, FODDER WITH HEADS, SUN-CURED	1-04-418	90	9.4	25.7	–	–	–	2.2	44.5	8.3	4.2	3.8	4.8
91	KAFIR, STOVER WITHOUT HEADS, SUN-CURED	1-04-419	82	8.1	26.9	–	–	–	1.5	40.7	4.6	1.6	0.9	2.1
92	MILO, FODDER WITH HEADS, SUN-CURED	1-04-433	89	8.5	21.8	–	–	–	2.9	49.4	6.5	2.5	2.3	3.4
93	MILO, STOVER WITHOUT HEADS, SUN-CURED	1-04-434	91	11.8	31.1	–	–	–	1.3	42.5	4.4	–0.8	0.4	1.8
94	SUDANGRASS, HAY, SUN-CURED	1-04-480	91	10.7	26.2	60.2	20.4	35.5	1.6	41.7	10.9	4.7	6.1	6.8
	SOYBEAN													
95	HAY, SUN-CURED	1-04-558	89	7.2	30.6	–	35.7	–	2.3	35.0	14.1	9.5	8.9	9.3
96	HULLS	1-04-560	91	4.6	36.2	59.4	42.4	1.8	2.0	37.0	10.8	6.5	3.4	6.7
97	STRAW	1-04-567	88	5.6	38.9	–	–	–	1.3	37.4	4.6	1.3	0.6	1.9
98	**STARGRASS,** HAY, SUN-CURED	1-13-407	90	8.6	28.8	–	–	–	1.5	44.4	7.7	2.0	–	2.8
99	**SUGARCANE (BAGASSE),** DEHY	1-04-686	91	2.9	42.3	78.8	54.5	12.8	0.7	43.8	1.4	–1.5	–2.3	–0.5
100	**SWEET POTATO,** HAY, SUN-CURED	1-04-779	90	10.2	21.3	–	23.4	–	3.2	43.5	11.9	4.6	7.0	7.6
101	**TIMOTHY,** HAY, SUN-CURED, ALL ANALYSES	1-04-893	91	4.6	30.3	63.7	36.4	–	2.4	47.3	6.8	2.8	2.4	3.3
102	**TIMOTHY-CLOVER,** HAY, SUN-CURED	1-04-973	89	5.1	31.7	–	–	–	2.0	42.7	7.8	4.1	3.4	4.4
103	**TREFOIL, BIRDSFOOT,** HAY, SUN-CURED	1-05-044	91	6.7	29.3	42.8	32.8	–	1.9	38.9	13.9	9.6	8.7	9.1
104	**VELVETBEAN,** HAY, SUN-CURED	1-05-080	93	7.4	27.5	–	–	–	3.1	38.4	16.4	9.5	10.8	11.0
105	**VETCH,** HAY, SUN-CURED	1-05-106	89	7.8	24.8	42.7	29.4	–	2.7	35.1	18.4	13.4	12.7	12.6
106	**WATERGRASS,** HAY, SUN-CURED	1-05-154	94	18.0	31.2	–	–	–	1.1	36.3	7.6	3.8	3.1	4.2
107	**WHEAT,** STRAW	1-05-175	90	6.9	37.4	70.3	47.7	8.4	1.8	40.4	3.2	0.3	–0.6	0.6
108	**WHEATGRASS, CRESTED,** HAY, SUN-CURED	1-05-418	92	6.4	30.8	–	33.2	5.1	2.1	42.2	10.3	6.4	5.5	6.3
109	**WHITE CLOVER,** HAY, SUN-CURED	1-01-464	90	9.0	21.9	–	–	–	2.4	40.1	16.9	12.4	11.4	11.5
110	**WOOD,** SAWDUST	1-07-714	90	0.7	71.5	–	–	–	0.8	16.7	0.3	–2.5	–3.2	–1.4

TABLE 19–1D SILAGES AND HAYLAGES, COMPOSITION OF FEEDS, DATA EXPRESSED **AS-FED**

Entry Number	Feed Name Description	International Feed Number	Proximate Analysis									Digestible Protein		
			Dry Matter	Ash	Crude Fiber	Neutral Det. Fib. (NDF)	Acid Det. Fib. (ADF)	Lignin	Ether Extract (Fat)	N-Free Extract	Crude Protein	Ruminant	Swine	Horse
			%	%	%	%	%	%	%	%	%	%	%	%
	ALFALFA (LUCERNE), SILAGE													
1	ALL ANALYSES	3-00-212	27	2.6	8.6	–	10.7	–	0.9	10.4	4.7	3.4	–	–
2	CORN GRAIN ADDED	3-00-226	23	2.3	7.0	–	–	–	0.9	9.0	4.1	3.0	–	–
3	MOLASSES ADDED	3-00-238	30	2.6	7.8	–	–	–	1.5	12.4	5.6	3.9	–	–
	ALFALFA-BROMEGRASS, SMOOTH, SILAGE													
4	ALL ANALYSES	3-00-268	26	2.0	9.4	–	–	–	1.1	10.4	3.6	2.3	–	–
5	WILTED	3-00-269	46	4.1	15.2	–	–	–	1.1	18.6	7.1	4.8	–	–
6	**ALFALFA-ORCHARDGRASS, SILAGE,** 30–50% DRY MATTER	3-08-144	37	4.0	11.2	–	16.8	3.7	1.5	13.7	6.7	4.7	–	–
7	**BEET, MANGEL, SILAGE,** TOPS WITH CROWNS	3-00-635	13	2.8	1.9	–	–	–	0.6	5.7	1.8	1.2	–	–
8	**CITRUS, SILAGE,** PULP	4-01-234	21	1.2	3.3	–	4.2	–	2.1	12.8	1.5	0.7	1.0	1.0
	CORN, SILAGE													
9	ALL ANALYSES	3-02-822	26	1.5	6.6	–	8.9	1.2	0.8	15.2	2.2	1.0	1.3	–
10	MOLASSES ADDED	3-02-834	29	1.9	6.8	–	–	–	0.8	17.2	2.3	1.1	–	–
11	EARS WITH HUSKS	4-02-839	43	1.5	4.8	–	–	–	1.6	31.5	4.0	2.2	2.8	2.7
12	HUSKS (HUSKLAGE)	3-26-074	78	2.7	26.8	–	–	–	0.7	45.0	2.9	–	–	–
13	**CORN, SWEET, SILAGE,** CANNERY RESIDUE	3-07-955	31	1.7	10.1	–	10.5	–	1.3	15.4	2.5	1.2	–	–
14	**CORN-SORGHUM, SILAGE,** MATURE	3-03-013	34	2.1	8.5	–	–	–	1.1	19.7	2.7	1.7	–	–
15	**CORN-SOYBEAN, SILAGE**	3-03-015	27	1.8	7.0	–	–	–	0.9	14.6	2.7	1.7	–	–
	GRASS, SILAGE													
16	EARLY BLOOM	3-02-218	23	2.1	7.3	–	–	–	0.6	10.6	2.8	1.7	–	–
17	MOLASSES ADDED	3-02-261	27	3.8	6.4	–	–	–	1.4	10.6	4.3	3.1	–	–
	GRASS-LEGUME, SILAGE													
18	ALL ANALYSES	3-02-303	31	2.5	10.1	18.1	12.4	3.1	1.1	13.6	3.7	1.9	–	–
19	BARLEY GRAIN ADDED	3-02-305	34	2.2	8.6	–	–	–	1.4	16.7	5.2	3.4	–	–
20	MOLASSES ADDED	3-02-309	28	2.0	9.1	–	–	–	1.1	12.8	3.4	2.0	–	–
21	**KAFIR SORGHUM, SILAGE**	3-04-425	30	2.2	8.1	–	–	–	1.0	16.1	2.1	0.9	–	–

DRY FORAGES

Entry Number	TDN Ruminant	TDN Swine	TDN Horse	DE Ruminant		DE Swine		DE Horse		ME Ruminant		ME Swine		ME Poultry ME_n		ME Horse		NE Ruminant NE_m		NE Ruminant NE_g		NE Lactating Cows NE_{lc}	
	%	%	%	Mcal lb	Mcal *kg*	kcal lb	kcal *kg*	Mcal lb	Mcal *kg*	Mcal lb	Mcal *kg*	kcal lb	kcal *kg*	kcal lb	kcal *kg*	Mcal lb	Mcal *kg*	Mcal lb	Mcal *kg*	Mcal lb	Mcal *kg*	Mcal lb	Mcal *kg*
76	48	39	32	0.95	*2.10*	–	–	0.64	*1.41*	0.78	*1.71*	–	–	–	–	0.52	*1.15*	0.39	*0.85*	0.16	*0.36*	0.45	*0.99*
77	50	24	30	0.94	*2.06*	–	–	0.61	*1.33*	0.77	*1.69*	–	–	–	–	0.50	*1.09*	0.38	*0.84*	0.17	*0.37*	0.44	*0.96*
78	55	56	53	1.12	*2.46*	–	–	0.99	*2.18*	0.94	*2.08*	–	–	–	–	0.81	*1.78*	0.57	*1.26*	0.33	*0.73*	0.58	*1.28*
79	46	31	34	0.94	*2.06*	–	–	0.67	*1.48*	0.76	*1.68*	–	–	–	–	0.55	*1.21*	0.44	*0.97*	0.21	*0.46*	0.48	*1.07*
80	50	41	37	1.02	*2.25*	–	–	0.73	*1.60*	0.84	*1.86*	–	–	–	–	0.60	*1.31*	0.50	*1.11*	0.27	*0.59*	0.53	*1.17*
81	44	36	43	0.93	*2.06*	–	–	0.82	*1.80*	0.76	*1.68*	–	–	–	–	0.67	*1.47*	0.46	*1.02*	0.24	*0.52*	0.50	*1.10*
82	11	–	12	0.27	*0.60*	–	–	0.30	*0.67*	0.16	*0.35*	–	–	36	*79*	0.25	*0.55*	−0.26	*−0.57*	−0.47	*−1.04*	0.08	*0.17*
83	40	12	22	0.80	*1.76*	–	–	0.47	*1.04*	0.62	*1.37*	–	–	–	–	0.39	*0.85*	0.34	*0.75*	0.12	*0.26*	0.41	*0.91*
84	39	–	32	0.78	*1.72*	–	–	0.64	*1.40*	0.60	*1.32*	–	–	–	–	0.52	*1.15*	0.25	*0.55*	0.03	*0.07*	0.35	*0.78*
85	53	37	40	1.06	*2.34*	–	–	0.76	*1.68*	0.89	*1.97*	–	–	–	–	0.62	*1.38*	0.54	*1.19*	0.31	*0.68*	0.55	*1.22*
86	32	–	–	0.65	*1.43*	–	–	–	–	0.47	*1.04*	–	–	–	–	–	–	–	–	–	–	–	–
87	29	–	–	0.58	*1.27*	–	–	–	–	0.40	*0.87*	–	–	–	–	–	–	–	–	–	–	–	–
88	51	34	36	1.02	*2.24*	–	–	0.71	*1.56*	0.84	*1.86*	–	–	–	–	0.58	*1.28*	0.51	*1.12*	0.28	*0.61*	0.53	*1.17*
89	47	27	33	0.92	*2.02*	–	–	0.66	*1.46*	0.74	*1.64*	–	–	–	–	0.54	*1.20*	0.40	*0.89*	0.18	*0.39*	0.42	*0.93*
90	52	37	37	1.01	*2.22*	–	–	0.71	*1.57*	0.83	*1.83*	–	–	–	–	0.59	*1.29*	0.45	*1.00*	0.23	*0.50*	0.49	*1.08*
91	47	24	30	0.93	*2.05*	–	–	0.59	*1.30*	0.77	*1.71*	–	–	–	–	0.48	*1.07*	0.45	*0.99*	0.24	*0.53*	0.47	*1.05*
92	57	43	36	1.07	*2.36*	–	–	0.70	*1.54*	0.90	*1.99*	–	–	–	–	0.57	*1.26*	0.49	*1.07*	0.26	*0.57*	0.51	*1.13*
93	48	19	29	0.96	*2.11*	–	–	0.59	*1.29*	0.78	*1.72*	–	–	–	–	0.48	*1.06*	0.44	*0.98*	0.21	*0.47*	0.49	*1.07*
94	51	35	41	1.00	*2.20*	–	–	0.79	*1.74*	0.82	*1.82*	–	–	–	–	0.65	*1.43*	0.48	*1.07*	0.25	*0.55*	0.52	*1.14*
95	49	40	41	1.11	*2.45*	–	–	0.79	*1.74*	0.86	*1.90*	–	–	–	–	0.65	*1.43*	0.52	*1.14*	0.29	*0.63*	0.55	*1.21*
96	69	47	39	1.20	*2.65*	851	*1877*	0.75	*1.65*	1.03	*2.26*	305	*671*	301	*665*	0.61	*1.35*	0.75	*1.66*	0.50	*1.09*	0.72	*1.59*
97	36	–	29	0.71	*1.57*	–	–	0.58	*1.27*	0.54	*1.19*	–	–	–	–	0.47	*1.04*	0.21	*0.46*	0.00	*0.00*	0.32	*0.71*
98	49	–	35	0.90	*1.98*	–	–	0.65	*1.44*	0.74	*1.62*	–	–	–	–	0.53	*1.17*	0.41	*0.90*	0.08	*0.18*	0.41	*0.90*
99	43	–	–	0.88	*1.93*	–	–	–	–	0.69	*1.53*	–	–	–	–	–	–	0.37	*0.82*	0.15	*0.32*	0.40	*0.88*
100	49	49	39	1.02	*2.25*	–	–	0.76	*1.67*	0.85	*1.86*	–	–	–	–	0.62	*1.37*	0.51	*1.13*	0.28	*0.62*	0.54	*1.18*
101	53	38	43	1.06	*2.35*	–	–	0.81	*1.79*	0.85	*1.88*	–	–	–	–	0.67	*1.47*	0.50	*1.10*	0.27	*0.58*	0.51	*1.13*
102	50	34	38	0.99	*2.18*	–	–	0.73	*1.60*	0.82	*1.80*	–	–	–	–	0.60	*1.31*	0.46	*1.01*	0.23	*0.51*	0.49	*1.09*
103	54	41	46	0.91	*2.01*	–	–	0.87	*1.91*	0.83	*1.84*	–	–	–	–	0.71	*1.56*	0.55	*1.22*	0.32	*0.69*	0.57	*1.25*
104	56	51	46	1.12	*2.47*	–	–	0.87	*1.91*	0.94	*2.08*	–	–	–	–	0.71	*1.57*	0.55	*1.22*	0.31	*0.69*	0.57	*1.26*
105	55	52	48	1.09	*2.41*	–	–	0.91	*2.00*	0.92	*2.04*	–	–	–	–	0.74	*1.64*	0.55	*1.22*	0.32	*0.71*	0.56	*1.24*
106	39	14	27	0.79	*1.75*	–	–	0.55	*1.22*	0.61	*1.34*	–	–	–	–	0.45	*1.00*	0.30	*0.66*	0.07	*0.16*	0.39	*0.87*
107	40	–	36	0.86	*1.90*	–	–	0.71	*1.56*	0.70	*1.54*	–	–	–	–	0.58	*1.28*	0.36	*0.79*	0.14	*0.30*	0.39	*0.86*
108	49	38	41	0.99	*2.19*	–	–	0.78	*1.73*	0.82	*1.80*	–	–	–	–	0.64	*1.42*	0.48	*1.06*	0.25	*0.55*	0.52	*1.14*
109	57	52	50	1.11	*2.45*	–	–	0.94	*2.07*	0.94	*2.07*	–	–	–	–	0.77	*1.70*	0.53	*1.18*	0.30	*0.66*	0.55	*1.22*
110	32	–	–	0.65	*1.43*	–	–	–	–	0.47	*1.04*	–	–	–	–	–	–	–	–	–	–	–	–

SILAGES AND HAYLAGES

Entry Number	TDN Ruminant	TDN Swine	TDN Horse	DE Ruminant		DE Swine		DE Horse		ME Ruminant		ME Swine		ME Poultry ME_n		ME Horse		NE Ruminant NE_m		NE Ruminant NE_g		NE Lactating Cows NE_{lc}	
	%	%	%	Mcal lb	Mcal *kg*	kcal lb	kcal *kg*	Mcal lb	Mcal *kg*	Mcal lb	Mcal *kg*	kcal lb	kcal *kg*	kcal lb	kcal *kg*	Mcal lb	Mcal *kg*	Mcal lb	Mcal *kg*	Mcal lb	Mcal *kg*	Mcal lb	Mcal *kg*
1	16	–	–	0.28	*0.61*	–	–	–	–	0.27	*0.59*	–	–	–	–	–	–	0.16	*0.36*	0.09	*0.20*	0.16	*0.35*
2	13	–	–	0.27	*0.59*	–	–	–	–	0.22	*0.49*	–	–	–	–	–	–	0.14	*0.31*	0.08	*0.17*	0.14	*0.32*
3	18	–	–	0.40	*0.88*	–	–	–	–	0.33	*0.72*	–	–	–	–	–	–	0.21	*0.46*	0.13	*0.28*	0.20	*0.44*
4	16	–	–	0.31	*0.69*	–	–	–	–	0.26	*0.57*	–	–	–	–	–	–	0.15	*0.34*	0.09	*0.19*	0.16	*0.35*
5	26	–	–	0.53	*1.16*	–	–	–	–	0.44	*0.97*	–	–	–	–	–	–	0.24	*0.54*	0.13	*0.28*	0.26	*0.58*
6	20	–	–	0.42	*0.93*	–	–	–	–	0.35	*0.77*	–	–	–	–	–	–	0.19	*0.41*	0.09	*0.20*	0.20	*0.45*
7	7	–	–	0.14	*0.32*	–	–	–	–	0.12	*0.26*	–	–	–	–	–	–	0.07	*0.15*	0.04	*0.08*	0.07	*0.16*
8	18	18	5	0.37	*0.80*	354	*781*	0.11	*0.24*	0.33	*0.72*	335	*739*	–	–	0.09	*0.19*	0.21	*0.45*	0.14	*0.31*	0.19	*0.42*
9	18	19	–	0.34	*0.75*	–	–	–	–	0.30	*0.66*	–	–	–	–	–	–	0.19	*0.42*	0.12	*0.26*	0.17	*0.37*
10	19	–	–	0.38	*0.85*	–	–	–	–	0.33	*0.73*	–	–	–	–	–	–	0.19	*0.42*	0.11	*0.25*	0.19	*0.42*
11	31	35	27	0.66	*1.45*	692	*1526*	0.49	*1.09*	0.57	*1.27*	653	*1439*	–	–	0.41	*0.89*	0.33	*0.74*	0.21	*0.47*	0.32	*0.71*
12	47	–	–	0.92	*2.03*	–	–	–	–	0.78	*1.72*	–	–	–	–	–	–	–	–	–	–	–	–
13	22	–	–	0.42	*0.92*	–	–	–	–	0.36	*0.79*	–	–	–	–	–	–	0.24	*0.53*	0.15	*0.34*	0.23	*0.51*
14	24	–	–	0.45	*1.00*	–	–	–	–	0.39	*0.86*	–	–	–	–	–	–	0.22	*0.49*	0.13	*0.29*	0.22	*0.49*
15	19	–	–	0.37	*0.82*	–	–	–	–	0.32	*0.70*	–	–	–	–	–	–	0.21	*0.46*	0.13	*0.29*	0.20	*0.44*
16	14	–	–	0.27	*0.61*	–	–	–	–	0.23	*0.51*	–	–	–	–	–	–	0.13	*0.28*	0.07	*0.15*	0.14	*0.30*
17	18	–	–	0.34	*0.75*	–	–	–	–	0.29	*0.64*	–	–	–	–	–	–	0.16	*0.36*	0.09	*0.20*	0.17	*0.37*
18	18	–	–	0.38	*0.84*	–	–	–	–	0.31	*0.68*	–	–	–	–	–	–	0.19	*0.42*	0.11	*0.24*	0.19	*0.41*
19	23	–	–	0.45	*1.00*	–	–	–	–	0.39	*0.86*	–	–	–	–	–	–	0.23	*0.51*	0.14	*0.31*	0.23	*0.50*
20	16	–	–	0.33	*0.72*	–	–	–	–	0.28	*0.61*	–	–	–	–	–	–	0.17	*0.37*	0.10	*0.22*	0.17	*0.37*
21	17	–	–	0.35	*0.77*	–	–	–	–	0.29	*0.64*	–	–	–	–	–	–	0.18	*0.40*	0.10	*0.23*	0.19	*0.41*

(Continued)

TABLE 19–1C DRY FORAGES, MINERAL AND VITAMIN COMPOSITION OF FEEDS, DATA EXPRESSED **AS-FED**—*(Continued)*

			Macro Minerals							Micro Minerals						
Entry Number	Feed Name Description	Dry Matter	Calcium (Ca)	Phos-phorus (P)	Sodium (Na)	Chlo-rine (Cl)	Mag-nesium (Mg)	Potas-sium (K)	Sulfur (S)	Cobalt (Co)	Copper (Cu)	Iodine (I)	Iron (Fe)	Man-ganese (Mn)	Sele-nium (Se)	Zinc (Zn)
		%	%	%	%	%	%	%	%	ppm or mg/kg	ppm or mg/kg	ppm or mg/kg	%	ppm or mg/kg	ppm or mg/kg	ppm or mg/kg
76	**PEANUT,** HAY, SUN-CURED	91	1.12	0.14	–	–	0.44	1.25	0.21	0.072	–	–	–	–	–	–
77	**PEARL MILLET,** HAY, SUN-CURED	87	–	–	–	–	–	–	–	–	–	–	–	–	–	–
78	**PEAVINE,** HAY, SUN-CURED	91	–	–	–	–	–	–	–	–	–	–	–	–	–	–
79	**PRAIRIE GRASS, MIDWEST (PRAIRIE HAY),** HAY, SUN-CURED	91	0.32	0.13	–	–	0.24	0.98	–	–	–	–	0.008	–	–	–
80	**REDTOP,** HAY, SUN-CURED	92	0.39	0.20	0.06	0.06	0.20	1.74	0.23	0.134	3.6	0.092	0.015	207.7	–	–
81	**REED CANARYGRASS,** HAY, SUN-CURED	89	0.32	0.21	0.01	–	0.19	2.60	–	–	10.6	–	0.014	82.5	–	–
	RICE															
82	HULLS	92	0.11	0.10	0.02	0.07	0.41	0.64	0.08	2.046	3.1	–	0.010	295.0	–	22.0
83	STRAW	91	0.19	0.07	0.28	–	0.10	1.20	–	–	–	–	–	313.9	–	–
84	**RYE,** STRAW	91	0.22	0.08	0.12	0.22	0.07	0.88	0.10	–	3.6	–	–	6.0	–	–
85	**RYEGRASS,** HAY, SUN-CURED	88	–	–	–	–	–	–	–	–	–	–	–	–	–	–
86	**SAWDUST,** WOOD	90	–	–	–	–	–	–	–	–	–	–	–	–	–	–
87	**SEAWEED (KELP),** WHOLE, DEHY	91	2.47	0.28	–	–	0.85	–	–	–	–	–	–	–	–	–
	SORGHUM															
88	FODDER WITH HEADS, SUN-CURED	90	0.56	0.17	0.02	–	0.27	1.12	–	–	–	–	–	–	–	–
89	STOVER WITHOUT HEADS, SUN-CURED	92	0.37	0.10	–	–	–	1.10	–	–	–	–	–	–	–	–
90	KAFIR, FODDER WITH HEADS, SUN-CURED	90	0.35	0.18	–	–	0.26	1.53	–	–	–	–	–	–	–	–
91	KAFIR, STOVER WITHOUT HEADS, SUN-CURED	82	0.50	0.08	–	–	–	–	–	–	–	–	–	–	–	–
92	MILO, FODDER WITH HEADS, SUN-CURED	89	0.35	0.18	–	–	–	–	–	–	–	–	–	–	–	–
93	MILO, STOVER WITHOUT HEADS, SUN-CURED	91	0.51	0.15	0.10	–	0.23	0.41	–	–	–	–	–	–	–	–
94	SUDANGRASS, HAY, SUN-CURED	91	0.47	0.28	0.01	–	0.34	1.90	0.06	0.116	28.6	–	0.015	69.5	–	34.6
	SOYBEAN															
95	HAY, SUN-CURED	89	1.13	0.22	0.10	0.13	0.71	0.92	0.25	0.083	8.0	0.216	0.026	94.3	–	21.5
96	HULLS	91	0.45	0.19	0.03	–	–	1.15	0.08	0.109	16.1	–	0.030	9.9	–	21.8
97	STRAW	88	1.40	0.05	0.11	–	0.81	0.49	0.23	–	–	–	0.027	44.9	–	–
98	**STARGRASS,** HAY, SUN-CURED	90	–	–	–	–	–	–	–	–	–	–	–	–	–	–
99	**SUGARCANE (BAGASSE),** DEHY	91	0.47	0.26	0.04	–	0.08	0.34	0.09	–	–	–	0.019	–	–	–
100	**SWEET POTATO,** HAY, SUN-CURED	90	0.08	0.32	–	–	–	–	–	–	–	–	–	–	–	–
101	**TIMOTHY,** HAY, SUN-CURED, ALL ANALYSES	91	0.38	0.17	0.03	0.49	0.11	1.43	0.11	0.071	4.3	0.034	0.010	45.2	–	15.5
102	**TIMOTHY-CLOVER,** HAY, SUN-CURED	89	0.62	0.17	0.17	0.48	0.17	1.33	0.13	–	6.3	–	0.011	48.7	–	–
103	**TREFOIL, BIRDSFOOT,** HAY, SUN-CURED	91	1.54	0.21	0.06	–	0.46	1.74	0.23	0.100	8.4	–	0.021	26.0	–	69.9
104	**VELVET BEAN,** HAY, SUN-CURED	93	–	0.24	–	–	–	2.20	–	–	–	–	–	–	–	–
105	**VETCH,** HAY, SUN-CURED	89	1.21	0.30	0.46	–	0.24	1.88	0.13	0.315	8.8	0.437	0.044	53.9	–	–
106	**WATERGRASS,** HAY, SUN-CURED	94	–	–	–	–	–	–	–	–	–	–	–	–	–	–
107	**WHEAT,** STRAW	90	0.16	0.05	0.13	0.29	0.11	1.27	0.17	0.041	3.2	–	0.015	36.7	–	5.8
108	**WHEATGRASS, CRESTED,** HAY, SUN-CURED	92	0.24	0.14	–	–	–	–	–	0.219	–	–	–	–	–	–
109	**WHITE CLOVER,** HAY, SUN-CURED	90	1.71	0.29	–	–	–	–	–	–	–	–	–	–	–	–
110	**WOOD,** SAWDUST	90	–	–	–	–	–	–	–	–	–	–	–	–	–	–

TABLE 19–1D SILAGES AND HAYLAGES, MINERAL AND VITAMIN COMPOSITION OF FEEDS, DATA EXPRESSED **AS-FED**

			Macro Minerals							Micro Minerals						
Entry Number	Feed Name Description	Dry Matter	Calcium (Ca)	Phos-phorus (P)	Sodium (Na)	Chlo-rine (Cl)	Mag-nesium (Mg)	Potas-sium (K)	Sulfur (S)	Cobalt (Co)	Copper (Cu)	Iodine (I)	Iron (Fe)	Man-ganese (Mn)	Sele-nium (Se)	Zinc (Zn)
		%	%	%	%	%	%	%	%	ppm or mg/kg	ppm or mg/kg	ppm or mg/kg	%	ppm or mg/kg	ppm or mg/kg	ppm or mg/kg
	ALFALFA (LUCERNE), SILAGE															
1	ALL ANALYSES	27	0.48	0.07	0.04	0.11	0.09	0.64	0.09	–	3.0	–	0.008	13.5	–	11.1
2	CORN GRAIN ADDED	23	–	–	–	–	–	–	–	–	–	–	–	–	–	–
3	MOLASSES ADDED	30	0.49	0.09	0.05	–	0.10	0.78	0.11	–	–	–	0.009	–	–	–
	ALFALFA-BROMEGRASS, SMOOTH, SILAGE															
4	ALL ANALYSES	26	0.17	0.05	–	–	0.05	0.49	–	0.030	3.0	–	0.004	8.0	–	–
5	WILTED	46	0.72	0.12	–	–	–	–	–	–	–	–	–	–	–	–
6	**ALFALFA-ORCHARDGRASS, SILAGE,** 30-50% DRY MATTER	37	0.32	0.11	0.04	–	0.12	1.07	0.10	–	–	–	0.010	–	–	–
7	**BEET, MANGEL, SILAGE,** TOPS WITH CROWNS	13	–	–	–	–	–	–	–	–	–	–	–	–	–	–
8	**CITRUS, SILAGE,** PULP	21	0.42	0.03	0.02	–	0.03	0.13	0.00	–	–	–	0.004	–	–	3.3
	CORN, SILAGE															
9	ALL ANALYSES	26	0.08	0.07	0.01	0.05	0.06	0.32	0.03	0.026	2.4	–	0.005	10.8	–	5.5
10	MOLASSES ADDED	29	0.07	0.09	–	–	–	–	–	–	–	–	–	–	–	–
11	EARS WITH HUSKS	43	0.04	0.12	0.00	–	0.05	0.21	0.06	–	–	–	0.004	–	–	–
12	HUSKS (HUSKLAGE)	78	–	–	–	–	–	–	–	–	–	–	–	–	–	–
13	**CORN, SWEET, SILAGE,** CANNERY RESIDUE	31	0.10	0.24	0.01	–	0.07	0.36	0.03	–	–	–	0.007	–	–	–
14	**CORN-SORGHUM, SILAGE,** MATURE	34	0.15	0.08	–	–	–	0.35	–	–	–	–	–	–	–	–
15	**CORN-SOYBEAN, SILAGE**	27	0.19	0.09	0.01	–	0.08	0.29	0.05	–	2.2	–	0.012	12.4	–	7.6
	GRASS, SILAGE															
16	EARLY BLOOM	23	–	–	–	–	–	–	–	–	–	–	–	–	–	–
17	MOLASSES ADDED	27	0.28	0.07	–	–	–	–	–	–	–	–	–	–	–	–
	GRASS-LEGUME, SILAGE															
18	ALL ANALYSES	31	0.26	0.08	0.02	0.33	0.08	0.57	0.17	0.040	1.9	–	0.013	18.0	–	8.7
19	BARLEY GRAIN ADDED	34	0.26	0.12	–	–	–	–	–	–	–	–	–	–	–	–
20	MOLASSES ADDED	28	0.30	0.10	0.04	–	0.09	0.55	0.07	–	–	–	0.015	–	–	–
21	**KAFIR SORGHUM, SILAGE,**	30	0.07	0.05	–	–	0.08	0.50	–	–	–	–	–	–	–	–

Entry Number	Fat-Soluble Vitamins: A (1 mg Carotene = 1667 IU Vit A)	Carotene (Provitamin A)	D	E	K	Water-Soluble Vitamins: B–12	Biotin	Choline	Folacin (Folic Acid)	Niacin	Pantothenic Acid (B–3)	(Pyridoxine) B–6	Riboflavin (B–2)	Thiamin (B–1)
	IU/g	ppm or mg/kg	IU/kg	ppm or mg/kg	ppm or mg/kg	ppb or mcg/kg	ppm or mg/kg	ppm or mg/kg	ppm or mg/kg	ppm or mg/kg	ppm or mg/kg	ppm or mg/kg	ppm or mg/kg	ppm or mg/kg
76	52.6	31.5	–	–	–	–	–	–	–	–	–	–	8.8	–
77	–	–	–	–	–	–	–	–	–	–	–	–	–	–
78	48.0	28.8	–	–	–	–	–	–	–	–	–	–	8.4	–
79	–	–	1	–	–	–	–	–	–	–	–	–	–	–
80	6.1	3.7	–	–	–	–	–	–	–	–	–	–	–	–
81	28.2	16.9	–	–	–	–	–	–	–	–	–	–	8.5	3.6
82	–	–	–	7.5	–	–	–	–	–	28	7.9	0.07	0.5	2.2
83	–	–	–	–	–	–	–	–	–	–	–	–	–	–
84	–	–	–	–	–	–	–	–	–	–	–	–	–	–
85	175.8	105.5	–	–	–	–	–	–	–	–	–	–	–	–
86	–	–	–	–	–	–	–	–	–	–	–	–	–	–
87	–	–	–	–	–	–	–	–	–	–	–	–	–	–
88	–	–	–	–	–	–	–	–	–	–	–	–	–	–
89	–	–	–	–	–	–	–	–	–	–	–	–	–	–
90	–	–	–	–	–	–	–	–	–	–	–	–	–	–
91	–	–	–	–	–	–	–	–	–	–	–	–	–	–
92	2.9	1.8	–	–	–	–	–	–	–	–	–	–	–	–
93	–	–	–	–	–	–	–	–	–	–	–	–	–	–
94	–	–	–	–	–	–	–	–	–	–	–	–	–	–
95	53.1	31.8	1	26.3	–	–	–	–	–	–	–	–	–	–
96	–	–	–	6.6	–	–	–	588	–	25	13.4	1.70	3.6	1.6
97	–	–	–	–	–	–	–	–	–	–	–	–	–	–
98	–	–	–	–	–	–	–	–	–	–	–	–	–	–
99	–	–	–	–	–	–	–	–	–	–	–	–	–	–
100	–	–	–	–	–	–	–	–	–	–	–	–	–	–
101	39.8	23.8	2	57.6	–	–	0.06	741	2.09	31	7.2	–	9.2	1.5
102	39.7	23.8	–	–	–	–	–	–	–	–	–	–	–	–
103	217.8	130.6	1	–	–	–	–	–	–	–	–	–	14.6	6.2
104	–	–	–	–	–	–	–	–	–	–	–	–	–	–
105	–	–	–	–	–	–	–	–	–	–	–	–	–	–
106	–	–	–	–	–	–	–	–	–	–	–	–	–	–
107	3.3	2.0	1	–	–	–	–	–	–	–	–	–	2.2	–
108	34.2	20.5	–	–	–	–	–	–	–	–	–	–	–	–
109	92.1	55.3	–	115.5	–	–	–	–	–	–	–	–	–	–
110	–	–	–	–	–	–	–	–	–	–	–	–	–	–

Entry Number	Fat-Soluble Vitamins: A (1 mg Carotene = 1667 IU Vit A)	Carotene (Provitamin A)	D	E	K	Water-Soluble Vitamins: B–12	Biotin	Choline	Folacin (Folic Acid)	Niacin	Pantothenic Acid (B–3)	(Pyridoxine) B–6	Riboflavin (B–2)	Thiamin (B–1)
	IU/g	ppm or mg/kg	IU/kg	ppm or mg/kg	ppm or mg/kg	ppb or mcg/kg	ppm or mg/kg	ppm or mg/kg	ppm or mg/kg	ppm or mg/kg	ppm or mg/kg	ppm or mg/kg	ppm or mg/kg	ppm or mg/kg
1	42.2	25.3	–	–	–	–	–	–	–	–	–	–	–	–
2	–	–	–	–	–	–	–	–	–	–	–	–	–	–
3	53.8	32.3	–	–	–	–	–	–	–	–	–	–	–	–
4	–	–	–	–	–	–	–	–	–	–	–	–	–	–
5	–	–	–	–	–	–	–	–	–	–	–	–	–	–
6	–	–	–	–	–	–	–	–	–	–	–	–	–	–
7	–	–	–	–	–	–	–	–	–	–	–	–	–	–
8	–	–	–	–	–	–	–	–	–	–	–	–	–	–
9	15.2	9.1	0	–	–	–	–	–	–	11	–	–	–	–
10	–	–	–	–	–	–	–	–	–	–	–	–	–	–
11	–	–	–	–	–	–	–	–	–	–	–	–	–	–
12	–	–	–	–	–	–	–	–	–	–	–	–	–	–
13	6.9	4.2	–	–	–	–	–	–	–	–	–	–	–	–
14	3.1	1.9	–	–	–	–	–	–	–	–	–	–	–	–
15	89.0	53.4	–	–	–	–	–	–	–	–	–	–	–	–
16	–	–	–	–	–	–	–	–	–	–	–	–	–	–
17	–	–	–	–	–	–	–	–	–	–	–	–	–	–
18	102.4	61.4	0	–	–	–	–	–	–	14	–	–	–	–
19	–	–	–	–	–	–	–	–	–	–	–	–	–	–
20	–	–	–	–	–	–	–	–	–	–	–	–	–	–
21	5.3	3.2	–	–	–	–	–	–	–	–	–	–	–	–

(Continued)

TABLE 19–1D SILAGES AND HAYLAGES, COMPOSITION OF FEEDS, DATA EXPRESSED AS-FED—*(Continued)*

Entry Number	Feed Name Description	International Feed Number	Proximate Analysis									Digestible Protein		
			Dry Matter	Ash	Crude Fiber	Neutral Det. Fib. (NDF)	Acid Det. Fib. (ADF)	Lignin	Ether Extract (Fat)	N-Free Extract	Crude Protein	Ruminant	Swine	Horse
			%	%	%	%	%	%	%	%	%	%	%	%
22	**LESPEDEZA, COMMON-KOREAN, SILAGE**	3-08-455	30	1.8	9.5	–	–	–	0.8	13.8	4.3	2.8	–	–
23	**LESPEDEZA, SERICEA (CHINESE), SILAGE**	3-02-614	30	1.7	9.5	–	–	–	0.9	14.0	4.3	2.7	–	–
24	**MILLET, JAPANESE (BARNYARD GRASS)-SOYBEAN, SILAGE**	3-26-066	21	2.9	7.3	–	–	–	1.1	6.9	2.9	1.7	–	–
25	**MILO SORGHUM, SILAGE**	3-04-437	31	2.5	5.5	–	–	–	0.5	20.0	2.3	1.0	–	–
	OATS, SILAGE													
26	DOUGH STAGE	3-03-296	35	2.4	11.6	–	–	–	1.4	16.1	3.5	2.0	–	–
27	MOLASSES ADDED	3-03-300	33	2.5	10.4	–	–	–	1.2	15.8	2.9	1.5	–	–
28	**OATS-PEA, SILAGE**	3-03-402	26	2.5	8.7	–	11.6	1.2	0.9	10.9	2.8	1.5	–	–
29	**OATS-VETCH, SILAGE,** MILK STAGE	3-03-408	27	2.2	8.0	–	–	–	1.2	12.4	3.4	2.1	–	–
30	**PEA, SILAGE,** VINES (WITHOUT SEEDS, WITH PODS)	3-03-596	25	2.2	7.3	–	–	–	0.8	11.0	3.2	1.9	–	–
31	**PEA, FIELD, SILAGE**	3-03-609	27	2.5	7.5	–	–	–	1.2	12.2	3.8	2.5	–	–
	POTATO, SILAGE													
32	TUBERS, ALFALFA HAY ADDED	3-03-770	35	2.2	7.1	–	–	–	0.5	20.7	4.2	2.6	–	–
33	VINES	3-03-765	15	2.8	3.4	–	–	–	0.5	5.7	2.3	1.7	–	–
	SORGHUM, SILAGE													
34	DOUGH STAGE	3-04-321	29	2.5	8.3	19.2	10.9	1.9	0.9	15.1	2.3	0.9	–	–
35	STOVER WITHOUT HEADS	3-04-326	62	5.6	16.2	–	–	–	1.2	36.3	2.7	–	–	0.1
36	KAFIR	3-04-425	30	2.2	8.1	–	–	–	1.0	16.1	2.1	0.9	–	–
37	MILO	3-04-437	31	2.5	5.5	–	–	–	0.5	20.0	2.3	1.0	–	–
38	**SOYBEAN, SILAGE**	3-04-581	30	3.0	9.0	–	12.0	–	0.8	12.2	5.2	3.2	–	–
39	**SUGAR BEET, SILAGE,** TOPS WITH CROWNS	3-00-660	25	8.6	3.2	–	–	–	0.7	9.3	3.4	2.2	–	–
40	**SUGARCANE, SILAGE,** ALL ANALYSES	3-04-693	22	1.4	7.5	–	–	–	0.4	11.6	1.1	0.3	–	–
41	**SUNFLOWER, SILAGE,** MILK STAGE	3-04-733	21	2.1	6.2	–	–	–	1.3	9.5	2.1	1.0	–	–
42	**SWEET CORN, SILAGE,** CANNERY RESIDUE	3-07-955	31	1.7	10.1	–	10.5	–	1.3	15.4	2.5	1.2	–	–
43	**SWEET POTATO, SILAGE,** VINES	3-04-785	13	1.5	3.6	–	–	–	0.5	5.9	1.7	0.7	–	–
44	**VETCH, SILAGE**	3-05-112	30	2.4	9.8	–	–	–	1.0	13.4	3.5	2.0	–	–

TABLE 19–1E PASTURE AND RANGE PLANTS, COMPOSITION OF FEEDS, DATA EXPRESSED AS-FED

Entry Number	Feed Name Description	International Feed Number	Proximate Analysis									Digestible Protein		
			Dry Matter	Ash	Crude Fiber	Neutral Det. Fib. (NDF)	Acid Det. Fib. (ADF)	Lignin	Ether Extract (Fat)	N-Free Extract	Crude Protein	Ruminant	Swine	Horse
			%	%	%	%	%	%	%	%	%	%	%	%
	ALFALFA (LUCERNE)													
1	FRESH, ALL ANALYSES	2-00-196	26	2.5	6.0	11.8	–	–	1.0	11.2	5.3	4.0	3.8	–
2	PREBLOOM, FRESH	2-00-181	20	2.1	4.9	6.4	5.2	1.5	0.6	8.5	4.3	3.3	3.0	–
3	EARLY BLOOM, FRESH	2-00-184	24	3.1	6.6	9.0	7.2	2.0	0.7	8.2	5.4	4.3	3.9	–
4	**ALFALFA-BROMEGRASS,** FRESH	2-08-328	23	2.2	5.3	–	–	–	0.8	9.4	4.8	3.3	3.5	–
5	**ALFALFA-BROMEGRASS, SMOOTH,** FRESH	2-00-262	22	2.1	5.5	–	–	–	0.8	9.0	4.2	3.2	3.0	–
6	**ALFALFA-ORCHARDGRASS,** FRESH	2-00-323	25	–	–	–	–	–	–	–	–	–	–	–
7	**ALFILERIA, REDSTEM (FILAREE),** FRESH	2-00-356	18	2.4	3.6	–	5.3	–	0.6	8.2	2.8	2.0	1.9	–
8	**ALTA (TALL) FESCUE,** FRESH	2-01-889	28	2.5	7.5	19.5	–	–	0.9	14.3	2.7	1.7	1.7	–
9	**ALYCE CLOVER,** FRESH	2-00-362	20	2.2	5.5	–	–	–	0.6	8.6	3.1	–	–	–
10	**BAHIAGRASS,** FRESH	2-00-464	30	3.3	9.0	20.4	11.4	–	0.5	14.2	2.6	1.5	1.6	–
11	**BEARDGRASS (BLUESTEM),** IMMATURE, FRESH	2-00-821	27	2.4	6.7	–	–	–	0.7	13.6	3.4	2.3	2.3	–
12	**BEET, SUGAR,** TOPS WITH CROWNS, FRESH	2-00-649	17	3.4	1.8	–	–	–	0.3	8.6	2.5	1.9	1.7	–
13	**BERMUDAGRASS,** FRESH	2-00-712	29	3.3	7.6	–	–	–	0.6	13.0	4.2	3.0	2.9	–
14	**BERMUDAGRASS, COASTAL,** FRESH	2-00-719	29	1.8	8.3	–	–	–	1.1	13.6	4.4	3.2	3.0	–
15	**BLUEGRASS, CANADA,** FRESH	2-00-764	31	2.8	8.3	–	–	–	1.2	13.8	5.3	3.9	3.7	–
	BLUEGRASS, KENTUCKY													
16	IMMATURE, FRESH	2-00-777	31	2.9	7.8	17.1	9.0	–	1.1	13.7	5.4	4.0	3.8	–
17	EARLY BLOOM, FRESH	2-00-779	35	2.5	9.6	22.8	11.2	–	1.4	15.7	5.8	4.2	4.1	–
18	**BLUEGRASS, KENTUCKY-CLOVER, WHITE,** FRESH	2-08-356	24	2.7	4.5	–	–	–	0.9	11.3	5.0	3.8	3.6	–
19	**BLUESTEM (BEARDGRASS),** IMMATURE, FRESH	2-00-821	27	2.4	6.7	–	–	–	0.7	13.6	3.4	2.3	2.3	–
20	**BRISTLEGRASS,** FRESH	2-00-876	26	3.2	8.3	–	–	–	0.5	10.9	3.1	2.0	2.0	–
21	**BROMEGRASS,** IMMATURE, FRESH	2-00-892	34	3.8	7.5	19.0	10.5	–	1.2	15.5	5.8	4.7	4.1	–
22	**BROMEGRASS, SMOOTH,** FRESH	2-00-963	27	–	7.7	–	–	–	0.8	13.1	3.1	2.0	2.0	–
23	**BUFFALOGRASS,** FRESH	2-01-010	46	5.6	12.7	33.9	16.7	2.9	0.9	22.0	4.7	2.5	3.0	–
24	**BUFFELGRASS,** PREBLOOM, FRESH	2-10-253	21	2.8	–	15.1	–	–	–	–	1.7	0.9	1.0	–
25	**BUR-CLOVER, CALIFORNIA,** FRESH	2-01-035	27	1.8	–	–	–	–	0.6	–	6.2	–	–	–
26	**CABBAGE,** HEADS, FRESH	2-01-046	9	0.9	1.0	–	–	–	0.2	5.3	1.9	1.6	1.6	1.4
27	**CANADA BLUEGRASS,** FRESH	2-00-764	31	2.8	8.3	–	–	–	1.2	13.8	5.3	3.9	3.7	–
28	**CANARYGRASS,** FRESH	2-01-093	26	2.4	6.9	–	–	–	1.0	12.1	3.4	2.3	2.3	–
29	**CANARYGRASS, REED,** FRESH	2-01-113	23	2.3	5.6	10.6	6.5	1.0	0.9	10.1	3.9	2.9	2.7	–
30	**CLOVER, ALSIKE,** FRESH	2-01-316	22	2.1	5.2	–	–	–	0.8	10.3	4.1	3.0	2.9	–
31	**CLOVER, CRIMSON,** FRESH	2-01-336	18	1.7	4.9	–	–	–	0.6	7.5	3.0	2.3	2.1	–
32	**CLOVER, LADINO,** FRESH	2-01-383	18	1.9	2.5	–	–	–	0.9	8.1	4.4	3.6	3.2	–
33	**CLOVER, RED,** EARLY BLOOM, FRESH	2-01-428	20	2.0	4.6	8.0	6.2	–	1.0	8.3	3.8	2.8	2.7	–
34	**CLOVER, STRAWBERRY,** FRESH	2-26-067	20	3.2	3.1	–	–	–	0.8	7.3	5.6	–	–	–
35	**CLOVER, SUBTERRANEAN,** FRESH	2-26-068	21	2.1	6.0	–	–	–	1.0	9.4	2.6	–	–	–
36	**CLOVER, WHITE,** FRESH	2-01-468	18	2.1	2.8	–	–	–	0.6	7.2	5.0	4.0	3.7	–

SILAGES AND HAYLAGES

Entry Number	TDN Ruminant	TDN Swine	TDN Horse	Digestible Energy Ruminant		Digestible Energy Swine		Digestible Energy Horse		Metabolizable Energy Ruminant		Metabolizable Energy Swine		Poultry ME_n		Metabolizable Energy Horse		Net Energy Ruminant NE_m		Net Energy Ruminant NE_g		Lactating Cows NE_{lc}	
	%	%	%	Mcal lb	Mcal *kg*	kcal lb	kcal *kg*	Mcal lb	Mcal *kg*	Mcal lb	Mcal *kg*	kcal lb	kcal *kg*	kcal lb	kcal *kg*	Mcal lb	Mcal *kg*	Mcal lb	Mcal *kg*	Mcal lb	Mcal *kg*	Mcal lb	Mcal *kg*
22	15	–	–	0.33	*0.73*	–	–	–	–	0.27	*0.60*	–	–	–	–	–	–	0.18	*0.40*	0.10	*0.22*	0.19	*0.41*
23	19	–	–	0.38	*0.84*	–	–	–	–	0.32	*0.71*	–	–	–	–	–	–	0.19	*0.41*	0.11	*0.24*	0.19	*0.42*
24	–	–	–	–	–	–	–	–	–	–	–	–	–	–	–	–	–	–	–	–	–	–	–
25	18	–	–	0.40	*0.88*	–	–	–	–	0.34	*0.75*	–	–	–	–	–	–	0.17	*0.38*	0.10	*0.21*	0.18	*0.40*
26	20	–	–	0.39	*0.85*	–	–	–	–	0.34	*0.75*	–	–	–	–	–	–	0.18	*0.40*	0.09	*0.20*	0.18	*0.40*
27	17	–	–	0.37	*0.81*	–	–	–	–	0.31	*0.68*	–	–	–	–	–	–	0.20	*0.43*	0.11	*0.24*	0.20	*0.44*
28	15	–	–	0.29	*0.65*	–	–	–	–	0.24	*0.54*	–	–	–	–	–	–	0.14	*0.31*	0.07	*0.16*	0.15	*0.33*
29	17	–	–	0.34	*0.75*	–	–	–	–	0.29	*0.64*	–	–	–	–	–	–	0.17	*0.38*	0.10	*0.22*	0.17	*0.38*
30	14	–	–	0.28	*0.61*	–	–	–	–	0.23	*0.51*	–	–	–	–	–	–	0.13	*0.29*	0.07	*0.15*	0.14	*0.31*
31	18	–	–	0.35	*0.76*	–	–	–	–	0.29	*0.65*	–	–	–	–	–	–	0.17	*0.38*	0.10	*0.22*	0.17	*0.38*
32	20	–	–	0.42	*0.93*	–	–	–	–	0.35	*0.78*	–	–	–	–	–	–	0.22	*0.48*	0.13	*0.28*	0.22	*0.49*
33	8	–	–	0.17	*0.36*	–	–	–	–	0.14	*0.30*	–	–	–	–	–	–	0.07	*0.16*	0.04	*0.08*	0.08	*0.18*
34	16	–	–	0.31	*0.68*	–	–	–	–	0.24	*0.53*	–	–	–	–	–	–	0.16	*0.35*	0.08	*0.18*	0.16	*0.35*
35	29	–	–	0.69	*1.53*	–	–	–	–	0.57	*1.25*	–	–	–	–	–	–	0.34	*0.74*	0.14	*0.31*	0.35	*0.78*
36	17	–	–	0.35	*0.77*	–	–	–	–	0.29	*0.64*	–	–	–	–	–	–	0.18	*0.40*	0.10	*0.23*	0.19	*0.41*
37	18	–	–	0.40	*0.88*	–	–	–	–	0.34	*0.75*	–	–	–	–	–	–	0.17	*0.38*	0.10	*0.21*	0.18	*0.40*
38	16	–	–	0.32	*0.71*	–	–	–	–	0.26	*0.58*	–	–	–	–	–	–	0.15	*0.33*	0.07	*0.16*	0.16	*0.36*
39	13	–	–	0.26	*0.58*	–	–	–	–	0.21	*0.47*	–	–	–	–	–	–	0.13	*0.28*	0.06	*0.14*	0.14	*0.30*
40	13	–	–	0.26	*0.57*	–	–	–	–	0.22	*0.48*	–	–	–	–	–	–	0.12	*0.26*	0.06	*0.14*	0.13	*0.28*
41	11	–	–	0.24	*0.53*	–	–	–	–	0.20	*0.44*	–	–	–	–	–	–	0.10	*0.21*	0.04	*0.09*	0.11	*0.24*
42	22	–	–	0.42	*0.92*	–	–	–	–	0.36	*0.79*	–	–	–	–	–	–	0.24	*0.53*	0.15	*0.34*	0.23	*0.51*
43	6	–	–	0.14	*0.31*	–	–	–	–	0.12	*0.26*	–	–	–	–	–	–	0.08	*0.17*	0.04	*0.10*	0.08	*0.18*
44	19	–	–	0.38	*0.83*	–	–	–	–	0.32	*0.71*	–	–	–	–	–	–	0.19	*0.42*	0.11	*0.25*	0.19	*0.43*

PASTURE AND RANGE PLANTS

Entry Number	TDN Ruminant	TDN Swine	TDN Horse	Digestible Energy Ruminant		Digestible Energy Swine		Digestible Energy Horse		Metabolizable Energy Ruminant		Metabolizable Energy Swine		Poultry ME_n		Metabolizable Energy Horse		Net Energy Ruminant NE_m		Net Energy Ruminant NE_g		Lactating Cows NE_{lc}	
	%	%	%	Mcal lb	Mcal *kg*	kcal lb	kcal *kg*	Mcal lb	Mcal *kg*	Mcal lb	Mcal *kg*	kcal lb	kcal *kg*	kcal lb	kcal *kg*	Mcal lb	Mcal *kg*	Mcal lb	Mcal *kg*	Mcal lb	Mcal *kg*	Mcal lb	Mcal *kg*
1	16	–	–	0.32	*0.70*	–	–	–	–	0.27	*0.59*	–	–	–	–	–	–	0.16	*0.35*	0.09	*0.20*	0.16	*0.36*
2	12	12	–	0.24	*0.54*	–	–	–	–	0.20	*0.44*	–	–	–	–	–	–	0.12	*0.27*	0.07	*0.15*	0.13	*0.28*
3	15	–	–	0.30	*0.66*	–	–	–	–	0.25	*0.56*	–	–	–	–	–	–	0.15	*0.34*	0.09	*0.20*	0.15	*0.34*
4	14	–	–	0.29	*0.64*	–	–	–	–	0.25	*0.55*	–	–	–	–	–	–	0.16	*0.35*	0.10	*0.22*	0.16	*0.35*
5	14	–	–	0.28	*0.62*	–	–	–	–	0.24	*0.53*	–	–	–	–	–	–	0.14	*0.30*	0.08	*0.18*	0.14	*0.31*
6	–	–	–	–	–	–	–	–	–	–	–	–	–	–	–	–	–	–	–	–	–	–	–
7	10	–	–	0.21	*0.46*	–	–	–	–	0.18	*0.39*	–	–	–	–	–	–	0.11	*0.25*	0.07	*0.15*	0.11	*0.25*
8	17	–	–	0.34	*0.76*	–	–	–	–	0.29	*0.64*	–	–	–	–	–	–	0.17	*0.38*	0.10	*0.22*	0.18	*0.39*
9	–	–	–	–	–	–	–	–	–	–	–	–	–	–	–	–	–	–	–	–	–	–	–
10	16	–	–	0.32	*0.70*	–	–	–	–	0.26	*0.57*	–	–	–	–	–	–	0.15	*0.33*	0.08	*0.17*	0.16	*0.36*
11	18	–	–	0.35	*0.78*	–	–	–	–	0.30	*0.67*	–	–	–	–	–	–	0.18	*0.39*	0.10	*0.23*	0.18	*0.39*
12	11	–	–	0.21	*0.46*	–	–	–	–	0.18	*0.39*	–	–	–	–	–	–	0.10	*0.22*	0.06	*0.13*	0.10	*0.23*
13	17	–	–	0.35	*0.77*	–	–	–	–	0.29	*0.65*	–	–	–	–	–	–	0.18	*0.39*	0.10	*0.22*	0.18	*0.40*
14	19	–	–	0.37	*0.82*	–	–	–	–	0.32	*0.70*	–	–	–	–	–	–	0.18	*0.40*	0.11	*0.23*	0.19	*0.41*
15	20	–	–	0.41	*0.90*	–	–	–	–	0.35	*0.77*	–	–	–	–	–	–	0.23	*0.51*	0.15	*0.32*	0.23	*0.50*
16	22	–	–	0.42	*0.93*	–	–	–	–	0.37	*0.81*	–	–	–	–	–	–	0.24	*0.52*	0.15	*0.33*	0.23	*0.51*
17	24	–	–	0.49	*1.07*	–	–	–	–	0.42	*0.93*	–	–	–	–	–	–	0.26	*0.57*	0.16	*0.36*	0.25	*0.55*
18	17	–	–	0.34	*0.74*	–	–	–	–	0.29	*0.64*	–	–	–	–	–	–	0.18	*0.39*	0.11	*0.24*	0.17	*0.38*
19	18	–	–	0.35	*0.78*	–	–	–	–	0.30	*0.67*	–	–	–	–	–	–	0.18	*0.39*	0.10	*0.23*	0.18	*0.39*
20	15	–	–	0.30	*0.66*	–	–	–	–	0.25	*0.55*	–	–	–	–	–	–	0.14	*0.31*	0.07	*0.16*	0.15	*0.33*
21	25	–	–	0.50	*1.11*	–	–	–	–	0.44	*0.97*	–	–	–	–	–	–	0.24	*0.53*	0.15	*0.33*	0.24	*0.52*
22	17	–	–	0.34	*0.74*	–	–	–	–	0.28	*0.62*	–	–	–	–	–	–	0.17	*0.37*	0.10	*0.22*	0.17	*0.38*
23	26	–	–	0.52	*1.15*	–	–	–	–	0.43	*0.95*	–	–	–	–	–	–	0.26	*0.57*	0.14	*0.31*	0.27	*0.59*
24	–	–	–	–	–	–	–	–	–	–	–	–	–	–	–	–	–	–	–	–	–	–	–
25	–	–	–	–	–	–	–	–	–	–	–	–	–	–	–	–	–	–	–	–	–	–	–
26	8	6	7	0.14	*0.32*	126	*279*	0.12	*0.27*	0.13	*0.28*	118	*261*	–	–	0.10	*0.22*	0.07	*0.15*	0.04	*0.10*	0.07	*0.15*
27	20	–	–	0.41	*0.90*	–	–	–	–	0.35	*0.77*	–	–	–	–	–	–	0.23	*0.51*	0.15	*0.32*	0.23	*0.50*
28	16	–	–	0.32	*0.71*	–	–	–	–	0.27	*0.60*	–	–	–	–	–	–	0.16	*0.36*	0.10	*0.21*	0.17	*0.37*
29	14	–	–	0.29	*0.64*	–	–	–	–	0.24	*0.54*	–	–	–	–	–	–	0.15	*0.34*	0.09	*0.20*	0.15	*0.34*
30	16	–	–	0.32	*0.71*	–	–	–	–	0.28	*0.61*	–	–	–	–	–	–	0.17	*0.37*	0.11	*0.24*	0.17	*0.36*
31	11	–	–	0.23	*0.50*	–	–	–	–	0.19	*0.42*	–	–	–	–	–	–	0.11	*0.25*	0.07	*0.15*	0.12	*0.25*
32	13	–	–	0.27	*0.60*	–	–	–	–	0.24	*0.52*	–	–	–	–	–	–	0.15	*0.33*	0.10	*0.22*	0.14	*0.32*
33	14	–	–	0.27	*0.58*	–	–	–	–	0.23	*0.50*	–	–	–	–	–	–	0.14	*0.32*	0.09	*0.20*	0.14	*0.31*
34	–	–	–	–	–	–	–	–	–	–	–	–	–	–	–	–	–	–	–	–	–	–	–
35	–	–	–	–	–	–	–	–	–	–	–	–	–	–	–	–	–	–	–	–	–	–	–
36	13	–	–	0.26	*0.57*	–	–	–	–	0.22	*0.49*	–	–	–	–	–	–	0.14	*0.30*	0.09	*0.19*	0.13	*0.29*

(Continued)

TABLE 19–1D SILAGES AND HAYLAGES, MINERAL AND VITAMIN COMPOSITION OF FEEDS, DATA EXPRESSED **AS-FED**—*(Continued)*

			Macro Minerals							Micro Minerals						
Entry Number	Feed Name Description	Dry Matter	Calcium (Ca)	Phosphorus (P)	Sodium (Na)	Chlorine (Cl)	Magnesium (Mg)	Potassium (K)	Sulfur (S)	Cobalt (Co)	Copper (Cu)	Iodine (I)	Iron (Fe)	Manganese (Mn)	Selenium (Se)	Zinc (Zn)
		%	%	%	%	%	%	%	%	ppm or mg/kg	ppm or mg/kg	ppm or mg/kg	%	ppm or mg/kg	ppm or mg/kg	ppm or mg/kg
22	**LESPEDEZA, COMMON-KOREAN, SILAGE**	30	–	–	–	–	–	–	–	–	–	–	–	–	–	–
23	**LESPEDEZA, SERICEA (CHINESE), SILAGE**	30	–	–	–	–	–	–	–	–	–	–	–	–	–	–
24	**MILLET, JAPANESE (BARNYARD GRASS)-SOYBEAN, SILAGE**	21	–	–	–	–	–	–	–	–	–	–	–	–	–	–
25	**MILO SORGHUM, SILAGE,**	31	0.11	0.06	–	–	–	–	–	–	–	–	–	–	–	–
	OATS, SILAGE															
26	DOUGH STAGE	35	0.17	0.12	–	–	–	–	–	–	–	–	–	–	–	–
27	MOLASSES ADDED	33	0.10	0.09	–	–	–	0.31	–	–	–	–	–	–	–	–
28	**OATS-PEA, SILAGE**	26	0.16	0.08	–	0.31	0.11	0.48	–	0.063	4.8	–	0.017	8.5	–	–
29	**OATS-VETCH, SILAGE,** MILK STAGE	27	–	–	–	–	–	–	–	–	–	–	–	–	–	–
30	**PEA, SILAGE,** VINES (WITHOUT SEEDS, WITH PODS)	25	0.32	0.06	0.00	–	0.10	0.34	0.06	–	–	–	0.003	–	–	–
31	**PEA, FIELD, SILAGE**	27	0.37	0.08	–	–	0.11	0.38	0.07	–	–	–	–	–	–	–
	POTATO, SILAGE															
32	TUBERS, ALFALFA HAY ADDED	35	–	–	–	–	–	–	–	–	–	–	–	–	–	–
33	VINES	15	0.31	0.03	–	0.06	0.02	0.59	0.06	–	–	–	–	–	–	–
	SORGHUM, SILAGE															
34	DOUGH STAGE	29	–	–	–	–	–	–	–	–	–	–	–	–	–	–
35	STOVER WITHOUT HEADS	62	0.25	0.07	–	–	–	–	–	–	–	–	–	91.7	–	–
36	KAFIR	30	0.07	0.05	–	–	0.08	0.50	–	–	–	–	–	–	–	–
37	MILO	31	0.11	0.06	–	–	–	–	–	–	–	–	–	–	–	–
38	**SOYBEAN, SILAGE**	30	0.40	0.13	0.03	–	0.12	0.39	0.09	–	2.9	–	0.010	42.7	–	10.3
39	**SUGAR BEET, SILAGE,** TOPS WITH CROWNS	25	0.39	0.07	0.14	–	0.27	1.45	0.14	–	–	–	0.006	–	–	–
40	**SUGARCANE, SILAGE,** ALL ANALYSES	22	0.08	0.04	–	–	0.05	–	–	–	–	–	–	–	–	–
41	**SUNFLOWER, SILAGE,** MILK STAGE	21	–	–	–	–	–	–	–	–	–	–	–	–	–	–
42	**SWEET CORN, SILAGE,** CANNERY RESIDUE	31	0.10	0.24	0.01	–	0.07	0.36	0.03	–	–	–	0.007	–	–	–
43	**SWEET POTATO, SILAGE,** VINES	13	–	–	–	–	–	–	–	–	–	–	–	–	–	–
44	**VETCH, SILAGE**	30	–	–	–	–	–	–	–	–	–	–	–	–	–	–

TABLE 19–1E PASTURE AND RANGE PLANTS, MINERAL AND VITAMIN COMPOSITION OF FEEDS, DATA EXPRESSED **AS-FED**

			Macro Minerals							Micro Minerals						
Entry Number	Feed Name Description	Dry Matter	Calcium (Ca)	Phosphorus (P)	Sodium (Na)	Chlorine (Cl)	Magnesium (Mg)	Potassium (K)	Sulfur (S)	Cobalt (Co)	Copper (Cu)	Iodine (I)	Iron (Fe)	Manganese (Mn)	Selenium (Se)	Zinc (Zn)
		%	%	%	%	%	%	%	%	ppm or mg/kg	ppm or mg/kg	ppm or mg/kg	%	ppm or mg/kg	ppm or mg/kg	ppm or mg/kg
	ALFALFA (LUCERNE)															
1	FRESH, ALL ANALYSES	26	0.40	0.07	0.05	0.12	0.09	0.83	0.10	0.092	3.2	–	0.009	24.1	–	9.4
2	PREBLOOM, FRESH	20	0.44	0.07	0.04	0.09	0.05	0.44	0.10	0.034	2.2	–	0.003	8.3	–	–
3	EARLY BLOOM, FRESH	24	0.39	0.07	0.04	–	0.12	0.88	–	0.107	4.4	–	0.008	33.2	–	9.6
4	**ALFALFA-BROMEGRASS,** FRESH	23	0.28	0.07	–	–	–	0.63	–	–	–	–	–	–	–	–
5	**ALFALFA-BROMEGRASS, SMOOTH,** FRESH	22	0.33	0.08	0.09	–	0.08	0.84	0.05	–	–	–	0.003	–	–	–
6	**ALFALFA-ORCHARDGRASS,** FRESH	25	0.10	0.13	–	–	0.06	–	–	–	–	–	–	–	–	–
7	**AFILERIA REDSTEM (FILAREE),** FRESH	18	0.35	0.08	–	–	–	0.59	–	–	–	–	–	–	–	–
8	**ALTA (TALL) FESCUE,** FRESH	28	0.13	0.05	0.03	–	0.07	0.70	–	0.113	1.0	–	0.003	18.0	–	5.9
9	**ALYCE CLOVER,** FRESH	20	–	–	–	–	–	–	–	0.018	–	–	–	–	–	–
10	**BAHIAGRASS,** FRESH	30	0.14	0.06	–	–	0.07	0.43	–	–	–	–	–	–	–	–
11	**BEARDGRASS (BLUESTEM),** IMMATURE, FRESH	27	0.17	0.05	–	–	–	0.46	–	–	12.6	–	0.024	28.5	–	–
12	**BEET, SUGAR,** TOPS WITH CROWNS, FRESH	17	0.17	0.04	0.09	0.09	0.18	0.96	0.09	–	2.3	–	0.003	9.0	–	–
13	**BERMUDAGRASS,** FRESH	29	0.16	0.06	0.13	–	0.07	0.55	–	0.022	1.6	–	0.033	28.6	–	–
14	**BERMUDAGRASS, COASTAL,** FRESH	29	0.14	0.08	–	–	–	–	–	–	–	–	–	–	–	–
15	**BLUEGRASS, CANADA,** FRESH	31	0.12	0.12	0.04	–	0.05	0.64	0.05	–	–	–	0.010	24.8	–	–
	BLUEGRASS, KENTUCKY															
16	IMMATURE, FRESH	31	0.15	0.14	0.04	–	0.05	0.70	0.05	–	–	–	0.010	–	–	–
17	EARLY BLOOM, FRESH	35	0.16	0.14	0.05	–	0.04	0.70	0.06	–	–	–	0.011	–	–	–
18	**BLUEGRASS, KENTUCKY-CLOVER, WHITE,** FRESH	24	0.31	0.11	–	–	–	–	–	–	–	–	–	–	–	–
19	**BLUESTEM (BEARDGRASS),** IMMATURE, FRESH	27	0.17	0.05	–	–	–	0.46	–	–	12.6	–	0.024	28.5	–	–
20	**BRISTLEGRASS,** FRESH	26	0.10	0.05	–	–	0.07	1.51	–	–	1.7	–	–	10.9	–	–
21	**BROMEGRASS,** IMMATURE, FRESH	34	0.20	0.13	0.01	–	0.06	1.46	0.07	–	–	–	0.007	–	–	–
22	**BROMEGRASS, SMOOTH,** FRESH	27	–	–	–	–	–	–	–	0.022	–	–	–	–	–	–
23	**BUFFALOGRASS,** FRESH	46	0.26	0.09	–	–	0.06	0.33	–	–	–	–	–	–	–	–
24	**BUFFELGRASS,** PREBLOOM, FRESH	21	0.19	0.03	0.03	–	0.12	0.89	–	0.063	2.0	–	0.014	27.7	–	9.5
25	**BUR-CLOVER, CALIFORNIA,** FRESH	27	–	–	–	–	–	–	–	–	–	–	–	–	–	–
26	**CABBAGE,** HEADS, FRESH	9	0.06	0.03	0.01	0.05	0.02	0.26	0.11	–	1.3	–	0.001	2.8	–	–
27	**CANADA BLUEGRASS,** FRESH	31	0.12	0.12	0.04	–	0.05	0.64	0.05	–	–	–	0.010	24.8	–	–
28	**CANARYGRASS,** FRESH	26	0.11	0.08	–	–	0.07	0.82	–	–	–	–	–	–	–	–
29	**CANARYGRASS, REED,** FRESH	23	0.08	0.08	–	–	–	0.83	–	–	–	–	–	–	–	–
30	**CLOVER, ALSIKE,** FRESH	22	0.31	0.06	0.10	0.17	0.07	0.61	0.05	–	1.3	–	0.010	26.3	–	–
31	**CLOVER, CRIMSON,** FRESH	18	0.24	0.05	0.07	0.11	0.05	0.55	0.05	–	–	–	–	43.1	–	–
32	**CLOVER, LADINO,** FRESH	18	0.22	0.07	0.02	–	0.09	0.33	0.02	–	–	–	0.007	12.7	–	–
33	**CLOVER, RED,** EARLY BLOOM, FRESH	20	0.45	0.08	0.04	–	0.10	0.49	0.03	–	–	–	0.006	–	–	–
34	**CLOVER, STRAWBERRY,** FRESH	20	0.37	0.09	–	–	–	–	–	–	–	–	–	–	–	–
35	**CLOVER, SUBTERRANEAN,** FRESH	21	0.31	0.07	–	–	–	–	–	–	–	–	–	–	–	–
36	**CLOVER, WHITE,** FRESH	18	0.25	0.09	0.07	0.11	0.08	0.38	0.06	–	–	–	0.006	54.4	–	–

SILAGES AND HAYLAGES

Entry Number	Fat-Soluble Vitamins					Water-Soluble Vitamins								
	A (1 mg Carotene = 1667 IU Vit A)	Carotene (Provitamin A)	D	E	K	B–12	Biotin	Choline	Folacin (Folic Acid)	Niacin	Pantothenic Acid (B–3)	(Pyridoxine) B–6	Riboflavin (B–2)	Thiamin (B–1)
	IU/g	ppm or mg/kg	IU/kg	ppm or mg/kg	ppm or mg/kg	ppb or mcg/kg	ppm or mg/kg	ppm or mg/kg	ppm or mg/kg	ppm or mg/kg	ppm or mg/kg	ppm or mg/kg	ppm or mg/kg	ppm or mg/kg
22	–	–	–	–	–	–	–	–	–	–	–	–	–	–
23	–	–	–	–	–	–	–	–	–	–	–	–	–	–
24	–	–	–	–	–	–	–	–	–	–	–	–	–	–
25	–	–	–	–	–	–	–	–	–	–	–	–	–	–
26	35.1	21.1	–	–	–	–	–	–	–	–	–	–	–	–
27	–	–	–	–	–	–	–	–	–	–	–	–	–	–
28	33.5	20.1	–	–	–	–	–	–	–	–	–	–	–	–
29	–	–	–	–	–	–	–	–	–	–	–	–	–	–
30	77.2	46.3	–	–	–	–	–	–	–	–	–	–	–	–
31	–	–	–	–	–	–	–	–	–	–	–	–	–	–
32	–	–	–	–	–	–	–	–	–	–	–	–	–	–
33	–	–	–	–	–	–	–	–	–	–	–	–	–	–
34	–	–	–	–	–	–	–	–	–	–	–	–	–	–
35	7.1	4.2	–	–	–	–	–	–	–	–	–	–	–	–
36	5.3	3.2	–	–	–	–	–	–	–	–	–	–	–	–
37	–	–	–	–	–	–	–	–	–	–	–	–	–	–
38	52.2	31.3	–	–	–	–	–	–	–	–	–	–	–	–
39	–	–	–	–	–	–	–	–	–	–	–	–	–	–
40	–	–	–	–	–	–	–	–	–	–	–	–	–	–
41	–	–	–	–	–	–	–	–	–	–	–	–	–	–
42	–	–	–	–	–	–	–	–	–	–	–	–	–	–
43	6.9	4.2	–	–	–	–	–	–	–	–	–	–	–	–
44	–	–	–	–	–	–	–	–	–	–	–	–	–	–

PASTURE AND RANGE PLANTS

Entry Number	Fat-Soluble Vitamins					Water-Soluble Vitamins								
	A (1 mg Carotene = 1667 IU Vit A)	Carotene (Provitamin A)	D	E	K	B–12	Biotin	Choline	Folacin (Folic Acid)	Niacin	Pantothenic Acid (B–3)	(Pyridoxine) B–6	Riboflavin (B–2)	Thiamin (B–1)
	IU/g	ppm or mg/kg	IU/kg	ppm or mg/kg	ppm or mg/kg	ppb or mcg/kg	ppm or mg/kg	ppm or mg/kg	ppm or mg/kg	ppm or mg/kg	ppm or mg/kg	ppm or mg/kg	ppm or mg/kg	ppm or mg/kg
1	101.3	60.8	0	–	–	–	0.13	374	0.64	15	8.9	1.66	4.6	1.7
2	–	–	0	34.8	–	–	–	–	–	–	–	–	–	–
3	69.9	41.9	–	–	–	–	–	–	–	–	–	–	–	–
4	–	–	–	–	–	–	–	–	–	–	–	–	–	–
5	–	–	–	–	–	–	–	–	–	–	–	–	–	–
6	–	–	–	–	–	–	–	–	–	–	–	–	–	–
7	–	–	–	–	–	–	–	–	–	–	–	–	–	–
8	–	–	–	–	–	–	–	–	–	–	–	–	–	–
9	35.0	21.0	–	–	–	–	–	–	–	–	–	–	–	–
10	89.7	53.8	–	–	–	–	–	–	–	–	–	–	–	–
11	97.9	58.7	–	–	–	–	–	–	–	–	–	–	–	–
12	9.6	5.8	–	–	–	–	–	–	–	–	–	–	1.1	–
13	147.8	88.7	–	–	–	–	–	–	–	–	–	–	–	–
14	160.3	96.1	–	–	–	–	–	–	–	–	–	–	–	–
15	199.9	119.9	–	–	–	–	–	–	–	–	–	–	–	–
16	247.6	148.5	–	47.8	–	–	–	–	–	–	–	–	–	–
17	163.4	98.0	–	–	–	–	–	–	–	–	–	–	–	–
18	–	–	–	–	–	–	–	–	–	–	–	–	–	–
19	97.9	58.7	–	–	–	–	–	–	–	–	–	–	–	–
20	–	–	–	–	–	–	–	–	–	–	–	–	–	–
21	259.6	155.7	–	–	–	–	–	–	–	–	–	–	–	–
22	142.0	85.2	0	–	–	–	–	–	–	–	–	–	2.1	0.8
23	71.6	42.9	–	–	–	–	–	–	–	–	–	–	–	–
24	–	–	–	–	–	–	–	–	–	–	–	–	–	–
25	–	–	–	–	–	–	–	–	–	–	–	–	–	–
26	0.7	0.4	–	–	–	–	–	249	0.63	3	–	–	0.5	0.6
27	199.9	119.9	–	–	–	–	–	–	–	–	–	–	–	–
28	–	–	–	–	–	–	–	–	–	–	–	–	–	–
29	–	–	–	–	–	–	–	–	–	–	–	–	–	–
30	–	–	–	–	–	–	–	–	–	–	–	–	4.4	2.0
31	–	–	–	–	–	–	–	–	–	–	–	–	–	–
32	96.2	57.7	–	–	–	–	–	–	–	–	–	–	4.2	–
33	81.5	48.9	–	–	–	–	–	–	–	–	–	–	–	–
34	–	–	–	–	–	–	–	–	–	–	–	–	–	–
35	–	–	–	–	–	–	–	–	–	–	–	–	–	–
36	44.0	26.4	–	54.6	–	–	–	–	–	11	–	–	16.0	2.5

(Continued)

SILAGES AND HAYLAGES

PASTURE AND RANGE PLANTS

TABLE 19-1E PASTURE AND RANGE PLANTS, COMPOSITION OF FEEDS, DATA EXPRESSED **AS-FED**—*(Continued)*

Entry Number	Feed Name Description	International Feed Number	Proximate Analysis									Digestible Protein		
			Dry Matter	Ash	Crude Fiber	Neutral Det. Fib. (NDF)	Acid Det. Fib. (ADF)	Lignin	Ether Extract (Fat)	N-Free Extract	Crude Protein	Ruminant	Swine	Horse
			%	%	%	%	%	%	%	%	%	%	%	%
37	**CORN,** FRESH, ALL ANALYSES	2-02-799	23	1.3	5.7	–	–	–	0.9	12.9	2.4	1.5	1.5	–
38	**CORN, SWEET,** STOVER WITHOUT EARS, WITHOUT HUSKS, FRESH	2-02-969	22	1.4	5.7	–	–	–	0.4	12.8	1.6	0.8	0.9	–
39	**COWPEA, COMMON,** FRESH	2-01-655	25	3.0	6.1	–	–	–	1.0	10.4	4.0	2.9	2.9	–
40	**CRESTED WHEATGRASS,** EARLY BLOOM, FRESH	2-05-422	41	–	8.9	–	–	–	1.7	–	4.8	3.1	3.1	–
	CURLY MESQUITE													
41	BROWSE, FRESH	2-01-728	35	5.1	9.8	–	–	–	0.7	16.4	3.0	1.7	1.8	–
42	BROWSE, MATURE, FRESH	2-01-729	50	7.6	14.3	32.0	–	–	1.1	23.9	2.8	1.1	1.4	–
43	**DALLISGRASS,** FRESH	2-01-741	25	3.0	7.3	–	–	–	0.6	11.1	3.0	2.0	2.0	–
44	**DROPSEED, SAND,** STEM CURED, FRESH	2-05-596	88	7.0	31.6	–	–	5.2	1.1	43.0	5.4	1.3	2.8	–
45	**FESCUE, TALL (ALTA),** FRESH	2-01-889	28	2.5	7.5	19.5	–	–	0.9	14.3	2.7	1.7	1.7	–
46	**FILAREE (ALFILERIA, REDSTEM),** FRESH	2-00-356	18	2.4	3.6	–	5.3	–	0.6	8.2	2.8	2.0	1.9	–
47	**FOXTAIL, MEADOW,** IMMATURE, FRESH	2-02-073	26	2.8	5.6	–	–	–	1.2	12.0	4.5	3.3	3.2	–
48	**GALLETA,** STEM CURED, FRESH	2-05-594	86	13.3	28.4	–	–	–	1.4	38.5	4.3	1.3	2.0	–
49	**GRAMA,** IMMATURE, FRESH	2-02-163	41	4.6	11.2	–	–	–	0.8	19.0	5.4	3.7	3.6	–
50	**GRASS-LEGUME,** FRESH	2-08-439	24	2.5	5.6	–	–	–	0.8	10.4	4.2	3.1	2.9	–
51	**GREASEWOOD,** BROWSE, FRESH	2-02-312	50	7.3	11.7	–	–	–	1.7	18.6	10.7	8.2	7.7	–
52	**INDIANGRASS,** FRESH	2-08-770	57	4.1	19.4	–	–	–	1.3	29.4	2.8	0.9	1.2	–
53	**INDIAN RICEGRASS,** FRESH	2-03-944	48	–	–	–	–	–	–	–	–	–	–	–
	JOHNSONGRASS SORGHUM–SEE SORGHUM													
54	**JUNEGRASS,** IMMATURE, FRESH	2-02-437	28	2.2	6.4	–	–	–	0.6	12.8	6.0	4.7	4.4	–
	KENTUCKY BLUEGRASS–SEE BLUEGRASS, KENTUCKY													
55	**KOA HAOLE (LEAD TREE, WHITE POPINAC),** BROWSE, FRESH	2-02-495	30	1.9	10.7	–	–	–	0.6	11.5	5.5	4.1	3.9	–
56	**LESPEDEZA, COMMON,** IMMATURE, FRESH	2-20-879	24	3.1	7.7	–	–	–	0.5	8.8	3.9	2.9	2.7	–
57	**LESPEDEZA, COMMON-KOREAN,** MATURE, FRESH	2-26-032	35	2.6	15.8	–	–	–	0.7	11.8	4.4	3.0	–	2.9
58	**LESPEDEZA, SERICEA (CHINESE LESPEDEZA),** FRESH	2-02-611	33	2.0	7.5	–	–	–	1.2	16.2	5.9	4.4	4.2	–
59	**LOVEGRASS,** IMMATURE, FRESH	2-02-647	43	2.8	13.1	–	–	–	1.3	20.2	5.4	3.6	3.6	–
60	**MEADOW FOXTAIL,** IMMATURE, FRESH	2-02-073	26	2.8	5.6	–	–	–	1.2	12.0	4.5	3.3	3.2	–
61	**MEDIC, BLACK (YELLOW TREFOIL),** FRESH	2-03-070	23	2.3	5.6	–	–	–	0.8	9.1	4.9	3.8	3.5	–
62	**MESQUITE, COMMON,** BROWSE, FRESH	2-03-081	35	2.1	9.6	–	–	–	1.2	14.8	7.4	5.7	5.3	–
63	**MILLET, FOXTAIL,** FRESH	2-03-101	29	2.5	9.2	–	–	–	0.9	13.4	2.8	1.7	1.7	–
64	**MILLET, JAPANESE (BARNYARD GRASS),** FRESH	2-03-108	22	1.6	6.8	–	–	–	0.6	11.0	1.7	1.0	1.0	–
65	**MILLET, PEARL (PEARL MILLET),** FRESH	2-03-115	21	1.9	6.5	–	–	–	0.6	9.7	2.1	1.3	1.3	–
66	**MILLET, PROSO (BROOMCORN; HOG MILLET),** FRESH	2-03-811	25	1.8	7.4	–	–	–	0.6	13.1	2.1	1.1	1.2	–
67	**MILO SORGHUM,** FRESH	2-04-436	30	1.8	7.6	19.5	12.0	–	0.5	17.2	2.6	1.5	1.5	–
68	**MUHLY, BUSH,** MIDBLOOM, FRESH	2-05-619	43	2.4	16.2	–	–	–	0.7	20.9	2.8	1.3	–	1.5
69	**NEEDLEGRASS,** MATURE, FRESH	2-03-205	42	2.9	17.1	–	–	–	0.5	18.4	3.2	1.7	1.8	–
70	**OATGRASS, TALL,** FRESH	2-03-267	30	2.0	10.5	–	–	–	0.9	14.3	2.6	1.5	1.6	–
71	**OATS,** IMMATURE, FRESH	2-03-286	16	1.7	4.0	–	–	–	0.4	7.4	2.5	1.8	1.7	–
72	**OATS-VETCH,** MILK STAGE, FRESH	2-03-407	33	2.5	9.1	–	–	–	1.0	16.3	3.5	2.2	2.3	–
73	**ORCHARDGRASS,** FRESH, ALL ANALYSES	2-03-451	26	2.6	6.4	13.9	–	–	1.6	11.3	3.9	2.7	2.7	–
74	**PANGOLAGRASS,** FRESH	2-03-493	20	1.5	6.6	–	7.5	1.0	0.5	9.8	1.8	0.9	0.9	–
75	**PANICUM,** FRESH	2-03-499	29	4.1	8.6	–	–	–	0.7	11.9	3.7	2.5	2.5	–
76	**PARAGRASS,** FRESH	2-03-525	26	3.0	9.1	–	–	–	0.4	11.9	2.0	1.2	1.1	–
77	**PEA, FIELD,** FRESH	2-03-603	18	1.7	4.6	–	–	–	0.6	7.6	3.6	3.0	2.6	–
78	**PEA-OATS,** FRESH	2-08-483	23	1.9	6.4	–	–	–	0.9	10.3	3.2	2.4	2.2	–
79	**PEARL MILLET (MILLET, PEARL),** FRESH	2-03-115	21	1.9	6.5	–	–	–	0.6	9.7	2.1	1.3	1.3	–
80	**PEAVINE,** FRESH	2-03-669	17	5.0	4.2	–	–	–	0.6	4.4	3.2	2.5	2.3	–
81	**RAPE (Canola),** FRESH	2-03-867	17	2.1	2.4	–	–	–	0.6	8.5	2.9	2.4	2.1	–
82	**REDTOP,** FULL BLOOM, FRESH	2-03-891	26	1.8	6.6	16.6	–	–	0.9	14.8	2.1	1.2	1.3	–
83	**REED CANARYGRASS,** FRESH	2-01-113	23	2.3	5.6	10.6	6.5	1.0	0.9	10.1	3.9	2.9	2.7	–
84	**RESCUEGRASS (BROMEGRASS, RESCUE),** FRESH	2-08-361	29	4.0	6.7	–	–	–	1.0	12.2	5.0	3.7	3.5	–
85	**RHODESGRASS,** FRESH	2-03-916	26	3.1	9.9	–	–	–	0.4	11.0	2.0	1.2	1.1	–
86	**RYE,** FRESH	2-04-018	20	1.9	5.9	–	–	–	0.8	8.1	3.6	2.8	2.5	–
87	**RYEGRASS,** FRESH	2-04-062	24	1.8	7.0	–	–	–	0.7	12.1	2.5	1.5	1.6	–
88	**RYEGRASS, ITALIAN,** FRESH	2-04-073	23	3.9	4.7	–	–	–	0.9	9.0	4.0	3.0	2.9	–
89	**SAGEBRUSH, BIG,** BROWSE, STEM CURED, FRESH	2-07-992	65	4.3	–	–	–	–	6.4	–	6.1	3.2	3.7	–
90	**SALTGRASS, DESERT,** FRESH	2-04-171	29	2.0	8.6	–	–	–	0.5	16.2	1.7	0.7	0.9	–
91	**SEDGE,** FRESH	2-04-195	25	2.2	–	15.4	–	1.0	–	–	3.0	2.0	2.0	–
	SORGHUM													
92	JOHNSONGRASS, IMMATURE, FRESH	2-04-409	20	2.1	5.6	–	–	–	0.6	8.4	3.1	2.2	2.1	–
93	JOHNSONGRASS, FULL BLOOM, FRESH	2-04-410	35	3.5	11.4	–	–	–	0.8	16.4	2.8	1.6	1.7	–
94	KAFIR, FRESH	2-04-424	24	1.9	6.6	–	–	–	0.7	12.0	2.4	1.5	1.5	–
95	MILO, FRESH	2-04-436	30	1.8	7.6	19.5	12.0	–	0.5	17.2	2.6	1.5	1.5	–
96	SUDANGRASS, MATURE, FRESH	2-04-487	30	2.4	10.6	–	–	–	0.5	14.5	1.6	0.6	0.8	–
97	**SOYBEAN,** FRESH	2-04-574	23	2.4	6.3	–	–	–	0.9	9.2	4.1	3.2	2.9	–
98	**STARGRASS,** FRESH	2-09-730	63	5.0	–	51.6	–	–	–	–	–	–	–	–
99	**SUGARCANE,** FRESH	2-04-689	28	1.6	8.8	–	–	–	0.6	15.2	1.4	0.8	0.7	–
100	**SUNFLOWER,** FRESH	2-04-723	15	1.7	4.7	–	–	–	0.4	7.2	1.4	0.8	0.8	–
101	**SWEET CLOVER, YELLOW,** FRESH	2-04-766	23	1.8	6.9	–	–	–	0.7	9.5	4.3	3.4	3.1	–
102	**SWITCHGRASS,** FRESH	2-04-800	55	3.5	19.2	41.7	–	5.3	1.3	27.8	3.5	1.6	1.9	–
103	**TALL (ALTA) FESCUE,** FRESH	2-01-889	28	2.5	7.5	19.5	–	–	0.9	14.3	2.7	1.7	1.7	–
104	**THREE-AWN (WIREGRASS),** FRESH	2-04-838	39	2.3	13.3	–	–	–	0.9	18.6	3.8	2.3	2.4	–
105	**TIMOTHY,** FRESH	2-04-912	28	2.3	7.5	19.4	–	–	1.1	13.4	3.4	2.1	2.3	–
106	**TOBOSA,** IMMATURE, FRESH	2-08-578	40	4.5	12.6	27.6	–	–	0.5	18.2	3.8	2.8	2.3	–
107	**TREFOIL, BIRDSFOOT (DEERVETCH, BIRDSFOOT),** FRESH	2-20-786	19	2.2	4.1	9.5	–	–	0.8	8.5	3.7	2.8	2.7	–
108	**VELVETBEAN,** DOUGH STAGE, FRESH	2-05-084	22	2.4	5.3	–	–	–	0.5	10.7	3.5	2.5	2.5	–
109	**VETCH,** FRESH	2-05-111	22	2.1	6.2	–	–	–	0.5	8.9	4.7	3.5	3.4	–
110	**VETCH-OATS,** FRESH	2-05-133	26	2.8	7.0	–	–	–	0.9	11.4	4.3	3.3	3.0	–
111	**WHEAT,** IMMATURE, FRESH	2-05-176	22	3.0	3.9	10.2	6.3	1.0	1.0	8.3	6.1	4.9	4.5	–
112	**WHEATGRASS, BLUEBUNCH,** PREBLOOM, FRESH	2-05-387	32	2.4	8.0	22.8	–	–	0.8	16.0	4.5	3.2	3.1	–
113	**WHEATGRASS, CRESTED,** IMMATURE, FRESH	2-05-420	28	2.9	6.2	–	–	–	0.6	12.9	6.0	5.1	4.3	–
114	**WHEATGRASS, SLENDER,** FRESH	2-05-439	32	3.0	10.4	23.1	–	–	1.4	14.5	3.1	1.8	1.9	–
115	**WHITE CLOVER,** FRESH	2-01-468	18	2.1	2.8	–	–	–	0.6	7.2	5.0	4.0	3.7	–
116	**WILD-RYE, RUSSIAN,** FRESH	2-05-469	35	3.1	7.8	25.6	–	–	0.9	18.9	4.2	2.8	2.8	–
117	**WINTERFAT, COMMON,** BROWSE, STEM CURED, FRESH	2-26-142	80	12.7	–	–	–	–	2.2	–	8.7	5.4	5.6	–

PASTURE AND RANGE PLANTS

Entry Number	TDN Ruminant	TDN Swine	TDN Horse	Digestible Energy Ruminant		Digestible Energy Swine		Digestible Energy Horse		Metabolizable Energy Ruminant		Metabolizable Energy Swine		Metabolizable Energy Poultry ME_n		Metabolizable Energy Horse		Net Energy Ruminant NE_m		Net Energy Ruminant NE_g		Net Energy Lactating Cows NE_{lc}	
	%	%	%	Mcal lb	Mcal *kg*	kcal lb	kcal *kg*	Mcal lb	Mcal *kg*	Mcal lb	Mcal *kg*	kcal lb	kcal *kg*	kcal lb	kcal *kg*	Mcal lb	Mcal *kg*	Mcal lb	Mcal *kg*	Mcal lb	Mcal *kg*	Mcal lb	Mcal *kg*
37	16	–	11	0.32	*0.70*	–	–	0.21	*0.46*	0.28	*0.61*	–	–	–	–	0.17	*0.38*	0.16	*0.35*	0.10	*0.22*	0.16	*0.35*
38	12	–	–	0.26	*0.58*	–	–	–	–	0.22	*0.49*	–	–	–	–	–	–	0.14	*0.31*	0.08	*0.19*	0.14	*0.32*
39	16	15	–	0.29	*0.65*	–	–	–	–	0.26	*0.58*	–	–	–	–	–	–	0.15	*0.32*	0.08	*0.18*	0.14	*0.32*
40	–	–	–	–	–	–	–	–	–	–	–	–	–	–	–	–	–	–	–	–	–	–	–
41	21	–	–	0.40	*0.88*	–	–	–	–	0.33	*0.73*	–	–	–	–	–	–	0.18	*0.40*	0.09	*0.20*	0.19	*0.43*
42	27	–	–	0.53	*1.17*	–	–	–	–	0.44	*0.96*	–	–	–	–	–	–	0.24	*0.52*	0.11	*0.25*	0.26	*0.58*
43	16	–	–	0.31	*0.67*	–	–	–	–	0.26	*0.57*	–	–	–	–	–	–	0.14	*0.31*	0.08	*0.17*	0.15	*0.33*
44	52	–	–	0.96	*2.12*	–	–	–	–	0.83	*1.82*	–	–	–	–	–	–	0.47	*1.04*	0.25	*0.54*	0.50	*1.11*
45	17	–	–	0.34	*0.76*	–	–	–	–	0.29	*0.64*	–	–	–	–	–	–	0.17	*0.38*	0.10	*0.22*	0.18	*0.39*
46	10	–	–	0.21	*0.46*	–	–	–	–	0.18	*0.39*	–	–	–	–	–	–	0.11	*0.25*	0.07	*0.15*	0.11	*0.25*
47	17	–	–	0.34	*0.75*	–	–	–	–	0.29	*0.64*	–	–	–	–	–	–	0.18	*0.40*	0.11	*0.24*	0.18	*0.39*
48	44	–	–	0.72	*1.58*	–	–	–	–	0.59	*1.29*	–	–	–	–	–	–	0.51	*1.13*	0.29	*0.63*	0.53	*1.16*
49	25	–	–	0.51	*1.12*	–	–	–	–	0.43	*0.94*	–	–	–	–	–	–	0.25	*0.54*	0.14	*0.31*	0.25	*0.56*
50	15	–	–	0.31	*0.68*	–	–	–	–	0.26	*0.58*	–	–	–	–	–	–	0.16	*0.35*	0.10	*0.21*	0.16	*0.35*
51	23	–	–	0.55	*1.20*	–	–	–	–	0.45	*0.99*	–	–	–	–	–	–	0.32	*0.70*	0.18	*0.41*	0.32	*0.71*
52	33	–	–	0.66	*1.45*	–	–	–	–	0.55	*1.21*	–	–	–	–	–	–	0.32	*0.70*	0.17	*0.38*	0.33	*0.74*
53	–	–	–	–	–	–	–	–	–	–	–	–	–	–	–	–	–	–	–	–	–	–	–
54	20	–	–	0.40	*0.88*	–	–	–	–	0.35	*0.76*	–	–	–	–	–	–	0.21	*0.45*	0.13	*0.29*	0.20	*0.44*
55	18	–	–	0.37	*0.82*	–	–	–	–	0.31	*0.69*	–	–	–	–	–	–	0.19	*0.43*	0.11	*0.25*	0.20	*0.43*
56	14	–	–	0.29	*0.63*	–	–	–	–	0.24	*0.53*	–	–	–	–	–	–	0.13	*0.29*	0.07	*0.16*	0.14	*0.31*
57	20	–	–	–	–	–	–	–	–	–	–	–	–	–	–	–	–	–	–	–	–	–	–
58	21	–	–	0.44	*0.96*	–	–	–	–	0.37	*0.83*	–	–	–	–	–	–	0.22	*0.47*	0.13	*0.28*	0.22	*0.48*
59	27	–	–	0.54	*1.19*	–	–	–	–	0.46	*1.01*	–	–	–	–	–	–	0.28	*0.61*	0.16	*0.36*	0.28	*0.61*
60	17	–	–	0.34	*0.75*	–	–	–	–	0.29	*0.64*	–	–	–	–	–	–	0.18	*0.40*	0.11	*0.24*	0.18	*0.39*
61	14	–	–	0.29	*0.64*	–	–	–	–	0.25	*0.54*	–	–	–	–	–	–	0.16	*0.35*	0.10	*0.21*	0.16	*0.34*
62	24	–	–	0.49	*1.08*	–	–	–	–	0.42	*0.93*	–	–	–	–	–	–	0.26	*0.56*	0.16	*0.35*	0.25	*0.55*
63	18	–	–	0.35	*0.77*	–	–	–	–	0.30	*0.65*	–	–	–	–	–	–	0.17	*0.37*	0.09	*0.21*	0.17	*0.38*
64	14	–	–	0.27	*0.60*	–	–	–	–	0.23	*0.51*	–	–	–	–	–	–	0.13	*0.29*	0.07	*0.16*	0.13	*0.29*
65	13	–	–	0.25	*0.56*	–	–	–	–	0.21	*0.47*	–	–	–	–	–	–	0.12	*0.27*	0.07	*0.15*	0.13	*0.28*
66	16	–	–	0.31	*0.68*	–	–	–	–	0.26	*0.58*	–	–	–	–	–	–	0.15	*0.34*	0.09	*0.19*	0.16	*0.34*
67	17	–	–	0.36	*0.80*	–	–	–	–	0.31	*0.67*	–	–	–	–	–	–	0.20	*0.43*	0.12	*0.26*	0.20	*0.44*
68	–	–	–	–	–	–	–	–	–	–	–	–	–	–	–	–	–	–	–	–	–	–	–
69	24	–	–	0.48	*1.06*	–	–	–	–	0.40	*0.88*	–	–	–	–	–	–	0.22	*0.48*	0.11	*0.25*	0.24	*0.52*
70	17	–	–	0.35	*0.78*	–	–	–	–	0.30	*0.65*	–	–	–	–	–	–	0.18	*0.39*	0.10	*0.22*	0.18	*0.41*
71	10	–	–	0.21	*0.46*	–	–	–	–	0.18	*0.39*	–	–	–	–	–	–	0.10	*0.23*	0.06	*0.14*	0.10	*0.23*
72	20	–	–	0.41	*0.90*	–	–	–	–	0.34	*0.76*	–	–	–	–	–	–	0.21	*0.46*	0.12	*0.27*	0.21	*0.46*
73	17	–	–	0.35	*0.77*	–	–	–	–	0.30	*0.66*	–	–	–	–	–	–	0.19	*0.42*	0.12	*0.26*	0.18	*0.41*
74	12	11	–	0.24	*0.54*	–	–	–	–	0.20	*0.45*	–	–	–	–	–	–	0.12	*0.26*	0.07	*0.14*	0.12	*0.26*
75	17	–	–	0.33	*0.74*	–	–	–	–	0.28	*0.61*	–	–	–	–	–	–	0.15	*0.34*	0.08	*0.18*	0.17	*0.36*
76	14	–	–	0.28	*0.63*	–	–	–	–	0.23	*0.51*	–	–	–	–	–	–	0.13	*0.29*	0.07	*0.15*	0.14	*0.32*
77	13	–	–	0.25	*0.55*	–	–	–	–	0.21	*0.47*	–	–	–	–	–	–	0.13	*0.28*	0.08	*0.17*	0.12	*0.27*
78	14	–	–	0.29	*0.64*	–	–	–	–	0.25	*0.54*	–	–	–	–	–	–	0.15	*0.32*	0.09	*0.19*	0.15	*0.33*
79	13	–	–	0.25	*0.56*	–	–	–	–	0.21	*0.47*	–	–	–	–	–	–	0.12	*0.27*	0.07	*0.15*	0.13	*0.28*
80	9	–	–	0.19	*0.42*	–	–	–	–	0.16	*0.34*	–	–	–	–	–	–	–	–	–	–	–	–
81	13	–	–	0.25	*0.54*	–	–	–	–	0.21	*0.47*	–	–	–	–	–	–	0.12	*0.26*	0.07	*0.16*	0.12	*0.26*
82	16	–	–	0.33	*0.72*	–	–	–	–	0.28	*0.61*	–	–	–	–	–	–	0.16	*0.36*	0.10	*0.21*	0.17	*0.37*
83	14	–	–	0.29	*0.64*	–	–	–	–	0.24	*0.54*	–	–	–	–	–	–	0.15	*0.34*	0.09	*0.20*	0.15	*0.34*
84	20	–	–	0.38	*0.84*	–	–	–	–	0.33	*0.72*	–	–	–	–	–	–	0.18	*0.40*	0.10	*0.23*	0.18	*0.40*
85	15	15	–	0.31	*0.68*	–	–	–	–	0.26	*0.58*	–	–	–	–	–	–	0.15	*0.33*	0.08	*0.18*	0.15	*0.33*
86	14	–	–	0.27	*0.60*	–	–	–	–	0.23	*0.51*	–	–	–	–	–	–	0.12	*0.27*	0.07	*0.15*	0.13	*0.28*
87	15	–	–	0.30	*0.66*	–	–	–	–	0.25	*0.56*	–	–	–	–	–	–	0.15	*0.34*	0.09	*0.20*	0.16	*0.34*
88	14	–	–	0.28	*0.61*	–	–	–	–	0.23	*0.51*	–	–	–	–	–	–	0.14	*0.31*	0.08	*0.18*	0.14	*0.32*
89	27	–	–	0.66	*1.46*	–	–	–	–	0.37	*0.81*	–	–	–	–	–	–	–	–	–	–	–	–
90	18	–	–	0.35	*0.78*	–	–	–	–	0.30	*0.66*	–	–	–	–	–	–	0.18	*0.39*	0.10	*0.22*	0.18	*0.40*
91	–	–	–	–	–	–	–	–	–	–	–	–	–	–	–	–	–	–	–	–	–	–	–
92	12	–	–	0.25	*0.54*	–	–	–	–	0.21	*0.46*	–	–	–	–	–	–	0.12	*0.27*	0.07	*0.15*	0.13	*0.28*
93	20	–	–	0.40	*0.89*	–	–	–	–	0.34	*0.74*	–	–	–	–	–	–	0.19	*0.42*	0.10	*0.22*	0.20	*0.45*
94	14	–	–	0.29	*0.64*	–	–	–	–	0.25	*0.54*	–	–	–	–	–	–	0.15	*0.33*	0.09	*0.19*	0.15	*0.33*
95	17	–	–	0.36	*0.80*	–	–	–	–	0.31	*0.67*	–	–	–	–	–	–	0.20	*0.43*	0.12	*0.26*	0.20	*0.44*
96	19	–	–	0.36	*0.79*	–	–	–	–	0.30	*0.67*	–	–	–	–	–	–	0.16	*0.35*	0.08	*0.18*	0.17	*0.37*
97	14	–	–	0.29	*0.64*	–	–	–	–	0.25	*0.54*	–	–	–	–	–	–	0.15	*0.33*	0.09	*0.19*	0.15	*0.33*
98	–	–	–	–	–	–	–	–	–	–	–	–	–	–	–	–	–	–	–	–	–	–	–
99	16	16	–	0.33	*0.73*	–	–	–	–	0.28	*0.62*	–	–	–	–	–	–	0.17	*0.36*	0.09	*0.21*	0.16	*0.36*
100	9	–	–	0.18	*0.39*	–	–	–	–	0.15	*0.33*	–	–	–	–	–	–	0.09	*0.19*	0.05	*0.10*	0.09	*0.20*
101	15	–	–	0.30	*0.66*	–	–	–	–	0.25	*0.56*	–	–	–	–	–	–	0.16	*0.34*	0.09	*0.21*	0.16	*0.34*
102	33	–	–	0.65	*1.44*	–	–	–	–	0.55	*1.20*	–	–	–	–	–	–	0.32	*0.70*	0.18	*0.39*	0.33	*0.73*
103	17	–	–	0.34	*0.76*	–	–	–	–	0.29	*0.64*	–	–	–	–	–	–	0.17	*0.38*	0.10	*0.22*	0.18	*0.39*
104	23	–	–	0.47	*1.04*	–	–	–	–	0.40	*0.87*	–	–	–	–	–	–	0.24	*0.52*	0.13	*0.30*	0.24	*0.53*
105	18	–	–	0.35	*0.78*	–	–	–	–	0.30	*0.66*	–	–	–	–	–	–	0.19	*0.42*	0.12	*0.26*	0.19	*0.42*
106	22	–	–	0.44	*0.97*	–	–	–	–	0.36	*0.80*	–	–	–	–	–	–	0.21	*0.47*	0.11	*0.25*	0.23	*0.50*
107	13	–	–	0.26	*0.58*	–	–	–	–	0.23	*0.50*	–	–	–	–	–	–	0.16	*0.35*	0.10	*0.22*	0.15	*0.33*
108	15	–	–	0.30	*0.67*	–	–	–	–	0.26	*0.58*	–	–	–	–	–	–	0.15	*0.32*	0.09	*0.19*	0.15	*0.33*
109	13	–	–	0.28	*0.62*	–	–	–	–	0.24	*0.52*	–	–	–	–	–	–	0.15	*0.33*	0.09	*0.20*	0.15	*0.33*
110	17	–	–	0.34	*0.74*	–	–	–	–	0.29	*0.63*	–	–	–	–	–	–	0.17	*0.37*	0.10	*0.22*	0.17	*0.38*
111	17	–	–	0.33	*0.73*	–	–	–	–	0.29	*0.64*	–	–	–	–	–	–	0.19	*0.42*	0.13	*0.28*	0.18	*0.40*
112	21	–	–	0.42	*0.92*	–	–	–	–	0.36	*0.79*	–	–	–	–	–	–	0.21	*0.47*	0.13	*0.29*	0.21	*0.47*
113	21	–	–	0.41	*0.90*	–	–	–	–	0.35	*0.78*	–	–	–	–	–	–	0.20	*0.45*	0.13	*0.28*	0.20	*0.44*
114	19	–	–	0.37	*0.83*	–	–	–	–	0.31	*0.69*	–	–	–	–	–	–	0.19	*0.41*	0.10	*0.23*	0.19	*0.43*
115	23	–	15	0.46	*1.02*	–	–	0.29	*0.64*	0.39	*0.87*	–	–	–	–	0.24	*0.52*	0.24	*0.52*	0.14	*0.31*	0.24	*0.52*
116	23	–	–	0.46	*1.00*	–	–	–	–	0.39	*0.86*	–	–	–	–	–	–	0.23	*0.52*	0.14	*0.31*	0.23	*0.52*
117	28	–	–	0.60	*1.33*	–	–	–	–	0.48	*1.05*	–	–	–	–	–	–	–	–	–	–	–	–

(Continued)

TABLE 19–1E PASTURE AND RANGE PLANTS, COMPOSITION OF FEEDS, DATA EXPRESSED **AS-FED**—*(Continued)*

			Macro Minerals							Micro Minerals						
Entry Number	Feed Name Description	Dry Matter	Calcium (Ca)	Phos-phorus (P)	Sodium (Na)	Chlo-rine (Cl)	Mag-nesium (Mg)	Potas-sium (K)	Sulfur (S)	Cobalt (Co)	Copper (Cu)	Iodine (I)	Iron (Fe)	Man-ganese (Mn)	Sele-nium (Se)	Zinc (Zn)
		%	%	%	%	%	%	%	%	ppm or mg/kg	ppm or mg/kg	ppm or mg/kg	%	ppm or mg/kg	ppm or mg/kg	ppm or mg/kg
37	**CORN,** FRESH, ALL ANALYSES	23	0.07	–	–	–	0.21	–	–	–	1.8	–	0.008	24.6	–	16.1
38	**CORN, SWEET,** STOVER WITHOUT EARS, WITHOUT HUSKS, FRESH	22	–	–	–	–	–	–	–	–	–	–	–	–	–	–
39	**COWPEA, COMMON,** FRESH	25	0.38	0.08	0.06	0.05	0.11	0.41	0.08	–	–	–	0.020	–	–	–
40	**CRESTED WHEATGRASS,** EARLY BLOOM, FRESH	41	0.09	0.07	–	–	–	–	–	–	–	–	–	–	–	–
	CURLY MESQUITE															
41	BROWSE, FRESH	35	0.18	0.05	–	–	0.06	0.23	–	–	3.5	–	–	16.4	–	–
42	BROWSE, MATURE, FRESH	50	0.27	0.04	–	–	0.08	0.19	–	–	–	–	–	–	–	–
43	**DALLISGRASS,** FRESH	25	0.14	0.05	0.09	–	0.10	0.43	–	0.019	–	–	0.005	–	–	–
44	**DROPSEED, SAND,** STEM CURED, FRESH	88	0.40	0.07	0.01	–	0.06	0.28	–	0.503	13.5	0.599	0.043	41.4	–	36.8
45	**FESCUE, TALL (ALTA),** FRESH	28	0.13	0.05	0.03	–	0.07	0.70	–	0.113	1.0	–	0.003	18.0	–	5.9
46	**FILAREE (ALFILERIA, REDSTEM),** FRESH	18	0.35	0.08	–	–	–	0.59	–	–	–	–	–	–	–	–
47	**FOXTAIL, MEADOW,** IMMATURE, FRESH	26	0.15	0.12	–	–	–	–	–	–	–	–	–	–	–	–
48	**GALLETA,** STEM CURED, FRESH	86	0.60	0.06	0.01	–	0.07	0.41	0.09	0.591	16.3	–	0.044	67.7	–	19.5
49	**GRAMA,** IMMATURE, FRESH	41	0.22	0.08	–	–	–	–	–	–	2.3	–	–	18.2	–	–
50	**GRASS-LEGUME,** FRESH	24	0.15	0.08	–	–	0.08	0.40	–	–	–	–	–	–	–	–
51	**GREASEWOOD,** BROWSE, FRESH	50	0.46	0.09	–	–	–	–	–	0.030	7.8	–	–	12.9	–	–
52	**INDIANGRASS,** FRESH	57	0.19	0.04	–	–	–	–	–	–	–	–	–	–	–	–
53	**INDIAN RICEGRASS,** FRESH	48	0.28	0.02	–	–	0.07	–	0.07	–	–	–	–	–	–	–
	JOHNSONGRASS SORGHUM–SEE SORGHUM															
54	**JUNEGRASS,** IMMATURE, FRESH	28	0.09	0.07	–	–	–	–	–	–	–	–	–	–	–	–
	KENTUCKY BLUEGRASS–SEE BLUEGRASS, KENTUCKY															
55	**KOA HAOLE (LEAD TREE, WHITE POPINAC),** BROWSE, FRESH	30	–	–	–	–	–	–	–	–	–	–	–	–	–	–
56	**LESPEDEZA, COMMON,** IMMATURE, FRESH	24	–	–	–	–	–	–	–	–	–	–	–	–	–	–
57	**LESPEDEZA, COMMON-KOREAN,** MATURE, FRESH	35	0.35	0.07	–	–	–	–	–	–	–	–	–	–	–	–
58	**LESPEDEZA, SERICEA (CHINESE LESPEDEZA),** FRESH	33	0.42	0.10	–	–	0.07	0.39	–	0.024	–	–	0.008	34.1	–	–
59	**LOVEGRASS,** IMMATURE, FRESH	43	0.20	0.10	–	–	–	–	–	–	–	–	–	–	–	–
60	**MEADOW FOXTAIL,** IMMATURE, FRESH	26	0.15	0.12	–	–	–	–	–	–	–	–	–	–	–	–
61	**MEDIC, BLACK (YELLOW TREFOIL),** FRESH	23	–	–	–	–	–	–	–	–	–	–	–	–	–	–
62	**MESQUITE, COMMON,** BROWSE, FRESH	35	0.68	0.07	–	–	0.08	0.49	–	–	–	–	–	–	–	–
63	**MILLET, FOXTAIL,** FRESH	29	0.09	0.05	–	–	–	0.56	–	–	–	–	–	–	–	–
64	**MILLET, JAPANESE (BARNYARD GRASS),** FRESH	22	0.11	0.07	–	–	–	0.52	–	–	–	–	–	–	–	–
65	**MILLET, PEARL (PEARL MILLET),** FRESH	21	–	–	–	–	–	–	–	–	–	–	–	–	–	–
66	**MILLET, PROSO (BROOMCORN; HOG MILLET),** FRESH	25	–	–	–	–	–	–	–	–	–	–	–	–	–	–
67	**MILO SORGHUM,** FRESH	30	0.09	0.05	–	–	–	0.81	–	–	–	–	–	–	–	–
68	**MUHLY BUSH,** MIDBLOOM, FRESH	43	0.13	0.03	0.00	–	0.02	0.22	–	0.215	0.2	–	0.006	5.2	–	5.2
69	**NEEDLEGRASS,** MATURE, FRESH	42	–	–	–	–	–	–	–	–	–	–	–	–	–	–
70	**OATGRASS, TALL,** FRESH	30	0.12	0.14	–	–	–	0.91	–	–	–	–	–	–	–	–
71	**OATS,** IMMATURE, FRESH	16	–	–	0.02	0.02	–	–	0.01	–	–	–	–	–	–	–
72	**OATS-VETCH,** MILK STAGE, FRESH	33	–	–	–	–	–	–	–	–	–	–	–	–	–	–
73	**ORCHARDGRASS,** FRESH, ALL ANALYSES	26	0.09	0.05	0.03	–	0.06	0.74	–	0.055	2.5	–	0.003	28.5	–	5.3
74	**PANGOLAGRASS,** FRESH	20	0.08	0.05	–	–	0.04	0.29	–	–	–	–	–	–	–	–
75	**PANICUM,** FRESH	29	0.14	0.05	–	–	0.10	0.93	–	–	–	–	–	–	–	–
76	**PARAGRASS,** FRESH	26	0.10	0.10	–	–	–	0.42	–	–	–	–	–	–	–	–
77	**PEA, FIELD,** FRESH	18	0.22	0.04	–	–	0.04	0.27	–	0.028	–	–	0.008	15.4	–	–
78	**PEA-OATS,** FRESH	23	0.17	0.07	–	–	–	0.38	–	–	–	–	–	–	–	–
79	**PEARL MILLET (MILLET, PEARL),** FRESH	21	–	–	–	–	–	–	–	–	–	–	–	–	–	–
80	**PEAVINE,** FRESH	17	–	–	–	–	–	–	–	–	–	–	–	–	–	–
81	**RAPE (CANOLA),** FRESH	17	0.25	0.07	–	–	0.01	0.56	0.11	–	1.4	–	0.004	7.7	–	–
82	**REDTOP,** FULL BLOOM, FRESH	26	0.16	0.10	0.01	–	0.07	0.62	0.04	–	–	–	0.006	–	–	–
83	**REED CANARYGRASS,** FRESH	23	0.08	0.08	–	–	–	0.83	–	–	–	–	–	–	–	–
84	**RESCUEGRASS (BROMEGRASS, RESCUE),** FRESH	29	0.15	0.08	–	–	–	–	–	–	–	–	–	–	–	–
85	**RHODESGRASS,** FRESH	26	0.13	0.10	–	–	–	0.61	–	–	–	–	–	–	–	–
86	**RYE,** FRESH	20	0.09	0.08	0.01	–	0.06	0.69	–	–	–	–	–	–	–	–
87	**RYEGRASS,** FRESH	24	–	–	–	–	–	–	–	–	–	–	–	–	–	–
88	**RYEGRASS, ITALIAN,** FRESH	23	0.15	0.09	0.00	–	0.08	0.45	0.02	–	–	–	0.023	–	–	–
89	**SAGEBRUSH, BIG,** BROWSE, STEM CURED, FRESH	65	0.46	0.12	–	–	–	–	–	–	–	–	–	–	–	–
90	**SALTGRASS, DESERT,** FRESH	29	0.05	0.03	–	–	–	–	–	–	–	–	–	–	–	–
91	**SEDGE,** FRESH	25	–	0.05	0.05	0.06	–	–	0.06	–	–	–	–	–	–	–
	SORGHUM															
92	JOHNSONGRASS, IMMATURE, FRESH	20	0.18	0.06	–	–	–	–	–	–	–	–	–	–	–	–
93	JOHNSONGRASS, FULL BLOOM, FRESH	35	0.29	0.06	–	–	–	–	–	–	–	–	–	–	–	–
94	KAFIR, FRESH	24	0.09	0.04	–	–	–	0.40	–	–	–	–	–	–	–	–
95	MILO, FRESH	30	0.09	0.05	–	–	–	0.81	–	–	–	–	–	–	–	–
96	SUDANGRASS, MATURE, FRESH	30	0.09	0.06	–	–	–	–	–	–	–	–	–	–	–	–
97	**SOYBEAN,** FRESH	23	0.25	0.07	–	–	0.12	0.21	–	–	2.1	–	0.005	27.3	–	–
98	**STARGRASS,** FRESH	63	0.39	0.19	0.02	–	0.20	2.07	–	0.133	6.7	–	0.012	50.6	–	38.1
99	**SUGARCANE,** FRESH	28	0.11	0.05	–	–	0.11	0.29	–	–	0.6	–	0.005	17.3	–	6.9
100	**SUNFLOWER,** FRESH	15	–	–	–	–	–	–	–	–	–	–	–	–	–	–
101	**SWEET CLOVER, YELLOW,** FRESH	23	0.31	0.06	0.02	0.09	0.08	0.38	0.11	–	2.3	–	0.004	29.0	–	–
102	**SWITCHGRASS,** FRESH	55	0.16	0.05	–	–	–	–	–	–	–	–	–	–	–	–
103	**TALL (ALTA) FESCUE,** FRESH	28	0.13	0.05	0.03	–	0.07	0.70	–	0.113	1.0	–	0.003	18.0	–	5.9
104	**THREE-AWN (WIREGRASS),** FRESH	39	–	–	–	–	–	–	–	–	–	–	–	–	–	–
105	**TIMOTHY,** FRESH	28	0.14	0.08	0.03	0.14	0.06	0.69	0.04	0.041	2.2	–	0.004	24.6	–	7.5
106	**TOBOSA,** IMMATURE, FRESH	40	0.18	0.05	0.01	–	0.04	0.21	–	0.277	5.6	–	0.023	32.3	–	12.3
107	**TREFOIL, BIRDSFOOT (DEERVETCH, BIRDSFOOT),** FRESH	19	0.34	0.05	0.02	–	0.08	0.63	0.05	0.094	2.5	–	0.006	16.0	–	6.0
108	**VELVET BEAN,** DOUGH STAGE, FRESH	22	–	–	–	–	–	–	–	–	–	–	–	–	–	–
109	**VETCH,** FRESH	22	–	–	0.11	0.42	–	–	0.03	0.068	–	–	–	–	–	–
110	**VETCH-OATS,** FRESH	26	0.18	0.08	–	–	0.06	0.45	–	–	–	–	–	–	–	–
111	**WHEAT,** IMMATURE, FRESH	22	0.09	0.09	0.04	–	0.05	0.78	0.05	–	–	–	0.003	–	–	–
112	**WHEATGRASS, BLUEBUNCH,** PREBLOOM, FRESH	32	0.12	0.08	0.02	–	0.06	1.03	–	0.063	2.6	–	0.008	15.7	–	9.2
113	**WHEATGRASS, CRESTED,** IMMATURE, FRESH	28	0.13	0.10	–	–	0.08	–	–	–	–	–	–	–	–	–
114	**WHEATGRASS, SLENDER,** FRESH	32	0.16	0.05	0.03	–	0.08	1.04	–	0.067	1.5	–	0.003	19.8	–	7.4
115	**WHITE CLOVER,** FRESH	18	0.25	0.09	0.07	0.11	0.08	0.38	0.06	–	–	–	0.006	54.4	–	–
116	**WILD-RYE, RUSSIAN,** FRESH	35	0.11	0.06	0.08	–	0.05	1.06	–	0.097	1.1	–	0.004	9.5	–	5.7
117	**WINTERFAT, COMMON,** BROWSE, STEM CURED, FRESH	80	1.58	0.09	–	–	–	–	–	–	–	–	–	–	–	–

Entry Number	Fat-Soluble Vitamins: A (1 mg Carotene = 1667 IU Vit A)	Carotene (Provitamin A)	D	E	K	Water-Soluble Vitamins: B-12	Biotin	Choline	Folacin (Folic Acid)	Niacin	Pantothenic Acid (B-3)	(Pyridoxine) B-6	Riboflavin (B-2)	Thiamin (B-1)
	IU/g	ppm or mg/kg	IU/kg	ppm or mg/kg	ppm or mg/kg	ppb or mcg/kg	ppm or mg/kg	ppm or mg/kg	ppm or mg/kg	ppm or mg/kg	ppm or mg/kg	ppm or mg/kg	ppm or mg/kg	ppm or mg/kg
37	–	–	–	–	–	–	–	–	–	–	–	–	–	–
38	–	–	–	–	–	–	–	–	–	–	–	–	–	–
39	–	–	–	–	–	–	–	–	–	–	–	–	–	–
40	–	–	–	–	–	–	–	–	–	–	–	–	–	–
41	–	–	–	–	–	–	–	–	–	–	–	–	–	–
42	–	–	–	–	–	–	–	–	–	–	–	–	–	–
43	126.0	75.6	–	–	–	–	–	–	–	–	–	–	–	–
44	14.0	8.4	–	–	–	–	–	–	–	–	–	–	–	–
45	160.0	96.0	–	46.9	–	–	–	–	–	–	–	–	2.4	3.4
46	–	–	–	–	–	–	–	–	–	–	–	–	–	–
47	–	–	–	–	–	–	–	–	–	–	–	–	–	–
48	0.3	0.2	–	–	–	–	–	–	–	–	–	–	–	–
49	–	–	–	–	–	–	–	–	–	–	–	–	–	–
50	–	–	–	–	–	–	–	–	–	–	–	–	–	–
51	36.2	21.7	–	–	–	–	–	–	–	–	–	–	–	–
52	92.5	55.5	–	–	–	–	–	–	–	–	–	–	–	–
53	0.4	0.2	–	–	–	–	–	–	–	–	–	–	–	–
54	–	–	–	–	–	–	–	–	–	–	–	–	–	–
55	–	–	–	–	–	–	–	–	–	–	–	–	–	–
56	–	–	–	–	–	–	–	–	–	–	–	–	–	–
57	–	–	–	–	–	–	–	–	–	–	–	–	–	–
58	–	–	–	–	–	–	–	–	–	–	–	–	–	–
59	–	–	–	–	–	–	–	–	–	–	–	–	–	–
60	–	–	–	–	–	–	–	–	–	–	–	–	–	–
61	–	–	–	–	–	–	–	–	–	–	–	–	–	–
62	–	–	–	–	–	–	–	–	–	–	–	–	–	–
63	–	–	–	–	–	–	–	–	–	–	–	–	–	–
64	–	–	–	–	–	–	–	–	–	–	–	–	–	–
65	63.0	37.8	–	–	–	–	–	–	–	–	–	–	–	–
66	–	–	–	–	–	–	–	–	–	–	–	–	–	–
67	–	–	–	–	–	–	–	–	–	–	–	–	–	–
68	–	–	–	–	–	–	–	–	–	–	–	–	–	–
69	–	–	–	–	–	–	–	–	–	–	–	–	–	–
70	–	–	–	–	–	–	–	–	–	–	–	–	–	–
71	150.0	90.0	–	–	–	–	–	–	–	–	–	–	–	–
72	–	–	–	–	–	–	–	–	–	–	–	–	–	–
73	137.1	82.2	–	112.3	–	–	–	–	–	–	–	–	–	1.9
74	–	–	–	–	–	–	–	–	–	–	–	–	–	–
75	–	–	–	–	–	–	–	–	–	–	–	–	–	–
76	–	–	–	–	–	–	–	–	–	–	–	–	–	–
77	–	–	–	–	–	–	–	–	–	–	–	–	–	–
78	–	–	–	–	–	–	–	–	–	–	–	–	–	–
79	63.0	37.8	–	–	–	–	–	–	–	–	–	–	–	–
80	29.7	17.8	–	–	–	–	–	–	–	–	–	–	2.2	–
81	–	–	–	–	–	–	–	–	–	–	–	–	–	–
82	66.9	40.1	–	–	–	–	–	–	–	–	–	–	–	–
83	–	–	–	–	–	–	–	–	–	–	–	–	–	–
84	–	–	–	–	–	–	–	–	–	–	–	–	–	–
85	–	–	–	–	–	–	–	–	–	–	–	–	–	–
86	115.0	69.0	–	–	–	–	–	–	–	–	–	–	–	–
87	–	–	–	–	–	–	–	–	–	–	–	–	–	–
88	–	–	–	–	–	–	–	–	–	–	–	–	–	–
89	17.3	10.4	–	–	–	–	–	–	–	–	–	–	–	–
90	–	–	–	–	–	–	–	–	–	–	–	–	–	–
91	–	–	–	–	–	–	–	–	–	–	–	–	–	–
92	–	–	–	–	–	–	–	–	–	–	–	–	–	–
93	–	–	–	–	–	–	–	–	–	–	–	–	–	–
94	6.9	4.2	–	–	–	–	–	–	–	9	3.3	1.41	1.0	–
95	–	–	–	–	–	–	–	–	–	–	–	–	–	–
96	–	–	–	–	–	–	–	–	–	–	–	–	–	–
97	121.6	73.0	–	64.2	–	–	–	–	–	–	–	–	–	–
98	–	–	–	–	–	–	–	–	–	–	–	–	–	–
99	–	–	–	–	–	–	–	–	–	–	–	–	–	–
100	–	–	–	–	–	–	–	–	–	–	–	–	–	–
101	102.5	61.5	–	–	–	–	–	–	–	8	–	–	19.4	1.2
102	83.0	49.8	–	–	–	–	–	–	–	–	–	–	–	–
103	–	–	–	–	–	–	–	–	–	–	–	–	–	–
104	–	–	–	–	–	–	–	–	–	–	–	–	–	–
105	103.2	61.9	–	42.6	–	–	–	–	–	–	–	–	3.2	0.8
106	–	–	–	–	–	–	–	–	–	–	–	–	–	–
107	–	–	–	–	–	–	–	–	–	–	–	–	–	–
108	–	–	–	–	–	–	–	–	–	–	–	–	–	–
109	–	–	–	–	–	–	–	–	–	–	–	–	–	–
110	–	–	–	–	–	–	–	–	–	–	–	–	–	–
111	192.5	115.4	–	–	–	–	–	–	–	13	4.7	–	6.1	–
112	173.6	104.2	–	–	–	–	–	–	–	–	–	–	–	–
113	205.8	123.4	–	–	–	–	–	–	–	–	–	–	–	–
114	–	–	–	–	–	–	–	–	–	–	–	–	3.4	1.6
115	44.0	26.4	–	54.6	–	–	–	–	–	11	–	–	16.0	2.5
116	–	–	–	–	–	–	–	–	–	–	–	–	–	–
117	24.1	14.5	–	–	–	–	–	–	–	–	–	–	–	–

TABLE 19-1F MINERAL SUPPLEMENTS, COMPOSITION, DATA EXPRESSED **AS-FED**

Entry Number	Feed Name Description	International Feed Number	Proximate Analysis: Dry Matter	Ash	Crude Fiber	Ether Extract (Fat)	N-Free Extract	Crude Protein (6.25 x N)	Digestible Protein: Ruminant	Non-Ruminant	Horse
			%	%	%	%	%	%	%	%	%
1	AMMONIUM PHOSPHATE, MONOBASIC	6-09-338	98	53.0	–	–	–	69.4	–	–	–
2	AMMONIUM PHOSPHATE, DIBASIC	6-00-370	98	35.5	–	–	–	112.9	–	–	–
3	AMMONIUM POLYPHOSPHATE SOLUTION	6-08-042	60	–	–	–	–	54.8	–	–	–
4	BONE, CHARCOAL	6-00-402	94	79.3	3.7	1.1	1.8	–	–	–	–
5	BONE MEAL	6-00-397	94	60.5	2.9	6.5	–	24.8	17.1	–	–
6	BONE MEAL, STEAMED*	6-00-400	95	67.3	1.9	3.6	3.8	18.6	–	–	–
7	CALCIUM CARBONATE*	6-01-069	100	97.1	–	–	–	–	–	–	–
8	CALCIUM OXIDE	6-14-003	97	–	–	–	–	–	–	–	–
9	CALCIUM PERIODATE*	6-09-335	–	–	–	–	–	–	–	–	–
10	CALCIUM PHOSPHATE, MONOBASIC, FROM DEFLUORINATED PHOSPHORIC ACID	6-01-082	99	87.1	–	–	–	–	–	–	–
11	CALCIUM PHOSPHATE, MONOBASIC, FROM FURNACED PHOSPHORIC ACID	6-26-334	96	–	–	–	–	–	–	–	–
12	CALCIUM PHOSPHATE, DIBASIC, FROM DEFLUORINATED PHOSPHORIC ACID*	6-01-080	97	89.7	–	–	–	–	–	–	–
13	CALCIUM PHOSPHATE, DIBASIC, FROM FURNACED PHOSPHORIC ACID*	6-26-335	97	85.6	–	–	–	–	–	–	–
14	CALCIUM PHOSPHATE, TRIBASIC, FROM FURNACED PHOSPHORIC ACID	6-01-084	98	92.1	–	–	–	–	–	–	–
15	CALCIUM SULFATE, ANHYDROUS	6-01-087	–	–	–	–	–	–	–	–	–
16	CALCIUM SULFATE (GYPSUM)	6-01-090	95	–	–	–	–	–	–	–	–
17	COBALT CARBONATE*	6-01-566	99	–	–	–	–	–	–	–	–
18	COBALT SULFATE*	6-01-564	–	–	–	–	–	–	–	–	–
19	COBALTOUS CHLORIDE	6-01-558	98	–	–	–	–	–	–	–	–
20	COBALTOUS OXIDE	6-01-560	99	–	–	–	–	–	–	–	–
21	COLLOIDAL CLAY (SOFT ROCK PHOSPHATE)	6-03-947	100	–	–	–	–	–	–	–	–
22	COPPER (CUPRIC) CARBONATE	6-01-703	98	–	–	–	–	–	–	–	–
23	COPPER (CUPRIC) CHLORIDE	6-01-705	99	–	–	–	–	–	–	–	–
24	COPPER (CUPRIC) GLUCONATE	6-01-707	99	–	–	–	–	–	–	–	–
25	COPPER (CUPRIC) HYDROXIDE	6-01-709	98	–	–	–	–	–	–	–	–
26	COPPER (CUPRIC) ORTHOPHOSPHATE	6-01-713	99	–	–	–	–	–	–	–	–
27	COPPER (CUPRIC) OXIDE	6-01-711	99	–	–	–	–	–	–	–	–
28	COPPER (CUPRIC) SULFATE, PENTAHYDRATE*	6-01-719	99	–	–	–	–	–	–	–	–
29	COPPER (CUPROUS) IODIDE	6-01-721	–	–	–	–	–	–	–	–	–
30	COPPER (CUPROUS) OXIDE*	6-28-224	99	–	–	–	–	–	–	–	–
31	CURACAO PHOSPHATE, GROUND	6-05-586	99	94.1	–	–	–	–	–	–	–
32	DIAMMONIUM PHOSPHATE*	6-00-370	98	35.5	–	–	–	112.9	–	–	–
33	DIIODOSALICYLIC ACID*	6-01-787	99	–	–	–	–	–	–	–	–
34	ETHYLENEDIAMINE DIHYDROIODIDE*	6-01-842	98	–	–	–	–	54.3	–	–	–
35	FERRIC (IRON) AMMONIUM CITRATE	6-01-857	99	–	–	–	–	42.1	–	–	–
36	FERRIC (IRON) CHLORIDE	6-01-865	98	–	–	–	–	–	–	–	–
37	FERRIC (IRON) OXIDE*	6-02-431	97	–	–	–	–	–	–	–	–
38	FERROUS (IRON) CARBONATE*	6-01-863	99	–	–	–	–	–	–	–	–
39	FERROUS (IRON) FUMARATE	6-08-097	99	–	–	–	–	–	–	–	–
40	FERROUS (IRON) GLUCONATE	6-01-867	99	–	–	–	–	–	–	–	–
41	FERROUS (IRON) OXIDE	6-20-728	97	–	–	–	–	–	–	–	–
42	FERROUS (IRON) SULFATE, MONOHYDRATE*	6-01-869	98	98.0	–	–	–	–	–	–	–
43	FERROUS (IRON) SULFATE, HEPTAHYDRATE	6-20-734	99	–	–	–	–	–	–	–	–
44	KELP (SEAWEED), WHOLE, DEHY	1-08-073	91	35.0	6.5	0.5	42.4	6.5	2.9	2.2	3.4
45	LIMESTONE, GROUND*	6-02-632	100	93.8	–	–	–	–	–	–	–
46	LIMESTONE, MAGNESIUM (DOLOMITE), GROUND*	6-02-633	100	–	–	–	–	–	–	–	–
47	MAGNESIUM CARBONATE	6-02-754	98	–	–	–	–	–	–	–	–
48	MAGNESIUM HYDROXIDE	6-26-012	98	–	–	–	–	–	–	–	–
49	MAGNESIUM OXIDE*	6-02-756	98	98.3	–	–	–	–	–	–	–
50	MAGNESIUM SULFATE (EPSOM SALTS)*	6-02-758	99	–	–	–	–	–	–	–	–
51	MANGANESE CHLORIDE	6-03-038	99	–	–	–	–	–	–	–	–
52	MANGANESE DIOXIDE	6-03-042	98	–	–	–	–	–	–	–	–
53	MANGANOUS (MANGANESE) OXIDE*	6-03-054	99	–	–	–	–	–	–	–	–
54	MANGANOUS (MANGANESE) SULFATE*	6-26-136	100	–	–	–	–	–	–	–	–
55	OYSTER SHELLS, GROUND (FLOUR)*	6-03-481	99	79.0	1.8	0.3	17.0	0.7	–	–	–
56	PHOSPHATE, DEFLUORINATED	6-01-780	100	99.3	–	–	–	–	–	–	–
57	PHOSPHATE ROCK, GROUND (RAW)	6-03-945	–	–	–	–	–	–	–	–	–
58	PHOSPHATE ROCK, LOW FLUORINE*	6-03-946	–	–	–	–	–	–	–	–	–
59	PHOSPHATE SOFT ROCK (COLLOIDAL CLAY)	6-03-947	100	–	–	–	–	–	–	–	–
60	PHOSPHORIC ACID, FEED GRADE (ORTHO)*	6-03-707	75	–	–	–	–	–	–	–	–
61	POTASSIUM CHLORIDE*	6-03-755	100	98.9	–	–	–	–	–	–	–
62	POTASSIUM IODIDE*	6-03-759	–	–	–	–	–	–	–	–	–
63	POTASSIUM MAGNESIUM SULFATE	6-06-177	98	–	–	–	–	–	–	–	–
64	POTASSIUM SULFATE	6-08-098	98	97.0	–	–	–	–	–	–	–
65	SEAWEED (KELP), WHOLE, DEHY	1-08-073	91	35.0	6.5	0.5	42.4	6.5	2.9	2.2	3.4
66	SODIUM BICARBONATE*	6-04-272	100	–	–	–	–	–	–	–	–
67	SODIUM CHLORIDE*	6-04-152	97	93.0	–	–	–	–	–	–	–
68	SODIUM IODIDE*	6-04-279	–	–	–	–	–	–	–	–	–
69	SODIUM PHOSPHATE, MONOBASIC*	6-04-288	97	96.9	–	–	–	–	–	–	–
70	SODIUM PHOSPHATE, DIBASIC	6-04-286	97	96.7	–	–	–	–	–	–	–
71	SODIUM SELENATE*	6-26-014	99	–	–	–	–	–	–	–	–
72	SODIUM SELENITE*	6-26-013	99	–	–	–	–	–	–	–	–
73	SODIUM SULFATE, DECAHYDRATE	6-04-291	97	–	–	–	–	–	–	–	–
74	SODIUM TRIPOLYPHOSPHATE*	6-08-076	97	89.7	–	–	–	–	–	–	–
75	SULFUR*	6-04-705	99	–	–	–	–	–	–	–	–
76	ZINC ACETATE	6-05-547	99	–	–	–	–	–	–	–	–
77	ZINC CARBONATE	6-05-549	99	–	–	–	–	–	–	–	–
78	ZINC CARBONATE, TETRAHYDRATE	6-29-585	98	–	–	–	–	–	–	–	–
79	ZINC CHLORIDE	6-05-551	98	–	–	–	–	–	–	–	–
80	ZINC OXIDE*	6-05-553	–	–	–	–	–	–	–	–	–
81	ZINC SULFATE, MONOHYDRATE*	6-05-555	99	–	–	–	–	–	–	–	–
82	ZINC SULFATE, HEPTAHYDRATE	6-20-729	98	–	–	–	–	–	–	–	–

*Sources most commonly used in commercial feeds.

MINERAL SUPPLEMENTS

Entry Number	Macro Minerals Calcium (Ca)	Phosphorus (P)	Sodium (Na)	Chlorine (Cl)	Magnesium (Mg)	Potassium (K)	Sulfur (S)	Micro Minerals Cobalt (Co)	Copper (Cu)	Fluorine (F)	Iodine (I)	Iron (Fe)	Manganese (Mn)	Selenium (Se)	Zinc (Zn)
	%	%	%	%	%	%	%	ppm or mg/kg	ppm or mg/kg	ppm or mg/kg	ppm or mg/kg	%	ppm or mg/kg	ppm or mg/kg	ppm or mg/kg
1	0.38	24.42	0.08	–	0.46	0.14	0.82	–	86	1833	–	0.991	462	–	640
2	0.50	20.09	0.04	–	0.45	–	2.47	–	81	1548	–	1.514	504	–	303
3	0.10	13.44	–	–	–	–	0.50	–	–	1341	–	0.505	–	–	–
4	31.92	14.84	–	–	0.55	0.15	–	–	–	–	–	–	–	–	–
5	24.52	11.43	0.61	0.22	0.35	0.14	0.12	–	19	2014	–	0.057	9	–	377
6	25.98	11.80	0.40	0.01	0.78	0.18	0.34	0	162	637	29	0.085	37	–	362
7	37.97	0.04	0.07	0.04	0.41	0.04	0.08	–	14	0	–	0.059	159	0.07	17
8	69.33	–	–	–	–	–	–	–	–	–	–	–	–	–	–
9	–	–	–	–	–	–	–	–	–	–	–	–	–	–	–
10	18.55	20.98	0.06	–	0.81	0.40	0.81	5	5	1410	–	1.007	201	–	419
11	22.00	23.00	–	–	–	–	–	–	–	300	–	–	–	–	–
12	22.00	18.43	1.56	–	0.51	0.10	0.69	8	9	940	–	0.844	253	–	122
13	23.00	18.50	0.08	–	0.60	0.07	–	–	80	1150	–	1.000	300	0.60	220
14	30.90	17.04	0.17	–	–	–	–	–	–	501	–	–	–	–	–
15	–	–	–	–	–	–	–	–	–	–	–	–	–	–	–
16	21.86	–	–	–	0.46	–	16.20	–	–	27	–	–	–	–	–
17	–	–	0.25	0.01	–	–	0.03	465000	15	–	–	0.020	100	–	15
18	–	–	–	–	–	–	–	–	–	–	–	–	–	–	–
19	–	–	–	29.20	–	–	0.07	242648	20	–	–	0.003	–	–	196
20	–	–	–	0.01	–	–	0.20	703494	–	–	–	0.050	–	–	–
21	16.01	9.00	0.10	–	0.38	–	–	–	–	12061	–	1.911	995	–	–
22	–	–	–	–	–	–	0.17	–	530000	–	–	0.147	–	–	196
23	–	–	–	41.17	–	–	0.03	–	368973	–	–	0.006	–	–	–
24	–	–	–	–	–	–	–	–	133353	–	–	–	–	–	–
25	–	–	–	–	–	–	–	–	602994	–	–		–	–	–
26	–	14.11	–	–	–	–	–	–	434214	–	–	–	–	–	–
27	0.01	–	–	–	0.00	–	–	–	753827	–	–	0.020	10	–	800
28	–	–	–	–	–	–	13.25	–	250976	–	–	0.010	2	–	9
29	–	–	–	–	–	–	–	–	–	–	–	–	–	–	–
30	–	–	–	–	–	–	–	–	879318	–	–	–	–	–	–
31	35.10	14.24	0.20	–	0.80	–	–	–	–	5445	–	0.347	–	–	–
32	0.50	20.09	0.04	–	0.45	–	2.47	–	81	1548	–	1.514	504	–	303
33	–	–	–	–	–	–	–	–	–	–	644391	–	–	–	–
34	–	–	–	–	–	–	–	–	–	–	787234	–	–	–	–
35	–	–	–	–	–	–	–	–	–	–	–	15.840	–	–	–
36	–	–	–	64.27	–	–	–	–	–	–	–	33.742	–	–	–
37	0.36	0.10	–	–	0.66	–	–	–	–	–	–	58.800	3600	–	–
38	1.24	0.01	–	–	0.33	–	1.77	200	3000	–	–	40.667	9000	–	–
39	–	–	–	–	–	–	–	–	–	–	–	32.542	–	–	–
40	–	–	–	–	–	–	–	–	–	–	–	11.465	–	–	–
41	–	–	–	–	–	–	–	–	–	–	–	75.369	–	–	–
42	–	–	–	–	0.50	–	17.80	–	–	–	–	31.000	–	–	–
43	–	–	–	–	0.21	–	11.00	–	100	–	–	20.899	0	–	100
44	2.47	0.28	–	–	0.85	–	–	–	–	–	–	–	–	–	–
45	37.12	0.21	0.06	0.03	1.13	0.11	0.04	–	11	–	–	0.357	269	–	19
46	20.61	0.02	0.38	0.12	10.37	0.27	0.01	–	20	–	–	0.053	–	–	–
47	0.02	–	–	–	30.19	–	–	–	–	–	–	0.020	–	–	–
48	–	–	–	–	40.86	–	–	–	–	–	–	–	–	–	–
49	1.66	–	–	–	55.19	–	0.10	501	5	251	–	1.048	80	0.35	9
50	0.02	–	–	0.01	9.60	–	13.00	–	–	–	–	–	–	10.12	–
51	–	–	–	35.47	–	–	–	–	–	–	–	–	274824	–	–
52	–	–	–	–	–	–	–	–	–	–	–	–	619262	–	–
53	0.16	0.10	0.06	–	0.70	0.58	0.01	300	724	–	–	3.436	620217	–	1349
54	–	–	–	–	0.30	–	19.01	–	–	–	–	0.040	250000	–	–
55	35.85	0.10	0.21	0.01	0.24	0.10	–	–	15	–	–	0.254	178	–	7
56	31.99	17.07	3.26	–	0.29	0.10	0.13	10	40	1794	–	0.840	496	–	90
57	–	–	–	–	–	–	–	–	–	–	–	–	–	–	–
58	–	–	–	–	–	–	–	–	–	–	–	–	–	–	–
59	16.01	9.00	0.10	–	0.38	–	–	–	–	12061	–	1.911	995	–	–
60	0.14	20.88	0.18	–	0.40	0.06	1.56	–	17	1900	–	0.913	500	–	210
61	0.05	–	1.00	46.88	0.23	51.31	0.32	–	7	–	–	0.061	7	–	9
62	–	–	–	–	–	–	–	–	–	–	–	–	–	–	–
63	0.06	–	0.75	1.24	11.58	18.45	21.97	–	2	10	–	0.010	10	–	9
64	0.15	–	0.09	1.50	0.59	43.04	17.64	–	3	–	–	0.069	9	–	4
65	2.47	0.28	–	–	0.85	–	–	–	–	–	–	–	–	–	–
66	–	–	26.87	–	–	0.01	–	–	–	450138	–	0.001	–	–	–
67	–	–	38.17	58.46	–	–	–	–	–	–	–	–	–	–	–
68	–	–	–	–	–	–	–	–	–	–	–	–	–	–	–
69	0.04	24.84	18.65	–	–	0.14	–	–	7	–	–	–	–	–	5
70	–	21.65	31.04	–	–	–	–	–	–	300	–	–	–	–	–
71	–	–	24.18	–	–	–	–	–	–	–	–	–	–	415898.96	–
72	–	–	26.40	–	0.01	–	–	–	10	–	–	0.031	–	452927.78	–
73	–	–	31.33	–	–	–	9.66	–	–	–	–	0.001	–	–	–
74	–	24.53	30.18	–	–	–	–	–	–	247	–	0.004	–	–	–
75	–	–	–	–	–	–	99.00	–	–	–	–	–	–	–	291951
76	–	–	–	–	–	–	0.07	–	–	–	–	0.001	–	–	294822
77	–	–	–	–	–	–	–	–	–	–	–	–	–	–	516285
78	–	–	–	–	–	–	–	–	–	–	–	–	–	–	534100
79	–	–	–	50.99	–	–	0.07	–	–	–	–	0.001	–	–	470008
80	–	–	–	–	–	–	–	–	–	–	–	–	–	–	
81	0.05	–	–	0.20	–	–	17.62	–	55	–	–	0.053	169	99.24	359073
82	–	–	–	–	–	–	10.93	–	–	–	–	–	–	–	222460

TABLE 19–1G VITAMIN SUPPLEMENTS, COMPOSITION OF FEEDS, DATA EXPRESSED **AS-FED**

Entry Number	Feed Name Description	International Feed Number	Proximate Analysis									Digestible Protein		
			Dry Matter	Ash	Crude Fiber	Neutral Det. Fib. (NDF)	Acid Det. Fib. (ADF)	Lignin	Ether Extract (Fat)	N-Free Extract	Crude Protein	Ruminant	Swine	Horse
			%	%	%	%	%	%	%	%	%	%	%	%
	ALFALFA (LUCERNE)													
1	IMMATURE, FRESH	2-00-177	21	2.3	4.3	–	–	–	0.7	8.6	5.4	4.5	4.0	–
2	HAY, SUN-CURED	1-00-078	90	8.6	28.2	35.4	30.9	8.9	1.7	35.9	16.0	11.2	7.5	11.9
3	HAY, SUN-CURED, PELLETED	1-00-124	92	10.2	25.4	–	31.9	6.1	1.9	39.1	15.7	11.5	10.2	10.4
4	LEAVES, SUN-CURED, GROUND	1-00-146	88	9.2	15.0	–	–	–	2.7	40.7	20.1	15.8	14.3	13.9
5	MEAL, DEHY, 17% PROTEIN	1-00-023	92	9.7	24.0	41.3	31.5	9.7	2.8	37.8	17.4	12.6	11.7	11.8
6	MEAL, DEHY, 20% PROTEIN	1-00-024	92	10.2	20.8	–	27.0	–	3.3	37.1	20.2	15.1	14.4	14.0
7	MEAL, DEHY, 22% PROTEIN	1-07-851	93	10.2	18.3	–	25.3	–	4.1	38.1	22.2	16.4	16.4	15.5
	ANIMAL													
8	LIVER-GLANDS, GROUND	5-00-390	93	5.9	1.8	–	–	–	15.8	3.3	66.5	58.2	64.8	66.5
9	LIVER, MEAL	5-00-389	93	6.3	1.4	–	–	–	15.7	3.2	66.1	57.9	64.4	66.1
10	MEAT SOLUBLES, DEHY	5-00-393	90	5.7	–	–	–	–	–	–	80.0	70.6	78.3	82.0
11	**BLUEGRASS, KENTUCKY,** IMMATURE, FRESH	2-00-777	31	2.9	7.8	–	–	–	1.1	13.7	5.4	4.0	3.8	–
12	**BREWERS GRAINS,** DEHY	5-02-141	92	3.6	13.0	38.7	23.9	4.6	6.6	41.6	27.3	20.1	21.8	21.0
13	**BUTTERMILK, CATTLE,** CONDENSED	5-01-159	29	3.6	0.1	–	–	–	2.4	12.4	10.8	9.2	10.0	9.5
14	**CARROT,** ROOTS, FRESH	4-01-145	11	1.0	1.1	–	–	–	0.2	8.1	1.2	0.8	0.8	0.8
	CATTLE													
15	BUTTERMILK, CONDENSED	5-01-159	29	3.6	0.1	–	–	–	2.4	12.4	10.8	9.2	10.0	9.5
16	LIVER, FRESH	5-01-166	28	1.4	0.2	–	–	–	5.1	1.9	19.5	17.0	19.0	19.4
17	WHEY, DEHY	4-01-182	93	8.8	0.2	0.3	0.2	–	0.8	70.2	13.3	8.9	13.1	9.6
	COD, FISH													
18	LIVER, MEAL	5-08-423	93	2.9	0.7	–	–	–	28.9	9.6	50.4	43.6	48.9	48.4
19	LIVER OIL	7-01-993	100	–	–	–	–	–	99.5	–	–	–	–	–
20	**CORN,** DISTILLERS GRAINS WITH SOLUBLES, DEHY	5-02-843	92	4.5	9.1	–	–	–	9.2	41.9	27.1	17.2	25.7	22.1
21	**CRAB,** CANNERY RESIDUE, MEAL (CRAB MEAL)	5-01-663	92	41.1	10.7	–	–	–	2.2	5.9	32.2	27.1	30.9	27.9
	DISTILLERS PRODUCTS (ALSO SEE CORN)													
22	GRAINS, DEHY	5-02-144	93	1.5	12.8	–	–	–	7.4	43.5	27.3	18.9	26.0	22.3
23	SOLUBLES, DEHY	5-02-147	92	6.2	3.4	–	–	–	8.9	44.8	28.8	24.0	27.4	24.0
	FATS AND OILS													
24	GERM OIL (WHEAT)	7-05-207	100	–	–	–	–	–	99.5	–	–	–	–	–
25	LIVER OIL (COD)	7-01-993	100	–	–	–	–	–	99.5	–	–	–	–	–
	FISH													
26	MEAL MECH EXTD	5-01-977	92	21.4	0.7	–	–	–	6.0	–	64.3	57.1	59.1	64.1
27	SOLUBLES, CONDENSED	5-01-969	50	10.1	0.5	–	–	–	6.1	2.2	31.5	28.0	29.3	30.9
28	SOLUBLES, DEHY	5-01-971	93	12.7	2.0	–	–	–	9.0	8.7	60.4	52.7	58.8	59.7
	FISH, COD													
29	LIVER, MEAL	5-08-423	93	2.9	0.7	–	–	–	28.9	9.6	50.4	43.6	48.9	48.4
30	LIVER OIL	7-01-993	100	–	–	–	–	–	99.5	–	–	–	–	–
	FISH, SARDINE													
31	MEAL MECH EXTD	5-02-015	93	15.8	1.0	–	–	–	5.0	6.1	65.2	53.5	63.6	65.1
32	SOLUBLES, CONDENSED	5-02-014	50	10.2	–	–	–	–	9.4	0.6	29.5	25.6	28.7	28.7
33	**KENTUCKY BLUEGRASS,** IMMATURE, FRESH	2-00-777	31	2.9	7.8	–	–	–	1.1	13.7	5.4	4.0	3.8	–
	LIVER													
34	CATTLE, FRESH	5-01-166	28	1.4	0.2	–	–	–	5.1	1.9	19.5	17.0	19.0	19.4
35	SHEEP, FRESH	5-08-116	29	1.4	–	–	–	–	3.9	2.9	21.0	18.4	20.5	21.0
36	SWINE, FRESH	5-04-792	30	1.6	0.1	–	–	–	5.0	2.8	20.8	18.2	20.3	20.7
37	MEAL	5-00-389	93	6.3	1.4	–	–	–	15.7	3.2	66.1	57.9	64.4	66.1
38	**LIVER-GLANDS,** MEAL	5-00-390	93	5.9	1.8	–	–	–	15.8	3.3	66.5	58.2	64.8	66.5
	LUCERNE–SEE ALFALFA													
39	**MAIZE (CORN),** DISTILLERS GRAINS WITH SOLUBLES, DEHY	5-02-843	92	4.5	9.1	–	–	–	9.2	41.9	27.1	17.2	25.7	22.1
40	**MEAT,** SOLUBLES, DEHY	5-00-393	90	5.7	–	–	–	–	–	–	80.0	70.6	78.3	82.0
41	**OATS,** IMMATURE, FRESH	2-03-286	16	1.7	4.0	–	–	–	0.4	7.4	2.5	1.8	1.7	–
	RICE													
42	BRAN WITH GERMS, MEAL SOLV EXTD (RICE BRAN, SOLV EXTD)	4-03-930	91	14.5	12.9	–	–	–	1.5	48.1	14.0	9.8	9.1	10.2
43	POLISHINGS	4-03-943	90	7.6	3.2	–	3.6	–	12.6	54.9	12.0	8.6	10.1	8.6
44	**SHEEP,** LIVER, FRESH	5-08-116	29	1.4	–	–	–	–	3.9	2.9	21.0	18.4	20.5	21.0
45	**SPINACH,** LEAVES, FRESH	2-08-125	9	2.2	0.7	–	–	–	0.4	3.0	3.1	2.5	2.3	–
46	**SWINE,** LIVER, FRESH	5-04-792	30	1.6	0.1	–	–	–	5.0	2.8	20.8	18.2	20.3	20.7
47	**TURNIP,** FRESH	2-05-603	13	2.1	1.4	–	–	–	0.3	6.8	2.9	1.1	2.1	–
48	**WHALE,** LIVER, DEHY	5-05-157	93	–	–	–	–	–	–	–	–	–	–	–
	WHEAT													
49	GERM MEAL	5-05-218	88	4.3	3.1	–	4.4	–	8.5	48.1	24.4	22.9	23.1	19.4
50	GERM OIL	7-05-207	100	–	–	–	–	–	99.5	–	–	–	–	–
51	**WHEY, CATTLE,** DEHY	4-01-182	93	8.8	0.2	0.3	0.2	–	0.8	70.2	13.3	8.9	13.1	9.6
52	**YEAST, BREWERS,** DEHY	7-05-527	93	6.5	3.0	–	3.7	–	0.9	38.8	43.8	39.0	–	–
53	**YEAST, PRIMARY,** DEHY	7-05-533	93	8.0	3.1	–	–	–	1.0	32.5	48.0	–	–	–
54	**YEAST, TORULA,** DEHY	7-05-534	93	8.0	2.5	–	3.7	–	1.6	31.5	49.6	45.1	40.6	–

Entry Number	TDN Ruminant	TDN Swine	TDN Horse	Digestible Energy Ruminant		Digestible Energy Swine		Digestible Energy Horse		Metabolizable Energy Ruminant		Metabolizable Energy Swine		Metabolizable Energy Poultry ME_n		Metabolizable Energy Horse		Net Energy Ruminant NE_m		Net Energy Ruminant NE_g		Net Energy Lactating Cows NE_{lc}	
	%	%	%	Mcal		kcal		Mcal		Mcal		kcal		kcal		Mcal		Mcal		Mcal		Mcal	
				lb	*kg*	lb	*kg*	lb	*kg*	lb	*kg*	lb	*kg*	lb	*kg*	lb	*kg*	lb	*kg*	lb	*kg*	lb	*kg*
1	15	–	–	0.30	*0.66*	–	–	–	–	0.26	*0.57*	–	–	–	–	–	–	0.16	*0.35*	0.10	*0.22*	0.15	*0.34*
2	51	32	48	1.03	*2.28*	–	–	0.90	*1.98*	0.90	*1.99*	–	–	–	–	0.74	*1.62*	0.48	*1.06*	0.25	*0.55*	0.49	*1.07*
3	54	44	48	1.01	*2.22*	–	–	0.90	*1.98*	0.84	*1.86*	–	–	–	–	0.74	*1.62*	0.49	*1.09*	0.26	*0.57*	0.53	*1.16*
4	56	63	56	1.13	*2.49*	–	–	1.03	*2.27*	0.96	*2.12*	–	–	–	–	0.85	*1.86*	0.58	*1.28*	0.35	*0.77*	0.58	*1.28*
5	55	44	47	1.12	*2.47*	643	*1418*	0.89	*1.96*	0.96	*2.12*	601	*1326*	682	*1504*	0.73	*1.61*	0.60	*1.32*	0.36	*0.79*	0.57	*1.25*
6	57	48	50	1.07	*2.36*	943	*2079*	0.94	*2.08*	0.93	*2.06*	872	*1923*	737	*1625*	0.77	*1.70*	0.58	*1.28*	0.34	*0.75*	0.59	*1.30*
7	60	49	52	1.20	*2.65*	991	*2186*	0.98	*2.15*	1.02	*2.26*	841	*1855*	768	*1692*	0.80	*1.76*	0.63	*1.39*	0.38	*0.84*	0.63	*1.38*
8	90	95	–	1.87	*4.11*	1711	*3771*	–	–	1.69	*3.73*	1503	*3314*	1323	*2917*	–	–	1.11	*2.44*	0.79	*1.73*	1.02	*2.25*
9	89	93	–	1.79	*3.94*	1867	*4116*	–	–	1.62	*3.57*	1645	*3627*	1306	*2878*	–	–	1.00	*2.21*	0.70	*1.55*	0.93	*2.04*
10	–	–	–	–	–	–	–	–	–	–	–	–	–	–	–	–	–	–	–	–	–	–	–
11	22	–	–	0.42	*0.93*	–	–	–	–	0.37	*0.81*	–	–	–	–	–	–	0.24	*0.52*	0.15	*0.33*	0.23	*0.51*
12	65	66	48	1.25	*2.76*	1045	*2303*	–	–	1.01	*2.22*	1038	*2288*	1047	*2308*	–	–	0.64	*1.41*	0.39	*0.86*	0.67	*1.48*
13	26	22	–	0.52	*1.14*	442	*974*	–	–	0.46	*1.02*	383	*844*	–	–	–	–	0.29	*0.63*	0.20	*0.43*	0.27	*0.59*
14	10	10	8	0.19	*0.43*	208	*458*	0.14	*0.31*	0.17	*0.38*	195	*430*	208	*458*	0.12	*0.26*	0.10	*0.23*	0.07	*0.16*	0.10	*0.22*
15	26	22	–	0.52	*1.14*	442	*974*	–	–	0.46	*1.02*	383	*844*	–	–	–	–	0.29	*0.63*	0.20	*0.43*	0.27	*0.59*
16	29	31	–	0.58	*1.28*	615	*1356*	–	–	0.53	*1.17*	544	*1200*	–	–	–	–	0.34	*0.76*	0.25	*0.54*	0.32	*0.70*
17	76	77	–	1.51	*3.33*	1444	*3183*	–	–	1.28	*2.83*	1411	*3110*	880	*1939*	–	–	0.87	*1.92*	0.59	*1.30*	0.78	*1.71*
18	109	118	–	2.18	*4.81*	2357	*5196*	–	–	2.01	*4.44*	2098	*4625*	–	–	–	–	1.27	*2.81*	0.92	*2.03*	1.17	*2.57*
19	–	–	–	–	–	–	–	–	–	–	–	–	–	–	–	–	–	–	–	–	–	–	–
20	81	79	–	1.55	*3.43*	1466	*3232*	–	–	1.34	*2.95*	1278	*2817*	1149	*2533*	–	–	0.92	*2.03*	0.63	*1.40*	0.86	*1.89*
21	27	–	–	0.54	*1.18*	686	*1511*	–	–	0.35	*0.78*	555	*1224*	827	*1823*	–	–	0.07	*0.16*	–	–	0.25	*0.54*
22	76	83	–	1.52	*3.36*	1669	*3680*	–	–	1.35	*2.97*	1463	*3225*	1138	*2509*	–	–	0.87	*1.92*	0.59	*1.31*	0.82	*1.80*
23	78	82	–	1.61	*3.54*	1637	*3609*	–	–	1.43	*3.16*	1433	*3160*	1316	*2901*	–	–	0.93	*2.06*	0.65	*1.43*	0.87	*1.92*
24	–	–	–	–	–	–	–	–	–	–	–	–	–	–	–	–	–	–	–	–	–	–	–
25	–	–	–	–	–	–	–	–	–	–	–	–	–	–	–	–	–	–	–	–	–	–	–
26	67	66	–	1.34	*2.95*	1317	*2903*	–	–	1.17	*2.57*	1138	*2508*	1174	*2587*	–	–	0.64	*1.40*	0.39	*0.86*	0.63	*1.39*
27	41	44	–	0.85	*1.87*	866	*1909*	–	–	0.78	*1.73*	736	*1623*	755	*1665*	–	–	0.54	*1.20*	0.38	*0.84*	0.44	*0.97*
28	77	66	–	1.50	*3.30*	1467	*3234*	–	–	1.21	*2.66*	1278	*2818*	1322	*2915*	–	–	0.81	*1.78*	0.54	*1.19*	0.77	*1.70*
29	109	118	–	2.18	*4.81*	2357	*5196*	–	–	2.01	*4.44*	2098	*4625*	–	–	–	–	1.27	*2.81*	0.92	*2.03*	1.17	*2.57*
30	–	–	–	–	–	–	–	–	–	–	–	–	–	–	–	–	–	–	–	–	–	–	–
31	70	67	–	1.40	*3.09*	1327	*2925*	–	–	1.23	*2.71*	1148	*2531*	1313	*2896*	–	–	0.76	*1.67*	0.49	*1.08*	0.73	*1.60*
32	–	–	–	–	–	–	–	–	–	–	–	–	–	–	–	–	–	–	–	–	–	–	–
33	22	–	–	0.42	*0.93*	–	–	–	–	0.37	*0.81*	–	–	–	–	–	–	0.24	*0.52*	0.15	*0.33*	0.23	*0.51*
34	29	31	–	0.58	*1.28*	615	*1356*	–	–	0.53	*1.17*	544	*1200*	–	–	–	–	0.34	*0.76*	0.25	*0.54*	0.32	*0.70*
35	–	–	–	–	–	–	–	–	–	–	–	–	–	–	–	–	–	–	–	–	–	–	–
36	31	32	–	0.62	*1.37*	650	*1433*	–	–	0.56	*1.24*	574	*1266*	–	–	–	–	0.37	*0.80*	0.26	*0.57*	0.34	*0.74*
37	89	93	–	1.79	*3.94*	1867	*4116*	–	–	1.62	*3.57*	1645	*3627*	1306	*2878*	–	–	1.00	*2.21*	0.70	*1.55*	0.93	*2.04*
38	90	95	–	1.87	*4.11*	1711	*3771*	–	–	1.69	*3.73*	1503	*3314*	1323	*2917*	–	–	1.11	*2.44*	0.79	*1.73*	1.02	*2.25*
39	81	79	–	1.55	*3.43*	1466	*3232*	–	–	1.34	*2.95*	1278	*2817*	1149	*2533*	–	–	0.92	*2.03*	0.63	*1.40*	0.86	*1.89*
40	–	–	–	–	–	–	–	–	–	–	–	–	–	–	–	–	–	–	–	–	–	–	–
41	10	–	–	0.21	*0.46*	–	–	–	–	0.18	*0.39*	–	–	–	–	–	–	0.10	*0.23*	0.06	*0.14*	0.10	*0.23*
42	55	72	–	1.15	*2.54*	1481	*3264*	–	–	0.98	*2.15*	1199	*2643*	909	*2003*	–	–	0.62	*1.36*	0.37	*0.82*	0.61	*1.35*
43	81	88	–	1.56	*3.45*	1684	*3713*	–	–	1.43	*3.16*	1555	*3428*	1367	*3015*	–	–	0.88	*1.93*	0.60	*1.32*	0.83	*1.82*
44	–	–	–	–	–	–	–	–	–	–	–	–	–	–	–	–	–	–	–	–	–	–	–
45	–	–	–	–	–	–	–	–	–	–	–	–	–	–	–	–	–	–	–	–	–	–	–
46	31	32	–	0.62	*1.37*	650	*1433*	–	–	0.56	*1.24*	574	*1266*	–	–	–	–	0.37	*0.80*	0.26	*0.57*	0.34	*0.74*
47	10	–	–	0.19	*0.41*	–	–	–	–	0.16	*0.36*	–	–	–	–	–	–	0.10	*0.21*	0.06	*0.13*	0.10	*0.21*
48	–	–	–	–	–	–	–	–	–	–	–	–	–	–	–	–	–	–	–	–	–	–	–
49	83	80	–	1.66	*3.66*	1727	*3807*	–	–	1.50	*3.30*	1522	*3357*	1223	*2696*	–	–	0.96	*2.11*	0.67	*1.48*	0.89	*1.95*
50	–	–	–	–	–	–	–	–	–	–	–	–	–	–	–	–	–	–	–	–	–	–	–
51	76	77	–	1.51	*3.33*	1444	*3183*	–	–	1.28	*2.83*	1411	*3110*	880	*1939*	–	–	0.87	*1.92*	0.59	*1.30*	0.78	*1.71*
52	73	70	–	1.46	*3.21*	–	–	–	–	1.28	*2.83*	1299	*2865*	928	*2047*	–	–	0.81	*1.79*	0.54	*1.19*	0.77	*1.69*
53	–	–	–	–	–	–	–	–	–	–	–	–	–	–	–	–	–	–	–	–	–	–	–
54	72	64	–	1.49	*3.29*	1287	*2837*	–	–	1.27	*2.81*	1096	*2416*	840	*1851*	–	–	0.82	*1.81*	0.55	*1.21*	0.78	*1.71*

(Continued)

TABLE 19–1G VITAMIN SUPPLEMENTS, MINERAL AND VITAMIN COMPOSITION OF FEEDS, DATA EXPRESSED **AS-FED**

Entry Number	Feed Name Description	Dry Matter	Macro Minerals							Micro Minerals						
			Calcium (Ca)	Phosphorus (P)	Sodium (Na)	Chlorine (Cl)	Magnesium (Mg)	Potassium (K)	Sulfur (S)	Cobalt (Co)	Copper (Cu)	Iodine (I)	Iron (Fe)	Manganese (Mn)	Selenium (Se)	Zinc (Zn)
		%	%	%	%	%	%	%	%	ppm or mg/kg	ppm or mg/kg	ppm or mg/kg	%	ppm or mg/kg	ppm or mg/kg	ppm or mg/kg
	ALFALFA (LUCERNE)															
1	IMMATURE, FRESH	21	0.50	0.09	0.04	0.08	0.05	0.48	0.13	–	–	–	0.006	6.7	–	–
2	HAY, SUN-CURED	90	1.28	0.24	0.07	0.33	0.30	1.85	0.25	0.250	10.0	–	0.019	41.4	–	21.9
3	HAY, SUN-CURED, PELLETED	92	1.48	0.20	–	–	0.25	–	0.24	–	0.6	–	–	–	–	–
4	LEAVES, SUN-CURED, GROUND	88	2.32	0.24	–	–	0.36	1.47	–	–	–	–	0.031	29.8	–	–
5	MEAL, DEHY, 17% PROTEIN	92	1.40	0.23	0.10	0.47	0.29	2.38	0.23	0.302	8.6	0.148	0.041	31.0	0.335	19.3
6	MEAL, DEHY, 20% PROTEIN	92	1.59	0.28	0.11	0.47	0.33	2.41	0.50	0.259	12.2	0.135	0.036	45.2	0.285	21.8
7	MEAL, DEHY, 22% PROTEIN	93	1.69	0.30	0.12	0.52	0.31	2.40	0.30	0.311	9.8	0.166	0.036	36.4	0.534	19.5
	ANIMAL															
8	LIVER-GLANDS, GROUND	93	0.63	1.18	0.10	–	–	0.40	–	0.170	94.2	–	0.056	8.1	–	50.1
9	LIVER, MEAL	93	0.56	1.26	–	–	0.10	–	–	0.135	89.4	–	0.064	8.8	–	61.8
10	MEAT SOLUBLES, DEHY	90	0.45	0.67	–	–	–	–	–	–	–	–	–	–	–	–
11	**BLUEGRASS, KENTUCKY,** IMMATURE, FRESH	31	0.15	0.14	0.04	–	0.05	0.70	0.05	–	–	–	0.010	–	–	–
12	**BREWERS' GRAINS,** DEHY	92	0.30	0.51	0.21	0.15	0.15	0.09	0.30	0.076	21.7	0.066	0.024	37.2	–	27.3
13	**BUTTERMILK, CATTLE,** CONDENSED	29	0.44	0.26	0.31	0.12	0.19	0.23	0.03	–	–	–	–	–	–	–
14	**CARROT,** ROOTS, FRESH	11	0.05	0.04	0.06	0.06	0.02	0.32	0.02	–	1.2	–	0.002	3.6	–	–
	CATTLE															
15	BUTTERMILK, CONDENSED	29	0.44	0.26	0.31	0.12	0.19	0.23	0.03	–	–	–	–	–	–	–
16	LIVER, FRESH	28	0.01	0.23	0.10	–	0.01	0.20	–	–	6.1	–	0.005	2.8	–	26.6
17	WHEY, DEHY	93	0.86	0.76	0.62	0.07	0.13	1.11	1.04	0.111	46.5	–	0.017	5.9	–	3.2
	COD, FISH															
18	LIVER, MEAL	93	0.16	0.69	–	–	–	–	–	–	–	–	–	–	–	–
19	LIVER OIL	100	–	–	–	–	–	–	–	–	–	–	–	–	–	–
20	**CORN,** DISTILLERS GRAINS WITH SOLUBLES, DEHY	92	0.16	0.69	0.47	0.17	0.18	0.47	0.31	0.152	52.6	0.051	0.024	24.0	0.331	80.7
21	**CRAB,** CANNERY RESIDUE, MEAL (CRAB MEAL)	92	14.46	1.58	0.88	1.51	0.94	0.45	0.25	–	32.7	0.557	0.435	132.8	–	–
	DISTILLERS PRODUCTS (ALSO SEE CORN)															
22	GRAINS, DEHY	93	0.12	0.54	0.05	0.05	0.09	0.20	0.46	0.092	47.9	–	0.027	35.0	–	–
23	SOLUBLES, DEHY	92	0.24	1.35	0.45	–	0.53	1.97	–	0.196	71.6	–	0.031	64.1	–	138.0
	FATS AND OILS															
24	GERM OIL (WHEAT)	100	–	–	–	–	–	–	–	–	–	–	–	–	–	–
25	LIVER OIL (COD)	100	–	–	–	–	–	–	–	–	–	–	–	–	–	–
	FISH															
26	MEAL MECH EXTD	92	6.63	3.61	1.11	1.25	0.21	0.40	0.25	0.110	15.1	–	0.038	23.6	–	99.1
27	SOLUBLES, CONDENSED	50	0.16	0.57	2.45	2.93	0.03	1.64	0.12	0.069	46.6	1.111	0.028	13.2	–	43.2
28	SOLUBLES, DEHY	93	0.40	1.27	1.70	–	0.30	2.50	0.45	–	20.0	–	0.095	50.4	2.692	76.7
	FISH, COD															
29	LIVER, MEAL	93	0.16	0.69	–	–	–	–	–	–	–	–	–	–	–	–
30	LIVER OIL	100	–	–	–	–	–	–	–	–	–	–	–	–	–	–
	FISH, SARDINE															
31	MEAL MECH EXTD	93	4.61	2.68	0.18	0.41	0.10	0.32	–	0.183	20.2	–	0.030	23.2	1.772	–
32	SOLUBLES, CONDENSED	50	0.14	0.83	0.18	0.28	–	0.18	0.11	–	25.8	4.934	0.002	24.9	–	–
33	**KENTUCKY BLUEGRASS,** IMMATURE, FRESH	31	0.15	0.14	0.04	–	0.05	0.70	0.05	–	–	–	0.010	–	–	–
	LIVER															
34	CATTLE, FRESH	28	0.01	0.23	0.10	–	0.01	0.20	–	–	6.1	–	0.005	2.8	–	26.6
35	SHEEP, FRESH	29	0.01	0.35	0.06	–	0.02	0.20	–	–	37.5	–	0.007	3.6	–	36.0
36	SWINE, FRESH	30	0.01	0.37	0.07	–	0.01	0.26	–	0.255	56.4	0.340	0.015	1.8	0.340	44.2
37	MEAL	93	0.56	1.26	–	–	0.10	–	–	0.135	89.4	–	0.064	8.8	–	61.8
38	**LIVER-GLANDS,** MEAL	93	0.63	1.18	0.10	–	–	0.40	–	0.170	94.2	–	0.056	8.1	–	50.1
	LUCERNE–SEE ALFALFA															
39	**MAIZE (CORN),** DISTILLERS GRAINS WITH SOLUBLES, DEHY	92	0.16	0.69	0.47	0.17	0.18	0.47	0.31	0.152	52.6	0.051	0.024	24.0	0.331	80.7
40	**MEAT,** SOLUBLES, DEHY	90	0.45	0.67	–	–	–	–	–	–	–	–	–	–	–	–
41	**OATS,** IMMATURE, FRESH	16	–	–	0.02	0.02	–	–	0.01	–	–	–	–	–	–	–
	RICE															
42	BRAN WITH GERMS, MEAL SOLV EXTD (RICE BRAN, SOLV EXTD)	91	0.11	1.37	–	–	–	1.48	0.18	0.111	13.0	0.045	0.019	232.2	–	30.0
43	POLISHINGS	90	0.05	1.34	0.04	0.11	0.60	1.28	0.17	3.890	8.0	–	0.009	126.8	–	63.2
44	**SHEEP,** LIVER, FRESH	29	0.01	0.35	0.06	–	0.02	0.20	–	–	37.5	–	0.007	3.6	–	36.0
45	**SPINACH,** LEAVES, FRESH	9	0.09	0.05	0.07	–	–	0.48	–	–	–	–	0.004	–	–	–
46	**SWINE,** LIVER, FRESH	30	0.01	0.37	0.07	–	0.01	0.26	–	0.255	56.4	0.340	0.015	1.8	0.340	44.2
47	**TURNIP,** FRESH	13	0.40	0.05	–	0.26	0.07	0.41	0.04	–	2.4	–	0.006	55.2	–	5.0
48	**WHALE,** LIVER, DEHY	93	–	–	–	1.99	–	–	–	–	–	–	–	–	–	–
	WHEAT															
49	GERM MEAL	88	0.06	0.95	0.02	0.06	0.25	0.94	0.27	0.120	9.2	–	0.006	132.5	0.463	119.4
50	GERM OIL	100	–	–	–	–	–	–	–	–	–	–	–	–	–	–
51	**WHEY, CATTLE,** DEHY	93	0.86	0.76	0.62	0.07	0.13	1.11	1.04	0.111	46.5	–	0.017	5.9	–	3.2
52	**YEAST, BREWERS,** DEHY	93	0.14	1.36	0.07	0.07	0.24	1.69	0.43	0.506	38.4	0.358	0.009	6.7	0.911	39.0
53	**YEAST, PRIMARY,** DEHY	93	0.36	1.72	–	0.02	0.36	–	0.57	–	–	–	0.030	3.7	–	–
54	**YEAST, TORULA,** DEHY	93	0.55	1.61	0.01	0.02	0.14	1.92	0.55	0.031	11.9	2.502	0.011	9.3	–	99.5

Entry Number	Fat-Soluble Vitamins					Water-Soluble Vitamins								
	A (1 mg Carotene = 1667 IU Vit A)	Carotene (Provitamin A)	D	E	K	B-12	Biotin	Choline	Folacin (Folic Acid)	Niacin	Pantothenic Acid (B-3)	(Pyridoxine) B-6	Riboflavin (B-2)	Thiamin (B-1)
	IU/g	ppm or mg/kg	IU/kg	ppm or mg/kg	ppm or mg/kg	ppb or mcg/kg	ppm or mg/kg	ppm or mg/kg	ppm or mg/kg	ppm or mg/kg	ppm or mg/kg	ppm or mg/kg	ppm or mg/kg	ppm or mg/kg
1	85.9	51.5	–	–	–	–	–	–	–	–	8.9	–	–	–
2	45.0	27.0	2	55.9	–	–	0.18	892	3.07	43	18.1	–	9.5	3.1
3	48.2	28.9	–	–	–	–	–	–	–	–	–	–	–	–
4	97.7	58.6	–	–	–	–	–	–	–	–	–	–	20.9	–
5	200.3	120.2	–	105.7	8.24	–	0.33	1369	4.37	37	29.7	7.18	12.9	3.4
6	265.4	159.2	–	143.3	14.19	–	0.35	1417	2.96	48	35.5	8.72	15.2	5.4
7	391.4	234.8	–	221.3	11.65	–	0.33	1605	5.15	50	39.0	8.28	17.6	5.9
8	–	–	–	–	–	440.5	0.41	10610	4.00	172	90.8	5.01	42.4	0.2
9	–	–	–	–	–	501.3	0.02	11370	5.56	205	29.2	–	36.2	0.2
10	–	–	–	–	–	881.6	–	–	–	–	–	–	–	–
11	247.6	148.5	–	47.8	–	–	–	–	–	–	–	–	–	–
12	0.8	0.5	–	26.7	–	3.6	0.44	1651	0.22	44	8.2	1.03	1.5	0.6
13	25.6	15.4	–	–	–	–	–	–	–	–	–	–	12.6	–
14	129.9	77.9	–	6.9	–	–	0.01	–	0.14	7	3.5	1.39	0.6	0.7
15	25.6	15.4	–	–	–	–	–	–	–	–	–	–	12.6	–
16	–	–	–	7.1	–	425.8	0.98	1424	2.33	75	46.1	5.03	25.8	1.8
17	–	–	–	0.2	–	18.9	0.35	1790	0.85	11	46.2	3.21	27.4	4.0
18	–	–	–	–	–	–	–	–	–	132	46.1	32.85	33.3	18.1
19	845.8	–	–	39.5	–	–	–	–	–	–	–	–	–	–
20	6.2	3.7	1	39.8	–	1.5	0.69	2582	0.91	73	13.8	4.74	8.5	3.0
21	–	–	–	–	–	437.6	0.07	2008	0.11	45	6.5	6.62	6.1	0.4
22	13.0	7.8	–	30.5	–	–	–	2645	–	47	11.9	6.00	6.6	2.5
23	1.9	1.1	–	–	–	2.9	2.84	4992	–	143	25.3	8.66	11.3	6.9
24	–	–	–	–	18.66	–	–	–	–	–	–	–	–	–
25	845.8	–	–	39.5	–	–	–	–	–	–	–	–	–	–
26	–	–	–	19.2	–	258.6	–	3644	–	75	15.0	14.68	5.6	0.8
27	2.2	1.3	–	–	–	506.6	0.14	3370	0.22	176	35.7	12.20	12.9	5.5
28	–	–	–	6.1	–	485.9	0.40	5525	0.57	256	50.4	19.71	13.5	7.4
29	–	–	–	–	–	–	–	–	–	132	46.1	32.85	33.3	18.1
30	845.8	–	–	39.5	–	–	–	–	–	–	–	–	–	–
31	–	–	–	–	–	238.0	0.10	3277	–	75	11.0	–	5.4	0.3
32	–	–	–	–	–	1041.0	0.13	3009	–	356	41.2	–	16.8	4.0
33	247.6	148.5	–	47.8	–	–	–	–	–	–	–	–	–	–
34	–	–	–	7.1	–	425.8	0.98	1424	2.33	75	46.1	5.03	25.8	1.8
35	–	–	–	–	–	–	–	–	–	169	–	–	32.8	4.0
36	–	–	–	–	–	282.7	0.75	–	2.07	165	23.6	3.02	27.3	2.3
37	–	–	–	–	–	501.3	0.02	11370	5.56	205	29.2	–	36.2	0.2
38	–	–	–	–	–	440.5	0.41	10610	4.00	172	90.8	5.01	42.4	0.2
39	6.2	3.7	1	39.8	–	1.5	0.69	2582	0.91	73	13.8	4.74	8.5	3.0
40	–	–	–	–	–	881.6	–	–	–	–	–	–	–	–
41	150.0	90.0	–	–	–	–	–	–	–	–	–	–	–	–
42	–	–	–	60.7	–	–	0.42	1128	2.21	284	23.0	29.11	2.9	22.6
43	–	–	–	90.2	–	–	0.62	1248	–	506	46.4	27.89	1.8	20.0
44	–	–	–	–	–	–	–	–	–	169	–	–	32.8	4.0
45	56.4	33.8	–	38.5	–	–	–	–	–	6	–	–	2.0	1.0
46	–	–	–	–	–	282.7	0.75	–	2.07	165	23.6	3.02	27.3	2.3
47	–	–	–	–	–	–	–	–	–	11	–	–	5.4	2.9
48	–	–	–	–	–	499.0	–	3351	–	200	36.4	9.04	79.1	2.6
49	–	–	–	141.2	–	–	0.22	3062	2.12	68	18.6	9.97	6.0	23.1
50	–	–	–	–	18.66	–	–	–	–	–	–	–	–	–
51	–	–	–	0.2	–	18.9	0.35	1790	0.85	11	46.2	3.21	27.4	4.0
52	–	–	–	2.1	–	1.1	1.04	3847	9.69	443	81.5	36.67	34.1	85.2
53	–	–	–	–	–	6.2	1.61	–	31.13	301	312.0	–	38.8	6.4
54	–	–	–	–	–	4.0	1.19	2981	25.66	512	107.5	34.48	47.7	6.8

TABLE 19-2 AMINO ACID, COMPOSITION OF FEEDS, DATA EXPRESSED **AS-FED**

Entry Number	Feed Name Description	International Feed Number	Dry Matter	Crude Protein	Arginine	Cystine	Glycine	Histidine	Isoleucine	Leucine	Lysine	Methionine	Phenylalanine	Serine	Threonine	Tryptophan	Tyrosine	Valine
					AMINO ACIDS													
			%	%	%	%	%	%	%	%	%	%	%	%	%	%	%	%
	ENERGY FEEDS																	
1	**BAKERY,** WASTE, DEHY (DRIED BAKERY PRODUCT)	4-00-466	91	10.1	0.47	0.17	0.69	0.16	0.45	0.77	0.31	0.17	0.45	0.65	0.46	0.10	0.36	0.47
	BARLEY																	
2	GRAIN	4-00-549	88	11.7	0.51	0.20	0.37	0.25	0.46	0.75	0.40	0.16	0.58	0.46	0.36	0.15	0.35	0.57
3	GRAIN, PACIFIC COAST	4-07-939	89	9.5	0.44	0.19	0.30	0.21	0.40	0.60	0.26	0.14	0.47	0.32	0.31	0.12	0.31	0.46
4	GRAIN SCREENINGS	4-00-542	89	11.5	–	–	–	–	–	–	–	–	–	–	0.36	–	–	–
5	MALT SPROUTS, DEHY	5-00-545	93	22.9	1.05	0.23	0.81	0.43	0.88	1.36	1.12	0.31	0.80	0.47	0.85	0.41	0.46	1.16
6	**BEAN, NAVY,** SEEDS	5-00-623	89	22.9	1.19	0.23	0.80	–	–	–	1.29	0.23	–	–	–	–	0.24	–
7	**BEET, SUGAR,** PULP, DEHY	4-00-669	91	8.8	0.30	0.01	–	0.20	0.30	0.60	0.60	0.01	0.30	–	0.40	0.10	0.40	0.40
8	**BROOMCORN (HOG MILLET; MILLET, PROSO),** GRAIN	4-03-120	90	11.6	0.34	0.20	0.25	0.20	0.44	1.13	0.22	0.26	0.54	0.63	0.37	0.16	0.23	0.55
9	**BUCKWHEAT, COMMON,** GRAIN	4-00-994	88	11.1	0.96	0.17	0.61	0.26	0.37	0.59	0.62	0.19	0.44	0.41	0.44	0.18	0.21	0.53
10	**CITRUS,** PULP WITHOUT FINES, DEHY (DRIED CITRUS PULP)	4-01-237	91	6.1	0.25	0.11	–	0.09	0.18	0.31	0.20	0.09	0.18	–	0.18	0.06	–	0.25
11	**CORN, DENT WHITE,** GRAIN	4-02-928	90	10.8	0.27	0.09	–	0.18	0.45	0.90	0.27	0.09	0.36	–	0.36	0.09	0.45	0.36
	CORN, DENT YELLOW																	
12	GRAIN, ALL ANALYSES	4-02-935	88	9.9	0.43	0.12	0.37	0.27	0.35	1.19	0.30	0.18	0.46	0.49	0.36	0.09	0.31	0.48
13	GRAIN, GRADE 2, 54 lb/bu *(69.5 kg/hl)*	4-02-931	87	8.9	0.45	0.11	0.45	0.20	0.40	1.00	0.19	0.11	0.45	–	0.35	0.09	0.43	0.35
14	GRAIN, FLAKED	4-28-244	89	9.9	0.44	0.25	0.36	0.28	0.34	1.24	0.25	0.15	0.44	0.48	0.35	–	0.39	0.47
15	DISTILLERS SOLUBLES, DEHY	5-28-237	93	27.4	0.99	0.44	1.12	0.67	1.32	2.38	0.92	0.55	1.47	1.22	1.01	0.25	0.88	1.53
16	EARS, GROUND (CORN-AND-COB MEAL)	4-28-238	87	7.8	0.36	0.12	0.31	0.16	0.35	0.86	0.17	0.14	0.39	–	0.28	0.07	0.32	0.31
17	GERM MEAL, WET MILLED, SOLV EXTD	5-28-240	92	20.7	1.31	0.40	1.10	0.70	0.70	1.81	0.90	0.58	0.90	1.00	1.09	0.20	0.70	1.20
18	GRITS (HOMINY GRITS)	4-03-011	90	10.3	0.47	0.15	0.35	0.20	0.39	0.85	0.38	0.16	0.33	–	0.39	0.11	0.50	0.49
19	**CORN, OPAQUE 2 (HIGH LYSINE),** GRAIN	4-11-445	90	10.1	0.64	0.19	0.48	0.35	0.33	0.98	0.42	0.16	0.43	0.46	0.37	0.12	0.40	0.48
20	**COWPEA, COMMON,** SEEDS	5-01-661	89	23.8	1.70	–	–	0.70	1.10	2.31	2.10	0.20	1.30	–	0.80	0.30	1.10	1.20
21	**DISTILLERS PRODUCTS** (ALSO SEE CORN; WHEAT), SOLUBLES, DEHY	5-02-147	92	28.8	1.06	0.40	1.20	0.66	1.21	2.35	0.95	0.50	1.24	0.93	1.00	0.24	0.93	1.40
22	**EMMER,** GRAIN	4-01-830	91	11.7	0.46	–	–	0.20	0.42	0.67	0.29	0.16	0.46	–	0.38	0.12	–	0.47
23	**GOOSEFOOT, LAMB'S QUARTER,** SEEDS	5-08-424	90	18.8	0.08	–	0.05	0.02	0.03	0.05	0.04	0.02	0.03	0.03	0.03	–	0.02	0.03
24	**GRAPE,** POMACE, DEHY (MARC)	1-02-208	90	12.1	0.67	0.17	0.89	0.26	0.55	1.63	0.50	0.18	0.55	–	0.38	0.07	0.16	1.09
25	**HOG MILLET (BROOMCORN; MILLET, PROSO),** GRAIN	4-03-120	90	11.6	0.34	0.20	0.25	0.20	0.44	1.13	0.22	0.26	0.54	0.63	0.37	0.16	0.23	0.55
26	**HOMINY GRITS (CORN, DENT YELLOW, GRITS)**	4-03-011	90	10.3	0.47	0.15	0.35	0.20	0.39	0.85	0.38	0.16	0.33	–	0.39	0.11	0.50	0.49
	MAIZE–SEE CORN																	
27	**MILLET,** GRAIN,	4-03-098	90	12.1	0.35	0.12	0.40	0.23	0.49	1.23	0.26	0.30	0.59	–	0.44	0.12	–	0.62
28	**MILLET, PROSO (BROOMCORN; HOG MILLET),** GRAIN	4-03-120	90	11.6	0.34	0.20	0.25	0.20	0.44	1.13	0.22	0.26	0.54	0.63	0.37	0.16	0.23	0.55
	OATS																	
29	GRAIN	4-03-309	89	11.9	0.71	0.19	0.51	0.17	0.48	0.87	0.40	0.18	0.57	0.50	0.38	0.15	0.45	0.62
30	GRAIN, GRADE 1, 34 lb/bu *(43.8 kg/hl)*	4-03-313	88	11.2	0.79	0.22	0.49	0.19	0.52	0.89	0.49	0.18	0.59	–	0.39	0.16	0.52	0.69
31	GRAIN, PACIFIC COAST	4-07-999	91	9.1	0.58	0.17	0.40	0.17	0.38	0.70	0.33	0.13	0.43	0.40	0.30	0.12	0.70	0.49
32	MIDDLINGS, LESS THAN 4% FIBER (FEEDING OAT MEAL)	4-03-303	91	14.8	0.81	0.25	0.62	0.30	0.56	1.05	0.53	0.21	0.69	0.70	0.48	0.20	0.72	0.74
33	GROATS	4-03-331	90	15.8	0.89	0.21	0.61	0.27	0.54	1.04	0.54	0.20	0.70	0.62	0.44	0.19	0.51	0.74
34	**PEA, GARDEN,** SEEDS	5-08-482	89	23.8	1.43	–	–	0.63	1.03	1.61	1.47	0.34	1.20	–	0.80	0.23	–	1.18
35	**PUMPKIN,** SEEDS	5-03-817	94	38.3	4.93	0.93	1.96	0.75	1.29	2.29	1.45	0.77	1.67	2.03	0.95	0.47	1.38	1.57
	RICE																	
36	GRAIN, GROUND (GROUND ROUGH RICE; GROUND PADDY RICE)	4-03-938	89	7.5	0.54	0.12	0.62	0.16	0.27	0.54	0.25	0.14	0.30	0.50	0.23	0.10	0.63	0.40
37	BRAN WITH GERMS (RICE BRAN)	4-03-928	91	13.0	0.82	0.16	0.81	0.29	0.50	0.84	0.54	0.26	0.53	0.73	0.44	0.10	0.59	0.75
38	GROATS, POLISHED (RICE, POLISHED)	4-03-942	89	7.0	0.48	0.09	0.42	0.18	0.28	0.47	0.24	0.17	0.31	0.29	0.25	0.09	0.23	0.40
39	POLISHINGS	4-03-943	90	12.0	0.57	0.14	0.65	0.19	0.37	0.73	0.51	0.22	0.43	0.49	0.35	0.11	0.45	0.68
40	**RYE,** GRAIN	4-04-047	87	12.0	0.52	0.19	0.44	0.25	0.48	0.68	0.42	0.17	0.56	0.61	0.35	0.12	0.26	0.58
41	**SAFFLOWER,** SEEDS	4-07-958	93	14.9	1.60	0.35	1.00	0.48	0.80	1.20	0.60	0.33	1.00	–	0.64	0.28	–	1.00
	SCREENINGS, GRAIN (CEREAL) (ALSO SEE WHEAT)																	
42	REFUSE	4-02-151	91	12.6	0.68	–	0.59	0.30	0.52	0.98	0.48	0.15	0.64	0.57	0.46	–	0.32	0.63
43	UNCLEANED	4-02-153	92	13.7	0.67	–	0.61	0.30	0.45	0.90	0.42	0.19	0.58	0.67	0.44	–	0.58	0.58
	SORGHUM																	
44	GRAIN, ALL ANALYSES	4-04-383	90	11.5	0.39	0.21	0.34	0.24	0.42	1.47	0.26	0.14	0.56	0.49	0.36	0.09	0.40	0.50
45	GRAIN, LESS THAN 9% PROTEIN	4-08-138	89	8.9	0.28	0.14	0.27	0.19	0.46	1.40	0.19	0.12	0.47	–	0.36	0.12	0.60	0.53
46	DARSO, GRAIN	4-04-357	90	10.1	0.36	–	–	0.18	0.45	1.23	0.19	0.11	0.48	–	0.31	0.11	–	0.48
47	FETERITA, GRAIN	4-04-369	89	11.7	0.46	–	–	0.26	0.58	1.78	0.20	0.18	0.67	–	0.46	0.17	–	0.67

V-3 APPARENT ILEAL DIGESTIBILITY OF CRUDE PROTEIN AND ESSENTIAL AMINO ACIDS IN FEEDSTUFFS FOR SWINE, AND

EEDSTUFF (FED BASIS)	**Obs	DM %	Crude Protein Total %	Crude Protein Coefficient %	Crude Protein Digestible %	Arginine Total %	Arginine Coefficient %	Arginine Digestible %	Cystine Total %	Cystine Coefficient %	Cystine Digestible %	Histidine Total %	Histidine Coefficient %	Histidine Digestible %	Isoleucine Total %	Isoleucine Coefficient %	Isoleucine Digestible %
LFA	3	89.85	16.95	39	6.61	0.66	59	0.39	0.18	15	0.03	0.32	46	0.15	0.66	55	0.36
LEY	33	87.74	10.59	70	7.41	0.52	75	0.39	0.22	74	0.16	0.24	71	0.17	0.37	73	0.27
LEY, NAKED	1	87.20	11.70	78	9.13	0.64	85	0.54	—	—	—	0.30	90	0.27	0.45	87	0.39
OD MEAL	3	90.57	87.87	87	76.45	3.68	94	3.46	—	—	—	5.44	94	5.11	0.72	70	0.50
E MEAL	2	94.11	28.10	72	20.23	1.78	79	1.41	0.20	38	0.08	0.26	71	0.18	0.54	71	0.38
WERS GRAINS	3	91.60	26.03	70	18.22	1.82	—	—	—	—	—	0.67	—	—	1.11	79	0.88
OLA MEAL	4	90.43	37.28	69	25.72	2.21	82	1.81	0.42	89	0.37	1.20	78	0.94	1.57	75	1.18
EIN	4	90.69	86.19	89	76.71	2.88	93	2.68	—	62	—	2.00	94	1.88	4.22	90	3.80
CONUT OIL MEAL	1	94.70	21.58	52	11.22	2.18	—	—	0.29	—	—	0.42	—	—	0.60	65	0.39
RN	10	88.34	9.19	77	7.08	0.41	82	0.34	0.17	74	0.13	0.25	82	0.21	0.33	80	0.26
RN GLUTEN FEED	4	90.30	19.58	54	10.57	1.07	72	0.77	0.47	—	—	0.66	58	0.38	0.66	62	0.41
RN GLUTEN MEAL	2	89.35	59.00	80	47.20	2.00	86	1.72	1.06	73	0.77	1.31	81	1.06	2.61	84	2.19
TTONSEED MEAL GLANDED	9	91.60	41.14	72	29.62	4.42	88	3.89	—	—	—	1.10	77	0.85	1.28	67	0.86
TTONSEED MEAL GLANDLESS	2	94.82	42.40	83	35.19	5.56	95	5.28	—	—	—	1.36	88	1.20	1.45	85	1.23
ED SKIM MILK	4	95.78	33.96	85	28.87	1.24	85	1.05	0.28	86	0.24	1.47	93	1.37	1.73	85	1.47
ATHER MEAL	3	94.30	83.20	71	59.07	5.09	79	4.02	3.85	72	2.77	0.57	45	0.26	3.69	78	2.88
SH MEAL	9	94.49	61.83	79	48.85	3.91	88	3.44	0.66	62	0.41	1.37	81	1.11	2.60	86	2.24
OUNDNUT MEAL	2	89.30	41.70	85	35.45	4.92	95	4.67	0.49	78	0.38	0.88	85	0.75	1.38	85	1.17
PIN MEAL	2	88.25	33.07	80	26.46	3.50	93	3.26	0.43	70	0.30	0.77	85	0.65	1.47	83	1.22
YSINE HCI	2	98.50	94.40	100	94.40												
EAT & BONE MEAL	29	93.38	55.07	67	36.90	3.44	77	2.65	0.63	51	0.32	0.93	71	0.66	1.53	68	1.04
ATS	1	87.60	13.00	61	7.93	0.65	—	—	0.29	—	—	0.21	—	—	0.39	—	—
AT GROATS	1	91.62	16.32	84	13.71	1.15	90	1.04	1.15	90	1.04	0.46	86	0.40	0.52	86	0.45
EANUT MEAL	4	93.87	44.23	73	32.29	5.20	90	4.68	—	—	—	1.04	73	0.76	1.46	76	1.11
EAS	3	89.03	20.87	73	15.24	1.83	86	1.57	0.37	61	0.23	0.57	81	0.46	0.90	74	0.67
OULTRY-BY-PRODUCT MEAL	5	92.76	63.71	72	45.87	4.34	88	3.82	2.04	—	—	1.23	81	1.00	2.52	79	1.99
APESEED MEAL	7	91.45	35.32	70	24.72	2.05	83	1.70	0.83	73	0.61	0.99	83	0.82	1.48	75	1.11
ICE	1	85.50	8.80	72	6.34	0.98	90	0.88	—	—	—	0.18	90	0.16	0.32	86	0.28
YE	3	88.27	10.03	68	6.82	0.48	74	0.36	0.20	72	0.14	0.21	70	0.15	0.33	70	0.23
AND EEL	1	89.42	72.56	79	57.32	3.45	—	—	0.70	—	—	1.63	54	0.88	2.93	88	2.58
ESAME MEAL	1	92.70	44.04	76	33.47	5.39	—	—	0.86	—	—	1.01	36	0.36	1.40	79	1.11
ORGHUM	2	89.99	9.83	81	7.96	0.45	85	0.38	—	—	—	0.24	81	0.19	0.44	88	0.39
OY FLOUR	1	94.44	50.68	82	41.56	4.02	91	3.66	0.74	78	0.58	1.39	88	1.22	2.42	83	2.01
OYBEANS EXTRUDED	2	92.68	35.11	74	25.98	2.72	84	2.23	0.57	64	0.36	0.99	80	0.79	1.69	74	1.25
OYBEAN MEAL 44%	9	89.62	44.27	80	35.42	3.29	90	2.96	0.69	74	0.51	1.18	86	1.01	2.03	83	1.68
SOYBEAN MEAL 48%	18	89.51	47.29	80	37.83	3.56	90	3.20	0.66	79	0.52	1.28	86	1.10	2.22	84	1.86
SOYFLAKES RAW	1	90.72	48.72	35	17.05	3.91	56	2.19	0.64	35	0.22	1.41	48	0.68	2.42	43	1.04
SOYFLAKES HEATED	1	91.96	49.29	79	38.94	3.91	88	3.44	0.63	74	0.47	1.40	82	1.15	2.48	78	1.93
SUNFLOWER MEAL	10	91.90	35.29	75	26.47	2.93	90	2.64	0.57	74	0.42	0.89	76	0.68	1.45	78	1.13
TRITICALE	5	89.14	13.31	78	10.38	0.55	87	0.48	—	—	—	0.25	84	0.21	0.36	83	0.30
WHEAT	24	87.93	12.68	82	10.40	0.57	83	0.47	0.20	80	0.16	0.25	80	0.20	0.43	83	0.36
WHEAT BRAN 7% FIBER	3	88.30	13.50	67	9.05	0.88	85	0.75	0.37	71	0.26	0.35	79	0.28	0.47	74	0.35
WHEAT FLOUR	1	88.00	13.46	91	12.25	0.43	91	0.39	—	—	—	0.26	91	0.24	0.50	94	0.47
WHEAT MIDDS 9% FIBER	5	89.14	16.63	67	11.14	1.18	84	0.99	—	—	—	0.45	79	0.36	0.55	70	0.39
WHEAT OFFAL 9% FIBER	1	88.00	15.84	70	11.09	0.97	95	0.92	—	—	—	0.39	79	0.31	0.51	73	0.37

*True Ileal Digestibility Values

**The number of observations for tryptophan may differ substantially from the value indicated.

DIGESTIBLE AMINO ACID RECOMMENDATIONS

Period Weight, lb.	Starting 10 - 22	Starting 22 - 55	Growing-Finishing[1] 55 - 110	Growing-Finishing[1] 110 - 220
DIGESTIBLE MINIMUMS, %				
CRUDE PROTEIN	16.50	14.50	12.50	11.00
LYSINE/CALORIE (g/Mcal ME)	3.60	3.30	2.50	2.00
LYSINE[2]	1.25	1.00	0.75	0.61
THREONINE	0.81	0.65	0.49	0.40
TRYPTOPHAN	0.23	0.18	0.14	0.11
ISOLEUCINE	0.75	0.60	0.45	0.37
METHIONINE & CYSTINE	0.69	0.55	0.41	0.34
HISTIDINE	0.41	0.33	0.25	0.20
LEUCINE	1.35	1.08	0.81	0.66
PHENYLALANINE & TYROSINE	1.50	1.20	0.90	0.73
VALINE	0.91	0.73	0.55	0.45

[1]Boar requirements are 10 to 15% higher than gilts or barrows

Heartland Lysine would like to express its appreciation o each of the authors who contributed data to this work. n particular our thanks are extended to Dr. Darrell Knabe Texas A&M University), Dr. Malcolm Fuller (Rowett Re earch Institute), and Dr. Michael Taverner (Animal Research Institute) for their contributions.

This chart reflects our interpretation of published literature on digestible amino acids for swine nutrition. It is the responsibility of the purchaser of our products to determine the best application of our products for their needs. Information and recommendations regarding our products and/or nutrient levels for swine feeding are to the best of our knowledge accurate. We do not warrant the accuracy or completeness of this information. Our making this information available does not relieve the purchaser or user of his obligation to verify the suitability of our products and recommendations for their intended application.

TABLE 19-2 AMINO ACID, COMPOSITION OF FEEDS, DATA EXPRESSED AS-FED—(Continued)

Entry Number	Feed Name Description	International Feed Number	Dry Matter %	Crude Protein %	Arginine %	Cystine %	Glycine %	Histidine %	Isoleucine %	Leucine %	Lysine %	Methionine %	Phenylalanine %	Serine %	Threonine %	Tryptophan %	Tyrosine %	Valine %
	SORGHUM (Continued)																	
48	HEGARI, GRAIN	4-04-398	89	10.4	0.29	–	–	0.18	0.47	1.40	0.17	0.12	0.54	–	0.36	0.11	–	0.55
49	KAFIR, GRAIN	4-04-428	89	10.8	0.38	0.17	0.30	0.27	0.55	1.62	0.26	0.19	0.64	–	0.45	0.15	–	0.62
50	MILO, GRAIN	4-04-444	89	10.1	0.37	0.13	0.35	0.24	0.44	1.32	0.23	0.16	0.49	0.49	0.35	0.10	0.37	0.53
51	MILO, GLUTEN WITH BRAN, MEAL	5-08-089	89	23.2	0.90	0.20	0.68	0.60	1.00	2.51	0.70	0.40	1.00	–	0.80	0.20	0.90	1.30
52	SHALLU, GRAIN	4-04-456	90	11.5	0.31	–	–	0.19	0.38	0.97	0.19	0.17	0.40	–	0.30	0.10	–	0.46
	SOYBEAN																	
53	SEEDS	5-04-610	92	38.4	2.63	0.42	1.42	0.92	1.62	2.72	2.32	0.48	1.76	1.99	1.46	0.56	1.29	1.61
54	SOYBEAN MILL FEED	4-04-594	90	12.6	0.70	0.13	0.47	0.18	0.41	0.58	0.59	0.12	0.37	–	0.30	0.13	0.23	0.37
55	**SPELT,** GRAIN	4-04-651	90	12.0	0.45	–	–	0.18	0.36	0.63	0.27	0.18	0.45	–	0.36	0.09	–	0.45
56	**TRITICALE,** GRAIN	4-20-362	89	15.4	0.85	0.27	0.68	0.38	0.58	1.11	0.52	0.22	0.77	0.73	0.53	0.18	0.49	0.78
	WHEAT																	
57	GRAIN, ALL ANALYSES	4-05-211	89	13.1	0.61	0.22	0.59	0.30	0.49	0.90	0.39	0.18	0.61	0.63	0.40	0.15	0.37	0.61
58	GRAIN, HARD RED SPRING	4-05-258	88	14.2	0.64	0.22	0.60	0.27	0.52	0.91	0.38	0.20	0.66	0.61	0.39	0.15	0.45	0.61
59	GRAIN, HARD RED WINTER	4-05-268	89	12.8	0.65	0.30	0.58	0.30	0.53	0.87	0.36	0.22	0.63	0.59	0.37	0.17	0.46	0.58
60	GRAIN, SOFT RED WINTER	4-05-294	88	11.4	0.65	0.36	0.55	0.32	0.45	0.90	0.36	0.22	0.64	0.65	0.39	0.27	0.37	0.58
61	GRAIN, SOFT WHITE WINTER	4-05-337	90	10.2	0.47	0.27	0.50	0.22	0.42	0.66	0.32	0.16	0.46	0.46	0.32	0.13	0.37	0.45
62	GRAIN, SOFT WHITE WINTER, PACIFIC COAST	4-08-555	89	10.0	0.45	0.24	0.50	0.20	0.40	0.75	0.30	0.14	0.48	0.49	0.31	0.12	0.36	0.46
63	BRAN	4-05-190	89	15.5	0.85	0.26	0.77	0.33	0.55	0.89	0.54	0.17	0.50	0.68	0.40	0.25	0.38	0.67
64	DISTILLERS GRAINS, DEHY	5-05-193	93	31.6	1.10	–	–	0.80	2.01	1.71	0.70	–	1.71	–	0.90	–	0.50	1.71
65	ENDOSPERM	4-05-197	88	11.1	0.60	0.30	–	0.30	1.10	1.70	0.40	0.20	0.60	–	0.40	0.30	–	0.60
66	FLOUR, LESS THAN 1.5% FIBER	4-05-199	88	13.7	0.42	0.25	0.46	0.28	0.56	0.89	0.27	0.13	0.62	0.51	0.30	0.10	0.25	0.51
67	GRAIN SCREENINGS	4-05-216	89	13.3	0.68	0.12	0.53	0.30	0.45	0.74	0.43	0.26	0.49	0.40	0.33	0.13	0.23	0.55
68	MIDDLINGS, LESS THAN 9.5% FIBER	4-05-205	89	16.4	0.98	0.22	0.96	0.40	0.68	1.11	0.68	0.19	0.66	0.80	0.57	0.19	0.43	0.80
69	MILL RUN, LESS THAN 9.5% FIBER	4-05-206	90	15.1	0.94	0.23	0.53	0.40	0.70	1.20	0.57	0.33	–	–	0.50	0.21	0.50	0.80
70	RED DOG, LESS THAN 4% FIBER	4-05-203	88	15.6	0.96	0.36	0.74	0.38	0.58	1.08	0.60	0.22	0.65	0.76	0.50	0.20	0.46	0.73
71	SHORTS, LESS THAN 7% FIBER	4-05-201	88	16.5	1.20	0.38	0.96	0.44	0.57	1.07	0.80	0.28	0.67	0.77	0.60	0.23	0.47	0.82
72	**WHEAT, DURUM,** GRAIN	4-05-224	88	13.8	0.58	0.13	0.46	0.27	0.48	1.40	1.05	0.14	0.53	0.45	0.37	0.26	0.29	0.54
	PROTEIN FEEDS																	
73	**ACACIA, SWEET,** SEEDS	5-09-110	87	47.9	4.40	–	1.63	1.10	1.67	3.58	2.25	0.44	1.67	1.97	1.20	–	1.34	1.86
	ANIMAL																	
74	BLOOD, MEAL	5-00-380	91	80.5	3.23	1.25	3.45	3.93	0.85	10.07	6.43	0.94	5.56	3.95	3.59	1.01	1.94	6.56
75	BLOOD, SPRAY DEHY	5-00-381	93	86.0	3.59	1.03	3.83	5.18	0.91	10.97	7.44	1.05	5.89	3.53	3.63	1.05	2.26	7.52
76	LIVER, MEAL	5-00-389	93	66.1	4.04	0.94	5.61	1.48	3.11	5.31	5.22	1.22	2.92	2.50	2.50	0.69	1.70	4.15
77	MEAT, MEAL RENDERED	5-00-385	94	50.7	3.58	0.60	7.23	0.87	1.63	3.11	3.00	0.69	1.74	2.31	1.67	0.35	1.09	2.42
78	MEAT WITH BONE, MEAL RENDERED	5-00-388	93	50.4	3.53	0.53	6.49	1.01	1.64	3.10	2.93	0.67	1.71	1.90	1.66	0.31	0.89	2.44
79	TANKAGE, MEAL RENDERED	5-00-386	92	60.5	3.60	0.48	6.45	2.06	1.82	5.10	3.89	0.75	2.56	2.81	2.34	0.65	1.38	3.83
80	TANKAGE WITH BONE, MEAL RENDERED	5-00-387	93	46.6	2.82	0.27	6.58	1.76	1.87	5.26	3.32	0.69	2.28	–	2.18	0.62	–	3.42
	BABASSU																	
81	KERNELS WITH COATS, MEAL MECH EXTD (BABASSU OIL MEAL)	5-00-454	92	22.3	2.87	–	–	0.40	1.04	1.34	0.89	0.30	0.89	–	0.60	0.20	0.40	1.09
82	KERNELS WITH COATS, MEAL SOLV EXTD (BABASSU OIL MEAL)	5-00-455	93	21.2	3.19	–	–	0.41	0.88	1.40	0.98	0.53	1.35	–	0.71	0.24	–	1.19
83	**BEAN, PINTO,** SEEDS	5-00-624	90	22.7	1.55	–	–	0.64	1.14	1.11	1.60	0.26	1.20	–	1.09	0.32	–	1.23
84	**BLOOD,** MEAL	5-00-380	91	80.5	3.23	1.25	3.45	3.93	0.85	10.07	6.43	0.94	5.56	3.95	3.59	1.01	1.94	6.56
85	**BUTTERMILK (CATTLE),** DEHY	5-01-160	92	31.7	1.08	0.39	0.47	0.85	2.42	3.21	2.28	0.71	1.46	1.50	1.52	0.49	1.00	2.58
86	**CASEIN,** ACID PRECIPITATED, DEHY	5-01-162	91	84.0	3.49	0.31	1.61	2.59	5.72	8.80	7.14	2.81	4.81	5.46	3.91	1.08	4.90	6.71
87	**CASTOR BEAN,** SEEDS WITHOUT TOXIN, MEAL	5-01-155	87	26.0	2.77	–	0.86	0.48	1.01	1.40	0.76	0.36	0.82	1.13	0.72	–	0.68	1.27
	CATTLE																	
88	BUTTERMILK, DEHY	5-01-160	92	31.7	1.08	0.39	0.47	0.85	2.42	3.21	2.28	0.71	1.46	1.50	1.52	0.49	1.00	2.58
89	MILK, FRESH	5-01-168	12	3.3	–	–	–	–	0.32	0.25	0.28	0.18	0.07	–	0.16	0.05	–	0.25
90	MILK, DEHY	5-01-167	95	25.3	0.92	–	–	0.71	1.32	2.54	2.24	0.61	1.32	–	1.02	0.41	1.32	1.73
91	SKIM MILK, DEHY	5-01-175	94	33.3	1.16	0.45	0.29	0.86	2.18	3.33	2.54	0.90	1.57	1.67	1.57	0.43	1.14	2.29
92	WHEY, DEHY	4-01-182	93	13.3	0.33	0.30	0.44	0.17	0.78	1.18	0.94	0.19	0.35	0.47	0.90	0.20	0.25	0.67
93	WHEY, LOW LACTOSE, DEHY	4-01-186	93	16.7	0.60	0.43	0.72	0.27	0.96	1.54	1.40	0.41	0.55	0.59	0.95	0.27	0.46	0.87
94	**CHICKPEA (GARBANZO; GRAM PEA),** SEEDS	5-01-218	89	19.1	1.52	–	0.69	0.40	0.76	1.32	1.25	0.24	1.14	0.90	0.61	–	0.57	0.80
95	**COCONUT,** KERNELS WITH COATS, MEAL MECH EXTD, (COPRA MEAL)	5-01-572	92	21.2	2.30	0.21	1.05	0.33	0.90	1.35	0.55	0.31	0.81	–	0.60	0.20	0.58	0.98

(Continued)

TABLE 19–2 AMINO ACID, COMPOSITION OF FEEDS, DATA EXPRESSED AS-FED—*(Continued)*

Entry Number	Feed Name Description	International Feed Number	Dry Matter	Crude Protein	Arginine	Cystine	Glycine	Histidine	Isoleucine	Leucine	Lysine	Methionine	Phenylalanine	Serine	Threonine	Tryptophan	Tyrosine	Valine
			%	%	%	%	%	%	%	%	%	%	%	%	%	%	%	%
	CORN																	
96	DISTILLERS GRAINS, DEHY	5-02-842	93	27.8	0.99	0.23	0.75	0.62	1.00	3.01	0.76	0.42	0.99	1.01	0.56	0.20	0.84	1.21
97	DISTILLERS SOLUBLES, DEHY	5-02-844	93	27.4	0.99	0.44	1.12	0.67	1.32	2.38	0.92	0.55	1.47	1.22	1.01	0.25	0.88	1.53
98	GLUTEN FEED	5-02-903	90	23.0	0.78	0.42	0.85	0.61	0.88	2.20	0.64	0.37	0.81	0.85	0.78	0.15	0.72	1.10
99	GLUTEN MEAL	5-02-900	91	43.2	1.40	0.67	1.51	0.97	2.25	7.38	0.80	1.03	2.85	1.70	1.43	0.21	1.01	2.23
	COTTON																	
100	SEEDS, MEAL MECH EXTD, 36% PROTEIN	5-01-625	92	37.2	3.55	0.79	1.83	0.91	1.32	—	1.22	0.55	1.88	—	1.12	0.46	—	2.84
101	SEEDS, MEAL MECH EXTD, 41% PROTEIN	5-01-617	93	41.0	4.20	0.71	1.87	1.07	1.42	2.30	1.60	0.57	2.19	1.70	1.33	0.52	0.97	1.89
102	SEEDS, MEAL, PREPRESSED, SOLV EXTD, 41% PROTEIN	5-07-872	90	41.3	4.32	0.78	1.89	1.14	1.42	2.42	1.80	0.56	2.05	1.80	1.34	0.50	1.14	1.97
103	SEEDS, MEAL, SOLV EXTD, 41% PROTEIN	5-01-621	91	41.2	4.24	0.76	1.95	1.10	1.50	2.46	1.69	0.58	2.23	1.76	1.37	0.55	1.04	1.97
104	SEEDS WITHOUT HULLS, MEAL, PREPRESSED, SOLV EXTD, 50% PROTEIN	5-07-874	93	50.3	4.83	1.05	2.82	1.21	1.86	2.82	1.93	0.76	2.62	—	1.66	0.62	0.81	2.16
105	**CRAB,** CANNERY RESIDUE, MEAL (CRAB MEAL)	5-01-663	92	32.2	1.66	0.24	1.74	0.48	1.16	1.54	1.38	0.52	1.16	1.38	1.00	0.29	1.17	1.47
106	**CRAMBE, ABYSSINIAN,** SEEDS WITHOUT HULLS, MEAL MECH EXTD	5-16-453	92	45.8	—	—	—	—	—	—	—	—	—	—	—	—	—	—
	DISTILLERS PRODUCTS (ALSO SEE CORN; RYE)																	
107	GRAINS, DEHY	5-02-144	93	27.3	1.04	0.42	0.56	0.53	1.16	2.66	0.81	0.46	1.03	0.70	0.81	0.21	0.73	1.22
108	SOLUBLES, DEHY	5-02-147	92	28.8	1.06	0.40	1.20	0.66	1.21	2.35	0.95	0.50	1.24	0.93	1.00	0.24	0.93	1.40
	FISH																	
109	MEAL MECH EXTD	5-01-977	92	64.3	31.28	0.62	3.99	1.46	3.27	4.90	5.26	1.63	2.60	2.42	2.59	0.75	1.79	3.14
110	SOLUBLES, CONDENSED	5-01-969	50	31.5	1.63	0.39	3.87	1.54	1.09	1.94	1.85	0.70	1.07	1.05	0.90	0.33	0.50	1.26
111	**FISH, ANCHOVY,** MEAL MECH EXTD	5-01-985	92	65.4	3.78	0.60	3.69	1.60	3.11	4.99	5.02	1.99	2.78	2.42	2.76	0.75	2.24	3.50
112	**FISH, MENHADEN,** MEAL MECH EXTD	5-02-009	92	61.2	3.74	0.58	4.19	1.44	2.85	4.48	4.74	1.75	2.46	2.25	2.51	0.65	1.93	3.19
113	**FISH, TUNA,** MEAL MECH EXTD	5-02-023	93	59.0	3.43	0.47	4.09	1.75	2.45	3.79	4.06	1.47	2.15	2.08	2.31	0.57	1.69	2.77
114	**FISH, WHITE,** MEAL MECH EXTD	5-02-025	91	62.6	4.26	0.77	5.15	1.38	2.85	4.65	4.70	1.79	2.44	3.44	2.56	0.67	2.27	3.25
	FLAX, COMMON																	
115	SEEDS, MEAL MECH EXTD (LINSEED MEAL)	5-02-048	90	34.6	2.94	0.61	1.74	0.69	1.68	2.02	1.16	0.54	1.46	1.93	1.22	0.51	1.09	1.74
116	SEEDS, MEAL SOLV EXTD (LINSEED MEAL)	5-02-045	91	34.3	2.81	0.61	1.64	0.65	1.69	1.92	1.18	0.58	1.38	1.90	1.14	0.51	0.96	1.61
117	**GARBANZO (CHICKPEA; GRAM PEA),** SEEDS	5-01-218	89	19.1	1.52	—	0.69	0.40	0.76	1.32	1.25	0.24	1.14	0.90	0.61	—	0.57	0.80
118	**HORSE BEAN,** SEEDS	5-02-407	88	25.5	—	—	—	—	—	—	—	—	—	—	—	—	—	—
119	**LIVER,** MEAL	5-00-389	93	66.1	4.04	0.94	5.61	1.48	3.11	5.31	5.22	1.22	2.92	2.50	2.50	0.69	1.70	4.15
120	**LOCUST, NEW MEXICO,** SEEDS	5-09-055	89	36.5	3.01	—	1.32	0.73	0.87	1.75	1.32	0.26	1.02	1.28	0.87	—	0.84	1.17
	MAIZE–SEE CORN																	
	MEAT																	
121	MEAL RENDERED	5-00-385	94	50.7	3.58	0.60	7.23	0.87	1.63	3.11	3.00	0.69	1.74	2.31	1.67	0.35	1.09	2.42
122	WITH BLOOD, MEAL RENDERED (TANKAGE)	5-00-386	92	60.5	3.60	0.48	6.45	2.06	1.82	5.10	3.89	0.75	2.56	2.81	2.34	0.65	1.38	3.83
123	WITH BLOOD, WITH BONE, MEAL RENDERED (TANKAGE)	5-00-387	93	46.6	2.82	0.27	6.58	1.76	1.87	5.26	3.32	0.69	2.28	—	2.18	0.62	—	3.42
124	WITH BONE, MEAL RENDERED	5-00-388	93	50.4	3.53	0.53	6.49	1.01	1.64	3.10	2.93	0.67	1.71	1.90	1.66	0.31	0.89	2.44
	MILK																	
125	FRESH (CATTLE)	5-01-168	12	3.3	—	—	—	—	0.32	0.25	0.28	0.18	0.07	—	0.16	0.05	—	0.25
126	DEHY (CATTLE)	5-01-167	95	25.3	0.92	—	—	0.71	1.32	2.54	2.24	0.61	1.32	—	1.02	0.41	1.32	1.73
127	SKIMMED, DEHY (CATTLE)	5-01-175	94	33.3	1.16	0.45	0.29	0.86	2.18	3.33	2.54	0.90	1.57	1.67	1.57	0.43	1.14	2.29
128	FRESH (HORSE)	5-02-401	17	4.2	—	—	—	0.11	0.25	0.34	0.25	0.07	0.18	—	0.16	0.05	—	0.29
129	FRESH (SHEEP)	5-08-510	19	4.6	—	—	—	0.18	0.39	0.60	0.51	0.17	0.32	—	0.30	0.09	—	0.48
130	FRESH (SWINE)	5-08-537	20	7.3	—	—	—	0.20	0.42	0.59	0.50	0.14	0.34	—	0.37	0.09	—	0.45
131	**MILKWEED, COMMON,** SEEDS	5-09-137	86	31.8	3.05	—	1.65	0.73	1.12	1.97	1.56	0.45	1.53	1.31	0.86	—	1.08	1.37
132	**PALM,** KERNELS WITH COATS, MEAL SOLV EXTD	5-03-486	90	18.2	2.52	—	—	0.31	0.76	1.22	0.66	0.41	0.81	—	0.60	0.20	—	1.02
133	**PEA,** SEEDS	5-03-600	89	23.2	1.40	0.21	1.09	0.60	1.20	1.81	1.53	0.27	1.25	—	0.94	0.21	—	1.25
134	**PEA, FIELD,** SEEDS	5-08-481	91	23.2	1.86	0.26	1.05	0.51	0.91	1.59	1.44	0.23	1.00	1.05	0.82	0.22	0.77	1.00
	PEANUT																	
135	PODS WITH SEEDS, MEAL SOLV EXTD	5-03-656	92	47.4	5.19	0.70	2.39	1.10	1.92	3.20	1.75	0.43	2.49	3.05	1.38	0.49	1.68	2.48
136	SEEDS WITHOUT HULLS, MEAL MECH EXTD (PEANUT MEAL)	5-03-649	93	49.2	5.08	0.96	2.49	1.03	1.78	3.13	1.69	0.50	2.38	1.44	1.27	0.46	1.59	2.29
137	SEEDS WITHOUT HULLS, MEAL SOLV EXTD (PEANUT MEAL)	5-03-650	93	49.0	5.82	0.54	2.88	1.46	1.84	3.27	1.45	0.44	2.12	3.12	1.37	0.48	—	2.16
	POULTRY																	
138	BY-PRODUCT, MEAL RENDERED	5-03-798	94	61.2	4.01	0.85	6.09	1.13	2.35	4.10	3.12	1.14	2.04	2.88	2.10	0.47	1.84	2.94
139	FEATHERS, HYDROLYZED, MEAL	5-03-795	93	83.8	5.33	3.21	6.32	0.47	3.51	6.42	1.55	0.54	3.59	9.16	3.63	0.52	2.35	5.85
140	**RAPE (CANOLA), SUMMER,** SEEDS, MEAL, PREPRESSED, SOLV EXTD	5-08-135	92	40.5	2.23	—	1.94	1.09	1.46	2.71	2.15	0.77	1.54	1.70	1.70	0.49	0.85	1.94

TABLE 19–2 AMINO ACID, COMPOSITION OF FEEDS, DATA EXPRESSED AS-FED—*(Continued)*

Entry Number	Feed Name Description	International Feed Number	Dry Matter	Crude Protein	Arginine	Cystine	Glycine	Histidine	Isoleucine	Leucine	Lysine	Methionine	Phenylalanine	Serine	Threonine	Tryptophan	Tyrosine	Valine
			%	%	%	%	%	%	%	%	%	%	%	%	%	%	%	%
	RYE																	
141	DISTILLERS GRAINS, DEHY	5-04-023	92	23.0	—	—	—	—	—	—	—	—	—	—	—	—	—	—
142	DISTILLERS GRAINS WITH SOLUBLES, DEHY	5-04-024	90	27.2	1.00	—	—	0.70	1.50	2.10	1.00	0.40	1.30	1.20	1.10	0.30	0.50	1.60
	SAFFLOWER																	
143	SEEDS WITHOUT HULLS, MEAL MECH EXTD	5-08-499	91	42.0	5.44	—	2.52	—	—	—	1.31	0.71	—	—	0.81	—	—	—
144	SEEDS WITHOUT HULLS, MEAL SOLV EXTD	5-07-959	91	42.8	3.67	0.71	2.36	0.97	1.58	2.42	1.26	0.67	1.73	—	1.30	0.59	1.01	2.17
145	**SESAME,** SEEDS, MEAL MECH EXTD	5-04-220	93	45.0	4.55	0.59	3.96	1.07	1.96	3.20	1.26	1.37	2.14	2.94	1.60	0.71	1.87	2.32
146	**SHRIMP,** CANNERY RESIDUE, MEAL (SHRIMP MEAL)	5-04-226	90	38.7	2.33	0.47	1.31	0.87	1.51	2.37	2.05	0.84	1.55	1.25	1.26	0.36	1.10	1.71
147	**SORGHUM,** GLUTEN MEAL	5-04-388	90	44.4	1.26	0.73	0.95	1.07	2.39	7.85	0.74	0.71	2.70	—	1.45	0.44	—	2.50
	SOYBEAN																	
148	FLOUR, SOLV EXTD	5-04-593	93	51.6	4.27	0.64	1.65	1.26	1.90	3.33	4.48	0.57	2.00	2.09	1.58	0.79	1.44	1.86
149	MEAL, SOLV EXTD	5-04-612	90	49.7	3.67	0.70	2.27	1.20	2.13	3.63	3.12	0.71	2.36	2.49	1.90	0.69	1.71	2.47
150	MEAL, SOLV EXTD, 44% PROTEIN	5-20-637	89	44.4	3.26	0.67	2.10	1.13	2.12	3.49	2.85	0.59	2.23	2.37	1.81	0.62	1.60	2.37
151	MEAL, SOLV EXTD, 49% PROTEIN	5-20-638	90	49.0	3.62	0.75	2.39	1.28	2.34	3.77	3.08	0.66	2.47	2.76	2.00	0.70	1.96	2.49
152	**WHALE,** MEAT, MEAL RENDERED	5-05-160	91	71.4	2.49	0.63	6.31	1.19	2.72	4.27	3.48	1.01	2.06	—	1.63	0.82	—	2.81
	WHEAT																	
153	GERM MEAL	5-05-218	88	24.4	1.83	0.47	1.46	0.62	0.95	1.47	1.53	0.41	0.93	1.12	0.94	0.30	0.74	1.16
154	GLUTEN	5-05-221	90	63.4	2.97	1.74	2.77	1.64	3.39	5.54	1.54	1.23	4.21	4.10	2.15	0.72	2.36	3.90
	WHEY–SEE CATTLE																	
155	**YEAST, IRRADIATED,** DEHY	7-05-529	94	48.1	2.46	—	—	1.00	2.94	3.56	3.70	1.00	2.77	—	2.41	0.73	—	3.06
156	**YEAST, TORULA,** DEHY	7-05-534	93	49.6	2.52	0.59	2.54	1.34	2.69	3.39	3.65	0.76	2.63	2.75	2.67	0.52	1.94	2.88
	DRY FORAGES																	
	ALFALFA (LUCERNE)																	
1157	HAY, SUN-CURED	1-00-078	90	16.0	0.81	—	—	0.28	0.87	1.12	1.00	0.12	0.71	—	0.62	0.18	0.50	0.69
158	HAY, SUN-CURED, EARLY BLOOM, MEAL	1-00-108	92	22.5	—	—	—	—	—	—	—	—	—	—	—	—	—	—
159	LEAVES, SUN-CURED, MEAL	1-00-146	88	20.1	—	—	—	—	—	—	—	—	—	—	—	—	—	—
160	LEAVES, MEAL, DEHY	1-00-137	92	20.6	0.97	0.27	0.99	0.39	0.92	1.45	0.95	0.32	0.94	0.89	0.86	0.43	0.60	1.05
161	MEAL, DEHY, 15% PROTEIN	1-00-022	90	15.6	0.59	0.17	0.70	0.27	0.64	1.02	0.59	0.22	0.62	0.60	0.56	0.38	0.41	0.75
162	MEAL, DEHY, 17% PROTEIN	1-00-023	92	17.4	0.77	0.29	0.84	0.33	0.81	1.28	0.85	0.27	0.80	0.71	0.71	0.34	0.54	0.88
163	MEAL, DEHY, 20% PROTEIN	1-00-024	92	20.2	0.95	0.32	0.99	0.38	0.89	1.43	0.89	0.32	0.94	0.90	0.82	0.41	0.60	1.05
164	MEAL, DEHY, 22% PROTEIN	1-07-851	93	22.2	0.96	0.30	1.09	0.44	1.06	1.63	0.97	0.34	1.13	0.97	0.97	0.49	0.64	1.29
165	**ALFALFA-GRASS,** HAY, SUN-CURED	1-08-331	91	14.5	—	—	—	—	—	—	—	—	—	—	—	—	—	—
166	**BEET, SUGAR,** LEAVES, SUN-CURED	1-00-641	91	23.2	1.00	—	—	0.27	1.00	1.55	1.27	0.36	0.91	—	0.91	0.27	—	1.18
167	**CLOVER, LADINO,** HAY, SUN-CURED	1-01-378	89	20.0	—	—	—	—	—	—	—	—	—	—	—	—	—	—
168	**COWPEA, COMMON,** HAY, SUN-CURED	1-01-645	90	17.7	1.11	—	—	0.45	1.27	2.01	1.08	0.51	1.26	—	1.06	0.52	—	1.44
169	**LESPEDEZA, COMMON,** HAY, SUN-CURED	1-08-591	89	13.8	—	—	—	—	—	—	—	—	—	—	—	—	—	—
170	**OATS,** HULLS	1-03-281	92	3.7	0.15	0.06	0.15	0.08	0.15	0.25	0.17	0.08	0.15	—	0.16	0.09	0.14	0.19
171	**SOYBEAN,** HAY, SUN-CURED	1-04-558	89	14.1	—	—	—	—	—	—	—	—	—	—	—	—	—	—
172	**VETCH,** HAY, SUN-CURED	1-05-106	89	18.4	—	—	—	—	—	—	—	—	—	—	—	—	—	—
	PASTURE AND RANGE PLANTS																	
173	**COWPEA, COMMON,** FRESH	2-01-655	25	4.0	—	—	—	—	—	—	—	—	—	—	—	—	—	—
174	**SPINACH,** LEAVES, FRESH	2-08-125	9	3.1	0.11	—	—	0.04	0.09	0.18	0.12	0.06	0.12	—	0.10	0.03	—	0.13
	VITAMIN SUPPLEMENTS																	
	ALFALFA (LUCERNE)																	
175	MEAL, DEHY, 20% PROTEIN	1-00-024	92	20.2	0.95	0.32	0.99	0.38	0.89	1.43	0.89	0.32	0.94	[illegible]	[illegible]	[illegible]	[illegible]	[illegible]
176	MEAL, DEHY, 22% PROTEIN	1-07-851	93	22.2	0.96	0.30	1.09	0.44	1.06	1.63	[illegible]	[illegible]	[illegible]	[illegible]	[illegible]	[illegible]	[illegible]	[illegible]
177	**BREWERS GRAINS,** DEHY	5-02-141	92	27.3	1.27	0.35	1.09	[illegible]	[illegible]	[illegible]	[illegible]	[illegible]	[illegible]	[illegible]	[illegible]	[illegible]	[illegible]	[illegible]
178	**CORN,** DISTILLERS GRAINS WITH SOLUBLES, DEHY	5-02-843	92	27.1	[illegible]	[illegible]	[illegible]	[illegible]	[illegible]	[illegible]	[illegible]	[illegible]	[illegible]	[illegible]	[illegible]	[illegible]	[illegible]	[illegible]
179	**FISH,** SOLUBLES, DEHY	[illegible]	[illegible]	[illegible]	[illegible]	[illegible]	[illegible]	[illegible]	[illegible]	[illegible]	[illegible]	[illegible]	[illegible]	[illegible]	[illegible]	[illegible]	[illegible]	[illegible]	[illegible]	[illegible]
	FISH, SARDINE																	
180	MEAL [illegible]	[illegible]	[illegible]	[illegible]	[illegible]	[illegible]	[illegible]	[illegible]	[illegible]	[illegible]	[illegible]	[illegible]	[illegible]	[illegible]	[illegible]	[illegible]	[illegible]	[illegible]	[illegible]

DIGESTIBLE AMINO ACID RECOMMENDATIONS FOR SWINE FEED FORMULATION[1]

Leucine			Lysine			Methionine			Phenylalanine			Threonine			Tryptophan			Valine		
Total %	Coefficient %	Digestible %	Total %	Coefficient %	Digestible %	Total %	Coefficient %	Digestible %	Total %	Coefficient %	Digestible %	Total %	Coefficient %	Digestible %	Total %	Coefficient %	Digestible %	Total %	Coefficient %	Digestible %
1.10	60	0.66	0.69	48	0.33	0.21	64	0.13	0.68	59	0.40	0.62	49	0.30	0.16	—	—	0.83	52	0.43
0.71	75	0.53	0.39	67	0.26	0.18	79	0.14	0.52	78	0.41	0.38	63	0.24	0.12	73	0.09	0.53	71	0.38
0.95	85	0.81	0.54	89	0.48	0.25	87	0.22	0.76	90	0.68	0.47	87	0.41	0.17	—	—	0.62	87	0.54
11.89	93	11.06	9.05	93	8.42	—	—	—	6.65	92	6.12	4.68	87	4.07	0.99	89	0.88	8.61	92	7.92
1.20	74	0.89	1.05	72	0.76	0.23	78	0.18	0.71	75	0.53	0.67	68	0.46	—	50	—	1.00	73	0.73
2.23	—	—	1.05	68	0.71	—	—	—	1.49	—	—	0.99	67	0.66	0.28	71	0.20	1.53	—	—
2.77	78	2.16	2.19	73	1.60	0.98	82	0.80	1.57	77	1.21	1.78	68	1.21	0.43	71	0.31	2.03	67	1.36
7.81	94	7.34	6.09	95	5.79	2.61	96	2.51	4.34	95	4.12	3.22	86	2.77	0.98	91	0.89	5.62	92	5.17
1.20	70	0.84	1.98	53	1.05	0.38	—	—	0.89	76	0.68	0.68	49	0.33	0.22	—	—	0.99	70	0.69
1.12	88	0.99	0.27	68	0.18	0.18	86	0.15	0.43	84	0.36	0.32	71	0.23	0.06	72	0.04	0.44	78	0.34
2.01	77	1.55	0.67	48	0.32	0.37	—	—	0.75	75	0.56	0.74	50	0.37	0.07	33	0.02	1.00	68	0.68
10.71	90	9.64	1.10	74	0.81	1.42	86	1.22	3.76	88	3.31	2.10	80	1.68	0.27	72	0.19	2.85	81	2.31
2.29	70	1.60	1.68	59	0.99	0.64	72	0.46	2.16	80	1.73	1.28	61	0.78	0.47	72	0.34	2.57	71	1.82
2.52	84	2.12	2.04	82	1.67	0.77	84	0.65	2.62	91	2.38	1.47	78	1.15	0.57	81	0.46	2.83	85	2.41
3.21	95	3.05	2.79	95	2.65	0.78	96	0.75	1.72	96	1.65	1.49	89	1.33	0.56	—	—	2.15	88	1.89
6.63	77	5.11	1.59	51	0.81	0.39	71	0.28	4.29	81	3.47	3.49	74	2.58	0.42	60	0.25	6.75	78	5.27
4.51	87	3.92	4.69	87	4.08	1.87	91	1.70	2.43	84	2.04	2.71	81	2.20	0.55	74	0.41	2.95	83	2.45
2.46	87	2.14	1.46	82	1.20	0.38	84	0.32	1.88	89	1.67	1.04	77	0.80	—	—	—	1.58	83	1.31
2.35	83	1.95	1.54	79	1.22	0.21	65	0.14	1.30	82	1.07	1.19	77	0.92	—	—	—	1.41	79	1.11
			78.80	100	78.80															
3.32	71	2.36	2.75	70	1.93	0.75	77	0.58	1.79	72	1.29	1.75	63	1.10	0.26	54	0.14	2.51	70	1.76
0.75	—	—	0.40	70	0.28	0.17	79	0.13	0.48	—	—	0.34	55	0.19	0.10	—	—	0.52	—	—
1.04	85	0.88	0.64	82	0.52	0.23	89	0.20	0.68	90	0.61	0.53	78	0.41	0.17	81	0.14	0.78	85	0.66
2.65	78	2.07	1.55	72	1.12	—	—	—	2.12	85	1.80	1.16	68	0.79	0.35	68	0.24	1.75	76	1.33
1.50	75	1.13	1.54	82	1.26	0.23	74	0.17	0.98	75	0.74	0.79	75	0.59	—	—	—	0.96	71	0.68
4.61	80	3.69	3.57	81	2.89	1.07	—	—	2.37	82	1.94	2.53	71	1.80	0.43	79	0.34	3.39	77	2.61
2.49	79	1.97	2.00	73	1.46	0.71	85	0.60	1.46	80	1.17	1.53	68	1.04	—	—	—	1.91	70	1.34
0.67	86	0.58	0.33	81	0.27	0.19	87	0.17	0.43	88	0.38	0.31	84	0.26	0.11	—	—	0.48	87	0.42
0.59	71	0.42	0.36	65	0.23	0.15	77	0.12	0.44	78	0.34	0.32	59	0.19	—	—	—	0.45	69	0.31
5.40	89	4.81	5.58	91	5.08	2.01	—	—	2.77	87	2.41	3.06	80	2.45	0.91	—	—	3.54	86	3.04
2.75	82	2.26	1.06	71	0.75	1.32	—	—	1.92	85	1.63	1.38	66	0.91	0.65	—	—	1.97	80	1.58
1.47	91	1.34	0.24	75	0.18	0.18	87	0.16	0.65	92	0.60	0.40	78	0.31	0.09	80	0.07	0.53	86	0.46
4.23	81	3.43	3.36	88	2.96	0.79	91	0.72	2.66	87	2.31	2.03	76	1.54	0.64	79	0.51	4.12	81	3.34
2.89	73	2.11	2.39	80	1.91	0.57	78	0.44	1.86	80	1.49	1.44	69	0.99	0.47	70	0.33	2.36	73	1.72
3.40	83	2.82	2.81	86	2.42	0.67	85	0.57	2.22	85	1.89	1.74	76	1.32	0.53	80	0.42	2.83	81	2.29
3.61	83	3.00	3.05	85	2.59	0.67	87	0.58	2.39	85	2.03	1.89	77	1.46	0.58	78	0.45	2.45	80	1.96
4.04	37	1.49	3.35	44	1.47	0.79	47	0.37	2.69	45	1.21	2.04	32	0.65	0.57	25	0.14	4.13	35	1.45
4.15	80	3.32	3.35	85	2.85	0.79	82	0.65	2.67	84	2.24	2.03	72	1.46	0.56	77	0.43	4.15	78	3.24
2.11	77	1.62	1.30	74	0.96	0.78	87	0.68	1.56	80	1.25	1.26	71	0.89	0.42	76	0.32	1.71	75	1.28
0.67	84	0.56	0.40	73	0.29	0.16	84	0.13	0.58	83	0.48	0.34	65	0.22	0.14	—	—	0.47	83	0.39
0.83	83	0.69	0.33	71	0.23	0.20	84	0.17	0.55	86	0.47	0.36	69	0.25	0.15	81	0.12	0.53	77	0.41
0.88	76	0.67	0.52	72	0.37	0.21	79	0.17	0.57	79	0.45	0.44	66	0.29	—	—	—	0.65	72	0.47
0.92	95	0.87	0.23	84	0.19	0.16	94	0.15	0.70	96	0.67	0.33	85	0.28	—	—	—	0.56	93	0.52
1.06	72	0.76	0.68	72	0.49	0.23	79	0.18	0.68	82	0.56	0.55	60	0.33	0.19	72	0.14	0.78	73	0.57
0.95	74	0.70	0.54	66	0.36	0.23	78	0.18	0.60	76	0.46	0.41	54	0.22	—	—	—	0.71	72	0.51

FOR SWINE FEED FORMULATION

	Sows		Mature
	Gestation	Lactation	Boar
DIGESTIBLE MINIMUMS, %			
CRUDE PROTEIN	9.50	12.00	12.00
LYSINE/CALORIE (g/Mcal ME)	1.60	2.00	2.00
LYSINE[2]	0.51	0.67	0.67
THREONINE	0.33	0.44	0.44
TRYPTOPHAN	0.09	0.12	0.12
ISOLEUCINE	0.30	0.40	0.40
METHIONINE & CYSTINE	0.28	0.37	0.37
HISTIDINE	0.17	0.22	0.22
LEUCINE	0.54	0.73	0.73
PHENYLALANINE & TYROSINE	0.60	0.80	0.80
VALINE	0.37	0.49	0.49

[2]For each 1% added fat digestible lysine should increase by .02%

[1]Table V-3 data was assembled by, and is presented through the courtesy of, Heartland Lysine, Inc., 8430 West Bryn Mawr Avenue, Suite 650, Chicago, IL 60631.

PRINCIPAL LIVESTOCK FEEDS

Pasture. (Courtesy, University of Missouri, Columbia)

Hay. (Courtesy, Ford New Holland, New Holland, PA)

Silage. (Courtesy, Harvestore Systems, DeKalb, IL)

Corn. (Courtesy, USDA)

Soybeans. (Courtesy, American Soybean Association, St. Louis, MO)

Corn stalklage, a crop residue. (Courtesy, Koehring Farm Division, Appleton, WI)

SCALES~
developed by
the ancient
Egyptians
to weigh
grain

Weights and measures are one of the most important parts of modern agriculture

SECTION

20

WEIGHTS AND MEASURES

Contents *Page*

BUTLER
ENSILMIXER
CAUTION

Fig. 20-1. Scale equipped, self-unloading truck conveying a precise (weighed) amount of feed into a fenceline bunk.

Weights and measures are the standards employed in arriving at weights, quantities, and volumes. Even among primitive people, such standards were necessary; and with the growing complexity of life, they become of greater and greater importance.

Weights and measures form one of the most important parts of modern agriculture. This section contains pertinent information relative to the most common standards used in the U.S. livestock industry.

METRIC SYSTEM[1,2]

The United States and a few other countries use standards that belong to the *customary*, or English, system of measurement. This system evolved in England from older measurement standards, beginning about the year 1200. All other countries—including England—now use a system of measurements called the *metric system*, which was created in France in the 1790s. Increasingly, the metric system is being used in the United States. Hence, everyone should have a working knowledge of it.

The basic metric units are the *meter* (length/distance), the *gram* (weight), and the *liter* (capacity). The units are then expanded in multiples of 10 or made smaller by 1/10. The prefixes, which are used in the same way with all basic metric units, follow:

"milli-"	=	1/1000	"deca-"	=	10
"centi-"	=	1/100	"hecto-"	=	100
"deci-"	=	1/10	"kilo-"	=	1,000

The following tables will facilitate conversion from metric units to U.S. customary, and vice versa:

Table 20–1 Weights and Measures—
- Weight
- Length
- Surface/Area
- Volume

Table 20–2 Temperature

[1]For further information on the federal government's metric activities contact: U.S. Dept. of Commerce, Office of Metric Programs, Room 4845, Washington, DC 20230, (202) 377-3036.

[2]For additional conversion factors, or for greater accuracy, see *Misc. Pub. 223*, the National Bureau of Standards.

TABLE 20–1
WEIGHTS AND MEASURES

Weight

Unit	Is Equal To	
Metric system:	**(metric)**	**(U.S. customary)**
1 microgram (mcg)	.001 mg	
1 milligram (mg)	.001 g	.015432356 grain
1 centigram (cg)	.01 g	.15432356 grain
1 decigram (dg)	.1 g	1.5432 grains
1 gram (g)	1,000 mg	.03527396 oz
1 decagram (dkg)	10 g	5.643833 dr
1 hectogram (hg)	100 g	3.527396 oz
1 kilogram (kg)	1,000 g	35.274 oz; 2.2046223 lb
1 ton	1,000 kg	2,204.6 lb; 1.102 tons (short or 0.984 ton (long)
U.S. customary:	**(U.S. customary)**	**(metric)**
1 grain	.037 dr	64.798918 mg; .064798918 g
1 dram (dr)	.063 oz	1.771845 g
1 ounce (oz)	16 dr	28.349527 g
1 pound (lb)	16 oz	453.5924 g or 0.4536 kg
1 hundredweight (cwt)	100 lb	
1 ton (short)	2,000 lb	907.18486 kg or 0.907 (metric) ton
1 ton (long)	2,200 lb	1,016.05 kg or 1.016 (metric) ton
1 part per million (ppm)	1 microgram/gram; 1 mg/l; 1 mg/kg	.4535924 mg/lb; .907 g/ton; .0001%; .00013 oz/gal
1 percent (%) (1 part in 100 parts)	10,000 ppm; 10 g/l	1.28 oz/gal; 8.34 lb/100 gal

Weight Conversions

U.S. Customary to Metric			Metric to U.S. Customary		
To Change		**Multiply By**	**To Change**		**Multiply By**
grains	to milligrams	64.799			
ounces	to grams	28.35	grams	to ounces	0.035
pounds	to grams	453.6			
pounds	to kg	0.454	kg	to pounds	2.205
tons	to metric tons	0.9	metric tons	to tons	1.102

Weight–Unit Conversion Factors

To Change		Multiply By	To Change		Multiply By
mg/lb	to g/ton	2	mg/g	to mg/lb	453.6
g/lb	to g/ton	2,000	mg/kg	to mg/lb	0.4536
lb/ton	to g/ton	453.6	mcg/kg	to g/lb	0.4536
ppm	to mg/lb	0.4536	g/ton	to g/lb	0.0005
ppm	to %	move decimal 4 places to left	g/ton	to lb/ton	0.0022
mg/lb	to ppm	2.2046	g/ton	to %	0.00011
			%	to g/ton	9,072.00
ppm	to g/ton	0.907	g/ton	to ppm	1.1

(Continued)

TABLE 20–1 *(Continued)*

Length

Unit	Is Equal To	
Metric system:	**(metric)**	**(U.S. customary)**
1 millimicron (m)	.000000001 m	.000000039 in.
1 micron ( )	.000001 m	.000039 in.
1 millimeter (mm)	.001 m	.0394 in.
1 centimeter (cm)	.01 m	.3937 in.
1 decimeter (dm)	.1 m	3.937 in.
1 meter (m)	1 m	39.37 in.; 3.281 ft;1.094 yd
1 hectometer (hm)	100 m	328.08 ft; 19.8338 rd
1 kilometer (km)	1,000 m	3,280.8 ft; 0.621 mi
U.S. customary:	**(U.S. customary)**	**(metric)**
1 inch (in.)	1 in.	25 mm; 2.54 cm
1 hand*	4 in.	10.16 cm
1 foot (ft)	12 in.	30.48 cm; .305 m
1 yard (yd)	3 ft	.914 m
1 fathom** (fath)	6.08 ft	1.829 m
1 rod (rd), pole, or perch	16½ ft; 5½ yd	5.029 m
1 chain	792 in.; 66 ft; 22 yd	20.116 m
1 furlong (fur.)	220 yd; 40 rd	201.168 m
1 mile (mi)	5,280 ft; 1,760 yd; 320 rd; 8 fur.	1,609.35 m; 1.609 km
1 knot or nautical mile	6,080 ft; 1.15 land miles	1.85 km
1 league (land)	3 mi (land)	4.827 km
1 league (nautical)	3 mi (nautical)	4.827 km

Length Conversions

U.S. Customary to Metric			Metric to U.S. Customary		
To Change		**Multiply By**	**To Change**		**Multiply By**
inches	to millimeters	25.4	millimeters	to inches	0.04
inches	to centimeters	2.54	centimeters	to inches	0.4
feet	to centimeters	30.5	centimeters	to feet	0.033
feet	to meters	0.305	meters	to feet	3.3
yards	to meters	0.914	meters	to yards	1.1
miles	to kilometers	1.609	kilometers	to miles	0.6

*Used in measuring height of horses.

**Used in measuring depth at sea.

(Continued)

TABLE 20–1 *(Continued)*

Surface/Area

Unit	Is Equal To (metric)	Is Equal To (U.S. customary)
Metric system:	**(metric)**	**(U.S. customary)**
1 square millimeter (mm^2)	.000001 m^2	.00155 $in.^2$
1 square centimeter (cm^2)	.0001 m^2	.155 $in.^2$
1 square decimeter (dm^2)	.01 m^2	15.50 $in.^2$
1 square meter (m^2)	1 centare (ca)	1,550 $in.^2$; 10.76 ft^2; 1.196 yd^2
1 are (a)	100 m^2	119.6 yd^2
1 hectare (ha)	10,000 m^2	2.47 acres
1 square kilometer (km^2)	1,000,000 m^2	247.1 acres; .386 mi^2
U.S. customary:	**(U.S. customary)**	**(metric)**
1 square inch ($in.^2$)	1 in. × 1 in.	6.452 cm^2
1 square foot (ft^2)	144 $in.^2$; 0.111 yd^2	.093 m^2
1 square yard (yd^2)	1,296 $in.^2$; 9 ft^2	.836 m^2
1 square rod (rd^2)	272.25 ft^2; 30.25 yd^2	25.29 m^2
1 rood	40 rd^2	10.117 a
1 acre	43,560 ft^2; 4,840 yd^2; 160 rd^2; 4 roods	4,046.87 m^2; 0.405 ha
1 square mile (mi^2)	640 acres; 1 section	2.59 km^2 or 259 ha
1 township	36 sections; 6 miles square	

Surface/Area Conversions

U.S. Customary to Metric			Metric to U.S. Customary		
To Change		**Multiply By**	**To Change**		**Multiply By**
sq in.	to cm^2	6.452	cm^2	to sq in.	0.155
sq ft	to cm^2	929.1	cm^2	to sq ft	0.001
sq ft	to m^2	0.09	m^2	to sq ft	10.764
sq yd	to m^2	0.836	m^2	to sq yd	1.196
sq mi	to km^2	2.6	km^2	to sq mi	0.4
acres	to ha	0.4	ha	to acres	2.5

Weights/Measures/Unit Area

Unit	Is Equal To
Volume per unit area:	
1 liter/hectare	0.107 gal/acre
1 gal/acre	9.354 liter/ha
Weight per unit area:	
1 kilogram/cm^2	14.22 lb/in^2
1 kilogram/hectare	0.892 lb/acre
1 lb/sq in.	0.0703 kg/cm^2
1 lb/acre	1.121 kg/ha
Area per unit weight:	
1 square centimeter/kilogram	0.0703 $in.^2$/lb
1 sq in./lb	14.22 cm^2/kg

(Continued)

TABLE 20-1 *(Continued)*

Volume

Unit	Is Equal To (U.S. customary) (liquid)	(U.S. customary) (dry)
Metric system liquid and dry:		
1 milliliter (ml) .001 liter	.271 dram (fl)	.061 in.^{3}
1 centiliter (cl) .01 liter	.338 oz (fl)	.610 in.^{3}
1 deciliter (dl) .1 liter	3.38 oz (fl)	
1 liter 1,000 cc	1.057 qt or 0.2642 gal (fl)	.908 qt
1 hectoliter (hl) 100 liters	26.418 gal	2.838 bu
1 kiloliter (kl) 1,000 liters	264.18 gal	1,308 yd^3

Unit	(ounces)	(cubic inches)	(metric)
U.S. customary liquid:			
1 teaspoon (t) 60 drops	1/8		5 ml
1 dessert spoon 2 t			
1 tablespoon (T) 3 t	1/2		15 ml
1 fl oz	1	1.805	29.57 ml
1 gill (gi) 1/2 c	4	7.22	118.29 ml
1 cup (c) 16 T	8	14.44	236.58 ml or 0.24 liter
1 pint (pt) 2 c	16	28.88	.47 liter
1 quart (qt) 2 pt	32	57.75	.95 liter
1 gallon (gal) 4 qt	8.34 lb	231	3.79 liters
1 barrel (bbl) 31 1/2 gal			
1 hogshead (hhd) 2 bbl			
Dry:			
1 pint (pt) 1/2 qt		33.6	.55 liter
1 quart (qt) 2 pt		67.20	1.10 liters
1 peck (pk) 8 qt		537.61	8.81 liters
1 bushel (bu) 4 pk		2,150.42	35.24 liters

Unit	Is Equal To (metric)	(U.S. customary)
Solid metric system:		
1 cubic millimeter (mm^3)	.001 cc	
1 cubic centimeter (cc)	1,000 mm^3	.061 cu in.
1 cubic decimeter (dm^3)	1,000 cc	61.023 cu in.
1 cubic meter (m^3)	1,000 dm^3	35.315 ft^3; 1.308 yd^3
U.S. customary:	**(U.S.customary)**	**(metric)**
1 cubic inch (in.^{3})		16.387 cc
1 board foot (fbm)	144 in.^{3}	2,359.8 cc
1 cubic foot (ft^3)	1,728 in.^{3}	.028 m^3
1 cubic yard (yd^3)	27 ft^3	.765 m^3
1 cord	128 ft^3	3.625 m^3

Volume Conversions

U.S. Customary to Metric			Metric to U.S. Customary		
To Change		**Multiply By**	**To Change**		**Multiply By**
ounces (fluid)	to cc	29.57	cc	to oz (fluid)	0.034
ounces	to ml	29.57	ml	to oz	0.034
qt	to liters	0.946	liters	to qt	1.057
cu in.	to cc	16.387	cc	to cu in.	0.061
cu yd	to cm	0.765	cm	to cu yd	1.308

TABLE 20-2
TEMPERATURE

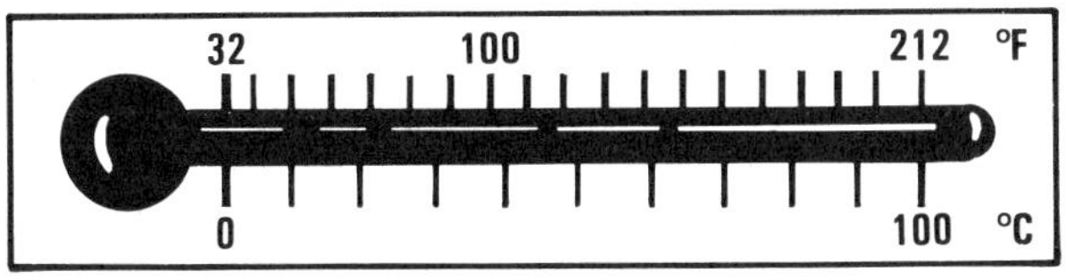

Fig. 20-2. Fahrenheit-Centigrade scale for direct conversion and reading.

One Fahrenheit (F) degree is 1/180 of the difference between the temperature of melting ice and that of water boiling at standard atmospheric pressure. One Fahrenheit degree equals 0.556° C.

One Centigrade (C) degree is 1/100 the difference between the temperature of melting ice and that of water boiling at standard atmospheric pressure. One Centigrade degree equals 1.8° F.

To Change	To	Do This
Degrees Fahrenheit	Degrees Centigrade	Subtract 32, then multiply by .556 (5/9)
Degrees Centigrade	Degrees Fahrenheit	Multiply by 1.8 (9/5) and add 32

WEIGHTS AND MEASURES OF COMMON FEEDS

In calculating rations and mixing concentrates, it is usually necessary to use weights rather than measures. However, in practical feeding operations it is often more convenient for the farmer or rancher to measure the concentrates by volume. Table 20-3 will serve as a guide in feeding by measure.

TABLE 20-3
WEIGHTS AND MEASURES OF COMMON FEEDS

Feed	Approximate Weight	
	Lb per Quart[1]	Lb per Bushel[1]
Alfalfa meal	0.6	19
Barley	1.5	48
Beet pulp (dried)	0.6	19
Brewers' grain (dried)	0.6	19
Buckwheat	1.6	51
Buckwheat bran	1.0	32
Corn, husked ear	—	70
Corn, cracked	1.6	51
Corn, shelled	1.8	58
Corn meal	1.6	51
Corn-and-cob meal	1.4	45
Cottonseed	0.9-1.0	29-32
Cottonseed meal	1.5	48
Cowpeas	1.9	61
Distillers' grain (dried)	0.6	19
Fish meal	1.0	32
Flax	1.7	54
Gluten feed	1.3	42
Linseed meal (old process)	1.1	35
Linseed meal (new process)	0.9	29
Meat scrap	1.3	42
Milo (grain sorghum)	1.7	54
Molasses feed	0.8	26
Oat middlings	1.5	48
Oats	1.0	32
Oats, ground	0.7	22
Peanut meal	1.0	32
Peas	1.9	61
Rice	1.4	45
Rice bran	0.8	26
Rye	1.7	54
Sorghum (grain)	1.7	54
Soybeans	1.8	58
Sunflower	0.7	22
Tankage	1.6	51
Velvet beans, shelled	1.8	58
Wheat	1.9	61
Wheat bran	0.5	16
Wheat middlings, standard	0.8	26
Wheat screenings	1.0	32

[1]32 qts per bushel.

GRAIN WEIGHT IN A BIN

Sometimes farmers need to estimate the weight of grain in storage. Such estimates are difficult to make because of differences in moisture content, depth of material stored, and other factors. However, the following procedure will enable one to figure feed quantities fairly closely.

1. **Corn (shelled) or small grain in rectangular cribs or bins.** Multiply the width by the length by the average depth (all in feet) and multiply by 0.8 to get the number of bushels (multiplying by 0.8 is the same as dividing by 1¼, the number of cubic feet in a bushel).

2. **Ear corn in rectangular cribs or bins.** Multiply the width by the length by the average depth (all in feet) and multiply by 0.4 to get the number of bushels (multiplying by 0.4 is the same as dividing by 2½, the number of cubic feet in a bushel of ear corn).

3. **Round bins or cribs.** To find the cubic feet in a cylindrical bin, multiply the squared radius by 3.1416 by the depth.

Thus, the volume of a round bin 20 ft in diameter and 10 ft deep is determined as follows:

a. The radius is half the diameter, or 10 ft.

b. 10 × 10 = 100

c. 100 × 3.1416 = 314.16

d. 314.16 × 10 = 3,141.6 cu ft

e. Where shelled corn or small grain is involved, one would multiply 3,141.6 × 0.8, which equals 2,513.28 bu of grain that it would hold if full.

f. Where ear corn is involved, one would multiply 3,141.6 × 0.4 which equals 1,256.64 bu of ear corn that it would hold if full.

HAY WEIGHT IN A BARN OR STACK

Livestock producers and hay dealers frequently buy and sell large quantities of hay in the stack or in the barn. This practice is especially prevalent in the western and Great Plains states where cattle and sheep are brought into the valleys to be wintered on hay bought from valley hay producers. Under such circumstances, the weight of hay is usually estimated, because (1) no scales are available, and/or (2) it is impractical to weigh the hay due to the time, labor, and wastage involved. In many such instances, the hay is fed directly from the stack or barn, in racks arranged about it. Under these and other circumstances, there is need for a simple and reasonably accurate method of estimating the weight of hay in a stack or in a barn.

In order to estimate the tonnage of hay in a stack or in a barn, it is necessary (1) to compute the volume of hay, and (2) to know the number of cubic feet per ton of hay. Table 20-4 gives the latter information.

TABLE 20-4
CUBIC FEET PER TON OF HAY

Feed	Settled 1-2 Months	Settled Over 3 Months
	(cu ft)	(cu ft)
Alfalfa	485	470
Clover	512	500
Hay, baled (closely stacked)	150-300	150-200
Hay, chopped	225	210
Straw, baled	200	200
Straw, loose	1,000	600-1,000
Timothy	640	625
Wild hay	600	450

In using Table 20–4, it should be recognized that many factors—other than kind of hay, form (loose, chopped, or baled), and period of settling—affect the density of hay in a barn or in a stack, including (1) moisture content at haying time, and (2) texture and foreign material.

It is relatively simple to compute the volume of hay in a mow, but it is more difficult to determine the volume of a stack. Although different rules or formulas may be and are used, the following are recommended by the U.S. Department of Agriculture.[3]

1. **Volume of hay in barns.** Multiply the width by the length by the height, all in feet, and divide by the cubic feet per ton as given in Table 20–4.

2. **Volume of hay in oblong stacks.** Three types of oblong stacks are common, as shown in Fig. 20–3.

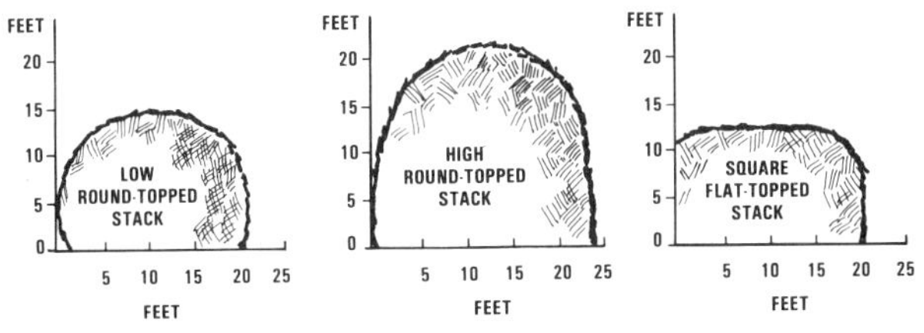

Fig. 20–3. Three common types of oblong stacks.

The volume of each type of oblong stack may be determined as follows:

a. For low, round-topped stacks—

$$(0.52 \times O) - (0.44 \times W) \times W \times L$$

b. For high, round-topped stacks—

$$(0.52 \times O) - (0.46 \times W) \times W \times L$$

c. For square, flat-topped stacks—

$$(0.56 \times O) - (0.55 \times W) \times W \times L$$

In these formulas "O" is the "over" or "overthrow," which is the distance in feet from the ground on one side of the stack, up and over the stack and down to the ground on the other side; W is the width; and L is the length.

The application of this formula is illustrated as follows:

Example. *It is desired to estimate the amount of alfalfa hay in a low, round-topped type of oblong stack that has settled for 4 months. The stack is 20 ft wide, 30 ft long, and has an over of 40 ft.*

The answer is secured as follows:

a. Volume = (0.52 × 40) – (0.44 × 20) × 20 × 30 = 7,200 cu ft.

b. Table A–6 shows that there are 470 cu ft per ton of settled alfalfa.

c. 7,200 ÷ 470 = 15 tons of hay.

[3]*Measuring Hay in Stacks*, USDA Leaflet No. 72.

3. **Volume of hay in round stacks.** The rules or formulas used for oblong stacks do not apply to round stacks. The volume of round stacks can be calculated by using the following formula:

$$\text{Volume} = (0.04 \times O) - (0.012 \times C) \times C^2$$

In this formula, C equals the circumference or distance around the stack at the ground, and O equals the over or distance from the ground on one side over the peak to the ground on the other side (usually it is best to take 2 over measurements at right angles to each other, and to average them).

Thus, the computation of the volume of a large round stack may be illustrated by the following example:

Example. *It is desired to determine the amount of alfalfa hay in a round stack that is 100 ft in circumference and has an average over of 60 ft.*

The answer is secured as follows:

a. Volume = (0.04 × 60) – (0.012 × 100) × $(100)^2$ = 12,000 cu ft.

b. Table 20–4 shows that there are 470 cu ft per ton of settled alfalfa.

c. 12,000 ÷ 470 = 25.5 tons of hay.

STORAGE SPACE REQUIREMENTS FOR FEED AND BEDDING

The space requirements for feed storage for the livestock enterprise—whether it be for cattle, sheep, hogs, or horses, or, as is more frequently the case, a combination of these—vary so widely that it is difficult to provide a standard method of calculating space requirements applicable to such diverse conditions. The amount of feed to be stored depends primarily upon (1) length of pasture season, (2) method of feeding and management, (3) kind of feed, (4) climate, and (5) the proportion of feeds produced on the farm or ranch in comparison with those purchased. Normally, the storage capacity should be sufficient to handle all feed grain and silage grown on the farm and to hold purchased supplies. Forage and bedding may or may not be stored under cover. In those areas where weather conditions permit, hay and straw are frequently stacked in the fields or near the barns in loose, baled, or chopped form. Sometimes poled, framed sheds or a cheap cover of waterproof paper, grass, or cereal straw grass are used for protection. Other forms of low-cost storage include temporary upright silos, trench silos, temporary grain bins, and open-walled buildings for hay.

Table 20–5 gives the storage space requirements for feed and bedding. This information may be helpful to the individual operator who desires to compute the barn storage space required for a specific livestock enterprise. This table provides a convenient means of estimating the amount of feed or bedding in storage.

TABLE 20-5
STORAGE SPACE REQUIREMENTS FOR FEED AND BEDDING

Kind of Feed or Bedding	Pounds per Cubic Foot	Cubic Feet per Ton	Pounds per Bushel of Grain
Hay-Straw:[1]			
1. Loose			
Alfalfa	4.4–4.0	450– 500	
Nonlegume	4.4–3.3	450– 600	
Straw	3.0–2.0	670–1,000	
2. Baled			
Alfalfa	10.0–6.0	200– 330	
Nonlegume	8.0–6.0	250– 330	
Straw	5.0–4.0	400– 500	
3. Chopped			
Alfalfa	7.0–5.5	285– 360	
Nonlegume	6.7–5.0	300– 400	
Straw	8.0–5.7	250– 350	
Corn:			
15½% moisture:			
Shelled	44.8		56.0
Ear	28.0		70.0
Shelled, ground	38.0		48.0
Ear, ground	36.0		45.0
30% moisture:			
Shelled	54.0		67.5
Ear, ground	35.8		89.6
Barley, 15% moisture	38.4		48.0
Ground	28.0		37.0
Flax, 11% moisture	44.8		56.0
Oats, 16% moisture	25.6		32.0
Ground	18.0		23.0
Rye, 16% moisture	44.8		56.0
Ground	38.0		48.0
Sorghum grain, 15% moisture	44.8		56.0
Soybeans, 14% moisture	48.0		60.0
Wheat, 14% moisture	48.0		60.0
Ground	43.0		50.0

[1]Many factors—other than kind of hay-straw, form (loose, baled, chopped), and period of settling—affect the density of hay-straw in a stack or in a barn, including (a) moisture content at haying time, and (b) texture and foreign material.

ANIMAL UNITS

An animal unit is a common animal denominator, based on feed consumption. It is assumed that 1 mature cow represents an animal unit. Then, the comparative (to a mature cow) feed consumption of other age groups or classes of animals determines the proportion of an animal unit which they represent. For example, it is generally estimated that the ration of one mature cow will feed 5 mature ewes, or that 5 mature ewes equal 1.0 animal unit.

The original concept of an animal unit included a weight stipulation—an animal unit referred to a 1,000-lb cow, with or without a calf at side. Unfortunately, in recent years, the 1,000-lb qualification has been dropped. Certainly, there is a wide difference in the daily feed requirements of a 900-lb cow and of a 1,500-lb cow. Both will consume dry matter on a daily basis at a level equivalent to about 2% of their body weight.

Hence, a 1,500-lb cow will consume 50% more feed than a 1,000-lb cow.

Also, the period of time to be grazed has an effect on the total carrying capacity. For example, if an animal is carried for 1 month only, it will take $\frac{1}{12}$ of the total feed required to carry the same animal 1 year. For this reason, the term *animal unit months* is becoming increasingly important. So, in addition to the weight factor, the time factor has a distinct bearing on the ultimate carrying capacity of a tract of land.

Table 20-6 gives the animal units of different classes and ages of livestock.

TABLE 20-6
ANIMAL UNITS

Type of Livestock	Animal Units
Cattle:	
Cow, with or without unweaned calf at side, or heifer 2 years old or older	1.0
Bull, 2 years old or older	1.3
Young cattle, 1 to 2 years	0.8
Weaned calves to yearlings	0.6
Horses:	
Horse, mature	1.3
Horse, yearling	1.0
Weanling colt or filly	0.75
Sheep:	
5 mature ewes, with or without unweaned lambs at side	1.0
5 rams, 2 years old or over	1.3
5 yearlings	0.8
5 weaned lambs to yearlings	0.6
Goats — 7	1.0
Swine:	
Sow	0.4
Boar	0.5
Pigs to 200 lb	0.2
Chickens:	
75 layers or breeders	1.0
325 replacement pullets to 6 months of age	1.0
650 8-week-old broilers	1.0
Turkeys:	
35 breeders	1.0
40 turkeys raised to maturity	1.0
75 turkeys to 6 months of age	1.0
Rabbits — 56	1.0
Fish — 259	1.0

Fig. 20-4. A cow, with an unweaned calf at side, equals one animal unit.

CALCULATING ANIMAL WEIGHTS

Feeders who finish large numbers of animals have scales in their feedyards for use in determining in-weights, out-weights, and interim weight gains of animals while they are on feed. Likewise, both purebred and commercial breeders usually have scales. However, those with only one animal, or a few head—such as 4–H Club and FFA members, and part-time farmers—may not have scales. As a result, rations cannot be accurately evaluated, rate of gain cannot be calculated, and an animal's "weight readiness" for a livestock show or for market cannot be determined. Under such circumstances, a simple but reasonably accurate method of estimating body weight is very useful. Fortunately, animal weights may be determined with reasonable accuracy by taking two body measurements (body length and heart girth), then applying an appropriate formula.

BEEF CATTLE WEIGHTS

Here is how to do it:

Step 1. Measure the circumference (heart girth), from a point slightly behind the shoulder blade, thence down over the foreribs and under the body, behind the elbow (distance C of Fig. 20–5).

Step 2. Measure the length of body, from the point of the shoulder to the point of the rump (pinbone), in inches (distance A-B of Fig. 20–5).

Step 3. Take the values obtained in Steps 1 and 2 and apply the following formula to calculate body weight:

Heart girth × heart girth × body length ÷ 300 = weight in pounds

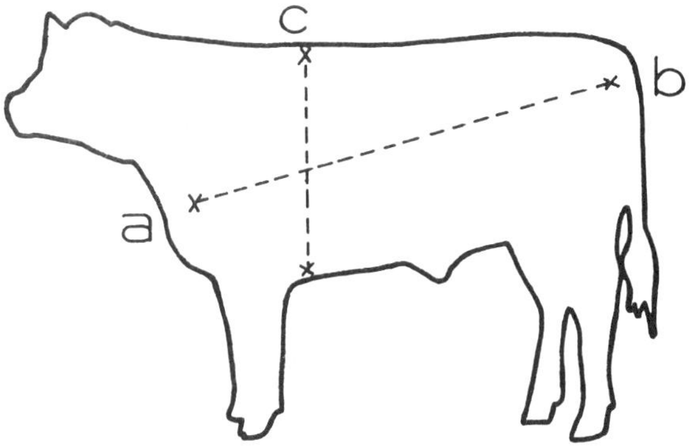

Fig. 20–5. How and where to measure beef cattle.

Example of a beef animal. *Assume that the heart girth measures 76 in. and the body length, 66 in. How much does the animal weigh?*

76 × 76 = 5,776
5,776 × 66 = 381,216
381,216 ÷ 300 = 1,270 lb

DAIRY HEIFER WEIGHTS

Weight for age is important in dairy heifers from the standpoint of determining the growth progress made by herd replacements.

Section 4, Feeding, p. 180, Table 4–23, of this book, shows the weight and heart girth measurements of dairy calves or heifers at monthly intervals up to 21 months of age. If producers do not have scales, they can measure the heart girth with a tape (see Fig. 20–6) and use Table 4–23 to estimate weight within 95% accuracy.

Fig. 20–6. How to tape measure a dairy heifer.

SHEEP AND GOAT WEIGHTS

The weight of sheep and goats is estimated in the same way as for beef cattle; hence, it involves making the measurements and applying the formula given for beef cattle. There is one important precaution, however; with unshorn sheep, be sure to part, or compress, the wool to ensure an accurate heart girth measurement.

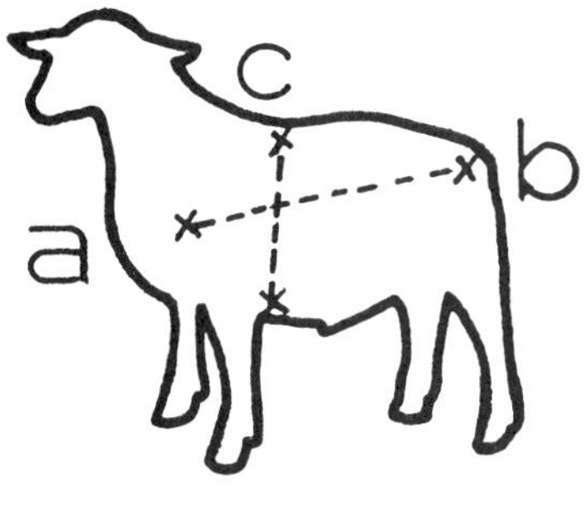

Fig. 20–7. How and where to measure sheep.

SWINE WEIGHTS

Hog weights can be calculated from body measurements, similar to beef cattle, but a different formula must be used. Here is how to estimate the weight of hogs:

Step 1. Measure the circumference (heart girth) of the animal (C in diagram).

Step 2. Measure the length of body (A-B in Fig. 20–8). With the animal standing or restrained in the position shown in Fig. 20–8, measure the distance from the poll (between the ears), over the backbone, to the base of the tail.

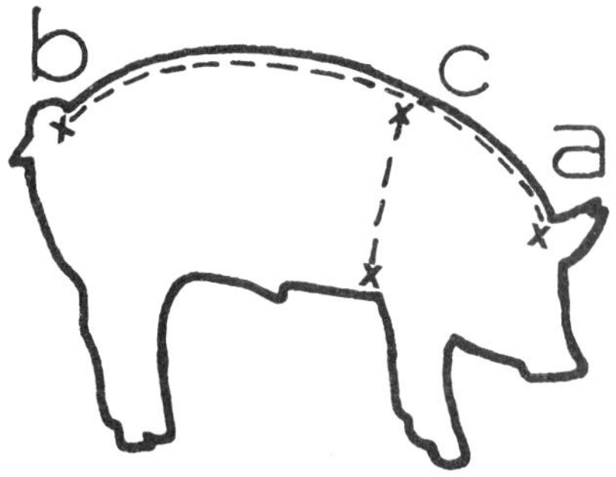

Fig. 20–8. How and where to measure hogs.

Step 3. Apply the following formula:

Heart girth × heart girth × length ÷ 400 = weight in pounds.

Note: For hogs weighing less than 150 lb, add 7 lb to the weight figure obtained from the formula. For animals weighing 151 to 400 lb, no adjustment is necessary.

HORSE WEIGHTS

It is easy to estimate the weight of a horse; and tests have shown that the results obtained this way are accurate within 3% of actual scale weight. This procedure is as follows:

Step 1. Measure the circumference (heart girth) of the body in inches (C in diagram).

Step 2. Measure the length of body from the point of the shoulder to the point of croup (A-B in the diagram).

Step 3. Apply the following formula to calculate the weight of the horse:

Heart girth × heart girth × **length ÷ 300 + 50 lb = weight of horse.**

Example. *Assume that the heart girth is 70 in. and the body length is 65 in. How much does the horse weigh?*

70 × 70 × 65 ÷ 300 + 50 lb = weight
4,900 × 65 = 318,500
318,500 ÷ 300 = 1,061 lb
1,061 + 50 = 1,111 lb body weight

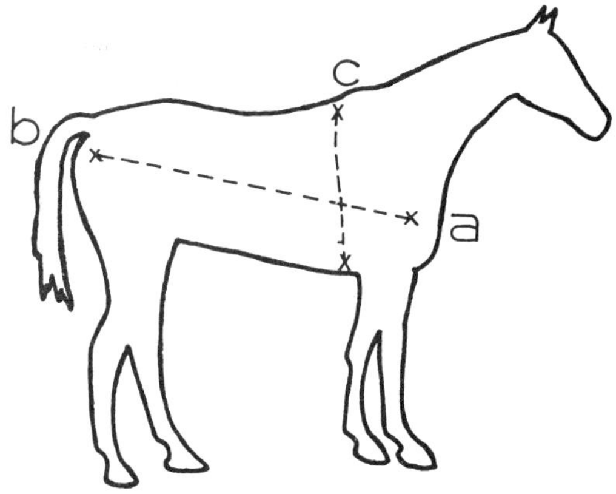

Fig. 20–9. How and where to measure horses.

A modern auction facility, with scales serving as the auction ring. As cattle enter the ring (scales), they are weighed and an electric scoreboard indicates the gross and average weights. (Courtesy, USDA, Washington, DC)

INDEX

A

B

C

D

E

F

G

H

I

J

Q

R

S

T

U

V

Upper Left: Hay making. (Courtesy, John Deere Co., Moline, IL)

Silos/buildings. (Courtesy, Harvestore Systems, DeKalb, IL)

Livestock Farming/Ranching In The 21st Century

Left: Branding. (Courtesy, USDA)

Animal health. (Courtesy, USDA)

Selecting and judging. (Courtesy, San Antonio Livestock Exposition, San Antonio, TX)

Fitting and showing. (Courtesy, San Antonio Livestock Exposition, San Antonio, TX)

Beef cattle/pond/farmstead. (Courtesy, H. Dietz, USDA, Soil Conservation Service, Ft. Worth, TX)

Jersey cow. (Courtesy, The American Jersey Cattle Club, Reynoldsburg, OH)

Livestock Farming/Ranching In The 21st Century

Above: Polypay ewe and lambs. (Courtesy, American Polypay Sheep Assn., Sidney, MT)

Right: Angora goat. (Courtesy, Mohair Council of America, San Angelo, TX)

Below: Horses. (Courtesy, Julie K. Smithson, London, OH)

Below: Smoked ham. (Courtesy, National Live Stock & Meat Board, Chicago, IL)

TABLE 19–2 AMINO ACID, COMPOSITION OF FEEDS, DATA EXPRESSED **AS-FED**—*(Continued)*

Entry Number	Feed Name Description	International Feed Number	Dry Matter	Crude Protein	Amino Acids: Arginine	Cystine	Glycine	Histidine	Isoleucine	Leucine	Lysine	Methionine	Phenylalanine	Serine	Threonine	Tryptophan	Tyrosine	Valine
			%	%	%	%	%	%	%	%	%	%	%	%	%	%	%	%
	CORN																	
96	DISTILLERS GRAINS, DEHY	5-02-842	93	27.8	0.99	0.23	0.75	0.62	1.00	3.01	0.76	0.42	0.99	1.01	0.56	0.20	0.84	1.21
97	DISTILLERS SOLUBLES, DEHY	5-02-844	93	27.4	0.99	0.44	1.12	0.67	1.32	2.38	0.92	0.55	1.47	1.22	1.01	0.25	0.88	1.53
98	GLUTEN FEED	5-02-903	90	23.0	0.78	0.42	0.85	0.61	0.88	2.20	0.64	0.37	0.81	0.85	0.78	0.15	0.72	1.10
99	GLUTEN MEAL	5-02-900	91	43.2	1.40	0.67	1.51	0.97	2.25	7.38	0.80	1.03	2.85	1.70	1.43	0.21	1.01	2.23
	COTTON																	
100	SEEDS, MEAL MECH EXTD, 36% PROTEIN	5-01-625	92	37.2	3.55	0.79	1.83	0.91	1.32	–	1.22	0.55	1.88	–	1.12	0.46	–	2.84
101	SEEDS, MEAL MECH EXTD, 41% PROTEIN	5-01-617	93	41.0	4.20	0.71	1.87	1.07	1.42	2.30	1.60	0.57	2.19	1.70	1.33	0.52	0.97	1.89
102	SEEDS, MEAL, PREPRESSED, SOLV EXTD, 41% PROTEIN	5-07-872	90	41.3	4.32	0.78	1.89	1.14	1.42	2.42	1.80	0.56	2.05	1.80	1.34	0.50	1.14	1.97
103	SEEDS, MEAL, SOLV EXTD, 41% PROTEIN	5-01-621	91	41.2	4.24	0.76	1.95	1.10	1.50	2.46	1.69	0.58	2.23	1.76	1.37	0.55	1.04	1.97
104	SEEDS WITHOUT HULLS, MEAL, PREPRESSED, SOLV EXTD, 50% PROTEIN	5-07-874	93	50.3	4.83	1.05	2.82	1.21	1.86	2.82	1.93	0.76	2.62	–	1.66	0.62	0.81	2.16
105	**CRAB,** CANNERY RESIDUE, MEAL (CRAB MEAL)	5-01-663	92	32.2	1.66	0.24	1.74	0.48	1.16	1.54	1.38	0.52	1.16	1.38	1.00	0.29	1.17	1.47
106	**CRAMBE, ABYSSINIAN,** SEEDS WITHOUT HULLS, MEAL MECH EXTD	5-16-453	92	45.8	–	–	–	–	–	–	–	–	–	–	–	–	–	–
	DISTILLERS PRODUCTS (ALSO SEE CORN; RYE)																	
107	GRAINS, DEHY	5-02-144	93	27.3	1.04	0.42	0.56	0.53	1.16	2.66	0.81	0.46	1.03	0.70	0.81	0.21	0.73	1.22
108	SOLUBLES, DEHY	5-02-147	92	28.8	1.06	0.40	1.20	0.66	1.21	2.35	0.95	0.50	1.24	0.93	1.00	0.24	0.93	1.40
	FISH																	
109	MEAL MECH EXTD	5-01-977	92	64.3	31.28	0.62	3.99	1.46	3.27	4.90	5.26	1.63	2.60	2.42	2.59	0.75	1.79	3.14
110	SOLUBLES, CONDENSED	5-01-969	50	31.5	1.63	0.39	3.87	1.54	1.09	1.94	1.85	0.70	1.07	1.05	0.90	0.33	0.50	1.26
111	**FISH, ANCHOVY,** MEAL MECH EXTD	5-01-985	92	65.4	3.78	0.60	3.69	1.60	3.11	4.99	5.02	1.99	2.78	2.42	2.76	0.75	2.24	3.50
112	**FISH, MENHADEN,** MEAL MECH EXTD	5-02-009	92	61.2	3.74	0.58	4.19	1.44	2.85	4.48	4.74	1.75	2.46	2.25	2.51	0.65	1.93	3.19
113	**FISH, TUNA,** MEAL MECH EXTD	5-02-023	93	59.0	3.43	0.47	4.09	1.75	2.45	3.79	4.06	1.47	2.15	2.08	2.31	0.57	1.69	2.77
114	**FISH, WHITE,** MEAL MECH EXTD	5-02-025	91	62.6	4.26	0.77	5.15	1.38	2.85	4.65	4.70	1.79	2.44	3.44	2.56	0.67	2.27	3.25
	FLAX, COMMON																	
115	SEEDS, MEAL MECH EXTD (LINSEED MEAL)	5-02-048	90	34.6	2.94	0.61	1.74	0.69	1.68	2.02	1.16	0.54	1.46	1.93	1.22	0.51	1.09	1.74
116	SEEDS, MEAL SOLV EXTD (LINSEED MEAL)	5-02-045	91	34.3	2.81	0.61	1.64	0.65	1.69	1.92	1.18	0.58	1.38	1.90	1.14	0.51	0.96	1.61
117	**GARBANZO (CHICKPEA; GRAM PEA),** SEEDS	5-01-218	89	19.1	1.52	–	0.69	0.40	0.76	1.32	1.25	0.24	1.14	0.90	0.61	–	0.57	0.80
118	**HORSE BEAN,** SEEDS	5-02-407	88	25.5	–	–	–	–	–	–	–	–	–	–	–	–	–	–
119	**LIVER,** MEAL	5-00-389	93	66.1	4.04	0.94	5.61	1.48	3.11	5.31	5.22	1.22	2.92	2.50	2.50	0.69	1.70	4.15
120	**LOCUST, NEW MEXICO,** SEEDS	5-09-055	89	36.5	3.01	–	1.32	0.73	0.87	1.75	1.32	0.26	1.02	1.28	0.87	–	0.84	1.17
	MAIZE–SEE CORN																	
	MEAT																	
121	MEAL RENDERED	5-00-385	94	50.7	3.58	0.60	7.23	0.87	1.63	3.11	3.00	0.69	1.74	2.31	1.67	0.35	1.09	2.42
122	WITH BLOOD, MEAL RENDERED (TANKAGE)	5-00-386	92	60.5	3.60	0.48	6.45	2.06	1.82	5.10	3.89	0.75	2.56	2.81	2.34	0.65	1.38	3.83
123	WITH BLOOD, WITH BONE, MEAL RENDERED (TANKAGE)	5-00-387	93	46.6	2.82	0.27	6.58	1.76	1.87	5.26	3.32	0.69	2.28	–	2.18	0.62	–	3.42
124	WITH BONE, MEAL RENDERED	5-00-388	93	50.4	3.53	0.53	6.49	1.01	1.64	3.10	2.93	0.67	1.71	1.90	1.66	0.31	0.89	2.44
	MILK																	
125	FRESH (CATTLE)	5-01-168	12	3.3	–	–	–	–	0.32	0.25	0.28	0.18	0.07	–	0.16	0.05	–	0.25
126	DEHY (CATTLE)	5-01-167	95	25.3	0.92	–	–	0.71	1.32	2.54	2.24	0.61	1.32	–	1.02	0.41	1.32	1.73
127	SKIMMED, DEHY (CATTLE)	5-01-175	94	33.3	1.16	0.45	0.29	0.86	2.18	3.33	2.54	0.90	1.57	1.67	1.57	0.43	1.14	2.29
128	FRESH (HORSE)	5-02-401	17	4.2	–	–	–	0.11	0.25	0.34	0.25	0.07	0.18	–	0.16	0.05	–	0.29
129	FRESH (SHEEP)	5-08-510	19	4.6	–	–	–	0.18	0.39	0.60	0.51	0.17	0.32	–	0.30	0.09	–	0.48
130	FRESH (SWINE)	5-08-537	20	7.3	–	–	–	0.20	0.42	0.59	0.50	0.14	0.34	–	0.37	0.09	–	0.45
131	**MILKWEED, COMMON,** SEEDS	5-09-137	86	31.8	3.05	–	1.65	0.73	1.12	1.97	1.56	0.45	1.53	1.31	0.86	–	1.08	1.37
132	**PALM,** KERNELS WITH COATS, MEAL SOLV EXTD	5-03-486	90	18.2	2.52	–	–	0.31	0.76	1.22	0.66	0.41	0.81	–	0.60	0.20	–	1.02
133	**PEA,** SEEDS	5-03-600	89	23.2	1.40	0.21	1.09	0.60	1.20	1.81	1.53	0.27	1.25	–	0.94	0.21	–	1.25
134	**PEA, FIELD,** SEEDS	5-08-481	91	23.2	1.86	0.26	1.05	0.51	0.91	1.59	1.44	0.23	1.00	1.05	0.82	0.22	0.77	1.00
	PEANUT																	
135	PODS WITH SEEDS, MEAL SOLV EXTD	5-03-656	92	47.4	5.19	0.70	2.39	1.10	1.92	3.20	1.75	0.43	2.49	3.05	1.38	0.49	1.68	2.48
136	SEEDS WITHOUT HULLS, MEAL MECH EXTD (PEANUT MEAL)	5-03-649	93	49.2	5.08	0.96	2.49	1.03	1.78	3.13	1.69	0.50	2.38	1.44	1.27	0.46	1.59	2.29
137	SEEDS WITHOUT HULLS, MEAL SOLV EXTD (PEANUT MEAL)	5-03-650	93	49.0	5.82	0.54	2.88	1.46	1.84	3.27	1.45	0.44	2.12	3.12	1.37	0.48	–	2.16
	POULTRY																	
138	BY-PRODUCT, MEAL RENDERED	5-03-798	94	61.2	4.01	0.85	6.09	1.13	2.35	4.10	3.12	1.14	2.04	2.88	2.10	0.47	1.84	2.94
139	FEATHERS, HYDROLYZED, MEAL	5-03-795	93	83.8	5.33	3.21	6.32	0.47	3.51	6.42	1.55	0.54	3.59	9.16	3.63	0.52	2.35	5.85
140	**RAPE (CANOLA), SUMMER,** SEEDS, MEAL, PREPRESSED, SOLV EXTD	5-08-135	92	40.5	2.23	–	1.94	1.09	1.46	2.71	2.15	0.77	1.54	1.70	1.70	0.49	0.85	1.94

TABLE 19–2 AMINO ACID, COMPOSITION OF FEEDS, DATA EXPRESSED **AS-FED**—*(Continued)*

Entry Number	Feed Name Description	International Feed Number	Dry Matter	Crude Protein	AMINO ACIDS Arginine	Cystine	Glycine	Histidine	Isoleucine	Leucine	Lysine	Methionine	Phenylalanine	Serine	Threonine	Tryptophan	Tyrosine	Valine
			%	%	%	%	%	%	%	%	%	%	%	%	%	%	%	%
	SORGHUM (Continued)																	
48	HEGARI, GRAIN	4-04-398	89	10.4	0.29	–	–	0.18	0.47	1.40	0.17	0.12	0.54	–	0.36	0.11	–	0.55
49	KAFIR, GRAIN	4-04-428	89	10.8	0.38	0.17	0.30	0.27	0.55	1.62	0.26	0.19	0.64	–	0.45	0.15	–	0.62
50	MILO, GRAIN	4-04-444	89	10.1	0.37	0.13	0.35	0.24	0.44	1.32	0.23	0.16	0.49	0.49	0.35	0.10	0.37	0.53
51	MILO, GLUTEN WITH BRAN, MEAL	5-08-089	89	23.2	0.90	0.20	0.68	0.60	1.00	2.51	0.70	0.40	1.00	–	0.80	0.20	0.90	1.30
52	SHALLU, GRAIN	4-04-456	90	11.5	0.31	–	–	0.19	0.38	0.97	0.19	0.17	0.40	–	0.30	0.10	–	0.46
	SOYBEAN																	
53	SEEDS	5-04-610	92	38.4	2.63	0.42	1.42	0.92	1.62	2.72	2.32	0.48	1.76	1.99	1.46	0.56	1.29	1.61
54	SOYBEAN MILL FEED	4-04-594	90	12.6	0.70	0.13	0.47	0.18	0.41	0.58	0.59	0.12	0.37	–	0.30	0.13	0.23	0.37
55	**SPELT,** GRAIN	4-04-651	90	12.0	0.45	–	–	0.18	0.36	0.63	0.27	0.18	0.45	–	0.36	0.09	–	0.45
56	**TRITICALE,** GRAIN	4-20-362	89	15.4	0.85	0.27	0.68	0.38	0.58	1.11	0.52	0.22	0.77	0.73	0.53	0.18	0.49	0.78
	WHEAT																	
57	GRAIN, ALL ANALYSES	4-05-211	89	13.1	0.61	0.22	0.59	0.30	0.49	0.90	0.39	0.18	0.61	0.63	0.40	0.15	0.37	0.61
58	GRAIN, HARD RED SPRING	4-05-258	88	14.2	0.64	0.22	0.60	0.27	0.52	0.91	0.38	0.20	0.66	0.61	0.39	0.15	0.45	0.61
59	GRAIN, HARD RED WINTER	4-05-268	89	12.8	0.65	0.30	0.58	0.30	0.53	0.87	0.36	0.22	0.63	0.59	0.37	0.17	0.46	0.58
60	GRAIN, SOFT RED WINTER	4-05-294	88	11.4	0.65	0.36	0.55	0.32	0.45	0.90	0.36	0.22	0.64	0.65	0.39	0.27	0.37	0.58
61	GRAIN, SOFT WHITE WINTER	4-05-337	90	10.2	0.47	0.27	0.50	0.22	0.42	0.66	0.32	0.16	0.46	0.46	0.32	0.13	0.37	0.45
62	GRAIN, SOFT WHITE WINTER, PACIFIC COAST	4-08-555	89	10.0	0.45	0.24	0.50	0.20	0.40	0.75	0.30	0.14	0.48	0.49	0.31	0.12	0.36	0.46
63	BRAN	4-05-190	89	15.5	0.85	0.26	0.77	0.33	0.55	0.89	0.54	0.17	0.50	0.68	0.40	0.25	0.38	0.67
64	DISTILLERS GRAINS, DEHY	5-05-193	93	31.6	1.10	–	–	0.80	2.01	1.71	0.70	–	1.71	–	0.90	–	0.50	1.71
65	ENDOSPERM	4-05-197	88	11.1	0.60	0.30	–	0.30	1.10	1.70	0.40	0.20	0.60	–	0.40	0.30	–	0.60
66	FLOUR, LESS THAN 1.5% FIBER	4-05-199	88	13.7	0.42	0.25	0.46	0.28	0.56	0.89	0.27	0.13	0.62	0.51	0.30	0.10	0.25	0.51
67	GRAIN SCREENINGS	4-05-216	89	13.3	0.68	0.12	0.53	0.30	0.45	0.74	0.43	0.26	0.49	0.40	0.33	0.13	0.23	0.55
68	MIDDLINGS, LESS THAN 9.5% FIBER	4-05-205	89	16.4	0.98	0.22	0.96	0.40	0.68	1.11	0.68	0.19	0.66	0.80	0.57	0.19	0.43	0.80
69	MILL RUN, LESS THAN 9.5% FIBER	4-05-206	90	15.1	0.94	0.23	0.53	0.40	0.70	1.20	0.57	0.33	–	–	0.50	0.21	0.50	0.80
70	RED DOG, LESS THAN 4% FIBER	4-05-203	88	15.6	0.96	0.36	0.74	0.38	0.58	1.08	0.60	0.22	0.65	0.76	0.50	0.20	0.46	0.73
71	SHORTS, LESS THAN 7% FIBER	4-05-201	88	16.5	1.20	0.38	0.96	0.44	0.57	1.07	0.80	0.28	0.67	0.77	0.60	0.23	0.47	0.82
72	**WHEAT, DURUM,** GRAIN	4-05-224	88	13.8	0.58	0.13	0.46	0.27	0.48	1.40	1.05	0.14	0.53	0.45	0.37	0.26	0.29	0.54
	PROTEIN FEEDS																	
73	**ACACIA, SWEET,** SEEDS	5-09-110	87	47.9	4.40	–	1.63	1.10	1.67	3.58	2.25	0.44	1.67	1.97	1.20	–	1.34	1.86
	ANIMAL																	
74	BLOOD, MEAL	5-00-380	91	80.5	3.23	1.25	3.45	3.93	0.85	10.07	6.43	0.94	5.56	3.95	3.59	1.01	1.94	6.56
75	BLOOD, SPRAY DEHY	5-00-381	93	86.0	3.59	1.03	3.83	5.18	0.91	10.97	7.44	1.05	5.89	3.53	3.63	1.05	2.26	7.52
76	LIVER, MEAL	5-00-389	93	66.1	4.04	0.94	5.61	1.48	3.11	5.31	5.22	1.22	2.92	2.50	2.50	0.69	1.70	4.15
77	MEAT, MEAL RENDERED	5-00-385	94	50.7	3.58	0.60	7.23	0.87	1.63	3.11	3.00	0.69	1.74	2.31	1.67	0.35	1.09	2.42
78	MEAT WITH BONE, MEAL RENDERED	5-00-388	93	50.4	3.53	0.53	6.49	1.01	1.64	3.10	2.93	0.67	1.71	1.90	1.66	0.31	0.89	2.44
79	TANKAGE, MEAL RENDERED	5-00-386	92	60.5	3.60	0.48	6.45	2.06	1.82	5.10	3.89	0.75	2.56	2.81	2.34	0.65	1.38	3.83
80	TANKAGE WITH BONE, MEAL RENDERED	5-00-387	93	46.6	2.82	0.27	6.58	1.76	1.87	5.26	3.32	0.69	2.28	–	2.18	0.62	–	3.42
	BABASSU																	
81	KERNELS WITH COATS, MEAL MECH EXTD (BABASSU OIL MEAL)	5-00-454	92	22.3	2.87	–	–	0.40	1.04	1.34	0.89	0.30	0.89	–	0.60	0.20	0.40	1.09
82	KERNELS WITH COATS, MEAL SOLV EXTD (BABASSU OIL MEAL)	5-00-455	93	21.2	3.19	–	–	0.41	0.88	1.40	0.98	0.53	1.35	–	0.71	0.24	–	1.19
83	**BEAN, PINTO,** SEEDS	5-00-624	90	22.7	1.55	–	–	0.64	1.14	1.11	1.60	0.26	1.20	–	1.09	0.32	–	1.23
84	**BLOOD,** MEAL	5-00-380	91	80.5	3.23	1.25	3.45	3.93	0.85	10.07	6.43	0.94	5.56	3.95	3.59	1.01	1.94	6.56
85	**BUTTERMILK (CATTLE),** DEHY	5-01-160	92	31.7	1.08	0.39	0.47	0.85	2.42	3.21	2.28	0.71	1.46	1.50	1.52	0.49	1.00	2.58
86	**CASEIN,** ACID PRECIPITATED, DEHY	5-01-162	91	84.0	3.49	0.31	1.61	2.59	5.72	8.80	7.14	2.81	4.81	5.46	3.91	1.08	4.90	6.71
87	**CASTOR BEAN,** SEEDS WITHOUT TOXIN, MEAL	5-01-155	87	26.0	2.77	–	0.86	0.48	1.01	1.40	0.76	0.36	0.82	1.13	0.72	–	0.68	1.27
	CATTLE																	
88	BUTTERMILK, DEHY	5-01-160	92	31.7	1.08	0.39	0.47	0.85	2.42	3.21	2.28	0.71	1.46	1.50	1.52	0.49	1.00	2.58
89	MILK, FRESH	5-01-168	12	3.3	–	–	–	–	0.32	0.25	0.28	0.18	0.07	–	0.16	0.05	–	0.25
90	MILK, DEHY	5-01-167	95	25.3	0.92	–	–	0.71	1.32	2.54	2.24	0.61	1.32	–	1.02	0.41	1.32	1.73
91	SKIM MILK, DEHY	5-01-175	94	33.3	1.16	0.45	0.29	0.86	2.18	3.33	2.54	0.90	1.57	1.67	1.57	0.43	1.14	2.29
92	WHEY, DEHY	4-01-182	93	13.3	0.33	0.30	0.44	0.17	0.78	1.18	0.94	0.19	0.35	0.47	0.90	0.20	0.25	0.67
93	WHEY, LOW LACTOSE, DEHY	4-01-186	93	16.7	0.60	0.43	0.72	0.27	0.96	1.54	1.40	0.41	0.55	0.59	0.95	0.27	0.46	0.87
94	**CHICKPEA (GARBANZO; GRAM PEA),** SEEDS	5-01-218	89	19.1	1.52	–	0.69	0.40	0.76	1.32	1.25	0.24	1.14	0.90	0.61	–	0.57	0.80
95	**COCONUT,** KERNELS WITH COATS, MEAL MECH EXTD, (COPRA MEAL)	5-01-572	92	21.2	2.30	0.21	1.05	0.33	0.90	1.35	0.55	0.31	0.81	–	0.60	0.20	0.58	0.98

(Continued)

TABLE 19–2 AMINO ACID, COMPOSITION OF FEEDS, DATA EXPRESSED **AS-FED**—*(Continued)*

Entry Number	Feed Name Description	International Feed Number	Dry Matter	Crude Protein	Arginine	Cystine	Glycine	Histidine	Isoleucine	Leucine	Lysine	Methionine	Phenylalanine	Serine	Threonine	Tryptophan	Tyrosine	Valine
			%	%	%	%	%	%	%	%	%	%	%	%	%	%	%	%
	RYE																	
141	DISTILLERS GRAINS, DEHY	5-04-023	92	23.0	–	–	–	–	–	–	–	–	–	–	–	–	–	–
142	DISTILLERS GRAINS WITH SOLUBLES, DEHY	5-04-024	90	27.2	1.00	–	–	0.70	1.50	2.10	1.00	0.40	1.30	1.20	1.10	0.30	0.50	1.60
	SAFFLOWER																	
143	SEEDS WITHOUT HULLS, MEAL MECH EXTD	5-08-499	91	42.0	5.44	–	2.52	–	–	–	1.31	0.71	–	–	0.81	–	–	–
144	SEEDS WITHOUT HULLS, MEAL SOLV EXTD	5-07-959	91	42.8	3.67	0.71	2.36	0.97	1.58	2.42	1.26	0.67	1.73	–	1.30	0.59	1.01	2.17
145	**SESAME,** SEEDS, MEAL MECH EXTD	5-04-220	93	45.0	4.55	0.59	3.96	1.07	1.96	3.20	1.26	1.37	2.14	2.94	1.60	0.71	1.87	2.32
146	**SHRIMP,** CANNERY RESIDUE, MEAL (SHRIMP MEAL)	5-04-226	90	38.7	2.33	0.47	1.31	0.87	1.51	2.37	2.05	0.84	1.55	1.25	1.26	0.36	1.10	1.71
147	**SORGHUM,** GLUTEN MEAL	5-04-388	90	44.4	1.26	0.73	0.95	1.07	2.39	7.85	0.74	0.71	2.70	–	1.45	0.44	–	2.50
	SOYBEAN																	
148	FLOUR, SOLV EXTD	5-04-593	93	51.6	4.27	0.64	1.65	1.26	1.90	3.33	4.48	0.57	2.00	2.09	1.58	0.79	1.44	1.86
149	MEAL, SOLV EXTD	5-04-612	90	49.7	3.67	0.70	2.27	1.20	2.13	3.63	3.12	0.71	2.36	2.49	1.90	0.69	1.71	2.47
150	MEAL, SOLV EXTD, 44% PROTEIN	5-20-637	89	44.4	3.26	0.67	2.10	1.13	2.12	3.49	2.85	0.59	2.23	2.37	1.81	0.62	1.60	2.37
151	MEAL, SOLV EXTD, 49% PROTEIN	5-20-638	90	49.0	3.62	0.75	2.39	1.28	2.34	3.77	3.08	0.66	2.47	2.76	2.00	0.70	1.96	2.49
152	**WHALE,** MEAT, MEAL RENDERED	5-05-160	91	71.4	2.49	0.63	6.31	1.19	2.72	4.27	3.48	1.01	2.06	–	1.63	0.82	–	2.81
	WHEAT																	
153	GERM MEAL	5-05-218	88	24.4	1.83	0.47	1.46	0.62	0.95	1.47	1.53	0.41	0.93	1.12	0.94	0.30	0.74	1.16
154	GLUTEN	5-05-221	90	63.4	2.97	1.74	2.77	1.64	3.39	5.54	1.54	1.23	4.21	4.10	2.15	0.72	2.36	3.90
	WHEY—SEE CATTLE																	
155	**YEAST, IRRADIATED,** DEHY	7-05-529	94	48.1	2.46	–	–	1.00	2.94	3.56	3.70	1.00	2.77	–	2.41	0.73	–	3.06
156	**YEAST, TORULA,** DEHY	7-05-534	93	49.6	2.52	0.59	2.54	1.34	2.69	3.39	3.65	0.76	2.63	2.75	2.67	0.52	1.94	2.88
	DRY FORAGES																	
	ALFALFA (LUCERNE)																	
1157	HAY, SUN-CURED	1-00-078	90	16.0	0.81	–	–	0.28	0.87	1.12	1.00	0.12	0.71	–	0.62	0.18	0.50	0.69
158	HAY, SUN-CURED, EARLY BLOOM, MEAL	1-00-108	92	22.5	–	–	–	–	–	–	–	–	–	–	–	–	–	–
159	LEAVES, SUN-CURED, MEAL	1-00-146	88	20.1	–	–	–	–	–	–	–	–	–	–	–	–	–	–
160	LEAVES, MEAL, DEHY	1-00-137	92	20.6	0.97	0.27	0.99	0.39	0.92	1.45	0.95	0.32	0.94	0.89	0.86	0.43	0.60	1.05
161	MEAL, DEHY, 15% PROTEIN	1-00-022	90	15.6	0.59	0.17	0.70	0.27	0.64	1.02	0.59	0.22	0.62	0.60	0.56	0.38	0.41	0.75
162	MEAL, DEHY, 17% PROTEIN	1-00-023	92	17.4	0.77	0.29	0.84	0.33	0.81	1.28	0.85	0.27	0.80	0.71	0.71	0.34	0.54	0.88
163	MEAL, DEHY, 20% PROTEIN	1-00-024	92	20.2	0.95	0.32	0.99	0.38	0.89	1.43	0.89	0.32	0.94	0.90	0.82	0.41	0.60	1.05
164	MEAL, DEHY, 22% PROTEIN	1-07-851	93	22.2	0.96	0.30	1.09	0.44	1.06	1.63	0.97	0.34	1.13	0.97	0.97	0.49	0.64	1.29
165	**ALFALFA-GRASS,** HAY, SUN-CURED	1-08-331	91	14.5	–	–	–	–	–	–	–	–	–	–	–	–	–	–
166	**BEET, SUGAR,** LEAVES, SUN-CURED	1-00-641	91	23.2	1.00	–	–	0.27	1.00	1.55	1.27	0.36	0.91	–	0.91	0.27	–	1.18
167	**CLOVER, LADINO,** HAY, SUN-CURED	1-01-378	89	20.0	–	–	–	–	–	–	–	–	–	–	–	–	–	–
168	**COWPEA, COMMON,** HAY, SUN-CURED	1-01-645	90	17.7	1.11	–	–	0.45	1.27	2.01	1.08	0.51	1.26	–	1.06	0.52	–	1.44
169	**LESPEDEZA, COMMON,** HAY, SUN-CURED	1-08-591	89	13.8	–	–	–	–	–	–	–	–	–	–	–	–	–	–
170	**OATS,** HULLS	1-03-281	92	3.7	0.15	0.06	0.15	0.08	0.15	0.25	0.17	0.08	0.15	–	0.16	0.09	0.14	0.19
171	**SOYBEAN,** HAY, SUN-CURED	1-04-558	89	14.1	–	–	–	–	–	–	–	–	–	–	–	–	–	–
172	**VETCH,** HAY, SUN-CURED	1-05-106	89	18.4	–	–	–	–	–	–	–	–	–	–	–	–	–	–
	PASTURE AND RANGE PLANTS																	
173	**COWPEA, COMMON,** FRESH	2-01-655	25	4.0	–	–	–	–	–	–	–	–	–	–	–	–	–	–
174	**SPINACH,** LEAVES, FRESH	2-08-125	9	3.1	0.11	–	–	0.04	0.09	0.18	0.12	0.06	0.12	–	0.10	0.03	–	0.13
	VITAMIN SUPPLEMENTS																	
	ALFALFA (LUCERNE)																	
175	MEAL, DEHY, 20% PROTEIN	1-00-024	92	20.2	0.95	0.32	0.99	0.38	0.89	1.43	0.89	0.32	0.94	0.90	0.82	0.41	0.60	1.05
176	MEAL, DEHY, 22% PROTEIN	1-07-851	93	22.2	0.96	0.30	1.09	0.44	1.06	1.63	0.97	0.34	1.13	0.97	0.97	0.49	0.64	1.29
177	**BREWERS GRAINS,** DEHY	5-02-141	92	27.3	1.27	0.35	1.09	0.53	1.57	2.53	0.88	0.46	1.46	1.30	0.93	0.37	1.16	1.58
178	**CORN,** DISTILLERS GRAINS WITH SOLUBLES, DEHY	5-02-843	92	27.1	0.97	0.31	0.59	0.64	1.33	2.31	0.70	0.50	1.47	1.21	0.93	0.18	0.72	1.47
179	**FISH,** SOLUBLES, DEHY	5-01-971	93	60.4	3.06	0.62	5.75	2.10	2.05	2.98	3.52	1.18	1.53	2.03	1.35	0.60	0.85	2.10
	FISH, SARDINE																	
180	MEAL MECH EXTD	5-02-015	93	65.2	2.70	0.80	4.50	1.80	3.34	–	5.91	2.01	2.00	–	2.60	0.50	2.79	4.10
181	SOLUBLES, CONDENSED	5-02-014	50	29.5	1.50	0.20	–	2.00	0.90	1.60	1.60	0.90	0.80	–	0.80	0.10	–	1.00
182	**RICE,** BRAN WITH GERM, MEAL SOLV EXTD (RICE BRAN, SOLV EXTD)	4-03-930	91	14.0	0.98	0.21	0.91	0.33	0.52	1.02	0.61	0.26	0.57	0.70	0.53	0.21	0.55	0.76
183	**YEAST, BREWERS,** DEHY	7-05-527	93	43.8	2.26	0.52	1.77	1.13	2.03	2.86	2.98	0.66	1.60	–	2.06	0.51	1.47	2.25
184	**YEAST, PRIMARY,** DEHY	7-05-533	93	48.0	2.60	0.50	–	5.60	3.60	3.70	3.80	1.00	2.50	–	2.50	0.40	–	3.20

TABLE V–3 APPARENT ILEAL DIGESTIBILITY OF CRUDE PROTEIN AND ESSENTIAL AMINO ACIDS IN FEEDSTUFFS FOR SWINE, AND

FEEDSTUFF (AS FED BASIS)	**Obs	DM %	Crude Protein			Arginine			Cystine			Histidine			Isoleucine		
			Total %	Coefficient %	Digestible %	Total %	Coefficient %	Digestible %	Total %	Coefficient %	Digestible %	Total %	Coefficient %	Digestible %	Total %	Coefficient %	Digestible %
ALFALFA	3	89.85	16.95	39	6.61	0.66	59	0.39	0.18	15	0.03	0.32	46	0.15	0.66	55	0.36
BARLEY	33	87.74	10.59	70	7.41	0.52	75	0.39	0.22	74	0.16	0.24	71	0.17	0.37	73	0.27
*BARLEY, NAKED	1	87.20	11.70	78	9.13	0.64	85	0.54	—	—	—	0.30	90	0.27	0.45	87	0.39
BLOOD MEAL	3	90.57	87.87	87	76.45	3.68	94	3.46	—	—	—	5.44	94	5.11	0.72	70	0.50
BONE MEAL	2	94.11	28.10	72	20.23	1.78	79	1.41	0.20	38	0.08	0.26	71	0.18	0.54	71	0.38
BREWERS GRAINS	3	91.60	26.03	70	18.22	1.82	—	—	—	—	—	0.67	—	—	1.11	79	0.88
CANOLA MEAL	4	90.43	37.28	69	25.72	2.21	82	1.81	0.42	89	0.37	1.20	78	0.94	1.57	75	1.18
CASEIN	4	90.69	86.19	89	76.71	2.88	93	2.68	—	62	—	2.00	94	1.88	4.22	90	3.80
COCONUT OIL MEAL	1	94.70	21.58	52	11.22	2.18	—	—	0.29	—	—	0.42	—	—	0.60	65	0.39
CORN	10	88.34	9.19	77	7.08	0.41	82	0.34	0.17	74	0.13	0.25	82	0.21	0.33	80	0.26
CORN GLUTEN FEED	4	90.30	19.58	54	10.57	1.07	72	0.77	0.47	—	—	0.66	58	0.38	0.66	62	0.41
CORN GLUTEN MEAL	2	89.35	59.00	80	47.20	2.00	86	1.72	1.06	73	0.77	1.31	81	1.06	2.61	84	2.19
COTTONSEED MEAL GLANDED	9	91.60	41.14	72	29.62	4.42	88	3.89	—	—	—	1.10	77	0.85	1.28	67	0.86
COTTONSEED MEAL GLANDLESS	2	94.82	42.40	83	35.19	5.56	95	5.28	—	—	—	1.36	88	1.20	1.45	85	1.23
DRIED SKIM MILK	4	95.78	33.96	85	28.87	1.24	85	1.05	0.28	86	0.24	1.47	93	1.37	1.73	85	1.47
FEATHER MEAL	3	94.30	83.20	71	59.07	5.09	79	4.02	3.85	72	2.77	0.57	45	0.26	3.69	78	2.88
FISH MEAL	9	94.49	61.83	79	48.85	3.91	88	3.44	0.66	62	0.41	1.37	81	1.11	2.60	86	2.24
GROUNDNUT MEAL	2	89.30	41.70	85	35.45	4.92	95	4.67	0.49	78	0.38	0.88	85	0.75	1.38	85	1.17
LUPIN MEAL	2	88.25	33.07	80	26.46	3.50	93	3.26	0.43	70	0.30	0.77	85	0.65	1.47	83	1.22
L-LYSINE HCl	2	98.50	94.40	100	94.40												
MEAT & BONE MEAL	29	93.38	55.07	67	36.90	3.44	77	2.65	0.63	51	0.32	0.93	71	0.66	1.53	68	1.04
OATS	1	87.60	13.00	61	7.93	0.65	—	—	0.29	—	—	0.21	—	—	0.39	—	—
OAT GROATS	1	91.62	16.32	84	13.71	1.15	90	1.04	1.15	90	1.04	0.46	86	0.40	0.52	86	0.45
PEANUT MEAL	4	93.87	44.23	73	32.29	5.20	90	4.68	—	—	—	1.04	73	0.76	1.46	76	1.11
PEAS	3	89.03	20.87	73	15.24	1.83	86	1.57	0.37	61	0.23	0.57	81	0.46	0.90	74	0.67
POULTRY-BY-PRODUCT MEAL	5	92.76	63.71	72	45.87	4.34	88	3.82	2.04	—	—	1.23	81	1.00	2.52	79	1.99
RAPESEED MEAL	7	91.45	35.32	70	24.72	2.05	83	1.70	0.83	73	0.61	0.99	83	0.82	1.48	75	1.11
*RICE	1	85.50	8.80	72	6.34	0.98	90	0.88	—	—	—	0.18	90	0.16	0.32	86	0.28
RYE	3	88.27	10.03	68	6.82	0.48	74	0.36	0.20	72	0.14	0.21	70	0.15	0.33	70	0.23
SAND EEL	1	89.42	72.56	79	57.32	3.45	—	—	0.70	—	—	1.63	54	0.88	2.93	88	2.58
SESAME MEAL	1	92.70	44.04	76	33.47	5.39	—	—	0.86	—	—	1.01	36	0.36	1.40	79	1.11
SORGHUM	2	89.99	9.83	81	7.96	0.45	85	0.38	—	—	—	0.24	81	0.19	0.44	88	0.39
SOY FLOUR	1	94.44	50.68	82	41.56	4.02	91	3.66	0.74	78	0.58	1.39	88	1.22	2.42	83	2.01
SOYBEANS EXTRUDED	2	92.68	35.11	74	25.98	2.72	84	2.28	0.57	64	0.36	0.99	80	0.79	1.69	74	1.25
SOYBEAN MEAL 44%	9	89.62	44.27	80	35.42	3.29	90	2.96	0.69	74	0.51	1.18	86	1.01	2.03	83	1.68
SOYBEAN MEAL 48%	18	89.51	47.29	80	37.83	3.56	90	3.20	0.66	79	0.52	1.28	86	1.10	2.22	84	1.86
SOYFLAKES RAW	1	90.72	48.72	35	17.05	3.91	56	2.19	0.64	35	0.22	1.41	48	0.68	2.42	43	1.04
SOYFLAKES HEATED	1	91.96	49.29	79	38.94	3.91	88	3.44	0.63	74	0.47	1.40	82	1.15	2.48	78	1.93
SUNFLOWER MEAL	10	91.90	35.29	75	26.47	2.93	90	2.64	0.57	74	0.42	0.89	76	0.68	1.45	78	1.13
TRITICALE	5	89.14	13.31	78	10.38	0.55	87	0.48	—	—	—	0.25	84	0.21	0.36	83	0.30
WHEAT	24	87.93	12.68	82	10.40	0.57	83	0.47	0.20	80	0.16	0.25	80	0.20	0.43	83	0.36
WHEAT BRAN 7% FIBER	3	88.30	13.50	67	9.05	0.88	85	0.75	0.37	71	0.26	0.35	79	0.28	0.47	74	0.35
WHEAT FLOUR	1	88.00	13.46	91	12.25	0.43	91	0.39	—	—	—	0.26	91	0.24	0.50	94	0.47
WHEAT MIDDS 9% FIBER	5	89.14	16.63	67	11.14	1.18	84	0.99	—	—	—	0.45	79	0.36	0.55	70	0.39
WHEAT OFFAL 9% FIBER	1	88.00	15.84	70	11.09	0.97	95	0.92	—	—	—	0.39	79	0.31	0.51	73	0.37

*True Ileal Digestibility Values

**The number of observations for tryptophan may differ substantially from the value indicated.

DIGESTIBLE AMINO ACID RECOMMENDATIONS

Heartland Lysine would like to express its appreciation to each of the authors who contributed data to this work. In particular our thanks are extended to Dr. Darrell Knabe (Texas A&M University), Dr. Malcolm Fuller (Rowett Research Institute), and Dr. Michael Taverner (Animal Research Institute) for their contributions.

Period	Starting		Growing-Finishing[1]	
Weight, lb.	10 - 22	22 - 55	55 - 110	110 - 220
DIGESTIBLE MINIMUMS, %				
CRUDE PROTEIN	16.50	14.50	12.50	11.00
LYSINE/CALORIE (g/Mcal ME)	3.60	3.30	2.50	2.00
LYSINE[2]	1.25	1.00	0.75	0.61
THREONINE	0.81	0.65	0.49	0.40
TRYPTOPHAN	0.23	0.18	0.14	0.11
ISOLEUCINE	0.75	0.60	0.45	0.37
METHIONINE & CYSTINE	0.69	0.55	0.41	0.34
HISTIDINE	0.41	0.33	0.25	0.20
LEUCINE	1.35	1.08	0.81	0.66
PHENYLALANINE & TYROSINE	1.50	1.20	0.90	0.73
VALINE	0.91	0.73	0.55	0.45

[1] Boar requirements are 10 to 15% higher than gilts or barrows